ELEMENTARY ALGEBRA OF THE REAL NUMBERS

$a + b = b + a$

$a + (b + c) = (a + b) + c$

$a(b + c) = ab + ac$

$a + 0 = a \cdot 1 = a$

$a - a = 0$

$-(-a) = a$

$-(a - b) = b - a$

$\dfrac{a}{b} + \dfrac{c}{d} = \dfrac{ad + bc}{bd} \quad (b, d \neq 0)$

$a^1 = a$

$1^n = 1$

$a^{m+n} = a^m a^n$

$a^{m-n} = a^m/a^n \quad (a \neq 0)$

$a^m b^m = (ab)^m$

$\sqrt[n]{-a} = -\sqrt[n]{a}, \quad (-a)^{m/n} = \sqrt[n]{(-a)^m}, \quad (a > 0, n \text{ odd})$

If $a < b$ then $a + c < b + c$.

If $a < b$, then $-b < -a$.

$|-a| = |a| \qquad |a| \cdot |b| = |ab| \qquad \dfrac{|a|}{|b|} = \left|\dfrac{a}{b}\right| \quad (b \neq 0).$

$ab = ba$

$a(bc) = (ab)c$

$a(-b) = (-a)b = -ab$

$a \cdot 0 = 0$

$a/a = 1 \quad (a \neq 0)$

$\dfrac{1}{1/a} = a \quad (a \neq 0)$

$\dfrac{1}{a/b} = \dfrac{b}{a} \quad (a, b \neq 0)$

$\dfrac{a}{b} \cdot \dfrac{c}{d} = \dfrac{ac}{bd} \quad (b, d \neq 0)$

$a^0 = 1 \quad (a \neq 0)$

$0^n = 0 \quad (n > 0)$

$a^{mn} = (a^m)^n$

$a^{m/n} = \sqrt[n]{a^m} \quad (a > 0)$

$\sqrt[m]{a}\sqrt[n]{b} = \sqrt[m]{ab} \quad (a, b > 0)$

If $a < b$ and $0 < c$, then $ac < bc$.

If $0 < a < b$, then $1/b < 1/a$.

The following terms are undefined:

$a/0, \quad 0^0, \quad 0^{-m}, \quad \sqrt[n]{-a} \qquad (a, m \text{ positive}, n \text{ even})$

Quadratic Formula: $ax^2 + bx + c = 0$ if and only if $x = \dfrac{-b \pm \sqrt{b^2 - 4ac}}{2a}$

ALGEBRA OF HYPERREAL NUMBERS

Notation:

ε, δ are positive infinitesimals

b, c are positive and finite but not infinitesimal

H, K are positive infinite.

The following are infinitesimal:

$-\varepsilon, \quad 1/H, \quad \varepsilon/b, \quad \varepsilon/H, \quad b/H, \quad \varepsilon + \delta, \quad \varepsilon - \delta, \quad \varepsilon \cdot \delta, \quad b \cdot \varepsilon, \quad \sqrt[n]{\varepsilon}$

The following are finite but not infinitesimal:

$-b, \quad 1/b, \quad b/c, \quad b + \varepsilon, \quad b \cdot c, \quad \sqrt[n]{b}, \quad b + c$

$b - c$ is finite (possibly infinitesimal).

The following are infinite:

$-H, \quad 1/\varepsilon, \quad b/\varepsilon, \quad H/\varepsilon, \quad H/b, \quad H + \varepsilon, \quad H + b, \quad H \cdot b, \quad H \cdot K, \quad \sqrt[n]{H}, \quad H + K$

The following can be infinitesimal, finite but not infinitesimal, or infinite:

$\varepsilon/\delta, \quad H/K, \quad H\varepsilon, \quad H - K$

Elementary Calculus

An Infinitesimal Approach

Elementary Calculus

An Infinitesimal Approach

THIRD EDITION

H. Jerome Keisler
University of Wisconsin

Dover Publications, Inc.
Mineola, New York

Bibliographical Note

This Dover edition, first published in 2012, is the first publication in book form of the Third Edition of the work originally published in 1976 by Prindle, Weber, and Schmidt, Boston.

Library of Congress Cataloging-in-Publication Data

Keisler, H. Jerome.
    Elementary calculus : an infinitesimal approach / H. Jerome Keisler. — 3rd ed.
      p. cm.
    "This Dover edition, first published in 2012, is the first publication in book form of the Third Edition of the work originally published in 1976 by Prindle, Weber, and Schmidt, Boston."
    Includes index.
    ISBN-13: 978-0-486-48452-5
    ISBN-10: 0-486-48452-1
    1. Calculus. I. Title.

QA303.2K44 2012
515—dc23

                                     2011033656

Manufactured in the United States by LSC Communications
48452108   2023
www.doverpublications.com

Dedicated to my sons, Randall, Jeffrey, and Thomas

PREFACE TO THE THIRD EDITION

This book contains all the usual topics in an elementary calculus course, plus one extra concept that makes the basic ideas more easily understood by beginning students. This concept is the infinitesimal, or infinitely small number. The calculus was originally developed in the 1670's by Leibniz and Newton using infinitesimals in an informal way, and infinitesimals were given a rigorous foundation by Abraham Robinson in 1960.

The work is designed as the main textbook for a three or four semester calculus course. It has been used at both the college and high school levels. Because of its unique approach with infinitesimals, it can also serve as a supplementary book for courses based on other textbooks, and as an interesting refresher for those who have had calculus courses in the past. In addition, it can be a starting point for mathematicians, scientists, and engineers who want to learn about the recent use of infinitesimals beyond calculus.

The First Edition of this book was published in 1976, and a considerably revised Second Edition was published in 1986, both by Prindle, Weber, and Schmidt. When the Second Edition went out of print in 1992, the copyright was returned to me as the author. Since 2002, the Second Edition with minor revisions has been available free in digital form under a Creative Commons License. It can be downloaded at http://www..math.wisc.edu/~keisler/calc. It is listed by the California Digital Textbook Initiative as meeting all content standards for calculus.

Many people find it easier and more convenient to learn mathematics from a printed book. With this Third Edition, Dover Publications has given people the option of purchasing a high-quality affordable printed version of the free digital book.

I have written a higher-level companion to this book, *Foundations of Infinitesimal Calculus* (2007). This companion is intended for instructors and advanced students. It is a self-contained treatment of the mathematical background for this book, and can be a bridge to more advanced topics. It is freely available online at http://www.Math.wisc.edu/~keisler/foundations.

Since 1960, modern infinitesimals have been applied to many areas of mathematics, as well as to the physical sciences and economics. For the most part, these applications have been reported in professional journal articles and in proceedings of conferences, but I will mention two books on the subject. *Nonstandard Methods in Stochastic Analysis and Mathematical Physics*, by Sergio Albeverio, et. al., is currently available from Dover Publications. The forthcoming book *Infinitesimal Methods in Mathematical Economics* by Robert M. Anderson is available online at http://www.econ.berkeley. edu/~anderson/Book.pdf.

H. Jerome Keisler
June, 2011

PREFACE TO THE SECOND EDITION

In this second edition, many changes have been made based on nine years of classroom experience. There are major revisions to the first six chapters and the Epilogue, and there is one completely new chapter, Chapter 14, on differential equations. In addition, the original Chapters 11 and 12 have been repackaged as three chapters: Chapter 11 on partial differentiation, Chapter 12 on multiple integration, and Chapter 13 on vector calculus.

Chapter 1 has been shortened, and much of the theoretical material from the first edition has been moved to the Epilogue. The calculus of transcendental functions has been fully integrated into the course beginning in Chapter 2 on derivatives. Chapter 3 focuses on applications of the derivative. The material on setting up word problems and on related rates has been moved from the first two chapters to the beginning of Chapter 3. The theoretical results on continuous functions, including the Intermediate, Extreme, and Mean Value Theorems, have been collected in a single section at the end of Chapter 3. The development of the integral in Chapter 4 has been streamlined. The Trapezoidal Rule has been moved from Chapter 5 to Chapter 4, and a discussion of Simpson's Rule has been added. The section on area between two curves has been moved from Chapter 6 to Chapter 4. Chapter 5 deals with limits, approximations, and analytic geometry. An extensive treatment of conic sections and a section on Newton's method have been added. Chapter 6 begins with new material on finding a volume by integrating areas of cross sections.

Only minor changes and corrections have been made to Chapters 7 through 13. The new Chapter 14 gives a first introduction to differential equations, with emphasis on solving first and second order linear differential equations. In Section 14.4, infinitesimals are used to give a simple proof that every differential equation $y' = f(t,y)$, where f is continuous, has a solution. The proof of this fact is beyond the scope of a traditional elementary calculus course, but is within reach with infinitesimals.

I wish to thank all my friends and colleagues who have suggested corrections and improvements to the first edition of the book.

H. Jerome Keisler

PREFACE TO THE
FIRST EDITION

The calculus was originally developed using the intuitive concept of an infinitesimal, or an infinitely small number. But for the past one hundred years infinitesimals have been banished from the calculus course for reasons of mathematical rigor. Students have had to learn the subject without the original intuition. This calculus book is based on the work of Abraham Robinson, who in 1960 found a way to make infinitesimals rigorous. While the traditional course begins with the difficult limit concept, this course begins with the more easily understood infinitesimals. It is aimed at the average beginning calculus student and covers the usual three or four semester sequence.

The infinitesimal approach has three important advantages for the student. First, it is closer to the intuition which originally led to the calculus. Second, the central concepts of derivative and integral become easier for the student to understand and use. Third, it teaches both the infinitesimal and traditional approaches, giving the student an extra tool which may become increasingly important in the future.

Before describing this book, I would like to put Robinson's work in historical perspective. In the 1670's, Leibniz and Newton developed the calculus based on the intuitive notion of infinitesimals. Infinitesimals were used for another two hundred years, until the first rigorous treatment of the calculus was perfected by Weierstrass in the 1870's. The standard calculus course of today is still based on the "ε, δ definition" of limit given by Weierstrass. In 1960 Robinson solved a three hundred year old problem by giving a precise treatment of the calculus using infinitesimals. Robinson's achievement will probably rank as one of the major mathematical advances of the twentieth century.

Recently, infinitesimals have had exciting applications outside mathematics, notably in the fields of economics and physics. Since it is quite natural to use infinitesimals in modelling physical and social processes, such applications seem certain to grow in variety and importance. This is a unique opportunity to find new uses for mathematics, but at present few people are prepared by training to take advantage of this opportunity.

Because the approach to calculus is new, some instructors may need additional background material. An instructor's volume, "Foundations of Infinitesimal

Calculus," gives the necessary background and develops the theory in detail. The instructor's volume is keyed to this book but is self-contained and is intended for the general mathematical public.

This book contains all the ordinary calculus topics, including the traditional limit definition, plus one extra tool—the infinitesimals. Thus the student will be prepared for more advanced courses as they are now taught. In Chapters 1 through 4 the basic concepts of derivative, continuity, and integral are developed quickly using infinitesimals. The traditional limit concept is put off until Chapter 5, where it is motivated by approximation problems. The later chapters develop transcendental functions, series, vectors, partial derivatives, and multiple integrals. The theory differs from the traditional course, but the notation and methods for solving practical problems are the same. There is a variety of applications to both natural and social sciences.

I have included the following innovation for instructors who wish to introduce the transcendental functions early. At the end of Chapter 2 on derivatives, there is a section beginning an alternate track on transcendental functions, and each of Chapters 3 through 6 have alternate track problem sets on transcendental functions. This alternate track can be used to provide greater variety in the early problems, or can be skipped in order to reach the integral as soon as possible. In Chapters 7 and 8 the transcendental functions are developed anew at a more leisurely pace.

The book is written for average students. The problems preceded by a square box go somewhat beyond the examples worked out in the text and are intended for the more adventuresome.

I was originally led to write this book when it became clear that Robinson's infinitesimal calculus could be made available to college freshmen. The theory is simply presented; for example, Robinson's work used mathematical logic, but this book does not. I first used an early draft of this book in a one-semester course at the University of Wisconsin in 1969. In 1971 a two-semester experimental version was published. It has been used at several colleges and at Nicolet High School near Milwaukee, and was tested at five schools in a controlled experiment by Sister Kathleen Sullivan in 1972–1974. The results (in her 1974 Ph.D. thesis at the University of Wisconsin) show the viability of the infinitesimal approach and will be summarized in an article in the American Mathematical Monthly.

I am indebted to many colleagues and students who have given me encouragement and advice, and have carefully read and used various stages of the manuscript. Special thanks are due to Jon Barwise, University of Wisconsin; G. R. Blakley, Texas A & M University; Kenneth A. Bowen, Syracuse University; William P. Francis, Michigan Technological University; A. W. M. Glass, Bowling Green University; Peter Loeb, University of Illinois at Urbana; Eugene Madison and Keith Stroyan, University of Iowa; Mark Nadel, Notre Dame University; Sister Kathleen Sullivan, Barat College; and Frank Wattenberg, University of Massachusetts.

H. Jerome Keisler

CONTENTS

INTRODUCTION

While arithmetic deals with sums, differences, products, and quotients, calculus deals with derivatives and integrals. The derivative and integral can be described in everyday language in terms of an automobile trip. An automobile instrument panel has a *speedometer* marked off in miles per hour with a needle indicating the speed. The instrument panel also has an *odometer* which tallies up the distance travelled in miles (the mileage).

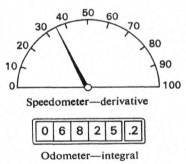

Speedometer—derivative

Odometer—integral

Both the speedometer reading and the odometer reading change with time; that is, they are both "functions of time." The speed shown on the speedometer is the rate of change, or *derivative*, of the distance. Speed is found by taking a very small interval of time and forming the ratio of the change in distance to the change in time. The distance shown on the odometer is the *integral* of the speed from time zero to the present. Distance is found by adding up the distance travelled from the first use of the car to the present.

The calculus has a great variety of applications in the natural and social sciences. Some of the possibilities are illustrated in the problems. However, future applications are hard to predict, and so the student should be able to apply the calculus himself in new situations. For this reason it is important to learn why the calculus works as well as what it can do. To explain why the calculus works, we present a large number of examples, and we develop the mathematical theory with great care.

1

REAL AND HYPERREAL NUMBERS

Chapter 1 takes the student on a direct route to the point where it is possible to study derivatives. Sections 1.1 through 1.3 are reviews of precalculus material and can be skipped in many calculus courses. Section 1.4 gives an intuitive explanation of the hyperreal numbers and how they can be used to find slopes of curves. This section has no problem set and is intended as the basis for an introductory lecture. The main content of Chapter 1 is in the last two sections, 1.5 and 1.6. In these sections, the student will learn how to work with the hyperreal numbers and in particular how to compute standard parts. Standard parts are used at the beginning of the next chapter to find derivatives of functions. Sections 1.5 and 1.6 take the place of the beginning chapter on limits found in traditional calculus texts.

For the benefit of the interested student, we have included an Epilogue at the end of the book that presents the theory underlying this chapter.

1.1 THE REAL LINE

Familiarity with the real number system is a prerequisite for this course. A review of the rules of algebra for the real numbers is given in the appendix. For convenience, these rules are also listed in a table inside the front cover. The letter R is used for the set of all real numbers. We think of the real numbers as arranged along a straight line with the integers (whole numbers) marked off at equal intervals, as shown in Figure 1.1.1. This line is called the *real line*.

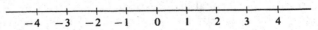

Figure 1.1.1 The real line.

In grade school and high school mathematics, the real number system is constructed gradually in several stages. Beginning with the positive integers, the systems of integers, rational numbers, and finally real numbers are built up. One way to construct the set of real numbers is as the set of all nonterminating decimals.

After constructing the real numbers, it is possible to prove the familiar rules for sums, differences, products, quotients, exponents, roots, and order. In this course, we take it for granted that these rules are familiar to the student, so that we can proceed as quickly as possible to the calculus.

Before going on, we pause to recall two special points that are important in the calculus. First, *division by zero is never allowed*. Expressions such as

$$\frac{2}{0}, \quad \frac{0}{0}, \quad \frac{x}{0}, \quad \frac{5}{1 + 3 - 4}$$

are always considered to be *undefined*.

Second, a positive real number c always has two square roots, $\sqrt{c}$ and $-\sqrt{c}$, and $\sqrt{c}$ always stands for the positive square root. Negative real numbers do not have real square roots. *For each positive real number c, $\sqrt{c}$ is positive and $\sqrt{-c}$ is undefined.*

On the other hand, every real number has one real cube root. If $c > 0$, c has the positive cube root $\sqrt[3]{c}$, and $-c$ has the negative cube root $\sqrt[3]{-c} = -\sqrt[3]{c}$.

In calculus, we often deal with sets of real numbers. By a *set S* of real numbers, we mean any collection of real numbers, called *members* of S, *elements* of S, or *points* in S.

A simple but important kind of set is an *interval*. Given two real numbers a and b with $a < b$, the *closed interval* $[a, b]$ is defined as the set of all real numbers x such that $a \leq x$ and $x \leq b$, or more concisely, $a \leq x \leq b$.

The *open interval* (a, b) is defined as the set of all real numbers x such that $a < x < b$. Closed and open intervals are illustrated in Figure 1.1.2.

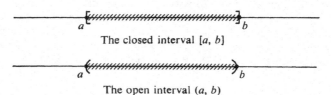

The closed interval $[a, b]$

Figure 1.1.2   The open interval (a, b)

For both open and closed intervals, the number a is called the *lower endpoint*, and b the *upper endpoint*. The difference between the closed interval $[a, b]$ and the open interval (a, b) is that the endpoints a and b are elements of $[a, b]$ but are not elements of (a, b). When $a \leq x \leq b$, we say that x is *between* a and b; when $a < x < b$, we say that x is *strictly between* a and b.

Three other types of sets are also counted as open intervals: the set (a, ∞) of all real numbers x greater than a; the set $(-\infty, b)$ of all real numbers x less than b, and the whole real line R. The real line R is sometimes denoted by $(-\infty, \infty)$. The symbols ∞ and $-\infty$, read "infinity" and "minus infinity," do not stand for numbers; they are only used to indicate an interval with no upper endpoint, or no lower endpoint.

Besides the open and closed intervals, there is one other kind of interval, called a *half-open interval*. The set of all real numbers x such that $a \leq x < b$ is a half-open interval denoted by $[a, b)$. The set of all real numbers x such that $a \leq x$ is also a half-open interval and is written $[a, \infty)$. Here is a table showing the various kinds of intervals.

Table 1.1.1   Kinds of Intervals

Type	Symbol	Defining Formula
Closed	$[a, b]$	$a \leq x \leq b$
Open	(a, b)	$a < x < b$
Open	(a, ∞)	$a < x$
Open	$(-\infty, b)$	$x < b$
Open	$(-\infty, \infty)$	
Half-open	$[a, b)$	$a \leq x < b$
Half-open	$[a, \infty)$	$a \leq x$
Half-open	$(a, b]$	$a < x \leq b$
Half-open	$(-\infty, b]$	$x \leq b$

We list some other important examples of sets of real numbers.

(1)   The empty set $\emptyset$, which has no elements.

(2)   The finite set $\{a_1, \ldots, a_n\}$, whose only elements are the numbers $a_1, a_2, \ldots, a_n$.

(3)   The set of all x such that $x \neq 0$.

(4)   The set $N = \{1, 2, 3, 4, \ldots\}$ of all positive integers.

(5)   The set $Z = \{\ldots, -3, -2, -1, 0, 1, 2, 3, \ldots\}$ of all integers.

(6)   The set Q of all rational numbers. A rational number is a quotient m/n where m and n are integers and $n \neq 0$.

While real numbers correspond to points on a line, ordered pairs of real numbers correspond to points on a plane. This correspondence gives us a way to draw pictures of calculus problems and to translate physical problems into the language of calculus. It is the starting point of the subject called *analytic geometry*.

An *ordered pair* of real numbers, (a, b), is given by the first number a and the second number b. For example, $(1, 3), (3, 1)$, and $(1, 1)$ are three different ordered pairs. Following tradition, we use the same symbol for the open interval (a, b) and the ordered pair (a, b). However the open interval and ordered pair are completely different things. It will always be quite obvious from the context whether (a, b) stands for the open interval or the ordered pair.

We now explain how ordered pairs of real numbers correspond to points in a plane. A system of *rectangular coordinates* in a plane is given by a horizontal and a vertical copy of the real line crossing at zero. The horizontal line is called the *horizontal axis*, or *x-axis*, while the vertical line is called the *vertical axis*, or *y-axis*. The point where the two axes meet is called the *origin* and corresponds to the ordered pair $(0, 0)$. Now consider any point P in the plane. A vertical line through P will cross the x-axis at a real number x_0, and a horizontal line through P will cross the y-axis at a real number y_0. The ordered pair (x_0, y_0) obtained in this way corresponds to the point P. (See Figure 1.1.3.) We sometimes call P the point (x_0, y_0) and sometimes write $P(x_0, y_0)$. x_0 is called the *x-coordinate* of P and y_0 the *y-coordinate* of P.

Conversely, given an ordered pair (x_0, y_0) of real numbers there is a corresponding point $P(x_0, y_0)$ in the plane. $P(x_0, y_0)$ is the point of intersection of the vertical line crossing the x-axis at x_0 and the horizontal line crossing the y-axis at y_0. We have described a one-to-one correspondence between all points in the plane and all ordered pairs of real numbers.

From now on, we shall simplify things by identifying points in the plane with ordered pairs of real numbers, as shown in Figure 1.1.4.

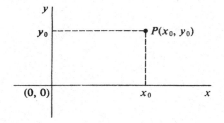

Figure 1.1.3

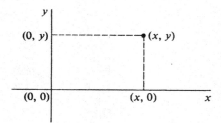

Figure 1.1.4

DEFINITION

*The (x, y) **plane** is the set of all ordered pairs (x, y) of real numbers. The **origin** is the point $(0, 0)$. The x-**axis** is the set of all points of the form $(x, 0)$, and the y-**axis** is the set of all points of the form $(0, y)$.*

The x- and y-axes divide the rest of the plane into four parts called *quadrants*. The quadrants are numbered I through IV, as shown in Figure 1.1.5.

In Figure 1.1.6, $P(x_1, y_1)$ and $Q(x_2, y_2)$ are two different points in the (x, y) plane. As we move from P to Q, the coordinates x and y will change by amounts that we denote by Δx and Δy. Thus

$$\text{change in } x = \Delta x = x_2 - x_1,$$
$$\text{change in } y = \Delta y = y_2 - y_1.$$

The quantities Δx and Δy may be positive, negative, or zero. For example, when $x_2 > x_1$, Δx is positive, and when $x_2 < x_1$, Δx is negative. Using Δx and Δy we define the basic notion of distance.

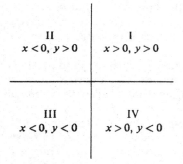

Figure 1.1.5   Quadrants

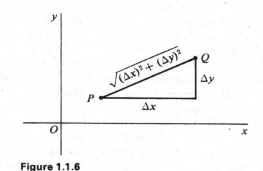

Figure 1.1.6

DEFINITION

*The **distance** between the points $P(x_1, y_1)$ and $Q(x_2, y_2)$ is the quantity*

$$\text{distance }(P, Q) = \sqrt{(\Delta x)^2 + (\Delta y)^2} = \sqrt{(x_2 - x_1)^2 + (y_2 - y_1)^2}.$$

When we square both sides of the distance formula, we obtain

$$[\text{distance }(P, Q)]^2 = (\Delta x)^2 + (\Delta y)^2.$$

One can also get this formula from the Theorem of Pythagoras in geometry: *The square of the hypotenuse of a right triangle is the sum of the squares of the sides.*

EXAMPLE 1 Find the distance between $P(7, 2)$ and $Q(4, 6)$ (see Figure 1.1.7).

$$\Delta x = 4 - 7 = -3, \qquad \Delta y = 6 - 2 = 4.$$

$$\text{distance } (P, Q) = \sqrt{(-3)^2 + 4^2} = 5.$$

We often deal with sets of points in the plane as well as on the line. One way to describe a set of points in the plane is by an equation or inequality in two variables, say x and y. A solution of an equation in x and y is a point (x_0, y_0) in the plane for which the equation is true. The set of all solutions is called the *locus,* or *graph,* of the equation. The circle is an important example of a set of points in the plane.

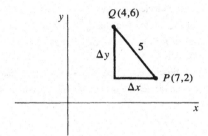

Figure 1.1.7

DEFINITION OF CIRCLE

*The set of all points in the plane at distance r from a point P is called the **circle** of **radius** r and **center** P.*

Using the distance formula, we see that the circle of radius r and center at the origin (Figure 1.1.8) is the locus of the equation

$$x^2 + y^2 = r^2.$$

The circle of radius r and center at $P(h, k)$ (Figure 1.1.8) is the locus of the equation

$$(x - h)^2 + (y - k)^2 = r^2.$$

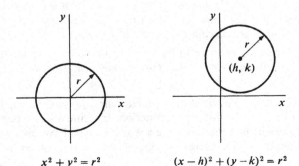

Figure 1.1.8 $x^2 + y^2 = r^2$ $(x - h)^2 + (y - k)^2 = r^2$

For example, the circle with radius 3 and center at $P(2, -4)$ has the equation

$$(x - 2)^2 + (y + 4)^2 = 9.$$

PROBLEMS FOR SECTION 1.1

In Problems 1–6, find the distance between the points P and Q.

1	$P(2, 9), Q(-1, 13)$		**2**	$P(1, -2), Q(2, 10)$
3	$P(0, 0), Q(-2, -3)$		**4**	$P(-1, -1), Q(4, 4)$
5	$P(6, 1), Q(-7, 1)$		**6**	$P(5, 10), Q(9, 10)$

Sketch the circles given in Problems 7–12.

7	$x^2 + y^2 = 4$		**8**	$x^2 + y^2 = \frac{1}{4}$
9	$(x - 1)^2 + (y + 2)^2 = 1$		**10**	$(x + 2)^2 + (y + 3)^2 = 9$
11	$(x - 1)^2 + (y - 1)^2 = 2$		**12**	$(x + 3)^2 + (y - 4)^2 = 25$

13   Find the equation of the circle of radius 2 with center at $(3, 0)$.

14   Find the equation of the circle of radius $\sqrt{3}$ with center at $(-1, -2)$.

☐ **15**   There are two circles of radius 2 that have centers on the line $x = 1$ and pass through the origin. Find their equations.

☐ **16**   Find the equation of the circle that passes through the three points $(0, 0), (0, 1), (2, 0)$.

☐ **17**   Find the equation of the circle one of whose diameters is the line segment from $(-1, 0)$ to $(5, 8)$.

1.2 FUNCTIONS OF REAL NUMBERS

The next two sections are about real numbers only. The calculus deals with problems in which one quantity depends on one or more others. For example, the area of a circle depends on its radius. The length of a day depends on both the latitude and the date. The price of an object depends on the supply and the demand. The way in which one quantity depends on one or more others can be described mathematically by a function of one or more variables.

DEFINITION

> A **real function of one variable** is a set f of ordered pairs of real numbers such that for every real number a one of the following two things happens:
>
> (i)   There is exactly one real number b for which the ordered pair (a, b) is a member of f. In this case we say that $f(a)$ is defined and we write $f(a) = b$. The number b is called the value of f at a.
>
> (ii) There is no real number b for which the ordered pair (a, b) is a member of f. In this case we say that $f(a)$ is undefined.

Thus $f(a) = b$ means that the ordered pair (a, b) is an element of f.

Here is one way to visualize a function. Imagine a black box labeled f as in Figure 1.2.1. Inside the box there is some apparatus, which we can't see. On both the left and right sides of the box there is a copy of the real line, called the input line and

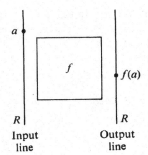

Figure 1.2.1

output line, respectively. Whenever we point to a number a on the input line, either one point b will light up on the output line to tell us that $f(a) = b$, or else nothing will happen, in which case $f(a)$ is undefined.

A second way to visualize a function is by drawing its graph. The *graph* of a real function f of one variable is the set of all points $P(x, y)$ in the plane such that $y = f(x)$. To draw the graph, we plot the value of x on the horizontal, or x-axis and the value of $f(x)$ on the vertical, or y-axis. How can we tell whether a set of points in the plane is the graph of some function? By reading the definition of a function again, we have an answer.

A set of points in the plane is the graph of some function f if and only if for each vertical line one of the following happens:

(1) Exactly one point on the line belongs to the set.
(2) No point on the line belongs to the set.

A vertical line crossing the x-axis at a point a will meet the set in exactly one point (a, b) if $f(a)$ is defined and $f(a) = b$, and the line will not meet the set at all if $f(a)$ is undefined. Try this rule out on the sets of points shown in Figure 1.2.2.

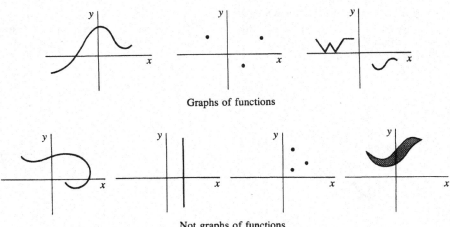

Graphs of functions

Not graphs of functions

Figure 1.2.2

Here are two examples of real functions of one variable. Each function will be described in two ways: the black box approach, where a rule is given for finding the value of the function at each real number, and the graph method, where an equation is given for the graph of the function.

EXAMPLE 1   The *square function*.

The square function is defined by the rule

$$f(x) = x^2$$

for each number x. The value of $f(a)$ is found by squaring a. For instance, the values of $f(0)$, $f(2)$, $f(-3)$, $f(r)$, $f(r + 1)$ are

$$f(0) = 0, \quad f(2) = 4, \quad f(-3) = 9,$$
$$f(r) = r^2, \quad f(r + 1) = r^2 + 2r + 1.$$

The graph of the square function is the parabola with the equation $y = x^2$. The graph of $y = x^2$, with several points marked in, is shown in Figure 1.2.3.

EXAMPLE 2   The *reciprocal function*.

The reciprocal function g is given by the rule

$$g(x) = \frac{1}{x}.$$

$g(x)$ is defined for all nonzero x, but is undefined at $x = 0$. Find the following values if they are defined: $g(0)$, $g(2)$, $g(-\tfrac{1}{3})$, $g(\tfrac{7}{4})$, $g(r + 1)$.

$$g(0) \text{ is undefined,} \quad g(2) = \tfrac{1}{2}, \quad g(-\tfrac{1}{3}) = -3,$$
$$g(\tfrac{7}{4}) = \tfrac{4}{7}, \quad g(r + 1) = \frac{1}{r + 1}.$$

The graph of the reciprocal function has the equation $y = 1/x$. This equation can also be written in the form $xy = 1$. The graph is shown in Figure 1.2.4.

In Examples 1 and 2 we have used the variables x and y in order to describe a function. A *variable* is a letter which stands for an arbitrary real number; that is, it "varies" over the real line. In the equation $y = x^2$, the value of y depends on the value of x; for this reason we say that x is the *independent variable* and y the *dependent variable* of the equation.

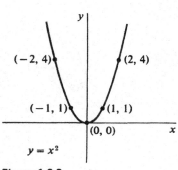

Figure 1.2.3

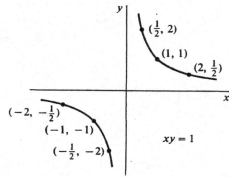

Figure 1.2.4

In describing a function, we do not always use x and y; sometimes other variables are more convenient, especially in problems involving several functions. The variable t is often used to denote time.

It is important to distinguish between the symbol f and the expression $f(x)$. f by itself stands for a *function*. $f(x)$ is called a *term* and stands for the value of the function at x. The need for this distinction is illustrated in the next example.

EXAMPLE 3   Let h be the function given by the rule

$$h(t) = t^3 + 1.$$

t is a variable, h is a function, and $h(t)$ is a term. The following expressions are also terms: $h(\frac{1}{2})$, $h(x)$, $h(t^3)$, $h(t^3) + 1$, $h(t^3 + 1)$, $h(x) - h(t)$, $h(t + \Delta t)$, $h(t + \Delta t) - h(t)$. Find the values of each of these terms.

The values are computed by careful substitution.

$$h(\tfrac{1}{2}) = (\tfrac{1}{2})^3 + 1 = 1\tfrac{1}{8}.$$
$$h(x) = x^3 + 1.$$
$$h(t^3) = (t^3)^3 + 1 = t^9 + 1.$$
$$h(t^3) + 1 = [(t^3)^3 + 1] + 1 = t^9 + 2.$$
$$h(t^3 + 1) = (t^3 + 1)^3 + 1 = t^9 + 3t^6 + 3t^3 + 2.$$
$$h(x) - h(t) = [x^3 + 1] - [t^3 + 1] = x^3 - t^3.$$
$$h(t + \Delta t) = (t + \Delta t)^3 + 1 = t^3 + 3t^2\,\Delta t + 3t\,\Delta t^2 + \Delta t^3 + 1.$$
$$h(t + \Delta t) - h(t) = [(t + \Delta t)^3 + 1] - [t^3 + 1]$$
$$= [t^3 + 3t^2\,\Delta t + 3t\,\Delta t^2 + \Delta t^3 + 1] - [t^3 + 1]$$
$$= 3t^2\,\Delta t + 3t\,\Delta t^2 + \Delta t^3.$$

The graph of h is given by the equation $x = t^3 + 1$. In this equation, t is the independent variable and x is the dependent variable. In Figure 1.2.5, the five points

$$h(-1) = 0, \qquad h(-\tfrac{1}{2}) = \tfrac{7}{8}, \qquad h(0) = 1, \qquad h(\tfrac{1}{2}) = 1\tfrac{1}{8}, \qquad h(1) = 2$$

are plotted and the graph is drawn.

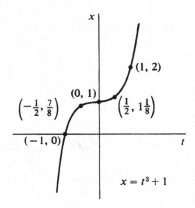

Figure 1.2.5

DEFINITION

*The **domain** of a real function f of one variable is the set of all real numbers x such that f(x) is defined.*

*The **range** of f is the set of all values f(x) where x is in the domain of f.*

EXAMPLE 1 (Continued)   The domain of the square function is the set R of all real numbers. The range is the interval $[0, \infty)$ of all nonnegative reals.

EXAMPLE 2 (Continued)   Both the domain and the range of the reciprocal function are equal to the set of all real x such that $x \neq 0$.

When a function is described by a rule, it is understood that the domain is the set of all real numbers for which the rule is meaningful.

EXAMPLE 3 (Continued)   The function h given by the rule

$$h(t) = t^3 + 1$$

has the whole real line as its domain and as its range.

EXAMPLE 4   Let f be the function given by the rule

$$f(x) = \sqrt{1 - x^2}.$$

Thus $f(x)$ is the positive square root of $1 - x^2$. The domain of f is the closed interval $[-1, 1]$. The range of f is $[0, 1]$.

For instance,

$$f(-2) \text{ is undefined,} \quad f(-1) = 0, \quad f(0) = 1,$$
$$f(\tfrac{1}{2}) = \sqrt{\tfrac{3}{4}}, \qquad f(1) = 0, \qquad f(2) \text{ is undefined.}$$

The graph of f is given by the equation $y = \sqrt{1 - x^2}$.

The equation can also be written in the form

$$x^2 + y^2 = 1, \quad y \geq 0.$$

The graph is just the upper half of the unit semicircle, shown in Figure 1.2.6.

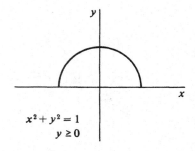

$$x^2 + y^2 = 1$$
$$y \geq 0$$

Figure 1.2.6

Sometimes a function is described by explicitly giving its domain in addition to a rule.

EXAMPLE 5   Let g be the function whose domain is the closed interval $[1, 2]$ with the rule

$$g(x) = x^2.$$

The domain and rule can be written in concise form with an equation and extra inequalities,

$$g(x) = x^2, \quad 1 \le x \le 2.$$

Note that

$$g(0) \text{ is undefined} \quad g(1) = 1$$
$$g(2) = 4 \quad g(3) \text{ is undefined.}$$

The graph is described by the formulas

$$y = x^2, \quad 1 \le x \le 2$$

and is drawn in Figure 1.2.7.

Some especially important functions are the *constant functions*, the *identity function*, and the *absolute value function*.

A real number is sometimes called a *constant*. This name is used to emphasize the difference between a fixed real number and a variable.

For a given real number c, the function f with the rule

$$f(x) = c$$

is called the *constant function* with value c. It has domain R and range $\{c\}$.

EXAMPLE 6   The constant function with value 5 is described by the rule

$$f(x) = 5.$$

Thus $f(0) = 5, \quad f(-3) = 5, \quad f(1{,}000{,}000) = 5.$

The graph (Figure 1.2.8) of the constant function with value 5 is given by the equation $y = 5$.

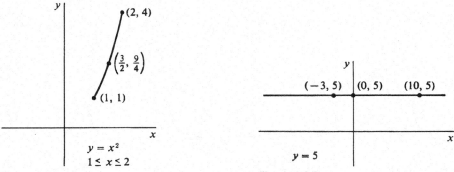

Figure 1.2.7

Figure 1.2.8

EXAMPLE 7   The function f given by the rule

$$f(x) = x$$

is called the *identity function.*

The graph (Figure 1.2.9) of the identity function is the straight line with the equation $y = x$.

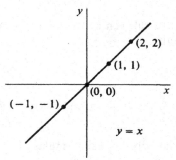

Figure 1.2.9

The *absolute value function* is defined by a rule which is divided into two cases.

DEFINITION

> *The* **absolute value function** $| \quad |$ *is defined by*
>
> $$|x| = \begin{cases} x & \text{if } x \geq 0, \\ -x & \text{if } x < 0. \end{cases}$$

The absolute value of x gives the distance between x and 0. It is always positive or zero. For example,

$$|3| = 3, \qquad |-3| = 3, \qquad |0| = 0.$$

The domain of the absolute value function is the whole real line R while its range is the interval $[0, \infty)$.

The absolute value function can also be described by the rule

$$|x| = \sqrt{x^2}.$$

Its graph is given by the equation $y = \sqrt{x^2}$. The graph is the V shown in Figure 1.2.10.

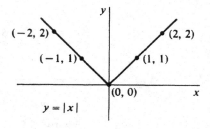

Figure 1.2.10

If a and b are two points on the real line, then from the definition of $|x|$ we see that

$$|a - b| = \begin{cases} a - b & \text{if } a \geq b, \\ b - a & \text{if } b \geq a. \end{cases}$$

Thus $|a - b|$ is the difference between the larger and the smaller of the two numbers. In other words, $|a - b|$ is the *distance* between the points a and b, as illustrated in Figure 1.2.11.

Figure 1.2.11

For example, $|2 - 5| = 3$, $|4 - (-4)| = 8$. Here are some useful facts about absolute values.

THEOREM 1

Let a and b be real numbers.

(i) $|-a| = |a|$.
(ii) $|ab| = |a| \cdot |b|$.
(iii) *If $b \neq 0$, $|a/b| = |a|/|b|$.*

PROOF   We use the equation $|x| = \sqrt{x^2}$.

(i) $|-a| = \sqrt{(-a)^2} = \sqrt{a^2} = |a|$.
(ii) $|ab| = \sqrt{(ab)^2} = \sqrt{a^2 b^2} = \sqrt{a^2}\sqrt{b^2} = |a| \cdot |b|$.
(iii) The proof is similar to (ii).

Warning   The equation $|a + b| = |a| + |b|$ is *false* in general. For example, $|2 + (-3)| = 1$, while $|2| + |(-3)| = 5$.

Functions arise in a great variety of situations. Here are some examples.

Geometry:

$$\pi r^2 = \text{area of a circle of radius } r$$
$$4\pi r^2 = \text{surface area of a sphere of radius } r$$
$$\tfrac{4}{3}\pi r^3 = \text{volume of a sphere of radius } r$$
$$\sin \theta = \text{the sine of the angle } \theta$$

Physics:

$$s(t) = \text{distance a particle travels from time 0 to } t$$
$$v(t) = \text{velocity of a particle at time } t$$
$$a(t) = \text{acceleration of a particle at time } t$$
$$p(y) = \text{water pressure at depth } y \text{ below the surface}$$
$$C = \tfrac{5}{9}(F - 32) = \text{Celsius temperature as a}$$
function of Fahrenheit temperature

Economics:

$$f(t) = \text{population at time } t$$
$$p(t) = \text{price of a commodity at time } t$$
$$c(x) = \text{cost of } x \text{ items of a commodity}$$
$$D(p) = \text{demand for a commodity at price } p, \text{ i.e., the}$$
$$\text{amount which can be sold at price } p$$

Functions of two or more variables can be dealt with in a similar way. Here is the precise definition of a function of two variables.

DEFINITION

*A **real function of two variables** is a set f of ordered triples of real numbers such that for every ordered pair of real numbers (a, b) one of the following two things occurs:*

(i)   *There is exactly one real number c for which the ordered triple (a, b, c) is a member of f. In this case, f(a, b) is defined and we write:*

$$f(a, b) = c.$$

(ii)   *There is no real number c for which the ordered triple (a, b, c) is a member of f. In this case f(a, b) is called undefined.*

If f is a real function of two variables, then the value of $f(x, y)$ depends on both the value of x and the value of y when $f(x, y)$ is defined.

A real function f of two variables can be visualized as a black box with *two* input lines and one output line, as in Figure 1.2.12.

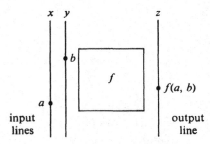

Figure 1.2.12

The *domain* of a real function f of two variables is the set of all pairs of real numbers (x, y) such that $f(x, y)$ is defined.

The most important examples of real functions of two variables are the sum, difference, product, and quotient functions:

$$f(x, y) = x + y, \qquad f(x, y) = xy,$$
$$f(x, y) = x - y, \qquad f(x, y) = x/y.$$

The sum, difference, and product functions have the whole plane as domain. The domain of the quotient function is the set of all ordered pairs (x, y) such that $y \neq 0$.

Here are some examples of functions of two or more variables arising in applications.

Geometry:

ab = area of a rectangle of sides a and b

abc = volume of a rectangular solid

$\frac{1}{2}bh$ = area of a triangle with base b and height h

$\pi r^2 h$ = volume of a cylinder with circular base of radius r and height h

$\frac{1}{3}\pi r^2 h$ = volume of a cone with circular base of radius r and height h

$\sqrt{x^2 + y^2}$ = distance from the origin to (x, y)

Physics:

$F = ma$ = force required to give a mass m an acceleration a

$\rho(x, y, z)$ = density of a three-dimensional object at the point (x, y, z)

$F = Gm_1 m_2/s^2$ = gravitational force between objects of mass m_1 and m_2 at distance s

$m = \dfrac{m_0}{\sqrt{1 - v^2/c^2}}$ = relativistic mass of an object with rest mass m_0 and velocity v

Economics:

$c(x, y)$ = cost of x items of one commodity and y items of another commodity

$D_1(p_1, p_2)$ = demand for commodity one when commodity one has price p_1 and commodity two has price p_2

PROBLEMS FOR SECTION 1.2

For each of the following functions (Problems 1–8), make a table showing the value of $f(x)$ when $x = -1, -\frac{1}{2}, 0, \frac{1}{2}, 1$. Put a * where $f(x)$ is undefined. Example:

$$f(x) = \frac{1}{x}$$

x	-1	$-\frac{1}{2}$	0	$\frac{1}{2}$	1
$f(x)$	-1	-2	*	2	1

1 $f(x) = x/3$ **2** $f(x) = 3$

3 $f(x) = 3x^3 - 5x^2 + 2$ **4** $f(x) = 1/(x - 1)$

5 $f(x) = \sqrt{-x}$ **6** $f(x) = |x|$

7 $f(x) = |x - \frac{1}{2}| + |x + \frac{1}{2}|$

8 $f(x) = \sqrt{x^2 - 1}$

9 Is the set of ordered pairs $\{(3, 2), (0, 1), (4, 2)\}$ a function?

10 Is the set of ordered pairs $\{(0, 2), (3, 6), (3, 4)\}$ a function?

11 If f is the function $f(x) = 1 + x + x^2$, find $f(2), f(t), f(t + \Delta t), f(1 + t + t^2), f(g(t))$.

12 If $f(x) = 1/x$, find $f(t), f(t + \Delta t), f(t^2), f(1/t), f(g(t))$.

13 If $f(x) = x\sqrt{x}$, find $f(t), f(t + \Delta t), f(t^2), f(\sqrt{t}), f(g(t))$.

14 If $f(x) = ax + b$, find $f(ct + d), f(t^2), f(1/t), f(t/a), f(g(t))$.

For each of the following functions (Problems 15–20), find $f(x + \Delta x) - f(x)$.

15 $f(x) = 4x + 1$ **16** $f(x) = x^2 - x$

17 $f(x) = x^{-2}$ **18** $f(x) = x^4$

19 $f(x) = \sqrt{x}$ **20** $f(x) = 4$

21   Find the domain of the function $f(x) = 1/(x^2 - 1)$.

22   Find the domain of the function $f(z) = \sqrt{z^2 - 1}$.

23   What is the domain of the function $f(x) = \sqrt{x}$?

24   What is the domain of the function $f(t) = \sqrt{|t|}$?

25   What is the domain of the function $f(x) = 1/\sqrt{1 - x^2}$?

☐ **26**   Show that if a and b have the same sign then $|a + b| = |a| + |b|$, and if a and b have opposite signs then $|a + b| < |a| + |b|$.

1.3 STRAIGHT LINES

DEFINITION

> Let $P(x_0, y_0)$ be a point and let m be a real number. The **line** through P with slope m is the set of all points $Q(x, y)$ with
>
> $$y - y_0 = m(x - x_0).$$
>
> This equation is called the **point-slope equation** of the line. (See Figure 1.3.1.)
> The **vertical line** through P is the set of all points $Q(x, y)$ with $x = x_0$. Vertical lines do not have slopes.

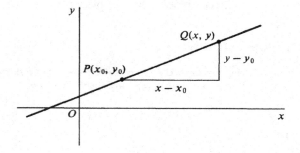

Figure 1.3.1

The slope is a measure of the direction of the line. Figure 1.3.2 shows lines with zero, positive, and negative slopes.

The line that crosses the y-axis at the point $(0, b)$ and has slope m has the simple equation.

$$y = mx + b.$$

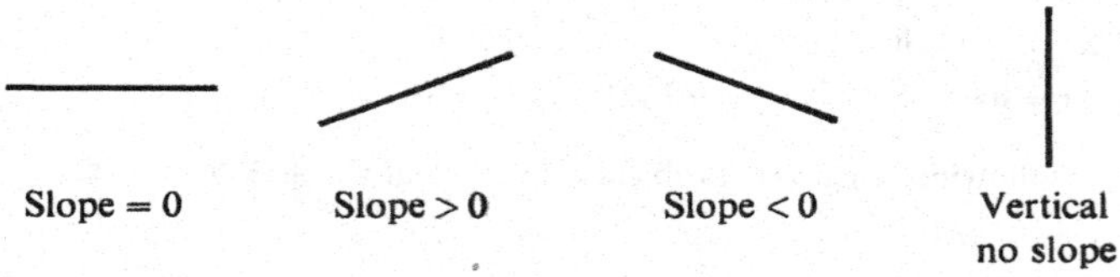

Slope $= 0$ Slope > 0 Slope < 0 Vertical no slope

Figure 1.3.2

This is called the *slope-intercept* equation for the line. We can get it from the point-slope equation by setting $x_0 = 0$ and $y_0 = b$.

EXAMPLE 1   The line through the point $P(-1, 2)$ with slope $m = -\frac{1}{2}$ (Figure 1.3.3) has the point-slope equation

$$y - 2 = (x - (-1)) \cdot (-\tfrac{1}{2}), \quad \text{or} \quad y - 2 = -\tfrac{1}{2}(x + 1).$$

The slope-intercept equation is

$$y = -\tfrac{1}{2}x + 1\tfrac{1}{2}.$$

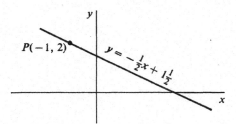

Figure 1.3.3

We now describe the functions whose graphs are nonvertical lines.

DEFINITION

> A **linear function** is a function f of the form
>
> $$f(x) = mx + b,$$
>
> where m and b are constants.

The graph of a linear function is just the line with slope-intercept equation

$$y = mx + b.$$

This is the line through $(0, b)$ with slope m.

If two points on a line are known, the slope can be found as follows.

THEOREM 1

> Suppose a line L passes through two distinct points $P(x_1, y_1)$ and $Q(x_2, y_2)$. If $x_1 = x_2$, then the line L is vertical. If $x_1 \neq x_2$, then the slope of the line L is equal to the change in y divided by the change in x,
>
> $$m = \frac{\Delta y}{\Delta x} = \frac{y_2 - y_1}{x_2 - x_1}.$$

PROOF Suppose $x_1 \neq x_2$, so L is not vertical. Let m be the slope of L. L has the point-slope formula

$$y - y_1 = m(x - x_1).$$

Substituting y_2 for y and x_2 for x, we see that $m = (y_2 - y_1)/(x_2 - x_1)$.

Theorem 1 shows why the slope of a line is a measure of its direction. Sometimes Δx is called the *run* and Δy the *rise*. Thus the slope is equal to the rise divided by the run. A large positive slope means that the line is rising steeply to the right, and a small positive slope means the line rises slowly to the right. A negative slope means that the line goes downward to the right. These cases are illustrated in Figure 1.3.4.

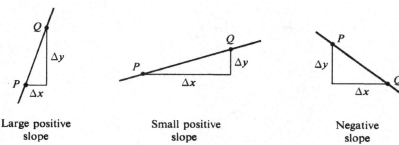

Large positive   Small positive   Negative
slope   slope   slope

Figure 1.3.4

There is exactly one line L passing through two distinct points $P(x_1, y_1)$ and $Q(x_2, y_2)$. If $x_1 \neq x_2$, we see from Theorem 1 that L has the equation

$$y - y_1 = \left(\frac{y_2 - y_1}{x_2 - x_1}\right)(x - x_1).$$

This is called the *two-point* equation for the line.

EXAMPLE 2   Given $P(3, 1)$ and $Q(1, 4)$, find the changes in x and y, the slope, and the equation of the line through P and Q. (See Figure 1.3.5.)

$$\Delta x = 1 - 3 = -2, \qquad \Delta y = 4 - 1 = 3.$$

The line through P and Q has slope $\Delta y/\Delta x = -\frac{3}{2}$, and its equation is

$$y - 1 = -\tfrac{3}{2}(x - 3).$$

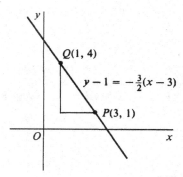

Figure 1.3.5

EXAMPLE 3   Given $P(1, -1)$ and $Q(1, 2)$, as in Figure 1.3.6,

$$\Delta x = 1 - 1 = 0, \qquad \Delta y = 2 - (-1) = 3.$$

The line through P and Q is the vertical line $x = 1$.

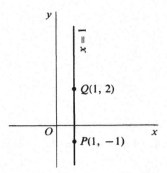

Figure 1.3.6

EXAMPLE 4 A particle moves along the y-axis with constant velocity. At time $t = 0$ sec, it is at the point $y = 3$ ft. At time $t = 2$ sec, it is at the point $y = 11$ ft. Find the velocity and the equation for the motion.

The velocity is defined as the distance moved divided by the time elapsed, so the velocity is

$$v = \frac{\Delta y}{\Delta t} = \frac{11 - 3}{2 - 0} = 4 \text{ ft/sec.}$$

If the motion of the particle is plotted in the (t, y) plane as in Figure 1.3.7,

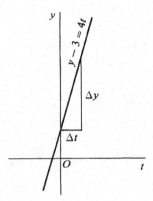

Figure 1.3.7

the result is a line through the points $P(0, 3)$ and $Q(2, 11)$. The velocity, being the ratio of Δy to Δt, is just the slope of this line. The line has the equation

$$y - 3 = 4t.$$

Suppose a particle moving with constant velocity is at the point $y = y_1$ at time $t = t_1$, and at the point $y = y_2$ at time $t = t_2$. Then the velocity is $v = \Delta y/\Delta t$. The motion of the particle plotted on the (t, y) plane is the line passing through the two points (t_1, y_1) and (t_2, y_2), and the velocity is the slope of this line.

An equation of the form

$$Ax + By + C = 0$$

where A and B are not both zero is called a *linear equation*. The reason for this name is explained by the next theorem.

THEOREM 2

Every linear equation determines a line.

PROOF

Case 1 $B = 0$. The equation $Ax + C = 0$ can be solved for x, $x = -C/A$. This is a vertical line.

Case 2 $B \neq 0$. In this case, we can solve the given equation for y, and the result is

$$y = \frac{-Ax - C}{B}, \qquad y = -\frac{A}{B}x - \frac{C}{B}.$$

This is a line with slope $-A/B$ crossing the y-axis at $-C/B$.

EXAMPLE 5 Find the slope of the line $6x - 2y + 7 = 0$.
The answer is $m = -A/B = -6/(-2) = 3$.

To draw the graph of a linear equation, find two points on the line and draw the line through them with a ruler.

EXAMPLE 6 Draw the graph of the line $4x + 2y + 3 = 0$.
First solve for y as a function of x:

$$y = -2x - \tfrac{3}{2}.$$

Next select any two values for x, say $x = 0$ and $x = 1$, and compute the corresponding values of y.

When $\qquad\qquad x = 0, \qquad y = -\tfrac{3}{2}$.

When $\qquad\qquad x = 1, \qquad y = -\tfrac{7}{2}$.

Finally, plot the two points $(0, -\tfrac{3}{2})$ and $(1, -\tfrac{7}{2})$, and draw the line through them. (See Figure 1.3.8.)

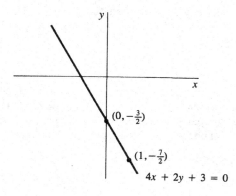

Figure 1.3.8

PROBLEMS FOR SECTION 1.3

In Problems 1–8, find the slope and equation of the line through P and Q.

1	$P(1, 2), \quad Q(3, 4)$	2	$P(1, -3), \quad Q(0, 2)$
3	$P(-4, 1), \quad Q(-4, 2)$	4	$P(2, 5), \quad Q(2, 7)$
5	$P(3, 0), \quad Q(0, 1)$	6	$P(0, 0), \quad Q(10, 4)$
7	$P(1, 3), \quad Q(3, 3)$	8	$P(6, -2), \quad Q(1, -2)$

In Problems 9–16, find the equation of the line with slope m through the point P.

9	$m = 2, \quad P(3, 3)$	10	$m = 3, \quad P(-2, 1)$
11	$m = -\frac{1}{2}, \quad P(1, -4)$	12	$m = -1, \quad P(2, 4)$
13	$m = 5, \quad P(0, 0)$	14	$m = -2, \quad P(0, 0)$
15	$m = 0, \quad P(7, 4)$	16	vertical line, $\quad P(4, 5)$

In Problems 17–22, a particle moves with constant velocity and has the given positions y at the given times t. Find the velocity and the equation of motion.

17    $y = 0$ at $t = 0, \quad y = 2$ at $t = 1$

18    $y = 3$ at $t = 0, \quad y = 1$ at $t = 2$

19    $y = 4$ at $t = 1, \quad y = 2$ at $t = 5$

20    $y = 1$ at $t = 2, \quad y = 3$ at $t = 3$

21    $y = 4$ at $t = 0, \quad y = 4$ at $t = 1$

22    $y = 1$ at $t = 3, \quad y = -2$ at $t = 6$

23   A particle moves with constant velocity 3, and at time $t = 2$ is at the point $y = 8$. Find the equation for its motion.

24   A particle moves with constant velocity $\frac{1}{4}$, and at time $t = 0$ is at $y = 1$. Find the equation for its motion.

In Problems 25–30, find the slope of the line with the given equation, and draw the line.

25	$3x - 2y + 5 = 0$	26	$x + y - 1 = 0$
27	$2x - y = 0$	28	$6x + 2y = 0$
29	$3x + 4y = 6$	30	$-2x + 4y = -1$

31   Show that the line that crosses the x-axis at $a \neq 0$ and the y-axis at $b \neq 0$ has the equation $(x/a) + (y/b) - 1 = 0$.

32   What is the equation of the line through the origin with slope m?

33   Find the points at which the line $ax + by + c = 0$ crosses the x- and y-axes. (Assume that $a \neq 0$ and $b \neq 0$.)

34   Let C denote Celsius temperature and F Fahrenheit temperature. Thus, $C = 0$ and $F = 32$ at the freezing point of water, while $C = 100$ and $F = 212$ at the boiling point of water. Use the two-point formula to find the linear equation relating C and F.

1.4 SLOPE AND VELOCITY; THE HYPERREAL LINE

In Section 1.3 the slope of the line through the points (x_1, y_1) and (x_2, y_2) is shown to be the ratio of the change in y to the change in x,

$$\text{slope} = \frac{\Delta y}{\Delta x} = \frac{y_2 - y_1}{x_2 - x_1}.$$

If the line has the equation

$$y = mx + b,$$

then the constant m is the slope.

What is meant by the slope of a *curve*? The differential calculus is needed to answer this question, as well as to provide a method of computing the value of the slope. We shall do this in the next chapter. However, to provide motivation, we now describe intuitively the method of finding the slope.

Consider the parabola

$$y = x^2.$$

The slope will measure the direction of a curve just as it measures the direction of a line. The slope of this curve will be different at different points on the x-axis, because the direction of the curve changes.

If (x_0, y_0) and $(x_0 + \Delta x, y_0 + \Delta y)$ are two points on the curve, then the "average slope" of the curve between these two points is defined as the ratio of the change in y to the change in x,

$$\text{average slope} = \frac{\Delta y}{\Delta x}.$$

This is exactly the same as the slope of the straight line through the points (x_0, y_0) and $(x_0 + \Delta x, y_0 + \Delta y)$, as shown in Figure 1.4.1.

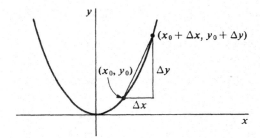

Figure 1.4.1

Let us compute the average slope. The two points (x_0, y_0) and $(x_0 + \Delta x, y_0 + \Delta y)$ are on the curve, so

$$y_0 = x_0^2,$$

$$y_0 + \Delta y = (x_0 + \Delta x)^2.$$

Subtracting, $\Delta y = (x_0 + \Delta x)^2 - x_0^2.$

Dividing by Δx, $\dfrac{\Delta y}{\Delta x} = \dfrac{(x_0 + \Delta x)^2 - x_0^2}{\Delta x}.$

This can be simplified,

$$\frac{\Delta y}{\Delta x} = \frac{x_0^2 + 2x_0\,\Delta x + (\Delta x)^2 - x_0^2}{\Delta x}$$

$$= \frac{2x_0\,\Delta x + (\Delta x)^2}{\Delta x} = 2x_0 + \Delta x.$$

Thus the average slope is

$$\frac{\Delta y}{\Delta x} = 2x_0 + \Delta x.$$

Notice that this computation can only be carried out when $\Delta x \neq 0$, because at $\Delta x = 0$ the quotient $\Delta y/\Delta x$ is undefined.

Reasoning in a nonrigorous way, the actual slope of the curve at the point (x_0, y_0) can be found thus. Let Δx be very small (but not zero). Then the point $(x_0 + \Delta x, y_0 + \Delta y)$ is close to (x_0, y_0), so the average slope between these two points is close to the slope of the curve at (x_0, y_0);

$$[\text{slope at } (x_0, y_0)] \text{ is close to } 2x_0 + \Delta x.$$

We neglect the term Δx because it is very small, and we are left with

$$[\text{slope at } (x_0, y_0)] = 2x_0.$$

For example, at the point $(0, 0)$ the slope is zero, at the point $(1, 1)$ the slope is 2, and at the point $(-3, 9)$ the slope is -6. (See Figure 1.4.2.)

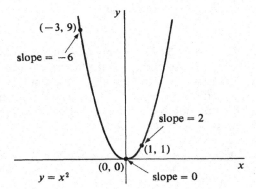

Figure 1.4.2 $y = x^2$

The whole process can also be visualized in another way. Let t represent time, and suppose a particle is moving along the y-axis according to the equation $y = t^2$. That is, at each time t the particle is at the point t^2 on the y-axis. We then ask: what is meant by the *velocity* of the particle at time t_0? Again we have the difficulty that the velocity is different at different times, and the calculus is needed to answer the question in a satisfactory way. Let us consider what happens to the particle between a time t_0 and a later time $t_0 + \Delta t$. The time elapsed is Δt, and the distance moved is $\Delta y = 2t_0 \Delta t + (\Delta t)^2$. If the velocity were constant during the entire interval of time, then it would just be the ratio $\Delta y/\Delta t$. However, the velocity is changing during the time interval. We shall call the ratio $\Delta y/\Delta t$ of the distance moved to the time elapsed the "average velocity" for the interval;

$$v_{\text{ave}} = \frac{\Delta y}{\Delta t} = 2t_0 + \Delta t.$$

The average velocity is not the same as the velocity at time t_0 which we are after. As a matter of fact, for $t_0 > 0$, the particle is speeding up; the velocity at time t_0 will be somewhat less than the average velocity for the interval of time between t_0 and $t_0 + \Delta t$, and the velocity at time $t_0 + \Delta t$ will be somewhat greater than the average.

But for a very small increment of time Δt, the velocity will change very little, and the average velocity $\Delta y/\Delta t$ will be close to the velocity at time t_0. To get the velocity v_0 at time t_0, we neglect the small term Δt in the formula

$$v_{\text{ave}} = 2t_0 + \Delta t,$$

and we are left with the value

$$v_0 = 2t_0.$$

When we plot y against t, the velocity is the same as the slope of the curve $y = t^2$, and the average velocity is the same as the average slope.

The trouble with the above intuitive argument, whether stated in terms of slope or velocity, is that it is not clear when something is to be "neglected." Nevertheless, the basic idea can be made into a useful and mathematically sound method of finding the slope of a curve or the velocity. What is needed is a sharp distinction between numbers which are small enough to be neglected and numbers which aren't. Actually, no real number except zero is small enough to be neglected. To get around this difficulty, we take the bold step of introducing a new kind of number, which is infinitely small and yet not equal to zero.

A number ε is said to be *infinitely small*, or *infinitesimal*, if

$$-a < \varepsilon < a$$

for every positive real number a. Then the only *real* number that is infinitesimal is zero. We shall use a new number system called the *hyperreal numbers*, which contains all the real numbers and also has infinitesimals that are not zero. Just as the real numbers can be constructed from the rational numbers, the hyperreal numbers can be constructed from the real numbers. This construction is sketched in the Epilogue at the end of the book. In this chapter, we shall simply list the properties of the hyperreal numbers needed for the calculus.

First we shall give an intuitive picture of the hyperreal numbers and show how they can be used to find the slope of a curve. The set of all hyperreal numbers is denoted by R^*. Every real number is a member of R^*, but R^* has other elements too. The infinitesimals in R^* are of three kinds: positive, negative, and the real number 0. The symbols $\Delta x, \Delta y, \ldots$ and the Greek letters ε (epsilon) and δ (delta) will be used for infinitesimals. If a and b are hyperreal numbers whose difference $a - b$ is infinitesimal, we say that a is *infinitely close to* b. For example, if Δx is infinitesimal then $x_0 + \Delta x$ is infinitely close to x_0. If ε is positive infinitesimal, then $-\varepsilon$ will be a negative infinitesimal. $1/\varepsilon$ will be an *infinite positive number*, that is, it will be greater than any real number. On the other hand, $-1/\varepsilon$ will be an *infinite negative number*, i.e., a number less than every real number. Hyperreal numbers which are not infinite numbers are called *finite numbers*. Figure 1.4.3 shows a drawing of the hyperreal line. The circles represent "infinitesimal microscopes" which are powerful enough to show an infinitely small portion of the hyperreal line. The set R of real numbers is scattered among the finite numbers. About each real number c is a portion of the hyperreal line composed of the numbers infinitely close to c (shown under an infinitesimal microscope for $c = 0$ and $c = 100$). The numbers infinitely close to 0 are the infinitesimals.

In Figure 1.4.3 the finite and infinite parts of the hyperreal line were separated from each other by a dotted line. Another way to represent the infinite parts of the hyperreal line is with an "infinite telescope" as in Figure 1.4.4. The field of view of an infinite telescope has the same scale as the finite portion of the hyperreal line, while the field of view of an infinitesimal microscope contains an infinitely small portion of the hyperreal line blown up.

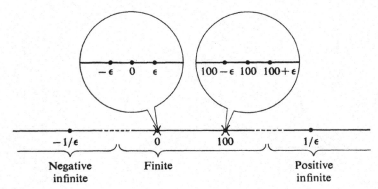

Figure 1.4.3

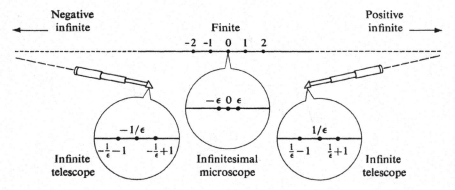

Figure 1.4.4

We have no way of knowing what a line in physical space is really like. It might be like the hyperreal line, the real line, or neither. However, in applications of the calculus it is helpful to imagine a line in physical space as a hyperreal line. The hyperreal line is, like the real line, a useful mathematical model for a line in physical space.

The hyperreal numbers can be algebraically manipulated just like the real numbers. Let us try to use them to find slopes of curves. We begin with the parabola $y = x^2$.

Consider a real point (x_0, y_0) on the curve $y = x^2$. Let Δx be either a positive or a negative infinitesimal (but not zero), and let Δy be the corresponding change in y. Then the slope at (x_0, y_0) is *defined* in the following way:

$$[\text{slope at } (x_0, y_0)] = \left[\text{the real number infinitely close to } \frac{\Delta y}{\Delta x} \right].$$

We compute $\dfrac{\Delta y}{\Delta x}$ as before: $\quad \dfrac{\Delta y}{\Delta x} = \dfrac{(x_0 + \Delta x)^2 - x_0^2}{\Delta x} = 2x_0 + \Delta x.$

This is a hyperreal number, not a real number. Since Δx is infinitesimal, the hyperreal number $2x_0 + \Delta x$ is infinitely close to the real number $2x_0$. We conclude that

$$[\text{slope at } (x_0, y_0)] = 2x_0.$$

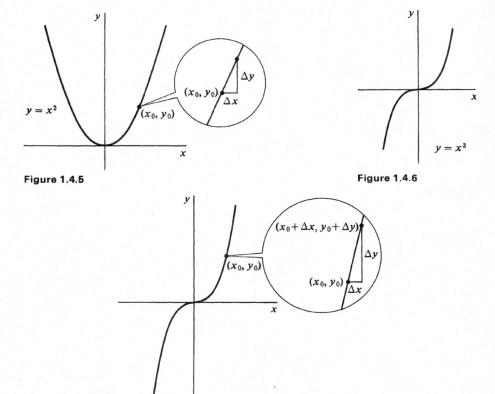

Figure 1.4.5

Figure 1.4.6

Figure 1.4.7

The process can be illustrated by the picture in Figure 1.4.5, with the infinitesimal changes Δx and Δy shown under a microscope.

The same method can be applied to other curves. The third degree curve $y = x^3$ is shown in Figure 1.4.6. Let (x_0, y_0) be any point on the curve $y = x^3$, and let Δx be a positive or a negative infinitesimal. Let Δy be the corresponding change in y along the curve. In Figure 1.4.7, Δx and Δy are shown under a microscope. We again define the slope at (x_0, y_0) by

$$[\text{slope at } (x_0, y_0)] = \left[\text{the real number infinitely close to } \frac{\Delta y}{\Delta x}\right].$$

We now compute the hyperreal number $\dfrac{\Delta y}{\Delta x}$.

$$y_0 = x_0^3,$$
$$y_0 + \Delta y = (x_0 + \Delta x)^3,$$
$$\Delta y = (x_0 + \Delta x)^3 - x_0^3,$$
$$\frac{\Delta y}{\Delta x} = \frac{(x_0 + \Delta x)^3 - x_0^3}{\Delta x}$$
$$= \frac{x_0^3 + 3x_0^2 \Delta x + 3x_0(\Delta x)^2 + (\Delta x)^3 - x_0^3}{\Delta x}$$
$$= \frac{3x_0^2 \Delta x + 3x_0(\Delta x)^2 + (\Delta x)^3}{\Delta x},$$

and finally
$$\frac{\Delta y}{\Delta x} = 3x_0^2 + 3x_0\,\Delta x + (\Delta x)^2.$$

In the next section we shall develop some rules about infinitesimals which will enable us to show that since Δx is infinitesimal,

$$3x_0\,\Delta x + (\Delta x)^2$$

is infinitesimal as well. Therefore the hyperreal number

$$3x_0^2 + 3x_0\,\Delta x + (\Delta x)^2$$

is infinitely close to the real number $3x_0^2$, whence

$$[\text{slope at } (x_0, y_0)] = 3x_0^2.$$

For example, at $(0, 0)$ the slope is zero, at $(1, 1)$ the slope is 3, and at $(2, 8)$ the slope is 12.

We shall return to the study of the slope of a curve in Chapter 2 after we have learned more about hyperreal numbers. From the last example it is evident that we need to know how to show that two numbers are infinitely close to each other. This is our next topic.

1.5 INFINITESIMAL, FINITE, AND INFINITE NUMBERS

Let us summarize our intuitive description of the hyperreal numbers from Section 1.4. The real line is a subset of the hyperreal line; that is, each real number belongs to the set of hyperreal numbers. Surrounding each real number r, we introduce a collection of hyperreal numbers infinitely close to r. The hyperreal numbers infinitely close to zero are called infinitesimals. The reciprocals of nonzero infinitesimals are infinite hyperreal numbers. The collection of all hyperreal numbers satisfies the same algebraic laws as the real numbers. In this section we describe the hyperreal numbers more precisely and develop a facility for computation with them.

This entire calculus course is developed from three basic principles relating the real and hyperreal numbers: the Extension Principle, the Transfer Principle, and the Standard Part Principle. The first two principles are presented in this section, and the third principle is in the next section.

We begin with the Extension Principle, which gives us new numbers called hyperreal numbers and extends all real functions to these numbers. The Extension Principle will deal with *hyperreal functions* as well as real functions. Our discussion of real functions in Section 1.2 can readily be carried over to hyperreal functions. Recall that for each real number a, a real function f of one variable either associates another real number $b = f(a)$ or is undefined. Now, for each hyperreal number H, a hyperreal function F of one variable either associates another hyperreal number $K = F(H)$ or is undefined. For each pair of hyperreal numbers H and J, a hyperreal function G of two variables either associates another hyperreal number $K = G(H, J)$ or is undefined. Hyperreal functions of three or more variables are defined in a similar way.

I. THE EXTENSION PRINCIPLE

(a) *The real numbers form a subset of the hyperreal numbers, and the order relation $x < y$ for the real numbers is a subset of the order relation for the hyperreal numbers.*

(b) *There is a hyperreal number that is greater than zero but less than every positive real number.*

(c) *For every real function f of one or more variables we are given a corresponding hyperreal function f* of the same number of variables. f* is called the natural extension of f.*

Part (a) of the Extension Principle says that the real line is a part of the hyperreal line. To explain part (b) of the Extension Principle, we give a careful definition of an infinitesimal.

DEFINITION

A hyperreal number b is said to be:

positive infinitesimal *if b is positive but less than every positive real number.*

negative infinitesimal *if b is negative but greater than every negative real number.*

infinitesimal *if b is either positive infinitesimal, negative infinitesimal, or zero.*

With this definition, part (b) of the Extension Principle says that there is at least one positive infinitesimal. We shall see later that there are infinitely many positive infinitesimals. A positive infinitesimal is a hyperreal number but cannot be a real number, so part (b) ensures that there are hyperreal numbers that are not real numbers.

Part (c) of the Extension Principle allows us to apply real functions to hyperreal numbers. Since the addition function $+$ is a real function of two variables, its natural extension $+^*$ is a hyperreal function of two variables. If x and y are hyperreal numbers, the sum of x and y is the number $x +^* y$ formed by using the natural extension of $+$. Similarly, the product of x and y is the number $x \cdot^* y$ formed by using the natural extension of the product function $\cdot$. To make things easier to read, we shall drop the asterisks and write simply $x + y$ and $x \cdot y$ for the sum and product of two hyperreal numbers x and y. Using the natural extensions of the sum and product functions, we will be able to develop algebra for hyperreal numbers. Part (c) of the Extension Principle also allows us to work with expressions such as $\cos(x)$ or $\sin(x + \cos(y))$, which involve one or more real functions. We call such expressions *real expressions*. These expressions can be used even when x and y are hyperreal numbers instead of real numbers. For example, when x and y are hyperreal, $\sin(x + \cos(y))$ will mean $\sin^*(x + \cos^*(y))$, where $\sin^*$ and $\cos^*$ are the natural extensions of sin and cos. The asterisks are dropped as before.

We now state the Transfer Principle, which allows us to carry out computations with the hyperreal numbers in the same way as we do for real numbers. Intuitively, the Transfer Principle says that the natural extension of each real function has the same properties as the original function.

II. TRANSFER PRINCIPLE

Every real statement that holds for one or more particular real functions holds for the hyperreal natural extensions of these functions.

Here are seven examples that illustrate what we mean by a *real statement*. In general, by a real statement we mean a combination of equations or inequalities about real expressions, and statements specifying whether a real expression is defined or undefined. A real statement will involve real variables and particular real functions.

(1) Closure law for addition: for any x and y, the sum $x + y$ is defined.
(2) Commutative law for addition: $x + y = y + x$.
(3) A rule for order: If $0 < x < y$, then $0 < 1/y < 1/x$.
(4) Division by zero is never allowed: $x/0$ is undefined.
(5) An algebraic identity: $(x - y)^2 = x^2 - 2xy + y^2$.
(6) A trigonometric identity: $\sin^2 x + \cos^2 x = 1$.
(7) A rule for logarithms: If $x > 0$ and $y > 0$, then $\log_{10}(xy) = \log_{10} x + \log_{10} y$.

Each example has two variables, x and y, and holds true whenever x and y are real numbers. The Transfer Principle tells us that each example also holds whenever x and y are hyperreal numbers. For instance, by Example (4), $x/0$ is undefined, even for hyperreal x. By Example (6), $\sin^2 x + \cos^2 x = 1$, even for hyperreal x.

Notice that the first five examples involve only the sum, difference, product, and quotient functions. However, the last two examples are real statements involving the transcendental functions sin, cos, and $\log_{10}$. The Transfer Principle extends all the familiar rules of trigonometry, exponents, and logarithms to the hyperreal numbers.

In calculus we frequently make a computation involving one or more unknown real numbers. The Transfer Principle allows us to compute in exactly the same way with hyperreal numbers. It "transfers" facts about the real numbers to facts about the hyperreal numbers. In particular, the Transfer Principle implies that a real function and its natural extension always give the same value when applied to a real number. This is why we are usually able to drop the asterisks when computing with hyperreal numbers.

A real statement is often used to define a new real function from old real functions. By the Transfer Principle, whenever a real statement defines a real function, the same real statement also defines the hyperreal natural extension function. Here are three more examples.

(8) The square root function is defined by the real statement $y = \sqrt{x}$ if, and only if, $y^2 = x$ and $y \geq 0$.
(9) The absolute value function is defined by the real statement $y = |x|$ if, and only if, $y = \sqrt{x^2}$.
(10) The common logarithm function is defined by the real statement $y = \log_{10} x$ if, and only if, $10^y = x$.

In each case, the hyperreal natural extension is the function defined by the given real statement when x and y vary over the hyperreal numbers. For example, the hyperreal natural extension of the square root function, $\sqrt{}$ *, is defined by Example (8) when x and y are hyperreal.

An important use of the Transfer Principle is to carry out computations with infinitesimals. For example, a computation with infinitesimals was used in the slope calculation in Section 1.4. The Extension Principle tells us that there is at least one positive infinitesimal hyperreal number, say ε. Starting from ε, we can use the Transfer Principle to construct infinitely many other positive infinitesimals. For example, ε^2 is a positive infinitesimal that is smaller than ε, $0 < \varepsilon^2 < \varepsilon$. (This follows from the Transfer Principle because $0 < x^2 < x$ for all real x between 0 and 1.) Here are several positive infinitesimals, listed in increasing order:

$$\varepsilon^3, \ \varepsilon^2, \ \varepsilon/100, \ \varepsilon, \ 75\varepsilon, \ \sqrt{\varepsilon}, \ \varepsilon + \sqrt{\varepsilon}.$$

We can also construct negative infinitesimals, such as $-\varepsilon$ and $-\varepsilon^2$, and other hyperreal numbers such as $1 + \sqrt{\varepsilon}$, $(10 - \varepsilon)^2$, and $1/\varepsilon$.

We shall now give a list of rules for deciding whether a given hyperreal number is infinitesimal, finite, or infinite. All these rules follow from the Transfer Principle alone. First, look at Figure 1.5.1, illustrating the hyperreal line.

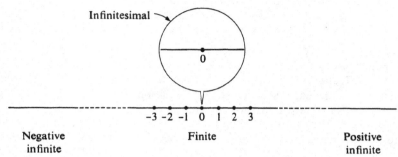

Infinitesimal

0

-3 -2 -1 0 1 2 3

Negative   Finite   Positive
infinite   infinite

Figure 1.5.1

DEFINITION

> *A hyperreal number b is said to be:*
>
> *finite if b is between two real numbers.*
>
> *positive infinite if b is greater than every real number.*
>
> *negative infinite if b is less than every real number.*

Notice that each infinitesimal number is finite. Before going through the whole list of rules, let us take a close look at two of them.

If ε is infinitesimal and a is finite, then the product $a \cdot \varepsilon$ is infinitesimal. For example, $\frac{1}{2}\varepsilon$, -6ε, 1000ε, $(5 - \varepsilon)\varepsilon$ are infinitesimal. This can be seen intuitively from Figure 1.5.2; an infinitely thin rectangle of length a has infinitesimal area.

If ε is positive infinitesimal, then $1/\varepsilon$ is positive infinite. From experience we know that reciprocals of small numbers are large, so we intuitively expect $1/\varepsilon$ to be positive infinite. We can use the Transfer Principle to prove $1/\varepsilon$ is positive infinite. Let r be any positive real number. Since ε is positive infinitesimal, $0 < \varepsilon < 1/r$. Applying the Transfer Principle, $1/\varepsilon > r > 0$. Therefore, $1/\varepsilon$ is positive infinite.

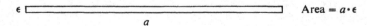

ε Area $= a \cdot \varepsilon$

a

Figure 1.5.2

RULES FOR INFINITESIMAL, FINITE, AND INFINITE NUMBERS   *Assume that ε, δ are infinitesimals; b, c are hyperreal numbers that are finite but not infinitesimal; and H, K are infinite hyperreal numbers.*

(i)   *Real numbers:*
The only infinitesimal real number is 0.
Every real number is finite.

(ii)   *Negatives:*
$-\varepsilon$ is infinitesimal.

$-b$ *is finite but not infinitesimal.*
$-H$ *is infinite.*

(iii) ***Reciprocals:***
If $\varepsilon \neq 0$, $1/\varepsilon$ *is infinite.*
$1/b$ *is finite but not infinitesimal.*
$1/H$ *is infinitesimal.*

(iv) ***Sums:***
$\varepsilon + \delta$ *is infinitesimal.*
$b + \varepsilon$ *is finite but not infinitesimal.*
$b + c$ *is finite (possibly infinitesimal).*
$H + \varepsilon$ *and* $H + b$ *are infinite.*

(v) ***Products:***
$\delta \cdot \varepsilon$ *and* $b \cdot \varepsilon$ *are infinitesimal.*
$b \cdot c$ *is finite but not infinitesimal.*
$H \cdot b$ *and* $H \cdot K$ *are infinite.*

(vi) ***Quotients:***
ε/b, ε/H, *and* b/H *are infinitesimal.*
b/c *is finite but not infinitesimal.*
b/ε, H/ε, *and* H/b *are infinite, provided that* $\varepsilon \neq 0$.

(vii) ***Roots:***
If $\varepsilon > 0$, $\sqrt[n]{\varepsilon}$ *is infinitesimal.*
If $b > 0$, $\sqrt[n]{b}$ *is finite but not infinitesimal.*
If $H > 0$, $\sqrt[n]{H}$ *is infinite.*

Notice that we have given no rule for the following combinations:

ε/δ, the quotient of two infinitesimals.
H/K, the quotient of two infinite numbers.
$H\varepsilon$, the product of an infinite number and an infinitesimal.
$H + K$, the sum of two infinite numbers.

Each of these can be either infinitesimal, finite but not infinitesimal, or infinite, depending on what ε, δ, H, and K are. For this reason, they are called *indeterminate forms.*

Here are three very different quotients of infinitesimals.

$\dfrac{\varepsilon^2}{\varepsilon}$ is infinitesimal (equal to ε).

$\dfrac{\varepsilon}{\varepsilon}$ is finite but not infinitesimal (equal to 1).

$\dfrac{\varepsilon}{\varepsilon^2}$ is infinite $\left(\text{equal to } \dfrac{1}{\varepsilon}\right)$.

Table 1.5.1 on the following page shows the three possibilities for each indeterminate form. Here are some examples which show how to use our rules.

EXAMPLE 1 Consider $(b - 3\varepsilon)/(c + 2\delta)$. ε is infinitesimal, so -3ε is infinitesimal, and $b - 3\varepsilon$ is finite but not infinitesimal. Similarly, $c + 2\delta$ is finite but not infinitesimal. Therefore the quotient

$$\frac{b - 3\varepsilon}{c + 2\delta}$$

is finite but not infinitesimal.

Table 1.5.1

indeterminate form	Examples		
	infinitesimal	finite (equal to 1)	infinite
$\dfrac{\varepsilon}{\delta}$	$\dfrac{\varepsilon^2}{\varepsilon}$	$\dfrac{\varepsilon}{\varepsilon}$	$\dfrac{\varepsilon}{\varepsilon^2}$
$\dfrac{H}{K}$	$\dfrac{H}{H^2}$	$\dfrac{H}{H}$	$\dfrac{H^2}{H}$
$H\varepsilon$	$H \cdot \dfrac{1}{H^2}$	$H \cdot \dfrac{1}{H}$	$H^2 \cdot \dfrac{1}{H}$
$H + K$	$H + (-H)$	$(H + 1) + (-H)$	$H + H$

The next three examples are quotients of infinitesimals.

EXAMPLE 2   The quotient

$$\frac{5\varepsilon^4 - 8\varepsilon^3 + \varepsilon^2}{3\varepsilon}$$

is infinitesimal, provided $\varepsilon \neq 0$.

The given number is equal to

(1)
$$\tfrac{5}{3}\varepsilon^3 - \tfrac{8}{3}\varepsilon^2 + \tfrac{1}{3}\varepsilon.$$

We see in turn that $\varepsilon, \varepsilon^2, \varepsilon^3, \tfrac{1}{3}\varepsilon, -\tfrac{8}{3}\varepsilon^2, \tfrac{5}{3}\varepsilon^3$ are infinitesimal; hence the sum (1) is infinitesimal.

EXAMPLE 3   If $\varepsilon \neq 0$, the quotient

$$\frac{3\varepsilon^3 + \varepsilon^2 - 6\varepsilon}{2\varepsilon^2 + \varepsilon}$$

is finite but not infinitesimal.

Cancelling an ε from numerator and denominator, we get

(2)
$$\frac{3\varepsilon^2 + \varepsilon - 6}{2\varepsilon + 1}.$$

Since $3\varepsilon^2 + \varepsilon$ is infinitesimal while -6 is finite but not infinitesimal, the numerator

$$3\varepsilon^2 + \varepsilon - 6$$

is finite but not infinitesimal. Similarly, the denominator $2\varepsilon + 1$, and hence the quotient (2) is finite but not infinitesimal.

EXAMPLE 4   If $\varepsilon \neq 0$, the quotient

$$\frac{\varepsilon^4 - \varepsilon^3 + 2\varepsilon^2}{5\varepsilon^4 + \varepsilon^3}$$

is infinite.

We first note that the denominator $5\varepsilon^4 + \varepsilon^3$ is not zero because it can be written as a product of nonzero factors,

$$5\varepsilon^4 + \varepsilon^3 = \varepsilon \cdot \varepsilon \cdot \varepsilon \cdot (5\varepsilon + 1).$$

When we cancel ε^2 from the numerator and denominator we get

$$\frac{\varepsilon^2 - \varepsilon + 2}{5\varepsilon^2 + \varepsilon}.$$

We see in turn that:

$\varepsilon^2 - \varepsilon + 2$ is finite but not infinitesimal,

$5\varepsilon^2 + \varepsilon$ is infinitesimal,

$\dfrac{\varepsilon^2 - \varepsilon + 2}{5\varepsilon^2 + \varepsilon}$ is infinite.

EXAMPLE 5 $\quad \dfrac{2H^2 + H}{H^2 - H + 2}$ is finite but not infinitesimal.

In this example the trick is to multiply both numerator and denominator by $1/H^2$. We get

$$\frac{2 + 1/H}{1 - 1/H + 2/H^2}.$$

Now $1/H$ and $1/H^2$ are infinitesimal. Therefore both the numerator and denominator are finite but not infinitesimal, and so is the quotient.

In the next theorem we list facts about the ordering of the hyperreals.

THEOREM 1

(i) *Every hyperreal number which is between two infinitesimals is infinitesimal.*

(ii) *Every hyperreal number which is between two finite hyperreal numbers is finite.*

(iii) *Every hyperreal number which is greater than some positive infinite number is positive infinite.*

(iv) *Every hyperreal number which is less than some negative infinite number is negative infinite.*

All the proofs are easy. We prove (iii), which is especially useful. Assume H is positive infinite and $H < K$. Then for any real number r, $r < H < K$. Therefore, $r < K$ and K is positive infinite.

EXAMPLE 6 If H and K are positive infinite hyperreal numbers, then $H + K$ is positive infinite. This is true because $H + K$ is greater than H.

Our last example concerns square roots.

EXAMPLE 7 If H is positive infinite then, surprisingly,

$$\sqrt{H+1} - \sqrt{H-1}$$

is infinitesimal.

This is shown using an algebraic trick.

$$\sqrt{H+1} - \sqrt{H-1} = \frac{(\sqrt{H+1} - \sqrt{H-1})(\sqrt{H+1} + \sqrt{H-1})}{\sqrt{H+1} + \sqrt{H-1}}$$

$$= \frac{(H+1) - (H-1)}{\sqrt{H+1} + \sqrt{H-1}} = \frac{2}{\sqrt{H+1} + \sqrt{H-1}}.$$

The numbers $H + 1, H - 1$, and their square roots are positive infinite, and thus the sum $\sqrt{H+1} + \sqrt{H-1}$ is positive infinite. Therefore the quotient

$$\sqrt{H+1} - \sqrt{H-1} = \frac{2}{\sqrt{H+1} + \sqrt{H-1}},$$

a finite number divided by an infinite number, is infinitesimal.

PROBLEMS FOR SECTION 1.5

In Problems 1–40, assume that: ε, δ are positive infinitesimal, H, K are positive infinite. Determine whether the given expression is infinitesimal, finite but not infinitesimal, or infinite.

1	$76{,}000{,}000\varepsilon$	**2**	$3\varepsilon + 4\delta$
3	$1 + 1/\varepsilon$	**4**	$3\varepsilon^3 - 2\varepsilon^2 + \varepsilon + 1$
5	$1/\sqrt{\varepsilon}$	**6**	ε/H
7	$H/1{,}000{,}000$	**8**	$(3 + \varepsilon)^2 - 9$
9	$(3 + \varepsilon)(4 + \delta) - 12$	**10**	$\dfrac{1 + \varepsilon + 3\varepsilon^2}{2 - \varepsilon - 8\varepsilon^3}$
11	$\dfrac{2\varepsilon^3 - \varepsilon^4}{4\varepsilon - \varepsilon^2 + \varepsilon^3}$	**12**	$\dfrac{2\varepsilon^3 - \varepsilon^4}{4\varepsilon^3 + \varepsilon^4}$
13	$\dfrac{3\varepsilon - 4\varepsilon^2}{\varepsilon^2 + 5\varepsilon^3}$	**14**	$\dfrac{\sqrt{\varepsilon} + \varepsilon}{\sqrt{\varepsilon} + 1}$
15	$\dfrac{1}{\sqrt{\varepsilon} - \varepsilon}$	**16**	$\dfrac{1}{\varepsilon} \cdot \sqrt{\varepsilon}$
17	$\dfrac{1}{\varepsilon} \cdot 5\varepsilon$	**18**	$\dfrac{1}{\varepsilon} \cdot \varepsilon^3$
19	$\dfrac{1}{\varepsilon}\left(\dfrac{1}{3 + \varepsilon} - \dfrac{1}{3}\right)$	**20**	$\dfrac{2H + 1}{3H + 2}$
21	$\dfrac{2H^4 + 3H - 6}{4H^3 + 5}$	**22**	$\dfrac{H + 4 + \varepsilon}{H^2 + 2\varepsilon}$
23	$\dfrac{H + K}{HK}$	**24**	$\dfrac{H - K}{H^2 + K^2}$
25	$H^2 - H$	**26**	$\sqrt{H+1} - \sqrt{H}$
27	$\left(H + \dfrac{1}{H}\right)^2 - \left(H - \dfrac{1}{H}\right)^2$	**28**	$\left(H + \dfrac{\varepsilon}{H}\right)^2 - \left(H - \dfrac{\varepsilon}{H}\right)^2$

29 $\dfrac{\sqrt{4 + \varepsilon} - 2}{\varepsilon}$ **30** $\dfrac{1}{\varepsilon}\left(1 - \dfrac{1}{\sqrt{1 + \varepsilon}}\right)$

31 $H\left(\sqrt{3 + \dfrac{1}{H}} - \sqrt{3}\right)$ **32** $\dfrac{\sqrt{H}}{\sqrt{H + 1} + \sqrt{H + 2}}$

33 $H(\sqrt{H + 2} - \sqrt{H})$ **34** $\dfrac{1 - \sqrt[3]{1 + \varepsilon}}{\varepsilon}$

35 $\sqrt[3]{H} - \sqrt[3]{H + 1}$ **36** $H - \sqrt{H + 1}\sqrt{H + 2}$

37 $\dfrac{(3 + \varepsilon)(4 + \delta) - 12}{\varepsilon\delta}$ **38** $\dfrac{5 + \varepsilon}{7 + \delta} - \dfrac{5}{7}$

39 $\dfrac{\varepsilon + \delta}{\sqrt{\varepsilon^2 + \delta^2}}$ **40** $\dfrac{H + K}{\sqrt{H^2 + K^2}}$

(*Hint:* Assume $\varepsilon \geq \delta$
and divide through by ε.)

41 In (a)–(f) below, determine which of the two numbers is greater.

  (a) ε or ε^2 (b) $\dfrac{1}{\varepsilon^3}$ or $\dfrac{1}{\varepsilon^4}$ (c) H or H^2

  (d) ε or $\sqrt{\varepsilon}$ (e) H or $\sqrt{H}$ (f) $\sqrt{H}$ or $\sqrt[3]{H}$

□ **42** Let x, y be positive hyperreal numbers. Can $\dfrac{x}{y} + \dfrac{y}{x}$ be infinite? Finite? Infinitesimal?

□ **43** Let a and b be real. When is $(3\varepsilon^2 - \varepsilon + a)/(4\varepsilon^2 + 2\varepsilon + b)$
  (a) infinitesimal?
  (b) finite but not infinitesimal?
  (c) infinite?

□ **44** Let a and b be real. When is $(aH^2 - 2H + 5)/(bH^2 + H - 2)$
  (a) infinitesimal?
  (b) finite but not infinitesimal?
  (c) infinite?

1.6 STANDARD PARTS

In this section we shall develop a method that will enable us to compute the slope of a curve by means of infinitesimals. We shall use the method to find slopes of curves in Chapter 2 and to find areas in Chapter 4. The key step will be to find the standard part of a given hyperreal number, that is, the real number that is infinitely close to it.

DEFINITION

*Two hyperreal numbers b and c are said to be **infinitely close** to each other, in symbols $b \approx c$, if their difference $b - c$ is infinitesimal. $b \not\approx c$ means that b is not infinitely close to c.*

Here are three simple remarks.

  (1) *If ε is infinitesimal, then $b \approx b + \varepsilon$.* This is true because the difference, $b - (b + \varepsilon) = -\varepsilon$, is infinitesimal.

(2)   *b is infinitesimal if and only if b $\approx$ 0.* The formula $b \approx 0$ will be used as a short way of writing "*b* is infinitesimal. "

(3)   *If b and c are real and b is infinitely close to c, then b equals c.*
$b - c$ is real and infinitesimal, hence zero; so $b = c$.

The relation $\approx$ between hyperreal numbers behaves somewhat like equality, but, of course, is not the same as equality. Here are three basic properties of $\approx$.

THEOREM 1

Let a, b and c be hyperreal numbers.

(i) $a \approx a$.
(ii)   *If $a \approx b$, then $b \approx a$.*
(iii)   *If $a \approx b$ and $b \approx c$, then $a \approx c$.*

These properties are useful when we wish to show that two numbers are infinitely close to each other.

The reason for (i) is that $a - a$ is an infinitesimal, namely zero. For (ii), we note that if $a - b$ is an infinitesimal ε, then $b - a = -\varepsilon$, which is also infinitesimal. Finally, (iii) is true because $a - c$ is the sum of two infinitesimals, namely $a - b$ and $b - c$.

THEOREM 2

Assume $a \approx b$. Then

(i)   *If a is infinitesimal, so is b.*
(ii)   *If a is finite, so is b.*
(iii)   *If a is infinite, so is b.*

The real numbers are sometimes called "standard" numbers, while the hyperreal numbers that are not real are called "nonstandard" numbers. For this reason, the real number that is infinitely close to *b* is called the "standard part" of *b*. An infinite number cannot have a standard part, because it can't be infinitely close to a finite number (Theorem 2). Our third principle (stated next) on hyperreal numbers is that every finite number has a standard part.

III. STANDARD PART PRINCIPLE

Every finite hyperreal number is infinitely close to exactly one real number.

DEFINITION

*Let b be a finite hyperreal number. The **standard part** of b, denoted by st(b), is the real number which is infinitely close to b. Infinite hyperreal numbers do not have standard parts.*

Here are some facts that follow at once from the definition.

Let b be a finite hyperreal number.

(1) *st(b) is a real number.*
(2) *b = st(b) + ε for some infinitesimal ε.*
(3) *If b is real, then b = st(b).*

Our next aim is to develop some skill in computing standard parts. This will be one of the basic methods throughout the Calculus course. The next theorem is the principal tool.

THEOREM 3

Let a and b be finite hyperreal numbers. Then

(i) $st(-a) = -st(a)$.

(ii) $st(a + b) = st(a) + st(b)$.

(iii) $st(a - b) = st(a) - st(b)$.

(iv) $st(ab) = st(a) \cdot st(b)$.

(v)   *If $st(b) \neq 0$, then $st(a/b) = st(a)/st(b)$.*

(vi) $st(a^n) = (st(a))^n$.

(vii) *If $a \geq 0$, then $st(\sqrt[n]{a}) = \sqrt[n]{st(a)}$.*

(viii) *If $a \leq b$, then $st(a) \leq st(b)$.*

This theorem gives formulas for the standard parts of the simplest expressions.

All of the rules in Theorem 3 follow from our three principles for hyperreal numbers. As an illustration, let us prove the formula (iv) for $st(ab)$. Let r be the standard part of a and s the standard part of b, so that

$$a = r + ε, \qquad b = s + δ,$$

where $ε$ and $δ$ are infinitesimal. Then

$$ab = (r + ε)(s + δ)$$
$$= rs + rδ + sε + εδ \approx rs.$$

Therefore $$st(ab) = rs = st(a) \cdot st(b).$$

Often the symbols $\Delta x, \Delta y$, etc. are used for infinitesimals. In the following examples we use the rules in Theorem 3 as a starting point for computing standard parts of more complicated expressions.

EXAMPLE 1 When Δx is an infinitesimal and x is real, compute the standard part of

$$3x^2 + 3x \, \Delta x + (\Delta x)^2.$$

Using the rules in Theorem 3, we can write

$$st(3x^2 + 3x \, \Delta x + (\Delta x)^2) = st(3x^2) + st(3x \, \Delta x) + st((\Delta x)^2)$$
$$= 3x^2 + st(3x) \cdot st(\Delta x) + st(\Delta x)^2$$
$$= 3x^2 + 3x \cdot 0 + 0^2 = 3x^2.$$

EXAMPLE 2   If $st(c) = 4$ and $c \neq 4$, find

$$st\left(\frac{c^2 + 2c - 24}{c^2 - 16}\right).$$

We note that the denominator has standard part 0,

$$st(c^2 - 16) = st(c)^2 - 16 = 4^2 - 16 = 0.$$

However, since $c \neq 4$ the fraction is defined, and it can be simplified by factoring the numerator and denominator,

$$\frac{c^2 + 2c - 24}{c^2 - 16} = \frac{(c + 6)(c - 4)}{(c + 4)(c - 4)} = \frac{c + 6}{c + 4}.$$

Then
$$st\left(\frac{c^2 + 2c - 24}{c^2 - 16}\right) = st\left(\frac{c + 6}{c + 4}\right) = \frac{st(c + 6)}{st(c + 4)}$$
$$= \frac{st(c) + 6}{st(c) + 4} = \frac{4 + 6}{4 + 4} = \frac{10}{8}.$$

We now have three kinds of computation available to us. First, there are computations involving hyperreal numbers. In Example 2, the two steps giving

$$\frac{c^2 + 2c - 24}{c^2 - 16} = \frac{c + 6}{c + 4}$$

are computations of this kind. The computations of this first kind are justified by the Transfer Principle.

Second, we have computations which involve standard parts. In Example 2, the three steps giving

$$st\frac{c^2 + 2c - 24}{c^2 - 16} = \frac{st(c) + 6}{st(c) + 4}$$

are of this kind. This second kind of computation depends on Theorem 3.

Third there are computations with ordinary real numbers. Sometimes the real numbers will appear as standard parts. In Example 2, the last two steps which give

$$\frac{st(c) + 6}{st(c) + 4} = \frac{10}{8}$$

are computations with ordinary real numbers.

Usually, in computing the standard part of a hyperreal number, we use the first kind of computation, then the second kind, and then the third kind, in that order. We shall give two more somewhat different examples and pick out these three stages in the computations.

EXAMPLE 3   If H is a positive infinite hyperreal number, compute the standard part of

$$c = \frac{2H^3 + 5H^2 - 3H}{7H^3 - 2H^2 + 4H}.$$

In this example both the numerator and denominator are infinite, and we have to use the first type of computation to get the equation into a different form before we can take standard parts.

First stage

$$c = \frac{2H^3 + 5H^2 - 3H}{7H^3 - 2H^2 + 4H} = \frac{H^{-3} \cdot (2H^3 + 5H^2 - 3H)}{H^{-3} \cdot (7H^3 - 2H^2 + 4H)} = \frac{2 + 5H^{-1} - 3H^{-2}}{7 - 2H^{-1} + 4H^{-2}}.$$

Second stage H^{-1} and H^{-2} are infinitesimal, so

$$st(c) = st\left(\frac{2 + 5H^{-1} - 3H^{-2}}{7 - 2H^{-1} + 4H^{-2}}\right) = \frac{st(2 + 5H^{-1} - 3H^{-2})}{st(7 - 2H^{-1} + 4H^{-2})}$$

$$= \frac{st(2) + st(5H^{-1}) - st(3H^{-2})}{st(7) - st(2H^{-1}) + st(4H^{-2})} = \frac{2 + 0 - 0}{7 - 0 + 0}.$$

Third stage
$$st(c) = \frac{2 + 0 - 0}{7 - 0 + 0} = \frac{2}{7}.$$

EXAMPLE 4   If ε is infinitesimal but not zero, find the standard part of

$$b = \frac{\varepsilon}{5 - \sqrt{25 + \varepsilon}}.$$

Both the numerator and denominator are nonzero infinitesimals.

First stage   We multiply both numerator and denominator by $5 + \sqrt{25 + \varepsilon}$.

$$b = \frac{\varepsilon}{5 - \sqrt{25 + \varepsilon}} = \frac{\varepsilon(5 + \sqrt{25 + \varepsilon})}{(5 - \sqrt{25 + \varepsilon})(5 + \sqrt{25 + \varepsilon})}$$

$$= \frac{\varepsilon(5 + \sqrt{25 + \varepsilon})}{25 - (25 + \varepsilon)} = \frac{\varepsilon(5 + \sqrt{25 + \varepsilon})}{-\varepsilon}$$

$$= -5 - \sqrt{25 + \varepsilon}.$$

Second stage $st(b) = st(-5 - \sqrt{25 + \varepsilon}) = st(-5) - st(\sqrt{25 + \varepsilon})$

$$= -5 - \sqrt{st(25 + \varepsilon)} = -5 - \sqrt{25}.$$

Third stage $st(b) = -5 - \sqrt{25} = -10.$

EXAMPLE 5   Remember that infinite hyperreal numbers do not have standard parts. Consider the infinite hyperreal number

$$\frac{3 + \varepsilon}{4\varepsilon + \varepsilon^2},$$

where ε is a nonzero infinitesimal. The numerator and denominator have standard parts

$$st(3 + \varepsilon) = 3, \qquad st(4\varepsilon + \varepsilon^2) = 0.$$

However, the quotient has no standard part. In other words,

$$st\left(\frac{3 + \varepsilon}{4\varepsilon + \varepsilon^2}\right) \text{ is undefined.}$$

PROBLEMS FOR SECTION 1.6

Compute the standard parts of the following.

1 $2 + \varepsilon + 3\varepsilon^2,$ $\qquad\qquad\qquad \varepsilon$ infinitesimal

2    $b + 2\varepsilon - \varepsilon^2,$ $st(b) = 5,$ ε infinitesimal

3    $\dfrac{2 - 3\varepsilon}{5 + 4\varepsilon},$ ε infinitesimal

4    $y^4 + 2y^2\,\Delta y + \Delta y^3,$ y real, Δy infinitesimal

5    $(x^2 + 3x\,\Delta x + \Delta x^2)^6,$ x real, Δx infinitesimal

6    $\sqrt{x + \Delta x} + \sqrt{x - \Delta x},$ x positive real, Δx infinitesimal

7    $\dfrac{\varepsilon^3 - \varepsilon^2 + 4\varepsilon}{3\varepsilon^2 + 2\varepsilon - 3},$ ε infinitesimal

8    $\dfrac{\varepsilon^4 - \varepsilon^3 + \varepsilon^2}{2\varepsilon^2},$ $\varepsilon \neq 0$ infinitesimal

9    $\dfrac{4\varepsilon^4 - 3\varepsilon^3 + 2\varepsilon^2}{3\varepsilon^4 - 2\varepsilon^3 + \varepsilon^2},$ $\varepsilon \neq 0$ infinitesimal

10    $(2 + \varepsilon + \delta)(3 - \varepsilon\delta),$ ε, δ infinitesimal

11    $\sqrt{a + \varepsilon}\sqrt{a + \delta},$ $st(a) = 3,$ ε, δ infinitesimal

12    $\dfrac{2H + 4}{3H - 6},$ H infinite

13    $\dfrac{6H - 7}{H^2 + 2},$ H infinite

14    $\dfrac{3H^2 - 5H + 2}{H^2 + 1},$ H infinite

15    $\dfrac{H + 1 + \varepsilon}{2H - 1 + 3\varepsilon},$ H infinite, ε infinitesimal

16    $\dfrac{H^4 + 3H^2 + 1}{4H^4 + 2H^2 - 1},$ H infinite

17    $\dfrac{2b^2 + c + 1}{3c^2 + 6b + 1},$ $st(b) = 2,$ $st(c) = -1$

18    $\sqrt{b^2 + bc + b - c},$ $st(b) = 3,$ $st(c) = 2$

19    $\dfrac{(x + \varepsilon)(y + \varepsilon) - xy}{\varepsilon},$ x, y real, $\varepsilon \neq 0$ infinitesimal

20    $\dfrac{(x + \Delta x)^2 - x^2}{\Delta x},$ x real, $\Delta x \neq 0$ infinitesimal

21    $\dfrac{(x + \Delta x)^3 - x^3}{\Delta x},$ x real, $\Delta x \neq 0$ infinitesimal

22    $\dfrac{1/(a + \varepsilon) - 1/a}{\varepsilon},$ $a \neq 0$ real, $\varepsilon \neq 0$ infinitesimal

23    $\dfrac{b^2 - 25}{b - 5},$ $b \neq 5$ and $st(b) = 5$

24    $\dfrac{4 - a}{2 - \sqrt{a}},$ $a \neq 4$ and $st(a) = 4$

25    $\dfrac{3 - \sqrt{c + 2}}{c - 7},$ $c \neq 7$ and $st(c) = 7$

26    $\dfrac{3 - \sqrt{c + 2}}{c - 7},$ $st(c) = 5$

27    $\dfrac{a^2 - 5a + 6}{a - 3},$ $a \neq 3$ and $st(a) = 3$

28    $\dfrac{2b^2 - b - 6}{b^2 - 3b + 2}$, $b \neq 2$ and $st(b) = 2$

29    $\dfrac{c^2 + 5c + 6}{c^2 + 4c + 3}$, $c \neq -3$ and $st(c) = -3$

30    $\dfrac{\sqrt{25 - \varepsilon} - 5}{\varepsilon}$, $\varepsilon \neq 0$ and ε infinitesimal

31    $\dfrac{1}{\varepsilon}\left(\dfrac{1}{\sqrt{4 + \varepsilon}} - \dfrac{1}{2} \right)$, $\varepsilon \neq 0$ and ε infinitesimal

32    $2H\left(\sqrt{1 + \dfrac{1}{H}} - 1 \right)$, H positive infinite

33    $\dfrac{\sqrt{H + 1}}{\sqrt{2H + \sqrt{H - 1}}}$, H positive infinite

34    $\sqrt{H^2 + H + 1} - H$, H positive infinite

In the following problems let a, b, a_1, b_1 be hyperreal numbers with $a \approx a_1, b \approx b_1$.

☐ 35   Show that $a + b \approx a_1 + b_1$.
Hint: Put $a_1 = a + \varepsilon$, $b_1 = b + \delta$, and compute the difference $(a_1 + b_1) - (a + b)$.

☐ 36   Show that if a, b are finite, then $ab \approx a_1 b_1$.

☐ 37   Show that if $a = b = H$, $a_1 = b_1 = H + 1/H$, then $ab \not\approx a_1 b_1$. (H positive infinite).

EXTRA PROBLEMS FOR CHAPTER 1

1   Find the distance between the points $P(2, 7)$ and $Q(1, -4)$.

2   Find the slope of the line through the points $P(2, -6)$ and $Q(3, 4)$.

3   Find the slope of the line through $P(3, 5)$ and $Q(6, 0)$.

4   Find the equation of the line through $P(4, 4)$ and $Q(5, 9)$.

5   Find the equation of the line through $P(4, 5)$ with slope $m = -2$.

6   Find the velocity and equation of motion of a particle which moves with constant velocity and has positions $y = 2$ at $t = 0$, $y = 5$ at $t = 2$.

7   Find the equation of the circle with radius $\sqrt{5}$ and center at $(1, 3)$.

8   Find the equation of the circle that has center $(1, 0)$ and passes through the point $(0, 1)$.

Let ε be positive infinitesimal. Determine whether the following are infinitesimal, finite but not infinitesimal, or infinite.

9    $(4\varepsilon + 5)(2\varepsilon + 6)$ 10 $(4\varepsilon + 5)(\varepsilon^2 - \varepsilon)$

11    $1/\varepsilon - 2/\varepsilon^2$ 12 $1 - \sqrt{1 - \varepsilon}$

Let H be positive infinite. Determine whether the following are infinitesimal, finite but not infinitesimal, or infinite.

13    $(H - 2)(2H + 5)$ 14 $\dfrac{H - 2}{2H + 5}$

15    $\dfrac{H + 6}{H^2 - 2}$ 16 $\sqrt{H^2 + 1} - H$

Compute the standard parts in Problems 17–22.

17    $(b + 2\varepsilon)(3b - 4\varepsilon)$, $st(b) = 4$, ε infinitesimal

18    $\sqrt{2 + \varepsilon + 3\varepsilon^2}$, ε infinitesimal

19    $\dfrac{4 + 5\varepsilon}{7 - 3\varepsilon^2}$, ε infinitesimal

20    $\left(\dfrac{1}{3} - \dfrac{1}{3 + \Delta x}\right)\left(\dfrac{1}{\Delta x}\right)$, $0 \neq \Delta x$ infinitesimal

21    $\dfrac{(3H + 4)(5K + 6)}{(H + 1)(1 - 4K)}$, H, K positive infinite

22    $(\sqrt{H^2 + 4} - H)H$, H positive infinite

23   If $f(x) = 1/\sqrt{x}$, find $f(x + \Delta x) - f(x)$.

24   What is the domain of the function $f(x) = \dfrac{1}{x(x + 1)(x + 2)}$?

☐ 25   Show that if $a < b$, then $(a + b)/2$ is between a and b; that is, $a < (a + b)/2 < b$.

☐ 26   Show that every open interval has infinitely many points.

☐ 27   The *union* of two sets X and Y, $X \cup Y$, is the set of all x such that x is either in X or Y or both. Prove that the union of two bounded sets is bounded.

28   The *intersection* of X and Y, $X \cap Y$, is the set of all x such that x is in both X and Y. Prove that the intersection of two closed intervals is either empty or is a closed interval.

☐ 29   Prove that the intersection of two open intervals is either empty or is an open interval.

☐ 30   Prove that two (real) straight lines with different slopes intersect.

☐ 31   Prove that if H is infinite, then $1/H$ is infinitesimal.

☐ 32   Prove that if H is infinite and b is finite, then $H + b$ is infinite.

☐ 33   Prove that if ε is positive infinitesimal, so is $\sqrt[n]{\varepsilon}$.

☐ 34   Prove that if a, b are not infinitesimal and $a \approx b$, then $1/a \approx 1/b$.

☐ 35   Prove that if a is finite, then $st(|a|) = |st(a)|$.

☐ 36   Suppose a is finite, r is real, and $st(a) < r$. Prove that $a < r$.

☐ 37   Suppose a and b are finite hyperreal numbers with $st(a) < st(b)$. Prove that there is a real number r with $a < r < b$.

☐ 38   Suppose that f is a real function.

Show that the set of real solutions of the equation $f(x) = 0$ is bounded if and only if every hyperreal solution of $f^*(x) = 0$ is finite.

2

DIFFERENTIATION

2.1 DERIVATIVES

We are now ready to explain what is meant by the slope of a curve or the velocity of a moving point. Consider a real function f and a real number a in the domain of f. When x has value a, $f(x)$ has value $f(a)$. Now suppose the value of x is changed from a to a hyperreal number $a + \Delta x$ which is infinitely close to but not equal to a. Then the new value of $f(x)$ will be $f(a + \Delta x)$. In this process the value of x will be changed by a nonzero infinitesimal amount Δx, while the value of $f(x)$ will be changed by the amount

$$f(a + \Delta x) - f(a).$$

The ratio of the change in the value of $f(x)$ to the change in the value of x is

$$\frac{f(a + \Delta x) - f(a)}{\Delta x}.$$

This ratio is used in the definition of the slope of f which we now give.

DEFINITION

S is said to be the **slope** of f at a if

$$S = st\left(\frac{f(a + \Delta x) - f(a)}{\Delta x}\right)$$

for every nonzero infinitesimal Δx.

The slope, when it exists, is infinitely close to the ratio of the change in $f(x)$ to an infinitely small change in x. Given a curve $y = f(x)$, the slope of f at a is also called the slope of the curve $y = f(x)$ at $x = a$. Figure 2.1.1 shows a nonzero infinitesimal Δx and a hyperreal straight line through the two points on the curve at a and $a + \Delta x$. The quantity

$$\frac{f(a + \Delta x) - f(a)}{\Delta x}$$

is the slope of this line, and its standard part is the slope of the curve.

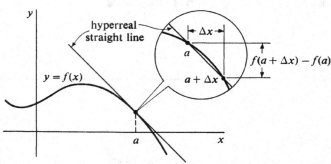

Figure 2.1.1

The slope of f at a does not always exist. Here is a list of all the possibilities.

(1) The slope of f at a exists if the ratio

$$\frac{f(a + \Delta x) - f(a)}{\Delta x}$$

is finite and has the same standard part for all infinitesimal $\Delta x \neq 0$. It has the value

$$S = st\left(\frac{f(a + \Delta x) - f(a)}{\Delta x}\right).$$

(2) The slope of f at a can fail to exist in any of four ways:

   (a) $f(a)$ is undefined.

   (b) $f(a + \Delta x)$ is undefined for some infinitesimal $\Delta x \neq 0$.

   (c) The term $\dfrac{f(a + \Delta x) - f(a)}{\Delta x}$ is infinite for some infinitesimal $\Delta x \neq 0$.

   (d) The term $\dfrac{f(a + \Delta x) - f(a)}{\Delta x}$ has different standard parts for different infinitesimals $\Delta x \neq 0$.

We can consider the slope of f at any point x, which gives us a new function of x.

DEFINITION

*Let f be a real function of one variable. The **derivative** of f is the new function f' whose value at x is the slope of f at x. In symbols,*

$$f'(x) = st\left(\frac{f(x + \Delta x) - f(x)}{\Delta x}\right)$$

whenever the slope exists.

The derivative $f'(x)$ is undefined if the slope of f does not exist at x.

For a given point a, the slope of f at a and the derivative of f at a are the same thing. We usually use the word "slope" to emphasize the geometric picture and "derivative" to emphasize the fact that f' is a function.

The process of finding the derivative of f is called *differentiation*. We say that f is *differentiable* at a if $f'(a)$ is defined; i.e., the slope of f at a exists.

Independent and dependent variables are useful in the study of derivatives. Let us briefly review what they are. A *system of formulas* is a finite set of equations and inequalities. If we are given a system of formulas which has the same graph as a simple equation $y = f(x)$, we say that y is *a function of x*, or that y *depends on x*, and we call x the *independent variable* and y the *dependent variable*.

When $y = f(x)$, we introduce a new independent variable Δx and a new dependent variable Δy, with the equation

(1) $$\Delta y = f(x + \Delta x) - f(x).$$

This equation determines Δy as a real function of the two variables x and Δx, when x and Δx vary over the real numbers. We shall usually want to use the Equation 1 for Δy when x is a real number and Δx is a nonzero infinitesimal. The Transfer Principle implies that Equation 1 also determines Δy as a hyperreal function of two variables when x and Δx are allowed to vary over the hyperreal numbers.

Δy is called the *increment* of y. Geometrically, the increment Δy is the change in y along the curve corresponding to the change Δx in x. The symbol y' is sometimes used for the derivative, $y' = f'(x)$. Thus the hyperreal equation

$$f'(x) = st\left(\frac{f(x + \Delta x) - f(x)}{\Delta x}\right)$$

now takes the short form

$$y' = st\left(\frac{\Delta y}{\Delta x}\right).$$

The infinitesimal Δx may be either positive or negative, but not zero. The various possibilities are illustrated in Figure 2.1.2 using an infinitesimal microscope. The signs of Δx and Δy are indicated in the captions.

Our rules for standard parts can be used in many cases to find the derivative of a function. There are two parts to the problem of finding the derivative f' of a function f:

(1) Find the domain of f'.
(2) Find the value of $f'(x)$ when it is defined.

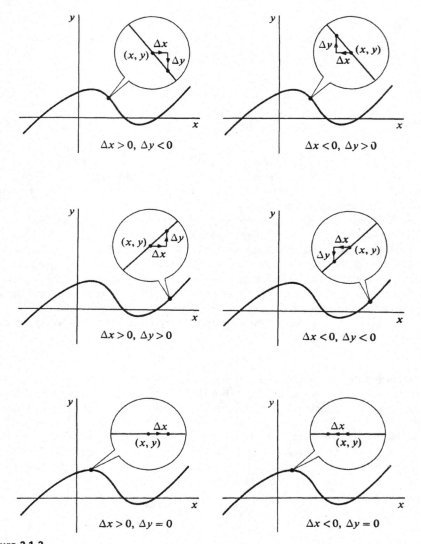

$\Delta x > 0, \ \Delta y < 0$

$\Delta x < 0, \ \Delta y > 0$

$\Delta x > 0, \ \Delta y > 0$

$\Delta x < 0, \ \Delta y < 0$

$\Delta x > 0, \ \Delta y = 0$

$\Delta x < 0, \ \Delta y = 0$

Figure 2.1.2

EXAMPLE 1   Find the derivative of the function

$$f(x) = x^3.$$

In this and the following examples we let x vary over the real numbers and Δx vary over the nonzero infinitesimals. Let us introduce the new variable y with the equation $y = x^3$. We first find $\Delta y / \Delta x$.

$$y = x^3,$$
$$y + \Delta y = (x + \Delta x)^3,$$
$$\Delta y = (x + \Delta x)^3 - x^3,$$
$$\frac{\Delta y}{\Delta x} = \frac{(x + \Delta x)^3 - x^3}{\Delta x}.$$

Next we simplify the expression for $\Delta y/\Delta x$.

$$\frac{\Delta y}{\Delta x} = \frac{(x^3 + 3x^2\,\Delta x + 3x(\Delta x)^2 + (\Delta x)^3) - x^3}{\Delta x}$$
$$= \frac{3x^2\,\Delta x + 3x(\Delta x)^2 + (\Delta x)^3}{\Delta x}$$
$$= 3x^2 + 3x\,\Delta x + (\Delta x)^2.$$

Then we take the standard part,

$$st\left(\frac{\Delta y}{\Delta x}\right) = st(3x^2 + 3x\,\Delta x + (\Delta x)^2)$$
$$= st(3x^2) + st(3x\,\Delta x) + st((\Delta x)^2)$$
$$= 3x^2 + 0 + 0 = 3x^2.$$

Therefore, $\quad\quad f'(x) = st\left(\frac{\Delta y}{\Delta x}\right) = 3x^2.$

We have shown that the derivative of the function

$$f(x) = x^3$$

is the function $\quad\quad f'(x) = 3x^2$

with the whole real line as domain. $f(x)$ and $f'(x)$ are shown in Figure 2.1.3.

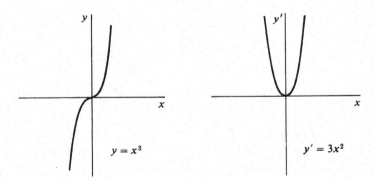

$y = x^3$

$y' = 3x^2$

Figure 2.1.3

EXAMPLE 2 Find $f'(x)$ given $f(x) = \sqrt{x}$.

Case 1 $x < 0$. Since $\sqrt{x}$ is not defined, $f'(x)$ does not exist.

Case 2 $x = 0$. When Δx is a negative infinitesimal, the term

$$\frac{\sqrt{x + \Delta x} - \sqrt{x}}{\Delta x} = \frac{\sqrt{0 + \Delta x} - \sqrt{0}}{\Delta x}$$

is not defined because $\sqrt{\Delta x}$ is undefined. When Δx is a positive infinitesimal, the term

$$\frac{\sqrt{x + \Delta x} - \sqrt{x}}{\Delta x} = \frac{\sqrt{\Delta x}}{\Delta x} = \frac{1}{\sqrt{\Delta x}}$$

is defined but its value is infinite. Thus for two reasons, $f'(x)$ does not exist.

Case 3 $x > 0$. Let $y = \sqrt{x}$. Then

$$y + \Delta y = \sqrt{x + \Delta x},$$
$$\Delta y = \sqrt{x + \Delta x} - \sqrt{x},$$
$$\frac{\Delta y}{\Delta x} = \frac{\sqrt{x + \Delta x} - \sqrt{x}}{\Delta x}.$$

We then make the computation

$$\frac{\Delta y}{\Delta x} = \frac{(\sqrt{x + \Delta x} - \sqrt{x})}{\Delta x} \cdot \frac{(\sqrt{x + \Delta x} + \sqrt{x})}{(\sqrt{x + \Delta x} + \sqrt{x})}$$

$$= \frac{(x + \Delta x) - x}{\Delta x(\sqrt{x + \Delta x} + \sqrt{x})}$$

$$= \frac{\Delta x}{\Delta x(\sqrt{x + \Delta x} + \sqrt{x})} = \frac{1}{\sqrt{x + \Delta x} + \sqrt{x}}.$$

Taking standard parts, $\quad st\left(\dfrac{\Delta y}{\Delta x}\right) = st\left(\dfrac{1}{\sqrt{x + \Delta x} + \sqrt{x}}\right)$

$$= \frac{1}{st(\sqrt{x + \Delta x} + \sqrt{x})}$$

$$= \frac{1}{st(\sqrt{x + \Delta x}) + st(\sqrt{x})}$$

$$= \frac{1}{\sqrt{x} + \sqrt{x}} = \frac{1}{2\sqrt{x}}.$$

Therefore, when $x > 0$, $\qquad f'(x) = \dfrac{1}{2\sqrt{x}}.$

So the derivative of $\qquad f(x) = \sqrt{x}$

is the function $\qquad f'(x) = \dfrac{1}{2\sqrt{x}},$

and the set of all $x > 0$ is its domain (see Figure 2.1.4).

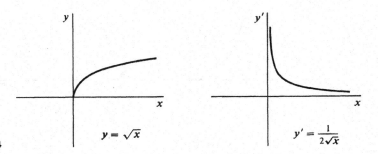

Figure 2.1.4

$y = \sqrt{x}$

$y' = \dfrac{1}{2\sqrt{x}}$

EXAMPLE 3 Find the derivative of $f(x) = 1/x$.

Case 1 $x = 0$. Then $1/x$ is undefined so $f'(x)$ is undefined.

Case 2 $x \neq 0$.

$$y = 1/x,$$

$$y + \Delta y = \frac{1}{x + \Delta x},$$

$$\Delta y = \frac{1}{x + \Delta x} - \frac{1}{x},$$

$$\frac{\Delta y}{\Delta x} = \frac{1/(x + \Delta x) - 1/x}{\Delta x}.$$

Simplifying,

$$\frac{1/(x + \Delta x) - 1/x}{\Delta x} = \frac{x - (x + \Delta x)}{x(x + \Delta x)\,\Delta x} = \frac{-\Delta x}{x(x + \Delta x)\,\Delta x}$$

$$= \frac{-1}{x(x + \Delta x)}.$$

Taking the standard part,

$$st\left(\frac{\Delta y}{\Delta x}\right) = st\left(-\frac{1}{x(x + \Delta x)}\right) = -\frac{1}{st(x(x + \Delta x))}$$

$$= -\frac{1}{st(x)st(x + \Delta x)} = -\frac{1}{x \cdot x} = -\frac{1}{x^2}.$$

Thus

$$f'(x) = -1/x^2.$$

The derivative of the function $f(x) = 1/x$ is the function $f'(x) = -1/x^2$ whose domain is the set of all $x \neq 0$. Both functions are graphed in Figure 2.1.5.

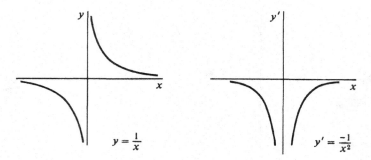

Figure 2.1.5 $y = \frac{1}{x}$ $y' = \frac{-1}{x^2}$

EXAMPLE 4   Find the derivative of $f(x) = |x|$.

Case 1 $x > 0$. In this case $|x| = x$, and we have

$$y = x,$$
$$y + \Delta y = x + \Delta x,$$
$$\Delta y = \Delta x,$$
$$\frac{\Delta y}{\Delta x} = 1, \qquad f'(x) = 1.$$

Case 2 $x < 0$. Now $|x| = -x$, and

$$y = -x,$$
$$y + \Delta y = -(x + \Delta x),$$
$$\Delta y = -(x + \Delta x) - (-x) = -\Delta x,$$
$$\frac{\Delta y}{\Delta x} = -\frac{\Delta x}{\Delta x} = -1, \qquad f'(x) = -1.$$

Case 3 $x = 0$. Then

$$y = 0,$$
$$y + \Delta y = |0 + \Delta x| = |\Delta x|,$$
$$\Delta y = |\Delta x|,$$

and
$$\frac{\Delta y}{\Delta x} = \frac{|\Delta x|}{\Delta x} = \begin{cases} 1 & \text{if } \Delta x > 0, \\ -1 & \text{if } \Delta x < 0. \end{cases}$$

The standard part of $\Delta y/\Delta x$ is then 1 for some values of Δx and -1 for others. Therefore $f'(x)$ does not exist when $x = 0$.

In summary,

$$f'(x) = \begin{cases} 1 & \text{if } x > 0, \\ -1 & \text{if } x < 0, \\ \text{undefined} & \text{if } x = 0. \end{cases}$$

Figure 2.1.6 shows $f(x)$ and $f'(x)$.

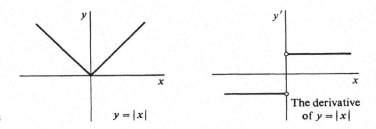

Figure 2.1.6 $y = |x|$ The derivative of $y = |x|$

The derivative has a variety of applications to the physical, life, and social sciences. It may come up in one of the following contexts.

Velocity: If an object moves according to the equation $s = f(t)$ where t is time and s is distance, the derivative $v = f'(t)$ is called the *velocity* of the object at time t.

Growth rates: A population y (of people, bacteria, molecules, etc.) grows according to the equation $y = f(t)$ where t is time. Then the derivative $y' = f'(t)$ is the *rate of growth* of the population y at time t.

Marginal values (economics): Suppose the total cost (or profit, etc.) of producing x items is $y = f(x)$ dollars. Then the cost of making one additional item is approximately the derivative $y' = f'(x)$ because y' is the change in y per unit change in x. This derivative is called the *marginal cost*.

EXAMPLE 5   A ball thrown upward with initial velocity b ft per sec will be at a height

$$y = bt - 16t^2$$

feet after t seconds. Find the velocity at time t. Let t be real and $\Delta t \neq 0$, infinitesimal.

$$y + \Delta y = b(t + \Delta t) - 16(t + \Delta t)^2,$$
$$\Delta y = [b(t + \Delta t) - 16(t + \Delta t)^2] - [bt - 16t^2],$$
$$\frac{\Delta y}{\Delta t} = \frac{[b(t + \Delta t) - 16(t + \Delta t)^2] - [bt - 16t^2]}{\Delta t}$$
$$= \frac{b \Delta t - 32t \Delta t - 16(\Delta t)^2}{\Delta t}$$
$$= b - 32t - 16 \Delta t.$$
$$st\left(\frac{\Delta y}{\Delta t}\right) = st(b - 32t - 16 \Delta t)$$
$$= st(b - 32t) - st(16 \Delta t)$$
$$= b - 32t - 0 = b - 32t.$$

At time t sec, $v = y' = b - 32t$ ft/sec.

Both functions are graphed in Figure 2.1.7.

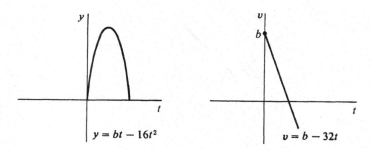

Figure 2.1.7 $y = bt - 16t^2$ $v = b - 32t$

EXAMPLE 6 Suppose a bacterial culture grows in such a way that at time t there are t^3 bacteria. Find the rate of growth at time $t = 1000$ sec.

$$y = t^3 \qquad y' = 3t^2 \quad \text{by Example 1.}$$

At $t = 1000$, $y' = 3{,}000{,}000$ bacteria/sec.

EXAMPLE 7 Suppose the cost of making x needles is $\sqrt{x}$ dollars. What is the marginal cost after 10,000 needles have been made?

$$y = \sqrt{x}, \qquad y' = \frac{1}{2\sqrt{x}} \qquad \text{by Example 2.}$$

At $x = 10{,}000$, $y' = \dfrac{1}{2\sqrt{10{,}000}} = \dfrac{1}{200}$ dollars per needle.

Thus the marginal cost is one half of a cent per needle.

PROBLEMS FOR SECTION 2.1

Find the derivative of the given function in Problems 1–21.

1	$f(x) = x^2$	2	$f(t) = t^2 + 3$
3	$f(x) = 1 - 2x^2$	4	$f(x) = 3x^2 + 2$
5	$f(t) = 4t$	6	$f(x) = 2 - 5x$
7	$f(t) = 4t^3$	8	$f(t) = -t^3$
9	$f(u) = 5\sqrt{u}$	10	$f(u) = \sqrt{u + 2}$
11	$g(x) = x\sqrt{x}$	12	$g(x) = 1/\sqrt{x}$
13	$g(t) = t^{-2}$	14	$g(t) = t^{-3}$
15	$f(y) = 3y^{-1} + 4y$	16	$f(y) = 2y^3 + 4y^2$
17	$f(x) = ax + b$	18	$f(x) = ax^2$
19	$f(x) = \sqrt{ax + b}$	20	$f(x) = 1/(x + 2)$
21	$f(x) = 1/(3 - 2x)$		

22   Find the derivative of $f(x) = 2x^2$ at the point $x = 3$.

23   Find the slope of the curve $f(x) = \sqrt{x - 1}$ at the point $x = 5$.

24   An object moves according to the equation $y = 1/(t + 2), t \geq 0$. Find the velocity as a function of t.

25   A particle moves according to the equation $y = t^4$. Find the velocity as a function of t.

26   Suppose the population of a town grows according to the equation $y = 100t + t^2$. Find the rate of growth at time $t = 100$ years.

27   Suppose a company makes a total profit of $1000x - x^2$ dollars on x items. Find the marginal profit in dollars per item when $x = 200$, $x = 500$, and $x = 1000$.

28   Find the derivative of the function $f(x) = |x + 1|$.

29   Find the derivative of the function $f(x) = |x^3|$.

30   Find the slope of the parabola $y = ax^2 + bx + c$ where a, b, c are constants.

2.2 DIFFERENTIALS AND TANGENT LINES

Suppose we are given a curve $y = f(x)$ and at a point (a, b) on the curve the slope $f'(a)$ is defined. Then the *tangent line* to the curve at the point (a, b), illustrated in Figure 2.2.1, is defined to be the straight line which passes through the point (a, b) and has the same slope as the curve at $x = a$. Thus the tangent line is given by the equation

$$l(x) - b = f'(a)(x - a),$$

or

$$l(x) = f'(a)(x - a) + b.$$

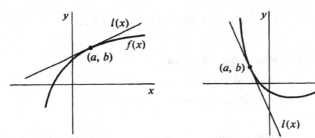

Figure 2.2.1   Tangent lines.

EXAMPLE 1   For the curve $y = x^3$, find the tangent lines at the points $(0, 0)$, $(1, 1)$, and $(-\frac{1}{2}, -\frac{1}{8})$ (Figure 2.2.2).

The slope is given by $f'(x) = 3x^2$. At $x = 0$, $f'(0) = 3 \cdot 0^2 = 0$. The tangent line has the equation

$$y = 0(x - 0) + 0, \quad \text{or} \quad y = 0.$$

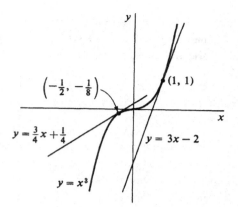

Figure 2.2.2

At $x = 1, f'(1) = 3$, whence the tangent line is

$$y = 3(x - 1) + 1, \quad \text{or} \quad y = 3x - 2.$$

At $x = -\frac{1}{2}, f'(-\frac{1}{2}) = 3 \cdot (-\frac{1}{2})^2 = \frac{3}{4}$, so the tangent line is

$$y = \tfrac{3}{4}(x - (-\tfrac{1}{2})) + (-\tfrac{1}{8}), \quad \text{or} \quad y = \tfrac{3}{4}x + \tfrac{1}{4}.$$

Given a curve $y = f(x)$, suppose that x starts out with the value a and then changes by an infinitesimal amount Δx. What happens to y? Along the curve, y will change by the amount

$$f(a + \Delta x) - f(a) = \Delta y.$$

But along the tangent line y will change by the amount

$$l(a + \Delta x) - l(a) = [f'(a)(a + \Delta x - a) + b] - [f'(a)(a - a) + b]$$
$$= f'(a)\,\Delta x.$$

When x changes from a to $a + \Delta x$, we see that:

$$\text{change in } y \text{ along curve} = f(a + \Delta x) - f(a),$$
$$\text{change in } y \text{ along tangent line} = f'(a)\,\Delta x.$$

In the last section we introduced the dependent variable Δy, the increment of y, with the equation

$$\Delta y = f(x + \Delta x) - f(x).$$

Δy is equal to the change in y along the curve as x changes to $x + \Delta x$.

The following theorem gives a simple but useful formula for the increment Δy.

INCREMENT THEOREM

Let $y = f(x)$. Suppose $f'(x)$ exists at a certain point x, and Δx is infinitesimal. Then Δy is infinitesimal, and

$$\Delta y = f'(x)\,\Delta x + \varepsilon\,\Delta x$$

for some infinitesimal ε, which depends on x and Δx.

PROOF

Case 1 $\Delta x = 0$. In this case, $\Delta y = f'(x)\,\Delta x = 0$, and we put $\varepsilon = 0$.

Case 2 $\Delta x \neq 0$. Then

$$\frac{\Delta y}{\Delta x} \approx f'(x);$$

so for some infinitesimal ε,

$$\frac{\Delta y}{\Delta x} = f'(x) + \varepsilon.$$

Multiplying both sides by Δx,

$$\Delta y = f'(x)\,\Delta x + \varepsilon\,\Delta x.$$

EXAMPLE 2 Let $y = x^3$, so that $y' = 3x^2$. According to the Increment Theorem,

$$\Delta y = 3x^2\,\Delta x + \varepsilon\,\Delta x$$

for some infinitesimal ε. Find ε in terms of x and Δx when $\Delta x \neq 0$. We have

$$\Delta y = 3x^2\,\Delta x + \varepsilon\,\Delta x,$$

$$\frac{\Delta y}{\Delta x} = 3x^2 + \varepsilon,$$

$$\varepsilon = \frac{\Delta y}{\Delta x} - 3x^2.$$

We must still eliminate Δy. From Example 1 in Section 2.1,

$$\Delta y = (x + \Delta x)^3 - x^3,$$

$$\frac{\Delta y}{\Delta x} = 3x^2 + 3x\,\Delta x + (\Delta x)^2.$$

Substituting, $\varepsilon = (3x^2 + 3x\,\Delta x + (\Delta x)^2) - 3x^2.$

Since $3x^2$ cancels,

$$\varepsilon = 3x\,\Delta x + (\Delta x)^2.$$

We shall now introduce a new dependent variable dy, called the differential of y, with the equation

$$dy = f'(x)\,\Delta x.$$

dy is equal to the change in y along the tangent line as x changes to $x + \Delta x$. In Figure 2.2.3 we see dy and Δy under the microscope.

$$\Delta y = \text{change in } y \text{ along curve}$$

$$dy = \text{change in } y \text{ along tangent line}$$

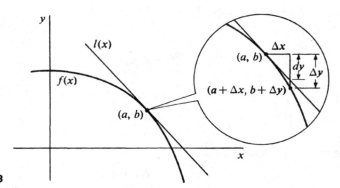

Figure 2.2.3

To keep our notation uniform we also introduce the symbol dx as another name for Δx. For an independent variable x, Δx and dx are the same, but for a dependent variable y, Δy and dy are different.

DEFINITION

Suppose y depends on x, $y = f(x)$.

(i) *The **differential of** x is the independent variable $dx = \Delta x$.*
(ii) *The **differential of** y is the dependent variable dy given by*

$$dy = f'(x)\, dx.$$

When $dx \neq 0$, the equation above may be rewritten as

$$\frac{dy}{dx} = f'(x).$$

Compare this equation with

$$\frac{\Delta y}{\Delta x} \approx f'(x).$$

The quotient dy/dx is a very convenient alternative symbol for the derivative $f'(x)$. In fact we shall write the derivative in the form dy/dx most of the time.

The differential dy depends on two independent variables x and dx. In functional notation,

$$dy = df(x, dx)$$

where df is the real function of two variables defined by

$$df(x, dx) = f'(x)\, dx.$$

When dx is substituted for Δx and dy for $f'(x)\,dx$, the Increment Theorem takes the short form

$$\Delta y = dy + \varepsilon\,dx.$$

The Increment Theorem can be explained graphically using an infinitesimal microscope. Under an infinitesimal microscope, a line of length Δx is magnified to a line of unit length, but a line of length $\varepsilon\,\Delta x$ is only magnified to an infinitesimal length ε. Thus the Increment Theorem shows that when $f'(x)$ exists:

(1) The differential dy and the increment $\Delta y = dy + \varepsilon\,dx$ are so close to each other that they cannot be distinguished under an infinitesimal microscope.

(2) The curve $y = f(x)$ and the tangent line at (x, y) are so close to each other that they cannot be distinguished under an infinitesimal microscope; both look like a straight line of slope $f'(x)$.

Figure 2.2.3 is not really accurate. The curvature had to be exaggerated in order to distinguish the curve and tangent line under the microscope. To give an accurate picture, we need a more complicated figure like Figure 2.2.4, which has a second infinitesimal microscope trained on the point $(a + \Delta x, b + \Delta y)$ in the field of view of the original microscope. This second microscope magnifies $\varepsilon\,dx$ to a unit length and magnifies Δx to an infinite length.

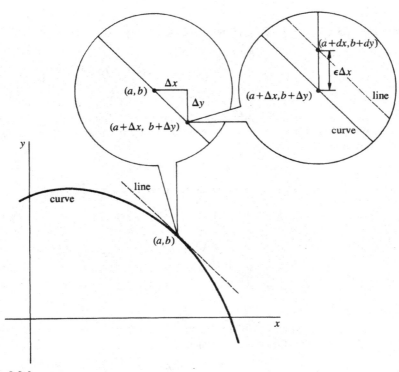

Figure 2.2.4

EXAMPLE 3 Whenever a derivative $f'(x)$ is known, we can find the differential dy at once by simply multiplying the derivative by dx, using the formula $dy = f'(x)\,dx$. The examples in the last section give the following differentials.

(a) $\qquad y = x^3, \qquad\qquad dy = 3x^2\,dx.$

(b) $\qquad y = \sqrt{x}, \qquad\qquad dy = \dfrac{dx}{2\sqrt{x}} \quad \text{where } x > 0.$

(c) $\qquad y = 1/x, \qquad\qquad dy = -dx/x^2 \quad \text{when } x \neq 0.$

(d) $\qquad y = |x|, \qquad\qquad dy = \begin{cases} dx & \text{when } x > 0, \\ -dx & \text{when } x < 0, \\ \text{undefined} & \text{when } x = 0. \end{cases}$

(e) $\qquad y = bt - 16t^2, \qquad dy = (b - 32t)\,dt.$

The differential notation may also be used when we are given a system of formulas in which two or more dependent variables depend on an independent variable. For example if y and z are functions of x,

$$y = f(x), \qquad z = g(x),$$

then $\Delta y, \Delta z, dy, dz$ are determined by

$$\Delta y = f(x + \Delta x) - f(x), \qquad \Delta z = g(x + \Delta x) - g(x),$$
$$dy = f'(x)\,dx, \qquad\qquad dz = g'(x)\,dx.$$

EXAMPLE 4 Given $y = \tfrac{1}{2}x, z = x^3$, with x as the independent variable, then

$$\Delta y = \tfrac{1}{2}(x + \Delta x) - \tfrac{1}{2}x = \tfrac{1}{2}\Delta x,$$
$$\Delta z = 3x^2\,\Delta x + 3x(\Delta x)^2 + (\Delta x)^3,$$
$$dy = \tfrac{1}{2}\,dx, \qquad dz = 3x^2\,dx.$$

The meaning of the symbols for increment and differential in this example will be different if we take y as the independent variable. Then x and z are functions of y.

$$x = 2y, \qquad z = 8y^3.$$

Now $\Delta y = dy$ is just an independent variable, while

$$\Delta x = 2(y + \Delta y) - 2y = 2\,\Delta y,$$
$$\Delta z = 8(y + \Delta y)^3 - 8y^3$$
$$= 8[3y^2\,\Delta y + 3y(\Delta y)^2 + (\Delta y)^3]$$
$$= 24y^2\,\Delta y + 24y(\Delta y)^2 + 8(\Delta y)^3.$$

Moreover, $\qquad dx = 2\,dy, \qquad dz = 24y^2\,dy.$

We may also apply the differential notation to terms. If $\tau(x)$ is a term with the variable x, then $\tau(x)$ determines a function f,

$$\tau(x) = f(x),$$

and the differential $d(\tau(x))$ has the meaning

$$d(\tau(x)) = f'(x)\, dx.$$

EXAMPLE 5

(a) $$d(x^3) = 3x^2\, dx.$$

(b) $$d(\sqrt{x}) = \frac{dx}{2\sqrt{x}}, \qquad x > 0.$$

(c) $$d(1/x) = -\frac{dx}{x^2}, \qquad x \neq 0.$$

(d) $$d(|x|) = \begin{cases} dx & \text{when } x > 0, \\ -dx & \text{when } x < 0, \\ \text{undefined} & \text{when } x = 0. \end{cases}$$

(e) Let $u = bt$ and $w = -16t^2$. Then

$$u + w = bt - 16t^2, \qquad d(u + w) = (b - 32t)\, dt.$$

PROBLEMS FOR SECTION 2.2

In Problems 1–8, express Δy and dy as functions of x and Δx, and for Δx infinitesimal find an infinitesimal ε such that $\Delta y = dy + \varepsilon\, \Delta x$.

1	$y = x^2$	2	$y = -5x^2$
3	$y = 2\sqrt{x}$	4	$y = x^4$
5	$y = 1/x$	6	$y = x^{-2}$
7	$y = x - 1/x$	8	$y = 4x + x^3$

9   If $y = 2x^2$ and $z = x^3$, find Δy, Δz, dy, and dz.

10   If $y = 1/(x + 1)$ and $z = 1/(x + 2)$, find Δy, Δz, dy, and dz.

11	Find $d(2x + 1)$	12	Find $d(x^2 - 3x)$
13	Find $d(\sqrt{x + 1})$	14	Find $d(\sqrt{2x + 1})$
15	Find $d(ax + b)$	16	Find $d(ax^2)$
17	Find $d(3 + 2/x)$	18	Find $d(x\sqrt{x})$
19	Find $d(1/\sqrt{x})$	20	Find $d(x^3 - x^2)$

21   Let $y = \sqrt{x}, z = 3x$. Find $d(y + z)$ and $d(y/z)$.

22   Let $y = x^{-1}$ and $z = x^3$. Find $d(y + z)$ and $d(yz)$.

In Problems 23–30 below, find the equation of the line tangent to the given curve at the given point.

23	$y = x^2$; $(2, 4)$	24	$y = 2x^2$; $(-1, 2)$
25	$y = -x^2$; $(0, 0)$	26	$y = \sqrt{x}$; $(1, 1)$

27	$y = 3x - 4$; $(1, -1)$	**28**	$y = \sqrt{t - 1}$; $(5, 2)$
29	$y = x^4$; $(-2, 16)$	**30**	$y = x^3 - x$; $(0, 0)$

31   Find the equation of the line tangent to the parabola $y = x^2$ at the point (x_0, x_0^2).

32   Find all points $P(x_0, x_0^2)$ on the parabola $y = x^2$ such that the tangent line at P passes through the point $(0, -4)$.

☐ **33**   Prove that the line tangent to the parabola $y = x^2$ at $P(x_0, x_0^2)$ does not meet the parabola at any point except P.

2.3 DERIVATIVES OF RATIONAL FUNCTIONS

A term of the form

$$a_1 x + a_0$$

where a_1, a_0 are real numbers, is called a *linear term* in x; if $a_1 \neq 0$, it is also called *polynomial of degree one* in x. A term of the form

$$a_2 x^2 + a_1 x + a_0, \qquad a_2 \neq 0$$

is called a *polynomial of degree two* in x, and, in general, a term of the form

$$a_n x^n + a_{n-1} x^{n-1} + \cdots + a_1 x + a_0, \qquad a_n \neq 0$$

is called a *polynomial of degree n* in x.

A *rational term* in x is any term which is built up from the variable x and real numbers using the operations of addition, multiplication, subtraction, and division. For example every polynomial is a rational term and so are the terms

$$\frac{(3x^2 - 5)(x + 2)^3}{5x - 11}, \qquad \frac{(1 + 1/x)^9}{x^3 + 1/(2 - x)}.$$

A *linear function, polynomial function,* or *rational function* is a function which is given by a linear term, polynomial, or rational term, respectively. In this section we shall establish a set of rules which enable us to quickly differentiate any rational function. The rules will also be useful later on in differentiating other functions.

THEOREM 1

The derivative of a linear function is equal to the coefficient of x. That is,

$$\frac{d(bx + c)}{dx} = b, \qquad d(bx + c) = b\,dx.$$

PROOF Let $y = bx + c$, and let $\Delta x \neq 0$ be infinitesimal. Then

$$y + \Delta y = b(x + \Delta x) + c,$$
$$\Delta y = (b(x + \Delta x) + c) - (bx + c) = b\,\Delta x,$$
$$\frac{\Delta y}{\Delta x} = \frac{b\,\Delta x}{\Delta x} = b.$$

Therefore $\dfrac{dy}{dx} = st(b) = b.$

Multiplying through by dx, we obtain at once

$$dy = b\,dx.$$

If in Theorem 1 we put $b = 1, c = 0$, we see that the derivative of the identity function $f(x) = x$ is $f'(x) = 1$; i.e.,

$$\frac{dx}{dx} = 1, \qquad dx = dx.$$

On the other hand, if we put $b = 0$ in Theorem 1 then the term $bx + c$ is just the constant c, and we find that the derivative of the constant function $f(x) = c$ is $f'(x) = 0$; i.e.,

$$\frac{dc}{dx} = 0, \qquad dc = 0.$$

THEOREM 2 (Sum Rule)

Suppose u and v depend on the independent variable x. Then for any value of x where du/dx and dv/dx exist,

$$\frac{d(u + v)}{dx} = \frac{du}{dx} + \frac{dv}{dx}, \qquad d(u + v) = du + dv.$$

In other words, the derivative of the sum is the sum of the derivatives.

PROOF Let $y = u + v$, and let $\Delta x \neq 0$ be infinitesimal. Then

$$y + \Delta y = (u + \Delta u) + (v + \Delta v),$$
$$\Delta y = [(u + \Delta u) + (v + \Delta v)] - [u + v] = \Delta u + \Delta v,$$
$$\frac{\Delta y}{\Delta x} = \frac{\Delta u + \Delta v}{\Delta x} = \frac{\Delta u}{\Delta x} + \frac{\Delta v}{\Delta x}.$$

Taking standard parts,

$$st\left(\frac{\Delta y}{\Delta x}\right) = st\left(\frac{\Delta u}{\Delta x} + \frac{\Delta v}{\Delta x}\right) = st\left(\frac{\Delta u}{\Delta x}\right) + st\left(\frac{\Delta v}{\Delta x}\right).$$

Thus $\qquad\qquad \dfrac{dy}{dx} = \dfrac{du}{dx} + \dfrac{dv}{dx}.$

By using the Sum Rule $n - 1$ times, we see that

$$\frac{d(u_1 + \cdots + u_n)}{dx} = \frac{du_1}{dx} + \cdots + \frac{du_n}{dx}, \quad \text{or} \quad d(u_1 + \cdots + u_n) = du_1 + \cdots + du_n.$$

THEOREM 3 (Constant Rule)

Suppose u depends on x, and c is a real number. Then for any value of x where du/dx exists,

$$\frac{d(cu)}{dx} = c\frac{du}{dx}, \qquad d(cu) = c\,du.$$

PROOF Let $y = cu$, and let $\Delta x \neq 0$ be infinitesimal. Then

$$y + \Delta y = c(u + \Delta u),$$

$$\Delta y = c(u + \Delta u) - cu = c\,\Delta u,$$

$$\frac{\Delta y}{\Delta x} = \frac{c\,\Delta u}{\Delta x} = c\frac{\Delta u}{\Delta x}.$$

Taking standard parts,

$$st\left(\frac{\Delta y}{\Delta x}\right) = st\left(c\frac{\Delta u}{\Delta x}\right) = c\,st\left(\frac{\Delta u}{\Delta x}\right)$$

whence

$$\frac{dy}{dx} = c\frac{du}{dx}.$$

The Constant Rule shows that in computing derivatives, a constant factor may be moved "outside" the derivative. It can only be used when c is a constant. For products of two functions of x, we have:

THEOREM 4 (Product Rule)

Suppose u and v depend on x. Then for any value of x where du/dx and dv/dx exist,

$$\frac{d(uv)}{dx} = u\frac{dv}{dx} + v\frac{du}{dx}, \qquad d(uv) = u\,dv + v\,du.$$

PROOF Let $y = uv$, and let $\Delta x \neq 0$ be infinitesimal.

$$y + \Delta y = (u + \Delta u)(v + \Delta v),$$

$$\Delta y = (u + \Delta u)(v + \Delta v) - uv = u\,\Delta v + v\,\Delta u + \Delta u\,\Delta v,$$

$$\frac{\Delta y}{\Delta x} = \frac{u\,\Delta v + v\,\Delta u + \Delta u\,\Delta v}{\Delta x} = u\frac{\Delta v}{\Delta x} + v\frac{\Delta u}{\Delta x} + \Delta u\frac{\Delta v}{\Delta x}.$$

Δu is infinitesimal by the Increment Theorem, whence

$$st\left(\frac{\Delta y}{\Delta x}\right) = st\left(u\frac{\Delta v}{\Delta x} + v\frac{\Delta u}{\Delta x} + \Delta u\frac{\Delta v}{\Delta x}\right)$$

$$= u\cdot st\left(\frac{\Delta v}{\Delta x}\right) + v\cdot st\left(\frac{\Delta u}{\Delta x}\right) + 0\cdot st\left(\frac{\Delta v}{\Delta x}\right).$$

So

$$\frac{dy}{dx} = u\frac{dv}{dx} + v\frac{du}{dx}.$$

The Constant Rule is really the special case of the Product Rule where v is a constant function of x, $v = c$. To check this we let v be the constant c and see what the Product Rule gives us:

$$\frac{d(u\cdot c)}{dx} = u\frac{dc}{dx} + c\frac{du}{dx} = u\cdot 0 + c\frac{du}{dx} = c\frac{du}{dx}.$$

This is the Constant Rule.

The Product Rule can also be used to find the derivative of a power of u.

THEOREM 5 (Power Rule)

Let u depend on x and let n be a positive integer. For any value of x where du/dx exists,

$$\frac{d(u^n)}{dx} = nu^{n-1}\frac{du}{dx}, \qquad d(u^n) = nu^{n-1}\, du.$$

PROOF To see what is going on we first prove the Power Rule for $n = 1, 2, 3, 4$.

$n = 1$: We have $u^n = u$ and $u^0 = 1$, whence

$$\frac{d(u^n)}{dx} = \frac{du}{dx} = 1 \cdot u^0 \cdot \frac{du}{dx}.$$

$n = 2$: We use the Product Rule,

$$\frac{d(u^2)}{dx} = \frac{d(u \cdot u)}{dx} = u\frac{du}{dx} + u\frac{du}{dx} = 2 \cdot u^1 \cdot \frac{du}{dx}.$$

$n = 3$: We write $u^3 = u \cdot u^2$, use the Product Rule again, and then use the result for $n = 2$.

$$\frac{d(u^3)}{dx} = \frac{d(u \cdot u^2)}{dx} = u\frac{d(u^2)}{dx} + u^2\frac{du}{dx}$$

$$= u \cdot 2u\frac{du}{dx} + u^2\frac{du}{dx} = 3u^2\frac{du}{dx}.$$

$n = 4$: Using the Product Rule and then the result for $n = 3$,

$$\frac{d(u^4)}{dx} = \frac{d(u \cdot u^3)}{dx} = u\frac{d(u^3)}{dx} + u^3\frac{du}{dx}$$

$$= u \cdot 3u^2\frac{du}{dx} + u^3\frac{du}{dx} = 4u^3\frac{du}{dx}.$$

We can continue this process indefinitely and prove the theorem for every positive integer n. To see this, assume that we have proved the theorem for m. That is, assume that

(1)
$$\frac{d(u^m)}{dx} = mu^{m-1}\frac{du}{dx}.$$

We then show that it is also true for $m + 1$. Using the Product Rule and the Equation 1,

$$\frac{d(u^{m+1})}{dx} = \frac{d(u \cdot u^m)}{dx} = u\frac{d(u^m)}{dx} + u^m\frac{du}{dx}$$

$$= u \cdot mu^{m-1}\frac{du}{dx} + u^m\frac{du}{dx} = (m+1)u^m\frac{du}{dx}.$$

Thus
$$\frac{d(u^{m+1})}{dx} = (m+1)u^m\frac{du}{dx}.$$

This shows that the theorem holds for $m + 1$.

We have shown the theorem is true for $1, 2, 3, 4$. Set $m = 4$; then the theorem holds for $m + 1 = 5$. Set $m = 5$; then it holds for $m + 1 = 6$. And so on. Hence the theorem is true for all positive integers n.

In the proof of the Power Rule, we used the following principle:

PRINCIPLE OF INDUCTION

Suppose a statement $P(n)$ about an arbitrary integer n is true when $n = 1$. Suppose further that for any positive integer m such that $P(m)$ is true, $P(m + 1)$ is also true. Then the statement $P(n)$ is true of every positive integer n.

In the previous proof, $P(n)$ was the Power Rule,

$$\frac{d(u^n)}{dx} = nu^{n-1}\frac{du}{dx}.$$

The Principle of Induction can be made plausible in the following way. Let a positive integer n be given. Set $m = 1$; since $P(1)$ is true, $P(2)$ is true. Now set $m = 2$; since $P(2)$ is true, $P(3)$ is true. We continue reasoning in this way for n steps and conclude that $P(n)$ is true.

The Power Rule also holds for $n = 0$ because when $u \neq 0, u^0 = 1$ and $d1/dx = 0$.

Using the Sum, Constant, and Power rules, we can compute the derivative of a polynomial function very easily. We have

$$\frac{d(x^n)}{dx} = nx^{n-1},$$

$$\frac{d(cx^n)}{dx} = cnx^{n-1},$$

and thus

$$\frac{d(a_nx^n + a_{n-1}x^{n-1} + \cdots + a_1x + a_0)}{dx} = a_n \cdot nx^{n-1} + a_{n-1}(n-1)x^{n-2} + \cdots + a_1.$$

EXAMPLE 1 $\dfrac{d(-3x^5)}{dx} = -3 \cdot 5x^4 = -15x^4.$

EXAMPLE 2 $\dfrac{d(6x^4 - 2x^3 + x - 1)}{dx} = 24x^3 - 6x^2 + 1.$

Two useful facts can be stated as corollaries.

COROLLARY 1

> The derivative of a polynomial of degree $n > 0$ is a polynomial of degree $n - 1$. (*A nonzero constant is counted as a polynomial of degree zero.*)

COROLLARY 2

> If u depends on x, then $\qquad \dfrac{d(u + c)}{dx} = \dfrac{du}{dx}$

> whenever du/dx exists. That is, adding a constant to a function does not change its derivative.

In Figure 2.3.1 we see that the effect of adding a constant is to move the curve up or down the y-axis without changing the slope.

For the last two rules in this section we need the formula for the derivative of $1/v$.

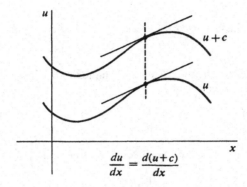

$$\frac{du}{dx} = \frac{d(u+c)}{dx}$$

Figure 2.3.1

LEMMA

> Suppose v depends on x. Then for any value of x where $v \neq 0$ and dv/dx exists,

$$\frac{d(1/v)}{dx} = -\frac{1}{v^2}\frac{dv}{dx}, \qquad d\left(\frac{1}{v}\right) = -\frac{1}{v^2}\,dv.$$

PROOF Let $y = 1/v$ and let $\Delta x \neq 0$ be infinitesimal.

$$y + \Delta y = \frac{1}{v + \Delta v},$$

$$\Delta y = \frac{1}{v + \Delta v} - \frac{1}{v},$$

$$\frac{\Delta y}{\Delta x} = \frac{1/(v + \Delta v) - 1/v}{\Delta x}$$

$$= \frac{v - (v + \Delta v)}{\Delta x v(v + \Delta v)}$$

$$= -\frac{1}{v(v + \Delta v)}\frac{\Delta v}{\Delta x}.$$

Taking standard parts,

$$st\left(\frac{\Delta y}{\Delta x}\right) = st\left(-\frac{1}{v(v + \Delta v)}\frac{\Delta v}{\Delta x}\right)$$

$$= st\left(-\frac{1}{v(v + \Delta v)}\right)st\left(\frac{\Delta v}{\Delta x}\right)$$

$$= -\frac{1}{v^2}st\left(\frac{\Delta v}{\Delta x}\right).$$

Therefore

$$\frac{dy}{dx} = -\frac{1}{v^2}\frac{dv}{dx}.$$

THEOREM 6 (Quotient Rule)

Suppose u, v depend on x. Then for any value of x where du/dx, dv/dx exist and v ≠ 0,

$$\frac{d(u/v)}{dx} = \frac{v\,du/dx - u\,dv/dx}{v^2}, \qquad d\left(\frac{u}{v}\right) = \frac{v\,du - u\,dv}{v^2}.$$

PROOF We combine the Product Rule and the formula for $d(1/v)$. Let $y = u/v$. We write y in the form

$$y = \frac{1}{v}\cdot u.$$

Then

$$dy = d\left(\frac{1}{v}u\right) = \frac{1}{v}du + u\,d\left(\frac{1}{v}\right)$$

$$= \frac{1}{v}du + u(-v^{-2})\,dv$$

$$= \frac{v\,du - u\,dv}{v^2}.$$

THEOREM 7 (Power Rule for Negative Exponents)

Suppose u depends on x and n is a negative integer. Then for any value of x where du/dx exists and u ≠ 0, $d(u^n)/dx$ exists and

$$\frac{d(u^n)}{dx} = nu^{n-1}\frac{du}{dx}, \qquad d(u^n) = nu^{n-1}\,du.$$

PROOF Since n is negative, $n = -m$ where m is positive. Let $y = u^n = u^{-m}$. Then $y = 1/u^m$. By the Lemma and the Power Rule,

$$\frac{dy}{dx} = -\frac{1}{(u^m)^2}\cdot\frac{d(u^m)}{dx}$$

$$= -\frac{1}{u^{2m}}\cdot mu^{m-1}\frac{du}{dx}$$

$$= (-m)u^{-2m}u^{m-1}\frac{du}{dx}$$

$$= (-m)u^{-m-1}\frac{du}{dx} = nu^{n-1}\frac{du}{dx}.$$

The Quotient Rule together with the Constant, Sum, Product, and Power Rules make it easy to differentiate any rational function.

EXAMPLE 3 Find dy when

$$y = \frac{1}{x^2 - 3x + 1}.$$

Introduce the new variable u with the equation

$$u = x^2 - 3x + 1.$$

Then $y = 1/u$, and $du = (2x - 3)\,dx$, so

$$dy = -\frac{1}{u^2}\,du = \frac{-(2x - 3)}{(x^2 - 3x + 1)^2}\,dx.$$

EXAMPLE 4 Let $y = \dfrac{(x^4 - 2)^3}{5x - 1}$ and find dy.

Let
$$u = (x^4 - 2)^3, \qquad v = 5x - 1.$$

Then
$$y = \frac{u}{v}, \qquad dy = \frac{v\,du - u\,dv}{v^2}.$$

Also,
$$du = 3 \cdot (x^4 - 2)^2 \cdot 4x^3\,dx = 12(x^4 - 2)^2 \cdot x^3\,dx,$$
$$dv = 5\,dx.$$

Therefore
$$dy = \frac{(5x - 1)12(x^4 - 2)^2 x^3\,dx - (x^4 - 2)^3 5\,dx}{(5x - 1)^2}$$

$$= \frac{(x^4 - 2)^2[12(5x - 1)x^3 - 5(x^4 - 2)]}{(5x - 1)^2}\,dx.$$

EXAMPLE 5 Let $y = 1/x^3 + 3/x^2 + 4/x + 5$.

Then
$$dy = \left(-\frac{3}{x^4} - \frac{6}{x^3} - \frac{4}{x^2}\right)dx.$$

EXAMPLE 6 Find dy where

$$y = \left(\frac{1}{x^2 + x} + 1\right)^2.$$

This problem can be worked by means of a double substitution. Let

$$u = x^2 + x, \qquad v = \frac{1}{u} + 1.$$

Then $$y = v^2.$$

We find dy, dv, and du,

$$dy = 2v \, dv,$$
$$dv = -u^{-2} \, du,$$
$$du = (2x + 1) \, dx.$$

Substituting, we get dy in terms of x and dx,

$$dy = 2v(-u^{-2} \, du)$$
$$= -2vu^{-2}(2x + 1) \, dx$$
$$= -2\left(\frac{1}{u} + 1\right)u^{-2}(2x + 1) \, dx$$
$$= -2\left(\frac{1}{x^2 + x} + 1\right)(x^2 + x)^{-2}(2x + 1) \, dx.$$

EXAMPLE 7 Assume that u and v depend on x. Given $y = (uv)^{-2}$, find dy/dx in terms of du/dx and dv/dx.

Let $s = uv$, whence $y = s^{-2}$. We have

$$dy = -2s^{-3} \, ds,$$
$$ds = u \, dv + v \, du.$$

Substituting, $$dy = -2(uv)^{-3}(u \, dv + v \, du),$$

and $$\frac{dy}{dx} = -2(uv)^{-3}\left(u\frac{dv}{dx} + v\frac{du}{dx}\right).$$

The six rules for differentiation which we have proved in this section are so useful that they should be memorized. We list them all together.

Table 2.3.1   Rules for Differentiation

(1)	$\dfrac{d(bx + c)}{dx} = b.$	$d(bx + c) = b \, dx.$
(2)	$\dfrac{d(u + v)}{dx} = \dfrac{du}{dx} + \dfrac{dv}{dx}.$	$d(u + v) = du + dv.$
(3)	$\dfrac{d(cu)}{dx} = c\dfrac{du}{dx}.$	$d(cu) = c \, du.$
(4)	$\dfrac{d(uv)}{dx} = u\dfrac{dv}{dx} + v\dfrac{du}{dx}.$	$d(uv) = u \, dv + v \, du.$
(5)	$\dfrac{d(u^n)}{dx} = nu^{n-1}\dfrac{du}{dx}.$	$d(u^n) = nu^{n-1} \, du$ (n is any integer).
(6)	$\dfrac{d(u/v)}{dx} = \dfrac{v \, du/dx - u \, dv/dx}{v^2}.$	$d(u/v) = \dfrac{v \, du - u \, dv}{v^2}.$

An easy way to remember the way the signs are in the Quotient Rule 6 is to put $u = 1$ and use the Power Rule 5 with $n = -1$,

$$d(1/v) = d(v^{-1}) = -1 \cdot v^{-2}\, dv = \frac{-1\, dv}{v^2}.$$

PROBLEMS FOR SECTION 2.3

In Problems 1–42 below, find the derivative.

1 $f(x) = 3x^2 + 5x - 4$

2 $s = \frac{1}{3}t^3 + \frac{1}{2}t^2 + t$

3 $y = (x + 8)^5$

4 $z = (2 + 3x)^4$

5 $f(t) = (4 - t)^3$

6 $g(x) = 3(2 - 5x)^6$

7 $y = (x^2 + 5)^3$

8 $u = (6 + 2x^2)^3$

9 $u = (6 - 2x^2)^3$

10 $w = (1 + 4x^3)^{-2}$

11 $w = (1 - 4x^3)^{-2}$

12 $y = 1 + x^{-1} + x^{-2} + x^{-3}$

13 $f(x) = 5(x + 1 - 1/x)$

14 $u = (x^2 + 3x + 1)^4$

15 $v = 4(2x^2 - x + 3)^{-2}$

16 $y = -(2x + 3 + 4x^{-1})^{-1}$

17 $y = \dfrac{1}{1 + 1/t}$

18 $y = \dfrac{1}{2x^2 + 1}$

19 $s = \dfrac{-3}{4t^2 - 2t + 1}$

20 $s = (2t + 1)(3t - 2)$

21 $h(x) = \frac{1}{2}(x^2 + 1)(5 - 2x)$

22 $y = (2x^3 + 4)(x^2 - 3x + 1)$

23 $v = (3t^2 + 1)(2t - 4)^3$

24 $z = (-2x + 4 + 3x^{-1})(x + 1 - 5x^{-1})$

25 $y = \dfrac{x + 1}{x - 1}$

26 $w = \dfrac{2 - 3x}{1 + 2x}$

27 $y = \dfrac{x^2 - 1}{x^2 + 1}$

28 $u = \dfrac{x}{x^2 + 1}$

29 $x = \dfrac{(s - 1)(s - 2)}{s - 3}$

30 $y = \dfrac{t}{1 + 1/t}$

31 $y = \dfrac{2x^{-1} - x^{-2}}{3x^{-1} - 4x^{-2}}$

32 $y = 4x - 5$

33 $y = 6$

34 $y = 2x(3x - 1)(4 - 2x)$

35 $y = 3(x^2 + 1)(2x^2 - 1)(2x + 3)$

36 $y = (4x + 3)^{-1} + (x - 4)^{-2}$

37 $z = \dfrac{1}{(2x + 1)(x - 3)}$

38 $y = (x^2 + 1)^{-1}(3x - 1)^{-2}$

39 $y = [(2x + 1)^{-1} + 3]^{-1}$

40 $s = [(t^2 + 1)^3 + t]^{-1}$

41 $y = (2x + 1)^3(x^2 + 1)^2$

42 $y = \left(\dfrac{2}{x - 1} - x^{-3}\right)^4$

In Problems 43–48, assume u and v depend on x and find dy/dx in terms of du/dx and dv/dx.

43 $y = u - v$

44 $y = u^2 v$

45 $y = 4u + v^2$

46 $y = 1/(u + v)$

47 $y = 1/uv$

48 $y = (u + v)(2u - v)$

49 Find the line tangent to the curve $y = 1 + x + x^2 + x^3$ at the point $(1, 4)$.

50   Find the line tangent to the curve $y = 9x^{-2}$ at the point $(3, 1)$.

☐ **51**   Consider the parabola $y = x^2 + bx + c$. Find values of b and c such that the line $y = 2x$ is tangent to the parabola at the point $x = 2$, $y = 4$.

☐ **52**   Show that if u, v, and w are differentiable functions of x and $y = uvw$, then

$$\frac{dy}{dx} = uv\frac{dw}{dx} + uw\frac{dv}{dx} + vw\frac{du}{dx}.$$

☐ **53**   Use the principle of induction to show that if n is a positive integer, $u_1, \ldots, u_n$ are differentiable functions of x, and $y = u_1 + \cdots + u_n$, then

$$\frac{dy}{dx} = \frac{du_1}{dx} + \cdots + \frac{du_n}{dx}.$$

☐ **54**   Use the principle of induction to prove that for every positive integer n,

$$1 + 2 + \cdots + n = \frac{n(n + 1)}{2}.$$

☐ **55**   Every rational function can be written as a quotient of two polynomials, $p(x)/q(x)$. Using this fact, show that the derivative of every rational function is a rational function.

2.4 INVERSE FUNCTIONS

Two real functions f and g are called *inverse functions* if the two equations

$$y = f(x), \qquad x = g(y)$$

have the same graphs in the (x, y) plane. That is, a point (x, y) is on the curve $y = f(x)$ if, and only if, it is on the curve $x = g(y)$. (In general, the graph of the equation $x = g(y)$ is different from the graph of $y = g(x)$, but is the same as the graph of $y = f(x)$; see Figure 2.4.1.)

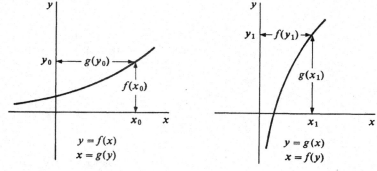

Figure 2.4.1   Inverse Functions

For example, the function $y = x^2$, $x \geq 0$, has the inverse function $x = \sqrt{y}$; the function $y = x^3$ has the inverse function $x = \sqrt[3]{y}$.

If we think of f as a black box operating on an input x to produce an output $f(x)$, the inverse function g is a black box operating on the output $f(x)$ to undo the work of f and produce the original input x (see Figure 2.4.2).

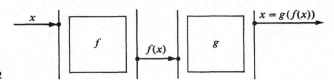

Figure 2.4.2

Many functions, such as $y = x^2$, do not have inverse functions. In Figure 2.4.3, we see that x is not a function of y because at $y = 1$, x has the two values $x = 1$ and $x = -1$.

Often one can tell whether a function f has an inverse by looking at its graph. If there is a horizontal line $y = c$ which cuts the graph at more than one point, the function f has no inverse. (See Figure 2.4.3.) If no horizontal line cuts the graph at more than one point, then f has an inverse function g. Using this rule, we can see in Figure 2.4.4 that the functions $y = |x|$ and $y = \sqrt{1 - x^2}$ do not have inverses.

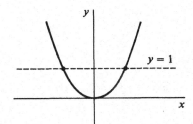

Figure 2.4.3

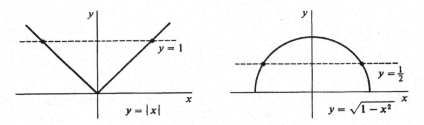

No inverse functions

Figure 2.4.4

Table 2.4.1 shows some familiar functions which do have inverses. Note that in each case, $\dfrac{dx}{dy} = \dfrac{1}{dy/dx}$.

Table 2.4.1

function $y = f(x)$	$\dfrac{dy}{dx}$	inverse function $x = g(y)$	$\dfrac{dx}{dy} = \dfrac{1}{dy/dx}$
$y = x + c$	1	$x = y - c$	1
$y = kx$	k	$x = y/k$	$1/k$
$y = x^2, \quad x \geq 0$	$2x$	$x = \sqrt{y}$	$\dfrac{1}{2\sqrt{y}} = \dfrac{1}{2x}$
$y = x^2, \quad x \leq 0$	$2x$	$x = -\sqrt{y}$	$-\dfrac{1}{2\sqrt{y}} = \dfrac{1}{2x}$
$y = 1/x$	$-\dfrac{1}{x^2}$	$x = 1/y$	$-\dfrac{1}{y^2} = -x^2$

Suppose the (x, y) plane is flipped over about the diagonal line $y = x$. This will make the x- and y-axes change places, forming the (y, x) plane. If f has an inverse function g, the graph of the function $y = f(x)$ will become the graph of the inverse function $x = g(y)$ in the (y, x) plane, as shown in Figure 2.4.5.

The following rule shows that the derivatives of inverse functions are always reciprocals of each other.

INVERSE FUNCTION RULE

Suppose f and g are inverse functions, so that the two equations

$$y = f(x) \quad and \quad x = g(y)$$

have the same graphs. If both derivatives $f'(x)$ and $g'(y)$ exist and are nonzero, then

$$f'(x) = \frac{1}{g'(y)};$$

that is,

$$\frac{dy}{dx} = \frac{1}{dx/dy}.$$

PROOF Let Δx be a nonzero infinitesimal and let Δy be the corresponding change in y. Then Δy is also infinitesimal because $f'(x)$ exists and is nonzero because $f(x)$ has an inverse function. By the rules for standard parts,

$$f'(x) \cdot g'(y) = st\left(\frac{\Delta y}{\Delta x}\right) \cdot st\left(\frac{\Delta x}{\Delta y}\right)$$

$$= st\left(\frac{\Delta y}{\Delta x} \cdot \frac{\Delta x}{\Delta y}\right) = st(1) = 1.$$

Therefore

$$f'(x) = \frac{1}{g'(y)}.$$

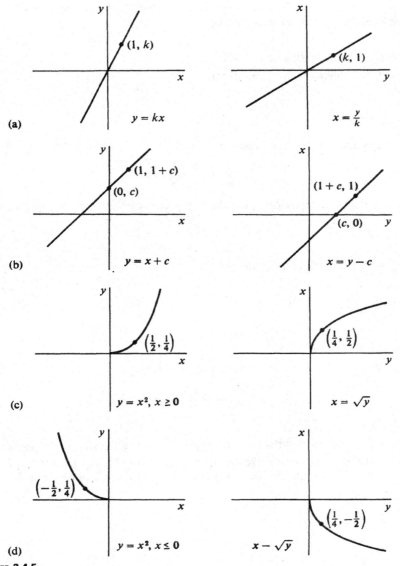

Figure 2.4.5

The formula

$$\frac{dy}{dx} = \frac{1}{dx/dy}$$

in the Inverse Function Rule is not as trivial as it looks. A more complete statement is

$\dfrac{dy}{dx}$ computed with x the independent variable

$$= \frac{1}{dx/dy} \text{ computed with } y \text{ the independent variable.}$$

Sometimes it is easier to compute dx/dy than dy/dx, and in such cases the Inverse Function Rule is a useful method.

EXAMPLE 1   Find dy/dx where $x = 1 + y^{-3}$.

Before solving the problem we note that

$$y = \frac{1}{\sqrt[3]{x-1}},$$

so x and y are inverse functions of each other. We want to find

$$\frac{dy}{dx} = \frac{d(1/\sqrt[3]{x-1})}{dx}$$

with x the independent variable. This looks hard, but it is easy to compute

$$\frac{dx}{dy} = \frac{d(1 + y^{-3})}{dy}$$

with y the independent variable.

SOLUTION $\dfrac{dx}{dy} = -3y^{-4}$,

$$\frac{dy}{dx} = \frac{1}{-3y^{-4}} = -\frac{1}{3}y^{4}.$$

We can write dy/dx in terms of x by substituting,

$$\frac{dy}{dx} = -\frac{1}{3}(x-1)^{-4/3}.$$

EXAMPLE 2   Find dy/dx where $x = y^5 + y^3 + y$. Compute dy/dx at the point $(3, 1)$.

Although we cannot solve the equation explicitly for y as a function of x, we can see from the graph in Figure 2.4.6 that there is an inverse function $y = f(x)$.

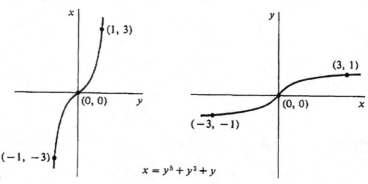

Figure 2.4.6

By the Inverse Function Rule,

$$\frac{dx}{dy} = 5y^4 + 3y^2 + 1,$$

$$\frac{dy}{dx} = \frac{1}{5y^4 + 3y^2 + 1}.$$

This time we must leave the answer in terms of y. At the point $(3,1)$, we substitute 1 for y and get $dy/dx = 1/9$.

For $y \geq 0$, the function $x = y^n$ has the inverse function $y = x^{1/n}$. In the next theorem, we use the Inverse Function Rule to find a new derivative, that of $y = x^{1/n}$.

THEOREM 1

If n is a positive integer and

$$y = x^{1/n},$$

then

$$\frac{dy}{dx} = \frac{1}{n}x^{(1/n)-1}.$$

Remember that $y = x^{1/n}$ is defined for all x if n is odd and for $x > 0$ if n is even. The derivative $\frac{1}{n}x^{(1/n)-1}$ is defined for $x \neq 0$ if n is odd and for $x > 0$ if n is even.

If we are willing to assume that dy/dx exists, then we can quickly find dy/dx by the Inverse Function Rule.

$$x = y^n,$$

$$\frac{dx}{dy} = ny^{n-1},$$

$$\frac{dy}{dx} = \frac{1}{dx/dy} = \frac{1}{ny^{n-1}} = \frac{1}{n}y^{1-n}$$

$$= \frac{1}{n}(x^{1/n})^{1-n} = \frac{1}{n}x^{(1-n)/n} = \frac{1}{n}x^{(1/n)-1}.$$

Here is a longer but complete proof which shows that dy/dx exists and computes its value.

PROOF OF THEOREM 1   Let $x \neq 0$ and let Δx be nonzero infinitesimal. We first show that

$$\Delta y = (x + \Delta x)^{1/n} - x^{1/n}$$

is a nonzero infinitesimal. $\Delta y \neq 0$ because $x + \Delta x \neq x$. The standard part of Δy is

$$st(\Delta y) = st((x + \Delta x)^{1/n}) - st(x^{1/n})$$
$$= x^{1/n} - x^{1/n} = 0.$$

Therefore Δy is nonzero infinitesimal.

Now
$$x = y^n,$$

$$\frac{dx}{dy} = ny^{n-1},$$

$$\frac{\Delta x}{\Delta y} \approx ny^{n-1}.$$

Therefore
$$\frac{\Delta y}{\Delta x} \approx \frac{1}{ny^{n-1}} = \frac{1}{n}x^{(1/n)-1},$$

$$\frac{dy}{dx} = \frac{1}{n}x^{(1/n)-1}.$$

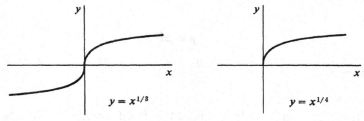

Figure 2.4.7

$y = x^{1/3}$ $y = x^{1/4}$

Figure 2.4.7 shows the graphs of $y = x^{1/3}$ and $y = x^{1/4}$. At $x = 0$, the curves are vertical and have no slope.

EXAMPLE 3   Find the derivatives of $y = x^{1/n}$ for $n = 2, 3, 4$.

$$\frac{d(x^{1/2})}{dx} = \frac{1}{2}x^{-1/2}, \qquad x > 0.$$

$$\frac{d(x^{1/3})}{dx} = \frac{1}{3}x^{-2/3}, \qquad x \neq 0.$$

$$\frac{d(x^{1/4})}{dx} = \frac{1}{4}x^{-3/4}, \qquad x > 0.$$

Using Theorem 1 we can show that the Power Rule holds when the exponent is any rational number.

POWER RULE FOR RATIONAL EXPONENTS

Let $y = x^r$ where r is a rational number. Then whenever $x > 0$,

$$\frac{dy}{dx} = rx^{r-1}.$$

PROOF Let $r = m/n$ where m and n are integers, $n > 0$. Let

$$u = x^{1/n}, \qquad y = u^m.$$

Then

$$\frac{du}{dx} = \frac{1}{n}x^{(1/n)-1}$$

and

$$\frac{dy}{dx} = mu^{m-1}\frac{du}{dx}$$

$$= m(x^{1/n})^{m-1}\left(\frac{1}{n}x^{(1/n)-1}\right)$$

$$= \frac{m}{n}x^{(m/n)-1} = rx^{r-1}.$$

EXAMPLE 4   Find dy/dx where

$$y = x^{-3/7}.$$
$$\frac{dy}{dx} = -\frac{3}{7}x^{(-3/7)-1} = -\frac{3}{7}x^{-10/7}.$$

EXAMPLE 5   Find dy/dx where

$$y = \frac{1}{2 + x^{3/2}}.$$

Let

$$u = 2 + x^{3/2}, \qquad y = u^{-1}.$$

Then

$$\frac{du}{dx} = \frac{3}{2}x^{1/2},$$

$$\frac{dy}{dx} = -u^{-2}\frac{du}{dx}$$

$$= -u^{-2}\left(\frac{3}{2}x^{1/2}\right) = -\frac{3}{2}\frac{x^{1/2}}{(2 + x^{3/2})^2}.$$

PROBLEMS FOR SECTION 2.4

In Problems 1–16, find dy/dx.

1	$x = 3y^3 + 2y$	**2**	$x = y^2 + 1, \quad y > 0$
3	$x = 1 - 2y^2, \quad y > 0$	**4**	$x = 2y^5 + y^3 + 4$
5	$x = (y^2 + 2)^{-1}, \quad y > 0$	**6**	$y = 1/\sqrt{x}$
7	$y = x^{4/3}$	**8**	$y = \sqrt{2x}$
9	$y = (\sqrt{x} + 1)/(\sqrt{x} - 1)$	**10**	$y = (2x^{1/3} + 1)^3$
11	$y = 1 + 2x^{1/3} + 4x^{2/3} + 6x$	**12**	$y = x^{-1/4} + 3x^{-3/4}$
13	$y = (x^{5/3} - x)^{-2}$	**14**	$x = y + 2\sqrt{y}$
15	$x = 3y^{1/3} + 2y, \quad y > 0$	**16**	$x = 1/(1 + \sqrt{y})$

In Problems 17–25, find the inverse function y and its derivative dy/dx as functions of x.

17	$x = ky + c, \quad k \neq 0$	**18**	$x = y^3 + 1$
19	$x = 2y^2 + 1, \quad y \geq 0$	**20**	$x = 2y^2 + 1, \quad y \leq 0$
21	$x = y^4 - 3, \quad y \geq 0$	**22**	$x = y^2 + 3y - 1, \quad y \geq -\frac{3}{2}$
23	$x = y^4 + y^2 + 1, \quad y \geq 0$	**24**	$x = 1/y^2 + 1/y - 1, \quad y > 0$
25	$x = \sqrt{y} + 2y, \quad y > 0$		

☐ **26** Show that no second degree polynomial $x = ay^2 + by + c$ has an inverse function.

☐ **27** Show that $x = ay^2 + by + c$, $y \geq -b/2a$, has an inverse function. What does its graph look like?

☐ **28** Prove that a function $y = f(x)$ has an inverse function if and only if whenever $x_1 \neq x_2$, $f(x_1) \neq f(x_2)$.

2.5 TRANSCENDENTAL FUNCTIONS

The transcendental functions include the trigonometric functions $\sin x$, $\cos x$, $\tan x$, the exponential function e^x, and the natural logarithm function $\ln x$. These functions are developed in detail in Chapters 7 and 8. This section contains a brief discussion.

1 TRIGONOMETRIC FUNCTIONS

The Greek letters θ (theta) and ϕ (phi) are often used for angles. In the calculus it is convenient to measure angles in *radians* instead of degrees. An angle θ in radians is defined as the length of the arc of the angle on a circle of radius one (Figure 2.5.1). Since a circle of radius one has circumference 2π,

$$360 \text{ degrees} = 2\pi \text{ radians}.$$

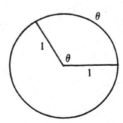

Figure 2.5.1

Thus a right angle is

$$90 \text{ degrees} = \pi/2 \text{ radians}.$$

To define the sine and cosine functions, we consider a point $P(x, y)$ on the unit circle $x^2 + y^2 = 1$. Let θ be the angle measured counterclockwise in radians from the point $(1, 0)$ to the point $P(x, y)$ as shown in Figure 2.5.2. Both coordinates

x and y depend on θ. The value of x is called the *cosine* of θ, and the value of y is the *sine* of θ. In symbols,

$$x = \cos\theta, \qquad y = \sin\theta.$$

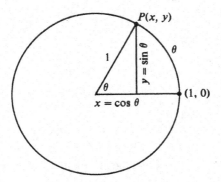

Figure 2.5.2

The *tangent* of θ is defined by

$$\tan\theta = \sin\theta/\cos\theta.$$

Negative angles and angles greater than 2π radians are also allowed.

The trigonometric functions can also be defined using the sides of a right triangle, but this method only works for θ between 0 and $\pi/2$. Let θ be one of the acute angles of a right triangle as shown in Figure 2.5.3.

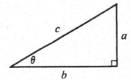

Figure 2.5.3

Then
$$\sin\theta = \frac{\text{opposite side}}{\text{hypotenuse}} = \frac{a}{c},$$
$$\cos\theta = \frac{\text{adjacent side}}{\text{hypotenuse}} = \frac{b}{c},$$
$$\tan\theta = \frac{\text{opposite side}}{\text{adjacent side}} = \frac{a}{b}.$$

The two definitions, with circles and right triangles, can be seen to be equivalent using similar triangles.

Table 2.5.1 gives the values of sin θ and cos θ for some important values of θ.

Table 2.5.1

θ in degrees	0°	30°	45°	60°	90°	180°	270°	360°
θ in radians	0	$\pi/6$	$\pi/4$	$\pi/3$	$\pi/2$	π	$3\pi/2$	2π
sin θ	0	1/2	$\sqrt{2}/2$	$\sqrt{3}/2$	1	0	-1	0
cos θ	1	$\sqrt{3}/2$	$\sqrt{2}/2$	1/2	0	-1	0	1

A useful identity which follows from the unit circle equation $x^2 + y^2 = 1$ is

$$\sin^2 \theta + \cos^2 \theta = 1.$$

Here $\sin^2 \theta$ means $(\sin \theta)^2$.

Figure 2.5.4 shows the graphs of sin θ and cos θ, which look like waves that oscillate between 1 and -1 and repeat every 2π radians.

The derivatives of the sine and cosine functions are:

$$\frac{d(\sin \theta)}{d\theta} = \cos \theta.$$

$$\frac{d(\cos \theta)}{d\theta} = -\sin \theta.$$

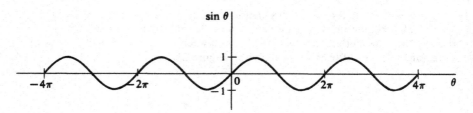

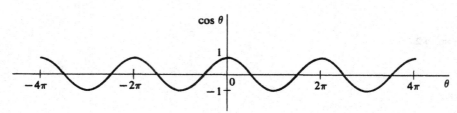

Figure 2.5.4

In both formulas θ is measured in radians. We can see intuitively why these are the derivatives in Figure 2.5.5.

In the triangle under the infinitesimal microscope,

$$\frac{\Delta(\sin \theta)}{\Delta\theta} \approx \frac{\text{adjacent side}}{\text{hypotenuse}} = \cos \theta;$$

$$\frac{\Delta(\cos \theta)}{\Delta\theta} \approx \frac{-\text{opposite side}}{\text{hypotenuse}} = -\sin \theta.$$

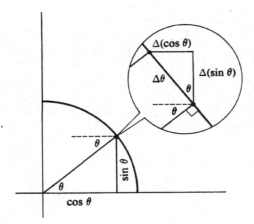

Figure 2.5.5

Notice that $\cos \theta$ decreases, and $\Delta(\cos \theta)$ is negative in the figure, so the derivative of $\cos \theta$ is $-\sin \theta$ instead of just $\sin \theta$.

Using the rules of differentiation we can find other derivatives.

EXAMPLE 1 Differentiate $y = \sin^2 \theta$. Let $u = \sin \theta$, $y = u^2$. Then

$$\frac{dy}{d\theta} = 2u\frac{du}{d\theta} = 2\sin \theta \cos \theta.$$

EXAMPLE 2 Differentiate $y = \sin \theta(1 - \cos \theta)$. Let $u = \sin \theta$, $v = 1 - \cos \theta$. Then $y = u \cdot v$, and

$$\frac{dy}{d\theta} = u\frac{dv}{d\theta} + v\frac{du}{d\theta} = \sin \theta(-(-\sin \theta)) + (1 - \cos \theta)\cos \theta$$

$$= \sin^2 \theta + \cos \theta - \cos^2 \theta.$$

The other trigonometric functions (the secant, cosecant, and cotangent functions) and the inverse trigonometric functions are discussed in Chapter 7.

2 EXPONENTIAL FUNCTIONS

Given a positive real number b and a rational number m/n, the rational power $b^{m/n}$ is defined as

$$b^{m/n} = \sqrt[n]{b^m},$$

the positive nth root of b^m. The negative power $b^{-m/n}$ is

$$b^{-m/n} = \frac{1}{b^{m/n}}.$$

As an example consider $b = 10$. Several values of $10^{m/n}$ are shown in Table 2.5.2.

Table 2.5.2

10^{-3}	$10^{-3/2}$	10^{-1}	$10^{-2/3}$	$10^{-1/3}$	10^0	$10^{1/3}$	$10^{2/3}$	10^1	$10^{3/2}$	10^3
$\dfrac{1}{1000}$	$\dfrac{1}{10\sqrt{10}}$	$\dfrac{1}{10}$	$\dfrac{1}{\sqrt[3]{100}}$	$\dfrac{1}{\sqrt[3]{10}}$	1	$\sqrt[3]{10}$	$\sqrt[3]{100}$	10	$10\sqrt{10}$	1000

If we plot all the rational powers $10^{m/n}$, we get a dotted line, with one value for each rational number m/n, as in Figure 2.5.6.

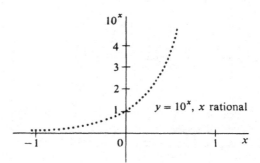

$y = 10^x$, x rational

Figure 2.5.6

By connecting the dots with a smooth curve, we obtain a function $y = 10^x$, where x varies over all real numbers instead of just the rationals. 10^x is called the *exponential function with base* 10. It is positive for all x and follows the rules

$$10^{a+b} = 10^a \cdot 10^b, \qquad 10^{a \cdot b} = (10^a)^b.$$

The derivative of 10^x is a constant times 10^x, approximately

$$\frac{d(10^x)}{dx} \sim (2.303)10^x.$$

To see this let Δx be a nonzero infinitesimal. Then

$$\frac{d(10^x)}{dx} = st\left[\frac{10^{x+\Delta x} - 10^x}{\Delta x}\right] = st\left[\frac{(10^{\Delta x} - 1)10^x}{\Delta x}\right] = st\left[\frac{10^{\Delta x} - 1}{\Delta x}\right]10^x.$$

The number $st[(10^{\Delta x} - 1)/\Delta x]$ is a constant which does not depend on x and can be shown to be approximately 2.303.

If we start with a given positive real number b instead of 10, we obtain the *exponential function with base* b, $y = b^x$. The derivative of b^x is equal to the constant $st[(b^{\Delta x} - 1)/\Delta x]$ times b^x. This constant depends on b. The derivative is computed as follows:

$$\frac{d(b^x)}{dx} = st\left[\frac{b^{x+\Delta x} - b^x}{\Delta x}\right] = st\left[\frac{(b^{\Delta x} - 1)b^x}{\Delta x}\right] = st\left[\frac{b^{\Delta x} - 1}{\Delta x}\right]b^x.$$

The most useful base for the calculus is the number e. e is defined as the real number such that the derivative of e^x is e^x itself,

$$\frac{d(e^x)}{dx} = e^x.$$

In other words, e is the real number such that the constant

$$st\left[\frac{e^{\Delta x} - 1}{\Delta x}\right] = 1$$

(where Δx is a nonzero infinitesimal). It will be shown in Section 8.3 that there is such a number e and that e has the approximate value

$$e \sim 2.71828.$$

The function $y = e^x$ is called the *exponential function*. e^x is always positive and follows the rules

$$e^{a+b} = e^a \cdot e^b, \qquad e^{a \cdot b} = (e^a)^b, \qquad e^0 = 1.$$

Figure 2.5.7 shows the graph of $y = e^x$.

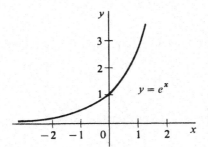

Figure 2.5.7

EXAMPLE 3 Find the derivative of $y = x^2 e^x$. By the Product Rule,

$$\frac{dy}{dx} = x^2 \frac{d(e^x)}{dx} + e^x \frac{d(x^2)}{dx} = x^2 e^x + 2x e^x.$$

3 THE NATURAL LOGARITHM

The inverse of the exponential function $x = e^y$ is the *natural logarithm function*, written

$$y = \ln x.$$

Verbally, $\ln x$ is the number y such that $e^y = x$. Since $y = \ln x$ is the inverse function of $x = e^y$, we have

$$e^{\ln a} = a, \qquad \ln(e^a) = a.$$

The simplest values of $y = \ln x$ are

$$\ln(1/e) = -1, \qquad \ln(1) = 0, \qquad \ln e = 1.$$

Figure 2.5.8 shows the graph of $y = \ln x$. It is defined only for $x > 0$.

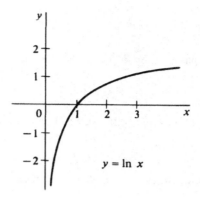

Figure 2.5.8

The most important rules for logarithms are

$$\ln(ab) = \ln a + \ln b,$$
$$\ln(a^b) = b \cdot \ln a.$$

The natural logarithm function is important in calculus because its derivative is simply $1/x$,

$$\frac{d(\ln x)}{dx} = \frac{1}{x}, \qquad (x > 0).$$

This can be derived from the Inverse Function Rule.

If

$$y = \ln x,$$

then

$$x = e^y,$$

$$\frac{dx}{dy} = e^y,$$

$$\frac{dy}{dx} = \frac{1}{dx/dy} = \frac{1}{e^y} = \frac{1}{x}.$$

The natural logarithm is also called the logarithm to the base e and is sometimes written $\log_e x$. Logarithms to other bases are discussed in Chapter 8.

EXAMPLE 4   Differentiate $y = \dfrac{1}{\ln x}$.

$$\frac{dy}{dx} = \frac{-1}{(\ln x)^2} \frac{d(\ln x)}{dx} = -\frac{1}{x(\ln x)^2}.$$

4 SUMMARY

Here is a list of the new derivatives given in this section.

$$\frac{d(\sin x)}{dx} = \cos x.$$

$$\frac{d(\cos x)}{dx} = -\sin x.$$

$$\frac{d(e^x)}{dx} = e^x.$$

$$\frac{d(\ln x)}{dx} = \frac{1}{x} \quad (x > 0).$$

Tables of values for $\sin x$, $\cos x$, e^x, and $\ln x$ can be found at the end of the book.

PROBLEMS FOR SECTION 2.5

In Problems 1–20, find the derivative.

1	$y = \cos^2 \theta$	**2**	$s = \tan^2 t$
3	$y = 2 \sin x + 3 \cos x$	**4**	$y = \sin x \cdot \cos x$
5	$w = \dfrac{1}{\cos z}$	**6**	$w = \dfrac{1}{\sin z}$
7	$y = \sin^n \theta$	**8**	$y = \tan^n \theta$
9	$s = t \sin t$	**10**	$s = \dfrac{\cos t}{t - 1}$
11	$y = xe^x$	**12**	$y = 1/(1 + e^x)$
13	$y = (\ln x)^2$	**14**	$y = x \ln x$
15	$y = e^x \cdot \ln x$	**16**	$y = e^x \cdot \sin x$
17	$u = \sqrt{v}(1 - e^v)$	**18**	$u = (1 + e^v)(1 - e^v)$
19	$y = x^n \ln x$	**20**	$y = (\ln x)^n$

In Problems 21–24, find the equation of the tangent line at the given point.

21	$y = \sin x$ at $(\pi/6, \frac{1}{2})$	**22**	$y = \cos x$ at $(\pi/4, \sqrt{2}/2)$
23	$y = x - \ln x$ at $(e, e - 1)$	**24**	$y = e^{-x}$ at $(0, 1)$

2.6 CHAIN RULE

The Chain Rule is more general than the Inverse Function Rule and deals with the case where x and y are both functions of a third variable t.

Suppose $\qquad\qquad x = f(t), \qquad y = G(x).$

Thus x depends on t, and y depends on x. But y is also a function of t,

$$y = g(t),$$

where g is defined by the rule

$$g(t) = G(f(t)).$$

The function g is sometimes called the *composition* of G and f (sometimes written $g = G \circ f$).

The composition of G and f may be described in terms of black boxes. The function $g = G \circ f$ is a large black box operating on the input t to produce $g(t) = G(f(t))$. If we look inside this black box (pictured in Figure 2.6.1), we see two smaller black boxes, f and G. First f operates on the input t to produce $f(t)$, and then G operates on $f(t)$ to produce the final output $g(t) = G(f(t))$.

The Chain Rule expresses the derivative of g in terms of the derivatives of f and G. It leads to the powerful method of "change of variables" in computing derivatives and, later on, integrals.

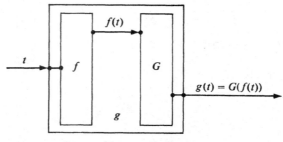

Figure 2.6.1   Composition $g = G \circ f$

CHAIN RULE

Let f, G be two real functions and define the new function g by the rule

$$g(t) = G(f(t)).$$

At any value of t where the derivatives $f'(t)$ and $G'(f(t))$ exist, $g'(t)$ also exists and has the value

$$g'(t) = G'(f(t))f'(t).$$

PROOF Let
$$x = f(t), \qquad y = g(t), \qquad y = G(x).$$
Take t as the independent variable, and let $\Delta t \neq 0$ be infinitesimal. Form the corresponding increments Δx and Δy. By the Increment Theorem for $x = f(t)$, Δx is infinitesimal. Using the Increment Theorem again but this time for $y = G(x)$, we have

$$\Delta y = G'(x)\,\Delta x + \varepsilon\,\Delta x$$

for some infinitesimal ε. Dividing by Δt,

$$\frac{\Delta y}{\Delta t} = G'(x)\frac{\Delta x}{\Delta t} + \varepsilon\frac{\Delta x}{\Delta t}.$$

Then taking standard parts,

$$st\left(\frac{\Delta y}{\Delta t}\right) = G'(x)st\left(\frac{\Delta x}{\Delta t}\right) + 0,$$

or
$$g'(t) = G'(x)f'(t) = G'(f(t))f'(t).$$

EXAMPLE 1   Find the derivative of $g(t) = \ln(\sin t)$. $g(t)$ is the natural logarithm of the sine of t. It can be written in the form

$$g(t) = G(f(t))$$

where $\qquad\qquad f(t) = \sin t, \qquad G(x) = \ln x.$

We have $\qquad\qquad f'(t) = \cos t, \qquad G'(x) = \dfrac{1}{x}.$

By the Chain Rule,

$$g'(t) = G'(f(t))f'(t)$$

$$= \frac{1}{\sin t} \cdot \cos t = \frac{\cos t}{\sin t}.$$

EXAMPLE 2   Find the derivative of $g(t) = \sqrt{3t + 1}$. $g(t)$ has the form

$$g(t) = G(f(t)),$$

where $\qquad\qquad f(t) = 3t + 1, \qquad G(x) = \sqrt{x}.$

We have $\qquad\qquad f'(t) = 3, \qquad G'(x) = \dfrac{1}{2}x^{-1/2}.$

Then $\qquad\qquad g'(t) = G'(f(t))f'(t)$

$$= \frac{1}{2}(3t + 1)^{-1/2}3 = \frac{3}{2\sqrt{3t + 1}}.$$

In practice it is more convenient to use the Chain Rule with dependent variables x and y instead of functions f and g.

CHAIN RULE WITH DEPENDENT VARIABLES

Let

$$x = f(t), \qquad y = g(t) = G(x).$$

Assume $g'(t)$ and $G'(x)$ exist. Then

$$\text{(i)} \quad \frac{dy}{dt} = \frac{dy}{dx}\frac{dx}{dt} \qquad \text{(ii)} \quad dy = \frac{dy}{dx}dx$$

where dx/dt, dy/dt are computed with t as the independent variable, and dy/dx is computed with x as the independent variable.

Let us work Examples 1 and 2 again with dependent variables.

EXAMPLE 1 (Continued)   Let $x = \sin t, \qquad y = \ln x$.

Find dy/dt using Chain Rule (i) and dy by using Chain Rule (ii).

$$\text{(i)} \qquad\qquad \frac{dx}{dt} = \cos t, \qquad \frac{dy}{dx} = \frac{1}{x},$$

$$\frac{dy}{dt} = \frac{dy}{dx} \cdot \frac{dx}{dt} = \frac{1}{x} \cdot \cos t = \frac{\cos t}{\sin t}.$$

(ii) $$dx = \cos t\, dt, \qquad \frac{dy}{dx} = \frac{1}{x},$$

$$dy = \frac{1}{x}\, dx = \frac{1}{x}\cos t\, dt = \frac{\cos t}{\sin t}\, dt.$$

EXAMPLE 2 (Continued)   Let $x = 3t + 1, \qquad y = \sqrt{x}.$

(i) $$\frac{dx}{dt} = 3, \qquad \frac{dy}{dx} = \frac{1}{2}x^{-1/2},$$

$$\frac{dy}{dt} = \frac{dy}{dx}\frac{dx}{dt} = \frac{3}{2}x^{-1/2} = \frac{3}{2}(3t + 1)^{-1/2}.$$

(ii) $$dx = 3\, dt, \qquad \frac{dy}{dx} = \frac{1}{2}x^{-1/2},$$

$$dy = \frac{1}{2}x^{-1/2}\, dx = \frac{1}{2}(3t + 1)^{-1/2}3\, dt = \frac{3}{2}(3t + 1)^{-1/2}\, dt.$$

The equation

$$\frac{dy}{dt} = \frac{dy}{dx}\frac{dx}{dt}$$

with t as the independent variable is trivial. We simply cancel the dx's. But when dy/dx is computed with x as the independent variable while dx/dt is computed with t as the independent variable, the two dx's have different meanings, and the equation is not trivial.

Similarly, the equation

$$dy = \frac{dy}{dx}\, dx$$

is trivial with x as the independent variable but not when t is the independent variable in dy and dx, while x is independent in dy/dx.

The Chain Rule shows that when we change independent variables the equations

$$\frac{dy}{dt} = \frac{dy}{dx}\frac{dx}{dt}, \qquad dy = \frac{dy}{dx}\, dx$$

remain true.

The Inverse Function Rule can be proved from the Chain Rule as follows. Let

$$y = f(x), \qquad x = g(y)$$

be inverse functions whose derivatives exist. Then

$$\frac{dy}{dx}\frac{dx}{dy} = \frac{dy}{dy} = 1,$$

whence $$\frac{dy}{dx} = \frac{1}{dx/dy}, \qquad f'(x) = \frac{1}{g'(y)}.$$

Using the Chain Rule we may write the Power Rule in a general form.

POWER RULE

Let r be a rational number, and let u depend on x. If $u > 0$ and du/dx exists, then

$$\frac{d(u^r)}{dx} = ru^{r-1}\frac{du}{dx}.$$

This is proved by letting $y = u^r$ and computing dy/dx by the Chain Rule,

$$\frac{dy}{dx} = \frac{dy}{du}\frac{du}{dx} = ru^{r-1}\frac{du}{dx}.$$

The Chain Rule has two types of applications.

(1) Given $x = f(t)$ and $y = G(x)$, find $\dfrac{dy}{dt}$. Use $\dfrac{dy}{dt} = \dfrac{dy}{dx}\dfrac{dx}{dt}$.

(2) Given $x = f(t)$ and $y = g(t)$, find $\dfrac{dy}{dx}$. Use $\dfrac{dy}{dx} = \dfrac{dy/dt}{dx/dt}$.

Applications of type (1) often arise when a new dependent variable x is introduced to help compute $\dfrac{dy}{dt}$. Applications of type (2) arise when two variables x and y both depend on a third variable t, for example, when x and y are the coordinates of a moving particle and t is time.

We give three examples of type (1) and then two of type (2).

EXAMPLE 3 Suppose that by investing t dollars a company can produce

$$x = \frac{t}{10} - 100, \qquad t \geq 1000,$$

items, and that it can sell x items for a total profit of

$$y = 5x - \frac{x^2}{100}.$$

Find $\dfrac{dy}{dt}$, which is the marginal profit with respect to the amount invested.

We have $\qquad \dfrac{dx}{dt} = \dfrac{1}{10}, \qquad \dfrac{dy}{dx} = 5 - \dfrac{x}{50}.$

By the Chain Rule,

$$\frac{dy}{dt} = \frac{dy}{dx}\frac{dx}{dt} = \left(5 - \frac{x}{50}\right)\frac{1}{10}$$

$$= \left(5 - \frac{\dfrac{t}{10} - 100}{50}\right)\frac{1}{10}$$

$$= 0.7 - \frac{t}{5000}.$$

Thus after t dollars have been invested, an additional dollar invested will bring $0.7 - t/5000$ dollars of additional profit.

EXAMPLE 4 Find dy/dt where $y = (5t^2 - 2)^{1/4}$.

Let
$$x = 5t^2 - 2, \qquad y = x^{1/4}.$$

Then
$$\frac{dx}{dt} = 10t, \qquad \frac{dy}{dx} = \frac{1}{4}x^{-3/4},$$

$$\frac{dy}{dt} = \frac{dy}{dx}\frac{dx}{dt} = \left(\frac{1}{4}x^{-3/4}\right)(10t)$$

$$= \frac{10}{4}(5t^2 - 2)^{-3/4}t.$$

EXAMPLE 5 Find dy/dx where $y = \sqrt{\sin(4x + 1) + \cos(4x - 1)}$. This problem requires three uses of the Chain Rule.

Let
$$u = \sin(4x + 1) + \cos(4x - 1), \qquad y = \sqrt{u}.$$

Then by the Chain Rule,

$$\frac{dy}{dx} = \frac{dy}{du}\cdot\frac{du}{dx} = \frac{1}{2\sqrt{u}}\cdot\frac{du}{dx}.$$

Now let $v = \sin(4x + 1)$, $\quad w = \cos(4x - 1)$, $\quad u = v + w$.

Then
$$\frac{du}{dx} = \frac{dv}{dx} + \frac{dw}{dx}.$$

We use the Chain Rule twice more to find dv/dx and dw/dx.

$$v = \sin(4x + 1).$$

$$\frac{dv}{dx} = \cos(4x + 1)\frac{d(4x + 1)}{dx} = 4\cos(4x + 1).$$

$$w = \cos(4x - 1).$$

$$\frac{dw}{dx} = -\sin(4x - 1)\frac{d(4x - 1)}{dx} = -4\sin(4x - 1).$$

Finally, we combine everything to get

$$\frac{dy}{dx} = \frac{1}{2\sqrt{u}}\cdot\frac{du}{dx} = \frac{1}{2\sqrt{u}}\left(\frac{dv}{dx} + \frac{dw}{dx}\right)$$

$$= \frac{4\cos(4x + 1) - 4\sin(4x - 1)}{2\sqrt{\sin(4x + 1) + \cos(4x - 1)}}.$$

If a particle is moving in the plane, its position (x, y) at time t will be given by a pair of equations

$$x = f(t), \qquad y = g(t).$$

These are called *parametric equations*. The slope of the curve traced out by this particle can be found by the Chain Rule,

$$\frac{dy}{dx} = \frac{dy/dt}{dx/dt} = \frac{g'(t)}{f'(t)},$$

whenever the derivatives exist and $f'(t) \neq 0$. This is a Chain Rule application of type (2).

EXAMPLE 6 A ball thrown horizontally from a 100 ft cliff at a velocity of 50 ft/sec will follow the parametric equations

$$x = 50t, \qquad y = 100 - 16t^2, \quad \text{in feet.}$$

Find the slope of its path at time t (Figure 2.6.2).

$$\frac{dx}{dt} = 50, \qquad \frac{dy}{dt} = -32t,$$

so

$$\frac{dy}{dx} = \frac{dy/dt}{dx/dt} = -\frac{32t}{50}.$$

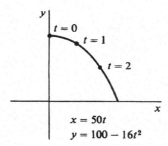

$$x = 50t$$
$$y = 100 - 16t^2$$

Figure 2.6.2

EXAMPLE 7 A particle moves according to the parametric equations

$$x = t^3 - t, \qquad y = t^2.$$

Find the slope of its path.

$$\frac{dx}{dt} = 3t^2 - 1, \qquad \frac{dy}{dt} = 2t,$$

so

$$\frac{dy}{dx} = \frac{dy/dt}{dx/dt} = \frac{2t}{3t^2 - 1}, \qquad t \neq \pm\sqrt{1/3}.$$

We see from Figure 2.6.3 that the path of this particle is not the graph of a function, and in fact contains a loop and crosses the point $(0, 1)$ twice, at $t = -1$ and $t = 1$. The path is vertical at the points $t = \pm\sqrt{1/3}$, where there is no slope. At the point $(0, 1)$, the two slopes of the path are $dy/dx = -1$ when $t = -1$, and $dy/dx = 1$ when $t = 1$.

EXAMPLE 8 A particle moving according to the parametric equations

$$x = \cos t, \qquad y = \sin t$$

will move counterclockwise around the unit circle at one radian per second

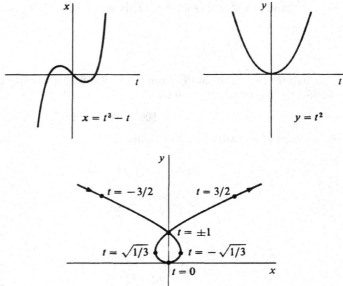

Figure 2.6.3

beginning at the point $(1, 0)$, as shown in Figure 2.6.4. Find the slope of its path at time t.

$$\frac{dx}{dt} = -\sin t, \qquad \frac{dy}{dt} = \cos t.$$

The slope is

$$\frac{dy}{dx} = \frac{dy/dt}{dx/dt} = -\frac{\cos t}{\sin t}.$$

In terms of x and y the slope is

$$\frac{dy}{dx} = -\frac{x}{y}.$$

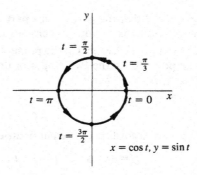

Figure 2.6.4

PROBLEMS FOR SECTION 2.6

In Problems 1–44, find dy/dx.

1 $y = \sqrt{x + 2}$ 2 $y = \sqrt{7 + 4x}$

3 $y = \sqrt{5 - x}$ 4 $y = \sqrt{1 - 10x}$

5 $y = \dfrac{1}{\sqrt{2 + 3x}}$ 6 $y = \dfrac{1}{\sqrt{4 - x}}$

7 $y = \sqrt[3]{6x + 1}$ 8 $y = \sqrt[5]{2 - 3x}$

9 $y = \sqrt{x^2 + 1}$ 10 $y = \sqrt{1 - x^2}$

11 $y = \sin(3x)$ 12 $y = \cos(4 - 2x)$

13 $y = \sin(x^{-2})$ 14 $y = \cos\sqrt{x}$

15 $y = e^{4x}$ 16 $y = e^{-x^2}$

17 $y = e^{\cos x}$ 18 $y = \ln(\ln x)$

19 $y = \cos u, \quad u = e^x$ 20 $y = \tan u, u = \ln x$

21 $y = u^{10}, \quad u = 1 - 4x$ 22 $y = u^{-10}, u = 1 - x^2$

23 $y = \sin u + \sin v, \quad u = 1 - x^2, \quad v = 2x - 1$

24 $y = e^u + e^v, \quad u = 1 - 3x, \quad v = 3 - 4x$

25 $y = e^u, \quad u = \sqrt{v}, \quad v = \sin x$

26 $y = \ln u, \quad u = \tan v, \quad v = 1/x$

27 $y = u^{-1/3}, \quad u = 1 + \sqrt{v}, \quad v = x^2 - 1$

28 $y = u^{-1}, \quad u = 3v + 4, \quad v = 1/(x + 1)$

29 $y = u^4, \quad u = 1 + 1/v, \quad v = x^3 + 1$

30 $y = u^2 + 1, \quad u = v^2 + 1, \quad v = x^2 + 1$

31 $y = (\sqrt{x^2 - 1} + \sqrt{x^2 + 1})^{1/3}$ 32 $y = (x + \sqrt{3 - 4x})^{-1/2}$

33 $y = 3x \sin(2x - 1)$ 34 $y = \sin(2x)\cos(3x)$

35 $x = \cos(3t), \quad y = \sin(3t)$ 36 $x = e^t, \quad y = \ln t$

37 $x = \sin t, \quad y = \sin(2t)$ 38 $x = \sin(e^t), \quad y = \cos(e^t)$

39 $x = \ln(t + 1), \quad y = t^2$ 40 $x = e^{\cos t}, \quad y = e^{\sin t}$

41 $x = \sqrt{t^2 - 4}, \quad y = \sqrt{t^2 + 4}$ 42 $x = 1 + \sqrt[3]{t}, \quad y = 2 + \sqrt[3]{t}$

43 $x = \sqrt[3]{t + 1}, \quad y = \sqrt[3]{t + 2}$ 44 $x = \dfrac{2t + 1}{t + 2}, \quad y = \dfrac{2t + 3}{t + 2}$

45 A particle moves in the plane according to the parametric equations
$$x = t^2 + 1, \qquad y = 3t^3.$$
Find the slope of its path.

46 An ant moves in the plane according to the equations
$$x = (1 - t^2)^{-1}, \qquad y = \sqrt{t}.$$
Find the slope of its path.

□ 47 Let y depend on u, u depend on v, and v depend on x. Assume the derivatives dy/du, du/dv, and dv/dx exist. Prove that
$$\frac{dy}{dx} = \frac{dy}{du}\frac{du}{dv}\frac{dv}{dx}.$$

□ 48 Let the function $f(x)$ be differentiable for all x, and let $g(x) = f(f(x))$. Show that $g'(x) = f'(f(x))f'(x)$.

2.7 HIGHER DERIVATIVES

DEFINITION

*The **second derivative** of a real function f is the derivative of the derivative of f, and is denoted by f''. The third derivative of f is the derivative of the second derivative, and is denoted by f''', or $f^{(3)}$. In general, the nth derivative of f is denoted by $f^{(n)}$.*

*If y depends on x, $y = f(x)$, then the **second differential** of y is defined to be*

$$d^2y = f''(x)\, dx^2.$$

In general the nth differential of y is defined by

$$d^ny = f^{(n)}(x)\, dx^n.$$

Here dx^2 means $(dx)^2$ and dx^n means $(dx)^n$.

We thus have the alternative notations

$$\frac{d^2y}{dx^2} = f''(x), \qquad \frac{d^ny}{dx^n} = f^{(n)}(x)$$

for the second and nth derivatives. The notation

$$y'' = f''(x), \qquad y^{(n)} = f^{(n)}(x)$$

is also used.

The definition of the second differential can be remembered in the following way. By definition,

$$dy = f'(x)\, dx.$$

Now hold dx constant and formally apply the Constant Rule for differentiation, obtaining

$$d(dy) = f''(x)\, dx\, dx,$$
or
$$d^2y = f''(x)\, dx^2.$$

(This is not a correct use of the Constant Rule because the rule applies to a real constant c, and dx is not a real number. It is only a mnemonic device to remember the definition of d^2y, not a proof.)

The third and higher differentials can be motivated in the same way. If we hold dx constant and formally use the Constant Rule again and again, we obtain

$$dy = f'(x)\, dx,$$
$$d^2y = f''(x)\, dx\, dx = f''(x)\, dx^2,$$
$$d^3y = f'''(x)\, dx^2\, dx = f'''(x)\, dx^3,$$
$$d^4y = f^{(4)}(x)\, dx^3\, dx = f^{(4)}(x)\, dx^4,$$

and so on.

The *acceleration* of a moving particle is defined to be the derivative of the velocity with respect to time,

$$a = dv/dt.$$

Thus the velocity is the first derivative of the distance and the acceleration is the second derivative of the distance. If s is distance, we have

$$v = \frac{ds}{dt}, \qquad a = \frac{d^2s}{dt^2}.$$

EXAMPLE 1   A ball thrown up with initial velocity b moves according to the equation

$$y = bt - 16t^2$$

with y in feet, t in seconds. Then the velocity is

$$v = b - 32t \text{ ft/sec},$$

and the acceleration (due to gravity) is a constant,

$$a = -32 \text{ ft/sec}^2.$$

EXAMPLE 2   Find the second derivative of $y = \sin(2\theta)$.

First derivative   Put $u = 2\theta$. Then

$$y = \sin u, \qquad \frac{dy}{du} = \cos u, \qquad \frac{du}{d\theta} = 2.$$

By the Chain Rule,

$$\frac{dy}{d\theta} = \frac{dy}{du} \cdot \frac{du}{d\theta} = \cos(2\theta) \cdot 2,$$

$$\frac{dy}{d\theta} = 2\cos(2\theta).$$

Second derivative   Let $v = 2\cos(2\theta)$. We must find $dv/d\theta$. Put $u = 2\theta$. Then

$$v = 2\cos u, \qquad \frac{dv}{du} = -2\sin u, \qquad \frac{du}{d\theta} = 2.$$

Using the Chain Rule again,

$$\frac{d^2y}{d\theta^2} = \frac{dv}{d\theta} = \frac{dv}{du} \cdot \frac{du}{d\theta} = (-2\sin(2\theta)) \cdot 2.$$

This simplifies to

$$\frac{d^2y}{d\theta^2} = -4\sin(2\theta).$$

EXAMPLE 3   A particle moves so that at time t it has gone a distance s along a straight line, its velocity is v, and its acceleration is a. Show that

$$a = v\frac{dv}{ds}.$$

By definition we have

$$v = \frac{ds}{dt}, \qquad a = \frac{dv}{dt},$$

so by the Chain Rule,

$$a = \frac{dv}{ds}\frac{ds}{dt} = v\frac{dv}{ds}.$$

EXAMPLE 4   If a polynomial of degree n is repeatedly differentiated, the kth derivative will be a polynomial of degree $n - k$ for $k \le n$, and the $(n + 1)$st derivative will be zero. For example,

$$y = 3x^5 - 10x^4 + x^2 - 7x + 4.$$
$$dy/dx = 15x^4 - 40x^3 + 2x - 7.$$
$$d^2y/dx^2 = 60x^3 - 120x^2 + 2.$$
$$d^3y/dx^3 = 180x^2 - 240x.$$
$$d^4y/dx^4 = 360x - 240.$$
$$d^5y/dx^5 = 360, \quad d^6y/dx^6 = 0.$$

Geometrically, the second derivative $f''(x)$ is the slope of the curve $y' = f'(x)$ and is also the rate of change of the slope of the curve $y = f(x)$.

PROBLEMS FOR SECTION 2.7

In Problems 1–23, find the second derivative.

1	$y = 1/x$	**2**	$y = x^5$	**3**	$y = \dfrac{-5}{x + 1}$
4	$f(x) = 3x^{-2}$	**5**	$f(x) = x^{1/2} + x^{-1/2}$	**6**	$f(t) = t^3 - 4t^2$
7	$f(t) = t\sqrt{t}$	**8**	$y = (3t - 1)^{10}$	**9**	$y = \sin x$
10	$y = \cos x$		**11**	$y = A \sin (Bx)$	
12	$y = A \cos (Bx)$		**13**	$y = e^{ax}$	
14	$y = e^{-ax}$		**15**	$y = \ln x$	
16	$y = x \ln x$		**17**	$y = \dfrac{1}{t^2 + 1}$	
18	$y = \sqrt{3t + 2}$	**19**	$z = \dfrac{x - 5}{x + 2}$	**20**	$z = \dfrac{2x - 1}{3x - 2}$
21	$z = x\sqrt{x + 1}$	**22**	$s = \left(\dfrac{t + 1}{t + 2}\right)^2$	**23**	$s = \sqrt{\dfrac{t}{t + 3}}$

24   Find the third derivative of $y = x^2 - 2/x$.

25   A particle moves according to the equation $s = 1 - 1/t^2$, $t > 0$. Find its acceleration.

26   An object moves in such a way that when it has moved a distance s its velocity is $v = \sqrt{s}$. Find its acceleration. (Use Example 3.)

27   Suppose u depends on x and d^2u/dx^2 exists. If $y = 3u$, find d^2y/dx^2.

28   If d^2u/dx^2 and d^2v/dx^2 exist and $y = u + v$, find d^2y/dx^2.

29   If d^2u/dx^2 exists and $y = u^2$, find d^2y/dx^2.

30   If d^2u/dx^2 and d^2v/dx^2 exist and $y = uv$, find d^2y/dx^2.

31   Let $y = ax^2 + bx + c$ be a polynomial of degree two. Show that dy/dx is a linear function and d^2y/dx^2 is a constant function.

☐ 32   Prove that the nth derivative of a polynomial of degree n is constant. (Use the fact that the derivative of a polynomial of degree k is a polynomial of degree $k - 1$.)

2.8 IMPLICIT FUNCTIONS

We now turn to the topic of implicit differentiation. We say that y is an *implicit function* of x if we are given an equation

$$\sigma(x, y) = \tau(x, y)$$

which determines y as a function of x. An example is $x + xy = 2y$. Implicit differentiation is a way of finding the derivative of y without actually solving for y as a function of x. Assume that dy/dx exists. The method has two steps:

Step 1   Differentiate both sides of the equation $\sigma(x, y) = \tau(x, y)$ to get a new equation

(1)
$$\frac{d(\sigma(x, y))}{dx} = \frac{d(\tau(x, y))}{dx}.$$

The Chain Rule is often used in this step.

Step 2   Solve the new Equation 1 for dy/dx. The answer will usually involve both x and y.

In each of the examples below, we assume that dy/dx exists and use implicit differentiation to find the value of dy/dx.

EXAMPLE 1   Given the equation $x + xy = 2y$, find dy/dx.

Step 1 $\dfrac{d(x + xy)}{dx} = \dfrac{d(2y)}{dx}$. We find each side by the Sum and Product Rules,

$$\frac{d(x + xy)}{dx} = \frac{dx + d(xy)}{dx} = \frac{dx + x\,dy + y\,dx}{dx}$$

$$= 1 + x\frac{dy}{dx} + y.$$

$$\frac{d(2y)}{dx} = 2\frac{dy}{dx}.$$

Thus our new equation is

$$1 + x\frac{dy}{dx} + y = 2\frac{dy}{dx}.$$

Step 2   Solve for dy/dx.

$$2\frac{dy}{dx} - x\frac{dy}{dx} = 1 + y.$$

$$\frac{dy}{dx} = \frac{1 + y}{2 - x}.$$

We can check our answer by solving the original equation for y and using ordinary differentiation:

$$x + xy = 2y.$$
$$2y - xy = x.$$
$$y = \frac{x}{2 - x}.$$

By the Quotient Rule,

$$\frac{dy}{dx} = \frac{(2-x)\cdot 1 - x(-1)}{(2-x)^2} = \frac{2}{(2-x)^2}.$$

A third way to find dy/dx is to solve the original equation for x, find dx/dy, and then use the Inverse Function Rule.

$$x + xy = 2y.$$

$$x = \frac{2y}{1+y}.$$

$$\frac{dx}{dy} = \frac{(1+y)\cdot 2 - 2y\cdot 1}{(1+y)^2} = \frac{2}{(1+y)^2}.$$

$$\frac{dy}{dx} = \frac{1}{2}(1+y)^2.$$

To see that our three answers

$$\frac{dy}{dx} = \frac{1+y}{2-x}, \qquad \frac{dy}{dx} = \frac{2}{(2-x)^2}, \qquad \frac{dy}{dx} = \frac{1}{2}(1+y)^2$$

are all the same we substitute $\dfrac{x}{2-x}$ for y:

$$\frac{dy}{dx} = \frac{1+y}{2-x} = \frac{1 + \dfrac{x}{2-x}}{2-x} = \frac{2}{(2-x)^2}.$$

$$\frac{dy}{dx} = \frac{1}{2}(1+y)^2 = \frac{1}{2}\left(1 + \frac{x}{2-x}\right)^2 = \frac{2}{(2-x)^2}.$$

In Example 1, we found dy/dx by three different methods.

(a) Implicit differentiation. We get dy/dx in terms of both x and y.

(b) Solve for y as a function of x and differentiate directly. This gives dy/dx in terms of x only.

(c) Solve for x as a function of y, find dx/dy directly, and use the Inverse Function Rule. This method gives dy/dx in terms of y only.

EXAMPLE 2   Given $y + \sqrt{y} = x^2$, find dy/dx.

$$\frac{d(y + \sqrt{y})}{dx} = \frac{d(x^2)}{dx}.$$

$$\frac{dy}{dx} + \frac{1}{2}y^{-1/2}\frac{dy}{dx} = 2x.$$

$$\frac{dy}{dx} = \frac{2x}{1 + \frac{1}{2}y^{-1/2}}.$$

This answer can be used to find the slope at any point on the curve. For example, at the point $(\sqrt{2}, 1)$ the slope is

$$\frac{2\sqrt{2}}{1 + \frac{1}{2} \cdot 1^{-1/2}} = \frac{2\sqrt{2}}{3/2} = \frac{4\sqrt{2}}{3}$$

while at the point $(-\sqrt{2}, 1)$ the slope is

$$\frac{2(-\sqrt{2})}{1 + \frac{1}{2} \cdot 1^{-1/2}} = \frac{-4\sqrt{2}}{3}.$$

To get dy/dx in terms of x, we solve the original equation for y using the quadratic formula:

$$y + \sqrt{y} - x^2 = 0,$$

$$\sqrt{y} = \frac{-1 \pm \sqrt{1 + 4x^2}}{2}.$$

Since $\sqrt{y} \geq 0$, only one solution may occur,

$$\sqrt{y} = \frac{-1 + \sqrt{1 + 4x^2}}{2}.$$

Then

$$y = \left(\frac{-1 + \sqrt{1 + 4x^2}}{2} \right)^2$$

The graph of this function is shown in Figure 2.8.1. By substitution we get

$$\frac{dy}{dx} = \frac{2x}{1 + \frac{1}{2}y^{-1/2}} = \frac{2x}{1 + (-1 + \sqrt{1 + 4x^2})^{-1}}.$$

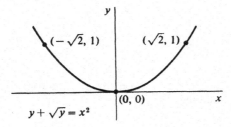

Figure 2.8.1

Often one side of an implicit function equation is constant and has derivative zero.

EXAMPLE 3 Given $x^2 - 2y^2 = 4$, $y \leq 0$, find dy/dx.

$$\frac{d(x^2 - 2y^2)}{dx} = \frac{d(4)}{dx}.$$

$$\frac{d(x^2 - 2y^2)}{dx} = 2x - 4y\frac{dy}{dx}.$$

$$\frac{d(4)}{dx} = 0.$$

$$2x - 4y\frac{dy}{dx} = 0.$$

$$\frac{dy}{dx} = \frac{-2x}{-4y} = \frac{x}{2y}.$$

Solving the original equation for y, we get

$$-2y^2 = 4 - x^2, \qquad y \le 0;$$

$$y^2 = \frac{x^2 - 4}{2}, \qquad y \le 0;$$

$$y = -\sqrt{\frac{x^2 - 4}{2}}.$$

Thus dy/dx in terms of x is

$$\frac{dy}{dx} = \frac{x}{2y} = \frac{x}{-2\sqrt{\dfrac{x^2 - 4}{2}}} = -\frac{x}{\sqrt{2(x^2 - 4)}}.$$

The graph of this function is shown in Figure 2.8.2.

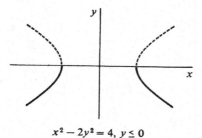

Figure 2.8.2 $x^2 - 2y^2 = 4, \ y \le 0$

Implicit differentiation can even be applied to an equation that does not by itself determine y as a function x. Sometimes extra inequalities must be assumed in order to make y a function of x.

EXAMPLE 4   Given

(2) $$x^2 + y^2 = 1,$$

find dy/dx. This equation does not determine y as a function of x; its graph is the unit circle. Nevertheless we differentiate both sides with respect to x and solve for dy/dx,

$$2x + 2y\frac{dy}{dx} = 0, \qquad \frac{dy}{dx} = -\frac{x}{y}.$$

We can conclude that for any system of formulas S which contains the Equation 2 and also determines y as a function of x, it is true that

(3) $$\frac{dy}{dx} = -\frac{x}{y}.$$

We can use Equation 3 to find the slope of the line tangent to the unit circle at any point on the circle. The following examples are illustrated in Figure 2.8.3.

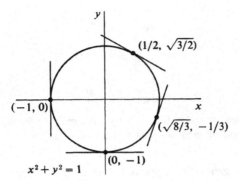

Figure 2.8.3

$x^2 + y^2 = 1$

$$\frac{dy}{dx} = 0 \text{ at } (0, -1), \qquad \frac{dy}{dx} = -\frac{1}{\sqrt{3}} \text{ at } \left(\frac{1}{2}, \frac{\sqrt{3}}{2}\right),$$

$$\frac{dy}{dx} = \sqrt{8} \text{ at } \left(\frac{\sqrt{8}}{3}, -\frac{1}{3}\right), \qquad \frac{dy}{dx} \text{ is undefined at } (-1, 0).$$

The system of formulas

$$x^2 + y^2 = 1, \qquad y \geq 0$$

gives us

$$y = \sqrt{1 - x^2}, \qquad \frac{dy}{dx} = -\frac{x}{y} = -\frac{x}{\sqrt{1 - x^2}}.$$

On the other hand the system

$$x^2 + y^2 = 1, \qquad y \leq 0$$

gives us

$$y = -\sqrt{1 - x^2}, \qquad \frac{dy}{dx} = -\frac{x}{y} = \frac{x}{\sqrt{1 - x^2}}.$$

EXAMPLE 5   Find the slope of the line tangent to the curve

(4)
$$x^5 y^3 + xy^6 = y + 1$$

at the points $(1, 1)$, $(1, -1)$, and $(0, -1)$.

The three points are all on the curve, and the first two points have the same x coordinate, so Equation 4 does not by itself determine y as a function of x. We differentiate with respect to x,

$$\frac{d(x^5 y^3 + xy^6)}{dx} = \frac{d(y + 1)}{dx},$$

$$5x^4 y^3 + x^5 \cdot 3y^2 \frac{dy}{dx} + y^6 + 6xy^5 \frac{dy}{dx} = \frac{dy}{dx},$$

and then solve for dy/dx,

$$5x^4 y^3 + y^6 + (3x^5 y^2 + 6xy^5 - 1)\frac{dy}{dx} = 0,$$

(5)
$$\frac{dy}{dx} = -\frac{5x^4 y^3 + y^6}{3x^5 y^2 + 6xy^5 - 1}.$$

Substituting, $$\frac{dy}{dx} = -\frac{6}{8} \quad \text{at } (1, 1).$$

$$\frac{dy}{dx} = -1 \quad \text{at } (1, -1),$$

$$\frac{dy}{dx} = +1 \quad \text{at } (0, -1).$$

Equation 5 for dy/dx is true of any system S of formulas which contains Equation 4 and determines y as a function of x.

Here is what generally happens in the method of implicit differentiation. Given an equation

(6) $$\tau(x, y) = \sigma(x, y)$$

between two terms which may involve the variables x and y, we differentiate both sides of the equation and obtain

(7) $$\frac{d(\tau(x, y))}{dx} = \frac{d(\sigma(x, y))}{dx}.$$

We then solve Equation 7 to get dy/dx equal to a term which typically involves both x and y. We can conclude that for any system of formulas which contains Equation 6 and determines y as a function of x, Equation 7 is true. Also, Equation 7 can be used to find the slope of the tangent line at any point on the curve $\tau(x, y) = \sigma(x, y)$.

PROBLEMS FOR SECTION 2.8

In Problems 1–26, find dy/dx by implicit differentiation. The answer may involve both x and y.

1	$xy = 1$	2	$2x^2 - 3y^2 = 4, \quad y \leq 0$
3	$x^3 + y^3 = 2$	4	$x^3 = y^5$
5	$y = 1/(x + y)$	6	$y^2 + 3y - 5 = x$
7	$x^{-2} + y^{-2} = 1$	8	$xy^3 = y + x$
9	$x^2 + 3xy + y^2 = 0$	10	$x/y + 3y = 2$
11	$x^5 = y^2 - y + 1$	12	$\sqrt{x} + \sqrt{y} = x + y$
13	$y = \sqrt{xy + 1}$	14	$x^4 + y^4 = 5$
15	$xy^2 - 3x^2y + x = 1$	16	$2xy^{-2} + x^{-2} = y$
17	$y = \sin(xy)$	18	$y = \cos(x + y)$
19	$x = \cos^2 y$	20	$x = \sin y + \cos y$
21	$y = e^{x+2y}$	22	$e^y = x^2 + y$
23	$e^x = \ln y$	24	$\ln y = \sin x$
25	$y^2 = \ln(2x + 3y)$	26	$\ln(\cos y) = 2x + 5$

In Problems 27–33, find the slope of the line tangent to the given curve at the given point or points.

27     $x^2 + xy + y^2 = 7$ at $(1, 2)$ and $(-1, 3)$

28     $x + y^3 = y$ at $(0, 0), (0, 1), (0, -1), (-6, 2)$

29     $x^2 - y^2 = 3$ at $(2, 1), (2, -1), (\sqrt{3}, 0)$

30     $\tan y = x^2$ at $(\pi/4, 1)$

31    $2 \sin^2 x = 3 \cos y$ at $(\pi/3, \pi/3)$

32    $y + e^y = 1 + \ln x$ at $(1, 0)$

33    $e^{\sin x} = \ln y$ at $(0, e)$

34   Given the equation $x^2 + y^2 = 1$, find dy/dx and d^2y/dx^2.

35   Given the equation $2x^2 - y^2 = 1$, find dy/dx and d^2y/dx^2.

36   Differentiating the equation $x^2 = y^2$ implicitly, we get $dy/dx = x/y$. This is undefined at the point $(0, 0)$. Sketch the graph of the equation to see what happens at the point $(0, 0)$.

EXTRA PROBLEMS FOR CHAPTER 2

1   Find the derivative of $f(x) = 4x^3 - 2x + 1$.

2   Find the derivative of $f(t) = 1/\sqrt{2t - 3}$.

3   Find the slope of the curve $y = x(2x + 4)$ at the point $(1, 6)$.

4   A particle moves according to the equation $y = 1/(t^2 - 4)$. Find the velocity as a function of t.

5   Given $y = 1/x^3$, express Δy and dy as functions of x and Δx.

6   Given $y = 1/\sqrt{x}$, express Δy and dy as functions of x and Δx.

7   Find $d(x^2 + 1/x^2)$.

8   Find $d(x - 1/x)$.

9   Find the equation of the line tangent to the curve $y = 1/(x - 2)$ at the point $(1, -1)$.

10   Find the equation of the line tangent to the curve $y = 1 + x\sqrt{x}$ at the point $(1, 2)$.

11   Find dy/dx where $y = -3x^3 - 5x + 2$.

12   Find dy/dx where $y = (2x - 5)^{-2}$.

13   Find ds/dt where $s = (3t + 4)(t^2 - 5)$.

14   Find ds/dt where $s = (4t^2 - 6)^{-1} + (1 - 2t)^{-2}$.

15   Find du/dv where $u = (2v^2 - 5v + 1)/(v^3 - 4)$.

16   Find du/dv where $u = (v + (1/v))/(v - (1/v))$.

17   Find dy/dx where $y = x^{1/2} + 4x^{3/2}$.

18   Find dy/dx where $y = (1 + \sqrt{x})^2$.

19   Find dy/dx where $y = x^{1/3} - x^{-1/4}$.

20   Find dy/dx where $y = e^x \cos^2 x$.

21   Find dy/dx where $x = \sqrt{y} + y^2, y > 0$.

22   Find dy/dx where $x = y^{-1/2} + y^{-1}, y > 0$.

23   Find dy/dx where $y = \sqrt{1 - 3x}$.

24   Find dy/dx where $y = \sin(2 + \sqrt{x})$.

25   Find dy/dx where $y = u^{-1/2}, u = 5x + 4$.

26   Find dy/dx where $y = u^5, u = 2 - x^3$.

27   Find the slope dy/dx of the path of a particle moving so that $y = 3t + \sqrt{t}, x = (1/t) - t^2$.

28   Find the slope dy/dx of the path of a particle moving so that $y = \sqrt{4t - 5}, x = \sqrt{3t + 6}$.

29   Find d^2y/dx^2 where $y = \sqrt{4x - 1}$.

30   Find d^2y/dx^2 where $y = x/(x^2 + 2)$.

31   An object moves so that $s = t\sqrt{t + 3}$. Find the velocity $v = ds/dt$ and the acceleration $a = d^2s/dt^2$.

32   Find dy/dx by implicit differentiation when $x + y + 2x^2 + 3y^3 = 2$.

33   Find dy/dx by implicit differentiation when $3xy^3 + 2x^3y = 1$.

34   Find the slope of the line tangent to the curve $2x\sqrt{y} - y^2 = \sqrt{x}$ at $(1, 1)$.

☐ **35**   Find the derivative of $f(x) = |x^2 - 1|$.

☐ **36**   Find the derivative of the function

$$f(x) = \begin{cases} 1 & \text{if } x \text{ is an integer,} \\ 0 & \text{otherwise.} \end{cases}$$

☐ **37**   Let $f(x) = (x - c)^{4/3}$. Show that $f'(x)$ exists for all real x but that $f''(c)$ does not exist.

☐ **38**   Let n be a positive integer and c a real number. Show that there is a function $g(x)$ which has an nth derivative at $x = c$ but does not have an $(n + 1)$st derivative at $x = c$. That is, $g^{(n)}(c)$ exists but $g^{(n+1)}(c)$ does not.

☐ **39**   (a) Let $u = |x|$, $y = u^2$. Show that at $x = 0$, dy/dx exists even though du/dx does not.
        (b) Let $u = x^4$, $y = |u|$. Show that at $x = 0$, dy/dx exists even though dy/du does not.

☐ **40**   Suppose $g(x)$ is differentiable at $x = c$ and $f(x) = |g(x)|$. Show that
        (a) $f'(c) = g'(c)$ if $g(c) > 0$,
        (b) $f'(c) = -g'(c)$ if $g(c) < 0$,
        (c) $f'(c) = 0$ if $g(c) = 0$ and $g'(c) = 0$,
        (d) $f'(c)$ does not exist if $g(c) = 0$ and $g'(c) \neq 0$.

☐ **41**   Prove by induction that for every positive integer n, $n < 2^n$.

☐ **42**   Prove by induction that the sum of the first n odd positive integers is equal to n^2,

$$1 + 3 + 5 + \cdots + (2n - 1) = n^2.$$

CONTINUOUS
FUNCTIONS

3.1 HOW TO SET UP A PROBLEM

In applications, a calculus problem is often presented verbally, and it is up to you
to set up the problem in mathematical terms. The problem can usually be described
mathematically by a list of equations and inequalities. The next two sections contain
several examples that illustrate the process of setting up a problem. The examples
in this section are from algebra and geometry, and those in the next section are from
calculus.

It is sometimes hard to see how to begin on a story problem. It is helpful
to break the process up into three steps:

Step 1 Draw a diagram if possible, and label all quantities involved.

Step 2 Write the given information as a set of equations and/or inequalities.

Step 3 Solve the mathematical problem, and interpret the mathematical solution
to answer the original story problem.

EXAMPLE 1 According to a treasure map, a buried treasure is located due east
of a cave and is 200 paces from a tree. The tree is 30 paces east and 40 paces
north of the cave. How far is the treasure from the cave?

The solution of this problem uses the quadratic formula, which will be
needed throughout the calculus course. We review it here.

QUADRATIC FORMULA If $a \neq 0$, then

$$ax^2 + bx + c = 0$$

if and only if

$$x = \frac{-b \pm \sqrt{b^2 - 4ac}}{2a}.$$

We solve Example 1 in three steps.

Step 1 Draw a diagram and label all quantities involved. In Figure 3.1.1, we put the cave at the origin and let x be the distance from the cave to the target along the x-axis. The tree is at the point $(30, 40)$, and the treasure is at the point $(x, 0)$.

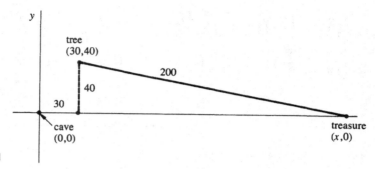

Figure 3.1.1

Step 2 Write the known information as a system of formulas. By the distance formula, we have

$$200 = \sqrt{(x - 30)^2 + (0 - 40)^2}, \quad x \geq 0.$$

The inequality $x \geq 0$ arises because the treasure is east of the cave.

Step 3 Solve for x. We square the Distance Formula.

$$40,000 = (x - 30)^2 + (0 - 40)^2 = x^2 - 60x + 900 + 1600$$
$$= x^2 - 60x + 2500$$
$$x^2 - 60x - 37,500 = 0$$

To find x we use the Quadratic Formula.

$$x = \frac{60 \pm \sqrt{(60)^2 - 4(-37,500)}}{2} = \frac{60 \pm \sqrt{153,600}}{2}$$
$$= 30 \pm \sqrt{38,400}$$

INTERPRET THE SOLUTION Since $x \geq 0$, we reject the negative solution. Thus $x = 30 + \sqrt{38,400} \sim 226$ paces. The treasure is approximately 226 paces from the cave.

Most calculus problems involve two or more variables.

EXAMPLE 2   A six-foot man stands near a ten-foot lamppost. Find the length of his shadow as a function of his distance from the lamppost.

Step 1 Draw a diagram and label all the quantities involved. In Figure 3.1.2, we let

$$x = \text{man's distance from lamppost,}$$
$$s = \text{length of his shadow.}$$

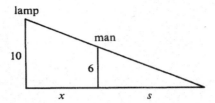

Figure 3.1.2

Step 2 Write the known information as a system of formulas. By similar triangles we have

$$\frac{s}{6} = \frac{s + x}{10}, \qquad x \geq 0.$$

The inequality $x \geq 0$ arises because the distance cannot be negative.

Step 3 Solve for s as a function of x.

$$10s = 6s + 6x,$$
$$4s = 6x,$$
$$s = \tfrac{3}{2}x.$$

INTERPRET THE SOLUTION $s = \tfrac{3}{2}x, \qquad x \geq 0.$

The domain of the function is $[0, \infty)$. The length of the shadow is $\tfrac{3}{2}$ times the distance from the lamppost. In this problem, x is the independent variable and s depends on x.

EXAMPLE 3 Two ships start at the same point at time $t = 0$. One ship moves north at 30 miles per hour, while the other ship moves east at 40 miles per hour. Find the distance between the two ships as a function of time.

Step 1 The ships start at the origin; the y-axis points north; and the x-axis points east. The diagram is shown in Figure 3.1.3. x and y are the distances of the east- and north-moving ships from the origin, and z is the distance between the ships, all in miles. t is the time in hours.

Step 2 $t \geq 0, \qquad y = 30t, \qquad x = 40t, \qquad z = \sqrt{x^2 + y^2}.$

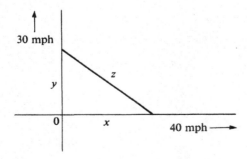

Figure 3.1.3

Step 3 $z = \sqrt{(30t)^2 + (40t)^2} = \sqrt{2500t^2} = 50t.$

INTERPRET THE SOLUTION $z = 50t,$ $t \geq 0.$

t is the independent variable, and x, y, z all depend on t. The distance between the ships is $50t$ miles, where t is the time in hours.

EXAMPLE 4   A brush fire starts along a straight line segment of length 20 ft and expands in all directions at the rate of 2 ft per second. Find the burned out area as a function of time.

Step 1 A = total burned out area
A_1 = area of left semicircle
A_2 = area of central rectangle
A_3 = area of right semicircle
s = distance of spread of fire
t = time

The diagram is shown in Figure 3.1.4.

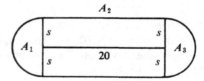

Figure 3.1.4

Step 2 $s = 2t,$ $t \geq 0.$
$A_1 = \frac{1}{2}\pi s^2,$ $A_2 = 20(2s),$ $A_3 = \frac{1}{2}\pi s^2.$
$A = A_1 + A_2 + A_3.$
Step 3 $A_1 = \frac{1}{2}\pi(2t)^2 = 2\pi t^2.$
$A_2 = 20 \cdot 2 \cdot 2t = 80t.$
$A_3 = \frac{1}{2}\pi(2t)^2 = 2\pi t^2.$
$A = 2\pi t^2 + 80t + 2\pi t^2 = 4\pi t^2 + 80t.$

INTERPRET THE SOLUTION   The burned out area is $A = 4\pi t^2 + 80t$ sq ft, $t \geq 0,$ where t is time in seconds.

An algebraic identity that comes up frequently in calculus problems is

$$(a - b)(a + b) = a^2 - b^2.$$

Sometimes it occurs in the form

$$(\sqrt{a} - \sqrt{b})(\sqrt{a} + \sqrt{b}) = a - b.$$

EXAMPLE 5   The area of square A is twelve square units greater than the area of square B, and the side of A is three units greater than the side of B. Find the areas of A and B.

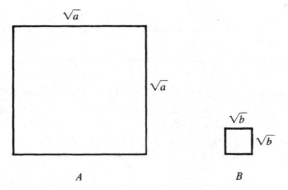

Figure 3.1.5   *A*   *B*

Step 1   Let a be the area of A and b the area of B. See Figure 3.1.5.

Step 2   The sides of the squares have length $\sqrt{a}$ and $\sqrt{b}$ respectively. Thus

$$a - b = 12, \qquad \sqrt{a} - \sqrt{b} = 3.$$

Step 3   We find $\sqrt{a} + \sqrt{b}$.

$$(\sqrt{a} - \sqrt{b})(\sqrt{a} + \sqrt{b}) = a - b,$$
$$3(\sqrt{a} + \sqrt{b}) = 12,$$
$$\sqrt{a} + \sqrt{b} = 4.$$

Adding the equations $\sqrt{a} + \sqrt{b} = 4$ and $\sqrt{a} - \sqrt{b} = 3$, we obtain $2\sqrt{a} = 7$, $\sqrt{a} = \frac{7}{2}$, $a = \frac{49}{4}$. Subtracting the equations gives $2\sqrt{b} = 1$, $\sqrt{b} = \frac{1}{2}, b = \frac{1}{4}$.

INTERPRET THE SOLUTION   The area of square A is $\frac{49}{4}$ square units, and the area of square B is $\frac{1}{4}$ square units.

PROBLEMS FOR SECTION 3.1

1   Find the perimeter p of a square as a function of its area A.

2   A piece of clay in the shape of a cube of side s is rolled into a sphere of radius r. Find r as a function of s.

3   Find the volume V of a sphere as a function of its surface area S.

4   Find the area A of a rectangle of perimeter 4 as a function of the length x.

5   Find the distance z between the origin and a point on the parabola $y = 1 - x^2$ as a function of x.

6   Express the perimeter p of a right triangle as a function of the base x and height y.

7   Four small squares of side x are cut from the corners of a large cardboard square of side s. The sides are then folded up to form an open top box. Find the volume of the box as a function of s and x.

8   A ladder of length L is propped up against a wall with its bottom at distance x from the wall. Find the height y of the top of the ladder as a function of x.

9   A man of height y stands 3 ft from a ten foot high lamp. Find the length s of his shadow as a function of y.

10   One ship traveling north at 30 mph passes the origin at time $t = 0$ hours. A second ship moving east at 30 mph passes the origin at $t = 1$. Find the distance z between them as a function of t.

11   A ball is thrown from ground level, and its path follows the equations

$$x = bt, \qquad y = t - 16t^2.$$

How far does it travel in the x direction before it hits the ground?

12   A circular weedpatch is initially 2 ft in radius. It grows so that its radius increases by 1 ft/day. Find its area after five days.

13   A rectangle originally has length l and width w. Its shape changes so that its length increases by one unit per second while its width decreases by 2 units per second. Find its area as a function of l, w and time t.

14   At p units of pollution per item, a product can be made at a cost of $2 + 1/p$ dollars per item. x items are to be produced with a total pollution of one unit. Find the cost.

15   In economics, the profit in producing and selling x items is equal to the revenue minus the cost,

$$P(x) = R(x) - C(x).$$

If a product can be manufactured at a cost of \$10 per item and x items can be sold at a price of $100 - \sqrt{x}$ per item, find the profit as a function of x.

16   Suppose the demand for a commodity at price p is $x = 1000/\sqrt{p}$, that is, $x = 1000/\sqrt{p}$ items can be sold at a price of p dollars per item. If it costs $100 + 10x$ dollars to produce x items, find the profit as a function of the selling price p.

3.2 RELATED RATES

In a related rates problem, we are given the rate of change of one quantity and wish to find the rate of change of another. Such problems can often be solved by implicit differentiation.

EXAMPLE 1   The point of a fountain pen is placed on an ink blotter, forming a circle of ink whose area increases at the constant rate of 0.03 in.²/sec. Find the rate at which the radius of the circle is changing when the circle has a radius of $\frac{1}{2}$ inch. We solve the problem in four steps.

Step 1   Label all quantities involved and draw a diagram.

$$t = \text{time} \qquad A = \text{area} \qquad r = \text{radius of circle}$$

Both A and r are functions of t. The diagram is shown in Figure 3.2.1.

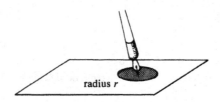

Figure 3.2.1

radius r

Step 2 Write the given information in the form of equations.

$$dA/dt = 0.03, \qquad A = \pi r^2.$$

The problem is to find dr/dt when $r = 1/2$.

Step 3 Differentiate both sides of the equation $A = \pi r^2$ with respect to t.

$$\frac{dA}{dt} = 2\pi r \frac{dr}{dt}, \quad \text{whence} \quad 0.03 = 2\pi r \frac{dr}{dt}.$$

Step 4 Set $r = \frac{1}{2}$ and solve for dr/dt.

$$0.03 = 2\pi \frac{1}{2} \cdot \frac{dr}{dt}, \quad \text{so} \quad \frac{dr}{dt} = \frac{0.03}{\pi} \text{in./sec.}$$

EXAMPLE 2 A 10 foot ladder is propped against a wall. The bottom end is being pulled along the floor away from the wall at the constant rate of 2 ft/sec. Find the rate at which the top of the ladder is sliding down the wall when the bottom end is 5 ft from the wall. *Warning*: although the bottom end of the ladder is being moved at a constant rate, the rate at which the top end moves will vary with time.

Step 1 t = time,
x = distance of the bottom end from the wall,
y = height of the top end above the floor.

The diagram is shown in Figure 3.2.2.

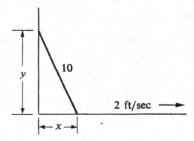

Figure 3.2.2

Step 2 $dx/dt = 2, \qquad x^2 + y^2 = 10^2 = 100.$

Step 3 We differentiate both sides of $x^2 + y^2 = 100$ with respect to t.

$$2x\frac{dx}{dt} + 2y\frac{dy}{dt} = 0, \quad \text{whence} \quad 4x + 2y\frac{dy}{dt} = 0.$$

Step 4 Set $x = 5$ ft and solve for dy/dt. We first find the value of y when $x = 5$.

$$x^2 + y^2 = 100, \qquad y = \sqrt{100 - x^2} = \sqrt{100 - 5^2} = \sqrt{75}.$$

Then we can solve for dy/dt,

$$4x + 2y\frac{dy}{dt} = 0,$$

$$4 \cdot 5 + 2\sqrt{75}\frac{dy}{dt} = 0,$$

$$\frac{dy}{dt} = -\frac{4 \cdot 5}{2\sqrt{75}} = -\frac{2}{\sqrt{3}}\text{ft/sec}.$$

The sign of dy/dt is negative because y is decreasing.

Related rates problems have the following form.

Given:

(a) Two quantities which depend on time, say x and y.
(b) The rate of change of one of them, say dx/dt.
(c) An equation showing the relationship between x and y.

(Usually this information is given in the form of a verbal description of a physical situation and part of the problem is to express it in the form of an equation.)

The problem: Find the rate of change of y, dy/dt, at a certain time t_0. (The time t_0 is sometimes specified by giving the value which x, or y, has at that time.)

Related rates problems can frequently be solved in four steps as we did in the examples.

Step 1 Label all quantities in the problem and draw a picture. If the labels are x, y, and t (time), the remaining steps are as follows:

Step 2 Write an equation for the given rate of change dx/dt. Write another equation for the given relation between x and y.

Step 3 Differentiate both sides of the equation relating x and y with respect to t. We choose the time t as the independent variable. The result is a new equation involving x, y, dx/dt, and dy/dt.

Step 4 Set $t = t_0$ and solve for dy/dt. It may be necessary to find the values of x, y, and dx/dt at $t = t_0$ first.

The hardest step is usually Step 2, because one has to start with the given verbal description of the problem and set it up as a system of formulas. Sometimes more than two quantities that depend on time are given. Here is an example with three.

EXAMPLE 3   One car moves north at 40 mph (miles per hour) and passes a point P at time 1:00. Another car moves east at 60 mph and passes the same point P at time 2:30. How fast is the distance between the two cars changing at the time 2:00?

It is not even easy to tell whether the two cars are getting closer or farther away at time 2:00. This is part of the problem.

Step 1 $t = $ time,
$y = $ position of the first car travelling north,
$x = $ position of the second car travelling east,
$z = $ distance between the two cars.

In the diagram in Figure 3.2.3, the point P is placed at the origin.

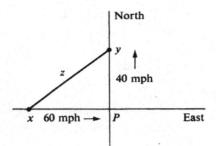

Figure 3.2.3

Step 2
$$\frac{dy}{dt} = 40, \quad \frac{dx}{dt} = 60, \quad x^2 + y^2 = z^2.$$

Step 3
$$2x\frac{dx}{dt} + 2y\frac{dy}{dt} = 2z\frac{dz}{dt}, \quad \text{whence} \quad 60x + 40y = z\frac{dz}{dt}.$$

Step 4 We first find the values of x, y, and z at the time $t = 2$ hrs. We are given that when $t = 1$, $y = 0$. In the next hour the car goes 40 miles, so at $t = 2$, $y = 40$. We are given that at time $t = 2\frac{1}{2}$, $x = 0$. One-half hour before that the car was 30 miles to the left of P, so at $t = 2$, $x = -30$. To sum up,

$$\text{at } t = 2, \quad y = 40 \quad \text{and} \quad x = -30.$$

We can now find the value of z at $t = 2$,

$$z = \sqrt{x^2 + y^2} = \sqrt{(-30)^2 + 40^2} = 50.$$

Finally, we solve for dz/dt at $t = 2$,

$$60 \cdot (-30) + 40 \cdot 40 = 50\frac{dz}{dt}, \quad \frac{dz}{dt} = \frac{-1800 + 1600}{50} = -4.$$

The negative sign shows that z is decreasing. Therefore at 2:00 the cars are getting closer to each other at the rate of 4 mph.

EXAMPLE 4 The population of a country is growing at the rate of one million people per year, while gasoline consumption is decreasing by one billion gallons per year. Find the rate of change of the per capita gasoline consumption when the population is 30 million and total gasoline consumption is 15 billion gallons per year.

By the per capita gasoline consumption we mean the total consumption divided by the population.

Step 1 t = time
x = population
y = gasoline consumption
z = per capita gasoline consumption.

Step 2   At $t = t_0$,

$$dx/dt = 1 \text{ million} = 10^6$$
$$dy/dt = -1 \text{ billion} = -10^9$$
$$z = y/x.$$

Step 3

$$\frac{dz}{dt} = \frac{x\,(dy/dt) - y\,(dx/dt)}{x^2},$$

$$\frac{dz}{dt} = \frac{-10^9 x - 10^6 y}{x^2}.$$

Step 4   At $t = t_0$, we are given

$$x = 30 \text{ million} = 30 \times 10^6,$$
$$y = 15 \text{ billion} = 15 \times 10^9.$$

Thus

$$\frac{dz}{dt} = \frac{-10^9 \cdot 30 \cdot 10^6 - 10^6 \cdot 15 \cdot 10^9}{(30 \cdot 10^6)^2}$$

$$= -\frac{45 \cdot 10^{15}}{900 \cdot 10^{12}} = -50.$$

The per capita gasoline consumption is decreasing at the annual rate of 50 gallons per person.

We conclude with another example from economics. In this example the independent variable is the quantity x of a commodity. The quantity x which can be sold at price p is called the *demand function* $D(p)$,

$$x = D(p).$$

When a quantity x is sold at price p, the *revenue* is the product

$$R = px.$$

The additional revenue from the sale of an additional unit of the commodity is called the *marginal revenue* and is given by the derivative

$$\text{marginal revenue} = dR/dx.$$

EXAMPLE 5   Suppose the demand for a product is equal to the inverse of the square of the price. Find the marginal revenue when the price is \$10 per unit.

Step 1 $p = $ price, $x = $ demand, $R = $ revenue.

Step 2 $x = 1/p^2, R = px.$

Step 3

$$\frac{dR}{dx} = p\frac{dx}{dx} + x\frac{dp}{dx} = p + x\frac{dp}{dx},$$

$$\frac{dx}{dp} = -2p^{-3},$$

so by the Inverse Function Rule,

$$\frac{dp}{dx} = \frac{1}{dx/dp} = -\frac{1}{2p^{-3}} = -\frac{1}{2}p^3.$$

Substituting,

$$\frac{dR}{dx} = p + \left(\frac{1}{p^2}\right)\left(-\frac{1}{2}p^3\right) = \frac{1}{2}p.$$

Step 4   We are given $p = \$10$. Therefore the marginal revenue is

$$dR/dx = \$5.$$

An additional unit sold would bring in an additional revenue of $5.

Here is a list of formulas from plane and solid geometry which will be useful in related rates problems. We always let $A = $ area and $V = $ volume.

Rectangle with sides a and b: $A = ab$, perimeter $= 2a + 2b$

Triangle with base b and height h: $A = \frac{1}{2}bh$

Circle of radius r: $A = \pi r^2$, circumference $= 2\pi r$

Sector (pie slice) of a circle of radius r and central angle θ (measured in radians): $A = \frac{1}{2}r^2\theta$

Rectangular solid with sides a, b, c: $V = abc$

Sphere of radius r: $V = \frac{4}{3}\pi r^3$, $A = 4\pi r^2$

Right circular cylinder, base of radius r, height of h: $V = \pi r^2 h$, $A = 2\pi rh$

Prism with base of area B and height h: $V = Bh$

Right circular cone, base of radius r, height h: $V = \pi r^2 h/3$, $A = \pi r\sqrt{r^2 + h^2}$

PROBLEMS FOR SECTION 3.2

1   Each side of a square is expanding at the rate of 5 cm/sec. How fast is the area changing when the length of each side is 10 cm?

2   The area of a square is decreasing at the constant rate of 2 sq cm/sec. How fast is the length of each side decreasing when the area is 1 sq cm?

3   The vertical side of a rectangle is expanding at the rate of 1 in./sec, while the horizontal side is contracting at the rate of 1 in./sec. At time $t = 1$ sec the rectangle is a square whose sides are 2 in. long. How fast is the area of the rectangle changing at time $t = 2$ sec?

4   Each edge of a cube is expanding at the rate of 1 in./sec. How fast is the volume of the cube changing when the volume is 27 cu in.?

5   Two cars pass point P at approximately the same time, one travelling north at 50 mph, the other travelling west at 60 mph. Find the rate of change of the distance between the two cars one hour after they pass the point P.

6   A cup in the form of a right circular cone with radius r and height h is being filled with water at the rate of 5 cu in./sec. How fast is the level of the water rising when the volume of the water is equal to one half the volume of the cup?

7   A spherical balloon is being inflated at the rate of 10 cu in./sec. Find the rate of change of the area when the balloon has radius 6 in.

8   A snowball melts at the rate equal to twice its surface area, with area in square inches and melting measured in cubic inches per hour. How fast is the radius shrinking?

9   A ball is dropped from a height of 100 ft, at which time its shadow is 500 ft from the ball. How fast is the shadow moving when the ball hits the ground? The ball falls with velocity 32 ft/sec, and the shadow is cast by the sun.

10   A 6 foot man walks away from a 10 foot high lamp at the rate of 3 ft/sec. How fast is the tip of his shadow moving?

11   A car is moving along a road at 60 mph. To the right of the road is a bush 10 ft away

and a parallel wall 30 ft away. Find the rate of motion of the shadow of the bush on the wall cast by the car headlights.

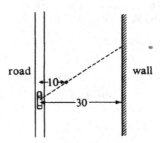

12   A car moves along a road at 60 mph. There is a bush 10 ft to the right of the road, and a wall 30 ft behind the bush is perpendicular to the road. Find the rate of motion of the shadow of the bush on the wall when the car is 26 ft from the bush.

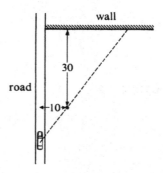

13   An airplane passes directly above a train at an altitude of 6 miles. If the airplane moves north at 500 mph and the train moves north at 100 mph, find the rate at which the distance between them is increasing two hours after the airplane passes over the train.

14   A rectangle has constant area, but its length is growing at the rate of 10 ft/sec. Find the rate at which the width is decreasing when the rectangle is 3 ft long and 1 ft wide.

15   A cylinder has constant volume, but its radius is growing at the rate of 1 ft/sec. Find the rate of change of its height when the radius and height are both 1 ft.

16   A country has constant national income, but its population is growing at the rate of one million people per year. Find the rate of change of the per capita income (national income divided by population) when the population is 20 million and the national income is 20 billion dollars.

17   If at time t a country has a birth rate of $1,000,000t$ births per year and a death rate of $300,000\sqrt{t}$ deaths per year, how fast is the population growing?

18   The population of a country is 10 million and is increasing at the rate of 500,000 people per year. The national income is $10 billion and is increasing at the rate of $100 million per year. Find the rate of change of the per capita income.

19   Work Problem 18 assuming that the population is decreasing by 500,000 per year.

20   Sand is poured at the rate of 4 cu in./sec and forms a conical pile whose height is equal to the radius of its base. Find the rate of increase of the height when the pile is 12 in. high.

21 A circular clock has radius 5 in. At time t minutes past noon, how fast is the area of the sector of the circle between the hour and minute hand increasing? ($t \le 60$).

22 The demand x for a commodity at price p is $x = 1/(1 + \sqrt{p})$. Find the *marginal revenue*, that is, the change in revenue per unit change in x, when the price is \$100 per unit.

23 x units of a commodity can be produced at a total cost of $y = 100 + 5x$. The *average cost* is defined as the total cost divided by x. Find the change in average cost per unit change in x (the marginal average cost) when $x = 100$.

24 The demand for a commodity at price p is $x = 1/(p + p^3)$. Find the change of the price per unit change in x, dp/dx, when the price is 3 dollars per unit.

25 In one day a company can produce x items at a total cost of $200 + 3x$ dollars and can sell x items at a price of $5 - x/1000$ dollars per item. *Profit* is defined as revenue minus cost. Find the change in profit per unit change in the number of items x (marginal profit).

26 In one day a company can produce x items at a total cost of $200 + 3x$ dollars and can sell $x = 1000/\sqrt{y}$ items at a price of y dollars per item.
(a) Find the change in profit per dollar change in the price y (the marginal profit with respect to price).
(b) Find the change in profit per unit change in x (the marginal profit).

27 An airplane P flies at 400 mph one mile above a line L on the surface. An observer is at the point O on L. Find the rate of change (in radians per hour) of the angle θ between the line L and the line OP from the observer to the airplane when $\theta = \pi/6$.

28 A train 20 ft wide is approaching an observer standing in the middle of the track at 100 ft/sec. Find the rate of increase of the angle subtended by the train (in radians per second) when the train is 20 ft from the observer.

29 Find the rate of increase of $e^{2x + 3y}$ when $x = 0$, $y = 0$, $dx/dt = 5$, and $dy/dt = 4$.

30 Find the rate of change of $\ln A$ where A is the area of a rectangle of sides x and y when $x = 1, y = 2, dx/dt = 3, dy/dt = -2$.

3.3 LIMITS

The notion of a limit is closely related to that of a derivative, but it is more general. In this chapter f will always be a real function of one variable. Let us recall the definition of the slope of f at a point a:

S is the slope of f at a if whenever Δx is infinitely close to but not equal to zero, the quotient

$$\frac{f(a + \Delta x) - f(a)}{\Delta x}$$

is infinitely close to S.

We now define the limit. c and L are real numbers.

DEFINITION

*L is the **limit** of $f(x)$ as x approaches c if whenever x is infinitely close to but not equal to c, $f(x)$ is infinitely close to L.*

In symbols,

$$\lim_{x \to c} f(x) = L$$

if whenever $x \approx c$ but $x \neq c$, $f(x) \approx L$. When there is no number L satisfying the above definition, we say that the limit of $f(x)$ as x approaches c *does not exist.*

Notice that the limit

$$\lim_{x \to c} f(x)$$

depends only on the values of $f(x)$ for x infinitely close but not equal to c. The value $f(c)$ itself has no influence at all on the limit. In fact, it very often happens that

$$\lim_{x \to c} f(x)$$

exists but $f(c)$ is undefined.

Figure 3.3.1(a) shows a typical limit. Looking at the point (c, L) through an infinitesimal microscope, we can see the entire portion of the curve with $x \approx c$ because $f(x)$ will be infinitely close to L and hence within the field of vision of the microscope.

In Figure 3.3.1(b), part of the curve with $x \approx c$ is outside the field of vision of the microscope, and the limit does not exist.

Our first example of a limit is the slope of a function.

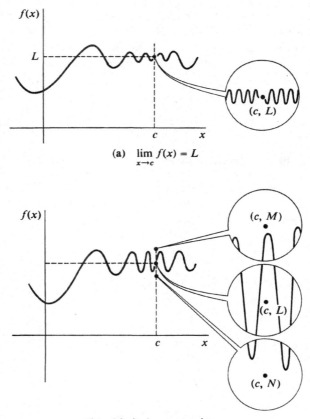

(a) $\lim_{x \to c} f(x) = L$

Figure 3.3.1   (b)   Limit does not exist

THEOREM 1

The slope of f at a is given by the limit

$$f'(a) = \lim_{\Delta x \to 0} \frac{f(a + \Delta x) - f(a)}{\Delta x}.$$

Verbally, the slope of f at a is the limit of the ratio of the change in $f(x)$ to the change in x as the change in x approaches zero. The theorem is seen by simply comparing the definitions of limit and slope. The slope exists exactly when the limit exists; and when they do exist they are equal. Notice that the ratio

$$\frac{f(a + \Delta x) - f(a)}{\Delta x}$$

is undefined when $\Delta x = 0$.

The slope of f at a is also equal to the limit

$$f'(a) = \lim_{x \to a} \frac{f(x) - f(a)}{x - a}.$$

This is seen by setting

$$\Delta x = x - a,$$
$$x = a + \Delta x.$$

Then when $x \approx a$ but $x \neq a$, we have $\Delta x \approx 0$ but $\Delta x \neq 0$; and

$$\frac{f(x) - f(a)}{x - a} = \frac{f(a + \Delta x) - f(a)}{\Delta x} \approx f'(a).$$

Sometimes a limit can be evaluated by recognizing it as a derivative and using Theorem 1 above.

EXAMPLE 1   Evaluate $\lim\limits_{\Delta x \to 0} \dfrac{(3 + \Delta x)^2 - 9}{\Delta x}$.

Let $F(x) = x^2$. The given limit is just $F'(3)$,

$$F'(3) = \lim_{\Delta x \to 0} \frac{F(3 + \Delta x) - F(3)}{\Delta x} = \lim_{\Delta x \to 0} \frac{(3 + \Delta x)^2 - 9}{\Delta x},$$

$$F'(3) = 2 \cdot 3 = 6.$$

Therefore

$$\lim_{\Delta x \to 0} \frac{(3 + \Delta x)^2 - 9}{\Delta x} = 6.$$

The symbol x in

$$\lim_{x \to c} f(x)$$

is an example of a "dummy variable." The value of the limit does not depend on x at all. However, it does depend on c. If we replace c by a variable u, we obtain a new function

$$L(u) = \lim_{x \to u} f(x).$$

A limit $\lim\limits_{x \to c} f(x)$ is usually computed as follows.

Step 1 Let x be infinitely close but not equal to c, and simplify $f(x)$.

Step 2 Compute the standard part $st(f(x))$.

CONCLUSION If the limit $\lim\limits_{x \to c} f(x)$ exists, it must equal $st(f(x))$.

EXAMPLE 1 (Continued) Instead of using the derivative, we can directly compute

$$\lim_{\Delta x \to 0} \frac{(3 + \Delta x)^2 - 9}{\Delta x}.$$

Step 1 Let $\Delta x \approx 0$, but $\Delta x \neq 0$. Then

$$\frac{(3 + \Delta x)^2 - 9}{\Delta x} = \frac{9 + 6\,\Delta x + \Delta x^2 - 9}{\Delta x} = \frac{6\,\Delta x + \Delta x^2}{\Delta x} = 6 + \Delta x.$$

Step 2 Taking standard parts,

$$st \frac{(3 + \Delta x)^2 - 9}{\Delta x} = st(6 + \Delta x) = 6.$$

Therefore the limit is equal to 6. (See Figure 3.3.2.)

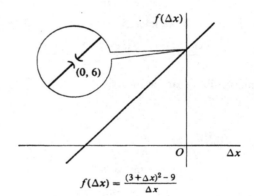

Figure 3.3.2

$$f(\Delta x) = \frac{(3 + \Delta x)^2 - 9}{\Delta x}$$

EXAMPLE 2 Find $\lim\limits_{t \to 4} (t^2 + 3t - 5)$.

Step 1 Let t be infinitely close to but not equal to 4.

Step 2 We take the standard part.

$$st(t^2 + 3t - 5) = 4^2 + 3 \cdot 4 - 5 = 23,$$

so the limit is 23.

EXAMPLE 3 Find $\lim\limits_{x \to 2} \dfrac{x^2 + 3x - 10}{x^2 - 4}$.

Step 1 This time the term inside the limit is undefined at $x = 2$. Taking $x \approx 2$ but $x \neq 2$, we have

$$\frac{x^2 + 3x - 10}{x^2 - 4} = \frac{(x-2)(x+5)}{(x-2)(x+2)} = \frac{x+5}{x+2}.$$

Step 2 $st\left(\dfrac{x^2 + 3x - 10}{x^2 - 4}\right) = st\left(\dfrac{x+5}{x+2}\right) = \dfrac{2+5}{2+2} = \dfrac{7}{4}.$

Thus $\displaystyle\lim_{x \to 2}\left(\frac{x^2 + 3x - 10}{x^2 - 4}\right) = \frac{7}{4}.$

EXAMPLE 4   Find $\displaystyle\lim_{x \to 0}\left(\frac{(2/x)+3}{(3/x)-1}\right)$

Step 1   Taking $x \approx 0$ but $x \neq 0$,

$$\frac{(2/x)+3}{(3/x)-1} = \frac{2+3x}{3-x}.$$

Step 2 $st\left(\dfrac{(2/x)+3}{(3/x)-1}\right) = st\left(\dfrac{2+3x}{3-x}\right) = \dfrac{2}{3}.$

Thus the limit exists and equals $\frac{2}{3}$.

EXAMPLE 5   Find $\displaystyle\lim_{x \to 9}\frac{\sqrt{x}-3}{x-9}.$

Step 1   Taking $x \approx 9$ and $x \neq 9$,

$$\frac{\sqrt{x}-3}{x-9} = \frac{(\sqrt{x}-3)(\sqrt{x}+3)}{(x-9)(\sqrt{x}+3)} = \frac{x-9}{(x-9)(\sqrt{x}+3)} = \frac{1}{\sqrt{x}+3}.$$

Step 2 $st\left(\dfrac{\sqrt{x}-3}{x-9}\right) = st\left(\dfrac{1}{\sqrt{x}+3}\right) = \dfrac{1}{\sqrt{9}+3} = \dfrac{1}{6},$

so the limit exists and is $\frac{1}{6}$.

Our rules for standard parts in Chapter 1 lead at once to rules for limits. We list these rules in Table 3.3.1. The limit rules can be applied whenever the two limits $\lim_{x \to c} f(x)$ and $\lim_{x \to c} g(x)$ exist.

Table 3.3.1

Standard Part Rule	Limit Rule
$st(kb) = k\, st(b)$, k real	$\displaystyle\lim_{x \to c} kf(x) = k \lim_{x \to c} f(x)$
$st(a+b) = st(a) + st(b)$	$\displaystyle\lim_{x \to c}(f(x) + g(x)) = \lim_{x \to c} f(x) + \lim_{x \to c} g(x)$
$st(ab) = st(a) \cdot st(b)$	$\displaystyle\lim_{x \to c}(f(x)g(x)) = \lim_{x \to c} f(x) \cdot \lim_{x \to c} g(x)$
$st(a/b) = st(a)/st(b)$, if $b \neq 0$	$\displaystyle\lim_{x \to c}(f(x)/g(x)) = \lim_{x \to c} f(x)/\lim_{x \to c} g(x)$, if $\lim_{x \to c} g(x) \neq 0$
$st(\sqrt[n]{a}) = \sqrt[n]{st(a)}$, if $a > 0$	$\displaystyle\lim_{x \to c} \sqrt[n]{f(x)} = \sqrt[n]{\lim_{x \to c} f(x)}$, if $\lim_{x \to c} f(x) > 0$

EXAMPLE 6   Find $\lim\limits_{x\to 1} (x^2 - 2x)\sqrt{(x^2 - 1)/(x - 1)}$.

All the limits involved exist, so we can use the limit rules to compute the limit as follows. First we find the limit of the expression inside the radical.

$$\lim_{x\to 1} \frac{x^2 - 1}{x - 1} = \lim_{x\to 1} \frac{(x - 1)(x + 1)}{x - 1} = \lim_{x\to 1} (x + 1) = 2.$$

Now we find the answer to the original problem.

$$\lim_{x\to 1} (x^2 - 2x)\sqrt{(x^2 - 1)/(x - 1)} = \lim_{x\to 1} (x^2 - 2x)\sqrt{\lim_{x\to 1} (x^2 - 1)/(x - 1)}$$
$$= (1 - 2)\sqrt{2} = -\sqrt{2}.$$

There are three ways in which a limit $\lim\limits_{x\to c} f(x)$ can fail to exist:

(1) $f(x)$ is undefined for some x which is infinitely close but not equal to c.
(2) $f(x)$ is infinite for some x which is infinitely close but not equal to c.
(3)   The standard part of $f(x)$ is different for different numbers x which are infinitely close but not equal to c.

EXAMPLE 7 $\lim\limits_{x\to 0} \sqrt{x}$ does not exist because $\sqrt{x}$ is undefined for negative infinitesimal x. (See Figure 3.3.3(a).)

EXAMPLE 8 $\lim\limits_{x\to 0} 1/x^2$ does not exist because $1/x^2$ is infinite for infinitesimal $x \neq 0$. (See Figure 3.3.3(b).)

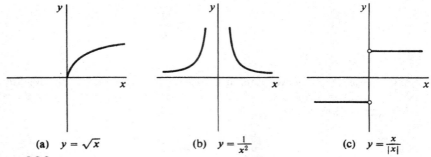

(a) $y = \sqrt{x}$ (b) $y = \dfrac{1}{x^2}$ (c) $y = \dfrac{x}{|x|}$

Figure 3.3.3

EXAMPLE 9 $\lim\limits_{x\to 0} x/|x|$ does not exist because

$$st\left(\frac{x}{|x|}\right) = \begin{cases} 1 & \text{if } x > 0, \\ -1 & \text{if } x < 0. \end{cases}$$

(See Figure 3.1.3(c).)

In the above examples the function behaves differently on one side of the point 0 than it does on the other side. For such functions, one-sided limits are useful.

We say that

$$\lim_{x \to c^+} f(x) = L$$

if whenever $x > c$ and $x \approx c$, $f(x) \approx L$.

$$\lim_{x \to c^-} f(x) = L$$

means that whenever $x < c$ and $x \approx c$, $f(x) \approx L$. These two kinds of limits, shown in Figure 3.3.4, are called the *limit from the right* and the *limit from the left*.

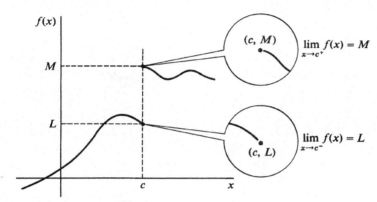

Figure 3.3.4 One-sided limits.

THEOREM 2

A limit has value L,

$$\lim_{x \to c} f(c) = L,$$

if and only if both one-sided limits exist and are equal to L,

$$\lim_{x \to c^-} f(x) = \lim_{x \to c^+} f(x) = L.$$

PROOF If $\lim_{x \to c} f(x) = L$, it follows at once from the definition that both one-sided limits are L.

Assume that both one-sided limits are equal to L. Let $x \approx c$, but $x \neq c$. Then either $x < c$ or $x > c$. If $x < c$, then because $\lim_{x \to c^-} f(x) = L$, we have $f(x) \approx L$. On the other hand if $x > c$, then $\lim_{x \to c^+} f(x) = L$ gives $f(x) \approx L$. Thus in either case $f(x) \approx L$. This shows that $\lim_{x \to c} f(x) = L$.

When a limit does not exist, it is possible that neither one-sided limit exists, that just one of them exists, or that both one-sided limits exist but have different values.

EXAMPLE 7 (Continued) $\lim_{x \to 0^+} \sqrt{x} = 0,$ and $\lim_{x \to 0^-} \sqrt{x}$ does not exist.

EXAMPLE 8 (Continued) Neither $\lim_{x \to 0^+} 1/x^2$ nor $\lim_{x \to 0^-} 1/x^2$ exists.

EXAMPLE 9 (Continued) $\lim_{x \to 0^+} x/|x| = 1,$ and $\lim_{x \to 0^-} x/|x| = -1.$

PROBLEMS FOR SECTION 3.3

In each problem below, determine whether or not the limit exists. When the limit exists, find its value. With a calculator, compute some values as x approaches its limit, and see what happens.

1    $\lim_{t \to 4} 3t^2 + t + 1$

2    $\lim_{\Delta x \to -1} \dfrac{\Delta x^2 + 2\Delta x + 1}{\Delta x + 1}$

3    $\lim_{x \to c} \sqrt{c - x}$

4    $\lim_{y \to 0} \dfrac{1}{y^3}$

5    $\lim_{x \to 2} \dfrac{x}{x^2 - 4}$

6    $\lim_{x \to 2} \dfrac{x^2 - 4}{x - 2}$

7    $\lim_{v \to 8} \dfrac{\sqrt{8} - \sqrt{v}}{v - 8}$

8    $\lim_{x \to 0} \dfrac{\sqrt{x + 1} - 1}{x}$

9    $\lim_{u \to 1} \dfrac{\sqrt[3]{u} - 1}{u - 1}$

10    $\lim_{t \to 0} \dfrac{t^3 - 2t^2 + 4}{3t^2 - 5t + 7}$

11    $\lim_{y \to 0} (\sqrt{1 + 1/y} - \sqrt{1/y}$

12    $\lim_{x \to 0} \dfrac{(a + x)^2 - a^2}{x}$

13    $\lim_{y \to -1} \dfrac{y^2 + 1}{y + 1}$

14    $\lim_{x \to 1} \dfrac{|x - 1|}{x - 1}$

15    $\lim_{x \to 1^+} \dfrac{|x - 1|}{x - 1}$

16    $\lim_{x \to c^-} \sqrt{c - x}$

17    $\lim_{z \to 1} \sqrt{z + \sqrt{z + \sqrt{z}}}$

18    $\lim_{x \to a} \sqrt{|a - x|}$

19    $\lim_{x \to 0^+} x\sqrt{1 + x^{-2}}$

20    $\lim_{x \to 0^-} x\sqrt{1 + x^{-2}}$

21    $\lim_{t \to 0} \dfrac{1 + 2t^{-1}}{3 - 4t^{-1}}$

22    $\lim_{x \to 0} \dfrac{3 + 4x^{-1} - 5x^{-2}}{6 - x^{-1} + 3x^{-2}}$

23    $\lim_{\Delta x \to 0} \dfrac{(x + \Delta x)^2 - x^2}{\Delta x}$

24    $\lim_{\Delta x \to 0} \dfrac{\dfrac{1}{x + \Delta x} - \dfrac{1}{x}}{\Delta x}$ $(x \neq 0)$

25    $\lim_{\Delta t \to 0} \dfrac{\sqrt{t + \Delta t} - \sqrt{t}}{\Delta t}$ $(t > 0)$

26    $\lim_{\Delta t \to 0} \dfrac{(t + \Delta t)^{1/5} - t^{1/5}}{\Delta t}$ $(t > 0)$

27    $\lim_{\Delta x \to 0} \dfrac{(x - \Delta x)^3 - x^3}{\Delta x}$

28    $\lim_{\Delta x \to 0} \dfrac{\dfrac{x + \Delta x}{x + \Delta x + 1} - \dfrac{x}{x + 1}}{\Delta x}$ $(x \neq -1)$

29    $\lim_{\Delta x \to 0^-} \dfrac{|(1 + \Delta x)^3 - (1 + \Delta x)|}{\Delta x}$

30    $\lim_{\Delta x \to 0^+} \dfrac{|(1 + \Delta x)^3 - (1 + \Delta x)|}{\Delta x}$

31    $\lim_{\Delta x \to 0^-} \dfrac{\sqrt{1 - (1 + \Delta x)^2}}{\Delta x}$

3.4 CONTINUITY

Intuitively, a curve $y = f(x)$ is continuous if it forms an unbroken line, that is, whenever x_1 is close to x_2, $f(x_1)$ is close to $f(x_2)$. To make this intuitive idea into a mathematical definition, we substitute "infinitely close" for "close."

DEFINITION

*f is said to be **continuous** at a point c if* :

(i) *f is defined at c*;

(ii) *whenever x is infinitely close to c, f(x) is infinitely close to f(c).*

If f is not continuous at c it is said to be *discontinuous* at c.

When f is continuous at c, the entire part of the curve where $x \approx c$ will be visible in an infinitesimal microscope aimed at the point $(c, f(c))$, as shown in Figure 3.4.1(a). But if f is discontinuous at c, some values of $f(x)$ where $x \approx c$ will either be undefined or outside the range of vision of the microscope, as in Figure 3.4.1(b).

Continuity, like the derivative, can be expressed in terms of limits. Again the proof is immediate from the definitions.

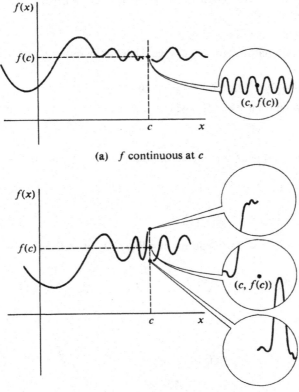

(a) f continuous at c

Figure 3.4.1   **(b)** f discontinuous at c

THEOREM 1

f is continuous at c if and only if

$$\lim_{x \to c} f(x) = f(c).$$

As an application, we have a set of rules for combining continuous functions. They can be proved either by the corresponding rules for limits (Table 3.3.1 in Section 3.3) or by computing standard parts.

THEOREM 2

Suppose f and g are continuous at c.

(i)   For any constant k, the function $k \cdot f(x)$ is continuous at c.

(ii) $f(x) + g(x)$ is continuous at c.

(iii) $f(x) \cdot g(x)$ is continuous at c.

(iv)   If $g(c) \neq 0$, then $f(x)/g(x)$ is continuous at c.

(v)   If $f(c)$ is positive and n is an integer, then $\sqrt[n]{f(x)}$ is continuous at c.

By repeated use of Theorem 2, we see that all of the following functions are continuous at c.

Every polynomial function.

Every rational function $f(x)/g(x)$, where $f(x)$ and $g(x)$ are polynomials and $g(c) \neq 0$.

The functions $f(x) = x^r$, r rational and x positive.

Sometimes a function $f(x)$ will be undefined at a point $x = c$ while the limit

$$L = \lim_{x \to c} f(x)$$

exists. When this happens, we can make the function continuous at c by defining $f(c) = L$.

EXAMPLE 1   Let $f(x) = \dfrac{x^2 + x - 2}{x - 1}$.

At any point $c \neq 1$, f is continuous. But $f(1)$ is undefined so f is discontinuous at 1. However,

$$\lim_{x \to 1} \frac{x^2 + x - 2}{x - 1} = \lim_{x \to 1} \frac{(x - 1)(x + 2)}{x - 1} = 3.$$

We can make f continuous at 1 by defining

$$f(x) = \begin{cases} \dfrac{x^2 + x - 2}{x - 1} & \text{if } x \neq 1, \\ 3 & \text{if } x = 1. \end{cases}$$

(See Figure 3.4.2.)

In terms of a dependent variable $y = f(x)$, the definition of continuity takes the following form, where $\Delta y = f(c + \Delta x) - f(c)$.

y is continuous at $x = c$ if :

(i) y is defined at $x = c$.

(ii)   Whenever Δx is infinitesimal, Δy is infinitesimal.

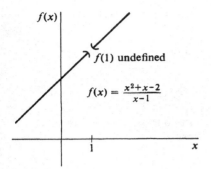

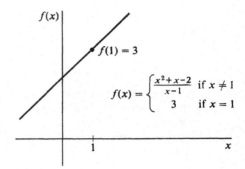

Figure 3.4.2

To summarize, given a function $y = f(x)$ defined at $x = c$, all the statements below are equivalent.

(1) f is continuous at c.
(2)   Whenever $x \approx c$, $f(x) \approx f(c)$.
(3)   Whenever $st(x) = c$, $st(f(x)) = f(c)$.
(4) $\lim_{x \to c} f(x) = f(c)$.
(5) y is continuous at $x = c$.
(6)   Whenever Δx is infinitesimal, Δy is infinitesimal.

Our next theorem is that differentiability implies continuity. That is, the set of differentiable functions at c is a subset of the set of continuous functions at c. (See Figure 3.4.3.)

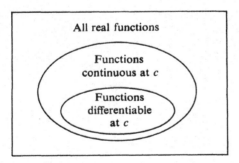

Figure 3.4.3

THEOREM 3

If f is differentiable at c then f is continuous at c.

PROOF Let $y = f(x)$, and let Δx be a nonzero infinitesimal. Then $\Delta y/\Delta x$ is infinitely close to $f'(c)$ and is therefore finite. Thus $\Delta y = \Delta x(\Delta y/\Delta x)$ is the product of an infinitesimal and a finite number, so Δy is infinitesimal.

For example, the transcendental functions $\sin x$, $\cos x$, e^x are continuous for all x, and $\ln x$ is continuous for $x > 0$. Theorem 3 can be used to show that combinations of these functions are continuous.

EXAMPLE 2 Find as large a set as you can on which the function

$$f(x) = \frac{\sin x \ln (x + 1)}{x^2 - 4}$$

is continuous.

$\sin x$ is continuous for all x. $\ln (x + 1)$ is continuous whenever $x + 1 > 0$, that is, $x > -1$. The numerator $\sin x \ln (x + 1)$ is thus continuous whenever $x > -1$. The denominator $x^2 - 4$ is continuous for all x but is zero when $x = \pm 2$. Therefore $f(x)$ is continuous whenever $x > -1$ and $x \neq 2$.

The next two examples give functions which are continuous but *not* differentiable at a point c.

EXAMPLE 3 The function $y = x^{1/3}$ is continuous but not differentiable at $x = 0$. (See Figure 3.4.4(a).) We have seen before that it is not differentiable at $x = 0$. It is continuous because if Δx is infinitesimal then so is

$$\Delta y = (0 + \Delta x)^{1/3} - 0^{1/3} = (\Delta x)^{1/3}.$$

EXAMPLE 4 The absolute value function $y = |x|$ is continuous but not differentiable at the point $x = 0$. (See Figure 3.4.4(b).)

We have already shown that the derivative does not exist at $x = 0$. To see that the function is continuous, we note that for any infinitesimal Δx,

$$\Delta y = |0 + \Delta x| - |0| = |\Delta x|$$

and thus Δy is infinitesimal.

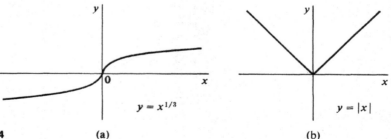

Figure 3.4.4   (a)   (b)

$y = x^{1/3}$ $y = |x|$

The path of a bouncing ball is a series of parabolas shown in Figure 3.4.5. The curve is continuous everywhere. At the points $a_1, a_2, a_3, \ldots$ where the ball bounces, the curve is continuous but not differentiable. At other points, the curve is both continuous and differentiable.

In the classical kinetic theory of gases, a gas molecule is assumed to be moving at a constant velocity in a straight line except at the instant of time when it collides with another molecule or the wall of the container. Its path is then a broken line in space, as in Figure 3.4.6.

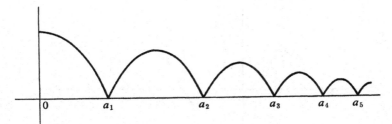

Figure 3.4.5   Path of Bouncing Ball

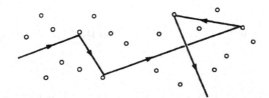

Figure 3.4.6

The position in three dimensional space at time t can be represented by three functions

$$x = f(t), \qquad y = g(t), \qquad z = h(t).$$

All three functions, f, g, and h are continuous for all values of t. At the time t of a collision, at least one and usually all three derivatives dx/dt, dy/dt, dz/dt will be undefined because the speed or direction of the molecule changes abruptly. At any other time t, when no collision is taking place, all three derivatives dx/dt, dy/dt, dz/dt will exist.

The functions we shall ordinarily encounter in this book will be defined and have a derivative at all but perhaps a finite number of points of an interval. The graph of such a function will be a smooth curve where the derivative exists. At points where the curve has a sharp corner (like 0 in $|x|$) or a vertical tangent line (like 0 in $x^{1/3}$), the function is continuous but not differentiable (see Figure 3.4.7). At points where the function is undefined or there is a jump, or the value approaches infinity or oscillates wildly, the function is discontinuous (see Figure 3.4.8).

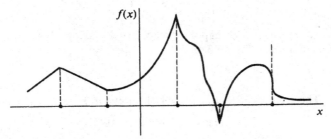

Figure 3.4.7   Points where f is continuous but nondifferentiable

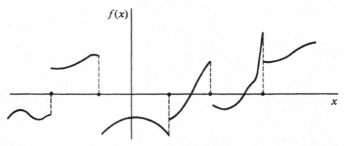

Figure 3.4.8   Points where f is discontinuous

The next theorem is similar to the Chain Rule for derivatives.

THEOREM 4

If f is continuous at c and G is continuous at $f(c)$, then the function

$$g(x) = G(f(x))$$

is also continuous at c. That is, a continuous function of a continuous function is continuous.

PROOF Let x be infinitely close to but not equal to c. Then

$$st(g(x)) = st(G(f(x))) = G(st(f(x))) = G(f(c)) = g(c).$$

For example, the following functions are continuous:

$$\begin{aligned}
f(x) &= \sqrt{x^2 + 1}, & \text{all } x \\
g(x) &= |x^3 - x|, & \text{all } x \\
h(x) &= (1 + \sqrt{x})^{1/3}, & x > 0 \\
j(x) &= e^{\sin x}, & \text{all } x \\
k(x) &= \ln|x|, & \text{all } x \neq 0
\end{aligned}$$

Here are two examples illustrating two types of discontinuities.

EXAMPLE 5   The function $g(x) = \dfrac{x^2 - 3x + 4}{4(x - 1)(x - 2)}$

is continuous at every real point except $x = 1$ and $x = 2$. At these two points $g(x)$ is undefined (Figure 3.4.9).

EXAMPLE 6   The *greatest integer function* $[x]$, shown in Figure 3.4.10, is defined by

$$[x] = \text{the greatest integer } n \text{ such that } n \leq x.$$

Thus $[x] = 0$ if $0 \leq x < 1$, $[x] = 1$ if $1 \leq x < 2$, $[x] = 2$ if $2 \leq x < 3$, and so on. For negative x, we have $[x] = -1$ if $-1 \leq x < 0$, $[x] = -2$

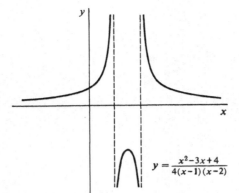

$$y = \frac{x^2 - 3x + 4}{4(x-1)(x-2)}$$

Figure 3.4.9

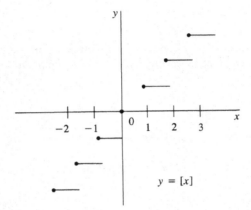

$$y = [x]$$

Figure 3.4.10

if $-2 \leq x < -1$, and so on. For example,

$$[7.362] = 7, \qquad [\pi] = 3, \qquad [-2.43] = -3.$$

For each integer n, $[n]$ is equal to n. The function $[x]$ is continuous when x is not an integer but is discontinuous when x is an integer n. At an integer n, both one-sided limits exist but are different,

$$\lim_{x \to n^-} f(x) = n - 1, \qquad \lim_{x \to n^+} f(x) = n.$$

The graph of $[x]$ looks like a staircase. It has a step, or jump discontinuity, at each integer n. The function $[x]$ will be useful in the last section of this chapter. Some hand calculators have a key for either the greatest integer function or for the similar function that gives $[x]$ for positive x and $[x] + 1$ for negative x.

Functions which are "continuous on an interval" will play an important role in this chapter. Intervals were discussed in Section 1.1. Recall that closed intervals have the form

$$[a, b],$$

open intervals have one of the forms

$$(a, b), \qquad (a, \infty), \qquad (-\infty, b), \qquad (-\infty, \infty),$$

and half-open intervals have one of the forms

$$[a, b), \qquad (a, b], \qquad [a, \infty), \qquad (-\infty, b].$$

In these intervals, a is called the *lower endpoint* and b, the *upper endpoint*. The symbol $-\infty$ indicates that there is no lower endpoint, while ∞ indicates that there is no upper endpoint.

DEFINITION

> We say that f is **continuous on an open interval** I if f is continuous at every point c in I. If in addition f has a derivative at every point of I, we say that f is **differentiable on** I.

To define what is meant by a function continuous on a closed interval, we introduce the notions of continuous from the right and continuous from the left, using one-sided limits.

DEFINITION

> f is **continuous from the right** at c if $\lim\limits_{x \to c^+} f(x) = f(c)$.
>
> f is **continuous from the left** at c if $\lim\limits_{x \to c^-} f(x) = f(c)$.

EXAMPLE 6 (Continued) The greatest integer function $f(x) = [x]$ is continuous from the right but not from the left at each integer n because

$$[n] = n, \qquad \lim_{x \to n^+} [x] = n, \qquad \lim_{y \to n^-} [x] = n - 1.$$

It is easy to check that f is continuous at c if and only if f is continuous from both the right and left at c.

DEFINITION

> f is said to be **continuous on the closed interval** $[a, b]$ if f is continuous at each point c where $a < c < b$, continuous from the right at a, and continuous from the left at b.

Figure 3.4.11 shows a function f continuous on $[a, b]$.

EXAMPLE 7 The semicircle

$$y = \sqrt{1 - x^2},$$

shown in Figure 3.4.12, is continuous on the closed interval $[-1, 1]$. It is

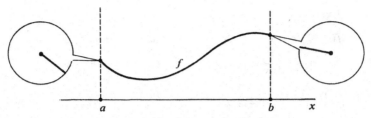

Figure 3.4.11   *f* is continuous on the interval [*a*, *b*]

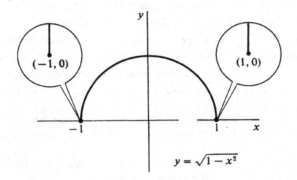

Figure 3.4.12

differentiable on the open interval $(-1, 1)$. To see that it is continuous from the right at $x = -1$, let Δx be positive infinitesimal. Then

$$y = \sqrt{1 - (-1)^2} = 0$$
$$y + \Delta y = \sqrt{1 - (-1 + \Delta x)^2} = \sqrt{1 - (1 - 2\Delta x + \Delta x^2)}$$
$$= \sqrt{2\Delta x - \Delta x^2} = \sqrt{(2 - \Delta x)\Delta x}.$$

Thus

$$\Delta y = \sqrt{(2 - \Delta x)\Delta x}.$$

The number inside the radical is positive infinitesimal, so Δy is infinitesimal. This shows that the function is continuous from the right at $x = -1$. Similar reasoning shows it is continuous from the left at $x = 1$.

PROBLEMS FOR SECTION 3.4

In Problems 1–17, find the set of all points at which the function is continuous.

1 $f(x) = 3x^2 + 5x + 4$

2 $f(x) = \dfrac{5x + 2}{x^2 + 1}$

3 $f(x) = \sqrt{x + 2}$

4 $f(x) = \dfrac{x}{x + 2}$

5 $f(x) = \sqrt{|x - 2| + 1}$

6 $f(x) = \dfrac{x + 3}{|x + 3|}$

7    $f(x) = \dfrac{x}{x^2 + x}$ 8 $f(x) = \dfrac{x + 2}{(x - 1)(x - 3)^{1/3}}$

9    $f(x) = \sqrt{4 - x^2}$ 10 $f(x) = \sqrt{x^2 - 4}$

11    $f(x) = \dfrac{1}{x - (1/(x + 1))}$ 12 $g(x) = \dfrac{1}{x} + \dfrac{1}{x - 1}$

13    $g(x) = \dfrac{x - 2}{x - 3} + \dfrac{x - 3}{x - 2}$ 14 $g(x) = \sqrt{x^3 - x}$

15    $g(x) = \sqrt[4]{x^2 - x^3}$ 16 $f(t) = \sqrt{t^{-2} - 1}$

17    $f(t) = \sqrt{t^{-1} - 1}$

18   Show that $f(x) = \sqrt{x}$ is continuous from the right at $x = 0$.

19   Show that $f(x) = \sqrt{1 - x}$ is continuous from the left at $x = 1$.

20   Show that $f(x) = \sqrt{1 - |x|}$ is continuous on the closed interval $[-1, 1]$.

21   Show that $f(x) = \sqrt{x} + \sqrt{2 - x}$ is continuous on the closed interval $[0, 2]$.

22   Show that $f(x) = \sqrt{9 - x^2}$ is continuous on the closed interval $[-3, 3]$.

23   Show that $f(x) = \sqrt{x^2 - 9}$ is continuous on the half-open intervals $(-\infty, -3]$ and $[3, \infty)$.

☐ 24   Suppose the function $f(x)$ is continuous on the closed interval $[a, b]$. Show that there is a function $g(x)$ which is continuous on the whole real line and has the value $g(x) = f(x)$ for x in $[a, b]$.

☐ 25   Suppose $\lim_{x \to c} f(x) = L$. Prove that the function $g(x)$, defined by $g(x) = f(x)$ for $x \neq c$ and $g(x) = L$ for $x = c$, is continuous at c.

26   In the curve $y = f(x)$ illustrated below, identify the points $x = c$ where each of the following happens:
(a) f is discontinuous at $x = c$
(b) f is continuous but not differentiable at $x = c$.

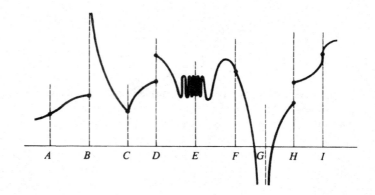

A   B   C   D   E   F   G   H   I

3.5 MAXIMA AND MINIMA

Let us assume throughout this section that f is a real function whose domain is an interval I, and furthermore that f is continuous on I. A problem that often arises is that of finding the point c where $f(c)$ has its largest value, and also the point c where $f(c)$ has its smallest value. The derivative turns out to be very useful in this problem. We begin by defining the concepts of maximum and minimum.

DEFINITION

Let c be a real number in the domain I of f.

(i) f has a **maximum** at c if $f(c) \geq f(x)$ for all real numbers x in I. In this case $f(c)$ is called the **maximum value** of f.

(ii) f has a **minimum** at c if $f(c) \leq f(x)$ for all real numbers x in I. $f(c)$ is then called the **minimum value** of f.

When we look at the graph of a continuous function f on I, the maximum will appear as the highest peak and the minimum as the lowest valley (Figure 3.5.1).

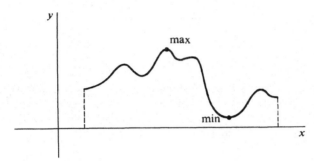

Figure 3.5.1   Maximum and Minimum

In general, all of the following possibilities can arise:

f has no maximum in its domain I.
f has a maximum at exactly one point in I.
f has a maximum at several different points in I.

However even if f has a maximum at several different points, f can have only one maximum value. Because if f has a maximum at c_1 and also at c_2, then $f(c_1) \geq f(c_2)$ and $f(c_2) \geq f(c_1)$, and therefore $f(c_1)$ and $f(c_2)$ are equal.

EXAMPLE 1   Each of the following functions, graphed in Figure 3.5.2, have no maximum and no minimum:

(a) $f(x) = 1/x$, $0 < x$.
(b) $f(x) = x^2$, $0 < x < 1$.
(c) $f(x) = 2x + 3$.

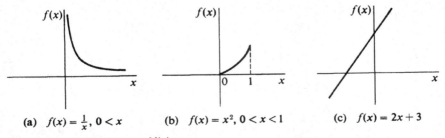

(a) $f(x) = \frac{1}{x}, 0 < x$ (b) $f(x) = x^2, 0 < x < 1$ (c) $f(x) = 2x + 3$

Figure 3.5.2   No Maximum or Minimum

EXAMPLE 2   The function $f(x) = x^2 + 1$ has no maximum. But f has a minimum at $x = 0$ with value 1, because for $x \neq 0$, we always have $x^2 > 0, x^2 + 1 > 1$. The graph is shown in Figure 3.5.3.

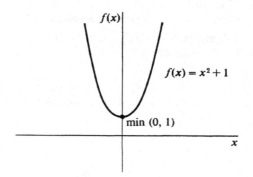

Figure 3.5.3

The use of the derivative in finding maxima and minima is based on the Critical Point Theorem. It shows that the maxima and minima of a function can only occur at certain points, called *critical points*. The theorem will be stated now, and its proof is given at the end of this section.

CRITICAL POINT THEOREM

Let f be continuous on its domain I. Suppose that c is a point in I and f has either a maximum or a minimum at c. Then one of the following three things must happen:

(i)   *c is an endpoint of I,*
(ii)   *$f'(c)$ is undefined,*
(iii)   *$f'(c) = 0$.*

We shall say that c is a *critical point* of f if either (i), (ii), or (iii) happens. The three types of critical points are shown in Figure 3.5.4. When I is an open interval, (i) cannot arise since the endpoints are not elements of I. But when I is a closed

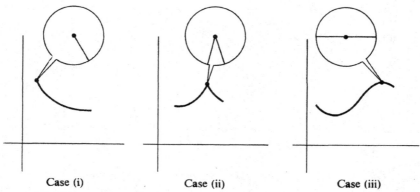

Case (i)   Case (ii)   Case (iii)

Figure 3.5.4   Critical Point Theorem

interval, the two endpoints of I will always be among the critical points. Geometrically the theorem says that if f has a maximum or minimum at c, then either c is an endpoint of the curve, or there is a sharp corner at c, or the curve has a horizontal slope at c. Thus at a maximum there is either an endpoint, a sharp peak, or a horizontal summit.

The Critical Point Theorem has some important applications to economics. Here is one example. Some other examples are described in the problem set.

EXAMPLE 3   Suppose a quantity x of a commodity can be produced at a total cost $C(x)$ and sold for a total revenue of $R(x)$, $0 < x < \infty$. The *profit* is defined as the difference between the revenue and the cost,

$$P(x) = R(x) - C(x).$$

Show that if the profit has a maximum at x_0, then the marginal cost is equal to the marginal revenue at x_0,

$$R'(x_0) = C'(x_0).$$

In this problem it is understood that $R(x)$ and $C(x)$ are differentiable functions, so that the marginal cost and marginal revenue always exist. Therefore $P'(x)$ exists and

$$P'(x) = R'(x) - C'(x).$$

Assume $P(x)$ has a maximum at x_0. Since $(0, \infty)$ has no endpoints and $P'(x_0)$ exists, the Critical Point Theorem shows that $P'(x_0) = 0$. Thus

$$P'(x_0) = R'(x_0) - C'(x_0) = 0$$
and $$R'(x_0) = C'(x_0).$$

DEFINITION

*An **interior point** of an interval I is an element of I which is not an endpoint of I.*

For example, if I is an open interval, then every point of I is an interior point of I. But if I is a closed interval $[a, b]$, then the set of all interior points of I is the open interval (a, b) (Figure 3.5.5).

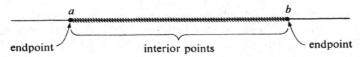

Figure 3.5.5

An interior point of I which is a critical point of f is called an *interior critical point*. There are a number of tests to determine whether or not f has a maximum at a given interior critical point. Here are two such tests. In both tests we assume that f is continuous on its domain I.

DIRECT TEST

Suppose c is the only interior critical point of f, and u, v are points in I with $u < c < v$.

(i)   *If $f(c) > f(u)$ and $f(c) > f(v)$, then f has a maximum at c and nowhere else.*

(ii) *If $f(c) < f(u)$ and $f(c) < f(v)$, then f has a minimum at c and nowhere else.*

(iii) *Otherwise, f has neither a maximum nor a minimum at c.*

The three cases in the Direct Test are shown in Figure 3.5.6. The advantage of the Direct Test is that one can determine whether f has a maximum or minimum at c by computing only the three values $f(u), f(v)$, and $f(c)$ instead of computing all values of $f(x)$.

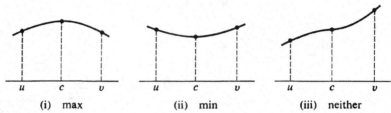

(i)   max   (ii)   min   (iii)   neither

Figure 3.5.6

PROOF OF THE DIRECT TEST   We must prove that if two points of I are on the same side of c, their values are on the same side of $f(c)$. Suppose, for instance, that $u_1 < u_2 < c$ (Figure 3.5.7). On the closed interval $[u_1, c]$ the only

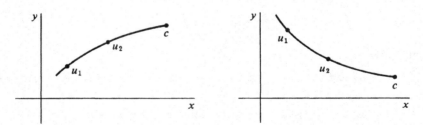

Figure 3.5.7

critical points are the endpoints. Thus when we restrict f to this interval, it has a maximum at one endpoint and a minimum at the other. If the maximum is at c, then $f(u_1)$ and $f(u_2)$ are both less than $f(c)$; if the minimum is at c, then $f(u_1)$ and $f(u_2)$ are both greater than $f(c)$. A similar proof works when $c < v_1 < v_2$.

SECOND DERIVATIVE TEST

Suppose c is the only interior critical point of f and that $f'(c) = 0$.

(i)   *If $f''(c) < 0$, f has a maximum at c and nowhere else.*

(ii) *If $f''(c) > 0$, f has a minimum at c and nowhere else.*

We omit the proof and give a simple intuitive argument instead. (See Figure 3.5.8.) Since $f'(c) = 0$, the curve is horizontal at c. If $f''(c)$ is negative the slope is decreasing. This means that the curve climbs up until it levels off at c and then falls down, so it has a maximum at c. On the other hand, if $f''(c)$ is positive, the slope is increasing, so the curve falls down until it reaches a minimum at c and then climbs up. This argument makes it easy to remember which way the inequalities go in the test.

The Second Derivative Test fails when $f''(c) = 0$ and when $f''(c)$ does not exist. When the Second Derivative Test fails any of the following things can still happen:

(1) f has a maximum at $x = c$.
(2) f has a minimum at $x = c$.
(3) f has neither a maximum nor a minimum at $x = c$.

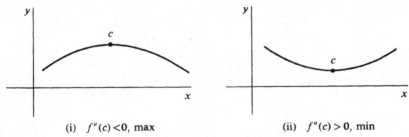

(i) $f''(c) < 0$, max   (ii) $f''(c) > 0$, min

Figure 3.5.8

In most maximum and minimum problems, there is only one critical point except for the endpoints of the interval. We develop a method for finding the maximum and minimum in that case.

METHOD FOR FINDING MAXIMA AND MINIMA

When to use: f is continuous on its domain I, and f has exactly one interior critical point.

Step 1 Differentiate f.

Step 2 Find the unique interior critical point c of f.

Step 3 Test to see whether f has a maximum or minimum at c. The Direct Test or the Second Derivative Test may be used.

This method can be applied to an open or half-open interval as well as a closed interval. The Second Derivative Test is more convenient because it requires only the single computation $f''(c)$, while the Direct Test requires the three computations $f(u)$, $f(v)$, and $f(c)$. However, the Direct Test always works while the Second Derivative Test sometimes fails.

We illustrate the use of both tests in the examples.

EXAMPLE 4   Find the point on the line $y = 2x + 3$ which is at minimum distance from the origin.

The distance is given by

$$z = \sqrt{x^2 + y^2},$$

and substituting $2x + 3$ for y,

$$z = \sqrt{x^2 + (2x + 3)^2} = \sqrt{5x^2 + 12x + 9}.$$

This is defined on the whole real line.

Step 1 $\dfrac{dz}{dx} = \dfrac{10x + 12}{2\sqrt{5x^2 + 12x + 9}} = \dfrac{5x + 6}{z}.$

Step 2 $\dfrac{dz}{dx} = 0$ only when $5x + 6 = 0$, or $x = -\frac{6}{5}$.

Step 3 $\dfrac{d^2z}{dx^2} = \dfrac{5z - (5x + 6)(dz/dx)}{z^2}.$

At $x = -\frac{6}{5}$, $5x + 6 = 0$ and $z > 0$ so $d^2z/dx^2 = 5/z > 0$. By the Second Derivative Test, z has a minimum at $x = -\frac{6}{5}$.

CONCLUSION   The distance is a minimum at $x = -\frac{6}{5}$, $y = 2x + 3 = \frac{3}{5}$. The minimum distance is $z = \sqrt{x^2 + y^2} = \sqrt{\frac{9}{5}}$. This is shown in Figure 3.5.9.

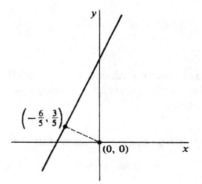

Figure 3.5.9

EXAMPLE 5   Find the minimum of $f(x) = x^6 + 10x^4 + 2$.

Step 1 $f'(x) = 6x^5 + 40x^3 = x^3(6x^2 + 40)$.

Step 2 $f'(x) = 0$ only when $x = 0$.

Step 3   The Second Derivative Test fails, because

$$f''(x) = 30x^4 + 120x^2, \qquad f''(0) = 0.$$

We use the Direct Test. Let $u = -1, v = 1$. Then

$$f(0) = 2, \qquad f(-1) = 13, \qquad f(1) = 13.$$

Hence f has a minimum at 0, as shown in Figure 3.5.10.

EXAMPLE 6   Find the maximum of $f(x) = 1 - x^{2/3}$.

Step 1 $f'(x) = -(\frac{2}{3})x^{-1/3}$.

Step 2 $f'(x)$ is undefined at $x = 0$, and this the only critical point.

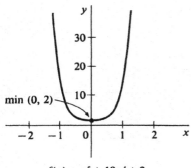

$$f(x) = x^6 + 10x^4 + 2$$

Figure 3.5.10

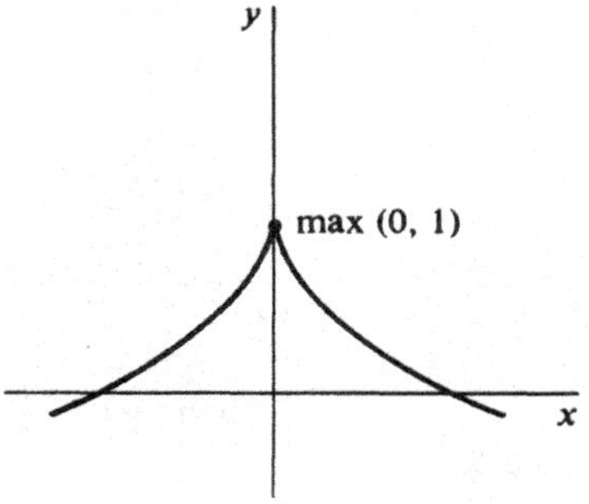

$$f(x) = 1 - x^{2/3}$$

Figure 3.5.11

Step 3 We use the Direct Test. Let $u = -1, v = 1$.

$$f(0) = 1, \qquad f(-1) = 0, \qquad f(1) = 0.$$

Thus f has a maximum at $x = 0$, as shown in Figure 3.5.11.

If f has more than one interior critical point, the maxima and minima can sometimes be found by dividing the interval into two or more parts.

EXAMPLE 7 Find the maximum and minimum of $f(x) = x/(x^2 + 1)$.

Step 1 $f'(x) = \dfrac{(x^2 + 1) - 2x^2}{(x^2 + 1)^2} = \dfrac{1 - x^2}{(x^2 + 1)^2}.$

Step 2 $f'(x) = 0$, when $x = -1$ and $x = 1$. There are two interior critical points. We divide the interval $(-\infty, \infty)$ on which f is defined into the two sub-intervals $(-\infty, 0]$ and $[0, \infty)$. On each of these subintervals, f has just one interior critical point.

Step 3 We shall use the direct test for the subinterval $(-\infty, 0]$. At the critical point -1, we have $f(-1) = -\frac{1}{2}$. By direct computation, we see that $f(-2) = -\frac{2}{5}$ and $f(0) = 0$. Both of these values are greater than $-\frac{1}{2}$. This shows that the restriction of f to the subinterval $(-\infty, 0]$ has a minimum at $x = -1$. Moreover, $f(x)$ is always ≥ 0 for x in the other subinterval $[0, \infty)$. Therefore f has a minimum at -1 for the whole interval $(-\infty, \infty)$. In a similar way, we can show that f has a maximum at $x = 1$.

CONCLUSION f has a minimum at $x = -1$ with value $f(-1) = -\frac{1}{2}$, and a maximum at $x = 1$ with value $f(1) = \frac{1}{2}$. (See Figure 3.5.12.)

The Critical Point Theorem can often be used to show that a curve has no maximum or minimum on an open interval $I = (a, b)$. The theorem shows that:

If $y = f(x)$ *has no critical points in* (a, b), *the curve has no maximum or minimum on* (a, b).

If $y = f(x)$ *has just one critical point* $x = c$ *in* (a, b) *and two points* x_1 *and*

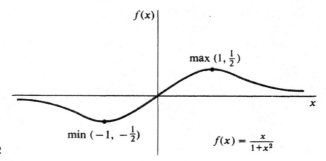

Figure 3.5.12

x_2 *are found where* $f(x_1) < f(c) < f(x_2)$, *then the curve has no maximum or minimum on* (a, b).

EXAMPLE 8 $f(x) = x^3 - 1$. Test for maxima and minima.

Step 1 $f'(x) = 3x^2$.

Step 2 $f'(x) = 0$ only when $x = 0$.

Step 3 The Second Derivative Test fails, because $f''(x) = 6x$, $f''(0) = 0$.
By direct computation, $f(0) = -1$, $f(-1) = -2$, $f(1) = 0$.
Therefore f has neither a minimum nor a maximum at $x = 0$.

CONCLUSION Since $x = 0$ is the only critical point of f and f doesn't have a maximum or minimum there, we conclude that f has no maximum and no minimum as shown in Figure 3.5.13.

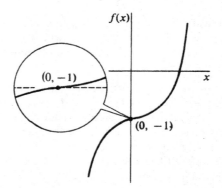

Figure 3.5.13

PROOF OF THE CRITICAL POINT THEOREM Assume that neither (i) nor (ii) holds; that is, assume that c is not an endpoint of I and $f'(c)$ exists. We must show that (iii) is true; i.e., $f'(c) = 0$. We give the proof for the case that f has a maximum at c. Let $x = c$, and let $\Delta x > 0$ be infinitesimal. Then

$$f(c + \Delta x) \le f(c),   \qquad f(c - \Delta x) \le f(c).$$

(See Figure 3.5.14.) Therefore

$$\frac{f(c + \Delta x) - f(c)}{\Delta x} \le 0 \le \frac{f(c - \Delta x) - f(c)}{-\Delta x}.$$

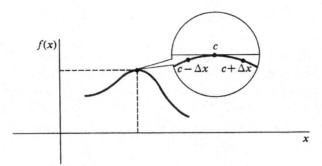

Figure 3.5.14 Proof of the Critical Point Theorem

Taking standard parts,

$$f'(c) = st\left(\frac{f(c + \Delta x) - f(c)}{\Delta x}\right) \le 0,$$

and also,

$$0 \le st\left(\frac{f(c - \Delta x) - f(c)}{-\Delta x}\right) = f'(c).$$

Therefore $f'(c) = 0$.

PROBLEMS FOR SECTION 3.5

In Problems 1–36, find the unique interior critical point and determine whether it is a maximum, a minimum, or neither.

1	$f(x) = x^2$	**2**	$f(x) = 1 - x^2$
3	$f(x) = x^4 + 2$	**4**	$f(x) = x^4 + 3x^2 + 5$
5	$f(x) = x^3 + 2$	**6**	$f(x) = x^3 - 3x^2 + 3x$
7	$f(x) = 3x^2 + 2x - 5$	**8**	$f(x) = 2(x - 1)^4 + (x - 1)^2 + 6$
9	$f(x) = x^{4/5}$	**10**	$f(x) = 2 - (x + 1)^{2/3}$
11	$f(x) = \dfrac{1}{x^2 - 1}, \quad -1 < x < 1$	**12**	$f(x) = \dfrac{1}{x^2 + 1}$
13	$f(x) = x^{1/3} + 1$	**14**	$f(x) = 4 - x^{1/5}$
15	$f(x) = x^2 - x^{-1}, \quad x < 0$	**16**	$f(x) = x^2 - x^{-1}, \quad x > 0$
17	$f(x) = x^{-1} - (x - 3)^{-1},$ $0 < x < 3$	**18**	$f(x) = x + x^{-1}, \quad 0 < x$
19	$f(x) = \sqrt{4 - x^2}, \quad -2 \le x \le 2$	**20**	$f(x) = (4 - x^2)^{-1/2}, \quad -2 < x < 2$
21	$y = \sin x + x, \quad 0 \le x \le 2\pi$	**22**	$y = \sin^2 x, \quad 0 < x < \pi$
23	$y = e^{-x^2}$	**24**	$y = e^{x^2 - 1}$
25	$y = \dfrac{1}{\cos x}, \quad -\dfrac{\pi}{2} < x < \dfrac{\pi}{2}$	**26**	$y = \ln(\sin x), \quad 0 < x < \pi$
27	$y = xe^x$	**28**	$y = x \ln x, \quad 0 < x < \infty$
29	$y = x - \ln x, \quad 0 < x < \infty$	**30**	$y = e^x - x$
31	$f(x) = |x - 3|$	**32**	$f(x) = 3 + |1 - x|$
33	$f(x) = 2 - |x|$	**34**	$f(x) = 2|x| - x$

35 $f(x) = \sqrt{x} + \sqrt{1 - x},$ **36** $f(x) = \sqrt{x} + \sqrt{9 - 3x}, \ 0 \le x \le 3$
$0 \le x \le 1$

37   Find the shortest distance between the line $y = 1 - 4x$ and the origin.

38   Find the shortest distance between the curve $y = 2/x$ and the origin.

39   Find the minimum of the curve $f(x) = x^m - mx, \ x > 0,$ where m is an integer ≥ 2.

40   Find the maximum of $f(x) = x^m - mx, \ x < 0,$ where m is an odd integer ≥ 2.

In Problems 41–44, find the maximum and minimum of the given curve.

41 $f(x) = \dfrac{x}{x^2 + 4}$ **42** $f(x) = \dfrac{3x + 4}{x^2 + 1}$

43 $f(x) = \dfrac{x}{x^4 + 1}$ **44** $f(x) = \dfrac{x^3}{x^4 + 1}$

3.6 MAXIMA AND MINIMA—APPLICATIONS

Maximum and minimum problems arise in both the physical and social sciences. We give three examples.

EXAMPLE 1   A woman wishes to rent a house. If she lives x miles from her work, her transportation cost will be cx dollars per year, while her rent will be $25c/(x + 1)$ dollars per year. How far should she live from work to minimize her rent and transportation expenses?

Let y be her expenses in dollars per year. Then

$$y = cx + \frac{25c}{x + 1}.$$

The problem is to find the minimum value of y in the interval $0 \le x < \infty$.

Step 1 $\dfrac{dy}{dx} = c - \dfrac{25c}{(x + 1)^2}.$

Step 2   To find x such that $dy/dx = 0$ we set $dy/dx = 0$ and solve for x.

$$c - \frac{25c}{(x + 1)^2} = 0, \quad c = \frac{25c}{(x + 1)^2}, \quad (x + 1)^2 = 25, \quad x + 1 = \pm 5.$$

Then $x = 4$ or $x = -6$. We reject $x = -6$ because $0 \le x$. The only interior critical point is $x = 4$.

Step 3   We use the Direct Test.

At $x = 0,$ $y = c \cdot 0 + 25c/(0 + 1) = 25c.$

At $x = 4,$ $y = 4c + 25c/(4 + 1) = 9c.$

At $x = 9,$ $y = 9c + 25c/(9 + 1) = 11.5c.$

CONCLUSION y has its minimum at $x = 4$ miles. So the woman should live four miles from work. (See Figure 3.6.1.)

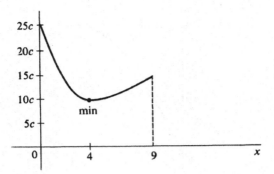

Figure 3.6.1

EXAMPLE 2   A farmer plans to use 1000 feet of fence to enclose a rectangular plot along the bank of a straight river. Find the dimensions which enclose the maximum area.

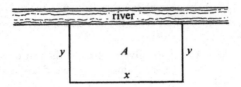

Figure 3.6.2

Let x be the dimension of the side along the river, and y be the other dimension, as in Figure 3.6.2. Call the area A.

No fencing is needed on the side of the plot bordering the river. The given information is expressed by the following system of formulas.

$$A = xy, \quad x + 2y = 1000, \quad 0 \le x \le 1000.$$

The problem is to find the values of x and y at which A is maximum. In this problem A is expressed in terms of two variables instead of one. However, we can select x as the independent variable, and then both y and A are functions of x. We find an equation for A as a function of x alone by eliminating y.

$$x + 2y = 1000, \qquad y = \frac{1000 - x}{2}.$$

$$A = xy = \frac{x(1000 - x)}{2} = 500x - \frac{1}{2}x^2.$$

We then find the maximum of A in the closed interval $0 \le x \le 1000$.

Step 1 $dA/dx = 500 - x.$

Step 2 $dA/dx = 0$ when $x = 500$. This is the unique interior critical point.

Step 3   We use the Second Derivative Test: $d^2A/dx^2 = -1$. Therefore A has a maximum at the critical point $x = 500$.

CONCLUSION   The maximum area occurs when the plot has dimensions $x = 500$ ft and $y = (1000 - x)/2 = 250$ ft (Figure 3.6.3).

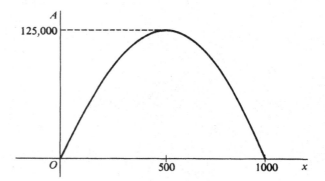

Figure 3.6.3

EXAMPLE 3   Find the shape of the cylinder of maximum volume which can be inscribed in a given sphere.

The *shape* of a right circular cylinder can be described by the ratio of the radius of its base to its height. This ratio for the inscribed cylinder of maximum volume should be a number which does not depend on the radius of the sphere. For example, we should get the same shape whether the radius of the sphere is given in inches or centimeters.

Let r be the radius of the given sphere, x the radius of the base of the cylinder, h its height, and V its volume. First, we draw a sketch of the problem in Figure 3.6.4.

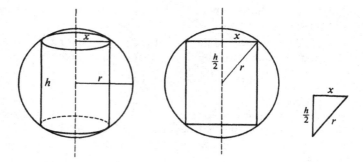

Figure 3.6.4

From the sketch we can read off the formulas

$$V = \pi x^2 h, \quad x^2 + (\tfrac{1}{2}h)^2 = r^2, \quad 0 \le x \le r.$$

r is a constant. We select x as the independent variable, while h and V are functions of x. To solve the problem we shall find the value of x where V is a maximum and then compute the ratio x/h at this point to describe the shape of the cylinder. The answer x/h should not depend on the constant r. We give two methods of solution.

FIRST SOLUTION   Express V as a function of x by eliminating h.

$$x^2 + (\tfrac{1}{2}h)^2 = r^2,$$
$$h = 2\sqrt{r^2 - x^2}.$$
$$V = \pi x^2 h = 2\pi x^2 \sqrt{r^2 - x^2}.$$

The problem is to find the maximum of V in the interval $0 \le x \le r$.

Step 1 $\dfrac{dV}{dx} = 4\pi x\sqrt{r^2 - x^2} - \dfrac{2\pi x^3}{\sqrt{r^2 - x^2}}, \quad (x < r).$

Step 2   There is one critical point at $x = r$, where dV/dx does not exist. We set $dV/dx = 0$ and solve for x to find the other critical points.

$$4\pi x\sqrt{r^2 - x^2} - \frac{2\pi x^3}{\sqrt{r^2 - x^2}} = 0, \quad 4\pi x(r^2 - x^2) - 2\pi x^3 = 0,$$

$$2\pi x(2r^2 - 3x^2) = 0, \qquad x = 0 \quad \text{or} \quad x = \pm r\sqrt{\tfrac{2}{3}}.$$

We reject $x = -r\sqrt{\tfrac{2}{3}}$ because $0 \le x \le r$. The only interior critical point is $x = r\sqrt{\tfrac{2}{3}}$.

Step 3   We use the Direct Test.

At $x = 0, \quad V = 0.$

At $x = r\sqrt{\tfrac{2}{3}}, \quad V = \dfrac{4\pi r^3}{3\sqrt{3}}.$

At $x = r, \quad V = 0.$

CONCLUSION   The maximum of V is at $x = r\sqrt{\tfrac{2}{3}}$ (see Figure 3.6.5). At that point, $h = 2\sqrt{r^2 - x^2} = 2r/\sqrt{3}$. Then the ratio of x to h is

$$x/h = 1/\sqrt{2}.$$

Notice that, as we expected, this number does not depend on r.

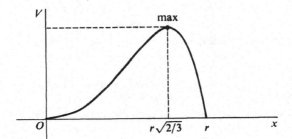

Figure 3.6.5

SECOND SOLUTION   Instead of eliminating h and expressing V as a function of x, we shall use the equations in their original form and find the critical points by implicit differentiation.

Step 1 $V = \pi x^2 h, \quad dV/dx = 2\pi x h + \pi x^2\, dh/dx.$

We find dh/dx by implicit differentiation.

$$x^2 + (\tfrac{1}{2}h)^2 = r^2, \quad 2x + \tfrac{1}{2}h\frac{dh}{dx} = 0, \quad \frac{dh}{dx} = -\frac{4x}{h}, \qquad (h \ne 0).$$

Then $\dfrac{dV}{dx} = 2\pi x h + \pi x^2\left(-\dfrac{4x}{h}\right) = 2\pi x h - \dfrac{4\pi x^3}{h}, \qquad (h \ne 0).$

Step 2   When $h = 0$ we have $x = r$, which is an endpoint. When $h \ne 0$ we set $dV/dx = 0$ and solve for x.

$$2\pi xh - 4\pi x^3/h = 0, \qquad xh - 2x^3/h = 0,$$
$$xh^2 = 2x^3, \qquad x = 0 \quad \text{or} \quad x = \pm h/\sqrt{2}.$$

We reject $x = -h/\sqrt{2}$ because x and $h \geq 0$. $x = 0$ is an endpoint. Thus $x = h/\sqrt{2}$ is the unique interior critical point.

Step 3   We use the Direct Test. At $x = 0$, $V = 0$. At $x = h/\sqrt{2}$, $V = \pi h^3/2$. At $x = r$, $V = 0$.

CONCLUSION   The maximum of V is at $x = h/\sqrt{2}$. At that point the ratio of x to h is $x/h = 1/\sqrt{2}$.

The second method of solution may be better in a problem where it is hard or impossible to find explicit equations for the dependent variables (like h and V) as functions of the independent variable.

PROBLEMS FOR SECTION 3.6

1   Split 20 into the sum of two numbers $x \geq 0$ and $y \geq 0$ such that the product of x and y^2 is a maximum.

2   Find two numbers $x \geq 0$ and $y \geq 0$ such that $x + y = 8$ and $x^2 + y^2$ is a minimum.

3   Find two numbers $x \geq 1$ and $y \geq 1$ such that $xy = 50$ and $2x + y$ is a maximum.

4   Find the rectangle with perimeter 8 which has maximum area.

5   Find the maximum value of $x^3 y$ if x and y belong to $[0, 1]$ and $x + y = 1$.

6   A rectangular box which is open at the top can be made from a 10 by 12 inch piece of metal by cutting a square from each corner and bending up the sides. Find the dimensions of the box with greatest volume.

7   A poster of total area 400 sq in. is to have a margin of 4 in. at the top and bottom and 3 in. at each side. Find the dimensions which give the largest printed area.

8   A man can travel 5 mph along the path AB and 3 mph off the path as shown in the figure. Find the quickest route APC from the point A to the point C.

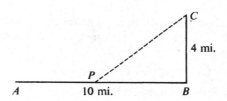

9   Find the dimensions of the right triangle of maximum area whose hypotenuse has length one.

10   Find the dimensions of the isosceles triangle of maximum area which has perimeter 3.

11   Find the five-sided figure of maximum area which has the shape of a square topped by an isosceles triangle, and such that the sum of the height of the figure and the perimeter of the square is 20 ft.

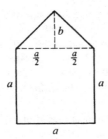

12 A wire of length L is to be divided into two parts; one part will be bent into a square and the other into a circle. How should the wire be divided to make the sum of the areas of the square and circle as large as possible? As small as possible?

13 Find the area of the largest rectangle which can be inscribed in a semicircle of radius r.

14 Find the dimensions of the rectangle of maximum area which can be inscribed in an equilateral triangle as shown in the figure.

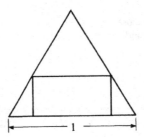

15 Find the shape of the right circular cylinder of maximum volume which can be inscribed in a right circular cone of height 3 and base of radius 1.

16 Find the shape of the right circular cone of maximum volume which can be inscribed in a given sphere.

17 Find the shape of the cylinder of maximum volume such that the sum of the height and the circumference of the base is equal to 4.

18 Find the shape of the largest trapezoid which can be inscribed in a semicircle as shown in the figure.

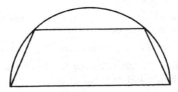

19 If a farmer plants x units of wheat in his field, $0 \le x \le 100$, the yield will be $10x - x^2/10$ units. How much wheat should he plant for the maximum yield?

20 In Problem 19 above, it costs the farmer \$100 for each unit of wheat he plants, and he is able to sell each unit he harvests for \$50. How much should he plant to maximize his profit?

21 A professional football team has a stadium which seats 60,000. It is found that x tickets can be sold at a price of $p = 10 - x/10,000$ dollars per ticket. Find the values of x and p at which the total money received will be a maximum.

22   In Problem 21 a tax of $1 per ticket is added onto the price. Find x and p so that the total revenue after taxes is a maximum.

23   A store can buy up to 300 seconds of advertising time daily on the radio at the rate of $2/sec for the first 100 sec, and $1/sec thereafter. x seconds on the radio increases daily sales by $32\sqrt{x}$ dollars. How many seconds on the radio will yield the maximum profit?

24   Work Problem 23 if the cost of advertising time is $1/sec for the first 100 sec and $2/sec thereafter.

25   Find the real number which most exceeds its square.

26   Find the rectangle of area 9 which has the smallest perimeter.

27   Find the right triangle of smallest area in which a 1 by 2 rectangle can be inscribed as shown in the figure.

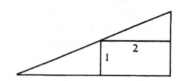

28   A farmer wishes to enclose 10,000 sq ft of land along a river by three sides of fence as shown in the figure. Find the dimensions which require the minimum length of fence.

29   Find the shortest distance between the line $y = 1 - 4x$ and the origin.

30   Find the shortest distance between the curve $y = 2/x$ and the origin.

31   A warehouse is to be built in the shape of a rectangular solid with a square base. The cost of the roof per unit area is three times the cost of the walls. Find the shape which will enclose the maximum volume for a given cost.

32   A rectangular box with volume 1 cu ft is to be made with a square base and no top. Find the dimensions which require the smallest amount of material.

33   Find the dimensions of the right circular cylinder of volume 1 cu ft which has the smallest surface area (top plus bottom plus sides).

34   Find the dimensions of the right circular cone of smallest volume which can be circumscribed about a sphere of radius r.

35   Given two real numbers a and b, find x such that $(x - a)^2 + (x - b)^2$ is a minimum.

36   The area of a sector of a circle with radius r and central angle θ is $A = \frac{1}{2}r^2\theta$, and its arc has length $s = r\theta$. Find r and θ so that $0 < \theta < 2\pi$, the sector has area 1, and the perimeter is a minimum.

37   Show that among all right circular cylinders of volume 1 cu ft which are open at both ends, there is no maximum or minimum surface area.

38   The population of a country at time $t = 0$ is 50 million and is increasing at the rate of one million people per year. The national income at time t is $(20{,}000 + t^2)$ million dollars per year. At what time $t \geq 0$ is the per capita income (= national income ÷ population) a minimum?

39   A man estimates that he can paint his house in x hours of his spare time if he buys equipment costing $200 + 2000/x^2$ dollars, and that his spare time is worth $2/hr. How many hours should he take?

40   An artisan can produce x items at a total cost of $100 + 5x$ dollars and sell x items at a price of $10 - x/100$ dollars per item. Find the value of x which gives the maximum profit.

41   A manufacturer can produce any number of buttons at a cost of two cents per button and can sell x buttons at a price of $1000/\sqrt{x}$ cents per button. How many buttons should be produced for maximum profit?

3.7 DERIVATIVES AND CURVE SKETCHING

If we compute n values of $f(x)$,

$$f(x_1), f(x_2), \ldots, f(x_n),$$

we obtain n points through which the curve $y = f(x)$ passes. The first and second derivatives tell us something about the shape of the curve in the intervals between these points and permit a much more accurate plot of the curve. It is especially helpful to know the signs of the first two derivatives.

When the first derivative is positive the curve is increasing from left to right, and when the first derivative is negative the curve is decreasing from left to right. When the first derivative is zero the curve is horizontal. These facts can be proved as a theorem if we define exactly what is meant by increasing and decreasing (see Figures 3.7.1 and 3.7.2).

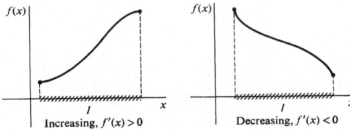

Figure 3.7.1

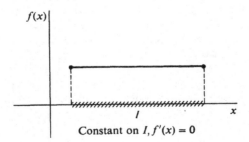

Figure 3.7.2

DEFINITION

*A function f is said to be **constant** on an interval I if :*

$$f(x_1) = f(x_2) \qquad \text{for all } x_1, x_2 \text{ in } I.$$

*f is **increasing** on I if :*

$$f(x_1) < f(x_2) \qquad \text{whenever } x_1 < x_2 \text{ in } I.$$

f is **decreasing** on *I* if :

$$f(x_1) > f(x_2) \qquad \text{whenever } x_1 < x_2 \text{ in } I.$$

THEOREM 1

Suppose f is continuous on I and has a derivative at every interior point of I.

(i) *If $f'(x) = 0$ for all interior points x of I, then f is constant on I.*

(ii) *If $f'(x) > 0$ for all interior points x of I, then f is increasing on I.*

(iii) *If $f'(x) < 0$ for all interior points x of I, then f is decreasing on I.*

A proof will be given in the next section.

EXAMPLE 1 The curve $y = x^3 + x - 1$ has derivative $dy/dx = 3x^2 + 1$. The derivative is always positive, so the curve is always increasing (Figure 3.7.3).

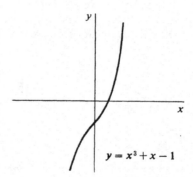

Figure 3.7.3

Let us now turn to the second derivative. It is the rate of change of the slope of the curve, so it has something to do with the way in which the curve is changing direction. When the second derivative is positive, the slope is increasing, and we would expect the curve to be concave upward, i.e., shaped like a ∪. When the second derivative is negative the slope is decreasing, so the curve should be shaped like ∩ (see Figure 3.7.4).

A precise definition of concave upward or downward can be given by comparing the curve with the chord (straight line segment) connecting two points on the curve.

DEFINITION

*Let f be defined on I. The curve $y = f(x)$ is **concave upward** on I if for any two points $x_1 < x_2$ in I and any value of x between x_1 and x_2, the curve at x is below the chord which meets the curve at x_1 and x_2.*

*The curve $y = f(x)$ is **concave downward** on I if for any two points $x_1 < x_2$ in I and any value of x between x_1 and x_2, the curve at x is above the chord which meets the curve at x_1 and x_2 (see Figure 3.7.5).*

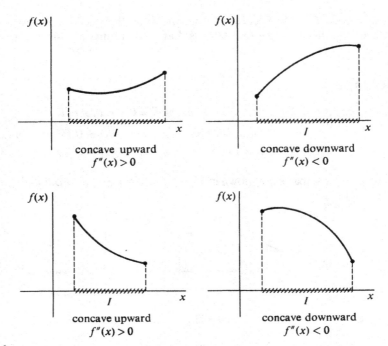

Figure 3.7.4

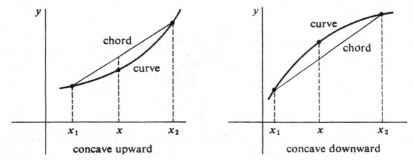

Figure 3.7.5

The next theorem gives the geometric meaning of the sign of the second derivative.

THEOREM 2

Suppose f is continuous on I and f has a second derivative at every interior point of I.

(i) *If $f''(x) > 0$ for all interior points x of I, then f is concave upward on I.*

(ii) *If $f''(x) < 0$ for all interior points x of I, then f is concave downward on I.*

We have already explained the intuitive reason for Theorem 2. The proof is omitted. Theorem 1 tells what happens when f' always has the same sign on an open interval I, while Theorem 2 does the same thing for f''. To use these results we need another theorem that tells us that certain functions always have the same sign on I.

THEOREM 3

Suppose g is continuous on I, and $g(x) \neq 0$ for all x in I.

(i) *If $g(c) > 0$ for at least one c in I, then $g(x) > 0$ for all x in I.*

(ii) *If $g(c) < 0$ for at least one c in I, then $g(x) < 0$ for all x in I.*

The two cases are shown in Figure 3.7.6. We give the proof in the next section.

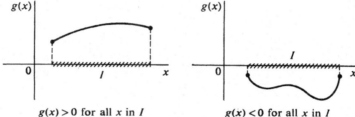

Figure 3.7.6 $g(x) > 0$ for all x in I $g(x) < 0$ for all x in I

Let us show with some simple examples how we can use the first and second derivatives in sketching curves. The three theorems above and the tests for minima and maxima are all helpful.

EXAMPLE 1(Continued) $y = x^3 + x - 1$. We have

$$\frac{dy}{dx} = 3x^2 + 1,$$

$$\frac{d^2y}{dx^2} = 6x.$$

dy/dx is always positive, while $d^2y/dx^2 = 0$ at $x = 0$. We make a table of values for y and its first two derivatives at $x = 0$ and at a point to the right and left side of 0.

x	y	$\dfrac{dy}{dx}$	$\dfrac{d^2y}{dx^2}$
-1	-3	4	-6
0	-1	1	0
1	1	4	6

With the aid of Theorems 1–3, we can draw the following conclusions:

(a) $dy/dx > 0$ and the curve is increasing for all x.

(b) $d^2y/dx^2 < 0$ for $x < 0$; concave downward.

(c) $d^2y/dx^2 > 0$ for $x > 0$; concave upward.

At the point $x = 0$, the curve changes from concave downward to concave upward. This is called a *point of inflection*.

To sketch the curve we first plot the three values of y shown in the table, then sketch the slope at these points as shown in Figure 3.7.7, then fill in a smooth curve, which is concave downward or upward as required.

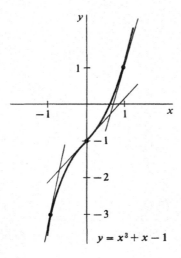

Figure 3.7.7

$$y = x^3 + x - 1$$

EXAMPLE 2   Sketch the curve $y = 2x - x^2$.

$$\frac{dy}{dx} = 2 - 2x, \qquad \frac{d^2y}{dx^2} = -2.$$

We see that $dy/dx = 0$ when $x = 1$, a critical point. d^2y/dx^2 is never zero because it is constant. We make a table of values including the critical point $x = 1$ and points to the right and left of it.

x	y	$\dfrac{dy}{dx}$	$\dfrac{d^2y}{dx^2}$
-1	-3	4	-2
0	0	2	-2
1	1	0	-2
2	0	-2	-2
3	-3	-4	-2

CONCLUSIONS

(a) $dy/dx > 0$ for $x < 1$; increasing.
(b) $dy/dx < 0$ for $x > 1$; decreasing.
(c) $d^2y/dx^2 < 0$ for all x; concave downward.
(d) $dy/dx = 0$, $d^2y/dx^2 < 0$ at $x = 1$; maximum.

The curve is shown in Figure 3.7.8.

   In general a curve $y = f(x)$ may go up and down several times. To sketch it we need to determine the intervals on which it is increasing or decreasing, and concave upward or downward. Here are some things which may happen at the endpoints of these intervals.

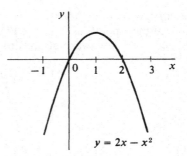

Figure 3.7.8

$y = 2x - x^2$

DEFINITION

Let c be an interior point of I.

*f has a **local maximum** at c if $f(c) \geq f(x)$ for all x in some open interval (a_0, b_0) containing c.*

*f has a **local minimum** at c if $f(c) \leq f(x)$ for all x in some open interval (a_0, b_0) containing c. (The interval (a_0, b_0) may be only a small subinterval of I.)*

*f has a **point of inflection** at c if f changes from one direction of concavity to the other at c.*

These definitions are illustrated in Figure 3.7.9.

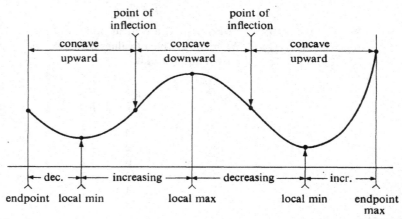

Figure 3.7.9

We may now describe the steps in sketching a curve. We shall stick to the simple case where f and its first two derivatives are continuous on a closed interval $[a, b]$, and either are never zero or are zero only finitely many times. (Curve plotting in a more general situation is discussed in Chapter 5 on limits.)

Step 1   Compute dy/dx and d^2y/dx^2.

Step 2   Find all points where $dy/dx = 0$ and all points where $d^2y/dx^2 = 0$.

Step 3   Pick a few points

$$a = x_0, x_1, x_2, \ldots, x_n = b$$

in the interval $[a, b]$. They should include both endpoints, all points where the first or second derivative is zero, and at least one point between any two consecutive zeros of dy/dx or d^2y/dx^2.

Step 4 At each of the points $x_0, \ldots, x_n$, compute the values of y and dy/dx and determine the sign of d^2y/dx^2. Make a table.

Step 5 From the table draw conclusions about where y is increasing or decreasing, where y has a local maximum or minimum, where the curve is concave upward or downward, and where it has a point of inflection. Use Theorems 1–3 of this section and the tests for maxima and minima.

Step 6 Plot the values of y and indicate slopes from the table. Then connect them with a smooth curve which agrees with the conclusions of Step 5.

EXAMPLE 3 $y = x^4/2 - x^2, \quad -2 \le x \le 2.$

Step 1 $dy/dx = 2x^3 - 2x. \quad d^2y/dx^2 = 6x^2 - 2.$

Step 2 $dy/dx = 0$ at $x = -1, 0, 1.$

Step 3 $d^2y/dx^2 = 0$ at $x \pm \sqrt{\frac{1}{3}}. \quad -2, -1, -\sqrt{\frac{1}{3}}, 0, \sqrt{\frac{1}{3}}, 1, 2.$

Step 4

x	y	$\dfrac{dy}{dx}$	$\dfrac{d^2y}{dx^2}$
-2	4	-12	$+$
-1	$-\frac{1}{2}$	0	$+$
$-\sqrt{\frac{1}{3}}$	$-\frac{5}{18}$	$4/(3\sqrt{3})$	0
0	0	0	$-$
$\sqrt{\frac{1}{3}}$	$-\frac{5}{18}$	$-4/(3\sqrt{3})$	0
1	$-\frac{1}{2}$	0	$+$
2	4	12	$+$

Step 5 We indicate the conclusions schematically in Figure 3.7.10.

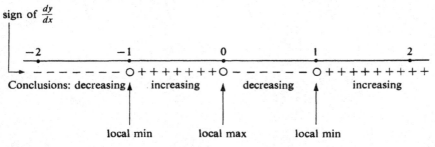

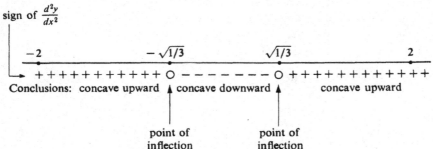

Figure 3.7.10

Step 6 The curve is W-shaped, as shown in Figure 3.7.11.

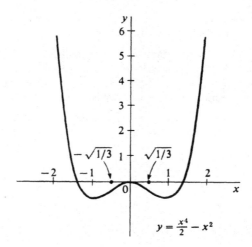

Figure 3.7.11

PROBLEMS FOR SECTION 3.7

Sketch each of the curves given below by the six-step process explained in the text. For each curve, give a table showing all the critical points, local maxima and minima, intervals on which the curve is increasing or decreasing, points of inflection, and intervals on which the curve is concave upward or downward.

1	$y = x^2 + 2,\quad -2 \le x \le 2$	**2**	$y = 1 - x^2,\quad -2 \le x \le 2$
3	$y = x^2 - 2x,\quad -2 \le x \le 2$	**4**	$y = \frac{1}{2}x^2 + x,\quad -2 \le x \le 2$
5	$y = 2x^2 - 4x + 3,\quad 0 \le x \le 2$	**6**	$y = -x^2 - 2x + 6,\quad -4 \le x \le 0$
7	$y = x^4,\quad -2 \le x \le 2$	**8**	$y = x^5,\quad -2 \le x \le 2$

9	$y = x^3 + x^2 + x,\quad -2 \le x \le 2$
10	$y = x^3 + x^2 - x,\quad -2 \le x \le 2$
11	$y = \frac{1}{3}x^3 + x^2 + x,\quad -2 \le x \le 2$
12	$y = -x^3 + 12x - 12,\quad -3 \le x \le 3$
13	$y = x^4 + 4x^3 + 2,\quad -4 \le x \le 2$
14	$y = \frac{1}{4}x^4 - x,\quad -2 \le x \le 2$
15	$y = x^2 - \frac{1}{2}x^4,\quad -2 \le x \le 2$
16	$y = x^2(x - 2)^2,\quad -1 \le x \le 3$
17	$y = 1/x,\quad -4 \le x \le -\frac{1}{4}$ and $\frac{1}{4} \le x \le 4$
18	$y = 1/x + x,\quad -4 \le x \le -\frac{1}{4}$ and $\frac{1}{4} \le x \le 4$
19	$y = x^{-2},\quad -2 \le x \le -\frac{1}{2}$ and $\frac{1}{2} \le x \le 2$
20	$y = x + x^{-2},\quad -2 \le x \le -\frac{1}{2}$ and $\frac{1}{2} \le x \le 2$

21	$y = \dfrac{x - 1}{x + 1},\quad 0 \le x \le 10$	**22**	$y = \dfrac{2x}{x + 1},\quad 0 \le x \le 10$
23	$y = \dfrac{1}{x^2 + 1},\quad -4 \le x \le 4$	**24**	$y = \dfrac{x}{x^2 + 1},\quad -4 \le x \le 4$
25	$y = \dfrac{x^2}{x^2 + 1},\quad -2 \le x \le 2$	**26**	$y = \dfrac{1}{x^2 - 1},\quad -\dfrac{9}{10} \le x \le \dfrac{9}{10}$

27	$y = \sqrt{x}, \ \frac{1}{4} \leq x \leq 4$	**28**	$y = 2\sqrt{x} - x, \ \frac{1}{4} \leq x \leq 4$
29	$y = 1/\sqrt{x}, \ \frac{1}{4} \leq x \leq 4$	**30**	$y = x^{1/2} + x^{-1/2}, \ \frac{1}{4} \leq x \leq 4$
31	$y = \sqrt{9 - x^2}, \ -2 \leq x \leq 2$	**32**	$y = \sqrt{9 + x^2}, \ -4 \leq x \leq 4$
33	$y = \sin x \cos x, \ 0 \leq x \leq 2\pi$	**34**	$y = \sin x + \cos x, \ 0 \leq x \leq 2\pi$
35	$y = 3\sin(\frac{1}{2}x), \ 0 \leq x \leq 2\pi$	**36**	$y = \sin^2 x, \ 0 \leq x \leq 2\pi$
37	$y = \tan x, \ -\pi/3 \leq x \leq \pi/3$	**38**	$y = 1/\cos x, \ -\pi/3 \leq x \leq \pi/3$
39	$y = e^{-x}, \ -2 \leq x \leq 2$	**40**	$y = e^{(1/2)x}, \ -2 \leq x \leq 2$
41	$y = \ln x, \ 1/e \leq x \leq e$	**42**	$y = (\ln x)^2, \ 1/e \leq x \leq e$
43	$y = xe^{-x}, \ -1 \leq x \leq 3$	**44**	$y = x - e^x, \ -2 \leq x \leq 2$
45	$y = x\ln x, \ e^{-2} \leq x \leq e$	**46**	$y = x - \ln x, \ e^{-2} \leq x \leq e$
47	$y = xe^x, \ -3 \leq x \leq 1$	**48**	$y = e^{-x^2}, \ -2 \leq x \leq 2$
49	$y = e^x/x, \ \frac{1}{4} \leq x \leq 4$	**50**	$y = \ln(1 + x^2), \ -3 \leq x \leq 3$

3.8 PROPERTIES OF CONTINUOUS FUNCTIONS

This section develops some theory that will be needed for integration in Chapter 4. We begin with a new concept, that of a *hyperinteger*. The hyperintegers are to the integers as the hyperreal numbers are to the real numbers. The hyperintegers consist of the ordinary finite integers, the positive infinite hyperintegers, and the negative infinite hyperintegers. The hyperintegers have the same algebraic properties as the integers and are spaced one apart all along the hyperreal line as in Figure 3.8.1.

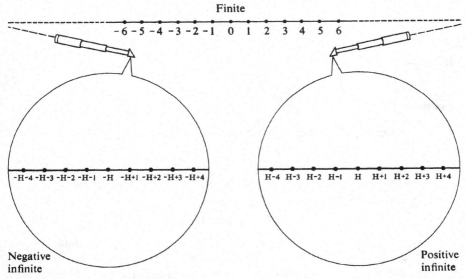

Figure 3.8.1 The Set of Hyperintegers

The rigorous definition of the hyperintegers uses the greatest integer function [x] introduced in Section 3.4, Example 6. Remember that for a real number x, [x] is the greatest integer n such that n ≤ x. A real number y is itself an integer if and only if y = [x] for some real x. To get the hyperintegers, we apply the function [x] to hyperreal numbers x (see Figure 3.8.2).

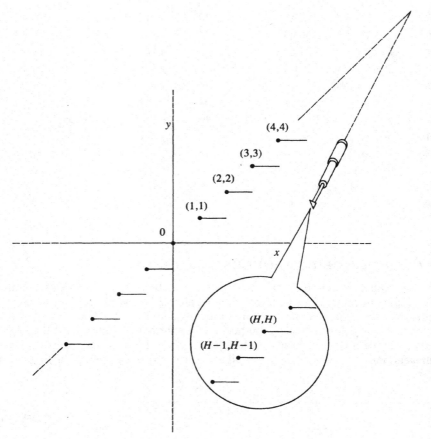

Figure 3.8.2

DEFINITION

*A **hyperinteger** is a hyperreal number y such that y = [x] for some hyperreal x.*

When x varies over the hyperreal numbers, $[x]$ is the greatest hyperinteger y such that $y \leq x$. Because of the Transfer Principle, every hyperreal number x is between two hyperintegers $[x]$ and $[x] + 1$,

$$[x] \leq x < [x] + 1.$$

Also, sums, differences, and products of hyperintegers are again hyperintegers.

We are now going to use the hyperintegers. In sketching curves we divided a closed interval $[a, b]$ into finitely many subintervals. For theoretical purposes in the calculus we often divide a closed interval into a finite or infinite number of equal subintervals. This is done as follows.

Given a closed real interval $[a, b]$, a *finite partition* is formed by choosing a positive integer n and dividing $[a, b]$ into n equal parts, as in Figure 3.8.3. Each part will be a subinterval of length $t = (b - a)/n$. The n subintervals are

$$[a, a + t], [a + t, a + 2t], \ldots, [a + (n - 1)t, b].$$

Figure 3.8.3

The endpoints

$$a, a + t, a + 2t, \ldots, a + (n - 1)t, a + nt = b$$

are called *partition points*.

The real interval $[a, b]$ is contained in the *hyperreal interval* $[a, b]^*$, which is the set of all hyperreal numbers x such that $a \le x \le b$. An infinite partition is applied to the hyperreal interval $[a, b]^*$ rather than the real interval. To form an infinite partition of $[a, b]^*$, choose a positive infinite hyperinteger H and divide $[a, b]^*$ into H equal parts as shown in Figure 3.8.4. Each subinterval will have the same infinitesimal length $\delta = (b - a)/H$. The H subintervals are

$$[a, a + \delta], [a + \delta, a + 2\delta], \ldots, [a + (K - 1)\delta, a + K\delta], \ldots, [a + (H - 1)\delta, b],$$

and the partition points are

$$a, a + \delta, a + 2\delta, \ldots, a + K\delta, \ldots, a + H\delta = b,$$

where K runs over the hyperintegers from 1 to H. Every hyperreal number x between a and b belongs to one of the infinitesimal subintervals,

$$a + (K - 1)\delta \le x < a + K\delta.$$

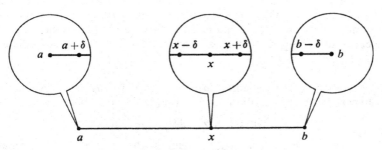

Figure 3.8.4   An infinite partition

We shall now use infinite partitions to sketch the proofs of three basic results, called the Intermediate Value Theorem, the Extreme Value Theorem, and Rolle's Theorem. The use of these results will be illustrated by studying zeros of continuous functions. By a *zero* of a function f we mean a point c where $f(c) = 0$. As we can see in Figure 3.8.5, the zeros of f are the points where the curve $y = f(x)$ intersects the x-axis.

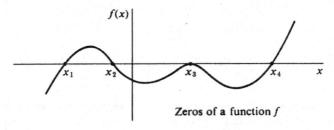

Zeros of a function f

Figure 3.8.5

INTERMEDIATE VALUE THEOREM

Suppose the real function f is continuous on the closed interval [a, b] and f(x) is positive at one endpoint and negative at the other endpoint. Then f has a zero in the interval (a, b); that is, f(c) = 0 for some real c in (a, b).

Discussion   There are two cases illustrated in Figure 3.8.6:

$$f(a) < 0 < f(b) \qquad \text{and} \qquad f(a) > 0 > f(b).$$

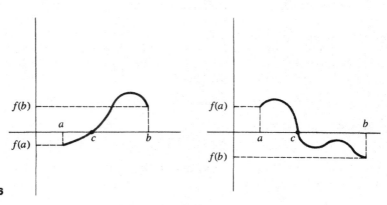

Figure 3.8.6

In the first case, the theorem says that if a continuous curve is below the x-axis at a and above it at b, then the curve must intersect the x-axis at some point c between a and b. Theorem 3 in the preceding Section 3.7 on curve sketching is simply a reformulation of the Intermediate Value Theorem.

SKETCH OF PROOF   We assume $f(a) < 0 < f(b)$. Let H be a positive infinite hyperinteger and partition the interval $[a, b]^*$ into H equal parts

$$a, a + \delta, a + 2\delta, \ldots, a + H\delta = b.$$

Let $a + K\delta$ be the last partition point at which $f(a + K\delta) < 0$. Thus

$$f(a + K\delta) < 0 \leq f(a + (K + 1)\delta).$$

Since f is continuous, $f(a + K\delta)$ is infinitely close to $f(a + (K + 1)\delta)$. We conclude that $f(a + K\delta) \approx 0$ (Figure 3.8.7). We take c to be the standard part of $a + K\delta$, so that

$$f(c) = st(f(a + K\delta)) = 0.$$

EXAMPLE 1   The function

$$f(x) = \frac{1}{1 + x} - x - \sqrt{x} - \sqrt[3]{x},$$

which is shown in Figure 3.8.8, is continuous for $0 \leq x \leq 1$. Moreover,

$$f(0) = 1, \qquad f(1) = \tfrac{1}{2} - 3 = -2\tfrac{1}{2}.$$

The Intermediate Value Theorem shows that $f(x)$ has a zero $f(c) = 0$ for some c between 0 and 1.

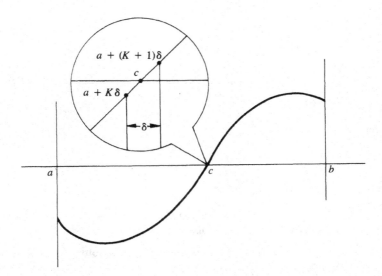

Figure 3.8.7

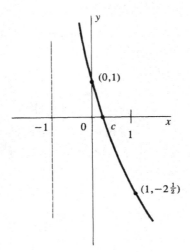

Figure 3.8.8

The Intermediate Value Theorem can be used to prove Theorem 3 of Section 3.7 on curve sketching:

Suppose g is a continuous function on an interval I, and $g(x) \neq 0$ for all x in I.

(i)   *If $g(c) > 0$ for at least one c in I, then $g(x) > 0$ for all x in I.*

(ii) *If $g(c) < 0$ for at least one c in I, then $g(x) < 0$ for all x in I.*

PROOF (i)   Let $g(c) > 0$ for some c in I. If $g(x_1) < 0$ for some other point x_1 in I, then by the Intermediate Value Theorem there is a point x_2 between c and x_1 such that $g(x_2) = 0$, contrary to hypothesis (Figure 3.8.9). Therefore we conclude that $g(x) > 0$ for all x in I.

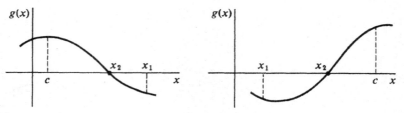

Figure 3.8.9

EXTREME VALUE THEOREM

Let f be continuous on its domain, which is a closed interval $[a, b]$. Then f has a maximum at some point in $[a, b]$, and a minimum at some point in $[a, b]$.

Discussion We have seen several examples of functions that do not have maxima on an *open* interval, such as $f(x) = 1/x$ on $(0, \infty)$, or $g(x) = 2x$ on $(0, 1)$. The Extreme Value Theorem says that on a closed interval a continuous function always has a maximum.

SKETCH OF PROOF Form an infinite partition of $[a, b]^*$,

$$a, a + \delta, a + 2\delta, \ldots, a + H\delta = b.$$

By the Transfer Principle, there is a partition point $a + K\delta$ at which $f(a + K\delta)$ has the largest value. Let c be the standard part of $a + K\delta$ (see Figure 3.8.10). Any point u of $[a, b]^*$ lies in a subinterval, say

$$a + L\delta \le u < a + (L + 1)\delta.$$

We have

$$f(a + K\delta) \ge f(a + L\delta),$$

and taking standard parts,

$$f(c) \ge f(u).$$

This shows that f has a maximum at c.

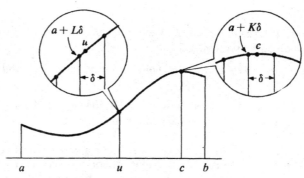

Figure 3.8.10   Proof of the Extreme Value Theorem

ROLLE'S THEOREM

Suppose that f is continuous on the closed interval [a, b] and differentiable on the open interval (a, b). If

$$f(a) = f(b) = 0,$$

then there is at least one point c strictly between a and b where f has derivative zero; i.e.,

$$f'(c) = 0 \quad \text{for some c in (a, b).}$$

Geometrically, the theorem says that a differentiable curve touching the x-axis at a and b must be horizontal for at least one point strictly between a and b.

PROOF We may assume that [a, b] is the domain of f. By the Extreme Value Theorem, f has a maximum value M and a minimum value m in [a, b]. Since f(a) = 0, $m \le 0$ and $M \ge 0$ (see Figure 3.8.11).

Case 1 $M = 0$ and $m = 0$. Then f is the constant function $f(x) = 0$, and therefore $f'(c) = 0$ for all points c in (a, b).

Case 2 $M > 0$. Let f have a maximum at c, $f(c) = M$. By the Critical Point Theorem, f has a critical point at c. c cannot be an endpoint because the value of $f(x)$ is zero at the endpoints and positive at $x = c$. By hypothesis, $f'(x)$ exists at $x = c$. It follows that c must be a critical point of the type $f'(c) = 0$.

Case 3 $m < 0$. We let f have a minimum at c. Then as in Case 2, c is in (a, b) and $f'(c) = 0$.

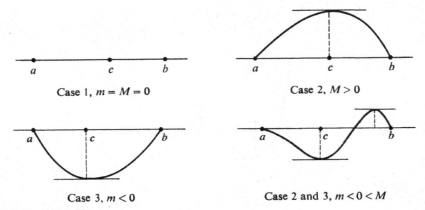

Case 1, $m = M = 0$

Case 2, $M > 0$

Case 3, $m < 0$

Case 2 and 3, $m < 0 < M$

Figure 3.8.11   Rolle's Theorem

EXAMPLE 2 $f(x) = (x - 1)^2(x - 2)^3$, $a = 1$, $b = 2$. The function f is continuous and differentiable everywhere (Figure 3.8.12). Moreover, $f(1) = f(2) = 0$. Therefore by Rolle's Theorem there is a point c in (1, 2) with $f'(c) = 0$.

Let us find such a point c. We have

$$f'(x) = 3(x - 1)^2(x - 2)^2 + 2(x - 1)(x - 2)^3 = (x - 1)(x - 2)^2(5x - 7).$$

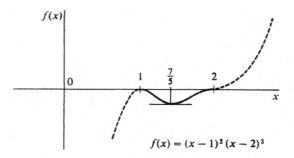

Figure 3.8.12 $f(x) = (x - 1)^2 (x - 2)^3$

Notice that $f'(1) = 0$ and $f'(2) = 0$. But Rolle's Theorem says that there is another point c which is in the *open* interval $(1, 2)$ where $f'(c) = 0$. The required value for c is $c = \frac{7}{5}$ because $f'(\frac{7}{5}) = 0$ and $1 < \frac{7}{5} < 2$.

EXAMPLE 3 Let $f(x) = \dfrac{x^4}{2} - x^2$, $a = -\sqrt{2}$, $b = \sqrt{2}$.

Then $f(a) = f(b) = 0$.

Rolle's Theorem says that there is at least one point c in $(-\sqrt{2}, \sqrt{2})$ at which $f'(c) = 0$. As a matter of fact there are three such points,

$$c = -1, \quad c = 0, \quad c = 1.$$

We can find these points as follows:

$$f'(x) = 2x^3 - 2x = 2x(x^2 - 1),$$
$$f'(x) = 0 \quad \text{when} \quad x = 0 \quad \text{or} \quad x = \pm 1.$$

The function is drawn in Figure 3.8.13.

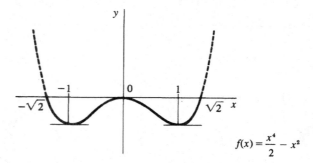

Figure 3.8.13 $f(x) = \dfrac{x^4}{2} - x^2$

EXAMPLE 4 $f(x) = \sqrt{1 - x^2}$, $a = -1$, $b = 1$. Then $f(-1) = f(1) = 0$. The function f is continuous on $[-1, 1]$ and has a derivative at each point of $(-1, 1)$, as Rolle's Theorem requires (Figure 3.8.14). Note, however, that $f'(x)$ does not exist at either endpoint, $x = -1$ or $x = 1$. By Rolle's Theorem there is a point c in $(-1, 1)$ such that $f'(c) = 0$, $c = 0$ is such a point, because

$$f'(x) = -\frac{x}{\sqrt{1 - x^2}}, \qquad f'(0) = 0.$$

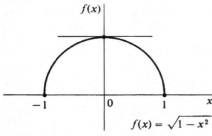

$$f(x) = \sqrt{1 - x^2}$$

Figure 3.8.14

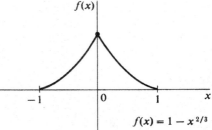

$$f(x) = 1 - x^{2/3}$$

Figure 3.8.15

EXAMPLE 5 $f(x) = 1 - x^{2/3}$, $a = -1$, $b = 1$. Then $f(-1) = f(1) = 0$, and $f'(x) = -\frac{2}{3}x^{-1/3}$ for $x \neq 0$. $f'(0)$ is undefined. There is no point c in $(-1, 1)$ at which $f'(c) = 0$. Rolle's Theorem does not apply in this case because $f'(x)$ does not exist at one of the points of the interval $(-1, 1)$, namely at $x = 0$. In Figure 3.8.15, we see that instead of being horizontal at a point in the interval, the curve has a sharp peak.

Rolle's Theorem is useful in finding the number of zeros of a differentiable function f. It shows that between any two zeros of f there must be one or more zeros of f'. It follows that if f' has no zeros in an interval I, then f cannot have more than one zero in I.

EXAMPLE 6 How many zeros does the function $f(x) = x^3 + x + 1$ have? We use both Rolle's Theorem and the Intermediate Value Theorem.

Using Rolle's Theorem: $f'(x) = 3x^2 + 1$. For all x, $x^2 \geq 0$, and hence $f'(x) \geq 1$. Therefore $f(x)$ has at most one zero.

Using the Intermediate Value Theorem: We have $f(-1) = -1$, $f(0) = 1$. Therefore f has at least one zero between -1 and 0.

CONCLUSION f has exactly one zero, and it lies between -1 and 0 (see Figure 3.8.16).

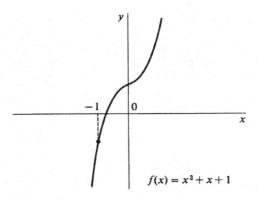

$$f(x) = x^3 + x + 1$$

Figure 3.8.16

Our method of sketching curves in Section 3.7 depends on a consequence of Rolle's Theorem called the Mean Value Theorem. It deals with the average slope of a curve between two points.

DEFINITION

Let f be defined on the closed interval $[a, b]$. The **average slope** of f between a and b is the quotient

$$average\ slope = \frac{f(b) - f(a)}{b - a}.$$

We can see in Figure 3.8.17 that the average slope of f between a and b is equal to the slope of the line passing through the points $(a, f(a))$ and $(b, f(b))$. This is shown by the two-point equation for a line (Section 1.3). In particular, if f is already a linear function $f(x) = mx + c$, then the average slope of f between a and b is equal to the slope m of the line $y = f(x)$.

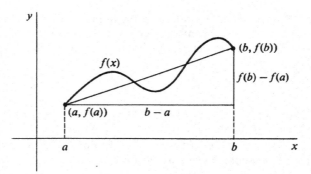

Figure 3.8.17   Average Slope

This is shown by the two-point equation for a straight line (Section 1.2). In particular, if f is already a linear function $f(x) = mx + c$, then the average slope of f between a and b is equal to the slope m of the straight line $y = f(x)$.

MEAN VALUE THEOREM

Assume that f is continuous on the closed interval $[a, b]$ and has a derivative at every point of the open interval (a, b). Then there is at least one point c in (a, b) where the slope $f'(c)$ is equal to the average slope of f between a and b,

$$f'(c) = \frac{f(b) - f(a)}{b - a}.$$

Remark In the special case that $f(a) = f(b) = 0$, the Mean Value Theorem becomes Rolle's Theorem:

$$f'(c) = \frac{f(b) - f(a)}{b - a} = \frac{0 - 0}{b - a} = 0.$$

On the other hand, we shall use Rolle's Theorem in the proof of the Mean Value Theorem. The Mean Value Theorem is illustrated in Figure 3.8.18.

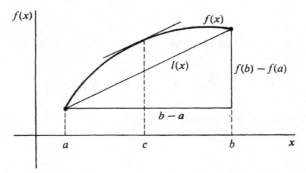

Figure 3.8.18   The Mean Value Theorem

PROOF OF THE MEAN VALUE THEOREM   Let m be the average slope, $m = (f(b) - f(a))/(b - a)$. The line through the points $(a, f(a))$ and $(b, f(b))$ has the equation

$$l(x) = f(a) + m(x - a).$$

Let $h(x)$ be the distance of $f(x)$ above $l(x)$,

$$h(x) = f(x) - l(x).$$

Then h is continuous on $[a, b]$ and has the derivative

$$h'(x) = f'(x) - l'(x) = f'(x) - m$$

at each point in (a, b). Since $f(x) = l(x)$ at the endpoints a and b, we have

$$h(a) = 0, \qquad h(b) = 0.$$

Therefore Rolle's Theorem can be applied to the function h, and there is a point c in (a, b) such that $h'(c) = 0$. Thus

$$0 = h'(c) = f'(c) - l'(c) = f'(c) - m,$$

whence $\qquad\qquad\qquad\qquad f'(c) = m.$

We can give a physical interpretation of the Mean Value Theorem in terms of velocity. Suppose a particle moves along the y-axis according to the equation $y = f(t)$. The *average velocity* of the particle between times a and b is the ratio

$$\frac{f(b) - f(a)}{b - a}$$

of the change in position to the time elapsed. The Mean Value Theorem states that there is a point of time c, $a < c < b$, when the velocity $f'(c)$ of the particle is equal to the average velocity between times a and b.

Theorems 1 and 2 in Section 3.7 on curve sketching are consequences of the Mean Value Theorem. As an illustration, we prove part (ii) of Theorem 1:

If $f'(x) > 0$ for all interior points x of I, then f is increasing on I.

PROOF Let $x_1 < x_2$ where x_1 and x_2 are points in I. By the Mean Value Theorem there is a point c strictly between x_1 and x_2 such that

$$f'(c) = \frac{f(x_2) - f(x_1)}{x_2 - x_1}.$$

Since c is an interior point of I, $f'(c) > 0$. Because $x_1 < x_2$, $x_2 - x_1 > 0$. Thus

$$\frac{f(x_2) - f(x_1)}{x_2 - x_1} > 0, \qquad f(x_2) - f(x_1) > 0, \qquad f(x_2) > f(x_1).$$

This shows that f is increasing on I.

PROBLEMS FOR SECTION 3.8

In Problems 1–16, use the Intermediate Value Theorem to show that the function has at least one zero in the given interval.

1 $f(x) = x^4 - 2x^3 - x^2 + 1, \quad 0 \le x \le 1$

2 $f(x) = x^2 + x - 3/x, \quad 1 \le x \le 2$

3 $f(x) = \sqrt{x} + \sqrt{x+1} - x, \quad 4 \le x \le 9$

4 $f(x) = \sqrt{x} + 1/x^2 - x^2, \quad 1 \le x \le 2$

5 $f(x) = \dfrac{2}{1 + x\sqrt{x}} - \sqrt{x^2 + 2}, \quad 0 \le x \le 1$

6 $f(x) = x^5 + x - \sqrt{x+1}, \quad 0 \le x \le 1$

7 $f(x) = x^3 + x^2 - 1, \quad 0 \le x \le 1$

8 $f(x) = x^2 + 1 - \dfrac{3}{x+1}, \quad 0 \le x \le 1$

9 $f(x) = 1 - 3x + x^3, \quad 0 \le x \le 1$

10 $f(x) = 1 - 3x + x^3, \quad 1 \le x \le 2$

11 $f(x) = x^2 + \sqrt{x} - 1, \quad 0 \le x \le 1$

12 $f(x) = x^2 - (x + 1)^{-1/2}, \quad 0 \le x \le 1$

13 $f(x) = \cos x - \frac{1}{10}, \quad 0 \le x \le \pi$

14 $f(x) = \sin x - 2\cos x, \quad 0 \le x \le \pi$

15 $f(x) = \ln x - \dfrac{1}{x}, \quad 1 \le x \le e$

16 $f(x) = e^x - 10x, \quad 1 \le x \le 10$

In Problems 17–30, determine whether or not f' has a zero in the interval (a, b). *Warning:* Rolle's Theorem may give a wrong answer unless all the hypotheses are met.

17 $f(x) = 5x^2 - 8x, \quad [a, b] = [0, \frac{8}{5}]$

18 $f(x) = 1 - x^{-2}, \quad [a, b] = [-1, 1]$

19 $f(x) = \sqrt{16 - x^4}, \quad [a, b] = [-2, 2]$

20 $f(x) = \sqrt{4 - x^{2/7}}, \quad [a, b] = [-128, 128]$

21 $f(x) = 1/x - x, \quad [a, b] = [-1, 1]$

22 $f(x) = (x - 1)^2(x - 2), \quad [a, b] = [1, 2]$

23 $f(x) = (x - 4)^3 x^4, \quad [a, b] = [0, 4]$

24    $f(x) = \dfrac{(x-2)(x-4)}{x^3 + x + 2}$, $[a, b] = [2, 4]$

25    $f(x) = |x| - 1$, $[a, b] = [-1, 1]$

26    $f(x) = \dfrac{x(x-2)}{x-1}$, $[a, b] = [0, 2]$

27    $f(x) = x \sin x$, $[a, b] = [0, \pi]$

28    $f(x) = e^x \cos x$, $[a, b] = [-\pi/2, \pi/2]$

29    $f(x) = \tan x$, $[a, b] = [0, \pi]$

30    $f(x) = \ln(1 - \sin x)$, $[a, b] = [0, \pi]$

31   Find the number of zeros of $x^4 + 3x + 1$ in $[-2, -1]$.

32   Find the number of zeros of $x^4 + 2x^3 - 2$ in $[0, 1]$.

33   Find the number of zeros of $x^4 - 8x - 4$.

34   Find the number of zeros of $2x + \sqrt{x} - 4$.

In Problems 35–42, find a point c in (a, b) such that $f(b) - f(a) = f'(c)(b - a)$.

35    $f(x) = x^2 + 2x - 1$, $[a, b] = [0, 1]$

36    $f(x) = x^3$, $[a, b] = [0, 3]$

37    $f(x) = x^{2/3}$, $[a, b] = [0, 1]$

38    $f(x) = \sqrt{x + 1}$, $[a, b] = [0, 2]$

39    $f(x) = x + \sqrt{x}$, $[a, b] = [0, 4]$

40    $f(x) = 2 + (1/x)$, $[a, b] = [1, 2]$

41    $f(x) = \dfrac{x-1}{x+1}$, $[a, b] = [0, 2]$

42    $f(x) = x\sqrt{x + 1}$, $[a, b] = [0, 3]$

43   Use Rolle's Theorem to show that the function $f(x) = x^3 - 3x + b$ cannot have more than one zero in the interval $[-1, 1]$, regardless of the value of the constant b.

44   Suppose f, f', and f'' are all continuous on the interval $[a, b]$, and suppose f has at least three distinct zeros in $[a, b]$. Use Rolle's Theorem to show that f'' has at least one zero in $[a, b]$.

□ 45   Suppose that $f''(x) > 0$ for all real numbers x, so that the curve $y = f(x)$ is concave upward on the whole real line as illustrated in the figure. Let L be the tangent line to the curve at $x = c$. Prove that the line L lies below the curve at every point $x \neq c$.

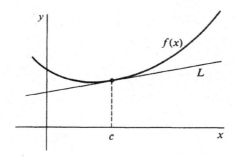

EXTRA PROBLEMS FOR CHAPTER 3

1   Find the surface area A of a cube as a function of its volume V.

2   Find the length of the diagonal d of a rectangle as a function of its length x and width y.

3   An airplane travels for t hours at a speed of 300 mph. Find the distance x of travel as a function of t.

4   An airplane travels x miles at 500 mph. Find the travelling time t as a function of x.

5   A 5 foot tall woman stands at a distance x from a 9 foot high lamp. Find the length of her shadow as a function of x.

6   The sides and bottom of a rectangular box are made of material costing \$1/sq ft. and the top of material costing \$2/sq ft. Find the cost of the box as a function of the length x, width y, and height z feet.

7   A piece of dough with a constant volume of 10 cu in. is being rolled in the shape of a right circular cylinder. Find the rate of increase of its length when the radius is $\frac{1}{2}$ inch and is decreasing at $\frac{1}{10}$ inch per second.

8   Car A travels north at 60 mph and passes the point P at 1:00. Car B travels east at 40 mph and passes the point P at 3:00. Find the rate of change of the distance between the two cars at 2:00.

9   A cup of water has the shape of a cone with the apex at the bottom, height 4 in., and a circular top of radius 2 in. The loss of water volume due to evaporation is $0.01A$ cu in./sec where A is the water surface area. Find the rate at which the water level drops due to evaporation.

10 A country has a constant national income and its population is decreasing by one million people per year. Find the rate of change of the per capita income when the population is 50 million and the national income is 100 billion dollars.

11 Evaluate $\lim\limits_{x \to 2} x^3 - 4x^2 + 3x - 1$

12 Evaluate $\lim\limits_{x \to 3} \dfrac{(x^2 - 9)^2}{(x - 3)^2}$ **13**   Evaluate $\lim\limits_{t \to 0^+} \dfrac{2 + t^{-1/2}}{3 - 4t^{-1/2}}$

14 Evaluate $\lim\limits_{\Delta x \to 0^+} \dfrac{\sqrt{0 + \Delta x} - \sqrt{0}}{\Delta x}$.

15 Find the set of all points at which $f(x) = \sqrt{1 + x} + \sqrt{1 - x}$ is continuous.

16 Find the set of all points at which

$$g(x) = \frac{x - 2}{(x - 3)(x - 4)(x - 5)}$$

is continuous.

17 Find the set of all points at which $f(x) = \sqrt{(4 - x^2)(x^2 - 1)}$ is continuous.

18 Assume $a < b$. Show that $f(x) = \sqrt{(x - a)(b - x)}$ is continuous on the closed interval $[a, b]$.

19 Show that $g(x) = (x - 1)^{1/3}$ is continuous at every real number $x = c$.

20 Find the maximum and minimum of
$$f(x) = 4x^3 - 3x^2 + 2, \qquad -1 \le x \le 1.$$

21 Find the maximum and minimum of
$$f(x) = x + \frac{4}{x^2}, \qquad 1 \le x \le 4.$$

22 Find the maximum and minimum of
$$f(x) = |2x - 5| + 3, \qquad 0 \le x \le 10.$$

23 Find the maximum and minimum of
$$f(x) = 4 - 3x^{2/3}, \qquad -1 \le x \le 1.$$

24 Find the maximum and minimum of
$$f(x) = (x - 1)^{1/3} - 2, \qquad 0 \le x \le 2.$$

25 Find the rectangle of maximum area which can be inscribed in a circle of radius 1.

26   A box with a square base and no top is to be made with 10 sq ft of material. Find the dimensions which will have the largest volume.

27   In one day a factory can produce x items at a total cost of $c_0 + ax$ dollars and can sell x items at a price of $bx^{-1/3}$ dollars per item. How many items should be produced for a maximum daily profit?

28   Test the curve $f(x) = x^3 - 5x + 4$ for maxima and minima.

29   Test the curve $f(x) = 3x^4 + 4x^3 - 12x^2$ for maxima and minima.

30   The light intensity from a light source is equal to S/D^2 where S is the strength of the source and D the distance from the source. Two light sources A and B have strengths $S_A = 2$ and $S_B = 1$ and are located on the x-axis at $x_A = 0$ and $x_B = 10$. Find the point $x, 0 < x < 10$, where the total light intensity is a minimum.

31   Find the right triangle of area $\frac{1}{2}$ with the smallest perimeter.

32   Find the points on the parabola $y = x^2$ which are closest to the point $(0, 2)$.

33   Find the number of zeros of $f(x) = x^3 - 8x^2 + 4x + 2$.

34   Find the number of zeros of $f(x) = x^3 - 2x^2 + 2x - 4$.

35   Sketch the curve $y = x^4 - x^3$, $-1 \le x \le 1$.

36   Sketch the curve $y = x^2 + x^{-2}$, $\frac{1}{2} \le x \le 2$.

37   Find all zeros of $f(x) = x^2 - 5x + 10$.

38   Show that the function $f(x) = x^6 - 5x^5 - 3x^2 + 4$ has at least one zero in the interval $[0, 1]$.

39   Show that the function $f(x) = \sqrt{x + 1} + \sqrt[3]{x + 8} - 2$ has at least one zero in the interval $[-1, 0]$.

40   Show that the equation $1 - x^2 = \sqrt{x}$ has at least one solution in the interval $[0, 1]$.

☐ **41**   Prove that $\lim_{x \to c} f(x)$ exists if and only if there is a function $g(x)$ such that
(a) $g(x)$ is continuous at $x = c$,
(b) $g(x) = f(x)$ whenever $x \ne c$.

☐ **42**   Let $S = \{a_1, \ldots, a_n\}$ be a finite set of real numbers. Show that the characteristic function of S,

$$f(x) = \begin{cases} 1 & \text{if } x \text{ is in } S, \\ 0 & \text{otherwise,} \end{cases}$$

is discontinuous for x in S and continuous for x not in S.

☐ **43**   Show that the function $f(x) = \sqrt{|x|}$ is continuous but not differentiable at $x = 0$.

☐ **44**   Let

$$f(x) = \begin{cases} 1 & \text{if } 1 \le |x| \\ 1/n & \text{if } 1/n \le |x| < 1/(n - 1), \quad n = 2, 3, 4, \ldots \\ 0 & \text{if } x = 0. \end{cases}$$

Show that f is continuous at $x = 0$ but discontinuous at $x = 1/n$ and $x = -1/n$, $n = 1, 2, 3, \ldots$.

☐ **45**   Let

$$f(x) = \begin{cases} 1 & \text{if } 1 \le |x| \\ 1/n^2 & \text{if } 1/n \le |x| < 1/(n - 1), \quad n = 2, 3, \ldots \\ 0 & \text{if } x = 0. \end{cases}$$

Prove that f is differentiable at $x = 0$ but discontinuous at $x = 1/n$ and $x = -1/n$, $n = 1, 2, 3, \ldots$.

☐ **46**   Suppose $f(x)$ is continuous on $[0, 1]$ and $f(0) = 1$, $f(1) = 0$. Prove that there is a point c in $(0, 1)$ such that $f(c) = c$.

☐ **47** Suppose $f(x)$ is continuous for all x, and $f(0) = 0$, $f(1) = 4$, $f(2) = 0$. Prove that there is a point c in $(0, 1)$ such that $f(c) = f(c + 1)$.

☐ **48** Prove that if $x = c$ is the only real solution of $f(x) = 0$, then $x = c$ is also the only hyperreal solution.

☐ **49** Prove that if n is odd, then the polynomial

$$x^n + a_{n-1}x^{n-1} + \cdots + a_1x + a_0$$

has no maximum and no minimum.

☐ **50** Prove that if n is even then the polynomial

$$x^n + a_{n-1}x^{n-1} + \cdots + a_1x + a_0$$

has no maximum.

☐ **51** Prove that if n is even then the polynomial

$$x^n + a_{n-1}x^{n-1} + \cdots + a_1x + a_0$$

has a minimum. You may use the fact that there are only finitely many critical points.

☐ **52** Prove the First Derivative Test: Assume $f(x)$ is continuous on an interval I.

If $f'(a) > 0$ for all $a < c$ and $f'(b) < 0$ for all $b > c$, then f has a maximum at $x = c$.

If $f'(a) < 0$ for all $a < c$ and $f'(b) > 0$ for all $b > c$, then f has a minimum at $x = c$.

☐ **53** Suppose f is differentiable and $f'(x) > 1$ for all x. If $f(0) = 0$, show that $f(x) > x$ for all positive x.

☐ **54** Suppose $f''(x) > 0$ for all x. Show that for any two points P and Q above the curve $y = f(x)$, every point on the line segment PQ is above the curve $y = f(x)$.

☐ **55** Suppose $f(0) = A$ and $f'(x)$ has the constant value B for all x. Use the Mean Value Theorem to show that f is the linear function $f(x) = A + Bx$.

☐ **56** Suppose $f'(x)$ is continuous for all real x. Use the Mean Value Theorem to show that for all finite hyperreal b and nonzero infinitesimal Δx,

$$f'(b) \approx \frac{f(b + \Delta x) - f(b)}{\Delta x}.$$

INTEGRATION

4.1 THE DEFINITE INTEGRAL

We shall begin our study of the integral calculus in the same way in which we began with the differential calculus—by asking a question about curves in the plane.

Suppose f is a real function continuous on an interval I and consider the curve $y = f(x)$. Let $a < b$ where a, b are two points in I, and let the curve be above the x-axis for x between a and b; that is, $f(x) \geq 0$. We then ask: What is meant by the *area* of the region bounded by the curve $y = f(x)$, the x-axis, and the lines $x = a$ and $x = b$? That is, what is meant by the area of the shaded region in Figure 4.1.1? We call this region the region under the curve $y = f(x)$ between a and b.

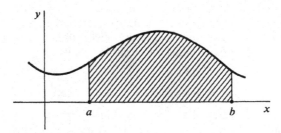

Figure 4.1.1 The Region under a Curve

The simplest possible case is where f is a constant function; that is, the curve is a horizontal line $f(x) = k$, where k is a constant and $k \geq 0$, shown in Figure 4.1.2. In this case the region under the curve is just a rectangle with height k and width $b - a$, so the area is defined as

$$\text{Area} = k \cdot (b - a).$$

The areas of certain other simple regions, such as triangles, trapezoids, and semi-circles, are given by formulas from plane geometry.

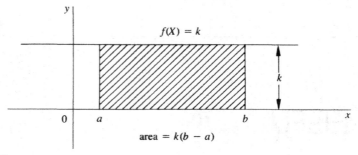

Figure 4.1.2

The area under any continuous curve $y = f(x)$ will be given by the definite integral, which is written

$$\int_a^b f(x)\,dx.$$

Before plunging into the detailed definition of the integral, we outline the main ideas.

First, the region under the curve is divided into infinitely many vertical strips of infinitesimal width dx. Next, each vertical strip is replaced by a vertical rectangle of height $f(x)$, base dx, and area $f(x)\,dx$. The next step is to form the sum of the areas of all these rectangles, called the infinite Riemann sum (look ahead to Figures 4.1.3 and 4.1.11). Finally, the integral $\int_a^b f(x)\,dx$ is defined as the standard part of the infinite Riemann sum.

The infinite Riemann sum, being a sum of rectangles, has an infinitesimal error. This error is removed by taking the standard part to form the integral.

It is often difficult to compute an infinite Riemann sum, since it is a sum of infinitely many infinitesimal rectangles. We shall first study finite Riemann sums, which can easily be computed on a hand calculator.

Suppose we slice the region under the curve between a and b into thin vertical strips of equal width. If there are n slices, each slice will have width $\Delta x = (b - a)/n$. The interval $[a, b]$ will be partitioned into n subintervals

$$[x_0, x_1], [x_1, x_2], \ldots, [x_{n-1}, x_n],$$

where $x_0 = a, x_1 = a + \Delta x, x_2 = a + 2\,\Delta x, \ldots, x_n = b.$

The points $x_0, x_1, \ldots, x_n$ are called *partition points*. On each subinterval $[x_{k-1}, x_k]$, we form the rectangle of height $f(x_{k-1})$. The kth rectangle will have area

$$f(x_{k-1}) \cdot \Delta x.$$

From Figure 4.1.3, we can see that the sum of the areas of all these rectangles will be fairly close to the area under the curve. This sum is called a *Riemann sum* and is equal to

$$f(x_0)\,\Delta x + f(x_1)\,\Delta x + \cdots + f(x_{n-1})\,\Delta x.$$

It is the area of the shaded region in the picture. A convenient way of writing Riemann sums is the "Σ-notation" (Σ is the capital Greek letter sigma),

$$\sum_a^b f(x)\,\Delta x = f(x_0)\,\Delta x + f(x_1)\,\Delta x + \cdots + f(x_{n-1})\,\Delta x.$$

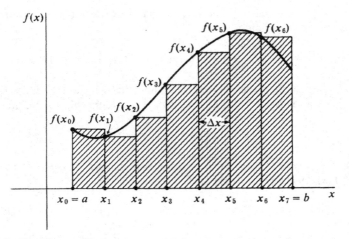

Figure 4.1.3 The Riemann Sum

The a and b indicate that the first subinterval begins at a and the last subinterval ends at b.

We can carry out the same process even when the subinterval length Δx does not divide evenly into the interval length $b - a$. But then, as Figure 4.1.4 shows, there will be a remainder left over at the end of the interval $[a, b]$, and the Riemann sum will have an extra rectangle whose width is this remainder. We let n be the largest integer such that

$$a + n\,\Delta x \le b,$$

and we consider the subintervals

$$[x_0, x_1], \ldots, [x_{n-1}, x_n], [x_n, b],$$

where the partition points are

$$x_0 = a, \quad x_1 = a + \Delta x, \quad x_2 = a + 2\,\Delta x, \ldots, \quad x_n = a + n\,\Delta x, \quad b.$$

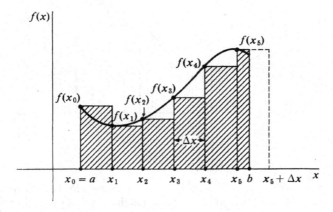

Figure 4.1.4

x_n will be less than or equal to b but $x_n + \Delta x$ will be greater than b. Then we define the *Riemann sum* to be the sum

$$\sum_a^b f(x)\,\Delta x = f(x_0)\,\Delta x + f(x_1)\,\Delta x + \cdots + f(x_{n-1})\,\Delta x + f(x_n)(b - x_n).$$

Thus given the function f, the interval $[a, b]$, and the real number $\Delta x > 0$, we have defined the Riemann sum $\sum_a^b f(x)\,\Delta x$. We repeat the definition more concisely.

DEFINITION

> Let $a < b$ and let Δx be a positive real number. Then the **Riemann sum** $\sum_a^b f(x)\,\Delta x$ is defined as the sum
>
> $$\sum_a^b f(x)\,\Delta x = f(x_0)\,\Delta x + f(x_1)\,\Delta x + \cdots + f(x_{n-1})\,\Delta x + f(x_n)(b - x_n)$$
>
> where n is the largest integer such that $a + n\,\Delta x \le b$, and
>
> $$x_0 = a, \quad x_1 = a + \Delta x, \cdots, \quad x_n = a + n\,\Delta x, \quad b$$
>
> are the partition points.

If $x_n = b$, the last term $f(x_n)(b - x_n)$ is zero. The Riemann sum $\sum_a^b f(x)\,\Delta x$ is a real function of three variables a, b, and Δx,

$$\sum_a^b f(x)\,\Delta x = S(a, b, \Delta x).$$

The symbol x which appears in the expression is called a *dummy variable* (or *bound variable*), because the value of $\sum_a^b f(x)\,\Delta x$ does not depend on x. The dummy variable allows us to use more compact notation, writing $f(x)\,\Delta x$ just once instead of writing $f(x_0)\,\Delta x, f(x_1)\,\Delta x, f(x_2)\,\Delta x$, and so on.

From Figure 4.1.5 it is plausible that by making Δx smaller we can get the Riemann sum as close to the area as we wish.

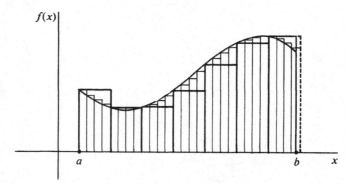

Figure 4.1.5

EXAMPLE 1 Let $f(x) = \frac{1}{2}x$. In Figure 4.1.6, the region under the curve from $x = 0$ to $x = 2$ is a triangle with base 2 and height 1, so its area should be

$$A = \tfrac{1}{2}bh = 1.$$

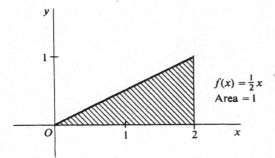

Figure 4.1.6

Let us compare this value for the area with some Riemann sums. In Figure 4.1.7, we take $\Delta x = \frac{1}{2}$. The interval $[0, 2]$ divides into four subintervals $[0, \frac{1}{2}]$, $[\frac{1}{2}, 1]$, $[1, \frac{3}{2}]$, and $[\frac{3}{2}, 2]$. We make a table of values of $f(x)$ at the *lower* endpoints.

x_k	0	$\frac{1}{2}$	1	$\frac{3}{2}$
$f(x_k)$	0	$\frac{1}{4}$	$\frac{1}{2}$	$\frac{3}{4}$

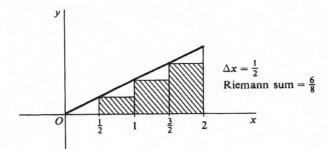

$\Delta x = \frac{1}{2}$
Riemann sum $= \frac{6}{8}$

Figure 4.1.7

The Riemann sum is then

$$\sum_{0}^{2} f(x)\,\Delta x = 0 \cdot \tfrac{1}{2} + \tfrac{1}{4} \cdot \tfrac{1}{2} + \tfrac{1}{2} \cdot \tfrac{1}{2} + \tfrac{3}{4} \cdot \tfrac{1}{2} = \tfrac{6}{8}.$$

In Figure 4.1.8, we take $\Delta x = \frac{1}{4}$. The table of values is as follows.

x_k	0	$\frac{1}{4}$	$\frac{2}{4}$	$\frac{3}{4}$	$\frac{4}{4}$	$\frac{5}{4}$	$\frac{6}{4}$	$\frac{7}{4}$
$f(x_k)$	0	$\frac{1}{8}$	$\frac{2}{8}$	$\frac{3}{8}$	$\frac{4}{8}$	$\frac{5}{8}$	$\frac{6}{8}$	$\frac{7}{8}$

The Riemann sum is

$$\sum_{0}^{2} f(x)\,\Delta x = 0 \cdot \tfrac{1}{4} + \tfrac{1}{8} \cdot \tfrac{1}{4} + \tfrac{2}{8} \cdot \tfrac{1}{4} + \tfrac{3}{8} \cdot \tfrac{1}{4} + \tfrac{4}{8} \cdot \tfrac{1}{4} + \tfrac{5}{8} \cdot \tfrac{1}{4} + \tfrac{6}{8} \cdot \tfrac{1}{4} + \tfrac{7}{8} \cdot \tfrac{1}{4} = \tfrac{7}{8}.$$

We see that the value is getting closer to one.

Finally, let us take a value of Δx that does not divide evenly into the interval length 2. Let $\Delta x = 0.6$. We see in Figure 4.1.9 that the interval then divides into three subintervals of length 0.6 and one of length 0.2, namely $[0, 0.6]$, $[0.6, 1.2]$, $[1.2, 1.8]$, $[1.8, 2.0]$.

x_k	0	0.6	1.2	1.8
$f(x_k)$	0	0.3	0.6	0.9

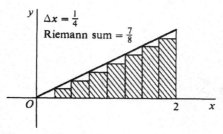

Figure 4.1.8   **Figure 4.1.9**

The Riemann sum is

$$\sum_{0}^{2} f(x)\,\Delta x = 0(.6) + (.3)(.6) + (.6)(.6) + (.9)(.2) = .72.$$

EXAMPLE 2   Let $f(x) = \sqrt{1 - x^2}$, defined on the closed interval $I = [-1, 1]$. The region under the curve is a semicircle of radius 1. We know from plane geometry that the area is $\pi/2$, or approximately $3.14/2 = 1.57$. Let us compute the values of some Riemann sums for this function to see how close they are to 1.57. First take $\Delta x = \frac{1}{2}$ as in Figure 4.1.10(a). We make a table of values.

x_k	-1	$-1/2$	0	$1/2$
$f(x_k)$	0	$\sqrt{3/4}$	1	$\sqrt{3/4}$

The Riemann sum is then

$$\sum_{-1}^{1} f(x)\,\Delta x = 0 \cdot 1/2 + \sqrt{3/4} \cdot 1/2 + 1 \cdot 1/2 + \sqrt{3/4} \cdot 1/2$$

$$= \frac{1 + \sqrt{3}}{2} \sim 1.37.$$

Next we take $\Delta x = \frac{1}{5}$. Then the interval $[-1, 1]$ is divided into ten subintervals as in Figure 4.1.10(b). Our table of values is as follows.

x_k	-1	$-\dfrac{4}{5}$	$-\dfrac{3}{5}$	$-\dfrac{2}{5}$	$-\dfrac{1}{5}$	0	$\dfrac{1}{5}$	$\dfrac{2}{5}$	$\dfrac{3}{5}$	$\dfrac{4}{5}$
$f(x_k)$	0	$\dfrac{3}{5}$	$\dfrac{4}{5}$	$\dfrac{\sqrt{21}}{5}$	$\dfrac{\sqrt{24}}{5}$	1	$\dfrac{\sqrt{24}}{5}$	$\dfrac{\sqrt{21}}{5}$	$\dfrac{4}{5}$	$\dfrac{3}{5}$

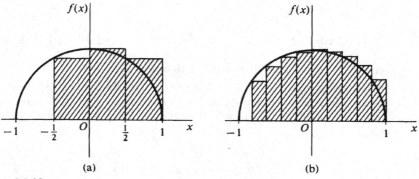

(a)   (b)

Figure 4.1.10

The Riemann sum is

$$\sum_{-1}^{1} f(x)\,\Delta x = \frac{1}{5}\left[0 + \frac{3}{5} + \frac{4}{5} + \frac{\sqrt{21}}{5} + \frac{\sqrt{24}}{5} + 1 + \frac{\sqrt{24}}{5} + \frac{\sqrt{21}}{5} + \frac{4}{5} + \frac{3}{5}\right]$$

$$= \frac{19 + 2\sqrt{21} + 2\sqrt{24}}{25} \sim 1.52.$$

Thus we are getting closer to the actual area $\pi/2 \sim 1.57$.

By taking Δx small we can get the Riemann sum to be as close to the area as we wish.

Our next step is to take Δx to be *infinitely small* and have an *infinite* Riemann sum. How can we do this? We observe that if the real numbers a and b are held fixed, then the Riemann sum

$$\sum_{a}^{b} f(x)\,\Delta x = S(\Delta x)$$

is a real function of the single variable Δx. (The symbol x which appears in the expression is a dummy variable, and the value of

$$\sum_{a}^{b} f(x)\,\Delta x$$

depends only on Δx and not on x.) Furthermore, the term

$$\sum_{a}^{b} f(x)\,\Delta x = S(\Delta x)$$

is defined for all real $\Delta x > 0$. Therefore by the Transfer Principle,

$$\sum_{a}^{b} f(x)\,dx = S(dx)$$

is defined for all hyperreal $dx > 0$. When $dx > 0$ is infinitesimal, there are infinitely many subintervals of length dx, and we call

$$\sum_{a}^{b} f(x)\,dx$$

an *infinite Riemann sum* (Figure 4.1.11).

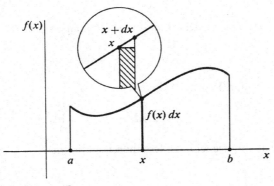

Figure 4.1.11   Infinite Riemann Sum

We may think intuitively of the Riemann sum

$$\sum_a^b f(x)\,dx$$

as the infinite sum

$$f(x_0)\,dx + f(x_1)\,dx + \cdots + f(x_{H-1})\,dx + f(x_H)(b - x_H)$$

where H is the greatest hyperinteger such that $a + H\,dx \le b$. (Hyperintegers are discussed in Section 3.8.) H is positive infinite, and there are $H + 2$ partition points $x_0, x_1, \ldots, x_H, b$. A typical term in this sum is the infinitely small quantity $f(x_K)\,dx$ where K is a hyperinteger, $0 \le K < H$, and $x_K = a + K\,dx$.

The infinite Riemann sum is a hyperreal number. We would next like to take the standard part of it. But first we must show that it is a finite hyperreal number and thus has a standard part.

THEOREM 1

Let f be a continuous function on an interval I, let $a < b$ be two points in I, and let dx be a positive infinitesimal. Then the infinite Riemann sum

$$\sum_a^b f(x)\,dx$$

is a finite hyperreal number.

PROOF Let B be a real number greater than the maximum value of f on $[a, b]$. Consider first a real number $\Delta x > 0$. We can see from Figure 4.1.12 that the

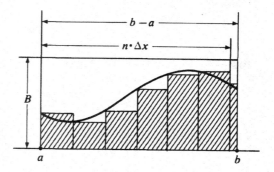

Figure 4.1.12

finite Riemann sum is less than the rectangular area $B \cdot (b - a)$;

$$\sum_a^b f(x)\,\Delta x < B \cdot (b - a).$$

Therefore by the Transfer Principle,

$$\sum_a^b f(x)\,dx < B \cdot (b - a).$$

In a similar way we let C be less than the minimum of f on $[a, b]$ and show that

$$\sum_a^b f(x)\, dx > C \cdot (b - a).$$

Thus the Riemann sum $\sum_a^b f(x)\, dx$ is finite.

We are now ready to define the central concept of this chapter, the definite integral. Recall that the derivative was defined as the standard part of the quotient $\Delta y / \Delta x$ and was written dy/dx. The "definite integral" will be defined as the standard part of the infinite Riemann sum

$$\sum_a^b f(x)\, dx,$$

and is written $\int_a^b f(x)\, dx$. Thus the Δx is changed to dx in analogy with our differential notation. The Σ is changed to the long thin S, i.e., $\int$, to remind us that the integral is obtained from an infinite sum. We now state the definition carefully.

DEFINITION

*Let f be a continuous function on an interval I and let $a < b$ be two points in I. Let dx be a positive infinitesimal. Then the **definite integral** of f from a to b with respect to dx is defined to be the standard part of the infinite Riemann sum with respect to dx, in symbols*

$$\int_a^b f(x)\, dx = st\left(\sum_a^b f(x)\, dx \right).$$

We also define
$$\int_a^a f(x)\, dx = 0,$$

$$\int_b^a f(x)\, dx = -\int_a^b f(x)\, dx.$$

By this definition, for each positive infinitesimal dx the definite integral

$$\int_u^w f(x)\, dx$$

is a real function of two variables defined for all pairs (u, w) of elements of I. The symbol x is a dummy variable since the value of

$$\int_u^w f(x)\, dx$$

does not depend on x.

In the notation $\sum_a^b f(x)\, dx$ for the Riemann sum and $\int_a^b f(x)\, dx$ for the integral, we always use matching symbols for the infinitesimal dx and the dummy variable x. Thus when there are two or more variables we can tell which one is the dummy variable in an integral. For example, $x^2 t$ can be integrated from 0 to 1 with respect to either x or t. With respect to x,

$$\sum_0^1 x^2 t\, dx = x_0^2 t\, dx + x_1^2 t\, dx + \cdots + x_{H-1}^2 t\, dx$$

(where $dx = 1/H$), and we shall see later that

$$\int_0^1 x^2 t \, dx = st(x_0^2 t \, dx + x_1^2 t \, dx + \cdots + x_{H-1}^2 t \, dx) = \tfrac{1}{3}t.$$

With respect to t, however,

$$\sum_0^1 x^2 t \, dt = x^2 t_0 \, dt + x^2 t_1 \, dt + \cdots + x^2 t_{K-1} \, dt,$$

and we shall see later that

$$\int_0^1 x^2 t \, dt = \tfrac{1}{2}x^2.$$

The next two examples evaluate the simplest definite integrals. These examples do it the hard way. A much better method will be developed in Section 4.2.

EXAMPLE 3   Given a constant $c > 0$, evaluate the integral $\int_a^b c \, dx$.

Figure 4.1.13 shows that for every positive real number Δx, the finite Riemann sum is

$$\sum_a^b c \, \Delta x = c(b - a).$$

By the Transfer Principle, the infinite Riemann sum in Figure 4.1.14 has the same value,

$$\sum_a^b c \, dx = c(b - a).$$

Taking standard parts,

$$\int_a^b c \, dx = c(b - a).$$

This is the familiar formula for the area of a rectangle.

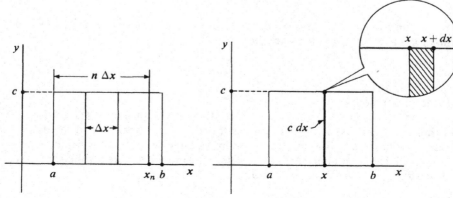

Figure 4.1.13   **Figure 4.1.14**

EXAMPLE 4   Given $b > 0$, evaluate the integral $\int_0^b x\, dx$.

The area under the line $y = x$ is divided into vertical strips of width dx. Study Figure 4.1.15. The area of the lower region A is the infinite Riemann sum

(1)
$$\text{area of } A = \sum_0^b x\, dx.$$

By symmetry, the upper region B has the same area as A;

(2)
$$\text{area of } A = \text{area of } B.$$

Call the remaining region C, formed by the infinitesimal squares along the diagonal. Thus

(3)
$$\text{area of } A + \text{area of } B + \text{area of } C = b^2.$$

Each square in C has height dx except the last one, which may be smaller, and the widths add up to b, so

(4)
$$0 \leq \text{area of } C \leq b\, dx.$$

Putting (1)–(4) together,

$$2\sum_0^b x\, dx \leq b^2 \leq \left(2\sum_0^b x\, dx\right) + b\, dx.$$

Since $b\, dx$ is infinitesimal,

$$2\sum_0^b x\, dx \approx b^2,$$

$$\sum_0^b x\, dx \approx \frac{b^2}{2}.$$

Taking standard parts, we have

$$\int_0^b x\, dx = \frac{b^2}{2}.$$

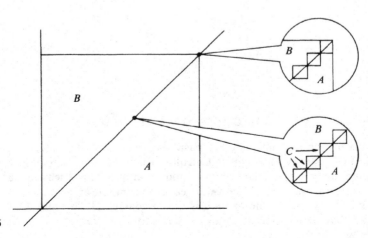

Figure 4.1.15

PROBLEMS FOR SECTION 4.1

Compute the following finite Riemann sums. If a hand calculator is available, the Riemann sums can also be computed with $\Delta x = \frac{1}{10}$.

1	$\sum_0^1 (3x + 1) \Delta x,$	$\Delta x = \frac{1}{3}$	**2**	$\sum_0^1 (3x + 1) \Delta x,$	$\Delta x = \frac{2}{5}$
3	$\sum_{-1}^1 (3x + 1) \Delta x,$	$\Delta x = \frac{1}{4}$	**4**	$\sum_0^1 2x^2 \Delta x,$	$\Delta x = \frac{1}{4}$
5	$\sum_{-1}^1 2x^2 \Delta x,$	$\Delta x = \frac{1}{4}$	**6**	$\sum_0^5 (2x - 1) \Delta x,$	$\Delta x = 1$
7	$\sum_0^5 (2x - 1) \Delta x,$	$\Delta x = 2$	**8**	$\sum_{-1}^1 (x^2 - 1) \Delta x,$	$\Delta x = \frac{1}{2}$
9	$\sum_0^2 (x^2 - 1) \Delta x,$	$\Delta x = \frac{1}{2}$	**10**	$\sum_{-1}^1 (x^2 - 1) \Delta x,$	$\Delta x = \frac{3}{10}$
11	$\sum_{-4}^3 (5x^2 - 12) \Delta x,$	$\Delta x = 2$	**12**	$\sum_{-4}^3 (5x^2 - 12) \Delta x,$	$\Delta x = 1$
13	$\sum_1^3 (1 + 1/x) \Delta x,$	$\Delta x = \frac{1}{3}$	**14**	$\sum_0^5 10^{-2x} \Delta x.$	$\Delta x = \frac{1}{2}$
15	$\sum_{-1}^0 x^4 \Delta x,$	$\Delta x = \frac{1}{4}$	**16**	$\sum_{-1}^1 2x^3 \Delta x,$	$\Delta x = \frac{1}{2}$
17	$\sum_0^\pi \sqrt{x} \Delta x,$	$\Delta x = 1$	**18**	$\sum_{-2}^9 |x - 4| \Delta x,$	$\Delta x = 2$
19	$\sum_0^\pi \sin x \Delta x,$	$\Delta x = \pi/4$	**20**	$\sum_0^\pi \sin^2 x \Delta x,$	$\Delta x = \pi/4$
21	$\sum_0^1 e^x \Delta x,$	$\Delta x = 1/5$	**22**	$\sum_0^1 xe^x \Delta x,$	$\Delta x = 1/5$
23	$\sum_1^5 \ln x \Delta x,$	$\Delta x = 1$	**24**	$\sum_1^5 \frac{\ln x}{x} \Delta x,$	$\Delta x = 1$

☐ **25** Let b be a positive real number and n a positive integer. Prove that if $\Delta x = b/n$,

$$\sum_0^b x \Delta x = (1 + 2 + \cdots + (n - 1)) \Delta x^2.$$

Using the formula $1 + 2 + \cdots + (n - 1) = \dfrac{n(n - 1)}{2}$, prove that

$$\sum_0^b x \Delta x = (1 - 1/n)b^2/2.$$

☐ **26** Let H be a positive infinite hyperinteger and $dx = b/H$. Using the Transfer Principle and Problem 25, prove that $\int_0^b x \, dx = b^2/2$.

☐ **27** Let b be a positive real number, n a positive integer, and $\Delta x = b/n$. Using the formula

$$1^2 + 2^2 + 3^2 + \cdots + (n - 1)^2 = \frac{n(n - 1)(2n - 1)}{6},$$

prove that

$$\sum_0^b x^2 \Delta x = \frac{n(n - 1)(2n - 1)}{6} \frac{b^3}{n^3}.$$

☐ **28** Use Problem 27 to show that $\int_0^b x^2 \, dx = b^3/3$.

4.2 FUNDAMENTAL THEOREM OF CALCULUS

In this section we shall state five basic theorems about the integral, culminating in the Fundamental Theorem of Calculus. Right now we can only approximate a definite integral by the laborious computation of a finite Riemann sum. At the end of this section we will be in a position easily to compute exact values for many definite integrals. The key to the method is the Fundamental Theorem. Our first theorem shows that we are free to choose any positive infinitesimal we wish for dx in the definite integral.

THEOREM 1

> *Given a continuous function f on $[a, b]$ and two positive infinitesimals dx and du, the definite integrals with respect to dx and du are the same,*
>
> $$\int_a^b f(x)\,dx = \int_a^b f(u)\,du.$$

From now on when we write a definite integral $\int_a^b f(x)\,dx$, it is understood that dx is a positive infinitesimal. By Theorem 1, it doesn't matter which infinitesimal.

The proof of Theorem 1 is based on the following intuitive idea. Figure 4.2.1 shows the two Riemann sums $\sum_a^b f(x)\,dx$ and $\sum_a^b f(u)\,du$. We see from the figure that the difference $\sum_a^b f(x)\,dx - \sum_a^b f(u)\,du$ is a sum of rectangles of infinitesimal height. These difference rectangles all lie between the horizontal lines $y = -\varepsilon$ and $y = \varepsilon$, where ε is the largest height. Thus $-\varepsilon(b - a) \le \sum_a^b f(x)\,dx - \sum_a^b f(u)\,du \le \varepsilon(b - a)$. Taking standard parts,

$$0 \le \int_a^b f(x)\,dx - \int_a^b f(u)\,du \le 0,$$

$$\int_a^b f(x)\,dx = \int_a^b f(u)\,du.$$

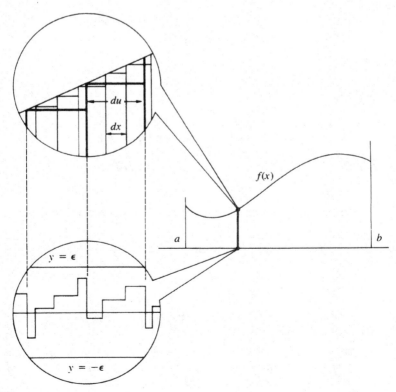

Figure 4.2.1

Theorem 1 shows that whenever Δx is positive infinitesimal, the Riemann sum is infinitely close to the definite integral,

$$\sum_{a}^{b} f(x)\,\Delta x \approx \int_{a}^{b} f(x)\,dx.$$

This fact can also be expressed in terms of limits. It shows that the Riemann sum approaches the definite integral as Δx approaches 0 from above, in symbols

$$\int_{a}^{b} f(x)\,dx = \lim_{\Delta x \to 0^{+}} \sum_{a}^{b} f(x)\,\Delta x.$$

Given a continuous function f on an interval I, Theorem 1 shows that the definite integral is a real function of two variables a and b,

$$A(a, b) = \int_{a}^{b} f(x)\,dx, \qquad a, b \text{ in } I.$$

We now formally define the area as the definite integral shown in Figure 4.2.2.

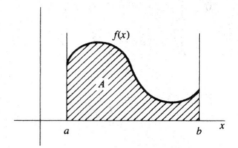

Figure 4.2.2

DEFINITION

If f is continuous and $f(x) \geq 0$ on $[a, b]$, the area of the region below the curve $y = f(x)$ from a to b is defined as the definite integral:

$$\text{Area} = \int_{a}^{b} f(x)\,dx.$$

The next two theorems give basic properties of the integral.

THEOREM 2 (The Rectangle Property)

Suppose f is continuous and has minimum value m and maximum value M on a closed interval $[a, b]$. Then

$$m(b - a) \leq \int_{a}^{b} f(x)\,dx \leq M(b - a).$$

That is, the area of the region under the curve is between the area of the rectangle whose height is the minimum value of f and the area of the rectangle whose height is the maximum value of f in the interval $[a, b]$.

The Extreme Value Theorem is needed to show that the minimum value m and maximum value M exist. The rectangle of height m is called the *inscribed rectangle* of the region, and the rectangle of height M is called the *circumscribed rectangle*. From Figure 4.2.3, we see that the inscribed rectangle is a subset of the region under the curve, which is in turn a subset of the circumscribed rectangle. The Rectangle Property says that the area of the region is between the areas of the inscribed and circumscribed rectangles.

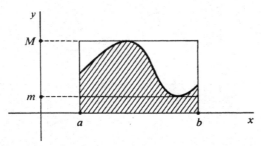

Figure 4.2.3 The Rectangle Property

PROOF By Theorem 1, any positive infinitesimal may be chosen for dx. Let us choose a positive infinite hyperinteger H and let $dx = (b - a)/H$. Then dx evenly divides $b - a$; that is, the interval $[a, b]$ is divided into H subintervals of exactly the same length dx. Then

$$\sum_a^b m\,dx = m \cdot H \cdot dx = m(b - a),$$

$$\sum_a^b M\,dx = M \cdot H \cdot dx = M(b - a).$$

For each x, we have $m \le f(x) \le M$. Adding up and taking standard parts, we obtain the required formula.

$$\sum_a^b m\,dx \le \sum_a^b f(x)\,dx \le \sum_a^b M\,dx,$$

$$m(b - a) \le \int_a^b f(x)\,dx \le M(b - a).$$

One useful consequence of the Rectangle Property is that the integral of a positive function is positive and the integral of a negative function is negative:

If $f(x) > 0$ on $[a, b]$,   then $0 < m(b - a) \le \int_a^b f(x)\,dx.$

If $f(x) < 0$ on $[a, b]$,   then $\int_a^b f(x)\,dx \le M(b - a) < 0.$

The definite integral of a negative function $f(x) = -g(x)$ from a to b is just the negative of the area of the region above the curve and below the x axis. This is because

$$f(x)\,dx = -g(x)\,dx,$$

$$\sum_a^b f(x)\,dx = -\sum_a^b g(x)\,dx,$$

$$\int_a^b f(x)\,dx = -\int_a^b g(x)\,dx.$$

(See Figure 4.2.4.)

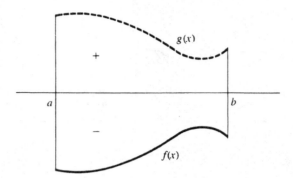

Figure 4.2.4

THEOREM 3 (The Addition Property)

Suppose f is continuous on an interval I. Then for all a, b, c in I,

$$\int_a^c f(x)\,dx = \int_a^b f(x)\,dx + \int_b^c f(x)\,dx.$$

This property is illustrated in Figure 4.2.5 for the case $a < b < c$. The Addition Property holds even if the points a, b, c are in some other order on the real line, such as $c < a < b$.

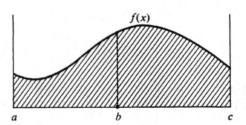

Figure 4.2.5

PROOF First suppose that $a < b < c$. Choose a dx that evenly divides the first interval length $b - a$. This simplifies our computation because it makes b a partition point, $b = a + H\,dx$. Then, as Figure 4.2.6 suggests,

$$\sum_a^c f(x)\,dx = \sum_a^b f(x)\,dx + \sum_b^c f(x)\,dx.$$

Taking standard parts we have the desired formula

$$\int_a^c f(x)\,dx = \int_a^b f(x)\,dx + \int_b^c f(x)\,dx.$$

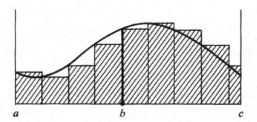

Figure 4.2.6

To illustrate the other cases, we prove the Addition Property when $c < a < b$. The previous case gives

$$\int_c^b f(x)\,dx = \int_c^a f(x)\,dx + \int_a^b f(x)\,dx.$$

Since reversing the endpoints changes the sign of the integral,

$$-\int_b^c f(x)\,dx = -\int_a^c f(x)\,dx + \int_a^b f(x)\,dx,$$

and the desired formula

$$\int_a^c f(x)\,dx = \int_a^b f(x)\,dx + \int_b^c f(x)\,dx$$

follows.

The definite integral of a curve can be thought of as area even if the curve crosses the x-axis. The curve in Figure 4.2.7 is positive from a to b and negative from b to c, crossing the x-axis at b. The integral $\int_a^b f(x)\,dx$ is a positive number and the integral $\int_b^c f(x)\,dx$ is a negative number. By the Addition Property, the integral

$$\int_a^c f(x)\,dx = \int_a^b f(x)\,dx + \int_b^c f(x)\,dx$$

is equal to the area from a to b minus the area from b to c. The definite integral $\int_a^c f(x)\,dx$ always gives the net area between the x-axis and the curve, counting areas above the x-axis as positive and areas below the x-axis as negative.

The definite integral $\int_u^v f(t)\,dt$ is a real function of two variables u and v and does not depend on the dummy variable t. If we replace u by a constant a and v by the variable x, we obtain a real function of one variable x, given by

$$F(x) = \int_a^x f(t)\,dt.$$

Our fourth theorem states that this new function is continuous.

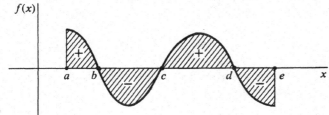

Figure 4.2.7

THEOREM 4

Let f be continuous on an interval I. Choose a point a in I. Then the function $F(x)$ defined by

$$F(x) = \int_a^x f(t)\, dt$$

is continuous on I.

SKETCH OF PROOF Let c be in I, and let x be infinitely close to c and between the endpoints of I. By the Addition Property,

$$\int_a^c f(t)\, dt = \int_a^x f(t)\, dt + \int_x^c f(t)\, dt,$$

$$\int_a^c f(t)\, dt - \int_a^x f(t)\, dt = \int_x^c f(t)\, dt,$$

and
$$F(c) - F(x) = \int_x^c f(t)\, dt.$$

This is the area of the infinitely thin strip under the curve $y = f(t)$ between $t = x$ and $t = c$ (see Figure 4.2.8). The strip has width $\Delta x = c - x$. By the Rectangle Property, its area is between $m\,\Delta x$ and $M\,\Delta x$ and hence is infinitely small. Therefore $F(x)$ is infinitely close to $F(c)$, and F is continuous on I.

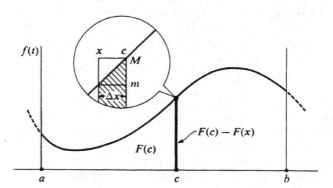

Figure 4.2.8

Our fifth theorem, the Fundamental Theorem of Calculus, shows ·that the definite integral can be evaluated by means of antiderivatives. The process of antidifferentiation is just the opposite of differentiation. To keep things simple, let I be an open interval, and assume that all functions mentioned have domain I.

DEFINITION

Let f and F be functions with domain I. If f is the derivative of F, then F is called an **antiderivative** of f.

For example, suppose a particle is moving upward along the y-axis with velocity $v = f(t)$ and position $y = F(t)$ at time t. The position $y = F(t)$ is an antiderivative of the velocity $v = f(t)$. We shall discuss antiderivatives in more detail in the next section. We are now ready for the Fundamental Theorem.

FUNDAMENTAL THEOREM OF CALCULUS

Suppose f is continuous on its domain, which is an open interval I.

(i) *For each point a in I, the definite integral of f from a to x considered as a function of x is an antiderivative of f. That is,*

$$d\left(\int_a^x f(t)\, dt \right) = f(x)\, dx.$$

(ii) *If F is any antiderivative of f, then for any two points (a, b) in I the definite integral of f from a to b is equal to the difference $F(b) - F(a)$,*

$$\int_a^b f(x)\, dx = F(b) - F(a).$$

The Fundamental Theorem of Calculus is important for two reasons. First, it shows the relation between the two main notions of calculus: the derivative, which corresponds to velocity, and the integral, which corresponds to area. It shows that differentiation and integration are "inverse" processes. Second, it gives a simple method for computing many definite integrals.

EXAMPLE 1

(a) Find $\int_a^b c\, dx$. Since cx is an antiderivative of c,

$$\int_a^b c\, dx = cb - ca = c(b - a).$$

(b) Find $\int_a^b x\, dx$. $\frac{1}{2}x^2$ is an antiderivative of x. Thus

$$\int_a^b x\, dx = \tfrac{1}{2}b^2 - \tfrac{1}{2}a^2.$$

The above example gives the same result that we got before but is much simpler. We can easily go further.

EXAMPLE 2   Find $\int_a^b x^2\, dx$. $x^3/3$ is an antiderivative of x^2 because

$$\frac{d(x^3/3)}{dx} = \frac{3x^2}{3} = x^2.$$

Therefore
$$\int_a^b x^2\, dx = \frac{b^3}{3} - \frac{a^3}{3}.$$

This gives the area of the region under the curve $y = x^2$ between a and b (Figure 4.2.9).

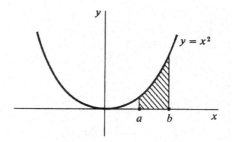

Figure 4.2.9

If a particle moves along the y-axis with continuous velocity $v = f(t)$, the position $y = F(t)$ is an antiderivative of the velocity, because $v = dy/dt$. The Fundamental Theorem of Calculus shows that the distance moved (the change in y) between times $t = a$ and $t = b$ is equal to the definite integral of the velocity,

$$\text{distance moved} = F(b) - F(a) = \int_a^b f(t)\, dt.$$

EXAMPLE 3   A particle moves along the y-axis with velocity $v = 8t^3$ cm/sec. How far does it move between times $t = -1$ and $t = 2$ sec? The function $G(t) = 2t^4$ is an antiderivative of the velocity $v = 8t^3$. Thus the definite integral is

$$\text{distance moved} = \int_{-1}^2 8t^3\, dt = 2 \cdot 2^4 - 2 \cdot (-1)^4 = 30 \text{ cm.}$$

EXAMPLE 4   Find $\int_0^4 \sqrt{t}\, dt$ (Figure 4.2.10). The function $\sqrt{t}$ is defined and continuous on the half-open interval $[0, \infty)$. But to apply the Fundamental Theorem we need a function continuous on an open interval that contains the limit points 0 and 4. We therefore define

$$f(t) = \begin{cases} 0 & \text{for } t < 0 \\ \sqrt{t} & \text{for } t \geq 0. \end{cases}$$

This function is continuous on the whole real line. In particular it is continuous at 0 because if $t \approx 0$ then $f(t) \approx 0$. The function

$$F(t) = \begin{cases} 0 & \text{for } t < 0 \\ \frac{2}{3}t^{3/2} & \text{for } t \geq 0 \end{cases}$$

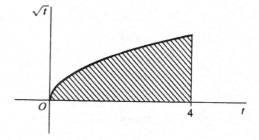

Figure 4.2.10

is an antiderivative of f. Then

$$\int_0^4 \sqrt{t}\, dt = F(4) - F(0) = (\tfrac{2}{3} \cdot 4^{3/2} - \tfrac{2}{3} \cdot 0^{3/2}) = \tfrac{16}{3}.$$

In the next section we shall develop some methods for finding antiderivatives. The antiderivative of a very simple function may turn out to be a "new" function which we have not yet given a name.

EXAMPLE 5 The only way we can show that the function $f(x) = \sqrt{1 + x^4}$ has an antiderivative is to take a definite integral

$$\int_0^x \sqrt{1 + t^4}\, dt.$$

This is a "new" function that cannot be expressed in terms of algebraic, trigonometric, and exponential functions without calculus.

The Fundamental Theorem can also be used to find the derivative of a function which is defined as a definite integral with a variable limit of integration. This can be done without actually evaluating the integral.

EXAMPLE 6 Let $y = \int_x^2 \sqrt{1 + t^2}\, dt$. Then $y = -\int_2^x \sqrt{1 + t^2}\, dt$,

and $$dy = -d\left(\int_2^x \sqrt{1 + t^2}\, dt \right) = -\sqrt{1 + x^2}\, dx.$$

EXAMPLE 7 Let $y = \int_3^{x^2 + x} \dfrac{1}{t^3 + 1}\, dt.$

Let $u = x^2 + x$. Then

$$\frac{du}{dx} = (2x + 1), \qquad y = \int_3^u \frac{1}{t^3 + 1}\, dt, \qquad \frac{dy}{du} = \frac{1}{u^3 + 1}.$$

By the Chain Rule,

$$\frac{dy}{dx} = \frac{dy}{du}\frac{du}{dx} = \frac{1}{u^3 + 1}(2x + 1) = \frac{2x + 1}{(x^2 + x)^3 + 1}.$$

We conclude this section with a proof of the Fundamental Theorem of Calculus.

PROOF (i) Let $F(x)$ be the area under the curve $y = f(t)$ from a to x,

$$F(x) = \int_a^x f(t)\, dt.$$

Imagine that the vertical line cutting the t-axis at x moves to the right as in Figure 4.2.11.

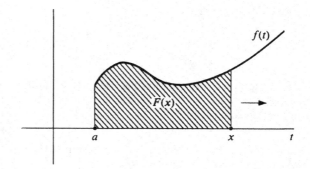

Figure 4.2.11

We show that the rate of change of $F(x)$ is equal to the length $f(x)$ of the moving vertical line.

Suppose x increases by an infinitesimal amount $\Delta x > 0$. Then

$$F(x + \Delta x) - F(x) = \int_{x}^{x+\Delta x} f(t)\, dt$$

is the area of an infinitely thin strip of width Δx and height infinitely close to $f(x)$. By the Rectangle Property the area of the strip is between the inscribed and circumscribed rectangles (Figure 4.2.12),

$$m\, \Delta x \le F(x + \Delta x) - F(x) \le M\, \Delta x.$$

Dividing by Δx, $$m \le \frac{F(x + \Delta x) - F(x)}{\Delta x} \le M.$$

Since f is continuous at x, the values m and M are both infinitely close to $f(x)$, and therefore

$$\frac{F(x + \Delta x) - F(x)}{\Delta x} \approx f(x).$$

The proof is similar when $\Delta x < 0$. Hence $F'(x) = f(x)$.

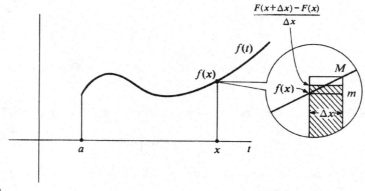

Figure 4.2.12

PROOF (ii)   Let $F(x)$ be any antiderivative of f. Then, by (i),

$$d\left(F(x) - \int_a^x f(t)\,dt\right) = f(x) - f(x) = 0.$$

In Section 3.7 on curve sketching, we saw that every function with derivative zero is constant. Thus

$$F(x) - \int_a^x f(t)\,dt = C_0, \qquad F(x) = \int_a^x f(t)\,dt + C_0$$

for some constant C_0. Then

$$F(b) - F(a) = \left(\int_a^b f(t)\,dt + C_0\right) - \left(\int_a^a f(t)\,dt + C_0\right)$$

$$= \int_a^b f(t)\,dt - 0 = \int_a^b f(t)\,dt,$$

so $$F(b) - F(a) = \int_a^b f(x)\,dx.$$

PROBLEMS FOR SECTION 4.2

In Problems 1–14, find an antiderivative of the given function.

1	$f(x) = 8\sqrt{x}$	2	$f(x) = 4/\sqrt{x}$
3	$f(t) = 3t^2 + 1$	4	$f(x) = 5x^3$
5	$f(t) = 4 - 3t^2$	6	$f(z) = 2/z^2$
7	$f(s) = 7s^{-3}$	8	$f(t) = t^2 + t^{-2}$
9	$f(x) = (x - 6)^2$	10	$f(u) = (5u + 1)^2$
11	$f(y) = y^{3/2}$	12	$f(x) = 2/x\sqrt{x}$
13	$f(x) = |x|$	14	$f(t) = |2t - 4|$

15   If $F'(x) = x + x^2$ for all x, find $F(1) - F(-1)$.

16   If $F'(x) = x^4$ for all x, find $F(2) - F(1)$.

17   If $F'(t) = t^{1/3}$ for all t, find $F(8) - F(0)$.

Evaluate the definite integrals in Problems 18–22.

18    $\displaystyle\int_{-1}^{1} 2x^2\,dx$ 19 $\displaystyle\int_{-2}^{2} x^3\,dx$

20    $\displaystyle\int_{-2}^{-1} t^{-2}\,dt$ 21 $\displaystyle\int_{0}^{4} 2\sqrt{x}\,dx$

22    $\displaystyle\int_{-3}^{-2} -5x^4\,dx$

In Problems 23–27 an object moves along the y-axis. Given the velocity v, find how far the object moves between the given times t_0 and t_1.

23    $v = 2t + 5$, $t_0 = 0$, $t_1 = 2$

24    $v = 4 - t$, $t_0 = 1$, $t_1 = 4$

25    $v = 3,$ $\qquad$ $t_0 = 2,$ $t_1 = 6$

26    $v = 3t^2,$ $\qquad$ $t_0 = 1,$ $t_1 = 3$

27    $v = 10t^{-2},$ $\qquad$ $t_0 = 1,$ $t_1 = 100$

In Problems 28–32, find the area of the region under the curve $y = f(x)$ from a to b.

28    $y = 4 - x^2,$ $\qquad$ $a = -2,$ $b = 2$

29    $y = \sqrt{x + 2},$ $\qquad$ $a = -2,$ $b = 2$

30    $y = 9x - x^2,$ $\qquad$ $a = 0,$ $b = 3$

31    $y = \sqrt{x} - x,$ $\qquad$ $a = 0,$ $b = 1$

32    $y = 3x^{1/3},$ $\qquad$ $a = 1,$ $b = 8$

33   If $F'(t) = t - 1$ for all t and $F(0) = 2$, find $F(2)$.

34   If $F'(x) = 1 - x^2$ for all x and $F(3) = 5$, find $F(-1)$.

☐ 35   Suppose $F(x)$ and $G(x)$ have continuous derivatives and $F'(x) + G'(x) = 0$ for all x. Prove that $F(x) + G(x)$ is constant.

☐ 36   Suppose $F(x)$ and $G(x)$ have continuous derivatives such that $F'(x) \le G'(x)$ for all x. Prove that $$F(b) - F(a) \le G(b) - G(a)$$ where $a < b$.

☐ 37   Prove that a function $F(x)$ has a constant derivative if and only if $F(x)$ is linear, i.e., of the form $F(x) = ax + b$.

☐ 38   Prove that a function $F(x)$ has a constant second derivative if and only if $F(x)$ has the form $F(x) = ax^2 + bx + c$.

☐ 39   Suppose that $F''(x) = G''(x)$ for all x. Prove that $F(x)$ and $G(x)$ differ by a linear function, that is, $G(x) = F(x) + ax + b$ for some real numbers a and b.

4.3 INDEFINITE INTEGRALS

The Fundamental Theorem of Calculus shows that every continuous function f has at least one antiderivative, namely $F(x) = \int_a^x f(t)\,dt$. Actually, f has infinitely many antiderivatives, but any two antiderivatives of f differ only by a constant. This is an important fact about antiderivatives, which we state as a theorem.

THEOREM 1

Let f be a real function whose domain is an open interval I.

(i)   *If $F(x)$ is an antiderivative of $f(x)$, then $F(x) + C$ is an antiderivative of $f(x)$ for every real number C.*

(ii)   *If $F(x)$ and $G(x)$ are two antiderivatives of $f(x)$, then $F(x) - G(x)$ is constant for all x in I. That is,*

$$G(x) = F(x) + C$$

for some real number C.

Discussion   Parts (i) and (ii) together show that if we can find one antiderivative $F(x)$ of $f(x)$, then the family of functions

$$F(x) + C, \qquad C = \text{a real number}$$

gives *all* antiderivatives of $f(x)$. We see from Figure 4.3.1 that the graph of $F(x) + C$ is just the graph of $F(x)$ moved vertically by a distance C. The graphs of $F(x)$ and $F(x) + C$ have the same slopes at every point x. For example, let $f(x) = 3x^2$. Then $F(x) = x^3$ is an antiderivative of $3x^2$ because

$$\frac{d(x^3)}{dx} = 3x^2.$$

But $x^3 + 6$ and $x^3 - \sqrt{2}$ are also antiderivatives of $3x^2$. In fact, $x^3 + C$ is an antiderivative of $3x^2$ for each real number C. Theorem 1 shows that $3x^2$ has no other antiderivatives.

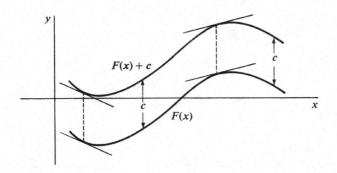

Figure 4.3.1

PROOF We prove (i) by differentiating,

$$\frac{d(F(x) + C)}{dx} = \frac{d(F(x))}{dx} + \frac{dC}{dx} = f(x) + 0 = f(x).$$

Part (ii) follows from a theorem in Section 3.7 on curve sketching. If a function has derivative zero on I, then the function is constant on I. The difference $F(x) - G(x)$ has derivative $f(x) - f(x) = 0$ and is therefore constant. We used this fact in the proof of the Fundamental Theorem of Calculus.

In computing integrals of f, we usually work with the family of all antiderivatives of f. We shall call this whole family of functions the *indefinite integral* of f. The symbol for the indefinite integral is $\int f(x)\, dx$. If $F(x)$ is one antiderivative of f, the indefinite integral is the set of all functions of the form $F(x) + C_0$, C_0 constant. We express this with the equation

$$\int f(x)\, dx = F(x) + C.$$

It is an equation between two families of functions rather than between two single functions. C is called the *constant of integration*. To illustrate the notation,

$$\int 3x^2\, dx = x^3 + C.$$

We repeat the above definitions in concise form.

DEFINITION

*Let the domain of f be an open interval I and suppose f has an antiderivative. The family of all antiderivatives of f is called the **indefinite integral** of f and is denoted by $\int f(x)\,dx$.*

Given a function F, the family of all functions which differ from F only by a constant is written $F(x) + C$. Thus if F is an antiderivative of f we write

$$\int f(x)\,dx = F(x) + C.$$

When working with indefinite integrals, it is convenient to use differentials and dependent variables. If we introduce the dependent variable u by $u = F(x)$, then

$$du = F'(x)\,dx = f(x)\,dx.$$

Thus the equation $\qquad\qquad \int f(x)\,dx = F(x) + C$

can be written in the form $\qquad\qquad \int du = u + C.$

The differential symbol d and the indefinite integral symbol $\int$ behave as inverses to each other. We can start with the family of functions $u + C$, form du, and then form $\int du = u + C$ to get back where we started. Some of the rules for differentiation given in Chapter 2 can be turned around to give a set of rules for indefinite integration.

THEOREM 2

Let u and v be functions of x whose domains are an open interval I and suppose du and dv exist for every x in I.

(i) $\quad \int du = u + C.$

(ii) **Constant Rule** $\quad \int c\,du = c\int du.$

(iii) **Sum Rule** $\quad \int du + dv = \int du + \int dv.$

(iv) **Power Rule** $\quad \int u^r\,du = \dfrac{u^{r+1}}{r+1} + C,$ *where r is rational, $r \neq -1$, and $u > 0$ on I.*

(v) $\quad \int \sin u\,du = -\cos u + C.$

(vi) $\quad \int \cos u\,du = \sin u + C.$

(vii) $\quad \int e^u\,du = e^u + C.$

(viii) $\displaystyle\int \frac{1}{u}\, du = \ln |u| + C$ $(u \neq 0)$.

Discussion The Power Rule gives the integral of u^r when $r \neq -1$, while Rule (viii) gives the integral of u^r when $r = -1$. When we put $u = f(x)$ and $v = g(x)$, the Constant and Sum Rules take the form

Constant Rule $\displaystyle\int cf(x)\, dx = c \int f(x)\, dx.$

Sum Rule $\displaystyle\int (f(x) + g(x))\, dx = \int f(x)\, dx + \int g(x)\, dx.$

In the Constant and Sum Rules we are multiplying a family of functions by a constant and adding two families of functions. If we do either of these two things to families of functions differing only by a constant, we get another family of functions differing only by a constant. For example,

$$7(3x^4 + C) = 21x^4 + 7C = 21x^4 + C'$$

is the family of all functions equal to $21x^4$ plus a constant. Similarly,

$$(3\sqrt{x} + C) + (5x - \sqrt{x} + D) = 5x + 2\sqrt{x} + (C + D) = 5x + 2\sqrt{x} + C'$$

is the family of all functions equal to $5x + 2\sqrt{x}$ plus a constant.

PROOF OF THEOREM 2

(i) This is just a short form of the theorem that $u + C$ is the family of all functions which have the same derivative as u.

(ii) We have $c\, du = d(cu)$, whence

$$\int c\, du = \int d(cu) = cu + C = c(u + C') = c \int du.$$

(iii) $du + dv = d(u + v)$,

$$\int du + dv = \int d(u + v) = u + v + C = \int du + \int dv.$$

(iv) $d\left(\dfrac{u^{r+1}}{r+1}\right) = \dfrac{(r+1)u^r}{r+1}\, du = u^r\, du,$

$$\int u^r\, du = \frac{u^{r+1}}{r+1} + C.$$

Rules (v)–(viii) are similar. Only the last formula, (viii), requires an explanation. The absolute value in $\ln |u|$ comes about by combining the two cases $u > 0$ and $u < 0$. When $u > 0$, $u = |u|$ and

$$d(\ln |u|) = d(\ln u) = \frac{1}{u}\, du.$$

When $u < 0$, $\ln u$ is undefined, but $|u| = -u$ and $\ln |u| = \ln(-u)$. Thus

$$d(\ln |u|) = d(\ln(-u)) = -\frac{1}{u}\, d(-u) = \frac{1}{u}\, du.$$

Thus, in both cases, when $u \neq 0$,

$$d(\ln |u|) = \frac{1}{u} du,$$

$$\int \frac{1}{u} du = \ln |u| + C.$$

EXAMPLE 1 $\int (2x^{-1} + 3 \sin x) \, dx = 2 \ln |x| - 3 \cos x + C.$

We can use the rules to write down at once the indefinite integral of any polynomial.

EXAMPLE 2 $\int (4x^3 - 6x^2 + 2x + 1) \, dx = x^4 - 2x^3 + x^2 + x + C.$

EXAMPLE 3 $\int \left(\frac{3}{x^2} + \sqrt{x} \right) dx = -\frac{3}{x} + \frac{2}{3} x^{3/2} + C.$

Indefinite integration is much harder than differentiation, because there are no rules for integrating the product or quotient of two functions. It often requires guesswork. The short list of rules in Theorem 1 will help, and as this course proceeds we shall add many more techniques for finding indefinite integrals.

EXAMPLE 4   Show that $\int \frac{dx}{(1 + x)^{1/2}(1 - x)^{3/2}} = \sqrt{\frac{1 + x}{1 - x}} + C.$

Our rules give no hint on finding this integral. However, once the answer is given to us we can easily prove that it is correct by differentiating,

$$\frac{d\sqrt{\frac{1 + x}{1 - x}}}{dx} = \frac{d((1 + x)^{1/2}(1 - x)^{-1/2})}{dx}$$

$$= (1 + x)^{1/2}(-1)(-\tfrac{1}{2})(1 - x)^{-3/2} + (1 - x)^{-1/2}(\tfrac{1}{2})(1 + x)^{-1/2}$$

$$= (1 + x)^{-1/2}(1 - x)^{-3/2}[\tfrac{1}{2}(1 + x) + \tfrac{1}{2}(1 - x)]$$

$$= \frac{1}{(1 + x)^{1/2}(1 - x)^{3/2}}.$$

Here is a warning that may prevent some common mistakes.
Warning: The integral of the product of two functions is *not* equal to the product of the integrals. The same goes for quotients. That is,

Wrong: $\int (uv) \, dx = \left(\int u \, dx \right) \left(\int v \, dx \right).$

For example,

Wrong: $\int x(x+1)\,dx = \left(\int x\,dx\right)\left(\int (x+1)\,dx\right) = \dfrac{x^2}{2}\left(\dfrac{x^2}{2} + x\right) + C$

$\qquad\qquad\qquad = \dfrac{x^4}{4} + \dfrac{x^3}{2} + C.$

Correct: $\int x(x+1)\,dx = \int (x^2 + x)\,dx = \dfrac{x^3}{3} + \dfrac{x^2}{2} + C.$

Wrong: $\int \dfrac{u}{v}\,dx = \dfrac{\int u\,dx}{\int v\,dx}.$

For example,

Wrong: $\int \dfrac{x+1}{\sqrt{x}}\,dx = \dfrac{\int (x+1)\,dx}{\int \sqrt{x}\,dx} = \dfrac{(\frac{1}{2})x^2 + x}{(\frac{2}{3})x^{3/2}} + C$

$\qquad\qquad\qquad = \dfrac{3\sqrt{x}}{4} + \dfrac{3}{2\sqrt{x}} + C.$

Correct: $\int \dfrac{x+1}{\sqrt{x}}\,dx = \int \left(\sqrt{x} + \dfrac{1}{\sqrt{x}}\right) dx = \frac{2}{3}x^{3/2} + 2\sqrt{x} + C.$

The indefinite integral can be used to solve problems of the following type. Given that a particle moves along the y-axis with velocity $v = f(t)$, and that at a certain time $t = t_0$ its position is $y = y_0$. Find the position y as a function of t.

EXAMPLE 5 A particle moves with velocity $v = 1/t^2$, $t > 0$. At time $t = 2$ it is at position $y = 1$. Find the position y as a function of t. We compute

$$\int v\,dt = \int \dfrac{1}{t^2}\,dt = -\dfrac{1}{t} + C.$$

Since $dy/dt = v$, y is one of the functions in the family $-1/t + C$. We can find the constant C by setting $t = 2$ and $y = 1$,

$$y = -\dfrac{1}{t} + C, \qquad 1 = -\dfrac{1}{2} + C, \qquad C = 1\tfrac{1}{2}.$$

Then the answer is

$$y = -\dfrac{1}{t} + 1\tfrac{1}{2}.$$

The next theorem shows that in such a problem we can always find the answer if we are given the position of the particle at just one point of time.

THEOREM 3

Suppose the domain of f is an open interval I and f has an antiderivative. Let $P(x_0, y_0)$ be any point with x_0 in I. Then f has exactly one antiderivative whose graph passes through P.

PROOF Let F be any antiderivative of f. Then $F(x) + C$ is the family of all anti-derivatives. We show that there is exactly one value of C such that the function $F(x) + C$ passes through $P(x_0, y_0)$ (Figure 4.3.2). We note that all of the following statements are equivalent:

(1) $F(x) + C$ passes through $P(x_0, y_0)$.
(2) $F(x_0) + C = y_0$.
(3) $C = y_0 - F(x_0)$.

Thus $y_0 - F(x_0)$ is the unique value of C which works.

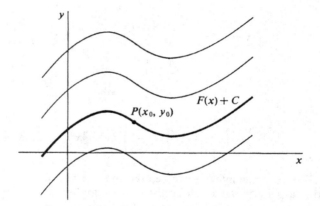

Figure 4.3.2

The Fundamental Theorem of Calculus, part (ii), may be expressed briefly as follows, where f is continuous on I.

If $\int f(x)\,dx = F(x) + C$, then

$$\int_a^b f(x)\,dx = F(b) - F(a).$$

For evaluating definite integrals we introduce the convenient notation

$$F(x)\Big]_a^b = F(b) - F(a).$$

It is read "$F(x)$ evaluated from a to b."

The Constant and Sum Rules hold for definite as well as indefinite integrals:

Constant Rule $$\int_a^b cf(x)\,dx = c\int_a^b f(x)\,dx.$$

Sum Rule $$\int_a^b (f(x) + g(x))\,dx = \int_a^b f(x)\,dx + \int_a^b g(x)\,dx.$$

The Constant Rule is shown by the computation

$$\int_a^b cf(x)\,dx = cF(b) - cF(a) = c(F(b) - F(a)) = c\int_a^b f(x)\,dx.$$

The Sum Rule is similar.

EXAMPLE 6 Evaluate the definite integral of $y = (1+t)/t^3$ from $t = 1$ to $t = 2$ (see Figure 4.3.3).

$$\int_1^2 \frac{1+t}{t^3}\,dt = \int_1^2 (t^{-3} + t^{-2})\,dt$$

$$= \int_1^2 t^{-3}\,dt + \int_1^2 t^{-2}\,dt = \frac{t^{-2}}{-2}\bigg]_1^2 + \frac{t^{-1}}{-1}\bigg]_1^2$$

$$= \left(\frac{1}{(-2)\cdot 4} - \frac{1}{(-2)\cdot 1}\right) + \left(\frac{1}{-2} - \frac{1}{-1}\right) = \frac{3}{8} + \frac{1}{2} = \frac{7}{8}.$$

Thus the area under the curve $y = (1+t)/t^3$ from $t = 1$ to $t = 2$ is $\frac{7}{8}$.

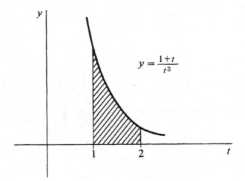

Figure 4.3.3

EXAMPLE 7 Find the area of the region under one arch of the curve $y = \sin x$ (see Figure 4.3.4).

One arch of the sine curve is between $x = 0$ and $x = \pi$. The area is the definite integral

$$\int_0^\pi \sin x\,dx = -\cos x\bigg]_0^\pi$$

$$= -\cos\pi - (-\cos 0) = -(-1) - (-1) = 2.$$

The area is exactly 2.

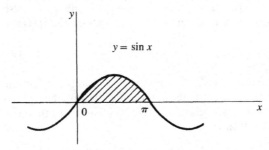

Figure 4.3.4

EXAMPLE 8   Find the area under the curve $y = -2x^{-1}$ from $x = -5$ to $x = -1$. (See Figure 4.3.5.)

The area is given by the definite integral

$$\int_{-5}^{-1} -2x^{-1}\, dx.$$

First compute the indefinite integral

$$\int -2x^{-1}\, dx = -2 \int x^{-1}\, dx = -2 \ln |x| + C.$$

Now compute the definite integral.

$$\int_{-5}^{-1} -2x^{-1}\, dx = -2 \ln |x| \Big]_{-5}^{-1}$$

$$= -2(\ln |-1| - \ln |-5|) = -2(\ln 1 - \ln 5)$$

$$= 2 \ln 5 \sim 3.219.$$

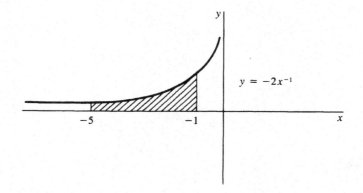

Figure 4.3.5

This example illustrates the need for the absolute value in the integration rule

$$\int x^{-1}\, dx = \ln |x| + C.$$

The natural logarithm $\ln x$ is undefined at $x = -5$ and $x = -1$, but $\ln |x|$ is defined for all $x \neq 0$. The absolute value sign is put in when integrating x^{-1} and removed when differentiating $\ln |x|$.

EXAMPLE 9   In computing definite integrals one must first make sure that the function to be integrated is continuous on the interval. For instance,

Incorrect: $\displaystyle \int_{-1}^{1} \frac{1}{x^2}\, dx = \frac{x^{-1}}{-1} \Big]_{-1}^{1} = -1 - (-(-1)) = -2.$

This is clearly wrong because $1/x^2 > 0$ so the area under the curve cannot be negative. The mistake is that $1/x^2$ is undefined at $x = 0$ and hence the function is discontinuous at $x = 0$. Therefore the area under the curve and the definite integral

$$\int_{-1}^{1} \frac{1}{x^2}\, dx$$

are undefined (Figure 4.3.6).

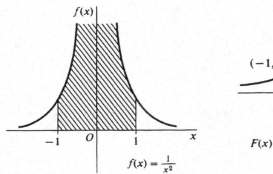
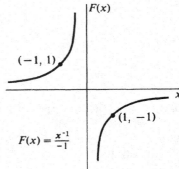

$$f(x) = \frac{1}{x^2}$$

$$F(x) = \frac{x^{-1}}{-1}$$

$(-1, 1)$

$(1, -1)$

Figure 4.3.6

PROBLEMS FOR SECTION 4.3

Evaluate the following integrals.

1 $\int (1 + 2x + 3x^2)\, dx$ **2** $\int (2x^2 - 6x + 9)\, dx$

3 $\int (12t^7 - 3t^5 + 2t^2 + 1)\, dt$ **4** $\int (5 + y^{-2} - 4y^{-3})\, dy$

5 $\int (t^{1/2} + t^{-1/2})\, dt$ **6** $\int (2y^{1/3} - 3y^{2/3})\, dy$

7 $\int (2x - 3)^2\, dx$ **8** $\int (x - 2)(2x + 1)\, dx$

9 $\int (z + 1/z)^2\, dz$ **10** $\int (z - 1/z)^2\, dz$

11 $\int 5 \cos x\, dx$ **12** $\int (\sin x + \cos x)\, dx$

13 $\int \frac{x + 1}{x}\, dx$ **14** $\int \frac{2x^2 - 3x + 6}{x^2}\, dx$

15 $\int (1 + x^{-1})^2\, dx$ **16** $\int 3e^x\, dx$

17 $\int (3 + \sqrt{t})(4 - 2\sqrt{t})\, dt$ **18** $\int \frac{3s + 1}{3\sqrt{s}}\, ds$

19 $\int \frac{4 + 3\sqrt{y} + y\sqrt{y}}{y^2}\, dy$ **20** $\int (3 - x^2)(1 + 4x^2)\, dx$

21 $\int (ax^2 + bx + c)\, dx$ **22** $\int (a_3x^3 + a_2x^2 + a_1x + a_0)\, dx$

23    $\displaystyle\int_{-2}^{2} (2x - 4x^3 + x^5)\,dx$ 24 $\displaystyle\int_{0}^{1} (1 + x^2 + 3x^4)\,dx$

25    $\displaystyle\int_{-1}^{1} (1 + x^2 + 3x^4)\,dx$ 26 $\displaystyle\int_{-1}^{1} e^x\,dx$

27    $\displaystyle\int_{0}^{\pi} \cos x\,dx$ 28 $\displaystyle\int_{0}^{\pi/2} \cos x\,dx$

29    $\displaystyle\int_{1}^{2} 3x^{-1}\,dx$ 30 $\displaystyle\int_{2}^{5} \frac{x - 1}{x}\,dx$

31    $\displaystyle\int_{-3}^{-1} \frac{1}{x}\,dx$

In Problems 32–36, find the position y as a function of t given the velocity $v = dy/dt$ and the value of y at one point of time.

32    $v = 2t + 3$, $y = 0$ when $t = 0$
33    $v = 4t^2 - 1$, $y = 2$ when $t = 0$
34    $v = 3t^4$, $y = 0$ when $t = -1$
35    $v = 2 \sin t$, $y = 10$ when $t = 0$
36    $v = 3t^{-1}$, $y = 1$ when $t = 1$

In Problems 37–42, find the position y and velocity v as a function of t given the acceleration a and the values of y and v at $t = 0$ or $t = 1$.

37    $a = t$, $v = 0$ and $y = 1$ when $t = 0$
38    $a = -32$, $v = 10$ and $y = 0$ when $t = 0$
39    $a = 3t^2$, $v = 1$ and $y = 2$ when $t = 0$
40    $a = 1 - \sqrt{t}$, $v = -2$ and $y = 1$ when $t = 0$
41    $a = t^{-3}$, $v = 1$ and $y = 0$ when $t = 1$
42    $a = -\sin t$, $v = 0$ and $y = 4$ when $t = 0$

43   Which of the following definite integrals are undefined?

(a) $\displaystyle\int_{-1}^{2} \frac{1}{x}\,dx$ (b) $\displaystyle\int_{1}^{2} \frac{1}{x}\,dx$

(c) $\displaystyle\int_{-1}^{0} \sqrt{-y}\,dy$ (d) $\displaystyle\int_{-1}^{0} \sqrt{y}\,dy$

(e) $\displaystyle\int_{-2}^{2} \sqrt{4 - x^2}\,dx$ (f) $\displaystyle\int_{-2}^{2} \sqrt{x^2 - 4}\,dx$

(g) $\displaystyle\int_{-1}^{1} \frac{1}{u^2 - 1}\,du$ (h) $\displaystyle\int_{-2}^{-1} \sqrt{t^2 - 1}\,dt$

(i) $\displaystyle\int_{-2}^{2} \sqrt{t^2 - 1}\,dt$ (j) $\displaystyle\int_{-3}^{3} |x - 1|\,dx$

(k) $\displaystyle\int_{-1}^{1} \tan x\,dx$ (l) $\displaystyle\int_{0}^{\pi} \tan x\,dx$

44   Find the function f such that f' is constant, $f(0) = f'(0)$ and $f(2) = f'(2)$.

45   An object moves with acceleration $a = 6t$. Find its position y as a function of t, given that $y = 1$ when $t = 0$ and $y = 4$ when $t = 1$.

46   Find the function h such that h'' is constant, $h(1) = 1$, $h(2) = 2$, and $h(3) = 3$.

☐ 47   Suppose that $F''(x)$ exists for all x, and let (x_0, y_0) and (x_1, y_1) be two given points. Prove that there is exactly one function $G(x)$ such that

$$G(x_0) = y_0$$
$$G'(x_1) = y_1$$
$$G''(x) = F''(x) \qquad \text{for all } x.$$

☐ **48**   Assume that $F''(x)$ exists for all x, and let (x_1, y_1) and (x_2, y_2) be two points with $x_1 \neq x_2$. Prove that there is exactly one function $G(x)$ such that $G''(x) = F''(x)$ for all x, and the graph of G passes through the two points (x_1, y_1) and (x_2, y_2).

4.4 INTEGRATION BY CHANGE OF VARIABLES

We have seen that the sum, constant, and power rules for differentiation can be turned around to give the sum, constant, and power rules for integration. In this section we shall show how to make use of the Chain Rule for differentiation in problems of integration. The Chain Rule will lead to the important method of *integration by change of variables*. The basic idea is to try to simplify the function to be integrated by changing from one independent variable to another.

If F is an antiderivative of f and we take u as the independent variable, then $\int f(u) \, du$ is a family of functions of u,

$$\int f(u) \, du = F(u) + C.$$

But if we take x as the independent variable and introduce u as a dependent variable $u = g(x)$, then du and $\int f(u) \, du$ mean the following:

$$du = g'(x) \, dx, \qquad \int f(u) \, du = \int f(g(x))g'(x) \, dx = H(x) + C.$$

The notation $\int f(u) \, du$ always stands for a family of functions of the independent variable, which in some cases is another variable such as x. The next theorem can be used as follows. To integrate a given function of x, properly choose a new variable $u = g(x)$ and integrate a new function with respect to u.

DEFINITION

> *Let I and J be intervals. We say that a function g **maps J into I** if for every point x in J, $g(x)$ is defined and belongs to I (Figure 4.4.1).*

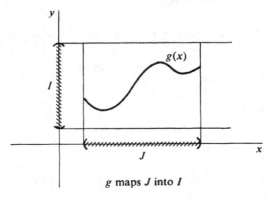

Figure 4.4.1   *g maps J into I*

THEOREM 1 (Indefinite Integration by Change of Variables)

Suppose I and J are open intervals, f has domain I, g maps J into I, and g is differentiable on J. Assume that when we take u as the independent variable,

$$\int f(u)\, du = F(u) + C.$$

Then when x is the independent variable and u = g(x),

$$\int f(u)\, du = F(g(x)) + C.$$

PROOF Let $H(x) = F(g(x))$. For any x in J, the derivatives $g'(x)$ and $F'(g(x)) = f(g(x))$ exist. Therefore by the Chain Rule,

$$H'(x) = F'(g(x))g'(x) = f(g(x))g'(x).$$

It follows that

$$\int f(g(x))g'(x)\, dx = H(x) + C = F(g(x)) + C.$$

So when $u = g(x)$, we have

$$f(u)\, du = f(g(x))g'(x)\, dx, \qquad \int f(u)\, du = F(g(x)) + C.$$

Theorem 1 gives another proof of the general power rule

$$\int u^n\, du = \frac{u^{n+1}}{n+1} + C, \qquad n \neq -1,$$

where u is given as a function of the independent variable x, from the simpler power rule

$$\int x^n\, dx = \frac{x^{n+1}}{n+1} + C, \qquad n \neq -1,$$

where x is the independent variable.

EXAMPLE 1   Find $\int (4x+1)^3 + (4x+1)^2 + (4x+1)\, dx$. Let $u = 4x + 1$. Then $du = 4\, dx$, $dx = \frac{1}{4}\, du$. Hence

$$\int (4x+1)^3 + (4x+1)^2 + (4x+1)\, dx$$

$$= \int (u^3 + u^2 + u) \cdot \frac{1}{4}\, du = \frac{1}{4}\left(\frac{u^4}{4} + \frac{u^3}{3} + \frac{u^2}{2} \right) + C$$

$$= \frac{1}{4}\left[\frac{(4x+1)^4}{4} + \frac{(4x+1)^3}{3} + \frac{(4x+1)^2}{2} \right] + C.$$

EXAMPLE 2   Find $\displaystyle\int \frac{-1}{x^2(1 + 1/x)^2}\, dx$.

Let $u = 1 + 1/x$. Then $du = -1/x^2\, dx$ and thus

$$\frac{-1}{x^2(1 + 1/x)^2} dx = \frac{1}{u^2} du.$$

So $\quad \int \frac{-1}{x^2(1 + 1/x)^2} dx = \int \frac{1}{u^2} du = \frac{u^{-1}}{-1} + C = -\frac{1}{1 + 1/x} + C.$

In a simple problem such as this example, we can save writing by using the term $1 + 1/x$ instead of introducing a new letter u,

$$\int \frac{-1}{x^2(1 + 1/x)^2} dx = \int \frac{1}{(1 + 1/x)^2} d\left(1 + \frac{1}{x}\right) = \frac{(1 + 1/x)^{-1}}{-1} + C.$$

In examples such as the above one, the trick is to find a new variable u such that the expression becomes simpler when we change variables. This usually must be done by an "educated" trial and error process.

One must be careful to express dx in terms of du before integrating with respect to u.

EXAMPLE 3 Find $\int(1 + 5x)^2 \, dx$. Let $u = 1 + 5x$. For emphasis we shall do it correctly and incorrectly.

$\quad$ *Correct:* $\qquad\qquad du = 5 \, dx, \qquad dx = \frac{1}{5} \, du,$

$$\int (1 + 5x)^2 \, dx = \int u^2 \cdot \frac{1}{5} \, du = \frac{u^3}{15} + C = \frac{(1 + 5x)^3}{15} + C.$$

$\quad$ *Incorrect:* $\qquad \int (1 + 5x)^2 \, dx = \int u^2 \, dx = \frac{u^3}{3} + C = \frac{(1 + 5x)^3}{3} + C.$

$\quad$ *Incorrect:* $\qquad \int (1 + 5x)^2 \, dx = \int u^2 \, du = \frac{u^3}{3} + C = \frac{(1 + 5x)^3}{3} + C.$

EXAMPLE 4 Find $\int x^3 \sqrt{2 - x^2} \, dx$. Let $u = 2 - x^2$, $du = -2x \, dx$, $dx = du/(-2x)$. We try to express the integral in terms of u.

$$\int x^3 \sqrt{2 - x^2} \, dx = \int x^3 \sqrt{u} \frac{du}{-2x} = \int -\frac{1}{2} x^2 \sqrt{u} \, du.$$

Since $u = 2 - x^2$, $x^2 = 2 - u$. Therefore

$$\int -\frac{1}{2} x^2 \sqrt{u} \, du = \int -\frac{1}{2}(2 - u)\sqrt{u} \, du = \int -\sqrt{u} + \frac{1}{2} u^{3/2} \, du$$

$$= -\frac{2}{3} u^{3/2} + \frac{1}{2} \cdot \frac{2}{5} u^{5/2} + C$$

$$= -\frac{2}{3}(2 - x^2)^{3/2} + \frac{1}{5}(2 - x^2)^{5/2} + C.$$

We next describe the method of *definite integration by change of variables*. In a definite integral

$$\int_a^b h(x) \, dx$$

it is always understood that x is the independent variable and we are integrating between the limits $x = a$ and $x = b$. Thus when we change to a new independent variable u, we must also change the limits of integration. The theorem below will show that if $u = c$ when $x = a$ and $u = d$ when $x = b$, then c and d will be the new limits of integration.

THEOREM 2 (Definite Integration by Change of Variables)

Suppose I and J are open intervals, f is continuous and has an antiderivative on I, g has a continuous derivative on J, and g maps J into I. Then for any two points a and b in J,

$$\int_a^b f(g(x))g'(x)\,dx = \int_{g(a)}^{g(b)} f(u)\,du.$$

PROOF Let F be an antiderivative of f. Then by Theorem 1, $H(x) = F(g(x))$ is an antiderivative of $h(x) = f(g(x))g'(x)$. Since f, g, and g' are continuous, h is continuous on J. Then by the Fundamental Theorem of Calculus,

$$\int_a^b f(g(x))g'(x)\,dx = H(b) - H(a) = F(g(b)) - F(g(a)) = \int_{g(a)}^{g(b)} f(u)\,du.$$

EXAMPLE 5   Find the area under the line $y = 1 + 3x$ from $x = 0$ to $x = 1$. This can be done either with or without a change of variables.

Without change of variable: $\int(1 + 3x)\,dx = x + 3x^2/2 + C$, so

$$\int_0^1 (1 + 3x)\,dx = x + \frac{3x^2}{2}\Bigg]_0^1 = \left(1 + \frac{3 \cdot 1^2}{2}\right) - \left(0 + \frac{3 \cdot 0^2}{2}\right) = \frac{5}{2}.$$

With change of variable: Let $u = 1 + 3x$. Then $du = 3\,dx$, $dx = \frac{1}{3}\,du$. When $x = 0$, $u = 1 + 3 \cdot 0 = 1$. When $x = 1$, $u = 1 + 3 \cdot 1 = 4$.

$$\int_0^1 (1 + 3x)\,dx = \int_1^4 u \cdot \frac{1}{3}\,du = \frac{u^2}{6}\Bigg]_1^4 = \frac{16}{6} - \frac{1}{6} = \frac{15}{6} = \frac{5}{2}.$$

Example 5 shows us that $\int_0^1 (1 + 3x)\,dx = \int_1^4 (u/3)\,du$; that is, the areas shown in Figure 4.4.2 are the same.

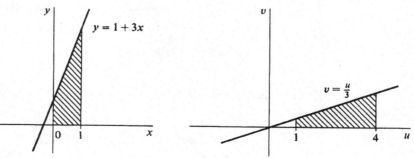

Figure 4.4.2

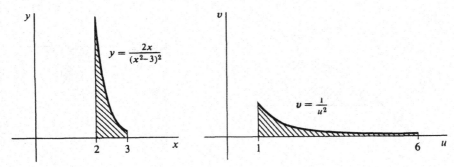

Figure 4.4.3

EXAMPLE 6 Find the area under the curve $y = 2x/(x^2 - 3)^2$ from $x = 2$ to $x = 3$ (Figure 4.4.3).

Let $u = x^2 - 3$. Then $du = 2x\,dx$. At $x = 2$, $u = 2^2 - 3 = 1$. At $x = 3$, $u = 3^2 - 3 = 6$. Then

$$\int_2^3 \frac{2x}{(x^2 - 3)^2}\,dx = \int_1^6 \frac{1}{u^2}\,du = -\frac{1}{u}\bigg]_1^6 = 1 - \frac{1}{6} = \frac{5}{6}.$$

EXAMPLE 7 Find $\int_0^1 \sqrt{1 - x^2}\,x\,dx$. The function $\sqrt{1 - x^2}\,x$ as given is only defined on the closed interval $[-1, 1]$. In order to use Theorem 2, we extend it to the open interval $J = (-\infty, \infty)$ by

$$h(x) = \begin{cases} 0 & \text{if } x < -1 \quad \text{or} \quad x > 1, \\ \sqrt{1 - x^2}\,x & \text{if } -1 \leq x \leq 1. \end{cases}$$

Let $u = 1 - x^2$. Then $du = -2x\,dx$, $dx = -du/2x$. At $x = 0$, $u = 1$. At $x = 1$, $u = 0$. Therefore

$$\int_0^1 \sqrt{1 - x^2}\,x\,dx = \int_1^0 \sqrt{u} \cdot \left(-\tfrac{1}{2}\,du\right) = \int_1^0 -\tfrac{1}{2}\sqrt{u}\,du$$

$$= \tfrac{1}{2}\int_0^1 \sqrt{u}\,du = \tfrac{1}{2} \cdot \tfrac{2}{3}u^{3/2}\bigg]_0^1 = \tfrac{1}{3} - 0 = \tfrac{1}{3}.$$

We see in Figure 4.4.4 that as x increases from 0 to 1, u decreases from 1 to 0, so the limits become reversed. The areas shown in Figure 4.4.5 are equal.

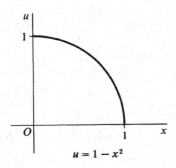

Figure 4.4.4

$u = 1 - x^2$

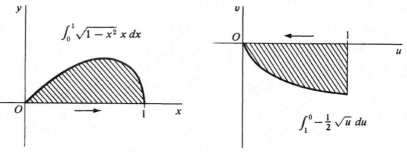

Figure 4.4.5

We can use integration by change of variables to derive the formula for the area of a circle, $A = \pi r^2$, where r is the radius. It is easier to work with a semicircle because the semicircle of radius r is just the region under the curve

$$y = \sqrt{r^2 - x^2}, \qquad -r \le x \le r.$$

To start with we need to give a rigorous definition of π. By definition, π is the area of a unit circle. Thus π is twice the area of the unit semicircle, which means:

DEFINITION

$$\pi = 2\int_{-1}^{1} \sqrt{1 - x^2}\, dx.$$

The area of a semicircle of radius r is the definite integral

$$\int_{-r}^{r} \sqrt{r^2 - x^2}\, dx.$$

To evaluate this integral we let $x = ru$. Then $dx = r\, du$. When $x = \pm r, u = \pm 1$. Thus

$$\int_{-r}^{r} \sqrt{r^2 - x^2}\, dx = \int_{-1}^{1} \sqrt{r^2 - (ru)^2}\, r\, du = \int_{-1}^{1} r^2\sqrt{1 - u^2}\, du$$

$$= r^2 \int_{-1}^{1} \sqrt{1 - u^2}\, du = r^2 \cdot \frac{\pi}{2}.$$

Therefore the semicircle has area $\pi r^2/2$ and the circle area πr^2 (Figure 4.4.6).

EXAMPLE 8   Find $\displaystyle \int_{0}^{1} \frac{3x^2 - 1}{1 + \sqrt{x - x^3}}\, dx.$

Let $u = x - x^3$. Then $du = (1 - 3x^2)\, dx$. When $x = 0, u = 0 - 0^3 = 0$. When $x = 1, u = 1 - 1^3 = 0$. Then

$$\int_{0}^{1} \frac{3x^2 - 1}{1 + \sqrt{x - x^3}}\, dx = \int_{0}^{0} -\frac{du}{1 + \sqrt{u}} = 0.$$

As x goes from 0 to 1, u starts at 0, increases for a time, then drops back to 0 (Figure 4.4.7).

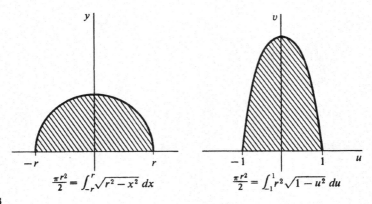

$$\frac{\pi r^2}{2} = \int_{-r}^{r}\sqrt{r^2 - x^2}\, dx \qquad\qquad \frac{\pi r^2}{2} = \int_{-1}^{1} r^2\sqrt{1 - u^2}\, du$$

Figure 4.4.6

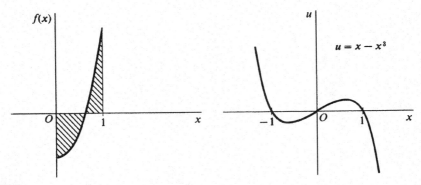

$$u = x - x^3$$

Figure 4.4.7

We do not know how to find the indefinite integrals in this example. Nevertheless the answer is 0 because on changing variables both limits of integration become the same. Using the Addition Property, we can also see that, for instance,

$$\int_{0}^{2/3} \frac{3x^2 - 1}{1 + \sqrt{x - x^3}}\, dx = -\int_{2/3}^{1} \frac{3x^2 - 1}{1 + \sqrt{x - x^3}}\, dx.$$

PROBLEMS FOR SECTION 4.4

In Problems 1–90, evaluate the integral.

1 $\displaystyle\int \frac{1}{(2x + 1)^2}\, dx$

2 $\displaystyle\int \sqrt{3y + 1}\, dy$

3 $\displaystyle\int (3 - 4z)^6\, dz$

4 $\displaystyle\int (1 - x)^{3/2}\, dx$

5 $\displaystyle\int 2t\sqrt{1 - t^2}\, dt$

6 $\displaystyle\int \frac{x}{\sqrt{2x^2 + 1}}\, dx$

7 $\displaystyle\int x(4 + 5x^2)^2\, dx$

8 $\displaystyle\int \frac{4y}{(2 + 3y^2)^2}\, dy$

9 $\displaystyle\int \sin(3x)\,dx$

10 $\displaystyle\int \cos(4 - 2x)\,dx$

11 $\displaystyle\int 6 \sin(4x - 1)\,dx$

12 $\displaystyle\int a \sin x + b \cos x\,dx$

13 $\displaystyle\int \sin\theta \cos\theta\,d\theta$

14 $\displaystyle\int \sin^2\theta \cos\theta\,d\theta$

15 $\displaystyle\int \cos^3\theta \sin\theta\,d\theta$

16 $\displaystyle\int \sin(2\theta) + \cos(3\theta)\,d\theta$

17 $\displaystyle\int x \sin(x^2 + 1)\,dx$

18 $\displaystyle\int x^2 \cos(x^3)\,dx$

19 $\displaystyle\int \frac{\sin(\ln x)}{x}\,dx$

20 $\displaystyle\int e^t \cos(e^t)\,dt$

21 $\displaystyle\int \sqrt{\sin t} \cos t\,dt$

22 $\displaystyle\int \sqrt{t} \cos(t\sqrt{t})\,dt$

23 $\displaystyle\int e^{2x}\,dx$

24 $\displaystyle\int 3e^{1-x}\,dx$

25 $\displaystyle\int ae^x + be^{-x}\,dx$

26 $\displaystyle\int (e^x + 1)^2\,dx$

27 $\displaystyle\int xe^{x^2}\,dx$

28 $\displaystyle\int xe^{1-x^2}\,dx$

29 $\displaystyle\int be^{ax}\,dx$

30 $\displaystyle\int e^{ax+b}\,dx$

31 $\displaystyle\int e^{\sin\theta}\cos\theta\,d\theta$

32 $\displaystyle\int e^t\sqrt{1 + e^t}\,dt$

33 $\displaystyle\int \frac{1}{x + 2}\,dx$

34 $\displaystyle\int \frac{2}{3 - 4x}\,dx$

35 $\displaystyle\int \frac{e^x}{e^x + 1}\,dx$

36 $\displaystyle\int \frac{x}{x^2 + 1}\,dx$

37 $\displaystyle\int \frac{x}{x + 1}\,dx$

38 $\displaystyle\int \frac{1}{x^2(1 + 1/x)}\,dx$

39 $\displaystyle\int \frac{1}{\sqrt{x}(1 + \sqrt{x})}\,dx$

40 $\displaystyle\int \frac{1 - 2t}{1 + 2t}\,dt$

41 $\displaystyle\int \frac{3t + 4}{5 - t}\,dt$

42 $\displaystyle\int \frac{x^2}{x^3 + 2}\,dx$

43 $\displaystyle\int x^3\sqrt{x^4 + 5}\,dx$

44 $\displaystyle\int \frac{1}{(2x - 1)\sqrt{1 - 2x}}\,dx$

45 $\displaystyle\int y\sqrt{2 + y^2}\,dy$

46 $\displaystyle\int \frac{t}{(2t^2 + 1)^3}\,dt$

47 $\displaystyle\int \frac{u}{\sqrt{1 - u^2}}\,du$

48 $\displaystyle\int (4x + 1)\sqrt{2x^2 + x + 5}\,dx$

49 $\displaystyle\int \frac{1}{\sqrt{3s + 2}}\,ds$

50 $\displaystyle\int \sqrt{1 - 5z}\,dz$

51 $\displaystyle\int \frac{1}{x^2\sqrt{1 + 1/x}}\,dx$

52 $\displaystyle\int \frac{1}{y^2(1 - 4/y)^2}\,dy$

53 $\displaystyle\int x^{-3}\sqrt{3 + 5x^{-2}}\,dx$

54 $\displaystyle\int \frac{1}{\sqrt{x}(1 + 2\sqrt{x})^2}\,dx$

55 $\displaystyle\int \frac{(3 - \sqrt{x})^2}{\sqrt{x}}\, dx$

56 $\displaystyle\int \frac{\sqrt{1 + \sqrt{t}}}{\sqrt{t}}\, dt$

57 $\displaystyle\int \frac{2 + 1/z}{z^2}\, dz$

58 $\displaystyle\int \frac{1}{(3y + 1)^3}\, dy$

59 $\displaystyle\int \frac{x^2}{\sqrt{x^3 + 4}}\, dx$

60 $\displaystyle\int \frac{x^2}{\sqrt{4x^3 + 1}}\, dx$

61 $\displaystyle\int \frac{x^3}{\sqrt{1 + x^4}}\, dx$

62 $\displaystyle\int x^3 \sqrt{2 - x^4}\, dx$

63 $\displaystyle\int t\sqrt{t + 1}\, dt$

64 $\displaystyle\int \frac{s}{(s + 2)^3}\, ds$

65 $\displaystyle\int (2s + 6)(1 - s)^{-4}\, ds$

66 $\displaystyle\int y^3 \sqrt{4 + y^2}\, dy$

67 $\displaystyle\int \frac{y^3}{(y^2 + 1)^3}\, dy$

68 $\displaystyle\int \frac{x^5}{\sqrt{1 + x^2}}\, dx$

69 $\displaystyle\int \frac{x}{\sqrt{4x + 1}}\, dx$

70 $\displaystyle\int \sqrt{2 + \sqrt{u}}\, du$

71 $\displaystyle\int u\sqrt{1 - 3u}\, du$

72 $\displaystyle\int \frac{1}{(2\sqrt{x} + 3)^3}\, dx$

73 $\displaystyle\int \frac{4x - 1}{\sqrt{4x + 1}}\, dx$

74 $\displaystyle\int \frac{x^2}{\sqrt{x - 1}}\, dx$

75 $\displaystyle\int \frac{x^3}{1 - x^4}\, dx$

76 $\displaystyle\int \frac{y^3}{2 - y^2}\, dy$

77 $\displaystyle\int \frac{y^5}{1 + y^2}\, dy$

78 $\displaystyle\int \frac{u}{(u + 4)^2}\, du$

79 $\displaystyle\int \frac{6u - 5}{(3u + 2)^2}\, du$

80 $\displaystyle\int \frac{1}{1 + \sqrt{x}}\, dx$

81 $\displaystyle\int \frac{\sqrt{x}}{2 + \sqrt{x}}\, dx$

82 $\displaystyle\int \frac{e^x + \cos x}{e^x + \sin x}\, dx$

83 $\displaystyle\int \frac{\cos \theta}{\sin \theta}\, d\theta$

84 $\displaystyle\int \tan \theta\, d\theta$

85 $\displaystyle\int \frac{1}{a + bx}\, dx$

86 $\displaystyle\int \frac{2x + 1}{x^2 + x + 1}\, dx$

87 $\displaystyle\int \frac{\sin \theta}{1 + \cos \theta}\, d\theta$

88 $\displaystyle\int \frac{\sin \theta - \cos \theta}{\sin \theta + \cos \theta}\, d\theta$

89 $\displaystyle\int \frac{\ln x}{x}\, dx$

90 $\displaystyle\int \frac{1}{x \ln x}\, dx$

In Problems 91–108, evaluate the definite integral.

91 $\displaystyle\int_0^{\pi/3} \sin \theta\, d\theta$

92 $\displaystyle\int_{-\pi/4}^{\pi/4} \cos(2\theta)\, d\theta$

93 $\displaystyle\int_{-1}^{1} e^x\, dx$

94 $\displaystyle\int_0^1 xe^{x^2}\, dx$

95 $\displaystyle\int_1^4 \frac{1}{2x}\, dx$

96 $\displaystyle\int_0^1 \frac{x}{x^2 + 1}\, dx$

97 $\displaystyle\int_0^{\pi/2} \sin \theta \cos \theta\, d\theta$

98 $\displaystyle\int_0^{2\pi} a \sin \theta + b \cos \theta\, d\theta$

99 $\displaystyle\int_0^2 \sqrt{x+1}\,dx$ **100** $\displaystyle\int_0^1 \frac{1}{(4x+3)^2}\,dx$

101 $\displaystyle\int_0^4 (2x+1)^{3/2}\,dx$ **102** $\displaystyle\int_0^1 t(t^2+3)^{-2}\,dt$

103 $\displaystyle\int_0^1 (1+6x)^3\,dx$ **104** $\displaystyle\int_1^5 \frac{2}{\sqrt{3t+1}}\,dt$

105 $\displaystyle\int_0^2 v\sqrt{2v^2+9}\,dv$ **106** $\displaystyle\int_{-1}^1 \frac{x^2}{(4-x^3)^2}\,dx$

107 $\displaystyle\int_{-1}^1 \frac{x}{2-x^2}\,dx$ **108** $\displaystyle\int_{\sqrt{6}}^5 x(x^2+2)^{1/3}\,dx$

109 Find the area of the region below the curve $y = 1/(10 - 3x)$ from $x = 1$ to $x = 2$.

110 Find the area of the region under one arch of the curve $y = \sin x \cos x$.

111 Find the area of the region under one arch of the curve $y = \cos(3x)$.

112 Find the area of the region below the curve $y = 4x\sqrt{4 - x^2}$ between $x = 0$ and $x = 2$.

113 Find the area below the curve $y = (1 + 7x)^{2/3}$ between $x = 0$ and $x = 1$.

114 Find the area below the curve $y = x/(x^2 + 1)^2$ between $x = 0$ and $x = 3$.

☐ **115** Evaluate: $\displaystyle\int_0^1 \frac{1 - 2x}{1 + \sqrt[3]{x - x^2}}\,dx$

☐ **116** Evaluate: $\displaystyle\int_{-1}^1 2x\sqrt{(1 - x^2)^3 + 1}\,dx$

☐ **117** Let f and g have continuous derivatives and evaluate $\int f'(g(x))g'(x)\,dx$.

☐ **118** A real function f is said to be *even* if $f(x) = f(-x)$ for all x. Show that if f is a continuous even function, then $\int_{-b}^0 f(x)\,dx = \int_0^b f(x)\,dx$.

☐ **119** An *odd* function is a real function g such that $g(-x) = -g(x)$ for all x. Prove that for a continuous odd function g, $\int_{-b}^b g(x)\,dx = 0$.

4.5 AREA BETWEEN TWO CURVES

A region in the plane can often be represented as the region between two curves. For example, the unit circle is the region between the curves

$$y = -\sqrt{1 - x^2}, \qquad y = \sqrt{1 - x^2}, \qquad -1 \le x \le 1$$

shown in Figure 4.5.1. Consider two continuous functions f and g on $[a, b]$ such that $f(x) \le g(x)$ for all x in $[a, b]$. The region R, bounded by the curves

$$y = f(x), \qquad y = g(x), \qquad x = a, \qquad x = b,$$

is called the region between $f(x)$ and $g(x)$ from a to b. If both curves are above the x-axis as in Figure 4.5.2, the area of the region R can be found by subtracting the area below f from the area below g:

$$\text{area of } R = \int_a^b g(x)\,dx - \int_a^b f(x)\,dx.$$

It is usually easier to work with a single integral and write

$$\text{area of } R = \int_a^b (g(x) - f(x))\,dx.$$

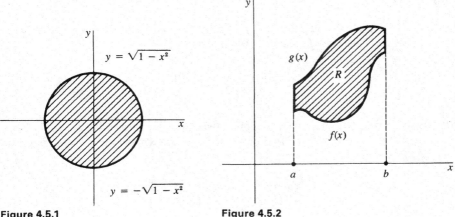

Figure 4.5.1

Figure 4.5.2

In the general case shown in Figure 4.5.3, we may move the region R above the x-axis by adding a constant c to both $f(x)$ and $g(x)$ without changing the area, and the same formula holds:

$$\text{area of } R = \int_a^b (g(x) + c)\, dx - \int_a^b (f(x) + c)\, dx$$
$$= \int_a^b (g(x) - f(x))\, dx.$$

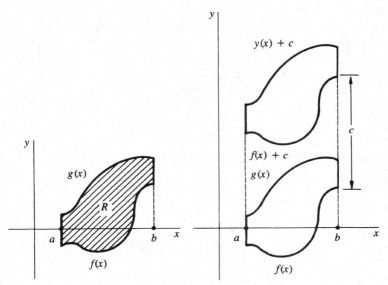

Figure 4.5.3

To sum up, we define the area between two curves as follows.

DEFINITION

*If f and g are continuous and $f(x) \le g(x)$ for $a \le x \le b$, then the **area** of the region R between $f(x)$ and $g(x)$ from a to b is defined as*

$$\int_a^b (g(x) - f(x))\, dx.$$

EXAMPLE 1   Find the area of the region between the curves $y = \frac{1}{2}x^2 - 1$ and $y = x$ from $x = 1$ to $x = 2$. In Figure 4.5.4, we sketch the curves to check that $\frac{1}{2}x^2 - 1 \le x$ for $1 \le x \le 2$. Then

$$A = \int_1^2 x - (\tfrac{1}{2}x^2 - 1)\, dx = \tfrac{1}{2}x^2 - \tfrac{1}{6}x^3 + x \Big]_1^2 = \tfrac{8}{6}.$$

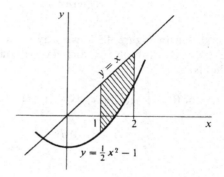

Figure 4.5.4

EXAMPLE 2   Find the area of the region bounded above by $y = x + 2$ and below by $y = x^2$.

Part of the problem is to find the limits of integration. First draw a sketch (Figure 4.5.5). The curves intersect at two points, which can be found by solving the equation $x + 2 = x^2$ for x.

$$x^2 - (x + 2) = 0, \qquad (x + 1)(x - 2) = 0,$$
$$x = -1 \quad \text{and} \quad x = 2.$$

Then $A = \displaystyle\int_{-1}^2 (x + 2 - x^2)\, dx = \tfrac{1}{2}x^2 + 2x - \tfrac{1}{3}x^3 \Big]_{-1}^2 = 4\tfrac{1}{2}.$

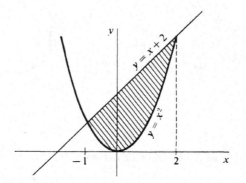

Figure 4.5.5

EXAMPLE 3 Find the area of the region R bounded below by the line $y = -1$ and above by the curves $y = x^3$ and $y = 2 - x$. The region is shown in Figure 4.5.6.

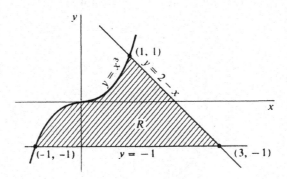

Figure 4.5.6

This problem can be solved in three ways. Each solution illustrates a different trick which is useful in other area problems. The three corners of the region are:

$(-1, -1)$,   where $y = x^3$ and $y = -1$ cross.
$(3, -1)$,   where $y = 2 - x$ and $y = -1$ cross.
$(1, 1)$,   where $y = x^3$ and $y = 2 - x$ cross.

Note that $y = x^3$ and $y = 2 - x$ can cross at only one point because x^3 is always increasing and $2 - x$ is always decreasing.

FIRST SOLUTION Break the region into the two parts shown in Figure 4.5.7: R_1 from $x = -1$ to $x = 1$, and R_2 from $x = 1$ to $x = 3$. Then

$$\text{area of } R = \text{area of } R_1 + \text{area of } R_2.$$

$$\text{area of } R_1 = \int_{-1}^{1} x^3 - (-1)\, dx = \tfrac{1}{4}x^4 + x \Big]_{-1}^{1} = 2.$$

$$\text{area of } R_2 = \int_{1}^{3} (2 - x) - (-1)\, dx = 3x - \tfrac{1}{2}x^2 \Big]_{1}^{3} = 2.$$

$$\text{area of } R = 2 + 2 = 4.$$

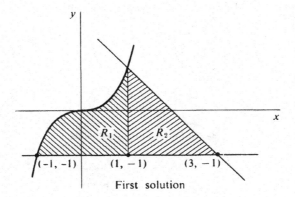

Figure 4.5.7   First solution

SECOND SOLUTION   Form the triangular region S between $y = -1$ and $y = 2 - x$ from -1 to 3. The region R is obtained by subtracting from S the region S_1 shown in Figure 4.5.8. Then

$$\text{area of } R = \text{area of } S - \text{area of } S_1.$$

$$\text{area of } S = \int_{-1}^{3} (2 - x) - (-1)\, dx = 3x - \tfrac{1}{2}x^2 \Big]_{-1}^{3} = 8.$$

$$\text{area of } S_1 = \int_{-1}^{1} (2 - x) - x^3\, dx = 2x - \tfrac{1}{2}x^2 - \tfrac{1}{4}x^4 \Big]_{-1}^{1} = 4.$$

$$\text{area of } R = 8 - 4 = 4.$$

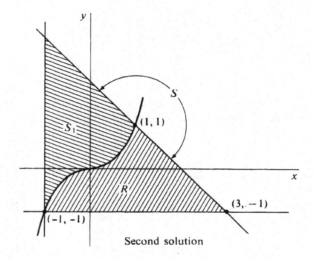

Figure 4.5.8   Second solution

THIRD SOLUTION   Use y as the independent variable and x as the dependent variable. Write the boundary curves with x as a function of y.

$$y = 2 - x \qquad \text{becomes } x = 2 - y.$$
$$y = x^3 \qquad \text{becomes } x = y^{1/3}.$$

The limits of integration are $y = -1$ and $y = 1$ (see Figure 4.5.9). Then

$$A = \int_{-1}^{1} (2 - y) - y^{1/3}\, dy = 2y - \tfrac{1}{2}y^2 - \tfrac{3}{4}y^{4/3} \Big]_{-1}^{1} = 4.$$

As expected, all three solutions gave the same answer.

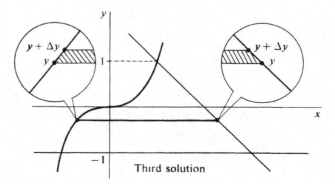

Figure 4.5.9   Third solution

PROBLEMS FOR SECTION 4.5

In Problems 1–43 below, sketch the given curves and find the area of the region bounded by them.

1      $f(x) = 0, \quad g(x) = 5x - x^2, \quad 0 \le x \le 4$

2      $f(x) = \sqrt{x}, \quad g(x) = x^2, \quad 1 \le x \le 4$

3      $f(x) = x\sqrt{1 - x^2}, \quad g(x) = 1, \quad -1 \le x \le 1$

4      $y = x - 2, \quad y = 3x^{1/3}, \quad 0 \le x \le 1$

5      $y = \sqrt{x}, \quad y = \sqrt{x + 1}, \quad 0 \le x \le 4$

6      $y = \sqrt{x^2 + 1} - x, \quad y = \sqrt{x^2 + 1} + x, \quad -0 \le x \le 1$

7   The x-axis and the curve $y = -5 + 6x - x^2$

8   The x-axis and the curve $y = 1 - x^4$

9   The y-axis and the curve $x = 25 - y^2$

10   The y-axis and the curve $x = y(8 - y)$

11     $y = \cos x, \quad y = 2 \cos x, \quad -\pi/2 \le x \le \pi/2$

12     $y = \sin x \cos x, \quad y = 1, \quad 0 \le x \le \pi/2$

13     $y = -\sin x, \quad y = \sin x, \quad 0 \le x \le \pi$

14     $y = \sin x, \quad y = \cos x, \quad 0 \le x \le \pi/4$

15     $y = \sin x \cos x, \quad y = \sin x, \quad 0 \le x \le \pi$

16     $y = \sin^2 x \cos x, \quad y = \sin x \cos x, \quad 0 \le x \le \pi/2$

17     $y = x, \quad y = e^x, \quad 0 \le x \le 2$

18     $y = e^{-x}, \quad y = e^x, \quad 0 \le x \le 2$

19     $y = -e^x, \quad y = e^x, \quad -1 \le x \le 1$

20     $y = xe^{x^2}, \quad y = e, \quad 0 \le x \le 1$

21     $y = \dfrac{1}{x + 1}, \quad y = 1, \quad 0 \le x \le 2$

22     $y = \dfrac{1}{2x + 1}, \quad y = \dfrac{1}{x + 1}, \quad 0 \le x \le 2$

23     $y = 1/x, \quad y = x, \quad 1 \le x \le 2$

24     $y = \dfrac{x}{x^2 + 1}, \quad y = \tfrac{1}{2}, \quad 0 \le x \le 1$

25     $f(x) = x^{3/2}, \quad g(x) = x^{2/3}$

26     $y = x^2 - 2x, \quad y = x - 2$

27     $y = x^4 - 2x^2, \quad y = 2x^2 + 12$

28     $y = x^4 - 1, \quad y = x^3 - x$

29     $y = x^4/(x^2 + 1), \quad y = 1/(x^2 + 1)$

30     $y = x^3\sqrt{1 - x^2}, \quad y = x\sqrt{1 - x^2}, \quad 0 \le x$

31     $y = 2x^2, \quad y = x^2 + 4$

32     $x = y^2, \quad x = 2 - y^2$

33     $\sqrt{x} + \sqrt{y} = 1$ and the x- and y-axes

34     $x^2y = 4, \quad x^2 + y = 5$ (first quadrant)

35     $y = x\sqrt{x + 1}, \quad y = 2x$

36     $y = 0, \quad y = x^3 + x + 2, \quad x = 2$

37     $y = 2x + 4, \quad y = 2 - 3x, \quad y = -x$

38     $y = x^2 - 1, \quad y = (x - 1)^2, \quad y = (x + 1)^2$

39     $y = \sqrt{x}, \quad y = 1, \quad y = 10 - 2x$

40     $y = x - 2, \quad y = 2 - x, \quad y = \sqrt{x}$

41     $y = -x, \quad y = \sqrt[3]{x}, \quad y = 3x - 2$

42     $y = -2, \quad y = x^3 + x, \quad x + y = 3$

43     $y = x^2, \quad y = 2x^{-2}, \quad y = 2x^{-3}$ (first quadrant)

44   Find the area of the ellipse $x^2/a^2 + y^2/b^2 = 1$. Use the fact that the unit circle has area π.

45   Sketch the four-sided region bounded by the lines $y = 1$, $y = x$, $y = 2x$, and $y = 6 - x$ and find its area.

46   Find the number $c > 0$ such that the region bounded by the curves $y = x$, $y = -2x$, and $x = c$ has area 6.

47   Find the number $c > 1$ such that the region bounded by the curves $y = 1$, $y = x^{-2}$, and $x = c$ has area 1.

48   Find the number c such that the region bounded by the curves $y = x^2$ and $y = c$ has area 36.

49   Find the number $c > 0$ such that the region bounded by the curves $y = x^2$ and $y = cx$ has area 9.

50   Find the value of c between -1 and 2 such that the area of the region bounded by the lines $y = -x$, $y = 2x$, and $y = 1 + cx$ is a minimum.

51   Find the value of c such that the line $y = c$ bisects the region bounded by the curves $y = x^2$ and $y = 1$.

52   Find the value of c such that the line $y = cx$ bisects the region bounded by the x-axis and the curve $y = x - x^2$.

4.6 NUMERICAL INTEGRATION

In numerical integration, one computes an approximate value for the definite integral rather than finding an exact value. In this section we shall present two methods of numerical integration, called the Trapezoidal Rule and Simpson's Rule.

The Fundamental Theorem of Calculus gives us a method of computing the definite integral of a given continuous function f from a to b. The method is to find, by trial and error, an antiderivative F of f and then to use the equation

$$\int_a^b f(t)\, dt = F(b) - F(a).$$

When the method works, it provides an exact value for the integral. However, the method succeeds only if the antiderivative happens to be a function that can be described in a simple way. For many integrals one cannot find a formula for the antiderivative, and the method fails. Such integrals can still be computed approximately using numerical integration.

The Trapezoidal Rule and Simpson's Rule can always be applied and do not use the antiderivative. They are easy to carry out on a computer or hand calculator. We already discussed one method of approximating the definite integral in Section 4.1, the Riemann sum. The Trapezoidal Rule is a modified form of the Riemann sum, which gives a much closer approximation for a given amount of effort. Simpson's Rule is a further modification that gives still better approximations.

Let f be a continuous function on an interval I, and let $a < b$ in I. By definition, for each positive infinitesimal dx the definite integral

$$\int_a^b f(x)\,dx$$

is the standard part of the infinite Riemann sum

$$\sum_a^b f(x)\,dx,$$

$$\int_a^b f(x)\,dx = st\left[\sum_a^b f(x)\,dx\right].$$

In Section 4.1, examples were worked out to show that the finite Riemann sums become very close to the definite integral when Δx is small; that is, the finite Riemann sums approximate the definite integral. In Section 4.2, we saw that the definite integral is the limit of the finite Riemann sums as $\Delta x \to 0^+$:

$$\int_a^b f(x)\,dx = \lim_{\Delta x \to 0^+} \sum_a^b f(x)\,\Delta x.$$

The Riemann sum, which is a sum of areas of rectangles, is a rather inefficient approximation of the definite integral. We can usually get a much closer approximation with the same amount of work by adding up areas of trapezoids instead of rectangles, forming the Trapezoidal Rule suggested by Figure 4.6.1. The Trapezoidal Rule also provides a formula, called an error estimate, which tells us how close the approximation is to the exact value of the definite integral.

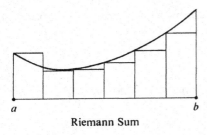

Riemann Sum   Trapezoidal Approximation

Figure 4.6.1

Choose a positive integer n and divide the interval $[a, b]$ into n subintervals of equal length $\Delta x = (b - a)/n$. The partition points are $a = x_0, x_1, x_2, \ldots, x_n = b$.

The trapezoidal approximation is the area of the region under the broken line connecting the points

$$(x_0, f(x_0)), (x_1, f(x_1)), \ldots, (x_n, f(x_n)).$$

Since all of these points lie on the curve $y = f(x)$, the broken line closely follows the curve. So one would expect the area of the region under the broken line to closely approximate the area under the curve.

Consider a single subinterval $[x_m, x_{m+1}]$ of width Δx. The region under the line segment connecting the two points

$$(x_m, f(x_m)), \qquad (x_{m+1}, f(x_{m+1}))$$

is a trapezoid and its area is

$$\frac{f(x_m) + f(x_{m+1})}{2} \Delta x.$$

The sum of the areas of the trapezoids is a modified Riemann sum

$$\sum_a^b \frac{f(x) + f(x + \Delta x)}{2} \Delta x$$

$$= \left(\frac{f(x_0) + f(x_1)}{2} + \frac{f(x_1) + f(x_2)}{2} + \cdots + \frac{f(x_{n-1}) + f(x_n)}{2} \right) \Delta x$$

$$= [\tfrac{1}{2}f(x_0) + f(x_1) + f(x_2) + \cdots + f(x_{n-1}) + \tfrac{1}{2}f(x_n)] \Delta x.$$

We thus make the definition:

DEFINITION

Let $\Delta x = (b - a)/n$ *evenly divide* $b - a$. *Then by the* **trapezoidal approximation** *to the definite integral* $\int_a^b f(x) \, dx$ *we mean the sum*

$$\sum_a^b \frac{f(x) + f(x + \Delta x)}{2} \Delta x = [\tfrac{1}{2}f(x_0) + f(x_1) + \cdots + f(x_{n-1}) + \tfrac{1}{2}f(x_n)] \Delta x.$$

The Trapezoidal Approximation of an integral $\int_a^b f(x) \, dx$ can be computed very efficiently on most hand calculators. First compute the sum

$$\tfrac{1}{2}f(x_0) + f(x_1) + f(x_2) + \cdots + \tfrac{1}{2}f(x_n)$$

by cumulative addition. Then multiply this sum by Δx to obtain the Trapezoidal Approximation.

THEOREM 1

For a continuous function f *on* $[a, b]$, *the trapezoidal approximation approaches the definite integral as* $\Delta x \to 0^+$, *that is,*

$$\int_a^b f(x) \, dx = \lim_{\Delta x \to 0^+} \sum_a^b \frac{f(x) + f(x + \Delta x)}{2} \Delta x.$$

PROOF Comparing the formulas for the trapezoidal approximation and the Riemann sum, we see that

$$\sum_a^b \frac{f(x) + f(x + \Delta x)}{2} \Delta x = \sum_a^b f(x) \Delta x + (\tfrac{1}{2}f(x_n) - \tfrac{1}{2}f(x_0)) \Delta x.$$

For dx positive infinitesimal, the extra term

$$(\tfrac{1}{2}f(x_H) - \tfrac{1}{2}f(x_0)) \, dx$$

is infinitely small. It follows that

$$\sum_a^b \frac{f(x) + f(x + dx)}{2} \, dx \approx \sum_a^b f(x) \, dx \approx \int_a^b f(x) \, dx.$$

From a practical standpoint, it is desirable to have a good estimate of error. We shall first work an example and then state a theorem which gives an error estimate for the trapezoidal approximation.

EXAMPLE 1   Approximate the definite integral

$$\int_0^1 \sqrt{1 + x^2}\, dx.$$

Use the trapezoidal approximation with $\Delta x = \frac{1}{5}$. We first make a table of values of $\sqrt{1 + x^2}$. The graph is drawn in Figure 4.6.2.

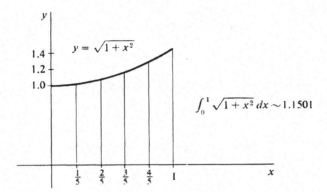

$$\int_0^1 \sqrt{1 + x^2}\, dx \sim 1.1501$$

Figure 4.6.2

x	$\sqrt{1 + x^2}$	$\sqrt{1 + x^2}$ to four places	term in trapezoidal approximation
$x_0 = 0$	1	1.0000	$0.5000 = \frac{1}{2}f(x_0)$
$x_1 = \frac{1}{5}$	$\sqrt{1.04}$	1.0198	$1.0198 = f(x_1)$
$x_2 = \frac{2}{5}$	$\sqrt{1.16}$	1.0770	$1.0770 = f(x_2)$
$x_3 = \frac{3}{5}$	$\sqrt{1.36}$	1.1662	$1.1662 = f(x_3)$
$x_4 = \frac{4}{5}$	$\sqrt{1.64}$	1.2806	$1.2806 = f(x_4)$
$x_5 = 1$	$\sqrt{2}$	1.4142	$0.7071 = \frac{1}{2}f(x_5)$
			$5.7507 = \text{total}$

Thus, $\frac{1}{2}f(x_0) + f(x_1) + f(x_2) + f(x_3) + f(x_4) + \frac{1}{2}f(x_5) = 5.7507$.   Since $\Delta x = \frac{1}{5}$, the trapezoidal approximation is

$$(5.7507) \cdot \tfrac{1}{5} = 1.1501,$$

$$\int_0^1 \sqrt{1 + x^2}\, dx \sim 1.1501.$$

The trapezoidal approximation can be made as close to the definite integral as we want by taking Δx small. From a practical standpoint, however, it is helpful to know how small we should take Δx in order to be sure of a given degree of accuracy. For instance, suppose we need to know the definite integral to three decimal places. How small must we take Δx in our trapezoidal approximation? The answer is given by the Trapezoidal Rule, which gives an error estimate for the trapezoidal approximation.

The *error* in the trapezoidal approximation is the absolute value of the difference between the trapezoidal sum and the definite integral,

$$\text{error} = \left| \sum_a^b \frac{f(x) + f(x + \Delta x)}{2} \Delta x - \int_a^b f(x)\, dx \right|.$$

An *error estimate* for the trapezoidal approximation is a function $E(\Delta x)$, which is known to be greater than or equal to the error.

Thus if $E(\Delta x)$ is an error estimate, the trapezoidal sum is within $E(\Delta x)$ of the definite integral. If we want to be sure that the trapezoidal approximation is accurate to three decimal places—i.e., the error is less than 0.0005—we choose Δx so that $E(\Delta x) \le 0.0005$. We are now ready to state the Trapezoidal Rule.

TRAPEZOIDAL RULE

Let f be a function whose second derivative f'' exists and has absolute value at most M on a closed interval $[a, b]$,

$$|f''(x)| \le M \qquad for\ a \le x \le b.$$

If Δx evenly divides $b - a$, then the trapezoidal approximation of the definite integral of f has the error estimate

$$\frac{b - a}{12} M(\Delta x)^2.$$

That is,

$$\left| \sum_a^b \frac{f(x) + f(x + \Delta x)}{2} \Delta x - \int_a^b f(x)\, dx \right| \le \frac{b - a}{12} M(\Delta x)^2.$$

The proof is omitted.

EXAMPLE 1 (Concluded)   We let $f(x) = \sqrt{1 + x^2}$.   Then

$$f'(x) = \frac{x}{\sqrt{1 + x^2}},$$

$$f''(x) = \frac{\sqrt{1 + x^2} - x^2/\sqrt{1 + x^2}}{1 + x^2} = \frac{1}{(1 + x^2)^{3/2}}.$$

Therefore $|f''(x)| \le 1$ for all x in $[0, 1]$. We take $M = 1$ and use the error estimate given by the Trapezoidal Rule,

$$\frac{b - a}{12} M(\Delta x)^2 = \frac{1}{12} \cdot 1 \cdot \left(\frac{1}{5}\right)^2 = \frac{1}{300}.$$

Thus our approximation is within an accuracy of 1/300,

$$\left| \int_0^1 \sqrt{1 + x^2}\, dx - 1.150 \right| \le 1/300 \sim 0.0033.$$

This shows that the integral is, at least, between 1.146 and 1.154.

In this particular example we can even conclude that the integral is between 1.146 and 1.150 (rounded off to three places). That is, the integral is less than its trapezoidal approximation. This is because the second derivative $f''(x) = (1 + x^2)^{-3/2}$ is always greater than 0, whence the curve is concave upwards and therefore $y = f(x)$ is always less than or equal to the broken line used in the trapezoidal approximation. Actually, the value to three places is 1.148. This can be found by taking $\Delta x = \frac{1}{10}$.

EXAMPLE 2 Consider the integral

$$\int_{-1}^{1} \sqrt{1 - x^2}\, dx = \pi/2.$$

Let

$$f(x) = \sqrt{1 - x^2}.$$

By Theorem 1, we have

$$\lim_{\Delta x \to 0^+} \sum_{-1}^{1} \frac{f(x) + f(x + \Delta x)}{2} \Delta x = \frac{\pi}{2}.$$

However, the Trapezoidal Rule fails to give an error estimate in this case because $f'(x)$ is discontinuous at $x = \pm 1$.

We now turn to Simpson's Rule, for which the number of subintervals n must be even. As before, we divide the interval $[a, b]$ into n subintervals of equal length Δx with the $n + 1$ partition points

$$a = x_0, x_1, \ldots, x_n = b.$$

We shall use subintervals of length $2\Delta x$ rather than Δx. On each of the $n/2$ subintervals

$$[x_0, x_2], [x_2, x_4], \ldots, [x_{n-2}, x_n],$$

of length $2\Delta x$ we approximate the curve $y = f(x)$ by a parabolic arc that meets the curve at both endpoints and the midpoint of the subinterval, as shown in Figure 4.6.3. We then add up the areas under each of the parabolic arcs to obtain an approximation to the area under the curve, which is the definite integral. We begin with a lemma that gives a formula for the area of the region under one parabolic arc.

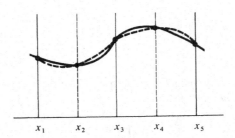

Figure 4.6.3

$x_1 \qquad x_2 \qquad x_3 \qquad x_4 \qquad x_5$

LEMMA

The area of the region under the parabola through three points (u, r), $(u + h, s)$, and $(u + 2h, t)$ (shown in Figure 4.6.4) is

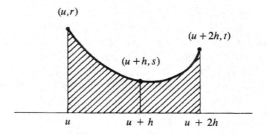

Figure 4.6.4

$$\frac{h}{3}(r + 4s + t).$$

The lemma is proved at the end of this section. Using the lemma, we find that the area of the region under one parabolic arc from x_k to x_{k+2} is

$$\frac{\Delta x}{3}[f(x_k) + 4f(x_{k+1}) + f(x_{k+2})].$$

It follows that the sum of the $n/2$ regions under the parabolic arcs is a modified Riemann sum,

$$\sum_a^b \left[\frac{f(x) + 4f(x + \Delta x) + f(x + 2\,\Delta x)}{3}\right]\Delta x$$

$$= \frac{\Delta x}{3}\{[f(x_0) + 4f(x_1) + f(x_2)] + [f(x_2) + 4f(x_3) + f(x_4)] + \cdots\}$$

$$= \frac{\Delta x}{3}[f(x_0) + 4f(x_1) + 2f(x_2) + 4f(x_3) + 2f(x_4) + \cdots + 4f(x_{n-1}) + f(x_n)].$$

This modified Riemann sum is *Simpson's approximation* to the definite integral. Note the sequence of coefficients,

$$1, 4, 2, 4, 2, \ldots, 2, 4, 1.$$

Like the trapezoidal approximation, it is easily computed on a computer or hand calculator.

THEOREM 2

For a continuous function f on $[a, b]$, Simpson's approximation approaches the definite integral as $\Delta x \to 0^+$,

$$\int_a^b f(x)\,dx = \lim_{\Delta x \to 0^+} \frac{\Delta x}{3}[f(x_0) + 4f(x_1) + 2f(x_2) + 4f(x_3) + \cdots + f(x_n)].$$

Simpson's approximation is almost as easy to calculate as the trapezoidal approximation, but is much more accurate. Simpson's Rule is an error estimate that involves the fourth derivative of the function and the fourth power of Δx.

SIMPSON'S RULE

Suppose the function f has a fourth derivative on the interval $[a, b]$ that has absolute value at most M,

$$|f^{(4)}(x)| \leq M \text{ for } a \leq x \leq b.$$

If $[a, b]$ is divided into an even number of subintervals of length Δx, then Simpson's approximation to the definite integral has the error estimate

$$\frac{b - a}{180} M(\Delta x)^4.$$

EXAMPLE 3   Use Simpson's Rule with $\Delta x = 0.25$ to approximate the integral

$$A = \int_0^1 e^{-x^2/2} \, dx$$

and find the error estimate.

The curve is the normal (bell-shaped) curve used in statistics, shown in Figure 4.6.5. We are to divide the interval $[0, 1]$ into four subintervals of equal length $\Delta x = 0.25$. The following table shows the values of x and y and the coefficient to be used in Simpson's approximation for each partition point.

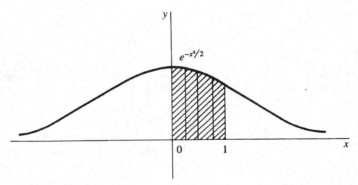

Figure 4.6.5   Example 3

x	$e^{-x^2/2}$	Coefficient
0.0	1.000000	1
0.25	0.969233	4
0.5	0.882496	2
0.75	0.754840	4
1.0	0.606531	1

The sum used in the Simpson approximation is then

$$[1.000000 + 4 \cdot (0.969233) + 2 \cdot (0.882496) + 4 \cdot (0.754840) + 0.606531]$$
$$= 10.267816$$

To get the Simpson approximation, we multiply this sum by $\Delta x/3$:

$$S = (10.267816) \cdot (0.25)/3 = 0.855651.$$

To find the error estimate we need the fourth derivative of

$$y = e^{-x^2/2}.$$

The fourth derivative can be computed as usual and turns out to be

$$y^{(4)} = (x^4 - 6x^2 + 3)e^{-x^2/2}.$$

On the interval $[0, 1]$, $y^{(4)}$ is decreasing because both $x^4 - 6x^2 + 3$ and $-x^2/2$ are decreasing, and therefore $y^{(4)}$ has its maximum value at $x = 0$ and its minimum value at $x = 1$,

$$\text{maximum:} \quad y^{(4)}(0) = 3$$
$$\text{minimum:} \quad y^{(4)}(1) = -1.213061$$

The maximum value of the absolute value $|y^{(4)}|$ is thus $M = 3$. The error estimate in Simpson's Rule is then

$$\frac{b-a}{180}(\Delta x)^4 M = \frac{1}{180} \cdot (0.25)^4 \cdot 3 = 0.000065.$$

This shows that the integral is within 0.000065 of the approximation; that is,

$$\int_0^1 e^{-x^2/2}\, dx = 0.855651 \pm 0.000065,$$

or using inequalities,

$$0.855586 \leq \int_0^1 e^{-x^2/2}\, dx \leq 0.855716.$$

For comparison, a more accurate computation with a smaller Δx shows that the actual value to six places is

$$\int_0^1 e^{-x^2/2}\, dx = 0.855624.$$

The Trapezoidal Rule for this integral and the same value of $\Delta x = 0.25$ give an approximate value of 0.85246 for the integral and an error estimate of 0.00521.

PROOF OF THE LEMMA   The algebra is simpler if the y-axis is drawn through the second point, so that $u + h = 0$, and the three points have coordinates

$$(-h, r), (0, s), (h, t).$$

Suppose the parabola has the equation $y = ax^2 + bx + c$. Then the area under the parabola is

$$A = \int_{-h}^{h} (ax^2 + bx + c)\, dx$$
$$= \frac{ax^3}{3} + \frac{bx^2}{2} + cx \Big]_{-h}^{h}$$
$$= \frac{2}{3}ah^3 + 2ch.$$

When we substitute the coordinates of the three points $(-h, r)$, $(0, s)$, (h, t) into the equation for the parabola, we obtain the three equations

$$r = ah^2 - bh + c,$$
$$s = c,$$
$$t = ah^2 + bh + c.$$

Add the first and third equations and solve for a:

$$r + t = 2ah^2 + 2c$$
$$a = \frac{r + t - 2c}{2h^2}$$

Finally, substitute the above expression for a and s for c in the equation for the area:

$$A = \frac{2}{3}ah^3 + 2ch$$
$$= \frac{r + t - 2c}{2h^2} \cdot \frac{2}{3} \cdot h^3 + 2ch$$
$$= \frac{r + t - 2c + 6c}{3} \cdot h$$
$$= \frac{h}{3}(r + 4c + t).$$
$$= \frac{h}{3}(r + 4s + t).$$

PROBLEMS FOR SECTION 4.6

Approximate the integrals in Problems 1–20 using (a) the Trapezoidal Rule and (b) Simpson's Rule. When possible, find error estimates. If a hand calculator is available, do the problems again with $\Delta x = 0.1$.

1 $\int_0^2 x \, dx$, $\Delta x = 0.5$

2 $\int_0^2 x^3 dx$, $\Delta x = 0.5$

3 $\int_1^2 \sqrt{x^2 - 1} \, dx$, $\Delta x = 0.25$

4 $\int_1^3 \frac{1}{x} dx$, $\Delta x = 0.5$

5 $\int_1^2 \frac{1}{1 + x^2} dx$, $\Delta x = 0.25$

6 $\int_0^1 x\sqrt{x + 1} \, dx$, $\Delta x = 0.25$

7 $\int_1^5 \frac{x}{x + 1} dx$, $\Delta x = 0.5$

8 $\int_0^1 \sqrt{x^3 + 1} \, dx$, $\Delta x = \frac{1}{4}$

9 $\int_0^1 \sqrt{x^4 + 1} \, dx$, $\Delta x = \frac{1}{4}$

10 $\int_1^4 \sqrt{1 + 1/x} \, dx$, $\Delta x = 0.5$

11 $\int_0^6 \frac{1}{x + 1} dx$, $\Delta x = 1$

12 $\int_0^{12} \frac{1}{2x + 3} dx$, $\Delta x = 2$

13 $\int_1^{13} \frac{1}{x + \sqrt{x}} dx$, $\Delta x = 3$

14 $\int_0^4 \frac{1}{2 + \sqrt{x}} dx$, $\Delta x = 1$

15 $\displaystyle\int_0^\pi \sin\theta\, d\theta, \quad \Delta x = \pi/2, \quad \pi/10$ **16** $\displaystyle\int_0^\pi \sin^2\theta\, d\theta, \quad \Delta x = \pi/2, \quad \pi/10$

17 $\displaystyle\int_0^1 e^x\, dx, \quad \Delta x = \frac{1}{4}$ **18** $\displaystyle\int_0^1 e^{x^2}\, dx, \quad \Delta x = \frac{1}{4}$

19 $\displaystyle\int_1^2 \ln x\, dx, \quad \Delta x = \frac{1}{4}$ **20** $\displaystyle\int_1^2 \ln(1/x)\, dx, \quad \Delta x = \frac{1}{4}$

☐ **21**   Let f be continuous on the interval $[a, b]$ and let $\Delta x = (b - a)/n$ where n is a positive integer. Prove that the trapezoidal sum is equal to the Riemann sum plus $\frac{1}{2}(f(b) - f(a))\,\Delta x$, that is,

$$\sum_a^b \tfrac{1}{2}(f(x) + f(x + \Delta x))\,\Delta x = \left(\sum_a^b f(x)\,\Delta x\right) + \tfrac{1}{2}(f(b) - f(a))\,\Delta x.$$

Show that if $f(a) = f(b)$ then the trapezoidal sum and Riemann sum are equal.

☐ **22**   Prove that for a linear function $f(x) = kx + c$, the trapezoidal sum is exactly equal to the integral.

☐ **23**   Show that if $f(x)$ is concave downward, $f''(x) > 0$, then the trapezoidal sum is less than the definite integral of $f(x)$.

☐ **24**   Show that for a quadratic function $f(x) = ax^2 + bx + c$, Simpson's approximation is equal to the definite integral.

☐ **25**   Show that for a cubic function $f(x) = ax^3 + bx^2 + cx + d$, Simpson's approximation is still equal to the definite integral.

EXTRA PROBLEMS FOR CHAPTER 4

1   Evaluate $\displaystyle\sum_1^2 \frac{1}{x^2}\,\Delta x, \quad \Delta x = 1/4$

2   Evaluate $\displaystyle\sum_1^{10} \frac{1}{x^2}\,\Delta x, \quad \Delta x = 2$

3   Evaluate $\displaystyle\sum_{-3}^3 2^x\,\Delta x, \quad \Delta x = 1$

4   Evaluate $\displaystyle\sum_0^2 x\sqrt{x + 1}\,\Delta x, \quad \Delta x = 1/2$

5   If $F'(x) = 1/(2x - 1)^2$ for all $x \neq 1/2$, find $F(2) - F(1)$.

6   If $G'(t) = \sqrt{4t + 1}$ for all $t > -1/4$, find $G(2) - G(0)$.

7   A particle moves with velocity $v = (3 + 2\sqrt{t})^2$. How far does it move from times $t_0 = 1$ to $t_1 = 5$?

8   A particle moves with velocity $v = t^2\sqrt{t^3 - 1}$. How far does it move from times $t_0 = 1$ to $t_1 = 4$?

9   A particle moves with velocity $v = (t + 1)(2t + 3)$. If it has position $y_0 = 0$ at time $t = 0$, find its position at time $t = 10$.

10   A particle moves with acceleration $a = 1/t^4$. If it has velocity $v_0 = 4$ and position $y_0 = 2$ at time $t = 1$, find its position at time $t = 3$.

11   Find the area of the region under the curve $y = 1/\sqrt{x}, 1 \leq x \leq 4$.

12   Find the area of the region under the curve $y = \sqrt{x} - x\sqrt{x}, 0 \leq x \leq 1$.

In Problems 13–30, evaluate the integral.

13 $\displaystyle\int (1 - x)(2 + 3x)\, dx$ **14** $\displaystyle\int \left(2 + \frac{1}{x}\right)\left(2 - \frac{1}{x}\right) dx$

15 $\displaystyle\int \frac{x}{(x^2-1)^3}\,dx$

16 $\displaystyle\int (4x+1)^{1/3}\,dx$

17 $\displaystyle\int (u/\sqrt{1-3u^2})\,du$

18 $\displaystyle\int x^{-2}\sqrt{2+x^{-1}}\,dx$

19 $\displaystyle\int (\sqrt{2t+1}-\sqrt{2t-1})\,dt$

20 $\displaystyle\int \frac{2x+1}{(x+4)^3}\,dx$

21 $\displaystyle\int y\sqrt{y+2}\,dy$

22 $\displaystyle\int (1-\sqrt{x})^{-4}\,dx$

23 $\displaystyle\int \cos\!\left(\frac{x}{2}\right) dx$

24 $\displaystyle\int \sqrt{x}\,\sin\sqrt{x}\,dx$

25 $\displaystyle\int e^{-t}\,dt$

26 $\displaystyle\int \frac{t+1}{t-1}\,dt$

27 $\displaystyle\int_0^4 (y+\sqrt{y})\,dy$

28 $\displaystyle\int_2^6 (x/\sqrt{x^2-1})\,dx$

29 $\displaystyle\int_0^1 e^{4x}\,dx$

30 $\displaystyle\int_0^1 x\sin(x^2)\,dx$

31   Differentiate $\displaystyle\int_1^x \sqrt{t^3+2}\,dt$

32   Differentiate $\displaystyle\int_0^{3x} (t^2/(t^2-1))\,dt$

33   Differentiate $\displaystyle\int_u^4 \sqrt{x}\sqrt{x-1}\,dx$

34   Differentiate $\displaystyle\int_y^{y^2} (1/(x+\sqrt{x}))\,dx$

35   Find the function F such that $F'(x) = x - 1$ for all x, and the minimum value of $F(x)$ is b.

36   Find the function F such that $F''(x) = x$ for all x, $F(0) = 1$, and $F(1) = 1$.

37   Find the function F such that $F''(x) = 6$ for all x, $F(x)$ has a minimum at $x = 1$, and the minimum value is 2.

38   Find all functions F such that $F''(x) = 1 + x^{-3}$ for all positive x.

☐ **39**   Find the function F such that

$$F'(x) = \begin{cases} 0 & \text{if } x < 0 \\ x & \text{if } x \ge 0 \end{cases}$$

and $F(0) = 1$.

☐ **40**   Find the value of b such that the area of the region under the curve $y = x(b - x)$, $0 \le x \le b$, is 1.

☐ **41**   Suppose f is increasing for $a \le x \le b$, and $\Delta x = (b - a)/n$ where n is a positive integer. Show that

$$\left| \sum_a^b f(x)\,\Delta x - \int_a^b f(x)\,dx \right| \le [f(b) - f(a)]\,\Delta x$$

☐ **42**   Suppose f is continuous for $a \le x \le b$. Show that

$$\left| \int_a^b f(x)\,dx \right| \le \int_a^b |f(x)|\,dx.$$

☐ **43**   Find the area of the top half of the ellipse $x^2/a^2 + y^2/b^2 = 1$ using the formula $\pi = 2\int_{-1}^{1} \sqrt{1-u^2}\,du$.

☐ **44**   Evaluate $\int_{-1}^1 (1-x)^{3/2}(1+x)^{1/2}\,dx$ using the formula $\pi = 2\int_{-1}^1 \sqrt{1-u^2}\,du$.

☐ **45**   Find dy/dx if $y = \int_0^x x f(t)\,dt$.

☐ **46**   Suppose $f(t)$ is continuous for all t and let $G(x) = \int_0^x (x-t)f(t)\,dt$. Prove that $G''(x) = f(x)$.

☐ **47**   Prove that for any continuous functions f and g,

$$2AB \int_a^b f(x)g(x)\,dx \le A^2 \int_a^b f(x)^2\,dx + B^2 \int_a^b g(x)^2\,dx.$$

☐ **48**   Prove Schwartz' Inequality,

$$\int_a^b f(x)g(x)\,dx \le \sqrt{\int_a^b f(x)^2\,dx \int_a^b g(x)^2\,dx}.$$

Hint: Use the preceding problem.

☐ **49**   Suppose f is continuous and dx is positive infinitesimal. Show that

$$\sum_a^b f(x + \tfrac{1}{2}\,dx)\,dx \approx \int_a^b f(x)\,dx.$$

Hint: For each positive real c,

$$f(x) - c < f\left(x + \frac{1}{2}\,dx\right) < f(x) + c.$$

Use this to show that

$$\int_a^b f(x)\,dx - c(b - a) < \sum_a^b f\left(x + \frac{1}{2}\,dx\right)dx < \int_a^b f(x)\,dx + c(b - a).$$

☐ **50**   Suppose f is continuous, n is an integer, and dx is positive infinitesimal. Prove that

$$\sum_a^b f(x + n\,dx)\,dx \approx \int_a^b f(x)\,dx.$$

LIMITS, ANALYTIC GEOMETRY, AND APPROXIMATIONS

5.1 INFINITE LIMITS

Up to this point we have studied three types of limits:

$$\lim_{x \to c} f(x) = L \quad \text{means} \quad f(x) \approx L \text{ whenever } x \approx c \text{ but } x \neq c.$$

$$\lim_{x \to c^+} f(x) = L \quad \text{means} \quad f(x) \approx L \text{ whenever } x \approx c \text{ but } x > c.$$

$$\lim_{x \to c^-} f(x) = L \quad \text{means} \quad f(x) \approx L \text{ whenever } x \approx c \text{ but } x < c.$$

The limit notation $\lim_{x \to \infty} f(x) = L$ means that whenever H is positive infinite, $f(H) \approx L$ (Figure 5.1.1(a)).

$\lim_{x \to c} f(x) = -\infty$ means that whenever $x \approx c$ and $x \neq c$, $f(x)$ is negative infinite (Figure 5.1.1(b)). The various other combinations have the meanings which one would expect.

EXAMPLE 1 $\lim\limits_{x \to 0} \dfrac{1}{x^2} = \infty$.

EXAMPLE 2 $\lim\limits_{x \to 0^+} \dfrac{1}{x} = \infty, \quad \lim\limits_{x \to 0^-} \dfrac{1}{x} = -\infty$.

EXAMPLE 3 Find $\lim\limits_{x \to \infty} \dfrac{3x^4 + 5x - 2}{2x^4 - 6x^3 + 7}$.

Let H be positive infinite. Then

$$\frac{3H^4 + 5H - 2}{2H^4 - 6H^3 + 7} = \frac{3 + 5H^{-3} - 2H^{-4}}{2 - 6H^{-1} + 7H^{-4}},$$

and therefore $\quad st\left(\dfrac{3H^4 + 5H - 2}{2H^4 - 6H^3 + 7}\right) = \dfrac{3 + 0 - 0}{2 - 0 + 0} = \dfrac{3}{2}.$

Thus the limit exists and is $\frac{3}{2}$.

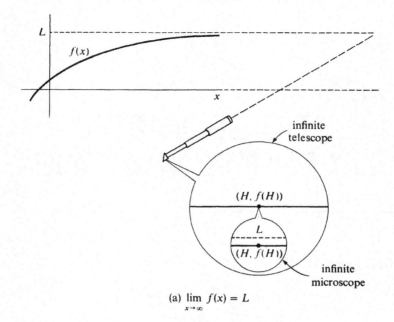

(a) $\lim\limits_{x \to \infty} f(x) = L$

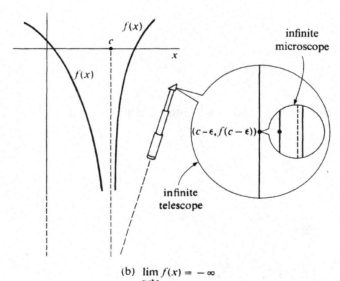

(b) $\lim\limits_{x \to c} f(x) = -\infty$

Figure 5.1.1

EXAMPLE 4 Find $\lim\limits_{x \to -\infty} (x^3 + 200x^2)$.

We have $x^3 + 200x^2 = x^2(x + 200)$. When H is negative infinite, H^2 is positive infinite and $(H + 200)$ is negative infinite, so their product is negative infinite. Thus

$$\lim_{x \to -\infty} (x^3 + 200x^2) = -\infty.$$

When
$$\lim_{x \to c} f(x) = \infty \quad \text{or} \quad -\infty,$$

the limit does not exist, because $f(x)$ has no standard part. The infinity symbol is only used to indicate the behavior of $f(x)$ and is not to be construed as a number.

EXAMPLE 5 A student can get a score of $100t/(t + 1)$ on his math exam if he studies t hours for it (Figure 5.1.2). If he studies infinitely long for the exam, his score will be infinitely close to 100, because if H is positive infinite,

$$st\left(\frac{100H}{H + 1}\right) = st\left(\frac{100}{1 + 1/H}\right) = \frac{100}{1 + 0} = 100.$$

In the notation of limits,

$$\lim_{t \to \infty} \frac{100t}{t + 1} = 100.$$

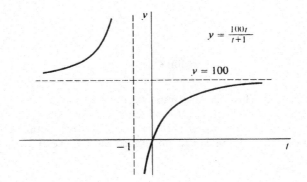

Figure 5.1.2

EXAMPLE 6 Given any polynomial

$$f(t) = a_n t^n + a_{n-1} t^{n-1} + \cdots + a_1 t + a_0$$

of degree $n > 0$, the limits as t approaches $-\infty$ or $+\infty$ are as follows. Suppose $a_n > 0$. When n is even, $\lim_{t \to -\infty} f(t) = \infty$, $\lim_{t \to \infty} f(t) = \infty$. When n is odd, $\lim_{t \to -\infty} f(t) = -\infty$, $\lim_{t \to \infty} f(t) = \infty$.

The signs are all reversed when $a_n < 0$.

All these limits can be computed from

$$f(t) = t^n\left(a_n + \frac{a^{n-1}}{t} + \cdots + \frac{a_1}{t^{n-1}} + \frac{a_0}{t^n}\right).$$

EXAMPLE 7 In the special theory of relativity, a body which is moving at constant velocity v, $-c < v < c$, will have mass

$$m = \frac{m_0}{\sqrt{1 - v^2/c^2}}$$

and its length in the direction of motion will be

$$l = l_0\sqrt{1 - v^2/c^2}.$$

Here m_0, l_0, and c are positive constants denoting the mass at rest (that is, the mass when $v = 0$), the length at rest, and the speed of light.

Suppose the velocity v is infinitely close to the speed of light c, that is,

$$v = c - \varepsilon, \qquad \varepsilon > 0 \text{ infinitesimal.}$$

Then

$$\sqrt{1 - \frac{v^2}{c^2}} = \sqrt{1 - \frac{(c - \varepsilon)^2}{c^2}} = \sqrt{\frac{c^2 - (c^2 - 2c\varepsilon + \varepsilon^2)}{c^2}}$$

$$= \sqrt{\frac{2\varepsilon}{c} - \frac{\varepsilon^2}{c^2}} = \sqrt{\varepsilon \cdot \left(\frac{2}{c} - \frac{\varepsilon}{c^2}\right)},$$

which is the square root of a positive infinitesimal. Thus $\sqrt{1 - v^2/c^2}$ is a positive infinitesimal. Therefore for v infinitely close to c, m is positive infinite and l is positive infinitesimal. That is, a body moving at velocity infinitely close to (but less than) the speed of light has infinite mass and infinitesimal length in the direction of motion. In the notation of limits this means that

$$\lim_{v \to c^-} \frac{m_0}{\sqrt{1 - v^2/c^2}} = +\infty,$$

$$\lim_{v \to c^-} l_0 \sqrt{1 - v^2/c^2} = 0.$$

Caution: This example must be understood in the light of our policy of speaking as if a line in physical space really is like the hyperreal line. Actually, there is no evidence one way or the other on whether a line in space is like the hyperreal line, but the hyperreal line is a useful model for the purpose of applications.

EXAMPLE 8   Evaluate $\lim\limits_{x \to \infty} \dfrac{\sin x}{x}$.

When H is positive infinite, $\sin H$ is between -1 and 1 and thus finite, so $(\sin H)/H$ is infinitesimal. The limit is therefore zero:

$$\lim_{x \to \infty} \frac{\sin x}{x} = 0.$$

EXAMPLE 9   Find $\lim\limits_{x \to \infty} \cos x$.

If H is any integer or hyperinteger, then

$$\cos(2\pi H) = 1, \qquad \cos(2\pi H + \pi) = -1.$$

In fact, $\cos x$ will keep oscillating between 1 and -1 even for infinite x. Therefore the limit does not exist.

Limits involving e^x and $\ln x$ will be studied in Chapter 8.

PROBLEMS FOR SECTION 5.1

Find the following limits. Your answer should be a real number, ∞, $-\infty$, or "does not exist."

1 $\displaystyle\lim_{x\to\infty} \frac{6x - 4}{2x + 5}$

2 $\displaystyle\lim_{x\to-\infty} \frac{3x}{4x - 10}$

3 $\displaystyle\lim_{t\to\infty} t^3 - 10t^2 - 6t - 2$

4 $\displaystyle\lim_{t\to-\infty} 4t^2 + 6t + 2$

5 $\displaystyle\lim_{x\to\infty} \frac{x^2 - x + 4}{3x^2 + 2x - 3}$

6 $\displaystyle\lim_{x\to\infty} \frac{2x^2 - 4x + 1}{3x^2 + 5x - 6}$

7 $\displaystyle\lim_{y\to-\infty} \frac{5y^3 + 3y^2 + 2}{3y^3 - 6y + 1}$

8 $\displaystyle\lim_{y\to\infty} \frac{y^4 - y^3 + 1}{2y^4 - 4y^2 + 5}$

9 $\displaystyle\lim_{x\to\infty} \frac{\sqrt{x + 2}}{\sqrt{3x + 1}}$

10 $\displaystyle\lim_{u\to\infty} \frac{3 + 2\sqrt{u}}{4 - \sqrt{u}}$

11 $\displaystyle\lim_{x\to\infty} x - \sqrt{x}$

12 $\displaystyle\lim_{x\to\infty} \sqrt{x} + \sqrt{x + 1}$

13 $\displaystyle\lim_{x\to\infty} \sqrt[3]{x + 2}$

14 $\displaystyle\lim_{x\to-\infty} \sqrt{2 - x}$

15 $\displaystyle\lim_{x\to\infty} \frac{1}{\sqrt[3]{x}}$

16 $\displaystyle\lim_{x\to 0^+} \frac{1}{\sqrt[3]{x}}$

17 $\displaystyle\lim_{x\to 0^-} 1 + \frac{1}{x}$

18 $\displaystyle\lim_{x\to 0^+} 1 + \frac{1}{x}$

19 $\displaystyle\lim_{x\to 0} \frac{1}{x^2} - \frac{1}{x}$

20 $\displaystyle\lim_{x\to 0^+} \frac{1}{\sqrt{x}} - \frac{1}{x}$

21 $\displaystyle\lim_{x\to\infty} \frac{5x + 6}{x^2 - 4}$

22 $\displaystyle\lim_{x\to\infty} \frac{10x^2 + x + 2}{x^3 - 4x^2 - 1}$

23 $\displaystyle\lim_{t\to\infty} \frac{t}{\sqrt{4t^2 + 1}}$

24 $\displaystyle\lim_{t\to-\infty} \frac{t}{\sqrt{4t^2 + 1}}$

25 $\displaystyle\lim_{t\to-\infty} \frac{\sqrt{t^2 + 2}}{4t + 2}$

26 $\displaystyle\lim_{t\to\infty} \frac{\sqrt{t^2 + 2}}{4t + 2}$

27 $\displaystyle\lim_{t\to\infty} \frac{5t + 2}{t^2 - 6t + 1}$

28 $\displaystyle\lim_{t\to\infty} \frac{t^3 - 6t^2 + 4}{2t^4 + t^3 - 5}$

29 $\displaystyle\lim_{t\to 0} \frac{1 - 5t^{-1}}{4 + 6t^{-1}}$

30 $\displaystyle\lim_{t\to 0} \frac{5 + 6t^{-1} + t^{-2}}{8 - 3t^{-1} + 2t^{-2}}$

31 $\displaystyle\lim_{t\to 0} \frac{1 + 2t^{-1}}{7 + t^{-1} - 5t^{-2}}$

32 $\displaystyle\lim_{t\to 0} \frac{1 - 2t^{-1} + t^{-2}}{3 - 4t^{-1}}$

33 $\displaystyle\lim_{x\to 2} \frac{1 - x}{2 - x}$

34 $\displaystyle\lim_{x\to 2^+} \frac{1 - x}{2 - x}$

35 $\displaystyle\lim_{y\to 3^+} \frac{y + 1}{(y - 2)(y - 3)}$

36 $\displaystyle\lim_{y\to 3^-} \frac{y + 1}{(y - 2)(y - 3)}$

37 $\displaystyle\lim_{y\to 3} \frac{y + 1}{(y - 2)(y - 3)}$

38 $\displaystyle\lim_{x\to 5} \frac{3x^2 + 4}{x^2 - 10x + 25}$

39 $\displaystyle\lim_{t\to 1} \frac{3t^3 + 4}{t^2 + t - 2}$

40 $\displaystyle\lim_{x\to 2} \frac{x^2 + 4}{x^2 - 4}$

41 $\displaystyle\lim_{x\to 2^+} \frac{x^2 + 4}{x^2 - 4}$

42 $\displaystyle\lim_{x\to 1^+} \frac{x - 1}{x - 2\sqrt{x} + 1}$

43 $\displaystyle\lim_{x\to\infty} \sqrt{x + 2} - \sqrt{x}$

44 $\displaystyle\lim_{x\to\infty} (x + 1)^{3/2} - x^{3/2}$

45 $\displaystyle\lim_{x\to\infty} \sqrt{3x + 1} - 2\sqrt{x}$

46 $\displaystyle\lim_{x\to\infty} \sqrt{2x + 3} - \sqrt{x}$

47 $\displaystyle\lim_{x\to\infty} \sqrt{x^2 + x} - x$ **48** $\displaystyle\lim_{x\to\infty} \sqrt{x^2 + 1} - x$

49 $\displaystyle\lim_{t\to\infty} t(\sqrt{t + 1} - \sqrt{t})$ **50** $\displaystyle\lim_{t\to\infty} \sqrt{t}(\sqrt{t + 2} - \sqrt{t + 1})$

51 $\displaystyle\lim_{u\to\infty} \sqrt{u^2 - 3u + 2} - \sqrt{u^2 + 1}$ **52** $\displaystyle\lim_{x\to\infty} \frac{\sqrt{x + 2} - \sqrt{x}}{\sqrt{x + 3} - \sqrt{x}}$

53 $\displaystyle\lim_{t\to\infty} \sqrt[3]{t + 4} - \sqrt[3]{t}$ **54** $\displaystyle\lim_{t\to\infty} \sqrt[3]{t^3 + 1} - t$

55 $\displaystyle\lim_{t\to\infty} \cos(1/t)$ **56** $\displaystyle\lim_{t\to\infty} \sin(1/t)$

57 $\displaystyle\lim_{t\to\infty} \frac{\sin t}{t}$ **58** $\displaystyle\lim_{t\to\infty} \frac{\cos t}{t^2}$

59 $\displaystyle\lim_{\theta\to\infty} \sin\theta$ **60** $\displaystyle\lim_{\theta\to\infty} \theta\cos\theta$

61 $\displaystyle\lim_{\theta\to 0} \tan\theta$ **62** $\displaystyle\lim_{\theta\to \pi/2} \tan\theta$

63 $\displaystyle\lim_{\theta\to \pi/2^+} \tan\theta$ **64** $\displaystyle\lim_{\theta\to \pi/2^-} \tan\theta$

65 $\displaystyle\lim_{x\to 0} \sin(1/x)$ **66** $\displaystyle\lim_{x\to 0} x\cos(1/x)$

67 $\displaystyle\lim_{x\to 0^+} \frac{\cos x}{x}$ **68** $\displaystyle\lim_{x\to 0^-} \frac{\cos x}{x}$

69   Prove that if $\lim_{x\to c} f(x) = \infty$ then $\lim_{x\to c} 1/f(x) = 0$.

71   Prove that if $\lim_{x\to c} f(x) = 0$ and $f(x) > 0$ for all x, then $\lim_{x\to c} 1/f(x) = \infty$.

72   Prove that if $\lim_{x\to 0^+} f(x)$ exists or is infinite, then

$$\lim_{x\to 0^+} f(x) = \lim_{t\to\infty} f(1/t).$$

73   Prove that if $\lim_{x\to\infty} f(x)$ exists or is infinite then

$$\lim_{x\to\infty} f(x) = \lim_{t\to 0^+} f(1/t).$$

5.2 L'HOSPITAL'S RULE

Suppose f and g are two real functions which are defined in an open interval containing a real number a, and we wish to compute the limit

$$\lim_{x\to a} \frac{f(x)}{g(x)}.$$

Sometimes the answer is easy. Assume that the limits of $f(x)$ and $g(x)$ exist as $x \to a$,

$$\lim_{x\to a} f(x) = L, \qquad \lim_{x\to a} g(x) = M.$$

If $M \neq 0$, then the limit of the quotient is simply the quotient of the limits,

$$\lim_{x\to a} \frac{f(x)}{g(x)} = \frac{L}{M}.$$

This is because for any infinitesimal $\Delta x \neq 0$,

$$\lim_{x\to a} \frac{f(x)}{g(x)} = st\left(\frac{f(a + \Delta x)}{g(a + \Delta x)}\right) = \frac{st(f(a + \Delta x))}{st(g(a + \Delta x))} = \frac{L}{M}.$$

If $L \neq 0$ and $M = 0$, then the limit

$$\lim_{x\to a} \frac{f(x)}{g(x)}$$

does not exist, because when $\Delta x \neq 0$ is infinitesimal, $f(a + \Delta x)$ has standard part $L \neq 0$ and $g(a + \Delta x)$ has standard part 0.

But what happens if both L and M are 0? In some cases a simple algebraic manipulation will enable us to compute the limit. For example,

$$\lim_{x \to -1} \frac{x^2 - 1}{x + 1} = \lim_{x \to -1} \frac{(x + 1)(x - 1)}{x + 1} = \lim_{x \to -1} (x - 1) = -2,$$

even though both the numerator $x^2 - 1$ and the denominator $x + 1$ approach 0 as x approaches -1.

In other cases l'Hospital's Rule is useful in computing limits of quotients where both L and M are 0. Before stating l'Hospital's Rule, we introduce the notion of a neighborhood of a point c (Figure 5.2.1).

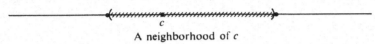

$$c$$

A neighborhood of c

Figure 5.2.1

DEFINITION

*By a **neighborhood** of a real number c we mean an interval which contains c as an interior point.*

*The set formed by removing the point c from a neighborhood I of c is called a **deleted neighborhood** of c. Thus a deleted neighborhood is the set of all points x in I such that $x \neq c$.*

L'HOSPITAL'S RULE FOR 0/0

Suppose that in some deleted neighborhood of a real number c, $f'(x)$ and $g'(x)$ exist and $g'(x) \neq 0$. Assume that

$$\lim_{x \to c} f(x) = 0, \qquad \lim_{x \to c} g(x) = 0.$$

If $\lim_{x \to c} \dfrac{f'(x)}{g'(x)}$ exists or is infinite, then

$$\lim_{x \to c} \frac{f(x)}{g(x)} = \lim_{x \to c} \frac{f'(x)}{g'(x)}.$$

(See Figure 5.2.2.) Usually the limit will be given by

$$\lim_{x \to c} \frac{f'(x)}{g'(x)} = \frac{f'(c)}{g'(c)},$$

and in this case the proof is very simple.

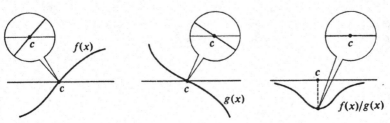

Figure 5.2.2 L'Hospital's Rule

PROOF IN THE CASE $$\lim_{x \to c} \frac{f'(x)}{g'(x)} = \frac{f'(c)}{g'(c)}$$

Let Δx be a nonzero infinitesimal. Then $f(c) = 0$, $g(c) = 0$, and

$$\frac{f(c + \Delta x)}{g(c + \Delta x)} = \frac{(f(c + \Delta x) - f(c))/\Delta x}{(g(c + \Delta x) - g(c))/\Delta x} \approx \frac{f'(c)}{g'(c)}.$$

Taking standard parts we get

$$\lim_{x \to c} \frac{f(x)}{g(x)} = \frac{f'(c)}{g'(c)} = \lim_{x \to c} \frac{f'(x)}{g'(x)}.$$

Intuitively, for $x \approx c$ the graphs of $f(x)$ and $g(x)$ are almost straight lines of slopes $f'(c), g'(c)$ passing through zero, so the graph of $f(x)/g(x)$ is almost the horizontal line through $f'(c)/g'(c)$ (Figure 5.2.3).

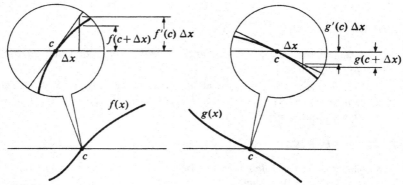

Figure 5.2.3

The equation

$$\lim_{x \to c} \frac{f'(x)}{g'(x)} = \frac{f'(c)}{g'(c)}$$

is not always true. For example, $g'(c)$ might be zero or undefined.

$$\lim_{x \to c} \frac{f'(x)}{g'(x)}$$

is sometimes another limit of type 0/0, that is,

$$\lim_{x \to c} f'(x) = 0 \quad \text{and} \quad \lim_{x \to c} g'(x) = 0.$$

When this happens, l'Hospital's Rule can often be reapplied to $\lim_{x \to c} f'(x)/g'(x)$. The proof of l'Hospital's Rule in general is fairly long and uses the Mean Value Theorem. It will not be given here.

Here are some examples showing how the rule can be applied.

EXAMPLE 1   Find $\lim\limits_{x \to 1} \dfrac{(1/x) - 1}{\sqrt{x - 1}}$.

Both $(1/x) - 1$ and $\sqrt{x} - 1$ approach 0 as x approaches 1. The limit is thus of the form 0/0. Using l'Hospital's Rule,

$$\lim_{x \to 1} \frac{(1/x) - 1}{\sqrt{x} - 1} = \lim_{x \to 1} \frac{-x^{-2}}{\frac{1}{2}x^{-1/2}} = \frac{-1}{\frac{1}{2}} = -2.$$

EXAMPLE 2 Find $\displaystyle\lim_{x \to 0} \frac{\sqrt{x + 1} - 1}{x^3}$.

The limit is of the form 0/0. The limit of $f'(x)/g'(x)$ as $x \to 0$ is ∞,

$$\lim_{x \to 0} \frac{d(\sqrt{x + 1} - 1)/dx}{d(x^3)/dx} = \lim_{x \to 0} \frac{\frac{1}{2}(x + 1)^{-1/2}}{3x^2} = \infty.$$

Thus by l'Hospital's Rule,

$$\lim_{x \to 0} \frac{\sqrt{x + 1} - 1}{x^3} = \infty.$$

EXAMPLE 3 Find $\displaystyle\lim_{x \to 3} \left(x + \frac{1}{x - 3}\right)(\sqrt{x + 1} - 2)$.

This limit is not in a form where we can apply l'Hospital's Rule. We must first use algebra to put it in another form,

$$\left(x + \frac{1}{x - 3}\right)(\sqrt{x + 1} - 2) = x(\sqrt{x + 1} - 2) + \frac{\sqrt{x + 1} - 2}{x - 3}.$$

By elementary computations, $\displaystyle\lim_{x \to 3} x(\sqrt{x + 1} - 2) = 3 \cdot 0 = 0$.

Using l'Hospital's Rule,

$$\lim_{x \to 3} \frac{\sqrt{x + 1} - 2}{x - 3} = \lim_{x \to 3} \frac{\frac{1}{2}(x + 1)^{-1/2}}{1} = \frac{1}{2} \cdot 4^{-1/2} = \frac{1}{4}.$$

We then add the limits to get the desired answer,

$$\lim_{x \to 3} \left(x + \frac{1}{x - 3}\right)(\sqrt{x + 1} - 2) = \lim_{x \to 3} x(\sqrt{x + 1} - 2) + \lim_{x \to 3} \frac{\sqrt{x + 1} - 2}{x - 3}$$
$$= 0 + \tfrac{1}{4} = \tfrac{1}{4}.$$

EXAMPLE 4 Find $\displaystyle\lim_{x \to 1} \frac{\dfrac{(x - 3)}{4} + \dfrac{1}{x + 1}}{(x - 1)^2}$.

This limit is of the form 0/0. When l'Hospital's Rule is used the limit is still of the form 0/0. But when it is used a second time we can compute the limit.

$$\lim_{x \to 1} \frac{\dfrac{x - 3}{4} + \dfrac{1}{x + 1}}{(x - 1)^2} = \lim_{x \to 1} \frac{\dfrac{1}{4} - \dfrac{1}{(x + 1)^2}}{2(x - 1)} = \lim_{x \to 1} \frac{2(x + 1)^{-3}}{2} = \frac{1}{8}.$$

L'Hospital's Rule also holds true for other types of limits. That is, it holds true if $x \to c$ is everywhere replaced by one of the following.

$$x \to c^+, \qquad x \to c^-, \qquad x \to \infty, \qquad x \to -\infty.$$

EXAMPLE 5   Find $\lim\limits_{x \to 0^+} \dfrac{\sqrt{x + 4} - 2}{\sqrt{x}}$.

The limit as $x \to 0$ does not exist because $\sqrt{x}$ is defined only for $x > 0$. However, the one-sided limit as $x \to 0^+$ has the form 0/0 and can be found by l'Hospital's Rule.

$$\lim_{x \to 0^+} \frac{\sqrt{x + 4} - 2}{\sqrt{x}} = \lim_{x \to 0^+} \frac{\frac{1}{2}(x + 4)^{-1/2}}{\frac{1}{2}x^{-1/2}} = \lim_{x \to 0^+} \frac{\sqrt{x}}{\sqrt{x + 4}} = 0.$$

A second form of l'Hospital's Rule deals with the case where both $f(x)$ and $g(x)$ approach ∞ as x approaches c.

L'HOSPITAL'S RULE FOR ∞/∞

Suppose c is a real number, and in some deleted neighborhood of c, $f'(x)$ and $g'(x)$ exist and $g'(x) \neq 0$. Assume that

$$\lim_{x \to c} f(x) = \infty, \qquad \lim_{x \to c} g(x) = \infty.$$

If $\lim\limits_{x \to c} \dfrac{f'(x)}{g'(x)}$ exists or is infinite, then

$$\lim_{x \to c} \frac{f(x)}{g(x)} = \lim_{x \to c} \frac{f'(x)}{g'(x)}.$$

The rule for ∞/∞ is exactly the same, word for word, as the rule for 0/0, except that 0 is replaced by ∞. We omit the proof, which is more difficult in the case ∞/∞. Actually, the assumption

$$\lim_{x \to c} f(x) = \infty$$

is not needed.

Again, l'Hospital's Rule for ∞/∞ also holds for the other types of limits,

$$x \to c^+, \qquad x \to c^-, \qquad x \to \infty, \qquad x \to -\infty.$$

EXAMPLE 6   Find $\lim\limits_{x \to \infty} \dfrac{x + \sqrt{x} + 1}{\sqrt{x} + \sqrt{x + 1}}$.

By l'Hospital's Rule for ∞/∞,

$$\lim_{x \to \infty} \frac{x + \sqrt{x} + 1}{\sqrt{x} + \sqrt{x + 1}} = \lim_{x \to \infty} \frac{1 + \dfrac{1}{2\sqrt{x}}}{\dfrac{1}{2\sqrt{x}} + \dfrac{1}{2\sqrt{x + 1}}} = \infty.$$

Warning: Before using l'Hospital's Rule, check to see whether the limit is of the form 0/0 or ∞/∞. A common mistake is to use the rule when the limit is not of one of these forms.

EXAMPLE 7 Find $\displaystyle\lim_{x \to 1} \frac{\sqrt{x} - (1/x)}{x}$.

The limit has the form 0/1, so l'Hospital's Rule does not apply.

Correct: $\displaystyle\lim_{x \to 1} \frac{\sqrt{x} - (1/x)}{x} = \frac{\displaystyle\lim_{x \to 1}(\sqrt{x} - (1/x))}{\displaystyle\lim_{x \to 1} x} = \frac{0}{1} = 0.$

Incorrect:

$$\lim_{x \to 1} \frac{\sqrt{x} - (1/x)}{x} = \lim_{x \to 1} \frac{d(\sqrt{x} - (1/x))/dx}{dx/dx} = \lim_{x \to 1} \left(\frac{1}{2\sqrt{x}} + \frac{1}{x^2} \right) = \frac{3}{2}.$$

PROBLEMS FOR SECTION 5.2

In Problems 1–34, evaluate the limit using l'Hospital's Rule.

1 $\displaystyle\lim_{x \to 0} \frac{\sqrt{9 + x} - 3}{x}$

2 $\displaystyle\lim_{t \to 1^+} \frac{1/t - 1}{t^2 - 2t + 1}$

3 $\displaystyle\lim_{x \to 2} \frac{2 - \sqrt{x + 2}}{4 - x^2}$

4 $\displaystyle\lim_{t \to \infty} \frac{t + 5 - 2t^{-1} - t^{-3}}{3t + 12 - t^{-2}}$

5 $\displaystyle\lim_{y \to \infty} \frac{\sqrt{y + 1} + \sqrt{y - 1}}{y}$

6 $\displaystyle\lim_{x \to 1} \frac{\sqrt{x} - 1}{\sqrt[3]{x} - 1}$

7 $\displaystyle\lim_{x \to 0} \frac{(1 - x)^{1/4} - 1}{x}$

8 $\displaystyle\lim_{t \to 0} \left(t + \frac{1}{t} \right)((4 - t)^{3/2} - 8)$

9 $\displaystyle\lim_{t \to 0^+} \left(\frac{1}{t} + \frac{1}{\sqrt{t}} \right)(\sqrt{t + 1} - 1)$

10 $\displaystyle\lim_{x \to 0} \frac{x^2}{\sqrt{2x + 1} - 1}$

11 $\displaystyle\lim_{u \to 1} \frac{(u - 1)^3}{u^{-1} - u^2 + 3u - 3}$

12 $\displaystyle\lim_{t \to 0} \frac{2 + 1/x}{3 - 2/x}$

13 $\displaystyle\lim_{u \to 0^+} \frac{1 + 5/\sqrt{u}}{2 + 1/\sqrt{u}}$

14 $\displaystyle\lim_{u \to 0^+} \frac{3 + u^{-1/2} + u^{-1}}{2 + 4u^{-1/2}}$

15 $\displaystyle\lim_{x \to \infty} \frac{x + x^{1/2} + x^{1/3}}{x^{2/3} + x^{1/4}}$

16 $\displaystyle\lim_{t \to \infty} \frac{1 - \sqrt{t/(t + 1)}}{2 - \sqrt{(4t + 1)/(t + 2)}}$

17 $\displaystyle\lim_{t \to \infty} \frac{1 - t/(t - 1)}{1 - \sqrt{t/(t - 1)}}$

18 $\displaystyle\lim_{y \to -\infty} \frac{y + y^{-1}}{1 + \sqrt{1 - y}}$

19 $\displaystyle\lim_{x \to 1} \frac{\int_1^x (1/t)\,dt}{\int_1^x 1/(2t + 1)\,dt}$

20 $\displaystyle\lim_{x \to \infty} \frac{\int_1^x \sqrt{t + t^{-1}}\,dt}{x\sqrt{x}}$

21 $\displaystyle\lim_{x \to 0} \frac{\sin x}{x}$

22 $\displaystyle\lim_{x \to 0} \frac{1 - \cos x}{x}$

23 $\displaystyle\lim_{x \to 0} \frac{\sin (2x)}{x}$

24 $\displaystyle\lim_{x \to 0} \frac{\sin^2 x}{x}$

25 $\displaystyle\lim_{\theta \to \pi/2} \frac{\cos \theta}{\pi/2 - \theta}$

26 $\displaystyle\lim_{\theta \to \pi/2} \frac{\cos (3\theta)}{\pi/2 - \theta}$

27 $\displaystyle\lim_{\theta \to 0} \frac{\tan \theta}{\theta}$

28 $\displaystyle\lim_{\theta \to 0} \frac{\sin (2\theta)}{\sin (5\theta)}$

29 $\displaystyle\lim_{t \to 0} \frac{e^t - 1}{t}$

30 $\displaystyle\lim_{t \to 0} \frac{t^2}{e^t - t - 1}$

31 $\displaystyle\lim_{t \to 1} \frac{\ln t}{t - 1}$

32 $\displaystyle\lim_{t \to 0} \frac{\ln (t^2 + 1)}{t}$

33 $\displaystyle\lim_{x \to 1} \frac{x \ln x}{x^2 - 1}$

34 $\displaystyle\lim_{x \to 0} \frac{\sin (2x)}{\ln (x + 1)}$

In Problems 35–52, evaluate the limit by l'Hospital's Rule or otherwise.

35 $\displaystyle\lim_{x \to 1} \frac{x^{1/4} - 1}{x}$

36 $\displaystyle\lim_{x \to 1^+} \frac{\sqrt{x}}{x - 1}$

37 $\displaystyle\lim_{x \to 1^+} \frac{\sqrt{x} - 1}{x - 1}$

38 $\displaystyle\lim_{x \to \infty} \frac{x^{-1} + x^{-1/2}}{x + x^{-1/2}}$

39 $\displaystyle\lim_{x \to \infty} \frac{x + x^{-2}}{2x + x^{-2}}$

40 $\displaystyle\lim_{x \to \infty} \frac{5 + x^{-1}}{1 + 2x^{-1}}$

41 $\displaystyle\lim_{x \to \infty} \frac{4x}{\sqrt{2x^2 + 1}}$

42 $\displaystyle\lim_{x \to 0} \frac{3x^2 + x + 2}{x - 4}$

43 $\displaystyle\lim_{x \to 0} \frac{\sqrt{x + 1} - 1}{\sqrt{x + 4} - 2}$

44 $\displaystyle\lim_{x \to 0} \frac{\sqrt{x + 1} - 1}{\sqrt{x + 2} - 1}$

45 $\displaystyle\lim_{x \to 0^+} \frac{\sqrt{x + 1} + 1}{\sqrt{x + 1} - 1}$

46 $\displaystyle\lim_{x \to 0} \frac{\sqrt{x^2 + 1} - 1}{\sqrt{x + 1} - 1}$

47 $\displaystyle\lim_{x \to \infty} (x + 5)\left(\frac{1}{2x} + \frac{1}{x + 2}\right)$

48 $\displaystyle\lim_{x \to 0^+} (x + 5)\left(\frac{1}{2x} + \frac{1}{x + 2}\right)$

49 $\displaystyle\lim_{x \to 1} (x + 5)\left(\frac{1}{2x} + \frac{1}{x + 2}\right)$

50 $\displaystyle\lim_{x \to 2} \frac{x^3 - 6x - 2}{x^3 + 4}$

51 $\displaystyle\lim_{x \to 2} \frac{x^3 - 6x - 2}{x^3 - 4x}$

52 $\displaystyle\lim_{x \to 1^+} \frac{x^3 + 4x + 8}{2x^3 - 2}$

☐ **53** Suppose f and g are continuous in a neighborhood of a and $g(a) \neq 0$. Show that

$$\lim_{x \to a} \frac{\int_a^x f(t)\,dt}{\int_a^x g(t)\,dt} = \frac{f(a)}{g(a)}.$$

5.3 LIMITS AND CURVE SKETCHING

By definition, $\lim_{x \to c} f(x) = L$ means that for every *hyperreal* number x which is infinitely close but not equal to c, $f(x)$ is infinitely close to L. What does $\lim_{x \to c} f(x) = L$ tell us about $f(x)$ for *real* numbers x? It turns out that if $\lim_{x \to c} f(x) = L$, then for every real number x which is close to but not equal to c, $f(x)$ is close to L.

In the next section we shall justify the above intuitive statement by a mathematical theorem. The main difficulty is to make the word "close" precise. For the time being we shall simply illustrate the idea with some examples.

EXAMPLE 1   Consider the limit $\displaystyle\lim_{x \to 0} \frac{2/x + 1}{1/x - 1} = 2$.

This limit is evaluated by letting $x \neq 0$ be infinitesimal:

$$\frac{2/x + 1}{1/x - 1} = \frac{2 + x}{1 - x},$$

$$\lim_{x \to 0} \frac{2/x + 1}{1/x - 1} = st\left(\frac{2 + x}{1 - x}\right) = \frac{st(2 + x)}{st(1 - x)} = \frac{2 + 0}{1 - 0} = 2.$$

Let us see what happens if instead of taking x to be infinitely small we take x to be a "small" real number. We shall make a table of values of

$$f(x) = \frac{2/x + 1}{1/x - 1}$$

for various small x.

x	$f(x) = \dfrac{2/x + 1}{1/x - 1}$	$f(x)$ to four places
0.1	21/9	2.3333
0.01	201/99	2.0303
0.001	2001/999	2.0030
0.0001	20001/9999	2.0003
−0.1	19/11	1.7364
−0.01	199/101	1.9703
−0.001	1999/1001	1.9970
−0.0001	19999/10001	1.9997

We see that as x gets closer and closer to zero, $f(x)$ gets closer and closer to 2.

The table helps us to draw the graph of the curve $y = f(x)$. Although the point $(0, 2)$ is not on the graph, we know that when x is close to 0, $f(x)$ is close to 2, and draw the graph accordingly. The graph is drawn in Figure 5.3.1.

Other types of limits also give information which is useful in drawing graphs. For instance, if $\lim_{x \to c} f(x) = \infty$, then for every number x which is close to but not equal to c, the value of $f(x)$ is large. And if $\lim_{x \to \infty} f(x) = L$, then for every large real number $x, f(x)$ is close to L.

In both the above statements, if we replace "close" by "infinitely close" and "large" by "infinitely large" we get our official definition of a limit. We give two more examples.

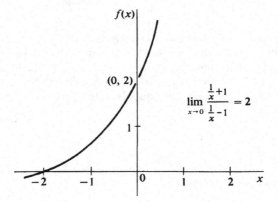

$$\lim_{x \to 0} \frac{\frac{1}{x} + 1}{\frac{1}{x} - 1} = 2$$

Figure 5.3.1

EXAMPLE 2 Consider the limit $\displaystyle\lim_{x \to 2} \frac{1}{(x - 2)^2} = \infty$.

For x infinitely close but not equal to 2, $1/(x - 2)^2$ is positive infinite. Let us make a table of values when x is a real number close to but not equal to 2.

x	$f(x)$
2.1	100
2.01	10000
2.001	1000000
1.9	100
1.99	10000
1.999	1000000

As x gets closer and closer to 2, $f(x)$ gets larger and larger.

EXAMPLE 3 $\lim\limits_{x \to \infty} \left(1 + \dfrac{1}{(x - 2)^2}\right) = 1.$

For infinitely large x, $1 + 1/(x - 2)^2$ is infinitely close to 1. Here is a table of values of $1 + 1/(x - 2)^2$ for large real x.

x	$1 + \dfrac{1}{(x - 2)^2}$
12	1.01
102	1.0001
1002	1.000001
10002	1.00000001

As x gets large, $1 + 1/(x - 2)^2$ gets close to 1. Also notice that

$$\lim\limits_{x \to -\infty} \left(1 + \frac{1}{(x - 2)^2}\right) = 1,$$

and for large negative x, $1 + 1/(x - 2)^2$ is close to 1.

In Chapter 3 we showed how to use the first and second derivatives to sketch the graph of a function which is continuous on a closed interval. In the next example we shall sketch the graph of the function $f(x) = 1 + 1/(x - 2)^2$. But this time the function is discontinuous at $x = 2$, and the domain is the whole real line except for the point $x = 2$. Our method uses not only the values but also the limits of the function and its first derivative.

EXAMPLE 4 Sketch the curve $f(x) = 1 + \dfrac{1}{(x - 2)^2}.$

The first two derivatives are

$$f'(x) = -2(x - 2)^{-3} \qquad f''(x) = 6(x - 2)^{-4}.$$

The first and second derivatives are never zero. $f(x)$ is undefined at $x = 2$. In our table we shall show the values of $f(x)$ and its first two derivatives at a point on each side of $x = 2$. We shall also show the limits of $f(x)$ and its first derivative as $x \to -\infty$, $x \to 2^-$, $x \to 2^+$, and $x \to \infty$. (We will not need the limits of $f''(x)$.)

	$f(x)$	$f'(x)$	$f''(x)$	Comments
$\lim_{x \to -\infty}$	1	0		horizontal
$x = 1$	2	2	6	increasing, $\cup$
$\lim_{x \to 2^-}$	∞	∞		vertical
$\lim_{x \to 2^+}$	∞	$-\infty$		vertical
$x = 3$	2	-2	6	decreasing, $\cup$
$\lim_{x \to \infty}$	1	0		horizontal

The first line of the table, $\lim_{x \to -\infty}$, shows that for large negative x the curve is close to 1 and its slope is nearly horizontal. The second line, $x = 1$, shows that the curve is increasing and concave upward in the interval $(-\infty, 2)$, and passes through the point $(1, 2)$ with a slope of 2. The third line, $\lim_{x \to 2^-}$, shows that just before $x = 2$ the curve is far above the x-axis and its slope is nearly vertical. Going through the table in this way, we are able to sketch the curve as in Figure 5.3.2.

The curve approaches the dotted horizontal line $y = 1$ and the dotted vertical line $x = 2$. These lines are called *asymptotes* of the curve.

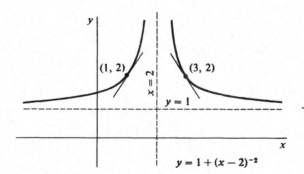

Figure 5.3.2

$$y = 1 + (x - 2)^{-2}$$

Suppose the function f and its derivative f' exist and are continuous at all but a finite number of points of an interval I. The following procedure can be used in sketching the curve $y = f(x)$.

Step 1 First carry out the procedure outlined in Section 3.9 concerning the first and second derivative.

Step 2 Compute $\lim_{x \to -\infty} f(x)$ and $\lim_{x \to \infty} f(x)$.
(They may either be real numbers, $+\infty$, $-\infty$, or may not exist.)

Step 3 At each point c of I where f is discontinuous, compute $f(c)$, $\lim_{x \to c^+} f(x)$ and $\lim_{x \to c^-} f(x)$.
(Some or all of these quantities may be undefined.)

Step 4 Compute $\lim_{x \to \infty} f'(x)$ and $\lim_{x \to -\infty} f'(x)$.

Step 5   At each point where f' is discontinuous, compute $f(c)$, $\lim_{x \to c^+} f'(x)$ and $\lim_{x \to c^-} f'(x)$.

We shall now work several more examples; the steps in computing the limits are left to the student.

EXAMPLE 5 $f(x) = x^{3/5}$.

Then
$$f'(x) = \tfrac{3}{5}x^{-2/5}, \qquad f''(x) = -\tfrac{6}{25}x^{-7/5}.$$

At the point $x = 0$, $f(x) = 0$ and $f'(x)$ does not exist. We first plot a few points, compute the necessary limits, and make a table.

	$f(x)$	$f'(x)$	$f''(x)$	Comments
$\lim\limits_{x \to -\infty}$	$-\infty$	0		horizontal
$x = -1$	-1	$3/5$	$6/25$	increasing, $\cup$
$\lim\limits_{x \to 0^-}$	0	∞		vertical
$x = 0$	0	undef.		
$\lim\limits_{x \to 0^+}$	0	∞		vertical
$x = 1$	1	$3/5$	$-6/25$	increasing, $\cap$
$\lim\limits_{x \to \infty}$	∞	0		horizontal

Figure 5.3.3 is a sketch of the curve.

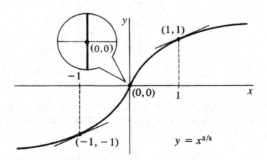

Figure 5.3.3

The behavior as x approaches $-\infty$, ∞, and zero are described by the limits we have computed. As x approaches either $-\infty$ or ∞, $f(x)$ gets large but the slope becomes more nearly horizontal. As x approaches zero the curve becomes nearly vertical, increasing from left to right, so we have a vertical tangent line at $x = 0$.

EXAMPLE 6 $f(x) = x^{4/5}$.

Then $\qquad f'(x) = \frac{4}{5}x^{-1/5}, \qquad f''(x) = -\frac{4}{25}x^{-6/5}.$

$f'(x)$ is undefined at $x = 0$. We make the table:

	$f(x)$	$f'(x)$	$f''(x)$	Comments
$\lim\limits_{x \to -\infty}$	∞	0		horizontal
$x = -1$	1	$-4/5$	$-4/25$	decreasing, $\cap$
$\lim\limits_{x \to 0^-}$	0	$-\infty$		vertical
$x = 0$	0	undef.		
$\lim\limits_{x \to 0^+}$	0	∞		vertical
$x = 1$	1	$4/5$	$-4/25$	increasing, $\cap$
$\lim\limits_{x \to \infty}$	∞	0		horizontal

With this information we can sketch the curve in Figure 5.3.4.

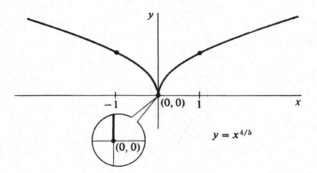

Figure 5.3.4

This time the limits of the derivative as x approaches zero show that there is a cusp at $x = 0$, with the curve decreasing when $x < 0$ and increasing when $x > 0$.

EXAMPLE 7 Sketch the curve $\quad f(x) = \dfrac{\cos x}{\sin x} \quad$ for $0 < x < 2\pi$.

$f(x)$ and $f'(x)$ are undefined at $x = \pi$ because the denominator $\sin \pi$ is zero. The first two derivatives are

$$f'(x) = -\frac{1}{\sin^2 x}, \qquad f''(x) = 2\frac{\cos x}{\sin^3 x}.$$

Thus $f'(x)$ is always negative, and $f''(x) = 0$ when $x = \pi/2, 3\pi/2$. Here is the table:

	$f(x)$	$f'(x)$	$f''(x)$	Comments
$\lim\limits_{x \to 0^+}$	∞	$-\infty$		vertical
$\pi/4$	1	$-1/2$	+	decreasing, $\cup$
$\pi/2$	0	-1	0	decreasing, inflection
$3\pi/4$	-1	$-1/2$	$-$	decreasing, $\cap$
$\lim\limits_{x \to \pi^-}$	$-\infty$	$-\infty$		vertical
$\lim\limits_{x \to \pi^+}$	∞	$-\infty$		vertical
$5\pi/4$	1	$-1/2$	+	decreasing, $\cup$
$3\pi/2$	0	-1	0	decreasing, inflection
$7\pi/4$	-1	$-1/2$	$-$	decreasing, $\cap$
$\lim\limits_{x \to 2\pi^-}$	$-\infty$	$-\infty$		vertical

Notice that the table from π to 2π is just a repeat of the table from 0 to π. This is because

$$\frac{\cos (x + \pi)}{\sin (x + \pi)} = \frac{-\cos x}{-\sin x} = \frac{\cos x}{\sin x}.$$

The curve is sketched in Figure 5.3.5.

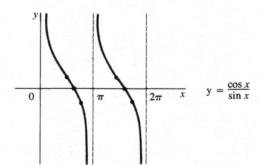

$$y = \frac{\cos x}{\sin x}$$

Figure 5.3.5

PROBLEMS FOR SECTION 5.3

1   This figure is a sketch of a curve $y = f(x)$. At which points $x = c$ do the following happen?

(a) f is discontinuous at c

(b) $\lim\limits_{x \to c^+} f(x)$ does not exist

(c) $\lim\limits_{x \to c^-} f(x)$ does not exist

(d) f is not differentiable at c

(e) $\lim\limits_{x \to c^+} f'(x)$ does not exist

(f) $\lim\limits_{x \to c^-} f'(x)$ does not exist.

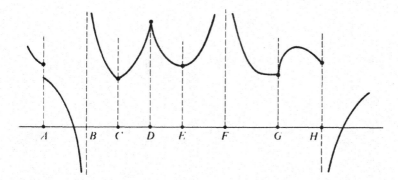

In Problems 2–42, sketch the graph of $f(x)$. Use a table of values of $f(x), f'(x), f''(x)$, and limits of $f(x)$ and $f'(x)$.

2 $f(x) = 2 - \frac{1}{2}x^2$

3 $f(x) = x^2 - 2x$

4 $f(x) = x^3 - x$

5 $f(x) = x^2 - \frac{1}{6}x^3$

6 $f(x) = \frac{1}{2}x^4 - x^2$

7 $f(x) = x^3 - \frac{1}{4}x^4$

8 $f(x) = 1 + \dfrac{1}{x}$

9 $f(x) = \dfrac{1}{2 - x}$

10 $f(x) = x^2 + \dfrac{2}{x}$

11 $f(x) = \dfrac{1}{x^2}$

12 $f(x) = x^2 + \dfrac{1}{x^2}$

13 $f(x) = \sqrt{x}$

14 $f(x) = \sqrt{2 - x}$

15 $f(x) = \dfrac{1}{\sqrt{x}}$

16 $f(x) = 1 - \dfrac{1}{\sqrt{x}}$

17 $f(x) = \sqrt[3]{x}$

18 $f(x) = 2 - (x - 1)^{1/3}$

19 $f(x) = \dfrac{x - 1}{x + 1}$

20 $f(x) = \dfrac{2x}{1 - x}$

21 $f(x) = \dfrac{1}{x^2 + 1}$

22 $f(x) = \dfrac{x}{x^2 + 1}$

23 $f(x) = \dfrac{x^2}{x^2 + 1}$

24 $f(x) = \dfrac{1}{x^2 - 1}$

25 $f(x) = \dfrac{x}{x^2 - 1}$

26 $f(x) = \dfrac{x^2}{x^2 - 1}$

27 $f(x) = x^{2/3}$

28 $f(x) = 2 + (x - 1)^{2/3}$

29 $f(x) = \sqrt{4 - x^2}$

30 $f(x) = 4\sqrt{1 - x^2}$

31 $f(x) = 1 - \sqrt{1 - x^2}$

32 $f(x) = \sqrt{x^2 - 1}$

33 $y = \dfrac{1}{\sin x}, \quad 0 < x < 2\pi$

34 $y = \dfrac{1}{\cos x}, \quad 0 < x < 2\pi$

35 $y = \tan x, \quad 0 \le x \le 2\pi$

36 $y = \tan^2 x, \quad -\pi \le x \le \pi$

37 $y = \dfrac{1}{\sin x + \cos x}, \quad 0 \le x \le 2\pi$

38 $y = \dfrac{1}{\sin x \cos x}, \quad 0 \le x \le 2\pi$

39 $f(x) = -\sqrt{x^2 - 4}$

40 $f(x) = 2 - \sqrt{x^2 + 4}$

41 $f(x) = \dfrac{1}{\sqrt{x^2 - 1}}$

42 $f(x) = \dfrac{1}{\sqrt{1 - x^2}}$

In Problems 43–55, graph the given function.

43 $f(x) = |x| - 1$ **44** $f(x) = 1 - |2x|$

45 $f(x) = |2x - 1|$ **46** $f(x) = 2 + \left|\dfrac{x}{2} - 3\right|$

47 $f(x) = 2x + |x - 2|$ **48** $f(x) = x^2 + |x|$

49 $f(x) = x^2 + |x + 1|$ **50** $f(x) = |x^2 - 1|$

51 $f(x) = \sqrt{|x|}$ **52** $f(x) = x/|x|$

53 $f(x) = x + \dfrac{x}{|x|}$ **54** $f(x) = \dfrac{x^3 - x}{|x|}$

☐ **55** $f(x) = x\sqrt{1 + 1/x^2}$

5.4 PARABOLAS

In this section we shall study the graph of the equation

$$y = ax^2 + bx + c,$$

which is a U-shaped curve called a vertical parabola. We begin with the general definition of a parabola in the plane.

Recall that the distance between a point P and a line L is the length of the perpendicular line from P to L, as in Figure 5.4.1. If we are given a line L and a point F not on L, the set of all points equidistant from L and F will form a U-shaped curve that passes midway between L and F. This curve is a parabola, shown in Figure 5.4.2.

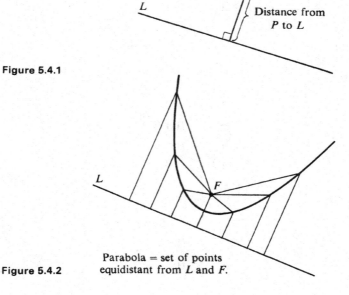

Figure 5.4.1

Figure 5.4.2   Parabola = set of points equidistant from L and F.

DEFINITION OF PARABOLA

*Given a line L and a point F not on the line, the set of all points equidistant from L and F is called the **parabola** with **directrix** L and **focus** F.*

The line through the focus perpendicular to the directrix is called the *axis* of the parabola. The point where the parabola crosses the axis is called the *vertex*. These are illustrated in Figure 5.4.3.

As we can see from the figure, the parabola is symmetric about its axis. That is, if we fold the page along the axis, the parabola will fold upon itself. The vertex is just the point halfway between the focus and directrix. It is the point on the parabola which is closest to the directrix and focus.

When a ball is thrown into the air, its path is the parabola shown in Figure 5.4.4, with the highest point at the vertex.

Telescope mirrors and radar antennae are in the shape of parabolas. This is done because all light rays coming from the direction of the axis will be reflected to a single point, the focus (see Figure 5.4.5). For the same reason, reflectors for searchlights and automobile headlights are shaped like parabolas, with the light at the focus.

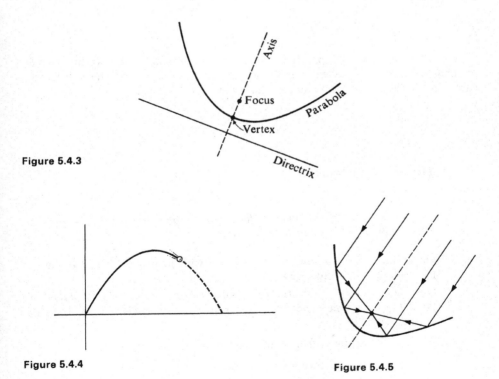

Figure 5.4.3

Figure 5.4.4

Figure 5.4.5

A parabola with a vertical axis (and horizontal directrix) is called a *vertical* parabola. The vertex of a vertical parabola is either the highest or lowest point, because it is the point closest to the directrix.

EXAMPLE 1   Find an equation for the vertical parabola with directrix $y = -1$ and focus $F(0, 1)$ (Figure 5.4.6).

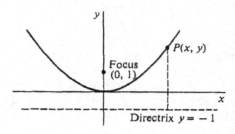

Figure 5.4.6

Given a point $P(x, y)$, the perpendicular from P to the directrix is a vertical line of length $\sqrt{(y + 1)^2}$. Thus

$$\text{distance from } P \text{ to directrix} = \sqrt{(y + 1)^2}.$$

Also, $\text{distance from } P \text{ to focus} = \sqrt{x^2 + (y - 1)^2}.$

The point P lies on the parabola exactly when these distances are equal,

$$\sqrt{(y + 1)^2} = \sqrt{x^2 + (y - 1)^2}.$$

The equation of a parabola is particularly simple if the coordinate axes are chosen so that the vertex is at the origin and the focus is on the y-axis. The parabola will then be vertical and have an equation of the form $y = ax^2$.

THEOREM 1

The graph of the equation

$$y = ax^2$$

(where $a \neq 0$) is the parabola with focus $F(0, 1/4a)$ and directrix $y = -1/(4a)$. Its vertex is $(0, 0)$, and its axis is the y-axis.

PROOF   Let us find the equation of the parabola with focus $F(0, d)$ and directrix $y = -d$, shown in Figure 5.4.7.

Our plan is to show that the equation is $y = ax^2$ where $d = 1/(4a)$. Given a point $P(x, y)$, the perpendicular from P to the directrix is a vertical line of length $\sqrt{(y + d)^2}$. Thus

$$\text{distance from } P \text{ to directrix} = \sqrt{(y + d)^2}.$$

Also, $\text{distance from } P \text{ to focus} = \sqrt{x^2 + (y - d)^2}.$

The point P lies on the parabola exactly when these distances are equal,

$$\sqrt{(y + d)^2} = \sqrt{x^2 + (y - d)^2}.$$

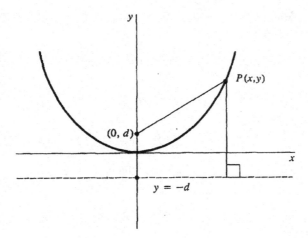

Figure 5.4.7

Simplifying we get

$$(y + d)^2 = x^2 + (y - d)^2$$
$$y^2 + 2yd + d^2 = x^2 + y^2 - 2yd + d^2$$
$$4yd = x^2$$
$$y = \frac{1}{4d}x^2.$$

Putting $a=1/4d$, we have $d=1/4a$ where $y = ax^2$ is the equation of the parabola.

Note that if a is negative, the focus will be below the x-axis and the directrix above the x-axis.

EXAMPLE 2   Find the focus and directrix of the parabola

$$y = -(1/2)\, x^2.$$

In Theorem 1, $a = -1/2$ and $d=1/4a = -\frac{1}{2}$. The focus is $F(0, -\frac{1}{2})$, and the directrix is $y = \frac{1}{2}$.

The next theorem shows that the graph of $y = ax^2 + bx + c$ is exactly like the graph of $y = ax^2$, except that its vertex is at the point (x_0, y_0) where the curve has slope zero. The focus and directrix are still at a distance $1/(4a)$ above and below the vertex.

THEOREM 2

The graph of the equation

$$y = ax^2 + bx + c$$

(where $a \neq 0$) is a vertical parabola. Its vertex is at the point (x_0, y_0) where

the curve has slope zero, the focus is $F(x_0, y_0 + 1/4a)$, *and the directrix is* $y = y_0 - 1/4a$.

PROOF We first compute x_0. The curve $y = ax^2 + bx + c$ has slope $dy/dx = 2ax + b$. The slope is zero when $2ax + b = 0$, $x = -b/2a$. Thus

$$x_0 = -b/2a.$$

Let p be the parabola with focus $F(x_0, y_0 + 1/4a)$ and directrix $y = y_0 - 1/4a$. Put $X = x - x_0$ and $Y = y - y_0$. In terms of X and Y, the focus and directrix are at

$$(X, Y) = (0, 1/4a), \qquad Y = -1/4a.$$

By Theorem 1, p has the equation

$$Y = aX^2,$$

or

$$y - y_0 = a(x - x_0)^2,$$

$$y = ax^2 - 2ax_0x + (ax_0^2 + y_0).$$

Substituting $-b/2a$ for x_0, we have

$$y = ax^2 + bx + (b^2/4a + y_0).$$

This shows that the parabola p and the curve $y = ax^2 + bx + c$ differ at most by a constant. Moreover, the point (x_0, y_0) lies on the curve. (x_0, y_0) is also the vertex of the parabola p, where $(X, Y) = (0, 0)$. Therefore the curve and the parabola are the same.

EXAMPLE 3   Find the vertex, focus and directrix of the parabola

$$y = 2x^2 - 5x + 4.$$

First find the point x_0 where the slope is 0.

$$\frac{dy}{dx} = 4x - 5.$$

Then

$$4x_0 - 5 = 0,$$

$$x_0 = \tfrac{5}{4}.$$

Substitute to find y_0.

$$y_0 = 2(x_0)^2 - 5x_0 + 4 = \tfrac{7}{8}.$$

The vertex is

$$(x_0, y_0) = (\tfrac{5}{4}, \tfrac{7}{8}).$$

We have $a = 2$, so $1/4a = \tfrac{1}{8}$. By Theorem 2, the focus is

$$\left(x_0, y_0 + \frac{1}{4a}\right) = \left(\frac{5}{4}, 1\right).$$

The directrix is

$$y = y_0 - \frac{1}{4a}, \qquad y = \frac{3}{4}.$$

The vertex, axis, focus, and directrix can be used to sketch quickly the graph of a vertical parabola.

GRAPHING A PARABOLA $y = ax^2 + bx + c$

Step 1 Make a table of values of x, y, dy/dx, and d^2y/dx^2 at $x \to -\infty$, $x = -b/2a$ (the vertex), and $x \to \infty$.

Step 2 Compute the axis, vertex, focus, and directrix, and draw them.

Step 3 Draw the two squares with sides along the axis and directrix and a corner at the focus. The two new corners level with the focus, P and Q, are on the parabola because they are equidistant from the focus and the directrix.

Step 4 Draw the diagonals of the squares through P and Q. These are the tangent lines to the parabola at P and Q. (The proof of this fact is left as a problem.)

Step 5 Draw the parabola through the vertex, P, and Q, using the table and tangent lines. The parabola should be symmetrical about the axis $x = -b/2a$. See Figure 5.4.8(a).

A *horizontal parabola* $x = ay^2 + by + c$ can be graphed by the same method with the roles of x and y interchanged, as in Figure 5.4.8(b).

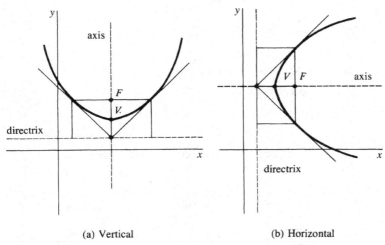

(a) Vertical   (b) Horizontal

Figure 5.4.8

EXAMPLE 2 (Continued) Sketch the parabola $y = -\frac{1}{2}x^2$.

The first two derivatives are

$$\frac{dy}{dx} = -x, \qquad \frac{d^2y}{dx^2} = -1.$$

The only critical point is at $x = 0$. The table of values follows.

x	y	dy/dx	d^2y/dx^2	Comments
$\lim\limits_{x \to -\infty}$	$-\infty$	∞		vertical
$x = 0$	0	0	-1	max, $\cap$
$\lim\limits_{x \to \infty}$	$-\infty$	$-\infty$		vertical

The parabola is drawn in Figure 5.4.9, using Steps 1–5.

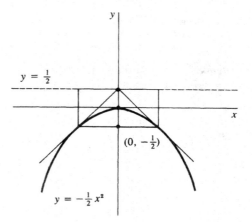

Figure 5.4.9

EXAMPLE 3 (Continued)   Sketch the parabola $y = 2x^2 - 5x + 4$.

The first two derivatives are

$$\frac{dy}{dx} = 4x - 5, \qquad \frac{d^2y}{dx^2} = 4.$$

The only critical point is at the vertex, where $x = \frac{5}{4}$. The table of values follows.

x	y	dy/dx	d^2y/dx^2	Comments
$\lim\limits_{x \to -\infty}$	∞	$-\infty$		vertical
$5/4$	$7/8$	0	$+$	min, $\cup$
$\lim\limits_{x \to \infty}$	∞	∞		vertical

The parabola is drawn in Figure 5.4.10, again using Steps 1–5.

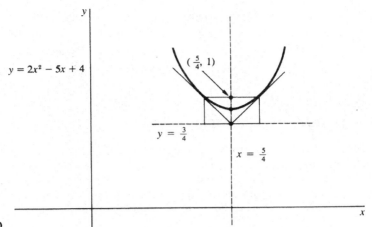

Figure 5.4.10

We can now sketch the graph of any equation of the form

$$Ax^2 + Dx + Ey + F = 0.$$

In the ordinary case where both A and E are different from zero, proceed as follows. First, solve the equation for y, obtaining the new equation

$$y = -\frac{A}{E}x^2 - \frac{D}{E}x - \frac{F}{E}.$$

Second, use the method in this section to sketch the graph, which will be a vertical parabola. There are also two *degenerate cases*. If $A = 0$, the graph is a straight line. If $E = 0$, then y does not appear at all, and the graph is either two vertical lines, one vertical line, or empty.

We can also sketch the graph of any equation of the form

$$Cy^2 + Dx + Ey + F = 0.$$

In the ordinary case where C and D are different from zero, the graph will be a horizontal parabola.

PROBLEMS FOR SECTION 5.4

In Problems 1–14, find the focus and directrix, and sketch the given parabola.

1	$y = 2x^2$	**2**	$y = \frac{1}{2}x^2$
3	$y = -x^2$	**4**	$y = 2 - x^2$
5	$y = x^2 - 2x$	**6**	$y = x^2 + 2x + 1$
7	$y = 2x^2 + x - 2$	**8**	$y = x^2 - x + 1$
9	$y = 3 + x - x^2$	**10**	$y = 1 - x - x^2$
11	$y = \frac{1}{2}x^2 + x - 1$	**12**	$y = \frac{1}{3}x^2 - x$
13	$y = (x - 2)^2$	**14**	$y = 2(x + 1)^2$
15	$x = y^2$	**16**	$x = 2y^2 - 4$
17	$x = -y^2 + y + 1$	**18**	$x = 3 - (y - 2)^2$

19   Find the equation of the parabola with directrix $y = 0$ and focus $F(2, 2)$.
20   Find the equation of the parabola with directrix $y = -1$ and focus $F(0, 0)$.
21   Find the focus of the parabola with directrix $y = 1$ and vertex $(1, 2)$.
22   Find the equation of the parabola with focus $(-1, -1)$ and vertex $(-1, 0)$.

5.5 ELLIPSES AND HYPERBOLAS

In this section we shall study two important types of curves, the ellipses and hyperbolas. The intersection of a circular cone and a plane will always be either a parabola, an ellipse, a hyperbola, or one of three degenerate cases—one line, two lines, or a point. For this reason, parabolas, ellipses, and hyperbolas are called *conic sections*. We begin with the definition of an ellipse in the plane.

DEFINITION OF ELLIPSE

> *Given two points, F_1 and F_2, and a constant, L, the **ellipse** with foci F_1 and F_2 and length L is the set of all points the sum of whose distances from F_1 and F_2 is equal to L.*

If the two foci F_1 and F_2 are the same, the ellipse is just the circle with center at the focus and diameter L. Circles are discussed in Section 1.1.

We shall concentrate on the case where the foci F_1 and F_2 are different. The ellipse will be an oval curve shown in Figure 5.5.1. The orbit of a planet is an ellipse with the sun at one focus. The eye sees a tilted circle as an ellipse.

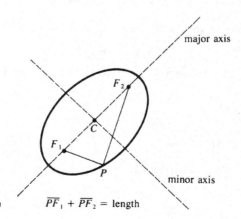

Figure 5.5.1   Ellipse $\overline{PF_1} + \overline{PF_2} = $ length

The line through the foci F_1 and F_2 is called the *major axis* of the ellipse. The point on the major axis halfway between the foci is called the *center*. The line through the center perpendicular to the major axis is called the *minor axis*.

An ellipse is symmetric about both its major and its minor axes. That is, for any point P on the ellipse, the mirror image of P on the other side of either axis is also on the ellipse. The equation of an ellipse has a simple form when the major and minor axes are chosen for the x-axis and y-axis.

THEOREM 1

For any positive a and b, the graph of the equation

$$\frac{x^2}{a^2} + \frac{y^2}{b^2} = 1$$

is an ellipse with its center at the origin. There are three cases:

(i) $a = b$. *The ellipse is a circle of radius a.*

(ii) $a > b$. *This is a horizontal ellipse, whose major axis is the x-axis, and whose minor axis is the y-axis. The length is 2a. The foci are at $(-c, 0)$ and $(c, 0)$, where c is found by*

$$c^2 = a^2 - b^2.$$

(iii) $a < b$. *This is a vertical ellipse whose major axis is the y-axis and whose minor axis is the x-axis. The length is 2b. The foci are at $(0, -c)$ and $(0, c)$, where c is found by*

$$c^2 = b^2 - a^2.$$

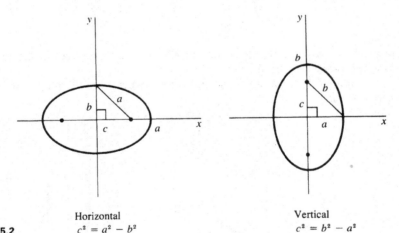

Horizontal
$c^2 = a^2 - b^2$

Vertical
$c^2 = b^2 - a^2$

Figure 5.5.2

· This theorem is illustrated by Figure 5.5.2. Here is the proof in case (ii), $a > b$. A point $P(x, y)$ is on the ellipse with foci $(-c, 0)$, $(c, 0)$ and length $2a$ if and only if the sum of the distances from P to the foci is $2a$. That is,

$$\sqrt{(x + c)^2 + y^2} + \sqrt{(x - c)^2 + y^2} = 2a.$$

Rewrite this as

$$\sqrt{(x - c)^2 + y^2} = 2a - \sqrt{(x + c)^2 + y^2}.$$

Square both sides:

$$x^2 - 2cx + c^2 + y^2 = 4a^2 - 4a\sqrt{(x + c)^2 + y^2} + x^2 + 2cx + c^2 + y^2.$$

Simplify:

$$a\sqrt{(x + c)^2 + y^2} = a^2 + cx.$$

Square both sides again:

$$a^2(x^2 + 2cx + c^2 + y^2) = a^4 + 2a^2cx + c^2x^2.$$

Collect the x^2 and y^2 terms and simplify.

$$x^2(a^2 - c^2) + y^2(a^2) = a^4 - a^2c^2 = a^2(a^2 - c^2).$$

Using the equation $b^2 = a^2 - c^2$, write this as

$$x^2b^2 + y^2a^2 = a^2b^2.$$

Finally, divide by a^2b^2 to obtain the required equation

$$\frac{x^2}{a^2} + \frac{y^2}{b^2} = 1.$$

Setting $x = 0$ we see that the ellipse meets the y-axis at the two points $y = \pm b$. Also, it meets the x-axis at $x = \pm a$. Since all terms are ≥ 0, at every point on the ellipse we have

$$\frac{x^2}{a^2} \leq 1, \qquad -a \leq x \leq a$$

and

$$\frac{y^2}{b^2} \leq 1, \qquad -b \leq y \leq b.$$

Using these facts we can easily sketch the ellipse. It is an oval curve inscribed in the rectangle bounded by the lines $x = \pm a, y = \pm b$.

Figure 5.5.3 shows a *horizontal* ellipse (where $a > b$) and a *vertical* ellipse (where $a < b$).

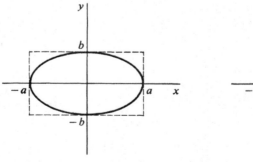

Horizontal ellipse   Vertical ellipse

Figure 5.5.3

EXAMPLE 1   Sketch the curve $\dfrac{x^2}{9} + y^2 = 1$.

The curve is an ellipse that cuts the x-axis at ± 3 and the y-axis at ± 1. To sketch the curve, we first draw the rectangle $x = \pm 3, y = \pm 1$ with dotted lines and then inscribe the ellipse in the rectangle. The ellipse, shown in Figure 5.5.4, is horizontal.

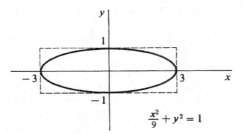

$$\frac{x^2}{9} + y^2 = 1$$

Figure 5.5.4

EXAMPLE 2 Sketch the curve $4x^2 + y^2 = 9$ and find the foci.

The equation may be rewritten as

$$\tfrac{4}{9}x^2 + \tfrac{1}{9}y^2 = 1.$$

The graph (Figure 5.5.5) is a vertical ellipse cutting the x-axis at $\pm\tfrac{3}{2}$ and the y-axis at ± 3.

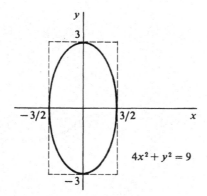

Figure 5.5.5

By Theorem 1, the foci are on the y-axis at $(0, \pm c)$. We compute c from the equation

$$c^2 = b^2 - a^2.$$

a and b are the x and y intercepts of the ellipse, $a = \tfrac{3}{2}$, $b = 3$. Thus

$$c^2 = 3^2 - (\tfrac{3}{2})^2 = \tfrac{27}{4}$$
$$c = \sqrt{\tfrac{27}{2}} \sim 2.598.$$

The foci are at $(0, \pm 2.598)$.

We turn next to the hyperbola. A hyperbola, like an ellipse, has two foci. However, the distances between the foci and a point on the hyperbola must have a constant difference instead of a constant sum.

DEFINITION OF HYPERBOLA

Given two distinct points, F_1 and F_2, and a constant, l, the **hyperbola** *with foci F_1 and F_2 and difference l is the set of all points the difference of whose distances from F_1 and F_2 is equal to l.*

In this definition, l must be a positive number less than the distance between the foci. A hyperbola will have two separate branches, each shaped like a rounded V. On one branch the points are closer to F_1 than F_2; and on the other branch they are closer to F_2 than F_1. Figure 5.5.6 shows a typical hyperbola. The path of a comet on an orbit that will escape the solar system is a hyperbola with the sun at one focus. The shadow of a cylindrical lampshade on a wall is a hyperbola (the section of the light cone cut by the wall).

The line through the foci is the *transverse axis* of the hyperbola, and the point on the axis midway between the foci is the *center*. The hyperbola crosses the transverse axis at two points called the *vertices*. The line through the center perpendicular to the transverse axis is the *conjugate axis*. The hyperbola never crosses its conjugate axis. A hyperbola is symmetric about both axes. A simple equation is obtained when the transverse and conjugate axes are chosen for the coordinate axes.

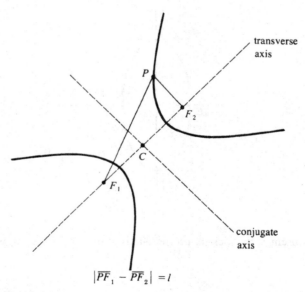

$$\left| \overline{PF}_1 - \overline{PF}_2 \right| = l$$

Figure 5.5.6 Hyperbola

THEOREM 2

For any positive a and b, the graph of the equation

$$\frac{y^2}{b^2} - \frac{x^2}{a^2} = 1$$

is a hyperbola with its center at the origin. Its transverse axis is the y-axis, and its conjugate axis is the x-axis. *The vertices are at* $(0, \pm b)$, *and the foci are at* $(0, \pm c)$, *where c is found by*

$$a^2 + b^2 = c^2.$$

The graph of the equation

$$\frac{x^2}{a^2} - \frac{y^2}{b^2} = 1$$

is a hyperbola with similar properties with the roles of x, a and y, b reversed. The proof of Theorem 2 uses a computation like the proof of Theorem 1 on ellipses and is omitted.

Using derivatives and limits, we can get additional information that is helpful in sketching the graph of a hyperbola. By solving the equation

$$\frac{y^2}{b^2} - \frac{x^2}{a^2} = 1$$

for y as a function of x, we see that the upper and lower branches have the equations

$$\text{upper branch:} \quad y = \frac{b}{a}\sqrt{a^2 + x^2},$$

$$\text{lower branch:} \quad y = -\frac{b}{a}\sqrt{a^2 + x^2}.$$

We concentrate on the upper branch. Its first two derivatives, after some algebraic simplification, come out to be

$$\frac{dy}{dx} = \frac{bx}{a\sqrt{a^2 + x^2}}, \qquad \frac{d^2y}{dx^2} = ab(a^2 + x^2)^{-3/2}.$$

Thus the first derivative is zero only at $x = 0$ (the vertex), and the second derivative is always positive. We have the following table of values for the upper branch.

x	y	dy/dx	d^2y/dx^2	Comments
$\lim\limits_{x \to -\infty}$	∞	$-b/a$	0	decreasing
0	b	0	b/a^2	minimum, $\cup$
$\lim\limits_{x \to \infty}$	∞	b/a	0	increasing

All the limit computations are easy except for dy/dx, which we work out for $x \to \infty$. Let H be positive infinite.

$$\lim_{x \to \infty} dy/dx = \lim_{x \to \infty} \frac{bx}{a\sqrt{a^2 + x^2}}$$

$$= st\left[\frac{bH}{a\sqrt{a^2 + H^2}}\right]$$

$$= st\left[\frac{b}{a\sqrt{a^2 H^{-2} + 1}}\right] = \frac{b}{a}.$$

We carry out a similar computation for the limit as $x \to -\infty$.

$$\lim_{x \to -\infty} \frac{dy}{dx} = \lim_{x \to -\infty} \frac{bx}{a\sqrt{a^2 + x^2}}$$

$$= st\left[\frac{b(-H)}{a\sqrt{a^2 + (-H)^2}}\right]$$

$$= st\left[\frac{-b}{a\sqrt{a^2 H^{-2} + 1}}\right] = -\frac{b}{a}.$$

The table shows that the upper branch is almost a straight line with slope $-b/a$ for large negative x and almost a straight line with slope b/a for large positive x. In fact, we shall show now that the lines

$$y = bx/a, \qquad y = -bx/a$$

are asymptotes of the hyperbola. That is, as x approaches ∞ or $-\infty$, the distance between the line and the hyperbola approaches zero. We show that the upper branch approaches the line $y = bx/a$ as $x \to \infty$; that is,

$$\lim_{x \to \infty} \left[\frac{b}{a}\sqrt{a^2 + x^2} - \frac{bx}{a}\right] = 0.$$

Let H be positive infinite. Then

$$\frac{b}{a}\sqrt{a^2 + H^2} - \frac{bH}{a} = \frac{b}{a}[\sqrt{a^2 + H^2} - H]$$

$$= \frac{b}{a}\left[\frac{(\sqrt{a^2 + H^2} - H)(\sqrt{a^2 + H^2} + H)}{\sqrt{a^2 + H^2} + H}\right]$$

$$= \frac{b}{a}\frac{a^2 + H^2 - H^2}{\sqrt{a^2 + H^2} + H}$$

$$= ab(\sqrt{a^2 + H^2} + H)^{-1}.$$

This is infinitesimal, so the limit is zero. Here are the steps for graphing a hyperbola $y^2/b^2 - x^2/a^2 = 1$.

GRAPHING A HYPERBOLA $\quad \dfrac{y^2}{b^2} - \dfrac{x^2}{a^2} = 1$

Step 1 Compute the values of a and b from the equation. Draw the rectangle with sides $x = \pm a, y = \pm b$.

Step 2 Draw the diagonals of the rectangle. They will be the asymptotes.

Step 3 Mark the vertices of the hyperbola at the points $(0, \pm b)$.

Step 4 Draw the upper and lower branches of the hyperbola. The upper branch has a minimum at the vertex $(0, b)$, is concave upward, and approaches the diagonal asymptotes from above. The lower branch is a mirror image. See Figure 5.5.7.

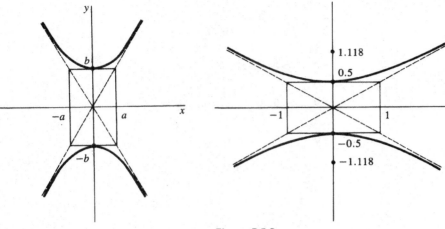

Figure 5.5.7   **Figure 5.5.8**

A hyperbola of the form

$$\frac{x^2}{a^2} - \frac{y^2}{b^2} = 1$$

is graphed in a similar manner, but with the roles of x and y reversed. There is a left branch and a right branch, which are vertical at the vertices $(\pm a, 0)$.

EXAMPLE 3 Sketch the hyperbola $4y^2 - x^2 = 1$ and find its foci.

First compute a and b.

$$4y^2 = y^2/b^2, \qquad b = \tfrac{1}{2}$$
$$x^2 = x^2/a^2, \qquad a = 1.$$

The rectangle has sides $x = \pm 1$, $y = \pm\tfrac{1}{2}$, and the vertices are at $(0, \pm\tfrac{1}{2})$. The hyperbola is sketched using Steps 1–4 in Figure 5.5.8. The foci are at $(0, \pm c)$ where

$$c^2 = a^2 + b^2 = 1^2 + (\tfrac{1}{2})^2 = 1.25$$
$$c = \sqrt{1.25} \sim 1.118.$$

Using the method of this section, we can sketch the graph of any equation of the form

$$Ax^2 + Cy^2 + F = 0.$$

In the ordinary case where A, C, and F are all different from zero, rewrite the equation as

$$A_1 x^2 + C_1 y^2 = 1,$$

where $A_1 = -A/F$, $C_1 = -C/F$. There are four cases depending on the signs of A_1 and C_1, which are listed in Table 5.5.1.

Table 5.5.1

A_1	C_1	Graph of $A_1x^2 + C_1y^2 = 1$	
>0	>0	ellipse	$\dfrac{x^2}{a^2} + \dfrac{y^2}{b^2} = 1$
>0	<0	hyperbola	$\dfrac{x^2}{a^2} - \dfrac{y^2}{b^2} = 1$
<0	>0	hyperbola	$\dfrac{y^2}{b^2} - \dfrac{x^2}{a^2} = 1$
<0	<0	empty	

If one or two of A, C, and F are zero, the graph will be degenerate (two lines, one line, a point, or empty).

PROBLEMS FOR SECTION 5.5

In Problems 1–12, find the foci and sketch the given ellipse or hyperbola.

1 $x^2 + 4y^2 = 1$ 2 $x^2 + \frac{1}{4}y^2 = 1$

3 $\frac{1}{4}x^2 + 4y^2 = 1$ 4 $\frac{1}{25}x^2 + \frac{1}{9}y^2 = 1$

5 $9x^2 + 4y^2 = 16$ 6 $x^2 + 9y^2 = 4$

7 $y^2 - 4x^2 = 1$ 8 $y^2 - x^2 = 4$

9 $9y^2 - x^2 = 4$ 10 $4y^2 - 4x^2 = 1$

11 $x^2 - y^2 = 1$ 12 $\dfrac{x^2}{9} - \dfrac{y^2}{4} = 1$

13 Prove that the hyperbola $x^2/a^2 - y^2/b^2 = 1$ has the two asymptotes $y = bx/a$ and $y = -bx/a$.

5.6 SECOND DEGREE CURVES

A *second degree equation* is an equation of the form

(1) $Ax^2 + Bxy + Cy^2 + Dx + Ey + F = 0.$

The graph of such an equation will be a conic section: a parabola, ellipse, hyperbola, or one of several degenerate cases. In Section 5.4 we saw that the graph of a second degree equation of one of the forms

(2) $Ax^2 + Dx + Ey + F = 0$

or

(3) $Cy^2 + Dx + Ey + F = 0$

is a parabola or degenerate. In Section 5.5 we saw that the graph of a second degree equation of the form

(4) $$Ax^2 + Cy^2 + F = 0$$

is an ellipse, a hyperbola, or degenerate.

In this and the next section we shall see how to describe and sketch the graph of any second degree equation. We will begin with the Discriminant Test, which shows at once whether a nondegenerate curve is a parabola, ellipse, or hyperbola. The next topic in this section will be translation of axes, which can change any second degree equation with no xy-term,

(5) $$Ax^2 + Cy^2 + Dx + Ey + F = 0,$$

into an equation of one of the simple forms (2), (3), or (4).

In the following section we will study rotation of axes, which can change any second degree equation into an equation of the form (5) with no xy-term. We will then be able to deal with any second degree equation by using first rotation and then translation of axes.

Here is the Discriminant Test.

DEFINITION

The quantity $B^2 - 4AC$ is called the **discriminant** of the equation

$$Ax^2 + Bxy + Cy^2 + Dx + Ey + F = 0.$$

DISCRIMINANT TEST

If we ignore the degenerate cases, the graph of a second degree equation is:

A parabola if the discriminant is zero.
An ellipse if the discriminant is negative.
A hyperbola if the discriminant is positive.

For example, the equation

$$xy - 1 = 0$$

has positive discriminant $1^2 - 4 \cdot 0 = 1$, and its graph is a hyperbola. The equation

$$2x^2 + xy + y^2 - 1 = 0$$

has negative discriminant $1^2 - 4 \cdot 2 \cdot 1 = -7$, and its graph is an ellipse.

The degenerate graphs that can arise are: two straight lines, one straight line, one point, and the empty graph. The Discriminant Test alone does not tell whether or not the graph is degenerate. However, a degenerate case can usually be recognized when one tries to sketch the graph. For the remainder of this section we shall ignore the degenerate cases.

We now turn to the method of Translation of Axes. This method is useful for graphing a second degree equation with no xy-term,

$$Ax^2 + Cy^2 + Dx + Ey + F = 0.$$

If A or C is zero, the graph will be a horizontal or vertical parabola, which can be graphed by the method of Section 5.4. If both A and C are nonzero, the graph turns out to be an ellipse or hyperbola with horizontal and vertical axes X and Y, as in

Figure 5.6.1. In the method of Translation of Axes, we take X and Y as a new pair of coordinate axes and get a new equation for the curve in the simple form

$$AX^2 + CY^2 + F_1 = 0.$$

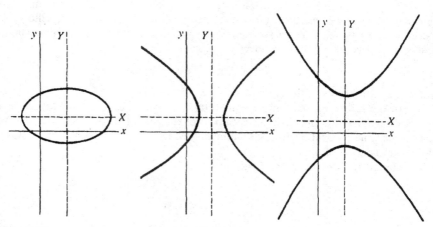

Figure 5.6.1

This curve can be sketched as in Section 5.5. The name "Translation of Axes" means that the original coordinate axes x and y are replaced by new coordinate axes X and Y, which are parallel to the original axes.

The new axes are found using a procedure from algebra called "completing the squares." This procedure changes an expression like $Ax^2 + Dx$ into a perfect square plus a constant.

FORMULA FOR COMPLETING THE SQUARES

Let A be different from zero. Then

$$Ax^2 + Dx = AX^2 + K,$$

where
$$X = x + \frac{D}{2A}, \qquad K = \frac{-D^2}{4A}.$$

For example,

$$4x^2 - 3x = 4X^2 - 9/16$$

where $X = x - \frac{3}{8}$.

We shall illustrate the method of Translation of Axes with an example and then describe the method in general.

EXAMPLE 1 Sketch the curve $4x^2 - y^2 - 16x - 2y + 11 = 0.$

Step 1 Apply the Discriminant Test to determine the type of curve.

$$B^2 - 4AC = 0^2 - 4 \cdot 4 \cdot (-1) = 16.$$

The discriminant is positive, so the graph is a hyperbola.

Step 2 Simplify by completing the squares. This is done by putting

$$X = x + \frac{D}{2A}, \qquad Y = y + \frac{E}{2C}$$

and writing the original equation in terms of X and Y.

$$X = x + \frac{-16}{2 \cdot 4} = x - 2, \qquad x = X + 2$$

$$Y = y + \frac{-2}{2 \cdot (-1)} = y + 1, \qquad y = Y - 1$$

$$4(X + 2)^2 - (Y - 1)^2 - 16(X + 2) - 2(Y - 1) + 11 = 0$$

$$4(X^2 + 4X + 4) - 16(X + 2) - (Y^2 - 2Y + 1) - 2(Y - 1) + 11 = 0.$$

The X and Y terms cancel, and

$$4X^2 + 16 - 32 - Y^2 - 1 + 2 + 11 = 0,$$

$$4X^2 - Y^2 - 4 = 0.$$

Step 3 Draw dotted lines for the X and Y axes, and sketch the curve as in Section 5.5. This is a hyperbola in the (X, Y)-plane. The X-axis is the line $Y = 0$, or $y = -1$. The Y-axis is the line $X = 0$, or $x = 2$. The graph is shown in Figure 5.6.2.

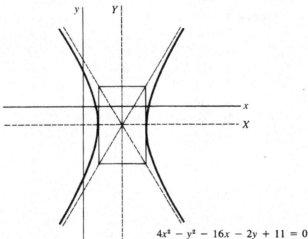

$$4x^2 - y^2 - 16x - 2y + 11 = 0$$

Figure 5.6.2 Example 1

METHOD OF TRANSLATION OF AXES

When to Use To graph an equation of the form $Ax^2 + Cy^2 + Dx + Ey + F = 0$ where A and C are both nonzero.

Step 1 Use the Discriminant Test to determine the type of curve.

Step 2 *Completing the Squares: Put*

$$X = x + \frac{D}{2A}, \qquad Y = y + \frac{E}{2C}$$

and rewrite the original equation in terms of X and Y. The new equation will have the simple form

$$Ax^2 + Cy^2 + F_1 = 0,$$

where F_1 is a new constant.

Step 3 *Draw dotted lines for the X and Y axes and sketch the curve as in Section 5.5.*

PROBLEMS FOR SECTION 5.6

In Problems 1–6, given that the graph is nondegenerate, use the Discriminant Test to determine whether the graph is a parabola, ellipse, or hyperbola.

1    $x^2 + 2xy - 3y^2 + 5x + 6y - 100 = 0$

2    $4x^2 - 8xy + 6y^2 + 10x - 2y - 20 = 0$

3    $4x^2 + 4xy + y^2 + 7x + 8y = 0$

4    $9x^2 + 6xy + y^2 + 6x - 22 = 0$

5    $x^2 + 5xy + 10y^2 - 16 = 0$

6    $4xy + 5x - 10y + 1 = 0$

In Problems 7–18, use the method of Translation of Axes to sketch the curve.

7	$x^2 + y^2 - 4x + 3 = 0$	8	$x^2 + y^2 + 2x - 6y + 6 = 0$
9	$x^2 - y^2 + 4x - 2y + 2 = 0$	10	$-x^2 + y^2 + 8x - 6y - 16 = 0$
11	$x^2 + 4y^2 - 4x + 24y + 36 = 0$	12	$4x^2 - 9y^2 + 8x + 18y - 41 = 0$
13	$9x^2 - 4y^2 - 36x - 24y - 36 = 0$	14	$-x^2 + 4y^2 + 16y + 12 = 0$
15	$-x^2 + 3y^2 + 8x + 30y + 56 = 0$	16	$5x^2 + 2y^2 + 10x + 12y + 28 = 0$
17	$16x^2 + 9y^2 - 320x - 108y + 1780 = 0$		
18	$25x^2 + 4y^2 + 250x - 40y + 625 = 0$		

5.7 ROTATION OF AXES

We have seen how to graph any second degree equation with no xy-term. These graphs are parabolas, ellipses, or hyperbolas with vertical and horizontal axes. When the equation has a nonzero xy-term, the graph will have diagonal axes. By rotating the axes, one can get new coordinate axes in the proper direction. The method will give us a new equation that has no xy-term and can be graphed by our previous method.

Suppose the x and y axes are rotated counterclockwise by an angle α, and the new coordinate axes are called X and Y, as in Figure 5.7.1. A point P in the plane will have a pair of coordinates (x, y) in the old coordinate system and (X, Y) in the new coordinate system. The old and new coordinates of P are related to each other by the equations for rotation of axes.

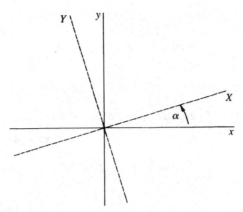

Figure 5.7.1 Rotation of Axes

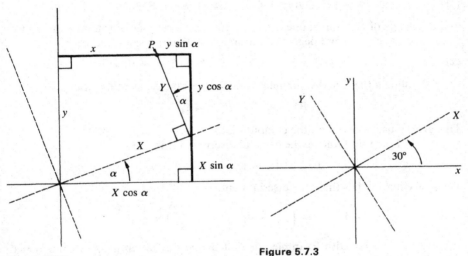

Figure 5.7.2

Figure 5.7.3

EQUATIONS FOR ROTATION OF AXES

$$x = X \cos \alpha - Y \sin \alpha, \qquad y = X \sin \alpha + Y \cos \alpha.$$

These equations can be seen directly from Figure 5.7.2. If we substitute the equations for rotation of axes into a second degree equation in x and y, we get a new second degree equation in the coordinates X and Y.

EXAMPLE 1 Find the equation of the curve

$$xy - 4 = 0,$$

with respect to the new coordinate axes X and Y formed by a counterclockwise rotation of 30 degrees (Figure 5.7.3).

In this example,

$$\alpha = 30°, \qquad \sin\alpha = \frac{1}{2}, \qquad \cos\alpha = \frac{\sqrt{3}}{2}.$$

Thus
$$x = -\frac{\sqrt{3}}{2}X - \frac{1}{2}Y, \qquad y = \frac{1}{2}X + \frac{\sqrt{3}}{2}Y.$$

Substitute into the original equation and collect terms.

$$xy - 4 = 0,$$

$$\left(\frac{\sqrt{3}}{2}X - \frac{1}{2}Y\right)\cdot\left(\frac{1}{2}X + \frac{\sqrt{3}}{2}Y\right) - 4 = 0,$$

$$\frac{\sqrt{3}}{4}X^2 + \frac{1}{2}XY - \frac{\sqrt{3}}{4}Y^2 - 4 = 0.$$

Given any second degree equation

(1) $$Ax^2 + Bxy + Cy^2 + Dx + Ey + F = 0$$

and any angle of rotation α, one can substitute the equations of rotation and collect terms to get a new second degree equation in the X and Y coordinates,

(2) $$A_1X^2 + B_1XY + C_1Y^2 + D_1X + E_1Y + F_1 = 0.$$

It can be shown that the discriminant is unchanged by the rotation; that is,

$$B^2 - 4AC = B_1^2 - 4A_1C_1.$$

This gives a useful check on the computations.

In Example 1 above, the original discriminant is

$$B^2 - 4AC = 1^2 - 4 \cdot 0 \cdot 0 = 1.$$

The new equation has the same discriminant,

$$B_1^2 - 4A_1C_1 = \left(\frac{1}{2}\right)^2 - 4\left(\frac{\sqrt{3}}{4}\right)\left(-\frac{\sqrt{3}}{4}\right) = \frac{1}{4} + \frac{3}{4} = 1.$$

The trouble with Example 1 is that the new equation is more complicated than the original equation, and in particular there is still a nonzero XY-term. We would like to be able to choose the angle of rotation α so that the new equation has no XY-term, because we could then sketch the curve. The next theorem tells us which angle of rotation is needed.

THEOREM 1

Given a second degree equation

$$Ax^2 + Bxy + Cy^2 + Dx + Ey + F = 0$$

with B nonzero. Rotate the coordinate axes counterclockwise through an angle α for which

$$\cot(2\alpha) = \frac{A - C}{B}.$$

Then the equation

$$A_1 X^2 + B_1 XY + C_1 Y^2 + D_1 X + E_1 Y + F_1 = 0$$

with respect to the new coordinate axes X and Y has XY-term $B_1 = 0$.

This theorem can be proved as follows. When the rotation equations are substituted and terms collected, the XY coefficient B_1 comes out to be

$$B_1 = B(\cos^2 \alpha - \sin^2 \alpha) - 2(A - C)\sin \alpha \cos \alpha.$$

From trigonometry,

$$\cos^2 \alpha - \sin^2 \alpha = \cos (2\alpha), \qquad 2 \sin \alpha \cos \alpha = \sin (2\alpha).$$

Thus $$B_1 = B \cos (2\alpha) - (A - C) \sin (2\alpha).$$

So $B_1 = 0$ if and only if

$$B \cos (2\alpha) - (A - C) \sin (2\alpha) = 0,$$

$$\frac{\cos (2\alpha)}{\sin (2\alpha)} - \frac{A - C}{B} = 0,$$

or $$\cot (2\alpha) = \frac{A - C}{B}.$$

As shown in Figure 5.7.4, α is the angle between the original coordinate axes and the axes of the parabola, ellipse, or hyperbola.

We are now ready to use rotation of axes to sketch a second degree curve. We illustrate the method for the curve introduced in Example 1.

EXAMPLE 2 Sketch the curve $xy - 4 = 0$.

Step 1 Apply the Discriminant Test to find the type of curve.

$$B^2 - 4AC = 1^2 - 4 \cdot 0 \cdot 0 = 1.$$

The discriminant is positive, so the curve is a hyperbola.

Step 2 Find an angle α with

$$\cot (2\alpha) = \frac{A - C}{B}.$$

$$\cot (2\alpha) = \frac{0 - 0}{1} = 0.$$

$$2\alpha = 90°, \qquad \alpha = 45°.$$

Step 3 Change coordinate axes using the rotation equations.

$$\cos \alpha = \frac{\sqrt{2}}{2}, \qquad \sin \alpha = \frac{\sqrt{2}}{2}.$$

$$x = X \cos \alpha - Y \sin \alpha = \frac{\sqrt{2}}{2} X - \frac{\sqrt{2}}{2} Y.$$

$$y = X \sin \alpha + Y \cos \alpha = \frac{\sqrt{2}}{2} X + \frac{\sqrt{2}}{2} Y.$$

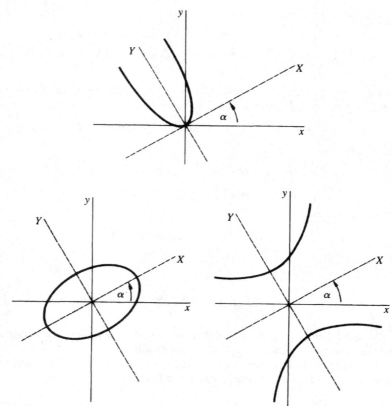

Figure 5.7.4

Substituting, we get

$$xy - 4 = 0,$$

$$\left(\frac{\sqrt{2}}{2}X - \frac{\sqrt{2}}{2}Y\right) \cdot \left(\frac{\sqrt{2}}{2}X + \frac{\sqrt{2}}{2}Y\right) - 4 = 0,$$

$$\frac{1}{2}X^2 - \frac{1}{2}Y^2 - 4 = 0.$$

As a check, the discriminant is still $0^2 - 4 \cdot (\frac{1}{2}) \cdot (-\frac{1}{2}) = 1$.

Step 4   Draw the X and Y axes as dotted lines and sketch the curve.

The new axes are found by rotating the old axes by $\alpha = 45°$. The curve is shown in Figure 5.7.5.

METHOD OF ROTATION OF AXES

When to Use   *To graph an equation of the form $Ax^2 + Bxy + Cy^2 + Dx + Ey + F = 0$ where B is nonzero.*

Step 1   *Use the Discriminant Test to determine the type of curve.*

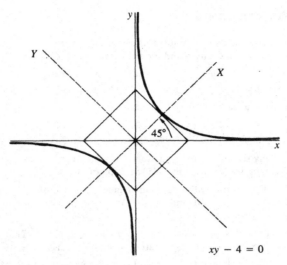

Figure 5.7.5 Example 2

Step 2 *Find an angle α with*

$$\cot(2\alpha) = \frac{A - C}{B}.$$

Step 3 *Change coordinate axes using the Rotation Equations. The new equation has the form*

$$A_1 X^2 + C_1 Y^2 + D_1 X + E_1 Y + F_1 = 0,$$

where $x = X \cos \alpha - Y \sin \alpha, \qquad y = X \sin \alpha + Y \cos \alpha.$

Step 4 *Draw the X- and Y-axes by rotating the old axes through the angle α. The curve can now be sketched by our previous method, using Translation of Axes if necessary.*

Here is an overall summary of the use of rotations and translations of axes. The problem is to graph an equation of the form

$$Ax^2 + Bxy + Cy^2 + Dx + Ey + F = 0.$$

By Rotation of Axes, we get a new equation of the simpler form

$$A_1 X^2 + C_1 Y^2 + D_1 X + E_1 Y + F_1 = 0.$$

If either $A_1 = 0$ or $C_1 = 0$, the curve is a parabola that can be sketched by the method of Section 5.4. If A_1 and C_1 are both nonzero, Translation of Axes gives us a new equation of the simpler form

$$A_2 U^2 + B_2 V^2 + F_2 = 0.$$

The graph of this equation is an ellipse or hyperbola, which can be sketched by the method of Section 5.5. The degenerate cases—two lines, one line, a point, or an empty graph—may also occur.

PROBLEMS FOR SECTION 5.7

In Problems 1–10, rotate the axes to transform the given equation into a new equation with no XY-term. Find the angle of rotation and the new equation.

1 $xy + 4 = 0$ **2** $x^2 + xy + y^2 = 2$

3 $x^2 - 4xy + y^2 = 1$ **4** $x^2 + 3xy + y^2 = 4$

5 $x^2 + 2\sqrt{3}xy - y^2 = 7$ **6** $5x^2 - \sqrt{3}xy + 4y^2 = 6$

7 $x^2 + xy = 3$ **8** $2x^2 - xy - y^2 = 1$

9 $4x^2 - \sqrt{3}xy + y^2 = 5$ **10** $2x^2 + \sqrt{3}xy - y^2 = -10$

11   Prove that any second degree Equation (1) in which $A = C$ can be transformed into an equation with no XY-term by a 45° rotation of axes.

☐ **12**   Prove that if we begin with a second degree equation with no first degree terms, $Ax^2 + Bxy + Cy^2 + F = 0$, and then rotate axes, the new equation will again have no first degree terms.

☐ **13**   Prove that the sum $A + C$ is not changed by rotation of axes. That is, if Equation (2) is obtained from Equation (1) by rotation of axes, then $A + C = A_1 + C_1$.

☐ **14**   Prove that the discriminant of a second degree equation is not changed by rotation of axes. That is, if Equation (2) is obtained from Equation (1) by rotation of axes, then $B^2 - 4AC = B_1^2 - 4A_1C_1$.

5.8 THE ε, δ CONDITION FOR LIMITS

The traditional calculus course is developed entirely without infinitesimals. The starting point is the concept of a limit. The intuitive idea of $\lim_{x \to c} f(x) = L$ is: *For every real number x which is close to but not equal to c, $f(x)$ is close to L.*

It is hard to make this idea into a rigorous definition, because one must clarify the word "close". Indeed, the whole point of our infinitesimal approach to calculus is that it is easier to define and explain limits using infinitesimals. The definition of limits in terms of real numbers is traditionally expressed using the Greek letters ε (epsilon) and δ (delta), and is therefore called the ε, δ condition for limits.

The ε, δ condition will be based on the notion of distance between two real numbers.

DEFINITION

The **distance** between two real numbers x and c is the absolute value of their difference,

$$distance = |x - c|.$$

x is within δ of c if $|x - c| \leq \delta$.
x is strictly within δ of c if $|x - c| < \delta$.

Notice that the distance $|x - c|$ is just the difference between the larger and the smaller of the two numbers x and c. This is a place where the absolute value sign is especially convenient. The following simple but helpful lemma is illustrated in Figure 5.8.1.

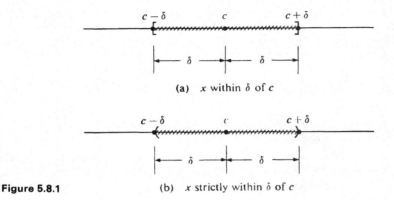

(a) x within δ of c

Figure 5.8.1   (b) x strictly within δ of c

LEMMA

(i)   *x is within δ of c if and only if*

$$c - \delta \le x \le c + \delta.$$

(ii)   *x is strictly within δ of c if and only if*

$$c - \delta < x < c + \delta.$$

PROOF (i)   Subtracting c from each term we see that

$$c - \delta \le x \le c + \delta$$

if and only if $-\delta \le x - c \le \delta,$

which is true if and only if $|x - c| \le \delta$.

The proof of (ii) is similar.

We shall repeat our infinitesimal definition of limit from Section 3.3 and then write down the ε, δ condition for limits. Later we shall prove that the two definitions of limit are equivalent to each other.

Suppose the real function f is defined for all real numbers $x \ne c$ in some neighborhood of c.

DEFINITION OF LIMIT (Repeated)

The equation

$$\lim_{x \to c} f(x) = L$$

means that whenever a hyperreal number x is infinitely close to but not equal to c, $f(x)$ is infinitely close to L.

ε, δ CONDITION FOR $\lim_{x \to c} f(x) = L$

For every real number $\varepsilon > 0$ there is a real number $\delta > 0$ which depends on ε such that whenever x is strictly within δ of c but not equal to c, $f(x)$ is strictly within ε of L. In symbols, if $0 < |x - c| < \delta$, then $|f(x) - L| < \varepsilon$.

In the ε, δ condition, the notion of being infinitely close to c is replaced by being strictly within δ of c, and being infinitely close to L is replaced by being strictly within ε of L. But why are there two numbers ε and δ, instead of just one? And why should δ depend on ε? Let us look at a simple example.

EXAMPLE 1   Consider the limit $\lim\limits_{x \to 0} \left(1 + \dfrac{10x^2}{x}\right) = 1.$

When $x = 0$, the function $f(x) = 1 + 10x^2/x$ is undefined. When x is a real number close to but not equal to 0, $f(x)$ is close to 1.

Now let us be more explicit. How should we choose x to get $f(x)$ strictly within $\frac{1}{5}$ of 1? To solve this problem we assume x is strictly within some distance δ of 0 and get inequalities for $f(x)$.

By the lemma, we must find a $\delta > 0$ such that whenever

$$-\delta < x < \delta \quad \text{and} \quad x \neq 0,$$

we have
$$1 - \tfrac{1}{5} < f(x) < 1 + \tfrac{1}{5}.$$

Assume $-\delta < x$ and $x < \delta.$

Then $-10\delta < 10x$ and $10x < 10\delta$

$$-10\delta < \frac{10x^2}{x} \quad \text{and} \quad \frac{10x^2}{x} < 10\delta \qquad \text{if } x \neq 0$$

$$1 - 10\delta < 1 + \frac{10x^2}{x} \quad \text{and} \quad 1 + \frac{10x^2}{x} < 1 + 10\delta$$

$$1 - 10\delta < f(x) < 1 + 10\delta.$$

If we set $\delta = \frac{1}{50}$, then

$$1 - \tfrac{1}{5} < f(x) < 1 + \tfrac{1}{5}.$$

This shows that

whenever $-\tfrac{1}{50} < x < \tfrac{1}{50}$ and $x \neq 0,$ $1 - \tfrac{1}{5} < f(x) < 1 + \tfrac{1}{5}.$

In other words,

whenever $0 < |x| < \tfrac{1}{50},$ $|f(x) - 1| < \tfrac{1}{5}.$

A similar computation shows that for each $\varepsilon > 0$, if $0 < |x| < \varepsilon/10$ then $|f(x) - 1| < \varepsilon$. Thus the ε, δ condition for $\lim_{x\to 0} (1 + 10x^2/x) = 1$ is true, and, for a given ε, a corresponding δ is $\delta = \varepsilon/10$.

EXAMPLE 2   In the limit

$$\lim_{x \to 2} x^2 = 4,$$

find a $\delta > 0$ such that whenever $0 < |x - 2| < \delta, |x^2 - 4| < \tfrac{1}{10}.$

By the Lemma, we must find $\delta > 0$ such that whenever

$$2 - \delta < x < 2 + \delta \quad \text{and} \quad x \neq 2,$$
$$4 - \tfrac{1}{10} < x^2 < 4 + \tfrac{1}{10}.$$

Assume that $2 - \delta < x$ and $x < 2 + \delta.$

As long as $2 - \delta$ and x are positive we may square both sides,

$$4 - 4\delta + \delta^2 < x^2 \quad \text{and} \quad x^2 < 4 + 4\delta + \delta^2$$

$$4 + (-4\delta + \delta^2) < x^2 \quad \text{and} \quad x^2 < 4 + (4\delta + \delta^2).$$

Now take δ small enough so that

$$-\tfrac{1}{10} \leq -4\delta + \delta^2 \quad \text{and} \quad 4\delta + \delta^2 \leq \tfrac{1}{10}.$$

For example, $\delta = \tfrac{1}{50}$ will do. Then

$$4 - \tfrac{1}{10} < x^2 < 4 + \tfrac{1}{10}.$$

Thus whenever $0 < |x - 2| < \tfrac{1}{50}$, $|x^2 - 4| < \tfrac{1}{10}$.

Notice that any smaller value of δ, such as $\delta = \tfrac{1}{100}$, will also work.

In geometric terms, the ϵ, δ condition says that for every horizontal strip (of width 2ϵ) centered at L, there exists a vertical strip (of width 2δ) centered at c such that whenever $x \neq c$ is in the vertical strip, $f(x)$ is in the horizontal strip. The graphs in Figure 5.8.2 indicate various horizontal strips and corresponding vertical strips. They should be examined closely.

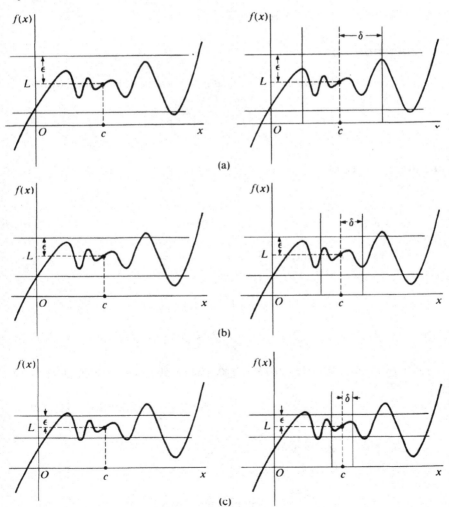

Figure 5.8.2

There are also ε, δ conditions for one-sided limits and infinite limits. The three cases below are typical.

ε, δ CONDITION FOR $\lim_{x \to c^+} f(x) = L$

For every real number $\varepsilon > 0$, there is a real number $\delta > 0$ which depends on ε such that whenever $c < x < c + \delta$, we have $|f(x) - L| < \varepsilon$.

Intuitively, when x is close to c but greater than c, $f(x)$ is close to L.

ε, δ CONDITION FOR $\lim_{x \to \infty} f(x) = L$

For every real number $\varepsilon > 0$ there is a real number $B > 0$ which depends on ε such that whenever $x > B$, we have $|f(x) - L| < \varepsilon$.

Intuitively, when x is large, $f(x)$ is close to L.

ε, δ CONDITION FOR $\lim_{x \to \infty} f(x) = \infty$

For every real number $A > 0$ there is a real number $B > 0$ which depends on A such that whenever $x > B$, we have $f(x) > A$.

Intuitively, when x is large, $f(x)$ is large.

EXAMPLE 3 In the limit

$$\lim_{t \to \infty} 2 + \frac{3}{t} = 2,$$

find a real number $B > 0$ such that whenever $t > B$, $(2 + 3/t)$ is strictly within $1/100$ of 2.

To find B, we assume $t > B$ and $t > 0$, and get inequalities for $2 + 3/t$.

$$0 < t, \qquad\qquad t > B$$

$$0 < \frac{3}{t}, \qquad\qquad \frac{3}{t} < \frac{3}{B}$$

$$2 < 2 + \frac{3}{t}, \qquad 2 + \frac{3}{t} < 2 + \frac{3}{B}$$

Now choose B so that $3/B \leq 1/100$. The number $B = 300$ will do. It follows that whenever $t > 300$,

$$2 < 2 + \frac{3}{t} < 2 + \frac{1}{100},$$

and $2 + \dfrac{3}{t}$ is strictly within $\varepsilon = \dfrac{1}{100}$ of 2.

EXAMPLE 4 In the limit

$$\lim_{x \to \infty} (x^2 - x) = \infty,$$

find a $B > 0$ such that whenever $x > B$, $x^2 - x > 10{,}000$.

This time we assume $x > B$ and get an inequality for $x^2 - x$. We may assume $B > 1$.

$$x > B > 1$$
$$x - 1 > B - 1 > 0$$
$$x(x - 1) > B(B - 1)$$
$$x^2 - x > B^2 - B.$$

Now take a B such that $B^2 - B > 10{,}000$. The number $B = 200$ will do, because $(200)^2 - 200 = 39800$. Thus whenever $x > 200$, $x^2 - x > 10{,}000$.

We conclude this section with the proof that the ε, δ condition is equivalent to the infinitesimal definition of a limit.

THEOREM 1

Let f be defined in some deleted neighborhood of c. Then the following are equivalent:

(i) $\lim_{x \to c} f(x) = L$.

(ii) *The ε, δ condition for* $\lim_{x \to c} f(x) = L$ *is true.*

PROOF We first assume the ε, δ condition and prove that

$$\lim_{x \to c} f(x) = L.$$

Let x be any hyperreal number which is infinitely close but not equal to c. To prove that $f(x)$ is infinitely close to L we must show that

for every real $\varepsilon > 0$, $|f(x) - L| < \varepsilon$.

Let ε be any positive real number, and let $\delta > 0$ be the corresponding number in the ε, δ condition. Since x is infinitely close to c and $\delta > 0$ is real, we have

$$0 < |x - c| < \delta.$$

By the ε, δ condition and the Transfer Principle,

$$|f(x) - L| < \varepsilon.$$

We conclude that $f(x)$ is infinitely close to L. This proves that

$$\lim_{x \to c} f(x) = L.$$

For the other half of the proof we assume that

$$\lim_{x \to c} f(x) = L,$$

and prove the ε, δ condition. This will be done by an indirect proof. Assume that the ε, δ condition is false for some real number $\varepsilon > 0$. That means that for every real $\delta > 0$ there is a real number $x = x(\delta)$ such that

(1) $x \neq c$, $|x - c| < \delta$, $|f(x) - L| \geq \varepsilon$.

Now let $\delta_1 > 0$ be a positive infinitesimal. By the Transfer Principle, Equation (1) holds for δ_1. Therefore $x_1 = x(\delta_1)$ is infinitely close but not equal to c. But since

$$|f(x_1) - L| \geq \varepsilon$$

and ε is a positive real number, $f(x_1)$ is not infinitely close to L. This contradicts the equation

$$\lim_{x \to c} f(x) = L.$$

We conclude that the ε, δ condition must be true after all.

The theorem is also true for the other types of limits.

The concept of continuity can be described in terms of limits, as we saw in Section 3.4. Therefore continuity can be defined in terms of the real number system only.

COROLLARY

The following are equivalent.

(i) *f is continuous at c.*

(ii) *For every real $\varepsilon > 0$ there is a real $\delta > 0$ depending on ε such that:*

whenever $|x - c| < \delta$, $|f(x) - f(c)| < \varepsilon$.

PROOF Both (i) and (ii) are equivalent to

$$\lim_{x \to c} f(x) = f(c).$$

Intuitively, this corollary says that *f is continuous at c* if and only if $f(x)$ is *close to $f(c)$ whenever x is close to c.*

PROBLEMS FOR SECTION 5.8

1   In the limit $\lim_{x \to 4} 10x = 40$, find a $\delta > 0$ such that whenever $0 < |x - 4| < \delta$, $|10x - 40| < 0.01$.

2   In the limit $\lim_{x \to 0} (x^2 - 4x)/2x = -2$, find a $\delta > 0$ such that whenever $0 < |x| < \delta$, $|(x^2 - 4x)/2x - (-2)| < 0.1$.

3   In the limit $\lim_{x \to 2} 1/x = 1/2$, find a $\delta > 0$ such that whenever $0 < |x - 2| < \delta$, $|1/x - 1/2| < 0.01$.

4   In the limit $\lim_{x \to -3} x^3 = -27$, find a $\delta > 0$ such that whenever $0 < |x - (-3)| < \delta$, $|x^3 - (-27)| < 0.01$.

5   In the limit $\lim_{x \to 0^+} \sqrt{x} = 0$, find a $\delta > 0$ such that whenever $0 < x < \delta$, $\sqrt{x} < 0.01$.

6   In the limit $\lim_{x \to 2^+} \sqrt{x^2 - 4} = 0$, find a $\delta > 0$ such that whenever $2 < x < 2 + \delta$, $\sqrt{x^2 - 4} < 0.1$.

7   In the limit $\lim_{x \to 1^-} \sqrt{1 - x^2} = 0$, find a $\delta > 0$ such that whenever $1 - \delta < x < 1$, $\sqrt{1 - x^2} < 0.001$.

8   In the limit $\lim_{x \to 2^-} \sqrt{6 - 3x} = 0$, find a $\delta > 0$ such that whenever $2 - \delta < x < 2$, $\sqrt{6 - 3x} < 0.01$.

9   In the limit $\lim_{x \to 0} x^{-2} = \infty$, find a $\delta > 0$ such that whenever $0 < |x| < \delta, x^{-2} > 10{,}000$.

10 In the limit $\lim_{x \to 0} 16/x^4 = \infty$, find a $\delta > 0$ such that whenever $0 < |x| < \delta$, $16/x^4 > 10,000$.

11 In the limit $\lim_{t \to 0^+} 1/10t = \infty$, find a $\delta > 0$ such that whenever $0 < t < \delta$, $1/10t > 100$.

12 In the limit $\lim_{t \to 4^+} 1/(4 - t) = -\infty$, find a $\delta > 0$ such that whenever $4 < t < 4 + \delta$, $1/(4 - t) < -100$.

13 In the limit $\lim_{x \to 0^+} 1/\sqrt{x} = \infty$, find a $\delta > 0$ such that whenever $0 < x < \delta$, $1/\sqrt{x} > 100$.

14 In the limit $\lim_{x \to 0^+} 1/x^3 = \infty$, find a $\delta > 0$ such that whenever $0 < x < \delta$, $1/x^3 > 1000$.

15 In the limit $\lim_{x \to 1^-} 1/(1 - x^2) = \infty$, find a $\delta > 0$ such that whenever $1 - \delta < x < 1$, $1/(1 - x^2) > 100$.

16 In the limit $\lim_{x \to 2^-} 5/\sqrt{2 - x} = \infty$, find a $\delta > 0$ such that whenever $2 - \delta < x < 2$, $5/\sqrt{2 - x} > 100$.

17 In the limit $\lim_{t \to \infty} 1/(1 + 4t) = 0$, find a $B > 0$ such that whenever $t > B$, $1/(1 + 4t) < 0.01$.

18 In the limit $\lim_{t \to \infty} 1/t^2 = 0$, find a $B > 0$ such that whenever $t > B$, $1/t^2 < 0.01$.

19 In the limit $\lim_{t \to \infty} 2t^2 - 5t = \infty$, find a $B > 0$ such that whenever $t > B$, $2t^2 - 5t > 1000$.

20 In the limit $\lim_{t \to \infty} t^3 + t^2 - 5 = \infty$, find a $B > 0$ such that whenever $t > B$, $t^3 + t^2 - 5 > 1000$.

21 In the limit $\lim_{x \to \infty} \sqrt{5x + 1} = \infty$, find a $B > 0$ such that whenever $x > B$, $\sqrt{5x + 1} > 100$.

22 In the limit $\lim_{x \to -\infty} \sqrt[3]{x - 1} = -\infty$, find a $B > 0$ such that whenever $x < -B$, $\sqrt[3]{x - 1} < -100$.

☐ **23** State the ε, δ condition for the limit $\lim_{x \to c^-} f(x) = L$.

☐ **24** State the ε, δ condition for the limit $\lim_{x \to c} f(x) = \infty$.

☐ **25** State the ε, δ condition for the limit $\lim_{x \to \infty} f(x) = -\infty$.

☐ **26** Prove that $\lim_{x \to \infty} f(x) = \infty$ if and only if the ε, δ condition for this limit holds: For every $A > 0$ there is a $B > 0$ such that whenever $x > B$, $f(x) > A$.

5.9 NEWTON'S METHOD

The Increment Theorem for derivatives shows that when $f'(c)$ exists and $x \approx c$, $f(x)$ is infinitely close to the tangent line $f(c) + f'(c)(x - c)$ even compared to $x - c$. Thus intuitively, when x is real and close to c, $f(x)$ is closely approximated by the tangent line $f(c) + f'(c)(x - c)$. Newton's method uses the tangent line to approximate a zero of $f(x)$. It is an iterative method that does not always work but usually gives a very good approximation.

Consider a real function f that crosses the x-axis as in Figure 5.9.1. From the graph we make a first rough approximation x_1 to the zero of $f(x)$. To get a better approximation, we take the tangent line at x_1 and compute the point x_2 where the tangent line intersects the x-axis. At x_2, the curve $f(x)$ is very close to zero, so we take x_2 as our new approximation. The tangent line has the equation

$$y = f(x_1) + f'(x_1)(x - x_1).$$

We get a formula for x_2 by setting $y = 0$ and $x = x_2$ and then solving for x_2.

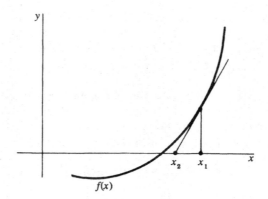

Figure 5.9.1 $f(x)$

$$0 = f(x_1) + f'(x_1)(x_2 - x_1)$$

$$x_2 = x_1 - \frac{f(x_1)}{f'(x_1)}.$$

We may then repeat the procedure starting from x_2 to get a still better approximation x_3 as in Figure 5.9.2,

$$x_3 = x_2 - \frac{f(x_2)}{f'(x_2)}.$$

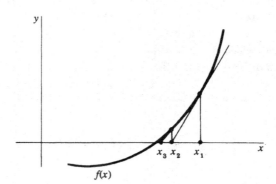

Figure 5.9.2 $f(x)$

NEWTON'S METHOD

When to Use   *We wish to approximate a zero of $f(x)$, where $f'(x)$ is continuous and not close to zero, as in Figure 5.9.1.*

Step 1   *Sketch the graph of $f(x)$, and choose a point x_1 near the zero of $f(x)$. x_1 is the first approximation.*

Step 2   *Compute $f'(x)$.*

Step 3   *Compute the second approximation*

$$x_2 = x_1 - \frac{f(x_1)}{f'(x_1)}.$$

Step 4   *For a closer approximation repeat Step 3. The $(n + 1)$st approximation is given by*

$$x_{n+1} = x_n - \frac{f(x_n)}{f'(x_n)}.$$

As a rough check on the accuracy, compute $f(x_n)$ and note how close it is to zero.

Steps 3 and 4 can be done conveniently on a hand calculator.

Warning: Since Newton's method involves division by $f'(x_1)$, avoid starting at a point where the slope is near zero. Figure 5.9.3 shows that when the slope is close to zero, the tangent line is nearly horizontal and the approximation may be poor.

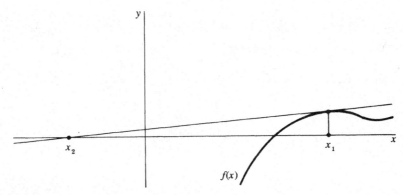

Figure 5.9.3

EXAMPLE 1 Approximate a zero of $f(x) = x^3 + 2x^2 - 5$ by Newton's method.

Step 1 The graph is shown in Figure 5.9.4. We choose $x_1 = 1$ as our first approximation.

Step 2 $f'(x) = 3x^2 + 4x$

Step 3 $x_2 = x_1 - \dfrac{f(x_1)}{f'(x_1)} = 1 - \dfrac{(-2)}{7} = \dfrac{9}{7} \sim 1.2857$

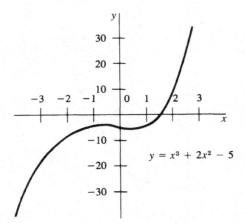

$$y = x^3 + 2x^2 - 5$$

Figure 5.9.4

Step 4 $x_3 = x_2 - \dfrac{f(x_2)}{f'(x_2)} = x_2 - \dfrac{x_2^3 + 2x_2^2 - 5}{3x_2^2 + 4x_2} \sim 1.2430$

As a check we compute

$$f(x_3) = x_3^3 + 2x_3^2 - 5 \sim 0.01$$

One more iteration gives much more accuracy:

$$x_4 = x_3 - \frac{f(x_3)}{f'(x_3)} = x_3 - \frac{x_3^3 + 2x_3^2 - 5}{3x_3^2 + 4x_3} \sim 1.241897$$
$$f(x_4) = x_4^3 + 2x_4^2 - 5 \sim 0.000007$$

EXAMPLE 2   Approximate the fifth root of 6 by Newton's method.

Step 1   We must find the zero of $f(x) = x^5 - 6$. The graph is shown in Figure 5.9.5. Choose $x_1 = 1.5$.

Step 2 $f'(x) = 5x^4$

Step 3 $x_2 = x_1 - \dfrac{x_1^5 - 6}{5x_1^4} \sim 1.437$

Step 4 $x_3 = x_2 - \dfrac{x_2^5 - 6}{5x_2^4} \sim 1.43102$

As a check we compute

$$(x_3)^5 \sim 6.001$$

In this example more iterations would be necessary if our first approximation had not been chosen as well. For instance, starting with $x_1 = 1$ we would not reach

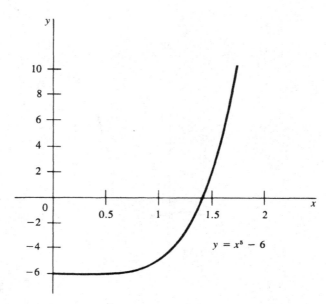

Figure 5.9.5

the approximation 1.431 until x_6, obtaining the successive approximations

$$x_1 = 1, \qquad x_2 = 2, \qquad x_3 = 1.675, \qquad x_4 = 1.49245,$$
$$x_5 = 1.43583, \qquad x_6 = 1.43100.$$

EXAMPLE 3 Approximate the point x where $\sin x = \ln x$.

As one can see from the graphs of $\sin x$ and $\ln x$ in Figure 5.9.6, $\sin x$ and $\ln x$ cross at one point x, which is somewhere between $x = 1$ (where $\ln x$ crosses the x-axis going up) and $x = \pi$ (where $\sin x$ crosses the x-axis going down). To apply Newton's method, we let $f(x)$ be the function

$$f(x) = \sin x - \ln x$$

shown in Figure 5.9.7. We wish to approximate the zero of $f(x)$.

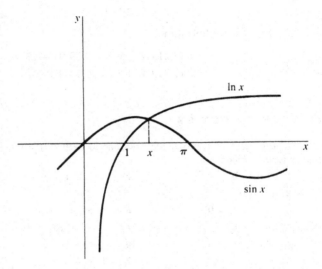

Figure 5.9.6

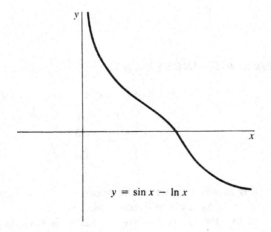

$y = \sin x - \ln x$

Figure 5.9.7

Step 1   Choose $x_1 = 2$ (since the zero of $f(x)$ is between 1 and π).

Step 2 $f'(x) = \cos x - 1/x$

Step 3 $x_2 = x_1 - \dfrac{\sin x_1 - \ln x_1}{\cos x_1 - 1/x_1} = 2 - \dfrac{\sin 2 - \ln 2}{\cos 2 - 1/2} \sim 2.23593$

Step 4   Repeat Step 3. The values of x_n, $f(x_n)$, and $f'(x_n)$ are shown in the table.

n	x_n	$f(x_n)$	$f'(x_n)$
1	2.000000000	0.216150246	-0.916146836
2	2.235934064	-0.017827280	-1.064407894
3	2.219185522	-0.000082645	-1.054519059
4	2.219107150	-0.000000001	-1.054472505

The answer is

$$x \sim 2.219107150.$$

On a calculator we find that

$$\sin (2.219107150) = 0.797104929$$
$$\ln (2.219107150) = 0.797104930.$$

PROBLEMS FOR SECTION 5.9

Use Newton's method to find approximate solutions to each of the following equations. (A hand calculator is recommended.)

1	$x^3 + 5x - 10 = 0$	2	$2x^3 + x + 4 = 0$
3	$x^5 + x^3 + x = 1$	4	$2x^5 + 3x = 2$
5	$x^4 = x + 1, \quad x > 0$	6	$x^4 = x + 1, \quad x < 0$
7	$x^3 - 10x + 4 = 0, \quad x > 1$	8	$x^3 - 10x + 4 = 0, \quad 0 < x < 1$
9	$x + \sqrt{x} = 1$	10	$x + 1/\sqrt{x} = 3$
11	$e^x = 1/x$	12	$e^x + x = 4$
13	$x + \sin x = 2$	14	$\cos x = x^2, \quad x > 0$
15	$\tan x = e^x, \quad 0 < x < \pi/2$	16	$e^x + \ln x = 0$

5.10 DERIVATIVES AND INCREMENTS

In Section 3.3 we found that the derivative of f is given by the limit

$$f'(c) = \lim_{\Delta x \to 0} \frac{f(c + \Delta x) - f(c)}{\Delta x}.$$

If $y = f(x)$, $$\frac{dy}{dx} = \lim_{\Delta x \to 0} \frac{\Delta y}{\Delta x}.$$

By definition this means that when the hyperreal number Δx is infinitely close to but not equal to zero, $\Delta y/\Delta x$ is infinitely close to dy/dx.

By contrast, the ε, δ condition for this limit says intuitively that when the real number Δx is close to but not equal to zero, $\Delta y/\Delta x$ is close to dy/dx.

The ε, δ condition for the derivative can be given a geometric interpretation, shown in Figure 5.10.1. Consider the curve $y = f(x)$, and suppose $f'(c)$ exists. Draw

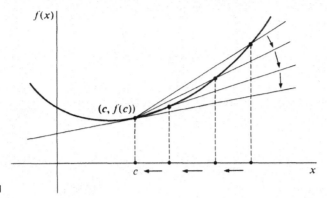

Figure 5.10.1

the line tangent to the curve at c. For $\Delta x \neq 0$, draw the *secant line* which intersects the curve at the points $(c, f(c))$ and $(c + \Delta x, f(c + \Delta x))$. Then the tangent line will have slope $f'(c)$ while the secant line will have slope

$$\frac{f(c + \Delta x) - f(c)}{\Delta x}.$$

The ε, δ condition shows that if we take values of Δx closer and closer to zero, then the slopes of the secant line will get closer and closer to the slope of the tangent line.

EXAMPLE 1   Consider the curve $f(x) = x^{1/3}$.

Then $f'(x) = \frac{1}{3}x^{-2/3}$.

At the point $x = 8$, we have

$$x = 8, \qquad f(x) = 2, \qquad f'(x) = \frac{1}{12} = 0.0833\ldots.$$

Thus $$\lim_{\Delta x \to 0} \frac{(8 + \Delta x)^{1/3} - 2}{\Delta x} = \frac{1}{12}.$$

This is the slope of the line tangent to the curve at the point $(8, 2)$. As Δx approaches zero, the slope of the secant line through the two points $(8, 2)$ and $(8 + \Delta x, (8 + \Delta x)^{1/3})$ will approach $\frac{1}{12}$. We make a table showing the slope of the secant line for various values of Δx.

Δx	$\Delta y = (8 + \Delta x)^{1/3} - 2$	$\dfrac{\Delta y}{\Delta x} = $ slope of secant line	$\left\lvert \dfrac{\Delta y}{\Delta x} - \dfrac{1}{12} \right\rvert$
10	0.6207	0.0621	0.0212
1	0.0801	0.0801	0.0032
$\dfrac{1}{10}$	0.00829	0.0830	0.0003
-10	-3.2599	0.3260	0.2427
-1	-0.0871	0.0871	0.0038
$-\dfrac{1}{10}$	-0.00837	0.0837	0.0004

The ε, δ condition for the derivative is of theoretical importance but does not give an error estimate for the limit. When the function f has a continuous second derivative, we can get a useful error estimate in a different way. It is more convenient to work with one-sided limits.

By an *error estimate* for a limit

$$\lim_{\Delta x \to 0^+} g(\Delta x) = L$$

we mean a real function $E(\Delta x)$, $0 < \Delta x \le b$, such that the approximation $g(\Delta x)$ is always within $E(\Delta x)$ of the limit L. In symbols,

$$|g(\Delta x) - L| \le E(\Delta x) \qquad \text{for} \quad 0 < \Delta x \le b.$$

THEOREM 1

Suppose f has a continuous second derivative and $|f''(t)| \le M$ for all t in the interval $[c, b]$. Then:

(i) *Whenever $c < c + \Delta x \le b, f(c + \Delta x)$ is within $\frac{1}{2}M \, \Delta x^2$ of $f(c) + f'(c) \, \Delta x$.*

(ii) *Whenever $c < c + \Delta x \le b$, $\dfrac{f(c + \Delta x) - f(c)}{\Delta x}$ is within $\frac{1}{2}M \, \Delta x$ of $f'(c)$. That is, $\frac{1}{2}M \, \Delta x$ is an error estimate for the right-sided limit*

$$\lim_{\Delta x \to 0^+} \frac{f(c + \Delta x) - f(c)}{\Delta x} = f'(c).$$

There is a similar theorem for the left-sided limit

$$\lim_{\Delta x \to 0^-} \frac{f(c + \Delta x) - f(x)}{\Delta x} = f'(c)$$

with the error estimate $\frac{1}{2}M|\Delta x|$.

PROOF Let $x = c + \Delta x$. Then

$$-M \le f''(t) \le M \qquad \text{for } c \le t \le x.$$

Integrating from c to t,

$$\int_c^t -M \, dt \le \int_c^t f''(t) \, dt \le \int_c^t M \, dt,$$

$$-M(t - c) \le f'(t) - f'(c) \le M(t - c).$$

Integrating again from c to x,

$$\int_c^x -M(t - c) \, dt \le \int_c^x f'(t) - f'(c) \, dt \le \int_c^x M(t - c) \, dt,$$

$$-M\frac{(x - c)^2}{2} \le f(t) - f'(c)t \Big]_c^x \le M\frac{(x - c)^2}{2},$$

or

$$-M\frac{\Delta x^2}{2} \le (f(x) - f(c)) - f'(c) \, \Delta x \le M\frac{\Delta x^2}{2},$$

$$-M\frac{\Delta x^2}{2} \le f(x) - (f(c) + f'(c) \, \Delta x) \le M\frac{\Delta x^2}{2}.$$

This proves part (i) (Figure 5.10.2). Dividing by Δx we get part (ii).

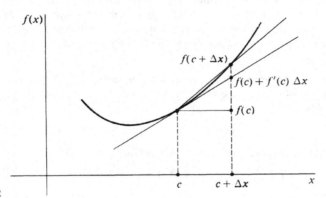

$f(x)$

$f(c + \Delta x)$

$f(c) + f'(c)\,\Delta x$

$f(c)$

c $c + \Delta x$ x

Figure 5.10.2

EXAMPLE 1 (Concluded) We consider once more the curve $f(x) = x^{1/3}$ at the point $x = 8$. The second derivative is

$$f''(x) = -\tfrac{2}{9}x^{-5/3}.$$

First consider the interval $[8, 9]$. In this interval $f''(x)$ has the maximum value

$$|f''(8)| = \tfrac{2}{9}(8)^{-5/3} = \tfrac{2}{9}2^{-5} = \tfrac{1}{144}.$$

Thus we may take $M = \tfrac{1}{144}$, and

$$\tfrac{1}{2}M\,\Delta x = \tfrac{1}{288}\Delta x \quad \text{is an error estimate for} \quad \lim_{\Delta x \to 0^+}\frac{\Delta y}{\Delta x} = \frac{1}{12}.$$

Thus   when $\Delta x = 1$, $\left|\dfrac{\Delta y}{\Delta x} - \dfrac{1}{12}\right| \le \dfrac{1}{288} = 0.0035$,

when $\Delta x = \dfrac{1}{10}$, $\left|\dfrac{\Delta y}{\Delta x} - \dfrac{1}{12}\right| \le \dfrac{1}{2880} = 0.00035$.

Next consider the interval $[7, 8]$. This time we take

$$M = |f''(7)| = \tfrac{2}{9}(7)^{-5/3} = 0.0087.$$

Then $\tfrac{1}{2}M|\Delta x| = 0.0044|\Delta x|$

is an error estimate for the limit

$$\lim_{\Delta x \to 0^-}\frac{\Delta y}{\Delta x} = \frac{1}{12}.$$

when $\Delta x = -1$, $\left|\dfrac{\Delta y}{\Delta x} - \dfrac{1}{12}\right| \le 0.0044$,

when $\Delta x = -\dfrac{1}{10}$, $\left|\dfrac{\Delta y}{\Delta x} - \dfrac{1}{12}\right| \le 0.00044$.

From the table in Example 1 we see that the error estimates are slightly greater than the actual values of $\left|\dfrac{\Delta y}{\Delta x} - \dfrac{1}{12}\right|$.

We shall now turn the problem around. Instead of using the increment Δy to approximate the derivative dy/dx, we shall use the derivative dy/dx to approximate the increment Δy. When Δx is small, $f(c + \Delta x)$ will be close to $f(c) + f'(c)\,\Delta x$ even compared to Δx. Part (i) of Theorem 1 gives the error estimate $\frac{1}{2}M\,\Delta x^2$ for this approximation. This method is especially useful for approximating $f(x)$ when there is a number c close to x such that both $f(c)$ and $f'(c)$ are known.

EXAMPLE 2   Find approximate values for $\sqrt[3]{9}$ and $\sqrt[3]{7.9}$. Both these numbers are close to 8, whose cube root 2 comes out even. Taking $f(x) = \sqrt[3]{x}$ and $c = 8$, we have

$$f(c) = 2, \qquad f'(c) = \tfrac{1}{12} = 0.0833\ldots.$$

From Theorem 1 the approximate values are

$$f(c + \Delta x) \sim f(c) + f'(c)\,\Delta x.$$

Thus

$$\sqrt[3]{9} \sim 2 + \tfrac{1}{12} = 2.0833,$$
$$\sqrt[3]{7.9} \sim 2 + \tfrac{1}{12}(-0.1) = 1.99167.$$

To get an error estimate for $\sqrt[3]{9}$, take the interval $[8, 9]$. From Example 1 we may take $M = \tfrac{1}{144}$. Therefore by Theorem 1,

$$\sqrt[3]{9} \sim 2.0833, \qquad \text{error} \le \tfrac{1}{2}\cdot\tfrac{1}{144}\cdot 1^2 = 0.0035.$$

Thus $$2.0798 \le \sqrt[3]{9} \le 2.0868.$$

To get an error estimate for $\sqrt[3]{7.9}$ take the interval $[7, 8]$ and $M = 0.0087$. By Theorem 1,

$$\sqrt[3]{7.9} \sim 1.991667, \qquad \text{error} \le \tfrac{1}{2}(0.0087)(0.1)^2 = 0.000044.$$

Thus $$1.991623 \le \sqrt[3]{7.9} \le 1.991711.$$

EXAMPLE 3   Find an approximate value for $(0.99)^5$.

Let $$f(x) = x^5, \quad c = 1.$$

Then $$f(c) = 1^5 = 1, \quad f'(c) = 5c^4 = 5.$$

We put $$0.99 = c + \Delta x, \qquad \Delta x = -0.01.$$

Then the approximate value is

$$f(c + \Delta x) \sim f(c) + f'(c)\,\Delta x,$$
$$(0.99)^5 \sim 1 + 5(-0.01) = 0.95.$$

To get an error estimate we see that $f''(u) = 20u^3$, so $|f''(u)| \le 20$ for u between 0.99 and 1. Then $M = 20$, and

$$(0.99)^5 \sim 0.95, \qquad \text{error} \le \frac{(0.01)^2}{2}(20) = 0.001,$$

or $$0.949 \le (0.99)^5 \le 0.951.$$

Theorem 1 is closely related to the Increment Theorem in Section 2.2. The relation between them can be seen when we write them next to each other.

INCREMENT THEOREM (Repeated)

Hypotheses $f'(c)$ *exists and* Δx *is infinitesimal.*

Conclusion $f(c + \Delta x) = f(c) + f'(c)\,\Delta x + \varepsilon\,\Delta x$ *for some infinitesimal* ε *which depends on c and* Δx.

THEOREM 1 OF THIS SECTION (in an equivalent form)

Hypotheses $f''(u)$ *exists and* $|f''(u)| \le M$ *for all u between the real numbers c and* $c + \Delta x$.

Conclusion $f(c + \Delta x) = f(c) + f'(c)\,\Delta x + \varepsilon\,\Delta x$ *for some real* ε *within* $\frac{1}{2}M|\Delta x|$ *of* 0.

Thus Theorem 1 has more hypotheses but also gives more specific information about ε in its conclusion.

PROBLEMS FOR SECTION 5.10

In Problems 1–6, find $f'(c)$ and an error estimate for the limit

$$f'(c) = \lim_{\Delta x \to 0^+} \frac{f(c + \Delta x) - f(c)}{\Delta x}$$

with $0 < \Delta x \le 1$.

1	$f(x) = x^2,\ \ c = 1$	**2**	$f(x) = x^3 - 5x,\ \ c = 10$
3	$f(x) = 2/\sqrt{x},\ \ c = 4$	**4**	$f(x) = x\sqrt{x},\ \ c = 4$
5	$f(x) = 1/x,\ \ c = 3$	**6**	$f(x) = 1/(x^2 + 1),\ \ c = 1$

7 $f(x) = \sin x,\ \ \ \ c = 0,\ \ (0 < \Delta x \le \pi)$

8 $f(x) = \tan x,\ \ \ \ c = 0,\ \ (0 < \Delta x \le \pi/6)$

9 $f(x) = \cos(2x),\ \ \ \ c = \pi/3,\ \ (0 < \Delta x \le \pi)$

10 $f(x) = \sin^2(2x),\ \ \ \ c = \pi/2,\ \ (0 < \Delta x \le \pi)$

11 $f(x) = \ln x,\ \ \ \ c = 1,\ \ (0 < \Delta x \le 1)$

12 $f(x) = x \ln x,\ \ \ \ c = 1,\ \ (0 < \Delta x \le 1)$

13 $f(x) = e^x,\ \ \ \ c = 1,\ \ (0 < \Delta x \le 1)$

14 $f(x) = e^{x^2},\ \ \ \ c = 0,\ \ (0 < \Delta x \le 1)$

In Problems 15–20, find $f'(c)$ and an error estimate for the limit

$$f'(c) = \lim_{\Delta x \to 0^-} \frac{f(c + \Delta x) - f(c)}{\Delta x}$$

with $-1 \le \Delta x < 0$.

15	$f(x) = \sqrt{x},\ \ c = 100$	**16**	$f(x) = 1/(3x + 6),\ \ c = 0$
17	$f(x) = \sqrt{x^2 + 1},\ \ c = 2$	**18**	$f(x) = 4x^3,\ \ c = 1$
19	$f(x) = x\sqrt{x + 1},\ \ c = 1$	**20**	$f(x) = x^{10},\ \ c = 2$

In Problems 21–38, approximate the given quantity and give an estimate of error.

21	$\sqrt{65}$	**22**	$1/\sqrt{50}$
23	$(0.301)^4$	**24**	$\sqrt[5]{30}$
25	$1/97$	**26**	$(99)^{3/2}$

27	$\sqrt{1.02} + \sqrt[3]{1.02}$	28	$(101 + \sqrt{101})^3$
29	$(1.003)^5$	30	$\sqrt[3]{0.9997}$
31	$\sin\left(\dfrac{\pi}{3} + 0.004\right)$	32	$\cos\left(\dfrac{\pi}{2} + 0.06\right)$
33	$\tan(0.005)$	34	$\sin(-0.003)$
35	$e^{0.002}$	36	$e^{-0.04}$
37	$\ln(1.006)$	38	$\ln(0.98)$

EXTRA PROBLEMS FOR CHAPTER 5

In Problems 1–10, find the limit.

1	$\displaystyle\lim_{x\to\infty} \dfrac{2x^2 - 3x + 2}{x^3 + 5x^2 - 1}$	2	$\displaystyle\lim_{x\to\infty} \dfrac{2x + 4}{5 - 3x}$
3	$\displaystyle\lim_{x\to-\infty} x^{-1/3}$	4	$\displaystyle\lim_{x\to\infty} (\sqrt[4]{x + 1} - \sqrt[4]{x})x^{3/4}$
5	$\displaystyle\lim_{x\to(2/3)^+} \dfrac{3x + 2}{3x - 2}$	6	$\displaystyle\lim_{x\to1} \dfrac{x - 1}{\sqrt{x - 1}}$
7	$\displaystyle\lim_{x\to1^+} \dfrac{x^{3/2} - 1}{\sqrt{x - 1}}$	8	$\displaystyle\lim_{x\to4} \dfrac{x^2 + 3x - 1}{x^2 - 16}$
9	$\displaystyle\lim_{x\to\infty} \dfrac{(x + 1)^{3/2} - x^{3/2}}{\sqrt{x}}$	10	$\displaystyle\lim_{x\to\infty} \dfrac{x + \sqrt{x + 1}}{x - \sqrt{x + 1}}$

11 Sketch the curve $y = x - 1/x$.

12 Sketch the curve $y = 1 - x^{1/3}$.

13 Sketch the curve $y = 1/((x - 1)(x - 2))$.

14 Sketch the curve $y^2 - 4x^2 = 9$.

15 Sketch the curve $y = |x - 1| + |x + 1|$.

16 Find the equation of the parabola with directrix $y = 1$ and focus $F(1, -1)$.

17 Sketch the curve $y = -x^2 + 2x + 4$.

18 Sketch the curve $y = (\frac{1}{4})x^2 + x$.

19 Find the foci and sketch the ellipse

$$\frac{x^2}{4} + \frac{y^2}{9} = 1.$$

20 Find the foci and sketch the hyperbola

$$\frac{x^2}{4} - \frac{y^2}{9} = 1.$$

21 Use Translation of Axes to sketch the curve

$$4x^2 + y^2 - 16x + 2y + 16 = 0.$$

22 Use Translation of Axes to sketch the curve

$$-x^2 + 4y^2 - 6x - 10 = 0.$$

23 Use Rotation of Axes to transform the equation $xy - 9 = 0$ into a second degree equation with no XY-term. Find the angle of rotation and the new equation.

24 Use Rotation of Axes to transform the equation $xy - y^2 = 5$ into a second degree equation with no XY-term. Find the angle of rotation and the new equation.

25 In the limit $\lim_{x\to4} 1/\sqrt{x} = 1/2$, find a $\delta > 0$ such that whenever $0 < |x - 4| < \delta$, $|1/\sqrt{x} - 1/2| < 0.01$.

26   In the limit $\lim_{x \to \infty} (x^2 - 1)^{1/2} = \infty$, find a $B > 0$ such that whenever $x > B$, $(x^2 - 1)^{1/2} > 10,000$.

27   Use Newton's method to find an approximate solution to the equation $x + x^{1/3} = 3$.

28   Use Newton's method to find an approximate solution to the equation $\cos x = \ln x$.

29   Find an error estimate for the limit
$$\lim_{\Delta x \to 0^+} \frac{(16 + \Delta x)^{1/4} - 2}{\Delta x} = \frac{1}{32}, \qquad 0 < \Delta x \le 1.$$

30   Find an error estimate for the limit
$$\lim_{\Delta x \to 0^+} \frac{(3 + \Delta x)^{-2} - \frac{1}{9}}{\Delta x} = -\frac{2}{27}, \qquad 0 < \Delta x \le 1.$$

31   Find an approximate value for $(124)^{2/3}$ and give an estimate of error.

32   Find an approximate value for $(0.9996)^6$ and give an estimate of error.

☐ 33   Prove that $\lim_{x \to \infty} f(x)$ exists if and only if whenever H and K are positive infinite, $f(H)$ is finite and $f(H) \approx f(K)$.

☐ 34   Prove that if $\lim_{t \to \infty} f(t) = L$ and $g(x)$ is continuous at $x = L$ then $\lim_{t \to \infty} g(f(t)) = g(L)$.

☐ 35   Prove that if $\lim_{t \to \infty} f(t) = \infty$ and $\lim_{x \to \infty} g(x) = \infty$ then $\lim_{t \to \infty} g(f(t)) = \infty$.

☐ 36   Suppose $\lim_{t \to \infty} f(t) = \infty$, c is a positive constant, and $cg(t) \ge f(t)$ for all t. Prove that $\lim_{t \to \infty} g(t) = \infty$.

☐ 37   Prove that $\lim_{x \to c} f(x) = L$ if and only if for every real $\varepsilon > 0$ there is a hyperreal $\delta > 0$ such that whenever $|x - c| < \delta$, $|f(x) - L| < \varepsilon$.

☐ 38   Let f be the function
$$f(x) = \begin{cases} 1 & \text{if } x \text{ is rational,} \\ 0 & \text{if } x \text{ is irrational.} \end{cases}$$
Using the ε, δ condition, prove that $f(x)$ is discontinuous at every real number $x = c$.

☐ 39   Let g be the function
$$g(x) = \begin{cases} x & \text{if } x \text{ is rational,} \\ 0 & \text{if } x \text{ is irrational.} \end{cases}$$
Prove that $g(x)$ is continuous at $x = 0$ but discontinuous everywhere else.

☐ 40   Prove that the function g in the preceding problem is not differentiable at $x = 0$.

☐ 41   Let
$$h(x) = \begin{cases} x^2 & \text{if } x \text{ is rational,} \\ 0 & \text{if } x \text{ is irrational.} \end{cases}$$
Prove that $h'(0)$ exists and equals 0.

☐ 42   Suppose $f(t)$ is continuous for all t and
$$\lim_{t \to -\infty} f(t) = A, \qquad \lim_{t \to \infty} f(t) = B.$$
If $A < C < B$, prove that there is a real number c with $f(c) = C$.

APPLICATIONS OF THE INTEGRAL

6.1 INFINITE SUM THEOREM

In Chapter 4 we obtained the formula

$$\text{Area} = \int_a^b f(x)\,dx$$

for the area of the region bounded by the x-axis, the curve $y = f(x)$, and the lines $x = a$ and $x = b$.

In this chapter we shall obtain integral formulas for several other quantities arising in geometry and physics, such as volumes, curve lengths, and work. We begin with the Infinite Sum Theorem, which will be useful in justifying these formulas. It tells when a given function $B(a, b)$ is equal to the definite integral $\int_a^b h(x)\,dx$.

Any two infinitesimals are infinitely close to each other. The following definition helps us to keep track of how close to each other they are.

DEFINITION

Let ε, δ be infinitesimals and let Δx be a nonzero infinitesimal. We say that ε is **infinitely close to** δ **compared to** Δx,

$$\varepsilon \approx \delta \qquad (\text{compared to } \Delta x), \quad \text{if} \quad \varepsilon/\Delta x \approx \delta/\Delta x.$$

In Figure 6.1.1, an infinitesimal microscope within an infinitesimal microscope is used to show $\varepsilon \approx \delta$ (compared to Δx).

For example, $3\,\Delta x + 5\,\Delta x^2 \approx 3\,\Delta x - \Delta x^2 + \Delta x^3$ (compared to Δx)
but $3\,\Delta x + 5\,\Delta x^2 \not\approx 2\,\Delta x$ (compared to Δx).

The Infinite Sum Theorem is used when we have a quantity $B(u, w)$ depending on two variables $u < w$ in $[a, b]$, and the total value $B(a, b)$ is the sum of infinitesimal pieces

$$\Delta B = B(x, x + \Delta x).$$

The theorem gives a method of expressing $B(a, b)$ as a definite integral.

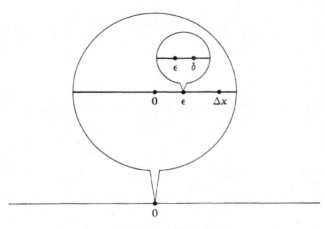

Figure 6.1.1 $\varepsilon \approx \delta$ (compared to Δx)

INFINITE SUM THEOREM

Let $B(u, w)$ be a real function of two variables that has the Addition Property in the interval $[a, b]$—i.e.,

$$B(u, w) = B(u, v) + B(v, w) \qquad \text{for } u < v < w \text{ in } [a, b].$$

Suppose $h(x)$ is a real function continuous on $[a, b]$ and for any infinitesimal subinterval $[x, x + \Delta x]$ of $[a, b]$,

$$\Delta B \approx h(x) \, \Delta x \qquad \text{(compared to } \Delta x\text{)}.$$

Then $B(a, b)$ is equal to the integral

$$B(a, b) = \int_a^b h(x) \, dx.$$

Intuitively, the theorem says that if each infinitely small piece ΔB is infinitely close to $h(x) \, \Delta x$ compared to Δx, then the sum $B(a, b)$ of all these pieces is infinitely close to $\sum_a^b h(x) \, \Delta x$ (Figure 6.1.2). This is why we call it the Infinite Sum Theorem.

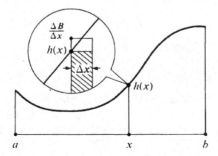

Figure 6.1.2 a x b

PROOF Divide the interval $[a, b]$ into subintervals of infinitesimal length Δx. Because $B(u, w)$ has the Addition Property, the sum of all the ΔB's is $B(a, b)$. Now let c be any positive real number. For each infinitesimal subinterval $[x, x + \Delta x]$ we have

$$h(x) \Delta x \approx \Delta B \qquad \text{(compared to } \Delta x\text{)}$$

$$h(x) \approx \frac{\Delta B}{\Delta x}$$

$$h(x) - c < \frac{\Delta B}{\Delta x} < h(x) + c$$

$$(h(x) - c) \Delta x < \Delta B < (h(x) + c) \Delta x.$$

Adding up, $\qquad \sum_{a}^{b} (h(x) - c) \Delta x < B(a, b) < \sum_{a}^{b} (h(x) + c) \Delta x.$

Now take standard parts,

$$\int_{a}^{b} (h(x) - c) \, dx \leq B(a, b) \leq \int_{a}^{b} (h(x) + c) \, dx$$

or $\qquad \int_{a}^{b} h(x) \, dx - c(b - a) \leq B(a, b) \leq \int_{a}^{b} h(x) \, dx + c(b - a).$

Since this holds for all positive real c, it follows that

$$B(a, b) = \int_{a}^{b} h(x) \, dx.$$

We shall use the Infinite Sum Theorem several times in this chapter. As a first illustration of the method, we derive again the formula from Chapter 4 for the area of the region between two curves, shown in Figure 6.1.3.

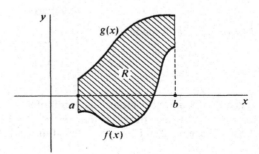

Figure 6.1.3

AREA BETWEEN TWO CURVES   Area $= \int_{a}^{b} [g(x) - f(x)] \, dx.$

where f and g are continuous and $f(x) \leq g(x)$ for $a \leq x \leq b$.

The justification of a definition resembles the proof of a theorem, but it shows that an intuitive concept is equivalent to a mathematical one. We shall now use the Infinitive Sum Theorem to give a justification of the formula for the area between two curves.

JUSTIFICATION We write $A(a, b)$ for the intuitive area of the region R between $f(x)$ and $g(x)$ from a to b. $A(u, w)$ has the Addition Property. Slice R into vertical strips of infinitesimal width Δx. Each strip is almost a rectangle of height $g(x) - f(x)$ and width Δx (Figure 6.1.4). The area $\Delta A = A(x, x + \Delta x)$ of the strip is infinitely close to the area of the rectangle compared to Δx,

$$\Delta A \approx [g(x) - f(x)] \Delta x \qquad \text{(compared to } \Delta x\text{)}.$$

The infinite sum theorem now shows that $A(a, b)$ is the integral of $g(x) - f(x)$ from a to b.

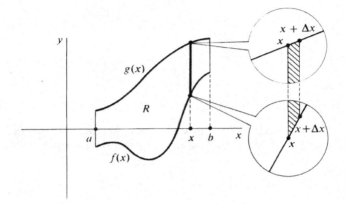

Figure 6.1.4

We now use the Infinite Sum Theorem to derive a formula for the volume of a solid when the area of each cross section is known. Suppose a solid S extends in the direction of the x-axis from $x = a$ to $x = b$, and for each x the plane perpendicular to the x-axis cuts the solid in a region of area $A(x)$, as shown in Figure 6.1.5. The area $A(x)$ is called the cross section of the solid at x. The volume is given by the formula:

VOLUME OF A SOLID $V = \displaystyle\int_a^b A(x)\, dx.$

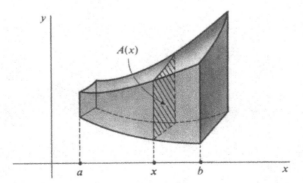

Figure 6.1.5

JUSTIFICATION Slice the solid S into vertical slabs of infinitesimal thickness Δx, as in Figure 6.1.6. Each slab, between x and $x + \Delta x$, has a face of area $A(x)$,

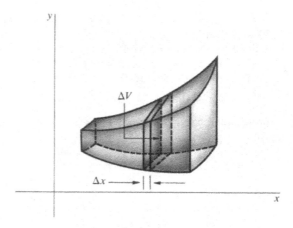

Figure 6.1.6

and thus its volume is given by

$$\Delta V \approx A(x)\, \Delta x \qquad \text{(compared to } \Delta x).$$

(The infinitesimal error arises because the area of the cross section changes slightly between x and $x + \Delta x$.) Then by the Infinite Sum Theorem,

$$V = \int_a^b A(x)\, dx.$$

The pattern used in justifying the two formulas in this section will be repeated again and again. First find a formula for an infinitesimal piece of volume ΔV. Then apply the Infinite Sum Theorem to get an integration formula for the total volume V.

EXAMPLE 1   Find the volume of a pyramid of height h whose base has area B, as in Figure 6.1.7.

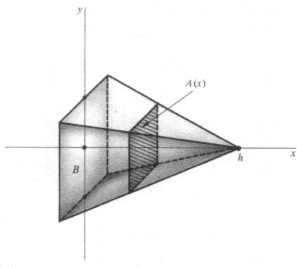

Figure 6.1.7   Example 1

Place the pyramid on its side with the apex at $x = 0$ and the base at $x = h$. We use the fact that at any point x between 0 and h, the cross section has area proportional to x^2, so that

$$\frac{A(x)}{x^2} = \frac{B}{h^2},$$

$$A(x) = \frac{Bx^2}{h^2}.$$

The volume is then

$$V = \int_0^h \frac{Bx^2}{h^2}\, dx = \frac{1}{3} \cdot \frac{Bx^3}{h^2}\Bigg]_0^h = \frac{1}{3} \cdot \frac{Bh^3}{h^2} = \frac{1}{3} Bh.$$

The solution is $V = (\tfrac{1}{3})Bh$.

EXAMPLE 2 A wedge is cut from a cylindrical tree trunk of radius 3 ft, by cutting the tree with two planes meeting on a line through the axis of the cylinder. The wedge is 1 ft thick at its thickest point. Find its volume.

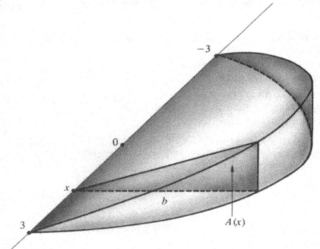

Figure 6.1.8 Example 2

The wedge is shown in Figure 6.1.8. The cross sections perpendicular to the x-axis are similar triangles. Place the edge along the x-axis with x from -3 to 3. At the thickest point, where $x = 0$, the cross section is a triangle with base 3 ft and altitude 1 ft. The base of the cross section triangle at x is

$$b = \sqrt{9 - x^2},$$

and the altitude is

$$\tfrac{1}{3}b = \tfrac{1}{3}\sqrt{9 - x^2}.$$

The area of the cross section is

$$A(x) = \tfrac{1}{2} \cdot \text{base} \cdot \text{altitude} = \tfrac{1}{2}b \cdot \tfrac{1}{3}b = \tfrac{1}{6}b^2 = \tfrac{1}{6}(9 - x^2).$$

The volume is thus

$$V = \int_{-3}^{3} A(x)\,dx = \int_{-3}^{3} \frac{1}{6}(9 - x^2)\,dx = \left(\frac{3}{2}x - \frac{1}{18}x^3\right)\Big]_{-3}^{3} = 6\,\text{ft}^3.$$

The solution is 6 cubic feet.

PROBLEMS FOR SECTION 6.1

1   The base of a solid is the triangle in the x, y-plane with vertices at $(0, 0)$, $(0, 1)$, and $(1, 0)$. The cross sections perpendicular to the x-axis are squares with one side on the base. Find the volume of the solid.

2   The base of a solid is the region in the x, y-plane bounded by the parabola $y = x^2$ and the line $y = 1$. The cross sections perpendicular to the x-axis are squares with one side on the base. Find the volume of the solid.

3   Find the volume of the solid in Problem 1 if the cross sections are equilateral triangles with one side on the base.

4   Find the volume of the solid in Problem 2 if the cross sections are equilateral triangles with one side on the base.

5   Find the volume of the solid in Problem 1 if the cross sections are semicircles with diameter on the base.

6   Find the volume of the solid in Problem 2 if the cross sections are semicircles with diameter on the base.

7   Find the volume of a wedge cut from a circular cylinder of radius r by two planes whose line of intersection passes through the axis of the cylinder, if the wedge has thickness c at its thickest point.

8   Find the volume of the smaller wedge cut from a circular cylinder of radius r by two planes whose line of intersection is a chord at distance b from the axis of the cylinder, if the greatest thickness is c.

6.2 VOLUMES OF SOLIDS OF REVOLUTION

Integrals are used in this section to find the volume of a solid of revolution. A solid of revolution is generated by taking a region in the first quadrant of the plane and rotating it in space about the x- or y-axis (Figure 6.2.1).

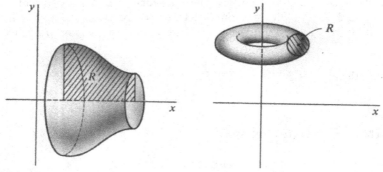

Figure 6.2.1   Solids of Revolution

We shall work with the region under a curve and the region between two curves. We use one method for rotating about the axis of the independent variable and another for rotating about the axis of the dependent variable.

For areas our starting point was the formula

$$\text{area} = \text{base} \times \text{height}$$

for the area of a rectangle. For volumes of a solid of revolution our starting point is the usual formula for the volume of a right circular cylinder (Figure 6.2.2).

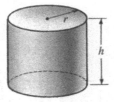

Figure 6.2.2

DEFINITION

*The **volume** of a right circular cylinder with height h and base of radius r is*

$$V = \pi r^2 h.$$

DISC METHOD: For rotations about the axis of the *independent variable.*

Let us first consider the region under a curve. Let R be the region under a curve $y = f(x)$ from $x = a$ to $x = b$, shown in Figure 6.2.3(a). x is the independent

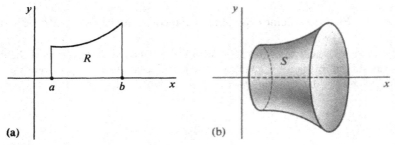

(a)   (b)

Figure 6.2.3

variable in this case. To keep R in the first quadrant we assume $0 \leq a < b$ and $0 \leq f(x)$. Rotate R about the x-axis, generating the solid of revolution S shown in Figure 6.2.3(b).

This volume is given by the formula below.

VOLUME BY DISC METHOD $V = \displaystyle\int_a^b \pi(f(x))^2 \, dx.$

To justify this formula we slice the region R into vertical strips of infinitesimal width Δx. This slices the solid S into discs of infinitesimal thickness Δx. Each disc is almost a cylinder of height Δx whose base is a circle of radius $f(x)$ (Figure 6.2.4). Therefore

$$\Delta V = \pi(f(x))^2 \, \Delta x \qquad \text{(compared to } \Delta x).$$

Then by the Infinite Sum Theorem we get the desired formula

$$V = \int_a^b \pi(f(x))^2 \, \Delta x.$$

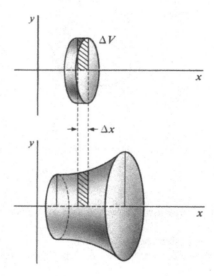

Figure 6.2.4 Disc Method

EXAMPLE 1   Find the volume of a right circular cone with height h and base of radius r.

It is convenient to center the cone on the x-axis with its vertex at the origin as shown in Figure 6.2.5. This cone is the solid generated by rotating about the x-axis the triangular region R under the line $y = (r/h)x, 0 \leq x \leq h$.

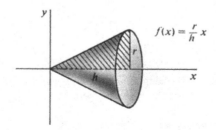

Figure 6.2.5

Since x is the independent variable we use the Disc Method. The volume formula gives

$$V = \int_0^h \pi \left(\frac{r}{h}x\right)^2 dx = \pi \frac{r^2}{h^2} \int_0^h x^2\, dx = \pi \frac{r^2}{h^2} \frac{x^3}{3}\bigg]_0^h = \frac{1}{3}\pi r^2 h,$$

or
$$V = \tfrac{1}{3}\pi r^2 h.$$

Now we consider the region R between two curves $y = f(x)$ and $y = g(x)$ from $x = a$ to $x = b$. Rotating R about the x-axis generates a solid of revolution S shown in Figure 6.2.6(c).

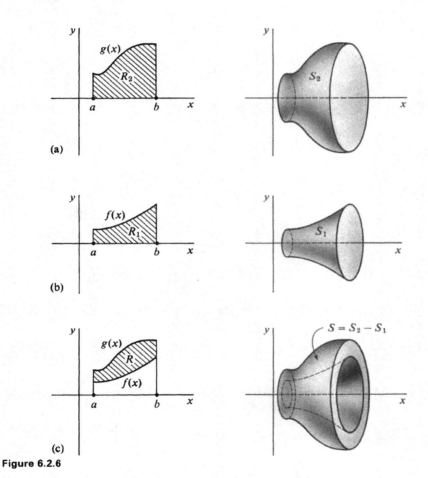

Figure 6.2.6

Let R_1 be the region under the curve $y = f(x)$ shown in Figure 6.2.6(b), and R_2, the region under the curve $y = g(x)$, shown in Figure 6.2.6(a). Then S can be found by removing the solid of revolution S_1 generated by R_1 from the solid of revolution S_2 generated by R_2. Therefore

volume of S = volume of S_2 − volume of S_1.

This justifies the formula

$$V = \int_a^b \pi(g(x))^2 \, dx - \int_a^b \pi(f(x))^2 \, dx.$$

We combine this into a single integral.

VOLUME BY DISC METHOD $V = \int_a^b \pi[(g(x))^2 - (f(x))^2] \, dx.$

Another way to see this formula is to divide the solid into annular discs (washers) with inner radius $f(x)$ and outer radius $g(x)$, as illustrated in Figure 6.2.7.

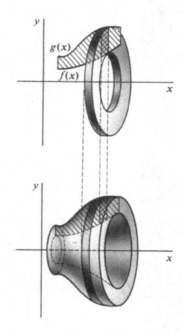

Figure 6.2.7

EXAMPLE 2   The region R between the curves $y = 2 - x^2$ and $y = x^2$ is rotated about the x-axis generating a solid S. Find the volume of S.

The curves $y = 2 - x^2$ and $y = x^2$ cross at $x = \pm 1$. The region is sketched in Figure 6.2.8. The volume is

$$V = \int_{-1}^1 \pi(2 - x^2)^2 \, dx - \int_{-1}^1 \pi(x^2)^2 \, dx$$

$$= \int_{-1}^1 \pi(2 - x^2)^2 - \pi x^4 \, dx$$

$$= \int_{-1}^1 \pi(4 - 4x^2) \, dx = \pi(4x - \tfrac{4}{3}x^3)\Big]_{-1}^1 = 16\pi/3.$$

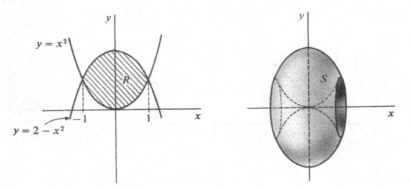

Figure 6.2.8

Warning: When using the disc method for a region between two curves, the correct formula is

$$V = \int_a^b \pi(g(x))^2\, dx - \int_a^b \pi(f(x))^2\, dx,$$

or

$$V = \int_a^b \pi[(g(x))^2 - (f(x))^2]\, dx.$$

A common mistake is to subtract $f(x)$ from $g(x)$ before squaring.

Wrong: $V = \int_a^b \pi(g(x) - f(x))^2\, dx.$

Wrong: (for Example 2):

$$V = \int_{-1}^1 \pi((2 - x^2) - x^2)^2\, dx = \int_{-1}^1 \pi(2 - 2x^2)^2\, dx$$

$$= \int_{-1}^1 \pi(4 - 8x^2 + 4x^4)\, dx = 64\pi/15.$$

CYLINDRICAL SHELL METHOD: For rotations about the axis of the *dependent variable.*

Let us again consider the region R under a curve $y = f(x)$ from $x = a$ to $x = b$, so that x is still the independent variable. This time rotate R about the y-axis to generate a solid of revolution S (Figure 6.2.9).

VOLUME BY CYLINDRICAL SHELL METHOD $V = \int_a^b 2\pi x f(x)\, dx.$

Let us justify this formula. Divide R into vertical strips of infinitesimal width Δx as shown in Figure 6.2.10. When a vertical strip is rotated about the y-axis it generates a *cylindrical shell* of thickness Δx and volume ΔV. This cylindrical shell is the difference between an outer cylinder of radius $x + \Delta x$ and an inner cylinder of radius Δx. Both cylinders have height infinitely close to $f(x)$. Thus compared to Δx,

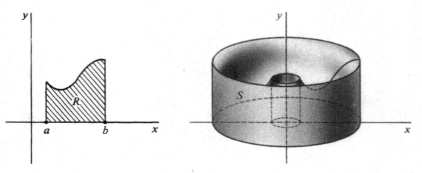

Figure 6.2.9

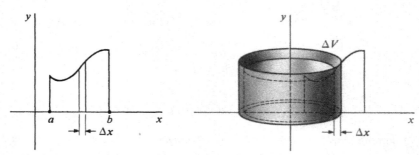

Figure 6.2.10   Cylindrical Shell Method

$$\Delta V \approx \text{outer cylinder} - \text{inner cylinder}$$
$$\approx \pi(x + \Delta x)^2 f(x) - \pi x^2 f(x)$$
$$= \pi(x^2 + 2x\,\Delta x + (\Delta x)^2 - x^2)f(x)$$
$$= \pi(2x\,\Delta x + (\Delta x)^2)f(x) \approx \pi 2x\,\Delta x f(x),$$

whence $\Delta V \approx 2\pi x f(x)\,\Delta x$ (compared to Δx).

By the Infinite Sum Theorem,

$$V = \int_a^b 2\pi x f(x)\,dx.$$

EXAMPLE 3   The region R between the line $y = 0$ and the curve $y = 2x - x^2$ is rotated about the y-axis to form a solid of revolution S. Find the volume of S.

We use the cylindrical shell method because y is the dependent variable. We see that the curve crosses the x-axis at $x = 0$ and $x = 2$, and sketch the region in Figure 6.2.11. The volume is

$$V = \int_0^2 2\pi x(2x - x^2)\,dx = 2\pi \int_0^2 2x^2 - x^3\,dx = 2\pi(\tfrac{2}{3}x^3 - \tfrac{1}{4}x^4)\Big]_0^2 = \tfrac{8}{3}\pi.$$

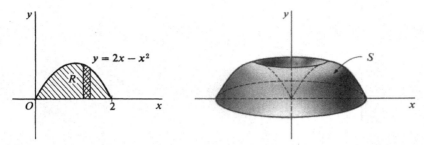

Figure 6.2.11

Now let R be the region between the curves $y = f(x)$ and $y = g(x)$ for $a \leq x \leq b$, and generate the solid S by rotating R about the y-axis. The volume of S can be found by subtracting the volume of the solid S_1 generated by the region under $y = f(x)$ from the volume of the solid S_2 generated by the region under $y = g(x)$ (Figure 6.2.12). The formula for the volume is

$$V = S_2 - S_1 = \int_a^b 2\pi x g(x)\, dx - \int_a^b 2\pi x f(x)\, dx.$$

Combining into one integral, we get

VOLUME BY CYLINDRICAL SHELL METHOD $\quad V = \int_a^b 2\pi x (g(x) - f(x))\, dx.$

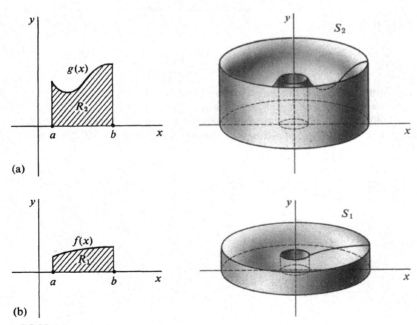

(a)

(b)

Figure 6.2.12

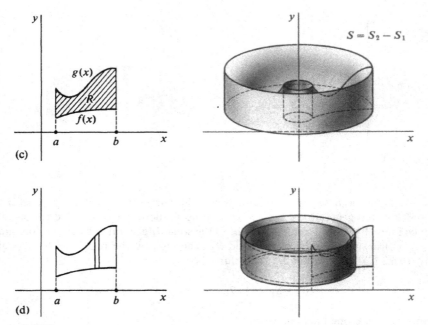

(c)

(d)

Figure 6.2.12

EXAMPLE 4   The region between the curves $y = x$ and $y = \sqrt{x}$ is rotated about the y-axis. Find the volume of the solid of revolution.

We make a sketch in Figure 6.2.13 and find that the curves cross at $x = 0$ and $x = 1$. We take x for the independent variable and use the Cylindrical Shell Method.

$$V = \int_0^1 2\pi x(\sqrt{x} - x)\, dx = \int_0^1 2\pi(x^{3/2} - x^2)\, dx = 2\pi(\tfrac{2}{5}x^{5/2} - \tfrac{1}{3}x^3)\Big]_0^1 = \tfrac{2}{15}\pi.$$

Some regions R are more easily described by taking y as the independent variable, so that R is the region between $x = f(y)$ and $x = g(y)$ for $c \le y \le d$. The volumes of the solids of revolution are then computed by integrating with respect to y. Often we have a choice of either x or y as the independent variable.

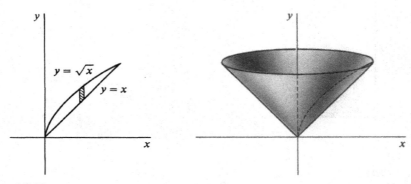

Figure 6.2.13

How can one decide whether to use the Disc or Cylindrical Shell Method? The answer depends on both the axis of rotation and the choice of independent variable. *Use the Disc Method when rotating about the axis of the independent variable. Use the Cylindrical Shell Method when rotating about the axis of the dependent variable.*

EXAMPLE 5   Derive the formula $V = \frac{4}{3}\pi r^3$ for the volume of a sphere by both the Disc Method and the Cylindrical Shell Method.

The circle of radius r and center at the origin has the equation

$$x^2 + y^2 = r^2.$$

The region R inside this circle in the first quadrant will generate a hemisphere of radius r when it is rotated about the x-axis (Figure 6.2.14).

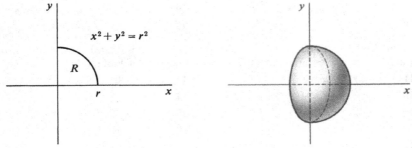

Figure 6.2.14

First take x as the independent variable and use the Disc Method. R is the region under the curve

$$y = \sqrt{r^2 - x^2}, \qquad 0 \le x \le r.$$

The hemisphere has volume

$$\frac{1}{2}V = \int_0^r \pi(f(x))^2 \, dx$$

$$= \int_0^r \pi(r^2 - x^2) \, dx = \pi r^2 x - \tfrac{1}{3}\pi x^3 \Big]_0^r$$

$$= \pi r^3 - \tfrac{1}{3}\pi r^3 = \tfrac{2}{3}\pi r^3.$$

Therefore the sphere has volume

$$V = \tfrac{4}{3}\pi r^3.$$

Now take y as the independent variable and use the Cylindrical Shell Method. R is the region under the curve

$$x = \sqrt{r^2 - y^2}, \qquad 0 \le y \le r.$$

The hemisphere has volume

$$\frac{1}{2}V = \int_0^r 2\pi y \sqrt{r^2 - y^2} \, dy.$$

Putting $u = r^2 - y^2$, $du = -2y\,dy$, we get

$$\tfrac{1}{2}V = \int_{r^2}^{0} 2\pi\sqrt{u}(-\tfrac{1}{2})\,du = \int_{r^2}^{0} -\pi\sqrt{u}\,du = -\tfrac{2}{3}\pi u^{3/2}\Big]_{r^2}^{0} = \tfrac{2}{3}\pi r^3.$$

Thus again $V = \tfrac{4}{3}\pi r^3$.

PROBLEMS FOR SECTION 6.2

In Problems 1–10 the region under the given curve is rotated about (a) the x-axis, (b) the y-axis. Sketch the region and find the volumes of the two solids of revolution.

1	$y = x^2$, $0 \le x \le 1$	2	$y = x^3$, $0 \le x \le 1$
3	$y = \sqrt{x}$, $0 \le x \le 4$	4	$y = \sqrt{2x - 4}$, $2 \le x \le 4$
5	$y = 1 - x$, $0 \le x \le 1$	6	$y = x$, $1 \le x \le 2$
7	$y = \sqrt{1 + x^2}$, $0 \le x \le 1$	8	$y = \sqrt{x^2 - 4}$, $2 \le x \le 4$
9	$y = x^{-3}$, $1 \le x \le 2$	10	$y = 1/x$, $1 \le x \le 2$

In Problems 11–22 the region bounded by the given curves is rotated about (a) the x-axis, (b) the y-axis. Sketch the region and find the volumes of the two solids of revolution.

11	$x, y \ge 0$, $y = x^2\sqrt{1 - x^4}$	12	$y = 0$, $y = x - x^2$
13	$y = x$, $y = 2x$, $0 \le x \le 3$	14	$y = x^2$, $y = x$
15	$y = x^3$, $y = x^2$	16	$y = 3/x$, $y = 4 - x$
17	$x = 0$, $x = y - y^4$	18	$x = y$, $x = 2y - y^2$
19	$x = 0$, $x = y + 1/y$, $1 \le y \le 2$		
20	$x \ge 0$, $y \ge 0$, $2x^2 + y^2 = 4$		
21	$y = 0$, $y = x - 2$, $y = \sqrt{x}$		
22	$y = \tfrac{3}{4}x$, $y = 1 - x$, $y = x - 1/x$ (first quadrant)		

In Problems 23–34 the region under the given curve is rotated about the x-axis. Find the volume of the solid of revolution.

23	$y = \sqrt{\sin x}$, $0 \le x \le \pi$		
24	$y = \cos x\sqrt{\sin x}$, $0 \le x \le \pi/2$		
25	$y = \cos x - \sin x$, $0 \le x \le \pi/4$		
26	$y = \sin(x/2) + \cos(x/2)$, $0 \le x \le \pi$		
27	$y = e^x$, $0 \le x \le 1$	28	$y = e^{1 - 2x}$, $0 \le x \le 2$
29	$y = xe^{x^3}$, $0 \le x \le 1$	30	$y = \sqrt{e^x + 1}$, $0 \le x \le 3$
31	$y = 1/\sqrt{x}$, $1 \le x \le 2$	32	$y = \dfrac{1}{\sqrt{2x + 1}}$, $0 \le x \le 1$
33	$y = \sqrt{\dfrac{x - 1}{x}}$, $1 \le x \le 4$	34	$y = \sqrt{\dfrac{2x}{x + 1}}$, $0 \le x \le 1$

In Problems 35–46 the region is rotated about the x-axis. Find the volume of the solid of revolution.

35	$y = \dfrac{\sin x}{x}$, $\pi/2 \le x \le \pi$	36	$y = \dfrac{\cos x}{x}$, $\pi/6 \le x \le \pi/2$
37	$y = \sin(x^2)$, $0 \le x \le \sqrt{\pi}$	38	$y = \cos(x^2)$, $0 \le x \le \sqrt{\pi/2}$
39	$y = e^{x^2}$, $0 \le x \le 1$	40	$y = e^x/x$, $1 \le x \le 2$
41	$y = 1/xe^x$, $1 \le x \le 4$	42	$y = xe^{x^3}$, $1 \le x \le 2$

43 $y = x^{-2}, \quad 1 \leq x \leq 2$ **44** $y = \dfrac{1}{x^2 + 1}, \quad 0 \leq x \leq 2$

45 $y = \dfrac{1}{2x^2 - 1}, \quad 1 \leq x \leq 2$ **46** $y = \dfrac{\ln x}{x^2}, \quad 1 \leq x \leq 2$

47   A hole of radius a is bored through the center of a sphere of radius r ($a < r$). Find the volume of the remaining part of the sphere.

48   A sphere of radius r is cut by a horizontal plane at a distance c above the center of the sphere. Find the volume of the part of the sphere above the plane ($c < r$).

49   A hole of radius a is bored along the axis of a cone of height h and base of radius r. Find the remaining volume ($a < r$).

50   Find the volume of the solid generated by rotating an ellipse $a^2x^2 + b^2y^2 = 1$ about the x-axis. *Hint:* The portion of the ellipse in the first quadrant will generate half the volume.

51   The sector of a circle shown in the figure is rotated about (a) the x-axis, (b) the y-axis. Find the volumes of the solids of revolution.

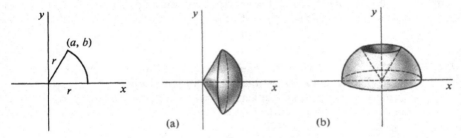

(a)   (b)

52   The region bounded by the curves $y = x^2$, $y = x$ is rotated about (a) the line $y = -1$, (b) the line $x = -2$. Find the volumes of the solids of revolution.

53   Find the volume of the torus (donut) generated by rotating the circle of radius r with center at $(c, 0)$ around the y-axis ($r < c$).

☐ **54**   (a) Find a general formula for the volume of the solid of revolution generated by rotating the region bounded by the curves $y = f(x), y = g(x), a \leq x \leq b$, about the line $y = -k$.
          (b) Do the same for a rotation about the line $x = -h$.

6.3 LENGTH OF A CURVE

A segment of a curve in the plane (Figure 6.3.1) is described by

$$y = f(x), \qquad a \leq x \leq b.$$

What is its length? As usual, we shall give a definition and then justify it. A curve $y = f(x)$ is said to be *smooth* if its derivative $f'(x)$ is continuous. Our definition will assign a length to a segment of a smooth curve.

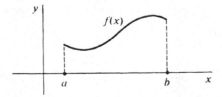

Figure 6.3.1

DEFINITION

Assume the function $y = f(x)$ has a continuous derivative for x in $[a, b]$, that is, the curve

$$y = f(x), \quad a \le x \le b$$

*is smooth. The **length** of the curve is defined as*

$$s = \int_a^b \sqrt{1 + (dy/dx)^2} \, dx.$$

Because $\sqrt{1 + (dy/dx)^2} \, dx = \sqrt{dx^2 + dy^2}$, the equation is sometimes written in the form

$$s = \int_a^b \sqrt{dx^2 + dy^2}$$

with the understanding that x is the independent variable. The length s is always greater than or equal to 0 because $a < b$ and

$$\sqrt{1 + (dy/dx)^2} > 0.$$

JUSTIFICATION   Let $s(u, w)$ be the intuitive length of the curve between $t = u$ and $t = w$. The function $s(u, w)$ has the Addition Property; the length of the curve from u to w equals the length from u to v plus the length from v to w. Figure 6.3.2 shows an infinitesimal piece of the curve from x to $x + \Delta x$. Its length is $\Delta s = s(x, x + \Delta x)$.

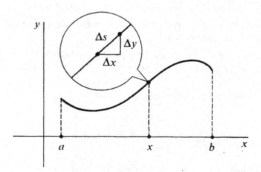

Figure 6.3.2

The slope dy/dx is a continuous function of x, and therefore changes only by an infinitesimal amount between x and $x + \Delta x$. Thus the infinitesimal piece of the curve is almost a straight line, the hypotenuse of a right triangle with sides Δx and Δy. Hence

$$\Delta s \approx \sqrt{\Delta x^2 + \Delta y^2} \quad \text{(compared to } \Delta x\text{)}.$$

Dividing by Δx,

$$\frac{\Delta s}{\Delta x} \approx \frac{\sqrt{\Delta x^2 + \Delta y^2}}{\Delta x} = \sqrt{\left(\frac{\Delta x}{\Delta x}\right)^2 + \left(\frac{\Delta y}{\Delta x}\right)^2} \approx \sqrt{1 + \left(\frac{dy}{dx}\right)^2}.$$

Then $\Delta s \approx \sqrt{1 + (dy/dx)^2} \, \Delta x$ (compared to Δx).

Using the Infinite Sum Theorem,

$$s(a, b) = \int_a^b \sqrt{1 + (dy/dx)^2} \, dx.$$

EXAMPLE 1   Find the length of the curve

$$y = 2x^{3/2}, \qquad 0 \le x \le 1$$

shown in Figure 6.3.3. We have

$$dy/dx = 3x^{1/2}, \qquad s = \int_0^1 \sqrt{1 + 9x} \, dx.$$

Put $u = 1 + 9x$. Then

$$s = \int_1^{10} \tfrac{1}{9}\sqrt{u} \, du = \tfrac{2}{3} \cdot \tfrac{1}{9} u^{3/2} \Big]_1^{10} = \tfrac{2}{27}(\sqrt{1000} - 1).$$

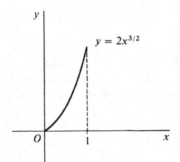

Figure 6.3.3

Sometimes a curve in the (x, y) plane is given by parametric equations

$$x = f(t), \qquad y = g(t), \qquad c \le t \le d.$$

A natural example is the path of a moving particle where t is time. We give a formula for the length of such a curve.

DEFINITION

Suppose the functions

$$x = f(t), \qquad y = g(t)$$

have continuous derivatives and the parametric curve does not retrace its path for t in $[a, b]$. *The length of the curve is defined by*

$$s = \int_a^b \sqrt{(dx/dt)^2 + (dy/dt)^2} \, dt.$$

JUSTIFICATION   The infinitesimal piece of the curve (Figure 6.3.4) from t to $t + \Delta t$

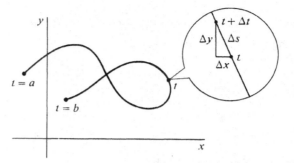

Figure 6.3.4

is almost a straight line, so its length Δs is given by

$$\Delta s \approx \sqrt{\Delta x^2 + \Delta y^2} \qquad \text{(compared to } \Delta t\text{),}$$

$$\Delta s \approx \sqrt{(dx/dt)^2 + (dy/dt)^2}\, \Delta t \qquad \text{(compared to } \Delta t\text{).}$$

By the Infinite Sum Theorem,

$$s = \int_a^b \sqrt{(dx/dt)^2 + (dy/dt)^2}\, dt.$$

The general formula for the length of a parametric curve reduces to our first formula when the curve is given by a simple equation $x = g(y)$ or $y = f(x)$.

If $y = f(x)$, $a \le x \le b$, we take $x = t$ and get

$$s = \int_a^b \sqrt{1 + (dy/dx)^2}\, dx.$$

If $x = g(y)$, $a \le y \le b$, we take $y = t$ and get

$$s = \int_a^b \sqrt{(dx/dy)^2 + 1}\, dy.$$

EXAMPLE 2   Find the length of the path of a ball whose motion is given by

$$x = 20t, \qquad y = 32t - 16t^2$$

from $t = 0$ until the ball hits the ground. (Ground level is $y = 0$, see Figure 6.3.5.) The ball is at ground level when

$$32t - 16t^2 = 0, \qquad t = 0 \quad \text{and} \quad t = 2.$$

We have
$$dx/dt = 20, \qquad dy/dt = 32 - 32t,$$

$$s = \int_0^2 \sqrt{20^2 + (32 - 32t)^2}\, dt.$$

We cannot evaluate this integral yet, so the answer is left in the above form. We can get an approximate answer by the Trapezoidal Rule. When $\Delta x = \frac{1}{5}$, the Trapezoidal Approximation is

$$s \sim 53.5 \qquad \text{error} \le 0.4.$$

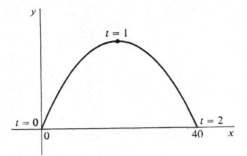

Figure 6.3.5

The following example shows what happens when a parametric curve does retrace its path.

EXAMPLE 3   Let

$$x = 1 - t^2, \qquad y = 1, \qquad -1 \le t \le 1.$$

As t goes from -1 to 1, the point (x, y) moves from $(0, 1)$ to $(1, 1)$ and then back along the same line to $(0, 1)$ again. The path is shown in Figure 6.3.6.

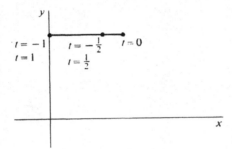

Figure 6.3.6

The path has length one. However, the point goes along the path twice for a total distance of two. The length formula gives the total distance the point moves.

$$s = \int_{-1}^{1} \sqrt{(dx/dt)^2 + (dy/dt)^2} \, dt = \int_{-1}^{1} \sqrt{(-2t)^2 + 0^2} \, dt$$

$$= \int_{-1}^{1} \sqrt{4t^2} \, dt = \int_{-1}^{1} 2|t| \, dt = 2.$$

We next prove a theorem which shows the connection between the length of an arc and the area of a sector of a circle. Given two points P and Q on a circle with center O, the *arc PQ* is the portion of the circle traced out by a point moving from P to Q in a counterclockwise direction. The *sector POQ* is the region bounded by the arc PQ and the radii OP and OQ as shown in Figure 6.3.7.

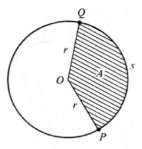

Figure 6.3.7

THEOREM

Let P and Q be two points on a circle with center O. The area A of the sector POQ is equal to one half the radius r times the length s of the arc PQ,

$$A = \tfrac{1}{2}rs.$$

DISCUSSION   The theorem is intuitively plausible because if we consider an infinitely small arc Δs of the circle as in Figure 6.3.8, then the corresponding sector is almost a triangle of height r and base Δs, so it has area

$$\Delta A \approx \tfrac{1}{2}r\,\Delta s \qquad \text{(compared to } \Delta s\text{)}.$$

Summing up, we expect that $A = \tfrac{1}{2}rs$.

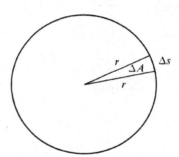

Figure 6.3.8

We can derive the formula $C = 2\pi r$ for the circumference of a circle using the theorem. By definition, π is the area of a circle of radius one,

$$\pi = \int_{-1}^{1} 2\sqrt{1 - y^2}\, dy.$$

Then a circle of radius r has area

$$A = \int_{-r}^{r} 2\sqrt{r^2 - y^2}\, dy = \int_{-1}^{1} 2r^2\sqrt{1 - (y/r)^2}\, d(y/r) = \pi r^2.$$

Therefore the circumference C is given by

$$A = \tfrac{1}{2}rC, \qquad \pi r^2 = \tfrac{1}{2}rC, \qquad C = 2\pi r.$$

PROOF OF THEOREM To simplify notation assume that the center O is at the origin, P is the point $(0, r)$ on the x-axis, and Q is a point (x, y) which varies along the circle (Figure 6.3.9). We may take y as the independent variable and

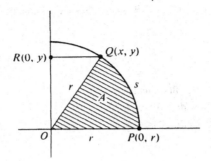

Figure 6.3.9

use the equation $x = \sqrt{r^2 - y^2}$ for the right half of the circle. Then A and s depend on y. Our plan is to show that

$$\frac{dA}{dy} = \frac{1}{2} r \frac{ds}{dy}.$$

First, we find dx/dy:

$$\frac{dx}{dy} = \frac{-y}{\sqrt{r^2 - y^2}} = -\frac{y}{x}.$$

Using the definition of arc length,

$$\frac{ds}{dy} = \sqrt{1 + \left(\frac{dx}{dy}\right)^2} = \sqrt{1 + \frac{y^2}{x^2}} = \sqrt{\frac{x^2 + y^2}{x^2}} = \frac{r}{x}.$$

The triangle OQR in the figure has area $\frac{1}{2}xy$, so the sector has area

$$A = \int_0^y x \, dy - \frac{1}{2} xy.$$

Then $$\frac{dA}{dy} = x - \frac{1}{2}\left(x + y\frac{dx}{dy}\right) = \frac{1}{2}x - \frac{1}{2}y\left(-\frac{y}{x}\right) = \frac{1}{2}\left(x + \frac{y^2}{x}\right)$$

$$= \frac{1}{2}\left(\frac{x^2 + y^2}{x}\right) = \frac{1}{2}\frac{r^2}{x}.$$

Thus $$\frac{dA}{dy} = \frac{1}{2}\frac{r^2}{x}, \qquad \frac{ds}{dy} = \frac{r}{x}, \qquad \frac{dA}{dy} = \frac{1}{2}r\frac{ds}{dy}.$$

So A and $\frac{1}{2}rs$ differ and only by a constant. But when $y = 0$, $A = \frac{1}{2}rs = 0$. Therefore $A = \frac{1}{2}rs$.

To prove the formula $A = \frac{1}{2}rs$ for arcs which are not within a single quadrant we simply cut the arc into four pieces each of which 'is within a single quadrant.

PROBLEMS FOR SECTION 6.3

Find the lengths of the following curves.

1   $y = \frac{2}{3}(x + 2)^{3/2}, \quad 0 \le x \le 3$ 2 $\bullet \, y = (x^2 + \frac{2}{3})^{3/2}, \quad -2 \le x \le 5$

3   $(3y - 1)^2 = x^3, \quad 0 \le x \le 2$ 4 $y = (4/5)x^{5/4}, \quad 0 \le x \le 1$

5   $y = (x - 1)^{2/3}, \quad 1 \le x \le 9$ *Hint:* Solve for x as a function of y.

6   $y = \dfrac{x^3}{12} + \dfrac{1}{x}, \quad 1 \le x \le 3$ 7 $x = \dfrac{y^4 + 3}{6y}, \quad 3 \le y \le 6$

8   $y = \frac{1}{3}x\sqrt{x} - \sqrt{x}, \quad 1 \le x \le 100$ 9 $y = \frac{3}{5}x^{5/3} - \frac{3}{4}x^{1/3}, \quad 1 \le x \le 8$

10   $8x = 2y^4 + y^{-2}, \quad 1 \le y \le 2$

11   $x^{2/3} + y^{2/3} = 1, \quad$ first quadrant

12   $y = \int_0^x \sqrt{t^2 + 2t} \, dt, \quad 0 \le x \le 10$

13   $y = 2\int_1^x \sqrt{t^2 + t} \, dt, \quad 2 \le x \le 6$

14   $y = \int_1^{2x} \sqrt{t^{-4} + t^{-2}} \, dt, \quad 1 \le x \le 3$

15   $x = \int_1^y \sqrt{\sqrt{t} - 1} \, dt, \quad 1 \le y \le 4$

16   $y = \int_0^{x^2} (\sqrt{t} + 1)^{-2} \, dt, \quad 0 \le x \le 1$

17   Find the distance travelled from $t = 0$ to $t = 1$ by an object whose motion is $x = t^{3/2}$, $y = (3 - t)^{3/2}$.

18   Find the distance moved from $t = 0$ to $t = 1$ by a particle whose motion is given by $x = 4(1 - t)^{3/2}, y = 2t^{3/2}$.

19   Find the distance travelled from $t = 1$ to $t = 4$ by an object whose motion is given by $x = t^{3/2}, y = 9t$.

20   Find the distance travelled from time $t = 0$ to $t = 3$ by a particle whose motion is given by the parametric equations $x = 5t^2, y = t^3$.

21   Find the distance moved from $t = 0$ to $t = 2\pi$ by an object whose motion is $x = \cos t$, $y = \sin t$.

22   Find the distance moved from $t = 0$ to $t = \pi$ by an object with motion $x = 3 \cos 2t$, $y = 3 \sin 2t$.

23   Find the distance moved from $t = 0$ to $t = 2\pi$ by an object with motion $x = \cos^2 t$, $y = \sin^2 t$.

24   Find the distance moved by an object with motion $x = e^t \cos t, y = e^t \sin t, 0 \le t \le 1$.

25   Let $A(t)$ and $L(t)$ be the area under the curve $y = x^2$ from $x = 0$ to $x = t$, and the length of the curve from $x = 0$ to $x = t$, respectively. Find $d(A(t))/d(L(t))$.

In Problems 26–30, find definite integrals for the lengths of the curves, but do not evaluate the integrals.

26   $y = x^3, \quad 0 \le x \le 1$

27   $y = 2x^2 - x + 1, \quad 0 \le x \le 4$

28   $x = 1/t, \quad y = t^2, \quad 1 \le t \le 5$

29   $x = 2t + 1, \quad y = \sqrt{t}, \quad 1 \le t \le 2$

30   The circumference of the ellipse $x^2 + 4y^2 = 1$.

31   Set up an integral for the length of the curve $y = \sqrt{x}, 1 \le x \le 2$, and find the Trapezoidal Approximation where $\Delta x = \frac{1}{4}$.

32   Set up an integral for the length of the curve $x = t^2 - t, y = \frac{4}{3}t^{3/2}, 0 \le t \le 1$, and find the Trapezoidal Approximation where $\Delta t = \frac{1}{4}$.

33   Set up an integral for the length of the curve $y = 1/x, 1 \le x \le 5$, and find the Trapezoidal Approximation where $\Delta x = 1$.

34   Set up an integral for the length of the curve $y = x^2$, $-1 \le x \le 1$, and find the Trapezoidal Approximation where $\Delta x = \frac{1}{2}$.

☐ **35**   Suppose the same curve is given in two ways, by a simple equation $y = F(x)$, $a \le x \le b$ and by parametric equations $x = f(t)$, $y = g(t)$, $c \le t \le d$. Assuming all derivatives are continuous and the parametric curve does not retrace its path, prove that the two formulas for curve length give the same values. *Hint:* Use integration by change of variables.

6.4 AREA OF A SURFACE OF REVOLUTION

When a curve in the plane is rotated about the x- or y-axis it forms a *surface of revolution*, as in Figure 6.4.1.

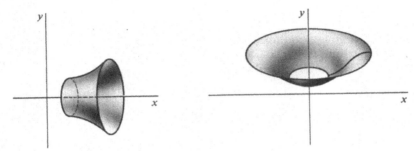

Figure 6.4.1   Surfaces of Revolution

The simplest surfaces of revolution are the right circular cylinders and cones. We can find their areas without calculus.

Figure 6.4.2 shows a right circular cylinder with height h and base of radius r. When the lateral surface is slit vertically and opened up it forms a rectangle with height h and base $2\pi r$. Therefore its area is

$$lateral\ area\ of\ cylinder = 2\pi hr.$$

Figure 6.4.3 shows a right circular cone with slant height l and base of radius r.

When the cone is slit vertically and opened up, it forms a circular sector with radius l and arc length $s = 2\pi r$. Using the formula $A = \frac{1}{2} sl$ for the area of a sector, we see that the lateral surface of the cone has area

$$lateral\ area\ of\ cone = \pi rl.$$

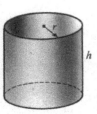

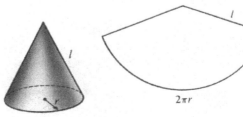

Figure 6.4.2   **Figure 6.4.3**

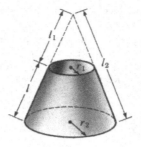

Figure 6.4.4   Cone frustum

Figure 6.4.4 shows the frustum of a cone with smaller radius r_1, larger radius r_2, and slant height l. The formula for the area of the lateral surface of a frustum of a cone is

$$lateral\ area\ of\ frustum = \pi(r_1 + r_2)l.$$

This formula is justified as follows. The frustum is formed by removing a cone of radius r_1 and slant height l_1 from a cone of radius r_2 and slant height l_2. The frustum therefore has lateral area

$$A = \pi r_2 l_2 - \pi r_1 l_1.$$

The slant heights are proportional to the radii,

$$\frac{l_1}{r_1} = \frac{l_2}{r_2}, \qquad so \quad r_1 l_2 = r_2 l_1.$$

The slant height l of the frustum is

$$l = l_2 - l_1.$$

Using the last two equations,

$$\begin{aligned}
\pi(r_1 + r_2)l &= \pi(r_2 + r_1)(l_2 - l_1) \\
&= \pi(r_2 l_2 + r_1 l_2 - r_2 l_1 - r_1 l_1) \\
&= \pi r_2 l_2 - \pi r_1 l_1 = A.
\end{aligned}$$

A surface of revolution can be sliced into frustums in the same way that a solid of revolution can be sliced into discs or cylindrical shells. Consider a smooth curve segment

$$y = f(x), \qquad a \le x \le b$$

in the first quadrant. When this curve segment is rotated about the y-axis it forms a surface of revolution (Figure 6.4.5).

Here is the formula for the area.

AREA OF SURFACE OF REVOLUTION

$$A = \int_a^b 2\pi x \sqrt{1 + (dy/dx)^2}\, dx \qquad (rotating\ about\ y\text{-}axis).$$

To justify this formula we begin by dividing the interval $[a, b]$ into infinitesimal subintervals of length Δx. This divides the curve into pieces of infinitesi-

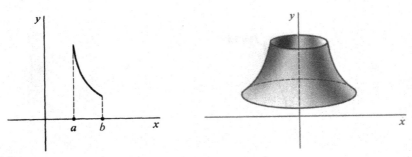

Figure 6.4.5

mal length Δs. When a piece Δs of the curve is rotated about the x-axis it sweeps out a piece of the surface, ΔA (Figure 6.4.6). Since Δs is almost a line segment, ΔA is almost a cone frustum of slant height Δs, and bases of radius x and $x + \Delta x$. Thus compared to Δx,

$$\Delta s \approx \sqrt{1 + (dy/dx)^2}\,\Delta x,$$
$$\Delta A \approx \pi(x + (x + \Delta x))\,\Delta s \approx 2\pi x\,\Delta s,$$
$$\Delta A \approx 2\pi x \sqrt{1 + (dy/dx)^2}\,\Delta x.$$

Then by the Infinite Sum Theorem,

$$A = \int_a^b 2\pi x \sqrt{1 + (dy/dx)^2}\,dx.$$

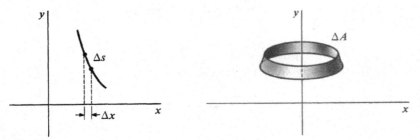

Figure 6.4.6

EXAMPLE 1   The line segment $y = 3x$, from $x = 1$ to $x = 4$, is rotated about the y-axis (Figure 6.4.7). Find the area of the surface of revolution.

FIRST SOLUTION   We use the integration formula. $dy/dx = 3$, so

$$A = \int_1^4 2\pi x \sqrt{1 + (dy/dx)^2}\,dx$$
$$= \int_1^4 2\pi x \sqrt{1 + 3^2}\,dx = 2\pi\sqrt{10}\int_1^4 x\,dx$$
$$= 2\pi\sqrt{10}\,\frac{x^2}{2}\Big]_1^4 = 2\pi\sqrt{10}\left(\frac{16 - 1}{2}\right) = 15\pi\sqrt{10}.$$

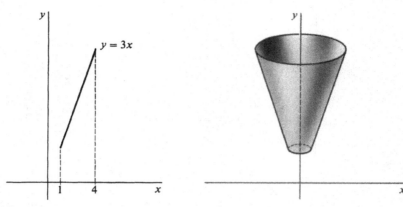

Figure 6.4.7

SECOND SOLUTION   This surface of revolution is a frustum of a cone, so the formula for the lateral area of a frustum can be used directly. From the diagram we see that the radii and slant height are:

$$r_1 = 1, \qquad r_2 = 4,$$
$$l = \text{distance from } (1, 3) \text{ to } (4, 12)$$
$$= \sqrt{(4 - 1)^2 + (12 - 3)^2} = \sqrt{3^2 + 9^2} = \sqrt{90} = 3\sqrt{10}.$$

Then $A = \pi(r_1 + r_2)l = \pi(1 + 4)3\sqrt{10} = 15\pi\sqrt{10}.$

EXAMPLE 2   The curve $y = \frac{1}{2}x^2$, $0 \le x \le 1$, is rotated about the y-axis (Figure 6.4.8). Find the area of the surface of revolution.

$$\frac{dy}{dx} = x,$$

$$A = \int_0^1 2\pi x \sqrt{1 + (dy/dx)^2}\, dx$$

$$= \int_0^1 2\pi x \sqrt{1 + x^2}\, dx = \int_1^2 \pi \sqrt{u}\, du \qquad (\text{where } u = 1 + x^2)$$

$$= \tfrac{2}{3}\pi u^{3/2}\Big]_1^2 = \tfrac{2}{3}\pi(2\sqrt{2} - 1).$$

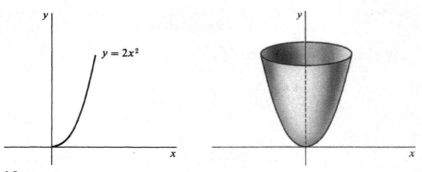

Figure 6.4.8

In finding a formula for surface area, why did we divide the surface into frustums of cones instead of into cylinders (as we did for volumes)? The reason is that to use the Infinite Sum Theorem we need something which is infinitely close to a small piece ΔA of area *compared to* Δx. The small frustum has area

$$(2x + \Delta x)\pi \, \Delta s$$

which is infinitely close to ΔA compared to Δx because it almost has the same shape as ΔA (Figure 6.4.9). The small cylinder has area $2x\pi \, \Delta y$. While this area is infinitesimal, it is not infinitely close to ΔA compared to Δx, because on dividing by Δx we get

$$\frac{\text{area of frustum}}{\Delta x} = 2x\pi \frac{\Delta s}{\Delta x} + \pi \, \Delta x \frac{\Delta s}{\Delta x} \approx 2x\pi \frac{ds}{dx},$$

$$\frac{\text{area of cylinder}}{\Delta x} = 2x\pi \frac{\Delta y}{\Delta x} \approx 2x\pi \frac{dy}{dx}.$$

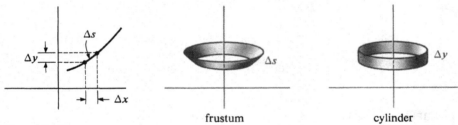

frustum   cylinder

Figure 6.4.9

Approximating the surface by small cylinders would give us the different and incorrect value $\int_a^b 2\pi x \frac{dy}{dx} \, dx$ for the surface area.

When a curve is given by parametric equations we get a formula for surface area of revolution analogous to the formula for lengths of parametric curves in Section 6.3.

Let
$$x = f(t), \qquad y = g(t), \qquad a \le t \le b$$

be a parametric curve in the first quadrant such that the derivatives are continuous and the curve does not retrace its path (Figure 6.4.10).

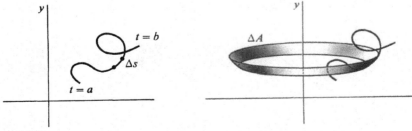

Figure 6.4.10

AREA OF SURFACE OF REVOLUTION

$$A = \int_a^b 2\pi x \sqrt{\left(\frac{dx}{dt}\right)^2 + \left(\frac{dy}{dt}\right)^2}\, dt \qquad (rotating\ about\ y\text{-}axis).$$

To justify this new formula we observe that an infinitesimal piece of the surface is almost a cone frustum of radii x, $x + \Delta x$ and slant height Δs. Thus compared to Δt,

$$\Delta s \approx \sqrt{(dx/dt)^2 + (dy/dt)^2}\, \Delta t,$$
$$\Delta A \approx \pi(x + (x + \Delta x))\,\Delta s \approx 2\pi x\, \Delta s,$$
$$\Delta A \approx 2\pi x\sqrt{(dx/dt)^2 + (dy/dt)^2}\, \Delta t.$$

The Infinite Sum Theorem gives the desired formula for area.

This new formula reduces to our first formula when the curve has the simple form $y = f(x)$. If $y = f(x)$, $a \le x \le b$, take $x = t$ and get

$$A = \int_a^b 2\pi x\sqrt{1 + (dy/dx)^2}\, dx \qquad (about\ y\text{-}axis).$$

Similarly, if $x = g(y)$, $a \le y \le b$, we take $y = t$ and get the formula

$$A = \int_a^b 2\pi x\sqrt{(dx/dy)^2 + 1}\, dy \qquad (about\ y\text{-}axis).$$

EXAMPLE 3 The curve $x = 2t^2$, $y = t^3$, $0 \le t \le 1$ is rotated about the y-axis. Find the area of the surface of revolution (Figure 6.4.11).

We first find dx/dt and dy/dt and then apply the formula for area.

$$\frac{dx}{dt} = 4t, \qquad \frac{dy}{dt} = 3t^2,$$

$$A = \int_0^1 2\pi x\sqrt{(dx/dt)^2 + (dy/dt)^2}\, dt$$

$$= \int_0^1 4\pi t^2\sqrt{(4t)^2 + (3t^2)^2}\, dt$$

$$= 4\pi\int_0^1 t^2\sqrt{16t^2 + 9t^4}\, dt$$

$$= 4\pi\int_0^1 t^3\sqrt{16 + 9t^2}\, dt.$$

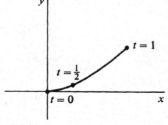

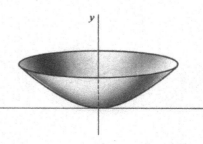

Figure 6.4.11

Let $u = 16 + 9t^2$, $du = 18t\,dt$, $dt = \dfrac{1}{18t}\,du$, $t^2 = \dfrac{u-16}{9}$. Then $u = 16$ at $t = 0$ and $u = 25$ at $t = 1$, so

$$A = 4\pi \int_{16}^{25} t^3 \sqrt{u}\,\frac{1}{18t}\,du = 4\pi \int_{16}^{25} \frac{1}{18}\left(\frac{u-16}{9}\right)\sqrt{u}\,du$$

$$= \frac{2\pi}{81} \int_{16}^{25} (u^{3/2} - 16\sqrt{u})\,du = \frac{5692}{1215}\pi \sim 4.7\pi.$$

EXAMPLE 4 Derive the formula $A = 4\pi r^2$ for the area of the surface of a sphere of radius r.

When the portion of the circle $x^2 + y^2 = r^2$ in the first quadrant is rotated about the y-axis it will form a hemisphere of radius r (Figure 6.4.12). The surface of the sphere has twice the area of this hemisphere.

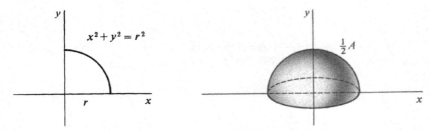

Figure 6.4.12

It is simpler to take y as the independent variable, so the curve has the equation

$$x = \sqrt{r^2 - y^2}, \qquad 0 \le y \le r.$$

Then
$$\frac{dx}{dy} = -\frac{y}{\sqrt{r^2 - y^2}}.$$

This derivative is undefined at $y = 0$. To get around this difficulty we let $0 < a < r$ and divide the surface into the two parts shown in Figure 6.4.13, the surface B generated by the curve from $y = 0$ to $y = a$ and the surface C generated by the curve from $y = a$ to $y = r$.

The area of C is

$$C = \int_a^r 2\pi x \sqrt{(dx/dy)^2 + 1}\,dy$$

$$= \int_a^r 2\pi \sqrt{r^2 - y^2}\sqrt{1 + y^2/(r^2 - y^2)}\,dy$$

$$= \int_a^r 2\pi \sqrt{r^2 - y^2}\sqrt{r^2/(r^2 - y^2)}\,dy$$

$$= \int_a^r 2\pi r\,dy = 2\pi r y \Big]_a^r = 2\pi r(r - a).$$

We could find the area of B by taking x as the independent variable. However,

Figure 6.4.13

it is simpler to let a be an infinitesimal ε. Then B is an infinitely thin ring-shaped surface, so its area is infinitesimal. Therefore the hemisphere has area

$$\tfrac{1}{2}A = B + C \approx 0 + 2\pi r(r - \varepsilon) \approx 2\pi r^2,$$

so

$$\tfrac{1}{2}A = 2\pi r^2,$$

and the sphere has area $A = 4\pi r^2$.

If a curve is rotated about the x-axis instead of the y-axis (Figure 6.4.14), we interchange x and y in the formulas for surface area,

$$A = \int_a^b 2\pi y \sqrt{(dx/dt)^2 + (dy/dt)^2}\, dt \qquad \text{(about x-axis),}$$

$$A = \int_a^b 2\pi y \sqrt{(dx/dy)^2 + 1}\, dy \qquad \text{(about x-axis),}$$

$$A = \int_a^b 2\pi y \sqrt{1 + (dy/dx)^2}\, dx \qquad \text{(about x-axis).}$$

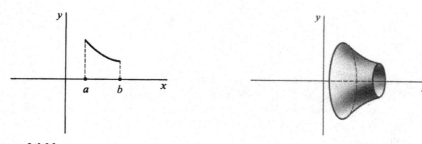

Figure 6.4.14

Most of the time the formula for surface area will give an integral which cannot be evaluated exactly but can only be approximated, for example by the Trapezoidal Rule.

EXAMPLE 5   Let C be the curve

$$y = x^4, \qquad 0 \le x \le 1. \qquad \text{(see Figure 6.4.15)}$$

Set up an integral for the surface area generated by rotating the curve C about (a) the y-axis, (b) the x-axis (see Figure 6.4.16).

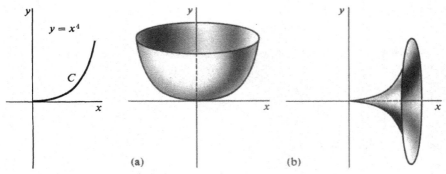

Figure 6.4.15   **Figure 6.4.16**

(a)
$$dy/dx = 4x^3$$

$$A = \int_0^1 2\pi x\sqrt{1 + (dy/dx)^2}\, dx$$

$$= \int_0^1 2\pi x\sqrt{1 + 16x^6}\, dx.$$

We cannot evaluate this integral, so we leave it in the above form. The Trapezoidal Rule can be used to get approximate values. When $\Delta x = \frac{1}{10}$ the Trapezoidal Approximation is

$$A \sim 6.42, \text{ error} \leq 0.26.$$

(b)
$$A = \int_0^1 2\pi y\sqrt{1 + (dy/dx)^2}\, dx$$

$$= \int_0^1 2\pi x^4\sqrt{1 + 16x^6}\, dx.$$

The Trapezoidal Approximation when $\Delta x = \frac{1}{10}$ is

$$A \sim 3.582 \qquad \text{error} \leq 0.9.$$

PROBLEMS FOR SECTION 6.4

In Problems 1–12, find the area of the surface generated by rotating the given curve about the y-axis.

1 $y = x^2, \ 0 \leq x \leq 2$ **2** $y = cx + d, \ a \leq x \leq b$

3 $y = 2x^{3/2}, \ 0 \leq x \leq 1$ **4** $y = \frac{1}{3}(x^2 + 2)^{3/2}, \ 1 \leq x \leq 2$

5 $y = \frac{1}{3}x\sqrt{x} - \sqrt{x}, \ 1 \leq x \leq 4$

6 $y = \frac{1}{4}x^4 + \frac{1}{8}x^{-2}, \ 1 \leq x \leq 2$

7 $y = \frac{3}{5}x^{5/3} - \frac{3}{4}x^{1/3}, \ 1 \leq x \leq 8$

8 $x = 2t + 1, y = 4 - t, \ 0 \leq t \leq 4$

9 $x = t + 1, y = \frac{1}{2}t^2 + t, \ 0 \leq t \leq 2$

10 $x = t^2, y = \frac{1}{3}t^3, \ 0 \leq t \leq 3$

11 $x = t^3, y = 3t + 1, \quad 0 \le t \le 1$

12 $x^{2/3} + y^{2/3} = 1, \quad \text{first quadrant}$

In Problems 13–20, find the area of the surface generated by rotating the given curve about the x-axis.

13 $y = \frac{1}{3}x^3, \quad 0 \le x \le 1$

14 $y = \sqrt{x}, \quad 1 \le x \le 2$

15 $y = \dfrac{x^3}{6} + \dfrac{1}{2x}, \quad 1 \le x \le 2$

16 $y = \frac{1}{4}x^4 + \frac{1}{8}x^{-2}, \quad 1 \le x \le 2$

17 $y = \frac{1}{3}x\sqrt{x} - \sqrt{x}, \quad 3 \le x \le 4$

18 $y = \frac{3}{5}x^{5/3} - \frac{3}{4}x^{1/3}, \quad 8 \le x \le 27$

19 $x = 2t + 1, y = 4 - t, \quad 0 \le t \le 4$

20 $x = t^2 + t, y = 2t + 1, \quad 0 \le t \le 1$

21   The part of the circle $x^2 + y^2 = r^2$ between $x = 0$ and $x = a$ in the first quadrant is rotated about the x-axis. Find the area of the resulting zone of the sphere $(0 < a < r)$.

22   Solve the above problem when the rotation is about the y-axis.

In Problems 23–26 set up integrals for the areas generated by rotating the given curve about (a) the y-axis, (b) the x-axis.

23 $y = x^5, \quad 0 \le x \le 1$

24 $x = y + \sqrt{y}, \quad 2 \le y \le 3$

25 $x = t^2 + t, y = t^2 - 1, \quad 1 \le t \le 10$

26 $x = t^4, y = t^3, \quad 2 \le t \le 4$

27   Set up an integral for the area generated by rotating the curve $y = \frac{1}{2}x^2, 0 \le x \le 1$ about the x-axis and find the Trapezoidal Approximation with $\Delta x = 0.2$.

28   Set up an integral for the area generated by rotating the curve $y = \frac{1}{3}x^3, 0 \le x \le 1$ about the y-axis and find the Trapezoidal Approximation with $\Delta x = 0.2$.

□ **29**   Show that the surface area of the torus generated by rotating the circle of radius r and center $(c, 0)$ about the y-axis $(r < c)$ is $A = 4\pi^2 rc$. *Hint:* Take y as the independent variable and use the formula $\int_a^b r\, dy/\sqrt{r^2 - y^2}$ for the length of the arc of the circle from $y = a$ to $y = b$.

6.5 AVERAGES

Given n numbers $y_1, \ldots, y_n$, their average value is defined as

$$y_{\text{ave}} = \frac{y_1 + \cdots + y_n}{n}.$$

If all the y_i are replaced by the average value y_{ave}, the sum will be unchanged,

$$y_1 + \cdots + y_n = y_{\text{ave}} + \cdots + y_{\text{ave}} = n y_{\text{ave}}.$$

If f is a continuous function on a closed interval $[a, b]$, what is meant by the average value of f between a and b (Figure 6.5.1)? Let us try to imitate the procedure for finding the average of n numbers. Take an infinite hyperreal number H and divide the interval $[a, b]$ into infinitesimal subintervals of length $dx = (b - a)/H$. Let

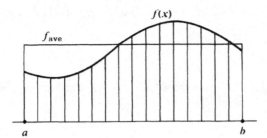

Figure 6.5.1

us "sample" the value of f at the H points $a, a + dx, a + 2\, dx, \ldots, a + (H - 1)\, dx$. Then the average value of f should be infinitely close to the sum of the values of f at $a, a + dx, \ldots, a + (H - 1)\, dx$, divided by H. Thus

$$f_{ave} \approx \frac{f(a) + f(a + dx) + f(a + 2\, dx) + \cdots + f(a + (H - 1)\, dx)}{H}.$$

Since $dx = \dfrac{b - a}{H}$, $\dfrac{1}{H} = \dfrac{dx}{b - a}$ and we have

$$f_{ave} \approx \frac{f(a)\, dx + f(a + dx)\, dx + \cdots + f(a + (H - 1)\, dx)\, dx}{b - a},$$

$$f_{ave} \approx \frac{\sum_a^b f(x)\, dx}{b - a}.$$

Taking standard parts, we are led to

DEFINITION

*Let f be continuous on $[a, b]$. The **average value** of f between a and b is*

$$f_{ave} = \frac{\int_a^b f(x)\, dx}{b - a}.$$

Geometrically, the area under the curve $y = f(x)$ is equal to the area under the constant curve $y = f_{ave}$ between a and b,

$$f_{ave} \cdot (b - a) = \int_a^b f(x)\, dx.$$

EXAMPLE 1   Find the average value of $y = \sqrt{x}$ from $x = 1$ to $x = 4$ (Figure 6.5.2).

$$y_{ave} = \frac{\int_1^4 \sqrt{x}\, dx}{(4 - 1)} = \frac{\frac{2}{3} x^{3/2} \big]_1^4}{3} = \frac{\frac{2}{3}(8 - 1)}{3} = \frac{14}{9}.$$

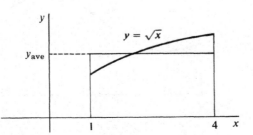

Figure 6.5.2

Recall that in Section 3.8, we defined the *average slope* of a function F between a and b as the quotient

$$\text{average slope} = \frac{F(b) - F(a)}{b - a}.$$

Using the Fundamental Theorem of Calculus we can find the connection between the average value of F' and the average slope of F.

THEOREM 1

Let F be an antiderivative of a continuous function f on an open interval I. Then for any $a < b$ in I, the average slope of F between a and b is equal to the average value of f between a and b,

$$\frac{F(b) - F(a)}{b - a} = \frac{\int_a^b f(x)\,dx}{b - a}.$$

PROOF   By the Fundamental Theorem,

$$F(b) - F(a) = \int_a^b f(x)\,dx.$$

THEOREM 2   (Mean Value Theorem for Integrals)

Let f be continuous on $[a, b]$. Then there is a point c strictly between a and b where the value of f is equal to its average value,

$$f(c) = \frac{\int_a^b f(x)\,dx}{b - a}.$$

PROOF Theorem 2 is illustrated in Figure 6.5.3. We can make f continuous on the whole real line by defining $f(x) = f(a)$ for $x < a$ and $f(x) = f(b)$ for $x > b$. By the Second Fundamental Theorem of Calculus, f has an antiderivative F. By the Mean Value Theorem there is a point c strictly between a and b at which $F'(c)$ is equal to the average slope of F,

$$F'(c) = \frac{F(b) - F(a)}{b - a}.$$

But $F'(c) = f(c)$ and $F(b) - F(a) = \int_a^b f(x)\,dx$, so

$$f(c) = \frac{\int_a^b f(x)\,dx}{b - a}.$$

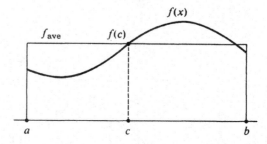

Figure 6.5.3

EXAMPLE 2 A car starts at rest and moves with velocity $v = 3t^2$. Find its average velocity between times $t = 0$ and $t = 5$. At what point of time is its velocity equal to the average velocity?

$$v_{\text{ave}} = \frac{\int_0^5 3t^2\, dt}{5 - 0} = \frac{t^3]_0^5}{5} = \frac{125}{5} = 25.$$

To find the value of t where $v = v_{\text{ave}}$, we put

$$3t^2 = 25, \qquad t = \sqrt{25/3} = 5/\sqrt{3}.$$

Suppose a car drives from city A to city B and back, a distance of 120 miles each way. From A to B it travels at a speed of 30 mph, and on the return trip it travels at 60 mph. What is the average speed?

If we choose distance as the independent variable we get one answer, and if we choose time we get another.

Average speed with respect to time: The car takes $120/30 = 4$ hours to go from A to B and $120/60 = 2$ hours to return to A. The total trip takes 6 hours.

$$v_{\text{ave}} = \frac{30 \cdot 4 + 60 \cdot 2}{6} = \frac{240}{6} = 40 \text{ mph.}$$

Average speed with respect to distance: The car goes 120 miles at 30 mph and 120 miles at 60 mph, with a total distance of 240 miles. Therefore

$$v_{\text{ave}} = \frac{30 \cdot 120 + 60 \cdot 120}{240} = 45 \text{ mph.}$$

From Figure 6.5.4 we see that the average with respect to time is smaller because most of the time was spent at the lower speed of 30 mph.

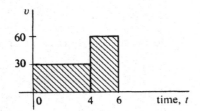 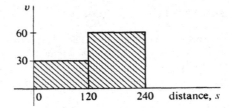

Figure 6.5.4

In general, if y is given both as a function of s and of t, $y = f(s) = g(t)$, then there is one average of y with respect to s, and another with respect to t.

EXAMPLE 3 A car travels with velocity $v = 4t + 10$, where t is time. Between times $t = 0$ and $t = 4$ find the average velocity with respect to (a) time, and (b) distance.

(a) $\qquad v_{\text{ave}} = \dfrac{\int_0^4 4t + 10\, dt}{4} = \dfrac{2t^2 + 10t]_0^4}{4} = 18$ (Figure 6.5.5(a)).

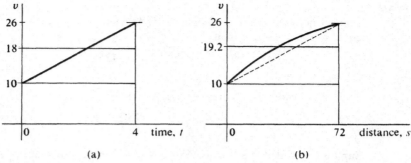

Figure 6.5.5

(b) Let s be the distance, and put $s = 0$ when $t = 0$. Since $ds = v\,dt = (4t + 10)\,dt$, at time $t = 4$ we have

$$s = \int_0^4 4t + 10\,dt = 2t^2 + 10t \Big]_0^4 = 72.$$

Then

$$v_{ave} = \frac{\int_0^{72} (4t + 10)\,ds}{72} = \frac{\int_0^4 (4t + 10)(4t + 10)\,dt}{72}$$

$$= \frac{\int_0^4 16t^2 + 80t + 100\,dt}{72} = \frac{\frac{16}{3}t^3 + 40t^2 + 100t]_0^4}{72}$$

$$= \frac{1024/3 + 640 + 400}{72} \sim 19.2 \text{ (Figure 6.5.5(b)).}$$

PROBLEMS FOR SECTION 6.5

In Problems 1–8, sketch the curve, find the average value of the function, and sketch the rectangle which has the same area as the region under the curve.

1	$f(x) = 1 + x, \quad -1 \leq x \leq 1$	**2**	$f(x) = 2 - \frac{1}{2}x, \quad 0 \leq x \leq 4$
3	$f(x) = 4 - x^2, \quad -2 \leq x \leq 2$	**4**	$f(x) = 1 + x^2, \quad -2 \leq x \leq 2$
5	$f(x) = \sqrt{2x - 1}, \quad 1 \leq x \leq 5$	**6**	$f(x) = x^3, \quad 0 \leq x \leq 2$
7	$f(x) = \sqrt[3]{x}, \quad 0 \leq x \leq 8$	**8**	$f(x) = 1 - x^4, \quad -1 \leq x \leq 1$

In Problems 9–22, find the average value of $f(x)$.

9	$f(x) = x^2 - \sqrt{x}, \quad 0 \leq x \leq 3$	**10**	$f(x) = \sqrt{x} + 1/\sqrt{x}, \quad 1 \leq x \leq 9$
11	$f(x) = 6x, \quad -4 \leq x \leq 2$	**12**	$f(x) = \dfrac{3x}{\sqrt{1 - x^2}}, \quad -\frac{1}{2} \leq x \leq \frac{1}{2}$
13	$f(x) = 2x\sqrt{1 + x^2}, \quad -3 \leq x \leq 3$	**14**	$f(x) = 5x^4 - 8x^3 + 10, \quad 0 \leq x \leq 10$
15	$f(x) = \sin x, \quad 0 \leq x \leq \pi$	**16**	$f(x) = \sin x, \quad 0 \leq x \leq 2\pi$
17	$f(x) = \sin x \cos x, \quad 0 \leq x \leq \pi/2$	**18**	$f(x) = x + \sin x, \quad 0 \leq x \leq 2\pi$
19	$f(x) = e^x, \quad -1 \leq x \leq 1$	**20**	$f(x) = e^x - 2x, \quad 0 \leq x \leq 2$
21	$f(x) = \dfrac{1}{x}, \quad 1 \leq x \leq 4$	**22**	$f(x) = \dfrac{x}{x + 1}, \quad 0 \leq x \leq 4$

In Problems 23–28, find a point c in the given interval such that $f(c)$ is equal to the average value of $f(x)$.

23 $f(x) = 2x, \quad -4 \le x \le 6$ **24** $f(x) = 3x^2, \quad 0 \le x \le 3$

25 $f(x) = \sqrt{2x}, \quad 0 \le x \le 2$ **26** $f(x) = x^2 - x, \quad -1 \le x \le 1$

27 $f(x) = x^{2/3}, \quad 0 \le x \le 2$ **28** $f(x) = |x - 3|, \quad 1 \le x \le 4$

29   What is the average distance between a point x in the interval $[5, 8]$ and the origin?

30   What is the average distance between a point in the interval $[-4, 3]$ and the origin?

31   Find the average distance from the origin to a point on the curve $y = x^{3/2}, 0 \le x \le 3$, with respect to x.

32   A particle moves with velocity $v = 6t$ from time $t = 0$ to $t = 10$. Find its average velocity with respect to (a) time, (b) distance.

33   An object moves with velocity $v = t^3$ from time $t = 0$ to $t = 2$. Find its average velocity with respect to (a) time, (b) distance.

☐ **34**   A particle moves with positive velocity $v = f(t)$ from $t = a$ to $t = b$. Thus its average velocity with respect to time is

$$\frac{\int_a^b f(t)\,dt}{(b - a)}.$$

Show that its average velocity with respect to distance is

$$\frac{\int_a^b (f(t))^2\,dt}{\int_a^b f(t)\,dt}.$$

6.6 SOME APPLICATIONS TO PHYSICS

The Infinite Sum Theorem can frequently be used to derive formulas in physics.

1 MASS AND DENSITY, ONE DIMENSION

Consider a one-dimensional object such as a length of wire. We ignore the atomic nature of matter and assume that it is distributed continuously along a line segment. If the density ρ per unit length is the same at each point of the wire, then the mass is the product of the density and the length, $m = \rho L$. If L is in centimeters and ρ in grams per centimeter, then m is in grams. (ρ is the Greek letter "rho".)

Now suppose that the density of the wire varies continuously with the position. Put the wire on the x-axis between the points $x = a$ and $x = b$, and let the density at the point x be $\rho(x)$. Consider the piece of the wire of infinitesimal length Δx and mass Δm shown in Figure 6.6.1. At each point between x and $x + \Delta x$, the density is infinitely close to $\rho(x)$, so

$$\Delta m \approx \rho(x)\,\Delta x \qquad \text{(compared to } \Delta x\text{)}.$$

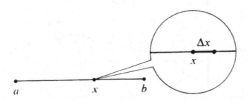

Figure 6.6.1

Therefore by the Infinite Sum Theorem, the total mass is

$$m = \int_a^b \rho(x)\, dx.$$

EXAMPLE 1   Find the mass of a wire 6 cm long whose density at distance x from the center is $9 - x^2$ gm/cm. In Figure 6.6.2, we put the center of the wire at the origin. Then

$$m = \int_{-3}^{3} 9 - x^2\, dx = 9x - \tfrac{1}{3}x^3 \Big]_{-3}^{3} = 36 \text{ gm.}$$

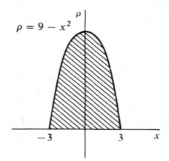

$$\rho = 9 - x^2$$

Figure 6.6.2

2 MASS AND DENSITY, TWO DIMENSIONS

Imagine a flat plate which occupies the region below the curve $y = f(x), f(x) \geq 0$, from $x = a$ to $x = b$. If its density per unit area is a constant ρ gm/cm^2, then its mass is the product of the density and area,

$$m = \rho A = \rho \int_a^b f(x)\, dx.$$

Suppose instead that the density depends on the value of x, $\rho(x)$. Consider a vertical strip of the plate of infinitesimal width Δx (Figure 6.6.3). On the strip between x and $x + \Delta x$, the density is everywhere infinitely close to $\rho(x)$, so

$$\Delta m \approx \rho(x)\, \Delta A \approx \rho(x) f(x)\, \Delta x \qquad \text{(compared to } \Delta x\text{).}$$

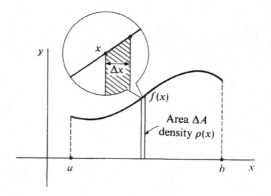

Figure 6.6.3

By the Infinite Sum Theorem,

$$m = \int_a^b \rho(x) f(x)\, dx.$$

EXAMPLE 2   A circular disc of radius r has density at each point equal to the distance of the point from the y-axis. Find its mass. (The center of the circle, shown in Figure 6.6.4, is at the origin.) The circle is the region between the curves $-\sqrt{r^2 - x^2}$ and $\sqrt{r^2 - x^2}$ from $-r$ to r. The density at a point (x, y) in the disc is $|x|$. By symmetry, all four quadrants have the same mass. We shall find the mass m_1 of the first quadrant and multiply by four.

$$m_1 = \int_0^r \sqrt{r^2 - x^2}\, x\, dx.$$

Put $u = r^2 - x^2$, $du = -2x\, dx$; $u = r^2$ when $x = 0$, and $u = 0$ when $x = r$.

$$m_1 = \int_{r^2}^0 -\tfrac{1}{2}\sqrt{u}\, du = \tfrac{1}{2} \int_0^{r^2} \sqrt{u}\, du = \tfrac{1}{2} \cdot \tfrac{2}{3} u^{3/2} \Big]_0^{r^2} = \tfrac{1}{3} r^3.$$

Then $m = 4m_1 = \tfrac{4}{3} r^3$.

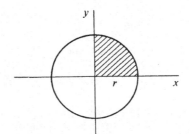

Figure 6.6.4

3 MOMENTS, ONE DIMENSION

Two children on a weightless seesaw will balance perfectly if the product of their masses and their distances from the fulcrum are equal, $m_1 d_1 = m_2 d_2$ (Figure 6.6.5).

Figure 6.6.5

For example, a 60 lb child 6 feet from the fulcrum will balance a 40 lb child 9 feet from the fulcrum, $60 \cdot 6 = 40 \cdot 9$. If the fulcrum is at the origin $x = 0$, the masses m_1 and m_2 have coordinates $x_1 = -d_1$ and $x_2 = d_2$. The equation for balancing becomes

$$m_1 x_1 + m_2 x_2 = 0.$$

Similarly, finitely many masses $m_1, \ldots, m_k$ at the points $x_1, \ldots, x_k$ will balance about the point $x = 0$ if

$$m_1 x_1 + \cdots + m_k x_k = 0.$$

Given a mass m at the point x, the quantity mx is called the *moment about the origin*.

The moment of a finite collection of point masses $m_1, \ldots, m_k$ at $x_1, \ldots, x_k$ about the origin is defined as the sum

$$M = m_1 x_1 + \cdots + m_k x_k.$$

Suppose the point masses are rigidly connected to a rod of mass zero. If the moment M is equal to zero, the masses will balance at the origin. In general they will balance at a point $\bar{x}$ called the *center of gravity* (Figure 6.6.6). $\bar{x}$ is equal to the moment divided by the total mass m,

$$\bar{x} = \frac{M}{m} = \frac{m_1 x_1 + \cdots + m_k x_k}{m_1 + \cdots + m_k}.$$

Since the mass m is positive, the moment M has the same sign as the center of gravity $\bar{x}$.

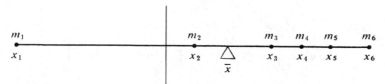

Figure 6.6.6

Now consider a length of wire between $x = a$ and $x = b$ whose density at x is $\rho(x)$. The *moment* of the wire about the origin is defined as the integral

$$M = \int_a^b x\rho(x)\,dx.$$

This formula is justified by considering a piece of the wire of infinitesimal length Δx. On the piece from x to $x + \Delta x$ the density remains infinitely close to $\rho(x)$. Thus if ΔM is the moment of the piece,

$$\Delta M \approx x\,\Delta m \approx x\rho(x)\,\Delta x \qquad \text{(compared to } \Delta x\text{)}.$$

The moment of an object is equal to the sum of the moments of its parts. Hence by the Infinite Sum Theorem,

$$M = \int_a^b x\rho(x)\,dx.$$

If the wire has moment M about the origin and mass m, the *center of mass* of the wire is defined as the point

$$\bar{x} = M/m.$$

A point of mass m located at $\bar{x}$ has the same moment about the origin as the whole wire, $M = \bar{x}m$. Physically, the wire will balance on a fulcrum placed at the center of mass.

EXAMPLE 3   A wire between $x = 0$ and $x = 1$ has density $\rho(x) = x^2$ (Figure 6.6.7). The moment is

$$M = \int_0^1 x^2 x\,dx = \frac{x^4}{4}\Big]_0^1 = \tfrac{1}{4}.$$

The mass and center of mass are

$$m = \int_0^1 x^2 \, dx = \tfrac{1}{3}, \qquad \bar{x} = M/m = \tfrac{3}{4}.$$

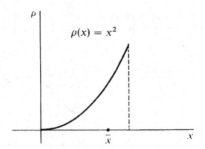

Figure 6.6.7

4 MOMENTS, TWO DIMENSIONS

A mass m at the point (x_0, y_0) in the (x, y) plane will have moments M_x about the x-axis and M_y about the y-axis (Figure 6.6.8). They are defined by

$$M_x = my_0, \qquad M_y = mx_0.$$

Consider a vertical length of wire of mass m and constant density which lies on the line $x = x_0$ from $y = a$ to $y = b$.

The wire has density

$$\rho = \frac{m}{b - a}.$$

The infinitesimal piece of the wire from y to $y + \Delta y$ shown in Figure 6.6.9 will have mass and moments

$$\Delta m = \rho \, \Delta y,$$

$$\Delta M_x \approx y \, \Delta m = y\rho \, \Delta y \qquad \text{(compared to } \Delta y\text{)},$$

$$\Delta M_y \approx x_0 \, \Delta m = x_0\rho \, \Delta y \qquad \text{(compared to } \Delta y\text{)}.$$

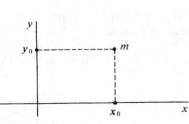

Figure 6.6.8

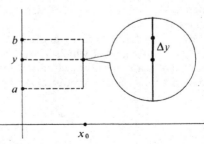

Figure 6.6.9

The Infinite Sum Theorem gives the moments for the whole wire,

$$M_x = \int_a^b y\rho \, dy = \rho\left(\frac{1}{2}b^2 - \frac{1}{2}a^2\right) = \frac{1}{2}(b + a)m,$$

$$M_y = \int_a^b x_0\rho \, dy = x_0\rho(b - a) = x_0 m.$$

We next take up the case of a flat plate which occupies the region R under the curve $y = f(x), f(x) \geq 0$, from $x = a$ to $x = b$ (Figure 6.6.10). Assume the density

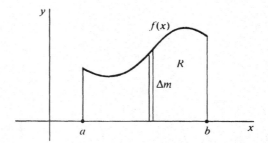

Figure 6.6.10

$\rho(x)$ depends only on the x-coordinate. A vertical slice of infinitesimal width Δx between x and $x + \Delta x$ is almost a vertical length of wire between 0 and $f(x)$ which has area ΔA and mass $\Delta m \approx \rho(x)\Delta A \approx \rho(x)f(x)\Delta x$ (compared to Δx). Putting the mass Δm into the vertical wire formulas, the moments are

$$\Delta M_y \approx x \, \Delta m \approx x\rho(x)f(x)\Delta x \qquad \text{(compared to } \Delta x),$$

$$\Delta M_x \approx \tfrac{1}{2}(f(x) + 0)\Delta m \approx \tfrac{1}{2}\rho(x)f(x)^2\Delta x \qquad \text{(compared to } \Delta x).$$

Then by the Infinite Sum Theorem, the total moments are

$$M_y = \int_a^b x\rho(x)f(x)\,dx,$$

$$M_x = \int_a^b \tfrac{1}{2}\rho(x)f(x)^2\,dx.$$

The *center of mass* of a two-dimensional object is defined as the point $(\bar{x}, \bar{y})$ with coordinates

$$\bar{x} = M_y/m, \qquad \bar{y} = M_x/m.$$

A single mass m at the point $(\bar{x}, \bar{y})$ will have the same moments as the two-dimensional body, $M_x = m\bar{y}$, $M_y = m\bar{x}$. The object will balance on a pin placed at the center of mass.

If a two-dimensional object has constant density, the center of mass depends only on the region R which it occupies. The *centroid* of a region R is defined as the center of mass of an object of constant density which occupies R. Thus if R is the region below the continuous curve $y = f(x)$ from $x = a$ to $x = b$, then the centroid has coordinates

$$\bar{x} = \int_a^b xf(x)\,dx/A, \qquad \bar{y} = \int_a^b \tfrac{1}{2}f(x)^2\,dx/A,$$

where A is the area $A = \int_a^b f(x)\,dx$.

EXAMPLE 4   Find the centroid of the triangular region R bounded by the x-axis, the y-axis, and the line $y = 1 - \frac{1}{2}x$ shown in Figure 6.6.11. R is the region under the curve $y = 1 - \frac{1}{2}x$ from $x = 0$ to $x = 2$. The area of R is

$$A = \int_0^2 1 - \tfrac{1}{2}x \, dx = x - \tfrac{1}{4}x^2 \Big]_0^2 = 1.$$

The centroid is $(\bar{x}, \bar{y})$ where

$$\bar{x} = \int_0^2 x(1 - \tfrac{1}{2}x) \, dx = \tfrac{1}{2}x^2 - \tfrac{1}{6}x^3 \Big]_0^2 = \tfrac{2}{3},$$

$$\bar{y} = \int_0^2 \tfrac{1}{2}(1 - \tfrac{1}{2}x)^2 \, dx = \int_0^2 \tfrac{1}{2} - \tfrac{1}{2}x + \tfrac{1}{8}x^2 \, dx$$

$$= \tfrac{1}{2}x - \tfrac{1}{4}x^2 + \tfrac{1}{24}x^3 \Big]_0^2 = \tfrac{1}{3}.$$

Thus the centroid is the point $(\tfrac{2}{3}, \tfrac{1}{3})$.

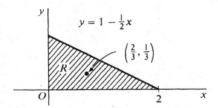

Figure 6.6.11

The following principle often simplifies a problem in moments.

If an object is symmetrical about an axis, then its moment about that axis is zero and its center of mass lies on the axis.

PROOF Consider the y-axis. Suppose a plane object occupies the region under the curve $y = f(y)$ from $-a$ to a and its density at a point (x, y) is $\rho(x)$ (Figure 6.6.12). The object is symmetric about the y-axis, so for all x between 0 and a,

$$f(-x) = f(x), \qquad \rho(-x) = \rho(x).$$

Then $M_y = \int_{-a}^a xf(x)\rho(x) \, dx = \int_{-a}^0 xf(x)\rho(x) \, dx + \int_0^a xf(x)\rho(x) \, dx$

$$= \int_a^0 (-x)f(-x)\rho(-x) \, d(-x) + \int_0^a xf(x)\rho(x) \, dx = 0.$$

Also, $\bar{x} = M_y/m = 0$.

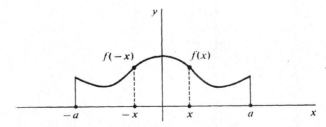

Figure 6.6.12   Symmetry about the y-axis

EXAMPLE 5   Find the centroid of the semicircle $y = \sqrt{1 - x^2}$ (Figure 6.6.13). By symmetry, the centroid is on the y-axis, $\bar{x} = 0$. The area of the semicircle is $\frac{1}{2}\pi$. Then

$$\frac{1}{2}\pi\bar{y} = \int_{-1}^{1} \frac{1}{2}(1 - x^2)\, dx = \frac{1}{2}x - \frac{1}{6}x^3 \Big]_{-1}^{1} = \frac{2}{3}, \qquad \bar{y} = \frac{2/3}{\frac{1}{2}\pi} = \frac{4}{3\pi}.$$

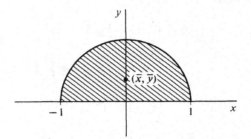

Figure 6.6.13

5 WORK

A constant force F acting along a straight line for a distance s requires the amount of work

$$W = Fs.$$

For example, the force of gravity on an object of mass m near the surface of the earth is very nearly a constant g times the mass, $F = gm$. Thus to lift an object of mass m a distance s against gravity requires the work $W = gms$. The following principle is useful in computing work done against gravity.

> *The amount of work done against gravity to move an object is the same as it would be if all the mass were concentrated at the center of mass. Moreover, the work against gravity depends only on the vertical change in position of the center of mass, not on the actual path of its motion.*
>
> *That is, $W = gms$ where s is the vertical change in the center of mass.*

EXAMPLE 6   A semicircular plate of radius one, constant density, and mass m lies flat on the table. (a) How much work is required to stand it up with the straight edge horizontal on the table (Figure 6.6.14(a))? (b) How much work is required to stand it up with the straight edge vertical and one corner on the table (Figure 6.6.14(b))? From the previous exercise, we know that the

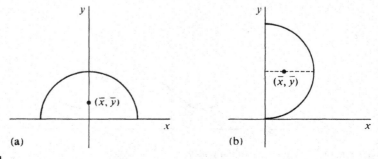

(a)   (b)

Figure 6.6.14

center of mass is on the central radius $4/3\pi$ from the center of the circle. Put the x-axis on the surface of the table.

(a) The center of mass is lifted a distance $4/3\pi$ above the table. Therefore $W = mg \cdot 4/(3\pi)$.

(b) The center of mass is lifted a distance 1 above the table, so $W = mg$.

Suppose a force $F(s)$ varies continuously with the position s and acts on an object to move it from $s = a$ to $s = b$. The work is then the definite integral of the force with respect to s,

$$W = \int_a^b F(s)\, ds.$$

To justify this formula we consider an infinitesimal length Δs. On the interval from s to $s + \Delta s$ the force is infinitely close to $F(s)$, so the work ΔW done on this interval satisfies

$$\Delta W \approx F(s)\, \Delta s \qquad \text{(compared to } \Delta s\text{)}.$$

By the Infinite Sum Theorem,

$$W = \int_a^b F(s)\, ds.$$

EXAMPLE 7   A spring, shown in Figure 6.6.15, of natural length L exerts a force $F = cx$ when compressed a distance x. Find the work done in compressing the spring from length $L - a$ to length $L - b$.

$$W = \int_a^b cx\, dx = \tfrac{1}{2}cx^2 \Big]_a^b = \tfrac{1}{2}c(b^2 - a^2).$$

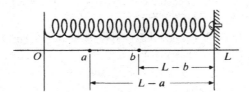

Figure 6.6.15

EXAMPLE 8   The force of gravity between two particles of mass m_1 and m_2 is

$$F = gm_1 m_2/s^2,$$

where g is a constant and s is the distance between the particles. Find the work required to move the particle m_2 from a distance a to a distance b from m_1 (Figure 6.6.16).

$$W = \int_a^b F\, ds = \int_a^b \frac{gm_1 m_2}{s^2}\, ds = gm_1 m_2(-s^{-1}) \Big]_a^b = gm_1 m_2\left(\frac{1}{a} - \frac{1}{b}\right).$$

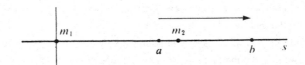

Figure 6.6.16

PROBLEMS FOR SECTION 6.6

In Problems 1–16 below, find (a) the mass, (b) the moments about the x- and y-axes, (c) the center of mass of the given object.

1   A wire on the x-axis, $0 \le x \le 2$ with density $\rho(x) = 2$.

2   A wire on the x-axis, $0 \le x \le 4$, with density $\rho(x) = x^3$.

3   A wire on the y-axis, $0 \le y \le 4$, whose density is twice the distance from the lower end of the wire times the square of the distance from the upper end.

4   A straight wire from the point $(0, 0)$ to the point $(1, 1)$ whose density at each point (x, x) is equal to $3x$.

5   A wire of length 6 and constant density k which is bent in the shape of an L covering the intervals $[0, 2]$ on the x-axis and $[0, 4]$ on the y-axis.

6   The plane object bounded by the x-axis and the curve $y = 4 - x^2$, with constant density k.

7   The plane object bounded by the x-axis and the curve $y = 4 - x^2$, with density $\rho(x) = x^2$.

8   The plane object bounded by the lines $x = 0, y = x, y = 4 - 3x$, with density $\rho(x) = 2x$.

9   The plane object between the x-axis and the curve $y = x^2$, $0 \le x \le 1$, with density $\rho(x) = 1/x$.

10   The object bounded by the x-axis and the curve $y = x^3$, $0 \le x \le 1$, with density $\rho(x) = 1 - x^2$.

11   The object bounded by the x-axis and the curve $y = 1/x$, $1 \le x \le 2$, with density $\rho(x) = \sqrt{x}$.

12   The disc bounded by $x^2 + y^2 = 4$ with density $\rho(x) = \sqrt{4 - x^2}$.

13   The object in the top half of the circle $x^2 + y^2 = 1$, with density $\rho(x) = 2|x|$.

14   The object between the x-axis and the curve $y = \sqrt{1 - x^4}$, with density equal to the cube of the distance from the y-axis.

15   The object bounded by the x-axis and the curve $y = 4x - x^2$, with density $\rho(x) = 2x$.

16   The object bounded by the curves $y = -f(x)$ and $y = f(x)$, $0 \le x \le 3$, with density $\rho(x) = 4/f(x)$. ($f(x)$ is always positive.)

In Problems 17–24, sketch and find the centroid of the region bounded by the given curves.

17 $y = 0, \quad y = 2, \quad -1 \le x \le 5$ **18** $y = 0, \quad x = 0, \quad 3x + 4y = 12$

19 $y = 0, \quad y = 1 - x^2$ **20** $y = 0, \quad y = 1 - x^2, \quad 0 \le x \le 1$

21 $y = 0, \quad y = \sqrt{9 - x^2}$ **22** $y = 0, \quad y = \sqrt{9 - x^2}, \quad 0 \le x \le 3$

23 $y = 0, \quad y = x^{1/3}, \quad 0 \le x \le 1$

24 $x = 0, \quad y = 0, \quad \sqrt{x} + \sqrt{y} = 1, \quad$ first quadrant

25   Find the mass of an object in the region under the curve $y = \sin x$, $0 \le x \le \pi$, with density $\rho(x) = \cos^2 x$.

26   Find the mass of an object in the region between the curves $y = \sin x \cos x$, $y = \sin x$, $0 \le x \le \pi/2$, with density $\rho(x) = \cos x$.

27   Find the mass of an object in the region under the curve $y = e^x$, $-1 \le x \le 1$, with density $e^{1 - 2x}$.

28   Find the mass of an object in the region under the curve $y = \ln x$, $1 \le x \le e$, with density $\rho(x) = 1/x$.

29   Find the centroid of the region under the curve $y = x^{-2}, 1 \le x \le 2$.

30   Find the centroid of the region under the curve $y = 1/\sqrt{x}, 1 \le x \le 4$.

31   Find the centroid of the region bounded by $y = 0, y = x(1 - x^2), 0 \le x \le 1$.

□ 32   Show that the moments of an object bounded by the two curves $y = f(x)$ and $y = g(x), a \le x \le b$, are

$$M_x = \int_a^b \tfrac{1}{2}\rho(x)(g(x)^2 - f(x)^2)\,dx, \qquad M_y = \int_a^b x\rho(x)(g(x) - f(x))\,dx.$$

33 Use the formulas in Problem 32 to find the centroid of the region between the curves $y = x^2$ and $y = x$.

34 A piece of metal weighing 50 lbs is in the shape of a triangle of sides 3, 4, and 5 ft. Find the amount of work required to stand the piece up on (a) the 3 ft side, (b) the 4 ft side.

35 A 4 ft chain lies flat on the ground and has constant density of 5 lbs/ft. How much work is required to lift one end 6 ft above the ground?

36 In Problem 35, how much work is required to lift the center of the chain 6 ft above the ground?

37 A 4 ft chain has a density of $4x$ lbs/ft at a point x ft from the left end. How much work is needed to lift the left end 6 ft above the ground?

38 In Problem 37, how much work is needed to lift both ends of the chain to the same point 6 ft above the ground?

39 A spring exerts a force of $4x$ lbs when compressed a distance x. How much work is needed to compress the spring 5 ft from its natural length?

40 A bucket of water weighs 10 lbs and is tied to a rope which has a density of $\tfrac{1}{10}$ lb/ft. How much work is needed to lift the bucket from the bottom of a 20 ft well?

41 The bucket in Problem 40 is leaking water at the rate of $\tfrac{1}{10}$ lb/sec and is raised from the well bottom at the rate of 4 ft/sec. How much work is expended in lifting the bucket?

42 Two electrons repel each other with a force inversely proportional to the square of the distance between them, $F = k/s^2$. If one electron is held fixed at the origin, find the work required to move a second electron along the x-axis from the point $(10, 0)$ to the point $(5, 0)$.

43 If one electron is held fixed at the point $(0, 0)$ and another at the point $(100, 0)$, find the work required to move a third electron along the x-axis from $(50, 0)$ to $(80, 0)$.

6.7 IMPROPER INTEGRALS

What is the area of the region under the curve $y = 1/\sqrt{x}$ from $x = 0$ to $x = 1$ (Figure 6.7.1(a))? The function $1/\sqrt{x}$ is not continuous at $x = 0$, and in fact $1/\sqrt{\varepsilon}$ is infinite for infinitesimal $\varepsilon > 0$. Thus our notion of a definite integral does not apply. Nevertheless we shall be able to assign an area to the region using improper integrals. We see from the figure that the region extends infinitely far up in the vertical direction. However, it becomes so thin that the area of the region turns out to be finite.

The region of Figure 6.7.1(b) under the curve $y = x^{-3}$ from $x = 1$ to $x = \infty$

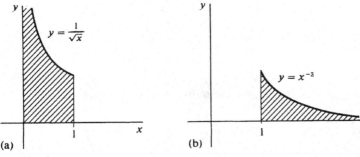

Figure 6.7.1

extends infinitely far in the horizontal direction. We shall see that this region, too, has a finite area which is given by an improper integral.

Improper integrals are defined as follows.

DEFINITION

*Suppose f is continuous on the half-open interval $(a, b]$. The **improper integral** of f from a to b is defined by the limit*

$$\int_a^b f(x)\, dx = \lim_{u \to a^+} \int_u^b f(x)\, dx.$$

If the limit exists the improper integral is said to converge. Otherwise the improper integral is said to diverge.

The improper integral can also be described in terms of definite integrals with hyperreal endpoints. We first recall that the definite integral

$$D(u, v) = \int_u^v f(x)\, dx$$

is a real function of two variables u and v. If u and v vary over the hyperreal numbers instead of the real numbers, the definite integral $\int_u^v f(x)\, dx$ stands for the natural extension of D evaluated at (u, v),

$$D^*(u, v) = \int_u^v f(x)\, dx.$$

Here is the description of the improper integral using definite integrals with hyperreal endpoints.

Let f be continuous on $(a, b]$.

(1) $\int_a^b f(x)\, dx = S$ *if and only if* $\int_{a+\varepsilon}^b f(x)\, dx \approx S$ *for all positive infinitesimal ε.*

(2) $\int_a^b f(x)\, dx = \infty$ *(or $-\infty$) if and only if* $\int_{a+\varepsilon}^b f(x)\, dx$ *is positive infinite (or negative infinite) for all positive infinitesimal ε.*

EXAMPLE 1 Find $\int_0^1 \dfrac{1}{\sqrt{x}}\, dx$. For $u > 0$,

$$\int_u^1 \frac{1}{\sqrt{x}}\, dx = 2\sqrt{x}\,\Big]_u^1 = 2 - 2\sqrt{u}.$$

Then $\displaystyle \int_0^1 \frac{1}{\sqrt{x}}\, dx = \lim_{u \to 0^+} \int_u^1 \frac{1}{\sqrt{x}}\, dx = \lim_{u \to 0^+} (2 - 2\sqrt{u}) = 2.$

Therefore the region under the curve $y = 1/\sqrt{x}$ from 0 to 1 shown in Figure 6.7.1(a) has area 2, and the improper integral converges.

EXAMPLE 2 Find $\int_0^1 x^{-2}\, dx$. For $u > 0$,

$$\int_u^1 x^{-2}\, dx = -x^{-1}\Big]_u^1 = -1 + \frac{1}{u}.$$

This time $$\lim_{u \to 0^+} \int_u^1 x^{-2}\, dx = \lim_{u \to 0^+} \left(-1 + \frac{1}{u}\right) = \infty.$$

The improper integral diverges. Since the limit goes to infinity we may write

$$\int_0^1 x^{-2}\, dx = \infty.$$

The region under the curve in Figure 6.7.2 is said to have *infinite area*.

Warning: We remind the reader once again that the symbols ∞ and $-\infty$ are not real or even hyperreal numbers. We use them only to indicate the behavior of a limit, or to indicate an interval without an upper or lower endpoint.

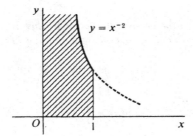

Figure 6.7.2

EXAMPLE 3 Find the length of the curve $y = x^{2/3}$, $0 \le x \le 8$. From Figure 6.7.3 the curve must have finite length. However, the derivative

$$\frac{dy}{dx} = \frac{2}{3}x^{-1/3}$$

is undefined at $x = 0$. Thus the length formula gives an improper integral,

$$s = \int_0^8 \sqrt{1 + (dy/dx)^2}\, dx = \int_0^8 \sqrt{1 + \tfrac{4}{9}x^{-2/3}}\, dx = \int_0^8 \sqrt{\frac{9x^{2/3} + 4}{9x^{2/3}}}\, dx$$

$$= \int_0^8 \frac{1}{3x^{1/3}}\sqrt{9x^{2/3} + 4}\, dx = \lim_{a \to 0^+} \int_a^8 \frac{1}{3x^{1/3}}\sqrt{9x^{2/3} + 4}\, dx.$$

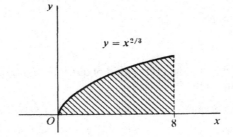

Figure 6.7.3

Let $u = 9x^{2/3} + 4$, $du = 6x^{-1/3}\, dx$. The indefinite integral is

$$\int \frac{1}{3x^{1/3}}\sqrt{9x^{2/3} + 4}\, dx = \int \frac{1}{18}\sqrt{u}\, du = \frac{1}{18} \cdot \frac{2}{3} u^{3/2} + C$$

$$= \frac{1}{27}(9x^{2/3} + 4)^{3/2} + C.$$

Therefore
$$s = \lim_{a \to 0^+} \frac{1}{27}(9x^{2/3} + 4)^{3/2}\Big]_a^8$$

$$= \frac{1}{27}((9 \cdot 4 + 4)^{3/2} - (9 \cdot 0 + 4)^{3/2}) = \frac{8}{27}(10\sqrt{10} - 1).$$

Notice that we use the same symbol for both the definite and the improper integral. The theorem below justifies this practice.

THEOREM 1

If f is continuous on the closed interval $[a, b]$ then the improper integral of f from a to b converges and equals the definite integral of f from a to b.

PROOF We have shown in Section 4.2 on the Fundamental Theorem that the function

$$F(u) = \int_u^b f(x)\, dx$$

is continuous on $[a, b]$. Therefore

$$\int_a^b f(x)\, dx = \lim_{u \to a^+} \int_u^b f(x)\, dx,$$

where $\int_a^b f(x)\, dx$ denotes the definite integral.

We now define a second kind of improper integral where the interval is infinite.

DEFINITION

*Let f be continuous on the half-open interval $[a, \infty)$. The **improper integral** of f from a to ∞ is defined by the limit*

$$\int_a^\infty f(x)\, dx = \lim_{u \to \infty} \int_a^u f(x)\, dx.$$

The improper integral is said to converge if the limit exists and to diverge otherwise.

Here is a description of this kind of improper integral using definite integrals with hyperreal endpoints.

Let f be continuous on $[a, \infty)$.

(1) $\int_a^\infty f(x)\, dx = S$ *if and only if* $\int_a^H f(x)\, dx \approx S$ *for all positive infinite H.*

(2) $\int_a^\infty f(x)\,dx = \infty$ (or $-\infty$) *if and only if* $\int_a^H f(x)\,dx$ *is positive infinite (or negative infinite) for all positive infinite H.*

EXAMPLE 4   Find the area under the curve $y = x^{-3}$ from 1 to ∞. The area is given by the improper integral

$$\int_1^\infty x^{-3}\,dx.$$

For $u > 0$, $\displaystyle\int_1^u x^{-3}\,dx = -\tfrac{1}{2}x^{-2}\Big]_1^u = -\tfrac{1}{2}u^{-2} + \tfrac{1}{2}$.

Thus $\displaystyle\int_1^\infty x^{-3}\,dx = \lim_{u\to\infty}\int_1^u x^{-3}\,dx = \lim_{u\to\infty}(-\tfrac{1}{2}u^{-2} + \tfrac{1}{2}) = \tfrac{1}{2}$.

So the improper integral converges and the region has area $\tfrac{1}{2}$. The region is shown in Figure 6.7.1(b) and extends infinitely far to the right.

EXAMPLE 5   Find the area under the curve $y = x^{-2/3}$, $1 \le x < \infty$.

$$A = \int_1^\infty x^{-2/3}\,dx = \lim_{u\to\infty}\int_1^u x^{-2/3}\,dx$$

$$= \lim_{u\to\infty} 3x^{1/3}\Big]_1^u = \lim_{u\to\infty} 3(u^{1/3} - 1) = \infty.$$

The region is shown in Figure 6.7.4 and has infinite area.

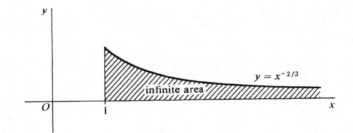

Figure 6.7.4

EXAMPLE 6   The region in Example 5 is rotated about the x-axis. Find the volume of the solid of revolution.

We use the Disc Method because the rotation is about the axis of the independent variable. The volume formula gives us an improper integral.

$$V = \int_1^\infty \pi(x^{-2/3})^2\,dx = \int_1^\infty \pi x^{-4/3}\,dx$$

$$= \lim_{u\to\infty}\int_1^u \pi x^{-4/3}\,dx = \lim_{u\to\infty} -3\pi x^{-1/3}\Big]_1^u$$

$$= \lim_{u\to\infty} 3\pi(-u^{-1/3} + 1) = 3\pi.$$

So the solid shown in Figure 6.7.5 has finite volume $V = 3\pi$.

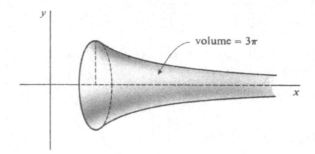

Figure 6.7.5

The last two examples give an unexpected result. A region with *infinite* area is rotated about the x-axis and generates a solid with *finite* volume! In terms of hyperreal numbers, the area of the region under the curve $y = x^{-2/3}$ from 1 to an infinite hyperreal number H is equal to $3(H^{1/3} - 1)$, which is positive infinite. But the volume of the solid of revolution from 1 to H is equal to

$$3\pi(1 - H^{-1/3}),$$

which is finite and has standard part 3π.

We can give a simpler example of this phenomenon. Let H be a positive infinite hyperinteger, and form a cylinder of radius $1/H$ and length H^2 (Figure 6.7.6). Then the cylinder is formed by rotating a rectangle of length H^2, width $1/H$, and infinite area $H^2/H = H$. But the volume of the cylinder is equal to π,

$$V = \pi r^2 h = \pi(1/H)^2(H^2) = \pi.$$

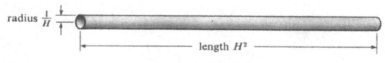

radius $\frac{1}{H}$

length H^2

Figure 6.7.6   Area $= H$, volume $= \pi$

Imagine a cylinder made out of modelling clay, with initial length and radius one. The volume is π. The clay is carefully stretched so that the cylinder gets longer and thinner. The volume stays the same, but the area of the cross section keeps getting bigger. When the length becomes infinite, the cylinder of clay still has finite volume $V = \pi$, but the area of the cross section has become infinite.

There are other types of improper integrals. If f is continuous on the half-open interval $[a, b)$ then we define

$$\int_a^b f(x)\, dx = \lim_{u \to b^-} \int_a^u f(x)\, dx.$$

If f is continuous on $(-\infty, b]$ we define

$$\int_{-\infty}^b f(x)\, dx = \lim_{u \to -\infty} \int_u^b f(x)\, dx.$$

We have introduced four types of improper integrals corresponding to the four types of half-open intervals

$$[a, b), \qquad [a, \infty), \qquad (a, b], \qquad (-\infty, b].$$

By piecing together improper integrals of these four types we can assign an improper integral to most functions which arise in calculus.

DEFINITION

*A function f is said to be **piecewise continuous** on an interval I if f is defined and continuous at all but perhaps finitely many points of I. In particular, every continuous function is piecewise continuous.*

We can introduce the improper integral $\int_a^b f(x)\,dx$ whenever f is piecewise continuous on I and a, b are either the endpoints of I or the appropriate infinity symbol. A few examples will show how this can be done.

Let f be continuous at every point of the closed interval $[a, b]$ except at one point c where $a < c < b$. We define

$$\int_a^b f(x)\,dx = \int_a^c f(x)\,dx + \int_c^b f(x)\,dx.$$

EXAMPLE 7   Find the improper integral $\int_{-8}^1 x^{-1/3}\,dx$. $x^{-1/3}$ is discontinuous at $x = 0$. The indefinite integral is

$$\int x^{-1/3}\,dx = \tfrac{3}{2}x^{2/3} + C.$$

Then
$$\int_{-8}^0 x^{-1/3}\,dx = \lim_{u \to 0^-} \int_{-8}^u x^{-1/3}\,dx = \lim_{u \to 0^-} \tfrac{3}{2}x^{2/3}\bigg]_{-8}^u$$

$$= \lim_{u \to 0^-} (\tfrac{3}{2}u^{2/3} - \tfrac{3}{2}(-8)^{2/3}) = -\tfrac{3}{2}\cdot 4 = -6.$$

Similarly,
$$\int_0^1 x^{-1/3}\,dx = \tfrac{3}{2}.$$

So
$$\int_{-8}^1 x^{-1/3}\,dx = -6 + \tfrac{3}{2} = -\tfrac{9}{2}$$

and the improper integral converges. Thus, the region shown in Figure 6.7.7 has finite area.

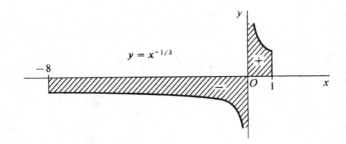

Figure 6.7.7

If f is continuous on the *open* interval (a, b), the improper integral is defined as the sum

$$\int_a^b f(x)\, dx = \int_a^c f(x)\, dx + \int_c^b f(x)\, dx,$$

where c is any point in the interval (a, b). The endpoints a and b may be finite or infinite. It does not matter which point c is chosen, because if e is any other point in (a, b), then

$$\int_a^c f(x)\, dx + \int_c^b f(x)\, dx = \int_a^c f(x)\, dx + \left(\int_c^e f(x)\, dx + \int_e^b f(x)\, dx \right)$$

$$= \left(\int_a^c f(x)\, dx + \int_c^e f(x)\, dx \right) + \int_e^b f(x)\, dx$$

$$= \int_a^e f(x)\, dx + \int_e^b f(x)\, dx.$$

EXAMPLE 8   Find $\displaystyle \int_0^2 \frac{2}{\sqrt{x}} + \frac{1}{\sqrt{2-x}}\, dx.$

The function $2/\sqrt{x} + 1/\sqrt{2-x}$ is continuous on the open interval $(0, 2)$ but discontinuous at both endpoints (Figure 6.7.8). Thus

$$\int_0^2 \frac{2}{\sqrt{x}} + \frac{1}{\sqrt{2-x}}\, dx = \int_0^1 \frac{2}{\sqrt{x}} + \frac{1}{\sqrt{2-x}}\, dx + \int_1^2 \frac{2}{\sqrt{x}} + \frac{1}{\sqrt{2-x}}\, dx.$$

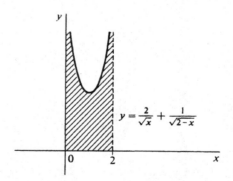

Figure 6.7.8

First we find the indefinite integral.

$$\int \frac{2}{\sqrt{x}} + \frac{1}{\sqrt{2-x}}\, dx = 4\sqrt{x} - 2\sqrt{2-x} + C.$$

Then $\displaystyle \int_0^1 \frac{2}{\sqrt{x}} + \frac{1}{\sqrt{2-x}}\, dx = \lim_{u \to 0^+} \int_u^1 \frac{2}{\sqrt{x}} + \frac{1}{\sqrt{2-x}}\, dx$

$$= \lim_{u \to 0^+} \left(4\sqrt{x} - 2\sqrt{2-x} \right) \Big]_u^1$$

$$= (4 - 2) - (0 - 2\sqrt{2}) = 2 + 2\sqrt{2}.$$

Also $\quad \displaystyle\int_1^2 \frac{2}{\sqrt{x}} + \frac{1}{\sqrt{2-x}}\, dx = \lim_{v \to 2^-} \int_1^v \frac{2}{\sqrt{x}} + \frac{1}{\sqrt{2-x}}\, dx$

$$= \lim_{v \to 2^-} \left(4\sqrt{x} - 2\sqrt{2-x}\,\right)\Big]_1^v$$

$$= (4\sqrt{2} - 0) - (4-2) = 4\sqrt{2} - 2.$$

Therefore $\quad \displaystyle\int_0^2 \frac{2}{\sqrt{x}} + \frac{1}{\sqrt{2-x}}\, dx = (2 + 2\sqrt{2}) + (4\sqrt{2} - 2) = 6\sqrt{2}.$

EXAMPLE 9 Find $\displaystyle\int_0^1 \frac{1}{x^2} + \frac{1}{(x-1)^2}\, dx.$

The function $1/x^2 + 1/(x-1)^2$ is continuous on the open interval $(0, 1)$ but discontinuous at both endpoints. The indefinite integral is

$$\int \frac{1}{x^2} + \frac{1}{(x-1)^2}\, dx = -\frac{1}{x} - \frac{1}{x-1} + C.$$

We have $\quad \displaystyle\int_0^{1/2} \frac{1}{x^2} + \frac{1}{(x-1)^2}\, dx = \lim_{u \to 0^+} \left(-\frac{1}{x} - \frac{1}{x-1}\right)\Big]_u^{1/2}$

$$= \lim_{u \to 0^+} \left(\frac{1}{u} + \frac{1}{u-1}\right) = \infty.$$

Similarly we find that

$$\int_{1/2}^1 \frac{1}{x^2} + \frac{1}{(x-1)^2}\, dx = \infty.$$

In this situation we may write

$$\int_0^1 \frac{1}{x^2} + \frac{1}{(x-1)^2}\, dx = \infty,$$

and we say that the region under the curve in Figure 6.7.9 has infinite area.

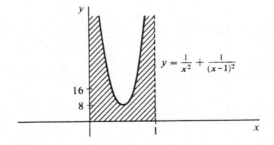

Figure 6.7.9

Remark In Example 9 we are faced with a sum of two infinite limits. Using the rules for adding infinite hyperreal numbers as a guide we can give rules for sums of infinite limits.

If H and K are positive infinite hyperreal numbers and c is finite, then

$H + K$ is positive infinite,

$H + c$ is positive infinite,

$-H - K$ is negative infinite,

$-H + c$ is negative infinite,

$H - K$ can be either finite, positive infinite, or negative infinite.

By analogy, we use the following rules for sums of two infinite limits or of a finite and an infinite limit. These rules tell us when such a sum can be considered to be positive or negative infinite. We use the infinity symbols as a convenient shorthand, keeping in mind that they are not even hyperreal numbers.

$$\infty + \infty = \infty,$$

$$\infty + c = \infty,$$

$$-\infty - \infty = -\infty,$$

$$-\infty + c = -\infty,$$

$$\infty - \infty \text{ is undefined.}$$

EXAMPLE 10   Find $\int_{-\infty}^{\infty} x \, dx$. We see that

$$\int_{-\infty}^{0} x \, dx = \lim_{u \to -\infty} \int_{u}^{0} x \, dx = \lim_{u \to -\infty} \tfrac{1}{2}x^2 \Big]_{u}^{0} = -\infty,$$

and

$$\int_{0}^{\infty} x \, dx = \infty.$$

Thus $\int_{-\infty}^{\infty} x \, dx$ diverges and has the form $\infty - \infty$. We do not assign it any value or either of the symbols ∞ or $-\infty$. The region under the curve $f(x) = x$ is shown in Figure 6.7.10.

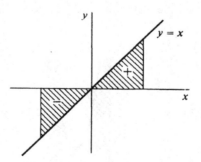

Figure 6.7.10

It is tempting to argue that the positive area to the right of the origin and the negative area to the left exactly cancel each other out so that the improper integral is zero. But this leads to a paradox.

Wrong: $\int_{-\infty}^{\infty} x \, dx = 0$. Let $v = x + 2$, $dv = dx$. Then

$$\int_{-\infty}^{\infty} (x + 2) \, dx = \int_{-\infty}^{\infty} v \, dv = 0.$$

Subtracting $\qquad \int_{-\infty}^{\infty} (x + 2) - x \, dx = 0 - 0 = 0, \qquad \int_{-\infty}^{\infty} 2 \, dx = 0.$

But $\qquad\qquad\qquad\qquad \int_{-\infty}^{\infty} 2 \, dx = \infty.$

So we do not give the integral $\int_{-\infty}^{\infty} x \, dx$ the value 0, and instead leave it undefined.

PROBLEMS FOR SECTION 6.7

In Problems 1–36, test the improper integral for convergence and evaluate when possible.

1 $\quad \int_{2}^{\infty} x^{-2} \, dx$

2 $\quad \int_{0}^{1} x^{-0.9} \, dx$

3 $\quad \int_{1}^{\infty} x^{-1/2} \, dx$

4 $\quad \int_{-\infty}^{0} (2x - 1)^{-3} \, dx$

5 $\quad \int_{0}^{1/2} (2x - 1)^{-3} \, dx$

6 $\quad \int_{-1}^{0} x^{-1/3} \, dx$

7 $\quad \int_{0}^{\infty} x^2 + 2x - 1 \, dx$

8 $\quad \int_{3}^{\infty} x^{-2} - x^{-3} \, dx$

9 $\quad \int_{0}^{\infty} x(1 + x^2)^{-2} \, dx$

10 $\quad \int_{1}^{\infty} x^{-1/2} + x^{-2} \, dx$

11 $\quad \int_{0}^{1} x^{-1/2} + x^{-2} \, dx$

12 $\quad \int_{0}^{1} (1 - x)^{-1/2} \, dx$

13 $\quad \int_{0}^{1} (x - 1)^{-2/3} \, dx$

14 $\quad \int_{-1}^{1} x^{-2} \, dx$

15 $\quad \int_{-1}^{1} x^{-2/3} \, dx$

16 $\quad \int_{0}^{1} \dfrac{x}{\sqrt{1 - x^2}} \, dx$

17 $\quad \int_{0}^{1} 2x(x^2 - 1)^{-1/3} \, dx$

18 $\quad \int_{-1}^{1} 2x^{-3} \, dx$

19 $\quad \int_{0}^{1} (2x - 1)^{-2/3} \, dx$

20 $\quad \int_{0}^{1} (3x - 1)^{-5} \, dx$

21 $\quad \int_{-\infty}^{\infty} x^2 \, dx$

22 $\quad \int_{-\infty}^{\infty} (2x - 1)^3 \, dx$

23 $\quad \int_{0}^{\infty} \dfrac{1}{\sqrt{x}} \, dx$

24 $\quad \int_{-\infty}^{\infty} x^{-1/3} \, dx$

25 $\quad \int_{-\infty}^{\infty} x^3 \, dx$

26 $\quad \int_{0}^{\infty} x^{-3/2} \, dx$

27 $\quad \int_{0}^{\infty} \dfrac{3x}{(x + 1)^4} \, dx$

28 $\quad \int_{-\infty}^{\infty} |x|(x^2 + 1)^{-3} \, dx$

29 $\quad \int_{-\infty}^{\infty} \dfrac{2x}{\sqrt{x^2 + 1}} \, dx$

30 $\quad \int_{1}^{3} (x - 1)^{-2} + (x - 3)^{-2} \, dx$

31 $\quad \int_{1}^{3} (x - 1)^{-1/2} + (3 - x)^{-1/2} \, dx$

32 $\quad \int_{-5}^{3} \dfrac{x}{|x|} \, dx$

33 $\quad \int_{0}^{\pi/2} \dfrac{\sin \theta}{\cos^2 \theta} \, d\theta$

34 $\quad \int_{0}^{\pi/2} \dfrac{\sin \theta}{\sqrt{\cos \theta}} \, d\theta$

35 $\displaystyle\int_0^{10} f(x)\,dx$ where $f(x) = \begin{cases} 1 & \text{for } 0 \le x < 1 \\ 6 & \text{for } 1 \le x < 5 \\ 3 & \text{for } 5 \le x \le 10 \end{cases}$

36 $\displaystyle\int_0^{\infty} f(x)\,dx$ where $f(x) = \begin{cases} 1/\sqrt{x} & \text{if } 0 < x < 1 \\ x^{-2} & \text{if } 1 \le x \end{cases}$

37   Show that if r is a rational number, the improper integral $\int_0^{100} x^{-r}\,dx$ converges when $r < 1$ and diverges when $r > 1$.

38   Show that if r is rational, the improper integral $\int_{1/2}^{\infty} x^{-r}\,dr$ converges when $r > 1$ and diverges when $r < 1$.

39   Find the area of the region under the curve $y = 4x^{-2}$ from $x = 1$ to $x = \infty$.

40   Find the area of the region under the curve $y = 1/\sqrt{2x - 1}$ from $x = \frac{1}{2}$ to $x = 1$.

41   Find the area of the region between the curves $y = x^{-1/4}$ and $y = x^{-1/2}$ from $x = 0$ to $x = 1$.

42   Find the area of the region between the curves $y = -x^{-3}$ and $y = x^{-2}$, $1 \le x < \infty$.

43   Find the volume of the solid generated by rotating the curve $y = 1/x$, $1 \le x < \infty$, about (a) the x-axis, (b) the y-axis.

44   Find the volume of the solid generated by rotating the curve $y = x^{-1/3}$, $0 < x \le 1$, about (a) the x-axis, (b) the y-axis.

45   Find the volume of the solid generated by rotating the curve $y = x^{-3/2}$, $0 < x \le 4$, about (a) the x-axis, (b) the y-axis.

46   Find the volume generated by rotating the curve $y = 4x^{-3}$, $-\infty < x \le -2$, about (a) the x-axis, (b) the y-axis.

47   Find the length of the curve $y = \sqrt{x} - \frac{1}{3}x\sqrt{x}$ from $x = 0$ to $x = 1$.

48   Find the length of the curve $y = \frac{3}{4}x^{1/3} - \frac{3}{8}x^{5/3}$ from $x = 0$ to $x = 1$.

49   Find the surface area generated when the curve $y = \sqrt{x} - \frac{1}{3}x\sqrt{x}$, $0 \le x \le 1$, is rotated about (a) the x-axis, (b) the y-axis.

50   Do the same for the curve $y = \frac{3}{4}x^{1/3} - \frac{3}{8}x^{5/3}$, $0 \le x \le 1$.

51   (a) Find the surface area generated by rotating the curve $y = \sqrt{x}$, $0 \le x \le 1$, about the x-axis.
(b) Set up an integral for the area generated about the y-axis.

52   Find the surface area generated by rotating the curve $y = x^{2/3}$, $0 \le x \le 8$, about the x-axis.

53   Find the surface area generated by rotating the curve $y = \sqrt{r^2 - x^2}$, $0 \le x \le a$, about (a) the x-axis, (b) the y-axis ($0 < a \le r$).

54   The force of gravity between particles of mass m_1 and m_2 is $F = gm_1m_2/s^2$ where s is the distance between them. If m_1 is held fixed at the origin, find the work done in moving m_2 from the point $(1, 0)$ all the way out the x-axis.

⫐ **55**   Show that the Rectangle and Addition Properties hold for improper integrals.

EXTRA PROBLEMS FOR CHAPTER 6

1   The skin is peeled off a spherical apple in four pieces in such a way that each horizontal cross section is a square whose corners are on the original surface of the apple. If the original apple had radius r, find the volume of the peeled apple.

2   Find the volume of a tetrahedron of height h and base a right triangle with legs of length a and b.

3   Find the volume of the wedge formed by cutting a right circular cylinder of radius r with two planes, meeting on a line crossing the axis, one plane perpendicular to the axis and the other at a $45°$ angle.

4   Find the volume of a solid whose base is the region between the x-axis and the curve $y = 1 - x^2$, and which intersects each plane perpendicular to the x-axis in a square.

In Problems 5–8, the region bounded by the given curves is rotated about (a) the x-axis, (b) the y-axis. Find the volumes of the two solids of revolution.

5 $y = 0, \quad y = \sqrt{4 - x^2}, \quad 0 \le x \le 1$

6 $y = 0, \quad y = x^{3/2}, \quad 0 \le x \le 1$

7 $y = x, \quad y = 4 - x, \quad 0 \le x \le 2$

8 $y = x^p, \quad y = x^q, \quad 0 \le x \le 1, \quad$ where $0 < q < p$

9   The region under the curve $y = \sqrt{1 - x^p}, 0 \le x \le 1$, where $0 < p$, is rotated about the x-axis. Find the volume of the solid of revolution.

10   The region under the curve $y = (x^2 + 4)^{1/3}, 0 \le x \le 2$, is rotated about the y-axis. Find the volume of the solid of revolution.

11   Find the length of the curve $y = (2x + 1)^{3/2}, 0 \le x \le 2$.

12   Find the length of the curve $y = 3x - 2, 0 \le x \le 4$.

13   Find the length of the curve $x = 3t + 1, y = 2 - 4t, 0 \le t \le 1$.

14   Find the length of the curve $x = f(t), y = f(t) + c, a \le t \le b$.

15   Find the length of the line $x = At + B, y = Ct + D, a \le t \le b$.

16   Find the area of the surface generated by rotating the curve $y = 3x^2 - 2, 0 \le x \le 1$, about the y-axis.

17   Find the area of the surface generated by rotating the curve $x = At^2 + Bt, y = 2At + B$, $0 \le t \le 1$, about the x-axis. $A > 0, B > 0$.

18   Find the average value of $f(x) = x/\sqrt{x^2 + 1}, 0 \le x \le 4$.

19   Find the average value of $f(x) = x^p, 1 \le x \le b, p \ne -1$.

20   Find the average distance from the origin of a point on the parabola $y = x^2, 0 \le x \le 4$, with respect to x.

21   Given that $f(x) = x^p, 0 \le x \le 1, p$ a positive constant, find a point c between 0 and 1 such that $f(c)$ equals the average value of $f(x)$.

22   Find the center of mass of a wire on the x-axis, $0 \le x \le 2$, whose density at a point x is equal to the square of the distance from $(x, 0)$ to $(0, 1)$.

23   Find the center of mass of a length of wire with constant density bent into three line segments covering the top, left, and right edges of the square with vertices $(0, 0), (0, 1),$ $(1, 1), (1, 0)$.

24   Find the center of mass of a plane object bounded by the lines $y = 0, y = x, x = 1$, with density $\rho(x) = 1/x$.

25   Find the center of mass of a plane object bounded by the curves $x = y^2, x = 1$, with density $\rho(y) = y^2$.

26   Find the centroid of the triangle bounded by the x- and y-axes and the line $ax + by = c$, where $a, b,$ and c are positive constants.

27   A spring exerts a force of $10x$ lbs when stretched a distance x beyond its natural length of 2 ft. Find the work required to stretch the spring from a length of 3 ft to 4 ft.

In Problems 28–36, test the improper integral for convergence and evaluate if it converges.

28 $\displaystyle\int_{-\infty}^{-2} x^{-3}\, dx$
        29 $\displaystyle\int_{0}^{\infty} (x + 2)^{-1/4}\, dx$

30 $\displaystyle\int_{-1}^{0} x^{-4}\,dx$

31 $\displaystyle\int_{-1}^{0} x^{-1/5}\,dx$

32 $\displaystyle\int_{-\infty}^{\infty} x^{1/5}\,dx$

33 $\displaystyle\int_{0}^{1} \frac{1}{x^2} + \frac{1}{(x-1)^2}\,dx$

34 $\displaystyle\int_{0}^{1} \frac{1}{\sqrt{x}} + \frac{1}{\sqrt{1-x}}\,dx$

35 $\displaystyle\int_{-4}^{4} \frac{1}{\sqrt{|x|}}\,dx$

36 $\displaystyle\int_{0}^{\infty} \sin x\,dx$

37 A wire has the shape of a curve $y = f(x)$, $a \le x \le b$, and has density $\rho(x)$ at value x. Justify the formulas below for the mass and moments of the wire.

$$m = \int_{a}^{b} \rho(x)\sqrt{1 + (f'(x))^2}\,dx,$$

$$M_x = \int_{a}^{b} f(x)\rho(x)\sqrt{1 + (f'(x))^2}\,dx,$$

$$M_y = \int_{a}^{b} x\rho(x)\sqrt{1 + (f'(x))^2}\,dx.$$

38 Find the mass, moments, and center of mass of a wire bent in the shape of a parabola $y = x^2$, $-1 \le x \le 1$, with density $\rho(x) = \sqrt{1 + 4x^2}$.

39 Find the mass, moments, and center of mass of a wire of constant density ρ bent in the shape of the semicircle $y = \sqrt{1 - x^2}$, $-1 \le x \le 1$.

☐ **40** An object fills the solid generated by rotating the region under the curve $y = f(x)$, $a \le x \le b$, about the x-axis. Its density per unit volume is $\rho(x)$. Justify the following formula for the mass of the object.

$$m = \int_{a}^{b} \rho(x)\pi(f(x))^2\,dx.$$

☐ **41** A container filled with water has the shape of a solid of revolution formed by rotating the curve $x = g(y)$, $a \le y \le b$, about the (vertical) y-axis. Water has constant density ρ per unit volume. Justify the formula below for the amount of work needed to pump all the water to the top of the container.

$$W = \int_{a}^{b} \rho\pi(g(y))^2(b - y)\,dy.$$

42 Find the work needed to pump all the water to the top of a water-filled container in the shape of a cylinder with height h and circular base of radius r.

43 Do Problem 46 if the container is in the shape of a hemispherical bowl of radius r.

44 Do Problem 46 if the container is in the shape of a cone with its vertex at the bottom, height h, and circular top of radius r.

☐ **45** The *pressure*, or force per unit area, exerted by water on the walls of a container is equal to $p = \rho(b - y)$ where ρ is the density of water and $b - y$ the water depth. Find the total force on a dam in the shape of a vertical rectangle of height b and width w, assuming the water comes to the top of the dam.

☐ **46** A water-filled container has the shape of a solid formed by rotating the curve $x = g(y)$, $a \le y \le b$ about the (vertical) y-axis. Justify the formula below for the total force on the walls of the container.

$$F = \int_{a}^{b} 2\pi\rho(b - y)x\sqrt{(dx/dy)^2 + 1}\,dy$$

7

TRIGONOMETRIC
FUNCTIONS

7.1 TRIGONOMETRY

In this chapter we shall study the *trigonometric functions*, i.e., the sine and cosine function and other functions that are built up from them. Let us start from the beginning and introduce the basic concepts of trigonometry.

The *unit circle* $x^2 + y^2 = 1$ has radius 1 and center at the origin.

Two points P and Q on the unit circle determine an *arc* $\overset{\frown}{PQ}$, an *angle* $\angle POQ$, and a *sector POQ*. The arc starts at P and goes counterclockwise to Q along the circle. The sector POQ is the region bounded by the arc $\overset{\frown}{PQ}$ and the lines OP and OQ. As Figure 7.1.1 shows, the arcs $\overset{\frown}{PQ}$ and $\overset{\frown}{QP}$ are different.

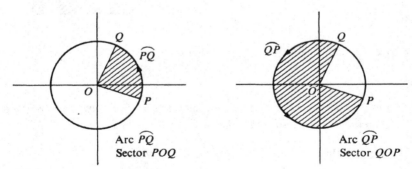

Arc $\overset{\frown}{PQ}$ Arc $\overset{\frown}{QP}$
Sector POQ Sector QOP

Figure 7.1.1

Trigonometry is based on the notion of the *length* of an arc. Lengths of curves were introduced in Section 6.3. Although that section provides a useful background, this chapter can also be studied independently of Chapter 6. As a starting point we shall give a formula for the length of an arc in terms of the area of a sector. (This formula was proved as a theorem in Section 6.3 but can also be taken as the definition of arc length.)

DEFINITION

The **length** of an arc $\stackrel{\frown}{PQ}$ on the unit circle is equal to twice the area of the sector POQ, $s = 2A$.

This formula can be seen intuitively as follows. Consider a small arc $\stackrel{\frown}{PQ}$ of length Δs (Figure 7.1.2). The sector POQ is a thin wedge which is almost a right triangle of altitude one and base Δs. Thus $\Delta A \sim \frac{1}{2}\Delta s$. Making Δs infinitesimal and adding up, we get $A = \frac{1}{2}s$.

The number $\pi \sim 3.14159$ is defined as the area of the unit circle. Thus the unit circle has circumference 2π.

The area of a sector POQ is a definite integral. For example, if P is the point $P(1, 0)$ and the point $Q(x, y)$ is in the first quadrant, then we see from Figure 7.1.3 that the area is

$$A(x) = \tfrac{1}{2}x\sqrt{1 - x^2} + \int_x^1 \sqrt{1 - t^2}\, dt.$$

Notice that $A(x)$ is a continuous function of x. The length of an arc has the following basic property.

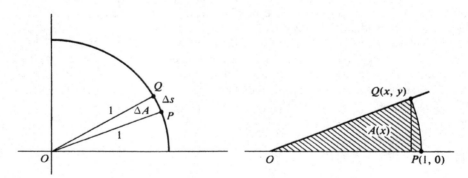

Figure 7.1.2   Figure 7.1.3

THEOREM 1

Let P be the point $P(1, 0)$. For every number s between 0 and 2π there is a point Q on the unit circle such that the arc $\stackrel{\frown}{PQ}$ has length s.

PROOF   We give the proof for s between 0 and $\pi/2$, whence

$$0 \le \tfrac{1}{2}s \le \pi/4.$$

Let $A(x)$ be the area of the sector POQ where $Q = Q(x, y)$ (Figure 7.1.4). Then $A(0) = \pi/4$, $A(1) = 0$ and the function $A(x)$ is continuous for $0 \le x \le 1$. By the Intermediate Value Theorem there is a point x_0 between 0 and 1 where the sector has area $\frac{1}{2}s$,

$$A(x_0) = \tfrac{1}{2}s.$$

Therefore the arc $\stackrel{\frown}{PQ}$ has length

$$2A(x_0) = s.$$

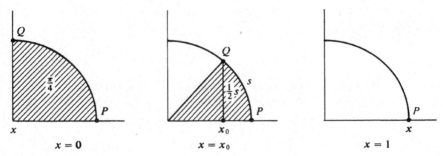

Figure 7.1.4

Arc lengths are used to measure angles. Two units of measurement for angles are radians (best for mathematics) and degrees (used in everyday life).

DEFINITION

> Let P and Q be two points on the unit circle. The measure of the angle $\angle POQ$ in **radians** is the length of the arc $\overset{\frown}{PQ}$. A **degree** is defined as
>
> $$1° = \pi/180 \text{ radians},$$
>
> whence the measure of $\angle POQ$ in degrees is $180/\pi$ times the length of $\overset{\frown}{PQ}$.

Approximately, $1° \sim 0.01745$ radians,
   1 radian $\sim 57°18' = (57\tfrac{18}{60})°$.

A complete revolution is $360°$ or 2π radians. A straight angle is $180°$ or π radians. A right angle is $90°$ or $\pi/2$ radians.

It is convenient to take the point $(1, 0)$ as a starting point and measure arc length around the unit circle in a counterclockwise direction. Imagine a particle which moves with speed one counterclockwise around the circle and is at the point $(1, 0)$ at time $t = 0$. It will complete a revolution once every 2π units of time. Thus if the particle is at the point P at time t, it will also be at P at all the times $t + 2k\pi$, k an integer. Another way to think of the process is to take a copy of the real line, place the origin at the point $(1, 0)$, and wrap the line around the circle infinitely many times with the positive direction going counterclockwise. Then each point on the circle will correspond to an infinite family of real numbers spaced 2π apart (Figure 7.1.5).

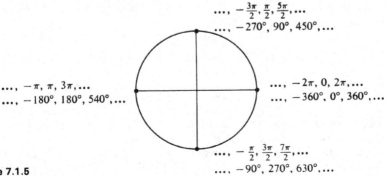

Figure 7.1.5

The Greek letters θ (theta) and ϕ (phi) are often used as variables for angles or circular arc lengths.

DEFINITION

*Let $P(x, y)$ be the point at counterclockwise distance θ around the unit circle starting from $(1, 0)$. x is called the **cosine** of θ and y the **sine** of θ,*

$$x = \cos \theta, \qquad y = \sin \theta.$$

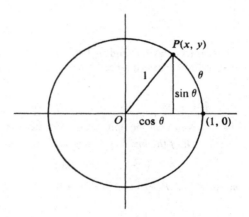

Figure 7.1.6

Cos θ·and sin θ are shown in Figure 7.1.6. Geometrically, if θ is between 0 and $\pi/2$ so that the point $P(x, y)$ is in the first quadrant, then the radius OP is the hypotenuse of a right triangle with a vertical side sin θ and horizontal side cos θ. By Theorem 1, sin θ and cos θ are real functions defined on the whole real line. We write $\sin^n \theta$ for $(\sin \theta)^n$, and $\cos^n \theta$ for $(\cos \theta)^n$. By definition $(\cos \theta, \sin \theta) = (x, y)$ is a point on the unit circle $x^2 + y^2 = 1$, so we always have

$$\sin^2 \theta + \cos^2 \theta = 1.$$

Also, $\qquad -1 \leq \sin \theta \leq 1, \qquad -1 \leq \cos \theta \leq 1.$

Sin θ and cos θ are *periodic functions* with period 2π. That is,

$$\sin (\theta + 2\pi n) = \sin \theta,$$
$$\cos (\theta + 2\pi n) = \cos \theta$$

for all integers n. The graphs of sin θ and cos θ are infinitely repeating waves which oscillate between -1 and $+1$ (Figure 7.1.7).

For infinite values of θ, the values of sin θ and cos θ continue to oscillate between -1 and 1. Thus the limits

$$\lim_{\theta \to \infty} \sin \theta, \qquad \lim_{\theta \to -\infty} \sin \theta,$$
$$\lim_{\theta \to \infty} \cos \theta, \qquad \lim_{\theta \to -\infty} \cos \theta,$$

do not exist. Figure 7.1.8 shows parts of the hyperreal graph of sin θ, for positive and negative infinite values of θ, through infinite telescopes.

The motion of our particle traveling around the unit circle with speed one

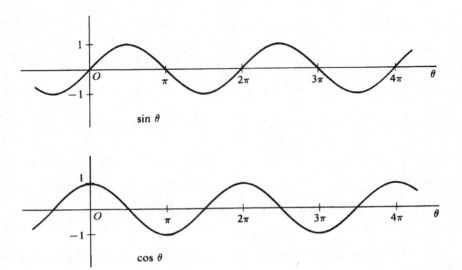

$\sin \theta$

$\cos \theta$

Figure 7.1.7

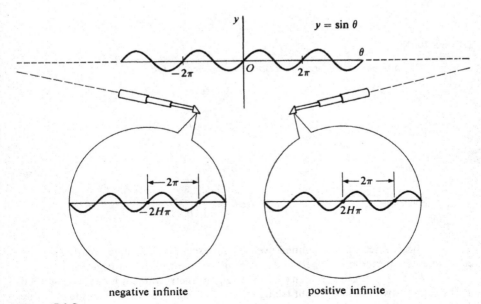

$y = \sin \theta$

negative infinite   positive infinite

Figure 7.1.8

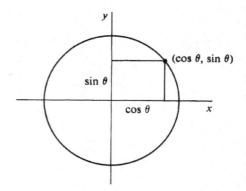

Figure 7.1.9

starting at $(1, 0)$ (Figure 7.1.9) has the parametric equations

$$x = \cos \theta, \qquad y = \sin \theta.$$

The following table shows a few values of $\sin \theta$ and $\cos \theta$, for θ in either radians or degrees.

Table 7.1.1

θ in radians	0	$\dfrac{\pi}{6}$	$\dfrac{\pi}{4}$	$\dfrac{\pi}{3}$	$\dfrac{\pi}{2}$	$\dfrac{3\pi}{4}$	π	$\dfrac{3\pi}{2}$	2π
θ in degrees	0°	30°	45°	60°	90°	135°	180°	270°	360°
$\sin \theta$	0	1/2	$\sqrt{2}/2$	$\sqrt{3}/2$	1	$\sqrt{2}/2$	0	-1	0
$\cos \theta$	1	$\sqrt{3}/2$	$\sqrt{2}/2$	1/2	0	$-\sqrt{2}/2$	-1	0	1

DEFINITION

The other trigonometric functions are defined as follows.

tangent: $$\tan \theta = \frac{\sin \theta}{\cos \theta}$$

cotangent: $$\cot \theta = \frac{\cos \theta}{\sin \theta}$$

secant: $$\sec \theta = \frac{1}{\cos \theta}$$

cosecant: $$\csc \theta = \frac{1}{\sin \theta}$$

These functions are defined everywhere except where there is a division by zero. They are periodic with period 2π. Their graphs are shown in Figure 7.1.10.

When θ is strictly between 0 and $\pi/2$, trigonometric functions can be described as the ratio of two sides of a right triangle with an angle θ. Let a be the side opposite θ, b the side adjacent to θ, c the hypotenuse as in Figure 7.1.11. Comparing this triangle

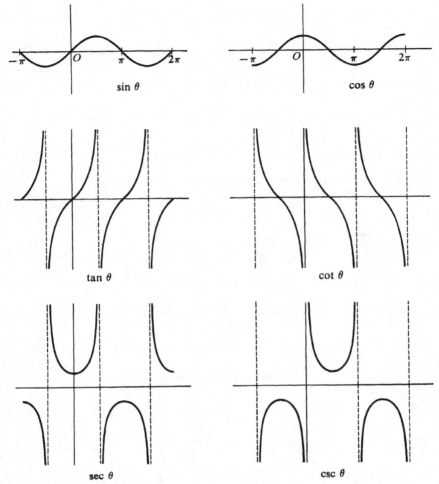

Figure 7.1.10

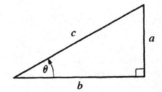

Figure 7.1.11

with a similar triangle whose hypotenuse is a radius of the unit circle, we see that

$$\sin \theta = \frac{a}{c}, \qquad \sec \theta = \frac{c}{b}, \qquad \tan \theta = \frac{a}{b},$$

$$\cos \theta = \frac{b}{c}, \qquad \csc \theta = \frac{c}{a}, \qquad \cot \theta = \frac{b}{a}.$$

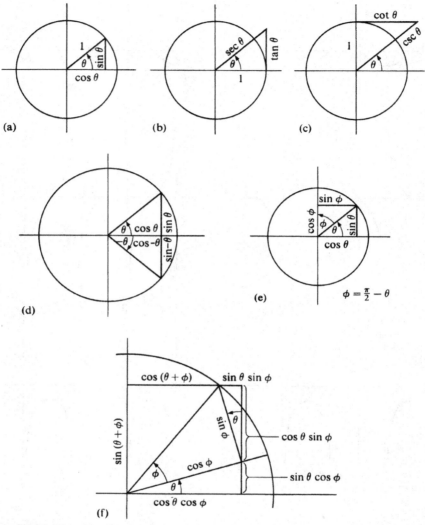

Figure 7.1.12 (Continued)

Here is a table of trigonometric identities. The diagrams in Figure 7.1.12 suggest possible proofs. ((6) and (7) are called the *addition formulas*.)

(1) $\sin^2 \theta + \cos^2 \theta = 1$ (Figure 7.1.12(a))

(2) $\tan^2 \theta + 1 = \sec^2 \theta$ (Figure 7.1.12(b))

(3) $\cot^2 \theta + 1 = \csc^2 \theta$ (Figure 7.1.12(c))

(4) $\sin(-\theta) = -\sin \theta,$ $\cos(-\theta) = \cos \theta$ (Figure 7.1.12(d))

(5) $\sin(\pi/2 - \theta) = \cos \theta,$ $\cos(\pi/2 - \theta) = \sin \theta$ (Figure 7.1.12(e))

(6) $\sin(\theta + \phi) = \sin \theta \cos \phi + \cos \theta \sin \phi$ (Figure 7.1.12(f))

(7) $\cos(\theta + \phi) = \cos \theta \cos \phi - \sin \theta \sin \phi$ (Figure 7.1.12(f))

PROBLEMS FOR SECTION 7.1

In Problems 1–6, derive the given identity using the formula $\sin^2 \theta + \cos^2 \theta = 1$ and the addition formulas for $\sin(\theta + \phi)$ and $\cos(\theta + \phi)$.

1	$\tan^2 \theta + 1 = \sec^2 \theta$	**2**	$\cos^2 \theta + \cos^2 \theta \cot^2 \theta = \cot^2 \theta$
3	$\sin 2\theta = 2 \sin \theta \cos \theta$	**4**	$\cos 2\theta = \cos^2 \theta - \sin^2 \theta$
5	$\sin^2 (\tfrac{1}{2}\theta) = \dfrac{1 - \cos \theta}{2}$	**6**	$\tan(\theta + \phi) = \dfrac{\tan \theta + \tan \phi}{1 - \tan \theta \tan \phi}$

In Problems 7–10, find all values of θ for which the given equation is true.

7	$\sin \theta = \cos \theta$	**8**	$\sin \theta \cos \theta = 0$
9	$\sec \theta = 0$	**10**	$5 \sin 3\theta = 0$

11   Find a value of θ where $\sin 2\theta$ is not equal to $2 \sin \theta$.

Determine whether the limits exist in Problems 12–17.

12	$\lim\limits_{x \to \infty} \sin x$	**13**	$\lim\limits_{x \to \infty} \dfrac{\sin x}{x}$
14	$\lim\limits_{x \to \infty} x \sin x$	**15**	$\lim\limits_{x \to 0} x \cos(1/x)$
16	$\lim\limits_{x \to 0} \cot x$	**17**	$\lim\limits_{x \to 0} \tan x$

18   Find all values of θ where $\tan \theta$ is undefined.

19   Find all values of θ where $\csc \theta$ is undefined.

7.2 DERIVATIVES OF TRIGONOMETRIC FUNCTIONS

THEOREM 1

The functions $x = \cos \theta$ and $y = \sin \theta$ are continuous for all θ.

PROOF We give the proof for θ in the first quadrant, $0 < \theta < \pi/2$. Let $\Delta\theta$ be infinitesimal and consider Figure 7.2.1.

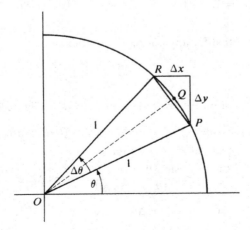

Figure 7.2.1

Let $\Delta s = \sqrt{\Delta x^2 + \Delta y^2}$ be the length of the line PR. Then

$$0 < \text{Area of quadrilateral } QPOR \leq \text{Area of sector } POR,$$
$$0 < \qquad\qquad \tfrac{1}{2}\Delta s \qquad\qquad \leq \qquad\qquad \tfrac{1}{2}\Delta\theta.$$

Thus Δs is infinitesimal. It follows that Δx and Δy are infinitesimal, whence the functions $x = \cos\theta$, $y = \sin\theta$ are continuous.

THEOREM 2

The functions $x = \cos\theta$ and $y = \sin\theta$ are differentiable for all θ, and

$$d(\sin\theta) = \cos\theta\, d\theta,$$
$$d(\cos\theta) = -\sin\theta\, d\theta.$$

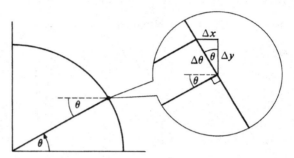

Figure 7.2.2

Discussion   Intuitively, the small triangle in Figure 7.2.2 is infinitely close to a right triangle with angle θ and hypotenuse $\Delta\theta$, whence

$$\frac{\Delta y}{\Delta\theta} \approx \cos\theta, \qquad \frac{\Delta x}{\Delta\theta} \approx -\sin\theta.$$

Notice that Δx is negative while Δy is positive when θ is in the first quadrant. The proof of Theorem 2 uses a lemma.

LEMMA

(i) $\displaystyle\lim_{\theta\to 0}\frac{\sin\theta}{\theta} = 1.$ (ii) $\displaystyle\lim_{\theta\to 0}\frac{\cos\theta - 1}{\theta} = 0.$

PROOF (i)   We show that for any nonzero infinitesimal $\Delta\theta$,

$$\frac{\sin\Delta\theta}{\Delta\theta} \approx 1.$$

When $\Delta\theta$ is positive we draw the figure shown in Figure 7.2.3. We have

$$\text{Area of triangle } QOR < \text{area of sector } QOR < \text{area of triangle } QOS,$$
$$\tfrac{1}{2}\sin\Delta\theta < \tfrac{1}{2}\Delta\theta < \tfrac{1}{2}\tan\Delta\theta.$$

Then $$\frac{\sin\Delta\theta}{\tan\Delta\theta} < \frac{\sin\Delta\theta}{\Delta\theta} < \frac{\sin\Delta\theta}{\sin\Delta\theta}, \qquad \cos\Delta\theta < \frac{\sin\Delta\theta}{\Delta\theta} < 1.$$

Since $\cos\theta$ is continuous, $\cos\Delta\theta \approx 1$, whence $\dfrac{\sin\Delta\theta}{\Delta\theta} \approx 1$. The case $\Delta\theta < 0$ is similar.

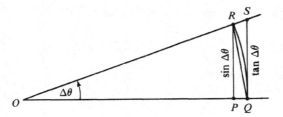

Figure 7.2.3

(ii) We compute the standard part of $(\cos \Delta\theta - 1)/\Delta\theta$.

$$st\left(\frac{\cos \Delta\theta - 1}{\Delta\theta}\right) = st\left(\frac{\cos^2 \Delta\theta - 1}{\Delta\theta(\cos \Delta\theta + 1)}\right) = st\left(\frac{-\sin^2 \Delta\theta}{\Delta\theta(\cos \Delta\theta + 1)}\right)$$

$$= -st\left(\frac{\sin \Delta\theta}{\Delta\theta}\right)\frac{st(\sin \Delta\theta)}{st(\cos \Delta\theta + 1)} = -1 \cdot \frac{0}{2} = 0.$$

PROOF OF THEOREM 2 Let $\Delta\theta$ be a nonzero infinitesimal. Then

$$\frac{d(\sin \theta)}{d\theta} = st\left(\frac{\sin (\theta + \Delta\theta) - \sin \theta}{\Delta\theta}\right)$$

$$= st\left(\frac{\sin \theta \cos \Delta\theta + \cos \theta \sin \Delta\theta - \sin \theta}{\Delta\theta}\right)$$

$$= st\left(\frac{\sin \theta(\cos \Delta\theta - 1) + \cos \theta \sin \Delta\theta}{\Delta\theta}\right)$$

$$= \sin \theta \, st\left(\frac{\cos \Delta\theta - 1}{\Delta\theta}\right) + \cos \theta \, st\left(\frac{\sin \Delta\theta}{\Delta\theta}\right)$$

$$= \sin \theta \cdot 0 + \cos \theta \cdot 1 = \cos \theta.$$

Here is a second proof that the derivative of the sine is the cosine. It uses the formula for the length of a curve in Section 6.3.

ALTERNATE PROOF OF THEOREM 2 (Optional) Let $0 \le \theta \le \pi/2$ and

$$x = \cos \theta, \qquad y = \sin \theta.$$

Then (x, y) is a point on the unit circle as shown in Figure 7.2.4.

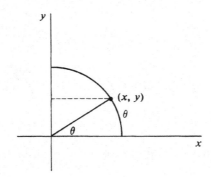

Figure 7.2.4

Take y as the independent variable. Then

$$x = \sqrt{1 - y^2}, \qquad \frac{dx}{dy} = \frac{-y}{\sqrt{1 - y^2}} = -\frac{y}{x}.$$

θ is the length of the arc from 0 to y, so

$$\theta = \int_0^y \sqrt{1 + (dx/dy)^2} \, dy.$$

By the Second Fundamental Theorem of Calculus,

$$\frac{d\theta}{dy} = \sqrt{1 + (dx/dy)^2} = \sqrt{1 + (y^2/x^2)} = \frac{\sqrt{x^2 + y^2}}{x} = \frac{1}{x}.$$

Then by the Chain Rule,

$$\frac{dy}{d\theta} = \frac{1}{d\theta/dy} = x,$$

and

$$\frac{dx}{d\theta} = \frac{dx}{dy}\frac{dy}{d\theta} = -\frac{y}{x} \cdot x = -y.$$

Substituting $\cos \theta$ for x and $\sin \theta$ for y,

$$\frac{d(\sin \theta)}{d\theta} = \cos \theta, \qquad \frac{d(\cos \theta)}{d\theta} = -\sin \theta.$$

We can now find the derivatives of all the trigonometric functions by using the Quotient Rule

$$d\left(\frac{u}{v}\right) = \frac{v \, du - u \, dv}{v^2}.$$

THEOREM 3

(i) $d(\sin \theta) = \cos \theta \, d\theta,$ $d(\cos \theta) = -\sin \theta \, d\theta,$

(ii) $d(\tan \theta) = \sec^2 \theta \, d\theta,$ $d(\cot \theta) = -\csc^2 \theta \, d\theta,$

(iii) $d(\sec \theta) = \sec \theta \tan \theta \, d\theta,$ $d(\csc \theta) = -\csc \theta \cot \theta \, d\theta.$

PROOF We prove the formula for $d(\tan \theta)$ and leave the rest as problems.

$$\tan \theta = \frac{\sin \theta}{\cos \theta},$$

$$d(\tan \theta) = d\left(\frac{\sin \theta}{\cos \theta}\right) = \frac{\cos \theta \, d(\sin \theta) - \sin \theta \, d(\cos \theta)}{\cos^2 \theta}$$

$$= \frac{\cos \theta \cos \theta - \sin \theta \, (-\sin \theta)}{\cos^2 \theta} d\theta = \frac{\cos^2 \theta + \sin^2 \theta}{\cos^2 \theta} d\theta$$

$$= \frac{1}{\cos^2 \theta} d\theta = \sec^2 \theta \, d\theta.$$

These formulas lead at once to new integration formulas.

THEOREM 4

(i) $\int \cos \theta \, d\theta = \sin \theta + C,$ $\int \sin \theta \, d\theta = -\cos \theta + C.$

(ii) $\int \sec^2 \theta \, d\theta = \tan \theta + C,$ $\int \csc^2 \theta \, d\theta = -\cot \theta + C.$

(iii) $\int \sec \theta \tan \theta \, d\theta = \sec \theta + C,$ $\int \csc \theta \cot \theta \, d\theta = -\csc \theta + C.$

We are not yet able to evaluate the integrals $\int \tan \theta \, d\theta$, $\int \cot \theta \, d\theta$, $\int \sec \theta \, d\theta$, $\int \csc \theta \, d\theta$. These integrals will be found in the next chapter.

EXAMPLE 1   Find the derivative of $y = \tan^2(3x)$.

$$dy = 2 \tan 3x \, d(\tan 3x) = 2 \tan 3x \sec^2 3x \, d(3x)$$
$$= 6 \tan 3x \sec^2 3x \, dx.$$

EXAMPLE 2   Evaluate $\lim\limits_{t \to \pi/2} \dfrac{\cos t}{t - \pi/2}.$

This is a limit of the form 0/0 because

$$\lim_{t \to \pi/2} \cos t = 0, \qquad \lim_{t \to \pi/2} \left(t - \frac{\pi}{2}\right) = 0.$$

By l'Hospital's Rule (Section 5.2),

$$\lim_{t \to \pi/2} \frac{\cos t}{t - \pi/2} = \lim_{t \to \pi/2} \frac{-\sin t}{1} = -\sin\left(\frac{\pi}{2}\right) = -1.$$

EXAMPLE 3   A particle travels around a vertical circle of radius r_0 with constant angular velocity $\omega = d\theta/dt$, beginning with $\theta = 0$ at time $t = 0$. If the sun is directly overhead, find the position, velocity, and acceleration of the shadow.

Let us center the circle at the origin in the (x, y) plane (Figure 7.2.5). Then

$$x = r_0 \cos \theta, \qquad y = r_0 \sin \theta.$$

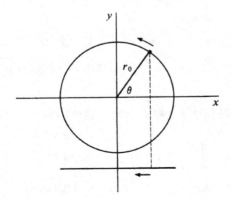

Figure 7.2.5

At time t, θ has the value $\theta = \omega t$. So the motion of the particle is given by the parametric equations

$$x = r_0 \cos(\omega t), \qquad y = r_0 \sin(\omega t).$$

The shadow is directly below the particle, and its position is given by the x-component

$$x = r_0 \cos(\omega t).$$

The velocity and acceleration of the shadow are

$$v = \frac{dx}{dt} = -r_0 \omega \sin(\omega t),$$

$$a = \frac{dv}{dt} = -r_0 \omega^2 \cos(\omega t).$$

EXAMPLE 4   A light beam on a 100 ft tower rotates in a vertical circle at the rate of one revolution per second. Find the speed of the spot of light moving along the ground at a point 1000 ft from the base of the tower.

We start by drawing the picture in Figure 7.2.6.

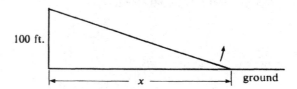

100 ft.

Figure 7.2.6 x ground

Assume the rotation is counterclockwise. Let t be time and let x and θ be as in the figure. Then

$$\frac{d\theta}{dt} = 2\pi \text{ radians/sec}, \qquad x = 100 \tan\theta \text{ ft}.$$

We wish to find dx/dt when $x = 1000$.

$$\frac{dx}{dt} = 100 \sec^2\theta \frac{d\theta}{dt} = 200\pi \sec^2\theta.$$

When $x = 1000$,

$$\sec^2\theta = 1 + \tan^2\theta = 1 + (x/100)^2 = 1 + 10^2 = 101.$$

Therefore $\dfrac{dx}{dt} = 20200\pi \sim 63,\!000 \text{ ft/sec}.$

EXAMPLE 5   Find $\int \sin^3 t \cos t \, dt$. Let $u = \sin t$, $du = \cos t \, dt$.

Then $\displaystyle \int \sin^3 t \cos t \, dt = \int u^3 \, du = \frac{u^4}{4} + C = \frac{\sin^4 t}{4} + C.$

EXAMPLE 6   Find the area under one arch of the curve $y = \cos x$.

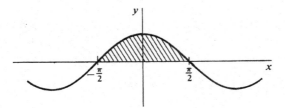

Figure 7.2.7

From Figure 7.2.7 we see that one arch lies between the limits $x = -\pi/2$ and $x = \pi/2$, therefore the area is

$$\int_{-\pi/2}^{\pi/2} \cos t \, dt = \sin t \Big]_{-\pi/2}^{\pi/2} = 1 - (-1) = 2.$$

Trigonometric identities can often be used to get an integral into a form which is easy to evaluate.

EXAMPLE 7 Evaluate $\int \sec^4 x \, dx$. Using the identity $\sec^2 x = 1 + \tan^2 x$, we have

$$\int \sec^4 x \, dx = \int (1 + \tan^2 x) \sec^2 x \, dx$$

$$= \int (1 + \tan^2 x) \, d(\tan x) = \tan x + \frac{\tan^3 x}{3} + C.$$

EXAMPLE 8 Find $\int \sqrt{1 - \cos x} \, dx$. Using the identity $\sin^2 x + \cos^2 x = 1$, we have

$$\sqrt{1 - \cos x} = \frac{\sqrt{1 - \cos x}\sqrt{1 + \cos x}}{\sqrt{1 + \cos x}} = \frac{\sqrt{1 - \cos^2 x}}{\sqrt{1 + \cos x}}$$

$$= \frac{\sqrt{\sin^2 x}}{\sqrt{1 + \cos x}} = \frac{|\sin x|}{\sqrt{1 + \cos x}}.$$

Case 1 In an interval where $\sin x \geq 0$,

$$\int \sqrt{1 - \cos x} \, dx = \int \frac{\sin x}{\sqrt{1 + \cos x}} dx = \int -\frac{1}{\sqrt{1 + \cos x}} d(1 + \cos x)$$

$$= -2\sqrt{1 + \cos x} + C.$$

Case 2 In an interval where $\sin x \leq 0$,

$$\int \sqrt{1 - \cos x} \, dx = 2\sqrt{1 + \cos x} + C.$$

PROBLEMS FOR SECTION 7.2

In Problems 1–14, find the derivative.

1	$y = \sin 5x$	2	$y = 3 \cos^2 x$
3	$x = \sin (3\theta^2)$	4	$y = \sec^3 x$

5	$x = \tan(4\theta - 3)$	6	$y = x \sin x$
7	$u = a \sin \theta + b \cos \theta$	8	$u = \sin(a\theta) + \cos(b\theta)$
9	$y = \cos \sqrt{x}$	10	$y = \sqrt{\cos x}$
11	$y = \tan(\sin \theta)$	12	$y = \sin \theta \tan \theta$
13	$u = \dfrac{1}{2 + \csc(3t)}$	14	$y = \cot(t^2 + 3t - 2)$

15   Find dy/dx where $x = \sin^2 y$

16   Find dy/dx where $y = \tan(xy)$

In Problems 17–24, evaluate the limit if it exists.

17	$\displaystyle\lim_{\theta \to \pi/3} 2 \sin^2 \theta$	18	$\displaystyle\lim_{x \to 0} \csc x$
19	$\displaystyle\lim_{x \to 0^+} \csc x$	20	$\displaystyle\lim_{t \to 0} \frac{\sin^2 t}{t}$
21	$\displaystyle\lim_{t \to 0} \frac{\sin(2t)}{t}$	22	$\displaystyle\lim_{\theta \to \pi} \frac{\sin \theta}{\pi - \theta}$
23	$\displaystyle\lim_{t \to 0} \frac{\sin(t^2)}{t \sin t}$	24	$\displaystyle\lim_{0 \to \pi/2} (\sec \theta - \tan \theta)$

In Problems 25–34, find the maxima, minima, inflection points, and limits when necessary, and sketch the curve for $0 \le x \le 2\pi$.

25	$y = 3 \sin x$	26	$y = \sin x \cos x$
27	$y = \sin^2 x$	28	$y = \cos(2x)$
29	$y = \sin\left(x - \dfrac{\pi}{4}\right)$	30	$y = \sec x$
31	$y = \tan x$	32	$y = 1 - \cos x$
33	$y = \csc^2 x$	34	$y = x + \sin x$

☐ 35   Show that at $\displaystyle\lim_{x \to 0} \sin(1/x)$ does not exist.

☐ 36   Let $f(x) = x \sin(1/x)$, with $f(0) = 0$. Show that f is continuous but not differentiable at $x = 0$.

In Problems 37–53, evaluate the integral.

37	$\displaystyle\int \sin(2t)\, dt$	38	$\displaystyle\int \sin x \cos x\, dx$
39	$\displaystyle\int \tan x \sec^3 x\, dx$	40	$\displaystyle\int \tan^2 \theta\, d\theta$
41	$\displaystyle\int \frac{1}{\sqrt{x}} \cos \sqrt{x}\, dx$	42	$\displaystyle\int t \sin(t^2 + 1)\, dt$
43	$\displaystyle\int \cot(5\theta) \csc(5\theta)\, d\theta$	44	$\displaystyle\int \sqrt{1 + \sin \theta}\, d\theta$
45	$\displaystyle\int \sec x \sqrt{\sec x - 1}\, dx$	46	$\displaystyle\int \frac{\sin \theta - \cos \theta}{(\sin \theta + \cos \theta)^2}\, d\theta$
47	$\displaystyle\int \frac{1}{1 + \sin \theta}\, d\theta$	48	$\displaystyle\int_0^\pi 3 \sin t\, dt$
49	$\displaystyle\int_{-\pi/4}^{\pi/4} \sec^2 \theta\, d\theta$	50	$\displaystyle\int_0^1 \sin(\pi x)\, dx$

51 $\displaystyle\int_{\pi/3}^{\pi/2} \sin\theta + \cos\theta\, d\theta$ **52** $\displaystyle\int_{0}^{\pi/2} \sec^2 x\, dx$

53 $\displaystyle\int_{0}^{\pi/2} \cot x \csc x\, dx$

54 A revolving light one mile from shore sweeps out eight revolutions per minute. Find the velocity of the beam of light along the shore at the instant when it makes an angle of 45° with the shoreline.

55 A ball is thrown vertically upward from a point P so that its height at time t is $y = 100t - 16t^2$ feet. Q is another point on the surface 100 ft from P. At time $t = 5$ find the rate of change of the angle between the horizontal line QP and the line from Q to the ball.

56 Two hallways of width a and b meet at right angles. Find the length of the longest rod which can be slid on the floor around the corner.

57 Find the area under one arch of the curve $y = 3\sin x$.

58 Find the area under one arch of the curve $y = \sin(3x)$.

59 Find the area of the region between the curves $y = \sin x \cos x$ and $y = \sin x, 0 \le x \le \pi/2$.

60 The region between the x-axis and the curve $y = \tan x, 0 \le x \le \pi/4$, is rotated about the x-axis. Find the volume of the solid of revolution.

61 The region between the x-axis and the curve $y = (\sin x)/x, \pi/2 \le x \le \pi$, is rotated about the y-axis. Find the volume of the solid of revolution.

62 Find the length of the parametric curve $x = 2\cos(3t), y = 2\sin(3t), 0 \le t \le 1$.

63 Find the length of the parametric curve $x = \cos^2 t, y = \sin^2 t, 0 \le t \le \pi/2$.

64 Find the length of the parametric curve $x = \cos^3 t, y = \sin^3 t, 0 \le t \le \pi/2$.

65 Find the area of the surface generated by rotating the curve in Problem 63 about the x-axis.

66 Find the area of the surface generated by rotating the curve in Problem 64 about the y-axis.

7.3 INVERSE TRIGONOMETRIC FUNCTIONS

Inverse functions were studied in Section 2.4. We now take up the topic again and apply it to trigonometric functions. A *binary relation* on the real numbers is any set of ordered pairs of real numbers. Thus a real function f of one variable is a binary relation such that for each x, either there is exactly one y with (x, y) in f or there is no y with (x, y) in f. (Other important relations are $x < y, x \le y, x \ne y, x = y$.)

DEFINITION

*Let S be a binary relation on the real numbers. The **inverse relation** of S is the set T of all ordered pairs (y, x) such that (x, y) is in S. If S and T are both functions they are called **inverse functions** of each other.*

The inverse of a function f may or may not be a function. For example, the inverse of $y = x^2$ is the relation $x = \pm\sqrt{y}$, which is not a function (Figure 7.3.1). But the inverse of $y = x^2, x \ge 0$, is the function $x = \sqrt{y}$ (Figure 7.3.2).

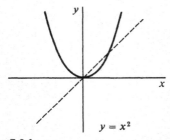

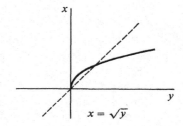

Figure 7.3.1

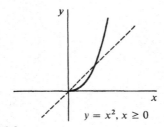

Figure 7.3.2

Geometrically, the graph of the inverse relation of $y = f(x)$ can be obtained by flipping the graph of $y = f(x)$ about the diagonal line $y = x$ (the dotted line in Figures 7.3.1 and 7.3.2). This flipping interchanges the x- and y-axes. This is because $f(x) = y$ means (x, y) is in f, and $g(y) = x$ means (y, x) is in g. It follows that:

> *If f and g are inverse functions then the range of f is the domain of g and vice versa.*

Which functions have inverse functions? We can answer this question with a definition and a simple theorem.

DEFINITION

> *A real function f with domain X is said to be one-to-one if f never takes the same value twice, that is, for all $x_1 \neq x_2$ in X we have $f(x_1) \neq f(x_2)$.*

THEOREM 1

> *f has an inverse function if and only if f is one-to-one.*

PROOF The following statements are equivalent.

   (1) f is a one-to-one function.
   (2)   For every y, either there is exactly one x with $f(x) = y$ or there is no x with $f(x) = y$.
   (3)   The equation $y = f(x)$ determines x as a function of y.
   (4) f has an inverse function.

COROLLARY

> *Every function which is increasing on its domain I has an inverse function. So does every function decreasing on its domain I.*

PROOF Let f be increasing on I. For any two points $x_1 \neq x_2$ in I, the value of f at the smaller of x_1, x_2 is less than the value of f at the greater, so $f(x_1) \neq f(x_2)$.

For example, the function $y = x^2$ is not one-to-one because $(-1)^2 = 1^2$, whence it has no inverse function. The function $y = x^2, x \geq 0$, is increasing on its domain $[0, \infty)$ and thus has an inverse.

Now let us examine the trigonometric functions. The function $y = \sin x$ is not one-to-one. For example, $\sin 0 = 0, \sin \pi = 0, \sin 2\pi = 0$, etc. We can see in Figure 7.3.3 that the inverse relation of $y = \sin x$ is not a function.

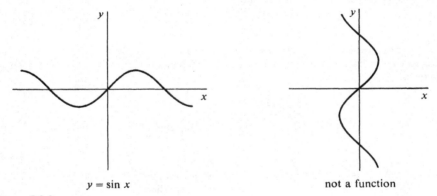

$y = \sin x$ not a function

Figure 7.3.3

However, the function $y = \sin x$ is increasing on the interval $[-\pi/2, \pi/2]$, because its derivative $\cos x$ is ≥ 0. So the sine function restricted to the interval $[-\pi/2, \pi/2]$,

$$y = \sin x, \qquad -\pi/2 \leq x \leq \pi/2,$$

has an inverse function shown in Figure 7.3.4. This inverse is called the *arcsine*

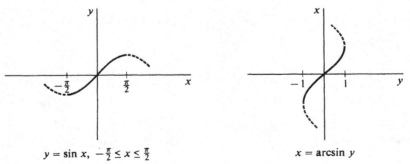

$y = \sin x, \; -\frac{\pi}{2} \leq x \leq \frac{\pi}{2}$ $x = \arcsin y$

Figure 7.3.4

function. It is written $x = \arcsin y$. Verbally, $\arcsin y$ is the angle x between $-\pi/2$ and $\pi/2$ whose sine is y.

The other trigonometric functions also are not one-to-one and thus do not have inverse functions. However, in each case we obtain a one-to-one function by restricting the domain to a suitable interval, either $[-\pi/2, \pi/2]$ or $[0, \pi]$. The resulting inverse functions are called the arccosine, arctangent, etc.

DEFINITION

The inverse trigonometric functions are defined as follows.

$x = \arcsin y$ *is the inverse of* $y = \sin x,$ $\quad -\pi/2 \le x \le \pi/2$

$x = \arccos y$ *is the inverse of* $y = \cos x,$ $\quad 0 \le x \le \pi$

$x = \arctan y$ *is the inverse of* $y = \tan x,$ $\quad -\pi/2 \le x \le \pi/2$

$x = \operatorname{arccot} y$ *is the inverse of* $y = \cot x,$ $\quad 0 \le x \le \pi$

$x = \operatorname{arcsec} y$ *is the inverse of* $y = \sec x,$ $\quad 0 \le x \le \pi$

$x = \operatorname{arccsc} y$ *is the inverse of* $y = \csc x,$ $\quad -\pi/2 \le x \le \pi/2$

The graphs of these functions are shown in Figure 7.3.5. The domains of the inverse trigonometric functions can be read off from the graphs, and are shown in the table below.

Table 7.3.1

Function	Domain
$\arcsin y$	$-1 \le y \le 1$
$\arccos y$	$-1 \le y \le 1$
$\arctan y$	whole real line
$\operatorname{arccot} y$	whole real line
$\operatorname{arcsec} y$	$y \le -1, \quad y \ge 1$
$\operatorname{arccsc} y$	$y \le -1, \quad y \ge 1$

We can prove the inverse trigonometric functions have these domains (i.e., the figures are correct) using the Intermediate Value Theorem. As an illustration we prove that $\arcsin y$ has domain $[-1, 1]$.

$\arcsin y$ is undefined outside $[-1, 1]$ because $-1 \le \sin x \le 1$ for all x. Suppose y_0 is in $[-1, 1]$. Then

$$\sin(-\pi/2) = -1 \le y_0 \le 1 = \sin(\pi/2).$$

$\sin x$ is continuous, so by the Intermediate Value Theorem there exists x_0 between $-\pi/2$ and $\pi/2$ such that $\sin x_0 = y_0$. Thus

$$\arcsin y_0 = x_0$$

and y_0 is in the domain of $\arcsin y$.

EXAMPLE 1 Find $\arccos(\sqrt{2}/2)$. From Table 7.1.1, $\cos(\pi/4) = \sqrt{2}/2$. Since $0 \le \pi/4 \le \pi$,

$$\arccos(\sqrt{2}/2) = \pi/4.$$

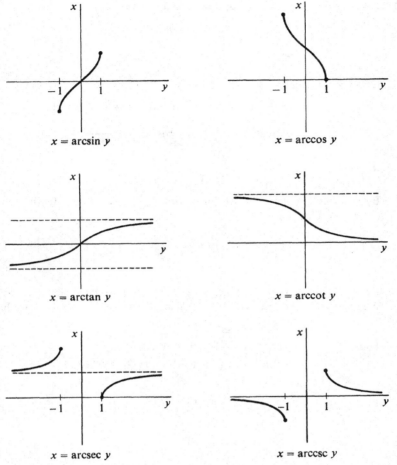

Figure 7.3.5

EXAMPLE 2   Find arcsin (-1). From Table 7.1.1, $\sin(3\pi/2) = -1$. But $3\pi/2$ is not in the interval $[-\pi/2, \pi/2]$. Using $\sin(\theta + 2n\pi) = \sin\theta$, we have

$$\sin(-\pi/2) = \sin(3\pi/2) = -1,$$

so $$\arcsin(-1) = -\pi/2.$$

EXAMPLE 3   Find arctan $(-\sqrt{3})$. We must find a θ in the interval $[-\pi/2, \pi/2]$ such that $\tan\theta = -\sqrt{3}$. From Table 7.1.1, $\sin(\pi/3) = \sqrt{3}/2$, $\cos(\pi/3) = 1/2$. Then $\sin(-\pi/3) = -\sqrt{3}/2$, $\cos(-\pi/3) = 1/2$. So

$$\tan(-\pi/3) = \frac{-\sqrt{3}/2}{1/2} = -\sqrt{3},$$

$$\arctan(-\sqrt{3}) = -\pi/3.$$

EXAMPLE 4. Find cos (arctan y). Let $\theta = \arctan y$. Thus $\tan \theta = y$. Using

$$\sin^2 \theta + \cos^2 \theta = 1$$

$$\frac{\sin \theta}{\cos \theta} = y,$$

we solve for $\cos \theta$.

$$\sin \theta = y \cos \theta, \qquad (y \cos \theta)^2 + \cos^2 \theta = 1,$$

$$\cos^2 \theta (y^2 + 1) = 1, \qquad \cos^2 \theta = \frac{1}{y^2 + 1}.$$

Thus $\cos \theta = \pm \dfrac{1}{\sqrt{y^2 + 1}}.$

By definition of arctan y, we know that $-\pi/2 \le \theta \le \pi/2$. In this interval, $\cos \theta \ge 0$. Therefore

$$\cos \theta = \frac{1}{\sqrt{y^2 + 1}}.$$

EXAMPLE 5   Show that $\arcsin y + \arccos y = \pi/2$ (Figure 7.3.6). Let $\theta = \arcsin y$. We have $y = \sin \theta = \cos(\pi/2 - \theta)$. Also, when $-\pi/2 \le \theta \le \pi/2$, we have

$$\pi/2 \ge -\theta \ge -\pi/2, \qquad \pi \ge \pi/2 - \theta \ge 0.$$

Thus $\pi/2 - \theta = \arccos y,$

$$\arcsin y + \arccos y = \theta + (\pi/2 - \theta) = \pi/2.$$

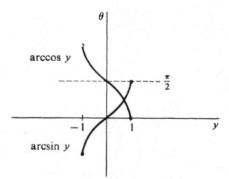

Figure 7.3.6

We shall next study the derivatives of the inverse trigonometric functions. Here is a general theorem which tells us when the derivative of the inverse function exists and gives a rule for computing its value.

INVERSE FUNCTION THEOREM

Suppose a real function f is differentiable on an open interval I and f has an inverse function g. Let x be a point in I where $f'(x) \ne 0$ and let $y = f(x)$. Then

(i) $g'(y)$ *exists,*

(ii) $g'(y) = \dfrac{1}{f'(x)}.$

We omit the proof that $g'(y)$ exists. Intuitively, the curve $y = f(x)$ has a non-horizontal tangent line, so the curve $x = g(y)$ should have a nonvertical tangent line and thus $g'(y)$ should exist.

The Inverse Function Rule from Chapter 2 says that (ii) is true if we assume (i). The proof of (ii) from (i) is an application of the Chain Rule:

$$g(f(x)) = x, \qquad g'(f(x))f'(x) = 1, \qquad g'(y)f'(x) = 1, \qquad g'(y) = \frac{1}{f'(x)}.$$

The Inverse Function Theorem shows that all the inverse trigonometric functions have derivatives. We now evaluate these derivatives.

THEOREM 2

(i) $d(\arcsin x) = \dfrac{dx}{\sqrt{1 - x^2}}$ (*where* $-1 < x < 1$).

    $d(\arccos x) = -\dfrac{dx}{\sqrt{1 - x^2}}$ (*where* $-1 < x < 1$).

(ii) $d(\arctan x) = \dfrac{dx}{1 + x^2}.$

    $d(\text{arccot } x) = -\dfrac{dx}{1 + x^2}.$

(iii) $d(\text{arcsec } x) = \dfrac{dx}{|x|\sqrt{x^2 - 1}}$ (*where* $|x| > 1$).

    $d(\text{arccsc } x) = -\dfrac{dx}{|x|\sqrt{x^2 - 1}}$ (*where* $|x| > 1$).

PROOF   We prove the first part of (i) and (iii). Since the derivatives exist we may use implicit differentiation.

(i)   Let $y = \arcsin x$. Then

$$x = \sin y, \qquad -\pi/2 \le y \le \pi/2,$$
$$dx = \cos y \, dy.$$

From $\sin^2 y + \cos^2 y = 1$ we get

$$\cos y = \pm\sqrt{1 - \sin^2 y} = \pm\sqrt{1 - x^2}.$$

Since $-\pi/2 \le y \le \pi/2$, $\cos y \ge 0$. Then

$$\cos y = \sqrt{1 - x^2}.$$

Substituting, $dx = \sqrt{1 - x^2} \, dy, \qquad dy = \dfrac{dx}{\sqrt{1 - x^2}}.$

(iii)   Let $y = \text{arcsec } x$.

Then
$$x = \sec y, \qquad 0 \le y \le \pi,$$
$$dx = \sec y \tan y \, dy.$$

From $\tan^2 y + 1 = \sec^2 y$ we get $\tan y = \pm\sqrt{\sec^2 y - 1} = \pm\sqrt{x^2 - 1}$.

Since $0 \le y \le \pi$, $\tan y$ and $\sec y = \dfrac{1}{\cos y}$ have the same sign.

Therefore
$$\sec y \tan y \ge 0$$

and
$$dx = |\sec y||\tan y| \, dy = |x|\sqrt{x^2 - 1} \, dy,$$
$$dy = \frac{dx}{|x|\sqrt{x^2 - 1}}.$$

When we turn these formulas for derivatives around we get some surprising new integration formulas.

THEOREM 3

(i) $\displaystyle\int \frac{1}{\sqrt{1 - x^2}} dx = \arcsin x + C = -\arccos x + C.$ (*Provided that* $|x| < 1$).

(ii) $\displaystyle\int \frac{dx}{1 + x^2} = \arctan x + C = -\operatorname{arccot} x + C.$

(iii) $\displaystyle\int \frac{dx}{|x|\sqrt{x^2 - 1}} = \operatorname{arcsec} x + C = -\operatorname{arccsc} x + C.$ (*Provided that* $|x| > 1$).

From part (i), $\arcsin x$ and $-\arccos x$ must differ only by a constant. We already knew this from Example 5,

$$\arcsin x = -\arccos x + \pi/2.$$

Before now we were not able to find the area of the regions under the curves

$$y = \frac{1}{\sqrt{1 - x^2}}, \qquad y = \frac{1}{1 + x^2}, \qquad y = \frac{1}{x\sqrt{x^2 - 1}}.$$

It is a remarkable and quite unexpected fact that these areas are given by inverse trigonometric functions.

EXAMPLE 6   (a)   Find the area of the region under the curve

$$y = \frac{1}{1 + x^2}$$

for $-1 \le x \le 1$.

(b)   Find the area of the region under the same curve for $-\infty < x < \infty$. The regions are shown in Figure 7.3.7.

(a) $A = \displaystyle\int_{-1}^{1} \frac{1}{1 + x^2} dx = \arctan x \Big]_{-1}^{1} = \frac{\pi}{4} - \left(-\frac{\pi}{4}\right) = \frac{\pi}{2}.$

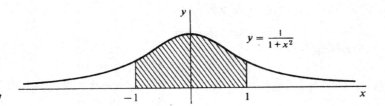

Figure 7.3.7

(b) $\quad A = \int_{-\infty}^{\infty} \frac{1}{1 + x^2}\, dx = \int_{-\infty}^{0} \frac{1}{1 + x^2}\, dx + \int_{0}^{\infty} \frac{1}{1 + x^2}\, dx$

$\qquad = \lim_{a \to -\infty} \int_{a}^{0} \frac{1}{1 + x^2}\, dx + \lim_{b \to \infty} \int_{0}^{b} \frac{1}{1 + x^2}\, dx$

$\qquad = \lim_{a \to -\infty} (\arctan 0 - \arctan a) + \lim_{b \to \infty} (\arctan b - \arctan 0)$

$\qquad = -\lim_{a \to -\infty} \arctan a + \lim_{b \to \infty} \arctan b.$

From the graph of arctan x we see that the first limit is $-\pi/2$ and the second limit is $\pi/2$, so

$$A = -\left(-\frac{\pi}{2}\right) + \frac{\pi}{2} = \pi.$$

Thus the region under $y = 1/(1 + x^2)$ has exactly the same area as the unit circle, and half of this area is between $x = -1$ and $x = 1$.

EXAMPLE 7   Find $\displaystyle\int_{-2}^{-\sqrt{2}} \frac{1}{x\sqrt{x^2 - 1}}\, dx.$

The region is shown in Figure 7.3.8. Since x is negative, $x = -|x|$. Thus

$\displaystyle\int_{-2}^{-\sqrt{2}} \frac{1}{x\sqrt{x^2 - 1}}\, dx = \int_{-2}^{-\sqrt{2}} -\frac{1}{|x|\sqrt{x^2 - 1}}\, dx$

$\qquad = -\operatorname{arcsec} x \Big]_{-2}^{-\sqrt{2}} = -(\operatorname{arcsec}(-\sqrt{2}) - \operatorname{arcsec}(-2))$

$\qquad = -\left(\frac{3\pi}{4} - \frac{2\pi}{3}\right) = -\frac{\pi}{12}.$

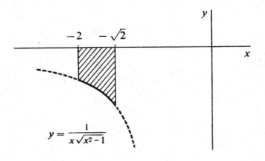

Figure 7.3.8

PROBLEMS FOR SECTION 7.3

In Problems 1–9, evaluate the given expression.

1 $\arcsin(\sqrt{3}/2)$

2 $\arcsin(-1/2)$

3 $\arctan(-1)$

4 $\sec(\arctan(-1))$

5 $\operatorname{arcsec} 2$

6 $\arcsin(\cos \pi)$

7 $\sin(\arccos x)$

8 $\cot(\operatorname{arcsec} x)$

9 $\arcsin(\cos x), \quad 0 \le x \le \pi$

10 Prove the identity $\arctan(-x) = -\arctan x$.

11 Prove $\arctan(1/x) = \operatorname{arccot} x$, for $0 < x$.

12 Prove $\arccos(-x) = \pi - \arccos x$.

13 Prove $\arctan x + \operatorname{arccot} x = \pi/2$.

Find the derivatives in Problems 14–25.

14 $y = \arcsin(x/2)$

15 $y = \operatorname{arcsec}(5x - 2)$

16 $y = (\arcsin x)^2$

17 $y = \arcsin(x^2)$

18 $y = \arctan\sqrt{x}$

19 $s = t\arcsin t$

20 $y = x\operatorname{arcsec} x$

21 $y = \arcsin x + \sqrt{1 - x^2}$

22 $y = \arccos x + (x/\sqrt{1 - x^2})$

23 $y = x\arcsin x + \sqrt{1 - x^2}$

24 $u = \operatorname{arcsec} t + \sqrt{t^2 - 1}$

25 $y = \arctan(1/\sqrt{x})$

26 Evaluate $\lim\limits_{x \to \infty} \operatorname{arccsc} x$.

27 Evaluate $\lim\limits_{x \to -\infty} \arctan x$.

28 Evaluate $\lim\limits_{x \to 0} \dfrac{\arcsin x}{x}$.

29 Evaluate $\lim\limits_{x \to \infty} \dfrac{\operatorname{arccot} x}{\operatorname{arccsc} x}$.

In Problems 30–47 evaluate the integrals.

30 $\displaystyle\int \frac{dx}{1 + 4x^2}$

31 $\displaystyle\int \frac{dx}{9 + x^2}$

32 $\displaystyle\int \frac{dx}{\sqrt{4 - x^2}}$

33 $\displaystyle\int \frac{dx}{\sqrt{x - x^2}}$

34 $\displaystyle\int \frac{\cos x}{1 + \sin^2 x}\,dx$

35 $\displaystyle\int \frac{dx}{x\sqrt{4x^2 - 1}}, \quad x > 1$

36 $\displaystyle\int \frac{x\,dx}{x^4 + 1}$

37 $\displaystyle\int \frac{x\,dx}{\sqrt{1 - x^4}}$

38 $\displaystyle\int \frac{dx}{(1 + x)\sqrt{x}}$

39 $\displaystyle\int \frac{dx}{x\sqrt{x - 1}}$

40 $\displaystyle\int \frac{\arctan x}{1 + x^2}\,dx$

41 $\displaystyle\int \frac{\arcsin x}{\sqrt{1 - x^2}}\,dx$

42 $\displaystyle\int_{-\sqrt{3}}^{\sqrt{3}} \frac{dx}{1 + x^2}$

43 $\displaystyle\int_{1}^{2} \frac{1}{x\sqrt{x^2 - 1}}\,dx$

44 $\displaystyle\int_{-\sqrt{2}}^{-1} \frac{1}{x\sqrt{x^2 - 1}}\,dx$

45 $\displaystyle\int_{0}^{1/2} \frac{1}{\sqrt{1 - x^2}}\,dx$

46 $\displaystyle\int_0^\infty \frac{dx}{25x^2 + 1}$ **47** $\displaystyle\int_{-\infty}^\infty \frac{dx}{a^2 + x^2}$

48 Find the area of the region bounded by the x-axis and the curve $y = 1/\sqrt{1 - x^2}$, $-1 < x < 1$.

49 Find the area of the region under the curve $y = 1/(x\sqrt{x^2 - 1})$, $1 \le x < \infty$.

50 Find the area of the region bounded below by the line $y = \frac{1}{2}$ and above by the curve $y = 1/(x^2 + 1)$.

7.4 INTEGRATION BY PARTS

One reason it is harder to integrate than differentiate is that for derivatives there is both a Sum Rule and a Product Rule,

$$d(u + v) = du + dv, \qquad d(uv) = u\,dv + v\,du$$

while for integrals there is only a Sum Rule,

$$\int du + dv = \int du + \int dv.$$

The Sum Rule for integrals is obtained in a simple way by reversing the sum rule for derivatives.

There is a way to turn the Product Rule for derivatives into a rule for integrals. It no longer looks like a product rule, and is called integration by parts. Integration by parts is a basic method which is needed for many integrals involving trigonometric functions (and later exponential functions).

INDEFINITE INTEGRATION BY PARTS

Suppose, for x in an open interval I, that u and v depend on x and that du and dv exist. Then

$$\int u\,dv = uv - \int v\,du.$$

PROOF We use the Product Rule

$$u\,dv + v\,du = d(uv), \qquad u\,dv = d(uv) - v\,du.$$

Integrating both sides with x as the independent variable,

$$\int u\,dv = \int (d(uv) - v\,du) = \int d(uv) - \int v\,du = uv - \int v\,du.$$

No constant of integration is needed because there are indefinite integrals on both sides of the equation.

Integration by parts is useful whenever $\int v\,du$ is easier to evaluate than a given integral $\int u\,dv$.

EXAMPLE 1   Evaluate $\int x \sin x \, dx$. Our plan is to break $x \sin x \, dx$ into a product of the form $u \, dv$, evaluate the integrals $\int dv$ and $\int v \, du$, and then use integration by parts to get $\int u \, dv$. There are several choices we might make for u and dv, and not all of them lead to a solution of the problem. Some guesswork is required.

First try: $u = \sin x, dv = x \, dx$. $\int dv = \int x \, dx = \frac{1}{2}x^2 + C$. Take $v = \frac{1}{2}x^2$. Next we find du and try to evaluate $\int v \, du$.

$$du = \cos x \, dx, \qquad \int v \, du = \int \frac{1}{2}x^2 \cos x \, dx.$$

This integral looks harder than the one we started with, so we shall start over with another choice of u and dv.

Second try: $u = x, dv = \sin x \, dx$.

$$\int dv = \int \sin x \, dx = -\cos x + C.$$

We take $v = -\cos x$. This time we find du and easily evaluate $\int v \, du$.

$$du = dx, \qquad \int v \, du = \int -\cos x \, dx = -\sin x + C_1.$$

Finally we use the rule

$$\int u \, dv = uv - \int v \, du,$$

$$\int x \sin x \, dx = x(-\cos x) - (-\sin x + C_1),$$

or
$$\int x \sin x \, dx = -x \cos x + \sin x + C.$$

EXAMPLE 2   Evaluate $\int \arcsin x \, dx$. A choice of u and dv which works is

$$u = \arcsin x, \qquad dv = dx.$$

We may take $v = x$. Then

$$du = \frac{dx}{\sqrt{1 - x^2}},$$

$$\int v \, du = \int \frac{x \, dx}{\sqrt{1 - x^2}} = -\sqrt{1 - x^2} + C_1.$$

Finally, $\int \arcsin x \, dx = x \arcsin x - (-\sqrt{1 - x^2} + C_1),$

$$\int \arcsin x \, dx = x \arcsin x + \sqrt{1 - x^2} + C.$$

This integral and the similar formula for $\int \arccos x \, dx$ are included in our table at the end of the book. We shall see how to integrate the other inverse trigonometric functions in the next chapter.

EXAMPLE 3 Evaluate $\int x^2 \sin x \, dx$. This requires two integrations by parts.

Step 1
$$u = x^2, \qquad dv = \sin x \, dx,$$
$$du = 2x \, dx, \qquad \int dv = \int \sin x \, dx = -\cos x + C.$$

We take $v = -\cos x$.

$$\int x^2 \sin x \, dx = uv - \int v \, du = -x^2 \cos x + \int 2x \cos x \, dx.$$

Step 2 Evaluate $\int 2x \cos x \, dx$.

$$u_1 = 2x, \qquad dv_1 = \cos x \, dx,$$
$$du_1 = 2 \, dx, \qquad \int dv_1 = \int \cos x \, dx = \sin x + C.$$

We take $v_1 = \sin x$.

$$\int 2x \cos x \, dx = u_1 v_1 - \int v_1 \, du_1$$

$$= 2x \sin x - \int 2 \sin x \, dx$$

$$= 2x \sin x + 2 \cos x + C.$$

Combining the two steps,

$$\int x^2 \sin x \, dx = -x^2 \cos x + 2x \sin x + 2 \cos x + C.$$

Sometimes integration by parts will yield an equation in which the given integral occurs on both sides. One can often solve for the answer.

EXAMPLE 4 Evaluate $\int \sin^2 \theta \, d\theta$. Let

$$u = \sin \theta, \qquad dv = \sin \theta \, d\theta.$$
Then
$$du = \cos \theta \, d\theta, \qquad v = -\cos \theta.$$

$$\int \sin^2 \theta \, d\theta = -\sin \theta \cos \theta - \int -\cos^2 \theta \, d\theta$$

$$= -\sin \theta \cos \theta + \int \cos^2 \theta \, d\theta$$

$$= -\sin \theta \cos \theta + \int (1 - \sin^2 \theta) \, d\theta$$

$$= -\sin \theta \cos \theta + \theta - \int \sin^2 \theta \, d\theta.$$

We solve this equation for $\int \sin^2 \theta \, d\theta$,

$$\int \sin^2 \theta \, d\theta = -\tfrac{1}{2} \sin \theta \cos \theta + \tfrac{1}{2}\theta + C.$$

Here is another way to evaluate $\int \sin^2 \theta \, d\theta$. Instead of using integration by parts, we can use the half-angle formula

$$\sin^2 \theta = \frac{1 - \cos(2\theta)}{2}.$$

This is derived from the addition formula,

$$\cos(\theta + \phi) = \cos \theta \cos \phi - \sin \theta \sin \phi,$$
$$\cos(2\theta) = \cos^2 \theta - \sin^2 \theta = 1 - 2 \sin^2 \theta,$$
$$\sin^2 \theta = \frac{1 - \cos(2\theta)}{2}.$$

Then
$$\int \sin^2 \theta \, d\theta = \int \frac{1 - \cos 2\theta}{2} \, d\theta = \frac{1}{2} \int d\theta - \frac{1}{2} \int \cos 2\theta \, d\theta$$
$$= \frac{1}{2} \int d\theta - \frac{1}{4} \int \cos 2\theta \, d(2\theta) = \frac{1}{2}\theta - \frac{1}{4} \sin 2\theta + C.$$

This answer agrees with Example 4 because

$$\sin 2\theta = \sin(\theta + \theta) = 2 \sin \theta \cos \theta,$$

so
$$\frac{1}{2}\theta - \frac{1}{4} \sin 2\theta = \frac{1}{2}\theta - \frac{1}{2} \sin \theta \cos \theta.$$

Integration by parts requires a great deal of guesswork. Given a problem $\int h(x) \, dx$ we try to find a way to split $h(x) \, dx$ into a product $f(x)g'(x) \, dx$ where we can evaluate both of the integrals $\int g'(x) \, dx$ and $\int g(x)f'(x) \, dx$.

Definite integrals take the following form when integration by parts is applied.

DEFINITE INTEGRATION BY PARTS

If $u = f(x)$ and $v = g(x)$ have continuous derivatives on an open interval I, then for a, b in I,

$$\int_a^b f(x)g'(x) \, dx = f(x)g(x) \Big]_a^b - \int_a^b g(x)f'(x) \, dx.$$

PROOF The Product Rule gives

$$f(x)g'(x) \, dx + g(x)f'(x) \, dx = d(f(x)g(x)).$$

Then by the Fundamental Theorem of Calculus,

$$\int_a^b (f(x)g'(x) + g(x)f'(x)) \, dx = f(x)g(x) \Big]_a^b,$$

and the desired result follows by the Sum Rule.

If we plot $u = f(x)$ on one axis and $v = g(x)$ on the other, we get a picture of definite integration by parts (Figure 7.4.1). The picture is easier to interpret if we change variables in the definite integrals and write the formula for integration by

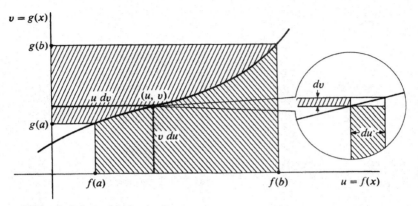

Figure 7.4.1   Definite Integration by Parts

parts in the form

$$\int_{g(a)}^{g(b)} u \, dv + \int_{f(a)}^{f(b)} v \, du = f(b)g(b) - f(a)g(a).$$

EXAMPLE 5   Evaluate $\int_0^\pi x \sin x \, dx$ (Figure 7.4.2). Take $u = x, dv = \sin x \, dx$ as in Example 1. Then $v = -\cos x$ and

$$\int_0^\pi x \sin x \, dx = -x \cos x \Big]_0^\pi - \int_0^\pi -\cos x \, dx$$

$$= -x \cos x \Big]_0^\pi + \sin x \Big]_0^\pi$$

$$= (-\pi(-1) + 0 \cdot 1) + (0 - 0) = \pi.$$

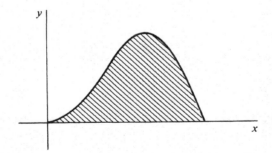

Figure 7.4.2

PROBLEMS FOR SECTION 7.4

Evaluate the integrals in Problems 1–35.

1   $\int x \cos x \, dx$ 2 $\int \arccos x \, dx$

3 $\int t^2 \cos t \, dt$ **4** $\int x \arctan x \, dx$

5 $\int t \sin (2t - 1) \, dt$ **6** $\int \arcsin (3t) \, dt$

7 $\int x^2 \sin (4x) \, dx$ **8** $\int x \, \mathrm{arcsec}\, x \, dx$

9 $\int x^3 \, \mathrm{arcsec}\, x \, dx$ **10** $\int x^3 \sin x \, dx$

11 $\int \sin \sqrt{x} \, dx$ **12** $\int \sin \theta \tan^2 \theta \, d\theta$

13 $\int \arctan \sqrt{x} \, dx$ **14** $\int x \tan x \sec^2 x \, dx$

15 $\int \dfrac{x^3}{\sqrt{x^2 - 1}} \, dx$ **16** $\int \cos^2 \theta \, d\theta$

17 $\int x \sin x \cos x \, dx$ **18** $\int t \sin^2 t \, dt$

19 $\int \sin \theta \sin (2\theta) \, d\theta$ **20** $\int \cos x \cos (3x) \, dx$

21 $\int \sin x \cos (5x) \, dx$ **22** $\int \cos x \cot^4 x \, dx$

23 $\int t^3 \sin (t^2) \, dt$ **24** $\int x^3 \cos (2x^2 - 1) \, dx$

25 $\int \dfrac{1}{x^3} \sin \left(\dfrac{1}{x} \right) dx$ **26** $\int \sin \theta \cos \theta \cos (\sin \theta) \, d\theta$

27 $\int t^3 \sqrt{t^2 + 4} \, dt$ **28** $\int \dfrac{1}{x^3} \sqrt{\dfrac{1}{x} - 1} \, dx$

29 $\int_0^{\pi/2} \theta \cos \theta \, d\theta$ **30** $\int_0^{1/2} \arcsin x \, dx$

31 $\int_0^{\pi} \sin^2 \theta \, d\theta$ **32** $\int_0^1 \arcsin x \, dx$

33 $\int_0^1 x \, \mathrm{arccot}\, x \, dx$ **34** $\int_0^{\infty} x \, \mathrm{arccot}\, x \, dx$

35 $\int_1^2 t \, \mathrm{arcsec}\, t \, dt$

36 Find the volume of the solid of revolution generated by rotating the region under the curve $y = \sin x, 0 \le x \le \pi$, about (a) the x-axis, (b) the y-axis.

☐ **37** Prove that if f is a differentiable function of x, then
$$\int f(x) \, dx = xf(x) - \int xf'(x) \, dx.$$

☐ **38** If u and v are differentiable functions of x, show that
$$\int u^2 \, dv = u^2 v - 2 \int uv \, du.$$

☐ **39** Show that if f' and g are differentiable for all x, then
$$\int g(x)g'(x)f''(g(x)) \, dx = f'(g(x))g(x) - f(g(x)) + C.$$

7.5 INTEGRALS OF POWERS OF TRIGONOMETRIC FUNCTIONS

It is often possible to transform an integral into one of the forms

$$\int \sin^n u \, du, \qquad \int \cos^n u \, du, \qquad \int \tan^n u \, du, \quad \text{etc.}$$

These integrals can be evaluated by means of *reduction formulas*, which express the integral of the nth power of a trigonometric function in terms of the $(n-2)$nd power. The easiest reduction formulas to prove are those for the tangent and cotangent, so we shall give them first.

THEOREM 1

Let $n \neq 1$. Then

(i) $\displaystyle \int \tan^n x \, dx = \frac{\tan^{n-1} x}{n-1} - \int \tan^{n-2} x \, dx.$

(ii) $\displaystyle \int \cot^n x \, dx = -\frac{\cot^{n-1} x}{n-1} - \int \cot^{n-2} x \, dx.$

PROOF We recall that

$$\tan^2 x = \sec^2 x - 1, \qquad d(\tan x) = \sec^2 x \, dx.$$

$$\text{Then} \qquad \int \tan^n x \, dx = \int \tan^{n-2} x \tan^2 x \, dx = \int \tan^{n-2} x \, (\sec^2 x - 1) \, dx$$

$$= \int \tan^{n-2} x \sec^2 x \, dx - \int \tan^{n-2} x \, dx$$

$$= \int \tan^{n-2} x \, d(\tan x) - \int \tan^{n-2} x \, dx$$

$$= \frac{\tan^{n-1} x}{n-1} - \int \tan^{n-2} x \, dx.$$

These reduction formulas are true for any rational number $n \neq 1$. They are most useful, however, when n is a positive integer.

EXAMPLE 1 $\displaystyle \int \tan^2 x \, dx = \frac{\tan x}{1} - \int \tan^0 x \, dx = \tan x - x + C.$

EXAMPLE 2 $\displaystyle \int \tan^4 x \, dx = \frac{\tan^3 x}{3} - \int \tan^2 x \, dx = \frac{\tan^3 x}{3} - \tan x + x + C.$

EXAMPLE 3 $\displaystyle \int \tan^3 x \, dx = \frac{\tan^2 x}{2} - \int \tan x \, dx.$

We will evaluate $\int \tan x \, dx$ in the next chapter.

Each time we use the reduction formula the exponent in the integral goes down by two. By repeated use of the reduction formulas we can integrate any even power of $\tan x$ or $\cot x$. We can also work the integral of any odd power of $\tan x$ or $\cot x$ down to an expression involving $\int \tan x$ or $\int \cot x$.

The reduction formulas for the other trigonometric functions are obtained by using integration by parts.

THEOREM 2

Let $n \neq 0$. Then

(i) $\int \sin^n x \, dx = -\dfrac{1}{n} \sin^{n-1} x \cos x + \dfrac{n-1}{n} \int \sin^{n-2} x \, dx.$

(ii) $\int \cos^n x \, dx = \dfrac{1}{n} \cos^{n-1} x \sin x + \dfrac{n-1}{n} \int \cos^{n-2} x \, dx.$

PROOF (i) Break the term $\sin^n x \, dx$ into two parts,

$$\sin^n x \, dx = \sin^{n-1} x (\sin x \, dx).$$

We shall let
$$u = \sin^{n-1} x, \qquad v = -\cos x,$$
$$du = (n-1)\sin^{n-2} x \cos x \, dx, \qquad dv = \sin x \, dx,$$

and use integration by parts. Then

$$\int \sin^n x \, dx = \int u \, dv = uv - \int v \, du$$

$$= -\sin^{n-1} x \cos x - \int (n-1)(-\cos x)\sin^{n-2} x \cos x \, dx$$

$$= -\sin^{n-1} x \cos x + (n-1)\int \sin^{n-2} x (1 - \sin^2 x) \, dx$$

$$= -\sin^{n-1} x \cos x + (n-1)\int \sin^{n-2} x \, dx - (n-1)\int \sin^n x \, dx.$$

We find that $\int \sin^n x \, dx$ appears on both sides of the equation, and we solve for it,

$$n \int \sin^n x \, dx = -\sin^{n-1} x \cos x + (n-1)\int \sin^{n-2} x \, dx,$$

$$\int \sin^n x \, dx = -\frac{1}{n} \sin^{n-1} x \cos x + \frac{n-1}{n} \int \sin^{n-2} x \, dx.$$

We already know the integrals

$$\int \sin x \, dx = -\cos x + C, \qquad \int \cos x \, dx = \sin x + C.$$

We can use the reduction formulas to integrate any positive power of $\sin x$ or $\cos x$. Again, the formulas are true where n is any rational number, $n \neq 0$.

EXAMPLE 4 $\int \sin^2 x \, dx = -\frac{1}{2}\sin x \cos x + \frac{1}{2}\int dx = -\frac{1}{2}\sin x \cos x + \frac{1}{2}x + C.$

$\int \cos^2 x \, dx = \frac{1}{2}\cos x \sin x + \frac{1}{2}\int dx = \frac{1}{2}\cos x \sin x + \frac{1}{2}x + C.$

EXAMPLE 5 $\int \cos^3 x \, dx = \frac{1}{3}\cos^2 x \sin x + \frac{2}{3}\int \cos x \, dx$

$= \frac{1}{3}\cos^2 x \sin x + \frac{2}{3}\sin x + C.$

THEOREM 3

Let $m \neq 1$. Then

(i) $\int \sec^m x \, dx = \frac{1}{m-1}\sec^{m-1} x \sin x + \frac{m-2}{m-1}\int \sec^{m-2} x \, dx.$

(ii) $\int \csc^m x \, dx = -\frac{1}{m-1}\csc^{m-1} x \cos x + \frac{m-2}{m-1}\int \csc^{m-2} x \, dx.$

PROOF (ii) This can be done by integration by parts, but it is easier to use Theorem 2. Let $n = 2 - m$. For $m \neq 2$, $n \neq 0$ and Theorem 2 gives

$\int \sin^{2-m} x \, dx = -\frac{1}{2-m}\sin^{1-m} x \cos x + \frac{1-m}{2-m}\int \sin^{-m} x \, dx,$

$\int \csc^{m-2} x \, dx = \frac{1}{m-2}\csc^{m-1} x \cos x + \frac{m-1}{m-2}\int \csc^m x \, dx,$

whence $\int \csc^m x \, dx = -\frac{1}{m-1}\csc^{m-1} x \cos x + \frac{m-2}{m-1}\int \csc^{m-2} x \, dx.$

For $m = 2$ the formula is already known,

$\int \csc^2 x \, dx = -\cot x + C = -\csc x \cos x + C.$

These reduction formulas can be used to integrate any even power of $\sec x$ or $\csc x$, and to get the integral of any odd power of $\sec x$ or $\csc x$ in terms of $\int \sec x$ or $\int \csc x$. We shall find $\int \sec x$ and $\int \csc x$ in the next chapter.

EXAMPLE 6 $\int \sec^3 x \, dx = \frac{1}{2}\sec^2 x \sin x + \frac{1}{2}\int \sec x \, dx.$

EXAMPLE 7 $\int \sec^4 x \, dx = \frac{1}{3}\sec^3 x \sin x + \frac{2}{3}\int \sec^2 x \, dx$

$= \frac{1}{3}\sec^3 x \sin x + \frac{2}{3}\tan x + C.$

By using the identity $\sin^2 x + \cos^2 x = 1$ we can evaluate any integral of the form $\int \sin^m x \cos^n x \, dx$ where m and n are positive integers. If either m or n is odd we let $u = \sin x$ or $u = \cos x$ and transform the integral into a polynomial in u.

EXAMPLE 8 $\int \sin^4 x \cos^3 x \, dx$. Let $u = \sin x$, $du = \cos x \, dx$.

$$\int \sin^4 x \cos^3 x \, dx = \int u^4 (1 - u^2) \, du$$

$$= \tfrac{1}{5} u^5 - \tfrac{1}{7} u^7 + C$$

$$= \tfrac{1}{5} \sin^5 x - \tfrac{1}{7} \sin^7 x + C.$$

This method also works for an odd power of $\sin x$ times any power of $\cos x$, and vice versa.

EXAMPLE 9 $\int \sqrt{\cos x} \, \sin^3 x \, dx$. Let $u = \cos x$, $du = -\sin x \, dx$.

$$\int \sqrt{\cos x} \, \sin^3 x \, dx = \int \sqrt{u} (1 - u^2)(-1) \, du$$

$$= \int -u^{1/2} + u^{5/2} \, du = -\tfrac{2}{3} u^{3/2} + \tfrac{2}{7} u^{7/2} + C$$

$$= -\tfrac{2}{3} (\cos x)^{3/2} + \tfrac{2}{7} (\cos x)^{7/2} + C.$$

EXAMPLE 10 $\int \sin^5 x \, dx$. Let $u = \cos x$, $du = -\sin x \, dx$.

$$\int \sin^5 x \, dx = \int (1 - u^2)^2 (-1) \, du$$

$$= -\int (1 - 2u^2 + u^4) \, du = -u + \tfrac{2}{3} u^3 - \tfrac{1}{5} u^5 + C$$

$$= -\cos x + \tfrac{2}{3} \cos^3 x - \tfrac{1}{5} \cos^5 x + C.$$

If m and n are both even, the integral $\int \sin^m x \cos^n x \, dx$ can be transformed into the integral of a sum of even powers of $\sin x$. Then the reduction formula can be used.

EXAMPLE 11 $\int \sin^4 x \cos^4 x \, dx = \int \sin^4 x (1 - \sin^2 x)^2 \, dx$

$$= \int \sin^4 x - 2 \sin^6 x + \sin^8 x \, dx.$$

We can also evaluate integrals of the form

$$\int \tan^m x \sec^n x \, dx,$$

$$\int \cot^m x \csc^n x \, dx.$$

EXAMPLE 12   When m is even use $\tan^2 x = \sec^2 x - 1$.

$$\int \tan^4 x \sec x \, dx = \int (\sec^2 x - 1)^2 \sec x \, dx$$

$$= \int \sec^5 x - 2 \sec^3 x + \sec x \, dx.$$

Now the reduction formula for $\sec x$ can be used.

EXAMPLE 13   When m is odd use the new variable $u = \sec x$ or $u = -\csc x$.

$$\int \cot^3 x \csc^3 x \, dx = \int \cot^2 x \csc^2 x \, (\cot x \csc x \, dx)$$

$$= \int (u^2 - 1)u^2 \, du = \frac{u^5}{5} - \frac{u^3}{3} + C$$

$$= -\frac{\csc^5 x}{5} + \frac{\csc^3 x}{3} + C.$$

PROBLEMS FOR SECTION 7.5

Evaluate the integrals in Problems 1–32.

1	$\int \dfrac{\sin^3 t}{\cos^2 t} \, dt$	**2**	$\int \sin^2(2t) \, dt$
3	$\int \cot^2 x \, dx$	**4**	$\int \sin^3 (5u) \, du$
5	$\int \cos^4 x \, dx$	**6**	$\int \dfrac{1}{\sin^4 x} \, dx$
7	$\int \tan^3 x \sec^4 x \, dx$	**8**	$\int \tan^6 \theta \, d\theta$
9	$\int \sin^2 x \cos^3 x \, dx$	**10**	$\int \cot \theta \csc^2 \theta \, d\theta$
11	$\int \cot^2 \theta \csc^2 \theta \, d\theta$	**12**	$\int \sin x (\cos x)^{3/2} \, dx$
13	$\int (\tan x)^{3/2} \sec^4 x \, dx$	**14**	$\int \sec^4 (3u - 1) \, du$
15	$\int \sec^2 \theta \csc^2 \theta \, d\theta$	**16**	$\int \dfrac{\sin^2 \theta}{1 - \cos \theta} \, d\theta$
17	$\int \dfrac{1 - \cos \theta}{\sin^2 \theta} \, d\theta$	**18**	$\int_0^{\pi/2} \sin^3 x \cos x \, dx$
19	$\int_0^{\pi/3} \tan^3 \theta \sec \theta \, d\theta$	**20**	$\int_0^{\pi/2} \sqrt{\cos x} \, \sin x \, dx$
21	$\int_0^{\pi/4} \tan^4 x \, dx$	**22**	$\int_0^{\pi/2} \tan^2 x \, dx$

23 $\displaystyle\int_0^\pi \sin^4\theta\, d\theta$

24 $\displaystyle\int \sin^3(2u)\cos^3(2u)\, du$

25 $\displaystyle\int \frac{\cos^2\sqrt{x}}{\sqrt{x}}\, dx$

26 $\displaystyle\int x\tan(x^2)\sec^2(x^2)\, dx$

27 $\displaystyle\int x\sin^3 x\, dx$

28 $\displaystyle\int x\tan^3 x\sec^2 x\, dx$

29 $\displaystyle\int x\sin^2 x\cos x\, dx$

30 $\displaystyle\int \sin^6\theta\cos^5\theta\, d\theta$

31 $\displaystyle\int \tan^4\theta\sec^6\theta\, d\theta$

32 $\displaystyle\int \sin^2 x\cos^2 x\, dx$

In Problems 33–39, express the given integral in terms of

$$\int \tan x\, dx, \quad \int \cot x\, dx, \quad \int \sec x\, dx, \quad \int \csc x\, dx.$$

33 $\displaystyle\int \sec^3 x\, dx$

34 $\displaystyle\int \cot^3 x\, dx$

35 $\displaystyle\int \tan^2 x\sec x\, dx$

36 $\displaystyle\int \csc^5 x\, dx$

37 $\displaystyle\int \cot^2 x\csc^3 x\, dx$

38 $\displaystyle\int \tan^4 x\sec x\, dx$

39 $\displaystyle\int \frac{\sin x + \cos x}{\sin x\cos x}\, dx$

40   Check the reduction formula for $\int \sin^n x\, dx$ by differentiating both sides of the equation. Do the same for $\int \tan^n x\, dx$ and $\int \sec^n x\, dx$.

☐ **41**   Find a reduction formula for $\int x^n \sin x\, dx$ using integration by parts.

42   Find the volume of the solid generated by rotating the region under the curve $y = \sin^2 x$, $0 \le x \le \pi$, about (a) the x-axis, (b) the y-axis.

43   Find the volume of the solid generated by rotating the region under the curve $y = \sin x\cos x$, $0 \le x \le \pi/2$, about (a) the x-axis, (b) the y-axis.

7.6 TRIGONOMETRIC SUBSTITUTIONS

Integrals containing one of the terms

$$\sqrt{a^2 + x^2}, \quad \sqrt{a^2 - x^2}, \quad \text{or} \quad \sqrt{x^2 - a^2}$$

can often be integrated by a *trigonometric substitution*. The idea is to take x, a, and the square root as the three sides of a right triangle and use one of its acute angles as a new variable θ. The three kinds of trigonometric substitutions are shown in Figure 7.6.1. These figures do not have to be memorized. Just remember that the sides must be labeled so that

$$(\text{opposite})^2 + (\text{adjacent})^2 = (\text{hypotenuse})^2.$$

These substitutions frequently give an integral of powers of trigonometric functions discussed in the preceding section.

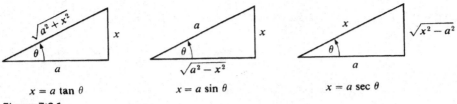

$x = a \tan \theta$ $x = a \sin \theta$ $x = a \sec \theta$

Figure 7.6.1

EXAMPLE 1   Find $\int (a^2 + x^2)^{-3/2} \, dx$.

Let $\theta = \arctan(x/a)$. Then from Figure 7.6.2,

$$x = a \tan \theta, \qquad dx = a \sec^2 \theta \, d\theta, \qquad \sqrt{a^2 + x^2} = a \sec \theta.$$

So

$$\int (a^2 + x^2)^{-3/2} \, dx = \int (a \sec \theta)^{-3} \, a \sec^2 \theta \, d\theta$$

$$= \frac{1}{a^2} \int (\sec \theta)^{-1} \, d\theta = \frac{1}{a^2} \int \cos \theta \, d\theta$$

$$= \frac{1}{a^2} \sin \theta + C = \frac{1}{a^2} \frac{\tan \theta}{\sec \theta} + C = \frac{x}{a^2 \sqrt{a^2 + x^2}} + C.$$

EXAMPLE 2   Find $\int \sqrt{x^2 - a^2} \, dx$.

Let $\theta = \operatorname{arcsec}(x/a)$ (Figure 7.6.3), so

$$x = a \sec \theta, \qquad dx = a \tan \theta \sec \theta \, d\theta, \qquad \sqrt{x^2 - a^2} = a \tan \theta.$$

So

$$\int \sqrt{x^2 - a^2} \, dx = \int a \tan \theta \, a \tan \theta \sec \theta \, d\theta = a^2 \int \tan^2 \theta \sec \theta \, d\theta$$

$$= a^2 \int (\sec^2 \theta - 1) \sec \theta \, d\theta$$

$$= a^2 \int \sec^3 \theta \, d\theta - a^2 \int \sec \theta \, d\theta$$

$$= \left(\tfrac{1}{2} a^2 \sec^2 \theta \sin \theta + \tfrac{1}{2} a^2 \int \sec \theta \, d\theta \right) - a^2 \int \sec \theta \, d\theta$$

$$= \tfrac{1}{2} a^2 \sec^2 \theta \sin \theta - \tfrac{1}{2} a^2 \int \sec \theta \, d\theta$$

$$= \tfrac{1}{2} x \sqrt{x^2 - a^2} - \tfrac{1}{2} a^2 \int \sec \theta \, d\theta.$$

This is as far as we can go on this problem until we find out how to integrate $\int \sec \theta \, d\theta$ in the next chapter.

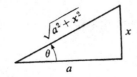

Figure 7.6.2

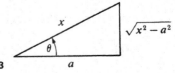

Figure 7.6.3

EXAMPLE 3 $\int \dfrac{1}{x^2\sqrt{a^2 - x^2}}\,dx$. Let $\theta = \arcsin(x/a)$ (Figure 7.6.4). Then

$$x = a\sin\theta, \qquad dx = a\cos\theta\,d\theta, \qquad \sqrt{a^2 - x^2} = a\cos\theta.$$

$$\int \frac{1}{x^2\sqrt{a^2 - x^2}}\,dx = \int \frac{1}{a^2\sin^2\theta\,a\cos\theta}\,a\cos\theta\,d\theta = \int \frac{1}{a^2\sin^2\theta}\,d\theta$$

$$= \frac{1}{a^2}\int \csc^2\theta\,d\theta = -\frac{1}{a^2}\cot\theta + C$$

$$= -\frac{1}{a^2}\frac{\sqrt{a^2 - x^2}}{x} + C.$$

EXAMPLE 4 $\int \dfrac{\sqrt{x^2 - a^2}}{x}\,dx$. Put $\theta = \text{arcsec}\,(x/a)$ (Figure 7.6.5). Then

$$x = a\sec\theta, \qquad dx = a\tan\theta\sec\theta\,d\theta, \qquad \sqrt{x^2 - a^2} = a\tan\theta.$$

$$\int \frac{\sqrt{x^2 - a^2}}{x}\,dx = \int \frac{a\tan\theta}{a\sec\theta}\,a\tan\theta\sec\theta\,d\theta = a\int \tan^2\theta\,d\theta$$

$$= a\int \sec^2\theta\,d\theta - a\int d\theta = a\tan\theta - a\theta + C$$

$$= \sqrt{x^2 - a^2} - a\,\text{arcsec}(x/a) + C.$$

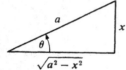

Figure 7.6.4

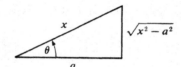

Figure 7.6.5

To keep track of a trigonometric substitution, it is a good idea to actually draw the triangle and label the sides.

EXAMPLE 5 The basic integrals:

(a) $$\int \frac{1}{\sqrt{1 - x^2}}\,dx = \arcsin x + C,$$

(b) $$\int \frac{dx}{1 + x^2}\,dx = \arctan x + C,$$

(c) $$\int \frac{dx}{x\sqrt{x^2 - 1}} = \text{arcsec}\,x + C, \qquad x > 1$$

can be evaluated very easily by a trigonometric substitution.

(a) $$\int \frac{1}{\sqrt{1 - x^2}}\,dx.$$

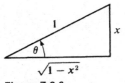

Figure 7.6.6

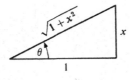

Figure 7.6.7

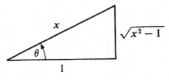

Figure 7.6.8

Let $\theta = \arcsin x$ (Figure 7.6.6). Then $x = \sin\theta$, $dx = \cos\theta\,d\theta$, $\sqrt{1 - x^2} = \cos\theta$.

$$\int \frac{1}{\sqrt{1 - x^2}}\,dx = \int \frac{\cos\theta\,d\theta}{\cos\theta} = \int d\theta = \theta + C,$$

$$\int \frac{1}{\sqrt{1 - x^2}}\,dx = \arcsin x + C.$$

(b)
$$\int \frac{dx}{1 + x^2}$$

Let $\theta = \arctan x$ (Figure 7.6.7). Then $x = \tan\theta$, $dx = \sec^2\theta\,d\theta$, $\sqrt{1 + x^2} = \sec\theta$.

$$\int \frac{dx}{1 + x^2} = \int \frac{\sec^2\theta}{\sec^2\theta}\,d\theta = \int d\theta = \theta + C,$$

$$\int \frac{dx}{1 + x^2} = \arctan x + C.$$

(c)
$$\int \frac{dx}{x\sqrt{x^2 - 1}}, \qquad x > 1.$$

Let $\theta = \text{arcsec } x$ (Figure 7.6.8). Then $x = \sec\theta$, $dx = \tan\theta\sec\theta\,d\theta$, $\sqrt{x^2 - 1} = \tan\theta$.

$$\int \frac{dx}{x\sqrt{x^2 - 1}} = \int \frac{\tan\theta\sec\theta}{\sec\theta\tan\theta}\,d\theta = \int d\theta = \theta + C,$$

$$\int \frac{dx}{x\sqrt{x^2 + 1}} = \text{arcsec } x + C, \qquad x > 1.$$

It is therefore more important to remember the method of trigonometric substitution than to remember the integration formulas (a), (b), (c).

PROBLEMS FOR SECTION 7.6

Draw the appropriate triangle and evaluate using trigonometric substitutions.

1 $\displaystyle\int \frac{dx}{\sqrt{1 - 4x^2}}$

2 $\displaystyle\int \sqrt{a^2 - x^2}\,dx$

3 $\displaystyle\int \frac{x^3\,dx}{\sqrt{9 + x^2}}$

4 $\displaystyle\int \frac{\sqrt{x^2 - 1}}{x}\,dx$

5 $\displaystyle\int (4 - x^2)^{-3/2}\, dx$ 6 $\displaystyle\int (1 - 3x^2)^{3/2}\, dx$

7 $\displaystyle\int \frac{\sin\theta\, d\theta}{\sqrt{2 - \cos^2\theta}}$ 8 $\displaystyle\int \frac{dx}{\sqrt{x}\sqrt{1 - x}}$

9 $\displaystyle\int \frac{dx}{x^2(1 + x^2)}$ 10 $\displaystyle\int \frac{x^4\, dx}{9 + x^2}$

11 $\displaystyle\int \frac{x^2\, dx}{\sqrt{4 - x^2}}$ 12 $\displaystyle\int \frac{\sqrt{9 - 4x^2}}{x^4}\, dx$

13 $\displaystyle\int x^2\sqrt{1 - x^2}\, dx$ 14 $\displaystyle\int \sqrt{x}\sqrt{1 - x}\, dx$

15 $\displaystyle\int \sqrt{4x - x^2}\, dx$ 16 $\displaystyle\int \frac{\sqrt{x}}{\sqrt{1 - x}}\, dx$

17 $\displaystyle\int \frac{x^3}{\sqrt{4x^2 - 1}}\, dx$ 18 $\displaystyle\int \frac{dx}{x^3\sqrt{x^2 - 3}}$

19 $\displaystyle\int \frac{x^3}{\sqrt{a^2 - x^2}}\, dx$ 20 $\displaystyle\int x^3\sqrt{1 + a^2 x^2}\, dx$

21 $\displaystyle\int \frac{\sqrt{x^4 - 1}}{x}\, dx$ 22 $\displaystyle\int x\sqrt{1 - x^4}\, dx$

23 $\displaystyle\int \frac{x^3}{(a^2 + x^2)^{3/2}}\, dx$ 24 $\displaystyle\int_0^2 \sqrt{4 - x^2}\, dx$

25 $\displaystyle\int_{-1}^{1} \frac{x^2}{\sqrt{1 - x^2}}\, dx$ 26 $\displaystyle\int_0^4 \frac{dx}{(9 + x^2)^{3/2}}$

27 $\displaystyle\int_0^\infty \frac{dx}{(9 + x^2)^{3/2}}$ 28 $\displaystyle\int_2^4 \frac{\sqrt{x^2 - 2}}{x}\, dx$

29 $\displaystyle\int_2^\infty \frac{\sqrt{x^2 - 2}}{x}\, dx$ 30 $\displaystyle\int_0^\infty \frac{x^3}{\sqrt{1 + x^2}}\, dx$

31 $\displaystyle\int x \arcsin x\, dx$ 32 $\displaystyle\int x \arccos x\, dx$

33 $\displaystyle\int x^2 \arcsin x\, dx$ 34 $\displaystyle\int x^3 \arctan x\, dx$

35 $\displaystyle\int x^{-3} \arcsin x\, dx$ 36 $\displaystyle\int x^{-3} \arctan x\, dx$

37   Find the surface area generated by rotating the ellipse $x^2 + 4y^2 = 1$ about the x-axis.

7.7 POLAR COORDINATES

The position of a point in the plane can be described by its distance and direction from the origin. In measuring direction we take the x-axis as the starting point. Let X be the point $(1, 0)$ on the x-axis and let P be a point in the plane as in Figure 7.7.1.

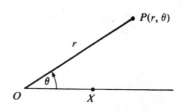

Figure 7.7.1

A pair of *polar coordinates* of P is given by (r, θ) where r is the distance from the origin to P and θ is the angle XOP.

Each pair of real numbers (r, θ) determines a point P in polar coordinates. To find P we first rotate the line OX through an angle θ, forming a new line OX', and then go out a distance r along the line OX'. If θ is negative then the rotation is in the negative, or clockwise direction. If r is negative the distance is measured along the line OX' in the direction away from X' (see Figure 7.7.2).

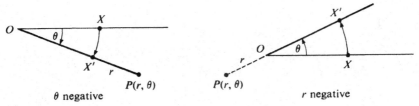

Figure 7.7.2

EXAMPLE 1 Plot the following points in polar coordinates.

$$(2, \pi/4), \qquad (-1, \pi/4), \qquad (3, 3\pi/4), \qquad (2, -\pi/4), \qquad (-4, -\pi/4).$$

The solution is shown in Figure 7.7.3.

Each point P has infinitely many different polar coordinate pairs. We see in Figure 7.7.4 that the point $P(3, \pi/2)$ has all the coordinates

$$\left.\begin{array}{l} (3, \pi/2 + 2n\pi), \\ (-3, 3\pi/2 + 2n\pi), \end{array}\right\} n \text{ an integer.}$$

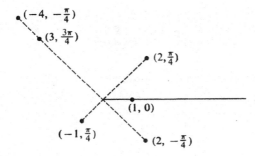

Figure 7.7.3

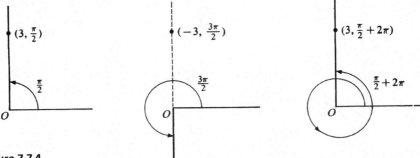

Figure 7.7.4

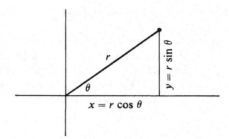

Figure 7.7.5

Any coordinate pair $(0, \theta)$ with $r = 0$ determines the origin. As we see in Figure 7.7.5, the coordinates of a point P in rectangular and in polar coordinates are related by the equations

$$x = r \cos\theta, \qquad y = r \sin\theta.$$

The *graph*, or *locus in polar coordinates* of a system of formulas in the variables r, θ is the set of all points $P(r, \theta)$ for which the formulas are true.

EXAMPLE 2   The graph of the equation $r = a$ is the circle of radius a centered at the origin (Figure 7.7.6(a)). The graph of the equation $\theta = b$ is a straight line through the origin (Figure 7.7.6(b)).

EXAMPLE 3   The graph of the system of formulas

$$r = \theta, \qquad 0 \le \theta$$

is the spiral of Archimedes formed by moving a pencil along the line OX while the line is rotating, with the pencil moving at the same speed as the point X. The graph is shown in Figure 7.7.6(c).

An equation in rectangular coordinates can readily be transformed into an equation in polar coordinates with the same graph by using $x = r \cos\theta$, $y = r \sin\theta$.

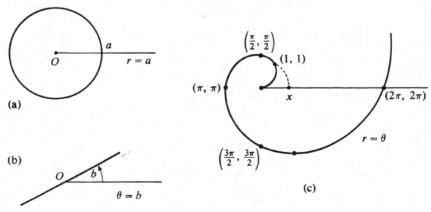

Figure 7.7.6

Here are the polar equations for various types of straight lines. Examples of their graphs are shown in Figure 7.7.7.

(1) Line through the origin (not vertical).
Rectangular equation: $y = mx.$
Polar equation: $r \sin \theta = mr \cos \theta,$
or: $\tan \theta = m.$

(2) Horizontal line (not through origin).
Rectangular equation: $y = b.$
Polar equation: $r \sin \theta = b,$
or: $r = b \csc \theta.$

(3) Vertical line (not through origin).
Rectangular equation: $x = a.$
Polar equation: $r \cos \theta = a,$
or: $r = a \sec \theta.$

(4) Vertical line through origin.
Rectangular equation: $x = 0.$
Polar equation: $r \cos \theta = 0,$
or: $\theta = \pi/2.$

(5) Other lines.
Rectangular equation: $y = mx + b.$
Polar equation: $r \sin \theta = mr \cos \theta + b,$

or: $r = \dfrac{b}{\sin \theta - m \cos \theta}.$

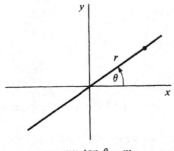

$y = mx, \tan \theta = m$

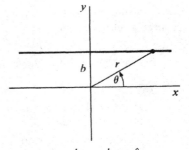

$y = b, r = b \csc \theta$

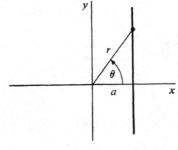

$x = a, r = a \sec \theta$

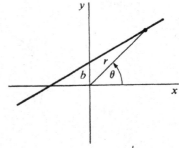

$y = mx + b, r = \dfrac{b}{\sin \theta - m \cos \theta}$

Figure 7.7.7

EXAMPLE 4   The parabola $y = x^2$ has the polar equation

$$r \sin \theta = (r \cos \theta)^2, \quad \text{or} \quad r = \frac{\sin \theta}{\cos^2 \theta} = \tan \theta \sec \theta.$$

EXAMPLE 5   The curve $y = 1/x$ has the polar equation

$$r \sin \theta = \frac{1}{r \cos \theta}, \quad \text{or} \quad r^2 = \sec \theta \csc \theta.$$

The graph is shown in Figure 7.7.8.

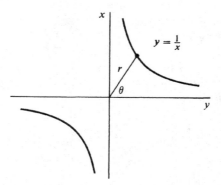

Figure 7.7.8

Some curves have much simpler equations in polar coordinates than in rectangular coordinates.

EXAMPLE 6   The graph of the equation

$$r = a \sin \theta$$

is the circle one of whose diameters is the line from the origin to a point a above the origin.

This can be seen from Figure 7.7.9, if we remember that a diameter and a point on the circle form a right triangle.

As θ increases, the point $(a \sin \theta, \theta)$ goes around this circle once for every π radians.

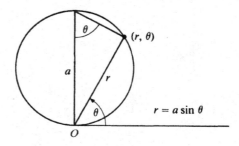

Figure 7.7.9

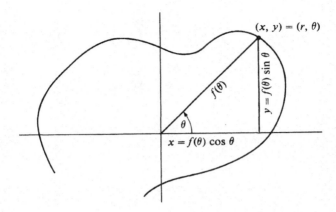

Figure 7.7.10

An equation $r = f(\theta)$ in polar coordinates has the same graph as the pair of parametric equations

$$x = f(\theta)\cos\theta, \qquad y = f(\theta)\sin\theta$$

in rectangular coordinates. This can be seen from Figure 7.7.10.

EXAMPLE 7

(a) The spiral $r = \theta$ has the parametric equations

$$x = \theta\cos\theta, \qquad y = \theta\sin\theta.$$

(b) The circle $r = a\sin\theta$ has the parametric equations

$$x = a\sin\theta\cos\theta, \qquad y = a\sin^2\theta.$$

PROBLEMS FOR SECTION 7.7

1   Plot the following points in polar coordinates:

(a) $(2, \pi/3)$ (b) $(-3, \pi/2)$ (c) $(1, 4\pi/3)$
(d) $(-2, -\pi/4)$ (e) $(\frac{1}{2}, \pi)$ (f) $(0, 3\pi/2)$

In Problems 2–12, find an equation in polar coordinates which has the same graph as the given equation in rectangular coordinates.

2    $y = 3x$ 3 $y = 5x + 2$
4    $y = -4$ 5 $x = 2$
6    $xy^2 = 1$ 7 $y = x^2 + 1$
8    $x^2 + y^2 = 5$ 9 $y = 3x^2 - 2x$
10   $y = x^3$ 11 $y = x^2 + y^2$
12   $y = \sin x$

In Problems 13–20, sketch the given curve in polar coordinates.

13   $r = \cos\theta$ 14 $r = -\sec\theta$
15   $r = \sin(\theta + \pi/4)$ 16 $r = \theta, \quad \theta \le 0$

17 $r = 1 + \theta^2/\pi^2$ **18** $r = \dfrac{1}{\sin\theta + \cos\theta}$

19 $r = \cot\theta \csc\theta$ **20** $r^2 = -2\sec\theta \csc\theta$

In Problems 21–24, find rectangular parametric equations for the given curves.

21 $r = \sin(3\theta)$ **22** $r = \sec\theta \csc\theta$

23 $r = \theta^2$ **24** $r = \tan\theta$

25   Prove that if $f(\theta) = f(-\theta)$ then the curve $r = f(\theta)$ is symmetric about the x-axis. That is, if (x, y) is on the curve then so is $(x, -y)$.

26   Prove that if $f(\theta) = f(\pi + \theta)$ then the curve $r = f(\theta)$ is symmetric about the origin. That is, if (x, y) is on the curve so is $(-x, -y)$.

27   Prove that if $f(\theta) = f(\pi - \theta)$ then the curve $r = f(\theta)$ is symmetric about the y-axis.

7.8 SLOPES AND CURVE SKETCHING IN POLAR COORDINATES

Derivatives can be used to measure direction in polar as well as in rectangular coordinates. We begin with two theorems, one about the direction of a curve at the origin (an unusual point in polar coordinates) and the other about the direction of a curve elsewhere. Then we shall use these theorems for sketching curves.

THEOREM 1

At any value θ_0 where the curve $r = f(\theta)$ passes through the origin, the curve is tangent to the line $\theta = \theta_0$.

More precisely, if $r = 0$ at $\theta = \theta_0$ but $r \neq 0$ for all $\theta \neq \theta_0$ in some neighborhood of θ_0, then

$$\lim_{\theta \to \theta_0} \frac{\Delta y}{\Delta x} = \tan\theta_0, \qquad \lim_{\theta \to \theta_0} \frac{\Delta x}{\Delta y} = \cot\theta_0.$$

PROOF Suppose $\cos\theta_0 \neq 0$, so $\tan\theta_0$ exists. Let $\Delta\theta$ be a nonzero infinitesimal. Then $\Delta r \neq 0$ and r changes from 0 to Δr. We compute $\Delta y/\Delta x$.

$$\Delta y = (0 + \Delta r)\sin(\theta_0 + \Delta\theta) - 0\sin\theta_0$$
$$= \Delta r \sin(\theta_0 + \Delta\theta),$$
$$\Delta x = \Delta r \cos(\theta_0 + \Delta\theta),$$
$$\frac{\Delta y}{\Delta x} = \frac{\Delta r \sin(\theta_0 + \Delta\theta)}{\Delta r \cos(\theta_0 + \Delta\theta)} = \tan(\theta_0 + \Delta\theta).$$

Taking standard parts,

$$\lim_{\theta \to \theta_0} \frac{\Delta y}{\Delta x} = \tan\theta_0.$$

Similarly, when $\sin\theta_0 \neq 0$,

$$\lim_{\theta \to \theta_0} \frac{\Delta x}{\Delta y} = \cot\theta_0.$$

Both limits were given in the theorem to cover the case where the curve is vertical and $\tan \theta_0$ is undefined.

The theorem tells us that if $r = 0$ at θ_0, the curve must approach the origin from the θ_0 direction. Figure 7.8.1 shows two cases.

(a) If r has a local maximum or minimum at θ_0, then r has the same sign on both sides of θ_0. In this case the curve has a cusp at θ_0.

(b) If r has no local maximum or minimum at θ_0, then r is positive on one side of θ_0 and negative on the other side. In this case the curve crosses the origin at θ_0.

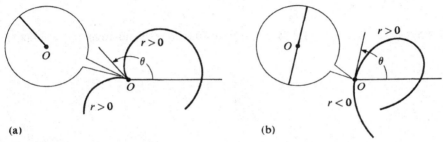

(a) (b)

Figure 7.8.1

We now consider points other than the origin. In rectangular coordinates, the slope of a curve $y = f(x)$ at a point P is $dy/dx = \tan \phi$ where ϕ is the angle between the x-axis and the line tangent to the curve at P as shown in Figure 7.8.2.

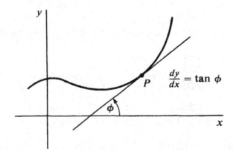

$$\frac{dy}{dx} = \tan \phi$$

Figure 7.8.2

When $r \neq 0$ in polar coordinates, a useful measure of the direction of the curve at a point P is $\tan \psi$, where ψ is the angle between the radius OP and the tangent line at P (see Figure 7.8.3).

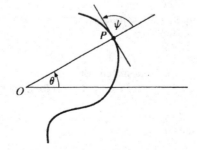

Figure 7.8.3

The following theorem gives a simple formula for $\tan \psi$ when $r \neq 0$.

THEOREM 2

Suppose $r = f(\theta)$ is a curve in polar coordinates and $dr/d\theta$ exists at a point P where $r \neq 0$. Let L be the line tangent to the curve at P and let ψ be the angle between OP and L. Then

$$\cot \psi = \frac{1}{r}\frac{dr}{d\theta}.$$

If $dr/d\theta \neq 0$,
$$\tan \psi = \frac{r}{dr/d\theta}$$

DISCUSSION When $r = 0$, P is the origin so the line OP and angle ψ are undefined.

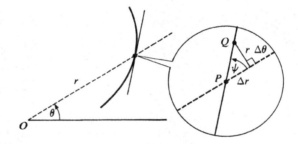

Figure 7.8.4

The formula can be seen intuitively in Figure 7.8.4.

$\Delta\theta$ is infinitesimal. As we move from the point $P(r, \theta)$ to the point $Q(r + \Delta r, \theta + \Delta\theta)$ on the curve, the change in the direction perpendicular to OP will be very close to $r\,\Delta\theta$, so we have

$$\frac{\Delta r}{r\,\Delta\theta} \approx \cot \psi, \qquad \frac{1}{r}\frac{dr}{d\theta} = \cot \psi.$$

We shall postpone the proof to the end of this section.
We can use Theorem 2 in curve sketching as follows.

(a) In an interval where $\tan \psi > 0$, the curve is going away from the origin as θ increases because $dr/d\theta$ has the same sign as r.

(b) Where $\tan \psi < 0$, the curve is going toward the origin as θ increases because $dr/d\theta$ has the opposite sign as r.

(c) Where r has either a local maximum or minimum and $dr/d\theta$ exists, the curve is going in a direction perpendicular to the radius. This is because $dr/d\theta = 0$ so $\cot \psi = 0$.

Each of these cases is shown in Figure 7.8.5.
Polar coordinates are best suited for trigonometric functions, which have the property that $f(\theta) = f(\theta + 2\pi)$. We shall therefore concentrate on the interval $0 \leq \theta < 2\pi$.

Suppose that the function $r = f(\theta)$ is differentiable for $0 \leq \theta \leq 2\pi$. The following steps may be used in sketching the curve.

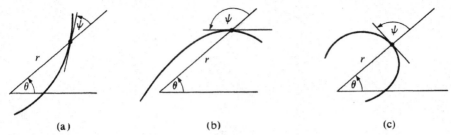

(a)   (b)   (c)

Figure 7.8.5

Step 1   Compute $dr/d\theta$.

Step 2   Find all points where $r = 0$ or $dr/d\theta = 0$.

Step 3   Sketch $y = f(x)$ in rectangular coordinates. (A method for doing this is given in Section 3.9.)

Step 4   Compute r, $dr/d\theta$, and $\tan\psi = r(dr/d\theta)$ at the points where $r = 0$ or $dr/d\theta = 0$ and at least one point between. Make a table, and test for local maxima or minima.

Step 5   Draw a smooth curve using the rectangular graph of step three and the table of step four.

EXAMPLE 1   Sketch the curve $r = 1 + \cos\theta$.

Step 1 $dr/d\theta = -\sin\theta$.

Step 2 $r = 0$ when $\theta = \pi$. $dr/d\theta = 0$ when $\theta = 0, \pi$.

Step 3   See Figure 7.8.6.

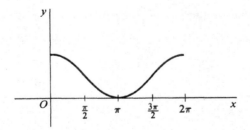

Figure 7.8.6

Step 4

θ	$r = 1 + \cos\theta$	$dr/d\theta$	$\tan\psi$	Comments
0	2	0	—	max
$\pi/2$	1	-1	-1	$\lvert r\rvert$ decreasing
π	0	0	—	min, cusp at 0
$3\pi/2$	1	1	1	$\lvert r\rvert$ increasing

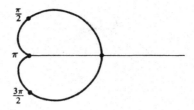

Figure 7.8.7

Step 5   We draw the curve in Figure 7.8.7. The curve is called a *cardioid* because of its heart shape.

EXAMPLE 2   Sketch the curve $r = \sin 2\theta$.

Step 1 $dr/d\theta = 2\cos 2\theta$.

Step 2 $r = 0$ at $\theta = 0, \dfrac{\pi}{2}, \pi, \dfrac{3\pi}{2}$. $dr/d\theta = 0$ at $\theta = \dfrac{\pi}{4}, \dfrac{3\pi}{4}, \dfrac{5\pi}{4}, \dfrac{7\pi}{4}$.

Step 3   See Figure 7.8.8.

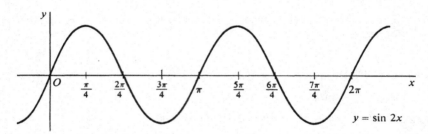

Figure 7.8.8

Step 4   We take values at intervals of $\dfrac{\pi}{8}$ beginning at $\theta = 0$. We can save some time by observing that the values from π to 2π are the same as those from 0 to π.

θ	$r = \sin 2\theta$	$dr/d\theta$	$\tan\psi$	Comments
0 and π	0	2	0	crosses origin
$\pi/8$ and $9\pi/8$	$\sqrt{2}/2$	$\sqrt{2}$	1/2	$|r|$ increasing
$2\pi/8$ and $10\pi/8$	1	0	—	max
$3\pi/8$ and $11\pi/8$	$\sqrt{2}/2$	$-\sqrt{2}$	$-1/2$	$|r|$ decreasing
$4\pi/8$ and $12\pi/8$	0	-2	0	crosses origin
$5\pi/8$ and $13\pi/8$	$-\sqrt{2}/2$	$-\sqrt{2}$	1/2	$|r|$ increasing
$6\pi/8$ and $14\pi/8$	-1	0	—	min
$7\pi/8$ and $15\pi/8$	$-\sqrt{2}/2$	$\sqrt{2}$	$-1/2$	$|r|$ decreasing

Step 5   We plot the points and trace out the curve as θ increases from 0 to 2π. Figure 7.8.9 shows the curve at various stages of development. The graph looks like a four-leaf clover.

If r approaches ∞ as θ approaches 0 or π, the curve may have a horizontal

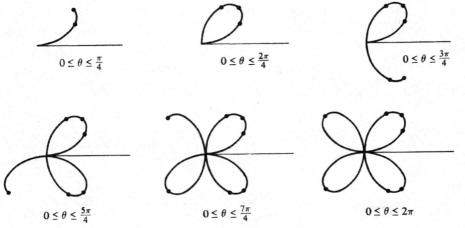

$0 \le \theta \le \frac{\pi}{4}$ $0 \le \theta \le \frac{2\pi}{4}$ $0 \le \theta \le \frac{3\pi}{4}$

$0 \le \theta \le \frac{5\pi}{4}$ $0 \le \theta \le \frac{7\pi}{4}$ $0 \le \theta \le 2\pi$

Figure 7.8.9

asymptote which can be found by computing the limit of y. At $\theta = \pi/2$ or $3\pi/2$ there may be vertical asymptotes. The method is illustrated in the following example.

EXAMPLE 3   Sketch $r = \tan(\frac{1}{2}\theta)$.

Step 1 $dr/d\theta = \frac{1}{2}\sec^2(\frac{1}{2}\theta)$.

$$y = r\sin\theta = \sin\tfrac{1}{2}\theta \, \sin\theta/\cos\tfrac{1}{2}\theta$$
$$= \sin\tfrac{1}{2}\theta(2\sin\tfrac{1}{2}\theta\cos\tfrac{1}{2}\theta)/\cos\tfrac{1}{2}\theta = 2\sin^2(\tfrac{1}{2}\theta).$$

Step 2 $r = 0$ at $\theta = 0$.
r is undefined at $\theta = \pi$.
$dr/d\theta$ is never 0.

Step 3   See Figure 7.8.10.

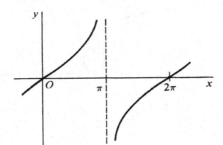

Figure 7.8.10

Step 4	θ	r or $\lim r$	$\lim y$	$dr/d\theta$	$\tan\psi$	Comments
	0	0		1/2		crosses origin
	$\pi/2$	1		1	1	$|r|$ increasing
	$\theta \to \pi^-$	∞	2			asymptote $y = 2$
	$\theta \to \pi^+$	$-\infty$	2			asymptote $y = 2$
	$3\pi/2$	-1		1	-1	$|r|$ decreasing

Step 5   The curve crosses itself at the point $x = 0$, $y = 1$, because this point has both polar coordinates

$$(r = 1, \theta = \pi/2), (r = -1, \theta = 3\pi/2).$$

Figure 7.8.11 shows the graph for various stages of development.

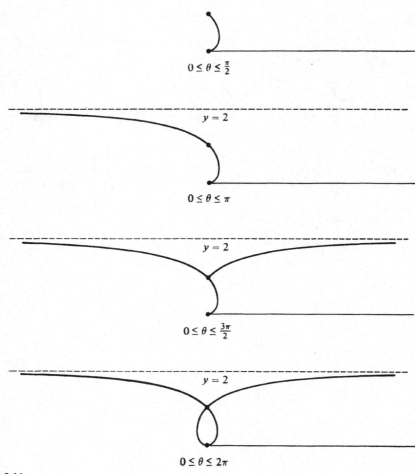

Figure 7.8.11

PROOF OF THEOREM 2   Assume the curve is not vertical at the point P, that is, $dx \neq 0$. Since

$$x = r\cos\theta, \qquad y = r\sin\theta,$$

we have $$\frac{dy}{dx} = \frac{dy/d\theta}{dx/d\theta} = \frac{r\cos\theta + (dr/d\theta)\sin\theta}{-r\sin\theta + (dr/d\theta)\cos\theta}.$$

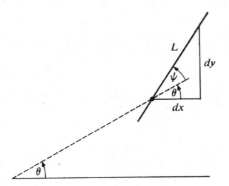

Figure 7.8.12

By the definition of the tangent line L (see Figure 7.8.12),

$$\frac{dy}{dx} = \frac{\text{change in } y \text{ along } L}{\text{change in } x \text{ along } L} = \frac{\sin(\theta + \psi)}{\cos(\theta + \psi)}.$$

Using the addition formulas,

$$\frac{dy}{dx} = \frac{\sin\theta\cos\psi + \cos\theta\sin\psi}{\cos\theta\cos\psi - \sin\theta\sin\psi}.$$

Thus $\dfrac{r\cos\theta + (dr/d\theta)\sin\theta}{-r\sin\theta + (dr/d\theta)\cos\theta} = \dfrac{\sin\psi\cos\theta + \cos\psi\sin\theta}{-\sin\psi\sin\theta + \cos\psi\cos\theta}.$

Multiplying out and canceling, we get

$$r\cos\psi(\sin^2\theta + \cos^2\theta) = \frac{dr}{d\theta}\sin\psi(\sin^2\theta + \cos^2\theta),$$

whence $r\cos\psi = \dfrac{dr}{d\theta}\sin\psi, \qquad \dfrac{1}{r}\dfrac{dr}{d\theta} = \cot\psi.$

If the curve is vertical at P we may use the same proof but with dx/dy instead of dy/dx.

PROBLEMS FOR SECTION 7.8

In Problems 1–6, find $\tan\psi$, where ψ is the angle between a line through the origin and the curve.

1	$r = \theta$	**2**	$r = \sin\theta$
3	$r = \cos\theta$	**4**	$r = \sec\theta$
5	$r = 1 + \cos\theta$	**6**	$r = \sin(2\theta)$

In Problems 7–25, sketch the given curve in polar coordinates by the method described in the text; $0 \le \theta \le 2\pi$ unless stated otherwise.

7	$r = \sin\theta + \cos\theta$	**8**	$r = 2 + 2\sin\theta$
9	$r = 1\frac{1}{2} + \sin\theta$	**10**	$r = 2 + \cos\theta$
11	$r = \frac{1}{2} + \cos\theta$	**12**	$r = \cos(\frac{1}{2}\theta), \quad 0 \le \theta \le 4\pi$
13	$r = \sin(\frac{1}{3}\theta), \quad 0 \le \theta \le 6\pi$	**14**	$r = \sin^2\theta$

15 $r = 1 + 3\cos^2(2\theta)$ **16** $r = \sin^2(3\theta)$

17 $r = \tan\theta$ **18** $r = \sec(\tfrac{1}{2}\theta), \quad 0 < \theta < 4\pi$

19 $r = 1 + \sec\theta$ **20** $r = \dfrac{1}{1 - \cos\theta}$

21 $r = \dfrac{1}{1 + \sin\theta}$ **22** $r = \cot(2\theta)$

23 $r = \pi/\theta, \quad 0 < \theta < \infty$ **24** $r = 1 + \pi/\theta, \quad 0 < \theta < \infty$

25 $r = \sqrt{\pi/\theta}, \quad 0 < \theta < \infty$

In Problems 26–29, find the points where x and y have maxima and minima.

26 $r = 1 + \cos\theta$ **27** $r = 1 + \sin^2\theta$

28 $r = \sin(2\theta)$ **29** $r = \tfrac{3}{2} + \cos\theta$

30 Find all points where the curves $r = 1 + \cos\theta$ and $r = 3\cos\theta$ intersect.

31 Find all points where the curves $r = \tfrac{1}{2}$ and $r = \sin(2\theta)$ intersect. *Warning:* The points (r, θ) and $(-r, \pi + \theta)$ are the same.

32 Find all points where the curves $r = \cos\theta$ and $r = \sin(2\theta)$ intersect.

7.9 AREA IN POLAR COORDINATES

In this section we derive a formula for the area of a region in polar coordinates. Section 6.3 on the length of a curve in rectangular coordinates should be studied before this and the following section.

Our starting point for areas in rectangular coordinates was the formula for the area of a rectangle. In polar coordinates our starting point is the formula for the area of a sector of a circle.

THEOREM 1

A sector of a circle with radius r and central angle θ has area

$$A = \tfrac{1}{2}r^2\theta.$$

An arc of a circle with radius r and central angle θ has length

$$s = r\theta.$$

PROOF Consider a sector POQ shown in Figure 7.9.1. To simplify notation let O be the origin, and put the sector POQ in the first quadrant with P on the

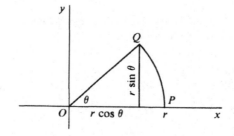

Figure 7.9.1

x-axis. Then

$$P = (r, 0), \qquad Q = (r \cos \theta, r \sin \theta).$$

The arc *QP* has the equation

$$y = \sqrt{r^2 - x^2}, \qquad r \cos \theta \leq x \leq r.$$

We see from the figure that

$$A = \tfrac{1}{2} r^2 \sin \theta \cos \theta + \int_{r \cos \theta}^{r} \sqrt{r^2 - x^2} \, dx.$$

Integrating by the trigonometric substitution $x = r \sin \phi$, we get

$$\int_{r \cos \theta}^{r} \sqrt{r^2 - x^2} \, dx = \tfrac{1}{2} r^2 \theta - \tfrac{1}{2} r^2 \sin \theta \cos \theta.$$

Therefore $A = \tfrac{1}{2} r^2 \theta$. By definition, $A = \tfrac{1}{2} rs$, so

$$s = \frac{2A}{r} = r\theta.$$

The next theorem gives the formula for area in polar coordinates.

THEOREM 2

Let $r = f(\theta)$ be continuous and $r \geq 0$ for $a \leq \theta \leq b$, where $b \leq a + 2\pi$. Then the region R bounded by the curve $r = f(\theta)$ and the lines $\theta = a$ and $\theta = b$ has area

$$A = \tfrac{1}{2} \int_{a}^{b} f(\theta)^2 \, d\theta.$$

Discussion   Imagine a point P moving along the curve $r = f(\theta)$ from $\theta = a$ to $\theta = b$. The line *OP* will sweep out the region R in Figure 7.9.2. Since $b \leq a + 2\pi$, the line will complete at most one revolution, so no point of R will be counted more than once.

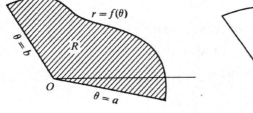

Figure 7.9.2

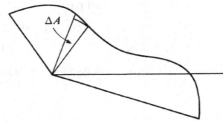

Figure 7.9.3

The formula for area can be seen intuitively by considering an infinitely small wedge ΔA of R between θ and $\theta + \Delta \theta$. (Figure 7.9.3). The wedge is almost a sector of a circle of radius $f(\theta)$ with central angle $\Delta\theta$, so

$$\Delta A \approx \tfrac{1}{2} f(\theta)^2 \, \Delta\theta \qquad \text{(compared to } \Delta\theta\text{)}.$$

By the Infinite Sum Theorem,

$$A = \tfrac{1}{2} \int_a^b f(\theta)^2 \, d\theta.$$

The actual proof follows this intuitive idea but the area of ΔA must be computed more carefully.

PROOF Let $\Delta\theta$ be positive infinitesimal and let θ be a hyperreal number between a and $b - \Delta\theta$. Consider the wedge of R with area ΔA between θ and $\theta + \Delta\theta$. Since $f(\theta)$ is continuous, it has a minimum value m and maximum value M between θ and $\theta + \Delta\theta$, and furthermore,

$$m \approx f(\theta), \qquad M \approx f(\theta).$$

The sector between θ and $\Delta\theta$ of radius m is inscribed in ΔA while the sector of radius M is circumscribed about ΔA.

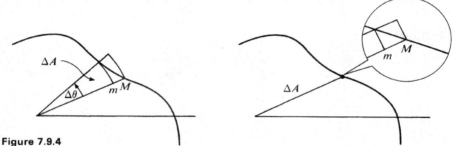

Figure 7.9.4

(Figure 7.9.4 shows the inscribed and circumscribed sectors for real $\Delta\theta$ and infinitesimal $\Delta\theta$.) By Theorem 1, the two sectors have areas $\tfrac{1}{2}m^2 \, \Delta\theta$ and $\tfrac{1}{2}M^2 \, \Delta\theta$. Moreover, ΔA is between those two areas,

$$\tfrac{1}{2}m^2 \, \Delta\theta \le \Delta A \le \tfrac{1}{2}M^2 \, \Delta\theta,$$
$$\tfrac{1}{2}m^2 \le \Delta A / \Delta\theta \le \tfrac{1}{2}M^2.$$

Taking standard parts,

$$\tfrac{1}{2} f(\theta)^2 \le st(\Delta A / \Delta\theta) \le \tfrac{1}{2} f(\theta)^2.$$

Therefore $\Delta A / \Delta\theta \approx \tfrac{1}{2} f(\theta)^2,$

and by the Infinite Sum Theorem,

$$A = \tfrac{1}{2} \int_a^b f(\theta)^2 \, d\theta.$$

Theorem 1 is also true in the case that $r = f(\theta)$ is continuous and $r \le 0$.

Since $A = \tfrac{1}{2} \int_a^b f(\theta)^2 \, d\theta = \tfrac{1}{2} \int_a^b (-f(\theta))^2 \, d\theta,$

the region R bounded by the curve $r = f(\theta)$ has the same area as the region S bounded by the curve $r = -f(\theta)$. Both areas are positive. As we can see from Figure 7.9.5, S looks exactly like R but is on the opposite side of the origin.

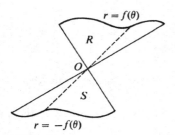

Figure 7.9.5

Figure 7.9.6

EXAMPLE 1 Find the area of one loop of the "four-leaf clover" $r = \sin 2\theta$. From Figure 7.9.6, we see that one loop is traced out when θ goes from 0 to $\pi/2$. Therefore the area is

$$A = \tfrac{1}{2}\int_0^{\pi/2} \sin^2(2\theta)\, d\theta = \tfrac{1}{2}\int_0^{\pi} \tfrac{1}{2}\sin^2 \phi\, d\phi$$

$$= \tfrac{1}{4}\int_0^{\pi} \sin^2 \phi\, d\phi = \tfrac{1}{4}\left(-\tfrac{1}{2}\sin\phi\cos\phi + \tfrac{1}{2}\phi\right)\Bigg]_0^{\pi} = \tfrac{1}{8}\pi.$$

As one would expect, all four loops have the same area.

On the loop from $\theta = \pi/2$ to $\theta = \pi$, the value of $r = \sin 2\theta$ is negative. However, the area is again

$$A = \tfrac{1}{2}\int_{\pi/2}^{\pi} \sin^2(2\theta)\, d\theta = \tfrac{1}{8}\pi.$$

Our next example shows why the hypothesis that r has the same sign for $a \le \theta \le b$ is needed in Theorem 2.

EXAMPLE 2 Find the area of the region inside the circle $r = \sin\theta$ (Figure 7.9.7).

The point (r, θ) goes around the circle once when $0 \le \theta \le \pi$ with r positive, and again when $\pi \le \theta \le 2\pi$ with r negative. The theorem says that we will get the correct area if we take either 0 and π, or π and 2π, as the limits of

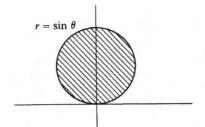

Figure 7.9.7

integration. Thus

$$A = \int_0^\pi \tfrac{1}{2}\sin^2\theta \, d\theta = \tfrac{1}{2}(-\tfrac{1}{2}\sin\theta\cos\theta + \tfrac{1}{2}\theta)\bigg]_0^\pi = \tfrac{1}{4}(\pi - 0) = \pi/4.$$

Alternatively,

$$A = \int_\pi^{2\pi} \tfrac{1}{2}\sin^2\theta \, d\theta = \tfrac{1}{2}(-\tfrac{1}{2}\sin\theta\cos\theta + \tfrac{1}{2}\theta)\bigg]_\pi^{2\pi} = \tfrac{1}{4}(2\pi - \pi) = \pi/4.$$

Since the curve is a circle of radius $\tfrac{1}{2}$, our answer $\pi/4$ agrees with the usual formula $A = \pi r^2$.

Integrating from 0 to 2π would count the area twice and give the wrong answer.

EXAMPLE 3   Find the area of the region inside both the circles $r = \sin\theta$ and $r = \cos\theta$.

The first thing to do is draw the graphs of both curves. The graphs are shown in Figure 7.9.8.

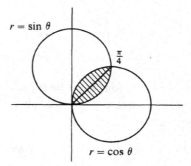

Figure 7.9.8

We see that the two circles intersect at the origin and at $\theta = \pi/4$. The region is divided into two parts, one bounded by $r = \sin\theta$ for $0 \le \theta \le \pi/4$ and the other bounded by $r = \cos\theta$ for $\pi/4 \le \theta \le \pi/2$. Thus

$$A = \int_0^{\pi/4} \frac{1}{2}\sin^2\theta \, d\theta + \int_{\pi/4}^{\pi/2} \frac{1}{2}\cos^2\theta \, d\theta$$

$$= \frac{1}{2}\left(-\frac{1}{2}\sin\theta\cos\theta + \frac{1}{2}\theta\right)\bigg]_0^{\pi/4} + \frac{1}{2}\left(\frac{1}{2}\sin\theta\cos\theta + \frac{1}{2}\theta\right)\bigg]_{\pi/4}^{\pi/2}$$

$$= \frac{1}{2}\left[\left(-\frac{1}{4} - 0\right) + \left(\frac{\pi}{8} - 0\right) + \left(0 - \frac{1}{4}\right) + \left(\frac{\pi}{4} - \frac{\pi}{8}\right)\right] = \frac{\pi}{8} - \frac{1}{4}.$$

PROBLEMS FOR SECTION 7.9

In Problems 1–13, find the area of the regions bounded by the following curves in polar coordinates.

1	$r = 2a\cos\theta$	**2**	$r = 1 + \cos\theta$
3	$r = \sqrt{\sin\theta}$	**4**	$r = 2 + \cos\theta$
5	The loop in $r = \tan(\tfrac{1}{2}\theta)$	**6**	One loop of $r = \cos(3\theta)$

7	One loop of $r = \sin^2\theta$	8	The large loop of $r = \frac{1}{2} + \cos\theta$
9	The small loop of $r = \frac{1}{2} + \cos\theta$	10	One loop of $r^2 = \cos(2\theta)$
11	$\theta = 0, \quad \theta = \pi/3, \quad r = \cos\theta$	12	$\theta = \pi/6, \quad \theta = \pi/3, \quad r = \sec\theta$
13	$r = \tan\theta, \quad r = \frac{1}{\sqrt{2}}\csc\theta$		

14   Find the area of the region inside the curve $r = 2\cos\theta$ and outside the curve $r = 1$.

15   Find the area of the region inside the curve $r = 2\sin\theta$ and above the line $r = \frac{3}{2}\csc\theta$.

16   Find the area of the region inside the spiral $r = \theta, 0 \le \theta \le 2\pi$.

17   Find the area of the region inside the spiral $r = \sqrt{\theta}, 0 \le \theta \le 2\pi$.

18   Find the area of the region inside both of the curves $r = \sqrt{3}\cos\theta, r = \sin\theta$.

19   Find the area of the region inside both of the curves $r = 1 - \cos\theta, r = \cos\theta$.

20   The center of a circle of radius one is on the circumference of a circle of radius two. Find the area of the region inside both circles.

21   Find a formula for the area of the region between the curves $r = f(\theta)$ and $r = g(\theta)$, $a \le \theta \le b$, when $0 \le f(\theta) \le g(\theta)$.

7.10 LENGTH OF A CURVE IN POLAR COORDINATES

Consider a curve

$$r = f(\theta), \qquad a \le \theta \le b$$

in polar coordinates. The curve is called *smooth* if $f'(\theta)$ is continuous for θ between a and b. In Chapter 6 we obtained a formula for the length of a smooth parametric curve in rectangular coordinates. We may now apply this to get a formula for the length of a smooth curve in polar coordinates.

THEOREM

The length of a smooth curve

$$r = f(\theta), \qquad a \le \theta \le b$$

in polar coordinates which does not retrace itself is

$$s = \int_a^b \sqrt{f(\theta)^2 + f'(\theta)^2}\, d\theta,$$

or equivalently $\qquad s = \int_a^b \sqrt{r^2 + (dr/d\theta)^2}\, d\theta.$

Discussion The formula can be seen intuitively as follows. We see from Figure 7.10.1 that

$$\Delta s \approx \sqrt{(r\,\Delta\theta)^2 + \Delta r^2} = \sqrt{r^2 + (\Delta r/\Delta\theta)^2}\,\Delta\theta$$
$$\approx \sqrt{r^2 + (dr/d\theta)^2}\,\Delta\theta \qquad \text{(compared to } \Delta\theta\text{).}$$

By the Infinite Sum Theorem,

$$s = \int_a^b \sqrt{r^2 + (dr/d\theta)^2}\, d\theta.$$

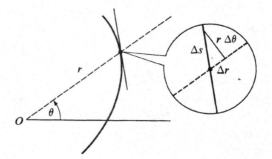

Figure 7.10.1

The length of a curve has already been defined using rectangular coordinates, and the theorem states that the new formula will give the same number for the length.

PROOF The curve is given in rectangular coordinates by the parametric equation

$$x = f(\theta) \cos \theta, \qquad y = f(\theta) \sin \theta.$$

The derivatives are

$$\frac{dx}{d\theta} = -f(\theta) \sin \theta + f'(\theta) \cos \theta,$$

$$\frac{dy}{d\theta} = f(\theta) \cos \theta + f'(\theta) \sin \theta.$$

Since $f(\theta)$ and $f'(\theta)$ are continuous, $dx/d\theta$ and $dy/d\theta$ are continuous. Recall the length formula for parametric equations:

$$s = \int_a^b \sqrt{\left(\frac{dx}{d\theta}\right)^2 + \left(\frac{dy}{d\theta}\right)^2}\, d\theta.$$

We compute

$$\left(\frac{dx}{d\theta}\right)^2 + \left(\frac{dy}{d\theta}\right)^2 = f(\theta)^2 \sin^2 \theta - 2f(\theta)f'(\theta) \sin \theta \cos \theta + f'(\theta) \cos^2 \theta$$
$$+ f(\theta)^2 \cos^2 \theta + 2f(\theta)f'(\theta) \sin \theta \cos \theta + f'(\theta)^2 \sin^2 \theta$$
$$= f(\theta)^2(\sin^2 \theta + \cos^2 \theta) + f'(\theta)^2(\cos^2 \theta + \sin^2 \theta)$$
$$= f(\theta)^2 + f'(\theta)^2.$$

The desired formula now follows by substitution.

EXAMPLE 1   Find the length of the spiral $r = \theta^2$ from $\theta = \pi$ to $\theta = 4\pi$, shown in Figure 7.10.2.

$$s = \int_\pi^{4\pi} \sqrt{r^2 + (dr/d\theta)^2}\, d\theta$$

$$= \int_\pi^{4\pi} \sqrt{\theta^4 + 4\theta^2}\, d\theta = \int_\pi^{4\pi} \sqrt{\theta^2 + 4}\, \theta\, d\theta.$$

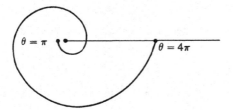

Figure 7.10.2

Let $u = \theta^2 + 4$, $du = 2\theta\, d\theta$. Then

$$s = \int_{\pi^2+4}^{16\pi^2+4} \tfrac{1}{2}\sqrt{u}\, du = \tfrac{1}{2}\cdot\tfrac{2}{3}u^{3/2}\Big]_{\pi^2+4}^{16\pi^2+4}$$
$$= \tfrac{1}{3}((16\pi^2 + 4)^{3/2} - (\pi^2 + 4)^{3/2}).$$

EXAMPLE 2   Find the length of the curve $r = \sin\theta$ from $\theta = \alpha$ to $\theta = \beta$, shown in Figure 7.10.3. $dr/d\theta = \cos\theta$, so

$$s = \int_\alpha^\beta \sqrt{\sin^2\theta + \cos^2\theta}\, d\theta = \int_\alpha^\beta d\theta = \beta - \alpha.$$

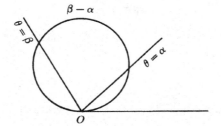

Figure 7.10.3

The graph of $r = \sin\theta$ is a circle of radius $\tfrac{1}{2}$ which passes through O. Example 2 proves that the length of an arc of the circle is equal to the angle formed by the ends of the arc and the origin. Note that if we take $\alpha = 0$ and $\beta = 2\pi$ we get an arc length of 2π, which is twice the circumference of the circle. This is because the point (r, θ) goes around the circle twice, once from $\theta = 0$ to $\theta = \pi$ and once from $\theta = \pi$ to $\theta = 2\pi$.

PROBLEMS FOR SECTION 7.10

In Problems 1–10, find the length in polar coordinates.

1	$r = 7$, $0 \le \theta \le 2\pi$	**2**	$r = \cos\theta$, $\pi/4 \le \theta \le \pi/3$
3	$r = \sec\theta$, $-\pi/4 \le \theta \le \pi/4$	**4**	$r = 6\theta^2$, $0 \le \theta \le \sqrt{5}$
5	$r = \theta^4$, $0 \le \theta \le 1$	**6**	$r = a\sin\theta + b\cos\theta$, $0 \le \theta \le \pi$
7	$r = 1 - \cos\theta$, $0 \le \theta \le \pi$	**8**	$r = 2 + 2\cos\theta$, $0 \le \theta \le 2\pi$
9	$r = \sin^2(\tfrac{1}{2}\theta)$, $0 \le \theta \le 2\pi$	**10**	$r = \sin^3(\tfrac{1}{3}\theta)$, $0 \le \theta \le 3\pi$

In Problems 11–14, set up an integral for the length of the curve.

11	$r = \sin(2\theta)$, $0 \le \theta \le 2\pi$	**12**	$r = \tan\theta$, $0 \le \theta \le \pi/4$

13 $\quad r = \theta, \quad 0 \le \theta \le b$ $\qquad\qquad$ **14** $\quad r = \theta^n, \quad 0 \le \theta \le b$

☐ **15** Show that the surface area generated by rotating the curve $r = f(\theta)$, $a \le \theta \le b$, about the y-axis is

$$A = \int_a^b 2\pi r \cos\theta\sqrt{r^2 + (dr/d\theta)^2}\, d\theta \qquad \text{(about } y\text{-axis).}$$

(Assume $0 \le a < b \le \pi/2$.) Show that the corresponding formula for a rotation about the x-axis is

$$A = \int_a^b 2\pi r \sin\theta\sqrt{r^2 + (dr/d\theta)^2}\, d\theta \qquad \text{(about } x\text{-axis).}$$

In Problems 16–21, find the surface area generated by rotating the curve about the given axis.

16 $\quad r = \sin\theta, \quad 0 \le \theta \le \pi/3, \quad$ about y-axis

17 $\quad r = a\sin\theta + b\cos\theta, \quad 0 \le \theta \le \pi/2, \quad$ about y-axis

18 $\quad r = 1 + \cos\theta, \quad 0 \le \theta \le \pi/2, \quad$ about x-axis

19 $\quad r = \sqrt{\cos(2\theta)}, \quad 0 \le \theta \le \pi/4, \quad$ about y-axis

20 $\quad r = \sqrt{\cos(2\theta)}, \quad 0 \le \theta \le \pi/4, \quad$ about x-axis

21 $\quad r = \cos^2(\tfrac{1}{2}\theta), \quad 0 \le \theta \le \pi/2, \quad$ about x-axis

EXTRA PROBLEMS FOR CHAPTER 7

1 $\quad$ Find dy/dx where $y = x + \sin x$. $\qquad$ **2** $\quad$ Find dy/dx where $y = \sin(1/x)$.

3 $\quad$ Find $dy/d\theta$ where $y = \sqrt{\theta}\cos\theta$. $\qquad$ **4** $\quad$ Find $dy/d\theta$ where $y = \sin(\tan\theta)$.

5 $\quad$ Evaluate $\lim\limits_{\theta\to 0} \dfrac{\sin(4\theta)}{\sin(3\theta)}$. $\qquad$ **6** $\quad$ Evaluate $\lim\limits_{u\to 0} \dfrac{\cos(6u) - 1}{u^2}$.

7 $\quad$ Evaluate $\displaystyle\int \cos(\cos\theta)\sin\theta\, d\theta$. $\qquad$ **8** $\quad$ Evaluate $\displaystyle\int 3\sqrt{\sin x}\cos x\, dx$

9 $\quad$ Evaluate $\displaystyle\int_{-\pi/2}^{\pi/2} 4\cos\theta\, d\theta$. $\qquad$ **10** $\quad$ Evaluate $\displaystyle\int_0^{\pi/2} \tan x\sec x\, dx$.

11 $\quad$ An airplane travels in a straight line at 600 mph at an altitude of 4 miles. Find the rate of change of the angle of elevation one minute after the airplane passes directly over an observer on the ground.

12 $\quad$ A 40 ft ladder is to be propped up against a 15 ft wall as shown in the figure. What angle should the ladder make with the ground if the horizontal distance the ladder extends beyond the wall is to be a maximum?

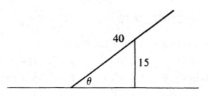

13 $\quad$ Find dy/dx where $y = \arccos\sqrt{x}$. $\quad$ **14** $\quad$ Find dy/dx where $y = \text{arcsec}\sqrt{x}$.

15 $\quad$ Find du/dt where $u = \arctan t - t$. $\quad$ **16** $\quad$ Find du/dt where $u = \arcsin(1/t)$.

17 $\quad$ Evaluate $\lim\limits_{x\to 0} \dfrac{\arctan x}{x}$. $\qquad$ **18** $\quad$ Evaluate $\displaystyle\int \dfrac{dt}{a^2 + b^2 t^2}$.

19   Evaluate $\displaystyle\int \frac{dx}{(x-1)\sqrt{x^2-2x}}$, $\quad x > 2$.

20   Evaluate $\displaystyle\int \frac{\sec^2 x}{\sqrt{1-\tan^2 x}}\, dx$.

21   Evaluate $\displaystyle\int \frac{x^3}{\sqrt{x^2+1}}\, dx$ in two ways, by change of variables and by parts.

22   Evaluate $\displaystyle\int x \sin(3x)\, dx$.

23   Evaluate $\displaystyle\int_0^1 \cos\sqrt{\theta}\, d\theta$.

24   Find the volume of the solid formed by rotating the region under the curve $y = x\sqrt{\sin x}$, $0 \le x \le \pi$, about the x-axis.

25   Find the volume of the solid generated by rotating the region under the curve $y = \tan x$, $0 \le x \le \pi/4$, about the x-axis.

26   Evaluate $\displaystyle\int \cot^4\theta\, d\theta$.

27   Evaluate $\displaystyle\int \tan^5\theta \sec^5\theta\, d\theta$.   **28**   Evaluate $\displaystyle\int (2x^2-1)^{-3/2}\, dx$.

29   Evaluate $\displaystyle\int \frac{\sqrt{2-x^2}}{x^2}\, dx$.   **30**   Evaluate $\displaystyle\int \frac{1}{(1+x^2)^2}\, dx$.

In Problems 31–34, sketch the given function in (a) rectangular coordinates, (b) polar coordinates. Let $0 \le \theta \le 2\pi$.

31 $r = 1 - \cos\theta$ **32** $r = \cos(3\theta)$

33 $r = \dfrac{1}{2 + \sin\theta}$ **34** $r^2 = \cos(2\theta)$

35   Find the area of the polar region bounded by $r = 1 + \sin^2\theta$.

36   Find the area of the polar region bounded by $r = \sin\theta + \cos\theta$.

37   Find the area of the polar region inside both the curves $r = 1 - \cos\theta$ and $r = 1 + \cos\theta$.

38   Find the length in polar coordinates of the curve

$$r = \sin^4(\tfrac{1}{4}\theta), \qquad 0 \le \theta \le \pi.$$

39   Find the surface area generated by rotating the polar curve

$$r = 1 - \cos\theta, \qquad 0 \le \theta \le \pi/2,$$

about the x-axis.

☐ **40**   Use the Intermediate Value Theorem to prove that $\arctan y$ has domain $(-\infty, \infty)$.

☐ **41**   Use the Intermediate Value Theorem to prove that the domain of $\operatorname{arcsec} y$ is the set of all y such that $y \le -1$ or $y \ge 1$.

☐ **42**   Prove that if f is a differentiable function of x then

$$\int f(x)\, dx = xf(x) - \int xf'(x)\, dx.$$

☐ **43**   If u and v are differentiable functions of x then

$$\int u^2\, dv = u^2 v - 2\int uv\, du.$$

☐ **44**   Show that if f' and g are differentiable for all x then

$$\int g(x)g'(x)f''(g(x))\, dx = f'(g(x))g(x) - f(g(x)) + C.$$

☐ **45**   Use integration by parts to prove the reduction formula

$$\int \frac{dx}{(1 + x^2)^{m+1}} = \frac{1}{2m} \frac{x}{(1 + x^2)^m} + \left(1 - \frac{1}{2m}\right) \int \frac{dx}{(1 + x^2)^m}.$$

Hint:
$$\frac{1}{(1 + x^2)^{m+1}} = \frac{1}{(1 + x^2)^m} - \frac{x^2}{(1 + x^2)^{m+1}}.$$

☐ **46**   Suppose $y = f(x)$, $a < x < b$ and $x = g(y)$, $c < y < d$ are inverse functions and are strictly increasing. Let $y_0 = f(x_0)$. Prove that:

(a) If f is continuous at x_0, g is continuous at y_0.
(b) If $f'(x_0)$ exists and $f'(x_0) \neq 0$, then $g'(y_0)$ exists.

☐ **47**   Justify the following formula for the area of the polar region bounded by the continuous curves

$$\theta = f(r), \qquad \theta = g(r), \qquad a \leq r \leq b,$$

where $0 \leq f(r) \leq g(r) \leq 2\pi$.

$$A = \int_a^b r(g(r) - f(r))\, dr.$$

☐ **48**   Justify the following formula for the mass of an object in the polar region $0 \leq r \leq f(\theta)$, $a \leq \theta \leq b$, with density $\rho(\theta)$ per unit area.

$$m = \int_a^b \tfrac{1}{2}\rho(\theta)(f(\theta))^2\, d\theta.$$

☐ **49**   Justify the following formulas for the centroid of the polar region $0 \leq r \leq f(\theta)$, $a \leq \theta \leq b$.

$$\bar{x} = \frac{\displaystyle\int_a^b \tfrac{1}{3}\cos\theta(f(\theta))^3\, d\theta}{\displaystyle\int_a^b \tfrac{1}{2}(f(\theta))^2\, d\theta}, \qquad \bar{y} = \frac{\displaystyle\int_a^b \tfrac{1}{3}\sin\theta(f(\theta))^3\, d\theta}{\displaystyle\int_a^b \tfrac{1}{2}(f(\theta))^2\, d\theta}.$$

Hint: The centroid of a triangle is located on a median $\tfrac{2}{3}$ of the way from a vertex to the opposite side.

☐ **50**   Find the centroid of the sector $0 \leq r \leq c, a \leq \theta \leq b$.

☐ **51**   Find the centroid of the region bounded by the cardioid $r = 1 + \cos\theta$.

EXPONENTIAL AND LOGARITHMIC FUNCTIONS

8.1 EXPONENTIAL FUNCTIONS

Any positive real number a can be raised to a rational exponent,

$$a^{m/n} = \sqrt[n]{a^m}, \qquad a > 0.$$

But what does a^b mean if b is an irrational number? For example, what are 2^π and $2^{\sqrt{3}}$?

We shall approach the problem of defining a^b by considering a^x as a function of x. Given a positive real number a, the function a^x is defined for all rational numbers x. Its graph may be·thought of as a "dotted" line as in Figure 8.1.1.

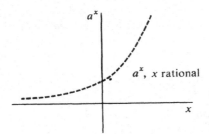

Figure 8.1.1

Our idea is to define a^x for all x by "connecting the dots." This will make a^x into a continuous function which agrees with the original dotted curve when x is rational. A number such as 2^π will thus be approximated by raising 2 to a rational exponent close to π. $2^{3.14}$ will be close to 2^π and $2^{3.14159}$ will be even closer.

To get the exact value of 2^π we use hyperrational numbers; if r is a hyperrational number infinitely close to π, then 2^r will be infinitely close to 2^π. The function $y = a^x$ will be called the *exponential function* with base a.

Hyperintegers were introduced in Section 3.8. To get an exact value of 2^π, we use hyperintegers. A quotient K/H of two hyperintegers is called a *hyperrational number*. Our idea is to take a hyperrational number K/H that is infinitely close to π and define 2^π to be the standard part of $2^{K/H}$.

In general, given a real number r, we can find a hyperrational number

K/H infinitely close to r as follows. Choose a positive infinite hyperinteger H. Let K be the greatest hyperinteger $\leq Hr$, $K = [Hr]$. Then

$$K \leq Hr < K + 1.$$

Dividing by H,

$$\frac{K}{H} \leq r < \frac{K}{H} + \frac{1}{H}, \qquad \frac{K}{H} \approx r.$$

Given a positive real number a, we then define a^r to be the standard part of $a^{K/H}$. It can be proved that the value for a^r obtained in this way does not depend on our choice of H. Thus the exponent a^x is defined for all real x. We summarize our procedure as a lemma and a definition.

LEMMA 1

Let a and r be real numbers, $a > 0$.

(i)   There is a hyperrational number K/H infinitely close to r.
(ii) The hyperrational exponent $a^{K/H}$ is defined and finite.
(iii) For any other hyperrational number $L/M \approx r$, $st(a^{K/H}) = st(a^{L/M})$.

DEFINITION

Let a and r be real, $a > 0$. We define $a^r = st(a^{K/H})$, where $K/H \approx r$.

The function $y = a^x$, also written $y = \exp_a x$, is called the *exponential function with base a*. If $a < 0$, we leave a^x undefined except when $x = m/n$, n odd.

The following rules for exponents should be familiar to the student when the exponents are rational, except for inequality (vii). They can be proved for real exponents by forming hyperrational exponents and taking standard parts.

RULES FOR EXPONENTS

Let a, b be positive real numbers.

(i) $1^x = 1, \qquad a^0 = 1$.
(ii) $a^{x+y} = a^x a^y, \qquad a^{x-y} = a^x/a^y$.
(iii) $a^{xy} = (a^x)^y$.
(iv) $a^x b^x = (ab)^x, \qquad (a^x/b^x) = (a/b)^x$.

INEQUALITIES FOR EXPONENTS

Let a, b be positive real numbers.

(v)   If $a < b$ and $x > 0$, then $a^x < b^x$.
(vi) If $1 < a$ and $x < y$, then $a^x < a^y$.
(vii) If $x \geq 1$, then $(a + 1)^x \geq ax + 1$.

PROOF (vii)   Since this inequality is probably new to the student, we give a proof for the case where x is a rational number $x = q$.

Replace a by the variable t. Let

$$y = (t + 1)^q - tq - 1.$$

We must show that $y \geq 0$. When $t = 0$, $y = 0$. For $t \geq 0$ and $q \geq 1$, we have

$$\frac{dy}{dt} = q(t + 1)^{q-1} - q \geq q \cdot 1^0 - q = 0.$$

Thus $dy/dt \geq 0$, so y is increasing and $y \geq 0$.

THEOREM 1

The exponential function $y = a^x$ is increasing if $a > 1$, constant if $a = 1$, and decreasing if $a < 1$.

PROOF Inequality (vi) shows that a^x is increasing if $a > 1$. If $a < 1$ and $q < r$ then

$$1/a > 1, \qquad (1/a)^q < (1/a)^r, \qquad a^q > a^r,$$

so a^x is decreasing. If $a = 1$ then $a^x = 1$ is constant. Figure 8.1.2 shows graphs of $y = a^x$ for different values of a.

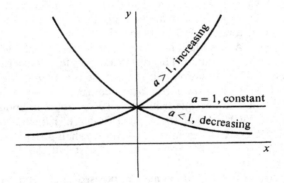

Figure 8.1.2

THEOREM 2

For each $a > 0$, the exponential function $y = a^x$ is continuous.

Consider the case $a > 1$. Suppose x_1 and x_2 are finite and $x_1 \approx x_2$. Say $x_1 < x_2$. Choose hyperrational numbers r_1 and r_2 infinitely close to x_1 and x_2 such that

$$r_1 < x_1 < x_2 < r_2.$$

The inequalities for exponents hold for hyperreal x by the Transfer Principle, so

$$a^{r_1} < a^{x_1} < a^{x_2} < a^{r_2}.$$

But $r_1 \approx r_2$, so $a^{r_1} \approx a^{r_2}$. Therefore $a^{x_1} \approx a^{x_2}$, and $y = a^x$ is continuous. The case $a \leq 1$ is similar.

An example of an exponential function is given by the growth of a population

$f(t)$ with a constant birth and death rate. It grows in such a way that the rate of change of the population is proportional to the population. Given an integer n, the population increase from time t to time $t + 1/n$ is a constant times $f(t)$.

$$f(t + 1/n) - f(t) = cf(t).$$

Then
$$f(t + 1/n) = kf(t)$$

where $k = c + 1$.

Let us set $f(0) = 1$; that is, we choose $f(0)$ for our unit of population. Then

$$f(0) = 1, f(1/n) = k, f(2/n) = k^2, \ldots, f(m/n) = k^m.$$

So if we put $f(1) = a = k^n$, we have

$$f(m/n) = a^{m/n}.$$

We conclude that for any rational number m/n, the population at time $t = m/n$ is $a^{m/n}$. In reality, of course, the population is not a continuous function of time because its value is always a whole number. However, it is convenient to approximate the population by the exponential function a^x, and to make a^x continuous by defining it for all real x.

If the birth rate of a population is greater than the death rate, the growth curve will be a^x where $a > 1$ and the population will increase. Similarly, if the birth and death rates are equal, $a = 1$ and the population is constant. When the death rate exceeds the birth rate, $a < 1$ and the population decreases.

Warning: A population grows exponentially only when the birth rate minus the death rate is constant. This rarely happens for long periods of time, because a large change in the population will itself cause the birth or death rate to change. For example, if the population of the earth quadrupled every century it would reach the impossible figure of one quadrillion, or 10^{15}, people in about 900 years. In the 20th century the birth rate of the United States has fluctuated wildly while the death rate has decreased. Later in this chapter we shall discuss more realistic growth functions which grow nearly exponentially at first but then level off at a limiting value.

The inequalities for exponents can be used to get approximate values for a^b and to evaluate limits.

EXAMPLE 1   Approximate $\sqrt{2}^\pi$. We have

$$\sqrt{2} \sim 1.4142, \qquad \pi \sim 3.14.$$

Thus
$$1.414 < \sqrt{2} < 1.415, \qquad 3.1 < \pi < 3.2.$$

By the inequalities for exponents,

$$(1.414)^{3.1} < \sqrt{2}^\pi < (1.415)^{3.2},$$

or
$$2.91 < \sqrt{2}^\pi < 3.06.$$

Thus $\sqrt{2}^\pi$ is within $\frac{1}{10}$ of 3.0.

EXAMPLE 2   If $a > 1$, evaluate the limit $\lim\limits_{x \to \infty} a^x$.

Let H be positive infinite and $a = b + 1$. Then $b > 0$ and by inequality (vii),

$$a^H = (b + 1)^H \geq bH + 1.$$

So a^H is positive infinite. Therefore

$$\lim_{x \to \infty} a^x = \infty.$$

EXAMPLE 3 Evaluate the limit $\lim_{x \to \infty} \dfrac{4^{x+1} + 5}{4^{x-1} - 3}$.

Let H be positive infinite. Then

$$\frac{4^{H+1} + 5}{4^{H-1} - 3} = \frac{4^{H+1} \cdot 4^{-H} + 5 \cdot 4^{-H}}{4^{H-1} \cdot 4^{-H} - 3 \cdot 4^{-H}} = \frac{4 + 5 \cdot 4^{-H}}{\frac{1}{4} - 3 \cdot 4^{-H}}.$$

By Example 2, 4^H is infinite, so $(\frac{1}{4})^H$ is infinitesimal. Thus

$$st\left(\frac{4^{H+1} + 5}{4^{H-1} - 3}\right) = st\left(\frac{4 + 5 \cdot 4^{-H}}{\frac{1}{4} - 3 \cdot 4^{-H}}\right) = \frac{4 + 5 \cdot 0}{\frac{1}{4} - 3 \cdot 0} = 16,$$

$$\lim_{x \to \infty} \frac{4^{x+1} + 5}{4^{x-1} - 3} = 16.$$

PROBLEMS FOR SECTION 8.1

In Problems 1–7, verify the inequalities.

1 $\quad 10\sqrt[3]{10} < 10^{\sqrt{2}} < 10\sqrt{10}$

2 $\quad 2\sqrt[3]{4} < 2^{\sqrt{3}} < 2\sqrt[4]{8}$

3 $\quad 10\sqrt{10} < \sqrt{10^{\pi}} < 10\sqrt[5]{1000}$

4 $\quad 3\sqrt[5]{9} < \pi^{\sqrt{2}} < 3.2\sqrt{3.2}$

5 $\quad \sqrt[10]{2} \geq 1.05 \quad$ (use inequality (vii))

6 $\quad (\pi - 2)^{\pi} \geq \pi^2 - 3\pi + 1$

7 $\quad \sqrt{2^{\sqrt{2}}} \geq 3 - \sqrt{2}$

In Problems 8–23 evaluate the limit.

8 $\quad \lim_{x \to \infty} a^x \quad$ if $0 < a < 1$

9 $\quad \lim_{x \to -\infty} a^x \quad$ if $a > 1$

10 $\quad \lim_{x \to 3} 5^{x^2 - 2x + 1}$

11 $\quad \lim_{x \to \infty} \dfrac{a^x}{b^x} \quad$ if $0 < b < a$

12 $\quad \lim_{t \to \infty} a^{1/t} \quad$ if $0 < a$

13 $\quad \lim_{t \to \infty} 10^{2t - t^2}$

14 $\quad \lim_{t \to \infty} 3^t - 2^t$

15 $\quad \lim_{t \to \infty} 2^{t+3} - 2^{t+1}$

16 $\quad \lim_{x \to \infty} \dfrac{3^x - 2^x + 1}{4 \cdot 3^x - 2^x - 1}$

17 $\quad \lim_{x \to \infty} \dfrac{3^{x+1} - 2^{x+4}}{3^{x-2} + 2^{x-1} + 6}$

18 $\quad \lim_{x \to \infty} \dfrac{3^{x+5} - 2^{2x+1}}{3^{x+1} - 2^{2x+4}}$

19 $\quad \lim_{x \to \infty} x^x$

20 $\quad \lim_{x \to \infty} x^{-x}$

21 $\quad \lim_{x \to 2} \dfrac{3 - \sqrt{3^x}}{9 - 3^x}$

22 $\quad \lim_{x \to 0} \dfrac{4^{1+x} - 4^{1-x}}{2^{1+x} - 2^{1-x}}$

23 $\quad \lim_{x \to 1} \dfrac{\pi^x - \pi}{\pi^{2x} - \pi^2}$

☐ **24** Prove that the function $y = x^x$, $x \geq 1$, is increasing.

☐ **25** Prove that if $a > 0$ and $\lim_{x \to c} f(x) = L$, then $\lim_{x \to c} a^{f(x)} = a^L$.

☐ **26** Prove that for each real number r, the function $y = x^r$, $x > 0$, is continuous.

8.2 LOGARITHMIC FUNCTIONS

The inverses of exponential functions are called logarithmic functions. Inverse functions were studied in Sections 2.4 and 7.3. Given a positive real number a different from one, the exponential function with base a is either increasing or decreasing. Therefore it has an inverse function.

DEFINITION

Let $a \neq 1$ be a positive real number. The **logarithmic function with base** a, denoted by
$$x = \log_a y,$$
is defined as the inverse of the exponential function with base a, $y = a^x$. That is, $\log_a y$ is defined as the exponent to which a must be raised to get y,

$$\log_a y = x \quad \text{if and only if} \quad y = a^x.$$

We see at once that

$$\log_a(a^x) = x, \qquad a^{\log_a y} = y$$

whenever $\log_a y$ is defined.

The logarithm of y to the base 10, written $\log y = \log_{10} y$, is called the *common logarithm* of y. Common logarithms are readily available in tables.

Logarithmic functions underlie such aids to computation as the slide rule and tables of logarithms. Some of the most basic integrals, such as the integrals of $1/x$ and $\tan x$, are functions that involve logarithms.

THEOREM 1

If $0 < a$ and $a \neq 1$, the function $x = \log_a y$ is defined and continuous for y in the interval $(0, \infty)$.

We skip the proof. $\log_a y$ is left undefined when either $a \leq 0$, $a = 1$, or $y \leq 0$.

THEOREM 2

The function $x = \log_a y$ is increasing if $a > 1$ and decreasing if $a < 1$.

PROOF

Case 1 $a > 1$. Let $0 < b < c$. Then

$$a^{\log_a b} = b < c = a^{\log_a c}.$$

We cannot have $\log_a b \geq \log_a c$ because the inequality (v) for exponents would then give $b \geq c$. We conclude that

$$\log_a b < \log_a c.$$

Case 2 $a < 1$ is similar.

In Figure 8.2.1 we have graphs of $y = a^x$ for $a > 1$ and for $a < 1$, and graphs of the inverse functions $x = \log_a y$.

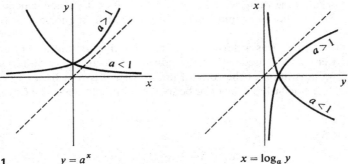

Figure 8.2.1 $y = a^x$ $x = \log_a y$

The rules for exponents can be turned around to give rules for logarithms.

RULES FOR LOGARITHMS

Let a, x, and y be positive real numbers, $a \neq 1$.

(i) $\log_a 1 = 0$, $\log_a a = 1$.

(ii) $\log_a (xy) = \log_a x + \log_a y$.

$\log_a \left(\dfrac{x}{y}\right) = \log_a x - \log_a y$.

(iii) $\log_a (x^r) = r \log_a x$.

These rules are useful because they reduce multiplication to addition and exponentiation to multiplication.

Let us make a quick check to see that these rules are correct for logarithms to the base 10. Here is a short table of common logarithms.

y	1	2	3	4	5	6	7	8	9	10
$\log_{10} y$	0	0.30	0.48	0.60	0.70	0.78	0.85	0.90	0.95	1

To find common logarithms of larger or smaller numbers we can use the rule

$$\log_{10} 10^n y = n + \log_{10} y.$$

We try a few cases to see if the answers agree, to one decimal place. We write $\log x$ for $\log_{10} x$ below.

$$
\begin{array}{ll}
\quad 2 & \log 2 \qquad \sim 0.30 \\
\times\, 3 & \log 3 \qquad \sim 0.48 \\
\hline
\quad 6 & \overline{\log 2 + \log 3 \sim 0.78} \\
 & \log 6 \qquad \sim 0.78
\end{array}
$$

$$
\begin{array}{ll}
\quad 700 & \log(7 \times 10^2) \qquad\qquad\qquad \sim \quad 2 + 0.85 \\
\times\, 0.3 & \log(3 \times 10^{-1}) \qquad\qquad\quad \sim -1 + 0.48 \\
\hline
\quad 210 & \overline{\log(7 \times 10^2) + \log(3 \times 10^{-1}) \sim \qquad 2.33} \\
\end{array}
$$

$$\log 210 \sim \log(2 \times 10^2) \qquad\qquad\qquad \sim \qquad 2.30$$

$$
\begin{array}{ll}
3^4 = 81 & \quad\log 3 \ \sim 0.48 \\
 & \ 4\log 3 \ \sim 1.92 \\
 & \log 81 \sim \log 80 \sim 1.90
\end{array}
$$

We could do the same thing with any other base. Base 10 is convenient because a number in decimal notation immediately can be put in the form $y = 10^n z$ where $1 \leq z \leq 10$.

The *slide rule* was a device for quickly looking up and adding logarithms. Slide rules were widely used before the advent of electronic calculators and give an interesting illustration of the rules of logarithms. If two ordinary rulers are slid together in slide rule fashion they can be used to compute the sum of two numbers, as shown in Figure 8.2.2.

In a slide rule, instead of marking off the distances 0, 1, 2, ..., 10, we mark off the distances

$$0 = \log 1, \log 2, \log 3, \ldots, \log 10.$$

The marks will be unevenly spaced, being closer together toward the right. We can then use the slide rule to compute the sum of two logarithms, and therefore the product of two numbers, as shown in Figure 8.2.3.

We know all the numbers are logarithms, so we can make a less cluttered slide rule by removing all the "log" symbols, as in Figure 8.2.4.

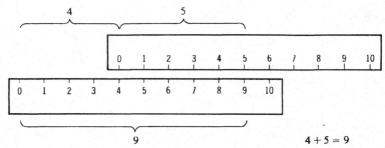

Figure 8.2.2

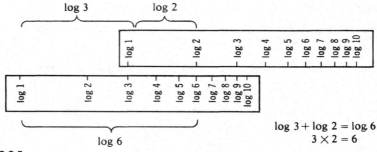

Figure 8.2.3

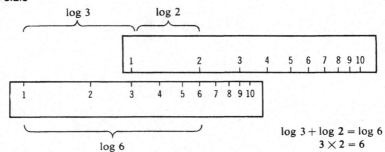

Figure 8.2.4

There is a simple relationship between logarithms with two different bases.

RULES FOR CHANGING BASES OF LOGARITHMS

Let a, b, and y be positive and a, b $\neq$ 1. Then

$$b^x = a^{x \log_a b}, \qquad \log_b y = \frac{\log_a y}{\log_a b}.$$

PROOF $a^{\log_a b} = b$, so

$$a^{x \log_a b} = (a^{\log_a b})^x = b^x,$$

$$(\log_a b)(\log_b y) = \log_a(b^{\log_b y}) = \log_a y,$$

whence $\qquad\qquad\qquad \log_b y = \dfrac{\log_a y}{\log_a b}.$

Setting $a = y$ we get the equation $\log_b a = 1/(\log_a b)$. If we hold the bases a and b fixed and let y vary, then the rule shows that $\log_a y$ and $\log_b y$ are proportional to each other, with the constant ratio

$$\frac{\log_a y}{\log_b y} = \log_a b.$$

Therefore a slide rule based on logarithms to the base 2, for example, would look exactly like a slide rule based on logarithms to the base 10 (common logarithms). If the same unit of length is used, all the distances would be multiplied by the constant factor

$$\log_2 10 = \frac{1}{\log_{10} 2} \sim 3.32.$$

So the slide rule would be similar but more than 3 times as big. Table 8.2.1 shows various logarithms with different bases.

Table 8.2.1

x	1	2	4	8	16	$\frac{1}{2}$	$\frac{1}{4}$	$\sqrt{2}$	$\frac{1}{2\sqrt{2}}$
$\log_2 x$	0	1	2	3	4	-1	-2	$\frac{1}{2}$	$-\frac{3}{2}$
$\log_4 x$	0	$\frac{1}{2}$	1	$1\frac{1}{2}$	2	$-\frac{1}{2}$	-1	$\frac{1}{4}$	$-\frac{3}{4}$
$\log_{1/2} x$	0	-1	-2	-3	-4	1	2	$-\frac{1}{2}$	$\frac{3}{2}$
$\log_{\sqrt{2}} x$	0	2	4	6	8	-2	-4	1	-3

Notice that for all $x > 0$,

$$\log_4 x = \frac{\log_2 x}{\log_2 4} = \frac{\log_2 x}{2},$$

$$\log_{1/2} x = \frac{\log_2 x}{\log_2 \frac{1}{2}} = -\log_2 x,$$

$$\log_{\sqrt{2}} x = \frac{\log_2 x}{\log_2 \sqrt{2}} = 2 \log_2 x.$$

Also, for each base a, $\log_a (1/x) = -\log_a x$.

EXAMPLE 1   Simplify the term $\log_a (\log_a(a^{a^x}))$.

$$\log_a (\log_a (a^{a^x})) = \log_a (a^x \log_a a) = \log_a (a^x) = x.$$

EXAMPLE 2   Express $\log_b \left(\dfrac{x^3 \sqrt{y}}{z}\right)$ in terms of $\log_b x$, $\log_b y$, and $\log_b z$.

$$\log_b \left(\frac{x^3 \sqrt{y}}{z}\right) = 3 \log_b x + \frac{1}{2} \log_b y - \log_b z.$$

EXAMPLE 3   Solve the equation below for x.

$$3^{x^2 - 2x} = \tfrac{1}{3}.$$

We take $\log_3$ of both sides of the equation.

$$(x^2 - 2x)\log_3 3 = \log_3 (3^{-1}),$$
$$x^2 - 2x = -1,$$
$$x^2 - 2x + 1 = 0,$$
$$x = 1.$$

The inequalities for exponents can be used to compute limits of logarithms.

EXAMPLE 4   Evaluate the limit $\lim\limits_{x \to \infty} \log_a x$, $a > 1$.

Let H be positive infinite. Then $0 = \log_a 1 < \log_a H$, so $\log_a H$ is positive. If $\log_a H$ is finite, say $\log_a H < n$, then

$$H = a^{\log_a H} < a^n,$$

which is impossible because H is infinite. Therefore $\log_a H$ is positive infinite, so

$$\lim_{x \to \infty} \log_a x = \infty.$$

PROBLEMS FOR SECTION 8.2

Simplify the following terms.

1	$a^{\log_a x}$	2	$\log_a (a^x)$
3	$\log_a(a^{-x^2})$	4	$a^{2 \log_a x}$
5	$a^{\log_a x - 2 \log_a y}$	6	$\log_a (\log_b(b^a))$

Express the following in terms of $\log_b x$, $\log_b y$, etc.

7	$\log_b(\sqrt[3]{x^2})$	8	$\log_b \left(\dfrac{xy}{z^3 w}\right)$
9	$\log_b \sqrt{xy}$	10	$\log_{1/b} x$

Evaluate the following.

11	$\log_3 9$	12	$\log_3\left(\dfrac{1}{27}\right)$
13	$\log_9 3$	14	$\log_{1/9} 27$

Solve the following equations for x.

15	$5^x = 3$	16	$x^5 = 3$
17	$2^{3x+5} = 8$	18	$\log_3 \sqrt{x} = 2$
19	$\log_x 5 = 3$	20	$\log_{10} x + \log_{10}(x + 3) = 1$
21	$2^{x^2+6} = 32^x$	22	$6^{x+1} = 7^x$
23	$\log_2 x = \log_3 x + 1$	24	$(\log_4 x)^2 + \log_4(x^{-3}) + 2 = 0$

25   Evaluate $\lim\limits_{x \to \infty} \log_a x$ when $0 < a < 1$.

26   Evaluate $\lim\limits_{x \to \infty} \log_x 2$.

27   Evaluate $\lim\limits_{x \to 0^+} \log_a x$ when $1 < a$.

28   Evaluate $\lim\limits_{x \to \infty} \log_{10}(\log_{10} x)$.

29   Evaluate $\lim\limits_{x \to \infty} \log_{10}\left(\dfrac{1}{3x + 1}\right)$.

☐ 30   Prove that for each $a > 0$, the function $y = \log_a x$ is continuous on $(0, \infty)$.

8.3 DERIVATIVES OF EXPONENTIAL FUNCTIONS AND THE NUMBER e

One of the most important constants in mathematics is the number e, whose value is approximately 2.71828. In this section we introduce e and show that it has the following remarkable properties.

(1)   The function $y = e^x$ is equal to its own derivative.

(2) e is the limit $\lim\limits_{x \to \infty}\left(1 + \dfrac{1}{x}\right)^x$.

Either property can be used as the definition of e. Because of property 1, it is convenient in the calculus to use exponential and logarithmic functions with the base e instead of 10. However, it is not at all easy to prove that such a number e exists. Before going into further detail we shall discuss these properties intuitively.

A function which equals its own derivative may be described as follows. Imagine a point moving on the (x, y) plane starting at $(0, 1)$. The point is equipped with a little man and a steering wheel which controls the direction of motion of the point. The man always steers directly away from the point $(x - 1, 0)$, so that

$$\frac{dy}{dx} = \frac{y - 0}{x - (x - 1)} = y.$$

Then the point will trace out a curve $y = f(x)$ which equals its own derivative, as in Figure 8.3.1.

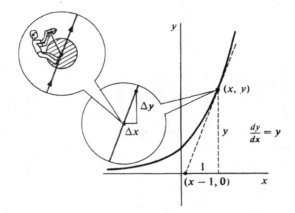

Figure 8.3.1

Another intuitive description is based on the example of the population growth function $y = a^t$. If the birth rate minus the death rate is equal to one, then the derivative of a^t is a^t, and a is the constant e. Imagine a country with one million people (one unit of population) at time $t = 0$ which has an annual birth rate of one million births per million people, and zero death rate. Then after one year the population will be approximately e million, or 2,718,282. (This high a growth rate is not recommended.)

The limit $e = \lim_{x \to \infty} (1 + 1/x)^x$ is suggested intuitively by the notion of continuously compounded interest. Suppose a bank gives interest at the annual rate of 100%, and we deposit one dollar in an account at time $t = 0$. If the interest is compounded annually, then after $t = 1$ year our account will have 2 dollars. If the interest is compounded quarterly (four times per year), then our account will grow to $1 + \frac{1}{4}$ dollars at time $t = \frac{1}{4}$, $(1 + \frac{1}{4})^2$ dollars at time $t = \frac{1}{2}$, and so on. After one year our account will have $(1 + \frac{1}{4})^4 \sim 2.44$ dollars. Similarly, if our account is compounded daily then after one year it will have $(1 + \frac{1}{365})^{365}$ dollars, and if it is compounded n times per year it will have $(1 + 1/n)^n$ dollars after one year.

Table 8.3.1 shows the value of $(1 + 1/n)^n$ for various values of n. (The last few values can be found with some hand calculators.)

Table 8.3.1

$n = 1$	$(1 + 1)^1 = 2$
$n = 2$	$(1 + \frac{1}{2})^2 = 2.25$
$n = 3$	$(1 + \frac{1}{3})^3 \sim 2.370$
$n = 4$	$(1 + \frac{1}{4})^4 \sim 2.441$
$n = 10$	$(1 + \frac{1}{10})^{10} \sim 2.594$
$n = 100$	$(1 + \frac{1}{100})^{100} \sim 2.705$
$n = 1000$	$(1 + \frac{1}{1000})^{1000} \sim 2.717$
$n = 10000$	$(1 + \frac{1}{10000})^{10000} \sim 2.718$

This table strongly suggests that the limit $e = \lim_{x \to \infty} (1 + 1/x)^x$ exists. A proof will be given later. Thus for H positive infinite,

$$\left(1 + \frac{1}{H}\right)^H \approx e.$$

If the interest is compounded H times per year, then in t years each dollar will grow to

$$\left(1 + \frac{1}{H}\right)^{Ht} = \left[\left(1 + \frac{1}{H}\right)^H\right]^t \approx e^t.$$

Thus if the 100% interest is continuously compounded, each dollar in the account grows to e^t dollars in t years. At the interest rate r, each dollar in a continuously compounded account will grow to e^{rt} dollars in t years. For more information, see Section 8.4. We now turn to a detailed discussion of e.

LEMMA

> *The limit* $\lim\limits_{x \to \infty} \left(1 + \dfrac{1}{x}\right)^x$ *exists.*

We shall save the proof of this lemma for the end of the section.

DEFINITION

> $e = \lim\limits_{x \to \infty} \left(1 + \dfrac{1}{x}\right)^x.$

As we have indicated before, e has the approximate value

$$e \sim 2.71828.$$

The function $y = e^x$ is called the *exponential function* and is sometimes written $y = \exp x$.

THEOREM 1

> e *is the unique real number such that*
> $$\frac{d(e^x)}{dx} = e^x.$$

PROOF Our plan is to show that whenever t and $t + \Delta t$ are finite and differ by a non-zero infinitesimal Δt,

$$\frac{e^{t + \Delta t} - e^t}{\Delta t} \approx e^t.$$

We may assume that t is the smaller of the two numbers, so that Δt is positive. By the rules of exponents,

(1) $$\frac{e^{t + \Delta t} - e^t}{\Delta t} = e^t \frac{e^{\Delta t} - 1}{\Delta t}.$$

Let $$b = \frac{e^{\Delta t} - 1}{\Delta t}.$$

(2)   Then $$b\,\Delta t = e^{\Delta t} - 1.$$

Since e^x is continuous and $e^0 = 1$, we see from Equation 2 that $b \, \Delta t$ is positive infinitesimal. Thus $H = 1/b \, \Delta t$ is positive infinite. From Equation 2,

$$\left(1 + \frac{1}{H}\right)^H = (1 + b \, \Delta t)^{1/b \, \Delta t} = (e^{\Delta t})^{1/b \, \Delta t} = e^{1/b}.$$

Taking standard parts,

$$e = st\left[\left(1 + \frac{1}{H}\right)^H\right] = st(e^{1/b}) = e^{1/st(b)}.$$

Therefore $st(b) = 1$, and by Equation 1,

$$\frac{e^{t + \Delta t} - e^t}{\Delta t} = e^t b \approx e^t.$$

We conclude that for real x,

$$\frac{d(e^x)}{dx} = e^x.$$

It remains to prove that e is the only real number with this property. Let a be any positive real number different from e, $a \neq e$. We may then differentiate a^x by the Chain Rule.

$$a^x = e^{x \log_e a}.$$

$$\frac{d(a^x)}{dx} = (\log_e a)e^{x \log_e a} = (\log_e a)a^x.$$

Since $a \neq e$, $\log_e a \neq 1$, so $(d(a^x))/dx \neq a^x$.

Since e^x is its own derivative, it is also its own antiderivative. We thus have a new differentiation formula and a new integration formula which should be memorized.

$$\frac{d(e^x)}{dx} = e^x, \qquad d(e^x) = e^x \, dx,$$

$$\int e^x \, dx = e^x + C.$$

We are now ready to plot the graph of the exponential curve $y = e^x$. Here is a short table. It gives both the value y and the slope y', because $y = y' = e^x$.

x	-2	-1	0	1	2
e^x	$1/e^2 \sim 0.14$	$1/e \sim 0.37$	1	$e \sim 2.7$	$e^2 \sim 7.3$

The number e^x is always positive, and y, y', and y'' all equal e^x. From this we can draw three conclusions.

$$y = e^x > 0 \qquad \text{the curve lies above the } x\text{-axis,}$$
$$y' = e^x > 0 \qquad \text{increasing,}$$
$$y'' = e^x > 0 \qquad \text{concave upward.}$$

If H is positive infinite, then by Rule (vii),

$$e^H \geq 1 + H(e - 1).$$

So $\qquad\qquad e^H$ is infinite, $\qquad e^{-H} = 1/e^H$ is infinitesimal.

Therefore, $\qquad\qquad\qquad \lim\limits_{x \to \infty} e^x = \infty, \qquad \lim\limits_{x \to -\infty} e^x = 0.$

We use this information to draw the curve in Figure 8.3.2.

EXAMPLE 1   Given $y = e^{\sin x}$, find d^2y/dx^2.

$$\frac{dy}{dx} = e^{\sin x} \cos x,$$

$$\frac{d^2y}{dx^2} = e^{\sin x} \cos^2 x - e^{\sin x} \sin x.$$

EXAMPLE 2   Find the area under the curve

$$y = \frac{e^{\arctan x}}{1 + x^2}, \qquad 0 \le x \le 1.$$

Let $u = \arctan x$, $du = \dfrac{1}{1 + x^2}\,dx$.

Then $\qquad \displaystyle\int_0^1 \frac{e^{\arctan x}}{1 + x^2}\,dx = \int_0^{\pi/4} e^u\,du = e^u \Big]_0^{\pi/4} = e^{\pi/4} - 1.$

EXAMPLE 3   Find $d(a^x)/dx$. We use the formula

$$a = e^{\log_e a}, \qquad a^x = e^{x \log_e a}.$$

Put $u = x \log_e a$. Then $a^x = e^u$, so

$$\frac{d(a^x)}{dx} = \frac{d(e^u)}{du}\frac{du}{dx} = e^u \frac{du}{dx} = (\log_e a) a^x,$$

$$\frac{d(a^x)}{dx} = (\log_e a) a^x.$$

This example shows that the derivative of a^x is equal to the constant $\log_e a$ times a^x itself. Figure 8.3.3 shows the graph of $y = a^x$ for various values of $a > 0$.

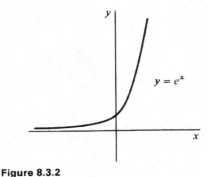

Figure 8.3.2   **Figure 8.3.3**

The slope of the curve $y = a^x$ at $x = 0$ is always equal to $\log_e a$. For all values of $a > 0$, a^x is positive for all x, so the derivative has the same sign as $\log_e a$. The three possibilities are shown:

$a > 1$	$\log_e a > 0$	a^x increasing for all x
$a = 1$	$\log_e a = 0$	$a^x = 1$ for all x
$0 < a < 1$	$\log_e a < 0$	a^x decreasing for all x

We conclude this section with the proof of the lemma that $\lim_{x \to \infty} (1 + 1/x)^x$ exists. We use the following formula from elementary algebra.

GEOMETRIC SERIES FORMULA

If $b \neq 1$, then $\qquad (1 + b + b^2 + \cdots + b^n) = \dfrac{b^{n+1} - 1}{b - 1}.$

This formula is proved by multiplying

$(1 + b + b^2 + \cdots + b^n)(b - 1)$

$\qquad = (b + b^2 + \cdots + b^n + b^{n+1}) - (1 + b + \cdots + b^{n-1} + b^n)$

$\qquad = b^{n+1} - 1.$

PROOF OF THE LEMMA The function $y = 2^t$ is continuous and positive. Therefore the integral

$$c = \int_0^1 2^t \, dt$$

is a positive real number. Our plan is to use the fact that the Riemann sums approach c to show that $(1 + 1/x)^x$ approaches the limit 2^c.

Let H be positive infinite. We wish to prove that

$$\left(1 + \frac{1}{H}\right)^H \approx 2^c.$$

It is easier to work with the logarithm

$$\log_2 \left[\left(1 + \frac{1}{H}\right)^H\right] = H \log_2 \left(1 + \frac{1}{H}\right).$$

Let $\qquad\qquad\qquad\qquad \Delta t = \log_2 \left(1 + \frac{1}{H}\right).$

Δt is positive and is infinitesimal because

$$\Delta t \approx \log_2 1 = 0.$$

Moreover, $\qquad\qquad 2^{\Delta t} = 1 + \frac{1}{H}, \qquad H = \frac{1}{2^{\Delta t} - 1}, \qquad$ so

(3)
$$H \log_2 \left(1 + \frac{1}{H} \right) = \frac{\Delta t}{2^{\Delta t} - 1}.$$

Let us form the Riemann sum

$$\sum_0^1 2^t \, \Delta t = (1 + 2^{\Delta t} + 2^{2 \, \Delta t} + \cdots + 2^{(K-1)\Delta t}) \, \Delta t.$$

For simplicity suppose Δt evenly divides 1, so $K \, \Delta t = 1$. By the Geometric Series Formula,

$$\sum_0^1 2^t \, \Delta t = \frac{2^{K \, \Delta t} - 1}{2^{\Delta t} - 1} \, \Delta t = \frac{2 - 1}{2^{\Delta t} - 1} \, \Delta t = \frac{\Delta t}{2^{\Delta t} - 1}.$$

By Equation 3, $\qquad\qquad \sum_0^1 2^t \, \Delta t = H \log_2 \left(1 + \frac{1}{H} \right)$

Taking standard parts we have

$$c \approx H \log_2 \left(1 + \frac{1}{H} \right)$$

Finally, $\qquad\qquad\qquad 2^c \approx \left(1 + \frac{1}{H} \right)^H$

The proof is the same when Δt does not evenly divide 1, except that $K \, \Delta t$ is infinitely close to 1 instead of equal to 1. Therefore

$$\lim_{x \to \infty} \left(1 + \frac{1}{x} \right)^x = 2^c.$$

We remark that in the above proof we could have used any other positive real number in place of 2. Notice that $2^c = e$, so the constant $c = \int_0^1 2^t \, dt$ is just $\log_2 e$.

PROBLEMS FOR SECTION 8.3

In Problems 1–12 find the derivative.

1	$y = e^{3x+4}$	2	$y = xe^x$
3	$y = 4^{-x}$	4	$s = 3^{t+1}$
5	$u = \sin(e^t)$	6	$y = e^{\arcsin x}$
7	$u = 2^{(t^2)}$	8	$y = e^{1/x}$
9	$u = e^{(e^t)}$	10	$y = (1 + e^t)^{-2}$
11	$y = 3^{\sqrt{x}}$	12	$y = \sqrt{3^x - 2^x}$

13   Find $\dfrac{dy}{dx}$ if $\cos y = e^{x+y}$.   14   Find $\dfrac{dy}{dx}$ if $x + y = e^{xy}$.

15   Find $\dfrac{dy}{dx}$ if $x = \dfrac{e^t}{t}$, $y = \sqrt{e^t}$.

16   Find $\dfrac{dy}{dx}$ if $x = e^{-t^2}$, $y = \sqrt{1 - t^2}$.

In Problems 17–26, evaluate the limit.

17    $\lim\limits_{t \to \infty} e^t/t$ 18 $\lim\limits_{t \to \infty} e^t/t^n$, n a fixed positive integer.

19    $\lim\limits_{t \to 0} \dfrac{e^t - 1}{\sin t}$ 20 $\lim\limits_{x \to \infty} e^x - x^2$

21    $\lim\limits_{x \to -\infty} e^x - x^2$ 22 $\lim\limits_{x \to 0} \dfrac{3^x - 2^x}{x}$

23    $\lim\limits_{x \to \infty} (1 + 1/x)^{cx}$ 24 $\lim\limits_{x \to \infty} (1 - 1/x)^x$ *Hint:* Let $u = -x$.

25    $\lim\limits_{x \to \infty} (1 + c/x)^x$ 26 $\lim\limits_{t \to 0} (1 + t)^{1/t}$

In Problems 27–34 use the first and second derivatives and limits to sketch the curve.

27    $y = 2^x$ 28 $y = 2^{-x}$

29    $y = xe^x$ 30 $y = e^{-x^2}$

31    $y = e^{x^3}$ 32 $y = e^x + e^{-x}$

33    $y = \dfrac{1}{1 + e^x}$ 34 $y = \dfrac{1}{1 + e^{-x}}$

In Problems 35–50 evaluate the integral.

35    $\displaystyle\int e^{2x}\, dx$ 36 $\displaystyle\int \dfrac{dx}{e^{3x}}$

37    $\displaystyle\int xe^{-x^2}\, dx$ 38 $\displaystyle\int 2^{-x}\, dx$

39    $\displaystyle\int e^{2x}\sqrt{1 + e^{2x}}\, dx$ 40 $\displaystyle\int \dfrac{e^x}{1 + e^{2x}}\, dx$ *Hint:* Try $u = e^x$.

41    $\displaystyle\int xe^x\, dx$ *Hint:* Use integration by parts.

42    $\displaystyle\int x^2 e^x\, dx$ 43 $\displaystyle\int e^x \sin x\, dx$

44    $\displaystyle\int e^{-x}\cos x\, dx$ 45 $\displaystyle\int_0^2 e^{5x}\, dx$

46    $\displaystyle\int_{-2}^2 e^{-x}\, dx$ 47 $\displaystyle\int_0^\infty e^x\, dx$

48    $\displaystyle\int_0^\infty e^{-rx}\, dx$ 49 $\displaystyle\int_0^\infty xe^{-rx}\, dx$

50    $\displaystyle\int_0^\infty x^2 e^{-rx}\, dx$

51   Find the volume generated by rotating the region under the curve $y = e^x, 0 \leq x \leq 1$, about (a) the x-axis, (b) the y-axis.

52   Find the volume generated by rotating the region under the curve $y = e^{-x}, 0 \leq x < \infty$, about (a) the x-axis, (b) the y-axis.

53   Find the length of the curve $x = e^t \cos t$, $y = e^t \sin t, 0 \leq t \leq 2\pi$.

54   A snail grows in the shape of an exponential spiral, $r = e^{a\theta}$ in polar coordinates.
     (a)   Find $\tan\psi$, the angle between a radius and the curve at θ.
     (b)   Sketch the curve for $a = 1$ and $a = 1/\sqrt{3}$.
     (c)   Find the length of the curve where $-\infty < \theta \leq b$.
     (d)   Find the area of the snail where $-\infty < \theta \leq b$. (To avoid overlap, one should integrate from $b - 2\pi$ to b.)

8.4 SOME USES OF EXPONENTIAL FUNCTIONS

In this section we shall discuss some functions involving exponentials which come up in physical and social sciences.

The hyperbolic functions are analogous to the trigonometric functions and are useful in physics and engineering.

The *hyperbolic sine*, sinh, and the *hyperbolic cosine*, cosh, are defined as follows.

$$\sinh x = \frac{e^x - e^{-x}}{2}, \qquad \cosh x = \frac{e^x + e^{-x}}{2}.$$

A chain fixed at both ends will hang in the shape of the curve $y = \cosh x$ (the *catenary*). The graphs of $y = \sinh x$ and $y = \cosh x$ are shown in Figure 8.4.1.

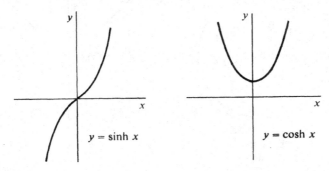

$y = \sinh x$ $y = \cosh x$

Figure 8.4.1

The hyperbolic functions have identities which are similar to, but different from, the trigonometric identities. We list some of them in Table 8.4.1.

Table 8.4.1

Trigonometric	Hyperbolic
$\sin^2 x + \cos^2 x = 1$	$\cosh^2 x - \sinh^2 x = 1$
$d(\sin x) = \cos x \, dx$	$d(\sinh x) = \cosh x \, dx$
$d(\cos x) = -\sin x \, dx$	$d(\cosh x) = \sinh x \, dx$
$\int \sin x \, dx = -\cos x + C$	$\int \sinh x \, dx = \cosh x + C$
$\int \cos x \, dx = \sin x + C$	$\int \cosh x \, dx = \sinh x + C$

These hyperbolic identities are easily verified. For example,

$$d(\sinh x) = d\left(\frac{e^x - e^{-x}}{2}\right) = \frac{d(e^x) - d(e^{-x})}{2}$$

$$= \left(\frac{e^x - (-e^{-x})}{2}\right) dx = \cosh x \, dx.$$

Notice that $\cosh x + \sinh x = \dfrac{e^x + e^{-x} + e^x - e^{-x}}{2} = e^x,$

$\cosh x - \sinh x = \dfrac{e^x + e^{-x} - e^x + e^{-x}}{2} = e^{-x}.$

When we multiply these we get the identity $\cosh^2 x - \sinh^2 x = 1$.

The other hyperbolic functions are defined like the other trigonometric functions,

$$\tanh x = \frac{\sinh x}{\cosh x}, \qquad \coth x = \frac{\cosh x}{\sinh x},$$

$$\operatorname{sech} x = \frac{1}{\cosh x}, \qquad \operatorname{csch} x = \frac{1}{\sinh x}.$$

The hyperbolic functions are related to the *unit hyperbola* $x^2 - y^2 = 1$ in the same way that the trigonometric functions are related to the unit circle $x^2 + y^2 = 1$ (Figure 8.4.2).

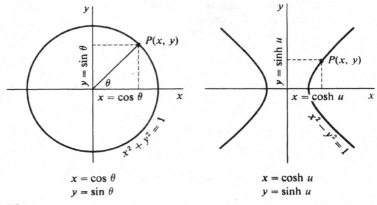

$x = \cos \theta$
$y = \sin \theta$

$x = \cosh u$
$y = \sinh u$

Figure 8.4.2

If we put

$$x = \cos\theta, \qquad y = \sin\theta,$$

we have

$$x^2 + y^2 = \cos^2\theta + \sin^2\theta = 1,$$

so the point $P(x, y)$ is on the unit circle $x^2 + y^2 = 1$.

On the other hand if we put

$$x = \cosh u, \qquad y = \sinh u,$$

we have

$$x^2 - y^2 = \cosh^2 u - \sinh^2 u = 1,$$

so the point $P(x, y)$ is on the unit hyperbola $x^2 - y^2 = 1$.

The hyperbolic functions differ from the trigonometric functions in some important ways. The most striking difference is that the hyperbolic functions are not periodic. In fact both $\sinh x$ and $\cosh x$ have infinite limits as x becomes infinite:

$$\lim_{x \to -\infty} \sinh x = -\infty, \qquad \lim_{x \to \infty} \sinh x = \infty,$$

$$\lim_{x \to -\infty} \cosh x = \infty, \qquad \lim_{x \to \infty} \cosh x = \infty.$$

Let us verify the last limit. If H is positive infinite, then

$$\cosh H = \frac{e^H + e^{-H}}{2} = \frac{1}{2} e^H + \frac{1}{2} e^{-H}$$

is the sum of a positive infinite number $\frac{1}{2} e^H$ and an infinitesimal $\frac{1}{2} e^{-H}$ and hence is positive infinite. Therefore $\lim_{x \to \infty} \cosh x = \infty$.

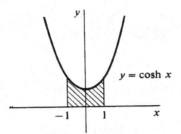

Figure 8.4.3

EXAMPLE 1   Find the area of the region under the catenary $y = \cosh x$ from $x = -1$ to $x = 1$, shown in Figure 8.4.3.

$$A = \int_{-1}^{1} \cosh x \, dx = \sinh x \Big]_{-1}^{1}$$
$$= \sinh 1 - \sinh(-1)$$
$$= \frac{e - e^{-1}}{2} - \frac{e^{-1} - e}{2} = e - \frac{1}{e}.$$

We now give an application of the exponential function to economics. Suppose money in the bank earns interest at the annual rate r, compounded continuously. (To keep our problem simple we assume r is constant with time, even though actual interest rates fluctuate with time.) Here is the problem: A person receives money continuously at the rate of $f(t)$ dollars per year and puts the money in the bank as he receives it. How much money will be accumulated during the time $a \le t \le b$? This is an integration problem.

We first consider a simpler problem. If a person puts y dollars in the bank at time $t = a$, how much will he have at time $t = b$? The answer is

$$ye^{r(b-a)} \text{ dollars.}$$

JUSTIFICATION   Divide the time interval $[a, b]$ into subintervals of infinitesimal length $\Delta t > 0$,

$$a, a + \Delta t, a + 2 \Delta t, \ldots, a + H \Delta t = b,$$

where $\Delta t = (b - a)/H$.

If the interest is compounded at time intervals of Δt, the account at the above times will be

$$y, y(1 + r \Delta t), y(1 + r \Delta t)^2, \ldots, y(1 + r \Delta t)^H.$$

Let $K = 1/(r \Delta t)$. Then $H = (b - a)/\Delta t = r(b - a)K$. At time b the account is

$$y(1 + r \Delta t)^H = y\left(1 + \frac{1}{K}\right)^H = y\left(1 + \frac{1}{K}\right)^{Kr(b-a)}$$

Since H, and hence K, is positive infinite,

$$\left(1 + \frac{1}{K}\right)^K \approx e, \qquad y\left(1 + \frac{1}{K}\right)^{Kr(b-a)} \approx ye^{r(b-a)}.$$

Thus when the interest is compounded infinitely often the account at time b

is infinitely close to $ye^{r(b-a)}$. So when the interest is compounded continuously the account at time b is

$$ye^{r(b-a)}.$$

Now we return to the original problem.

CAPITAL ACCUMULATION FORMULA

If money is received continuously at the rate of $f(t)$ dollars per year and earns interest at the annual rate r, the amount of capital accumulated between times $t = a$ and $t = b$ is

$$C = \int_a^b f(t)e^{r(b-t)}\, dt.$$

JUSTIFICATION   During an infinitesimal time interval $[t, t + \Delta t]$, of length Δt, the amount received is

$$\Delta y \approx f(t)\,\Delta t \qquad \text{(compared to } \Delta t\text{).}$$

This amount Δy will earn interest from time t to b, so its contribution to the total capital at time b will be

$$\Delta C \approx \Delta y e^{r(b-t)} = f(t)e^{r(b-t)}\,\Delta t \qquad \text{(compared to } \Delta t\text{).}$$

Therefore by the Infinite Sum Theorem, the total capital accumulated from $t = a$ to $t = b$ is the integral

$$C = \int_a^b f(t)e^{r(b-t)}\, dt.$$

EXAMPLE 2   If money is received at the rate $f(t) = 2t$ dollars per year, and earns interest at the annual rate of 7%, how much will be accumulated from times $t = 0$ to $t = 10$?

The formula gives

$$C = \int_0^{10} 2te^{0.07(10-t)}\, dt.$$

We first find the indefinite integral.

$$\int 2t\, e^{0.07(10-t)}\, dt = \int 2te^{0.7}\, e^{-0.07t}\, dt$$

$$= 2e^{0.7} \int te^{-0.07t}\, dt.$$

Let $u = -0.07t$, $du = -0.07\, dt$. Then

$$\int 2te^{0.07(10-t)}\, dt = 2e^{0.7} \int \frac{u}{-0.07}\, e^u\, \frac{1}{-0.07}\, du$$

$$= 2e^{0.7}(0.07)^{-2} \int ue^u\, du.$$

Using integration by parts,

$$\int ue^u \, du = ue^u - \int e^u \, du = ue^u - e^u + \text{Constant}.$$

Therefore $\quad \int 2te^{0.07(10-t)} \, dt = 2e^{0.7}(0.07)^{-2}(ue^u - e^u) + \text{Constant}.$

When $t = 0$, $u = 0$ and when $t = 10$, $u = -0.7$. Thus

$$
\begin{aligned}
C &= [2e^{0.7}(0.07)^{-2}(ue^u - e^u)]_0^{-0.7} \\
&= 2e^{0.7}(0.07)^{-2}(-0.7e^{-0.7} - e^{-0.7} + e^0) \\
&= 2(0.07)^{-2}(e^{0.7} - 1.7) \sim 128.08.
\end{aligned}
$$

The answer is \$128.08.

Notice that if the money were placed under a mattress and earned no interest, the capital accumulated between times $t = 0$ and $t = 10$ would be

$$\int_0^{10} 2t \, dt = \$100.$$

The formula for capital accumulation also has a meaning when $f(t)$ is negative part or all of the time. A negative value of $f(t)$ means that money is being paid out instead of received. When $f(t)$ is negative, money must be either withdrawn from the bank account or else borrowed from the bank at interest rate r. The formula

$$C = \int_a^b f(t) \, e^{r(b-t)} \, dt$$

then represents the net gain or loss of capital from times $t = a$ to $t = b$, provided that the bank pays interest on savings and charges interest on loans at the same rate r.

PROBLEMS FOR SECTION 8.4

In Problems 1–4, find the derivative.

1 $y = \sinh(3x)$

2 $y = \cosh^2 x$

3 $y = \text{sech}\, x$

4 $y = \tanh x$

5 Evaluate $\lim\limits_{x \to \infty} \tanh x.$

6 Evaluate $\lim\limits_{x \to 0} \dfrac{\sinh x}{x}.$

7 Evaluate $\lim\limits_{x \to 0} \dfrac{1 - \cosh x}{x}.$

8 Evaluate $\lim\limits_{x \to \infty} (\cosh x - \sinh x).$

In Problems 9–12 use the first and second derivatives to sketch the curve.

9 $y = \tanh x$

10 $y = \coth x$

11 $y = \text{sech}\, x$

12 $y = \text{csch}\, x$

In Problems 13–20 evaluate the integral.

13 $\displaystyle\int \sinh x \cosh x \, dx$

14 $\displaystyle\int x^{-2} \cosh(1/x) \, dx$

15 $\displaystyle\int x \sinh x \, dx$ **16** $\displaystyle\int \sinh^2 x \, dx$

17 $\displaystyle\int x \cosh^2 x \, dx$ **18** $\displaystyle\int_0^1 \sinh x \, dx$

19 $\displaystyle\int_{-\infty}^{\infty} \cosh x \, dx$ **20** $\displaystyle\int_{-\infty}^{\infty} \operatorname{sech}^2 x \, dx$

21 Prove the identity $\tanh^2 x + \operatorname{sech}^2 x = 1$.

22 Find the length of the curve $y = \cosh x$, $-1 \le x \le 1$.

23 Find the volume of the solid formed by rotating the curve $y = \cosh x$, $0 \le x \le 1$, about (a) the x-axis, (b) the y-axis.

24 Find the surface area generated by rotating the curve $y = \cosh x$, $0 \le x \le 1$, about (a) the x-axis, (b) the y-axis.

25 Money is received at the constant rate of 5000 dollars per year and earns interest at the annual rate of 10%. How much is accumulated in 20 years?

26 Money is received at the rate of $20 - 2t$ dollars per year and earns interest at the annual rate of 8%. How much capital is accumulated between times $t = 0$ and $t = 10$?

27 A firm initially loses (and borrows) money but later makes a profit, and its net rate of profit is

$$f(t) = 10^6(t - 1)$$

dollars per year. All interest rates are at 10%. Starting at $t = 0$, find the net capital accumulated after (a) 2 years, (b) 3 years.

28 A firm in a fluctuating economy receives or loses money at the rate $f(t) = \sin t$. Find the net capital accumulated between times $t = 0$ and $t = 2\pi$ if all interest is at 10%. .

☐ **29** The *present value* of z dollars t years in the future is the quantity $y = ze^{-rt}$, where r is the interest rate. This is because $y = ze^{-rt}$ dollars today will grow to $ye^{rt} = z$ dollars in t years. Use the Infinite Sum Theorem to justify the following formula for the present value V of all future profits where $f(t)$ is the profit per unit time.

$$V = \int_0^{\infty} f(t)e^{-rt} \, dt.$$

8.5 NATURAL LOGARITHMS

DEFINITION

Given $x > 0$, the **natural logarithm** of x is defined as the logarithm of x to the base e. The symbol ln is used for natural logarithm; thus

$$\ln x = \log_e x,$$

and
$$y = \ln x \text{ if and only if } x = e^y.$$

Natural logarithms are particularly convenient for problems involving derivatives and integrals. When we write $\ln x$ instead of $\log_e x$, the rules for logarithms take the following form:

(i) $\ln 1 = 0,$ $\ln e = 1.$

(ii) $\ln(xy) = \ln x + \ln y$,

$\ln(x/y) = \ln x - \ln y$.

(iii) $\ln(x^r) = r \ln x$.

The rules for changing the base become

$$b^x = e^{x \ln b}, \qquad \log_b y = \frac{\ln y}{\ln b}.$$

Using the above equations, the formulas for the derivative and integral of b^x take the form

$$\frac{d(b^x)}{dx} = (\ln b)b^x,$$

$$\int b^x \, dx = \frac{1}{\ln b} b^x + C, \qquad (b \neq 1).$$

Recall the Power Rule for integrals,

$$\int x^n \, dx = \frac{x^{n+1}}{n+1} + C, \qquad n \neq -1.$$

It shows how to integrate x^n for $n \neq -1$. Now, at long last, we are about to determine the integral of x^{-1}. It turns out to be the natural logarithm of x.

THEOREM 1

(i)   *On the interval* $(0, \infty)$,

$$d(\ln x) = \frac{1}{x}, \qquad \int \frac{1}{x} dx = \ln x + C.$$

(ii)   *On both the intervals* $(-\infty, 0)$ *and* $(0, \infty)$,

$$d(\ln|x|) = \frac{1}{x} dx, \qquad \int \frac{1}{x} dx = \ln|x| + C.$$

PROOF   (i)   Let $y = \ln x$. Then $x = e^y$, $dx/dy = e^y$. By the Inverse Function Theorem,

$$\frac{dy}{dx} = \frac{1}{dx/dy} = \frac{1}{e^y} = \frac{1}{x}.$$

(ii)   Let $x < 0$ and let $y = \ln|x|$. For $x < 0$, $|x| = -x$ so

$$\frac{d|x|}{dx} = -1.$$

Then $\dfrac{d(\ln|x|)}{dx} = \dfrac{d(\ln|x|)}{d|x|} \dfrac{d|x|}{dx} = \dfrac{1}{|x|}(-1) = \dfrac{1}{-x}(-1) = \dfrac{1}{x}.$

In the above theorem we had to be careful because $1/x$ is defined for all $x \neq 0$ but $\ln x$ is only defined for $x > 0$. Thus on the negative interval $(-\infty, 0)$ the antiderivative of $1/x$ cannot be $\ln x$. Since $|x| > 0$ for both positive and negative x, $\ln|x|$ is defined for all $x \neq 0$. Fortunately, it turns out to be the antiderivative of $1/x$ in all

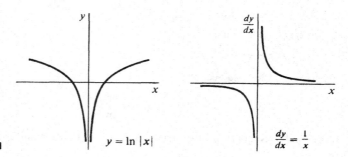

Figure 8.5.1 $y = \ln |x|$ $\dfrac{dy}{dx} = \dfrac{1}{x}$

cases. For $x > 0$, $1/x > 0$ and $\ln|x|$ is increasing, while for $x < 0$, $1/x < 0$ and $\ln|x|$ is decreasing (see Figure 8.5.1).

We now evaluate the integral of $\ln x$. This integral can be found in the table at the end of the book.

THEOREM 2

$$\int \ln x \, dx = x \ln x - x + C.$$

PROOF We use integration by parts. Let

$$u = \ln x, \quad du = \frac{1}{x} dx, \quad dv = dx, \quad v = x.$$

Then

$$\int \ln x \, dx = uv - \int v \, du$$

$$= x \ln x - \int \frac{x}{x} dx$$

$$= x \ln x - x + C.$$

Let us study the graph of $y = \ln x$. Here are a few values of y and dy/dx.

x	$\frac{1}{4}$	$\frac{1}{2}$	1	2	4
$y = \ln x$	-1.4	-0.7	0	0.7	1.4
$dy/dx = 1/x$	4	2	1	$\frac{1}{2}$	$\frac{1}{4}$

The limits as $x \to 0^+$ and $x \to \infty$ (see Example 4, Section 8.2) are:

$$\lim_{x \to 0^+} (\ln x) = -\infty, \qquad \lim_{x \to \infty} (\ln x) = \infty,$$

$$\lim_{x \to 0^+} (1/x) = \infty, \qquad \lim_{x \to \infty} (1/x) = 0.$$

From the sign of dy/dx and d^2y/dx^2 we get the following information.

$$\frac{dy}{dx} = \frac{1}{x} > 0, \qquad \text{increasing}$$

$$\frac{d^2y}{dx^2} = \frac{-1}{x^2} < 0, \qquad \text{concave downward.}$$

We use this information to draw the curve in Figure 8.5.2.

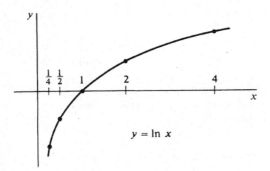

Figure 8.5.2

There are two bases for logarithms which are especially useful for different purposes, base 10 and base e. The student should be careful not to confuse the two.

Table 8.5.1

Name	Common Logarithms	Natural Logarithms
base	base 10	base e
symbols	$\log_{10} x, \log x$	$\log_e x, \ln x$
use	numerical computation	derivatives and integrals

To pass back and forth between common and natural logarithm we need the constants

$$\log_{10} e \sim 0.4343, \qquad \ln 10 \sim 2.3026.$$

Then

$$\log_{10} x = \frac{\ln x}{\ln 10}, \qquad \ln x \sim 2.3026 \log_{10} x$$

and

$$\ln x = \frac{\log_{10} x}{\log_{10} e}, \qquad \log_{10} x \sim 0.4343 \ln x.$$

Warning: Do not make the mistake of using common logarithms instead of natural logarithms in differentiating and integrating.

EXAMPLE 1 Find $\dfrac{d(\log_{10} x)}{dx}$.

Right: $\dfrac{d(\log_{10} x)}{dx} \sim \dfrac{d(0.4343 \ln x)}{dx} = \dfrac{0.4343}{x}$.

Wrong: $\dfrac{d(\log_{10} x)}{dx} = \dfrac{1}{x}$.

EXAMPLE 2 Find $\displaystyle\int_1^{10} \frac{1}{x}\,dx$.

Right: $\displaystyle\int_1^{10} \frac{1}{x}\,dx = \ln x\,\Big]_1^{10} = \ln 10 - \ln 1 \sim 2.3026$.

Wrong: $\displaystyle\int_1^{10} \frac{1}{x}\,dx = \log_{10} x\,\Big]_1^{10} = \log_{10} 10 - \log_{10} 1 = 1$.

EXAMPLE 3 Find $\displaystyle\int_{-e}^{-1} \frac{1}{x}\, dx$.

$$\int_{-e}^{-1} \frac{1}{x}\, dx = \ln|x|\ \Big]_{-e}^{-1} = \ln 1 - \ln e = -1.$$

Note that $\ln x$ is undefined at -1 and $-e$ but $\ln|x|$ is defined there. The absolute value sign is put in when integrating $1/x$ and removed when differentiating $\ln|x|$.

EXAMPLE 4 Find dy/dx where $y = \ln[(3 - 2x)^2]$.

We have $(3 - 2x)^2 = |3 - 2x|^2$, and by the rules of logarithms,

$$y = 2\ln|3 - 2x|.$$

By Theorem 1, $\displaystyle\frac{dy}{dx} = \frac{2}{3 - 2x}\ \frac{d(3 - 2x)}{dx} = \frac{-4}{3 - 2x}.$

This answer is correct when $3 - 2x$ is negative as well as positive.

EXAMPLE 5 Find $d(\log_a x)/dx$.

$$\log_a x = \frac{\ln x}{\ln a},$$

$$\frac{d(\log_a x)}{dx} = \frac{1}{\ln a}\ \frac{d(\ln x)}{dx} = \frac{1}{x \ln a}.$$

EXAMPLE 6 Find $\displaystyle\int \frac{1}{2x - 5}\, dx$. Let $u = 2x - 5$, $du = 2\, dx$.

$$\int \frac{1}{2x - 5}\, dx = \frac{1}{2} \int \frac{1}{u}\, du = \frac{1}{2}\ln|u| + C = \frac{1}{2}\ln|2x - 5| + C.$$

EXAMPLE 7 Find the improper integral $\displaystyle\int_{1}^{\infty} \frac{1}{x}\, dx$.

$$\int_{1}^{\infty} \frac{1}{x}\, dx = \lim_{b\to\infty} \int_{1}^{b} \frac{1}{x}\, dx = \lim_{b\to\infty}\left(\ln x\ \Big]_{1}^{b}\right) = \lim_{b\to\infty} \ln b = \infty.$$

Thus the region under the curve $y = 1/x$ from 1 to ∞, shown in Figure 8.5.3, has infinite area.

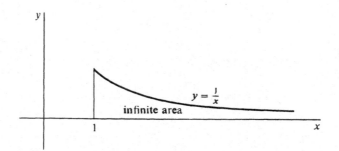

Figure 8.5.3

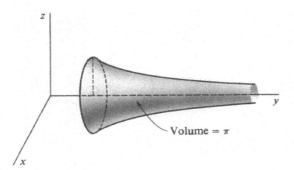

Figure 8.5.4

EXAMPLE 8   The region R under the curve $y = 1/x$ from 1 to ∞ is rotated about the x-axis, forming a solid of revolution. Find the volume of this solid (Figure 8.5.4).

The volume is given by the improper integral

$$V = \int_1^\infty \pi\left(\frac{1}{x}\right)^2 dx = \pi \int_1^\infty x^{-2}\, dx.$$

Then
$$V = \pi \lim_{b \to \infty} \int_1^b x^{-2}\, dx = \pi \lim_{b \to \infty}\left(-\frac{1}{x}\Big]_1^b\right) = \pi \lim_{b \to \infty}\left(1 - \frac{1}{b}\right) = \pi.$$

Thus the solid has volume π.

In the last two examples a region of infinite area was rotated about the x-axis to form a solid of finite volume. We saw another example of this kind in Section 6.7 on improper integrals.

PROBLEMS FOR SECTION 8.5

In Problems 1–12 find the derivatives.

1	$y = (\ln x)^3$	2	$y = \ln|3x + 4|$
3	$y = \ln(\cos x)$	4	$y = \ln(x^4 + x - 1)$
5	$s = t \ln t - t$	6	$s = \ln(t^{-1})$
7	$s = \ln(\sqrt[t]{t})$	8	$y = \ln(\ln x)$
9	$y = \log_2(3x)$	10	$y = \log_x a$
11	$z = \ln(y\sqrt{3y + 1})$	12	$z = \ln\left(\dfrac{(y^4 + 1)^2}{(y - 1)^3}\right)$

13   Find dy/dx where $x = \ln(xy)$.

14   Find dy/dx where $y = \ln(x^2 y)$.

15   Find dy/dx where $y = \ln(x + y)$.

In Problems 16–25 evaluate the limit.

16	$\displaystyle\lim_{x \to \infty} \frac{\ln x}{x}$	17	$\displaystyle\lim_{x \to \infty} \frac{(\ln x)^2}{\sqrt{x}}$

18 $\displaystyle\lim_{t\to 1^-} \frac{\ln t}{\sqrt{1-t}}$

19 $\displaystyle\lim_{x\to 0^+} x\ln x$

20 $\displaystyle\lim_{t\to\infty} \ln(\ln t)$

21 $\displaystyle\lim_{t\to 0} \frac{a^t - 1}{t}, \quad a > 0$

22 $\displaystyle\lim_{x\to\infty} x(a^{1/x} - 1), \quad a > 0$

23 $\displaystyle\lim_{x\to\infty} \sqrt[x]{x}$ *Hint:* Find the limit of the logarithm.

24 $\displaystyle\lim_{x\to 0^+} \sqrt[x]{x}$

25 $\displaystyle\lim_{x\to 0^+} x^x$

26 Sketch the curve $y = x - \ln x$.

27 Sketch the curve $y = \ln(x(2 - x))$.

28 Sketch the curve $y = x \ln x$.

In Problems 29–51 evaluate the integral.

29 $\displaystyle\int \frac{dx}{2x + 3}$

30 $\displaystyle\int \frac{x \, dx}{5x^2 - 2}$

31 $\displaystyle\int \frac{2x}{x - 1} dx$

32 $\displaystyle\int \frac{x - 1}{x + 1} dx$

33 $\displaystyle\int \frac{\ln x}{x} dx$

34 $\displaystyle\int \frac{e^t}{e^t + 1} dt$

35 $\displaystyle\int \frac{dt}{t \ln t}$

36 $\displaystyle\int \frac{\cos\theta}{1 + \sin\theta} d\theta$

37 $\displaystyle\int \frac{\ln(\ln x)}{x} dx$

38 $\displaystyle\int x\ln x \, dx$ *Hint:* Integrate by parts.

39 $\displaystyle\int x^n \ln x \, dx, \quad n \neq -1$

40 $\displaystyle\int (\ln x)^2 \, dx$

41 $\displaystyle\int (\ln x)^3 \, dx$

42 $\displaystyle\int x(\ln x)^2 \, dx$

43 $\displaystyle\int \frac{1}{x} \cos(\ln x) \, dx$

44 $\displaystyle\int \cos(\ln x) \, dx$

45 $\displaystyle\int_0^{10} \frac{x}{x + 1} dx$

46 $\displaystyle\int_{-3}^{-2} \frac{1}{x} dx$

47 $\displaystyle\int_1^e \ln x \, dx$

48 $\displaystyle\int_0^1 \frac{1}{x} dx$

49 $\displaystyle\int_{-\infty}^{-1} \frac{1}{x} dx$

50 $\displaystyle\int_0^1 \ln x \, dx$

51 $\displaystyle\int_1^{\infty} \ln x \, dx$

52 The region bounded by the curve $y = 1/\sqrt{x}$, $1 \le x \le 4$, is rotated about the x-axis. Find the volume of the solid of revolution.

53 Find the volume generated by rotating the region under the curve $y = \ln x$, $1 \le x \le e$, about (a) the x-axis, (b) the y-axis.

54 Find the volume generated by rotating the region under the curve $y = -\ln x$, $0 < x \le 1$, about (a) the x-axis, (b) the y-axis.

55 Find the length of the curve $y = \ln x$, $1 \le x \le e$.

56 Find the surface area generated by rotating the curve $y = \ln x$, $0 \le x \le 1$, about the y-axis.

☐ **57**   The *inverse hyperbolic sine* is defined by

$$\text{arcsinh}\, x = \ln(x + \sqrt{x^2 + 1}).$$

Show that this is the inverse of the hyperbolic sine function by solving the equation below for y:

$$x = \sinh y = \frac{e^y - e^{-y}}{2}.$$

☐ **58**   Show that $d(\text{arcsinh}\, x) = 1/\sqrt{x^2 + 1}$.

☐ **59**   Show that

$$\text{arctanh}\, x = \frac{1}{2} \ln\left(\frac{1 + x}{1 - x}\right), \quad |x| < 1$$

is the inverse function of $\tanh y$, and that $d(\text{arctanh}\, x) = 1/(1 - x^2)$.

8.6 SOME DIFFERENTIAL EQUATIONS

This section contains a brief preview of differential equations. They are studied in more detail in Chapter 14.

A *first order differential equation* is an equation that involves x, y, and dy/dx. If d^2y/dx^2 also appears in the equation it is called a *second order differential equation*. The simplest differential equation is

(1) $dy/dx = f(x)$

where the function f is continuous on an open interval I.

To solve such an equation we must find a function $y = F(x)$ such that $dy/dx = f(x)$. Differential Equation 1 arises from problems such as the following. Given the velocity $v = dy/dt$ at each time t, find the position y as a function of t. Given the slope dy/dx of a curve at each x, find the curve.

Any antiderivative $y = F(x)$ of $f(x)$ is a solution of this differential equation. Remember that all the antiderivatives of $f(x)$ form a family of functions which differ from each other by a constant.

This family is just the indefinite integral of f,

(1') $\displaystyle\int f(x)\, dx = F(x) + C.$

The family of functions (Equation 1') is the *general solution* of the Differential Equation 1.

In this chapter we have solved the problem of finding a nonzero function which is equal to its own derivative. This problem may be set up as another differential equation,

(2) $dy/dx = y.$

We found one solution, namely $y = e^x$. Are there any other solutions?

THEOREM 1

The general solution of the differential equation

$$dy/dx = y$$

is $y = Ce^x.$

That is, the only functions which are equal to their own derivatives are

$$y = Ce^x.$$

PROOF Assume y is a differentiable function of x. The following are equivalent, where x is the independent variable.

$$\frac{dy}{dx} = y,$$

$$\frac{1}{y} dy = dx,$$

$$\int \frac{1}{y} dy = \int dx,$$

$$\begin{array}{ll} \ln |y| = x + C_1 & \text{for some } C_1, \\ |y| = e^{x+C_1} & \text{for some } C_1, \\ y = Ce^x & \text{for some } C. \end{array}$$

In the last step, $C = e^{C_1}$ if y is positive and $C = -e^{C_1}$ if y is negative.

It can be shown in a similar way that the general solution of the differential equation

(3) $$dy/dx = ky,$$

where k is constant, is

(3′) $$y = Ce^{kx}.$$

The constant C is just the value of y at $x = 0$,

$$Ce^{k \cdot 0} = C.$$

In applications we often find a differential equation (3) plus an *initial condition* which gives the value of y at $x = 0$. The problem can be solved by writing down the general solution of the differential equation and then putting in the value of C given by the initial condition.

EXAMPLE 1 A country has a population of ten million at time $t = 0$, and constant annual birth rate $b = 0.020$ and death rate $d = 0.015$ per person. Find the population at time t.

The population satisfies the differential equation

$$\frac{dy}{dt} = (b - d)y = 0.005\, y.$$

The initial condition is

$$y = 10^7 \quad \text{at } t = 0.$$

The general solution is

$$y = Ce^{0.005t}.$$

Since at $t = 0$, $10^7 = Ce^0 = C$, the actual solution is

$$y = 10^7 e^{0.005t}.$$

EXAMPLE 2 A radioactive element has a half-life of N years, that is, half of the substance will decay every N years. Given ten pounds of the element at time $t = 0$, how much will remain at time t?

In radioactive decay the amount y of the element is decreasing at a rate proportional to y, so the differential equation has the form

$$dy/dt = ky.$$

The general solution is

$$y = Ce^{kt}.$$

Since y is decreasing, k will be negative. We must find the constants C and k. To find C we use the initial condition

$$y = 10 \quad \text{at } t = 0, \qquad C = 10.$$

To find k we use the given half-life. It tells us that

$$y = \tfrac{1}{2} \cdot 10 = 5 \quad \text{at } t = N.$$

Therefore
$$10e^{kN} = 5,$$
$$e^k = (\tfrac{1}{2})^{1/N},$$
$$k = \ln\left((\tfrac{1}{2})^{1/N}\right) = -\frac{\ln 2}{N}.$$

The solution is
$$y = 10e^{-(t \ln 2)/N}.$$

As we mentioned at the beginning of this chapter, the exponential growth function $y = Ce^{kt}$ is unrealistic for populations except for short periods of time. Here is a more realistic, but still quite simple, population growth function.

A population often has a limiting value L at which overcrowding will overcome reproduction. It is reasonable to suppose that the growth rate dy/dt is proportional to both the population y and the difference $L - y$. That is, the population satisfies the differential equation

$$\frac{dy}{dt} = ky(L - y)$$

for some constant k. The spread of an epidemic also satisfies this differential equation, where y is the number of victims and L is the total population. That is, the rate of increase of the number of victims is proportional to the product of the number of victims and the remaining population.

THEOREM 2

The general solution of the differential equation

$$\frac{dy}{dx} = ky(L - y)$$

is
$$y = \frac{L}{1 + Ce^{-kLx}}.$$

PROOF The constant functions $y = L$, $y = 0$ are trivial solutions. Suppose $y \neq L$, $y \neq 0$. The following are equivalent.

$$\frac{dy}{dx} = ky(L - y),$$

$$\frac{dy}{y(L - y)} = k\,dx,$$

$$\frac{L - y + y}{Ly(L - y)}\,dy = k\,dx,$$

$$\frac{1}{L}\left(\frac{1}{y} + \frac{1}{L - y}\right)dy = k\,dx,$$

$$\left(\frac{1}{y} + \frac{1}{L - y}\right)dy = kL\,dx,$$

$$\ln|y| - \ln|L - y| = kLx + C_1 \qquad \text{for some } C_1,$$

$$\ln\left|\frac{y}{L - y}\right| = kLx + C_1,$$

$$\left|\frac{y}{L - y}\right| = e^{kLx + C_1},$$

$$\frac{y}{L - y} = C_2 e^{kLx} \qquad \text{for some } C_2 \neq 0,$$

$$y(1 + C_2 e^{kLx}) = C_2 L e^{kLx},$$

$$y = \frac{C_2 I\, e^{kLx}}{1 + C_2 e^{kLx}} = \frac{L}{1 + (1/C_2)e^{-kLx}},$$

$$y = \frac{L}{1 + Ce^{-kLx}} \qquad \text{for some } C \neq 0.$$

The important case of this function is where C, k, and L are positive constants. In this case the function is called a *logistic function*. As the graph in Figure 8.6.1 shows, the value of the function approaches zero as $t \to -\infty$ and L as $t \to \infty$; that is,

$$\lim_{t \to -\infty} y = 0, \qquad \lim_{t \to \infty} y = L.$$

A population given by this function will approach but never quite reach the limiting value L.

It is easy to see intuitively that a differential equation

$$\frac{dy}{dx} = g(x, y)$$

will have a solution if the function $g(x, y)$ behaves reasonably. We return to our picture of a moving point controlled by a little man with a steering wheel (Figure 8.6.2). At $x = 0$ the point starts at $y = C$. (This is the initial condition.) At each value of x, the little man computes the value of $g(x, y)$ and turns the steering wheel so that the slope will be $dy/dx = g(x, y)$. Then the curve traced out by the point will be a solution of the differential equation. In general, there will always be a family of solutions which depend on the constant C of the initial condition.

Using indefinite integrals we can solve any differential equation where dy/dx is equal to a product of a function of x and a function of y,

$$\text{(4)} \qquad \frac{dy}{dx} = f(x)\,h(y), \qquad h(y) \neq 0.$$

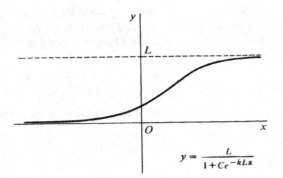

Figure 8.6.1

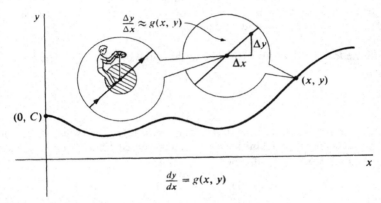

Figure 8.6.2

.We simply separate the x and y terms and integrate,

$$\frac{dy}{h(y)} = f(x)\,dx,$$

$$\int \frac{dy}{h(y)} = \int f(x)\,dx.$$

In an equation of the form (Equation 4) the variables are said to be *separable*.

EXAMPLE 3   Solve $dy/dx = e^y \sin x$.

$$e^{-y}\,dy = \sin x\,dx,$$
$$-e^{-y} = -\cos x - C,$$
$$e^{-y} = \cos x + C,$$
$$-y = \ln(\cos x + C),$$
$$y = -\ln(\cos x + C).$$

Second order differential equations also arise frequently in applications. As a rule, the general solution of a second order differential equation will involve two constants, and two initial conditions are needed to determine a particular solution.

EXAMPLE 4   Newton's law, $F = ma$, states that force equals mass times acceleration. Suppose a constant force F is applied along the y-axis to an object of constant mass m. Then the position y of the object is governed by the second order differential equation

$$m\frac{d^2y}{dt^2} = F, \qquad \frac{d^2y}{dt^2} = \frac{F}{m}.$$

The general solution of this equation is found by integrating twice,

$$\frac{dy}{dt} = \frac{Ft}{m} + v_0,$$

$$y = \frac{Ft^2}{2m} + v_0 t + y_0.$$

Setting $t = 0$ we see that the constants v_0 and y_0 are just the velocity and position at time $t = 0$. Thus the motion of the object is known if we know its initial position y_0 and velocity v_0.

If the force $F(t)$ varies with time we have the differential equation

$$\frac{d^2y}{dt^2} = \frac{F(t)}{m}.$$

The general solution can still be found by integrating twice, and the motion will still be determined by the initial position and velocity. Suppose for example that $F(t) = t^2$, and $y_0 = 5$, $v_0 = 1$ at time $t = 0$. Then

$$\frac{d^2y}{dt^2} = \frac{t^2}{m},$$

$$\frac{dy}{dt} = \frac{t^3}{3m} + 1,$$

$$y = \frac{t^4}{12m} + t + 5.$$

We shall now discuss an important second order differential equation whose solution involves sines and cosines.

The general solution of the equation

$$\frac{d^2y}{dt^2} = -y$$

is $y = a \cos t + b \sin t.$

We have $\dfrac{d(\sin t)}{dt} = \cos t, \qquad \dfrac{d^2(\sin t)}{dt^2} = -\sin t,$

$\dfrac{d(\cos t)}{dt} = -\sin t, \qquad \dfrac{d^2(\cos t)}{dt^2} = -\cos t.$

Therefore both $y = \sin t$ and $y = \cos t$ are solutions. It then follows easily that every function $a \cos t + b \sin t$ is a solution. Notice also that if

$$y = a \cos t + b \sin t$$

then at time $t = 0$, $y = a$ and $dy/dt = b$.

It can be proved that there are no other solutions, but we shall not give the proof here.

More generally, given a constant ω the equation

$$\frac{d^2y}{dt^2} = -\omega^2 y$$

has the general solution

$$y = a\cos\omega t + b\sin\omega t.$$

EXAMPLE 5   When a spring of natural length L is compressed a distance x it exerts a force $F = -kx$. The negative sign indicates that the force is in the opposite direction from x (Figure 8.6.3).

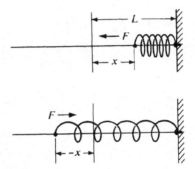

Figure 8.6.3

When x is negative the spring is expanded and the equation $F = -kx$ still holds.

Suppose a mass m is attached to the end of the spring and at time $t = 0$ is at position x_0 and has velocity v_0. The motion of the mass follows the differential equation

$$F = ma, \quad -kx = m\frac{d^2x}{dt^2}, \quad \frac{d^2x}{dt^2} = -\frac{k}{m}x.$$

The general solution is

$$x = a\cos\omega t + b\sin\omega t$$

where $\omega = \sqrt{k/m}$. Using the initial conditions, the motion of the mass is

$$x = x_0\cos\omega t + \frac{v_0}{\omega}\sin\omega t.$$

This function is periodic with period $2\pi/\omega$, so as expected the mass oscillates back and forth.

In the following second order equation, hyperbolic sines and cosines arise. *The general solution of the differential equation*

$$d^2y/dx^2 = y$$

is
$$y = a\cosh x + b\sinh x.$$

We see that cosh x and sinh x are solutions because

$$\frac{d(\cosh x)}{dx} = \sinh x, \qquad \frac{d^2(\cosh x)}{dx} = \cosh x,$$

$$\frac{d(\sinh x)}{dx} = \cosh x, \qquad \frac{d^2(\sinh x)}{dx} = \sinh x.$$

Another solution is e^x. Note that

$$e^x = \frac{(e^x + e^{-x}) + (e^x - e^{-x})}{2} = \cosh x + \sinh x.$$

PROBLEMS FOR SECTION 8.6

In Problems 1–16, find all solutions of the differential equation.

1 $\dfrac{dy}{dx} = xy^2$

2 $\dfrac{dy}{dx} = 2y - 5$

3 $\dfrac{dy}{dx} = \dfrac{x^2}{y}$

4 $\dfrac{dy}{dx} = x^2 y^2 + x^2$

5 $\dfrac{dy}{dx} = xe^y$

6 $\dfrac{dy}{dx} = xy + x + y + 1$

7 $\dfrac{dy}{dx} = e^{x-y}$

8 $\dfrac{dy}{dx} = \sqrt{xy}$

9 $\dfrac{dy}{dx} = y^3 \sin x$

10 $\dfrac{dy}{dx} = B + ky$

11 $\dfrac{d^2y}{dx^2} = 2x + 1$

12 $\dfrac{d^2y}{dx^2} = x^{-2}$

13 $\dfrac{d^2y}{dx^2} = 0$

14 $\dfrac{d^3y}{dx^3} = e^x$

15 $\dfrac{d^2y}{dx^2} = 3y$

16 $\dfrac{d^2y}{dx^2} = -4y$

17   A country has a population of 10 million at time $t = 0$ and constant annual birth rate $b = 0.025$ and death rate $d = 0.015$ per person. Find the population as a function of time.

18   Suppose a tree grows at a yearly rate equal to $\frac{1}{10}$ of its height. If the tree is 10 ft tall now, how tall will it be in 5 years?

19   A bacteria culture is found to double in size every minute. How long will it take to increase by a factor of one million?

20   If a bacteria culture has a population of B at time $t = 0$ and $2B$ at time $t = 10$, what will be its population at time $t = 25$?

21   A city had a population of 100,000 ten years ago and its current population is 115,000. If the growth is exponential, what will its population be in 30 years?

22   A radioactive element has a half-life of 100 years. In how many years will 99% of the original material decay?

23   What is the half-life of a radioactive substance if 10 grams decay to 9 grams in one year?

24   A body of mass m moving in a straight line is slowed down by a force due to air resistance which is proportional to its velocity, $F = -kv$. If the velocity at time $t = 0$ is v_0, find its velocity as a function of time. Use Newton's law, $F = ma = m\,dv/dt$.

25   A particle is accelerated at a rate equal to its position on the y-axis, $d^2y/dt^2 = y$. At time $t = 0$ it has position $y = 2$ and velocity $dy/dt = 0$. Find y as a function of t.

26   A mass of m grams at the end of a certain spring oscillates at the rate of one cycle every 10 seconds. How fast would a mass of $2m$ grams oscillate?

27   A particle is accelerated at a rate equal to its position on the y-axis but in the opposite direction, $d^2y/dt^2 = -y$. At time $t = 0$ it has position $y = 1$ and velocity $dy/dt = -2$. Find y as a function of t.

28   In Problem 27, suppose that at time $t = 0$ the position is $y = -3$ and at time $t = \pi/2$ the position is $y = 2$. Find y as a function of t.

29   Suppose the birth rate of a country is declining so that its population satisfies a differential equation of the form $dy/dt = ky/t$. If $y = 10,000,000$ at time $t = 10$ and $y = 20,000,000$ at time $t = 40$, find y as a function of t.

30   Work Problem 29 under the assumption that the population satisfies a differential equation of the form $dy/dt = ky/t^2$.

31   Suppose a population satisfies the differential equation $dy/dt = 10^{-8}y(10^8 - y)$ and $y_0 = 10^7$ at time $t_0 = 0$ years. Find the population y at time $t = 1$ year.

32   Suppose a population satisfies a differential equation of the form $dy/dt = ky(10^8 - y)$. At time $t_0 = 0$ years the population is $y_0 = 10^7$, and at time $t_1 = 1$ year the population is $y_1 = 2 \cdot 10^7$. Find y as a function of t.

☐ **33**   Suppose a population grows according to the differential equation $dy/dt = ky(L - y)$, and $0 < y < L, 0 < k$.
   (a)   Show that there is a single inflection point t_0, and the growth curve is concave upward when $t < t_0$ and concave downward when $t > t_0$.
   (b)   Find the population y_0 at the inflection point t_0.

☐ **34**   A population with a constant annual birth rate b and death rate d per person, and a constant annual immigration rate I, grows according to the differential equation $dy/dt = (b - d)y + I$. Suppose $b = 0.025$, $d = 0.015$, $I = 10^4$ people per year, and the population at time $t = 0$ is ten million people. Find the population as a function of time.

☐ **35**   Suppose the population of a country has a rate of growth proportional to the difference between 10,000,000 and the population, $dy/dt = k(10,000,000 - y)$. Find y as a function of t assuming that:
   (a) $y = 4,000,000$ at $t = 0$ and $y = 7,000,000$ at $t = 1$.
   (b) $y = 13,000,000$ at $t = 0$ and $y = 11,000,000$ at $t = 1$.

36   Find all curves with the property that the slope of the curve through each point P is equal to twice the slope of the line through P and the origin.

37   Find all curves whose slope at each point P is the reciprocal of the slope of the line through P and the origin.

38   Find all curves whose tangent line at each point (x, y) meets the x-axis at $(x - 4, 0)$.

8.7 DERIVATIVES AND INTEGRALS INVOLVING ln x

Sometimes it is easier to differentiate the natural logarithm of a function $y = f(x)$ than to differentiate the function itself. The method of computing the derivative of a function by differentiating its natural logarithm is called *logarithmic differentiation*.

THEOREM 1 (Logarithmic Differentiation)

Suppose the function $y = f(x)$ is differentiable and not zero at x. Then

$$\frac{dy}{dx} = y\frac{d(\ln|y|)}{dx}.$$

PROOF $$\frac{d(\ln|y|)}{dx} = \frac{d(\ln|y|)}{dy}\frac{dy}{dx} = \frac{1}{y}\frac{dy}{dx}.$$

Logarithmic differentiation is useful when the function is a product or involves an exponent, because logarithms turn exponents into products and products into sums.

EXAMPLE 1   Find dy/dx where $y = (2x + 1)(3x - 1)(4 - x)$.

$$\ln|y| = \ln|2x + 1| + \ln|3x - 1| + \ln|4 - x|,$$

$$\frac{dy}{dx} = y\left(\frac{2}{2x + 1} + \frac{3}{3x - 1} - \frac{1}{4 - x}\right)$$

$$= (2x + 1)(3x - 1)(4 - x)\left(\frac{2}{2x + 1} + \frac{3}{3x - 1} - \frac{1}{4 - x}\right).$$

EXAMPLE 2   Find dy/dx where $y = x^x$.

$$\ln y = x \ln x,$$

$$\frac{dy}{dx} = y\frac{d(x \ln x)}{dx} = x^x\left(\frac{x}{x} + \ln x\right) = x^x(1 + \ln x).$$

In this example, $\ln y = \ln|y|$ because $y > 0$.

EXAMPLE 3   Find dy/dx where $y = \dfrac{(x^2 + 1)^3(x^3 + x + 2)}{(x - 1)\sqrt{x + 4}}$.

$$\ln|y| = 3\ln|x^2 + 1| + \ln|x^3 + x + 2| - \ln|x - 1| - \tfrac{1}{2}\ln|x + 4|.$$

$$\frac{dy}{dx} = y\left(\frac{6x}{x^2 + 1} + \frac{3x^2 + 1}{x^3 + x + 2} - \frac{1}{x - 1} - \frac{1}{2(x + 4)}\right)$$

$$= \frac{(x^2 + 1)^3(x^3 + x + 2)}{(x - 1)\sqrt{x + 4}}\left(\frac{6x}{x^2 + 1} + \frac{3x^2 + 1}{x^3 + x + 2} - \frac{1}{x - 1} - \frac{1}{2(x + 4)}\right).$$

This derivative could have been found using the Product and Quotient Rules but it would take a great deal of work.

The Power Rule $d(x^r) = rx^{r-1}\,dx$ was proved in Chapter 2 only when the exponent r is rational. We can use logarithmic differentiation to show that the Power Rule holds even for irrational r.

THEOREM 2 (Power Rule)

Let b be any real number. Then

$$d(x^b) = bx^{b-1}\, dx,$$

$$\int x^b\, dx = \frac{x^{b+1}}{b+1} + C, \qquad (b \neq 1).$$

PROOF Let $y = x^b$. Then ln $y = b$ ln x, and

$$\frac{dy}{dx} = y\frac{d(\ln y)}{dx} = y\frac{d(b \ln x)}{dx} = x^b \cdot b \cdot \frac{1}{x} = bx^{b-1}.$$

The formula $\int 1/x\, dx = \ln |x| + C$ allows us to integrate a number of basic functions which we could not handle before.

EXAMPLE 4 Find $\int \tan \theta\, d\theta$. We have $\tan \theta = (\sin \theta/\cos \theta)$. Let $u = \cos \theta, du = -\sin \theta\, d\theta$. Then

$$\int \tan \theta\, d\theta = -\int 1/u\, du = -\ln |u| + C = -\ln |\cos \theta| + C.$$

Remember the absolute value sign inside the logarithm. It is needed because $\cos \theta$ may be negative.

EXAMPLE 5 Find $\int \sec \theta\, d\theta$.

$$\int \sec \theta\, d\theta = \int \frac{\sec \theta\,(\sec \theta + \tan \theta)}{\sec \theta + \tan \theta}\, d\theta$$

$$= \int \frac{d(\sec \theta + \tan \theta)}{\sec \theta + \tan \theta} = \ln |\sec \theta + \tan \theta| + C.$$

With the above two examples and the reduction formulas from Section 7.5 we can integrate any power of $\tan \theta$ or $\sec \theta$.

These integrals often arise in trigonometric substitutions.

EXAMPLE 6 Find $\int \sec^3 \theta\, d\theta$. From the reduction formula in Section 7.5,

$$\int \sec^3 \theta\, d\theta = \tfrac{1}{2} \sec^2 \theta \sin \theta + \tfrac{1}{2} \int \sec \theta\, d\theta.$$

Therefore $\int \sec^3 \theta\, d\theta = \tfrac{1}{2} \sec^2 \theta \sin \theta + \tfrac{1}{2} \ln |\sec \theta + \tan \theta| + C.$

EXAMPLE 7 Find $\int \frac{x\, dx}{a^2 + x^2}$.

Let $u = a^2 + x^2$. Then $du = 2x\, dx$,

$$\int \frac{x\, dx}{a^2 + x^2} = \frac{1}{2} \int \frac{du}{u} = \frac{1}{2} \ln |u| + C = \frac{1}{2} \ln |a^2 + x^2| + C.$$

Since $a^2 + x^2$ is always positive

$$\int \frac{x \, dx}{a^2 + x^2} = \frac{1}{2} \ln (a^2 + x^2) + C$$

is equally correct.

EXAMPLE 8 Find $\int \dfrac{dx}{\sqrt{x^2 - a^2}}$.

Assume $a > 0$. We make the trigonometric substitution $x = a \sec \theta$, illustrated in Figure 8.7.1.

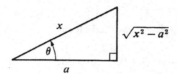

Figure 8.7.1

Then $dx = a \tan \theta \sec \theta \, d\theta$, $\sqrt{x^2 - a^2} = a \tan \theta$.

$$\int \frac{dx}{\sqrt{x^2 - a^2}} d\theta = \int \frac{a \tan \theta \sec \theta}{a \tan \theta} d\theta = \int \sec \theta \, d\theta$$

$$= \ln |\sec \theta + \tan \theta| + C' \qquad \text{(by Example 5)}$$

$$= \ln \left| \frac{x}{a} + \frac{\sqrt{x^2 - a^2}}{a} \right| + C' = \ln |x + \sqrt{x^2 - a^2}| - \ln a + C'.$$

Therefore
$$\int \frac{dx}{\sqrt{x^2 - a^2}} = \ln |x + \sqrt{x^2 - a^2}| + C.$$

The formula $\int \dfrac{dx}{\sqrt{a^2 + x^2}} = \ln |x + \sqrt{a^2 + x^2}| + C$

can be derived in a similar way and is left as an exercise.

The integrals $\qquad \displaystyle\int \arctan x \, dx, \qquad \int \operatorname{arcsec} x \, dx$

can now be evaluated using integration by parts,

$$\int u \, dv = uv - \int v \, du.$$

EXAMPLE 9 Find $\int \arctan x \, dx$.

Let $u = \arctan x$, $\quad du = dx/(1 + x^2)$, $\quad v = x$, $\quad dv = dx$.

Then
$$\int \arctan x \, dx = \int u \, dv = uv - \int v \, du$$

$$= x \arctan x - \int \frac{x}{1 + x^2} dx.$$

From Example 7,

$$\int \frac{x\,dx}{1 + x^2} = \frac{1}{2}\ln(1 + x^2) + C.$$

Therefore $\int \arctan x\,dx = x \arctan x - \frac{1}{2}\ln(1 + x^2) + C.$

$\int \text{arccot } x\,dx$ can be evaluated in a similar way.

EXAMPLE 10 Find $\int \text{arcsec } x\,dx$, when $x > 1$.

Let $u = \text{arcsec } x,\quad du = \dfrac{1}{|x|\sqrt{x^2 - 1}}dx = \dfrac{1}{x\sqrt{x^2 - 1}}dx,\quad v = x,\quad dv = dx.$

Then $\displaystyle\int \text{arcsec } x\,dx = \int u\,dv = uv - \int v\,du = x\,\text{arcsec } x - \int \frac{x}{x\sqrt{x^2 - 1}}dx$

$$= x\,\text{arcsec } x - \int \frac{1}{\sqrt{x^2 - 1}}dx.$$

From Example 8,

$$\int \frac{1}{\sqrt{x^2 - 1}}dx = \ln|x + \sqrt{x^2 - 1}| + C.$$

Therefore

$$\int \text{arcsec } x\,dx = x\,\text{arcsec } x - \ln|x + \sqrt{x^2 - 1}| + C.$$

PROBLEMS FOR SECTION 8.7

In Problems 1–10 find the derivatives by logarithmic differentiation.

1 $y = \dfrac{3x - 2}{4x + 3}$

2 $y = (5x - 2)^3(6x + 1)^2$

3 $y = \dfrac{(x^2 + 1)\sqrt{3x + 4}}{(2x - 3)\sqrt{x^2 - 4}}$

4 $y = x^{2x}$

5 $y = (x - 1)^{x^2 + 1}$

6 $y = (\sin\theta)^{\tan\theta}$

7 $y = e^{(e^x)}$

8 $y = (2x + 1)^e$

9 $s = \sqrt[t]{t}$

10 $y = x^{(x^x)}$

11 Using derivatives and limits, sketch the curve $y = x^x, x > 0$.

12 Using derivatives and limits, sketch the curve $y = \sqrt[x]{x}, x > 0$.

13 Prove the differentiation rule $d(u^v) = u^v(v/u\,du + \ln u\,dv), (u > 0)$.

In Problems 14–38 evaluate the integral.

14 $\displaystyle\int \tan^3\theta\,d\theta$

15 $\displaystyle\int \cot\theta\,d\theta$

16 $\displaystyle\int \csc\theta\,d\theta$

17 $\displaystyle\int \tan(3\theta)\,d\theta$

18 $\displaystyle\int \text{sech } x\,dx$

19 $\displaystyle\int \sec^5 x\,dx$

20    $\displaystyle\int \tan^5 x \, dx$

21    $\displaystyle\int \frac{\sec^2 x}{\tan x} \, dx$

22    $\displaystyle\int \frac{\sec^3 x}{\tan x} \, dx$

23    $\displaystyle\int \frac{\tan^2 x}{\sec x} \, dx$

24    $\displaystyle\int \frac{dx}{\sqrt{a^2 + x^2}}$

25    $\displaystyle\int \sqrt{x^2 - 1} \, dx$

26    $\displaystyle\int \frac{1}{x\sqrt{x^2 + 1}} \, dx$

27    $\displaystyle\int \frac{1}{x\sqrt{4 - x^2}} \, dx$

28    $\displaystyle\int \sqrt{4 + x^2} \, dx$

29    $\displaystyle\int \frac{1}{x^2 - 4} \, dx$

30    $\displaystyle\int \sqrt{1 + \frac{1}{x^2}} \, dx$

31    $\displaystyle\int \sqrt{1 - \frac{1}{x^2}} \, dx$

32    $\displaystyle\int (x^2 - 1)^{3/2} \, dx$

33    $\displaystyle\int x^2 \sqrt{1 + x^2} \, dx$

34    $\displaystyle\int x^2 \operatorname{arcsec} x \, dx$

35    $\displaystyle\int x^{-2} \arcsin x \, dx$

36    $\displaystyle\int x \sec^2 x \, dx$

37    $\displaystyle\int \operatorname{arccsc} x \, dx$

38    $\displaystyle\int \operatorname{arccot} x \, dx$

39   Find the length of the parabola $y = x^2$, $-1 \le x \le 1$.

40   Find the surface area generated by rotating the parabola $y = x^2, 0 \le x \le 1$ about the x-axis.

41   Find the length of the spiral of Archimedes $r = \theta, 0 \le \theta \le a$, in polar coordinates.

42   Find the volume of the solid generated by rotating the region under the curve $y = \sec^2 x, 0 \le x \le \pi/3$, about (a) the x-axis, (b) the y-axis.

8.8 INTEGRATION OF RATIONAL FUNCTIONS

A rational function is a quotient of two polynomials,

$$f(x) = \frac{F(x)}{G(x)}.$$

Using the Quotient, Constant, Sum, and Power Rules, one can easily find the derivative of any rational function. We shall now show how to find the integral of any rational function. This is fairly easy to do if the degree of the denominator $G(x)$ is only two or three, but becomes more difficult as the degree of $G(x)$ gets larger. Let us work some examples and then formulate a general procedure.

Our first example shows how to integrate when the denominator $G(x)$ has degree one.

EXAMPLE 1 $\displaystyle\int \frac{x^3 + 4x^2 - 1}{x + 2} \, dx.$

The first step is to divide the denominator into the numerator by long division.

$$\frac{x^3 + 4x^2 - 1}{x + 2} = x^2 + 2x - 4 + \frac{7}{x + 2}.$$

We now easily integrate each term in the sum.

$$\int \frac{x^3 + 4x^2 - 1}{x + 2} dx = \int \left(x^2 + 2x - 4 + \frac{7}{x + 2} \right) dx$$

$$= \frac{x^3}{3} + x^2 - 4x + 7 \ln |x + 2| + C.$$

EXAMPLE 2 $\int \dfrac{x^3 + 2x^2 - 20x - 33}{x^2 - 3x - 10} dx.$

Step 1 By long division, divide the denominator into the numerator. The result is

$$\frac{x^3 + 2x^2 - 20x - 33}{x^2 - 3x - 10} = x + 5 + \frac{5x + 17}{x^2 - 3x - 10}.$$

Step 2 Break up the remainder $\dfrac{5x + 17}{x^2 - 3x - 10}$ into a sum,

(1)
$$\frac{5x + 17}{x^2 - 3x - 10} = \frac{-1}{x + 2} + \frac{6}{x - 5}.$$

One can readily check that Equation 1 is true,

$$\frac{-1}{x + 2} + \frac{6}{x - 5} = \frac{-(x - 5) + 6(x + 2)}{(x + 2)(x - 5)} = \frac{5x + 17}{x^2 - 3x - 10}.$$

The terms $\dfrac{-1}{x + 2}$ and $\dfrac{6}{x - 5}$ are called *partial fractions*. Later on we shall explain how they were found. Notice that the denominators of the partial fractions are factors of the denominator of the rational function,

$$(x + 2)(x - 5) = x^2 - 3x - 10.$$

Step 3 We now have

$$\int \frac{x^3 + 2x^2 - 20x - 33}{x^2 - 3x - 10} dx = \int x \, dx + \int 5 \, dx + \int -\frac{1}{x + 2} dx + \int \frac{6}{x - 5} dx$$

$$= \frac{x^2}{2} + 5x - \ln |x + 2| + 6 \ln |x - 5| + C.$$

EXAMPLE 3 $\int \dfrac{x^2}{x^3 + 3x^2 + 3x + 1} dx.$

Step 1 This time the numerator already has smaller degree than the denominator, so no long division is needed.

Step 2 Break the rational function into a sum of partial fractions. The denominator can be factored as

$$x^3 + 3x^2 + 3x + 1 = (x + 1)^3.$$

It turns out that

$$\frac{x^2}{(x+1)^3} = \frac{1}{x+1} - \frac{2}{(x+1)^2} + \frac{1}{(x+1)^3}.$$

This can again be readily checked.

Step 3 $\displaystyle\int \frac{x^2}{(x+1)^3}dx = \int \frac{1}{x+1}dx + \int -\frac{2}{(x+1)^2}dx + \int \frac{1}{(x+1)^3}dx$

$$= \ln|x+1| + \frac{2}{x+1} - \frac{1}{2(x+1)^2} + C.$$

EXAMPLE 4 $\displaystyle\int \frac{2x+3}{x^2+x+1}dx.$

Step 1 No long division is needed.

Step 2 The denominator $x^2 + x + 1$ cannot be factored, i.e., it is irreducible. In this case no sum of partial fractions is needed.

Step 3 To integrate $\displaystyle\int \frac{2x+3}{x^2+x+1}dx$

we use the method of *completing the square*. We have

$$x^2 + x + 1 = (x + \tfrac{1}{2})^2 + \tfrac{3}{4}.$$

Let $u = x + \tfrac{1}{2}$. Then $du = dx$ and

$$\int \frac{2x+3}{x^2+x+1}dx = \int \frac{2(u-\tfrac{1}{2})+3}{u^2+\tfrac{3}{4}}du = \int \frac{2u+2}{u^2+\tfrac{3}{4}}du$$

$$= \int \frac{2u}{u^2+\tfrac{3}{4}}du + \int \frac{2}{u^2+\tfrac{3}{4}}du$$

$$= \int \frac{d(u^2+\tfrac{3}{4})}{u^2+\tfrac{3}{4}} + 2\int \frac{1}{u^2+(\sqrt{3}/2)^2}du$$

$$= \ln\left|u^2 + \frac{3}{4}\right| + \frac{4}{\sqrt{3}}\arctan\left(\frac{2}{\sqrt{3}}u\right) + C$$

$$= \ln|x^2 + x + 1| + \frac{4}{\sqrt{3}}\arctan\left(\frac{2}{\sqrt{3}}\left(x + \frac{1}{2}\right)\right) + C.$$

We used the trigonometric substitution illustrated in Figure 8.8.1.

$$u = \left(\frac{\sqrt{3}}{2}\right)\tan\theta, \qquad \sqrt{u^2 + \left(\sqrt{\frac{3}{2}}\right)^2} = \left(\frac{\sqrt{3}}{2}\right)\sec\theta.$$

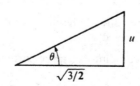

Figure 8.8.1

In all four examples the idea was to break the rational function into a sum of simpler functions which can easily be integrated. Here are three steps in the method.

METHOD FOR INTEGRATING A RATIONAL FUNCTION $f(x) = \dfrac{F(x)}{G(x)}$

Step 1   If the degree of $F(x)$ is $\geq$ the degree of $G(x)$, apply long division. This puts the quotient $F(x)/G(x)$ in the form

$$\frac{F(x)}{G(x)} = Q(x) + \frac{R(x)}{G(x)}$$

where the degree of the polynomial $R(x)$ is less than that of $G(x)$.

Step 2   Break the quotient $R(x)/G(x)$ into a sum of *partial fractions*.

Step 3   Integrate the polynomial $Q(x)$ and each of the partial fractions separately.

Sometimes Step 1 or 2 will be unnecessary.

How to do Step 2: We wish to break a quotient $R(x)/G(x)$ into a sum of partial fractions. First, factor the denominator $G(x)$ into a product of linear terms of the form $ax + b$, and irreducible quadratic terms of the form $ax^2 + bx + c$. It can be proved that every polynomial can be so factored, but we shall not give the proof here. Two theorems from elementary algebra are useful for factoring a given polynomial.

FACTOR THEOREM

  $x - r$ *is a factor of a polynomial* $G(x)$ *if and only if* r *is a root of* $G(x) = 0$.

QUADRATIC FORMULA

  Let $a \neq 0.$ x *is a root of* $ax^2 + bx + c = 0$ *if and only if*

$$x = \frac{-b \pm \sqrt{b^2 - 4ac}}{2a}.$$

If $(ax + b)^n$ appears in the factorization of $G(x)$, the sum of partial fractions will contain the following terms:

$$\frac{A_1}{ax + b} + \frac{A_2}{(ax + b)^2} + \cdots + \frac{A_n}{(ax + b)^n}.$$

If $(ax^2 + bx + c)^n$ appears, the sum of partial fractions will include

$$\frac{B_1 x + C_1}{ax^2 + bx + c} + \frac{B_2 x + C_2}{(ax^2 + bx + c)^2} + \cdots + \frac{B_n x + C_n}{(ax^2 + bx + c)^n}.$$

To find the partial fractions we must solve for the unknown constants A_i, B_i, and C_i. We show how this is done in the examples.

EXAMPLE 2 (Continued)   From Step 1 we obtained the remainder $\dfrac{5x + 17}{x^2 - 3x - 10}$.

We first factor the denominator $x^2 - 3x - 10$. Since it has degree two we can find its roots from the quadratic formula.

$$x = \frac{-(-3) \pm \sqrt{(-3)^2 - 4 \cdot 1(-10)}}{2 \cdot 1} = \frac{3 \pm \sqrt{49}}{2} = \frac{3 \pm 7}{2},$$

$$x = 5 \text{ and } x = -2.$$

By the Factor Theorem, $x^2 - 3x - 10$ has the two factors $x - 5$ and $x + 2$, whence

$$x^2 - 3x - 10 = (x + 2)(x - 5).$$

Now we find the sum of partial fractions. It must have the form

$$\frac{5x + 17}{(x + 2)(x - 5)} = \frac{A}{x + 2} + \frac{B}{x - 5}.$$

The way we find A and B is to use $(x + 2)(x - 5)$ as a common denominator so the numerators of both sides of the equation are equal.

$$\frac{5x + 17}{(x + 2)(x - 5)} = \frac{A(x - 5) + B(x + 2)}{(x + 2)(x - 5)},$$

$$5x + 17 = A(x - 5) + B(x + 2),$$

$$5x + 17 = (A + B)x + (-5A + 2B).$$

The x terms and the constant terms must be equal, so we get two equations in the unknowns A and B.

$$5 = A + B, \qquad 17 = -5A + 2B.$$

Solving for A and B we have

$$A = -1, \qquad B = 6,$$

$$\frac{5x + 17}{x^2 - 3x - 10} = \frac{-1}{x + 2} + \frac{6}{x - 5}.$$

EXAMPLE 3 (Continued)   We have $\dfrac{x^2}{x^3 + 3x^2 + 3x + 1}.$

One might recognize $x^3 + 3x^2 + 3x + 1$ at once as $(x + 1)^3$. Alternatively, one can see easily that $x = -1$ is a root of $x^3 + 3x^2 + 3x + 1$. Therefore $x + 1$ is a factor of it. Dividing by $x + 1$ we get the quotient $x^2 + 2x + 1 = (x + 1)^2$.

The sum of partial fractions has the form

$$\frac{x^2}{(x + 1)^3} = \frac{A}{x + 1} + \frac{B}{(x + 1)^2} + \frac{C}{(x + 1)^3}.$$

Then
$$\frac{x^2}{(x + 1)^3} = \frac{A(x + 1)^2 + B(x + 1) + C}{(x + 1)^3},$$

$$x^2 = A(x + 1)^2 + B(x + 1) + C,$$

$$x^2 = Ax^2 + (2A + B)x + (A + B + C).$$

$$A = 1, \qquad 2A + B = 0, \qquad A + B + C = 0.$$

Solving these three equations for A, B, and C we have

$$A = 1, \qquad B = -2, \qquad C = 1.$$

Therefore

$$\frac{x^2}{(x + 1)^3} = \frac{1}{x + 1} - \frac{2}{(x + 1)^2} + \frac{1}{(x + 1)^3}.$$

EXAMPLE 4 (Continued) We are given $\dfrac{2x + 3}{x^2 + x + 1}$.

The denominator $x^2 + x + 1$ has no real roots because the quadratic formula gives

$$x = \frac{-1 \pm \sqrt{1 - 4}}{2} = \frac{-1 \pm \sqrt{-3}}{2}.$$

We therefore proceed immediately to Step 3.

How to do Step 3: The rational function has been broken up into a sum of a polynomial and partial fractions of the two types

(1)
$$\frac{A}{(ax + b)^n},$$

(2)
$$\frac{Bx + C}{(ax^2 + bx + c)^n}, \qquad \text{where } ax^2 + bx + c \text{ is irreducible.}$$

Polynomials and fractions of type (1) are easily integrated using the Power Rule,

$$\int u^n \, du = \frac{u^{n+1}}{n + 1} + C, \qquad n \neq -1,$$

and the rule,

$$\int \frac{du}{u} = \ln |u| + C.$$

Partial fractions of type (2) can be integrated as follows.
First divide the denominator by a^n so the fraction has the simpler form

$$\frac{Bx + C}{a^n(x^2 + b_1 x + c_1)^n}.$$

When we make the substitution $u = x + \dfrac{b_1}{2}$, we find that

$$x^2 + b_1 x + c_1 = u^2 + \left(c_1 - \frac{b_1^2}{4} \right) = u^2 + k^2.$$

This substitution is called the method of *completing the square*. Now the integral takes the even simpler form

$$\frac{1}{a^n} \int \frac{Bu + C}{(u^2 + k^2)^n} du = \frac{1}{a^n} \int \frac{Bu}{(u^2 + k^2)^n} du + \frac{1}{a^n} \int \frac{C \, du}{(u^2 + k^2)^n}.$$

The first integral can be evaluated by putting $w = u^2 + k^2$, $dw = 2u \, du$. The second integral can be evaluated by the trigonometric substitution shown in Figure 8.8.2, $u = k \tan \theta$.

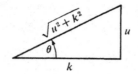

Figure 8.8.2

Example 4 is an integral of the form

$$\int \frac{Bx + C}{ax^2 + bx + c} dx$$

and was worked out in this way.

PROBLEMS FOR SECTION 8.8

Evaluate the following integrals.

1 $\displaystyle\int \frac{dx}{2x - 7}$

2 $\displaystyle\int \frac{dx}{(2x - 1)(x + 2)}$

3 $\displaystyle\int \frac{dx}{3x(x - 4)}$

4 $\displaystyle\int \frac{x + 5}{3x - 1} dx$

5 $\displaystyle\int \frac{2x - 3}{(x - 1)(x + 4)} dx$

6 $\displaystyle\int \frac{3x^2 - 4x + 2}{x - 5} dx$

7 $\displaystyle\int \frac{x^3 + x^2 + x + 1}{x(x + 4)} dx$

8 $\displaystyle\int \frac{2x^2 + x - 5}{(x - 3)(x + 2)} dx$

9 $\displaystyle\int \frac{dx}{(x + 1)^3}$

10 $\displaystyle\int \frac{x \, dx}{(2x - 1)^2}$

11 $\displaystyle\int \frac{x^2 - x + 1}{(x - 1)^3} dx$

12 $\displaystyle\int \frac{x^4}{x^2 - 1} dx$

13 $\displaystyle\int \frac{1}{x^3 - x} dx$

14 $\displaystyle\int \frac{dx}{(x + 1)(x + 3)(x + 5)}$

15 $\displaystyle\int \frac{x^3 - 1}{x^3 - x^2} dx$

16 $\displaystyle\int \frac{dx}{4 + x^2}$

17 $\displaystyle\int \frac{x^2 \, dx}{16 + x^2}$

18 $\displaystyle\int \frac{x \, dx}{x^2 + 4x + 5}$

19 $\displaystyle\int \frac{x + 2}{x^2 - 2x - 3} dx$

20 $\displaystyle\int \frac{dx}{x^3 + x}$

21 $\displaystyle\int \frac{x^4}{1 + x^2} dx$

22 $\displaystyle\int \frac{x^2}{(4 + x^2)^2} dx$

23 $\displaystyle\int \frac{1}{(1 + x^2)(2 + x^2)} dx$

24 $\displaystyle\int \frac{x^4 + 3x + 1}{x^2 + x + 1} dx$

25 $\displaystyle\int \frac{dx}{x^4 + x^2}$

26 $\displaystyle\int \frac{dx}{x^4 - 16}$

27 $\displaystyle\int \frac{dx}{x^3 + 1}$

28 $\displaystyle\int \frac{3x + 6}{x^4 - 2x^2 + 1} dx$

29 $\displaystyle\int \frac{x^5 + 3x^2 + 1}{x^4 - 1} dx$

30 $\displaystyle\int \frac{dx}{x^4 + 1}$

31 $\displaystyle\int \frac{\arctan x}{x^2} dx$

32 $\displaystyle\int x^2 \arctan x \, dx$

8.9 METHODS OF INTEGRATION

During this course we have developed several methods for evaluating indefinite integrals, such as the Sum and Constant Rules, change of variables, integration by parts, and partial fractions. In the integration problems up to this point, the method to be used was usually given. But in a real life integration problem, one will have to decide which method to use on his own.

This section has two purposes. First, to review all the methods of integration. Second, to explain how one might decide which method to use for a given problem.

Almost all the examples and problems in this book involve what are called elementary functions. A real function $f(x)$ is called an *elementary function* if $f(x)$ is given by a term $\tau(x)$ which is built up from constants, sums, differences, products, quotients, powers, roots, exponential functions, logarithmic functions, and trigonometric functions and their inverses. These are the functions for which we have introduced names. Given an elementary function $f(x)$, an indefinite integral $\int f(x)\,dx$ may or may not be an elementary function. For example, it turns out that the integrals

$$\int e^{-x^2}\,dx, \qquad \int \sqrt{1-x^4}\,dx$$

are not elementary functions.

What is meant by the problem "evaluate the indefinite integral $\int f(x)\,dx$"? The problem is really the following.

Given an elementary function $f(x)$, find another elementary function $F(x)$ (if there is one) such that

$$\int f(x)\,dx = F(x) + C.$$

This is a hard problem. Sometimes the integral is not an elementary function at all. Sometimes the integral is an elementary function but it can be found only by guesswork. There is no routine way to evaluate an indefinite integral. However, one can often find clues which will cut down on the guesswork. We shall point out some of these clues here.

The corresponding problem for differentiation is much easier. Given an elementary function $f(x)$, the derivative $f'(x)$ is always another elementary function. It can be found in a routine way using the rules for differentiation and the Chain Rule.

The starting point for evaluating indefinite integrals is a list of twelve basic formulas which should be memorized.

A. BASIC FORMULAS

Let u and v be differentiable functions of x.

I. $du = \dfrac{du}{dx}\,dx,$ $\qquad\qquad \displaystyle\int du = u + C$

II. $d(ku) = k\,du,$ $\qquad\qquad \displaystyle\int k\,du = k\int du$

III. $d(u+v) = du + dv,$ $\qquad\qquad \displaystyle\int du + dv = \int du + \int dv$

IV. $d(u^r) = ru^{r-1}\,du,$ $\qquad\qquad \displaystyle\int u^r\,du = \dfrac{u^{r+1}}{r+1} + C, \qquad r \neq -1$

$\text{V.} \quad d(\ln u) = \dfrac{du}{u}, \qquad\qquad \displaystyle\int \dfrac{du}{u} = \ln |u| + C$

$\text{VI.} \quad d(e^u) = e^u\, du, \qquad\qquad \displaystyle\int e^u\, du = e^u + C$

$\text{VII.} \quad d(\sin u) = \cos u\, du, \qquad\qquad \displaystyle\int \cos u\, du = \sin u + C$

$\text{VIII.} \quad d(\cos u) = -\sin u\, du, \qquad\qquad \displaystyle\int \sin u\, du = -\cos u + C$

$\text{IX.} \quad d(\tan u) = \sec^2 u\, du, \qquad\qquad \displaystyle\int \sec^2 u\, du = \tan u + C$

$\text{X.} \quad d(\cot u) = -\csc^2 u\, du, \qquad\qquad \displaystyle\int \csc^2 u\, du = -\cot u + C$

$\text{XI.} \quad d(\sec u) = \tan u \sec u\, du, \qquad\qquad \displaystyle\int \tan u \sec u\, du = \sec u + C$

$\text{XII.} \quad d(\csc u) = -\cot u \csc u\, du, \qquad \displaystyle\int \cot u \csc u\, du = -\csc u + C$

We shall see later, when we discuss the method of integration by change of variables, why it is important to actually *memorize* these formulas.

B. TABLES OF INTEGRALS

The integrals of the following functions were computed in Chapters 7 and 8; they can be found in the table at the end of the book. These integrals are more complicated and need not be memorized. Instead, one should remember that their integrals are elementary functions which can be looked up in a table.

$$\int \tan x\, dx \qquad \int \cot x\, dx$$

$$\int \sec x\, dx \qquad \int \csc x\, dx$$

$$\int \arcsin x\, dx \qquad \int \arccos x\, dx$$

$$\int \arctan x\, dx \qquad \int \operatorname{arccot} x\, dx$$

$$\int \operatorname{arcsec} x\, dx \qquad \int \operatorname{arccsc} x\, dx$$

$$\int \ln x\, dx$$

The following integrals of powers of trigonometric functions are given by reduction formulas in terms of smaller powers.

$$\int \sin^n x\, dx \qquad \int \cos^n x\, dx$$

$$\int \tan^n x \, dx \qquad \int \cot^n x \, dx$$

$$\int \sec^n x \, dx \qquad \int \csc^n x \, dx$$

C. INTEGRALS OF RATIONAL FUNCTIONS

In Section 8.8 we explained how to integrate any rational function. The only part of the procedure which requires guesswork is factoring the denominator into linear and quadratic terms. Once that is done, any rational function can be integrated in a routine manner.

The integrals in lists A and B (which can be found in tables) and the rational integrals are easily recognized. Now we come to grips with the real problem. Given an integral which cannot be found in a table, we wish to transform it into either a rational integral or an integral which can be found in a table. We have three main methods for transforming integrals: using the Sum Rule, integration by change of variables, and integration by parts.

D. USING THE SUM RULE

Sometimes we can break an integral into a sum of two or more easier integrals. We may use algebraic identities, trigonometric identities, or rules of logarithms to do this.

EXAMPLE 1 $\displaystyle\int \frac{dx}{\sqrt{x+1} - \sqrt{x}}.$

By multiplying the numerator and denominator by $\sqrt{x+1} + \sqrt{x}$ (i.e., rationalizing the denominator), we get the sum

$$\int \frac{dx}{\sqrt{x+1} - \sqrt{x}} = \int \frac{\sqrt{x+1} + \sqrt{x}}{(x+1) - x} dx = \int (\sqrt{x+1} + \sqrt{x}) \, dx$$

$$= \int \sqrt{x+1} \, dx + \int \sqrt{x} \, dx.$$

EXAMPLE 2 $\int \tan^3 x \sec^2 x \, dx.$ Using the identity $\sec^2 x = 1 + \tan^2 x$, we obtain a sum of integrals of powers of $\tan x$:

$$\int \tan^3 x \sec^2 x \, dx = \int \tan^3 x (1 + \tan^2 x) \, dx = \int \tan^3 x \, dx + \int \tan^5 x \, dx.$$

EXAMPLE 3 $\ln\left(\dfrac{x^2}{x+1}\right) dx.$ Using the rules of logarithms we have

$$\int \ln\left(\frac{x^2}{x+1}\right) dx = \int (2\ln x - \ln(x+1)) \, dx = 2\int \ln x \, dx - \int \ln(x+1) \, dx.$$

EXAMPLE 4 $\int \sin(x + a) \sin(x - a)\,dx$. Using the addition formulas,

$$\sin(x + a) = \sin x \cos a + \cos x \sin a,$$
$$\sin(x - a) = \sin x \cos a - \cos x \sin a,$$

we have

$$\int \sin(x + a) \sin(x - a)$$

$$= \int (\sin x \cos a + \cos x \sin a)(\sin x \cos a - \cos x \sin a)\,dx$$

$$= \int (\sin^2 x \cos^2 a - \cos^2 x \sin^2 a)\,dx$$

$$= \cos^2 a \int \sin^2 x\,dx - \sin^2 a \int \cos^2 x\,dx.$$

The method of partial fractions also makes use of the Sum Rule.

EXAMPLE 5 $\displaystyle\int \frac{x}{(x - a)(x - b)}\,dx,$ $a \neq 0,\quad b \neq 0.$ We have

$$\frac{x}{(x - a)(x - b)} = \frac{A}{x - a} + \frac{B}{x - b},$$

$$A = \frac{a}{a - b},\qquad B = \frac{b}{b - a},$$

$$\int \frac{x}{(x - a)(x - b)}\,dx = \frac{a}{a - b}\int \frac{dx}{x - a} + \frac{b}{b - a}\int \frac{dx}{x - b}.$$

E.   INTEGRATION BY CHANGE OF VARIABLES (Integration by Substitution)

Suppose an integral has the form

$$\int f(g(x))g'(x)\,dx.$$

When we make the substitution $u = g(x),\ du = g'(x)\,dx$, the integral becomes $\int f(u)\,du$. This new integral is often simpler than the original one.

EXAMPLE 6 $\int \sqrt{2x + 1}\,dx$. Let $u = 2x + 1,\ du = 2\,dx$. Then

$$\int \sqrt{2x + 1}\,dx = \int \sqrt{u} \cdot \tfrac{1}{2}\,du.$$

This can be integrated using the Constant and Power Rules,

$$\int \sqrt{u} \cdot \frac{1}{2}\,du = \frac{\tfrac{1}{2}u^{3/2}}{\tfrac{3}{2}} = \frac{1}{3}u^{3/2} = \frac{1}{3}(2x + 1)^{3/2}.$$

Clue   *If an integral has the form $f(ax + b)\,dx$, try the substitution $u = ax + b$, $du = a\,dx$.*

EXAMPLE 7 $\int \dfrac{1}{\sqrt{x}+1}dx$. Let $u = \sqrt{x}$. Then $du = \dfrac{1}{2\sqrt{x}}dx$, $dx = 2u\,du$. We get the rational integral

$$\int \frac{1}{\sqrt{x}+1}\,dx = \int \frac{2u}{u+1}\,du.$$

Clue   *If an integral involves $\sqrt{x}$, try the substitution $u = \sqrt{x}$, $dx = 2u\,du$. If an integral involves $\sqrt[n]{x}$, try $u = \sqrt[n]{x}$, $dx = nu^{n-1}\,du$.*

EXAMPLE 8 $\int \sin{(3x^2 - 1)}x\,dx$. Let $u = 3x^2 - 1$, $du = 6x\,dx$. Then

$$\int \sin{(3x^2 - 1)}x\,dx = \int (\sin u)\tfrac{1}{6}\,du.$$

Clue   *If an integral has the form $\int f(ax^2 + b)x\,dx$, try $u = ax^2 + b$, $du = ax\,dx$.*

If the derivatives in formulas I–XII are solidly memorized, then one can often recognize integrals of the form $\int f(g(x))g'(x)\,dx$ and find the right substitution. Here are three more clues.

Clue   *Given $\int f(a^x)a^x\,dx$, put $a^x = e^{x\ln a}$ and try the substitution $u = a^x$, $du = (\ln a)a^x\,dx$.*

Clue   *Given $\int f(\sin x)\cos x\,dx$, try $u = \sin x$, $du = \cos x\,dx$.*

Clue   *Given $\int f(\sin x, \cos x)\,dx$, try the substitution $u = \tan{(x/2)}$. It can be shown using trigonometric identities that*

$$\cos x = \frac{1 - u^2}{1 + u^2}, \quad \sin x = \frac{2u}{1 + u^2}, \quad dx = \frac{2\,du}{1 + u^2}.$$

EXAMPLE 9 $\int \dfrac{1}{2\sin x + \cos x}\,dx$. Putting $u = \tan\dfrac{x}{2}$, we obtain the rational integral

$$\int \frac{1}{\dfrac{4u}{1 + u^2} + \dfrac{1 - u^2}{1 + u^2}} \cdot \frac{2}{1 + u^2}\,du = \int \frac{2}{1 + 4u - u^2}\,du.$$

F. TRIGONOMETRIC SUBSTITUTIONS

If the simple substitutions corresponding to the basic formulas I–XII do not work, look for a trigonometric substitution. Trigonometric substitutions correspond to the formulas for derivatives of the inverse trigonometric functions. We have not asked you to memorize these formulas, because it is easier to remember the method of trigonometric substitution. The three trigonometric substitutions can be remembered by drawing right triangles. They are shown once more in Figure 8.9.1 They often result in an integral of powers of trigonometric functions of θ.

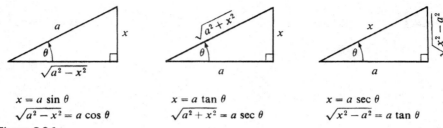

$$x = a \sin \theta$$
$$\sqrt{a^2 - x^2} = a \cos \theta$$

$$x = a \tan \theta$$
$$\sqrt{a^2 + x^2} = a \sec \theta$$

$$x = a \sec \theta$$
$$\sqrt{x^2 - a^2} = a \tan \theta$$

Figure 8.9.1

Clue   *If an integral contains $\sqrt{a^2 - x^2}$, $\sqrt{a^2 + x^2}$, or $\sqrt{x^2 - a^2}$, draw a triangle and label its sides so that it can be used to find the appropriate trigonometric substitution.*

EXAMPLE 10 $\int x^2 \sqrt{x^2 - 6^2} \, dx$. We draw the triangle shown in Figure 8.9.2 and use the substitution $x = 6 \sec \theta$.

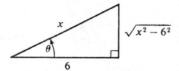

Figure 8.9.2

Then $\sqrt{x^2 - 6^2} = 6 \tan \theta$, $dx = 6 \tan \theta \sec \theta \, d\theta$, and the integral becomes

$$\int 6^2 \sec^2 \theta \cdot 6 \tan \theta \cdot 6 \tan \theta \sec \theta \, d\theta$$

$$= \int 6^4 \tan^2 \theta \sec^3 \theta \, d\theta = 6^4 \int (\sec^2 \theta - 1) \sec^3 \theta \, d\theta$$

$$= 6^4 \int \sec^5 \theta \, d\theta - 6^4 \int \sec^3 \theta \, d\theta.$$

G. INTEGRATION BY PARTS

When all else fails, try integration by parts. If u and v are differentiable functions of x, then

$$\int u \, dv = uv - \int v \, du.$$

To use the method on a given integral $\int f(x) \, dx$, we must break $f(x) \, dx$ into a product of the form $u \, dv$. u and dv are chosen by guesswork. The method works when we are able to evaluate both the integrals

$$\int dv, \qquad \int v \, du.$$

One should therefore look for a dv whose integral is known.

EXAMPLE 11 $\int x \ln x \, dx$. Try $u = \ln x$, $dv = x \, dx$. Then

$$du = 1/x \, dx, \qquad v = x^2/2,$$

$$\int x \ln x \, dx = \frac{x^2}{2} \ln x - \int \frac{x^2}{2} \frac{1}{x} dx = \frac{x^2}{2} \ln x - \frac{x^2}{4} + C.$$

We give two more clues and illustrate them with examples.

EXAMPLE 12 $\int (\ln x)^2 \, dx$. Put $u = (\ln x)^2$, $dv = dx$. Then

$$du = \frac{2 \ln x}{x} dx, \qquad v = x,$$

$$\int (\ln x)^2 \, dx = x(\ln x)^2 - 2 \int \ln x \, dx.$$

Clue   *Sometimes $u = f(x)$, $dv = dx$ can be used to evaluate an integral $\int f(x) \, dx$ by parts.*

Clue   *Sometimes one can perform two integrations by parts and solve for the desired integral.*

EXAMPLE 13 $\int \sin (\ln x) \, dx$. Let $u = \sin (\ln x)$, $dv = dx$. Then

$$du = \frac{\cos (\ln x)}{x} dx, \qquad v = x.$$

Integrating by parts,

$$\int \sin (\ln x) \, dx = x \sin (\ln x) - \int \cos (\ln x) \, dx.$$

Integrating by parts again,

$$\int \cos (\ln x) \, dx = x \cos (\ln x) + \int \sin (\ln x) \, dx.$$

Then $\int \sin (\ln x) \, dx = x \sin (\ln x) - x \cos (\ln x) - \int \sin (\ln x) \, dx,$

$$\int \sin (\ln x) \, dx = \tfrac{1}{2} x \sin (\ln x) - \tfrac{1}{2} x \cos (\ln x) + C.$$

PROBLEMS FOR SECTION 8.9

Evaluate the following integrals.

1    $\int 3 \sin x + 4 \cos x \, dx$
       2    $\int \tan (3x - 5) \, dx$

3    $\int \dfrac{x}{\sqrt[3]{x^2 - 1}} dx$
       4    $\int x e^{-x} \, dx$

5 $\displaystyle\int \frac{dx}{\sqrt{x+2}-\sqrt{x}}$

6 $\displaystyle\int \frac{x^3-4}{x+1}dx$

7 $\displaystyle\int \frac{1}{1+\sqrt{x}}dx$

8 $\displaystyle\int \frac{\sec^2 x}{1+\tan x}dx$

9 $\displaystyle\int \frac{1}{x^2\sqrt{4x^2+1}}dx$

10 $\displaystyle\int \frac{e^{1/x}}{x^2}dx$

11 $\displaystyle\int \frac{\sin\sqrt{x}}{\sqrt{x}}dx$

12 $\displaystyle\int x^{-3}\ln x\,dx$

13 $\displaystyle\int x^2\sqrt{x-3}\,dx$ ·

14 $\displaystyle\int \frac{x^3}{\sqrt{1-9x^2}}dx$

15 $\displaystyle\int \ln(3x+4)\,dx$

16 $\displaystyle\int \frac{x-2}{3x(x+4)}dx$

17 $\displaystyle\int x\tan^2 x\,dx$

18 $\displaystyle\int \frac{1}{x(x-1)^3}dx$

19 $\displaystyle\int x\sin(3x^2+1)\,dx$

20 $\displaystyle\int \frac{dx}{\sqrt[3]{x}+1}$

21 $\displaystyle\int \ln(x^2+x^3)\,dx$

22 $\displaystyle\int x^2 e^x\,dx$

23 $\displaystyle\int \sin\theta\ln(\cos\theta)\,d\theta$

24 $\displaystyle\int e^{\sqrt{x}}\,dx$

25 $\displaystyle\int \frac{\sin\theta}{2-\cos\theta}d\theta$

26 $\displaystyle\int \sqrt{2x+3}\,dx$

27 $\displaystyle\int e^x(e^x+1)^3\,dx$

28 $\displaystyle\int \frac{1}{x\sqrt{1-x^2}}dx$

29 $\displaystyle\int \sin\sqrt{x}\,dx$

30 $\displaystyle\int \frac{2^x}{2^x+1}dx$

31 $\displaystyle\int \frac{\sinh x}{1+\cosh x}dx$

32 $\displaystyle\int \frac{x\,dx}{\sqrt{x-1}-\sqrt{x}}$

33 $\displaystyle\int \frac{\ln x}{(1+x)^2}dx$

34 $\displaystyle\int \cos^3 x\sqrt{\sin x}\,dx$

35 $\displaystyle\int \sin^3 x\sqrt{1-\cos x}\,dx$

36 $\displaystyle\int \sqrt{e^x+1}\,dx$

37 $\displaystyle\int \frac{1}{x\sqrt{2+x^2}}dx$

38 $\displaystyle\int \frac{x-2}{3x(x+4)}dx$

39 $\displaystyle\int \sqrt{x}\sqrt{4-x}\,dx$

40 $\displaystyle\int \cos^3 x\sin^3 x\,dx$

41 $\displaystyle\int \arcsin(5x-2)\,dx$

42 $\displaystyle\int \frac{1}{\sin\theta+\cos\theta}d\theta$

43 $\displaystyle\int e^x\cos x\,dx$

44 $\displaystyle\int \frac{1}{x(1+(\ln x)^2)}dx$

45 $\displaystyle\int \frac{x^4+1}{x^2+1}dx$

46 $\displaystyle\int \frac{x^2}{\sqrt{x^2-1}}dx$

47 $\displaystyle\int x\operatorname{arcsec}(x^2)\,dx$

48 $\displaystyle\int x\sqrt{x-2}\,dx$

49 $\displaystyle\int \ln(x^2\sqrt{4x-1})\,dx$

50 $\displaystyle\int 4^x\sin(4^x)\,dx$

51 $\displaystyle\int \frac{\sqrt{4x^2 - 1}}{x^2}\, dx$

52 $\displaystyle\int x^3 e^{x^2}\, dx$

53 $\displaystyle\int \arctan \sqrt{x}\, dx$

54 $\displaystyle\int \frac{x}{\sqrt{4x + 1}}\, dx$

55 $\displaystyle\int x \sec (4x^2 + 7)\, dx$

56 $\displaystyle\int \frac{1}{4 + \sin \theta}\, d\theta$

57 $\displaystyle\int \frac{x^3}{\sqrt{1 - 9x^2}}\, dx$

58 $\displaystyle\int \tan \theta \ln (\sin \theta)\, d\theta$

59 $\displaystyle\int \frac{dx}{x^2 \sqrt{x^2 - 3}}$

60 $\displaystyle\int \frac{\sqrt{4x^2 + 1}}{x^2}\, dx$

61 $\displaystyle\int \cos (\sqrt[3]{x})\, dx$

62 $\displaystyle\int \ln (1 + x^2)\, dx$

63 $\displaystyle\int \frac{dx}{(1 - x^2)^{5/2}}$

64 $\displaystyle\int \frac{dx}{1 - \cos 3x}$

65 $\displaystyle\int \cos^2 (\ln x)\, dx$

EXTRA PROBLEMS FOR CHAPTER 8

1 Evaluate $\displaystyle\lim_{x \to \infty} 8^{\sqrt{x}} - 2^x$.

2 Evaluate $\displaystyle\lim_{x \to \infty} 2^{3x-1} - 3^{2x}$.

3 Find $\dfrac{dy}{d\theta}$ where $y = e^{\cos \theta}$.

4 Find $\dfrac{dy}{dx}$ where $y = x^2 e^{-x^2}$.

5 Find $\dfrac{dy}{dx}$ where $y = \operatorname{csch}^3 x$.

6 Sketch the curve $y = \operatorname{csch} x$.

7 Evaluate $\displaystyle\int 3^x \sin (3^x)\, dx$.

8 Evaluate $\displaystyle\int \frac{e^{\arcsin x}}{\sqrt{1 - x^2}}\, dx$.

9 Evaluate $\displaystyle\int e^x \sqrt{1 - e^{2x}}\, dx$.

10 Evaluate $\displaystyle\int \frac{dx}{\sqrt{e^{2x} - 1}}$.

11 Evaluate $\displaystyle\int e^x \sinh x\, dx$.

12 Evaluate $\displaystyle\int x^2 \sinh x\, dx$.

13 Find $\dfrac{dy}{dx}$ where $y = \ln [(x^2 - 1)^4]$.

14 Find $\dfrac{ds}{dt}$ where $s = e^t \ln t$.

15 Find $\dfrac{dy}{dx}$ where $y = \ln \left| \dfrac{(3x + 2)(5x - 4)}{(2x - 1)(x^2 + 1)} \right|$.

16 Evaluate $\displaystyle\lim_{t \to \infty} \frac{\ln t}{\ln (\ln t)}$.

17 Evaluate $\displaystyle\lim_{t \to 0} (1 + t)^{2/t}$.

18 Evaluate $\displaystyle\int \frac{\sec^2 \theta}{1 + \tan \theta}\, d\theta$.

19 Evaluate $\displaystyle\int \frac{1}{x(a + bx)}\, dx$.

20 Evaluate $\displaystyle\int_{-1}^{0} \frac{1}{x}\, dx$.

21 Evaluate $\displaystyle\int_{0}^{\infty} \frac{1}{x}\, dx$.

22 Find all solutions of $dy/dx = ay^2$.

23 Find all solutions of $dy/dx = ax/y$.

24 A falling object of mass m is subject to a force due to gravity of $-mg$ and a force due to air resistance of $-kv$, where v is its velocity. If $v = 0$ at time $t = 0$, find v as a function of time.

25   The pressure P and volume V of a gas in an adiabatic process (a process with no heat transfer) are related by the differential equation

$$P + kV\frac{dP}{dV} = 0,$$

where k is constant. Solve for P as a function of V.

26   An electrical condenser discharges at a rate proportional to its charge Q, so that $dQ/dt = -kQ$ for some constant k. If the charge at time $t = 0$ is Q_0, find Q as a function of t.

27   Newton's Law of Cooling states that a hot object cools down at a rate proportional to the difference between the temperature of the object and the air temperature. If the object has temperature $140°$ at $t = 0$, $100°$ at $t = 10$, and $80°$ at $t = 20$, find the temperature y of the object as a function of t, and find the air temperature.

28   Find $\dfrac{dy}{dx}$ where $y = x^{(2^x)}$.   29   Find $\dfrac{dy}{dt}$ where $y = (4t + 1)^t(t - 3)^{2t+1}$.

30   Evaluate $\displaystyle\int \sec{(5\theta)}\, d\theta$.   31   Evaluate $\displaystyle\int \tanh x\, dx$.

32   Evaluate $\displaystyle\int \sqrt{(1/x^2) - 1}\, dx$.   33   Evaluate $\displaystyle\int (x + 1)^{3/2}\, dx$.

34   Evaluate $\displaystyle\int \theta \tan^2 \theta\, d\theta$.

35   Find the surface area generated by rotating the curve $y = \sin x, 0 \le x \le \pi$, about the x-axis.

36   Find the surface area generated by rotating the parabola $y = x^2, 0 \le x \le 1$, about the y-axis.

37   Approximate $e^{0.03}$ and give an error estimate.

38   Approximate $\ln{(0.996)}$ and give an error estimate.

39   Use the trapezoidal rule with $\Delta x = 1$ to approximate $\ln 6$ and give an error estimate.

40   Find the centroid of the region under the curve $y = e^x, 0 \le x \le 1$.

41   Find the centroid of the region under the curve $y = \ln x, 1 \le x \le 2$.

42   Find the length of the curve $y = e^x, 0 \le x \le 1$.

43   Find the surface area generated by rotating the curve $y = e^x, 0 \le x \le 1$, about the x-axis.

☐ 44   Obtain a reduction formula for $\int x^n e^x\, dx$.

☐ 45   Prove that the function $y = x^x, x > 0$, is continuous, using the continuity of $\ln x$ and e^x.

☐ 46   Let $y = f(x)$ be a function which is continuous on the whole real line and such that for all u and v, $f(u + v) = f(u)f(v)$. Prove that $f(x) = a^x$ where $a = f(1)$. *Hint:* First prove it for x rational.

☐ 47   Prove that for all $x > 0$,

$$\ln\left(1 + \frac{1}{x}\right) > \frac{1}{x + 1}.$$

Hint: Use the formula

$$\ln\left(1 + \frac{1}{x}\right) = \int_1^{1 + 1/x} \frac{1}{t}\, dt.$$

☐ 48   Prove that the function $f(x) = (1 + 1/x)^x$ is increasing for $x > 0$.

□ **49**   Show that the improper integral $\int_0^\infty \sqrt{x}\, e^{-x}\, dx$ converges. *Hint*: Show that the definite integrals $\int_0^H \sqrt{x}\, e^{-x}\, dx$ are finite and have the same standard part for all positive infinite H.

□ **50**   Show that $\int_{-\infty}^\infty e^{-x^2}\, dx$ converges.

□ **51**   The inverse square law for gravity shows that an object projected vertically from the earth's surface will rise according to the differential equation

(1) $$\frac{d^2y}{dt^2} = -\frac{k}{y^2}, \qquad t \geq 0.$$

Here y is the height above the earth's center. If $v = dy/dt$ is the velocity at time t, then

$$\frac{d^2y}{dt^2} = \frac{dv}{dt} = \frac{dv}{dy}\frac{dy}{dt} = v\frac{dv}{dy},$$

so Equation 1 may be written as

(2) $$v\frac{dv}{dy} = -\frac{k}{y^2}.$$

Assume that at time $t = 0$, $y = 4000$ miles (the radius of the earth) and $v = v_0$ (the initial velocity). Solve for velocity as a function of y. Find the *escape velocity*, i.e., the smallest initial velocity v_0 such that the velocity v never drops to zero.

INFINITE SERIES

9.1 SEQUENCES

DEFINITION

*An **infinite sequence** is a real function whose domain is the set of all positive integers.*

A sequence a can be displayed in the form

$$a(1), a(2), \ldots, a(n), \ldots.$$

The value $a(n)$ is called the nth *term* of the sequence and is usually written a_n. The whole sequence is denoted by

$$\langle a_n \rangle = a_1, a_2, \ldots, a_n, \ldots.$$

Hyperintegers, which were introduced in Section 3.8, are a basic tool in this chapter. Since a_n is defined for every positive integer n, a_H is defined for every positive infinite hyperinteger H.

EXAMPLE 1   If the sequence is simple enough one can look at the first few terms and guess the general rule for computing the nth term. For instance:

$$1, 1, 1, 1, 1, \ldots \qquad\qquad a_n = 1$$
$$-1, 0, 1, 2, 3, \ldots \qquad\qquad a_n = n - 2$$
$$-2, -4, -6, -8, -10, \ldots \qquad a_n = -2n$$
$$1, -1, 1, -1, 1, \ldots \qquad\qquad a_n = (-1)^{n-1}$$
$$1, \frac{1}{2}, \frac{1}{3}, \frac{1}{4}, \frac{1}{5}, \ldots \qquad\qquad a_n = \frac{1}{n}$$

The graph of a sequence will look like a collection of dots whose x-coordinates are spaced one apart. Some examples of graphs of sequences are shown in Figure 9.1.1.

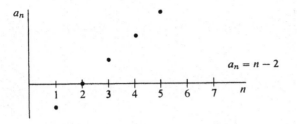

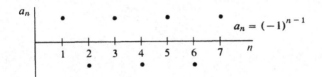

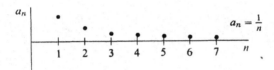

Figure 9.1.1

EXAMPLE 2   The sequence

$$3.1, 3.14, 3.141, 3.1415, 3.14159, \ldots$$

is defined by the rule

$$a_n = \pi \text{ to } n \text{ decimal places},$$

that is, $a_n = \dfrac{m}{10^n}$ where m is the integer such that $\dfrac{m}{10^n} \leq \pi < \dfrac{m+1}{10^n}$.

EXAMPLE 3   The number $n!$, read n factorial, is defined as the product of the first n positive integers;

$$n! = 1 \cdot 2 \cdot \cdots \cdot n$$

$\langle n! \rangle$ is an important sequence. Its first few terms are

$$1, 2, 6, 24, 120, 720, \ldots.$$

By convention, $0!$ is defined by $0! = 1.$

DEFINITION

*An infinite sequence $\langle a_n \rangle$ is said to **converge** to a real number L if a_H is infinitely close to L for all positive infinite hyperintegers H (Figure 9.1.2). L is called the* ***limit*** *of the sequence and is written*

$$L = \lim_{n \to \infty} a_n.$$

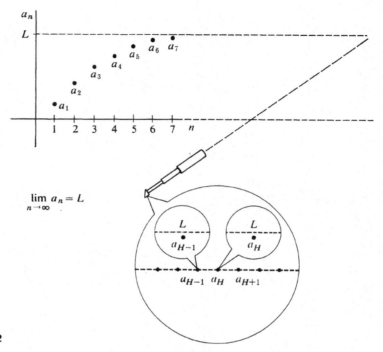

Figure 9.1.2

*A sequence which does not converge to any real number is said to **diverge**. If a_H is positive infinite for all positive infinite hyperintegers H, the sequence is said to **diverge to** ∞, and we write*

$$\lim_{n \to \infty} a_n = \infty.$$

Sequences can diverge to $-\infty$, and also diverge without diverging to ∞ or to $-\infty$.

Throughout this chapter, H and K will always be used for positive infinite hyperintegers. One can often determine whether or not a sequence converges by examining the values of a_H for infinite H. The definition gives us some convenient working rules.

(1)   If a_H is infinitely close to L for all H, the sequence converges to L.

(2)   If we can find a_H and a_K which are not infinitely close to each other, the sequence diverges.

(3)   If at least one a_H is infinite, the sequence diverges.

(4)   If all the a_H are positive infinite, the sequence diverges to ∞.

EXAMPLE 1 (Continued)

$\lim_{n \to \infty} 1 = 1$,   converges,   because $a_H = 1$ for all H.

$\lim_{n \to \infty} n - 2 = \infty$,   diverges,   because $H - 2$ is positive infinite for all H.

$$\lim_{n\to\infty} (-2n) = -\infty, \quad \text{diverges,} \quad \text{because } -2H \text{ is negative infinite for all } H.$$

$$\lim_{n\to\infty} (-1)^n \text{ is undefined,} \quad \text{diverges,} \quad \text{because } (-1)^{2H} = 1 \text{ but } (-1)^{2H+1} = -1.$$

$$\lim_{n\to\infty} \frac{1}{n} = 0, \quad \text{converges,} \quad \text{because } \frac{1}{H} \text{ has standard part zero.}$$

EXAMPLE 2 (Continued)   The sequence

$$3.1, \ 3.14, \ 3.141, \ 3.1415, \ 3.14159, \dots, a_n, \dots$$

where $a_n = (\pi$ to n decimal places), converges to π. That is,

$$\lim_{n\to\infty} a_n = \pi.$$

PROOF Let H be positive infinite. For some K,

$$\frac{K}{10^H} \le \pi < \frac{K+1}{10^H}.$$

Then $\qquad\qquad a_H = \dfrac{K}{10^H}, \qquad a_H \le \pi \le a_H + \dfrac{1}{10^H}.$

But $1/10^H$ is infinitesimal, so $a_H \approx \pi$.

EXAMPLE 3 (Continued) $\lim\limits_{n\to\infty} n! = \infty.$

PROOF For any $n > 1$, we have

$$(n-1)! \ge 1, \qquad n! = n\cdot(n-1)! \ge n.$$

Therefore for positive infinite H, $H! \ge H$ is positive infinite.

Given a function $f(x)$ defined for all $x \ge 1$, we can form the sequence

$$f(1), f(2), \dots, f(n), \dots.$$

The graph of the sequence $\langle f(n) \rangle$ is the collection of dots on the curve $y = f(x)$ where the x-coordinate is a positive integer (Figure 9.1.3).

If $\lim\limits_{x\to\infty} f(x) = L$, then $\lim\limits_{n\to\infty} f(n) = L$ because $f(H) \approx L$ for any positive infinite H.

EXAMPLE 4 $\lim\limits_{n\to\infty} \dfrac{4n^2+1}{n^2+3n} = \lim\limits_{x\to\infty} \dfrac{4x^2+1}{x^2+3x} = 4.$

Similarly, if $\lim\limits_{x\to\infty} f(x) = \infty$ then $\lim\limits_{n\to\infty} f(n) = \infty.$

EXAMPLE 5 $\lim\limits_{n\to\infty} \ln(n) = \lim\limits_{x\to\infty} \ln(x) = \infty.$

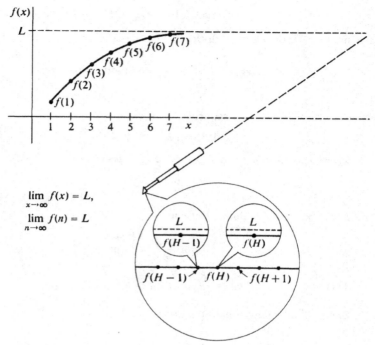

Figure 9.1.3

If $\lim_{x \to 0^+} f(x) = L$, then $\lim_{n \to \infty} f(1/n) = L$. If H is positive infinite, then $\varepsilon = 1/H$ is infinitesimal and

$$f(1/H) = f(\varepsilon) \approx L.$$

EXAMPLE 6 $\lim_{n \to \infty} c^{1/n} = \lim_{x \to 0^+} c^x = c^0 = 1$, if $c > 0$.

EXAMPLE 7   Evaluate the limits

(a) $\lim_{n \to \infty} \left(1 + \dfrac{1}{c}\right)^n$ where $c > 0$,

(b) $\lim_{n \to \infty} \left(1 + \dfrac{1}{n}\right)^c$ where $c > 0$,

(c) $\lim_{n \to \infty} \left(1 + \dfrac{1}{n}\right)^n$.

The answers are

(a) $\lim_{n \to \infty} \left(1 + \dfrac{1}{c}\right)^n = \lim_{x \to \infty} \left(1 + \dfrac{1}{c}\right)^x = \infty$.

(b) $\lim_{n \to \infty} \left(1 + \dfrac{1}{n}\right)^c = \lim_{x \to 0^+} (1 + x)^c = 1$.

(c) $\lim_{n \to \infty} \left(1 + \dfrac{1}{n}\right)^n = \lim_{x \to \infty} \left(1 + \dfrac{1}{x}\right)^x = e$.

The limit $\lim_{n\to\infty}(1 + 1/n)^n = e$ is closely related to compound interest. Suppose a bank pays interest on one dollar at the rate of 100% per year. If the interest is compounded n times per year the dollar will grow to $(1 + 1/n)$ after $1/n$ years, to $(1 + 1/n)^k$ after k/n years, and thus to $(1 + 1/n)^n$ after one year. Since $\lim_{n\to\infty}(1 + 1/n)^n = e$, one dollar will grow to e dollars if the interest is compounded continuously for one year, and to e^t dollars after t years.

More generally, suppose the account initially has a dollars and the bank pays interest at the rate of $b\%$ per year. If the interest is compounded n times per year, the account will grow as follows:

$$
\begin{array}{ll}
0 \text{ years} & a \\[2mm]
\dfrac{1}{n}\text{ years} & a\left(1 + \dfrac{b}{100}\cdot\dfrac{1}{n}\right) \\[4mm]
\dfrac{k}{n}\text{ years} & a\left(1 + \dfrac{b}{100}\cdot\dfrac{1}{n}\right)^k \\[4mm]
1\text{ year} & a\left(1 + \dfrac{b}{100}\cdot\dfrac{1}{n}\right)^n
\end{array}
$$

If the interest is compounded continuously the account will grow in one year to

$$\lim_{n\to\infty} a\left(1 + \frac{b}{100}\cdot\frac{1}{n}\right)^n.$$

We can evaluate this limit by setting $x = \dfrac{100}{b}n,\ n = \dfrac{b}{100}x.$

$$\lim_{n\to\infty} a\left(1 + \frac{b}{100}\cdot\frac{1}{n}\right)^n = \lim_{x\to\infty} a\left(1 + \frac{1}{x}\right)^{bx/100} = ae^{b/100}.$$

Thus the account grows to $ae^{b/100}$ dollars after one year and to $ae^{bt/100}$ dollars after t years.

Sometimes we may wish to know how rapidly a sequence grows. If two sequences approach ∞ and their quotient also approaches ∞,

$$\lim_{n\to\infty} a_n = \infty, \qquad \lim_{n\to\infty} b_n = \infty, \qquad \lim_{n\to\infty} \frac{a_n}{b_n} = \infty,$$

the sequence $\langle a_n \rangle$ is said to *grow faster* than the sequence $\langle b_n \rangle$. For each infinite H, both a_H and b_H are infinite. But a_H/b_H is still infinite, so a_H is infinite even compared to b_H.

THEOREM 1

Each of the following sequences approaches ∞.

$$\lim_{n\to\infty} n! = \infty,$$

$$\lim_{n\to\infty} b^n = \infty \qquad \text{if } b > 1,$$

$$\lim_{n\to\infty} n^c = \infty \qquad \text{if } c > 0,$$

$$\lim_{n\to\infty} \ln(n) = \infty.$$

Moreover, each sequence in the list grows faster than the next one,

(i) $\lim\limits_{n \to \infty} \dfrac{n!}{b^n} = \infty$ $(b > 1),$

(ii) $\lim\limits_{n \to \infty} \dfrac{b^n}{n^c} = \infty$ $(b > 1, \;\; c > 0),$

(iii) $\lim\limits_{n \to \infty} \dfrac{n^c}{\ln (n)} = \infty$ $(c > 0).$

PROOF Let H be positive infinite. We already know that $\ln H$ is positive infinite. We must show that each of the following are also positive infinite.

$$\frac{H!}{b^{H'}} \qquad \frac{b^H}{H^{c'}} \qquad \frac{H^c}{\ln H}.$$

It is easier to show that their logarithms are positive infinite. We need the fact that, by l'Hospital's rule for ∞/∞,

$$\lim_{x \to \infty} \frac{\ln x}{x} = \lim_{x \to \infty} \frac{1/x}{1} = 0,$$

so $\dfrac{\ln K}{K} \approx 0$ for all infinite K.

(i) $\dfrac{H!}{b^H}.$ Let $m > b$. Then

$$\ln\left(\frac{H!}{b^H}\right) = \ln 1 + \cdots + \ln(m - 1) + \ln m + \cdots + \ln H - H \ln b$$

$$> (H - m) \ln m - H \ln b = H(\ln m - \ln b) - m \ln m.$$

Since $m > b$, $\ln m > \ln b$, and $\ln(H!/b^H)$ is positive infinite.

(ii) $\dfrac{b^H}{H^c}.$

$$\ln\left(\frac{b^H}{H^c}\right) = H \ln b - c \ln H = H\left(\ln b - c\,\frac{\ln H}{H}\right).$$

Since $b > 1$, $\ln b > 0$. $\dfrac{\ln H}{H}$ is infinitesimal. Therefore $\ln(b^H/H^c)$ is positive infinite.

(iii) $\dfrac{H^c}{\ln H}.$ Let $K = \ln H.$

$$\ln\left(\frac{H^c}{\ln H}\right) = c \ln H - \ln(\ln H) = K\left(c - \frac{\ln K}{K}\right).$$

Since $c > 0$, K is infinite, and $(\ln K)/K$ is infinitesimal, $\ln(H^c/\ln H))$ is positive infinite.

Note: For $n = 1$, the term $n^c/(\ln n)$ is undefined, so we should start the sequence with $n = 2$.

EXAMPLE 8   From Theorem 1, the following sequences all approach ∞.

$$\frac{\sqrt{2}}{\ln 2}, \frac{\sqrt{3}}{\ln 3}, \frac{\sqrt{4}}{\ln 4}, \cdots, \frac{\sqrt{n}}{\ln(n)}, \cdots$$

$$\frac{2^1}{1^{10}}, \frac{2^2}{2^{10}}, \frac{2^3}{3^{10}}, \frac{2^4}{4^{10}}, \cdots, \frac{2^n}{n^{10}}, \cdots$$

$$\frac{1!}{100^1}, \frac{2!}{100^2}, \frac{3!}{100^3}, \frac{4!}{100^4}, \cdots, \frac{n!}{100^n}, \cdots$$

If $\lim_{n \to \infty} a_n = \infty$, then $\lim_{n \to \infty} 1/a_n = 0$ because $1/a_H$ will be infinitesimal.

COROLLARY

(i) $\lim_{n \to \infty} b^{-n} = 0$ if $b > 1$.

(ii) $\lim_{n \to \infty} n^{-c} = 0$ if $c > 0$.

Like other types of limits, limits of sequences have an ε, N condition. It will be used later to prove theorems on series.

THEOREM 2 (ε, N Condition for Limits of Sequences)

$$\lim_{n \to \infty} a_n = L$$

if and only if for every real number $\varepsilon > 0$ there is a positive integer N such that the numbers

$$a_N, a_{N+1}, a_{N+2}, \ldots, a_{N+m}, \cdots$$

are all within ε of L.

The proof is similar to that of the ε, δ condition for limits of functions. The ε, N condition says intuitively that a_n gets close to L as the integer n gets large.

A similar condition can be formulated for $\lim_{n \to \infty} a_n = \infty$.

THEOREM 3 (ε, N Condition for Infinite Limits)

$$\lim_{n \to \infty} a_n = \infty$$

if and only if for every real number B, there is a positive integer N such that the numbers

$$a_N, a_{N+1}, a_{N+2}, \ldots, a_{N+m}, \cdots$$

are all greater than B.

We conclude this section with another useful criterion for convergence.

CAUCHY CONVERGENCE TEST FOR SEQUENCES

> *A sequence $\langle a_n \rangle$ converges if and only if*

(1) $a_H \approx a_K$ *for all infinite H and K.*

PROOF First suppose $\langle a_n \rangle$ converges, say $\lim_{n \to \infty} a_n = L$. Then for all infinite H and K,

$$a_H \approx L \approx a_K.$$

Now assume Equation 1 and let H be infinite. There are three cases to consider.

Case 1 a_H is finite. Then for all infinite K,

$$st(a_K) = st(a_H),$$

so the sequence converges to $st(a_H)$.

Case 2 a_H is positive infinite. For each finite m, $a_H \geq a_m + 1$. Among the hyperintegers $\{1, 2, \ldots, H - 1\}$, there must be a largest element M such that $a_H \geq a_M + 1$. But this largest M cannot be finite, and since $a_M \not\approx a_H$, M cannot be infinite. Therefore Case 2 cannot arise.

Case 3 a_H is negative infinite. By a similar argument this case cannot arise.

Therefore, only Case 1 is possible, whence $\langle a_n \rangle$ converges.

PROBLEMS FOR SECTION 9.1

In Problems 1–8, find the nth term of the sequence.

1	$\frac{1}{2}, \frac{1}{4}, \frac{1}{8}, \frac{1}{16}, \ldots$	**2**	$\frac{1}{2}, \frac{2}{3}, \frac{3}{4}, \frac{4}{5}, \ldots$
3	$-1, 2, -3, 4, -5, 6, \ldots$	**4**	$2, 5, 10, 17, 26, 37, \ldots$
5	$1, 1\frac{1}{2}, 1\frac{3}{4}, 1\frac{7}{8}, \ldots$	**6**	$1, 3, 6, 10, 15, \ldots$
7	$2, 4, 16, 256, \ldots$	**8**	$0.6, 0.61, 0.616, 0.6161, \ldots$

Determine whether the following sequences converge, and find the limits when they exist.

9	$a_n = \sqrt{n}$	**10**	$a_n = \dfrac{n + 2}{n}$
11	$a_n = n - \dfrac{n^2}{n + 1}$	**12**	$a_n = n(-1)^n$
13	$a_n = \dfrac{(-1)^n}{\sqrt{n}}$	**14**	$a_n = \dfrac{n!}{n^3}$
15	$a_n = \dfrac{n}{(\ln(n))^2}$	**16**	$a_n = \sqrt[n]{n}$
17	$a_n = \ln(\ln(n))$	**18**	$a_n = \sqrt{n^2 + n} - n$
19	$a_n = \left(\dfrac{n - 1}{n}\right)^n$	**20**	$a_n = \dfrac{3n^2 - 2n + 4}{2n^2 - n + 1}$
21	$a_n = \dfrac{n^2 + 1}{n^3 + 4}$	**22**	$a_n = \dfrac{n^3 - 2}{n^2 + 5}$

23 $\quad a_n = \dfrac{2^n + 3^n}{2^n - 3^n}$ 　　　　　　　　24 $\quad a_n = 2^n - n^2$

25 $\quad a_n = n! - 10^n$ 　　　　　　　　　　26 $\quad a_n = \dfrac{n! + 2}{(n+1)! + 1}$

27 $\quad a_n = \dfrac{\ln(n)}{\ln(\ln(n))}$ 　　　　　　　　28 $\quad a_n = (n!)^{1/n}$

29 $\quad a_n = \dfrac{(n+1)^n}{n^{n+1}}$

☐ 30　Formulate an ε, N condition for $\lim_{n \to \infty} a_n = -\infty$.

☐ 31　Show that if $\lim_{n \to \infty} a_n = L$ and $\lim_{n \to \infty} b_n = M$ then $\lim_{n \to \infty} (a_n + b_n) = L + M$.

☐ 32　Show that if $\lim_{n \to \infty} a_n = L$ then $\lim_{n \to \infty} ca_n = cL$.

9.2 SERIES

The sum of finitely many real numbers $a_1, a_2, \ldots, a_n$ is again a real number $a_1 + a_2 + \cdots + a_n$. Sometimes we wish to form the sum of an infinite sequence of real numbers,

$$a_1 + a_2 + \cdots + a_n + \cdots.$$

For example, if a man walks halfway across a room of unit width, then half of the remaining distance, then half the remaining distance again, and so forth, the total distance he will travel is an infinite sum

$$\frac{1}{2} + \frac{1}{4} + \frac{1}{8} + \frac{1}{16} + \cdots + \frac{1}{2^n} + \cdots.$$

In n steps he will travel $1 - \dfrac{1}{2^n}$ units,

$$\frac{1}{2} + \frac{1}{4} + \frac{1}{8} + \cdots + \frac{1}{2^n} = 1 - \frac{1}{2^n}.$$

Thus he will get closer and closer to the other side of the room, and we have the limit

$$\lim_{n \to \infty} \left(\frac{1}{2} + \frac{1}{4} + \frac{1}{8} + \cdots + \frac{1}{2^n} \right) = 1.$$

It is natural to call this limit the infinite sum,

$$1 = \frac{1}{2} + \frac{1}{4} + \frac{1}{8} + \frac{1}{16} + \cdots + \frac{1}{2^n} + \cdots.$$

We can go from this example to the general notion of an infinite sum. When we wish to find the sum of an infinite sequence $\langle a_n \rangle$ we call it an *infinite series* and write it in the form

$$a_1 + a_2 + \cdots + a_n + \cdots.$$

Given an infinite sequence $\langle a_n \rangle$, each finite sum

$$a_1 + \cdots + a_n$$

is defined. This sum is called the nth *partial sum* of the series. Thus, with each infinite series

$$a_1 + a_2 + \cdots + a_n + \cdots,$$

there are associated two sequences, the *sequence of terms*,

$$a_1, a_2, \ldots, a_n, \ldots,$$

and the *sequence of partial sums*,

$$S_1, S_2, \ldots, S_n, \ldots \qquad \text{where } S_n = a_1 + \cdots + a_n.$$

For each positive hyperreal number H, the *infinite partial sum*

$$S_H = a_1 + \cdots + a_H$$

is also defined, by the Extension Principle.

The sum of an infinite series will be a real number which is close to the nth partial sum for large n, and infinitely close to the infinite partial sums. Before stating the definition precisely, let us examine some infinite series and their partial sum sequences, and guess at their sums.

Table 9.2.1

Series	Partial sums	Sum
$1 + 0.1 + 0.01 + 0.001 + \cdots$	$1, 1.1, 1.11, 1.111, \ldots$	$1\frac{1}{9}$
$1 + \frac{1}{2} + \frac{1}{4} + \frac{1}{8} + \frac{1}{16} + \cdots$	$1, 1\frac{1}{2}, 1\frac{3}{4}, 1\frac{7}{8}, 1\frac{15}{16}, \ldots$	2
$1 - 1 + 1 - 1 + 1 - 1 + \cdots$	$1, 0, 1, 0, 1, 0, \ldots$	?
$1 + 1 + 1 + 1 + 1 + \cdots$	$1, 2, 3, 4, 5, \ldots$	∞
$1 + \frac{1}{2} + \frac{1}{3} + \frac{1}{4} + \frac{1}{5} + \cdots$	$1, \frac{3}{2}, \frac{11}{6}, \frac{25}{12}, \frac{137}{60}, \ldots$	?
$3 + 0.1 + 0.04 + 0.001 + \cdots$	$3, 3.1, 3.14, 3.141, \ldots$	π

DEFINITION

> The **sum** of an infinite series is defined as the limit of the sequence of partial sums if the limit exists,
>
> $$a_1 + a_2 + \cdots + a_n + \cdots = \lim_{n \to \infty} (a_1 + \cdots + a_n).$$
>
> The series is said to **converge** to a real number S, **diverge**, or **diverge to** ∞, if the sequence of partial sums converges to S, diverges, or diverges to ∞, respectively.

The sum of an infinite series can often be found by looking at the infinite partial sums $a_1 + \cdots + a_H$. Corresponding to our working rules for limits of sequences, we have the following rules for sums of series.

(1)   If the value of every infinite partial sum is finite with standard part S, then the series converges to S,

$$a_1 + \cdots + a_n + \cdots = S.$$

(2)   If there are two infinite partial sums which are not infinitely close to each other, the series diverges.

(3) If there is an infinite partial sum whose value is infinite, then the series diverges.

(4) If all infinite partial sums have positive infinite values, the series diverges to ∞,

$$a_1 + \cdots + a_n + \cdots = \infty.$$

Given an infinite series, we often wish to answer two questions. Does the series converge? What is the sum of the series? Our next theorem gives a formula for the sum of an important kind of series, the geometric series.

For each constant c, the series

$$1 + c + c^2 + \cdots + c^n + \cdots$$

is called the *geometric series* for c.

THEOREM 1

If $|c| < 1$, *the geometric series converges and*

$$1 + c + c^2 + \cdots + c^n + \cdots = \frac{1}{1 - c}.$$

PROOF For each n we have

$$(1 - c)(1 + c + c^2 + \cdots + c^n)$$
$$= (1 + c + c^2 + \cdots + c^n) - (c + c^2 + \cdots + c^n + c^{n+1})$$
$$= 1 - c^{n+1}.$$

The nth partial sum is therefore

$$1 + c + c^2 + \cdots + c^n = \frac{1 - c^{n+1}}{1 - c}.$$

The infinite partial sum up to H is

$$1 + c + \cdots + c^H = \frac{1 - c^{H+1}}{1 - c}.$$

Since $|c| < 1$, c^{H+1} is infinitesimal, so

$$1 + c + \cdots + c^H \approx \frac{1}{1 - c}.$$

EXAMPLE 1 $1 + 0.1 + 0.01 + 0.001 + \cdots = \dfrac{1}{1 - 1/10} = 1\dfrac{1}{9}.$

$$1 - \frac{1}{2} + \frac{1}{4} - \frac{1}{8} + \cdots = \frac{1}{1 - (-1/2)} = \frac{2}{3}.$$

EXAMPLE 2 Every sequence $\qquad S_1, S_2, S_3, \ldots, S_n, \ldots$

is the partial sum sequence of an infinite series, namely

$$S_1 + (S_2 - S_1) + (S_3 - S_2) + \cdots + (S_{n+1} - S_n) + \cdots.$$

For example,
$$1, \frac{1}{2}, \frac{1}{3}, \frac{1}{4}, \ldots, \frac{1}{n}, \ldots$$

is the partial sum sequence of

$$1 + \left(\frac{1}{2} - 1\right) + \left(\frac{1}{3} - \frac{1}{2}\right) + \cdots + \left(\frac{1}{n+1} - \frac{1}{n}\right) + \cdots$$

or
$$1 - \frac{1}{2} - \frac{1}{6} - \cdots - \frac{1}{n(n+1)} - \cdots.$$

The Cauchy Convergence Test from the preceding section takes on the following form for series.

CAUCHY CONVERGENCE TEST FOR SERIES

$a_1 + a_2 + \cdots + a_n + \cdots$ *converges if and only if*

(1)   *for all infinite $H < K$, $a_{H+1} + a_{H+2} + \cdots + a_K \approx 0$.*

DISCUSSION   The sum in (1) is just the difference in partial sums,

$$a_{H+1} + a_{H+2} + \cdots + a_K = S_K - S_H.$$

A very important consequence of the Cauchy Convergence Criterion is that all the infinite terms of a convergent series must be infinitesimal. We state this consequence as a corollary, which is illustrated in Figure 9.2.1.

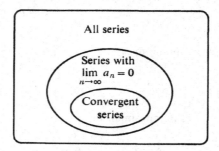

Figure 9.2.1

COROLLARY

If the series $a_1 + a_2 + \cdots + a_n + \cdots$ converges, then $\lim_{n \to \infty} a_n = 0$. That is, $a_K \approx 0$ for every infinite K.

PROOF   This is true by the Cauchy Criterion, with $K = H + 1$.

Warning: The converse of this corollary is false. It is possible for a sequence to have $\lim_{n \to \infty} a_n = 0$ and yet diverge. We shall give an example later (Example 3).

The Cauchy Convergence Criterion and its corollary can often be used to show that a series diverges. Table 9.2.2 sums up the various possibilities. In this table it is understood that

$$a_1 + a_2 + \cdots + a_n + \cdots$$

is an infinite series and H, K are positive infinite hyperintegers with $H < K$.

Table 9.2.2 Cauchy Convergence and Divergence Tests

Hypothesis	Conclusion
all $a_{H+1} + \cdots + a_K \approx 0$	Converges
all $a_K \approx 0$	none
some $a_{H+1} + \cdots + a_K \not\approx 0$	Diverges
some $a_K \not\approx 0$	Diverges

We shall give many other convergence tests later on in this chapter. For convenience there is a summary of all these tests at the end of Section 9.6.

THEOREM 2

(i)   *If $|c| \geq 1$ the geometric series $1 + c + c^2 + \cdots + c^n + \cdots$ diverges.*

(ii)   **The harmonic series** $1 + \dfrac{1}{2} + \dfrac{1}{3} + \cdots + \dfrac{1}{n} + \cdots$ *diverges.*

PROOF   (i)   For infinite H the term c^H is not infinitesimal, so the series diverges.

(ii)   Intuitively this can be seen by writing

$$1 + \tfrac{1}{2} + (\tfrac{1}{3} + \tfrac{1}{4}) + (\tfrac{1}{5} + \tfrac{1}{6} + \tfrac{1}{7} + \tfrac{1}{8}) + \cdots$$
$$\geq 1 + \tfrac{1}{2} + (\tfrac{1}{4} + \tfrac{1}{4}) + (\tfrac{1}{8} + \tfrac{1}{8} + \tfrac{1}{8} + \tfrac{1}{8}) + \cdots$$
$$= 1 + \tfrac{1}{2} + \tfrac{1}{2} + \tfrac{1}{2} + \cdots = \infty.$$

Instead we can use the Cauchy Test. We see that for each n,

$$\frac{1}{2^n + 1} + \frac{1}{2^n + 2} + \cdots + \frac{1}{2^{n+1}} \geq 2^n \cdot \frac{1}{2^{n+1}} = \frac{1}{2}.$$

Therefore for infinite H,

$$\frac{1}{2^H + 1} + \frac{1}{2^H + 2} + \cdots + \frac{1}{2^{H+1}} \geq \frac{1}{2}.$$

Since the above sum is not infinitesimal the series diverges.

EXAMPLE 3   The harmonic series

$$1 + \frac{1}{2} + \frac{1}{3} + \cdots + \frac{1}{n} + \cdots$$

is the example promised in our warning. It has the property that

$$\lim_{n \to \infty} a_n = \lim_{n \to \infty} \frac{1}{n} = 0$$

and yet the series diverges.

PROBLEMS FOR SECTION 9.2

In Problems 1–13 find the nth partial sum, determine whether the series converges, and find the sum when it exists.

1 $1 + \frac{1}{3} + \frac{1}{9} + \cdots + (\frac{1}{3})^n + \cdots$

2 $1 - \frac{1}{3} + \frac{1}{9} - \cdots + (-\frac{1}{3})^n + \cdots$

3 $1 + \frac{3}{4} + \frac{9}{16} + \cdots + (\frac{3}{4})^n + \cdots$

4 $1 - 2 + 4 - 8 + \cdots + (-2)^n + \cdots$

5 $\left(1 - \frac{1}{2}\right) + \left(\frac{1}{2} - \frac{1}{6}\right) + \left(\frac{1}{6} - \frac{1}{24}\right) + \cdots + \left(\frac{1}{n!} - \frac{1}{(n+1)!}\right) + \cdots$

6 $(a_1 - a_2) + (a_2 - a_3) + \cdots + (a_n - a_{n+1}) + \cdots$ where $\lim_{n \to \infty} a_n = 0$. This is called a *telescoping series*.

7 $\frac{1}{1 \cdot 2} + \frac{1}{2 \cdot 3} + \cdots + \frac{1}{n(n+1)} + \cdots.$ *Hint*: $\frac{1}{n(n+1)} = \frac{1}{n} - \frac{1}{n+1}.$

8 $\ln\frac{1}{2} + \ln\frac{2}{3} + \ln\frac{3}{4} + \cdots + \ln\frac{n}{n+1} + \cdots$

9 $1 - 2 + 3 - 4 + \cdots + n(-1)^{n-1} + \cdots$

10 $\frac{1}{1 \cdot 3} + \frac{1}{2 \cdot 4} + \frac{1}{3 \cdot 5} + \cdots + \frac{1}{n(n+2)} + \cdots$

11 $\frac{3}{1^2 \cdot 2^2} + \frac{5}{2^2 \cdot 3^2} + \cdots + \frac{2n+1}{n^2(n+1)^2} + \cdots.$ *Hint*: $\frac{2n+1}{n^2(n+1)^2} = \frac{1}{n^2} - \frac{1}{(n+1)^2}$

12 $\ln\frac{4}{3} + \ln\frac{9}{8} + \ln\frac{16}{15} + \cdots + \ln\frac{n^2}{n^2-1} + \cdots$

13 $\frac{1}{1 \cdot 3} + \frac{1}{3 \cdot 5} + \frac{1}{5 \cdot 7} + \cdots + \frac{1}{(2n-1)(2n+1)} + \cdots$

In Problems 14–19, show that the series diverges.

14 $\frac{1}{2} + \frac{2}{3} + \frac{3}{4} + \cdots + \frac{n}{n+1} + \cdots$

15 $\frac{1}{3} - \frac{2}{5} + \frac{3}{7} - \frac{4}{9} + \cdots + \frac{(-1)^{n-1}n}{2n+1} + \cdots$

16 $1 + \frac{1}{3} + \frac{1}{5} + \cdots + \frac{1}{2n-1} + \cdots$

17 $\frac{1}{4} + \frac{1}{7} + \frac{1}{10} + \cdots + \frac{1}{3n+1} + \cdots$

18 $1 + \sqrt{2} + \sqrt[3]{3} + \cdots + \sqrt[n]{n} + \cdots$

19 $\ln 1 + \ln 2 + \ln 3 + \cdots + \ln n + \cdots$

20   A ball bounces along a street. On each bounce it goes $\frac{4}{5}$ as far as it did on the previous bounce. If the first bounce is one foot long, how far will the ball go before it stops bouncing?

21   Two students are sharing a loaf of bread. Student A eats half of the loaf, then student B eats half of what's left, then A eats half of what's left, and so on. How much of the loaf will each student eat?

22   In the Problem 21, how much will each student eat if only $\frac{1}{3}$ of the remaining loaf is eaten at each turn?

23   Three students A, B, C take turns eating a loaf of bread, taking $\frac{1}{3}$ of the remaining loaf at each turn. How much will each student eat?

9.3 PROPERTIES OF INFINITE SERIES

It is convenient to use capital sigmas, $\sum$, for partial sums and infinite series, as we did for finite and infinite Riemann sums. We write

$$S_m = \sum_{n=1}^{m} a_n = a_1 + a_2 + \cdots + a_m$$

for the mth partial sum,

$$S_H = \sum_{n=1}^{H} a_n = a_1 + a_2 + \cdots + a_H$$

for an infinite partial sum, and

$$S = \sum_{n=1}^{\infty} a_n = a_1 + a_2 + \cdots + a_n + \cdots$$

for the infinite series. Thus S is the standard part of S_H,

$$\sum_{n=1}^{\infty} a_n = st\left(\sum_{n=1}^{H} a_n\right).$$

Sometimes we start counting from zero instead of one. For example, the formula for the sum of a geometric series can be written

$$\sum_{n=0}^{\infty} c^n = \frac{1}{1-c}, \qquad \text{where } |c| < 1.$$

Infinite series are similar to definite integrals. Table 9.3.1 compares and contrasts the two notions.

Table 9.3.1

Infinite series	Definite integral
$\sum_{n=1}^{\infty} a_n$	$\int_a^b f(x)\, dx$
Finite partial sum	Finite Riemann sum
$\sum_{n=1}^{m} a_n = a_1 + \cdots + a_m$	$\sum_a^b f(x)\,\Delta x = f(x_1)\,\Delta x + \cdots + f(x_m)\,\Delta x$
Infinite partial sum	Infinite Riemann sum
$\sum_{n=1}^{H} a_n = a_1 + \cdots + a_H$	$\sum_a^b f(x)\, dx = f(x_1)\, dx + \cdots + f(x_H)\, dx$
$\sum_{n=1}^{\infty} a_n = st\left(\sum_{n=1}^{H} a_n\right)$	$\int_a^b f(x)\, dx = st\left(\sum_a^b f(x)\, dx\right)$
$\sum_{n=1}^{\infty} a_n = \lim_{m\to\infty}\left(\sum_{n=1}^{m} a_n\right)$	$\int_a^b f(x)\, dx = \lim_{\Delta x\to 0^+}\left(\sum_a^b f(x)\,\Delta x\right)$

The difference between them is that the infinite series is formed by adding up the terms of an infinite sequence, while the definite integral is formed by adding up the values of $f(x)\, dx$ for x between a and b. The definite integral of a continuous function always exists. But the problem of whether an improper integral converges is similar to the problem of whether an infinite series converges.

Here are some basic theorems about infinite series which are like theorems about integrals.

THEOREM 1

Suppose $\sum_{n=1}^{\infty} a_n$ and $\sum_{n=1}^{\infty} b_n$ are convergent.

(i) **Constant Rule**   For any constant c, $\sum_{n=1}^{\infty} c a_n = c \sum_{n=1}^{\infty} a_n$.

(ii) **Sum Rule** $\sum_{n=1}^{\infty} (a_n + b_n) = \sum_{n=1}^{\infty} a_n + \sum_{n=1}^{\infty} b_n$.

(iii) **Inequality Rule**   If $a_n \leq b_n$ for all n then $\sum_{n=1}^{\infty} a_n \leq \sum_{n=1}^{\infty} b_n$.

PROOF   To illustrate we prove (ii). For any H,

$$(a_1 + b_1) + \cdots + (a_H + b_H) = (a_1 + \cdots + a_H) + (b_1 + \cdots + b_H).$$

Taking standard parts we get the Sum Rule.

EXAMPLE 1   For any constant b, and any $|c| < 1$,

$$b + bc + bc^2 + \cdots + bc^n + \cdots = b(1 + c + c^2 + \cdots + c^n + \cdots)$$

$$= \frac{b}{1 - c}.$$

The next theorem corresponds to the Addition Property for integrals,

$$\int_a^c f(x)\,dx = \int_a^b f(x)\,dx + \int_b^c f(x)\,dx.$$

DEFINITION

The series
$$\sum_{n=m+1}^{\infty} a_n = a_{m+1} + a_{m+2} + \cdots + a_{m+n} + \cdots$$

is defined as
$$\sum_{n=1}^{\infty} b_n = b_1 + b_2 + \cdots + b_n + \cdots$$

where $b_n = a_{m+n}$. This series is called a **tail** of the original series $\sum_{n=1}^{\infty} a_n$.

THEOREM 2

A series $\sum_{n=1}^{\infty} a_n$ converges if and only if its tail $\sum_{n=m+1}^{\infty} a_n$ converges for any m. The sum of a convergent series is equal to the mth partial sum plus the remaining tail,

$$\sum_{n=1}^{\infty} a_n = \sum_{n=1}^{m} a_n + \sum_{n=m+1}^{\infty} a_n,$$

or

$$a_1 + \cdots + a_n + \cdots = (a_1 + \cdots + a_m) + (a_{m+1} + \cdots + a_{m+n} + \cdots).$$

PROOF First assume the tail converges. For any infinite H, we have

$$a_1 + \cdots + a_H = (a_1 + \cdots + a_m) + (a_{m+1} + \cdots + a_H),$$

or

$$\sum_{n=1}^{H} a_n = \sum_{n=1}^{m} a_n + \sum_{n=m+1}^{H} a_n.$$

Taking standard parts,

$$st\left(\sum_{n=1}^{H} a_n\right) = \sum_{n=1}^{m} a_n + \sum_{n=m+1}^{\infty} a_n.$$

Therefore the series converges and

$$\sum_{n=1}^{\infty} a_n = \sum_{n=1}^{m} a_n + \sum_{n=m+1}^{\infty} a_n.$$

If we assume the series converges we can prove the tail converges in a similar way.

EXAMPLE 2   The series $\dfrac{1}{5^3} + \dfrac{1}{5^4} + \dfrac{1}{5^5} + \cdots = \displaystyle\sum_{n=3}^{\infty} \left(\dfrac{1}{5}\right)^n$

is a tail of the geometric series

$$\sum_{n=0}^{\infty} \left(\frac{1}{5}\right)^n.$$

Its sum can be found in two ways.

(a) $\displaystyle\sum_{n=3}^{\infty} \left(\frac{1}{5}\right)^n = \sum_{n=0}^{\infty} \left(\frac{1}{5}\right)^n - \sum_{n=0}^{2} \left(\frac{1}{5}\right)^n = \frac{1}{1 - \frac{1}{5}} - \left(1 + \frac{1}{5} + \frac{1}{25}\right) = \frac{1}{100}.$

(b) $\displaystyle\sum_{n=3}^{\infty} \left(\frac{1}{5}\right)^n = \frac{1}{5^3} \sum_{n=0}^{\infty} \left(\frac{1}{5}\right)^n = \frac{1}{125} \cdot \frac{1}{1 - \frac{1}{5}} = \frac{1}{125} \cdot \frac{5}{4} = \frac{1}{100}.$

COROLLARY 1

If $\sum_{n=1}^{\infty} a_n$ converges, then the tails $\sum_{n=m}^{\infty} a_n$ approach zero as m approaches ∞,

$$\lim_{m \to \infty} \left(\sum_{n=m}^{\infty} a_n\right) = 0.$$

PROOF If H is infinite, then

$$\sum_{n=1}^{H} a_n \approx \sum_{n=1}^{\infty} a_n,$$

so

$$\sum_{n=H+1}^{\infty} a_n = \sum_{n=1}^{\infty} a_n - \sum_{n=1}^{H} a_n \approx 0.$$

COROLLARY 2

If a series $\sum_{n=1}^{\infty} a_n$ converges, then it remains convergent if finitely many terms are added, deleted, or changed.

PROOF If a_m is the last term changed, then the tail

$$\sum_{n=m+1}^{\infty} a_n$$

is left unchanged, so it still converges.

Warning: Although the convergence properties of a series are not affected by changing finitely many terms, the value of the sum, if finite, is affected.

EXAMPLE 3   Here is a convergent geometric series.

$$\frac{1}{5^0} + \frac{1}{5^1} + \frac{1}{5^2} + \frac{1}{5^3} + \frac{1}{5^4} + \frac{1}{5^5} + \cdots = \frac{1}{1 - \frac{1}{5}} = \frac{5}{4} = 1.25.$$

The following series still converges by Corollary 2. Find its sum.

$$3 - 8 + \frac{1}{5^3} + \frac{1}{5^4} + \frac{1}{5^5} + \cdots.$$

We have

$$3 - 8 + \frac{1}{5^3} + \frac{1}{5^4} + \frac{1}{5^5} + \cdots = 3 - 8 + \frac{1}{5^3}\left(\frac{1}{5^0} + \frac{1}{5^1} + \frac{1}{5^2} + \cdots\right)$$

$$= (3 - 8) + \frac{1}{5^3} \cdot \frac{5}{4} = -5 + \frac{1}{100}$$

$$= -4.99.$$

PROBLEMS FOR SECTION 9.3

Find the sum of the following series.

1 $\dfrac{1}{7^2} + \dfrac{1}{7^3} + \cdots + \dfrac{1}{7^{n+2}} + \cdots$ $\qquad$ **2** $\dfrac{2}{1} + \dfrac{4}{3} + \dfrac{8}{9} + \dfrac{16}{27} + \cdots + \dfrac{2^{n+1}}{3^n} + \cdots$

3 $(1 + 1) + (\frac{1}{3} + \frac{1}{5}) + (\frac{1}{9} + \frac{1}{25}) + \cdots + (3^{-n} + 5^{-n}) + \cdots$

4 $\displaystyle\sum_{n=0}^{\infty} \left(-\frac{2}{7}\right)^n$ $\qquad\qquad$ **5** $\displaystyle\sum_{n=3}^{\infty} 5 \cdot 4^{-n}$

6 $1 + \dfrac{1}{5} + \dfrac{1}{7^2} + \dfrac{1}{7^3} + \dfrac{1}{7^9} + \cdots + \dfrac{1}{7^n} + \cdots$

7 $6^2 + 6 + 1 + 6^{-1} + 6^{-2} + \cdots + 6^{-n} + \cdots$

8 $\displaystyle\sum_{n=0}^{\infty} \dfrac{3^n + 4^n}{5^n}$

9 $8.88888\ldots = 8 + 8 \cdot 10^{-1} + 8 \cdot 10^{-2} + \cdots + 8 \cdot 10^{-n} + \cdots$

10 $2.36666\ldots = 2.3 + 6 \cdot 10^{-2} + 6 \cdot 10^{-3} + 6 \cdot 10^{-4} + \ldots$

11 $5.434343\ldots = 5 + 43 \cdot 100^{-1} + 43 \cdot 100^{-2} + 43 \cdot 100^{-3} + \cdots$

12 $0.286286286\ldots$ $\qquad\qquad$ **13** $492.315041041041041\ldots$

14   Prove the Constant Rule $\sum_{n=1}^{\infty} ca_n = c\sum_{n=1}^{\infty} a_n$.

15   Prove that the repeating decimal $0.142857142857142857\ldots$ is a rational number.

9.4 SERIES WITH POSITIVE TERMS

By a *positive term series*, we mean a series in which every term is greater than zero. For example, the geometric series

$$1 + c + c^2 + \cdots + c^n + \cdots$$

is a positive term series if $c > 0$ but not if $c \leq 0$. We call a sequence $S_1, S_2, \ldots, S_n, \ldots$ *increasing* if $S_m < S_n$ whenever $m < n$. It is easy to see that

$$a_1 + a_2 + \cdots + a_n + \cdots$$

is a positive term series if and only if its partial sum sequence is increasing. We are going to give several tests for the convergence of a positive term series. The starting point is the following theorem.

THEOREM 1

An increasing sequence $\langle S_n \rangle$ either converges or diverges to ∞.

Geometrically, this says that, as n gets large, the graph of the sequence either levels out at a limit L or the value of S_n gets large (Figure 9.4.1). We omit the proof. (The proof is given in the Epilogue at the end of the book.)

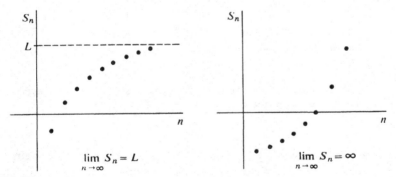

Figure 9.4.1

Theorem 1 has an equivalent form for positive term series because the partial sum sequence of a positive term series is increasing.

THEOREM 1 (Second Form)

A positive term series either converges or diverges to ∞.

EXAMPLE 1   The harmonic series diverges to ∞,

$$1 + \frac{1}{2} + \frac{1}{3} + \frac{1}{4} + \cdots + \frac{1}{n} + \cdots = \infty.$$

This is because it is a positive term series and we have shown that it diverges.

EXAMPLE 2   If $0 < a$ the geometric series

$$1 + a + a^2 + \cdots + a^n + \cdots$$

is a positive term series. It converges when $a < 1$ and diverges to ∞ when $a \geq 1$.

Remark   Theorem 1 shows that to determine whether a positive term series converges, we need only look at one infinite partial sum. If it is finite the series converges and if it is infinite the series diverges to ∞.

COMPARISON TEST

Let c be a positive constant. Suppose $\sum_{n=1}^{\infty} a_n$ and $\sum_{n=1}^{\infty} b_n$ are positive term series and $a_n \leq cb_n$ for all n.

(i)   If $\sum_{n=1}^{\infty} b_n$ converges then $\sum_{n=1}^{\infty} a_n$ converges.

(ii)   If $\sum_{n=1}^{\infty} a_n$ diverges then $\sum_{n=1}^{\infty} b_n$ diverges.

PROOF   (i)   Suppose $\sum_{n=1}^{\infty} b_n$ converges to S. The Constant Rule gives $cS = \sum_{n=1}^{\infty} cb_n$. Each finite partial sum of $\sum_{n=1}^{\infty} a_n$ is less than cS,

$$\sum_{n=1}^{m} a_n \leq \sum_{n=1}^{m} c\,b_n < cS.$$

Therefore, an infinite partial sum $\sum_{n=1}^{H} a_n$ is less than cS and hence finite. It follows that $\sum_{n=1}^{\infty} a_n$ converges.

(ii)   If $\sum_{n=1}^{\infty} a_n$ diverges then $\sum_{n=1}^{\infty} b_n$ cannot converge by part (i).

To use the Comparison Test we compare a series whose convergence or divergence is unknown with one which is known.

EXAMPLE 3   Test the series $\sum_{n=1}^{\infty} 6^n/(7^n - 5^n)$ for convergence. Intuitively, the 7^n should overcome the -5^n, so we shall compare with $6^n/7^n$. The simplest approach is to factor out 7^n. We have

$$\frac{6^n}{7^n - 5^n} = \frac{6^n}{7^n(1 - (5/7)^n)} \leq \frac{6^n}{7^n(2/7)} = \frac{7}{2}\left(\frac{6}{7}\right)^n.$$

The geometric series $\sum_{n=1}^{\infty} (6/7)^n$ is convergent, so the given series converges.

EXAMPLE 4   Test for convergence: $\sum_{n=1}^{\infty} n^2/(n^3 + 1)$. We have $n^3 + 1 \leq 2n^3$, so

$$\frac{n^2}{n^3 + 1} \geq \frac{n^2}{2n^3} = \frac{1}{2} \cdot \frac{1}{n}.$$

The harmonic series $\sum_{n=1}^{\infty} 1/n$ diverges, whence the given series diverges.

Sometimes the following comparison test is easier to use.

LIMIT COMPARISON TEST

Let $\sum_{n=1}^{\infty} a_n$ and $\sum_{n=1}^{\infty} b_n$ be positive term series and c a positive real number. Suppose that

$$a_K \leq cb_K \qquad \text{for all infinite } K.$$

Then:

(i)   If $\sum_{n=1}^{\infty} b_n$ converges then $\sum_{n=1}^{\infty} a_n$ converges.

(ii) If $\sum_{n=1}^{\infty} a_n$ diverges then $\sum_{n=1}^{\infty} b_n$ diverges.

PROOF Assume $\sum_{n=1}^{\infty} b_n$ converges. Let H and K be infinite. By the Cauchy Convergence Test (Section 9.2),

$$b_{H+1} + b_{H+2} + \cdots + b_K \approx 0.$$

Hence
$$0 \leq a_{H+1} + \cdots + a_K \leq cb_{H+1} + \cdots + cb_K$$
$$= c(b_{H+1} + \cdots + b_K) \approx 0.$$

It follows that
$$a_{H+1} + \cdots + a_K \approx 0$$

and $\sum_{n=1}^{\infty} a_n$ converges.

EXAMPLE 5   Test $\sum_{n=2}^{\infty} \dfrac{1}{(\ln n)^p}$ where p is a positive constant.

We compare this series with the divergent series

$$\sum_{n=2}^{\infty} \frac{1}{n}.$$

Let H be positive infinite. Then by Theorem 1 in Section 9.1,

$$\ln H < H^{1/p},$$
$$(\ln H)^p < H,$$
$$\frac{1}{(\ln H)^p} > \frac{1}{H}.$$

By the Limit Comparison Test, the given series $\sum_{n=2}^{\infty} 1/(\ln n)^p$ diverges.

For our last test we need another theorem which is similar to Theorem 1.

THEOREM 2

If the function $F(x)$ increases for $x \geq 1$, then $\lim_{x \to \infty} F(x)$ either exists or is infinite.

This says that the curve $y = F(x)$ is either asymptotic to some horizontal line $y = L$ or increases indefinitely, as illustrated in Figure 9.4.2.

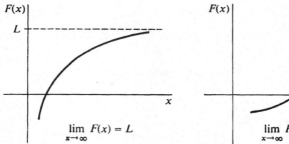

$$\lim_{x \to \infty} F(x) = L \qquad\qquad \lim_{x \to \infty} F(x) = \infty$$

Figure 9.4.2

INTEGRAL TEST

Suppose f is a continuous decreasing function and $f(x) > 0$ for all $x \geq 1$. Then the improper integral

$$\int_1^\infty f(x)\, dx$$

and the infinite series

$$\sum_{n=1}^\infty f(n)$$

either both converge or both diverge to ∞.

Discussion   Figure 9.4.3 suggests that

$$\sum_{n=2}^\infty f(n) < \int_1^\infty f(x)\, dx < \sum_{n=1}^\infty f(n)$$

so the series and the integral should both converge or both diverge to ∞. The Integral Test shows that the integral $\int_1^\infty f(x)\, dx$ and the series $\sum_{n=1}^\infty f(n)$ have the same convergence properties. However, their values, when finite, are different. In fact, we can see from Figure 9.4.3(c) that the integral is less than the series sum,

$$\int_1^\infty f(x)\, dx < \sum_{n=1}^\infty f(n).$$

PROOF As we can see from Figure 9.4.3, for each m we have

$$\sum_{n=2}^m f(n) \leq \int_1^m f(x)\, dx \leq \sum_{n=1}^{m-1} f(n).$$

The improper integral is defined by

$$\int_1^\infty f(x)\, dx = \lim_{u \to \infty} \int_1^u f(x)\, dx.$$

Since $f(x)$ is always positive, the function $F(u) = \int_1^u f(x)\, dx$ is increasing, so by Theorem 2, the limit either exists or is infinite. Hence the improper integral either converges or diverges to ∞.

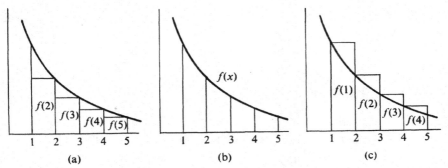

Figure 9.4.3   The Integral Test

Case 1 $\int_1^\infty f(x)\,dx = S$ converges. For infinite H we have

$$\sum_{n=2}^{H} f(n) \le \int_1^H f(x)\,dx \approx S;$$

thus the infinite partial sum is finite. Hence the tail $\sum_{n=2}^{\infty} f(n)$ and the series $\sum_{n=1}^{\infty} f(n)$ converge.

Case 2 $\int_1^\infty f(x)\,dx$ diverges to ∞. Since $\int_1^H f(x)\,dx \le \sum_{n=1}^{H-1} f(n)$, the infinite partial sum has infinite value, whence the series $\sum_{n=1}^{\infty} f(n)$ diverges to ∞.

The series $\sum_{n=1}^{\infty} 1/n^p$, where p is constant, is called the *p series*.

COROLLARY

The p series $\sum_{n=1}^{\infty} 1/n^p$ converges if $p > 1$ and diverges if $p \le 1$.

PROOF

Case 1 $p = 1$. The p series is just $\sum_{n=1}^{\infty} 1/n = \infty$.

Case 2 $p > 1$. The improper integral converges,

$$\int_1^\infty \frac{1}{x^p}\,dx = \lim_{u \to \infty} \int_1^u x^{-p}\,dx$$

$$= \lim_{u \to \infty} \frac{u^{1-p}-1}{1-p} = -\frac{1}{1-p}.$$

Therefore the p series converges.

Case 3 $p < 1$. The improper integral diverges to ∞, $\int_1^\infty (1/x^p)\,dx = \lim_{u \to \infty} \int_1^u x^{-p}\,dx = \lim_{u \to \infty} (u^{1-p} - 1)/(1 - p) = \infty$.
Therefore the p series diverges to ∞.

EXAMPLE 6   The p series $\displaystyle \sum_{n=1}^{\infty} \frac{1}{n\sqrt[3]{n}} = \sum_{n=1}^{\infty} \frac{1}{n^{4/3}}$

converges because $4/3 > 1$. The p series $\sum_{n=1}^{\infty} 1/\sqrt{n}$ diverges to ∞ because *
$1/2 < 1$.

The p series is often used in the Comparison Tests.

EXAMPLE 7   Test the series
$$\sum_{n=1}^{\infty} \frac{\ln n}{n^2}$$

for convergence.

If H is positive infinite then by Theorem 1 in Section 9.1,

$$\ln H < H^c,$$

$$\frac{\ln H}{H^2} < \frac{H^c}{H^2} = \frac{1}{H^{2-c}}, \qquad \text{for real } c > 0.$$

Now take c so that $0 < c < 1$. Then $2 - c > 1$ so the p series $\sum_{n=1}^{\infty} 1/n^{2-c}$
converges. By the Limit Comparison Test, the given series $\sum_{n=1}^{\infty} (\ln n)/n^2$
converges.

EXAMPLE 8   Use the Integral Test to test the improper integral $\int_3^{\infty} ((\ln x)/x^2)\, dx$ for
convergence.

By Example 7 the series $\sum_{n=3}^{\infty} (\ln n)/n^2$ converges. For $x > 1$ the function
$f(x) = (\ln x)/x^2$ is continuous, positive, and has derivative

$$f'(x) = x^{-3}(1 - 2 \ln x).$$

Thus for $x > \sqrt{e}$, $f'(x) < 0$ and $f(x)$ is decreasing. Therefore the Integral
Test applies and the improper integral converges.

PROBLEMS FOR SECTION 9.4

Test the following series for convergence.

1 $\sum_{n=0}^{\infty} \frac{n}{n + 4}$

2 $\sum_{n=1}^{\infty} \frac{2}{4n - 3}$

3 $\sum_{n=1}^{\infty} \frac{n + 1}{n^3}$

4 $\sum_{n=0}^{\infty} \frac{n}{n^2 + 2}$

5 $\sum_{n=0}^{\infty} \frac{1}{(n + 1)(n + 2)}$

6 $\sum_{n=1}^{\infty} \frac{1}{\sqrt{n(n + 1)}}$

7 $\sum_{n=3}^{\infty} \frac{n + 3}{n(n + 1)(n - 2)}$

8 $\sum_{n=0}^{\infty} \frac{n}{(n + 1)(n^2 + 1)}$

9 $\sum_{n=0}^{\infty} \frac{n}{(n + 1)(n + 2)}$

10 $\sum_{n=1}^{\infty} \frac{1}{n\sqrt{n}}$

11 $\sum_{n=0}^{\infty} \sqrt{n + 1} - \sqrt{n}$

12 $\sum_{n=1}^{\infty} \frac{\sqrt{n + 2} - \sqrt{n}}{n}$

13 $\sum_{n=1}^{\infty} \frac{3n^2 + 1}{2n^4 - 1}$

14 $\sum_{n=0}^{\infty} \frac{1}{\sqrt{n^3 + 4}}$

15 $\displaystyle\sum_{n=0}^{\infty} \frac{n}{\sqrt{n^3 + 1}}$

16 $\displaystyle\sum_{n=0}^{\infty} \frac{\sqrt{n}}{3n + 2}$

17 $\displaystyle\sum_{n=1}^{\infty} n^{-n}$

18 $\displaystyle\sum_{n=1}^{\infty} \frac{1}{2^n - n}$

19 $\displaystyle\sum_{n=1}^{\infty} \frac{\sin^2 n}{n^2}$

20 $\displaystyle\sum_{n=0}^{\infty} \frac{n^2}{2^n}$

21 $\displaystyle\sum_{n=1}^{\infty} \frac{\sqrt[n]{n}}{n}$

22 $\displaystyle\sum_{n=1}^{\infty} \frac{\sqrt[n]{n}}{n^2}$

23 $\displaystyle\sum_{n=0}^{\infty} \frac{1}{n!}$

24 $\displaystyle\sum_{n=0}^{\infty} \frac{5^n}{3^n + 4^n}$

25 $\displaystyle\sum_{n=0}^{\infty} \frac{5^n + 6^n}{2^n + 7^n}$

26 $\displaystyle\sum_{n=1}^{\infty} \frac{1}{2 + \ln n}$

27 $\displaystyle\sum_{n=1}^{\infty} \frac{\ln n}{n}$

28 $\displaystyle\sum_{n=1}^{\infty} \frac{\ln n}{n\sqrt{n}}$

29 $\displaystyle\sum_{n=2}^{\infty} \frac{1}{n \ln n}$

30 $\displaystyle\sum_{n=1}^{\infty} \frac{1}{\ln(n^2 + 1)}$

31 $\displaystyle\sum_{n=2}^{\infty} \frac{1}{n(\ln n)^2}$

32 $\displaystyle\sum_{n=0}^{\infty} \left(\frac{\pi}{2} - \arctan n\right)$

33 $\displaystyle\sum_{n=2}^{\infty} \frac{1}{\ln(n!)}$

Use the Integral Test to determine whether the following improper integrals converge or diverge.

34 $\displaystyle\int_2^{\infty} \frac{dx}{\ln x}$

35 $\displaystyle\int_2^{\infty} \frac{dx}{x^2 + \ln x}$

36 $\displaystyle\int_2^{\infty} \frac{1}{x + \ln x}\, dx$

37 $\displaystyle\int_1^{\infty} \frac{x + 1}{x^3 + x^2 + 1}\, dx$

38 $\displaystyle\int_3^{\infty} \frac{\ln x\, dx}{x\sqrt{x}}$

39 $\displaystyle\int_0^{\infty} e^{-x^2}\, dx$

40 $\displaystyle\int_1^{\infty} x^{-x}\, dx$

☐ **41** Prove that if each a_n is positive and $\sum_{n=1}^{\infty} a_n$ converges, then $\sum_{n=1}^{\infty} a_n^2$ converges.

☐ **42** Using Theorem 1 (page 539), prove that a negative term series either converges or diverges to $-\infty$.

9.5 ALTERNATING SERIES

An *alternating series* is a series in which the odd numbered terms are positive and the even numbered terms are negative, or vice versa. An example is the geometric series

$$\sum_{n=1}^{\infty} a^n, \qquad a < 0.$$

Given any positive term series

$$\sum_{n=1}^{\infty} a_n = a_1 + a_2 + a_3 + a_4 + \cdots,$$

the series
$$\sum_{n=1}^{\infty} (-1)^{n+1} a_n = a_1 - a_2 + a_3 - a_4 + \cdots$$

and
$$\sum_{n=1}^{\infty} (-1)^n a_n = -a_1 + a_2 - a_3 + a_4 - \cdots.$$

are alternating series. Here is a test for convergence of alternating series.

ALTERNATING SERIES TEST

Assume that

(i) $\sum_{n=1}^{\infty} (-1)^{n+1} a_n$ *is an alternating series.*

(ii) *The terms a_n are decreasing, $a_1 > a_2 > \cdots > a_n > \cdots$.*

(iii) *The terms approach zero, $\lim_{n \to \infty} a_n = 0$.*

Then the series converges to a sum $\sum_{n=1}^{\infty} (-1)^{n+1} a_n = S$. Moreover, the sum S is between any two consecutive partial sums,

$$S_{2n} < S < S_{2n+1}.$$

Discussion   We see from the graph in Figure 9.5.1 that the partial sums S_n alternately increase and decrease, but the change is less each time. The value of S_n "vibrates" back and forth and the vibration damps down around the limit S.

PROOF The sequence of even partial sums is increasing.

$$S_2 < S_4 < \cdots < S_{2n} < \cdots,$$

because $S_4 = S_2 + (a_3 - a_4),$ $S_6 = S_4 + (a_5 - a_6),$ etc.

The sequence of odd partial sums is decreasing,

$$S_1 > S_3 > S_5 > \cdots,$$

for $S_3 = S_1 - (a_2 - a_3),$ $S_5 = S_3 - (a_4 - a_5),$ etc.

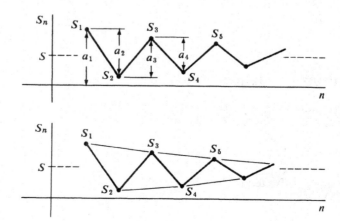

Figure 9.5.1

It follows that each even partial sum is less than S_1,

$$S_1 > S_1 - a_2 = S_2, \qquad S_1 > S_3 - a_4 = S_4, \qquad S_1 > S_5 - a_6 = S_6, \quad \text{etc.}$$

Theorem 1 (Section 9.4) shows that the increasing sequence of even partial sums converges,

$$\lim_{n \to \infty} S_{2n} = S.$$

Given any infinite H, $a_{2H+1} \approx 0$ and $S_{2H} \approx S$, so

$$S_{2H+1} = S_{2H} + a_{2H+1} \approx S.$$

Therefore the sequence of all partial sums converges to S, and

$$\sum_{n=1}^{\infty} (-1)^{n+1} a_n = \lim_{n \to \infty} S_n = S.$$

Finally, since the even partial sums are increasing and the odd partial sums are decreasing, we have the estimate

$$S_{2n} < S < S_{2n+1}.$$

Figure 9.5.2 shows a graph of the partial sums.

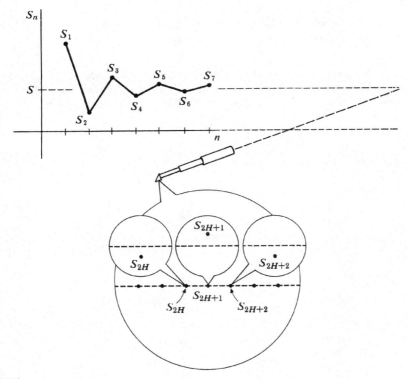

Figure 9.5.2

EXAMPLE 1 The *alternating harmonic series*

$$1 - \frac{1}{2} + \frac{1}{3} - \frac{1}{4} + \frac{1}{5} - \cdots + \frac{(-1)^{n+1}}{n} + \cdots$$

converges by the Alternating Series Test, because $\frac{1}{n}$ is decreasing and approaches zero as $n \to \infty$. The partial sums are

$$1, \tfrac{1}{2}, \tfrac{5}{6}, \tfrac{7}{12}, \tfrac{47}{60}, \tfrac{37}{60}, \ldots$$

or
$$\tfrac{60}{60}, \tfrac{30}{60}, \tfrac{50}{60}, \tfrac{35}{60}, \tfrac{47}{60}, \tfrac{37}{60}, \ldots.$$

The sum S is between any two consecutive partial sums, for example

$$\tfrac{37}{60} < S < \tfrac{47}{60}.$$

EXAMPLE 2 The alternating series

$$2 - \frac{3}{2} + \frac{4}{3} - \frac{5}{4} + \frac{6}{5} - \cdots + (-1)^{n+1}\frac{n+1}{n} + \cdots$$

diverges. The terms $(n + 1)/n$ are decreasing, but their limit is one instead of zero,

$$\lim_{n \to \infty} \frac{n+1}{n} = 1.$$

The Cauchy Test for Divergence in Section 9.2 shows that if the terms a_n do not converge to zero the series diverges.

We have now built up quite a long list of convergence tests. The next section contains one more important test, the Ratio Test. At the end of that section is a summary of all the convergence tests with hints on when to use them.

PROBLEMS FOR SECTION 9.5

Test the following alternating series for convergence.

1   $\displaystyle\sum_{n=1}^{\infty} (-1)^n \sqrt{n}$
     2   $\displaystyle\sum_{n=1}^{\infty} \frac{(-1)^{n+1}}{\sqrt{n}}$

3   $\displaystyle\sum_{n=1}^{\infty} (-1)^{n+1}\frac{n}{10n+5}$
     4   $\displaystyle\sum_{n=1}^{\infty} (-1)^n \frac{\sqrt{n}}{n+1}$

5   $\displaystyle\sum_{n=1}^{\infty} (-1)^n n^{-2}$
     6   $\displaystyle\sum_{n=1}^{\infty} (-1)^n n^{-1/3}$

7   $\displaystyle\sum_{n=1}^{\infty} (-1)^n \sqrt[n]{\tfrac{1}{2}}$
     8   $\displaystyle\sum_{n=2}^{\infty} \frac{(-1)^n}{\ln n}$

9   $\displaystyle\sum_{n=2}^{\infty} \frac{n(-1)^{n+1}}{\ln n}$
     10   $\displaystyle\sum_{n=1}^{\infty} (-1)^n$

11   $\displaystyle\sum_{n=0}^{\infty} \frac{(-1)^{n+1} n!}{2^n}$
     12   $\displaystyle\sum_{n=0}^{\infty} \frac{(-1)^{n+1} 2^n}{n!}$

13 $\displaystyle\sum_{n=3}^{\infty} \frac{(-1)^n}{\ln(\ln n)}$

14 $\displaystyle\sum_{n=2}^{\infty} \frac{(-1)^n}{\sqrt[n]{n}}$

15 $\displaystyle\sum_{n=1}^{\infty} \frac{\cos(\pi n)}{n^2}$

16 $\displaystyle\sum_{n=1}^{\infty} (-1)^{n+1}\left(1 - \frac{1}{n}\right)$

17 $\displaystyle\sum_{n=1}^{\infty} (-1)^n(\sqrt{n+1} - \sqrt{n})$

18 $\displaystyle\sum_{n=0}^{\infty} (-1)^n \frac{2^n + 1}{3^n - 2}$

19 $\displaystyle\sum_{n=0}^{\infty} (-1)^n \frac{2^{n-2} + 1}{2^{n+3} + 5}$

20 $\displaystyle\sum_{n=1}^{\infty} (-1)^{n+1}\left(1 + \frac{1}{n}\right)^{-n}$

21 Approximate the series $\sum_{n=1}^{\infty} (-1)^{n+1} n^{-3}$ to two decimal places.

22 Approximate the series $1 - \frac{2}{10} + \frac{3}{100} - \frac{4}{1000} + \cdots$ to four decimal places.

23 Approximate $\sum_{n=0}^{\infty} (-1)^n/n!$ to two decimal places.

24 Approximate $\sum_{n=1}^{\infty} (-n)^{-n}$ to three decimal places.

9.6 ABSOLUTE AND CONDITIONAL CONVERGENCE

Consider a series $\sum_{n=1}^{\infty} a_n$ which has both positive and negative terms. We may form a new series $\sum_{n=1}^{\infty} |a_n|$ whose terms are the absolute values of the terms of the given series. If all the terms a_n are nonzero, then $|a_n| > 0$ so $\sum_{n=1}^{\infty} |a_n|$ is a positive term series.

If $\sum_{n=1}^{\infty} a_n$ is already a positive term series, then $|a_n| = a_n$ and the series is identical to its absolute value series $\sum_{n=1}^{\infty} |a_n|$.

Sometimes it is simpler to study the convergence of the absolute value series $\sum_{n=1}^{\infty} |a_n|$ than of the given series $\sum_{n=1}^{\infty} a_n$. This is because we have at our disposal all the convergence tests for positive term series from the preceding sections.

DEFINITION

*A series $\sum_{n=1}^{\infty} a_n$ is said to be **absolutely convergent** if its absolute value series $\sum_{n=1}^{\infty} |a_n|$ is convergent. A series which is convergent but not absolutely convergent is called **conditionally convergent**.*

THEOREM 1

Every absolutely convergent series is convergent. That is, if the absolute value series $\sum_{n=1}^{\infty} |a_n|$ converges, then $\sum_{n=1}^{\infty} a_n$ converges.

Discussion This theorem shows that if a positive term series $\sum_{n=1}^{\infty} b_n$ is convergent, then it remains convergent if we make some or all of the terms b_n negative, because the new series will still be absolutely convergent.

Given an arbitrary series $\sum_{n=1}^{\infty} a_n$, the theorem shows that exactly one of the following three things can happen:

The series is absolutely convergent.
The series is conditionally convergent.
The series is divergent.

PROOF OF THEOREM 1   We use the Sum Rule. Assume $\sum_{n=1}^{\infty} |a_n|$ converges and let

$$b_n = a_n + |a_n|.$$

Then $a_n = b_n - |a_n|$ and

$$b_n = \begin{cases} 2|a_n| & \text{if } a_n > 0, \\ 0 & \text{if } a_n < 0. \end{cases}$$

(See Figure 9.6.1). Both $\sum_{n=1}^{\infty} |a_n|$ and $\sum_{n=1}^{\infty} b_n$ have nonnegative terms. Moreover, $\sum_{n=1}^{\infty} |a_n|$ converges and $b_n \le 2|a_n|$. By the Comparison Test, $\sum_{n=1}^{\infty} b_n$ converges. Then using the Sum and Constant Rules,

$$\sum_{n=1}^{\infty} a_n = \sum_{n=1}^{\infty} b_n - \sum_{n=1}^{\infty} |a_n|$$

converges.

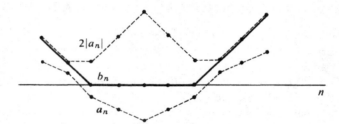

Figure 9.6.1

EXAMPLE 1   The alternating series

$$1 - \frac{1}{2^2} + \frac{1}{3^2} - \frac{1}{4^2} + \frac{1}{5^2} - \cdots,$$

is absolutely convergent, because its absolute value series

$$1 + \frac{1}{2^2} + \frac{1}{3^2} + \frac{1}{4^2} + \cdots$$

is convergent.

EXAMPLE 2   The alternating harmonic series

$$1 - \tfrac{1}{2} + \tfrac{1}{3} - \tfrac{1}{4} + \tfrac{1}{5} - \cdots$$

is conditionally convergent. It converges by the Alternating Series Test. But its absolute value series

$$1 + \tfrac{1}{2} + \tfrac{1}{3} + \tfrac{1}{4} + \tfrac{1}{5} + \cdots$$

diverges.

Given a series

$$a_1 + a_2 + a_3 + a_4 + \cdots + a_k + \cdots,$$

one can form a new series by listing the terms in a different order, for example

$$a_1 + a_3 + a_2 + a_5 + a_4 + \cdots.$$

Such a series is called a *rearrangement* of $\sum_{n=1}^{\infty} a_n$. The difference between absolute convergence and conditional convergence is shown emphatically by the following pair of theorems.

THEOREM 2

A. *Every rearrangement of an absolutely convergent series is also convergent and has the same sum.*

B. *Let $\sum_{n=1}^{\infty} a_n$ be a conditionally convergent series.*

   (i) *The series has a rearrangement which diverges to ∞.*

   (ii) *The series has another rearrangement which diverges to $-\infty$.*

   (iii) *For each real number r, the series has a rearrangement which converges to r.*

We shall not prove these theorems. Instead we give a pair of rearrangements of the conditionally convergent series

$$1 - \tfrac{1}{2} + \tfrac{1}{3} - \tfrac{1}{4} + \cdots,$$

one diverging to ∞ and the other converging to -1.

The alternating series

$$1 - \tfrac{1}{2} + \tfrac{1}{3} - \tfrac{1}{4} + \cdots$$

conditionally converges to a number between $\tfrac{1}{2}$ and 1.

To get a rearrangement which diverges to ∞, we write down terms in the following order:

1st positive term,	1st negative term,
next 2 positive terms,	2nd negative term,
next 4 positive terms,	3rd negative term,

$$\vdots$$

next 2^m positive terms,	mth negative term,

$$\vdots$$

We thus obtain the series

$$1 - \tfrac{1}{2} + \tfrac{1}{3} + \tfrac{1}{5} - \tfrac{1}{4} + \tfrac{1}{7} + \tfrac{1}{9} + \tfrac{1}{11} + \tfrac{1}{13} - \tfrac{1}{6} + \cdots.$$

Each block of 2^m positive terms adds up to at least $\tfrac{1}{4}$,

$$1 \geq \tfrac{1}{4},$$

$$\tfrac{1}{3} + \tfrac{1}{5} \geq 2 \times \tfrac{1}{8} = \tfrac{1}{4},$$

$$\tfrac{1}{7} + \tfrac{1}{9} + \tfrac{1}{11} + \tfrac{1}{13} \geq 4 \times \tfrac{1}{16} = \tfrac{1}{4},$$

$$\tfrac{1}{15} + \cdots + \tfrac{1}{29} \geq 8 \times \tfrac{1}{32} = \tfrac{1}{4}.$$

However, all the negative terms except $-\tfrac{1}{2}$ and $-\tfrac{1}{4}$ have absolute value $\leq \tfrac{1}{6}$. Hence after the mth negative term the partial sum is more than

$$\frac{m}{4} - \frac{m}{6} - \frac{1}{2} - \frac{1}{4} = \frac{m}{12} - \frac{3}{4}.$$

Therefore the partial sums, and hence the series, diverge to ∞.

To get a rearrangement which converges conditionally to -1 we proceed as follows:

Write down negative terms until the partial sum is below -1, then positive terms until the partial sum is above -1, then negative terms until the partial sum is below -1, and so on.

The mth time the partial sum goes above -1, it must be between -1 and $-1 + (1/m)$. The mth time it goes below -1 it must be between -1 and $-1 - (1/m)$. Therefore the series converges to -1.

The comparison tests for positive term series give us tests for absolute convergence.

COMPARISON TEST

If $|a_n| \leq c|b_n|$ and $\sum_{n=1}^{\infty} b_n$ is absolutely convergent then $\sum_{n=1}^{\infty} a_n$ is absolutely convergent.

LIMIT COMPARISON TEST

Let c be a positive real number. If

$$|a_K| \leq c|b_K| \qquad \text{for all infinite } K$$

and $\sum_{n=1}^{\infty} b_n$ is absolutely convergent then $\sum_{n=1}^{\infty} a_n$ is absolutely convergent.

The above tests do not help to distinguish between conditional convergence and divergence. Theorem 2 in Section 9.2 is often useful as a test for divergence.

There is another test which can be used either to show that a series is absolutely convergent or that a series is divergent.

RATIO TEST

Suppose the limit of the ratio $|a_{n+1}|/|a_n|$ exists or is ∞,

$$\lim_{n \to \infty} \frac{|a_{n+1}|}{|a_n|} = L.$$

(i)   *If $L < 1$, the series $\sum_{n=1}^{\infty} a_n$ converges absolutely.*

(ii) *If $L > 1$, or $L = \infty$, the series diverges.*

(iii) *If $L = 1$, the test gives no information and the series may converge absolutely, converge conditionally, or diverge.*

PROOF (i)   Choose b with $L < b < 1$. By the ε, N condition, there is an N such that all the ratios

$$\frac{|a_{N+1}|}{|a_N|}, \frac{|a_{N+2}|}{|a_{N+1}|}, \ldots, \frac{|a_{N+k+1}|}{|a_{N+k}|}, \ldots$$

are less than b. Therefore with $c = |a_N|$,

$$|a_{N+1}| < cb, \qquad |a_{N+2}| < cb^2, \ldots, |a_{N+n}| < cb^n, \ldots$$

The geometric series $\sum_{n=1}^{\infty} b^n$ converges, so by the Comparison Test, the tail $\sum_{n=N}^{\infty} |a_n|$ converges. Therefore the absolute value series $\sum_{n=1}^{\infty} |a_n|$ converges.

(ii)   By the ε, N condition there is an N such that the ratios

$$\frac{|a_{N+1}|}{|a_N|}, \ldots, \frac{|a_{N+n+1}|}{|a_{N+n}|}, \ldots$$

are all greater than one. Therefore

$$|a_N| < |a_{N+1}| < \cdots < |a_{N+n}| < \cdots.$$

It follows that the terms a_n do not converge to zero, so the series $\sum_{n=1}^{\infty} a_n$ diverges.

The Ratio Test is useful even for positive term series, and is often effective for series involving $n!$ and a^n.

EXAMPLE 3   Test the series $\sum_{n=1}^{\infty} \dfrac{1}{n!}$.

$$\lim_{n \to \infty} \frac{1/(n+1)!}{1/n!} = \lim_{n \to \infty} \frac{1}{n+1} = 0,$$

so by the Ratio Test the series converges.

EXAMPLE 4   Test $\displaystyle\sum_{n=1}^{\infty} \frac{(-1)^n n^n}{n!}$.

$$\lim_{n \to \infty} \left(\frac{(n+1)^{n+1}/(n+1)!}{n^n/n!} \right) = \lim_{n \to \infty} \left(\frac{(n+1)}{n} \right)^n = \lim_{n \to \infty} \left(1 + \frac{1}{n} \right)^n = e.$$

e is greater than one, so by the Ratio Test the series diverges.

EXAMPLE 5   The Ratio Test does not apply to either of the series

$$\sum_{n=1}^{\infty} \frac{1}{n}, \qquad \sum_{n=1}^{\infty} \frac{1}{n^2},$$

since

$$\lim_{n \to \infty} \frac{1/(n+1)}{1/n} = 1, \qquad \lim_{n \to \infty} \frac{1/(n+1)^2}{1/n^2} = 1.$$

SUMMARY OF SERIES CONVERGENCE TESTS

A.   *Particular Series*

(1)   *Geometric Series*

$$\sum_{n=0}^{\infty} c^n \text{ converges to } \frac{1}{1-c} \text{ if } |c| < 1,$$

diverges if $|c| \geq 1$.

(2)   *Harmonic Series*

$$\sum_{n=1}^{\infty} \frac{1}{n} \text{ diverges.}$$

(3)   *p Series*

$$\sum_{n=1}^{\infty} \frac{1}{n^p} \text{ converges if } p > 1,$$

$$\text{diverges if } p \leq 1.$$

B.   *Tests for Positive and Alternating Series*

In the tests below, assume $a_n \geq 0$ for all n.

(1)   *Convergence versus Divergence to ∞*

Let H be infinite.

$\sum_{n=1}^{\infty} a_n$ converges if $\sum_{n=1}^{H} a_n$ is finite,

diverges to ∞ if $\sum_{n=1}^{H} a_n$ is infinite.

(2)   *Comparison Test*

Suppose $a_n \leq cb_n$ for all n.

If $\sum_{n=1}^{\infty} b_n$ converges then $\sum_{n=1}^{\infty} a_n$ converges.

If $\sum_{n=1}^{\infty} a_n$ diverges then $\sum_{n=1}^{\infty} b_n$ diverges.

Hint: Often a series can be compared with one of the particular series above: a geometric, harmonic, or p series.

(3)   *Limit Comparison Test*

Suppose $a_K \leq cb_K$ for all infinite K.

If $\sum_{n=1}^{\infty} b_n$ converges then $\sum_{n=1}^{\infty} a_n$ converges.

If $\sum_{n=1}^{\infty} a_n$ diverges then $\sum_{n=1}^{\infty} b_n$ diverges.

Hint: Try this test if the Comparison Test almost works.

(4)   *Integral Test*

Suppose f is continuous, decreasing, and positive for $x \geq 1$.

If $\int_1^{\infty} f(x)\, dx$ converges, then $\sum_{n=1}^{\infty} f(n)$ converges.

If $\int_1^{\infty} f(x)\, dx$ diverges, then $\sum_{n=1}^{\infty} f(n)$ diverges.

Hint: This test may be useful if a_n comes from a continuous function $f(x)$.

(5)   *Alternating Series Test*

$\sum_{n=1}^{\infty} (-1)^n a_n$ converges if the a_n are decreasing and approach 0.

Hint: This is usually the simplest test if you see a $(-1)^n$ in the expression.

C.   *Tests for General Series*

(1)   *Definition of Convergence*

$\sum_{n=1}^{\infty} a_n$ converges if and only if the partial sum series $\sum_{n=1}^{k} a_n = S_k$ converges.

(2)   *Cauchy Convergence Test*

$\sum_{n=1}^{\infty} a_n$ converges if for all infinite H and $K > H$,

$$a_{H+1} + \cdots + a_K \approx 0,$$

diverges if for some infinite H and $K > H$,
$$a_{H+1} + \cdots + a_K \not\approx 0,$$
diverges if $\lim_{n \to \infty} a_n \neq 0$.

Hint: This test is useful for showing a series diverges.

(3) *Constant and Sum Rules*

Sums and constant multiples of convergent series converge.

(4) *Tail Rule*

$\sum_{n=1}^{\infty} a_n$ converges if and only if $\sum_{n=m}^{\infty} a_n$ converges.

(5) *Absolute Convergence*

If $\sum_{n=1}^{\infty} |a_n|$ converges then $\sum_{n=1}^{\infty} a_n$ converges.

Hint: Remember that $\sum_{n=1}^{\infty} |a_n|$ is a positive term series. Thus tests in group B may be applied to $\sum_{n=1}^{\infty} |a_n|$.

(6) *Ratio Test*

Suppose $\lim_{n \to \infty} \dfrac{|a_{n+1}|}{|a_n|} = L$.

$\sum_{n=1}^{\infty} a_n$ converges absolutely if $L < 1$,

$\qquad\qquad$ diverges if $L > 1$.

Hint: This is useful if a_n involves a factorial. Watch for $\left(\dfrac{n+1}{n}\right)^n$ in the

ratio because $\quad \lim_{n \to \infty} \left(\dfrac{n+1}{n}\right)^n = \lim_{n \to \infty} \left(1 + \dfrac{1}{n}\right)^n = e$.

If the limit L is one, try another test because the Ratio Test gives no information.

PROBLEMS FOR SECTION 9.6

Problems 1-20: For each of the first 20 problems in Section 9.5, determine whether the alternating series is absolutely convergent, conditionally convergent, or divergent.

Problems 21-44: Apply the Ratio Test to the given series. Possible answers are "convergent," "divergent," or "Ratio Test gives no information."

21 $\displaystyle\sum_{n=1}^{\infty} 3^n$
$\qquad\qquad\qquad$ 22 $\displaystyle\sum_{n=1}^{\infty} \frac{1}{2^n}$

23 $\displaystyle\sum_{n=1}^{\infty} \frac{1}{\sqrt{n}}$
$\qquad\qquad\qquad$ 24 $\displaystyle\sum_{n=2}^{\infty} n^2$

25 $\displaystyle\sum_{n=1}^{\infty} \frac{2^n}{n!}$
$\qquad\qquad\qquad$ 26 $\displaystyle\sum_{n=1}^{\infty} \frac{1}{n^3}$

27 $\displaystyle\sum_{n=1}^{\infty} \frac{5^n}{3^n + 4^n}$
$\qquad\qquad\qquad$ 28 $\displaystyle\sum_{n=1}^{\infty} \frac{5^n}{6^n - 5^n}$

29 $\displaystyle\sum_{n=1}^{\infty} \frac{n!}{n^n}$
$\qquad\qquad\qquad$ 30 $\displaystyle\sum_{n=1}^{\infty} \frac{n^n}{(2n)!}$

31 $\displaystyle\sum_{n=0}^{\infty} \frac{(n!)^2}{(2n)!}$

32 $\displaystyle\sum_{n=1}^{\infty} \frac{3^n(n!)}{n^n}$

33 $\displaystyle\sum_{n=1}^{\infty} \frac{e^n(n!)}{n^n}$

34 $\displaystyle\sum_{n=1}^{\infty} \frac{4^n(n!)^2}{(2n)!}$

35 $\displaystyle\sum_{n=1}^{\infty} \frac{3^n(n!)^2}{(2n)!}$

36 $\displaystyle\sum_{n=2}^{\infty} \frac{1}{(\ln n)^n}$

37 $\displaystyle\sum_{n=2}^{\infty} \frac{10^n}{(\ln n)^n}$

38 $\displaystyle\sum_{n=3}^{\infty} \frac{1}{(\ln (\ln n))^n}$

39 $\displaystyle\sum_{n=1}^{\infty} \frac{n!}{1\cdot 3\cdot 5\cdot\cdots\cdot(2n-1)}$

40 $\displaystyle\sum_{n=1}^{\infty} \frac{1\cdot 3\cdot 5\cdot\cdots\cdot(2n-1)}{(n!)^2}$

41 $\displaystyle\sum_{n=1}^{\infty} \frac{1\cdot 3\cdot 5\cdot\cdots\cdot(2n-1)}{2\cdot 4\cdot 6\cdot\cdots\cdot(2n)}$

42 $\displaystyle\sum_{n=1}^{\infty} \frac{1\cdot 3\cdot 5\cdot\cdots\cdot(2n-1)}{1\cdot 4\cdot 7\cdot\cdots\cdot(3n-2)}$

43 $\displaystyle\sum_{n=1}^{\infty} \frac{(n!)^3}{(3n)!}$

44 $\displaystyle\sum_{n=1}^{\infty} \frac{1}{\sqrt{n!}}$

9.7 POWER SERIES

So far we have studied series of constants,

$$\sum_{n=0}^{\infty} a_n = a_0 + a_1 + \cdots + a_n + \cdots.$$

One can also form a *series of functions*

$$\sum_{n=0}^{\infty} f_n(x) = f_0(x) + f_1(x) + \cdots + f_n(x) + \cdots.$$

Such a series will converge for some values of x and diverge for others. The *sum* of the series is a new function

$$f(x) = \sum_{n=0}^{\infty} f_n(x)$$

which is defined at each point x_0 where the series converges. We shall concentrate on a particular kind of series of functions called a power series. Its importance will be evident in the next section where we show that many familiar functions are sums of power series.

DEFINITION

A **power series** in x is a series of functions of the form

$$\sum_{n=0}^{\infty} a_n x^n = a_0 + a_1 x + a_2 x^2 + \cdots + a_n x^n + \cdots.$$

The nth finite partial sum of a power series is just a polynomial of degree n,

$$\sum_{k=0}^{n} a_k x^k = a_0 + a_1 x + \cdots + a_n x^n.$$

The infinite partial sums are polynomials of infinite degree,

$$\sum_{n=0}^{H} a_n x^n = a_0 + a_1 x + \cdots + a_H x^H.$$

At $x = 0$ every power series converges absolutely,

$$\sum_{n=0}^{\infty} a_n x^n = a_0 + a_1 0 + a_2 0^2 + \cdots = a_0.$$

(In a power series we use the convention $a_0 x^0 = a_0$.) If a power series converges absolutely at $x = u$, it also converges absolutely at $x = -u$, because the absolute value series $\sum_{n=0}^{\infty} |a_n u^n|$ and $\sum_{n=0}^{\infty} |a_n(-u)^n|$ are the same.

Intuitively, the smaller the absolute value $|x|$, the more likely the power series is to converge at x. This intuition is borne out in the following theorem.

THEOREM 1

(i) *If a power series*

$$\sum_{n=0}^{\infty} a_n x^n = a_0 + a_1 x + \cdots + a_n x^n + \cdots$$

converges when $x = u$, then it converges absolutely whenever $|x| < |u|$.

(ii) *If a power series diverges when $x = v$, then it diverges whenever $|x| > |v|$.*

PROOF (i) Suppose the series $\sum_{n=0}^{\infty} a_n u^n$ converges. Then for any positive infinite H, $a_H u^H$ is infinitesimal. Let $|v| < |u|$. The ratio $b = |v|/|u|$ is then less than one. It follows that:

(1) The positive term geometric series $\sum_{n=0}^{\infty} b^n$ converges,

(2) $|a_H v^H| = \left| a_H u^H \left(\dfrac{v}{u}\right)^H \right| = |a_H u^H| b^H \le b^H.$

Now by the Limit Comparison Test, $\sum_{n=0}^{\infty} a_n v^n$ converges absolutely.

(ii) This follows trivially from (i). Let $\sum_{n=0}^{\infty} a_n v^n$ diverge and $|u| > |v|$. $\sum_{n=0}^{\infty} a_n u^n$ cannot converge because if it did $\sum_{n=0}^{\infty} a_n v^n$ would converge absolutely. Therefore $\sum_{n=0}^{\infty} a_n u^n$ diverges.

Theorem 1 shows that if a power series converges at $x = u$ and at $x = v$, then it converges absolutely at every point strictly between u and v. We conclude that the set of points where the power series converges is an interval, called the *interval of convergence*. (A rigorous proof that the set is an interval is given in the Epilogue.) The next corollary summarizes what we know about the interval of convergence.

COROLLARY

For each power series $\sum_{n=0}^{\infty} a_n x^n$, one of the following happens.

(i) *The series converges absolutely at $x = 0$ and diverges everywhere else.*

(ii) *The series converges absolutely on the whole real line $(-\infty, \infty)$.*

(iii)   *The series converges absolutely at every point in an open interval* $(-r, r)$ *and diverges at every point outside the closed interval* $[-r, r]$. *At the endpoints* $-r$ *and* r *the series may converge or diverge, so the interval of convergence is one of the sets*

$$(-r, r), \quad [-r, r), \quad (-r, r], \quad [-r, r].$$

Figure 9.7.1 illustrates part (iii) of the Corollary. The number r is called the *radius of convergence* of the power series. In case (i) the radius of convergence is zero, and in case (ii) it is ∞. Once the radius of convergence is determined, we need only test the series at $x = r$ and $x = -r$ to find the interval of convergence.

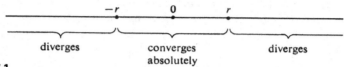

Figure 9.7.1

EXAMPLE 1   Find the interval of convergence of the power series

$$\sum_{n=0}^{\infty} b^n x^n, \qquad \text{where } b > 0.$$

This is just the geometric series

$$1 + bx + (bx)^2 + \cdots + (bx)^n + \cdots.$$

It converges absolutely when $|bx| < 1$, $|x| < 1/b$, and diverges when $|bx| > 1$, $|x| > 1/b$. So the radius of convergence is $r = 1/b$. At $x = r$ and at $x = -r$ the series diverges, because $b^n r^n = 1$. Thus the interval of convergence is $(-1/b, 1/b)$.

The Ratio Test can often be used to find the radius of convergence of a power series.

EXAMPLE 2   Find the interval of convergence of

$$\sum_{n=1}^{\infty} \frac{x^n}{n} = x + \frac{x^2}{2} + \frac{x^3}{3} + \cdots + \frac{x^n}{n} + \cdots.$$

We compute the limit

$$\lim_{n \to \infty} \frac{|x^{n+1}/(n+1)|}{|x^n/n|} = |x| \lim_{n \to \infty} \frac{n}{n+1} = |x|.$$

By the Ratio Test the series converges for $|x| < 1$ and diverges for $|x| > 1$, so the radius of convergence is $r = 1$.

At $x = 1$ the series is

$$1 + \frac{1}{2} + \frac{1}{3} + \cdots + \frac{1}{n} + \cdots$$

which is divergent. At $x = -1$ the series is

$$-1 + \frac{1}{2} - \frac{1}{3} + \frac{1}{4} + \cdots + \frac{(-1)^n}{n} + \cdots$$

which converges by the Alternating Series Test. The interval of convergence is $[-1, 1)$.

EXAMPLE 3 Find the interval of convergence of

$$\sum_{n=0}^{\infty} \frac{x^n}{n!} = 1 + x + \frac{x^2}{2} + \frac{x^3}{6} + \cdots + \frac{x^n}{n!} + \cdots.$$

For all x we have

$$\lim_{n \to \infty} \frac{|x^{n+1}/(n+1)!|}{|x^n/n!|} = \lim_{n \to \infty} \frac{|x|}{n} = 0.$$

Therefore by the Ratio Test the series converges for all x. It has radius of convergence ∞, and interval of convergence $(-\infty, \infty)$.

EXAMPLE 4 Find the radius of convergence of

$$\sum_{n=0}^{\infty} n! x^n = 1 + x + 2x^2 + 6x^3 + \cdots.$$

For $x \neq 0$, $\quad \lim_{n \to \infty} \frac{|(n+1)! x^{n+1}|}{|n! x^n|} = \lim_{n \to \infty} n|x| = \infty.$

By the Ratio Test the series diverges for $x \neq 0$ and the radius of convergence is $r = 0$.

If we replace x by $x - c$ we obtain a *power series in $x - c$*,

$$\sum_{n=0}^{\infty} a_n(x - c)^n = a_0 + a_1(x - c) + a_2(x - c)^2 + \cdots.$$

The power series $\sum_{n=0}^{\infty} a_n(x - c)^n$ has the same radius of convergence as $\sum_{n=0}^{\infty} a_n x^n$, and the interval of convergence is simply moved over so that its center is c instead of 0. For example, if $\sum_{n=0}^{\infty} a_n x^n$ has interval of convergence $(-r, r]$, then

$$\sum_{n=0}^{\infty} a_n(x - c)^n$$

has interval of convergence $(c - r, c + r]$, illustrated in Figure 9.7.2.

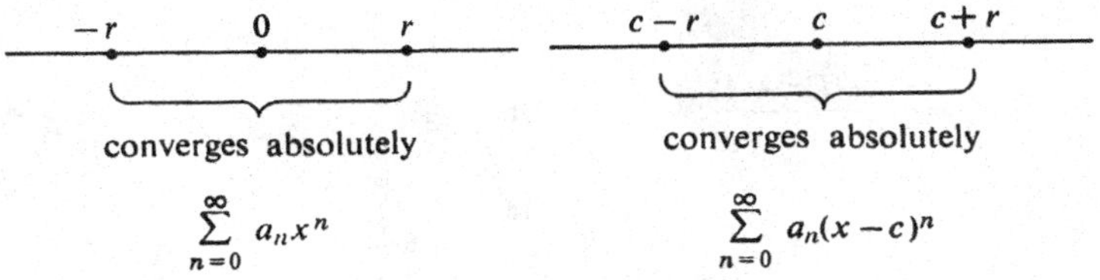

Figure 9.7.2

EXAMPLE 5   Find the interval of convergence of

$$\sum_{n=0}^{\infty} \frac{(n!)^2}{(2n)!} (x + 5)^n = 1 + \frac{1}{2}(x + 5) + \frac{(2!)^2}{4!}(x + 5)^2 + \cdots.$$

We have

$$\lim_{n \to \infty} \left| \frac{(n + 1)! \, (n + 1)! \, (x + 5)^{n+1}/(2n + 2)!}{(n!)(n!)(x + 5)^n/(2n)!} \right|$$

$$= \lim_{n \to \infty} \left| \frac{(n + 1)^2(x + 5)}{(2n + 1)(2n + 2)} \right| = \frac{|x + 5|}{4}.$$

By the Ratio Test the series converges for $|x + 5| < 4$ and diverges for $|x + 5| > 4$. The radius of convergence is $r = 4$, and the interval of convergence is centered at -5. We note that

$$\frac{(k!)^2}{(2k)!} = \frac{1}{1} \cdot \frac{1}{2} \cdot \frac{2}{3} \cdot \frac{2}{4} \cdots \cdot \frac{n}{(2n - 1)} \cdot \frac{n}{2n} > \left(\frac{1}{2} \cdot \frac{1}{2}\right)^n = \left(\frac{1}{4}\right)^n.$$

Therefore at $|x + 5| = 4$,

$$\left| \frac{(n!)^2}{(2n)!}(x + 5)^n \right| > \left(\frac{1}{4}\right)^n 4^n = 1.$$

Thus at $x + 5 = 4$ and $x + 5 = -4$ the terms do not approach zero and the series diverges. The interval of convergence is $(-9, -1)$.

PROBLEMS FOR SECTION 9.7

In Problems 1–25, find the radius of convergence.

1    $\displaystyle\sum_{n=0}^{\infty} 5x^n$

2    $\displaystyle\sum_{n=0}^{\infty} \frac{x^n}{3^n}$

3    $\displaystyle\sum_{n=1}^{\infty} n^n x^n$

4    $\displaystyle\sum_{n=1}^{\infty} \sqrt[n]{n}\, x^n$

5    $\displaystyle\sum_{n=1}^{\infty} \frac{n!}{n^n} x^n$

6    $\displaystyle\sum_{n=1}^{\infty} \frac{n^n}{n!} x^n$

7    $\displaystyle\sum_{n=1}^{\infty} \frac{n^{2n}}{(2n)!} x^n$

8    $\displaystyle\sum_{n=1}^{\infty} \frac{(3n)!}{(n!)^3} x^n$

9    $\displaystyle\sum_{n=1}^{\infty} \left(1 - \frac{1}{n}\right) x^n$

10   $\displaystyle\sum_{n=1}^{\infty} \left(1 + \frac{1}{n}\right)^n x^n$

11   $\displaystyle\sum_{n=2}^{\infty} \frac{x^n}{\ln n}$

12   $\displaystyle\sum_{n=2}^{\infty} \frac{x^n}{(\ln n)^n}$

13   $\displaystyle\sum_{n=2}^{\infty} \frac{n^n x^n}{(\ln n)^n}$

14   $\displaystyle\sum_{n=3}^{\infty} \frac{x^n}{(\ln (\ln n))^n}$

15   $\displaystyle\sum_{n=0}^{\infty} \frac{1 \cdot 4 \cdot 7 \cdot \cdots \cdot (3n + 1)}{n!} x^n$

16   $\displaystyle\sum_{n=0}^{\infty} \frac{n!}{1 \cdot 3 \cdot 5 \cdot \cdots \cdot (2n + 1)} x^n$

17   $\displaystyle\sum_{n=0}^{\infty} \frac{x^n}{3^{(n^2)}}$

18   $\displaystyle\sum_{n=0}^{\infty} \frac{x^n}{5^{\sqrt{n}}}$

19   $\displaystyle\sum_{n=0}^{\infty} \frac{x^n}{5^n \sqrt{n}}$

20   $\displaystyle\sum_{n=1}^{\infty} \frac{x^n}{\sqrt{n^n}}$

21 $\displaystyle\sum_{n=1}^{\infty} \frac{n!x^n}{\sqrt{n^n}}$

22 $\displaystyle\sum_{n=1}^{\infty} 3^n x^{2n}$

23 $\displaystyle\sum_{n=1}^{\infty} \frac{x^{3n}}{5^n}$

24 $\displaystyle\sum_{n=1}^{\infty} \frac{x^{6n}}{n!}$

25 $\displaystyle\sum_{n=1}^{\infty} \frac{1}{n!} x^{(n^2)}$

In Problems 26–45, find the interval of convergence.

26 $\displaystyle\sum_{n=1}^{\infty} \frac{x^n}{n^2}$

27 $\displaystyle\sum_{n=0}^{\infty} 2x^n$

28 $\displaystyle\sum_{n=0}^{\infty} n3^n x^n$

29 $\displaystyle\sum_{n=1}^{\infty} (-1)^n \frac{x^n}{n}$

30 $\displaystyle\sum_{n=1}^{\infty} \frac{x^n}{6\sqrt{n}}$

31 $\displaystyle\sum_{n=2}^{\infty} \frac{2^n x^n}{\ln n}$

32 $\displaystyle\sum_{n=1}^{\infty} \frac{(-1)^n x^n}{3^n n^2}$

33 $\displaystyle\sum_{n=0}^{\infty} \frac{(-1)^n x^n}{n!}$

34 $\displaystyle\sum_{n=1}^{\infty} \frac{(x+2)^n}{n\sqrt{n}}$

35 $\displaystyle\sum_{n=1}^{\infty} \frac{(-1)^n (x+2)^n}{n}$

36 $\displaystyle\sum_{n=0}^{\infty} n!(x-3)^n$

37 $\displaystyle\sum_{n=0}^{\infty} \frac{(x-5)^n}{n!}$

38 $\displaystyle\sum_{n=0}^{\infty} \frac{(x+8)^n}{2^n}$

39 $\displaystyle\sum_{n=1}^{\infty} \left(1 - \frac{1}{n}\right) x^n$

40 $\displaystyle\sum_{n=0}^{\infty} (3^n + 4^n) x^n$

41 $\displaystyle\sum_{n=0}^{\infty} \frac{4^n}{3^n + 5^n} x^n$

42 $\displaystyle\sum_{n=0}^{\infty} 3^n x^{2n}$

43 $\displaystyle\sum_{n=1}^{\infty} \frac{x^{2n}}{n \cdot 5^n}$

44 $\displaystyle\sum_{n=1}^{\infty} \frac{e^n (x-4)^{2n}}{n^2}$

45 $\displaystyle\sum_{n=0}^{\infty} \frac{x^{2n}}{n!}$

9.8 DERIVATIVES AND INTEGRALS OF POWER SERIES

In the last section we concentrated on the problem of finding the interval of convergence of a power series. We shall now find the sums of some important power series. Our general plan will be as follows.

First, find the sums of two basic power series:

$$\frac{1}{1-x} = 1 + x + x^2 + \cdots + x^n + \cdots,$$

$$e^x = 1 + x + \frac{x^2}{2!} + \cdots + \frac{x^n}{n!} + \cdots.$$

Then, starting with these basic power series, find the sums of other power series by differentiation and integration. (Based on Theorem 1.)

An especially useful property of power series is that they can be differentiated and integrated like polynomials. If we have a power series for a function $f(x)$, we can use Theorem 1 to immediately write down the power series for the derivative $f'(x)$ and integral $\int_0^x f(t)\, dt$.

THEOREM 1

Suppose $f(x)$ is the sum of a power series

$$f(x) = \sum_{n=0}^{\infty} a_n x^n$$

with radius of convergence $r > 0$, and let $-r < x < r$. Then:

(i) *f has the derivative*

$$f'(x) = \sum_{n=1}^{\infty} n a_n x^{n-1}$$

(ii) *f has the integral*

$$\int_0^x f(t)\, dt = \sum_{n=0}^{\infty} \frac{a_n}{n+1} x^{n+1}.$$

(iii)   *The power series in* (i) *and* (ii) *both have radius of convergence r.*

Discussion   This theorem says that a power series can be differentiated and integrated term by term. Also, the radius of convergence remains the same. To differentiate or integrate each term of a power series we simply use the Power Rule.

$$n\text{th term of } f(x) = a_n x^n$$

$$\text{derivative} = \begin{cases} n a_n x^{n-1} & n \neq 0 \\ 0 & n = 0 \end{cases}$$

$$\text{integral} = \frac{a_n}{n+1} x^{n+1}$$

We postpone the proof of Theorem 1 until later.

EXAMPLE 1   Differentiate and integrate the power series $\sum_{n=0}^{\infty} n^2 x^n$, and find the radii of convergence.

By the Ratio Test this power series has radius of convergence $r = 1$, for

$$\lim_{n \to \infty} \frac{|(n+1)^2 x^{n+1}|}{|n^2 x^n|} = |x| \lim_{n \to \infty} \frac{(n+1)^2}{n^2} = |x|.$$

Derivative: $\dfrac{d}{dx}\left(\sum_{n=0}^{\infty} n^2 x^n \right) = \sum_{n=1}^{\infty} n^3 x^{n-1} = \sum_{m=0}^{\infty} (m+1)^3 x^m.$

Integral: $\displaystyle \int_0^x \left(\sum_{n=0}^{\infty} n^2 t^n \right) dt = \sum_{n=0}^{\infty} \frac{n^2}{n+1} x^{n+1} = \sum_{m=1}^{\infty} \frac{(m-1)^2}{m} x^m.$

For convenience we rewrote the derivative as a power series in x^m where $m = n - 1$, and the integral as a power series in x^m where $m = n + 1$. Both the derivative and integral also have radius of convergence $r = 1$.

We are now ready to prove the power series formulas for $1/(1 - x)$ and e^x.

THEOREM 2

(i) $\dfrac{1}{1-x} = 1 + x + x^2 + \cdots + x^n + \cdots,$ $r = 1.$

(ii) $e^x = 1 + x + \dfrac{x^2}{2!} + \cdots + \dfrac{x^n}{n!} + \cdots,$ $r = \infty.$

PROOF (i) is just the geometric series for x. We proved in Section 9.2 that it converges to $1/(1-x)$ for $|x| < 1$ and diverges for $|x| \geq 1$.

(ii)   Let

$$y = \sum_{n=0}^{\infty} \frac{x^n}{n!} = 1 + x + \frac{x^2}{2!} + \cdots + \frac{x^n}{n!} + \cdots.$$

At $x = 0$ we have $y = 1$. We can find dy/dx by Theorem 1.

$$\frac{dy}{dx} = \sum_{n=1}^{\infty} \frac{nx^{n-1}}{n!} = \sum_{n=1}^{\infty} \frac{x^{n-1}}{(n-1)!} = \sum_{m=0}^{\infty} \frac{x^m}{m!} = y.$$

The radius of convergence is ∞, so for all x,

$$\frac{dy}{dx} = y.$$

The general solution of this differential equation (see Section 8.6) is

$$y = Ce^x.$$

At $x = 0$, $1 = Ce^0 = C$. Therefore $y = e^x$.

We shall now get several new power series formulas starting from the power series for $1/(1-x)$. We shall use the following methods:

A.   Differentiate a power series.
B.   Integrate a power series.
C.   Substitute bu for x.
D.   Substitute u^p for x.
E.   Multiply a power series by a constant.
F.   Multiply a power series by x^p.
G.   Add two power series.

Methods C, D, and G may change the radius of convergence.
We start with

(1) $\dfrac{1}{1-x} = 1 + x + x^2 + \cdots + x^n + \cdots,$ $r = 1.$

Substitute $-u$ for x in Equation 1.

(2) $\dfrac{1}{1+u} = 1 - u + u^2 - \cdots + (-1)^n u^n + \cdots,$ $r = 1.$

The radius of convergence is still $r = 1$ because when $|-u| < 1$, $|u| < 1$. Let us instead substitute $2u$ for x in Equation 1 and see what happens to the radius of convergence.

(3) $\dfrac{1}{1-2u} = 1 + 2u + 2^2u^2 + \cdots + 2^n u^n + \cdots,$ $r = \dfrac{1}{2}.$

The radius of convergence in Equation 3 is $r = \frac{1}{2}$ because when $|2u| < 1$, $|u| < \frac{1}{2}$. For convenience we rewrite Equations 2 and 3 with x's instead of u's. Thus

(2) $$\frac{1}{1 + x} = 1 - x + x^2 - \cdots + (-1)^n x^n + \cdots, \qquad r = 1.$$

(3) $$\frac{1}{1 - 2x} = 1 + 2x + 2^2 x^2 + \cdots + 2^n x^n + \cdots, \qquad r = \frac{1}{2}.$$

By integrating $1/(1 - x)$ and multiplying by -1 we get a power series for $\ln(1 - x)$.

$$\int_0^x \frac{1}{1 - t}\,dt = -\ln(1 - x).$$

(4) $$\ln(1 - x) = -x - \frac{x^2}{2} - \frac{x^3}{3} - \cdots - \frac{x^n}{n} - \cdots, \qquad r = 1.$$

We next use the power series Equation 2 for $1/(1 + x)$. Substitute x^2 for x in Equation 2.

(5) $$\frac{1}{1 + x^2} = 1 - x^2 + x^4 - \cdots + (-1)^n x^{2n} + \cdots, \qquad r = 1.$$

r is still 1 because if $|x^2| < 1$, $|x| < 1$. We obtain a power series for $\arctan x$ by integrating (5).

$$\int_0^x \frac{1}{1 + t^2}\,dt = \arctan x.$$

(6) $$\arctan x = x - \frac{x^3}{3} + \frac{x^5}{5} - \cdots + \frac{(-1)^n x^{2n+1}}{2n + 1} + \cdots, \qquad r = 1.$$

Finally let us differentiate the series (1) for $1/(1 - x)$.

$$\frac{d}{dx}\left(\frac{1}{1 - x}\right) = \frac{1}{(1 - x)^2}.$$

(7) $$\frac{1}{(1 - x)^2} = 1 + 2x + 3x^2 + \cdots + (n + 1)x^n + \cdots, \qquad r = 1.$$

Let us begin again, this time with

(8) $$e^x = 1 + x + \frac{x^2}{2!} + \cdots + \frac{x^n}{n!} + \cdots, \qquad r = \infty.$$

Substitute $-x$ for x in Equation 8.

(9) $$e^{-x} = 1 - x + \frac{x^2}{2!} - \cdots + \frac{(-1)^n x^n}{n!} + \cdots, \qquad r = \infty.$$

Using the formulas

$$\cosh x = \frac{e^x + e^{-x}}{2}, \qquad \sinh x = \frac{e^x - e^{-x}}{2}$$

we can obtain power series for $\cosh x$ and $\sinh x$. This is our first chance to use the method of adding power series.

(10) $$\cosh x = 1 + \frac{x^2}{2!} + \frac{x^4}{4!} + \cdots + \frac{x^{2n}}{(2n)!} + \cdots, \qquad r = \infty.$$

$$(11) \qquad \sinh x = x + \frac{x^3}{3!} + \frac{x^5}{5!} + \cdots + \frac{x^{2n+1}}{(2n+1)!} + \cdots, \qquad r = \infty.$$

Notice that the odd terms cancel out for $\cosh x$ and the even terms cancel out for $\sinh x$.

In Section 9.11 we shall obtain power series for $\sin x$ and $\cos x$ by another method.

We can easily get new power series by multiplying by x^p. For example, starting with the power series for $\ln(1-x)$, we obtain

$$\ln(1-x) = -x - \frac{x^2}{2} - \frac{x^3}{3} - \cdots, \qquad r = 1,$$

$$x \ln(1-x) = -x^2 - \frac{x^3}{2} - \frac{x^4}{3} - \cdots, \qquad r = 1,$$

$$x^2 \ln(1-x) = -x^3 - \frac{x^4}{2} - \frac{x^5}{3} - \cdots, \qquad r = 1,$$

and so on. Since the series for $\ln(1-x)$ has no constant term, we may also divide by x to get a new power series. To cover the case $x = 0$, we let

$$f(x) = \begin{cases} \dfrac{\ln(1-x)}{x} & \text{if } x \neq 0, \\ -1 & \text{if } x = 0. \end{cases}$$

Then $$f(x) = -1 - \frac{x}{2} - \frac{x^2}{3} - \frac{x^3}{4} - \cdots, \qquad r = 1.$$

We can often get a power series formula for an indefinite integral which cannot be evaluated in other ways. For example, the integral

$$\int_0^x e^{-t^2}\, dt$$

is of central importance in probability theory. It is the area under the normal (bell-shaped) curve $y = e^{-x^2}$. This integral is not an elementary function at all, so the methods of integration in Chapter 9 will fail. However, we can easily find a power series for this integral. First substitute x^2 for x in Equation 9.

$$(12) \qquad e^{-x^2} = 1 - x^2 + \frac{x^4}{2!} - \frac{x^6}{3!} + \cdots + \frac{(-1)^n x^{2n}}{n!} + \cdots, \qquad r = \infty.$$

Then integrate.

$$(13) \qquad \int_0^x e^{-t^2}\, dt = x - \frac{x^3}{3} + \frac{x^5}{5\cdot 2!} - \frac{x^7}{7\cdot 3!} + \cdots + \frac{(-1)^{n+1} x^{2n+1}}{(2n+1)n!} + \cdots, \qquad r = \infty.$$

PROOF OF THEOREM 1   It is easiest to prove (iii), then (ii), and finally (i).

(iii) The series $\sum_{n=1}^{\infty} na_n x^{n-1}$ and $\sum_{n=0}^{\infty} (a_n/(n+1))x^{n+1}$ have radius of convergence r.

Let $|x| < r$. We may choose c with $|x| < c < r$. Then $\sum_{n=0}^{\infty} a_n c^n$ converges absolutely. For positive infinite H, Theorem 1 in Section 9.1 (page 526) shows that $|c/x|^H/H$ is positive infinite, so $H|x/c|^H \approx 0$. Therefore

$$\left| Ha_H x^{H-1} \right| = H \left| \frac{x}{c} \right|^H \cdot \left| \frac{1}{x} \cdot a_H c^H \right| < \left| a_H c^H \right|.$$

Then by the Limit Comparison Test, $\sum_{n=1}^{\infty} na_n x^{n-1}$ converges absolutely. Similarly $\sum_{n=0}^{\infty} (a_n/(n+1))x^{n+1}$ converges absolutely.

Now let $|x| > r$. Using the same test we can show that $\sum_{n=1}^{\infty} na_n x^{n-1}$ and $\sum_{n=0}^{\infty} (a_n/(n+1))x^{n+1}$ diverge. Therefore both series have radius of convergence r.

(ii) $\displaystyle \int_0^x f(t)\, dt = \sum_{n=0}^{\infty} \frac{a_n}{n+1} x^{n+1}.$

Let $0 < c < r$. Our proof has three main steps. First, get an error estimate for the difference between $f(t)$ and the mth partial sum. Second, show that $f(t)$ is continuous for $-c \le t \le c$. Third, show that $f(t)$ has the required integral.

The series $\sum_{n=0}^{\infty} a_n c^n$ converges absolutely. Let E_m be the tail

$$E_m = \sum_{n=m+1}^{\infty} |a_n| c^n.$$

Then $$\lim_{m \to \infty} E_m = 0.$$

Moreover, for $-c \le t \le c$,

$$\left| \sum_{n=m+1}^{\infty} a_n t^n \right| \le \sum_{n=m+1}^{\infty} |a_n t^n| \le E_m.$$

Therefore E_m is an error estimate for $f(t)$ minus the partial sum,

(14) $$-E_m \le f(t) - \sum_{n=0}^{m} a_n t^n \le E_m.$$

We now prove f is continuous on $[-c, c]$. Since c was chosen arbitrarily between 0 and r, it will follow that f is continuous on $(-r, r)$. Let $t \approx u$ in $[-c, c]$. For each finite m,

$$\left| f(t) - \sum_{n=0}^{m} a_n t^n \right| \le E_m,$$

$$\left| \sum_{n=0}^{m} a_n t^n - \sum_{n=0}^{m} a_n u^n \right| \approx 0,$$

$$\left| \sum_{n=0}^{m} a_n u^n - f(u) \right| \le E_m.$$

Therefore $$st|f(t) - f(u)| \le E_m + 0 + E_m.$$

Since the E_m's approach zero, it follows that $f(t) \approx f(u)$. Hence f is continuous on $[-c, c]$.

To prove the integral formula we integrate both sides of Equation 14 from 0 to x. Let $0 < x$.

$$-E_m x \le \int_0^x f(t)\, dt - \sum_{n=0}^{m} \frac{a_n}{n+1} x^{n+1} \le E_m x.$$

Again since E_m approaches zero, we conclude that

$$\int_0^x f(t)\, dt = \sum_{n=0}^{\infty} \frac{a_n}{n+1} x^{n+1}.$$

The case $x < 0$ is similar.

(i) $f'(x) = \sum_{n=1}^{\infty} n a_n x^{n-1}$.

Let
$$g(t) = \sum_{n=1}^{\infty} n a_n t^{n-1}.$$

Integrating term by term,

$$\int_0^x g(t)\, dt = \sum_{n=1}^{\infty} a_n x^n.$$

Thus
$$\int_0^x g(t)\, dt = f(x) - a_0 = f(x) - f(0).$$

By the Fundamental Theorem of Calculus, $g(x) = f'(x)$.

In part (i) of the proof we needed part (iii) to be sure that the series for $g(t)$ converges for $-r < t < r$, and part (ii) to justify the term by term integration.

PROBLEMS FOR SECTION 9.8

In Problems 1–10 find power series for $f'(x)$ and for $\int_0^x f(t)\, dt$.

1 $f(x) = \sum_{n=0}^{\infty} 10^n x^n$

2 $f(x) = \sum_{n=1}^{\infty} n^{-n} x^n$

3 $f(x) = \sum_{n=1}^{\infty} n^{-3} x^n$

4 $f(x) = \sum_{n=2}^{\infty} \frac{x^n}{\ln n}$.

5 $f(x) = \sum_{n=1}^{\infty} \frac{n+1}{n} x^n$

6 $f(x) = \sum_{n=1}^{\infty} \sqrt{n}\sqrt{n+1}\, x^n$

7 $f(x) = \sum_{n=1}^{\infty} \frac{n!}{n^n} x^n$

8 $f(x) = \sum_{n=0}^{\infty} (-1)^n \frac{n+1}{3^n} x^n$

9 $f(x) = \sum_{n=0}^{\infty} x^{2n}$

10 $f(x) = \sum_{n=1}^{\infty} \frac{1}{n^2} x^{2n}$

In Problems 11–34 find a power series for the given function and determine its radius of convergence.

11 $f(x) = \dfrac{1}{1 + 3x}$

12 $f(x) = \dfrac{1}{1 - x^2}$

13 $f(x) = \arctan(4x^2)$

14 $f(x) = \ln(1 - 3x^2)$

15 $f(x) = x \ln(1 + 2x)$

16 $f(x) = \dfrac{\arctan x}{x}$ if $x \neq 0,\ f(0) = 1$

17 $f(x) = e^{-4x}$

18 $f(x) = x^2 e^x$

19 $f(x) = \sinh(3x)$

20 $f(x) = \cosh(x^2)$

21 $f(x) = \displaystyle\int_0^x \ln(1 + 2t^2)\, dt$

22 $f(x) = \displaystyle\int_0^x \arctan(t^3)\, dt$

23 $f(x) = \displaystyle\int_0^x e^{t^3}\, dt$

24 $f(x) = \displaystyle\int_0^x \sinh(t^2)\, dt$

25 $\quad f(x) = \int_0^x t \ln(1 - t)\, dt$ $\qquad$ 26 $\quad f(x) = \int_0^x \dfrac{t^2}{1 + t^2}\, dt$

27 $\quad f(x) = \int_0^x \dfrac{\ln(1 + t)}{t}\, dt$ $\qquad$ 28 $\quad f(x) = \int_0^x \dfrac{e^t - 1}{t}\, dt$

29 $\quad f(x) = \dfrac{2x}{(1 + x^2)^2}$ $\qquad$ Hint: $f(x) = \dfrac{d}{dx} \dfrac{-1}{1 + x^2}$.

30 $\quad f(x) = \dfrac{1}{(1 + x^2)^2}$

31 $\quad f(x) = \dfrac{2x}{1 + x^4}$ $\qquad$ Hint: $f(x) = \dfrac{d}{dx} \arctan(x^2)$.

32 $\quad f(x) = \dfrac{1}{(1 - x)^3}$

33 $\quad f(x) = \arctan x + \arctan(2x)$

34 $\quad f(x) = \sinh x + x \cosh x$

35 $\quad$ Check the formulas $d(\sinh x)/dx = \cosh x$, $d(\cosh x)/dx = \sinh x$ by differentiating the power series.

☐ 36 $\quad$ Prove that if the power series $f(x) = \sum_{n=0}^{\infty} a_n x^n$ has finite radius of convergence r, then the power series

$$f(bx) = \sum_{n=0}^{\infty} a_n(bx)^n$$

has radius of convergence r/b $(b > 0)$.

☐ 37 $\quad$ Prove that if $f(x) = \sum_{n=0}^{\infty} a_n x^n$ has finite radius of convergence r, then

$$f(x^2) = \sum_{n=0}^{\infty} a_n x^{2n}$$

has radius of convergence $\sqrt{r}$.

☐ 38 $\quad$ Prove that if

$$f(x) = \sum_{n=0}^{\infty} a_n x^n, \qquad g(x) = \sum_{n=0}^{\infty} b_n x^n$$

have radii of convergence r and s respectively and $r \le s$, then $f(x) + g(x)$ has a radius of convergence of at least r.

☐ 39 $\quad$ Show that if $f(x) = \sum_{n=0}^{\infty} a_n x^n$ has radius of convergence r, then for any positive integer p,

$$x^p f(x) = \sum_{n=0}^{\infty} a_n x^{n+p}$$

has radius of convergence r.

☐ 40 $\quad$ Evaluate $\sum_{n=1}^{\infty} n x^n$, $|x| < 1$, using the derivative of the power series $\sum_{n=0}^{\infty} x^n$.

☐ 41 $\quad$ Evaluate $\sum_{n=1}^{\infty} n^2 x^n$, $|x| < 1$, using the first and second derivatives of $\sum_{n=0}^{\infty} x^n$.

9.9 APPROXIMATIONS BY POWER SERIES

Power series are one of the most important methods of approximation in mathematics. Consider a power series

$$f(x) = a_0 + a_1 x + a_2 x^2 + \cdots + a_n x^n + \cdots.$$

The partial sums give approximate values for the function,

$$f(x) \sim a_0 + a_1 x + a_2 x^2 + \cdots + a_n x^n,$$

and the tails E_n give the error in the approximation,

$$f(x) = a_0 + a_1 x + a_2 x^2 + \cdots + a_n x^n + E_n.$$

If we can estimate the error E_n we can compute approximate values for $f(x)$ to any desired degree of accuracy.

In this section we shall give two simple methods of estimating the error. A more general method will be given in the next section. Our first method is to use the Alternating Series Test. It can be applied whenever a power series is alternating.

EXAMPLE 1 Approximate $\ln\left(1\frac{1}{2}\right)$ within 0.01.

We use the power series for $\ln(1 - x)$,

$$\ln(1 - x) = -x - \frac{x^2}{2} - \frac{x^3}{3} - \frac{x^4}{4} - \frac{x^5}{5} - \cdots, \qquad r = 1.$$

Setting $1 - x = 1\frac{1}{2}$, $x = -\frac{1}{2}$,

$$\ln\left(1\frac{1}{2}\right) = \frac{1}{2} - \frac{1}{2\cdot 4} + \frac{1}{3\cdot 8} - \frac{1}{4\cdot 16} + \frac{1}{5\cdot 32} - \cdots.$$

This is an alternating series. The last term shown is less than 0.01,

$$\frac{1}{5\cdot 32} = \frac{1}{160} \sim 0.006.$$

By the Alternating Series Test, the error in each partial sum is less than the next term. So

$$\ln\left(1\frac{1}{2}\right) \sim \frac{1}{2} - \frac{1}{2\cdot 4} + \frac{1}{3\cdot 8} - \frac{1}{4\cdot 16}, \qquad \text{error} \le \frac{1}{5\cdot 32},$$

or

$$\ln\left(1\frac{1}{2}\right) \sim 0.401, \qquad \text{error} \le 0.006.$$

The actual value is $\ln\left(1\frac{1}{2}\right) \sim 0.405$.

EXAMPLE 2 Approximate $\arctan\frac{1}{2}$ within 0.001.

The power series for $\arctan x$ is

$$\arctan x = x - \frac{x^3}{3} + \frac{x^5}{5} - \frac{x^7}{7} + \frac{x^9}{9} - \cdots, \qquad r = 1.$$

Setting $x = \frac{1}{2}$,

$$\arctan\frac{1}{2} = \frac{1}{2} - \frac{1}{3\cdot 8} + \frac{1}{5\cdot 32} - \frac{1}{7\cdot 128} + \frac{1}{9\cdot 512} - \cdots.$$

This an an alternating series. The last term is less than 0.001,

$$\frac{1}{9\cdot 512} \sim 0.0002.$$

Therefore

$$\arctan\frac{1}{2} \sim \frac{1}{2} - \frac{1}{3\cdot 8} + \frac{1}{5\cdot 32} - \frac{1}{7\cdot 128}, \qquad \text{error} \le 0.0002.$$

Adding up, $\qquad \arctan\frac{1}{2} \sim 0.4635, \qquad \text{error} \le 0.0002.$

The series

$$\arctan x = x - \frac{x^3}{3} + \frac{x^5}{5} - \frac{x^7}{7} + \cdots, \qquad r = 1$$

can be used to approximate π. We start with

$$\tan \frac{\pi}{6} = \frac{1}{\sqrt{3}}, \qquad \arctan \frac{1}{\sqrt{3}} = \frac{\pi}{6}.$$

Setting $x = 1/\sqrt{3}$ in the series,

$$\frac{\pi}{6} = \frac{1}{\sqrt{3}} - \frac{1}{3}\left(\frac{1}{\sqrt{3}}\right)^3 + \frac{1}{5}\left(\frac{1}{\sqrt{3}}\right)^5 - \frac{1}{7}\left(\frac{1}{\sqrt{3}}\right)^7 + \cdots,$$

or

$$\frac{\sqrt{3}}{6}\pi = 1 - \frac{1}{3}\left(\frac{1}{3}\right) + \frac{1}{5}\left(\frac{1}{3}\right)^2 - \frac{1}{7}\left(\frac{1}{3}\right)^3 + \frac{1}{9}\left(\frac{1}{3}\right)^4 - \cdots.$$

This is an alternating series, so

$$\frac{\sqrt{3}}{6}\pi \sim 1 - \frac{1}{9} + \frac{1}{45} - \frac{1}{189} + \frac{1}{729}, \qquad \text{error} \le \frac{1}{11}\left(\frac{1}{3}\right)^5,$$

$$\frac{\sqrt{3}}{6}\pi \sim 0.9072, \qquad \text{error} \le 0.0004.$$

Dividing everything by $\sqrt{3}/6$ we get

$$\pi \sim 3.1426, \qquad \text{error} \le 0.0013.$$

EXAMPLE 3   Approximate e^{-1} within 0.001.

The power series for e^x is

$$e^x = 1 + x + \frac{x^2}{2!} + \frac{x^3}{3!} + \frac{x^4}{4!} + \frac{x^5}{5!} + \cdots, \qquad r = \infty.$$

Setting $x = -1$,

$$e^{-1} = 1 - 1 + \tfrac{1}{2} - \tfrac{1}{6} + \tfrac{1}{24} - \tfrac{1}{120} + \tfrac{1}{720} - \tfrac{1}{5040} + \cdots.$$

The series alternates and the last term is less than 0.001, so

$$e^{-1} \sim 1 - 1 + \tfrac{1}{2} - \tfrac{1}{6} + \tfrac{1}{24} - \tfrac{1}{120} + \tfrac{1}{720}, \qquad \text{error} \le \tfrac{1}{5040} \sim 0.0002.$$

Adding up, $\qquad e^{-1} \sim 0.36806, \qquad \text{error} \le 0.0002.$

The actual value is $e^{-1} \sim 0.36788$.

Our second method of approximation is to start with a known error estimate for the geometric series and carefully keep track of the error each time we make a new series.

We recall the formula for the partial sum of a geometric series.

$$1 + x + x^2 + \cdots + x^n = \frac{1 - x^{n+1}}{1 - x} = \frac{1}{1 - x} - \frac{x^{n+1}}{1 - x}.$$

Thus $\dfrac{1}{1-x} = 1 + x + x^2 + \cdots + x^n + E_n, \qquad E_n = \dfrac{x^{n+1}}{1-x}.$

This formula is valid for all x, but the error E_n approaches zero only when x is within the interval of convergence $(-1, 1)$.

EXAMPLE 4 Approximate $1/(1 - 0.02)$ to six decimal places. Take $x = 0.02$.

$$\frac{1}{1 - 0.02} = 1 + 0.02 + (0.02)^2 + (0.02)^3 + E_4$$

$$= 1 + 0.02 + 0.0004 + 0.000008 + E_4$$

$$= 1.020408 + E_4.$$

The error E_4 after four terms is

$$E_4 = \frac{(0.02)^4}{1 - 0.02} = \frac{0.00000016}{0.98} < \frac{0.00000016}{0.8} = 0.00000020.$$

So $1/(1 - 0.02) \sim 1.020408$ to six places.

Suppose we wish to approximate $\ln \frac{1}{2}$ within 0.01. If in the series

$$\ln(1 - x) = -x - \frac{x^2}{2} - \frac{x^3}{3} - \frac{x^4}{4} - \cdots, \qquad r = 1$$

we set $1 - x = \frac{1}{2}$, $x = \frac{1}{2}$, we get

$$\ln\frac{1}{2} = -\frac{1}{2} - \frac{1}{2\cdot 4} - \frac{1}{3\cdot 8} - \frac{1}{4\cdot 16} - \cdots.$$

We know this series converges, but to be sure of an approximation within 0.01 we need an error estimate. The next example shows how to get such an error estimate.

EXAMPLE 5 Given a constant c where $-1 < c < 1$, find a simple error estimate for the power series

$$\ln(1 - x) = -x - \frac{x^2}{2} - \frac{x^3}{3} - \cdots - \frac{x^n}{n} - \cdots$$

valid for $-1 < x \le c$.

We start with the equation

(1) $$\frac{1}{1 - t} = (1 + t + t^2 + \cdots + t^n) + E_n, \qquad E_n = \frac{t^{n+1}}{1 - t}.$$

For $-1 < t \le c$ we have

$$1 - t \ge 1 - c, \qquad |E_n| \le \frac{|t|^{n+1}}{1 - c}.$$

Integrating Equation 1 from 0 to x we have

(2) $$-\ln(1 - x) = \left(x + \frac{x^2}{2} + \frac{x^3}{3} + \cdots + \frac{x^{n+1}}{n + 1}\right) + \int_0^x E_n \, dt$$

and
$$\left| \int_0^x E_n \, dt \right| \leq \int_0^x \frac{|t|^{n+1}}{1-c} \, dt = \frac{|x|^{n+2}}{(1-c)(n+2)}.$$

Multiplying Equation 2 by -1 and setting $m = n + 1$ we have the following error estimate for $\ln(1-x)$, valid for $-1 < x \leq c$.

(3)
$$\ln(1-x) \sim \left(-x - \frac{x^2}{2} - \frac{x^3}{3} - \cdots - \frac{x^m}{m} \right),$$

$$\text{error} \leq \frac{|x|^{m+1}}{(1-c)(m+1)}.$$

EXAMPLE 6   Use Example 5 to approximate $\ln\frac{1}{2}$ within 0.01. We set $c = x = \frac{1}{2}$ in Equation 3.

$$\ln\frac{1}{2} \sim -\frac{1}{2} - \frac{1}{2\cdot 4} - \frac{1}{3\cdot 8} - \frac{1}{4\cdot 16} - \cdots - \frac{1}{m\cdot 2^m},$$

$$|\text{error}| \leq \frac{(1/2)^{m+1}}{\frac{1}{2}(m+1)} = \frac{1}{(m+1)2^m}.$$

Table 9.9.1 shows approximate values and error estimates.

Table 9.9.1

m	$\dfrac{1}{m\cdot 2^m}$	Approximate value for $\ln\frac{1}{2}$ $-\dfrac{1}{2} - \dfrac{1}{2\cdot 4} - \cdots - \dfrac{1}{m\cdot 2^m}$	Error estimate $\dfrac{1}{(m+1)2^m}$
1	0.5000	-0.5000	0.2500
2	0.1250	-0.6250	0.0833
3	0.04167	-0.6667	0.0313
4	0.01563	-0.6823	0.0125
5	0.00625	-0.6886	0.0052

We see that the error estimate drops below 0.01 when $m = 5$.

So $\ln\frac{1}{2} \sim -0.689$,   error ≤ 0.01.

Since $\ln\frac{1}{2} = -\ln 2$, we also have

$$\ln 2 \sim 0.689, \qquad \text{error} \leq 0.01.$$

A more rapidly converging series for $\ln 2$ can be obtained in the following way. Any number $a > 1$ can be put in the form

$$a = \frac{1+x}{1-x}, \qquad 0 < x < 1.$$

We simply take
$$x = \frac{a-1}{a+1}.$$

By the rules of logarithms,

$$\ln\left(\frac{1+x}{1-x}\right) = \ln(1+x) - \ln(1-x).$$

We can subtract two series by the Sum Rule, whence

$$\ln(1+x) = x - \frac{x^2}{2} + \frac{x^3}{3} - \frac{x^4}{4} + \frac{x^5}{5} - \cdots, \qquad r = 1,$$

$$\ln(1-x) = -x - \frac{x^2}{2} - \frac{x^3}{3} - \frac{x^4}{4} - \frac{x^5}{5} - \cdots, \qquad r = 1,$$

$$\ln\left(\frac{1+x}{1-x}\right) = 2x + \frac{2x^3}{3} + \frac{2x^5}{5} + \cdots, \qquad r = 1.$$

This power series is convenient because half of the terms are zero.

EXAMPLE 7 Find an error estimate for the power series for $\ln((1+x)/(1-x))$ valid for $-c \le x \le c$. Use it to approximate $\ln 2$ within 0.00001.

From Example 5 we have the following error estimates for $\ln(1+x)$ and $-\ln(1-x)$ valid for $-c \le x \le c$.

$$\ln(1+x) \sim x - \frac{x^2}{2} + \frac{x^3}{3} - \cdots + (-1)^{m+1}\frac{x^m}{m},$$

$$\text{error} \le \frac{|x|^{m+1}}{(1-c)(m+1)}.$$

$$-\ln(1-x) \sim x + \frac{x^2}{2} + \frac{x^3}{3} + \cdots + \frac{x^m}{m},$$

$$\text{error} \le \frac{|x|^{m+1}}{(1-c)(m+1)}.$$

We add the two sums and error estimates,

$$\ln\left(\frac{1+x}{1-x}\right) \sim 2x + \frac{2x^3}{3} + \frac{2x^5}{5} + \cdots + \frac{2x^{2m-1}}{2m-1},$$

$$\text{error} \le \frac{2|x|^{2m+1}}{(1-c)(2m+1)}.$$

We wish to choose x so that $(1+x)/(1-x) = 2$. Solving for x we get $x = \frac{1}{3}$. Now set $c = \frac{1}{3}$ and $x = \frac{1}{3}$. The error estimate for $x = \frac{1}{3}$ is

$$\frac{2|x|^{2m+1}}{(1-c)(2m+1)} = \frac{1}{(2m+1)3^{2m}}.$$

Table 9.9.2

m	$\dfrac{2}{(2m-1)3^{2m-1}}$	Approximate value for $\ln 2$ $\dfrac{2}{1\cdot 3} + \dfrac{2}{3\cdot 27} + \cdots + \dfrac{2}{(2m-1)3^{2m-1}}$	Error estimate $\dfrac{1}{(2m+1)3^{2m}}$
1	0.666667	0.666667	0.037037
2	0.024691	0.691358	0.002469
3	0.001646	0.693004	0.000196
4	0.000131	0.693134	0.000017
5	0.000011	0.693146	0.000002

The error estimate drops below 0.00001 when $m = 5$. Thus

$$\ln 2 \sim 0.693146, \qquad \text{error} \le 0.00001.$$

EXAMPLE 8   Find the sum of the alternating harmonic series

$$1 - \tfrac{1}{2} + \tfrac{1}{3} - \tfrac{1}{4} + \cdots.$$

Our first guess is to set $x = -1$ in the power series

$$\ln(1 - x) = -x - \frac{x^2}{2} - \frac{x^3}{3} - \frac{x^4}{4} - \cdots, \qquad r = 1.$$

This suggests to us the sum

$$\ln 2 = 1 - \tfrac{1}{2} + \tfrac{1}{3} - \tfrac{1}{4} + \cdots.$$

We know the series converges to something by the Alternating Series Test. For $-1 < x < 1$ the series converges to $\ln(1 - x)$. But $x = -1$ is an endpoint of the interval of convergence and the general theorem on integrating a power series does not apply. So we must go back to the beginning and use the equation

$$\frac{1}{1 - t} = (1 + t + \cdots + t^n) + \frac{t^{n+1}}{1 - t}.$$

For $t \le 0$, $|t^{n+1}/(1 - t)| \le |t^{n+1}|$, whence

$$\frac{1}{1 - t} = (1 + t + \cdots + t^n) + E_n, \qquad |E_n| \le |t^{n+1}|.$$

Integrating from 0 to x,

$$-\ln(1 - x) = \left(x + \frac{x^2}{2} + \cdots + \frac{x^{n+1}}{n + 1}\right) + F_n, \qquad |F_n| \le \left|\frac{x^{n+2}}{n + 2}\right|.$$

This holds for all $x \le 0$.

Now we set $x = -1$ and see that the error term $|F_n| \le 1/(n + 2)$ approaches zero. This proves that $\ln 2$ really is the sum of the alternating harmonic series,

$$\ln 2 = 1 - \tfrac{1}{2} + \tfrac{1}{3} - \tfrac{1}{4} + \cdots.$$

The alternating harmonic series converges very slowly, because after n terms the error estimate is only $1/(n + 1)$.

PROBLEMS FOR SECTION 9.9

Problems 1–12 below are to be done using a power series with an error estimate. If a hand calculator is available they can be worked with the errors reduced by an additional factor of 1000.

1   Approximate $\ln(1.2)$ within 0.01.

2   Approximate $\arctan\left(\tfrac{1}{10}\right)$ within 10^{-7}.

3   Approximate $e^{-1/4}$ within 0.00001.

4   Approximate $\displaystyle\int_0^1 e^{-t^2}\, dt$ within 0.01.

5   Approximate $\displaystyle\int_0^{1/2} \frac{1}{1 + t^3}\, dt$ within 0.0001.

6   Approximate $\displaystyle\int_0^{1/2} \ln(1 + t^2)\, dt$ within 0.001.

7   Approximate $\displaystyle\int_0^{1/3} \frac{1}{t} \arctan(t)\, dt$ within 0.0001.

8   Approximate $\displaystyle\int_0^{1/2} \arctan(t^2)\, dt$ within 0.00001.

9   Approximate $1/(1 - 0.003)$ within 0.0001.

10   Approximate $\ln 3$ within 0.1 by the method of Example 6. *Hint:* $\ln 3 = -\ln(1 - x)$ where $x = \frac{2}{3}$.

11   Approximate $\ln 3$ within 0.001 by the method of Example 7.

12   (a) Approximate $\ln(1\frac{1}{2})$ within 0.00001 by the method of Example 7.
(b) Approximate $\ln 3$ within 0.00002 using the formula $\ln 3 = \ln 2 + \ln(1\frac{1}{2})$.

In Problems 13–18 find a power series approximation with an error estimate for $f(x)$ valid for $-\frac{1}{2} \le x \le \frac{1}{2}$. Then approximate $f(\frac{1}{2})$ within 0.01.

13    $f(x) = \dfrac{x}{1 - x}$ 14 $f(x) = \displaystyle\int_0^x \ln(1 - t)\, dt$

15    $f(x) = \dfrac{1}{1 - x^2}.$ 16 $f(x) = \displaystyle\int_0^x \dfrac{1}{1 - t^3}\, dt$
Hint: $x^2 = \frac{1}{4}$ when $x = \frac{1}{2}$.

17    $f(x) = \displaystyle\int_0^x \dfrac{\ln(1 - t)}{t}\, dt$ 18 $f(x) = \displaystyle\int_0^x \ln(1 - t^2)\, dt$

☐ 19   Using the power series for $\arctan x$ at $x = 1$, show that
$$\frac{\pi}{4} = 1 - \frac{1}{3} + \frac{1}{5} - \frac{1}{7} + \frac{1}{9} - \frac{1}{11} + \cdots.$$

☐ 20   Using the power series for $\int_0^x \ln(1 + t)\, dt$ at $x = 1$, show that
$$2\ln 2 - 1 = \frac{1}{1 \cdot 2} - \frac{1}{2 \cdot 3} + \frac{1}{3 \cdot 4} - \frac{1}{4 \cdot 5} + \frac{1}{5 \cdot 6} - \cdots.$$

9.10 TAYLOR'S FORMULA

If we wish to express $f(x)$ as a power series in $x - c$, we need two things:

(1) A sequence of polynomials which approximate $f(x)$ near $x = c$,

(1) $a_0, a_0 + a_1(x - c), \ldots, a_0 + a_1(x - c) + \cdots + a_n(x - c)^n, \ldots.$

(2) An estimate for the error E_n between $f(x)$ and the nth polynomial,

(2) $f(x) = a_0 + a_1(x - c) + \cdots + a_n(x - c)^n + E_n.$

In the last section the formula
$$\frac{1}{1 - x} = 1 + x + \cdots + x^n + E_n, \qquad E_n = \frac{x^{n+1}}{1 - x}$$

was used to obtain power series approximations. A much more general formula of this type is Taylor's Formula. In Taylor's Formula the nth polynomial $P_n(x)$ is chosen so that its value and first n derivatives agree with $f(x)$ at $x = c$.

The tangent line at $x = c$,
$$P_1(x) = f(c) + f'(c)(x - c),$$

has the same value and first derivative as $f(x)$ at $x = c$. A polynomial of degree two with the same value and first two derivatives as $f(x)$ at c is

$$P_2(x) = f(c) + f'(c)(x - c) + \frac{f''(c)}{2}(x - c)^2.$$

$P_1(x)$ and $P_2(x)$ are the first and second *Taylor polynomials* of $f(x)$ (see Figure 9.10.1).

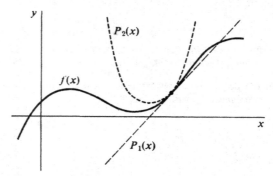

Figure 9.10.1   First and Second Taylor Polynomials

To continue the procedure we need a formula for the nth derivative of a polynomial.

LEMMA 1

Let $P(x)$ be a polynomial in $x - c$ of degree n.

$$P(x) = a_0 + a_1(x - c) + a_2(x - c) + \cdots + a_n(x - c)^n.$$

For each $m \leq n$, the mth *derivative* of $P(x)$ at $x = c$ divided by $m!$ is equal to the coefficient a_m,

$$\frac{P^{(m)}(c)}{m!} = a_m.$$

PROOF Consider one term $a_k(x - c)^k$. Its mth derivative is

$$
\begin{aligned}
k(k - 1) \cdots (k - m + 1)a_k(x - c)^{k-m} \quad &\text{if } m < k,\\
m!\, a_m \quad &\text{if } m = k,\\
0 \quad &\text{if } m > k.
\end{aligned}
$$

At $x = c$, the mth derivative of $a_k(x - c)^k$ is:

$$0 \text{ if } m < k, \qquad m!\, a_m \text{ if } m = k, \qquad 0 \text{ if } m > k.$$

It follows that $P^{(m)}(c) = m!\, a_m$.

This lemma shows us how to find a polynomial $P(x)$ whose value and first n derivatives agree with $f(x)$ at $x = c$. The mth coefficient of $P(x)$ must be

$$a_m = \frac{f^{(m)}(c)}{m!}.$$

DEFINITION

Let $f(x)$ have derivatives of all orders at $x = c$. The nth **Taylor polynomial** of $f(x)$ at $x = c$ is the polynomial

$$P_n(x) = f(c) + f'(c)(x - c) + \frac{f''(c)}{2!}(x - c)^2 + \cdots + \frac{f^{(n)}(c)}{n!}(x - c)^n.$$

By Lemma 1, $P_n(x)$ is the unique polynomial of degree n whose value and first n derivatives at $x = c$ agree with $f(x)$,

$$P_n(c) = f(c), P_n'(c) = f'(c), \ldots, P_n^{(n)}(c) = f^{(n)}(c).$$

The difference between $f(x)$ and the nth Taylor polynomial is called the nth **Taylor remainder**,

$$R_n(x) = f(x) - P_n(x).$$

Thus

$$f(x) = f(c) + f'(c)(x - c) + \frac{f''(c)}{2}(x - c)^2 + \cdots + \frac{f^{(n)}(c)}{n!}(x - c)^n + R_n(x).$$

EXAMPLE 1   Find the first five Taylor polynomials of $\sin x$ at $x = 0$. We work them out in Table 9.10.1.

Table 9.10.1

k	$f^{(k)}(x)$	$f^{(k)}(0)$	$P_k(x)$
0	$\sin x$	0	0
1	$\cos x$	1	x
2	$-\sin x$	0	x
3	$-\cos x$	-1	$x - x^3/3!$
4	$\sin x$	0	$x - x^3/3!$
5	$\cos x$	1	$x - x^3/3! + x^5/5!$

Since the even degree terms are zero, the 2nth Taylor polynomial is the same as the $(2n - 1)$st. Figure 9.10.2 compares the first and third Taylor polynomials with $\sin x$.

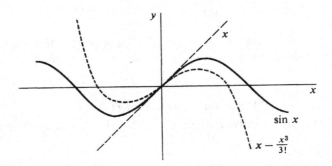

$\sin x$

$x - \dfrac{x^3}{3!}$

Figure 9.10.2

We can easily find the Taylor polynomials of $f(x)$ by differentiating. Let us now try to find a formula for the Taylor remainders. The Mean Value Theorem gives a formula for the Taylor remainder, $R_0(x)$.

MEAN VALUE THEOREM (Repeated)

Suppose $f(t)$ is differentiable at all t between c and d. Then

$$f'(t_0) = \frac{f(d) - f(c)}{d - c}$$

for some point t_0 strictly between c and d.

When we replace d by x, this gives the formula

$$f(x) = f(c) + R_0(x), \qquad R_0(x) = f'(t_0)(x - c).$$

Taylor's Formula is a generalization of the Mean Value Theorem which gives the nth Taylor remainder.

TAYLOR'S FORMULA

Suppose the $(n + 1)$st derivative $f^{(n+1)}(t)$ exists for all t between c and x. Then

$$f(x) = f(c) + f'(c)(x - c) + \frac{f''(c)}{2!}(x - c)^2 + \cdots + \frac{f^{(n)}(c)}{n!}(x - c)^n + R_n(x)$$

where

$$R_n(x) = \frac{f^{(n+1)}(t_n)}{(n + 1)!}(x - c)^{n+1}$$

for some point t_n strictly between c and x.

Notice that the remainder term looks just like the $(n + 1)$st term of a Taylor polynomial except that $f^{(n+1)}(c)$ is replaced by $f^{(n+1)}(t_n)$.

When $c = 0$ Taylor's Formula is sometimes called *MacLaurin's Formula*.

Taylor's Formula can be used to get an estimate of the error $R_n(x)$ between $f(x)$ and the Taylor polynomial $P_n(x)$. For example if

$$|f^{(n+1)}(t)| \leq M_{n+1}$$

for all t between c and x, then we obtain the error estimate

$$|R_n(x)| \leq \frac{M_{n+1}}{(n + 1)!}|x - c|^{n+1}.$$

Taylor polynomials with the error estimate are of great practical value in obtaining approximations. In the next example we use Taylor's Formula to approximate the value of e.

EXAMPLE 2   Find MacLaurin's Formula for $f(x) = e^x$.

The nth derivative is $f^{(n)}(x) = e^x$, $f^{(n)}(0) = 1$. MacLaurin's Formula is

$$e^x = 1 + x + \frac{x^2}{2!} + \frac{x^3}{3!} + \cdots + \frac{x^n}{n!} + R_n(x), \qquad R_n(x) = e^{t_n}\frac{x^{n+1}}{(n + 1)!}$$

for some t_n between 0 and x. For t between 0 and x the value of e^t is always less than or equal to $3^{|x|}$, for

$$e^t \le e^{|x|} \le 3^{|x|}.$$

We therefore have the formula

(3) $$e^x = 1 + x + \frac{x^2}{2!} + \cdots + \frac{x^n}{n!} + R_n(x), \qquad |R_n(x)| \le 3^{|x|} \cdot \frac{|x|^{n+1}}{(n+1)!}.$$

The formula (3) can be used to approximate e^x. Let us set $x = 1$ and approximate e. The error estimate is now

$$3^{|x|} \cdot \frac{|x|^{n+1}}{(n+1)!} = \frac{3}{(n+1)!}.$$

n	$1/n!$	Approximate value for e $1 + 1 + \frac{1}{2!} + \frac{1}{3!} + \cdots + \frac{1}{n!}$	Error estimate $\frac{3}{(n+1)!}$
2	0.500000	2.500000	0.500000
3	0.166667	2.666667	0.125000
4	0.041667	2.708333	0.025000
5	0.008333	2.716667	0.004167
6	0.001389	2.718056	0.000594
7	0.000198	2.718254	0.000075
8	0.000025		

This compares with $e = 2.718282$.

EXAMPLE 3   Find MacLaurin's Formula for $f(x) = \sin x$. The derivatives are

$$\begin{aligned} f(x) &= \sin x & f(0) &= 0 \\ f'(x) &= \cos x & f'(0) &= 1 \\ f''(x) &= -\sin x & f''(0) &= 0 \\ f^{(3)}(x) &= -\cos x & f^{(3)}(0) &= -1 \\ f^{(4)}(x) &= \sin x & f^{(4)}(0) &= 0 \\ f^{(5)}(x) &= \cos x & f^{(5)}(0) &= 1 \\ &\ \ \vdots & &\ \ \vdots \end{aligned}$$

MacLaurin's Formula for $2n$ terms is

$$\sin x = x - \frac{x^3}{3!} + \frac{x^5}{5!} - \frac{x^7}{7!} + \cdots + (-1)^{n-1}\frac{x^{2n-1}}{(2n-1)!} + R_{2n}(x),$$

$$R_{2n}(x) = (-1)^n \cos t \frac{x^{2n+1}}{(2n+1)!}.$$

For all t, $|\cos t| \le 1$, so we have the error estimate

$$|R_{2n}(x)| \le \frac{|x|^{2n+1}}{(2n+1)!}.$$

MacLaurin's Formula can be used to approximate $\sin x$ (with x in radians) when x is close to zero. We approximate $\sin(18°)$ as follows.

$$x = 18° = \frac{\pi}{10} \sim 0.31415927 \text{ radians.}$$

n	$(-1)^{n-1}\dfrac{x^{2n-1}}{(2n-1)!}$	Approximate value of $P_{2n}(x)$	Error estimate $|x|^{2n+1}/(2n+1)!$
1	0.31415927	0.31415927	0.00516771
2	−0.00516771	0.30899156	0.00002550
3	0.00002550	0.30901706	0.00000006

Thus $\sin(18°) \sim 0.3090171$ to seven places.

The proof of Taylor's Formula uses the following generalized form of the Mean Value Theorem.

GENERALIZED MEAN VALUE THEOREM

Suppose f and g are differentiable at all t between c and d, and that $g'(t) \neq 0$ for t strictly between c and d. Then

$$\frac{f'(t_0)}{g'(t_0)} = \frac{f(d) - f(c)}{g(d) - g(c)}$$

for some point t_0 strictly between c and d.

This theorem can be illustrated graphically by plotting the parametric equations $x = g(t)$, $y = f(t)$ in the (x, y) plane, as in Figure 9.10.3.

If $f(c) = 0$ and $g(c) = 0$, the formula in the theorem takes on the simpler form

$$\frac{f'(t_0)}{g'(t_0)} = \frac{f(d)}{g(d)}.$$

This is the form which will be used in the proof of Taylor's Formula.

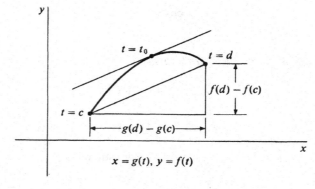

Figure 9.10.3

PROOF OF THE GENERALIZED MEAN VALUE THEOREM Introduce the new function

$$h(t) = f(t)(g(d) - g(c)) - g(t)(f(d) - f(c)).$$

Then $h(t)$ is also differentiable at all points between c and d. Furthermore, at the endpoints c and d we have

$$h(c) = f(c)g(d) - f(d)g(c) = h(d).$$

We may therefore apply Rolle's Theorem, whence there is a point t_0 strictly between c and d such that $h'(t_0) = 0$. Differentiating $h(t)$, we get

$$h'(t) = f'(t)(g(d) - g(c)) - g'(t)(f(d) - f(c)).$$

Therefore at $t = t_0$,

$$0 = f'(t_0)(g(d) - g(c)) - g'(t_0)(f(d) - f(c)).$$

$g'(t)$ is never zero. Also, $g(c) \neq g(d)$ because otherwise Rolle's Theorem would give a t with $g'(t) = 0$. We may therefore divide out and obtain the desired formula

$$\frac{f'(t_0)}{g'(t_0)} = \frac{f(d) - f(c)}{g(d) - g(c)}.$$

PROOF OF TAYLOR'S FORMULA Let $F(x) = R_n(x)$, $G(x) = (x - c)^{n+1}$.

Then $F(x) = f(x) - P_n(x)$. $f(x)$ and the nth Taylor polynomial $P_n(x)$ have the same value and first n derivatives at $x = c$. Therefore

$$F(c) = F'(c) = F''(c) = \cdots = F^{(n)}(c) = 0.$$

We also see that

$$G(c) = G'(c) = G''(c) = \cdots = G^{(n)}(c) = 0.$$

Using the Generalized Mean Value Theorem $n + 1$ times, we have

$$\frac{F'(t_0)}{G'(t_0)} = \frac{F(x)}{G(x)} \qquad \text{for some } t_0 \text{ strictly between } c \text{ and } x;$$

$$\frac{F''(t_1)}{G''(t_1)} = \frac{F'(t_0)}{G'(t_0)} \qquad \text{for some } t_1 \text{ strictly between } c \text{ and } t_0;$$

$$\vdots$$

$$\frac{F^{(n+1)}(t_n)}{G^{(n+1)}(t_n)} = \frac{F^{(n)}(t_{n-1})}{G^{(n)}(t_{n-1})} \qquad \text{for some } t_n \text{ strictly between } c \text{ and } t_{n-1}.$$

It follows that

$$\frac{F^{(n+1)}(t_n)}{G^{(n+1)}(t_n)} = \frac{F(x)}{G(x)}.$$

Either

$$x < t_0 < t_1 < \cdots < t_n < c \quad \text{or} \quad x > t_0 > t_1 > \cdots > t_n > c,$$

so t_n is strictly between c and x. The $(n + 1)$st derivatives of $F(t)$ and $G(t)$ are

$$F^{(n+1)}(t) = f^{(n+1)}(t) - 0, \qquad G^{(n+1)}(t) = (n + 1)!$$

Substituting, we have

$$\frac{f^{(n+1)}(t_n)}{(n+1)!} = \frac{R_n(x)}{(x - c)^{n+1}},$$

and Taylor's Formula follows at once.

PROBLEMS FOR SECTION 9.10

In Problems 1–8, find MacLaurin's Formula for $f(x)$, and use it to approximate $f(\frac{1}{2})$ within 0.01. (If a hand calculator is available, the approximations should be found within 0.0001.)

1	$f(x) = \cos x$	**2**	$f(x) = \sinh x$
3	$f(x) = \sin(2x)$	**4**	$f(x) = 100e^x$
5	$f(x) = \sin x \cos x$	**6**	$f(x) = \sqrt{1 + x}$
7	$f(x) = (4 + x)^{-3/2}$	**8**	$f(x) = (1 - x)^{1/3}$

In Problems 9–18 find the first two nonzero terms in MacLaurin's Formula and use it to approximate $f(\frac{1}{3})$.

9	$f(x) = \tan x$	**10**	$f(x) = \sec x$
11	$f(x) = \arcsin x$	**12**	$f(x) = \sin(e^x)$
13	$f(x) = \ln(1 + \sin x)$	**14**	$f(x) = \sqrt{x^2 + 1}$
15	$\int_0^x e^{t^2}\, dt$	**16**	$\int_0^x \sin(t^2)\, dt$
17	$\int_0^x \sin(\ln(1 + t))\, dt$	**18**	$\int_0^x \arcsin(t^2)\, dt$

19   Find Taylor's Formula for $f(x) = e^x$ in powers of $x - 2$.

20   Find Taylor's Formula for $f(x) = \ln x$ in powers of $x - 10$.

21   Find Taylor's Formula for $f(x) = x^p$ in powers of $x - 1$, where p is a constant real number.

22   Find Taylor's Formula for $f(x) = \sin x$ in powers of $x - \pi$.

9.11 TAYLOR SERIES

DEFINITION

If we continue the Taylor polynomial (by adding three dots at the end) we obtain a power series

$$f(c) + f'(c)(x - c) + \frac{f''(c)}{2!}(x - c)^2 + \cdots + \frac{f^{(n)}(c)}{n!}(x - c)^n + \cdots$$

$$= \sum_{n=0}^{\infty} \frac{f^{(n)}(c)}{n!}(x - c)^n.$$

*This series is called the **Taylor series** for the function $f(x)$ about the point $x = c$.*

*The Taylor series about the point $x = 0$ is called the **MacLaurin series**,*

$$f(0) + f'(0)x + \frac{f''(0)}{2!}x^2 + \cdots + \frac{f^{(n)}(0)}{n!}x^n + \cdots.$$

At $x = c$ the Taylor series about the point c converges to $f(c)$. But we have no assurance that the Taylor series converges to $f(x)$ at any other point x. There are three possibilities and all of them arise:

(1)   *The Taylor series diverges at x.*

(2)   *The Taylor series converges but to a value different than $f(x)$. (For an example, see Problem 28 at the end of this section.)*

(3)   *The Taylor series converges to $f(x)$; i.e., $f(x)$ is equal to the sum of its Taylor series.*

Theorem 1 shows that if we already know a function $f(x)$ is the sum of a power series, then that power series must be the Taylor series of $f(x)$.

THEOREM 1

Suppose $f(x)$ is equal to the sum of a power series with radius of convergence $r > 0$,

$$f(x) = \sum_{n=0}^{\infty} a_n(x - c)^n.$$

Then the power series is the same as the Taylor series for f about c. In other words, a_n is just $f^{(n)}(c)/n!$ for $n = 0, 1, 2, \ldots$.

Discussion   A function which is equal to the sum of a power series in $x - c$ (with nonzero radius of convergence) is called *analytic* at c. The theorem shows that every analytic function is equal to the sum of its Taylor series.

PROOF   Since power series can be differentiated term by term within its interval of convergence, all the nth derivatives $f^{(n)}(c)$ exist. Let us compute $f^{(n)}(x)$ and set $x = c$.

$$f(x) = \sum_{n=0}^{\infty} a_n(x - c)^n, \qquad\qquad f(c) = a_0$$

$$f'(x) = \sum_{n=1}^{\infty} na_n(x - c)^{n-1}, \qquad\qquad f'(c) = a_1$$

$$f''(x) = \sum_{n=2}^{\infty} n(n - 1)a_n(x - c)^{n-2}, \qquad\qquad f''(c) = 2!a_2$$

$$f'''(x) = \sum_{n=3}^{\infty} n(n - 1)(n - 2)a_n(x - c)^{n-3}, \qquad\qquad f'''(c) = 3!a_3$$

$$f^{(k)}(x) = \sum_{n=k}^{\infty} n(n - 1)\cdots(n - k + 1)a_n(x - c)^{n-k}, \qquad f^{(k)}(c) = k!a_k.$$

Thus for each n,

$$a_n = f^{(n)}(c)/n!,$$

and the original power series is the same as the Taylor series of $f(x)$,

$$\sum_{n=0}^{\infty} a_n(x - c)^n = \sum_{n=0}^{\infty} \frac{f^{(n)}(c)}{n!}(x - c)^n.$$

EXAMPLE 1   Let $f(x)$ be a polynomial in $x - c$,

$$f(x) = a_0 + a_1(x - c) + \cdots + a_n(x - c)^n.$$

This is just a power series with all but the first $n + 1$ coefficients equal to zero. So by Theorem 1, the Taylor series of the polynomial is just the polynomial itself followed by infinitely many zeros,

$$a_0 + a_1(x - c) + \cdots + a_n(x - c)^n + 0 + 0 + \cdots.$$

We can also see this directly from Lemma 1 of the last section, namely

$$\frac{f^{(m)}(c)}{m!} = a_m \qquad \text{for } m \leq n.$$

Here is a review of the power series obtained earlier in this chapter. By Theorem 1, they are all MacLaurin series.

(1) $\dfrac{1}{1 - x} = 1 + x + x^2 + x^3 + x^4 + \cdots, \qquad |x| < 1$

(2) $\dfrac{1}{1 + x} = 1 - x + x^2 - x^3 + x^4 - \cdots, \qquad |x| < 1$

(3) $\dfrac{1}{1 - 2x} = 1 + 2x + 2^2 x^2 + 2^3 x^3 + 2^4 x^4 + \cdots, \qquad |x| < \frac{1}{2}$

(4) $\ln(1 - x) = -x - \dfrac{x^2}{2} - \dfrac{x^3}{3} - \dfrac{x^4}{4} - \cdots, \qquad |x| < 1$

(5) $\dfrac{1}{1 + x^2} = 1 - x^2 + x^4 - x^6 + x^8 - \cdots, \qquad |x| < 1$

(6) $\arctan x = x - \dfrac{x^3}{3} + \dfrac{x^5}{5} - \dfrac{x^7}{7} + \dfrac{x^9}{9} - \cdots, \qquad |x| < 1$

(7) $\dfrac{1}{(1 - x)^2} = 1 + 2x + 3x^2 + 4x^3 + 5x^4 + \cdots, \qquad |x| < 1$

(8) $e^x = 1 + x + \dfrac{x^2}{2!} + \dfrac{x^3}{3!} + \dfrac{x^4}{4!} + \cdots$

(9) $e^{-x} = 1 - x + \dfrac{x^2}{2!} - \dfrac{x^3}{3!} + \dfrac{x^4}{4!} - \cdots$

(10) $\cosh x = 1 + \dfrac{x^2}{2!} + \dfrac{x^4}{4!} + \dfrac{x^6}{6!} + \dfrac{x^8}{8!} + \cdots$

(11) $\sinh x = x + \dfrac{x^3}{3!} + \dfrac{x^5}{5!} + \dfrac{x^7}{7!} + \dfrac{x^9}{9!} + \cdots$

(12) $e^{-x^2} = 1 - x^2 + \dfrac{x^4}{2!} - \dfrac{x^6}{3!} + \dfrac{x^8}{4!} - \cdots$

(13) $\displaystyle\int_0^x e^{-t^2}\,dt = x - \frac{x^3}{3} + \frac{x^5}{5\cdot 2!} - \frac{x^7}{7\cdot 3!} + \frac{x^9}{9\cdot 4!} - \cdots$

(14) $\displaystyle\ln\left(\frac{1+x}{1-x}\right) = 2x + \frac{2x^3}{3} + \frac{2x^5}{5} + \frac{2x^7}{7} + \frac{2x^9}{9} + \cdots, \qquad |x| < 1$

At the end of this section we shall add three important power series to our list:

(15) $\displaystyle\sin x = x - \frac{x^3}{3!} + \frac{x^5}{5!} - \frac{x^7}{7!} + \frac{x^9}{9!} - \cdots$

(16) $\displaystyle\cos x = 1 - \frac{x^2}{2!} + \frac{x^4}{4!} - \frac{x^6}{6!} + \frac{x^8}{8!} - \cdots$

(17) $\displaystyle(1 + x)^p = 1 + px + \frac{p(p-1)}{2!}x^2 + \frac{p(p-1)(p-2)}{3!}x^3 + \cdots, \qquad |x| < 1,$

where p is constant.

The last series is called the *binomial series*.

It is interesting to observe that the derivatives of an analytic function at zero can be read directly from the MacLaurin series. Sometimes it is quite hard to compute the derivative directly but easy to take it from the MacLaurin series.

EXAMPLE 2 Find the sixth derivative of $f(x) = 1/(1 + x^2)$ at $x = 0$.

If we try to differentiate directly we will be hopelessly bogged down at about the third derivative. But from the MacLaurin series we see that

$$\frac{1}{1 + x^2} = 1 - x^2 + x^4 - x^6 + \cdots,$$

$$\frac{f^{(6)}(0)}{6!}x^6 = -x^6,$$

$$\frac{f^{(6)}(0)}{6!} = -1,$$

$$f^{(6)}(0) = -6! = -720.$$

Suppose we are given a function $f(x)$ and a point c, and we wish to represent $f(x)$ as the sum of a power series in $x - c$. This will be possible for some functions (the analytic functions), but not for all. Theorem 1 shows that if there is such a power series it is the Taylor series for $f(x)$. Thus we use the following steps to represent $f(x)$ as a power series.

Step 1 Compute all the derivatives $f^{(n)}(c), n = 0, 1, 2, \ldots$. If these derivatives do not all exist, $f(x)$ is not the sum of a power series in powers of $x - c$.

Step 2 Write down the Taylor series of $f(x)$ at $x = c$ and find its radius of convergence r.

Step 3 If possible, show that $f(x)$ is equal to the sum of its Taylor series for $c - r < x < c + r$.

We shall now use Steps 1–3 to obtain the power series for $\sin x$, $\cos x$, and $(1 + x)^p$.

THE POWER SERIES FOR sin x

Step 1   This step was carried out in the preceding section. The values of $f^{(n)}(0)$ for $n = 0, 1, 2, \ldots$ are

$$0, 1, 0, -1, 0, 1, 0, -1, \ldots.$$

Step 2   The MacLaurin series for sin x is

$$x - \frac{x^3}{3!} + \frac{x^5}{5!} - \frac{x^7}{7!} + \frac{x^9}{9!} - \cdots + (-1)^{n-1}\frac{x^{2n-1}}{(2n-1)!} + \cdots.$$

Let $b_n = (-1)^{n-1}\dfrac{x^{2n-1}}{(2n-1)!}$. We use the Ratio Test,

$$\lim_{n\to\infty}\left|\frac{b_{n+1}}{b_n}\right| = \lim_{n\to\infty}\frac{x^2}{2n(2n+1)} = 0.$$

Therefore the series converges for all x and has radius of convergence ∞.

Step 3   We use MacLaurin's Formula,

$$\sin x = x - \frac{x^3}{3!} + \frac{x^5}{5!} - \frac{x^7}{7!} + \cdots + (-1)^{n-1}\frac{x^{2n-1}}{(2n-1)!} + R_{2n}(x),$$

$$|R_{2n}(x)| \le \frac{|x|^{2n+1}}{(2n+1)!}$$

Let us show that the remainders approach zero. We have

$$\lim_{n\to\infty}\frac{|x|^{2n+1}}{(2n+1)!} = 0, \qquad \lim_{n\to\infty}R_{2n}(x) = 0.$$

Since the even terms are zero, $R_{2n-1}(x) = R_{2n}(x)$. Therefore

$$\lim_{n\to\infty}R_n(x) = 0.$$

Conclusion: Since the remainders approach zero, the MacLaurin polynomials approach sin x. So for all x,

$$\sin x = x - \frac{x^3}{3!} + \frac{x^5}{5!} - \frac{x^7}{7!} + \frac{x^9}{9!} - \cdots.$$

THE POWER SERIES FOR cos x

This power series can be found by the same method as was used for sin x. However, it is simpler to differentiate the power series for sin x.

$$\frac{d(\sin x)}{dx} = \cos x.$$

$$\cos x = 1 - \frac{3x^2}{3!} + \frac{5x^4}{5!} - \frac{7x^6}{7!} + \frac{9x^8}{9!} - \cdots,$$

$$\cos x = 1 - \frac{x^2}{2!} + \frac{x^4}{4!} - \frac{x^6}{6!} + \frac{x^8}{8!} - \cdots.$$

THE BINOMIAL SERIES FOR $(1+x)^p$

Let us first consider the case where p is a nonnegative integer m, whence $(1 + x)^m$ is a polynomial. The *Binomial Theorem* states that for nonnegative integers m,

$$(a + b)^m = a^m + ma^{m-1}b + \frac{m(m-1)}{2!}a^{m-2}b^2$$

$$+ \cdots + \frac{m(m-1)\cdots(m-k+1)}{k!}a^{m-k}b^k + \cdots + b^m.$$

Setting $a = 1$ and $b = x$ we obtain a finite power series for $(1 + x)^m$,

$$(1 + x)^m = 1 + mx + \frac{m(m-1)}{2!}x^2$$

$$+ \cdots + \frac{m(m-1)\cdots(m-k+1)}{k!}x^k + \cdots + x^m.$$

When $p < 0$, and when $p > 0$ but p is not an integer, we shall see that $(1 + x)^p$ is the sum of a similar power series but with infinitely many terms. Let $g(x) = (1 + x)^p$.

Step 1 By differentiation we see that

$$g'(x) = p(1 + x)^{p-1},$$
$$g''(x) = p(p-1)(1 + x)^{p-2},$$
$$g^{(n)}(x) = p(p-1)\cdots(p-n+1)(1 + x)^{p-n}.$$

Thus at $x = 0$, $\quad g(0) = 1,$
$$g'(0) = p,$$
$$g''(0) = p(p-1),$$
$$g^{(n)}(0) = p(p-1)\cdots(p-n+1).$$

Step 2 The MacLaurin series is

$$f(x) = 1 + px + \frac{p(p-1)}{2!}x^2 + \cdots + \frac{p(p-1)\cdots(p-n+1)}{n!}x^n + \cdots.$$

We use the Ratio Test.

$$\left|\frac{a_{n+1}}{a_n}\right| = \left|\frac{p(p-1)\cdots(p-n)/(n+1)!}{p(p-1)\cdots(p-n+1)/n!}\right||x| = \frac{|p-n|}{n+1}|x|.$$

$$\lim_{n\to\infty}\left|\frac{a_{n+1}}{a_n}\right| = \lim_{n\to\infty}\frac{|p-n|}{n+1}|x| = |x|.$$

Therefore the series converges for $|x| < 1$, diverges for $|x| > 1$, and has radius of convergence $r = 1$. We denote the sum by $f(x)$.

Step 3 We wish to show that the sum $f(x)$ is equal to $(1 + x)^p$ for $|x| < 1$. In this case, the MacLaurin Formula does not give the needed information (see Problem 27 at the end of this section). Instead we show that the quotient $f(x)/(1 + x)^p$ has derivative zero for $|x| < 1$. We have

$$\frac{d}{dx}[f(x)(1 + x)^{-p}] = \frac{f'(x)(1 + x) - pf(x)}{(1 + x)^{p+1}}.$$

It suffices to show that

$$f'(x)(1 + x) = pf(x) \quad \text{or} \quad f'(x) + xf'(x) = pf(x).$$

Let us compute $f'(x)$ and $xf'(x)$.

$$f'(x) = p + p(p - 1)x + \frac{p(p - 1)(p - 2)}{2!}x^2$$

$$+ \frac{p(p - 1)(p - 2)(p - 3)}{3!}x^3 + \cdots,$$

$$xf'(x) = px + p(p - 1)x^2 + \frac{p(p - 1)(p - 2)}{2!}x^3 + \cdots.$$

Adding the power series, we have

$$f'(x) + xf'(x) = p + p[(p - 1) + 1]x + \frac{p(p - 1)}{2!}[(p - 2) + 2]x^2$$

$$+ \frac{p(p - 1)(p - 2)}{3!}[(p - 3) + 3]x^3 + \cdots$$

$$= p\left[1 + px + \frac{p(p - 1)}{2!}x^2 + \frac{p(p - 1)(p - 2)}{3!}x^3 + \cdots\right]$$

$$= pf(x).$$

Thus $f'(x) + xf'(x) = pf(x)$, $\dfrac{d}{dx}[f(x)(1 + x)^{-p}] = 0$.

We conclude that for some constant C,

$$f(x)(1 + x)^{-p} = C.$$

At $x = 0$, $f(x) = 1 = (1 + x)^{-p}$. Hence $C = 1$. This shows that $(1 + x)^p = f(x)$ for $|x| < 1$.

Thus we have the binomial series

$$(1 + x)^p = 1 + px + \frac{p(p - 1)}{2!}x^2 + \frac{p(p - 1)(p - 2)}{3!'}x^3 + \cdots, \qquad |x| < 1.$$

EXAMPLE 3   Find the power series for arcsin x.

Recall that for $|x| < 1$,

$$\arcsin x = \int_0^x \frac{dt}{\sqrt{1 - t^2}} = \int_0^x (1 - t^2)^{-1/2}\, dt.$$

We start with the binomial series with $p = -\frac{1}{2}$ and obtain the following power series by substitution and integration. They are valid for $|x| < 1$.

$$(1 + x)^{-1/2} = 1 - \frac{1}{2}x + \left(-\frac{1}{2}\right)\left(-\frac{3}{2}\right)\frac{1}{2!}x^2 - \cdots$$

$$+ (-1)^n \frac{1 \cdot 3 \cdot \cdots \cdot (2n - 1)}{2^n n!}x^n + \cdots$$

$$(1 - x)^{-1/2} = 1 + \frac{1}{2}x + \frac{3}{8}x^2 + \cdots + \frac{1 \cdot 3 \cdot \cdots \cdot (2n - 1)}{2^n n!}x^n + \cdots$$

$$(1 - x^2)^{-1/2} = 1 + \frac{1}{2}x^2 + \frac{3}{8}x^4 + \cdots + \frac{1 \cdot 3 \cdot \cdots \cdot (2n - 1)}{2^n n!}x^{2n} + \cdots$$

$$\arcsin x = x + \frac{1}{6}x^3 + \frac{3}{40}x^5 + \cdots + \frac{1 \cdot 3 \cdot \cdots \cdot (2n - 1)}{2^n n!(2n + 1)}x^{2n+1} + \cdots.$$

PROBLEMS FOR SECTION 9.11

1   Find $f^{(4)}(0)$ where $f(x) = 1/(1 - 2x^2)$.

2   Find $f^{(5)}(0)$ where $f(x) = x/(1 + x^2)$.

3   Find $f^{(6)}(0)$ where $f(x) = xe^x$.

4   Find $f^{(8)}(0)$ where $f(x) = \cos(x^2)$.

5   Find $f^{(7)}(0)$ where $f(x) = x^2 \ln(1 + x)$.

6   Find $f^{(6)}(0)$ where $f(x) = (\arctan x)/x$ if $x \neq 0$, and $f(0) = 1$.

In Problems 7–24, find a power series converging to $f(x)$ and determine the radius of convergence.

7     $f(x) = e^{x/2}$ 8 $f(x) = x^2 e^x$

9     $f(x) = \sqrt{1 + 2x}$ 10 $f(x) = (1 - 4x)^{-1/3}$

11     $f(x) = \cos\sqrt{x}$ 12 $f(x) = \arcsin(x^3)$

13     $f(x) = \dfrac{\sin x}{x}$ if $x \neq 0$, $f(0) = 1$

14     $f(x) = \dfrac{1 - \cos x}{x^2}$ if $x \neq 0$, $f(0) = \frac{1}{2}$.

15     $f(x) = \sqrt{1 - x^2}$ 16 $f(x) = \dfrac{x}{(1 + x)^4}$

17     $f(x) = \displaystyle\int_0^x \sin(t^3)\, dt$ 18 $f(x) = \displaystyle\int_0^x t^{-1} \sin t\, dt$

19     $f(x) = \displaystyle\int_0^x t^{-2} \sinh(t^2)\, dt$ 20 $f(x) = \displaystyle\int_0^x \ln(1 + t^2)\, dt$

21     $f(x) = \displaystyle\int_0^x (1 + t^2)^{1/3}\, dt$ 22 $f(x) = \displaystyle\int_0^x \sqrt{1 - t^3}\, dt$

23     $f(x) = \displaystyle\int_0^x \dfrac{\arcsin t}{t}\, dt$ 24 $f(x) = \displaystyle\int_0^x \arcsin(t^2)\, dt$

25   Find the Taylor series for $\ln x$ in powers of $x - 1$.

26   Find the Taylor series for $\sin x$ in powers of $x - \pi/4$.

☐ 27   Use Taylor's Formula to prove that the binomial series converges to $(1 + x)^p$ when $-\frac{1}{2} \leq x < 1$. (The proof in the text shows that it actually converges to $(1 + x)^p$ for $-1 < x < 1$.)

☐ 28   Let
$$f(x) = \begin{cases} 0 & \text{if } x = 0, \\ e^{-1/x^2} & \text{if } x \neq 0, \end{cases}$$

Show that $f^{(n)}(0) = 0$ for all integers n; so for $x \neq 0$ the MacLaurin series converges but to zero instead of to $f(x)$.

EXTRA PROBLEMS FOR CHAPTER 9

Determine whether the sequences 1–5 converge and find the limits when they exist.

1     $a_n = \left(1 + \dfrac{1}{n^2}\right)^n$ 2 $a_n = \left(1 + \dfrac{1}{\sqrt{n}}\right)^n$

3     $a_n = (1 + n)^{1/n}$ 4 $a_n = n! - 10^n$

☐ 5 $a_n = n^n/n!$ (*Hint*: Show that $a_{n+1} \geq 2a_n$.)

Determine whether the series 6–12 converge and find the sums when they exist.

6 $1 + \frac{3}{7} + \frac{9}{49} + \cdots + (\frac{3}{7})^n + \cdots$

7 $1 - 1.1 + 1.11 - 1.111 + 1.1111 - \cdots$

8 $\left(1 - \frac{1}{8}\right) + \left(\frac{1}{8} - \frac{1}{27}\right) + \left(\frac{1}{27} - \frac{1}{64}\right) + \cdots + \left(\frac{1}{n^3} - \frac{1}{(n+1)^3}\right) + \cdots$

9 $6 + 19 + 3 + \frac{4}{25} + \frac{8}{125} + \frac{16}{625} + \cdots + (\frac{2}{5})^n + \cdots$

10 $\displaystyle\sum_{n=0}^{\infty} \frac{7^n - 6^n}{5^n}$ **11** $\displaystyle\sum_{n=0}^{\infty} \frac{7 \cdot 5^n}{6^n}$

12 $\displaystyle\sum_{n=0}^{\infty} \frac{2n - 3}{5n + 6}$

Test the series 13–23 for convergence.

13 $\displaystyle\sum_{n=0}^{\infty} \frac{3n - 7}{10n + 9}$ **14** $\displaystyle\sum_{n=1}^{\infty} \frac{5}{6n^2 + n - 1}$

15 $\displaystyle\sum_{n=1}^{\infty} \frac{\sqrt{n}}{1 + 2\sqrt{n} + 3n}$ **16** $\displaystyle\sum_{n=1}^{\infty} ne^{-n}$

17 $\displaystyle\sum_{n=2}^{\infty} (\ln (n))^{-n}$ **18** $\displaystyle\sum_{n=2}^{\infty} n^{-\ln n}$

19 $\displaystyle\sum_{n=2}^{\infty} \ln n^{-\ln n}$ **20** $\displaystyle\sum_{n=3}^{\infty} (-1)^n / \sqrt{\ln n}$

21 $\displaystyle\sum_{n=1}^{\infty} (-1)^n \left(1 - \frac{1}{n^2}\right)$ **22** $\displaystyle\sum_{n=1}^{\infty} (-1)^n n^{-1/n}$

23 $\displaystyle\sum_{n=1}^{\infty} (-1)^n \frac{n^{1/n}}{n}$

24   Test the integral $\int_0^\infty e^{-\sqrt{x}} \, dx$ for convergence.

25   Test the integral $\int_2^\infty (\ln x)^{-x} \, dx$ for convergence.

26   Approximate the series $\sum_{n=1}^{\infty} (-1)^n (1/n^3)$ to three decimal places.

Test the series 27–30 by the Ratio Test.

27 $\displaystyle\sum_{n=1}^{\infty} \frac{n^n}{(n!)^2}$ **28** $\displaystyle\sum_{n=1}^{\infty} \frac{2^n(n!)}{n^n}$

29 $\displaystyle\sum_{n=2}^{\infty} \frac{n^n}{(\ln n)^n}$ **30** $\displaystyle\sum_{n=1}^{\infty} \frac{100^n(n!)^3}{(3n)!}$

Find the radius of convergence of the power series in Problems 31–35.

31 $\displaystyle\sum_{n=0}^{\infty} 2^n n^3 x^n$ **32** $\displaystyle\sum_{n=1}^{\infty} n^{-n} x^n$

33 $\displaystyle\sum_{n=1}^{\infty} \frac{x^n}{(n!)^{1/n}}$ **34** $\displaystyle\sum_{n=1}^{\infty} x^n \sqrt{n!/n^n}$

35 $\displaystyle\sum_{n=1}^{\infty} \frac{n!}{n^n} x^{2n}$

36   Find the interval of convergence of $\sum_{n=2}^{\infty} (x + 10)^n/(\ln n)$.

37   Find the power series and radius of convergence for $f'(x)$ and $\int_0^x f(t) \, dt$ where

$$f(x) = \sum_{n=1}^{\infty} n^a(n + 1)^b 2^n x^n.$$

38   Find a power series for $f(x) = 1/(1 + 2x^3)$ and determine its radius of convergence.

39   Find a power series for

$$f(x) = \int_0^x \frac{\arctan (t^2)}{t^2} dt$$

and determine its radius of convergence.

40   Approximate $\int_0^{1/2} \frac{\arctan (t^2)}{t^2} dt$ within 0.0001.

41   Approximate $\int_0^{1/4} t \ln (1 - t) dt$ within 0.001.

42   Approximate $e^{1/5}$ within 10^{-7}.

43   Approximate $\int_0^{1/2} e^{\sin t} dt$ within 0.01.

44   Find a power series for $(1 + x^3)^{-3/2}$ and give its radius of convergence.

45   Find a power series for $\int_0^x (1 + 2t^2)^{-2/3} dt$ and determine its radius of convergence.

☐ **46**   Prove that any repeating decimal

$$0. b_1 b_2 \ldots b_n b_1 b_2 \ldots b_n b_1 b_2 \ldots b_n \ldots$$

(where each of $b_1, \ldots, b_n$ is a digit from the set $\{0, 1, \ldots, 9\}$) is equal to a rational number.

☐ **47**   Approximately how many terms of the harmonic series $1 + \frac{1}{2} + \frac{1}{3} + \cdots + 1/n + \cdots$ are needed to reach a partial sum of at least 50? *Hint:* Compare with $\int_1^n (1/t) dt$.

☐ **48**   Suppose $\sum_{n=1}^\infty a_n = \infty$ and $\sum_{n=1}^\infty b_n$ is either finite or ∞. Prove that $\sum_{n=1}^\infty (a_n + b_n) = \infty$.

☐ **49**   Suppose $\sum_{n=1}^\infty a_n$ is a convergent positive term series and $\sum_{n=1}^\infty b_n$ is a rearrangement of $\sum_{n=1}^\infty a_n$. Prove that $\sum_{n=1}^\infty b_n$ converges and has the same sum. *Hint:* Show that each finite partial sum of $\sum_{n=1}^\infty a_n$ is less than or equal to each infinite partial sum of $\sum_{n=1}^\infty b_n$, and vice versa.

☐ **50**   Give a rearrangement of the series $1 - \frac{1}{2} + \frac{1}{3} - \frac{1}{4} + \cdots$ which diverges to $-\infty$.

☐ **51**   Suppose $\sum_{n=1}^\infty a_n = \sum_{n=1}^\infty b_n = S$, and $a_n \le c_n \le b_n$ for all n. Prove that $\sum_{n=1}^\infty c_n = S$.

☐ **52**   Prove the following result using the Limit Comparison Test.
Let $\sum_{n=1}^\infty a_n$ and $\sum_{n=1}^\infty b_n$ be positive term series and suppose $\lim_{n \to \infty} (a_n/b_n)$ exists. If $\sum_{n=1}^\infty b_n$ converges then $\sum_{n=1}^\infty a_n$ converges. If $\sum_{n=1}^\infty a_n$ diverges then $\sum_{n=1}^\infty b_n$ diverges.

☐ **53**   Multiplication of Power Series.
Prove that if $f(x) = \sum_{n=0}^\infty a_n x^n$ and $g(x) = \sum_{n=0}^\infty b_n x^n$ then $f(x)g(x) = \sum_{n=0}^\infty c_n x^n$ where

$$c_n = a_0 b_n + a_1 b_{n-1} + \cdots + a_{n-1} b_1 + a_n b_0.$$

Hint: First prove the corresponding formula for partial sums, then take the standard part of an infinite partial sum.

☐ **54**   Suppose $f(x)$ is the sum of a power series for $|x| < r$ and let $g(x) = f(x^2)$. Prove that for each n,

$$g^{(n)}(0) = \begin{cases} 0 & \text{if } n \text{ is odd}, \\ \dfrac{n!}{(n/2)!} f^{(n/2)}(0) & \text{if } n \text{ is even}. \end{cases}$$

☐ **55**   Show that if $p \le -1$ then the binomial series

$$1 + px + \frac{p(p - 1)}{2!} x^2 + \frac{p(p - 1)(p - 2)}{3!} x^3 + \cdots$$

diverges at $x = 1$ and $x = -1$. *Hint:* Cauchy Test.

If $p \ge 1$, the series converges at $x = 1$ and $x = -1$. *Hint:* Compare with $\sum_{n=1}^\infty 1/n^2$. *Note:* The cases $-1 < p < 1$ are more difficult. It turns out that if $-1 < p < 0$ the series converges at $x = 1$ and diverges at $x = -1$. If $p \ge 0$ the series converges at $x = 1$ and $x = -1$.

☐ **56**   Prove that e is irrational, that is, $e \ne a/b$ for all integers a, b. *Hint:* Suppose $e = a/b$, $e^{-1} = b/a$. Let $c = e^{-1} - \sum_{n=0}^a (-1)^n/n!$. Then $|c| \ge 1/a!$ but $|c| \le 1/(a + 1)!$.

VECTORS

10.1 VECTOR ALGEBRA

Figure 10.1.1 shows a directed line segment from the point P to the point Q.

Figure 10.1.1

We pictorially represent a directed line segment as an arrow from P to Q, and use the symbol $\overrightarrow{PQ}$. Mathematically, a directed line segment is most easily represented as an ordered pair of points.

The *directed line segment* from P to Q, in symbols $\overrightarrow{PQ}$, is the ordered pair of points (P, Q). P is called the *initial point*, and Q, the *terminal point*, of the directed line segment.

The directed line segments $\overrightarrow{PQ}$ and $\overrightarrow{QP}$ are considered to be different. $\overrightarrow{QP}$ has initial point Q and terminal point P. If $P(p_1, p_2)$ and $Q(q_1, q_2)$ are two points in the plane, the *x-component* of $\overrightarrow{PQ}$ is the increment $q_1 - p_1$ of x from P to Q, and the *y-component* is the increment $q_2 - p_2$ of y, as shown in Figure 10.1.2.

$$x\text{-component of } \overrightarrow{PQ} = q_1 - p_1.$$
$$y\text{-component of } \overrightarrow{PQ} = q_2 - p_2.$$

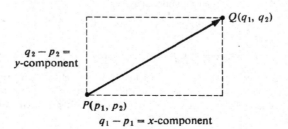

Figure 10.1.2

Usually we are not really interested in the exact placement of a directed line segment $\overrightarrow{PQ}$ on the (x, y) plane, but in the length and direction of $\overrightarrow{PQ}$. These can be determined by the x and y components of $\overrightarrow{PQ}$. We are thus led to the notion of a vector.

DEFINITION

*The family of all directed line segments with the same components as $\overrightarrow{PQ}$ will be called the **vector** from P to Q. We say that $\overrightarrow{PQ}$ **represents** this vector.*

Since all directed line segments with the same components have the same length and direction, a vector may be regarded as a quantity which has length and direction.

Vectors arise quite naturally in both physics and economics. Here are some examples of vector quantities.

Position If an object is at the point (p_1, p_2) in the plane, its position vector is the vector with components p_1 and p_2.

Velocity If a particle is moving in the plane according to the parametric equations

$$x = f(t), \qquad y = g(t),$$

the velocity vector is the vector with x and y components dx/dt and dy/dt.

Acceleration The acceleration vector of a moving particle has the x and y components d^2x/dt^2 and d^2y/dt^2.

Force In physics, force is a vector quantity which will accelerate a free particle in the direction of the force vector at a rate proportional to the length of the force vector

Displacement (change in position) If an object moves from the point P to the point Q, its displacement vector is the vector from P to Q.

Commodity vector In economics, one often compares two or more commodities (such as guns and butter). If a trader in a market has a quantity a_1 of one commodity and a_2 of another, his commodity vector has the x and y components (a_1, a_2).

Price vector If two commodities have prices p_1 and p_2 respectively, the price vector has components (p_1, p_2). The components of a commodity or price vector are always greater than or equal to zero.

EXAMPLE 1 Find the components of the vectors represented by the given directed line segments.

(a) $\overrightarrow{(3, 2), (5, 1)}$.

x-component $= 5 - 3 = 2$, $\qquad$ y-component $= 1 - 2 = -1$.

(b) $\overrightarrow{(0, -2), (2, -3)}$.

x-component $= 2 - 0 = 2$, $\qquad$ y-component $= -3 - (-2) = -1$.

Notice that both of these directed line segments represent the same vector. They are shown in Figure 10.1.3.

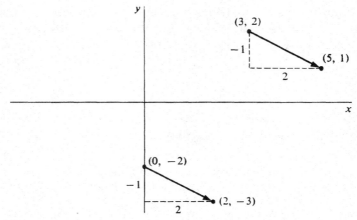

Figure 10.1.3

Vectors are denoted by boldface letters, **A**. A vector is represented by a whole family of directed line segments. However, given a vector **A** and an initial point P, there is exactly one point Q such that the directed line segment $\vec{PQ}$ represents **A**. To find the x-coordinate of Q we add the x-coordinate of P and the x component of **A**; similarly for the y-coordinate.

EXAMPLE 2   Let **A** be the vector with components -4 and 1, and let P be the point $(1, 2)$. Find Q so that $\vec{PQ}$ represents **A**.

Q has the x-coordinate $1 + (-4) = -3$ and the y-coordinate $2 + 1 = 3$. Thus $Q = (-3, 3)$, as shown in Figure 10.1.4.

We shall now begin the algebra of vectors. In vector algebra, real numbers are called *scalars*. We study two different kinds of quantities, scalars and vectors.

The *length* (or norm) of a vector **A** is the distance between P and Q where $\vec{PQ}$ represents **A**. The length is a scalar, denoted by $|\mathbf{A}|$. If **A** has components a_1 and a_2, then the length, shown in Figure 10.1.5, is given by the distance formula,

$$|\mathbf{A}| = \sqrt{a_1^2 + a_2^2}.$$

The length of a position vector is the *distance from the origin*. The length of a

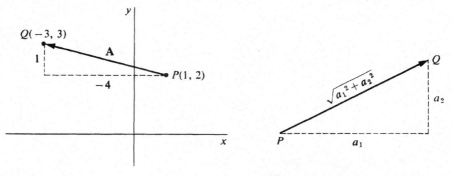

Figure 10.1.4   **Figure 10.1.5**   Length of a Vector

velocity vector is the *speed* of a particle. The length of a force vector is the *magnitude* of the force. The length of a displacement vector is the *distance moved*. For price or commodity vectors, the notion of length does not arise in a natural way.

EXAMPLE 3 The vector **A** with components 3 and −4 has length $|A| = \sqrt{3^2 + (-4)^2}$ = 5.

The vector with components $(0, 0)$ is called the *zero vector*, denoted by **0**. The zero vector is represented by the degenerate line segments $\overrightarrow{PP}$. It has no direction. The length of the zero vector is zero, while the length of every other vector is a positive scalar.

The sum **A** + **B** of vectors **A** and **B** is defined as follows. Let $\overrightarrow{PQ}$ represent **A** and let QR represent **B**. Then **A** + **B** is the vector represented by $\overrightarrow{PR}$. More briefly, if **A** is the vector from P to Q and **B** is the vector from Q to R, then **A** + **B** is the vector from P to R. Figure 10.1.6 shows two ways of drawing the sum **A** + **B**.

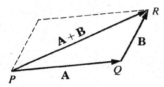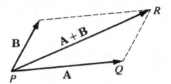

Figure 10.1.6   Sum of Two Vectors

If an object in the plane originally has the position vector **P** and is moved by a displacement vector **D**, its new position vector will be the vector sum **P** + **D**. If an object is moved twice, first by a displacement vector **D** and then by a displacement vector **E**, the total displacement vector is the sum **D** + **E**.

If two forces **F** and **G** are acting simultaneously on an object, their combined effect is the vector sum **F** + **G** (Figure 10.1.7). The combined effect of three or more forces acting on an object is also the vector sum, e.g., (**F** + **G**) + **H**. *Newton's first law of motion* states that if an object is at rest, the vector sum of all forces acting on the object is the zero vector.

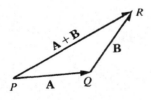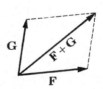

Changes in position   Forces

Figure 10.1.7

In economics, if a trader initially has a commodity vector **A** and buys a commodity vector **B** (i.e., he buys a quantity b_1 of commodity one and b_2 of commodity two), his new commodity vector will be the vector sum **A** + **B**.

The vector sum is also useful in discussing an exchange between two or more traders. Suppose traders A and B initially have commodity vectors $\mathbf{A_1}$ and $\mathbf{B_1}$. After exchanging goods, they have new commodity vectors $\mathbf{A_2}$ and $\mathbf{B_2}$. Since the total amount of each good remains unchanged, we see that $\mathbf{A_1} + \mathbf{B_1} = \mathbf{A_2} + \mathbf{B_2}$.

Vector sums obey rules similar to the rules for sums of real numbers.

THEOREM 1

Let **A**, **B**, *and* **C** *be vectors.*

(i)   Identity Law $\mathbf{A} + \mathbf{0} = \mathbf{0} + \mathbf{A} = \mathbf{A}$.

(ii) Commutative Law $\mathbf{A} + \mathbf{B} = \mathbf{B} + \mathbf{A}$.

(iii) Associative Law $(\mathbf{A} + \mathbf{B}) + \mathbf{C} = \mathbf{A} + (\mathbf{B} + \mathbf{C})$.

(iv) Triangle Inequality $|\mathbf{A} + \mathbf{B}| \le |\mathbf{A}| + |\mathbf{B}|$.

We shall skip the proofs, which use the corresponding laws for real numbers. The Commutative and Associative Laws are illustrated by Figure 10.1.8.

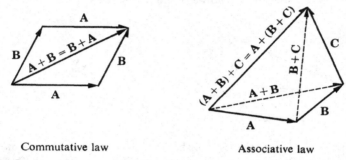

Commutative law   Associative law

Figure 10.1.8

The Triangle Inequality says that the length of one side of a triangle is at most the sum of the lengths of the other two sides. This is because the vectors **A**, **B**, and **A** + **B** are represented by sides of a triangle. The proof of the Triangle Inequality is left as a problem (with a hint). It is illustrated in Figure 10.1.9.

The sum of three or more vectors is formed in the same way as the sum of two vectors, as in Figure 10.1.10.

Triangle inequality

Figure 10.1.9

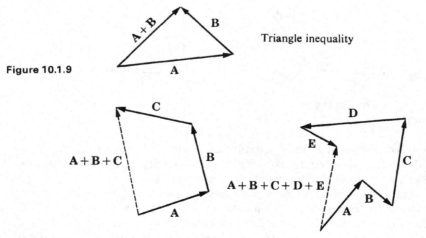

Figure 10.1.10   Sum of Vectors

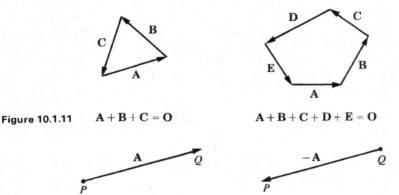

Figure 10.1.11 $\mathbf{A} + \mathbf{B} + \mathbf{C} = \mathbf{0}$ $\mathbf{A} + \mathbf{B} + \mathbf{C} + \mathbf{D} + \mathbf{E} = \mathbf{0}$

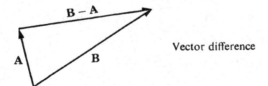

Vector negative

Figure 10.1.12

The sum of the vectors (clockwise or counterclockwise) around the perimeter of a triangle or polygon is always the zero vector (Figure 10.1.11).

We next define the *vector negative*, $-\mathbf{A}$, and the *vector difference*, $\mathbf{B} - \mathbf{A}$. If $\mathbf{A}$ is the vector from P to Q, then $-\mathbf{A}$ is the vector from Q to P (Figure 10.1.12). $\mathbf{B} - \mathbf{A}$ is the vector which, when added to $\mathbf{A}$, gives $\mathbf{B}$; i.e.,

$$\mathbf{A} + (\mathbf{B} - \mathbf{A}) = \mathbf{B}.$$

Thus if $\mathbf{A}$ is the vector from P to Q and $\mathbf{B}$ is the vector from P to R, then $\mathbf{B} - \mathbf{A}$ is the vector from Q to R (Figure 10.1.13).

Vector difference

Figure 10.1.13

If a trader initially has a commodity vector $\mathbf{A}$ and sells a quantity b_1 of the first commodity and b_2 of the second, his new commodity vector will be the vector difference $\mathbf{A} - \mathbf{B}$.

Given a force vector $\mathbf{F}$, $-\mathbf{F}$ is the force vector of the same magnitude but exactly the opposite direction.

If an object initially has position vector $\mathbf{P}$, then $\mathbf{Q} - \mathbf{P}$ is the displacement vector which will change its position to $\mathbf{Q}$.

THEOREM 2

Let $\mathbf{A}$ and $\mathbf{B}$ be vectors.

(i) $-\mathbf{0} = \mathbf{0}$,
(ii) $-(-\mathbf{A}) = \mathbf{A}$,
(iii) $\mathbf{A} - \mathbf{A} = \mathbf{0}$,
(iv) $\mathbf{B} - \mathbf{A} = \mathbf{B} + (-\mathbf{A})$.

Rule (iv) is illustrated in Figure 10.1.14.

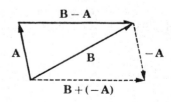

Figure 10.1.14

If **A** is a vector with components a_1 and a_2 and c is a scalar, then the *scalar multiple* c**A** is the vector with components ca_1, ca_2. Notice that the product of a scalar and a vector is a vector. Geometrically, for positive c, c**A** is the vector in the same direction as **A** whose length is c times the length of **A** (Figure 10.1.15). $(-c)$**A** is the vector in the opposite direction from **A** whose length is $c|$**A**$|$. We sometimes write **A**c for c**A**, and **A**$/c$ for $(1/c)$**A**.

Figure 10.1.15   Scalar Multiples

In physics, *Newton's second law of motion* states that

$$\mathbf{F} = m\mathbf{A}$$

where **F** is the force vector acting on an object, **A** is the acceleration vector, and the scalar m is the mass of the object.

In economics, if all prices are increased by the same factor c due to inflation, then the new price vector **Q** will be a scalar multiple of the initial price vector **P**,

$$\mathbf{Q} = c\mathbf{P}.$$

THEOREM 3

Let **A** and **B** *be vectors and* s, t *be scalars.*

(i) $0\mathbf{A} = \mathbf{0}$, $1\mathbf{A} = \mathbf{A}$, $(-s)\mathbf{A} = -(s\mathbf{A})$.

(ii)   Scalar Associative Law $s(t\mathbf{A}) = (st)\mathbf{A}$.

(iii)   Distributive Laws $(s + t)\mathbf{A} = s\mathbf{A} + t\mathbf{A}$,

$$s(\mathbf{A} + \mathbf{B}) = s\mathbf{A} + s\mathbf{B}.$$

(iv) $|s\mathbf{A}| = |s|\,|\mathbf{A}|$.

We shall prove only part (iv) which says that the length of s**A** is $|s|$ times the length of **A**.

Let **A** have components a_1, a_2. Then s**A** has components sa_1, sa_2.

Thus
$$|s\mathbf{A}| = \sqrt{(sa_1)^2 + (sa_2)^2} = \sqrt{s^2 a_1^2 + s^2 a_2^2}$$
$$= \sqrt{s^2}\sqrt{a_1^2 + a_2^2} = |s|\,|\mathbf{A}|.$$

A *unit vector* is a vector **U** of length one. The two most important unit vectors are the *basis vectors* **i** and **j**. **i**, the unit vector along the x-axis, has components $(1, 0)$. **j**, the unit vector along the y-axis, has components $(0, 1)$. Figure 10.1.16 shows **i** and **j**.

Basis vectors   **j**

Figure 10.1.16   **i**

A vector can be conveniently expressed in terms of the basis vectors.

COROLLARY 1

*The vector with components a and b is a**i** + b**j**.*

PROOF $a\mathbf{i}$ is the vector from $(0, 0)$ to $(a, 0)$, $b\mathbf{j}$ is the vector from $(0, 0)$ to $(0, b)$. Therefore the sum $a\mathbf{i} + b\mathbf{j}$ is the vector from $(0, 0)$ to (a, b) (Figure 10.1.17).

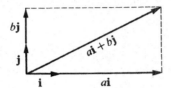

Figure 10.1.17

Sums, differences, scalar multiples, and lengths of vectors can easily be computed using the basis vectors and components. The necessary formulas are given in the next corollary.

COROLLARY 2

Let $\mathbf{A} = a_1\mathbf{i} + a_2\mathbf{j}$ and $\mathbf{B} = b_1\mathbf{i} + b_2\mathbf{j}$ be vectors and let c be a scalar.

(i) $\mathbf{A} + \mathbf{B} = (a_1 + b_1)\mathbf{i} + (a_2 + b_2)\mathbf{j}.$

(ii) $\mathbf{A} - \mathbf{B} = (a_1 - b_1)\mathbf{i} + (a_2 - b_2)\mathbf{j}.$

(iii) $c\mathbf{A} = (ca_1)\mathbf{i} + (ca_2)\mathbf{j}.$

(iv) $|\mathbf{A}| = \sqrt{a_1^2 + a_2^2}.$

For example, (i) is shown by the computation

$$\mathbf{A} + \mathbf{B} = (a_1\mathbf{i} + a_2\mathbf{j}) + (b_1\mathbf{i} + b_2\mathbf{j})$$
$$= (a_1\mathbf{i} + b_1\mathbf{i}) + (a_2\mathbf{j} + b_2\mathbf{j}) = (a_1 + b_1)\mathbf{i} + (a_2 + b_2)\mathbf{j}.$$

It is illustrated in Figure 10.1.18.

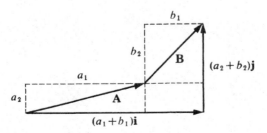

Figure 10.1.18

EXAMPLE 4   Let $\mathbf{A} = 2\mathbf{i} - 5\mathbf{j}$, $\mathbf{B} = \mathbf{i} + 3\mathbf{j}$.

(a) Find $\mathbf{A} + \mathbf{B}$, $\mathbf{A} - \mathbf{B}$, $-\mathbf{A}$,   and $6\mathbf{B}$.

$$A + B = (2 + 1)i + (-5 + 3)j = 3i - 2j,$$
$$A - B = (2 - 1)i + (-5 - 3)j = i - 8j,$$
$$-A = (-1)A = (-1)2i + (-1)(-5)j = -2i + 5j,$$
$$6B = 6(i + 3j) = 6i + 18j.$$

(b) Find the vector **D** such that $3A + 5D = B$.

$$5D = -3A + B,$$
$$D = \tfrac{1}{5}(-3A + B),$$
$$= \tfrac{1}{5}(-3 \cdot 2 + 1)i + \tfrac{1}{5}(-3(-5) + 3)j,$$
$$= -i + \tfrac{18}{5}j.$$

EXAMPLE 5   A triangle has vertices $(0, 0)$, $(2, -1)$, and $(3, 1)$ (Figure 10.1.19). Find the vectors counterclockwise around the perimeter of the triangle and check that their sum is the zero vector.

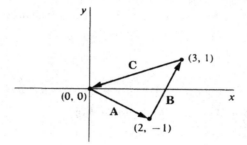

Figure 10.1.19

The three vectors are

$$A = (2 - 0)i + (-1 - 0)j = 2i - j,$$
$$B = (3 - 2)i + (1 - (-1))j = i + 2j,$$
$$C = (0 - 3)i + (0 - 1)j = -3i - j.$$

Their sum is

$$A + B + C = (2 + 1 - 3)i + (-1 + 2 + (-1))j = 0i + 0j.$$

We need a convenient way of describing the direction as well as the magnitude of a vector. First we define the angle between two vectors (Figure 10.1.20).

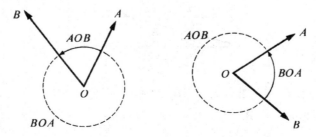

Figure 10.1.20

DEFINITION

Let **A** and **B** be two nonzero vectors in the plane, and let O be the origin.

The **angle** between **A** and **B** is either the angle AOB or the angle BOA, whichever is in the interval $[0, \pi]$. (The angle between **A** and the zero vector is undefined.)

Notice that if AOB is between 0 and π, then BOA is between π and 2π, and vice versa. So exactly one is between 0 and π. The angle θ between **A** and **B** can be computed by using the Law of Cosines from trigonometry, illustrated in Figure 10.1.21.

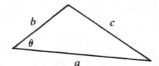

Figure 10.1.21

LAW OF COSINES

In a triangle with sides a, b, c, and angle θ opposite side c,
$$c^2 = a^2 + b^2 - 2ab \cos \theta.$$

Notice that when $\theta = \pi/2$, $\cos \theta = 0$ and the Law of Cosines reduces to the familiar Theorem of Pythagoras, $c^2 = a^2 + b^2$.

Given vectors **A** and **B** with angle θ between them, we form a triangle with sides $|\mathbf{A}|$, $|\mathbf{B}|$, and $|\mathbf{B} - \mathbf{A}|$. Then by the Law of Cosines,
$$|\mathbf{B} - \mathbf{A}|^2 = |\mathbf{A}|^2 + |\mathbf{B}|^2 - 2|\mathbf{A}||\mathbf{B}| \cos \theta.$$

Solving for $\cos \theta$,
$$\cos \theta = \frac{|\mathbf{A}|^2 + |\mathbf{B}|^2 - |\mathbf{B} - \mathbf{A}|^2}{2|\mathbf{A}||\mathbf{B}|}.$$

Since the arccosine is always between 0 and π,
$$\theta = \arccos \left(\frac{|\mathbf{A}|^2 + |\mathbf{B}|^2 - |\mathbf{B} - \mathbf{A}|^2}{2|\mathbf{A}||\mathbf{B}|} \right).$$

EXAMPLE 6   Find the angle between $\mathbf{A} = 3\mathbf{i} - 4\mathbf{j}$ and $\mathbf{B} = \mathbf{i} + \mathbf{j}$.

$$|\mathbf{A}| = \sqrt{3^2 + (-4)^2} = \sqrt{25} = 5,$$
$$|\mathbf{B}| = \sqrt{1^2 + 1^2} = \sqrt{2},$$
$$|\mathbf{B} - \mathbf{A}| = \sqrt{(3 - 1)^2 + (-4 - 1)^2} = \sqrt{4 + 25} = \sqrt{29},$$
$$\cos \theta = \frac{|\mathbf{A}|^2 + |\mathbf{B}|^2 - |\mathbf{B} - \mathbf{A}|^2}{2|\mathbf{A}||\mathbf{B}|} = \frac{25 + 2 - 29}{2 \cdot 5 \cdot \sqrt{2}}$$
$$= -\frac{2}{10\sqrt{2}} = -\frac{\sqrt{2}}{10}.$$
$$\theta = \arccos \left(-\frac{\sqrt{2}}{10} \right).$$

The direction of a vector can be described in one of three closely related ways: by its direction angles, its direction cosines, or its unit vector.

Let **A** be a nonzero vector. The angles α between **A** and **i**, and β between **A** and **j**, are called the *direction angles* of **A**. The cosines of these angles, $\cos \alpha$ and $\cos \beta$, are called the *direction cosines of* **A**.

The vector **U** = **A**/|**A**| is called the *unit vector* of **A**. **U** has length one, |**U**| = |**A**|/|**A**| = 1.

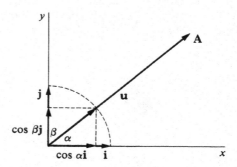

Figure 10.1.22

We can see from Figure 10.1.22 that the components of **U** are the direction cosines of **A**,

$$\mathbf{U} = \cos \alpha \mathbf{i} + \cos \beta \mathbf{j}.$$

A vector **A** is determined by its length and its direction cosines,

$$\mathbf{A} = |\mathbf{A}|\, \mathbf{U} = |\mathbf{A}| \cos \alpha \mathbf{i} + |\mathbf{A}| \cos \beta \mathbf{j}.$$

The sum of the squares of the direction cosines is always one, for

$$|\mathbf{U}| = \cos^2 \alpha + \cos^2 \beta = 1.$$

EXAMPLE 7 Find the unit vector and direction cosines of the given vector.

First find the length, then the unit vector, and then the direction cosines.

(a) $\mathbf{A} = 2\mathbf{i} + \mathbf{j}$ $|\mathbf{A}| = \sqrt{2^2 + 1^2} = \sqrt{5}$

Unit vector $= \dfrac{\mathbf{A}}{|\mathbf{A}|} = \dfrac{2\mathbf{i} + \mathbf{j}}{\sqrt{5}}$

Direction cosines $= \left(\dfrac{2}{\sqrt{5}}, \dfrac{1}{\sqrt{5}} \right)$

(b) $\mathbf{B} = 5\mathbf{i} - 12\mathbf{j}$ $|\mathbf{B}| = \sqrt{5^2 + (-12)^2} = \sqrt{169} = 13$

Unit vector $= \dfrac{\mathbf{B}}{|\mathbf{B}|} = \dfrac{5\mathbf{i} - 12\mathbf{j}}{13}$

Direction cosines $= (\tfrac{5}{13}, -\tfrac{12}{13})$.

(c) $\mathbf{C} = \tfrac{1}{4}\mathbf{j}$ $|\mathbf{C}| = \sqrt{0^2 + (\tfrac{1}{4})^2} = \tfrac{1}{4}$

Unit vector $= \dfrac{\tfrac{1}{4}\mathbf{j}}{\tfrac{1}{4}} = \mathbf{j}$

Direction cosines $= (0, 1)$.

EXAMPLE 8 Find the vector **A** which has length 6 and direction cosines $(-1/2, \sqrt{3}/2)$.

$$\mathbf{A} = 6(-1/2)\mathbf{i} + 6(\sqrt{3}/2)\mathbf{j} = -3\mathbf{i} + 3\sqrt{3}\mathbf{j}.$$

PROBLEMS FOR SECTION 10.1

In Problems 1–4 find the vector represented by the directed line segment $\overrightarrow{PQ}$.

1	$P = (3, 1), \quad Q = (4, 3)$	**2**	$P = (-1, -1), \quad Q = (2, -2)$
3	$P = (3, 4), \quad Q = (0, 0)$	**4**	$P = (0, 0), \quad Q = (0, 3)$

In Problems 5–8 find the point Q such that **A** is the vector from P to Q.

5	$P = (1, -1), \quad \mathbf{A} = \mathbf{i} - 3\mathbf{j}$	**6**	$P = (0, 0), \quad \mathbf{A} = 3\mathbf{i} - 5\mathbf{j}$
7	$P = (4, 6), \quad \mathbf{A} = -5\mathbf{i} + 6\mathbf{j}$	**8**	$P = (3, 3), \quad \mathbf{A} = 2\mathbf{j}$

In Problems 9–32, find the given vector or scalar, where

$$\mathbf{A} = \mathbf{i} - 2\mathbf{j}, \qquad \mathbf{B} = -4\mathbf{i} + 3\mathbf{j}, \qquad \mathbf{C} = 3\mathbf{i}.$$

9	$\mathbf{A} + \mathbf{B}$	**10**	$\mathbf{A} + \mathbf{C}$
11	$\mathbf{A} + \mathbf{B} + \mathbf{C}$	**12**	$-\mathbf{A}$
13	$3\mathbf{A}$	**14**	$\mathbf{A} - \mathbf{B}$
15	$\mathbf{B} - \mathbf{A}$	**16**	$3\mathbf{A} + 4\mathbf{B}$
17	$\mathbf{A} - 2\mathbf{B} + 3\mathbf{C}$	**18**	$|\mathbf{A}|$
19	$|\mathbf{B}|$	**20**	$|\mathbf{A} + \mathbf{B}|$
21	$|\mathbf{A} - \mathbf{B}|$	**22**	$|\mathbf{A}| + |\mathbf{B}|$

23 $\quad |6\mathbf{A}|$

24 $\quad$ The vector **D** such that $\mathbf{A} + 2\mathbf{D} = \mathbf{B}$.

25 $\quad$ The vector **D** such that $2\mathbf{A} + 4\mathbf{D} = \mathbf{C} - 3\mathbf{B}$.

26 $\quad$ The unit vector and direction cosines of **A**.

27 $\quad$ The unit vector and direction cosines of **B**.

28 $\quad$ The unit vector and direction cosines of **C**.

29 $\quad$ The angle between **A** and **B**.

30 $\quad$ The angle between **A** and **C**.

31 $\quad$ The angle between **B** and **C**.

32 $\quad$ The angle between $-\mathbf{B}$ and **C**.

33 $\quad$ An object initially has position vector $\mathbf{P} = 3\mathbf{i} + 5\mathbf{j}$ and is displaced by the vector $\mathbf{A} = 4\mathbf{i} - 2\mathbf{j}$. Find its new position vector.

34 $\quad$ An object is displaced first by the vector $\mathbf{A} = -\mathbf{i} - 2\mathbf{j}$ and then by the vector $\mathbf{B} = 4\mathbf{i} - \mathbf{j}$. Find the total displacement vector.

35 $\quad$ Find the displacement vector necessary to change the position vector of an object from $\mathbf{P} = -3\mathbf{i} + 6\mathbf{j}$ to $\mathbf{Q} = 5\mathbf{i} + 4\mathbf{j}$.

36 $\quad$ Three forces are acting on an object, with vectors

$$\mathbf{F} = \mathbf{i} + 3\mathbf{j}, \qquad \mathbf{G} = 2\mathbf{i}, \qquad \mathbf{H} = -2\mathbf{i} - \mathbf{j}.$$

Find the total force on the object.

37   Three forces are acting on an object which is at rest. The first two forces are
$$\mathbf{F}_1 = -6\mathbf{i} + 9\mathbf{j}, \qquad \mathbf{F}_2 = 10\mathbf{i} - 3\mathbf{j}.$$
Find the third force $\mathbf{F}_3$.

38   An object of mass 10 is being accelerated so that its acceleration vector is $\mathbf{A} = 5\mathbf{i} - 6\mathbf{j}$. Find the total force acting on the object.

39   An object is displaced by the vector $3\mathbf{i} - 4\mathbf{j}$. Find the distance it is moved.

40   An object has the velocity vector $\mathbf{V} = \mathbf{i} - \mathbf{j}$. Find its speed.

41   A trader initially has the commodity vector $\mathbf{A} = 3\mathbf{i} + \mathbf{j}$ and buys the commodity vector $\mathbf{B} = \mathbf{i} + 2\mathbf{j}$. Find his new total commodity vector.

42   Two traders initially have commodity vectors $\mathbf{A}_0 = 4\mathbf{i} + \mathbf{j}$, $\mathbf{B}_0 = 3\mathbf{i} + 6\mathbf{j}$. After trading with each other, trader A has the commodity vector $\mathbf{A}_1 = 3\mathbf{i} + 3\mathbf{j}$. Find the new commodity vector $\mathbf{B}_1$ of trader B.

43   A trader initially has the commodity vector $\mathbf{A} = 15\mathbf{i} + 12\mathbf{j}$ and sells the commodity vector $5\mathbf{i} + 10\mathbf{j}$. Find his new commodity vector.

44   A pair of commodities initially has the price vector $\mathbf{P} = 6\mathbf{i} + 9\mathbf{j}$. Due to inflation all prices are increased by 10%. Find the new price vector.

45   Find the vector with length 4 and direction cosines $(-\sqrt{2}/2, \sqrt{2}/2)$.

46   Find the vector with length 4 and direction cosines $(-1, 0)$.

47   Find the vector with length 10 and direction cosines $(\frac{3}{5}, \frac{4}{5})$.

In Problems 48–50 find the vectors counterclockwise around the perimeter of the polygon with the given vertices.

48 $(0, 0), (1, 0), (0, 1)$.

49 $(1, 1), (3, 0), (5, 2), (0, 4)$.

50   The regular hexagon inscribed in the unit circle $x^2 + y^2 = 1$ with the initial vertex $(1, 0)$.

51   Use the Triangle Inequality to prove the following.
$$|\mathbf{A} - \mathbf{B}| \le |\mathbf{A}| + |\mathbf{B}|,$$
$$|\mathbf{A}| - |\mathbf{B}| \le |\mathbf{A} + \mathbf{B}|,$$
$$|\mathbf{A} + \mathbf{B} + \mathbf{C}| \le |\mathbf{A}| + |\mathbf{B}| + |\mathbf{C}|.$$

☐ **52**   Prove that for every nonzero vector $\mathbf{A}$ and positive scalar s, there are exactly two scalar multiples $t\mathbf{A}$ of length s.

☐ **53**   Prove that two nonzero vectors $\mathbf{A}$ and $\mathbf{B}$ have the same direction cosines if and only if $\mathbf{B} = t\mathbf{A}$ for some positive t.

☐ **54**   Prove the Commutative Law for vector addition.

☐ **55**   Prove the Distributive Laws for scalar multiples.

☐ **56**   Prove the Triangle Inequality. *Hint*: Assume
$$\sqrt{(a_1 + b_1)^2 + (a_2 + b_2)^2} > \sqrt{a_1^2 + a_2^2} + \sqrt{b_1^2 + b_2^2}$$
and get a contradiction. This is done by squaring both sides, simplifying, and then squaring and simplifying again.

10.2 VECTORS AND PLANE GEOMETRY

In this section we apply the algebra of two-dimensional vectors to plane geometry.

Given a point $P(p_1, p_2)$ in the plane, the *position vector* of P is the vector $\mathbf{P}$ from the origin to P (Figure 10.2.1). $\mathbf{P}$ has components p_1 and p_2, so

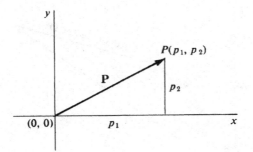

Figure 10.2.1   The position vector

$$\mathbf{P} = p_1\mathbf{i} + p_2\mathbf{j}.$$

If A and B are two points in the plane with position vectors $\mathbf{A}$ and $\mathbf{B}$, then the vector from A to B is the vector difference $\mathbf{B} - \mathbf{A}$. This can be seen from Figure 10.2.2.

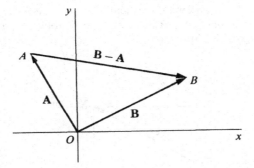

Figure 10.2.2

In Section 1.3, we saw that a line in the plane may be defined as the graph of a linear equation

$$ax + by = c$$

where a and b are not both zero (Figure 10.2.3). We shall call the above equation a *scalar equation* of the line.

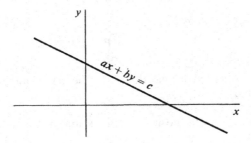

Figure 10.2.3

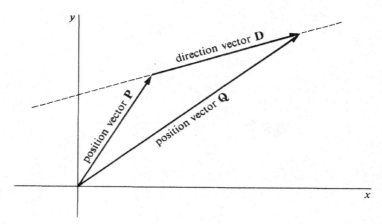

Figure 10.2.4

    The position vector of any point P on a line L is called a *position vector* of L. If P and Q are two distinct points on L, the vector $\mathbf{D}$ from P to Q is called a *direction vector* of L. Thus $\mathbf{D} = \mathbf{Q} - \mathbf{P}$ (Figure 10.2.4).

    Theorem 1 will show how to represent a line by a *vector equation*. Let us use the symbol X for the *variable point* $X(x, y)$, and the symbol $\mathbf{X}$ for the *variable vector* $\mathbf{X} = x\mathbf{i} + y\mathbf{j}$. (see Figure 10.2.5).

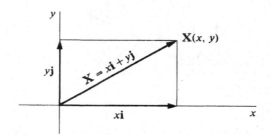

Figure 10.2.5

THEOREM 1

    A line L is uniquely determined by a position vector $\mathbf{P}$ and a direction vector $\mathbf{D}$. L has the scalar equation

$$xd_2 - yd_1 = p_1d_2 - p_2d_1$$

and the vector equation

$$\mathbf{X} = \mathbf{P} + t\mathbf{D}.$$

    The vector equation means that L is the set of all points X such that $\mathbf{X} = \mathbf{P} + t\mathbf{D}$ for some t.

PROOF Let L be any line with position vector $\mathbf{P}$ and direction vector $\mathbf{D}$. We must show that:

(i) L has the scalar equation given in Theorem 1.
(ii) If $\mathbf{X} = \mathbf{P} + t\mathbf{D}$ for some t then X is a point of L.
(iii) If X is a point of L then $\mathbf{X} = \mathbf{P} + t\mathbf{D}$ for some t.

(i) $\mathbf{D}$ is the vector from A to B where A and B are points on L. Since L is the line through A and B, it has the scalar equation

$$(x - a_1)(b_2 - a_2) = (y - a_2)(b_1 - a_1),$$
$$(x - a_1)d_2 = (y - a_2)d_1,$$
$$xd_2 - yd_1 = a_1d_2 - a_2d_1.$$

This equation holds for the point P of L,

$$p_1d_2 - p_2d_1 = a_1d_2 - a_2d_1.$$

Combining the last two equations we get the required equation:

(1)
$$xd_2 - yd_1 = p_1d_2 - p_2d_1.$$

(ii) Let X be a point such that $\mathbf{X} = \mathbf{P} + t\mathbf{D}$ for some t. Then

$$x = p_1 + td_1, \qquad y = p_2 + td_2,$$
$$d_2x - d_1y = d_2p_1 + d_2td_1 - d_1p_2 - d_1td_2 = d_2p_1 - d_1p_2,$$

so X is a point of L.

(iii) Let X be a point of L. If $d_1 \neq 0$ we set $t = (x - p_1)/d_1$, and using Equation 1 we get $\mathbf{X} = \mathbf{P} + t\mathbf{D}$. The case $d_2 \neq 0$ is similar. Therefore $\mathbf{X} = \mathbf{P} + t\mathbf{D}$ is a vector equation for L (Figure 10.2.6).

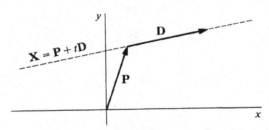

The line with position vector $\mathbf{P}$
and direction vector $\mathbf{D}$

Figure 10.2.6

The vector equation $\mathbf{X} = \mathbf{P} + t\mathbf{D}$ can be put in the form

$$x\mathbf{i} + y\mathbf{j} = (p_1 + td_1)\mathbf{i} + (p_2 + td_2)\mathbf{j}.$$

It can also be written as a pair of *parametric equations*

$$x = p_1 + td_1, \qquad y = p_2 + td_2.$$

EXAMPLE 1   Find a vector equation for the line through the two points $A(2, 1)$ and $B(-4, 0)$, shown in Figure 10.2.7.

The vector $\mathbf{D} = \mathbf{B} - \mathbf{A}$ from A to B is given by

$$\mathbf{D} = (-4 - 2)\mathbf{i} + (0 - 1)\mathbf{j} = -6\mathbf{i} - \mathbf{j}.$$

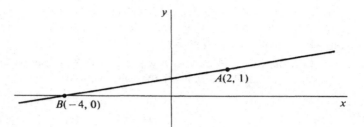

Figure 10.2.7

Since **A** is a position vector and **D** a direction vector of the line, the line has the vector equation

$$\mathbf{X} = \mathbf{A} + t\mathbf{D}$$
$$= 2\mathbf{i} + \mathbf{j} + t(-6\mathbf{i} - \mathbf{j}).$$

In general, the line L through points A and B has the vector equation $\mathbf{X} = \mathbf{A} + t(\mathbf{B} - \mathbf{A})$ because **A** is a position vector and $\mathbf{B} - \mathbf{A}$ is a direction vector of L.

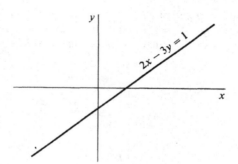

Figure 10.2.8

EXAMPLE 2   Find a vector equation for the line in Figure 10.2.8:

$$2x - 3y = 1.$$

Step 1   Find two points on the line by taking two values of x and solving for y.

$$x = 0, \qquad 0 - 3y = 1, \qquad y = -\tfrac{1}{3}, \qquad (0, -\tfrac{1}{3}).$$
$$x = 1, \qquad 2 - 3y = 1, \qquad y = \tfrac{1}{3}, \qquad (1, \tfrac{1}{3}).$$

Step 2   Find a position and direction vector.

$$\mathbf{P} = 0\mathbf{i} + (-\tfrac{1}{3})\mathbf{j} = -\tfrac{1}{3}\mathbf{j}.$$
$$\mathbf{D} = (1 - 0)\mathbf{i} + (\tfrac{1}{3} - (-\tfrac{1}{3}))\mathbf{j} = \mathbf{i} + \tfrac{2}{3}\mathbf{j}.$$

Step 3   Use Theorem 1. The vector equation is

$$\mathbf{X} = \mathbf{P} + t\mathbf{D}$$
$$= -\tfrac{1}{3}\mathbf{j} + t(\mathbf{i} + \tfrac{2}{3}\mathbf{j}).$$

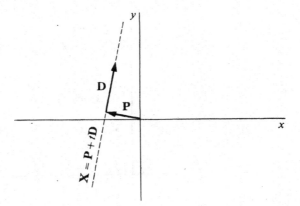

Figure 10.2.9

EXAMPLE 3 Find a scalar equation for the line in Figure 10.2.9:

$$X = -4i + j + t(i + 6j).$$

First method By Theorem 1, the line has the equation

$$xd_2 - yd_1 = p_1d_2 - p_2d_1,$$
$$6x - y = (-4)\cdot 6 - 1\cdot 1,$$
$$6x - y = -25.$$

Second method We convert the vector equation to parametric equations and then eliminate t.

$$x = -4 + t, \qquad y = 1 + 6t,$$
$$t = x + 4, \qquad y = 1 + 6(x + 4),$$
$$y = 25 + 6x.$$

This is equivalent to the first solution.

EXAMPLE 4 Determine whether the three points

$$A(1, 3), \qquad B(2, 5), \qquad C(3, 10)$$

are on the same line.

The line L through A and B has the vector equation

$$X = A + t(B - A)$$
$$= i + 3j + t(i + 2j) = (1 + t)i + (3 + 2t)j.$$

The only point on L with x component 3 is given by

$$3 = 1 + t, \qquad t = 2, \qquad P = 3i + 7j.$$

Since C is another point with x component 3, C is not on L. Therefore A, B, and C are not on the same line, as we see in Figure 10.2.10.

Some applications of vectors to plane geometry follow.

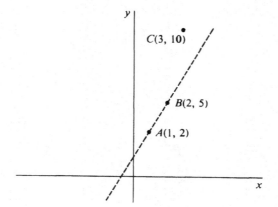

Figure 10.2.10

EXAMPLE 5 Let A and B be two distinct points. Prove that the midpoint of the line segment AB is the point P with position vector $\mathbf{P} = \frac{1}{2}\mathbf{A} + \frac{1}{2}\mathbf{B}$.

PROOF We shall prove that the point P is on the line AB and is equidistant from A and B (see Figure 10.2.11). The line through A and B has the direction

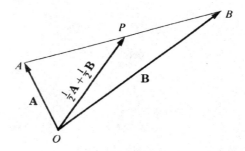

Figure 10.2.11

vector $\mathbf{D} = \mathbf{B} - \mathbf{A}$. The vector $\mathbf{P}$ has the form

$$\mathbf{P} = \tfrac{1}{2}\mathbf{A} + \tfrac{1}{2}\mathbf{B} = \mathbf{A} + \tfrac{1}{2}(\mathbf{B} - \mathbf{A}) = \mathbf{A} + \tfrac{1}{2}\mathbf{D}.$$

Therefore by Theorem 1, P is on the line AB. To prove that P is equidistant, we show that the vector from A to P is the same as the vector from P to B

$$\mathbf{P} - \mathbf{A} = \tfrac{1}{2}\mathbf{A} + \tfrac{1}{2}\mathbf{B} - \mathbf{A} = \tfrac{1}{2}\mathbf{B} - \tfrac{1}{2}\mathbf{A}.$$
$$\mathbf{B} - \mathbf{P} = \mathbf{B} - \tfrac{1}{2}\mathbf{A} - \tfrac{1}{2}\mathbf{B} = \tfrac{1}{2}\mathbf{B} - \tfrac{1}{2}\mathbf{A}.$$

EXAMPLE 6 Find the midpoint of the line segment from $A(-1, 2)$ to $B(3, 3)$ (Figure 10.2.12).

The points have position vectors

$$\mathbf{A} = -\mathbf{i} + 2\mathbf{j}, \qquad \mathbf{B} = 3\mathbf{i} + 3\mathbf{j}.$$

The midpoint $\mathbf{P}$ has the position vector

$$\mathbf{P} = \tfrac{1}{2}\mathbf{A} + \tfrac{1}{2}\mathbf{B} = \tfrac{1}{2}(-\mathbf{i} + 2\mathbf{j}) + \tfrac{1}{2}(3\mathbf{i} + 3\mathbf{j}) = \mathbf{i} + \tfrac{5}{2}\mathbf{j}.$$

Therefore P is the point $(1, \frac{5}{2})$.

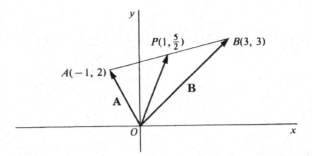

Figure 10.2.12

A four-sided figure whose opposite sides represent equal vectors is called a *parallelogram*.

EXAMPLE 7   Prove that the diagonals of a parallelogram bisect each other.

PROOF We are given a parallelogram *ABCD*, shown in Figure 10.2.13.

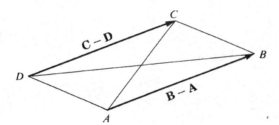

Figure 10.2.13

Since the opposite sides represent equal vectors, we have

(2) $$\mathbf{B} - \mathbf{A} = \mathbf{C} - \mathbf{D}.$$

The diagonal *AC* has midpoint $\frac{1}{2}\mathbf{A} + \frac{1}{2}\mathbf{C}$ and the other diagonal *BD* has midpoint $\frac{1}{2}\mathbf{B} + \frac{1}{2}\mathbf{D}$. We show that these two midpoints are equal. The Equation 2 gives

$$\mathbf{C} = \mathbf{B} - \mathbf{A} + \mathbf{D}.$$

Then $\frac{1}{2}\mathbf{A} + \frac{1}{2}\mathbf{C} = \frac{1}{2}\mathbf{A} + \frac{1}{2}(\mathbf{B} - \mathbf{A} + \mathbf{D}) = \frac{1}{2}\mathbf{B} + \frac{1}{2}\mathbf{D}.$

Thus the two diagonals meet at their midpoints.

EXAMPLE 8   Prove that the lines from the vertices of a triangle *ABC* to the midpoints of the opposite sides all meet at the single point *P* given by

$$\mathbf{P} = \tfrac{1}{3}\mathbf{A} + \tfrac{1}{3}\mathbf{B} + \tfrac{1}{3}\mathbf{C}.$$

PROOF We are given triangle *ABC*, shown in Figure 10.2.14. Let *A'*, *B'*, *C'* be the midpoints of the opposite sides. We prove that all three lines *AA'*, *BB'*, *CC'* pass through the point *P*.

The point *A'* has position vector

$$\mathbf{A}' = \tfrac{1}{2}\mathbf{B} + \tfrac{1}{2}\mathbf{C}.$$

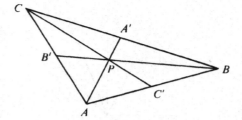

Figure 10.2.14

The line AA' has the direction vector $\mathbf{A}' - \mathbf{A}$. AA' has the vector equation

$$\mathbf{X} = \mathbf{A} + t(\mathbf{A}' - \mathbf{A}).$$

The computation below shows that P is on the line AA'

$$\mathbf{P} = \tfrac{1}{3}\mathbf{A} + (\tfrac{1}{3}\mathbf{B} + \tfrac{1}{3}\mathbf{C}) = \tfrac{1}{3}\mathbf{A} + \tfrac{2}{3}\mathbf{A}' = \mathbf{A} + \tfrac{2}{3}(\mathbf{A}' - \mathbf{A}).$$

A similar proof shows that P is on BB' and CC'.

PROBLEMS FOR SECTION 10.2

In Problems 1–14, find a vector equation for the given line.

1   The line through $P(3, -1)$ with direction vector $\mathbf{D} = -\mathbf{i} + \mathbf{j}$.

2   The line through $P(0, 0)$ with direction vector $\mathbf{D} = \mathbf{i} + 2\mathbf{j}$.

3   The line with parametric equations $x = 3 - 2t, y = 4 + 5t$.

4   The line with parametric equations $x = 4t, y = 1 + t$.

5   The line through the points $P(1, 4)$ and $Q(2, -1)$.

6   The line through the points $P(5, 5)$ and $Q(-6, 6)$.

7   The vertical line through $P(2, 5)$.

8   The horizontal line through $P(4, 1)$.

9   The line $y = 2 + 5x$.

10   The line $x + y = 3$.

11   The line $y = 3$.

12   The line $x = y$.

13   The line through $P(6, 5)$ with slope -3.

14   The line through $P(1, 2)$ with slope 4.

15   Find a scalar equation for the line $\mathbf{X} = 3\mathbf{i} - 4\mathbf{j} + t(\mathbf{i} - 2\mathbf{j})$.

16   Find a scalar equation for the line $\mathbf{X} = 2\mathbf{i} + t(-\mathbf{i} + 4\mathbf{j})$.

17   Find a scalar equation for the line $\mathbf{X} = \mathbf{i} + 3\mathbf{j} + 4t\mathbf{i}$.

18   Find a scalar equation for the line with parametric equations $x = 3 - 4t, y = 1 + 2t$.

In Problems 19–24, determine whether the given three points are on a line.

19    $A(1, 1)$, $B(2, 4)$, $C(-1, -2)$.   20 $A(1, 3)$, $B(2, 5)$, $C(-1, -1)$.

21    $A(4, 0)$, $B(0, 1)$, $C(12, -2)$.   22 $A(6, 3)$, $B(5, 7)$, $C(4, 10)$.

23    $A(5, -1)$, $B(5, 2)$, $C(5, 6)$.   24 $A(-3, 2)$, $B(-3, 3)$, $C(0, 0)$.

25   Find the midpoint of the line AB where $A = (2, 5)$, $B = (-6, 1)$.

26   Find the midpoint of the line AB where $A = (-1, -4)$, $B = (9, 16)$.

27   Find the midpoint of the line AB where $A = (5, 10)$, $B = (-1, 10)$.

28   Find the point of intersection of the diagonals of the parallelogram $A(1, 4)$, $B(6, 4)$, $C(6, 6)$, $D(1, 6)$.

29   Find the point of intersection of the diagonals of the parallelogram $A(2, 0)$, $B(5, 1)$, $C(6, 6)$, $D(3, 5)$.

30   Find the point of intersection of the lines from the vertices to the midpoints of the opposite sides of the triangle ABC, where $A = (1, 4)$, $B = (2, -1)$, $C = (6, 3)$.

31   Prove that the slope of a line with direction vector $\mathbf{D} = d_1\mathbf{i} + d_2\mathbf{j}$ is $m = d_2/d_1$ (vertical if $d_1 = 0$).

☐ 32   Prove that if the diagonals of a four-sided figure bisect each other then the figure is a parallelogram. (Converse of Example 7.)

☐ 33   Prove that if the opposite sides of a four-sided figure are scalar multiples of each other then the figure is a parallelogram (i.e., the opposite sides are equal as vectors).

☐ 34   Let ABC be a triangle and let A_1, B_1, C_1 be the midpoints of the sides opposite A, B, C respectively. Show that the line AA_1 bisects the line B_1C_1.

☐ 35   Show that the midpoints of the sides of any four-sided figure are the vertices of a parallelogram.

☐ 36   Given a triangle ABC, let D be the midpoint of AB and E the midpoint of AC. Show that DE is parallel to BC and DE has half the length of BC. *Hint*: Show that $\mathbf{E} - \mathbf{D} = \frac{1}{2}(\mathbf{C} - \mathbf{B})$.

10.3 VECTORS AND LINES IN SPACE

In the preceding section, we used the algebra of vectors to prove some facts from plane geometry. This approach really comes into its own in solid geometry. Without using vectors, it is quite hard to define such basic concepts as a straight line, or the angle between two lines, in space. In this section we shall develop geometry in three-dimensional space with vectors as our starting point. The notions of a straight line and an angle in space will be defined using vectors, and we shall use vector algebra to solve problems about lines and angles. Later on in this chapter we shall continue our development of solid geometry, using vectors to study planes in space.

Vectors in space are developed in the same way as vectors in the plane. Three-dimensional space has three perpendicular coordinate axes, x, y, and z, as shown in Figure 10.3.1. This is called a *right-handed coordinate system*, because the

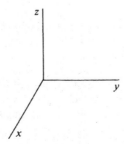

Figure 10.3.1

right thumb, forefinger, and middle finger can point in the direction of the positive x, y, and z axes respectively.

A point in space has three coordinates, one along each axis. We thus identify a point in space with an ordered triple of real numbers, as in Figure 10.3.2.

Given two points $P(p_1, p_2, p_3)$ and $Q(q_1, q_2, q_3)$ in space, the directed line segment PQ has the x-component $q_1 - p_1$, y-component $q_2 - p_2$, and z-component $q_3 - p_3$ (Figure 10.3.3).

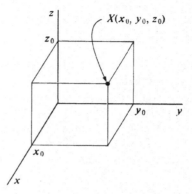

Figure 10.3.2

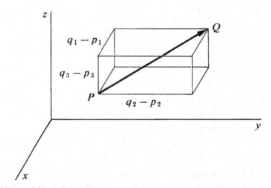

Figure 10.3.3 A Directed Line Segment

The family of all directed line segments in space which have the same three components as $\overrightarrow{PQ}$ is called the *vector in three dimensions*, or the *vector in space*, represented by $\overrightarrow{PQ}$.

The examples of vectors which we discussed in the plane also arise naturally in space. In space, position, velocity, acceleration, force, and displacement are vector quantities with three dimensions. In an economic model with three commodities, the commodity and price vectors have three dimensions.

Vectors in n dimensions arise quite naturally in economics, as commodity and price vectors in an economic model with n commodities. They also arise in more advanced parts of physics, such as quantum mechanics.

Sums, negatives, differences, and scalar multiples of vectors in three dimensions are defined exactly as in two dimensions. The *length*, or *norm*, of a vector **A** with components a_1, a_2, a_3 is defined by

$$|\mathbf{A}| = \sqrt{a_1^2 + a_2^2 + a_3^2}.$$

THEOREM 1

> *All the rules for vector algebra given in Section* 10.1 *hold for vectors in three dimensions.*

These rules are in Theorems 1, 2, and 3 of Section 10.1, and include the Triangle Inequality.

In space, there are three *basis vectors*, denoted by **i**, **j**, and **k**. **i** has components $(1, 0, 0)$, **j** has components $(0, 1, 0)$, and **k** has components $(0, 0, 1)$. **i**, **j**, and **k** are shown in Figure 10.3.4. As in the case of two dimensions, we see that

$a\mathbf{i} + b\mathbf{j} + c\mathbf{k}$ *is the vector in three dimensions with components a, b, and c.*

The rules for computing vectors by their components take the following form in three dimensions.

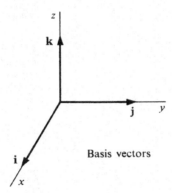

Basis vectors

Figure 10.3.4

COROLLARY 1 (Three Dimensions)

> Let $\mathbf{A} = a_1\mathbf{i} + a_2\mathbf{j} + a_3\mathbf{k}$ *and* $\mathbf{B} = b_1\mathbf{i} + b_2\mathbf{j} + b_3\mathbf{k}$ *be vectors in three dimensions and let c be a scalar.*
>
> (i) $\mathbf{A} + \mathbf{B} = (a_1 + b_1)\mathbf{i} + (a_2 + b_2)\mathbf{j} + (a_3 + b_3)\mathbf{k}.$
>
> (ii) $\mathbf{A} - \mathbf{B} = (a_1 - b_1)\mathbf{i} + (a_2 - b_2)\mathbf{j} + (a_3 - b_3)\mathbf{k}.$
>
> (iii) $c\mathbf{A} = (ca_1)\mathbf{i} + (ca_2)\mathbf{j} + (ca_3)\mathbf{k}.$

EXAMPLE 1   Given $\mathbf{A} = \mathbf{i} - \mathbf{j} + 2\mathbf{k}$ and $\mathbf{B} = 2\mathbf{i} - 2\mathbf{k}$, find $\mathbf{A} + \mathbf{B}$, $\mathbf{A} - \mathbf{B}$, $|\mathbf{A}|$, and
3A.

$$\mathbf{A} + \mathbf{B} = (1 + 2)\mathbf{i} + (-1 + 0)\mathbf{j} + (2 - 2)\mathbf{k} = 3\mathbf{i} - \mathbf{j}.$$
$$\mathbf{A} - \mathbf{B} = (1 - 2)\mathbf{i} + (-1 - 0)\mathbf{j} + (2 - (-2))\mathbf{k} = -\mathbf{i} - \mathbf{j} + 4\mathbf{k}.$$
$$|\mathbf{A}| = \sqrt{1^2 + (-1)^2 + (2^2)} = \sqrt{6}.$$
$$3\mathbf{A} = 3\mathbf{i} - 3\mathbf{j} + 6\mathbf{k}.$$

The Law of Cosines gives us a formula for the angle between two vectors in space. In fact, we shall use the Law of Cosines to *define* the angle between two vectors.

DEFINITION

Let **A** and **B** be two nonzero vectors in space. The **angle** between **A** and **B** is the angle θ between 0 and π such that

$$\cos \theta = \frac{|\mathbf{A}|^2 + |\mathbf{B}|^2 - |\mathbf{B} - \mathbf{A}|^2}{2|\mathbf{A}||\mathbf{B}|}.$$

One can prove from the Triangle Inequality that the above quantity is always between -1 and 1, and therefore is the cosine of some angle θ (Problem 42 at the end of this section).

EXAMPLE 2   Find the angle between $\mathbf{A} = \mathbf{i} - \mathbf{j} - \mathbf{k}$ and $\mathbf{B} = 2\mathbf{i} + \mathbf{j} + \mathbf{k}$.

$$|\mathbf{A}| = \sqrt{1^2 + (-1)^2 + (-1)^2} = \sqrt{3}.$$
$$|\mathbf{B}| = \sqrt{2^2 + 1^2 + 1^2} = \sqrt{6}.$$
$$|\mathbf{B} - \mathbf{A}| = \sqrt{(2-1)^2 + (1-(-1))^2 + (1-(-1))^2}$$
$$= \sqrt{1^2 + 2^2 + 2^2} = 3.$$
$$\cos \theta = \frac{3 + 6 - 9}{2\sqrt{3}\sqrt{6}} = 0. \qquad \theta = \arccos 0 = \frac{\pi}{2}.$$

The *direction angles* of a nonzero vector **A** in space are the three angles α, β, γ between **A** and **i**, **j**, **k** respectively. The cosines of the direction angles are called the *direction cosines* of **A**. Let us compute the direction cosines in terms of the components of **A**.

$$\cos \alpha = \frac{|\mathbf{A}|^2 + |\mathbf{i}|^2 - |\mathbf{i} - \mathbf{A}|^2}{2|\mathbf{A}||\mathbf{i}|}$$
$$= \frac{a_1^2 + a_2^2 + a_3^2 + 1 - ((1 - a_1)^2 + a_2^2 + a_3^2)}{2|\mathbf{A}|}$$
$$= \frac{a_1^2 + a_2^2 + a_3^2 + 1 - 1 + 2a_1 - a_1^2 - a_2^2 - a_3^2}{2|\mathbf{A}|}$$
$$= \frac{a_1}{|\mathbf{A}|}.$$

The computations for β and γ are similar. Thus

$$\cos \alpha = \frac{a_1}{|\mathbf{A}|}, \qquad \cos \beta = \frac{a_2}{|\mathbf{A}|}, \qquad \cos \gamma = \frac{a_3}{|\mathbf{A}|}.$$

The *unit vector* of **A** is defined as

$$\mathbf{U} = \frac{\mathbf{A}}{|\mathbf{A}|} = \frac{a_1}{|\mathbf{A}|}\mathbf{i} + \frac{a_2}{|\mathbf{A}|}\mathbf{j} + \frac{a_3}{|\mathbf{A}|}\mathbf{k}.$$

As in the two-dimensional case, the components of **U**, shown in Figure 10.3.5, are

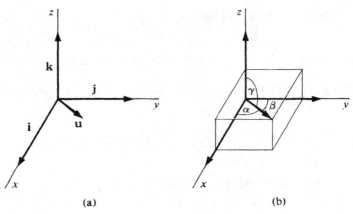

Figure 10.3.5

the direction cosines of **A**,

$$\mathbf{U} = \cos \alpha \mathbf{i} + \cos \beta \mathbf{j} + \cos \gamma \mathbf{k}.$$

Again, **A** is determined by its length and direction cosines, and the sum of the squares of the direction cosines is one,

$$\mathbf{A} = |\mathbf{A}| \cos \alpha \mathbf{i} + |\mathbf{A}| \cos \beta \mathbf{j} + |\mathbf{A}| \cos \gamma \mathbf{k},$$
$$\cos^2 \alpha + \cos^2 \beta + \cos^2 \gamma = 1.$$

EXAMPLE 3   Find the unit vectors and direction cosines of the vector $\mathbf{A} = 2\mathbf{i} + \mathbf{j} - 2\mathbf{k}$.

We first find the length, then the unit vector, then the direction cosines.

$$|\mathbf{A}| = \sqrt{2^2 + 1^2 + (-2)^2} = \sqrt{9} = 3.$$
$$\mathbf{U} = \frac{\mathbf{A}}{|\mathbf{A}|} = \frac{2\mathbf{i} + \mathbf{j} - 2\mathbf{k}}{3}.$$

Direction cosines $= (\frac{2}{3}, \frac{1}{3}, -\frac{2}{3}).$

The *position vector* of a point $P(p_1, p_2, p_3)$ in space is the vector

$$\mathbf{P} = p_1 \mathbf{i} + p_2 \mathbf{j} + p_3 \mathbf{k}.$$

X denotes the *variable vector*

$$\mathbf{X} = x\mathbf{i} + y\mathbf{j} + z\mathbf{k}.$$

We shall now define the notion of a line in space. The simplest way to describe a line in space is by a vector equation.

DEFINITION

Let **P** be a vector and **D** a nonzero vector in space. The **line** with the vector equation $\mathbf{X} = \mathbf{P} + t\mathbf{D}$ is the set of all points X such that $\mathbf{X} = \mathbf{P} + t\mathbf{D}$ for some scalar t.

The vector equation can also be written as a set of parametric equations

$$x = p_1 + d_1 t, \qquad y = p_2 + d_2 t, \qquad z = p_3 + d_3 t.$$

If t is time, the line is the path of a moving particle in space given by these parametric equations.

The three coordinate axes are lines with the following vector equations.

$$x\text{-axis:} \quad \mathbf{X} = t\mathbf{i},$$
$$y\text{-axis:} \quad \mathbf{X} = t\mathbf{j},$$
$$z\text{-axis:} \quad \mathbf{X} = t\mathbf{k}.$$

EXAMPLE 4   Find a vector equation for the line L with the parametric equations

$$x = 3t + 2, \qquad y = 0t - 4, \qquad z = t + 0.$$

Let $\mathbf{P} = 2\mathbf{i} - 4\mathbf{j}, \qquad \mathbf{D} = 3\mathbf{i} + \mathbf{k},$

then L has the vector equation

$$\mathbf{X} = \mathbf{P} + t\mathbf{D}, \quad \text{or} \quad \mathbf{X} = (2\mathbf{i} - 4\mathbf{j}) + t(3\mathbf{i} + \mathbf{k}).$$

L is shown in Figure 10.3.6.

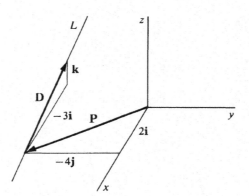

Figure 10.3.6

If A is a point on L, let us call $\mathbf{A}$ a *position vector* of L. A vector $\mathbf{D}$ is said to be a *direction vector* of L if $\mathbf{D}$ is the vector from one point of L to another point of L. Thus if $\mathbf{A}$ and $\mathbf{B}$ are distinct position vectors of L, then $\mathbf{B} - \mathbf{A}$ is a direction vector of L (Figure 10.3.7).

The next theorem shows that a line in space is uniquely determined by a position vector and a direction vector. That is, if two lines L and M have a position vector and direction vector in common, then L and M must be the same line.

THEOREM 2

Given a vector $\mathbf{P}$ and a nonzero vector $\mathbf{D}$, the line $\mathbf{X} = \mathbf{P} + t\mathbf{D}$ is the unique line with position vector $\mathbf{P}$ and direction vector $\mathbf{D}$.

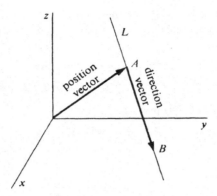

Figure 10.3.7

PROOF Let L be the line $\mathbf{X} = \mathbf{P} + t\mathbf{D}$. Setting $t = 0$ and $t = 1$ we see that $\mathbf{P}$ and $\mathbf{P} + \mathbf{D}$ are position vectors of L, so $\mathbf{D}$ is a direction vector of L.

Let $\mathbf{X} = \mathbf{Q} + s\mathbf{E}$ be any line M with position vector $\mathbf{P}$ and direction vector $\mathbf{D}$. We show $\mathbf{X} = \mathbf{Q} + s\mathbf{E}$ is another vector equation for L. For some s_0,

$$\mathbf{P} = \mathbf{Q} + s_0\mathbf{E}.$$

Also, $\mathbf{D} = \mathbf{B} - \mathbf{A}$ for some position vectors of M,

$$\mathbf{A} = \mathbf{Q} + s_1\mathbf{E}, \qquad \mathbf{B} = \mathbf{Q} + s_2\mathbf{E}.$$

Thus $\qquad\qquad \mathbf{D} = (\mathbf{Q} + s_2\mathbf{E}) - (\mathbf{Q} + s_1\mathbf{E}) = (s_2 - s_1)\mathbf{E}.$

Since $\mathbf{D} \neq 0$, $s_2 - s_1 \neq 0$. Thus the following are equivalent:

$$\mathbf{X} = \mathbf{P} + t\mathbf{D} \qquad\qquad \text{for some } t,$$
$$\mathbf{X} = \mathbf{Q} + s_0\mathbf{E} + t(s_2 - s_1)\mathbf{E} \qquad \text{for some } t,$$
$$\mathbf{X} = \mathbf{Q} + (s_0 + ts_2 - ts_1)\mathbf{E} \qquad \text{for some } t,$$
$$\mathbf{X} = \mathbf{Q} + s\mathbf{E} \qquad\qquad \text{for some } s.$$

COROLLARY 2

Two points in space determine a line. The line through A and B has the vector equation

$$\mathbf{X} = \mathbf{A} + t(\mathbf{B} - \mathbf{A}).$$

PROOF A line L passes through A and B if and only if $\mathbf{A}$ is a position vector of L and $\mathbf{B} - \mathbf{A}$ is a direction vector of L. By Theorem 2, this happens if and only if L is the line with the vector equation $\mathbf{X} = \mathbf{A} + t(\mathbf{B} - \mathbf{A})$.

EXAMPLE 5 Find a vector equation of the line through the points

$$A(3, -4, 2), \qquad B(0, 8, 1).$$

The line has the equation

$$\mathbf{X} = 3\mathbf{i} - 4\mathbf{j} + 2\mathbf{k} + t((0 - 3)\mathbf{i} + (8 - (-4))\mathbf{j} + (1 - 2)\mathbf{k}),$$
$$\mathbf{X} = 3\mathbf{i} - 4\mathbf{j} + 2\mathbf{k} + t(-3\mathbf{i} + 12\mathbf{j} - \mathbf{k}).$$

The formula $\frac{1}{2}(\mathbf{A} + \mathbf{B})$ for the midpoint of the line segment AB holds for three as well as two dimensions.

EXAMPLE 6   Find the midpoint of the line segment AB where

$$A = (1, 4, -6), \qquad B = (2, 6, 0).$$

The midpoint C has position vector

$$\mathbf{C} = \tfrac{1}{2}[(\mathbf{i} + 4\mathbf{j} - 6\mathbf{k}) + (2\mathbf{i} + 6\mathbf{j})] = \tfrac{3}{2}\mathbf{i} + 5\mathbf{j} - 3\mathbf{k}.$$

Thus $C = (\tfrac{3}{2}, 5, -3)$.

PROBLEMS FOR SECTION 10.3

In Problems 1–3, find the vector represented by the directed line segment $\vec{PQ}$.

1    $P = (0, 0, 1), \quad Q = (5, -1, 8)$
2    $P = (5, 10, 0), \quad Q = (4, 10, 1)$
3    $P = (7, -2, 4), \quad Q = (7, -2, 3)$

In Problems 4–6, find the point Q such that $\mathbf{A}$ is the vector from P to Q.

4    $P = (4, 6, -4), \quad \mathbf{A} = \mathbf{i} + \mathbf{j} + \mathbf{k}$
5    $P = (1, -2, 3), \quad \mathbf{A} = -\mathbf{i} + 2\mathbf{j} - 3\mathbf{k}$
6    $P = (0, 0, 0), \quad \mathbf{A} = -3\mathbf{i} - 4\mathbf{j} + 2\mathbf{k}$

In Problems 7–22, find the given vector or scalar where

$$\mathbf{A} = \mathbf{i} - 2\mathbf{j} + 2\mathbf{k}, \qquad \mathbf{B} = 2\mathbf{i} + 3\mathbf{j} - 6\mathbf{k}$$

7	$\mathbf{A} + \mathbf{B}$	8	$\mathbf{A} - \mathbf{B}$
9	$\mathbf{B} - \mathbf{A}$	10	$4\mathbf{A}$
11	$-\mathbf{B}$	12	$-3\mathbf{A} + 4\mathbf{B}$
13	$|\mathbf{A}|$	14	$|\mathbf{B}|$
15	$|\mathbf{A} + \mathbf{B}|$	16	$|\mathbf{B} - \mathbf{A}|$

17   The angle between $\mathbf{A}$ and $\mathbf{B}$
18   The angle between $\mathbf{A}$ and $\mathbf{A} + \mathbf{B}$
19   The angle between $\mathbf{A}$ and $3\mathbf{A}$
20   The angle between $\mathbf{A}$ and $-2\mathbf{A}$
21   The unit vector and direction cosines of $\mathbf{A}$
22   The unit vector and direction cosines of $\mathbf{B}$
23   Find the vector with length 6 and direction cosines $(-1/2, 1/2, 1/\sqrt{2})$.
24   Find the vector with length $\sqrt{3}$ and direction cosines $(1/\sqrt{3}, 1/\sqrt{3}, 1/\sqrt{3})$.
25   If $\frac{1}{3}$ and $\frac{2}{3}$ are two of the direction cosines of a vector, what are the two possible values for the third direction cosine?
26   If the three forces

$$\mathbf{F}_1 = \mathbf{i} + 2\mathbf{j} + 3\mathbf{k}, \qquad \mathbf{F}_2 = 3\mathbf{i} - \mathbf{j} - \mathbf{k}, \qquad \mathbf{F}_3 = 4\mathbf{k}$$

are acting on an object, find the total force.

27   If a force $\mathbf{F} = 6\mathbf{i} - 10\mathbf{j} + 2\mathbf{k}$ is acting on an object of mass 20, find its acceleration vector.

28   If a trader has the initial commodity vector $\mathbf{A} = 15\mathbf{i} + 20\mathbf{j} + 30\mathbf{k}$ and buys the commodity vector $\mathbf{B} = 2\mathbf{i} + \mathbf{k}$, find his new commodity vector.

29   If three commodities have the original price vector $\mathbf{P} = 100\mathbf{i} + 200\mathbf{j} + 500\mathbf{k}$ and all prices increase 25%, find the new price vector.

In Problems 30–35, find a vector equation for the given line.

30   The line with parametric equations $x = -t$, $y = 1 + \sqrt{2}t$, $z = 6 - 8t$.

31   The line with parametric equations $x = 1 + t$, $y = 3$, $z = 1 - t$.

32   The line through the points $P(0, 0, 0)$, $Q(1, 2, 3)$.

33   The line through the points $P(-1, 4, 3)$, $Q(-2, -3, 6)$.

34   The line through the point $P(4, 4, 5)$ with direction cosines $(1/\sqrt{6}, \sqrt{2}/\sqrt{6}, \sqrt{3}/\sqrt{6})$.

35   The line through the origin with direction cosines $(-\frac{3}{5}, 0, \frac{4}{5})$.

36   Find the midpoint of the line segment AB where $A = (-6, 3, 1)$, $B = (0, -4, 0)$.

37   Find the midpoint of AB where $A = (1, 2, 3)$, $B = (-1, 2, 7)$.

38   Find the midpoint of AB where $A = (6, 8, 10)$, $B = (-6, -8, -10)$.

39   Prove that if two sides of a triangle in space have equal lengths, then the angles opposite them are equal.

☐ 40   Prove that if θ is the angle between $\mathbf{A}$ and $\mathbf{B}$ then $\pi - \theta$ is the angle between $\mathbf{A}$ and $-\mathbf{B}$. *Hint*: Show that the sum of the cosines is zero.

☐ 41   Prove the Triangle Inequality for three dimensions.

☐ 42   Use the Triangle Inequality to prove that if $\mathbf{A}$ and $\mathbf{B}$ are two nonzero vectors then

$$-1 \leq \frac{|\mathbf{A}|^2 + |\mathbf{B}|^2 - |\mathbf{B} - \mathbf{A}|^2}{2|\mathbf{A}||\mathbf{B}|} \leq 1.$$

0.4 PRODUCTS OF VECTORS

In the preceding sections we studied the sum of two vectors and the product of a scalar and a vector. We shall now define the inner product (or scalar or dot product) of two vectors $\mathbf{A}$ and $\mathbf{B}$, denoted by $\mathbf{A} \cdot \mathbf{B}$.

The inner product arises in quite different ways in physics and economics. We first discuss an example from economics.

If the price per unit of a commodity is p, the cost of a units of the commodity is the product pa. Similarly, if a pair of commodities has price vector

$$\mathbf{P} = p_1\mathbf{i} + p_2\mathbf{j},$$

the cost of a commodity vector

$$\mathbf{A} = a_1\mathbf{i} + a_2\mathbf{j}$$

is found by adding the products of the prices and quantities,

$$\text{cost} = p_1a_1 + p_2a_2.$$

If three commodities have price vector

$$\mathbf{P} = p_1\mathbf{i} + p_2\mathbf{j} + p_3\mathbf{k},$$

the cost of a commodity vector

$$\mathbf{A} = a_1\mathbf{i} + a_2\mathbf{j} + a_3\mathbf{k}$$

is the sum of products,

$$\text{cost} = p_1 a_1 + p_2 a_2 + p_3 a_3.$$

Notice that the cost is always a scalar. The quantity

$$p_1 a_1 + p_2 a_2 + p_3 a_3$$

is the inner product of the vectors **P** and **A**.

DEFINITION

Two Dimensions   *The **inner product** of* $\mathbf{A} = a_1\mathbf{i} + a_2\mathbf{j}$ *and* $\mathbf{B} = b_1\mathbf{i} + b_2\mathbf{j}$ *is the scalar*

$$\mathbf{A} \cdot \mathbf{B} = a_1 b_1 + a_2 b_2.$$

Three Dimensions   *The **inner product** of* $\mathbf{A} = a_1\mathbf{i} + a_2\mathbf{j} + a_3\mathbf{k}$ *and* $\mathbf{B} = b_1\mathbf{i} + b_2\mathbf{j} + b_3\mathbf{k}$ *is the scalar*

$$\mathbf{A} \cdot \mathbf{B} = a_1 b_1 + a_2 b_2 + a_3 b_3.$$

Thus the cost of a commodity vector **A** at the price vector **P** is equal to the inner product of **P** and **A**, cost = **P** · **A**.

EXAMPLE 1   Compute the inner product of $\mathbf{i} - \mathbf{j} + 3\mathbf{k}$ and $\mathbf{j} + \mathbf{k}$.

$$(\mathbf{i} - \mathbf{j} + 3\mathbf{k}) \cdot (\mathbf{j} + \mathbf{k}) = 1 \cdot 0 + (-1) \cdot 1 + 3 \cdot 1 = 2.$$

EXAMPLE 2   Find the cost of one unit of commodity a, 3 units of commodity b, and 2 units of commodity c if the prices per unit are 6, 4, and 10 respectively.

$$\text{cost} = (6\mathbf{i} + 4\mathbf{j} + 10\mathbf{k}) \cdot (\mathbf{i} + 3\mathbf{j} + 2\mathbf{k})$$
$$= 6 \cdot 1 + 4 \cdot 3 + 10 \cdot 2 = 38.$$

EXAMPLE 3   Suppose a trader buys a commodity vector

$$\mathbf{A} = 40\mathbf{i} + 60\mathbf{j} + 100\mathbf{k}$$

at the price vector

$$\mathbf{P} = 3\mathbf{i} + 2\mathbf{j} + 4\mathbf{k}$$

and then sells it at the new price vector

$$\mathbf{Q} = 2\mathbf{i} + 5\mathbf{j} + 3\mathbf{k}.$$

Find his profit (or loss).

Since the trader pays **P** · **A** and receives **Q** · **A**, his profit is given by

$$\text{profit} = \mathbf{Q} \cdot \mathbf{A} - \mathbf{P} \cdot \mathbf{A}.$$

Thus $\text{profit} = (2 \cdot 40 + 5 \cdot 60 + 3 \cdot 100) - (3 \cdot 40 + 2 \cdot 60 + 4 \cdot 100)$
$$= 40.$$

A positive number indicates a profit and a negative number indicates a loss.

EXAMPLE 4   A buyer has \$7500 and plans to buy a commodity vector **B** in the direction of the unit vector

$$\mathbf{U} = \tfrac{2}{3}\mathbf{i} + \tfrac{2}{3}\mathbf{j} + \tfrac{1}{3}\mathbf{k}.$$

Find the largest such commodity vector **B** which he can buy if the price vector is

$$\mathbf{P} = 2\mathbf{i} + 5\mathbf{j} + \mathbf{k}.$$

We must have $\mathbf{B} = t\mathbf{U}$ for some positive t, and also

$$\mathbf{P} \cdot \mathbf{B} = 7500.$$

We solve for t.

$$7500 = \mathbf{P} \cdot \mathbf{B} = \mathbf{P} \cdot t\mathbf{U} = t(\mathbf{P} \cdot \mathbf{U}).$$

$$t = \frac{7500}{\mathbf{P} \cdot \mathbf{U}} = \frac{7500}{2 \cdot \tfrac{2}{3} + 5 \cdot \tfrac{2}{3} + 1 \cdot \tfrac{1}{3}} = \frac{7500}{5} = 1500.$$

Thus $\mathbf{B} = t\mathbf{U} = 1000\mathbf{i} + 1000\mathbf{j} + 500\mathbf{k}.$

Another illustration of an inner product is the notion of *work* in physics. Suppose a force vector **F** acts on an object which moves in a straight line with a displacement vector **S**. If the force **F** has the same direction as the displacement **S**, i.e., the angle θ between **F** and **S** is zero, work is simply the product of the magnitudes of **F** and **S**;

$$W = |\mathbf{F}||\mathbf{S}| \qquad \text{if } \theta = 0.$$

In general, work depends on the *component* of the force in the direction of the displacement, that is, the product $|\mathbf{F}| \cos \theta$. The geometric meaning of this component is shown in Figure 10.4.1.

Work is defined as the product of the component of force in the direction of **S** and the length of **S**, so

$$W = |\mathbf{F}||\mathbf{S}| \cos \theta.$$

Work is thus a scalar quantity. It is positive if the angle θ is less than 90°, zero if $\theta = 90°$, and negative if $\theta > 90°$. Our first theorem shows that work is equal to the inner product of **F** and **S**,

$$W = \mathbf{F} \cdot \mathbf{S}.$$

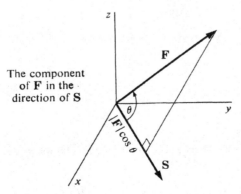

The component of **F** in the direction of **S**

Figure 10.4.1

THEOREM 1

If θ is the angle between two nonzero vectors **A** *and* **B**, *then* $\mathbf{A} \cdot \mathbf{B} = |\mathbf{A}|\,|\mathbf{B}|\cos\theta$.

PROOF We give the proof in two dimensions. By the Law of Cosines,

$$\cos\theta = \frac{|\mathbf{A}|^2 + |\mathbf{B}|^2 - |\mathbf{B} - \mathbf{A}|^2}{2|\mathbf{A}|\,|\mathbf{B}|}$$

$$= \frac{(a_1^2 + a_2^2) + (b_1^2 + b_2^2) - [(b_1 - a_1)^2 + (b_2 - a_2)^2]}{2|\mathbf{A}|\,|\mathbf{B}|}$$

$$= \frac{2b_1a_1 + 2b_2a_2}{2|\mathbf{A}|\,|\mathbf{B}|} = \frac{\mathbf{A}\cdot\mathbf{B}}{|\mathbf{A}|\,|\mathbf{B}|}.$$

Multiplying through by $|\mathbf{A}|\,|\mathbf{B}|$, we have

$$\mathbf{A}\cdot\mathbf{B} = |\mathbf{A}|\,|\mathbf{B}|\cos\theta.$$

EXAMPLE 5   A lawnmower is moved horizontally (in the x direction) a distance of 10 feet. Find the work done if the lawnmower is pushed by a force **F** where

(a) $|\mathbf{F}| = 15$ pounds, $\theta = 30°$. (See Figure 10.4.2a.)
(b) $\mathbf{F} = 8\mathbf{i} - 5\mathbf{j}$, in pounds. (See Figure 10.4.2b.)

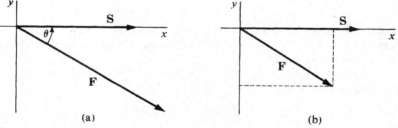

(a)   (b)

Figure 10.4.2

(a) $\cos\theta = \frac{1}{2}\sqrt{3}.\ |\mathbf{S}| = 10$.

$$W = |\mathbf{F}|\,|\mathbf{S}|\cos\theta = 15\cdot 10\cdot\tfrac{1}{2}\sqrt{3} = 75\sqrt{3}\ \text{ft lbs}.$$

(b) $W = \mathbf{F}\cdot\mathbf{S} = 8\cdot 10 + (-5)\cdot 0 = 80$ ft lbs.

The angle between two vectors can be easily computed using the inner product.

COROLLARY

If **A** *and* **B** *are nonzero vectors, the angle* θ *between them has cosine*

$$\cos\theta = \frac{\mathbf{A}\cdot\mathbf{B}}{|\mathbf{A}|\,|\mathbf{B}|}.$$

EXAMPLE 6 Find the angle between the vectors
$$A = 3i + j - k, \qquad B = -i + 5j + k.$$
$$\cos \theta = \frac{A \cdot B}{|A||B|} = \frac{3(-1) + 1 \cdot 5 + (-1) \cdot 1}{\sqrt{3^2 + 1^2 + 1^2}\sqrt{1^2 + 5^2 + 1^2}} = \frac{1}{\sqrt{11 \cdot 27}}.$$
$$\theta = \arccos \frac{1}{\sqrt{11 \cdot 27}}.$$

Here is a list of algebraic rules for inner products. All the rules are easy to prove in either two or three dimensions.

THEOREM 2 (Algebraic Rules for Inner Products)

    (i) $A \cdot i = a_1, \quad A \cdot j = a_2, \quad A \cdot k = a_3.$

    (ii) $A \cdot 0 = 0 \cdot A = 0.$

    (iii) $A \cdot B = B \cdot A$ (*Commutative Law*).

    (iv) $A \cdot (B + C) = A \cdot B + A \cdot C$ (*Distributive Law*).

    (v) $(tA) \cdot B = t(A \cdot B)$ (*Associative Law*).

    (vi) $A \cdot A = |A|^2.$

PROOF Rule (vi) is proved as follows in three dimensions.
$$A \cdot A = a_1 a_1 + a_2 a_2 + a_3 a_3 = a_1^2 + a_2^2 + a_3^2 = |A|^2.$$

Inner products are useful in the study of perpendicular and parallel vectors.

DEFINITION

*Two nonzero vectors A and B are said to be **perpendicular** (or **orthogonal**), A ⊥ B, if the angle between them is π/2. A and B are said to be **parallel**, A ∥ B, if the angle between them is either 0 or π.*

TEST FOR PERPENDICULARS

Let A and B be nonzero vectors. Then A ⊥ B if and only if A · B = 0.

PROOF The following are equivalent:
$$A \cdot B = 0, \qquad \frac{A \cdot B}{|A||B|} = 0, \qquad \cos \theta = 0, \qquad \theta = \pi/2.$$

TEST FOR PARALLELS

Given two nonzero vectors A and B, the following are equivalent:

    (i) $A \parallel B.$

    (ii) $|A \cdot B| = |A||B|.$

    (iii) A *is a scalar multiple of* $B.$

PROOF To show that (i) is equivalent to (ii), we note that the following are equivalent.

$$\mathbf{A} \parallel \mathbf{B}.$$
$$\cos \theta = \pm 1.$$
$$\frac{\mathbf{A} \cdot \mathbf{B}}{|\mathbf{A}||\mathbf{B}|} = \pm 1.$$
$$\mathbf{A} \cdot \mathbf{B} = \pm |\mathbf{A}||\mathbf{B}|.$$
$$|\mathbf{A} \cdot \mathbf{B}| = |\mathbf{A}||\mathbf{B}|.$$

We now show that (i) implies (iii), and (iii) implies (i). Assume (i), $\mathbf{A} \parallel \mathbf{B}$.

Case 1 $\theta = 0$. Let $\mathbf{U}$ and $\mathbf{V}$ be the unit vectors of $\mathbf{A}$ and $\mathbf{B}$. By the Law of Cosines,

$$\cos \theta = \frac{|\mathbf{U}|^2 + |\mathbf{V}|^2 - |\mathbf{V} - \mathbf{U}|^2}{2|\mathbf{U}||\mathbf{V}|},$$

$$1 = \frac{2 - |\mathbf{V} - \mathbf{U}|^2}{2},$$

$$|\mathbf{V} - \mathbf{U}| = 0,$$

$$\frac{\mathbf{A}}{|\mathbf{A}|} = \mathbf{U} = \mathbf{V} = \frac{\mathbf{B}}{|\mathbf{B}|},$$

$$\mathbf{A} = \frac{|\mathbf{A}|}{|\mathbf{B}|}\mathbf{B}.$$

Case 2 $\theta = \pi$. We see, by a similar proof, that $\mathbf{A} = -\dfrac{|\mathbf{A}|}{|\mathbf{B}|}\mathbf{B}$. In either case, $\mathbf{A}$ is a scalar multiple of $\mathbf{B}$.

Finally, assume (iii), say $\mathbf{A} = t\mathbf{B}$. Then

$$\cos \theta = \frac{\mathbf{A} \cdot \mathbf{B}}{|\mathbf{A}||\mathbf{B}|} = \frac{t\mathbf{B} \cdot \mathbf{B}}{|t\mathbf{B}||\mathbf{B}|} = \frac{t(\mathbf{B} \cdot \mathbf{B})}{|t||\mathbf{B}||\mathbf{B}|} = \frac{t|\mathbf{B}|^2}{|t||\mathbf{B}|^2} = \pm 1.$$

Therefore $\theta = 0$ or $\theta = \pi$, so $\mathbf{A} \parallel \mathbf{B}$.

EXAMPLE 7   Test for $\mathbf{A} \perp \mathbf{B}$ and $\mathbf{A} \parallel \mathbf{B}$ using the inner product.

(a) $\mathbf{A} = 3\mathbf{i} + \mathbf{j} - \mathbf{k}$, $\mathbf{B} = \mathbf{i} - 3\mathbf{j} + \mathbf{k}$.

We compute $\mathbf{A} \cdot \mathbf{B}$ and $|\mathbf{A}||\mathbf{B}|$.

$$\mathbf{A} \cdot \mathbf{B} = -1, \qquad |\mathbf{A}||\mathbf{B}| = 11.$$

Since $\mathbf{A} \cdot \mathbf{B} \neq 0$, not $\mathbf{A} \perp \mathbf{B}$.

Since $\mathbf{A} \cdot \mathbf{B} \neq \pm |\mathbf{A}||\mathbf{B}|$, not $\mathbf{A} \parallel \mathbf{B}$

(b) $\mathbf{A} = 2\mathbf{i} - \sqrt{3}\mathbf{j} + \mathbf{k}$, $\mathbf{B} = -\sqrt{8}\mathbf{i} + \sqrt{6}\mathbf{j} - \sqrt{2}\mathbf{k}$.

$$\mathbf{A} \cdot \mathbf{B} = -8\sqrt{2}, \qquad |\mathbf{A}||\mathbf{B}| = 8\sqrt{2}.$$

Therefore $\mathbf{A} \parallel \mathbf{B}$.

(c) $\mathbf{A} = 3\mathbf{i} + \mathbf{j} - \mathbf{k}$, $\mathbf{B} = \mathbf{i} - 3\mathbf{j}$.

$\mathbf{A} \cdot \mathbf{B} = 0$. Therefore $\mathbf{A} \perp \mathbf{B}$.

Figure 10.4.3 illustrates this example.

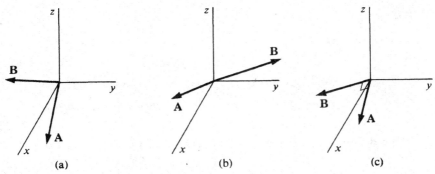

Figure 10.4.3

We conclude this section with a theorem about perpendicular vectors, first in the plane and then in space.

THEOREM 3

Let $\mathbf{A} = a_1\mathbf{i} + a_2\mathbf{j}$ be a nonzero vector in the plane.

(i)   The vector $\mathbf{B} = a_2\mathbf{i} - a_1\mathbf{j}$ is perpendicular to $\mathbf{A}$.

(ii) Any vector perpendicular to $\mathbf{A}$ is parallel to $\mathbf{B}$.

PROOF   (i)   We compute $\mathbf{A} \cdot \mathbf{B} = a_1 a_2 + a_2(-a_1) = 0$.

(ii) If $\mathbf{C} \perp \mathbf{A}$, then both $\mathbf{B}$ and $\mathbf{C}$ make angles of $\pi/2$ with $\mathbf{A}$, so the angle between $\mathbf{B}$ and $\mathbf{C}$ is either 0 or π. Therefore $\mathbf{B} \parallel \mathbf{C}$.

EXAMPLE 8   Find a vector perpendicular to $\mathbf{A} = 4\mathbf{i} - 7\mathbf{j}$.

Answer $\mathbf{B} = -7\mathbf{i} - 4\mathbf{j}$ (Figure 10.4.4).

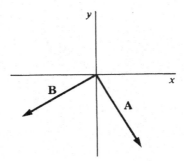

Figure 10.4.4

Theorem 3 raises the following problem about vectors in space. *Given two vectors $\mathbf{A}$ and $\mathbf{B}$ which are neither zero nor parallel, find a third vector $\mathbf{C}$ which is perpendicular to both $\mathbf{A}$ and $\mathbf{B}$.* For example, if $\mathbf{A}$ is $\mathbf{i}$ and $\mathbf{B}$ is $\mathbf{j}$, then the vector $\mathbf{k}$ is perpendicular to both $\mathbf{i}$ and $\mathbf{j}$. So is any scalar multiple of $\mathbf{k}$. In general it is not easy to see how to find a vector perpendicular to both $\mathbf{A}$ and $\mathbf{B}$. In fact, to solve the problem we need a new kind of product of vectors, the vector product

$$\mathbf{A} \times \mathbf{B}.$$

DEFINITION

Given two vectors

$$\mathbf{A} = a_1\mathbf{i} + a_2\mathbf{j} + a_3\mathbf{k}, \qquad \mathbf{B} = b_1\mathbf{i} + b_2\mathbf{j} + b_3\mathbf{k}$$

*in space, the **vector product** (or **cross product**) is the new vector*

$$\mathbf{A} \times \mathbf{B} = (a_2b_3 - a_3b_2)\mathbf{i} + (a_3b_1 - a_1b_3)\mathbf{j} + (a_1b_2 - a_2b_1)\mathbf{k}.$$

This definition can be remembered by writing down the determinant

$$\mathbf{A} \times \mathbf{B} = \begin{vmatrix} \mathbf{i} & \mathbf{j} & \mathbf{k} \\ a_1 & a_2 & a_3 \\ b_1 & b_2 & b_3 \end{vmatrix}.$$

The positive and negative terms of $\mathbf{A} \times \mathbf{B}$ are the products of the diagonals shown in Figure 10.4.5.

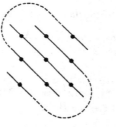

Figure 10.4.5   Positive terms   Negative terms

EXAMPLE 9   Find $\mathbf{A} \times \mathbf{B}$ where

$$\mathbf{A} = 4\mathbf{i} - \mathbf{j} + \mathbf{k}, \qquad \mathbf{B} = 2\mathbf{j} - \mathbf{k}.$$

$$\mathbf{A} \times \mathbf{B} = \begin{vmatrix} \mathbf{i} & \mathbf{j} & \mathbf{k} \\ 4 & -1 & 1 \\ 0 & 2 & -1 \end{vmatrix}$$
$$= ((-1)(-1) - 1 \cdot 2)\mathbf{i} + (1 \cdot 0 - 4(-1))\mathbf{j} + (4 \cdot 2 - (-1) \cdot 0)\mathbf{k}$$
$$= -\mathbf{i} + 4\mathbf{j} + 8\mathbf{k}.$$

$\mathbf{A}$, $\mathbf{B}$ and $\mathbf{A} \times \mathbf{B}$ are shown in Figure 10.4.6.

The vector products of the unit vectors $\mathbf{i}$, $\mathbf{j}$, and $\mathbf{k}$ are

$$\mathbf{i} \times \mathbf{i} = \mathbf{0}, \qquad \mathbf{j} \times \mathbf{j} = \mathbf{0}, \qquad \mathbf{k} \times \mathbf{k} = \mathbf{0},$$
$$\mathbf{i} \times \mathbf{j} = \mathbf{k}, \qquad \mathbf{j} \times \mathbf{k} = \mathbf{i}, \qquad \mathbf{k} \times \mathbf{i} = \mathbf{j},$$
$$\mathbf{j} \times \mathbf{i} = -\mathbf{k}, \qquad \mathbf{k} \times \mathbf{j} = -\mathbf{i}, \qquad \mathbf{i} \times \mathbf{k} = -\mathbf{j}.$$

Notice that $\mathbf{A} \cdot \mathbf{B}$ is a scalar but $\mathbf{A} \times \mathbf{B}$ is a vector.

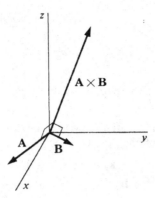

Figure 10.4.6

THEOREM 4

*Let **A** and **B** be two vectors in space which are not zero and not parallel.*

(i) **A** × **B** *is perpendicular to both **A** and **B**.*

(ii) *Any vector perpendicular to both **A** and **B** is parallel to **A** × **B**.*

PROOF (i) We compute the inner products.

$$\mathbf{A} \cdot (\mathbf{A} \times \mathbf{B}) = a_1(a_2b_3 - a_3b_2) + a_2(a_3b_1 - a_1b_3) + a_3(a_1b_2 - a_2b_1)$$
$$= a_1a_2b_3 - a_1a_3b_2 + a_2a_3b_1 - a_1a_2b_3 + a_1a_3b_2 - a_2a_3b_1$$
$$= 0.$$

Similarly $\mathbf{B} \cdot (\mathbf{A} \times \mathbf{B}) = 0$.

It remains to prove that $\mathbf{A} \times \mathbf{B} \neq \mathbf{0}$. At least one component of $\mathbf{A}$, say a_1, is nonzero. Let $t = b_1/a_1$ and let $\mathbf{C} = \mathbf{A} \times \mathbf{B}$. When we solve the equations

$$c_3 = a_1b_2 - a_2b_1, \qquad c_2 = a_3b_1 - a_1b_3$$

for b_2 and b_3, we get

$$b_1 = ta_1, \qquad b_2 = \frac{c_3}{a_1} + ta_2, \qquad b_3 = \frac{c_2}{a_1} + ta_3.$$

Since $\mathbf{B}$ is not parallel to $\mathbf{A}$, $\mathbf{B} \neq t\mathbf{A}$. Therefore at least one of c_2, c_3 is nonzero, so $\mathbf{C} \neq \mathbf{0}$.

(ii) Let $\mathbf{C} = \mathbf{A} \times \mathbf{B}$ and let $\mathbf{D}$ be any other vector perpendicular to both $\mathbf{A}$ and $\mathbf{B}$. Then

$$\mathbf{A} \cdot \mathbf{D} = a_1d_1 + a_2d_2 + a_3d_3 = 0.$$
$$\mathbf{B} \cdot \mathbf{D} = b_1d_1 + b_2d_2 + b_3d_3 = 0.$$

At least one component of $\mathbf{D}$, say d_1, is nonzero. We may then solve the above equations for a_1 and b_1,

$$a_1 = -\frac{a_2d_2 + a_3d_3}{d_1}, \qquad b_1 = -\frac{b_2d_2 + b_3d_3}{d_1}.$$

Let $s = c_1/d_1$. Then $c_1 = sd_1$. Also,

$$c_2 = a_3b_1 - a_1b_3 = -\frac{a_3(b_2d_2 + b_3d_3)}{d_1} + \frac{b_3(a_2d_2 + a_3d_3)}{d_1}$$

$$= \frac{(a_2b_3 - a_3b_2)d_2}{d_1} = \frac{c_1d_2}{d_1} = sd_2.$$

Similarly, $c_3 = sd_3$. Therefore $\mathbf{C} = s\mathbf{D}$, and $\mathbf{C}$ is parallel to $\mathbf{D}$.

Warning: The Commutative Law and the Associative Law do *not* hold for the vector product. For example,

$$\mathbf{i} \times \mathbf{j} = \mathbf{k}, \qquad \mathbf{j} \times \mathbf{i} = -\mathbf{k}$$
$$\mathbf{i} \times (\mathbf{j} \times \mathbf{j}) = \mathbf{0}, \qquad (\mathbf{i} \times \mathbf{j}) \times \mathbf{j} = -\mathbf{i}.$$

However, vector products do satisfy the Distributive Laws

$$(s\mathbf{A} + t\mathbf{B}) \times \mathbf{C} = s(\mathbf{A} \times \mathbf{C}) + t(\mathbf{B} \times \mathbf{C}),$$
$$\mathbf{C} \times (s\mathbf{A} + t\mathbf{B}) = s(\mathbf{C} \times \mathbf{A}) + t(\mathbf{C} \times \mathbf{B}).$$

The proof is left as an exercise (Problem 36 at the end of this section).

Here is a brief summary of the operations on scalars and vectors.

Addition: $s + t$ is a scalar
$\mathbf{A} + \mathbf{B}$ is a vector
$s + \mathbf{A}$ is undefined

Multiplication: st is a scalar
$s\mathbf{A}$ is a vector
$\mathbf{A} \cdot \mathbf{B}$ is a scalar
$\mathbf{A} \times \mathbf{B}$ is a vector

Division: s/t is a scalar
$\mathbf{A}/t$ is a vector
$s/\mathbf{B}$ and $\mathbf{A}/\mathbf{B}$ are undefined

Absolute value and length : $|s|$ is a scalar
$|\mathbf{A}|$ is a scalar

One must be careful in forming longer expressions. For example,

$$\mathbf{A} \cdot (\mathbf{B} + \mathbf{C}) \text{ is a scalar,}$$
$$(\mathbf{A} \cdot \mathbf{B}) + \mathbf{C} \text{ is undefined,}$$
$$(\mathbf{A} \cdot \mathbf{B})\mathbf{C} \text{ is a vector.}$$

PROBLEMS FOR SECTION 10.4

In Problems 1–11, (a) compute $\mathbf{A} \cdot \mathbf{B}$, (b) test whether $\mathbf{A}$ is perpendicular or parallel to $\mathbf{B}$, and (c) find the cosine of the angle between $\mathbf{A}$ and $\mathbf{B}$ using $\mathbf{A} \cdot \mathbf{B}$.

1    $\mathbf{A} = \mathbf{i} - 3\mathbf{j}, \quad \mathbf{B} = 2\mathbf{i} - 6\mathbf{j}$

2    $\mathbf{A} = 4\mathbf{i} - \mathbf{j}, \quad \mathbf{B} = \mathbf{i} - \mathbf{j}$

3    $\mathbf{A} = \mathbf{i} + \mathbf{j} + \mathbf{k}, \quad \mathbf{B} = \mathbf{i} - \mathbf{k}$

4	$A = i + k, \quad B = j$
5	$A = 6i - j, \quad B = 2i - 12j$
6	$A = 5i - j + k, \quad B = -5i + j - k$
7	$A = i + 4j - 10k, \quad B = 4i + j + 10k$
8	$A = i + 2j + 3k, \quad B = 2i - 3j + k$
9	$A = \sqrt{2}i + \sqrt{3}j + k, \quad B = \sqrt{2}i - \sqrt{3}j + k$
10	$A = \sqrt{3}i + \sqrt{6}j - k, \quad B = \sqrt{3}i - \sqrt{6}j - 3k$
11	$A = i - \sqrt{5}j + \sqrt{2}k, \quad B = \sqrt{5}i - 5j + \sqrt{10}k$
12	Which of the following are vectors, which are scalars, and which are undefined?

(a) $s(A + B)$ (b) $(s + t)A$

(c) $(sA) \cdot (tB)$ (d) $s + (tA)$

(e) $s(A \cdot B)$ (f) $A + (B \cdot C)$

(g) $A \times (B + C)$ (h) $A \times (B \cdot C)$

(i) $A \cdot (B \times C)$ (j) $(A \times B) + C$

13   Find the cost of the commodity vector $A = 15i + 4j + 6k$ at the price vector $P = i + 2j + 3k$.

14   Find the profit or loss if a trader buys the commodity vector $A = 3i + 16j + 4k$ at the price vector $P = 2i + 4j + 6k$ and sells it at the price vector $3i + 2j + 10k$.

15   A trader initially has the commodity vector $A = i + 3j + 6k$. He sells his whole commodity vector at the price $P = 3i + j + 2k$ and uses the revenue from this sale to buy an equal amount of each commodity. Find his new commodity vector.

16   Find the amount of work done by the force vector $F = 3i - j - 4k$ acting along the displacement vector $S = 5i + 3j + k$.

17   Find the work done by a force vector of magnitude 10 acting along a displacement of length 40 if the angle between the force and displacement is 45°.

18   Prove that the basis vectors i, j, k are perpendicular.

19   Find a vector in the plane perpendicular to $A = i + j$.

20   Find a vector in the plane perpendicular to $A = 2i - 9j$.

21   Compute $A \times B$ where $A = i - 3j + k, B = -i - j + k$.

22   Compute $A \times B$ where $A = i - j + k, B = i + j + k$.

23   Compute $A \times B$ where $A = i + k, B = j - k$.

24   Find a vector perpendicular to both $A = i + j - k, B = i - j + k$.

25   Find a vector perpendicular to both $A = i + 2j + 3k, B = i + 3j + 4k$.

26   Find a vector perpendicular to both $A = -i - 4j + k, B = j - 2k$.

27   Find a unit vector in the plane perpendicular to $A = 3i - 4j$.

28   Find a unit vector in the plane perpendicular to $A = 2i - j$.

29   Find a unit vector perpendicular to both $A = i + j, B = k$.

30   Find a unit vector perpendicular to both $A = 2i + 3k, B = -i + j - k$.

31   Find the angle between two long diagonals of a cube.

32   Find the angle between a long diagonal and a diagonal along a face of a cube.

33   Find the angle between the diagonals of two adjacent faces of a cube.

34   Show that the inner product of two unit vectors is equal to the cosine of the angle between them.

35   Use inner products to prove that the diagonals of a rhombus (a parallelogram whose sides have equal lengths) are perpendicular.

☐ **36**   Prove the Distributive Law for vector products,

$$(sA + tB) \times C = s(A \times C) + t(B \times C).$$

☐ **37**   Prove the Anticommutative Law for vector products,

$$B \times A = -(A \times B).$$

☐ **38**   Prove that $A \parallel B$ if and only if $A \times B = 0$ (where A, B are nonzero).

☐ **39**   Show that the length of $A \times B$ is equal to the area of the parallelogram with sides A and B, in symbols

$$|A \times B| = |A||B| \sin \theta.$$

☐ **40**   Prove that the "scalar triple product" $A \cdot (B \times C)$ is equal to the volume of a parallelopiped with edges A, B, and C.

10.5 PLANES IN SPACE

DEFINITION

> A **plane** in space is the graph of an equation of the form
>
> $$ax + by + cz = d$$
>
> where a, b, c are not all zero.

The simplest planes are those where two of the numbers a, b, c are zero.

The plane $x = d$ is parallel to the yz-plane.
The plane $y = d$ is parallel to the xz-plane.
The plane $z = d$ is parallel to the xy-plane.

These three cases are illustrated in Figure 10.5.1.

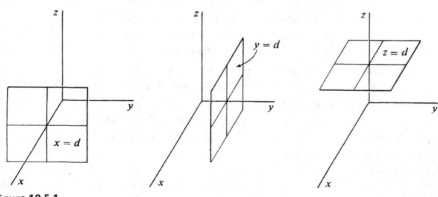

Figure 10.5.1

The examples below show how to draw sketches to help visualize other planes. The idea is to use the points where the plane cuts the coordinate axes, and to draw a triangular or rectangular portion of the plane.

EXAMPLE 1 (For sketching a plane where a, b, c and d are nonzero.) Sketch the plane $x + 2y + z = 2$.

Step 1 Find the points where the plane crosses the coordinate axes.

$$x\text{-axis}: \quad \text{When } y = z = 0, \quad x = 2.$$

The plane crosses the x-axis at $(2, 0, 0)$.

$$y\text{-axis}: \quad \text{When } x = z = 0, \quad y = 1.$$

The plane crosses the y-axis at $(0, 1, 0)$.

$$z\text{-axis}: \quad \text{When } x = y = 0, \quad z = 2.$$

The plane crosses the z-axis at $(0, 0, 2)$.

Step 2 Draw the triangle connecting these three points, as shown in Figure 10.5.2. This triangle lies in the plane.

EXAMPLE 2 (For sketching a plane where two of a, b, c are nonzero and $d \neq 0$.) Sketch the plane $2x + z = 4$.

Step 1 Find the points where the plane crosses the x- and z-axes.
The plane crosses the x-axis at $(2, 0, 0)$.
The plane crosses the z-axis at $(0, 0, 4)$.

Step 2 The plane is parallel to the y-axis. Draw a rectangle with two sides parallel to the y-axis and two sides parallel to the line segment from $(2, 0, 0)$ to $(0, 0, 4)$, as in Figure 10.5.3. This rectangle lies in the plane.

EXAMPLE 3 (For sketching a plane with $d = 0$.) Sketch the plane $x + 2y - z = 0$.

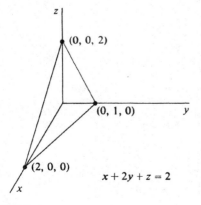

$$x + 2y + z = 2$$

Figure 10.5.2

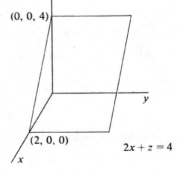

$$2x + z = 4$$

Figure 10.5.3

Step 1 The plane passes through the origin because $(0, 0, 0)$ is a solution of the equation. Find another point where $x = 0$ and a third point where y or $z = 0$,

$$x = 0, \qquad y = 1, \qquad z = 2,$$
$$x = 1, \qquad y = 0, \qquad z = 1.$$

Step 2 Connect the points $(0, 0, 0)$, $(0, 1, 2)$, $(1, 0, 1)$ to form a triangle which lies in the plane, as in Figure 10.5.4.

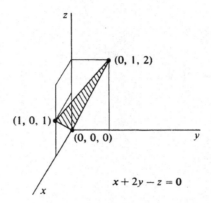

Figure 10.5.4

A *position vector* of a plane p is a vector $\mathbf{P}$ such that P is a point on the plane. A *direction vector* of p is a *vector* $\mathbf{D}$ from one point of p to another. A *normal vector* of p is a vector $\mathbf{N}$ which is perpendicular to every direction vector of p. These vectors are illustrated in Figure 10.5.5.

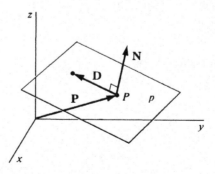

Figure 10.5.5   Position, Direction, and Normal Vectors

We shall often find it convenient to write a scalar equation

$$ax + by + cz = d$$

for a plane in vector form,

$$(a\mathbf{i} + b\mathbf{j} + c\mathbf{k}) \cdot \mathbf{X} = d.$$

We call this a *vector equation* for the plane.

THEOREM 1

    (i)   *A vector is normal to a plane* $\mathbf{N} \cdot \mathbf{X} = d$ *if and only if it is parallel to* $\mathbf{N}$.

    (ii) *There is a unique plane with a given normal vector* $\mathbf{N}$ *and position vector* $\mathbf{P}$, *and it has the vector equation*

$$\mathbf{N} \cdot \mathbf{X} = \mathbf{N} \cdot \mathbf{P}.$$

PROOF   (i) Call the plane p. For any direction vector $\mathbf{D} = \mathbf{Q} - \mathbf{P}$ of p, we have

$$\mathbf{N} \cdot \mathbf{D} = \mathbf{N} \cdot (\mathbf{Q} - \mathbf{P}) = \mathbf{N} \cdot \mathbf{Q} - \mathbf{N} \cdot \mathbf{P} = d - d = 0.$$

Let $\mathbf{M}$ be parallel to $\mathbf{N}$, say $\mathbf{M} = s\mathbf{N}$.

$$\mathbf{M} \cdot \mathbf{D} = (s\mathbf{N}) \cdot \mathbf{D} = s(\mathbf{N} \cdot \mathbf{D}) = 0.$$

Hence $\mathbf{M} \perp \mathbf{D}$ and $\mathbf{M}$ is normal to p.

Now suppose $\mathbf{M}$ is normal to p. Let $\mathbf{C}$ and $\mathbf{D}$ be two nonparallel direction vectors of p. Then both $\mathbf{M}$ and $\mathbf{N}$ are perpendicular to $\mathbf{C}$ and $\mathbf{D}$. Therefore $\mathbf{M}$ and $\mathbf{N}$ are parallel to $\mathbf{C} \times \mathbf{D}$ and hence parallel to each other.

    (ii) Set $d = \mathbf{N} \cdot \mathbf{P}$. The plane p with the equation $\mathbf{N} \cdot \mathbf{X} = d$ has position vector $\mathbf{P}$ and normal vector $\mathbf{N}$ by (i).

To show p is unique let q be any plane with position vector $\mathbf{P}$ and normal vector $\mathbf{N}$. q has a vector equation $\mathbf{M} \cdot \mathbf{X} = e$. By (i), $\mathbf{N}$ is parallel to $\mathbf{M}$, say $\mathbf{N} = s\mathbf{M}$. Then the following equations are equivalent for all $\mathbf{X}$:

$$\mathbf{N} \cdot \mathbf{X} = \mathbf{N} \cdot \mathbf{P} = d.$$
$$(s\mathbf{M}) \cdot \mathbf{X} = (s\mathbf{M}) \cdot \mathbf{P}.$$
$$s(\mathbf{M} \cdot \mathbf{X}) = s(\mathbf{M} \cdot \mathbf{P}).$$
$$\mathbf{M} \cdot \mathbf{X} = \mathbf{M} \cdot \mathbf{P} = e.$$

It follows that q equals p.

EXAMPLE 4   The plane $2x + 3y - z = 5$ has the normal vector

$$\mathbf{N} = 2\mathbf{i} + 3\mathbf{j} - \mathbf{k}$$

and the vector equation

$$(2\mathbf{i} + 3\mathbf{j} - \mathbf{k}) \cdot \mathbf{X} = 5.$$

EXAMPLE 5   Find the vector and scalar equations for the plane with position and normal vectors

$$\mathbf{P} = 3\mathbf{i} - \mathbf{j} - 2\mathbf{k}, \qquad \mathbf{N} = \mathbf{i} + \mathbf{j} + 4\mathbf{k}.$$

We first compute $\mathbf{N} \cdot \mathbf{P}$,

$$\mathbf{N} \cdot \mathbf{P} = 1 \cdot 3 + 1 \cdot (-1) + 4 \cdot (-2) = -6.$$

A vector equation is $(\mathbf{i} + \mathbf{j} + 4\mathbf{k}) \cdot \mathbf{X} = -6.$

A scalar equation is $x + y + 4z = -6.$

The plane is shown in Figure 10.5.6.

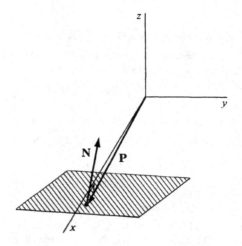

Figure 10.5.6

The vector product can sometimes be used to find a normal vector of a plane.

COROLLARY

> If **C** and **D** *are two nonparallel direction vectors of a plane p, then* **C** × **D** *is a normal vector of p.*

PROOF p has some normal vector **N**. **N** is perpendicular to both **C** and **D**, and hence parallel to **C** × **D**, so **C** × **D** is a normal vector of p.

EXAMPLE 6   Find the plane with position vector **P** = **k** and direction vectors **C** = −2**i** + **j** + **k**,   **D** = −**j**.

First we find a normal vector of the plane,

$$\mathbf{N} = \mathbf{C} \times \mathbf{D} = (1 \cdot 0 - 1 \cdot (-1))\mathbf{i} + (1 \cdot 0 - (-2) \cdot 0)\mathbf{j} + ((-2)(-1) - 1 \cdot 0)\mathbf{k}$$
$$= \mathbf{i} + 2\mathbf{k}.$$

Then $\mathbf{N} \cdot \mathbf{P} = 1 \cdot 0 + 0 \cdot 0 + 2 \cdot 1 = 2.$

The plane has the vector equation $(\mathbf{i} + 2\mathbf{k}) \cdot \mathbf{X} = 2$

and the scalar equation $x + 2z = 2.$

The plane is shown in Figure 10.5.7.

EXAMPLE 7   Find the plane through the three points

$$P(-1, 3, 1), \qquad Q(1, 2, 3), \qquad S(-1, -1, 0).$$

The plane has position vector

$$\mathbf{P} = -\mathbf{i} + 3\mathbf{j} + \mathbf{k}$$

and the two direction vectors

$$\mathbf{C} = \mathbf{Q} - \mathbf{P} = 2\mathbf{i} - \mathbf{j} + 2\mathbf{k},$$
$$\mathbf{D} = \mathbf{S} - \mathbf{P} = -4\mathbf{j} - \mathbf{k}.$$

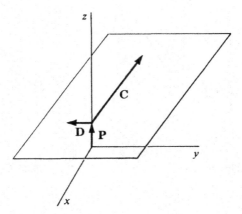

Figure 10.5.7

A normal vector of the plane is

$$\mathbf{N} = \mathbf{C} \times \mathbf{D} = ((-1)(-1) - 2(-4))\mathbf{i} + (2 \cdot 0 - 2(-1))\mathbf{j} + (2(-4) - (-1) \cdot 0)\mathbf{k}$$
$$= 9\mathbf{i} + 2\mathbf{j} - 8\mathbf{k}.$$

Then $\mathbf{N} \cdot \mathbf{P} = 9(-1) + 2 \cdot 3 + (-8) \cdot 1 = -11.$

The plane has the vector equation $(9\mathbf{i} + 2\mathbf{j} - 8\mathbf{k}) \cdot \mathbf{X} = -11$

and the scalar equation $9x + 2y - 8z = -11.$

The plane is shown in Figure 10.5.8.

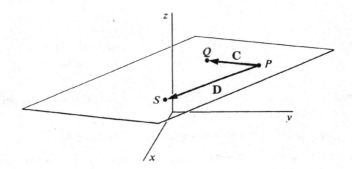

Figure 10.5.8

Two planes are said to be *parallel* if their normal vectors are parallel. A line L is said to be *parallel* to a plane p if the direction vectors of L are perpendicular to the normal vectors of p.

Two planes are said to be *perpendicular* if their normal vectors are perpendicular. A line L is said to be *perpendicular* to a plane p if the direction vectors of L are normal to p. Figure 10.5.9 illustrates these definitions.

EXAMPLE 8 Determine whether the plane $3x - 2y + z = 4$ and the line $\mathbf{X} = (3\mathbf{i} - \mathbf{j} + \mathbf{k}) + t(\mathbf{i} + \mathbf{j} - \mathbf{k})$ are parallel.

The plane has the normal vector $\mathbf{N} = 3\mathbf{i} - 2\mathbf{j} + \mathbf{k}.$

The line has the direction vector $\mathbf{D} = \mathbf{i} + \mathbf{j} - \mathbf{k}.$

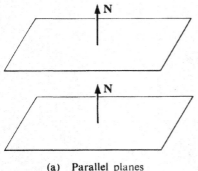

(a) Parallel planes

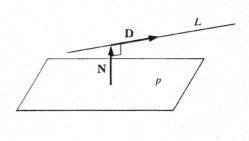

(b) A line L parallel to a plane p

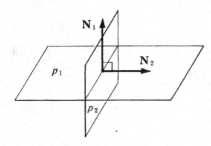

(c) Perpendicular planes

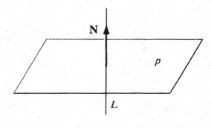

(d) A line L perpendicular to a plane p

Figure 10.5.9

We compute $\quad \mathbf{N} \cdot \mathbf{D} = 3 \cdot 1 + (-2) \cdot 1 + 1(-1) = 0.$

Therefore the plane and line are parallel (Figure 10.5.10).

EXAMPLE 9 Find the line L through the point $P(1, 2, 3)$ which is perpendicular to the plane $3x - 4y + z = 10$.

The plane has the normal vector $\mathbf{N} = 3\mathbf{i} - 4\mathbf{j} + \mathbf{k}$.

Therefore $\mathbf{N}$ is a direction vector of L, and L has the vector equation

$$\mathbf{X} = \mathbf{P} + t\mathbf{N},$$
$$= \mathbf{i} + 2\mathbf{j} + 3\mathbf{k} + t(3\mathbf{i} - 4\mathbf{j} + \mathbf{k})$$

(see Figure 10.5.11).

EXAMPLE 10 Find the plane p containing the line $\mathbf{X} = \mathbf{i} + t(\mathbf{j} + \mathbf{k})$ which is perpendicular to the plane $x + 3y - 2z = 0$.

The given plane q has the normal vector $\quad \mathbf{M} = \mathbf{i} + 3\mathbf{j} - 2\mathbf{k}$,

and the given line L has the direction vector $\quad \mathbf{D} = \mathbf{j} + \mathbf{k}$.

The required plane p must have a normal vector $\mathbf{N}$ which is perpendicular to both $\mathbf{M}$ and $\mathbf{D}$, so we take

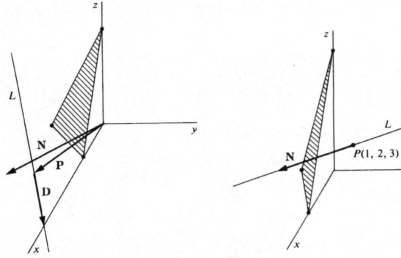

Figure 10.5.10   Figure 10.5.11

$$\mathbf{N} = \mathbf{M} \times \mathbf{D} = \begin{vmatrix} \mathbf{i} & \mathbf{j} & \mathbf{k} \\ 1 & 3 & -2 \\ 0 & 1 & 1 \end{vmatrix}, \qquad \mathbf{N} = 5\mathbf{i} - \mathbf{j} + \mathbf{k}.$$

The vector $\mathbf{P} = \mathbf{i}$ is a position vector of L and therefore a position vector of p. So p has the vector equation

$$\mathbf{N} \cdot \mathbf{X} = \mathbf{N} \cdot \mathbf{P},$$
$$(5\mathbf{i} - \mathbf{j} + \mathbf{k}) \cdot \mathbf{X} = 5,$$

and the scalar equation $\qquad 5x - y + z = 5 \qquad$ (see Figure 10.5.12).

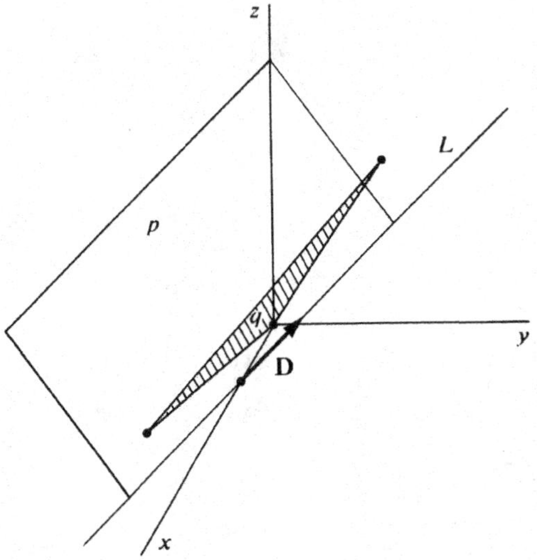

Figure 10.5.12

A line which is not parallel to a plane will intersect the plane at exactly one point.

EXAMPLE 11 Find the point at which the line $\mathbf{X} = \mathbf{i} - \mathbf{j} + \mathbf{k} + t(3\mathbf{i} - \mathbf{j} - \mathbf{k})$ intersects the plane $3x - 2y + z = 4$.

The line has the parametric equations

$$x = 1 + 3t, \qquad y = -1 - t, \qquad z = 1 - t.$$

We substitute these in the equation for the plane and solve for t.

$$3(1 + 3t) - 2(-1 - t) + (1 - t) = 4,$$
$$6 + 10t = 4,$$
$$t = -\tfrac{1}{5}.$$

Therefore the point of intersection is given by the parametric equations for the line at $t = -\tfrac{1}{5}$;

$$x = \tfrac{2}{5}, \qquad y = -\tfrac{4}{5}, \qquad z = \tfrac{6}{5},$$

(see Figure 10.5.13).

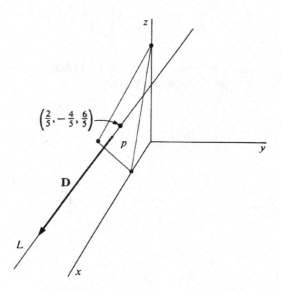

Figure 10.5.13

Two planes which are not parallel intersect at a line.

EXAMPLE 12 Find the line L of intersection of the planes

$$4x - 5y + z = 2,$$
$$x + 2z = 0.$$

Step 1   To get a position vector of L, we find any point on both planes. Setting $z = 0$ and solving for x and y, we obtain the point $S(0, -\tfrac{2}{5}, 0)$ on both planes. Thus $\mathbf{S} = -\tfrac{2}{5}\mathbf{j}$ is a position vector of L.

Step 2   To get a direction vector **D** of L we need a vector perpendicular to the normal
vectors of both planes. The normal vectors are

$$\mathbf{M} = 4\mathbf{i} - 5\mathbf{j} + \mathbf{k}, \qquad \mathbf{N} = \mathbf{i} + 2\mathbf{k}.$$

We take
$$\mathbf{D} = \mathbf{M} \times \mathbf{N} = \begin{vmatrix} \mathbf{i} & \mathbf{j} & \mathbf{k} \\ 4 & -5 & 1 \\ 1 & 0 & 2 \end{vmatrix}$$

$$= -10\mathbf{i} - 7\mathbf{j} + 5\mathbf{k}.$$

Thus L is the line $\mathbf{X} = -\tfrac{2}{5}\mathbf{j} + t(-10\mathbf{i} - 7\mathbf{j} + 5\mathbf{k}).$

PROBLEMS FOR SECTION 10.5

In Problems 1–12 find the points, if any, where the plane meets the x, y, and z axes. and sketch
the plane.

1	$x + y + z = 1$	2	$10x + 5y + z = 10$
3	$2x - 2y + z = 2$	4	$-x + 3y + 3z = -3$
5	$x + y = 1$	6	$y + 3z = 1$
7	$x - z = 2$	8	$x + y + z = 0$
9	$4x - y + 2z = 0$	10	$\tfrac{1}{2}x - y - z = 0$
11	$x - y = 0$	12	$2y + z = 0$

13   Find a normal vector to the following planes.

(a) $x - 3y + 6z = 4$ (b) $x + 2y = 0$
(c) $-3x + 4y + z = 0$ (d) $x + 6z = 8$
(e) $y = 4$ (f) $-y + z = 5$

Find a scalar equation for the plane described in Problems 14–32

14   The plane with normal vector $\mathbf{N} = \mathbf{i} + \mathbf{j} - \mathbf{k}$ and position vector $\mathbf{P} = 2\mathbf{i} + \mathbf{k}$.
15   The plane with normal vector $\mathbf{N} = \mathbf{j} + 2\mathbf{k}$ and position vector $\mathbf{P} = \mathbf{i} + 3\mathbf{j} - 6\mathbf{k}$.
16   The plane through the point $(1, 5, 8)$ with normal vector $\mathbf{N} = 5\mathbf{i} + \mathbf{j} - \mathbf{k}$.
17   The plane through the origin with normal vector $\mathbf{N} = \mathbf{i} + \mathbf{j} + 2\mathbf{k}$.
18   The plane with position vector $\mathbf{P} = \mathbf{i} - \mathbf{j}$ and direction vectors $\mathbf{C} = \mathbf{i} + \mathbf{j} + \mathbf{k}$,
  $\mathbf{D} = \mathbf{i} - \mathbf{j} - \mathbf{k}$.
19   The plane through the point $(1, 2, 3)$ with direction vectors $\mathbf{C} = \mathbf{i}, \mathbf{D} = \mathbf{j} + \mathbf{k}$.
20   The plane through the points $A(0, 4, 6)$, $B(5, 1, -1)$, $C(2, 6, 0)$.
21   The plane through the points $A(5, 0, 0)$, $B(0, 1, 0)$, $C(0, 0, -4)$.
22   The plane through the points $A(4, 9, -6)$, $B(6, 6, 6)$, $C(1, 10, 0)$.
23   The plane through the point $A(1, 2, 4)$ containing the line
$$\mathbf{X} = 2\mathbf{i} + 3\mathbf{j} + \mathbf{k} + t(\mathbf{i} - \mathbf{k}).$$
24   The plane through the point $A(0, 5, 1)$ containing the line
$$\mathbf{X} = \mathbf{i} + t(3\mathbf{i} - \mathbf{j} + \mathbf{k}).$$
25   The plane through the point $A(5, 0, 1)$ perpendicular to the line
$$\mathbf{X} = \mathbf{i} + \mathbf{j} + \mathbf{k} + t(2\mathbf{i} + \mathbf{j} + 3\mathbf{k}).$$

26   The plane through the origin perpendicular to the line
$$\mathbf{X} = t(5\mathbf{i} - \mathbf{j} + 6\mathbf{k}).$$

27   The plane through the point $A(4, 10, -3)$ parallel to the plane $x + y - 2z = 1$.

28   The plane through the origin parallel to the plane $4x + y + z = 6$.

29   The plane containing the line $\mathbf{X} = \mathbf{i} + \mathbf{j} - \mathbf{k} + t(3\mathbf{i} + \mathbf{k})$ and perpendicular to the plane $2x - y + z = 3$.

30   The plane containing the line $\mathbf{X} = 3\mathbf{j} + t(5\mathbf{i} + \mathbf{j} - 6\mathbf{k})$ and perpendicular to the plane $x + y + z = 0$.

31   The plane containing the line $\mathbf{X} = 3\mathbf{i} + \mathbf{j} + \mathbf{k} + t(\mathbf{i} - 6\mathbf{k})$ and parallel to the line $\mathbf{X} = \mathbf{i} + \mathbf{j} + t(3\mathbf{i} + 4\mathbf{j} + \mathbf{k})$.

32   The plane containing the x-axis and parallel to the line $\mathbf{X} = t(\mathbf{i} + 2\mathbf{j} - \mathbf{k})$.

In Problems 33–36, test for perpendiculars and parallels.

33   The planes $x - 3y + 2z = 4$, $-2x + 6y - 4z = 0$.

34   The planes $4x + 3y - z = 6$, $x + y + 7z = 4$.

35   The plane $-x + y - 2z = 8$ and the line $\mathbf{X} = 2\mathbf{i} + \mathbf{k} + t(3\mathbf{i} - \mathbf{j} + \mathbf{k})$.

36   The plane $x + y + 3z = 10$ and the line $\mathbf{X} = 3\mathbf{j} + t(\mathbf{i} + 2\mathbf{j} - \mathbf{k})$.

In Problems 37–42 find a vector equation for the given line.

37   The line through $P(5, 3, -1)$, perpendicular to the plane $x - y + 3z = 1$.

38   The line through the origin, perpendicular to the plane $x - y + z = 0$.

39   The line of intersection of the planes $x + y + z = 0$, $x - y + 2z = 1$.

40   The line of intersection of the planes $2x + 3y - 4z = 1$, $x + z = 4$.

41   The line of intersection of the planes $x + y = 1$, $y - z = 2$.

42   The line of intersection of the planes $x - 2y + 3z = 0$, $z = -2$.

In Problems 43–49, find the coordinates of the given point.

43   The point where the line $\mathbf{X} = 3\mathbf{i} + \mathbf{j} + \mathbf{k} + t(-\mathbf{i} + 3\mathbf{j} - \mathbf{k})$ intersects the plane $x + 2y - z = 4$.

44   The point where the line $\mathbf{X} = \mathbf{i} + \mathbf{k} + t(\mathbf{j} + \mathbf{k})$ intersects the plane $x + 2y = -3$.

45   The point where the line $\mathbf{X} = t(\mathbf{i} - 2\mathbf{k})$ intersects the plane $x - 3y + 2z = 4$.

46   The point P on the plane $x + 3y + 6z = 6$, nearest to the origin. *Hint:* The line from the origin to P must be perpendicular to the plane.

47   The point P on the plane $x + y + z = 1$, nearest to the point $A(-1, 2, 3)$.

48   The point P on the line $\mathbf{X} = \mathbf{i} + 2\mathbf{j} + 3\mathbf{k} + t(\mathbf{i} - \mathbf{j} + \mathbf{k})$ nearest to the origin. *Hint:* P must be on the plane through the origin perpendicular to the line.

49   The point P on the line $\mathbf{X} = \mathbf{j} + t(\mathbf{i} + 3\mathbf{k})$ nearest to the point $A(1, 2, 3)$.

☐ 50   Prove that any three points which are not all on a line determine a plane.

☐ 51   Prove that if a line and plane are parallel and have at least one point in common then the line is a subset of the plane.

☐ 52   Prove that if two parallel planes have at least one point in common then they are equal.

☐ 53   Let p be a plane with normal vector $\mathbf{N}$. Prove that every vector $\mathbf{D}$ perpendicular to $\mathbf{N}$ is a direction vector of p.

☐ 54   Given a plane p and a line L not perpendicular to p, prove that there is a unique plane q which contains L and is perpendicular to p.

0.6 VECTOR VALUED FUNCTIONS

A vector valued function is a function **F** which maps real numbers to vectors. We shall study vector valued functions in either two or three dimensions. Here is the exact definition.

DEFINITION

A **vector valued function** in two dimensions is a set **F** of ordered pairs $(t, \mathbf{X})$ such that for every real number t one of the following occurs.

(i)   There is exactly one vector **X** in two dimensions for which the ordered pair $(t, \mathbf{X})$ belongs to **F**. In this case $\mathbf{F}(t)$ is defined and $\mathbf{F}(t) = \mathbf{X}$.

(ii) There is no **X** for which $(t, \mathbf{X})$ belongs to **F**. In this case $\mathbf{F}(t)$ is said to be undefined.

The definition of a three-dimensional vector valued function is similar.
A vector valued function in two dimensions can be written as a sum

$$\mathbf{F}(t) = f_1(t)\mathbf{i} + f_2(t)\mathbf{j}.$$

The functions f_1 and f_2 are real functions of one variable, called the *components* of **F**. The *vector equation* $\mathbf{X} = \mathbf{F}(t)$ can also be written as a pair of parametric equations

$$x = f_1(t), \qquad y = f_2(t).$$

As t varies over the real numbers, the point $X(x, y)$ traces out a *parametric curve* in the plane. The vector valued function $\mathbf{F}(t)$ is called the *position vector* of the curve.
The line with the vector equation $\mathbf{X} = \mathbf{P} + t\mathbf{C}$ is a parametric curve with position vector $\mathbf{F}(t) = \mathbf{P} + t\mathbf{C}$ and components

$$f_1(t) = p_1 + tc_1, \qquad f_2(t) = p_2 + tc_2.$$

EXAMPLE 1   Find the vector equation for a particle which moves counterclockwise around the unit circle, and is at the point $(1, 0)$ at time $t = 0$, shown in Figure 10.6.1.

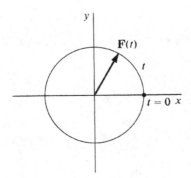

Figure 10.6.1

The motion is given by the parametric equations

$$x = \cos t, \qquad y = \sin t,$$

and the vector equation $\qquad \mathbf{X} = \cos t\mathbf{i} + \sin t\mathbf{j}.$

EXAMPLE 2   A ball thrown at time $t = 0$ with initial velocity of v_1 in the x direction and v_2 in the y direction will follow the parabolic curve

$$x = v_1 t, \qquad y = v_2 t - 16t^2.$$

The curve (Figure 10.6.2) has the vector equation $\mathbf{X} = v_1 t\mathbf{i} + (v_2 t - 16t^2)\mathbf{j}.$

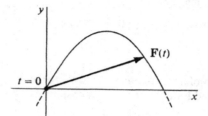

Figure 10.6.2

EXAMPLE 3   A point on the rim of a wheel rolling along a line traces out a curve called *a cycloid*. Find the vector equation for the cycloid if the wheel has radius one, rolls at one radian per second along the x-axis, and starts at $t = 0$ with the point at the origin.

As we can see from the close-up in Figure 10.6.3, the parametric equations are

$$x = t - \sin t, \qquad y = 1 - \cos t.$$

The vector equation is $\mathbf{X} = (t - \sin t)\mathbf{i} + (1 - \cos t)\mathbf{j}.$

A vector valued function in three dimensions can be written in the form

$$\mathbf{F}(t) = f_1(t)\mathbf{i} + f_2(t)\mathbf{j} + f_3(t)\mathbf{k}$$

and has the three components f_1, f_2, and f_3. The equation $\mathbf{X} = \mathbf{F}(t)$ can be written as three parametric equations

$$x = f_1(t), \qquad y = f_2(t), \qquad z = f_3(t),$$

and as t varies over the reals we get a parametric curve in space.

EXAMPLE 4   The space curve

$$\mathbf{X} = \cos t\mathbf{i} + \sin t\mathbf{j} + t\mathbf{k}$$

is a *circular helix*. The point (x, y) goes around a horizontal circle of radius one whose center is rising vertically at a constant rate (Figure 10.6.4).

EXAMPLE 5   In economics the price vector may change with time and thus be a vector valued function of t. Find the price vector function $\mathbf{P}(t)$ for three

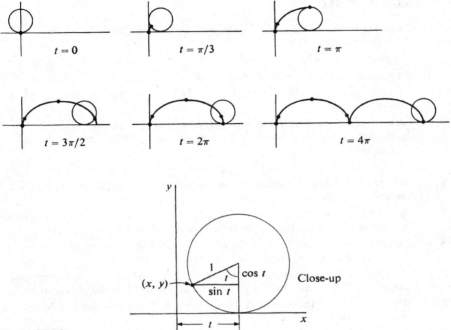

Figure 10.6.3

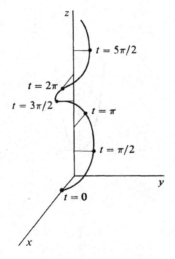

Figure 10.6.4

commodities such that the first commodity has price t^2, the second has price $t + 1$, and the price of the third commodity is the sum of the other two $(t \geq 0)$. The answer is

$$\mathbf{P}(t) = t^2\mathbf{i} + (t + 1)\mathbf{j} + (t^2 + t + 1)\mathbf{k}.$$

PROBLEMS FOR SECTION 10.6

Find the vector equations for the motion of the given point in the plane. The positions at $t = 0$ and $t = 1$ are as shown in the figures.

1   A point moving along the parabola $y = x^2$ in such a way that $x = 3t$.

2   A point moving along $y = x^2$ so that $xy = t$.

3   A point moving upward along the line $y = 2x$ so that its distance from the origin at time t is t^3.

4   A wheel of radius one is turning at the rate of one radian per second. At the same time its center is moving along the x-axis at one unit per second. Find the motion of a point on the circumference of the wheel.

5   The point at distance one from the origin in the direction of the point $(t, 1)$.

6   The point where the parabola $y = x^2$ intersects the line through the origin which makes an angle t with the x-axis.

7   The point halfway between a point P going around the circle $x^2 + y^2 = 1$ at one radian per second and a point Q going around the same circle at 3 radians per second.

8   A wheel of radius one rolls along the x-axis at one radian per second. Find the motion of a point on the circumference of a concentric axle of radius $\frac{1}{2}$.

9   A circle of radius one rolls around the outside of the circle $x^2 + y^2 = 9$ at one radian per second. Find the motion of a point on the circumference of the smaller circle.

10   Find the motion of the point in Problem 9 if the small circle rolls around the inside of the large circle.

11   A string is unwound from a circular reel of radius one at one radian per second. The string is held taut and forms a line tangent to the reel. Find the motion of the end of the string.

1.

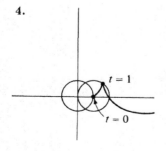

2.

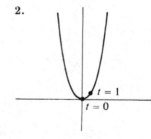

3.

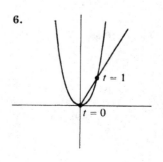

4.

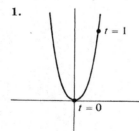

5.

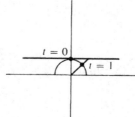

6.

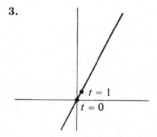

7.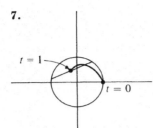

8.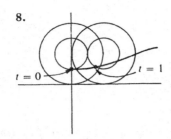

9.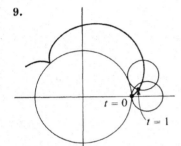

10.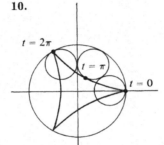

11.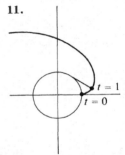

In Problems 12–23, find the vector equation for the motion of the given point in space.

12 A point moving so that at time t its position vector has length t^2 and direction cosines $(\frac{1}{3}, \frac{2}{3}, \frac{2}{3})$.

13 A point X moving at one radian per second counterclockwise around a horizontal unit circle whose center is at $(0, 0, t^2)$ at time t. (At $t = 0$, $X = i$.)

14 The point which at time t is at distance one from the origin in the direction of the vector $ti + j + t^2k$.

15 The point at distance one from the point $P(1, 2, 1)$ in the direction of the vector $t^2j + (t^2 - 1)k$.

16 The point where the line through the origin in the direction of $i + tj + t^2k$ intersects the plane $x + 2y + 3z = 1$.

17 The point halfway between a point P going around the circle $x^2 + y^2 = 1$ in the (x, y) plane at one radian per second and a point Q going around the circle $x^2 + z^2 = 1$ in the (x, z) plane at 2 radians per second. (At $t = 0$, $P = Q = i$. Both motions are counterclockwise.)

18 The point at distance $f(t)$ from the point $P(t)$ in the direction of the vector $D(t)$.

19 The point on the plane $x + y + z = 1$ which is nearest to the point

$$\cos ti + \sin tj + 6k.$$

20 The point where the rotating plane $x \cos t + y \sin t = 0$ intersects the line through $(1, 1, 1)$ and $(2, 3, 4)$.

21 The point on the rotating plane $x \cos t + y \sin t = 0$ which is nearest to the point $ti + 2tj + 3tk$.

22 Find the price vector $P(t)$ for three commodities such that the first has price $1/t$, the second has double the price of the first, and the sum of the prices is 4 $(t \geq 1)$.

23 Find the price vector $P(t)$ of three commodities such that the product of the three prices is one, the first commodity has price $2t$, and the third commodity has price $t + 1$ $(t \geq 1)$.

10.7 VECTOR DERIVATIVES

The derivative of a vector valued function is defined in terms of its components. We shall state the definitions for three dimensions. The two-dimensional case is similar.

DEFINITION

> *Given a vector valued function*
>
> $$\mathbf{F}(t) = f_1(t)\mathbf{i} + f_2(t)\mathbf{j} + f_3(t)\mathbf{k},$$
>
> *the **derivative** $\mathbf{F}'(t)$ is defined by*
>
> $$\mathbf{F}'(t) = f_1'(t)\mathbf{i} + f_2'(t)\mathbf{j} + f_3'(t)\mathbf{k}.$$
>
> $\mathbf{F}'(t)$ *exists if and only if* $f_1'(t)$, $f_2'(t)$, *and* $f_3'(t)$ *all exist.*

When we use the notation $\mathbf{X} = x\mathbf{i} + y\mathbf{j} + z\mathbf{k}$, the derivative is written

$$\frac{d\mathbf{X}}{dt} = \frac{dx}{dt}\mathbf{i} + \frac{dy}{dt}\mathbf{j} + \frac{dz}{dt}\mathbf{k}.$$

EXAMPLE 1   Find $d\mathbf{X}/dt$ where

$$\mathbf{X} = t^{1/3}\mathbf{i} + \frac{1}{t+1}\mathbf{j} + 2t\mathbf{k}, \qquad t \neq -1.$$

$$d\mathbf{X}/dt = \tfrac{1}{3}t^{-2/3}\mathbf{i} - (t+1)^{-2}\mathbf{j} + 2\mathbf{k}.$$

$d\mathbf{X}/dt$ is undefined at $t = 0$ and $t = -1$.

If $\mathbf{X}$ is the position vector of a line L, $\mathbf{X} = \mathbf{P} + t\mathbf{C}$, then the derivative of $\mathbf{X}$ is the constant direction vector $\mathbf{C}$, $d\mathbf{X}/dt = \mathbf{C}$. For

$$\frac{d\mathbf{X}}{dt} = \frac{d(p_1 + c_1 t)}{dt}\mathbf{i} + \frac{d(p_2 + c_2 t)}{dt}\mathbf{j} + \frac{d(p_3 + c_3 t)}{dt}\mathbf{k}$$

$$= c_1\mathbf{i} + c_2\mathbf{j} + c_3\mathbf{k} = \mathbf{C}.$$

The next two theorems show the geometric meaning of the vector derivative.

THEOREM 1

> *Given a curve $\mathbf{X} = \mathbf{F}(t)$ in the plane, if $\mathbf{F}'(t_0) \neq \mathbf{0}$ then $\mathbf{F}'(t_0)$ is a direction vector of the line tangent to the curve at t_0.*

PROOF

Case 1   The curve is not vertical at t_0. The tangent line has slope

$$\frac{dy}{dx} = \frac{dy/dt}{dx/dt} = \frac{f_2'(t_0)}{f_1'(t_0)}$$

at t. Therefore the vector

$$\mathbf{F}'(t_0) = f_1'(t_0)\mathbf{i} + f_2'(t_0)\mathbf{j}$$

is a direction vector of the tangent line.

Case 2   The curve is vertical at t_0. Then $f'_1(t_0) = 0$, so $\mathbf{F}'(t_0) = f'_2(t_0)\mathbf{j}$ is a direction vector of the vertical tangent line. $\mathbf{F}'(t_0)$ is shown in Figure 10.7.1 for a curve $\mathbf{X} = \mathbf{F}(t)$.

For curves in space we can use the vector derivative to *define* the tangent line.

DEFINITION

> *If* $\mathbf{X} = \mathbf{F}(t)$ *is a curve in space and* $\mathbf{F}'(t_0) \neq \mathbf{0}$, *the **tangent line** of the curve at* t_0 *is the line with position vector* $\mathbf{F}(t_0)$ *and direction vector* $\mathbf{F}'(t_0)$.
>
> *A vector parallel to* $\mathbf{F}'(t_0)$ *is said to be a **tangent vector** of the curve at* t_0.

EXAMPLE 2   Find the vector equation of the tangent line for the spiral

$$\mathbf{F}(t) = \cos t\,\mathbf{i} + \sin t\,\mathbf{j} + \tfrac{1}{4}t\mathbf{k}$$

at the point $t = \pi/3$.

The derivative is $\mathbf{F}'(t) = -\sin t\,\mathbf{i} + \cos t\,\mathbf{j} + \tfrac{1}{4}\mathbf{k}.$

At $t = \pi/3$ the tangent line has the equation

$$\mathbf{X} = \mathbf{F}(\pi/3) + t\mathbf{F}'(\pi/3)$$

or $$\mathbf{X} = \left(\frac{1}{2}\mathbf{i} + \frac{\sqrt{3}}{2}\mathbf{j} + \frac{\pi}{12}\mathbf{k}\right) + t\left(-\frac{\sqrt{3}}{2}\mathbf{i} + \frac{1}{2}\mathbf{j} + \frac{1}{4}\mathbf{k}\right).$$

The tangent line is shown in Figure 10.7.2.

We have seen that the direction of the vector derivative is tangent to the curve. We next discuss the length of the vector derivative.

Suppose all the derivatives dx/dt, dy/dt, and dz/dt are continuous on an interval $a \leq t \leq b$. Recall that in two dimensions the length of the curve is defined as

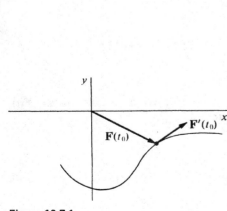

Figure 10.7.1

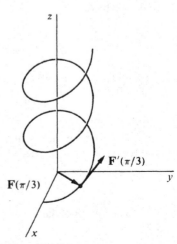

Figure 10.7.2

the integral

$$s = \int_a^b \sqrt{(dx/dt)^2 + (dy/dt)^2} \, dt.$$

The *length of a curve in space* is defined in a similar way,

$$s = \int_a^b \sqrt{(dx/dt)^2 + (dy/dt)^2 + (dz/dt)^2} \, dt.$$

EXAMPLE 3   Find the length of the helix

$$\mathbf{X} = \cos t\mathbf{i} + \sin t\mathbf{j} + \tfrac{1}{4}t\mathbf{k},$$

from $t = a$ to $t = b$.

$$\frac{dx}{dt} = -\sin t, \qquad \frac{dy}{dt} = \cos t, \qquad \frac{dz}{dt} = \frac{1}{4}.$$

$$s = \int_a^b \sqrt{\sin^2 t + \cos^2 t + \tfrac{1}{16}} \, dt$$

$$= \int_a^b \sqrt{1 + \tfrac{1}{16}} \, dt = \int_a^b \frac{\sqrt{17}}{4} \, dt = \frac{\sqrt{17}}{4}(b - a).$$

THEOREM 2

Let $\mathbf{X} = \mathbf{F}(t)$ be a curve in space such that all the derivatives dx/dt, dy/dt, and dz/dt are continuous for $a \le t \le b$. Then the vector derivative $d\mathbf{X}/dt$ has length ds/dt, where s is the length of the curve from a to t. That is,

$$|d\mathbf{X}/dt| = ds/dt.$$

PROOF We have

$$s = \int_a^t \sqrt{\left(\frac{dx}{dt}\right)^2 + \left(\frac{dy}{dt}\right)^2 + \left(\frac{dz}{dt}\right)^2} \, dt,$$

$$\frac{d\mathbf{X}}{dt} = \frac{dx}{dt}\mathbf{i} + \frac{dy}{dt}\mathbf{j} + \frac{dz}{dt}\mathbf{k}.$$

Therefore

$$\frac{ds}{dt} = \sqrt{\left(\frac{dx}{dt}\right)^2 + \left(\frac{dy}{dt}\right)^2 + \left(\frac{dz}{dt}\right)^2} = \left|\frac{d\mathbf{X}}{dt}\right|.$$

If a particle moves in space so that its position vector at time t is $\mathbf{S} = \mathbf{F}(t)$, the vector derivative is called the *velocity vector*,

$$\mathbf{V} = \frac{d\mathbf{S}}{dt}.$$

The length of the velocity vector is called the *speed* of the particle. Theorems 1 and 2 show that:

The velocity vector **V** *is tangent to the curve.*

The speed |**V**| *is equal to the rate of change of the length of the curve,*

$$|\mathbf{V}| = \frac{ds}{dt}.$$

The second derivative is called the *acceleration vector*

$$\mathbf{A} = \frac{d\mathbf{V}}{dt} = \frac{d^2\mathbf{S}}{dt^2}.$$

EXAMPLE 4 Find the velocity, speed, and acceleration of a particle which moves around the unit circle with position vector

$$\mathbf{S} = \cos t\mathbf{i} + \sin t\mathbf{j}.$$

Velocity: $\mathbf{V} = -\sin t\mathbf{i} + \cos t\mathbf{j}.$

Speed: $|\mathbf{V}| = \sqrt{\sin^2 t + \cos^2 t} = 1.$

Acceleration: $\mathbf{A} = -\cos t\mathbf{i} - \sin t\mathbf{j}.$

As Figure 10.7.3 shows, the velocity **V** is tangent to the circle and the acceleration **A** points to the center of the circle.

EXAMPLE 5 Find the velocity, speed, and acceleration of a ball moving on the parabolic curve

$$\mathbf{S} = v_1 t\mathbf{i} + (v_2 t - 16t^2)\mathbf{j}.$$

Velocity: $\mathbf{V} = v_1\mathbf{i} + (v_2 - 32t)\mathbf{j}.$

Speed: $|\mathbf{V}| = \sqrt{v_1^2 + (v_2 - 32t)^2}.$

Acceleration: $\mathbf{A} = -32\mathbf{j}.$

We see in Figure 10.7.4 that the velocity vector is tangent to the parabola, while the acceleration vector points straight down.

EXAMPLE 6 Find the position vector of a particle which moves with velocity

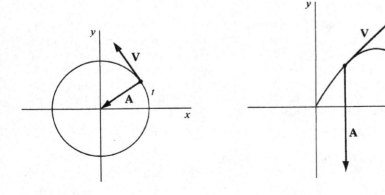

Figure 10.7.3   **Figure 10.7.4**

$$V = -\sin t\mathbf{i} + \cos t\mathbf{j} + \sin t \cos t\mathbf{k}$$

and at time $t = 0$ has position $\mathbf{F}(0) = \mathbf{i} + 2\mathbf{k}$.

We find each component separately by integration.

$$f_1'(t) = -\sin t, \qquad f_1(0) = 1.$$
$$f_1(t) = \cos t + C_1.$$
$$1 = \cos 0 + C_1, \qquad C_1 = 0.$$
$$f_1(t) = \cos t.$$

$$f_2'(t) = \cos t, \qquad f_2(0) = 0.$$
$$f_2(t) = \sin t + C_2.$$
$$0 = \sin 0 + C_2, \qquad C_2 = 0.$$
$$f_2(t) = \sin t.$$

$$f_3'(t) = \sin t \cos t, \qquad f_3(0) = 2.$$
$$f_3(t) = \tfrac{1}{2}\sin^2 t + C_3.$$
$$2 = \tfrac{1}{2}\sin^2 0 + C_3, \qquad C_3 = 2.$$
$$f_3(t) = \tfrac{1}{2}\sin^2 t + 2.$$

$$\mathbf{F}(t) = \cos t\mathbf{i} + \sin t\mathbf{j} + (\tfrac{1}{2}\sin^2 t + 2)\mathbf{k}.$$

The path of the particle is shown in Figure 10.7.5.

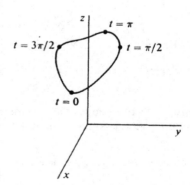

Figure 10.7.5

Let us briefly consider the derivative $\mathbf{P}'(t)$ of a price vector $\mathbf{P}(t)$. $\mathbf{P}'(t)$ is the *marginal price vector with respect to time*. It represents the rates at which the prices of all the commodities are changing. In a time of pure inflation, prices will increase but the ratios between prices of different commodities will stay the same, hence $\mathbf{P}'(t)$ will have the same direction as $\mathbf{P}(t)$. In a time of pure deflation $\mathbf{P}'(t)$ will have the opposite direction from $\mathbf{P}(t)$. Usually $\mathbf{P}'(t)$ is not parallel to $\mathbf{P}(t)$ at all, because the prices of some commodities are changing relative to others.

THEOREM 3 (Rules for Vector Derivatives)

Let $u = h(t)$ be a real function and let $\mathbf{X} = \mathbf{F}(t)$, $\mathbf{Y} = \mathbf{G}(t)$ be vector valued functions whose derivatives at t exist.

(i) **Constant Rules** $\dfrac{d(c\mathbf{X})}{dt} = c\dfrac{d\mathbf{X}}{dt},$ $\dfrac{d(C u)}{dt} = \mathbf{C}\dfrac{du}{dt}.$

(ii) **Sum Rule** $\dfrac{d(\mathbf{X} + \mathbf{Y})}{dt} = \dfrac{d\mathbf{X}}{dt} + \dfrac{d\mathbf{Y}}{dt}.$

(iii) **Inner Product Rule** $\dfrac{d(\mathbf{X}\cdot\mathbf{Y})}{dt} = \mathbf{X}\cdot\dfrac{d\mathbf{Y}}{dt} + \dfrac{d\mathbf{X}}{dt}\cdot\mathbf{Y}.$

PROOF We prove (iii).

$$\mathbf{X}\cdot\mathbf{Y} = f_1(t)g_1(t) + f_2(t)g_2(t) + f_3(t)g_3(t).$$

$$\frac{d(\mathbf{X}\cdot\mathbf{Y})}{dt} = f_1(t)g_1'(t) + f_2(t)g_2'(t) + f_3(t)g_3'(t)$$

$$+ f_1'(t)g_1(t) + f_2'(t)g_2(t) + f_3'(t)g_3(t)$$

$$= \mathbf{F}(t)\cdot\mathbf{G}'(t) + \mathbf{F}'(t)\cdot\mathbf{G}(t) = \mathbf{X}\cdot\frac{d\mathbf{Y}}{dt} + \frac{d\mathbf{X}}{dt}\cdot\mathbf{Y}.$$

COROLLARY

Suppose $\mathbf{X} = \mathbf{F}(t)$ is a curve whose distance $|\mathbf{F}(t)|$ from the origin is a constant r_0. Then the derivative $\mathbf{F}'(t)$ is perpendicular to $\mathbf{F}(t)$ whenever $\mathbf{F}'(t) \neq \mathbf{0}$.

PROOF We use the Inner Product Rule. For all t,

$$r_0^2 = \mathbf{F}(t)\cdot\mathbf{F}(t).$$

$$0 = \frac{d(\mathbf{F}(t)\cdot\mathbf{F}(t))}{dt} = \mathbf{F}(t)\cdot\mathbf{F}'(t) + \mathbf{F}'(t)\cdot\mathbf{F}(t)$$

$$= 2\mathbf{F}(t)\cdot\mathbf{F}'(t).$$

Therefore $\mathbf{F}(t)\cdot\mathbf{F}'(t) = 0$, so $\mathbf{F}(t) \perp \mathbf{F}'(t)$, as shown in Figure 10.7.6.

We see from the corollary that if a particle moves with constant speed $|\mathbf{V}| = v_0$, then its acceleration vector is always perpendicular to the velocity vector (Figure 10.7.7).

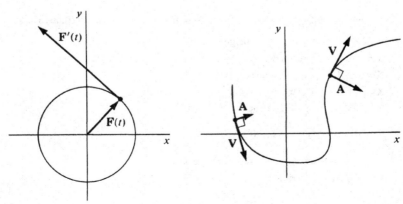

Figure 10.7.6   **Figure 10.7.7**   Motion with Constant Speed

PROBLEMS FOR SECTION 10.7

In Problems 1–9 find the derivative.

1	$\mathbf{X} = 5(\sin t\mathbf{i} + \cos t\mathbf{j})$	**2**	$\mathbf{X} = \cos(2t)\mathbf{i} + \sin(3t)\mathbf{j}$
3	$\mathbf{X} = \cos(e^t)\mathbf{i} + \sin(e^t)\mathbf{j}$	**4**	$\mathbf{X} = t^2\mathbf{i} + t^3\mathbf{j} - t\mathbf{k}$
5	$\mathbf{X} = -6(t\mathbf{i} - \ln t\mathbf{j} + e^t\mathbf{k})$	**6**	$\mathbf{X} = 4t^3(2\mathbf{i} - 3\mathbf{j} + \mathbf{k})$

7 $\quad u = (\cos t\mathbf{i} + \sin t\mathbf{j}) \cdot (\sin t\mathbf{i} + \cos t\mathbf{j})$

8 $\quad u = (2^t\mathbf{i} + 3^t\mathbf{j} + 4^t\mathbf{k}) \cdot (2t\mathbf{i} + 3t\mathbf{j} + 4t\mathbf{k})$

9 $\quad u = |\cos t\mathbf{i} + \sin t\mathbf{j} + t\mathbf{k}|$

10 $\quad$ Find the line tangent to the curve $\mathbf{X} = \sin^2 t\mathbf{i} + \cos^2 t\mathbf{j} + \sin t \cos t\mathbf{k}$ at $t = \pi/3$.

11 $\quad$ Find the line tangent to the curve $\mathbf{X} = t\mathbf{i} + t^2\mathbf{j} + t^3\mathbf{k}$ at the point $(1, 1, 1)$.

12 $\quad$ Find the line tangent to the cycloid $\mathbf{X} = (t - \sin t)\mathbf{i} + (1 - \cos t)\mathbf{j}$ at $t = \pi/4$.

In Problems 13–25 find the velocity, speed, and acceleration.

13 $\quad \mathbf{S} = 2t\mathbf{i} + 3t\mathbf{j} - 4t\mathbf{k}$

14 $\quad \mathbf{S} = t^2\mathbf{i} + t\mathbf{j} + \mathbf{k}$

15 $\quad \mathbf{S} = \cos t\mathbf{i} + \sin t\mathbf{j} + t\mathbf{k}$

16 $\quad$ The cycloid $\mathbf{S} = (t - \sin t)\mathbf{i} + (1 - \cos t)\mathbf{j}$

17 $\quad \mathbf{S} = \cos(e^t)\mathbf{i} + \sin(e^t)\mathbf{j}$

18 $\quad \mathbf{S} = \cos t\mathbf{i} + \sin t\mathbf{j} + t^2\mathbf{k}$

19 $\quad \mathbf{S} = (t^2 + 1)\mathbf{i} + (2t^2 + 1)\mathbf{j} + (-t^2 + 1)\mathbf{k}$

20 $\quad$ A point on the rim of a wheel of radius one in the (x, y) plane which is spinning counterclockwise at one radian per second and whose center at time t is at $(t, 0)$. (At $t = 0$, $\mathbf{S} = \mathbf{i}$.)

21 $\quad$ A bug which is crawling outward along a spoke of a wheel at one unit per second while the wheel is spinning at one radian per second. The center of the wheel is at the origin, and at $t = 0$, the bug is at the origin and the spoke is along the x-axis. (A spiral of Archimedes.)

22 $\quad$ The point at distance one from the origin in the direction of the vector

$$\mathbf{i} + t\mathbf{j} + \sqrt{2t}\mathbf{k}, \qquad t > 0.$$

23 $\quad$ A car going counterclockwise around a circular track $x^2 + y^2 = 1$ with speed $|2t|$ at time t. At $t = 0$ the car is at $(1, 0)$.

24 $\quad$ A point moving at speed one along the parabola $y = x^2$, going from left to right. ($\mathbf{S} = \mathbf{0}$ at $t = 0$.)

25 $\quad$ A point moving at speed y along the curve $y = e^x$ going from left to right. ($\mathbf{S} = \mathbf{j}$ at $t = 0$.)

In Problems 26–33 find the length of the given curve.

26 $\quad \mathbf{X} = \cos t\mathbf{i} + \sin t\mathbf{j} + t^2\mathbf{k}, \;\; 0 \le t \le 2$

27 $\quad \mathbf{X} = \cos t\mathbf{i} + \frac{3}{5}\sin t\mathbf{j} - \frac{4}{5}\sin t\mathbf{k}, \;\; 0 \le t \le 2\pi$

28 $\quad \mathbf{X} = 6t\mathbf{i} - 8t\mathbf{j} + t^2\mathbf{k}, \;\; 0 \le t \le 5$

29 $\quad \mathbf{X} = 2t\mathbf{i} + 3t^2\mathbf{j} + 3t^3\mathbf{k}, \;\; 0 \le t \le 1$

30 $\quad \mathbf{X} = t\mathbf{i} + \dfrac{1}{\sqrt{2}}t^2\mathbf{j} + \dfrac{1}{3}t^3\mathbf{k}, \;\; 0 \le t \le 1$

31 $\quad \mathbf{X} = \cos^2 t\mathbf{i} + \sin^2 t\mathbf{j} + 2\sin t\mathbf{k}, \;\; 0 \le t \le \pi$

32 $\quad \mathbf{X} = \cosh t\mathbf{i} + \sinh t\mathbf{j} + t\mathbf{k}, \;\; 0 \le t \le 1$

33 $\quad \mathbf{X} = \ln t\mathbf{i} + \sqrt{2}t\mathbf{j} + \tfrac{1}{2}t^2\mathbf{k}, \;\; 1 \le t \le 2$

In Problems 34–37 find the position vector of a particle with the given velocity vector and initial position.

34     $\mathbf{V} = e^t\mathbf{i} + e^{2t}\mathbf{j} + e^{3t}\mathbf{k},\ \ \mathbf{F}(0) = \mathbf{0}$

35     $\mathbf{V} = t\mathbf{i} + t^2\mathbf{j} + t^3\mathbf{k},\ \ \mathbf{F}(1) = \mathbf{i} + 2\mathbf{j} + 3\mathbf{k}$

36     $\mathbf{V} = \dfrac{\mathbf{i}}{t^2 + 1} + \dfrac{\mathbf{j}}{t^2 - 1} + \mathbf{k},\ \ \mathbf{F}(0) = \mathbf{i} + \mathbf{j} + \mathbf{k}$

37     $\mathbf{V} = \dfrac{\mathbf{i}}{t - 1} + \dfrac{\mathbf{j}}{t - 2} + \dfrac{\mathbf{k}}{t - 3},\ \ \mathbf{F}(0) = \mathbf{0}$

38   Find the position vector of a particle whose acceleration vector at time t is $\mathbf{A} = \mathbf{i} + t\mathbf{j} + e^t\mathbf{k}$, if at $t = 0$ the velocity and position vectors are both zero.

39   Find the position vector $\mathbf{S}$ if $\mathbf{A} = \sin t\mathbf{i} + \cos t\mathbf{j} + \mathbf{k}$, and at $t = 0$, $\mathbf{V} = \mathbf{0}$ and $\mathbf{S} = \mathbf{0}$.

☐ 40   Show that if $\mathbf{U}$ is the unit vector of $\mathbf{X}$, then

$$\frac{d|\mathbf{X}|}{dt} = \frac{d\mathbf{X}}{dt}\cdot\mathbf{U}.$$

☐ 41   Show using the Chain Rule that if $\mathbf{X}$ is the position vector of a curve and s is the length from 0 to t, then $d\mathbf{X}/ds$ is a unit vector tangent to the curve.

☐ 42   Suppose a particle moves so that its speed is constant and its distance from the origin at time t is e^t. Show that the angle between the position and velocity vectors is constant.

☐ 43   Prove that if $\mathbf{F}(t)$ is perpendicular to a constant vector $\mathbf{C}$ for all t, then $\mathbf{F}'(t)$ is also perpendicular to $\mathbf{C}$.

☐ 44   Prove that if $\mathbf{F}(t)$ is parallel to a constant vector $\mathbf{C}$ for all t, then $\mathbf{F}'(t)$ is also parallel to $\mathbf{C}$.

☐ 45   Prove the following differentiation rule for scalar multiples:

$$\frac{d(u\mathbf{X})}{dt} = u\frac{d\mathbf{X}}{dt} + \frac{du}{dt}\mathbf{X}.$$

☐ 46   Prove the vector product rule $\dfrac{d(\mathbf{X} \times \mathbf{Y})}{dt} = \mathbf{X} \times \dfrac{d\mathbf{Y}}{dt} + \dfrac{d\mathbf{X}}{dt} \times \mathbf{Y}.$

0.8 HYPERREAL VECTORS

This section may be skipped without affecting the rest of the course. We introduce hyperreal vectors and use them to give an infinitesimal treatment of vector derivatives. We shall concentrate on three dimensions; the theory for two dimensions is similar.

A *hyperreal vector* in three dimensions is a vector

$$\mathbf{A} = a_1\mathbf{i} + a_2\mathbf{j} + a_3\mathbf{k}$$

whose components a_1, a_2, and a_3 are hyperreal numbers. The algebra of hyperreal vectors is in many ways similar to the algebra of hyperreal numbers. It begins with the notions of infinitesimal, finite, and infinite hyperreal vectors.

A hyperreal vector $\mathbf{A}$ is said to be *infinitesimal, finite,* or *infinite* if its length $|\mathbf{A}|$ is an infinitesimal, finite, or infinite number, respectively. Two hyperreal vectors $\mathbf{A}$ and $\mathbf{B}$ are said to be *infinitely close*, $\mathbf{A} \approx \mathbf{B}$, if their difference $\mathbf{B} - \mathbf{A}$ is infinitesimal (Figure 10.8.1).

EXAMPLE 1 Let ε be a positive infinitesimal and H be a positive infinite hyperreal

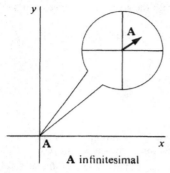 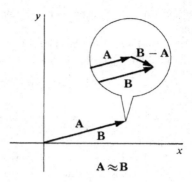

A infinitesimal   **A** ≈ **B**

Figure 10.8.1

number. The vector $5\varepsilon\mathbf{i} + \varepsilon^2\mathbf{k}$ is infinitesimal. Its length is

$$\sqrt{25\varepsilon^2 + 0 + \varepsilon^4} = \varepsilon\sqrt{25 + \varepsilon^2} \approx 0.$$

The vector $\varepsilon\mathbf{i} + \mathbf{j} + \mathbf{k}$ is finite but not infinitesimal. Its length is

$$\sqrt{\varepsilon^2 + 1^2 + 1^2} = \sqrt{\varepsilon^2 + 2} \approx \sqrt{2}.$$

The vector $\mathbf{i} + \varepsilon\mathbf{j} + H\mathbf{k}$ is infinite. Its length is

$$\sqrt{1 + \varepsilon^2 + H^2} > H.$$

Our first theorem shows how these notions depend on the components of the vectors.

THEOREM 1

> Let **A** and **B** be hyperreal vectors.
>
> (i)   **A** is infinitesimal if and only if all of its components are infinitesimal.
> (ii) **A** is finite if and only if all of its components are finite.
> (iii) **A** is infinite if and only if at least one of its components is infinite.
> (iv) **A** ≈ **B** if and only if $a_1 \approx b_1$, $a_2 \approx b_2$, and $a_3 \approx b_3$.

PROOF (i), (ii), and (iii) are proved using the inequalities

$$|a_1| \leq \sqrt{a_1^2 + a_2^2 + a_3^2}, \qquad |a_2| \leq \sqrt{a_1^2 + a_2^2 + a_3^2}, \qquad |a_3| \leq \sqrt{a_1^2 + a_2^2 + a_3^2},$$

$$\sqrt{a_1^2 + a_2^2 + a_3^2} \leq |a_1| + |a_2| + |a_3|,$$

and (iv) follows easily from (i). We prove (i). Suppose **A** is infinitesimal. This means that its length

$$|\mathbf{A}| = \sqrt{a_1^2 + a_2^2 + a_3^2}$$

is infinitesimal. The inequalities show that $|a_1|$, $|a_2|$, and $|a_3|$ are all between 0 and $|\mathbf{A}|$. Therefore all the components a_1, a_2, and a_3 are infinitesimal. On the other hand, if all the components are infinitesimal, then $|a_1| + |a_2| + |a_3|$ is infinitesimal, and by the last inequality, the length $|\mathbf{A}|$ is infinitesimal.

The following facts are obvious from the definitions.

The only infinitesimal real vector is **0**.

Every real vector is finite.

Every infinitesimal vector is finite.

A *is infinitesimal if and only if* **A** $\approx$ **0**.

Here is a list of algebraic rules for hyperreal vectors. Suppose the scalars and vectors ε, δ, are infinitesimal, c, **A**, are finite but not infinitesimal, and H, **K** are infinite.

Negatives:

$-\delta$ *is infinitesimal.*

$-$**A** *is finite but not infinitesimal.*

$-$**K** *is infinite.*

Sums:

$\delta_1 + \delta_2$ *is infinitesimal.*

A $+ \delta$ *is finite but not infinitesimal.*

$\mathbf{A}_1 + \mathbf{A}_2$ *is finite (possibly infinitesimal).*

K $+ \delta$ *and* **K** $+$ **A** *are infinite.*

Scalar multiples:

$\varepsilon\delta$, $c\delta$, and ε**A** *are infinitesimal.*

c**A** *is finite but not infinitesimal.*

c**K**, H**A**, *and* H**K** *are infinite.*

Inner products:

$\delta_1 \cdot \delta_2$ *and* $\delta \cdot$ **A** *are infinitesimal.*

$\mathbf{A}_1 \cdot \mathbf{A}_2$ *is finite (possibly infinitesimal).*

Each of these rules can be proved using Theorem 1. For example ε**A** is infinitesimal because each of its components εa_1, εa_2, and εa_3 is infinitesimal.

Other combinations, such as ε**K** and $H\delta$, can be either infinitesimal, finite, or infinite.

As in the case of hyperreal numbers, our next step is to introduce the standard part. If **A** is a finite hyperreal vector, the *standard part* of **A** is the real vector

$$st(\mathbf{A}) = st(a_1)\mathbf{i} + st(a_2)\mathbf{j} + st(a_3)\mathbf{k}.$$

Since each component of **A** is infinitely close to its standard part, **A** is infinitely close to its standard part. Thus

$st(\mathbf{A})$ *is the real vector infinitely close to* **A**.

The standard part of an infinite hyperreal vector is undefined.

Here is a list of rules for standard parts of vectors. **A** and **B** are finite hyperreal vectors and c is a finite hyperreal number.

$$st(-\mathbf{A}) = -st(\mathbf{A})$$
$$st(\mathbf{A} + \mathbf{B}) = st(\mathbf{A}) + st(\mathbf{B})$$
$$st(c\mathbf{A}) = st(c)st(\mathbf{A})$$
$$st(\mathbf{A} \cdot \mathbf{B}) = st(\mathbf{A}) \cdot st(\mathbf{B})$$

$$st(\mathbf{A} \times \mathbf{B}) = st(\mathbf{A}) \times st(\mathbf{B})$$
$$st(|\mathbf{A}|) = |st(\mathbf{A})|$$

As an example we prove the equation for inner products,

$$st(\mathbf{A} \cdot \mathbf{B}) = st(a_1 b_1 + a_2 b_2 + a_3 b_3)$$
$$= st(a_1)st(b_1) + st(a_2)st(b_2) + st(a_3)st(b_3)$$
$$= st(\mathbf{A}) \cdot st(\mathbf{B}).$$

Given a nonzero hyperreal vector $\mathbf{A}$, we may form its *unit vector* $\mathbf{U} = \mathbf{A}/|\mathbf{A}|$. The three components of $\mathbf{U}$ are the *direction cosines* of $\mathbf{A}$. As in the case of real vectors, $\mathbf{U}$ has length one and is parallel to $\mathbf{A}$.

Two new concepts which arise in the study of hyperreal vectors are vectors with real length and vectors with real direction. We say that $\mathbf{A}$ has *real length* if $|\mathbf{A}|$ is a real number. We say that $\mathbf{A}$ has *real direction* if the unit vector of $\mathbf{A}$ is real, or equivalently, the direction cosines of $\mathbf{A}$ are real.

There are four types of hyperreal vectors:

(a) Vectors with real length and real direction.
(b) Vectors with real length but nonreal direction.
(c) Vectors with nonreal length but real direction.
(d) Vectors with nonreal length and nonreal direction.

THEOREM 2

A vector is real if and only if it has both real length and real direction.

PROOF $\mathbf{A}$ has real length and direction if and only if $|\mathbf{A}|$ and $\mathbf{U} = \mathbf{A}/|\mathbf{A}|$ are both real if and only if $\mathbf{A} = |\mathbf{A}|\mathbf{U}$ is real.

EXAMPLE 2   Here are some vectors of type (b), (c), and (d), illustrated in Figure 10.8.2.

(b) The vector $\mathbf{B} = \sin\varepsilon\, \mathbf{i} + \cos\varepsilon\, \mathbf{j}$ has real length but nonreal direction (where ε is a positive infinitesimal). $\mathbf{B}$ has length one,

$$|\mathbf{B}| = \sqrt{\sin^2\varepsilon + \cos^2\varepsilon} = 1.$$

However, $\mathbf{B}$ is its own unit vector and is not real, so it has nonreal direction.

(c) The following vectors have nonreal lengths but real directions.

$3\varepsilon\mathbf{i} + 4\varepsilon\mathbf{j}$,   infinitesimal length 5ε,
$(6 + 3\varepsilon)\mathbf{i} + (8 + 4\varepsilon)\mathbf{j}$,   finite length $5(2 + \varepsilon)$,
$3H\mathbf{i} + 4H\mathbf{j}$,   infinite length $5H$.

All three of these vectors are parallel and have the same real unit vector

$$\mathbf{U} = \tfrac{3}{5}\mathbf{i} + \tfrac{4}{5}\mathbf{j}.$$

(d) The vector $\mathbf{D} = \mathbf{i} + \varepsilon\mathbf{j}$ has nonreal length and nonreal direction. Its length is $\sqrt{1 + \varepsilon^2}$, and its unit vector is

$$\mathbf{U} = \frac{1}{\sqrt{1 + \varepsilon^2}}\mathbf{i} + \frac{\varepsilon}{\sqrt{1 + \varepsilon^2}}\mathbf{j}.$$

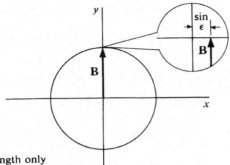

(b) real length only
$\mathbf{B} = \sin \epsilon \mathbf{i} + \cos \epsilon \mathbf{j}$

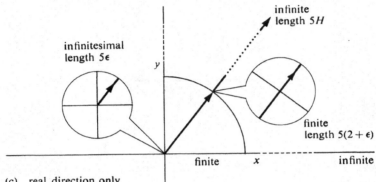

(c) real direction only
$3\epsilon \mathbf{i} + 4\epsilon \mathbf{j}$
$3H\mathbf{i} + 4H\mathbf{j}$
$(6 + 3\epsilon)\mathbf{i} + (8 + 4\epsilon)\mathbf{j}$

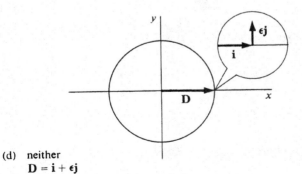

(d) neither
$\mathbf{D} = \mathbf{i} + \epsilon \mathbf{j}$

Figure 10.8.2

Two hyperreal vectors **A** and **B** with unit vectors **U** and **V** are said to be *almost parallel* if either $\mathbf{U} \approx \mathbf{V}$ or $\mathbf{U} \approx -\mathbf{V}$.

EXAMPLE 3   The vectors

$$\mathbf{A} = 2\mathbf{i}, \qquad \mathbf{B} = 2\mathbf{i} + \varepsilon\mathbf{j}, \qquad \mathbf{C} = -\varepsilon\mathbf{i} + \varepsilon^2\mathbf{j}$$

are almost parallel to each other (Figure 10.8.3). Their unit vectors are

$$\mathbf{i}, \qquad \frac{2}{\sqrt{4 + \varepsilon^2}}\mathbf{i} + \frac{\varepsilon}{\sqrt{4 + \varepsilon^2}}\mathbf{j} \approx \mathbf{i}, \qquad \frac{-\varepsilon}{\sqrt{\varepsilon^2 + \varepsilon^4}}\mathbf{i} + \frac{\varepsilon^2}{\sqrt{\varepsilon^2 + \varepsilon^4}}\mathbf{j} \approx -\mathbf{i}.$$

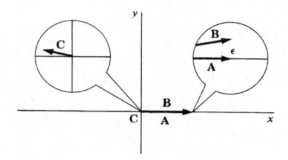

Figure 10.8.3

Let $\mathbf{A} \neq 0$ be a hyperreal vector with unit vector $\mathbf{U} = \mathbf{A}/|\mathbf{A}|$. **A** is almost parallel to the real unit vector $st(\mathbf{U})$. Thus every nonzero hyperreal vector is almost parallel to a real vector.

Now let us consider a vector valued function

$$\mathbf{F}(t) = f_1(t)\mathbf{i} + f_2(t)\mathbf{j} + f_3(t)\mathbf{k}.$$

Each of the real functions f_1, f_2, f_3 has a natural extension to a hyperreal function. Thus the real vector valued function **F** can be extended to a hyperreal vector valued function. When t is a hyperreal number, $\mathbf{F}(t)$ is defined if and only if all of $f_1(t), f_2(t)$, and $f_3(t)$ are defined, and its value is

$$\mathbf{F}(t) = f_1(t)\mathbf{i} + f_2(t)\mathbf{j} + f_3(t)\mathbf{k}.$$

We shall now return to the study of vector derivatives.

THEOREM 3

The vector valued function $\mathbf{F}(t)$ has derivative $\mathbf{V}$ at t if and only if

$$\mathbf{V} = st\left(\frac{\mathbf{F}(t + \Delta t) - \mathbf{F}(t)}{\Delta t}\right)$$

for every nonzero infinitesimal Δt.

This theorem is exactly like the definition of the derivative of a real function in Chapter 2, except that it applies to a vector valued function.

PROOF OF THEOREM 3   Suppose first that $\mathbf{F}'(t) = \mathbf{V}$. This means that

$$f_1'(t)\mathbf{i} + f_2'(t)\mathbf{j} + f_3'(t)\mathbf{k} = v_1\mathbf{i} + v_2\mathbf{j} + v_3\mathbf{k}.$$

Then $f_1'(t) = v_1,$ $f_2'(t) = v_2,$ $f_3'(t) = v_3.$

Let Δt be a nonzero infinitesimal. Then

$$v_1 = st\left(\frac{f_1(t + \Delta t) - f_1(t)}{\Delta t}\right)$$

and similarly for v_2, v_3. It follows that

$$\mathbf{V} = st\left(\frac{\mathbf{F}(t + \Delta t) - \mathbf{F}(t)}{\Delta t}\right).$$

By reversing the steps we see that if the above equation holds for all nonzero infinitesimal Δt, then $\mathbf{V} = \mathbf{F}'(t)$.

We shall now discuss the increment and differential of a vector function. Given a curve

$$\mathbf{X} = \mathbf{F}(t),$$

t is a scalar independent variable and $\mathbf{X}$ a vector dependent variable. We introduce a new scalar independent variable Δt and a new vector dependent variable $\Delta \mathbf{X}$ with the equation

$$\Delta \mathbf{X} = \mathbf{F}(t + \Delta t) - \mathbf{F}(t).$$

$\Delta \mathbf{X}$ is called the *increment* of $\mathbf{X}$. $\Delta \mathbf{X}$ depends on both t and Δt, and is the vector from the point on the curve at t to the point on the curve at $t + \Delta t$ (Figure 10.8.4).

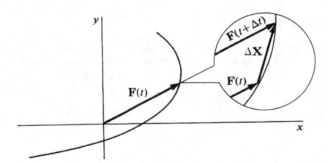

Figure 10.8.4   The Increment of $\mathbf{X}$

Now suppose the vector derivative $\mathbf{F}'(t)$ exists. We introduce another vector dependent variable $d\mathbf{X}$ with the equation

$$d\mathbf{X} = \mathbf{F}'(t)\,\Delta t.$$

$d\mathbf{X}$ is called the *differential* of $\mathbf{X}$. It is customary to write dt for Δt, so we get the familiar quotient formulas

$$d\mathbf{X} = \mathbf{F}'(t)\,dt, \qquad \frac{d\mathbf{X}}{dt} = \mathbf{F}'(t).$$

The relationship between the vector increment and differential may be summarized as follows. At each value t where $\mathbf{F}'(t)$ exists and is not zero, and for each nonzero infinitesimal Δt, we have:

$d\mathbf{X}$ is an infinitesimal vector tangent to the curve $\mathbf{X} = \mathbf{F}(t)$.
$\Delta\mathbf{X}$ is an infinitesimal vector which is almost parallel to $d\mathbf{X}$.

$d\mathbf{X}$ and $\Delta\mathbf{X}$ are infinitely close compared to Δt, i.e,

$$\frac{\Delta\mathbf{X}}{\Delta t} \approx \frac{d\mathbf{X}}{dt}$$

as shown in Figure 10.8.5.

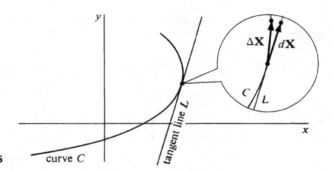

Figure 10.8.5   curve C

PROBLEMS FOR SECTION 10.8

In Problems 1–20, determine whether the given vector or scalar is infinitesimal, finite but not infinitesimal, or infinite. (ε, δ are infinitesimal but not zero and H, $\mathbf{K}$ are infinite.)

1	$2\delta_1 - 5\delta_2$	**2**	$5\delta - 3\mathbf{K}$
3	$H(2\mathbf{i} - \mathbf{j})$	**4**	$\varepsilon(5\mathbf{i} + \mathbf{j})$
5	$(2 + \varepsilon)\mathbf{i} + (3 - \varepsilon)\mathbf{j}$	**6**	$\dfrac{5\mathbf{i} + 6\mathbf{j} + \mathbf{k}}{\varepsilon}$
7	$\varepsilon\mathbf{i} - 4\mathbf{j} + H\mathbf{k}$	**8**	$\mathbf{K}/|\mathbf{K}|$
9	$\mathbf{K} \cdot \mathbf{K}$	**10**	$5\delta/|\delta|$
11	$(H_1\mathbf{i}) \cdot (H_2\mathbf{j})$	**12**	$(\mathbf{i} + \delta_1) \cdot (\mathbf{j} + \delta_2)$
13	$(H\mathbf{i} + \mathbf{j}) \cdot (\mathbf{i} + \varepsilon\mathbf{j})$		
14	$(\sqrt{H + 1}\,i + \sqrt{H}\mathbf{j}) \cdot (\sqrt{H + 1}\,i - \sqrt{H}\mathbf{j})$		
15	$(H\mathbf{i} + H^2\mathbf{j}) \cdot (H^{-2}\mathbf{i} + H^{-1}\mathbf{j})$	**16**	$\delta_1 \times \delta_2$
17	$(2\mathbf{i} + 3\mathbf{j} - \mathbf{k}) \times \delta$	**18**	$(\mathbf{i} + \delta_1) \times (\mathbf{j} + \delta_2)$
19	$(H_1\mathbf{i}) \times (H_2\mathbf{j})$	**20**	$(H\mathbf{i}) \times (\mathbf{j} + \delta)$

In Problems 21–30, compute the standard part. Assume $\mathbf{A}$, $\mathbf{B}$ are real.

21 $\dfrac{\cos(x + \Delta x)\mathbf{i} + \sin(x + \Delta x)\mathbf{j} - (\cos x\mathbf{i} + \sin x\mathbf{j})}{\Delta x}$

22 $(2 + \varepsilon)\mathbf{i} + (3 - \varepsilon)\mathbf{j}$ **23** $(5 + 6\varepsilon)(2\mathbf{i} - 4\mathbf{j} + \mathbf{k})$

24 $\dfrac{2\varepsilon\mathbf{i} + 4\varepsilon^2\mathbf{j} + 6\varepsilon^3\mathbf{k}}{\varepsilon + \varepsilon^2 + \varepsilon^3}$ **25** $\dfrac{(2H + 1)\mathbf{i} + (3H - 1)\mathbf{j} - H\mathbf{k}}{H + 4}$

26 $\dfrac{(\mathbf{A} + \boldsymbol{\delta}) \cdot (\mathbf{B} + \boldsymbol{\delta}) - \mathbf{A} \cdot \mathbf{B}}{|\boldsymbol{\delta}|}$ where $st\left(\dfrac{\boldsymbol{\delta}}{|\boldsymbol{\delta}|}\right) = \mathbf{U}$

27 $|H\mathbf{i} + \mathbf{j}| - H$

28 $|H\mathbf{i} + \sqrt{H}\mathbf{j}| - H$

29 $\dfrac{|\mathbf{A} + \boldsymbol{\delta}| - |\mathbf{A}|}{|\boldsymbol{\delta}|}$ where $st\left(\dfrac{\boldsymbol{\delta}}{|\boldsymbol{\delta}|}\right) = \mathbf{U}$

30 $\dfrac{(x + \Delta x)^2(\mathbf{A} + \Delta x\mathbf{B}) - x^2\mathbf{A}}{\Delta x}$

In Problems 31–40 determine whether or not the vector has (a) real length, (b) real direction.

31 $H\mathbf{i} + \sqrt{H}\mathbf{j}$ **32** $\mathbf{i} + \varepsilon\mathbf{j} + \varepsilon^2\mathbf{k}$

33 $(2\mathbf{i} + 2\sqrt{H}\mathbf{j} + H\mathbf{k})/(H + 2)$ **34** $2H\mathbf{i} - 3H\mathbf{j}$

35 $\cos(2 + \varepsilon)\mathbf{i} + \sin(2 + \varepsilon)\mathbf{j}$ **36** $\dfrac{\sqrt{5}}{\sqrt{1 + \varepsilon^2}}\mathbf{i} + \dfrac{\varepsilon}{\sqrt{1 + \varepsilon^2}}\mathbf{j} - \dfrac{2\varepsilon}{\sqrt{1 + \varepsilon^2}}\mathbf{k}$

37 $\dfrac{1}{\sqrt{1 + \varepsilon^2}}\mathbf{i} + \dfrac{\varepsilon}{\sqrt{1 + \varepsilon^2}}\mathbf{j} + \dfrac{\varepsilon}{\sqrt{1 + \varepsilon^2}}\mathbf{k}$

38 $\dfrac{\varepsilon}{\sqrt{1 + \varepsilon^2}}\mathbf{i} + \dfrac{2\varepsilon}{\sqrt{1 + \varepsilon^2}}\mathbf{j} - \dfrac{3\varepsilon}{\sqrt{1 + \varepsilon^2}}\mathbf{k}$

39 $5\cos(1 + \varepsilon)\mathbf{i} + 3\sin(1 + \varepsilon)\mathbf{j} + 4\sin(1 + \varepsilon)\mathbf{k}$

40 $5\cos(1 + \varepsilon)\mathbf{i} + 3\cos(1 + \varepsilon)\mathbf{j} + 4\cos(1 + \varepsilon)\mathbf{k}$

41   Prove that $st(\mathbf{A} + \mathbf{B}) = st(\mathbf{A}) + st(\mathbf{B})$.

42   Prove that $st(\mathbf{A} \times \mathbf{B}) = st(\mathbf{A}) \times st(\mathbf{B})$.

43   Prove that if $\mathbf{A}$ is infinite and $\mathbf{A} - \mathbf{B}$ is finite, then $\mathbf{A}$ is almost parallel to $\mathbf{B}$.

44   Prove that if $\mathbf{A}$ is finite but not infinitesimal and $\mathbf{A} - \mathbf{B}$ is infinitesimal, then $\mathbf{A}$ is almost parallel to $\mathbf{B}$.

45   Prove that a vector which is parallel to a real vector has a real direction.

The following problems use the notion of a continuous vector valued function. $\mathbf{F}(t)$ is said to be *continuous* at t_0 if each of the components $f_1(t)$, $f_2(t)$, and $f_3(t)$ is continuous at t_0.

☐ **46**   Prove that $\mathbf{F}(t)$ is continuous at t_0 if and only if whenever $t \approx t_0$, $\mathbf{F}(t) \approx \mathbf{F}(t_0)$.

☐ **47**   Assume $\mathbf{F}(t)$ and $\mathbf{G}(t)$ are continuous at t_0. Prove that the following functions are continuous at t_0.

$$\mathbf{F}(t) + \mathbf{G}(t), \quad\quad \mathbf{F}(t) \cdot \mathbf{G}(t), \quad\quad |\mathbf{F}(t)|, \quad\quad \mathbf{F}(t) \times \mathbf{G}(t).$$

☐ **48**   Prove that if $\mathbf{F}(t)$ and $h(t)$ are continuous at t_0, so is $h(t)\mathbf{F}(t)$.

EXTRA PROBLEMS FOR CHAPTER 10

1   Find the vector represented by the directed line segment $\overrightarrow{PQ}$ where $P = (4, 7)$, $Q = (9, -5)$.

2   Find the vector $\mathbf{A}/|\mathbf{B}|$ where $\mathbf{A} = 5\mathbf{i} - 10\mathbf{j}$, $\mathbf{B} = 3\mathbf{i} - 4\mathbf{j}$.

3   If $\mathbf{A} = 7\mathbf{i} + 2\mathbf{j}$, $\mathbf{B} = -4\mathbf{i} + \mathbf{j}$, find a vector $\mathbf{C}$ such that $\mathbf{A} + \mathbf{B} + \mathbf{C} = \mathbf{0}$.

4   An object originally has position vector $\mathbf{P} = 12\mathbf{i} - 5\mathbf{j}$ and is displaced twice, once by the vector $\mathbf{A} = 3\mathbf{i} + 3\mathbf{j}$ and once by the vector $\mathbf{B} = 6\mathbf{j}$. Find the new position vector.

5   Two traders initially have commodity vectors $\mathbf{A}_0 = 18\mathbf{i} + 2\mathbf{j}$, $\mathbf{B}_0 = 20\mathbf{j}$. They exchange in such a way that their new commodity vectors are equal, $\mathbf{A}_1 = \mathbf{B}_1$. Find their new commodity vectors.

6   Find a vector equation for the line through $P(2, 4)$ with direction vector $\mathbf{D} = \mathbf{i}$.

7   Find a vector equation for the line $3x + 4y = -1$.

8   Find the midpoint of the line AB where $A = (0, 0)$, $B = (-4, 2)$.

9   Find the point of intersection of the diagonals of the parallelogram $A(-1, -3)$, $B(0, -3)$, $C(5, 8)$, $D(4, 8)$.

10 Find the vector represented by $\overrightarrow{PQ}$ where $P = (4, 2, 1)$, $Q = (9, 6, 0)$.

11 Find the direction cosines of $\mathbf{A} = \mathbf{i} - 10\mathbf{j} + 2\mathbf{k}$.

12 If an object at rest has three forces acting on it and two of the forces are $\mathbf{F}_1 = \mathbf{i} + 3\mathbf{j} - \mathbf{k}$, $\mathbf{F}_2 = 4\mathbf{i} - 3\mathbf{j} + 2\mathbf{k}$, find the third force $\mathbf{F}_3$.

13 Find the force required to cause an object of mass 100 to accelerate with the acceleration vector $\mathbf{A} = \mathbf{i} - 5\mathbf{j} + 3\mathbf{k}$.

14 If a trader has the commodity vector $\mathbf{A} = 5\mathbf{i} + 10\mathbf{j} + 15\mathbf{k}$ and sells the commodity vector $\mathbf{B} = 5\mathbf{i} + 5\mathbf{j} + 5\mathbf{k}$, find his new commodity vector.

15 Find the vector equation of the line through $P(1, 4, 3)$ and $Q(1, 4, 4)$.

16 Find the vector equation of the line through $P(1, 1, 1)$ with direction cosines $(1/2, -1/2, 1/\sqrt{2})$.

17 Determine whether the vectors $\mathbf{A} = 3\mathbf{i} - 4\mathbf{j} + 5\mathbf{k}$, $\mathbf{B} = 10\mathbf{i} + 5\mathbf{j} - 2\mathbf{k}$, are perpendicular.

18 Find the cost of the commodity vector $\mathbf{A} = 8\mathbf{i} + 20\mathbf{j} + 10\mathbf{k}$ at the price vector $\mathbf{P} = 6\mathbf{i} + 12\mathbf{j} + 15\mathbf{k}$.

19 Find the amount of work done by a force vector $\mathbf{F} = 10\mathbf{i} - 20\mathbf{j} + 5\mathbf{k}$ acting along the displacement vector $\mathbf{S} = 2\mathbf{i} + 3\mathbf{j} + 4\mathbf{k}$.

20 Find a vector in the plane perpendicular to $\mathbf{A} = -2\mathbf{i} + 3\mathbf{j}$.

21 Find a vector in space perpendicular to both

$$\mathbf{A} = \mathbf{i} + \mathbf{j} + 2\mathbf{k}, \qquad \mathbf{B} = 2\mathbf{i} + \mathbf{j} + \mathbf{k}.$$

22 Find two vectors in space perpendicular to each other and to $\mathbf{A} = \mathbf{i} + \mathbf{j} + \mathbf{k}$.

23 Sketch the plane $x + 2y + 3z = 6$.

24 Sketch the plane $3x - z = 0$.

25 Find a scalar equation for the plane through the point $(1, 3, 2)$ with normal vector $\mathbf{N} = -\mathbf{i} - \mathbf{j} + 2\mathbf{k}$.

26 Find a scalar equation for the plane through the points $A(4, 1, 1)$, $B(2, 3, 4)$, $C(5, 1, 6)$.

27 Find the point where the line $\mathbf{X} = 2\mathbf{i} - 2\mathbf{j} + 4\mathbf{k} + t\mathbf{i}$ intersects the plane $x + y + z = 1$.

28 A bug is crawling along a spoke of a wheel towards the rim at a inches per second. At the same time the wheel is rotating counterclockwise at b radians per second. The center of the wheel is at $(0, 0)$ and at time $t = 0$, the bug is at $(0, 0)$. Find the vector equation for the motion of the bug, $0 \le t \le 1/a$.

29 The sphere $x^2 + y^2 + z^2 = 1$ is rotating about the z-axis counterclockwise at one radian per second. A bug crawls south at one inch per second along a great circle. At time $t = 0$ the bug is at $(0, 0, 1)$ and the great circle is in the (x, z) plane. Find the vector equation for the motion of the bug, $0 \leq t \leq \pi$. (There are two possible answers.)

30 Find the velocity, speed, and acceleration of the bug in Problem 28.

31 Find the velocity, speed, and acceleration of the bug in Problem 29.

32 Find the derivative of $\mathbf{X} = (\cosh t)\mathbf{i} + (\sinh t)\mathbf{j}$.

33 Find the line tangent to the curve

$$\mathbf{X} = \frac{\mathbf{i}}{t+1} + \frac{\mathbf{j}}{t+2} + \frac{\mathbf{k}}{t+3} \quad \text{at } t = 0.$$

34 Find the length of the curve

$$\mathbf{X} = (\cosh^2 t)\mathbf{i} + (\sinh^2 t)\mathbf{j} + (\sqrt{8}\sinh t)\mathbf{k}, \qquad 0 \leq t \leq 1.$$

35 Find the position vector of a particle which moves with velocity

$$\mathbf{V} = (e^t \sin e^t)\mathbf{i} + (e^t \cos e^t)\mathbf{j} + e^t \mathbf{k},$$

if the particle is at the origin at $t = 0$.

36 If $\varepsilon > 0$ is infinitesimal, determine whether or not the vector $(\sin \varepsilon)\mathbf{i} + (1 - \cos \varepsilon)\mathbf{j}$ is infinitesimal.

37 Determine whether or not the vector in Problem 36 has real direction.

38 If $\varepsilon > 0$ is infinitesimal, find the standard part of the vector

$$\frac{(\sin \varepsilon)\mathbf{i} + \varepsilon^2 \mathbf{j} + (e^\varepsilon - 1)\mathbf{k}}{\varepsilon}.$$

☐ **39** Let $\mathbf{D}$ be a direction vector of a line L in the (x, y) plane. Prove that the set of all direction vectors of L is equal to the set of all scalar multiples of $\mathbf{D}$.

☐ **40** Let $\mathbf{U}$ and $\mathbf{V}$ be perpendicular unit vectors in the plane. Prove that for any vector $\mathbf{A}$,

$$|\mathbf{A}|^2 = (\mathbf{A} \cdot \mathbf{U})^2 + (\mathbf{A} \cdot \mathbf{V})^2.$$

☐ **41** Let $\mathbf{U}$ and $\mathbf{V}$ be perpendicular unit vectors in the plane. Prove that for any vector $\mathbf{A}$,

$$\mathbf{A} = (\mathbf{A} \cdot \mathbf{U})\mathbf{U} + (\mathbf{A} \cdot \mathbf{V})\mathbf{V}.$$

Hint: Let $\mathbf{B} = (\mathbf{A} \cdot \mathbf{U})\mathbf{U} + (\mathbf{A} \cdot \mathbf{V})\mathbf{V}$ and show that $\mathbf{B} \cdot \mathbf{U} = \mathbf{A} \cdot \mathbf{U}$ and $\mathbf{B} \cdot \mathbf{V} = \mathbf{A} \cdot \mathbf{V}$. $\mathbf{A} \cdot \mathbf{U}$ and $\mathbf{A} \cdot \mathbf{V}$ are called the $\mathbf{U}$ and $\mathbf{V}$ *components* of $\mathbf{A}$.

☐ **42** Let $\mathbf{A}$ and $\mathbf{B}$ be two vectors in the plane which are not parallel. Prove that every vector $\mathbf{C}$ in the plane can be expressed uniquely in the form $\mathbf{C} = s\mathbf{A} + t\mathbf{B}$.

☐ **43** Prove the Schwartz inequality $|\mathbf{A} \cdot \mathbf{B}| \leq |\mathbf{A}||\mathbf{B}|$ for vectors $\mathbf{A}, \mathbf{B}$ in space.

☐ **44** Prove that if s and t are positive scalars, then the angle between two vectors $\mathbf{A}$ and $\mathbf{B}$ in space is equal to the angle between $s\mathbf{A}$ and $t\mathbf{B}$.

☐ **45** Let p be a plane in space with position vector $\mathbf{P}$ and nonparallel direction vectors $\mathbf{C}$ and $\mathbf{D}$. Prove that $\mathbf{Q}$ is a position vector of p if and only if $\mathbf{Q} = \mathbf{P} + s\mathbf{C} + t\mathbf{D}$ for some scalars s and t.

Hint: If $\mathbf{E}$ is a direction vector of p, then $\mathbf{E} \times \mathbf{D}$ is zero or parallel to $\mathbf{C} \times \mathbf{D}$, so $\mathbf{E} \times \mathbf{D} = s(\mathbf{C} \times \mathbf{D})$ for some s, $(\mathbf{E} - s\mathbf{C}) \times \mathbf{D} = 0$, and hence $\mathbf{E} - s\mathbf{C}$ is parallel to $\mathbf{D}$.

☐ **46** Let A, B, C be three distinct points in space whose plane does not pass through the origin. Prove that any vector $\mathbf{P}$ may be expressed uniquely in the form $\mathbf{P} = s\mathbf{A} + t\mathbf{B} + u\mathbf{C}$.

Hint: Consider the point where the line $\mathbf{X} = s\mathbf{A}$ intersects the plane with position vector $\mathbf{P}$ and direction vectors $\mathbf{B}$ and $\mathbf{C}$.

☐ **47** Let C be a curve represented by the vector equation $\mathbf{X} = \mathbf{F}(s)$, $0 \leq s \leq b$. Assume that the length of the curve from $\mathbf{F}(0)$ to $\mathbf{F}(s)$ equals s, and that no tangent line crosses the curve. A string is stretched along the curve, attached at the end b, and carefully un-

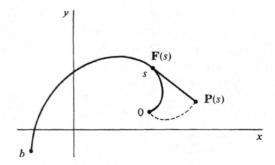

wrapped starting at 0 as shown in the figure. Show that the point at the end of the string has the position vector $\mathbf{P}(s) = \mathbf{F}(s) + s\mathbf{F}'(s)$.

☐ **48** A ball is thrown with initial velocity vector $\mathbf{V}_0 = b(\cos \alpha \mathbf{i} + \sin \alpha \mathbf{j})$ and position vector $\mathbf{S}_0 = \mathbf{0}$ at time $t = 0$. Its acceleration at time t is $\mathbf{A} = -32\mathbf{j}$. Find its position at time t, its maximum height, and the point where it hits the ground.

PARTIAL DIFFERENTIATION

11.1 SURFACES

The rectangular coordinate axes in (x, y, z) space are drawn as in Figure 11.1.1. Points in real space are identified with triples (x, y, z) of real numbers, and points in hyperreal space with triples (x, y, z) of hyperreal numbers. The set of all points for which an equation is true is called the *graph*, or *locus*, of the equation. The graph of an equation in the three variables x, y, and z is a surface in space. We have seen in the last chapter that the graph of a linear equation

$$ax + by + cz = d$$

is a plane. The graphs of other equations are often curved surfaces. The simplest planes are:

The vertical planes $x = x_0$ perpendicular to the x-axis. The plane $x = 0$ is called the (y, z) *plane*.

The vertical planes $y = y_0$ perpendicular to the y-axis. The plane $y = 0$ is called the (x, z) *plane*.

The horizontal planes $z = z_0$ perpendicular to the z-axis. The plane $z = 0$ is called the (x, y) *plane*.

Examples of the planes $x = x_0$, $y = y_0$ and $z = z_0$ are pictured in Figure 11.1.2.

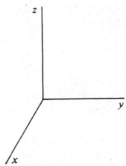

Figure 11.1.1

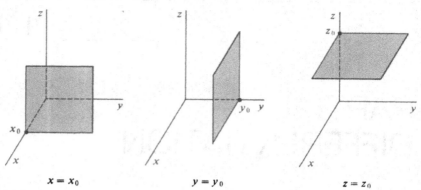

$x = x_0$ $y = y_0$ $z = z_0$

Figure 11.1.2

By the *graph* of a function f of two variables we mean the graph of the equation $z = f(x, y)$. Recall that a real function of two variables is a set of ordered triples (x, y, z) such that for each (x, y) there is at most one z with $z = f(x, y)$. Geometrically this means that the graph of a function intersects each vertical line through (x, y) in at most one point (x, y, z). The value of z is the height of the surface above (x, y). Figure 11.1.3 shows part of a surface $z = f(x, y)$.

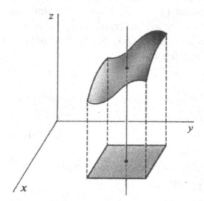

Figure 11.1.3

Whenever one quantity depends on two others we have a function of two variables. The height of a surface above (x, y) is one example. A few other examples are: the density of a plane object at (x, y), the area of a rectangle of length x and width y, the size of a wheat crop in a season with rainfall r and average temperature t, the number of items which can be sold if the price is p and the advertising budget is a, and the force of the sun's gravity on an object of mass m at distance d.

A rough sketch of the graph can be very helpful in understanding a function of two variables or an equation in three variables. In this section we do two things. First we describe a class of surfaces whose equations are simple and easily recognized, the quadric surfaces. After that we shall give a general method for sketching the graph of an equation. Graph paper with lines in the x, y, and z directions is available in many bookstores.

The graph of a second degree equation in x, y, and z is called a *quadric surface*. These surfaces correspond to the conic sections in the plane. There are several types of quadric surfaces. We shall present each of them in its simplest form.

Quadric Cylinders   If z does not appear in an equation, its graph will be a cylinder parallel to the z-axis. The cylinder is generated by a line parallel to the z-axis moving along a curve in the plane $z = 0$.

The graph (in space) of $\dfrac{x^2}{a^2} + \dfrac{y^2}{b^2} = 1$ is an *elliptic cylinder*.

It intersects any horizontal plane $z = z_0$ in an ellipse.

The graph of $y = ax^2 + bx + c$ is a *parabolic cylinder*.

The graph of $\dfrac{x^2}{a^2} - \dfrac{y^2}{b^2} = c$ is a *hyperbolic cylinder*.

Cylinders parallel to other axes are similar. The three types of quadric cylinders are shown in Figure 11.1.4.

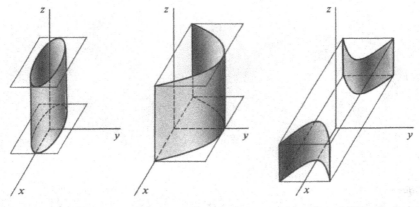

(a) Elliptic cylinder   (b) Parabolic cylinder   (c) Hyperbolic cylinder

Figure 11.1.4

The Sphere   The sphere of radius r and center $P(a, b, c)$ has the equation

$$(x - a)^2 + (y - b)^2 + (z - c)^2 = r^2.$$

It is the set of all points at distance r from P (Figure 11.1.5). A sphere intersects any plane in a circle (possibly a single point or no intersection).

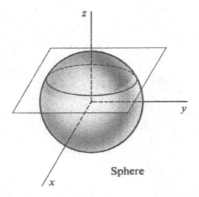

Sphere

Figure 11.1.5

The Ellipsoid $\dfrac{x^2}{a^2} + \dfrac{y^2}{b^2} + \dfrac{z^2}{c^2} = 1.$

This egg-shaped surface intersects a plane perpendicular to any coordinate axis in an ellipse (Figure 11.1.6). It is inscribed in the rectangular solid

$$-a \leq x \leq a, \qquad -b \leq y \leq b, \qquad -c \leq z \leq c.$$

The Elliptic Cone $\dfrac{x^2}{a^2} + \dfrac{y^2}{b^2} = \dfrac{z^2}{c^2}.$

This surface intersects a horizontal plane $z = z_0$ in an ellipse, and the vertical planes $x = 0$ and $y = 0$ in two intersecting lines (Figure 11.1.7).

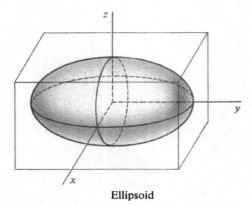

Ellipsoid

Figure 11.1.6

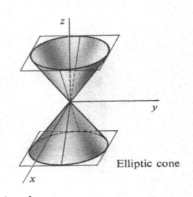

Elliptic cone

Figure 11.1.7

The Elliptic Paraboloid $\dfrac{x^2}{a^2} + \dfrac{y^2}{b^2} = \dfrac{z}{c}.$

This surface intersects a horizontal plane $z = z_0$ in an ellipse and a vertical plane $x = x_0$ or $y = y_0$ in a parabola. It is shaped like a bowl if c is positive and a mound if c is negative (Figure 11.1.8).

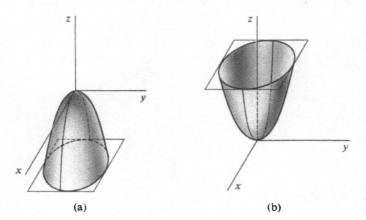

Figure 11.1.8   (a)   (b)

The Hyperboloid of One Sheet $\dfrac{x^2}{a^2} + \dfrac{y^2}{b^2} - \dfrac{z^2}{c^2} = 1.$

The intersection of this surface with a horizontal plane $z = z_0$ is an ellipse. The intersection with a vertical plane $x = x_0$ or $y = y_0$ is a hyperbola (Figure 11.1.9).

The Hyperboloid of Two Sheets $-\dfrac{x^2}{a^2} - \dfrac{y^2}{b^2} + \dfrac{z^2}{c^2} = 1.$

The surface has an upper sheet with $z \geq c$ and a lower sheet with $z \leq -c$. It intersects a horizontal plane $z = z_0$ in an ellipse if $|z_0| > c$. It intersects a vertical plane $x = x_0$ or $y = y_0$ in a hyperbola (Figure 11.1.10).

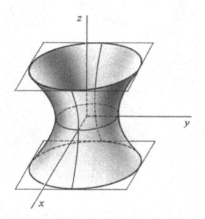

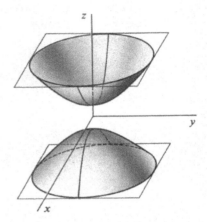

Hyperboloid of one sheet

Figure 11.1.9

Hyperboloid of two sheets

Figure 11.1.10

The Hyperbolic Paraboloid $\dfrac{x^2}{a^2} - \dfrac{y^2}{b^2} = \dfrac{z}{c}.$

This surface has the shape of a saddle. It intersects a horizontal plane $z = z_0$ in a hyperbola, and a vertical plane $x = x_0$ or $y = y_0$ in a parabola (Figure 11.1.11).

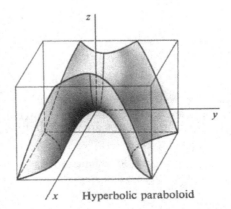

Figure 11.1.11

Hyperbolic paraboloid

We shall describe a method for sketching cylinders and then other graphs in space. We concentrate on a finite portion of (x, y, z) space.

EXAMPLE 1 Sketch the portion of the cylinder $x^2 + y^2 = 1$ where $1 \le z \le 2$ (Figure 11.1.12).

Step 1 Draw the curve $x^2 + y^2 = 1$ in the (x, y) plane. The curve is a circle of radius one.

Step 2 Draw the three coordinate axes and the horizontal planes $z = 1$, $z = 2$.

Step 3 Draw the circles $x^2 + y^2 = 1$ where the surface intersects the two planes $z = 1$, $z = 2$.

Step 4 Complete the sketch by drawing heavy lines for all edges which would be visible on an "opaque" model of the given surface. This surface is called a *circular cylinder*.

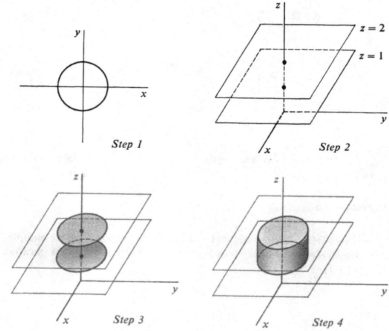

Figure 11.1.12

EXAMPLE 2 Sketch the part of the cylinder $z = x^2$ where $0 \le y \le 2$, $0 \le z \le 1$. This is a parabolic cylinder parallel to the y-axis, because y does not appear in the equation. The four steps are shown in Figure 11.1.13.

For sketching the graph of a function $z = f(x, y)$, a *topographic map*, or *contour map*, can often be used as a first step. It is a method of representing a surface which is often found in atlases. In a topographic map, the curves $f(x, y) = z_0$ are

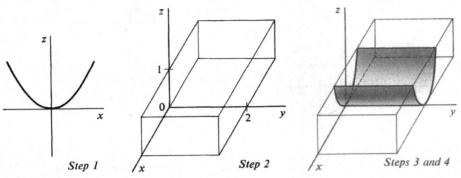

Step 1   *Step 2*   *Steps 3 and 4*

Figure 11.1.13

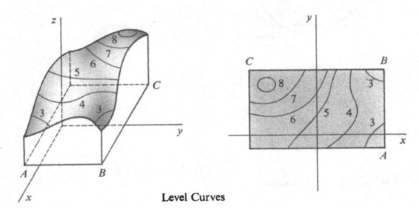

Level Curves

Figure 11.1.14

sketched in the (x, y) plane for several different constants z_0, and each curve is labeled (Figure 11.1.14). These curves are called *level curves*, or *contours*.

EXAMPLE 3 Sketch the part of the surface $z = x^2 + y^2$ where $-1 \leq z \leq 1$. This is an elliptic paraboloid (Figure 11.1.15).

Step 1 Draw the topographic map. The level curves are circles.

Step 2 Draw the axes and the planes $z = -1$, $z = 1$.

Step 3 Draw the intersections of the surface with the planes $z = -1$, $z = 1$ and also the planes $x = 0$ and $y = 0$.

$$
\begin{aligned}
z = -1: &\quad \text{No intersection.} \\
z = 1: &\quad \text{The circle } x^2 + y^2 = 1. \\
x = 0: &\quad \text{The parabola } z = y^2. \\
y = 0: &\quad \text{The parabola } z = x^2.
\end{aligned}
$$

Step 4 Complete the figure with heavy lines for visible edges.

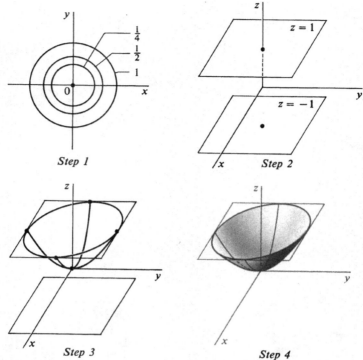

Figure 11.1.15

EXAMPLE 4 Graph the function

$$\frac{x^2}{4} - y^2 = z,$$

where $-3 \le x \le 3$, $-2 \le y \le 2$, $-1 \le z \le 1$. This is a hyperbolic paraboloid (Figure 11.1.16).

Step 1 Draw a topographic map. The level curves are hyperbolas.

Step 2 Draw the axes and rectangular solid.

Step 3 Draw the curves where the surface intersects the faces and also the planes $x = 0$, $y = 0$. The topographic map gives the curves on $z = -1$, $z = 0$, and $z = 1$. The curves on $x = 0$ and $y = 0$ are parabolas.

Step 4 Complete Figure 11.1.16.

EXAMPLE 5 Sketch the surface

$$-x^2 - \frac{y^2}{4} + z^2 = 1$$

where $-2 \le z \le 2$. This is a hyperboloid of two sheets (Figure 11.1.17).

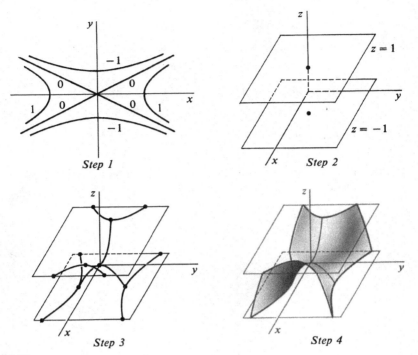

Step 1

Step 2

Step 3

Step 4

Figure 11.1.16

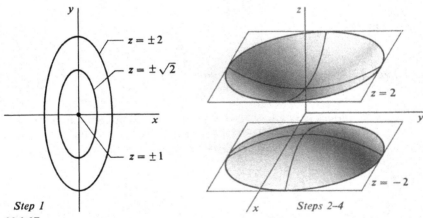

$z = \pm 2$

$z = \pm \sqrt{2}$

$z = \pm 1$

Step 1

Figure 11.1.17

$z = 2$

$z = -2$

Steps 2–4

Although it is not a function, it can be broken up into two functions

$$z = \sqrt{1 + x^2 + \frac{y^2}{4}}, \qquad z = -\sqrt{1 + x^2 + \frac{y^2}{4}}.$$

Step 1 Draw topographic maps for $z = \sqrt{1 + x^2 + y^2/4}$ and $z = -\sqrt{1 + x^2 + y^2/4}$. The level curves are ellipses.

Step 2 Draw the axes and the planes $z = 2, z = -2$.

Step 3 Draw the intersections of the surface with the planes

$$z = -2, \qquad z = 2, \qquad x = 0, \qquad y = 0.$$

The surface intersects $x = 0$ and $y = 0$ in the hyperbolas

$$-\tfrac{1}{4}y^2 + z^2 = 1, \qquad -x^2 + z^2 = 1.$$

Step 4 Complete Figure 11.1.17.

EXAMPLE 6 Graph the *sum function* $z = x + y$. The graph is a plane. A topographic map and sketch of the surface are shown in Figure 11.1.18.

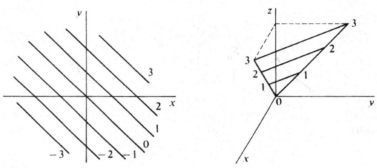

Figure 11.1.18

EXAMPLE 7 Sketch the graph of the *product function* $z = xy$, where

$$-2 \le x \le 2, \qquad -2 \le y \le 2, \qquad -1 \le z \le 1.$$

The surface is saddle shaped. It intersects the horizontal plane $z = z_0$ in the curve $y = z_0/x$. It intersects the vertical planes $x = x_0$ and $y = y_0$ in the lines $z = x_0 y$ and $z = xy_0$. The surface is shown in Figure 11.1.19.

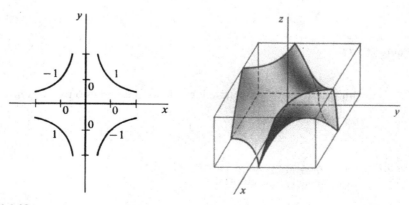

Figure 11.1.19

EXAMPLE 8   Graph the function $z = \sqrt{x} + y^2$ where

$$0 \leq x \leq 1, \qquad -1 \leq y \leq 1, \qquad 0 \leq z \leq 1.$$

Step 1   The topographic map has level curves

$$\sqrt{x} + y^2 = c, \qquad x = (c - y^2)^2 \quad \text{with} \quad y^2 \leq c.$$

The derivative $dx/dy = 4y(c - y^2)$ has zeros at $y = 0$ and $y = \pm\sqrt{c}$. The table shows that the curves are bell shaped.

y	x	dx/dy	
$-\sqrt{c}$	0	0	Min
0	c^2	0	Max
$\sqrt{c}$	0	0	Min

Step 2   Draw the rectangular solid.

Step 3   The surface intersects the plane $x = 0$ in the parabola $z = y^2$, and intersects the plane $y = 0$ in the curve $z = \sqrt{x}$. It intersects the plane $z = 1$ in the curve $x = (1 - y^2)^2$.

Step 4   The surface, shown in Figure 11.1.20, is shaped like a beaker spout.

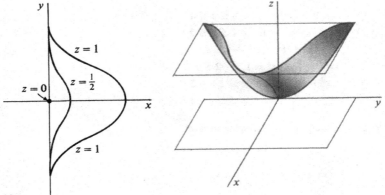

Figure 11.1.20

PROBLEMS FOR SECTION 11.1

Sketch the following graphs in (x, y, z) space.

1   $x^2 + y^2 = 4, \quad -1 \leq z \leq 1$

2   $x^2 + z^2 = 1, \quad 0 \leq y \leq 2$

3   $(x - 2)^2 + (y - 1)^2 = 1, \quad -1 \leq z \leq 1$

4   $(y - 1)^2 + (z + 1)^2 = 1, \quad 0 \leq x \leq 3$

5   $y^2 + z = 1, \quad 0 \leq z, \quad 0 \leq x \leq 2$

6 $y = x^2 - x, \quad y \le 0, \quad 0 \le z \le 2$

7 $x^2 + \frac{1}{4}y^2 = 1, \quad 0 \le z \le 3$

8 $y^2 - x^2 = 1, \quad -2 \le x \le 2, \quad 0 \le z \le 4$

9 $x = \sin y, \quad 0 \le y \le \pi, \quad 0 \le z \le 2$

10 $z = e^{-x}, \quad 0 \le x \le 2, \quad 0 \le y \le 2$

11 $x^2 + y^2 + z^2 = 4$

12 $(x - 1)^2 + y^2 + (z + 1)^2 = 1$

13 $(x - 1)^2 + (y - 2)^2 + (z - 3)^2 = 9$

14 $x^2 + (y + 4)^2 + (z - 2)^2 = 4$

15 $x^2 + \frac{1}{4}y^2 + \frac{1}{9}z^2 = 1$

16 $\frac{1}{4}x^2 + \frac{1}{25}y^2 + z^2 = 1$

Make contour maps and sketch the following surfaces.

17 $x^2 + y^2 = z^2, \quad -4 \le z \le 4$

18 $x^2 + y^2 = 4z^2, \quad -1 \le z \le 1$

19 $x^2 + \frac{1}{4}y^2 = z^2, \quad -4 \le z \le 4$

20 $z = \frac{1}{9}x^2 + y^2, \quad -4 \le z \le 4$

21 $z = -x^2 - y^2, \quad -4 \le z \le 4$

22 $2z = -x^2 - \frac{1}{4}y^2, \quad -4 \le z \le 4$

23 $x^2 + \frac{1}{4}y^2 - z^2 = 1, \quad -4 \le z \le 4$

24 $x^2 + y^2 - 9z^2 = 1, \quad -2 \le z \le 2$

25 $-x^2 - y^2 + z^2 = 1, \quad -4 \le z \le 4$

26 $-4x^2 - y^2 + 4z^2 = 1, \quad -2 \le z \le 2$

27 $z = x^2 - y^2, \quad -2 \le x \le 2, \quad -2 \le y \le 2$

28 $z = y^2 - x^2, \quad -2 \le x \le 2, \quad -2 \le y \le 2$

Make contour maps of the following surfaces.

29 $z = x - y$

30 $z = y - 2x$

31 $z = (x^2 + y^2 + 1)^{-1}, \quad -4 \le x \le 4, \quad -4 \le y \le 4$

32 $z = \dfrac{x^2 + y^2}{x^2 + y^2 + 1}, \quad -4 \le x \le 4, \quad -4 \le y \le 4$

33 $z = x + y^2, \quad -2 \le x \le 2, \quad -2 \le y \le 2$

34 $z = xy^2, \quad -2 \le x \le 2, \quad -2 \le y \le 2$

35 $z = x\sqrt{y}, \quad -2 \le x \le 2, \quad 0 \le y \le 4$

36 $z = \sqrt{x} + \sqrt{y}, \quad 0 \le x \le 4, \quad 0 \le y \le 4$

37 $z = \dfrac{x}{y}, \quad -2 \le x \le 2, \quad -2 \le y \le 2, \quad -4 \le z \le 4$

38 $z = (x + y)^{-1}, \quad -2 \le x \le 2, \quad -2 \le y \le 2, \quad -4 \le z \le 4$

39 $z = \cos x + \sin y, \quad -\pi/2 \le x \le \pi/2, \quad 0 \le y \le \pi$

40 $z = \cos x \cdot \sin y, \quad -\pi/2 \le x \le \pi/2, \quad 0 \le y \le \pi$

41 $z = e^{x+y}, \quad -2 \le x \le 2, \quad -2 \le y \le 2$

42 $z = e^{-x^2 - y^2}, \quad -2 \le x \le 2, \quad -2 \le y \le 2$

43 $z = x^y, \quad 0 < x \le 4, \quad -2 \le y \le 2, \quad -4 \le z \le 4$

44 $z = \log_x y, \quad 0 < x \le 4, \quad 0 < y \le 4, \quad -4 \le z \le 4$

11.2 CONTINUOUS FUNCTIONS OF TWO OR MORE VARIABLES

Two points (x_1, y_1) and (x_2, y_2) in the hyperreal plane are said to be *infinitely close*, $(x_1, y_1) \approx (x_2, y_2)$, if both $x_1 \approx x_2$ and $y_1 \approx y_2$. If

$$\Delta x = x_2 - x_1, \qquad \Delta y = y_2 - y_1,$$

then the distance between (x_1, y_1) and (x_2, y_2) is

$$\Delta s = \sqrt{\Delta x^2 + \Delta y^2}.$$

LEMMA 1

> *Two points are infinitely close to each other if and only if the distance between them is infinitesimal.*

This lemma can be seen from Figure 11.2.1. (An easy proof of the lemma in terms of vectors was given in Section 10.8.)

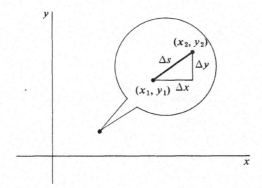

Figure 11.2.1

The definition of a continuous function in two variables is similar to the definition in one variable.

DEFINITION

> *A real function $f(x, y)$ is said to be **continuous** at a real point (a, b) if whenever (x, y) is infinitely close to (a, b), $f(x, y)$ is infinitely close to $f(a, b)$. In other words,*
>
> *if $st(x) = a$ and $st(y) = b$, then $st(f(x, y)) = f(a, b)$.*

Figure 11.2.2 shows (a, b) and $f(a, b)$ under the microscope.

Remark   It follows from the definition that if $f(x, y)$ is continuous at (a, b), then $f(x, y)$ is defined at every hyperreal point infinitely close to (a, b). In fact, it can even be proved that $f(x, y)$ is defined at every point in some real rectangle $a_1 < x < a_2, b_1 < y < b_2$ containing (a, b).

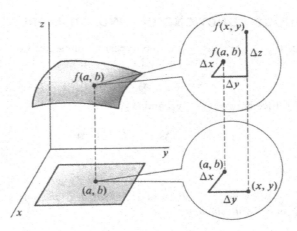

Figure 11.2.2

EXAMPLE 1   Show that $f(x, y) = 2x + xy^2$ is continuous for all (a, b). Let $st(x) = a$ and $st(y) = b$. Then

$$st(2x + xy^2) = st(2x) + st(xy^2) = 2st(x) + st(x)st(y^2) = 2a + ab^2.$$

Here is a list of important continuous functions of two variables.

THEOREM 1

The following are continuous at all real points (x, y) as indicated.

(i)   The Sum Function $f(x, y) = x + y$.

(ii)   The Difference Function $f(x, y) = x - y$.

(iii)   The Product Function $f(x, y) = xy$.

(iv)   The Quotient Function $f(x, y) = x/y$, $(y \neq 0)$.

(v)   The Exponential Function $f(x, y) = x^y$, $(x > 0)$.

(i)–(iv) follow at once from the corresponding rules for standard parts,

$$st(x + y) = st(x) + st(y),$$
$$st(x - y) = st(x) - st(y),$$
$$st(xy) = st(x)st(y),$$
$$st\left(\frac{x}{y}\right) = \frac{st(x)}{st(y)} \qquad \text{if } st(y) \neq 0.$$

(v) is equivalent to the new standard parts rule

$$st(x^y) = st(x)^{st(y)} \qquad \text{if } st(x) > 0.$$

We prove this rule using the fact that e^u and $\ln u$ are continuous functions of one variable.

$$st(x^y) = st(e^{y \ln x}) = e^{st(y \ln x)} = e^{st(y)st(\ln x)} = e^{st(y) \ln st(x)} = st(x)^{st(y)}.$$

The next theorem shows that most functions we deal with are continuous.

THEOREM 2

(i) *If $f(x, y)$ is continuous at (a, b) and $g(u)$ is continuous at $f(a, b)$, then*

$$h(x, y) = g(f(x, y))$$

is continuous at (a, b).

(ii) *Sums, differences, products, quotients, and exponents of continuous functions are continuous.*

PROOF (i) If $(x, y) \approx (a, b)$ then $f(x, y) \approx f(a, b)$, hence $g(f(x, y)) \approx g(f(a, b))$, and thus $h(x, y) \approx h(a, b)$.

(ii) Let $f(x, y)$ and $g(x, y)$ be continuous at (a, b). As an illustration we show that if $f(x, y) > 0$ then

$$h(x, y) = f(x, y)^{g(x, y)}$$

is continuous at (a, b). Let $(x, y) \approx (a, b)$. Then

$$st(h(x, y)) = st(f(x, y)^{g(x, y)}) = st(f(x, y))^{st(g(x, y))} = f(a, b)^{g(a, b)} = h(a, b).$$

EXAMPLE 2   By (i), $h(x, y) = \sin(x + y)$ is continuous for all (x, y).

EXAMPLE 3   By (ii), $h(x, y) = \sin x \cos y$ is continuous for all (x, y).

A function is said to be *continuous on a set S* of points in the plane if it is continuous at every point in S. Thus the quotient function $f(x, y) = x/y$ is continuous on the set of all (x, y) such that $y \neq 0$. The function $f(x, y) = x^y$ is continuous on the set of all (x, y) such that $x > 0$.

EXAMPLE 4   Find a set on which $h(x, y) = \ln(x + y)$ is continuous.

By Theorems 1 and 2,

$$x + y \text{ is continuous for all } (x, y),$$
$$\ln u \text{ is continuous for } u > 0,$$
$$\ln(x + y) \text{ is continuous for } x + y > 0.$$

Answer $\ln(x + y)$ is continuous on the set of all (x, y) such that $x + y > 0$, shown in Figure 11.2.3.

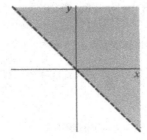

Figure 11.2.3

EXAMPLE 5   Find a set on which $h(x, y) = x^y + \cos\sqrt{x^2 - y}$ is continuous.

x^y is continuous for $x > 0$.
x^2 is continuous for all x.
$x^2 - y$ is continuous for all (x, y).

$\sqrt{x^2 - y}$ is continuous for $x^2 - y > 0$.

$\cos\sqrt{x^2 - y}$ is continuous for $x^2 - y > 0$.

$x^y + \cos\sqrt{x^2 - y}$ is continuous for $x > 0$ and $x^2 - y > 0$.

Answer $h(x, y)$ is continuous on the set of all (x, y) such that $x > 0$ and $x^2 - y > 0$. The set is shown in Figure 11.2.4.

EXAMPLE 6   Find a set on which $h(x, y) = \log_x y$ is continuous.

We use the identity $$\log_x y = \frac{\ln y}{\ln x}.$$

$\ln y$ is continuous for $y > 0$.
$\ln x$ is continuous for $x > 0$.
$\ln y/\ln x$ is continuous for $x > 0$, $\ln x \neq 0$, $y > 0$,
that is, $x > 0$, $x \neq 1$, $y > 0$.
$\log_x y$ is continuous for $x > 0$, $x \neq 1$, $y > 0$.

Answer $\log_x y$ is continuous on the set of all (x, y) such that $x > 0$, $x \neq 1$, $y > 0$ (Figure 11.2.5).

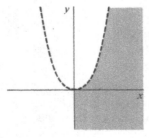

Figure 11.2.4   **Figure 11.2.5**

Continuous functions of three or more variables are defined in the natural way, and Theorem 2 holds for such functions.

EXAMPLE 7   Find a set where the function

$$h(x, y, z) = \frac{x^2 y}{x + y + z}$$

is continuous.

x^2 is continuous for all x.
$x^2 y$ is continuous for all (x, y).

$x + y$ is continuous for all (x, y).

$(x + y) + z$ is continuous for all (x, y, z).

$\dfrac{x^2 y}{x + y + z}$ is continuous for $x + y + z \neq 0$.

Answer $h(x, y, z)$ is continuous on the set of all (x, y, z) such that $x + y + z \neq 0$.

PROBLEMS FOR SECTION 11.2

Find the largest set you can in which the following functions are continuous.

1 $f(x, y) = 2x - 3y$

2 $f(x, y) = \dfrac{1}{1 + x^2 + y^2}$

3 $f(x, y) = e^{x^2 - y}$

4 $f(x, y) = \dfrac{1}{2 + \sin(xy)}$

5 $f(x, y) = \dfrac{xy}{x + y}$

6 $f(x, y) = \dfrac{1}{x^2 + y^2}$

7 $f(x, y) = \dfrac{x^3}{y + 2}$

8 $f(x, y) = \dfrac{x + y}{xy}$

9 $f(x, y) = \dfrac{1}{(x - 2)(y + 1)}$

10 $f(x, y) = \sqrt{x} + \sqrt{y}$

11 $f(x, y) = \sqrt{x + y}$

12 $f(x, y) = \sqrt{x^2 + y^2}$

13 $f(x, y) = \sqrt{x - y}$

14 $f(x, y) = \dfrac{\sqrt{y}}{\sqrt{x + 2y}}$

15 $f(x, y) = x^{x + y}$

16 $f(x, y) = y^{\sin x}$

17 $f(x, y) = (x^2 - y)^x$

18 $f(x, y) = y^{1/x}$

19 $f(x, y) = x^{y^x}$

20 $f(x, y) = \dfrac{1}{1 - x^y}$

21 $f(x, y) = \ln(x^2 - y)$

22 $f(x, y) = \ln(xy)$

23 $f(x, y) = \dfrac{1}{\ln x + \ln y}$

24 $f(x, y) = \ln(\ln(x - y))$

25 $f(x, y) = \log_{x + y}(xy)$

26 $f(x, y) = \log_{2x - y}(x + 3y)$

27 $\dfrac{\sqrt{y^2 - x}}{x - 4y}$

28 $\dfrac{1}{\sin x \cos y}$

29 $\sqrt{\cos x + y}$

30 $\sqrt{|x| + |y|}$

31 $\ln|x - y|$

32 $x^y + y^z$

33 $\dfrac{\sqrt{x - y}}{y - z}$

34 $\dfrac{1}{x + 2y + 3z}$

35 $\dfrac{1}{x^2 + y^2 + z^2}$

36 $\log_x(y + z)$

37 $(x + y)^{1/z}$

☐ **38**   Let $f(x, y) = \begin{cases} 0 & \text{if } xy = 0, \\ 1 & \text{if } xy \neq 0. \end{cases}$

Show that f is not continuous at $(0, 0)$.

☐ 39   Suppose $f(x, y)$ is continuous at (a, b). Prove that $g(x) = f(x, b)$ is continuous at $x = a$.

☐ 40   Prove that if $f(x)$ and $g(x)$ are continuous at $x = a$ and if $h(u, v)$ is continuous at $(f(a), g(a))$, then

$$k(x) = h(f(x), g(x))$$

is continuous at $x = a$.

☐ 41   Prove that if $f(x, y)$ and $g(x, y)$ are both continuous at (a, b) and if $h(u, v)$ is continuous at $(f(a, b), g(a, b))$, then

$$k(x, y) = h(f(x, y), g(x, y))$$

is continuous at (a, b).

The notation
$$\lim_{(x,y)\to(a,b)} f(x, y) = L$$
means that whenever (x, y) is infinitely close to but not equal to (a, b), $f(x, y)$ is infinitely close to L.

☐ 42   Evaluate $\lim_{(x,y)\to(0,0)} \dfrac{x^2 + y^2}{|x| + |y|}$.

☐ 43   Evaluate $\lim_{(x,y)\to(0,0)} (1 + x^2 + y^2)^{1/(x^2 + y^2)}$.

☐ 44   Evaluate $\lim_{(x,y)\to(0,0)} \dfrac{xy}{\sqrt{x^2 + y^2}}$.

☐ 45   Evaluate $\lim_{(x,y)\to(0,0)} \dfrac{1}{\sqrt{x^2 + y^2}}$.

☐ 46   Show that $\lim_{(x,y)\to(0,0)} \dfrac{x}{\sqrt{x^2 + y^2}}$ does not exist.

11.3 PARTIAL DERIVATIVES

Partial derivatives are used to study the rates of change of functions of two or more variables. In general the rate of change of $z = f(x, y)$ will depend both on the rate of change of x and the rate of change of y. Partial derivatives deal with the simplest case, where only one of the independent variables is changing and the other is held constant.

Given a function $z = f(x, y)$, if we hold y fixed at some constant value b we obtain a function

$$g(x) = f(x, b)$$

of x only. Geometrically the curve $z = g(x)$ is the intersection of the surface $z = f(x, y)$ with the vertical plane $y = b$. The rate of change of z with respect to x with y held constant is the slope of the curve $z = g(x)$. This slope is called the partial derivative of $f(x, y)$ with respect to x (Figure 11.3.1(a)). There is also a partial derivative with respect to y (Figure 11.3.1(b)).

Here is a precise definition.

DEFINITION

*The **partial derivatives** of $f(x, y)$ at the point (a, b) are the limits*

$$f_x(a, b) = \lim_{\Delta x\to 0} \frac{f(a + \Delta x, b) - f(a, b)}{\Delta x},$$

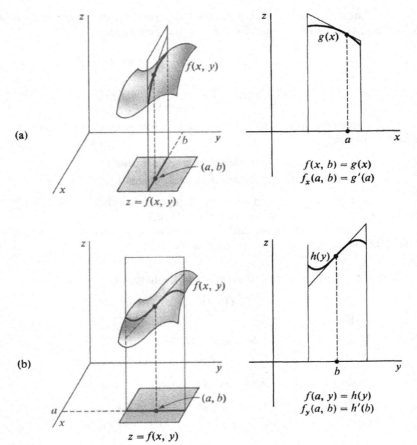

Figure 11.3.1 Partial Derivatives

$$f_y(a, b) = \lim_{\Delta y \to 0} \frac{f(a, b + \Delta y) - f(a, b)}{\Delta y}.$$

A partial derivative is undefined if the limit does not exist.

When $f_x(a, b)$ exists, it is equal to the standard part

$$f_x(a, b) = st\left(\frac{f(a + \Delta x, b) - f(a, b)}{\Delta x}\right)$$

for any nonzero infinitesimal Δx. Similarly when $f_y(a, b)$ exists,

$$f_y(a, b) = st\left(\frac{f(a, b + \Delta y) - f(a, b)}{\Delta y}\right)$$

for any nonzero infinitesimal Δy.

Just as the one-variable derivative $f'(x)$ is a function of x, the partial derivatives $f_x(x, y)$ and $f_y(x, y)$ are again functions of x and y. At each point (x, y), the partial derivative $f_x(x, y)$ either has exactly one value or is undefined.

Another convenient notation for the partial derivatives uses the Cyrillic lower case D, ∂, called a "round d". If $z = f(x, y)$, we use:

$$\frac{\partial z}{\partial x}(x, y), \quad \frac{\partial z}{\partial x}, \quad \text{or} \quad \frac{\partial f}{\partial x} \quad \text{for } f_x(x, y),$$

$$\frac{\partial z}{\partial y}(x, y), \quad \frac{\partial z}{\partial y}, \quad \text{or} \quad \frac{\partial f}{\partial y} \quad \text{for } f_y(x, y).$$

Partial derivatives, like ordinary derivatives, may be represented as quotients of infinitesimals.

In $\partial z/\partial x$, ∂x means Δx and ∂z means $f_x(x, y)\,\Delta x$.

In $\partial z/\partial y$, ∂y means Δy and ∂z means $f_y(x, y)\,\Delta y$.

Notice that ∂z has a different meaning in $\partial z/\partial x$ than it has in $\partial z/\partial y$. For this reason we shall avoid using the symbol ∂z alone.

Partial derivatives are easily computed using the ordinary rules of differentiation with all but one variable treated as a constant.

EXAMPLE 1   Find the partial derivatives of the function

$$f(x, y) = x^2 + 3xy - 8y$$

at the point $(2, -1)$.

To find $f_x(x, y)$, we treat y as a constant,

$$f_x(x, y) = 2x + 3y.$$

To find $f_y(x, y)$, we treat x as a constant,

$$f_y(x, y) = 3x - 8.$$

Thus $f_x(2, -1) = 2 \cdot 2 + 3(-1) = 1, \qquad f_y(2, -1) = 3 \cdot 2 - 8 = -2.$

Figure 11.3.2 shows the surface $z = f(x, y)$ and the tangent lines at the point $(2, -1)$.

EXAMPLE 2   A point $P(x, y)$ has distance $z = \sqrt{x^2 + y^2}$ from the origin (Figure 11.3.3). Find the rate of change of z at $P(3, 4)$ if:

(a) P moves at unit speed in the x direction.
(b) P moves at unit speed in the y direction.

In this problem the round d notation is convenient.

(a) $\dfrac{\partial z}{\partial x}(x, y) = \dfrac{x}{\sqrt{x^2 + y^2}},$

$\dfrac{\partial z}{\partial x}(3, 4) = \dfrac{3}{\sqrt{3^2 + 4^2}} = \dfrac{3}{5}.$

(b) $\dfrac{\partial z}{\partial y}(x, y) = \dfrac{y}{\sqrt{x^2 + y^2}},$

$\dfrac{\partial z}{\partial y}(3, 4) = \dfrac{4}{\sqrt{3^2 + 4^2}} = \dfrac{4}{5}.$

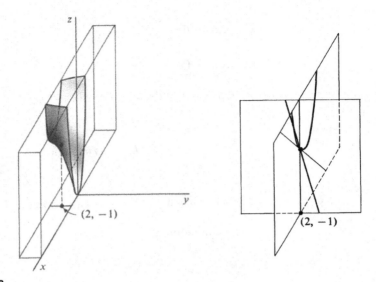

Figure 11.3.2

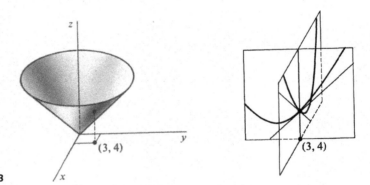

Figure 11.3.3

Functions of three or more variables cannot easily be represented graphically. However, they can be given other physical interpretations. For example, $w = f(x, y, t)$ may be pictured as a moving surface in (x, y, w) space where t is time. Alternatively, $w = f(x, y, z)$ may be thought of as assigning a number to each point of (x, y, z) space where it is defined; for example, w could be the density of a three-dimensional object at the point (x, y, z).

Partial derivatives of functions of three or more variables are defined in a manner analogous to the two-variable case.

DEFINITION

The partial derivatives of $f(x, y, z)$ at the point (a, b, c) are the limits

$$f_x(a, b, c) = \lim_{\Delta x \to 0} \frac{f(a + \Delta x, b, c) - f(a, b, c)}{\Delta x},$$

$$f_y(a, b, c) = \lim_{\Delta y \to 0} \frac{f(a, b + \Delta y, c) - f(a, b, c)}{\Delta y},$$

$$f_z(a, b, c) = \lim_{\Delta z \to 0} \frac{f(a, b, c + \Delta z) - f(a, b, c)}{\Delta z}.$$

A partial derivative is undefined if the limit does not exist.

When $f_x(a, b, c)$ exists we have

$$f_x(a, b, c) = st\left(\frac{f(a + \Delta x, b, c) - f(a, b, c)}{\Delta x}\right)$$

for nonzero infinitesimal Δx.

Thus $f_x(x, y, z)$ is the rate of change of $f(x, y, z)$ with respect to x when y and z are held constant.

We also use the round d notation. If $w = f(x, y, z)$, we use:

$$\frac{\partial w}{\partial x}(x, y, z), \quad \frac{\partial w}{\partial x}, \quad \text{or} \quad \frac{\partial f}{\partial x} \quad \text{for } f_x(x, y, z),$$

$$\frac{\partial w}{\partial y}(x, y, z), \quad \frac{\partial w}{\partial y}, \quad \text{or} \quad \frac{\partial f}{\partial y} \quad \text{for } f_y(x, y, z),$$

$$\frac{\partial w}{\partial z}(x, y, z), \quad \frac{\partial w}{\partial z}, \quad \text{or} \quad \frac{\partial f}{\partial z} \quad \text{for } f_z(x, y, z).$$

EXAMPLE 3   Find the partial derivatives of

$$f(x, y, z) = \sin(x^2 y - z)$$

at the point $(1, 0, 0)$.

To find $f_x(x, y, z)$ we treat y and z as constants.

$$f_x(x, y, z) = 2xy \cos(x^2 y - z).$$
$$f_y(x, y, z) = x^2 \cos(x^2 y - z).$$
$$f_z(x, y, z) = -\cos(x^2 y - z).$$

Thus
$$f_x(1, 0, 0) = 2 \cdot 1 \cdot 0 \cos(1^2 \cdot 0 - 0) = 0.$$
$$f_y(1, 0, 0) = 1^2 \cos(1^2 \cdot 0 - 0) = 1.$$
$$f_z(1, 0, 0) = -\cos(1^2 \cdot 0 - 0) = -1.$$

PROBLEMS FOR SECTION 11.3

In Problems 1–28, find the partial derivatives.

1	$z = 4x - 3y$	2	$z = 1 + 3x + 5y$
3	$z = xy^2 + x^3 y$	4	$z = x^3 y^2$
5	$z = \dfrac{1}{x^2 + y^2}$	6	$z = \dfrac{1}{xy + 1}$
7	$f(x, y) = xy$	8	$f(x, y) = x/y$

9	$f(x, y) = ax + by$	10	$f(x, y) = e^{ax+by}$
11	$f(x, y) = e^{x^2 - y^2}$	12	$f(x, y) = \sin x \cos y$
13	$f(x, y) = \sqrt{x + 2y}$	14	$f(x, y) = \sqrt{xy} + \sqrt{x} + \sqrt{y}$
15	$z = x^y$	16	$z = x^{1/y}$
17	$z = \ln(xy)$	18	$z = \ln(ax + by)$
19	$z = \log_x y$	20	$z = \tan x \arctan y$
21	$z = \arcsin(x^2 y)$	22	$w = xyz$
23	$w = \sqrt{x^2 + y^2 + z^2}$	24	$f(x, y, z) = xe^{y-z}$
25	$f(x, y, z) = ax + by + cz$	26	$f(x, y, z) = x^a y^b z^c$
27	$w = z \cos x + z \sin y$	28	$w = z \cosh x + z \sinh y$

In Problems 29–40 find the partial derivatives at the given point.

29   $f(x, y) = xy^2$, $x = 1$, $y = 2$

30   $f(x, y) = x\sqrt{y}$, $x = 2$, $y = 4$

31   $f(x, y) = 1/xy$, $x = -1$, $y = 1$

32   $f(x, y) = \dfrac{1}{x} + \dfrac{1}{y}$, $x = 3$, $y = 4$

33   $z = e^{xy}$, $x = 0$, $y = 2$

34   $z = e^{x+y}$, $x = 0$, $y = 2$

35   $z = e^x \cos y$, $x = 1$, $y = 0$

36   $z = e^x \sin y$, $x = 1$, $y = 0$

37   $z = \dfrac{1}{x^2 + y^3}$, $x = 2$, $y = 3$

38   $z = \sqrt{x^2 + xy + 2y^2}$, $x = 1$, $y = 1$

39   $f(x, y, z) = x^2 + y^2 + z^2$, $x = 1$, $y = 2$, $z = 3$

40   $f(x, y, z) = \dfrac{x}{y} - \dfrac{x}{z}$, $x = 1$, $y = 1$, $z = 1$

41   A point $P(x, y)$ at $(1, 2)$ is moving at unit speed in the x direction. Find the rate of change of the distance from P to the origin.

42   A point $P(x, y)$ at $(1, 2)$ is moving at unit speed in the y direction. Find the rate of change of the distance from P to the point $(5, -1)$.

43   A point $P(x, y, z)$ is moving at unit speed in the x direction. Find the rate of change of the distance from P to the origin when P is at $(1, 2, 2)$.

44   A point $P(x, y, z)$ is moving at unit speed in the z direction. Find the rate of change of the distance from P to the origin when P is at $(3, \sqrt{3}, 2)$.

45   Find b and c if for all x and y,

$$z = x^2 + bxy + cy^2 \quad \text{and} \quad \frac{\partial z}{\partial x} = \frac{\partial z}{\partial y}.$$

46   Find b if for all x and y

$$z = \sin x \sin y + b \cos x \cos y \quad \text{and} \quad \frac{\partial z}{\partial x} = \frac{\partial z}{\partial y}.$$

47   It is found that the cost of producing x units of commodity one and y units of commodity two is

$$C(x, y) = 100 + 3x + 4y - \sqrt{xy}.$$

Find the partial marginal costs with respect to x and y, $\partial C/\partial x$ and $\partial C/\partial y$.

48   When a certain three commodities are produced in quantities x, y, and z respectively, it is found that they can be sold at a profit of

$$P(x, y, z) = 100x + 100y + yz - xy - z^2.$$

Find the marginal profits with respect to x, y and z; i.e., $\partial P/\partial x$, $\partial P/\partial y$, and $\partial P/\partial z$.

11.4 TOTAL DIFFERENTIALS AND TANGENT PLANES

Most of the functions we encounter have continuous partial derivatives. To keep our theory simple we shall concentrate on such functions in this chapter.

DEFINITION

*A function $f(x, y)$ is said to be **smooth** at (a, b) if both of its partial derivatives exist and are continuous at (a, b).*

The definition for three or more variables is similar.

The Increment Theorem for a differentiable function of one variable shows that the increment Δz is very close to the differential dz, and leads to the notion of a tangent line. In this section we introduce the increment and total differential for a function of two variables. Then we state an Increment Theorem for a smooth function of two variables, which leads to the notion of a tangent plane.

Let z depend on the two independent variables x and y, $z = f(x, y)$. Let Δx and Δy be two new independent variables, called the *increments* of x and y. Usually Δx and Δy are taken to be infinitesimals.

We now introduce two new dependent variables, the increment Δz and the total differential dz.

DEFINITION

*When $z = f(x, y)$, the **increment** of z is the dependent variable Δz given by*

$$\Delta z = f(x + \Delta x, y + \Delta y) - f(x, y).$$

The increment Δz depends on the four independent variables x, y, Δx, Δy, and is equal to the change in z as x changes by Δx and y changes by Δy. Thus

$$\Delta z = \Delta f(x, y, \Delta x, \Delta y),$$

where Δf is the function

$$\Delta f(x, y, \Delta x, \Delta y) = f(x + \Delta x, y + \Delta y) - f(x, y).$$

DEFINITION

*When $z = f(x, y)$, the **total differential** of z is the dependent variable dz given by*

$$dz = f_x(x, y)\, dx + f_y(x, y)\, dy,$$

or equivalently
$$dz = \frac{\partial z}{\partial x}\, dx + \frac{\partial z}{\partial y}\, dy.$$

When x and y are independent variables, dx and dy are the same as Δx and Δy. The total differential dz depends on the four independent variables x, y, dx, and dy. Thus

$$dz = df(x, y, dx, dy),$$

where df is the function

$$df(x, y, dx, dy) = f_x(x, y)\,dx + f_y(x, y)\,dy.$$

Figure 11.4.1 shows Δz under the microscope.

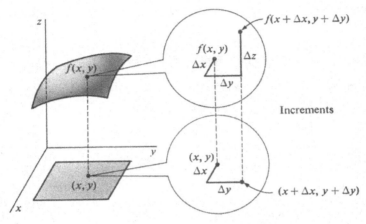

Figure 11.4.1

EXAMPLE 1   Find the increment and total differential of the product function $z = xy$ (Figure 11.4.2).

Increment: $\Delta z = (x + \Delta x)(y + \Delta y) - xy = y\,\Delta x + x\,\Delta y + \Delta x\,\Delta y.$

Total differential: $dz = \dfrac{\partial z}{\partial x}\,dx + \dfrac{\partial z}{\partial y}\,dy = y\,dx + x\,dy.$

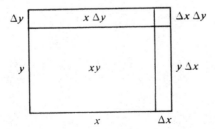

Figure 11.4.2

EXAMPLE 2   Find the increment and total differential of $z = x^2 - 3xy^2$.

Increment:

$$\Delta z = [(x + \Delta x)^2 - 3(x + \Delta x)(y + \Delta y)^2] - [x^2 - 3xy^2]$$
$$= [x^2 + 2x\,\Delta x + \Delta x^2 - 3xy^2 - 6xy\,\Delta y - 3x\,\Delta y^2 - 3\,\Delta x y^2$$
$$-6\,\Delta x y\,\Delta y - 3\,\Delta x\,\Delta y^2] - [x^2 - 3xy^2]$$

$$= 2x\,\Delta x + \Delta x^2 - 6xy\,\Delta y - 3x\,\Delta y^2 - 3\,\Delta xy^2 - 6\,\Delta xy\,\Delta y - 3\,\Delta x\,\Delta y^2$$
$$= (2x - 3y^2)\,\Delta x - 6xy\,\Delta y + \Delta x^2 - 3x\,\Delta y^2 - 6y\,\Delta x\,\Delta y - 3\,\Delta x\,\Delta y^2.$$

Total differential:

$$\frac{\partial z}{\partial x} = 2x - 3y^2, \qquad \frac{\partial z}{\partial y} = -6xy.$$

$$dz = \frac{\partial z}{\partial x}\,dx + \frac{\partial z}{\partial y}\,dy = (2x - 3y^2)\,dx - 6xy\,dy.$$

We shall now state the Increment Theorem. It shows that Δz is very close to dz.

INCREMENT THEOREM FOR TWO VARIABLES

Suppose $z = f(x, y)$ is smooth at (a, b). Let Δx and Δy be infinitesimal. Then

$$\Delta z = dz + \varepsilon_1\,\Delta x + \varepsilon_2\,\Delta y$$

for some infinitesimals ε_1 and ε_2 which depend on Δx and Δy.

Before proving the Increment Theorem, let us check it for Examples 1 and 2.

EXAMPLE 1 (Continued)   The product function $z = xy$ is smooth for all (x, y). Express Δz in the form

$$\Delta z = dz + \varepsilon_1\,\Delta x + \varepsilon_2\,\Delta y.$$

We have
$$\Delta z = y\,\Delta x + x\,\Delta y + \Delta x\,\Delta y,$$
$$dz = y\,\Delta x + x\,\Delta y.$$

Thus
$$\Delta z = dz + \Delta x \cdot \Delta y.$$

The problem has more than one correct answer. One answer is $\varepsilon_1 = 0$ and $\varepsilon_2 = \Delta x$, so that

$$\Delta z = dz + 0 \cdot \Delta x + \Delta x \cdot \Delta y = dz + \varepsilon_1\,\Delta x + \varepsilon_2\,\Delta y.$$

Another answer is $\varepsilon_1 = \Delta y$ and $\varepsilon_2 = 0$, so that

$$\Delta z = dz + \Delta y \cdot \Delta x + 0 \cdot \Delta y = dz + \varepsilon_1\,\Delta x + \varepsilon_2\,\Delta y.$$

EXAMPLE 2 (Continued)   The function $z = x^2 - 3xy^3$ is smooth for all (x, y). Express Δz in the form

$$\Delta z = dz + \varepsilon_1\,\Delta x + \varepsilon_2\,\Delta y$$

at an arbitrary point (x, y) and at the point $(5, 4)$. We have

$$\Delta z = (2x - 3y^2)\,\Delta x - 6xy\,\Delta y + \Delta x^2 - 3x\,\Delta y^2 - 6y\,\Delta x\,\Delta y - 3\,\Delta x\,\Delta y^2,$$
$$dz = (2x - 3y^2)\,\Delta x - 6xy\,\Delta y.$$

Then
$$\Delta z = dz + \Delta x^2 - 3x\,\Delta y^2 - 6y\,\Delta x\,\Delta y - 3\,\Delta x\,\Delta y^2.$$

Each term after the dz has either a Δx or a Δy or both. Factor Δx from all the terms where Δx appears and Δy from the remaining terms.

$$\Delta z = dz + (\Delta x - 6y\,\Delta y - 3\,\Delta y^2)\,\Delta x + (-3x\,\Delta y)\,\Delta y.$$

Then
$$\Delta z = dz + \varepsilon_1\,\Delta x + \varepsilon_2\,\Delta y,$$

where
$$\varepsilon_1 = \Delta x - 6y\,\Delta y - 3\,\Delta y^2, \qquad \varepsilon_2 = -3x\,\Delta y.$$

At the point $(5, 4)$,

$$\Delta z = dz + \varepsilon_1\,\Delta x + \varepsilon_2\,\Delta y,$$

where
$$\varepsilon_1 = \Delta x - 24\,\Delta y - 3\,\Delta y^2, \qquad \varepsilon_2 = -15\,\Delta y.$$

PROOF OF THE INCREMENT THEOREM We break Δz into two parts by going first from (a, b) to $(a + \Delta x, b)$ and then from $(a + \Delta x, b)$ to $(a + \Delta x, b + \Delta y)$, as shown in Figure 11.4.3,

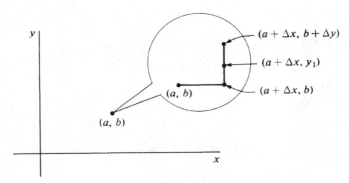

Figure 11.4.3

$$\Delta z = [f(a + \Delta x, b + \Delta y) - f(a + \Delta x, b)] + [f(a + \Delta x, b) - f(a, b)].$$

Our plan is as follows. First, we regard $f(a, b)$ as a one-variable function of a and show that

(1) $\qquad f(a + \Delta x, b) - f(a, b) = f_x(a, b)\,\Delta x + \varepsilon_1\,\Delta x$ for some infinitesimal ε_1.

Second, we regard $f(a + \Delta x, b)$ as a one-variable function of b and show that

(2) $\qquad f(a + \Delta x, b + \Delta y) - f(a + \Delta x, b) = f_y(a, b)\,\Delta y + \varepsilon_2\,\Delta y$

$$\text{for some infinitesimal } \varepsilon_2.$$

Once Equations 1 and 2 are established the proof will be complete because by adding Equations 1 and 2 we get the desired result

$$\Delta z = f_x(a, b)\,\Delta x + f_y(a, b)\,\Delta y + \varepsilon_1\,\Delta x + \varepsilon_2\,\Delta y = dz + \varepsilon_1\,\Delta x + \varepsilon_2\,\Delta y.$$

Equation 1 follows at once from the one-variable Increment Theorem since $f_x(a, b)$ exists.

We now prove Equation 2. We regard $f(a + \Delta x, y)$ as a one-variable function of y. For all y between b and $b + \Delta y$, the point $(a + \Delta x, y)$ is infinitely close to (a, b), so $f_y(a + \Delta x, y)$ is defined. By the one-variable Mean Value Theorem on the interval $[b, b + \Delta y]$, there is a y_1 between b and $b + \Delta y$ such that

$$f_y(a + \Delta x, y_1) = \frac{f(a + \Delta x, b + \Delta y) - f(a + \Delta x, b)}{\Delta y}.$$

Since f_y is continuous at (a, b),

$$f_y(a + \Delta x, y_1) = f_y(a, b) + \varepsilon_2,$$

where ε_2 is infinitesimal. Then

$$\frac{f(a + \Delta x, b + \Delta y) - f(a + \Delta x, b)}{\Delta y} = f_y(a, b) + \varepsilon_2,$$

and Equation 2 follows.

The following corollary is analogous to the theorem that a differentiable function of one variable is continuous.

COROLLARY 1

If a function $z = f(x, y)$ is smooth at (a, b) then it is continuous at (a, b).

PROOF Let (x, y) be infinitely close to (a, b) and let

$$\Delta x = x - a, \qquad \Delta y = y - b.$$

Then
$$\Delta z = dz + \varepsilon_1 \Delta x + \varepsilon_2 \Delta y$$

$$= \frac{\partial z}{\partial x} \Delta x + \frac{\partial z}{\partial y} \Delta y + \varepsilon_1 \Delta x + \varepsilon_2 \Delta y.$$

Since Δx and Δy are infinitesimal, Δz is infinitesimal, so $f(x, y) \approx f(a, b)$.

Some examples of what can happen when the function is not smooth are given in the problem set.

If a function $z = f(x, y)$ is smooth at (a, b), the curve $z = f(x, b)$ has a tangent line L_1 on the plane $y = b$, and the curve $z = f(a, y)$ has a tangent line L_2 on the plane $x = a$.

L_1 has the equation $z - f(a, b) = f_x(a, b)(x - a)$

and L_2 has the equation $z - f(a, b) = f_y(a, b)(y - b)$.

The plane determined by the lines L_1 and L_2 is called the *tangent plane*. It has the equation

$$z - f(a, b) = f_x(a, b)(x - a) + f_y(a, b)(y - b),$$

because the graph p of this equation is a plane and intersects the plane $y = b$ in L_1 and the plane $x = a$ in L_2 (Figure 11.4.4).

DEFINITION

The **tangent plane** of a smooth function $z = f(x, y)$ at (a, b) is the plane with the equation

$$z - f(a, b) = f_x(a, b)(x - a) + f_y(a, b)(y - b).$$

If we set $x = a$ and $y = b$ in this equation we get $z = f(a, b)$. If we set $x - a = dx$ and $y - b = dy$ we get $z - f(a, b) = dz$. Therefore:

The tangent plane touches the surface at (a, b).

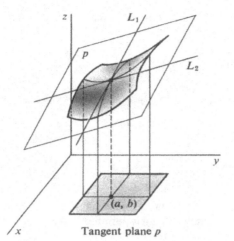

Figure 11.4.4   Tangent plane p

$\Delta z = $ *change in z on the surface.*
$dz = $ *change in z on the tangent plane.*

Figure 11.4.5 shows Δz and dz.

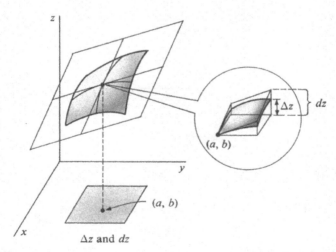

Δz and dz

Figure 11.4.5

Our second corollary to the Increment Theorem shows that the tangent plane closely follows the surface.

COROLLARY 2

Suppose $z = f(x, y)$ is smooth at (a, b). Then for every point (x, y) at an infinitesimal distance

$$\Delta s = \sqrt{\Delta x^2 + \Delta y^2}$$

from (a, b), the change in z on the tangent plane is infinitely close to the change in z along the surface compared to Δs, i.e.,

$$\frac{\Delta z}{\Delta s} \approx \frac{dz}{\Delta s}.$$

PROOF We have $\Delta z = dz + \varepsilon_1 \Delta x + \varepsilon_2 \Delta y$. Both $\Delta x/\Delta s$ and $\Delta y/\Delta s$ are finite, so

$$\frac{\Delta z}{\Delta s} = \frac{dz}{\Delta s} + \varepsilon_1 \frac{\Delta x}{\Delta s} + \varepsilon_2 \frac{\Delta y}{\Delta s},$$

$$\frac{\Delta z}{\Delta s} - \frac{dz}{\Delta s} = \varepsilon_1 \frac{\Delta x}{\Delta s} + \varepsilon_2 \frac{\Delta y}{\Delta s} \approx 0.$$

In Figure 11.4.6, we see that the piece of the surface seen through an infinitesimal microscope aimed at $(a, b, f(a, b))$ is infinitely close to a piece of the tangent plane, compared to the field of view of the microscope.

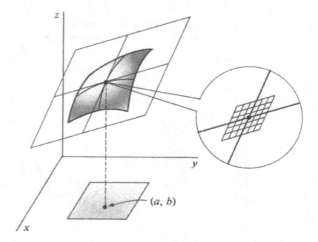

Figure 11.4.6

EXAMPLE 3 Find the equation of the tangent plane to

$$z = 1 + \sin(2x + 3y)$$

at the point $(0, 0)$.

We have

$$\frac{\partial z}{\partial x}(x, y) = 2\cos(2x + 3y), \qquad \frac{\partial z}{\partial y}(x, y) = 3\cos(2x + 3y).$$

At the point $(0, 0)$, $z = 1 + \sin(0 + 0) = 1$,

$$\frac{\partial z}{\partial x}(0, 0) = 2\cos(0 + 0) = 2, \qquad \frac{\partial z}{\partial y}(0, 0) = 3\cos(0 + 0) = 3.$$

The equation of the tangent plane is $z - 1 = 2(x - 0) + 3(y - 0)$, or $z = 2x + 3y + 1$.

EXAMPLE 4   Find the tangent plane to the sphere

$$x^2 + y^2 + z^2 = 14$$

at the point $(1, 2, 3)$ (Figure 11.4.7).

The top hemisphere has the equation $z = \sqrt{14 - x^2 - y^2}$.

Then

$$\frac{\partial z}{\partial x}(x, y) = -\frac{x}{\sqrt{14 - x^2 - y^2}} = -\frac{x}{z},$$

$$\frac{\partial z}{\partial y}(x, y) = -\frac{y}{\sqrt{14 - x^2 - y^2}} = -\frac{y}{z}.$$

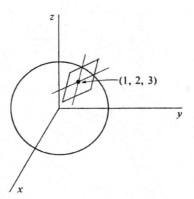

Figure 11.4.7

At $(1, 2)$, $z = 3$, $\dfrac{\partial z}{\partial x}(1, 2) = -\dfrac{1}{3}$, $\dfrac{\partial z}{\partial y}(1, 2) = -\dfrac{2}{3}$.

Then the tangent plane has the equation

$$z - 3 = \frac{\partial z}{\partial x}(x - 1) + \frac{\partial z}{\partial y}(y - 2),$$

$$z - 3 = -\frac{1}{3}(x - 1) + \left(-\frac{2}{3}\right)(y - 2),$$

or $$x + 2y + 3z = 14.$$

The *total differential* of a function $w = f(x, y, z)$ of three variables is the dependent variable dw given by

$$dw = f_x(x, y, z)\, dx + f_y(x, y, z)\, dy + f_z(x, y, z)\, dz,$$

or equivalently $$dw = \frac{\partial w}{\partial x}\, dx + \frac{\partial w}{\partial y}\, dy + \frac{\partial w}{\partial z}\, dz.$$

The following Increment Theorem has a proof like the Increment Theorem for two variables.

INCREMENT THEOREM FOR THREE VARIABLES

Suppose $w = f(x, y, z)$ is smooth at (a, b, c). Let Δx, Δy, and Δz be infinitesimal. Then the increment Δw is equal to

$$\Delta w = dw + \varepsilon_1 \Delta x + \varepsilon_2 \Delta y + \varepsilon_3 \Delta z$$

for some infinitesimals ε_1, ε_2, ε_3 which depend on Δx, Δy, and Δz.

EXAMPLE 5   Given $w = xyz$, express the increment Δw in the form

$$\Delta w = dw + \varepsilon_1 \Delta x + \varepsilon_2 \Delta y + \varepsilon_3 \Delta z.$$

We first find Δw and dw,

$$\Delta w = (x + \Delta x)(y + \Delta y)(z + \Delta z) - xyz$$
$$= yz\,\Delta x + xz\,\Delta y + xy\,\Delta z + x\,\Delta y\,\Delta z + y\,\Delta x\,\Delta z + z\,\Delta x\,\Delta y + \Delta x\,\Delta y\,\Delta z$$

$$\frac{\partial w}{\partial x} = yz, \qquad \frac{\partial w}{\partial y} = xz, \qquad \frac{\partial w}{\partial z} = xy.$$

$$dw = yz\,\Delta x + xz\,\Delta y + xy\,\Delta z.$$

Thus $\Delta w = dw + (y\,\Delta z + z\,\Delta y)\,\Delta x + (x\,\Delta z)\,\Delta y + (\Delta x\,\Delta y)\,\Delta z.$

Figure 11.4.8 pictures dw and Δw.

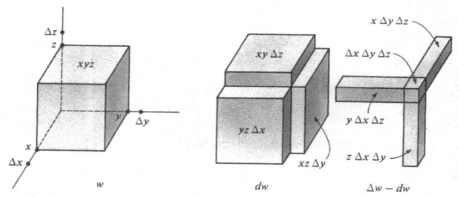

Figure 11.4.8

PROBLEMS FOR SECTION 11.4

In Problems 1–16, find the increment and total differential.

1	$z = 1 + 3x - 2y$		**2**	$z = x^2 - y^2$
3	$z = x^2 y^2$		**4**	$z = x^3 y$
5	$z = 1/xy$		**6**	$z = e^{x+y}$
7	$z = e^{3x-4y}$		**8**	$z = \cos x + \sin y$
9	$z = \cos x \sin y$		**10**	$z = \ln(x + 2y)$
11	$z = x \ln y$		**12**	$z = \sqrt{xy}$
13	$w = x + 2y + 3z$		**14**	$w = x^2 + y^2 + z^2$
15	$w = xy + yz$		**16**	$w = \sqrt{x} + \sqrt{y} + \sqrt{z}$

In Problems 17–22, express Δz in the form $\Delta z = dz + \varepsilon_1\, \Delta x + \varepsilon_2\, \Delta y$.

17   $z = x^2 + y^2$ 18 $z = x^3 + y^3$

19   $z = x^2 y$ 20 $z = 3xy - 2x^2 + y^2$

21   $z = \dfrac{x}{y}$ 22 $z = y\sqrt{x}$

In Problems 23–40 find the tangent plane at the given point.

23   $z = 2x^2 + y^2$ at $(1, 2)$ 24 $z = x^2 - 4y^2$ at $(-2, 1)$

25   $z = 2x^2 y + y^2 + 3$ at $(1, 1)$ 26 $z = x^2 y^2 + xy^3 + 2$ at $(-1, 2)$

27   $z = \sqrt{xy} + 1$ at $(1, 1)$ 28 $z = \sqrt{x} - 2\sqrt{y}$ at $(4, 1)$

29   $z = e^{x^2 y}$ at $(1, 3)$ 30 $z = e^{x^2 + y^3}$ at $(-1, -1)$

31   $z = \sin x \sin y$ at $(\pi/3, \pi/4)$ 32 $z = \tan(xy)$ at $(\pi, 1/4)$

33   $z = xy^2 - 2$ at $(0, 1)$ 34 $z = x^2 y^2 + 2$ at $(0, 0)$

35   $z = \cos x \cos y$ at $(0, 0)$ 36 $z = \arctan(2x - y)$ at $(1, 4)$

37   $x^2 + y^2 + z^2 = 9$ at $(1, -2, 2)$

38   $x^2 + 2y^2 + 3z^2 = 6$ at $(-1, 1, -1)$

39   $x^2 + y^2 - z^2 = 1$ at $(1, 1, 1)$ 40 $-x^2 - y^2 + z^2 = 1$ at $(2, -2, 3)$

☐ 41   Show that if z is a linear function of x and y, $z = ax + by + c$, then $\Delta z = dz$ at every point (x, y).

☐ 42   Let $f(x, y) = \begin{cases} 0 & \text{if } xy = 0 \\ 1 & \text{if } xy \neq 0. \end{cases}$

Show that at $(0, 0)$
(a) $f(x, y)$ is not continuous;
(b) $f_x(0, 0)$ and $f_y(0, 0)$ exist;
(c) $f(x, y)$ is not smooth.

☐ 43   Let $f(x, y) = \sqrt{xy}$. Prove that at the point $(0, 0)$,
(a) $f(x, y)$ is continuous;
(b) $f_x(0, 0)$ and $f_y(0, 0)$ exist;
(c) $f(x, y)$ is not smooth;
(d) Δz is not infinitely close to dz compared to $\Delta s = \sqrt{\Delta x^2 + \Delta y^2}$.

☐ 44   Let $f(x, y) = |xy|$. Show that at $(0, 0)$,
(a) $f(x, y)$ is continuous;
(b) $f_x(0, 0)$ and $f_y(0, 0)$ exist;
(c) $f(x, y)$ is not smooth;
(d) Δz is infinitely close to dz compared to Δs.

☐ 45   Let $f(x, y) = |x| + |y|$. Show that at $(0, 0)$,
(a) $f(x, y)$ is continuous;
(b) $f_x(0, 0)$ and $f_y(0, 0)$ do not exist.

11.5 CHAIN RULE

The Chain Rule is useful when several variables depend on each other. A typical case is where z depends on x and y, while x and y depend on another variable t. We shall call t the *independent variable*, x and y the *intermediate variables*, and z the *dependent variable*. Figure 11.5.1 shows which variables depend on which.

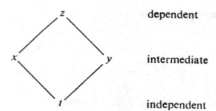

dependent

intermediate

Figure 11.5.1

independent

CHAIN RULE

If z is a smooth function of x and y, while x and y are differentiable functions of t, then dz/dt exists and

$$\frac{dz}{dt} = \frac{\partial z}{\partial x}\frac{dx}{dt} + \frac{\partial z}{\partial y}\frac{dy}{dt}.$$

Discussion   If $z = F(x, y)$ and $x = g(t)$, $y = h(t)$, then z as a function of t is

$$z = f(t) = F(g(t), h(t)).$$

We can give a more precise statement of the Chain Rule using functional notation:

If $g(t)$ and $h(t)$ are differentiable at t_0, and $F(x, y)$ is smooth at (x_0, y_0) where $x_0 = g(t_0)$ and $y_0 = h(t_0)$, then $f'(t_0)$ exists and

$$f'(t_0) = F_x(x_0, y_0)g'(t_0) + F_y(x_0, y_0)h'(t_0).$$

We shall give some examples and then prove the Chain Rule.

EXAMPLE 1   A particle moves in such a way that

$$\frac{dx}{dt} = 6, \qquad \frac{dy}{dt} = -2.$$

Find the rate of change of the distance from the particle to the origin when the particle is at the point $(3, -4)$ (Figure 11.5.2).

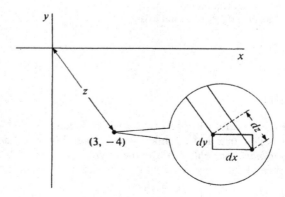

Figure 11.5.2

$$z = \sqrt{x^2 + y^2},$$

$$\frac{\partial z}{\partial x} = \frac{x}{\sqrt{x^2 + y^2}}, \qquad \frac{\partial z}{\partial y} = \frac{y}{\sqrt{x^2 + y^2}}.$$

$$\frac{dz}{dt} = \frac{\partial z}{\partial x}\frac{dx}{dt} + \frac{\partial z}{\partial y}\frac{dy}{dt} = \frac{6x}{\sqrt{x^2 + y^2}} - \frac{2y}{\sqrt{x^2 + y^2}}$$

$$= \frac{6 \cdot 3}{\sqrt{3^2 + 4^2}} - \frac{2 \cdot (-4)}{\sqrt{3^2 + 4^2}} = \frac{26}{5}.$$

EXAMPLE 2   Find the derivative of $z = \sqrt[t]{\sin t}$, using the Chain Rule. (This can also be done by logarithmic differentiation.)

Let
$$x = \sin t, \qquad y = \frac{1}{t}.$$

Then
$$z = x^y.$$

$$\frac{\partial z}{\partial x} = yx^{y-1}, \qquad \frac{\partial z}{\partial y} = (\ln x)x^y,$$

$$\frac{dx}{dt} = \cos t, \qquad \frac{dy}{dt} = -\frac{1}{t^2}.$$

$$\frac{dz}{dt} = \frac{\partial z}{\partial x}\frac{dx}{dt} + \frac{\partial z}{\partial y}\frac{dy}{dt}$$

$$= yx^{y-1}\cos t + (\ln x)x^y\left(-\frac{1}{t^2}\right)$$

$$= \frac{\sqrt[t]{\sin t}\,\cos t}{t \sin t} - \frac{\ln(\sin t)\sqrt[t]{\sin t}}{t^2}.$$

EXAMPLE 3   Suppose the price z of steel is proportional to the population x divided by the supply y,

$$z = \frac{cx}{y}.$$

x and y depend on time in such a way that

$$\frac{dx}{dt} = 0.01x, \qquad \frac{dy}{dt} = -\sqrt{x}.$$

Find the rate of increase in the price z when $x = 1{,}000{,}000$, $y = 10{,}000$.

$$\frac{\partial z}{\partial x} = \frac{c}{y}, \qquad \frac{\partial z}{\partial y} = -\frac{cx}{y^2}.$$

$$\frac{dz}{dt} = \frac{\partial z}{\partial x}\frac{dx}{dt} + \frac{\partial z}{\partial y}\frac{dy}{dt} = \frac{c}{y}(0.01x) + \left(-\frac{cx}{y^2}\right)(-\sqrt{x})$$

$$= c \cdot 10^{-4} \cdot 10^{-2} \cdot 10^6 + c \cdot 10^6 \cdot (10^{-4})^2 \cdot (10^6)^{1/2}$$

$$= c(1 + 10) = 11c.$$

PROOF OF THE CHAIN RULE   We use the Increment Theorem. Let Δt be a non-zero infinitesimal, and let Δx, Δy, and Δz be the corresponding increments of x, y and z. Then Δx and Δy are infinitesimal, and

$$\Delta z = \frac{\partial z}{\partial x}\Delta x + \frac{\partial z}{\partial y}\Delta y + \varepsilon_1 \Delta x + \varepsilon_2 \Delta y,$$

where ε_1 and ε_2 are infinitesimal. Dividing by Δt,

$$\frac{\Delta z}{\Delta t} = \frac{\partial z}{\partial x}\frac{\Delta x}{\Delta t} + \frac{\partial z}{\partial y}\frac{\Delta y}{\Delta t} + \varepsilon_1 \frac{\Delta x}{\Delta t} + \varepsilon_2 \frac{\Delta y}{\Delta t}.$$

Taking standard parts, we see that

$$\frac{dz}{dt} = \frac{\partial z}{\partial x}\frac{dx}{dt} + \frac{\partial z}{\partial y}\frac{dy}{dt}.$$

There is a Chain Rule for any number of independent and intermediate variables. We state the simplest cases here.

The Chain Rules for two or more independent variables follow from the Chain Rules for one independent variable.

If z depends on x and x depends on s and t, we have the diagram in Figure 11.5.3. The Chain Rule for this case is:

If z is a differentiable function of x and x is a smooth function of s and t, then

$$\frac{\partial z}{\partial s} = \frac{dz}{dx}\frac{\partial x}{\partial s}, \qquad \frac{\partial z}{\partial t} = \frac{dz}{dx}\frac{\partial x}{\partial t}.$$

This follows from the ordinary Chain Rule in Chapter 2 by holding s or t constant.

If z depends on x and y while x and y depend on s and t, we have the diagram in Figure 11.5.4. The Chain Rule for this case is:

If z is a smooth function of x and y while x and y are smooth functions of s and t, then

$$\frac{\partial z}{\partial s} = \frac{\partial z}{\partial x}\frac{\partial x}{\partial s} + \frac{\partial z}{\partial y}\frac{\partial y}{\partial s}, \qquad \frac{\partial z}{\partial t} = \frac{\partial z}{\partial x}\frac{\partial x}{\partial t} + \frac{\partial z}{\partial y}\frac{\partial y}{\partial t}.$$

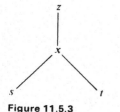

dependent

intermediate

independent

Figure 11.5.3

dependent

intermediate

independent

Figure 11.5.4

The Chain Rule for three intermediate variables is proved like the Chain Rule for two intermediate variables. We have the diagram in Figure 11.5.5.

If w is a smooth function of x, y, and z, which are in turn differentiable functions of t, then

$$\frac{dw}{dt} = \frac{\partial w}{\partial x}\frac{dx}{dt} + \frac{\partial w}{\partial y}\frac{dy}{dt} + \frac{\partial w}{\partial z}\frac{dz}{dt}.$$

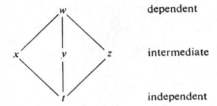

dependent

intermediate

Figure 11.5.5

independent

EXAMPLE 4 Use the Chain Rule to compute $\partial z/\partial s$ and $\partial z/\partial t$ where

$$z = \frac{x^2}{y}, \qquad x = st, \qquad y = s^2 - t^2.$$

$$\frac{\partial z}{\partial x} = \frac{2x}{y}, \qquad \frac{\partial z}{\partial y} = -\frac{x^2}{y^2}.$$

$$\frac{\partial x}{\partial s} = t, \qquad \frac{\partial y}{\partial s} = 2s.$$

$$\frac{\partial x}{\partial t} = s, \qquad \frac{\partial y}{\partial t} = -2t.$$

$$\frac{\partial z}{\partial s} = \frac{\partial z}{\partial x}\frac{\partial x}{\partial s} + \frac{\partial z}{\partial y}\frac{\partial y}{\partial s} = \frac{2x}{y}t - \frac{x^2}{y^2}2s$$

$$= \frac{2st^2}{s^2 - t^2} - \frac{2s^3t^2}{(s^2 - t^2)^2}.$$

$$\frac{\partial z}{\partial t} = \frac{\partial z}{\partial x}\frac{\partial x}{\partial t} + \frac{\partial z}{\partial y}\frac{\partial y}{\partial t} = \frac{2x}{y}s - \frac{x^2}{y^2}(-2t)$$

$$= \frac{2s^2t}{s^2 - t^2} + \frac{2s^2t^3}{(s^2 - t^2)^2}.$$

As a check, we compute $\partial z/\partial s$ and $\partial z/\partial t$ directly without the Chain Rule.

$$z = \frac{x^2}{y} = \frac{s^2t^2}{s^2 - t^2}.$$

$$\frac{\partial z}{\partial s} = \frac{(s^2 - t^2)2st^2 - (2s)s^2t^2}{(s^2 - t^2)^2} = \frac{2st^2}{s^2 - t^2} - \frac{2s^3t^2}{(s^2 - t^2)^2}.$$

$$\frac{\partial z}{\partial t} = \frac{(s^2 - t^2)(2s^2t) - (-2t)s^2t^2}{(s^2 - t^2)^2} = \frac{2s^2t}{s^2 - t^2} + \frac{2s^2t^3}{(s^2 - t^2)^2}.$$

EXAMPLE 5   Let z depend on x and y and let $x = r\cos\theta$, $y = r\sin\theta$. Use the Chain Rule to obtain formulas for $\partial z/\partial r$ and $\partial z/\partial\theta$.

$$\frac{\partial x}{\partial r} = \cos\theta, \qquad \frac{\partial y}{\partial r} = \sin\theta.$$

$$\frac{\partial x}{\partial\theta} = -r\sin\theta, \qquad \frac{\partial y}{\partial\theta} = r\cos\theta.$$

$$\frac{\partial z}{\partial r} = \frac{\partial z}{\partial x}\frac{\partial x}{\partial r} + \frac{\partial z}{\partial y}\frac{\partial y}{\partial r} = \frac{\partial z}{\partial x}\cos\theta + \frac{\partial z}{\partial y}\sin\theta.$$

$$\frac{\partial z}{\partial\theta} = \frac{\partial z}{\partial x}\frac{\partial x}{\partial\theta} + \frac{\partial z}{\partial y}\frac{\partial y}{\partial\theta} = -\frac{\partial z}{\partial x}r\sin\theta + \frac{\partial z}{\partial y}r\cos\theta.$$

EXAMPLE 6   A rectangular solid has sides x, y, and z. Find the rate of change of the volume $V = xyz$ if

$$x = 1, \qquad y = 2, \qquad z = 3 \qquad \text{(in feet)},$$

$$\frac{dx}{dt} = 1, \qquad \frac{dy}{dt} = -5, \qquad \frac{dz}{dt} = 2 \qquad \text{(in feet per second)}.$$

We have
$$\frac{\partial V}{\partial x} = yz, \qquad \frac{\partial V}{\partial y} = xz, \qquad \frac{\partial V}{\partial z} = xy,$$

so
$$\frac{dV}{dt} = \frac{\partial V}{\partial x}\frac{dx}{dt} + \frac{\partial V}{\partial y}\frac{dy}{dt} + \frac{\partial V}{\partial z}\frac{dz}{dt}$$

$$= 2\cdot3\frac{dx}{dt} + 1\cdot3\frac{dy}{dt} + 1\cdot2\frac{dz}{dt}$$

$$= 2\cdot3\cdot1 + 1\cdot3\cdot(-5) + 1\cdot2\cdot2 = -5.$$

Thus the volume is decreasing at -5 cubic feet per second (Figure 11.5.6).

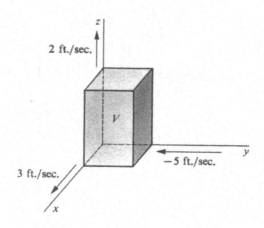

Figure 11.5.6

PROBLEMS FOR SECTION 11.5

In Problems 1–6, calculate dz/dt by the Chain Rule and check by a direct calculation.

1      $z = x^2 - y^2,\quad x = e^t,\quad y = e^{-t}$

2      $z = x^2 y^2,\quad x = \cos t,\quad y = \sin t$

3      $z = \dfrac{1}{ax + by},\quad x = \sin\left(\dfrac{t}{a}\right),\quad y = \sin\left(\dfrac{t}{b}\right)$

4      $z = e^{ax+by},\quad \dot{x} = \sqrt{t},\quad y = \dfrac{1}{\sqrt{t}}$

5      $z = \dfrac{\ln x}{\ln y},\quad x = \cosh(2t),\quad y = \sinh(2t)$

6      $z = \ln x \ln y,\quad x = \tan(3t),\quad y = \sec(3t)$

In Problems 7–14, calculate dz/dt by the Chain Rule.

7      $z = (t + 1)^{1/t}$ 8 $z = \left(1 + \dfrac{1}{t}\right)^t$

9      $z = \sin t^{\cos t}$ 10 $z = \sqrt{t}^{\sqrt{t}}$

11      $z = \log_{(t^2+1)}(t^2 - 1)$ 12 $z = \log_{\sin t}(\cos t)$

13      $z = 3x - 2y,\quad \dfrac{dx}{dt} = \sqrt{1 - t^4},\quad \dfrac{dy}{dt} = \sqrt{1 - t^3}$

14      $z = x + 2y + 3,\quad \dfrac{dx}{dt} = \cos\dfrac{1}{t},\quad \dfrac{dy}{dt} = \sin\dfrac{1}{t}$

In Problems 15–20, find $\partial z/\partial s$ and $\partial z/\partial t$ by the Chain Rule.

15      $z = y^3,\quad y = s\cos t$ 16 $z = \sin y,\quad y = st^2$

17      $z = \ln x,\quad x = s^2 - t^2$ 18 $z = e^x,\quad x = \cos(2s) + \sin(3t)$

19      $z = ax + by,\quad x = \dfrac{1}{s + t},\quad y = \dfrac{1}{s - t}$

20      $z = x^2 - y^2,\quad x = s\cos t,\quad y = s\sin t$

21   If $z = f(ax + by)$ and f is differentiable, show that $b\dfrac{\partial z}{\partial x} = a\dfrac{\partial z}{\partial y}$.

22   If $z = f(x + at, y + bt)$ and f is smooth, show that

$$\frac{\partial z}{\partial t} = a\frac{\partial z}{\partial x} + b\frac{\partial z}{\partial y}.$$

23   If $z = f(x, y)$, $x = r\cos\theta$, $y = r\sin\theta$, and f is smooth, show that

$$\left(\frac{\partial z}{\partial r}\right)^2 + \frac{1}{r^2}\left(\frac{\partial z}{\partial \theta}\right)^2 = \left(\frac{\partial f}{\partial x}\right)^2 + \left(\frac{\partial f}{\partial y}\right)^2.$$

24   If $z = f\left(\dfrac{xy}{x^2 + y^2}\right)$, where f is differentiable, show that

$$x\frac{\partial z}{\partial x} + y\frac{\partial z}{\partial y} = 0.$$

25   Find dw/dt where $w = x\cos z + y\sin z$, $x = e^t$, $y = e^{-t}$, $z = \sqrt{t}$.

26   Find dw/dt where $w = xy^2 z^3$, $x = 2t + 1$, $y = 3t - 2$, $z = 1 - 4t$.

In Problems 27–30, find formulas for dz/dt.

27      $z = \sqrt{x^2 + y^2},\quad x = f(t),\quad y = g(t)$

28 $z = x^a y^b$, $x = f(t)$, $y = g(t)$

29 $z = x^y$, $x = f(t)$, $y = g(t)$

30 $z = \log_x y$, $x = f(t)$, $y = g(t)$

In Problems 31–36, find formulas for $\partial z/\partial s$ and $\partial z/\partial t$.

31 $z = f(u)$, $u = as + bt$ **32** $z = f(u)$, $u = st$

33 $z = e^u$, $u = f(s, t)$ **34** $z = f(u)$, $u = g(s) + h(t)$

35 $z = g(s)h(t)$ **36** $z = f(x, y)$, $x = g(s)$, $y = h(t)$

37   A particle moves in the (x, y) plane so that $dx/dt = 2$, $dy/dt = -4$. Find dz/dt, where z is the distance from the origin, when the particle is at the point $(3, 4)$.

38   A particle moves in the (x, y) plane so that

$$\frac{dx}{dt} = \frac{1}{x} + \frac{1}{y}, \qquad \frac{dy}{dt} = 2x + y.$$

Find dz/dt, where z is the distance of the particle from the point $(1, 2)$, when the particle is at $(2, 3)$.

39   A particle moves in space so that

$$\frac{dx}{dt} = 3, \qquad \frac{dy}{dt} = 4, \qquad \frac{dz}{dt} = -2.$$

Find the rate of change of the distance from the origin when $x = 1, y = -2, z = 2$.

40   Find the rate of change of the area of a rectangle when the sides have lengths $x = 5$ and $y = 6$ and are changing at rates $dx/dt = 3$, $dy/dt = -4$.

41   Find the rate of change of the perimeter of a rectangle when the sides are $x = 2$, $y = 4$ and are changing at the rates $dx/dt = -2$, $dy/dt = 3$.

42   The per capita income of a country is equal to the national income x divided by the population y. Find the rate of change in per capita income when $x = \$10$ billion, $y = 10$ million, $dx/dt = \$10$ million per year, $dy/dt = 50,000$ people per year.

43   The profit of a manufacturer is equal to the total revenue x minus the total cost y. As the number of items produced, u, is increased, the revenue and cost increase at the rates $dx/du = 500/u$ and $dy/du = 1/\sqrt{u}$. Find the rate of increase of profit with respect to u when $u = 10,000$.

44   When commodities one and two have prices p and q respectively, their respective demands are $D_1(p, q)$ and $D_2(p, q)$. The revenue at prices p and q is the quantity

$$R(p, q) = pD_1(p, q) + qD_2(p, q),$$

since a quantity $D_1(p, q)$ can be sold at price p and a quantity $D_2(p, q)$ at price q. Find formulas for the partial marginal revenues with respect to price, $\partial R/\partial p$ and $\partial R/\partial q$.

11.6 IMPLICIT FUNCTIONS

In many applications of the Chain Rule, one or more of the independent variables is also used as an intermediate variable. The simplest case where this occurs is when z depends on x and y while y depends on x,

$$y = g(x), \qquad z = F(x, y).$$

Figure 11.6.1 shows which variables depend on which.

Assuming $F(x, y)$ is smooth and $dy/dx = g'(x)$ exists, the Chain Rule gives

$$\frac{dz}{dx} = \frac{\partial z}{\partial x}\frac{dx}{dx} + \frac{\partial z}{\partial y}\frac{dy}{dx}, \quad \text{or} \quad \frac{dz}{dx} = \frac{\partial z}{\partial x} + \frac{\partial z}{\partial y}\frac{dy}{dx}.$$

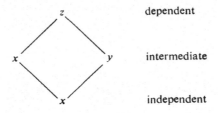

dependent

intermediate

independent

Figure 11.6.1

Here dz/dx stands for $f'(x)$ where $z = f(x) = F(x, g(x))$, and $\partial z/\partial x$ stands for $F_x(x, y)$. The round d, ∂, is useful in telling the two apart.

EXAMPLE 1   If $\qquad\qquad z = 2x + 3y, \qquad y = \sin x,$

find $\partial z/\partial x$ and dz/dx when $x = 0$.

$$\frac{\partial z}{\partial x} = 2.$$

$$\frac{dz}{dx} = \frac{\partial z}{\partial x} + \frac{\partial z}{\partial y}\frac{dy}{dx} = 2 + 3\cos x.$$

When $x = 0$, $\qquad\qquad \dfrac{dz}{dx} = 2 + 3\cos 0 = 5.$

As a check, we find dz/dx directly.

$$z = 2x + 3y = 2x + 3\sin x.$$

$$\frac{dz}{dx} = 2 + 3\cos x.$$

When $x = 0$, $\qquad\qquad \dfrac{dz}{dx} = 2 + 3\cos 0 = 5.$

EXAMPLE 2   Use the Chain Rule to obtain a formula for dz/dx where $z = x^y$ and y depends on x.

$$\frac{\partial z}{\partial x} = yx^{y-1}, \qquad \frac{\partial z}{\partial y} = (\ln x)x^y.$$

$$\frac{dz}{dx} = \frac{\partial z}{\partial x} + \frac{\partial z}{\partial y}\frac{dy}{dx} = yx^{y-1} + (\ln x)x^y\frac{dy}{dx}.$$

The Chain Rule can also be used in problems where dz/dx is known and dy/dx is to be found.

In many problems we are given a relationship between x and y which can be expressed by an equation of the form $F(x, y) = 0$, and we wish to find dy/dx. The graph of $F(x, y) = 0$ is usually a curve in the (x, y) plane. If we put $z = F(x, y) = 0$, then $dz/dx = 0$ while dy/dx is the slope of the curve. Ordinarily such a curve can be divided into finitely many pieces each of which is the graph of a function $y = g(x)$.

For example, the top and bottom halves of the circle

$$x^2 + y^2 - 1 = 0$$

are the functions

$$y = \sqrt{1 - x^2}, \qquad y = -\sqrt{1 - x^2}$$

shown in Figure 11.6.2.

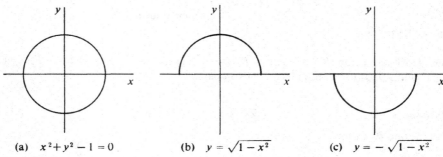

(a) $x^2 + y^2 - 1 = 0$ (b) $y = \sqrt{1 - x^2}$ (c) $y = -\sqrt{1 - x^2}$

Figure 11.6.2

DEFINITION

*An **implicit function** of the curve $F(x, y) = 0$ at (a, b) is a function $y = g(x)$ such that:*

(i)   *$g(a) = b$;*

(ii) *The domain of $g(x)$ is an open interval containing a;*

(iii) *The graph of $y = g(x)$ is a subset of the graph of $F(x, y) = 0$.*

*If every implicit function of $F(x, y) = 0$ has the same slope S at (a, b), we call S the **slope** of the curve.*

Figure 11.6.3 shows an implicit function $y = g(x)$ of a curve $F(x, y) = 0$. It is often hard or impossible to express an implicit function in terms of known (or elementary) functions. However, the next theorem gives an easy test for showing that there is an implicit function and finding its slope.

IMPLICIT FUNCTION THEOREM

Suppose that at the point (a, b), $z = F(x, y)$ is smooth, $F(a, b) = 0$, and $\partial z / \partial y \neq 0$.

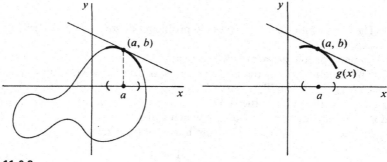

Figure 11.6.3

Then the curve $F(x, y) = 0$ at (a, b) has an implicit function and the slope

$$\frac{dy}{dx} = -\frac{\partial z / \partial x}{\partial z / \partial y}.$$

There are three things to prove:

(1) There exists an implicit function $y = g(x)$ at (a, b).
(2) The slope $dy/dx = g'(a)$ exists.
(3) dy/dx has the required value.

Instead of proving the whole theorem, we give an intuitive argument for (1) and (2) and then prove (3). The surface $z = F(x, y)$ has a tangent plane at $(a, b, 0)$. If we intersect the surface and tangent plane with the plane $z = 0$ we get the curve $0 = F(x, y)$ and a line L. Through an infinitesimal microscope aimed at the point (a, b), the curve looks like the graph of a function $y = g(x)$ which has the tangent line L and thus has a slope at (a, b) (Figure 11.6.4).

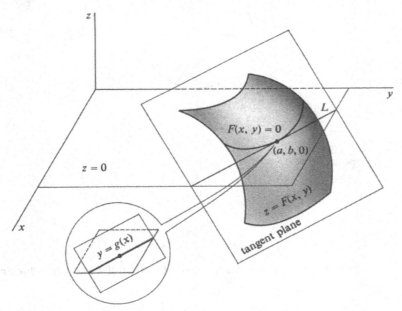

Figure 11.6.4

PROOF OF (3) Given that the slope $\dfrac{dy}{dx}$ exists, we compute its value.

By the Chain Rule, $\qquad \dfrac{dz}{dx} = \dfrac{\partial z}{\partial x} + \dfrac{\partial z}{\partial y}\dfrac{dy}{dx}.$

But $F(x, g(x))$ is identically zero, so $dz/dx = 0$ and

$$0 = \frac{\partial z}{\partial x} + \frac{\partial z}{\partial y}\frac{dy}{dx}.$$

Since $\dfrac{\partial z}{\partial y} \neq 0,$ $\qquad \dfrac{dy}{dx} = -\dfrac{\partial z / \partial x}{\partial z / \partial y}.$

The best way to remember the minus sign in the above equation is to derive the equation yourself. Start with the Chain Rule for $dz/dx = 0$ and solve for dy/dx. One way to understand the minus sign is as follows: if $\partial z/\partial x$ and $\partial z/\partial y$ are positive, an increase in x must be offset by a decrease in y to keep z constant, so dy/dx should be negative.

Warning: The two ∂z's have different meanings and cannot be cancelled.

EXAMPLE 3   Find the slope dy/dx of the circle

$$x^2 + y^2 - 4 = 0$$

at the point $(1, \sqrt{3})$ (see Figure 11.6.5).

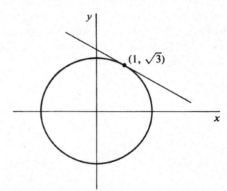

Figure 11.6.5

Put
$$z = x^2 + y^2 - 4 = 0.$$

At a point (x, y),

$$\frac{\partial z}{\partial x} = 2x, \qquad \frac{\partial z}{\partial y} = 2y, \qquad \frac{dy}{dx} = -\frac{\partial z/\partial x}{\partial z/\partial y} = -\frac{x}{y}.$$

At the given point $(1, \sqrt{3})$, $\qquad \dfrac{dy}{dx} = -\dfrac{1}{\sqrt{3}}.$

In this problem we can solve for y as a function of x and check the answer directly.

$$y = \sqrt{4 - x^2}.$$

$$\frac{dy}{dx} = \frac{-2x}{2\sqrt{4 - x^2}} = \frac{-2}{2\sqrt{4 - 1}} = -\frac{1}{\sqrt{3}}.$$

The Implicit Function Theorem gives us a convenient equation for the tangent line to the curve $F(x, y) = 0$ at (a, b).

$$y - b = \frac{dy}{dx}(x - a),$$

$$y - b = -\frac{\partial z/\partial x}{\partial z/\partial y}(x - a),$$

and finally

$$\text{Tangent Line:} \quad \frac{\partial z}{\partial x}(x - a) + \frac{\partial z}{\partial y}(y - b) = 0.$$

EXAMPLE 3 (Continued) Find the equation for the tangent line in Example 3.
At the point $(1, \sqrt{3})$,

$$\frac{\partial z}{\partial x} = 2x = 2, \quad \frac{\partial z}{\partial y} = 2y = 2\sqrt{3},$$

and the tangent line is

$$2(x - 1) + 2\sqrt{3}(y - \sqrt{3}) = 0.$$

EXAMPLE 4 Find the tangent line and slope of the curve

$$y + \ln y + x^3 = 0$$

at the point $(-1, 1)$ (Figure 11.6.6).

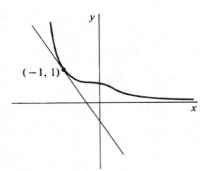

Figure 11.6.6

Put $$z = y + \ln y + x^3.$$

Then $$\frac{\partial z}{\partial x} = 3x^2, \quad \frac{\partial z}{\partial y} = 1 + \frac{1}{y}.$$

At $(-1, 1)$, $$\frac{\partial z}{\partial x} = 3, \quad \frac{\partial z}{\partial y} = 2.$$

Tangent Line: $$3(x + 1) + 2(y - 1) = 0.$$

Slope: $$\frac{dy}{dx} = -\frac{3}{2}.$$

EXAMPLE 5 Find the tangent line and slope of the level curve of the hyperbolic
paraboloid

$$z = x^2 - y^2$$

at the point (a, b) (where $b \neq 0$) (Figure 11.6.7).

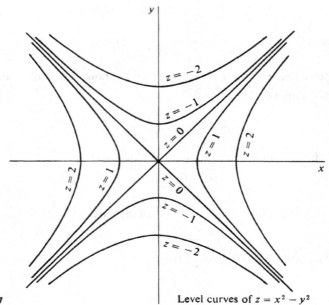

Figure 11.6.7   Level curves of $z = x^2 - y^2$

The level curve has the equation

$$x^2 - y^2 = a^2 - b^2,$$
$$x^2 - y^2 - (a^2 - b^2) = 0.$$

Put $w = x^2 - y^2 - (a^2 - b^2) = 0.$

Then $\dfrac{\partial w}{\partial x} = 2x, \qquad \dfrac{\partial w}{\partial y} = -2y.$

At (a, b), $\dfrac{\partial w}{\partial x} = 2a, \qquad \dfrac{\partial w}{\partial y} = -2b.$

Tangent Line: $2a(x - a) - 2b(y - b) = 0.$

Slope: $\dfrac{dy}{dx} = -\dfrac{2a}{-2b} = \dfrac{a}{b}.$

Let us next consider the case where w depends on x, y, and z, while z depends on x and y,

$$w = F(x, y, z), \qquad z = g(x, y).$$

Figure 11.6.8 shows which variables depend on which.

If $F(x, y, z)$ is smooth and $\partial z/\partial x$, $\partial z/\partial y$ exist, the Chain Rule gives

$$\frac{\partial w}{\partial x}(x, y) = \frac{\partial w}{\partial x}(x, y, z)\frac{\partial x}{\partial x} + \frac{\partial w}{\partial y}(x, y, z)\frac{\partial y}{\partial x} + \frac{\partial w}{\partial z}(x, y, z)\frac{\partial z}{\partial x},$$

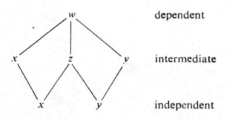

dependent

intermediate

independent

Figure 11.6.8

or
$$\frac{\partial w}{\partial x}(x, y) = \frac{\partial w}{\partial x}(x, y, z) + \frac{\partial w}{\partial z}(x, y, z)\frac{\partial z}{\partial x}.$$

Similarly,
$$\frac{\partial w}{\partial y}(x, y) = \frac{\partial w}{\partial y}(x, y, z) + \frac{\partial w}{\partial z}(x, y, z)\frac{\partial z}{\partial y}.$$

We used the fact that for the independent variables x and y,

$$\frac{\partial x}{\partial x} = \frac{\partial y}{\partial y} = 1, \qquad \frac{\partial x}{\partial y} = \frac{\partial y}{\partial x} = 0.$$

Notice that in this case $\partial w/\partial x$ alone is ambiguous so we had to use the more complete notation

$$\frac{\partial w}{\partial x}(x, y, z) \qquad \text{for } F_x(x, y, z),$$

$$\frac{\partial w}{\partial x}(x, y) \qquad \text{for } f_x(x, y), \quad \text{where } f(x, y) = F(x, y, g(x, y)).$$

EXAMPLE 6 Find $\dfrac{\partial w}{\partial x}(x, y)$ and $\dfrac{\partial w}{\partial y}(x, y)$ where

$$w = x^2 + 2y^2 + 3z^2, \qquad z = e^{5x+y}.$$

$$\frac{\partial w}{\partial x}(x, y, z) = 2x, \qquad \frac{\partial w}{\partial y}(x, y, z) = 4y, \qquad \frac{\partial w}{\partial z}(x, y, z) = 6z.$$

$$\frac{\partial z}{\partial x} = 5e^{5x+y}, \qquad \frac{\partial z}{\partial y} = e^{5x+y}.$$

Then
$$\frac{\partial w}{\partial x}(x, y) = 2x + 6z \cdot 5e^{5x+y} = 2x + 30ze^{5x+y}$$

$$= 2x + 30e^{2(5x+y)}.$$

$$\frac{\partial w}{\partial y}(x, y) = 4y + 6z \cdot e^{5x+y} = 4y + 6e^{2(5x+y)}.$$

The graph of an equation

$$F(x, y, z) = 0$$

is a surface in space. The Implicit Function Theorem can be generalized to this case.

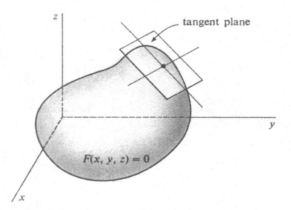

Figure 11.6.9

We shall skip the details, but the end result is an equation for the tangent plane of the surface, pictured in Figure 11.6.9.

THEOREM

Suppose the function $w = F(x, y, z)$ is smooth at the point (a, b, c), and $F_z(a, b, c) \neq 0$. Then the implicit surface $F(x, y, z) = 0$ has the partial derivatives

$$\frac{\partial z}{\partial x} = -\frac{F_x(a, b, c)}{F_z(a, b, c)}, \qquad \frac{\partial z}{\partial y} = -\frac{F_y(a, b, c)}{F_z(a, b, c)}$$

and the tangent plane

$$F_x(a, b, c)(x - a) + F_y(a, b, c)(y - b) + F_z(a, b, c)(z - c) = 0.$$

The equation for the tangent plane is obtained as follows.

$$z - c = \frac{\partial z}{\partial x}(x - a) + \frac{\partial z}{\partial y}(y - b),$$

$$z - c = -\frac{F_x(a, b, c)}{F_z(a, b, c)}(x - a) - \frac{F_y(a, b, c)}{F_z(a, b, c)}(y - b),$$

and finally $F_x(a, b, c)(x - a) + F_y(a, b, c)(y - b) + F_z(a, b, c)(z - c) = 0$.

EXAMPLE 7   Find the tangent plane to the ellipsoid

$$x^2 + 2y^2 + 3z^2 = 6$$

at the point $(1, 1, 1)$ (see Figure 11.6.10).

Put $F(x, y, z) = x^2 + 2y^2 + 3z^2 - 6.$

Then $F_x(x, y, z) = 2x,$ $F_y(x, y, z) = 4y,$ $F_z(x, y, z) = 6z.$
          $F_x(1, 1, 1) = 2,$ $F_y(1, 1, 1) = 4,$ $F_z(1, 1, 1) = 6.$

The tangent plane has the equation

$$2(x - 1) + 4(y - 1) + 6(z - 1) = 0.$$

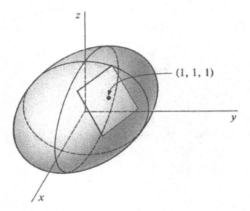

Figure 11.6.10

Find $\partial z/\partial x$ and $\partial z/\partial y$ at $(1, 1, 1)$.

$$\frac{\partial z}{\partial x} = -\frac{F_x(1, 1, 1)}{F_z(1, 1, 1)} = -\frac{2x}{6z} = -\frac{x}{3z},$$

$$\frac{\partial z}{\partial y} = -\frac{F_y(1, 1, 1)}{F_z(1, 1, 1)} = -\frac{4y}{6z} = -\frac{2y}{3z}.$$

At $(1, 1, 1)$, $\dfrac{\partial z}{\partial x} = -\dfrac{1}{3}$, $\dfrac{\partial y}{\partial y} = -\dfrac{2}{3}$.

PROBLEMS FOR SECTION 11.6

In Problems 1–8, find $\partial z/\partial x$ and dz/dx by the Chain Rule.

1	$z = 3x - 4y$, $y = e^x$	**2**	$z = xy$, $y = \ln x$
3	$z = \cos x + \sin y$, $y = 3x$	**4**	$z = \dfrac{1}{2x + 3y}$, $y = \sqrt{x}$
5	$z = x^y$, $y = x$	**6**	$z = x^y$, $y = \sqrt{x}$
7	$z = \arctan(xy)$, $y = e^{-x}$	**8**	$z = \sin x \sin y$, $y = 2x$

In Problems 9–14, find dy/dx.

9	$x^2 + 2xy - y^2 = 2$	**10**	$\sqrt{x} + \sqrt{xy} + \sqrt{y} = 1$
11	$x^2 + 2xy^3 + y = 2$	**12**	$e^{xy} + 3x + 2y^2 = 1$
13	$\sin xy + x + 2 = 0$	**14**	$\ln x + 2 \ln y + xy = 1$

In Problems 15–22, find the tangent line and the slope of the curve at the given point.

15	$(x + 1)^2 + (y + 2)^2 = 25$ at $(2, 2)$		
16	$x^2 + 4y^2 = 4$ at $(\sqrt{3}, \frac{1}{2})$	**17**	$x^2 - 3xy - y^2 = 3$ at $(1, -1)$
18	$\sqrt{x} + \sqrt{y} = 2$ at $(1, 1)$	**19**	$x^3 + y^3 = 2$ at $(1, 1)$
20	$x + \sqrt{xy} - 2y = 8$ at $(8, 2)$	**21**	$\cos x \sin y = \frac{1}{2}$ at $(\pi/4, \pi/4)$
22	$y + e^x \ln y = 1$ at $(2, 1)$		

In Problems 23–26 find $\frac{\partial w}{\partial x}(x, y)$ and $\frac{\partial w}{\partial y}(x, y)$.

23      $w = 3x - 4y + 6z, \quad z = 2x - 5y$

24      $w = z\cos x + z\sin y, \quad z = \sqrt{x^2 + y^2}$

25      $w = \sqrt{x^2 + y^2 + z^2}, \quad z = -3x + 2y$

26      $w = z/xy, \quad z = y\ln x$

In Problems 27–32, find the tangent plane to the surface at the given point.

27      $3x^2 + 5y^2 + 4z^2 = 21$ at $(-2, 1, -1)$

28      $2x^2 - 4y^2 + z^2 = 2$ at $(1, 1, -2)$

29      $xyz + x^2 + y^2 + z^2 = 4$ at $(1, 1, 1)$

30      $xy + xz + yz = 3$ at $(1, 1, 1)$

31      $xe^y + ye^z + ze^x = 0$ at $(0, 0, 0)$

32      $\sin x \cos y \tan z = 1$ at $(\pi/2, 0, \pi/4)$

In Problems 33–38, find $\partial z/\partial x$ and $\partial z/\partial y$.

33      $x^2 y + z^2 = 1$ 34 $x^2 + 2y^2 - 3z^2 = 4$

35      $\sin xy + \cos yz = 1$ 36 $e^x + e^y + e^z = 1$

37      $xy^2 z^3 + 2 = 0$ 38 $x^2 + y^3 + \ln z = 2$

39   Suppose that x items can be bought at a price of y dollars per item, where y depends on x in such a way that $dy/dx = -1/(1 + \sqrt{x})$. Find the rate of change of the total cost $z = xy$ with respect to x.

40   A point moves along the parabola $y = x^2$. Find the rate of change with respect to x of the distance from the origin.

41   Suppose w depends on x, y, and z, and both y and z depend on x. Find a formula for dw/dx using the Chain Rule.

42   Suppose z depends on x and y, while y depends on x and t. Use the Chain Rule to find a formula for $\frac{\partial z}{\partial x}(x, t)$.

11.7 MAXIMA AND MINIMA

The theory of maxima and minima for functions of two variables is similar to the theory for one variable. The student should review the one-variable case at this time.

DEFINITION

Let $z = f(x, y)$ be a function with domain D. f is said to have a **maximum** at a point (x_0, y_0) in D if

$$f(x_0, y_0) \geq f(x, y)$$

for all (x, y) in D. The value $f(x_0, y_0)$ is called the **maximum value** of f.

A **minimum** and the **minimum value** of f are defined analogously.

We shall first study functions defined on closed regions, which correspond to closed intervals. By a *closed region* in the plane we mean a set D defined by inequalities

$$a \leq x \leq b, \qquad f(x) \leq y \leq g(x),$$

where f and g are continuous and $f(x) \leq g(x)$ on $[a, b]$. *D is called the region between* $f(x)$ *and* $g(x)$ *for* $a \leq x \leq b$ (Figure 11.7.1).

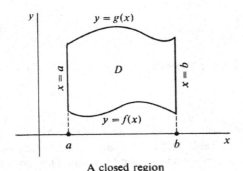

Figure 11.7.1   A closed region

The points of D on the four curves

$$x = a, \qquad x = b, \qquad y = f(x), \qquad y = g(x)$$

are called *boundary points*. All other points of D are called *interior points*.

EXTREME VALUE THEOREM

Suppose $z = f(x, y)$ *is continuous at every point of a closed region D. Then the function f with its domain restricted to D has a maximum and a minimum.*

The proof is similar to the corresponding proof for one variable.

CRITICAL POINT THEOREM

Suppose the domain of $z = f(x, y)$ *is a closed region D and f is smooth at every interior point of D. If f has a maximum or minimum at* (x_0, y_0), *then either*

(i) $f_x(x_0, y_0) = 0$ *and* $f_y(x_0, y_0) = 0$, *or*

(ii) (x_0, y_0) *is a boundary point of D.*

Figure 11.7.2 illustrates the two cases when f has a maximum at (x_0, y_0). An interior point where both partial derivatives are zero is called a *critical point*. Thus a critical point is a point where the tangent plane is horizontal. On the graph of a surface, an interior point looks like a mountain summit if it is a maximum and a valley bottom if it is a minimum. The theorem states that every interior maximum or minimum is a critical point. An interesting kind of critical point which is neither

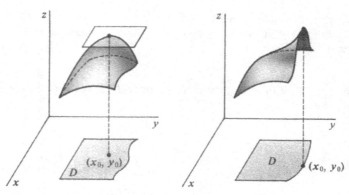

Case (i) Interior Maximum   *Case (ii)* Boundary Maximum

Figure 11.7.2 Critical Point Theorem

a maximum nor a minimum is a *saddle point*, which looks like the summit of a pass between two mountains. Table 11.7.1 gives three simple examples of critical points, one maximum, one minimum, and one saddle point. They are illustrated in Figure 11.7.3.

Table 11.7.1

Function	Partials	Critical Point	Type
$z = -(x^2 + y^2)$	$\dfrac{\partial z}{\partial x} = -2x, \quad \dfrac{\partial z}{\partial y} = -2y$	$(0, 0)$	Maximum
$z = x^2 + y^2$	$\dfrac{\partial z}{\partial x} = 2x, \quad \dfrac{\partial z}{\partial y} = 2y$	$(0, 0)$	Minimum
$z = x^2 - y^2$	$\dfrac{\partial z}{\partial x} = 2x, \quad \dfrac{\partial z}{\partial y} = -2y$	$(0, 0)$	Saddle Point

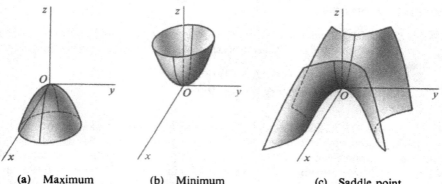

(a) Maximum   (b) Minimum   (c) Saddle point

Figure 11.7.3

PROOF OF THE CRITICAL POINT THEOREM Suppose f has a maximum at an interior point (x_0, y_0) of D. (x_0, y_0) is not a boundary point so we must prove (i). The function

$$g(x) = f(x, y_0)$$

is differentiable and has a maximum at x_0. By the Critical Point Theorem for one variable, $g'(x_0) = f_x(x_0, y_0) = 0$. Similarly $f_y(x_0, y_0) = 0$.

METHOD FOR FINDING MAXIMA AND MINIMA ON A CLOSED REGION

When to Use $z = f(x, y)$ *is continuous on a closed region D and smooth on the interior of D.*

Step 1   *Set the problem up and sketch D.*

Step 2   *Compute $\partial z / \partial x$ and $\partial z / \partial y$.*

Step 3   *Find the critical points of f, if any, and the value of f at each critical point.*

Step 4   *Find the maximum and minimum of f on the boundary of D. This can be done by solving for z as a function of x or y alone and using the method for one variable.*

CONCLUSION   *The largest of the values from Steps 3 and 4 is the maximum value, and the smallest is the minimum value.*

It is convenient to record the results of Steps 3 and 4 on the sketch of D.

EXAMPLE 1   Find the maximum and minimum of $z = x^2 + y^2 - xy - x$ on the closed rectangle $0 \le x \le 1, 0 \le y \le 1$.

Step 1   The region D is sketched in Figure 11.7.4.

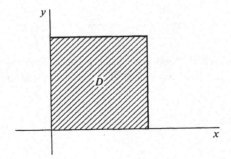

Figure 11.7.4

Step 2 $\dfrac{\partial z}{\partial x} = 2x - y - 1, \qquad \dfrac{\partial z}{\partial y} = 2y - x.$

Step 3 $2x - y - 1 = 0, \qquad 2y - x = 0.$

Solving for x and y we get one critical point

$$y = \tfrac{1}{3}, \qquad x = \tfrac{2}{3}.$$

The value of z at that point is

$$z = (\tfrac{2}{3})^2 + (\tfrac{1}{3})^2 - \tfrac{2}{3} \cdot \tfrac{1}{3} - \tfrac{2}{3} = -\tfrac{1}{3}.$$

Step 4   We make a table.

Boundary Line	z	Maximum	Minimum
$x = 0, \quad 0 \le y \le 1$	y^2	1 at $(0, 1)$	0 at $(0, 0)$
$x = 1, \quad 0 \le y \le 1$	$y^2 - y$	0 at corners	$-\tfrac{1}{4}$ at $(1, \tfrac{1}{2})$
$y = 0, \quad 0 \le x \le 1$	$x^2 - x$	0 at corners	$-\tfrac{1}{4}$ at $(\tfrac{1}{2}, 0)$
$y = 1, \quad 0 \le x \le 1$	$x^2 + 1 - 2x$	1 at $(0, 1)$	0 at $(1, 1)$

The values from Steps 3 and 4 are also shown on the sketch of D in Figure 11.7.5.

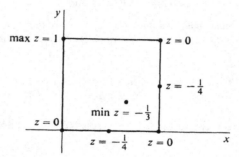

Figure 11.7.5

CONCLUSION

> *Maximum:* $z = 1$ at $(0, 1)$
> *Minimum:* $z = -\tfrac{1}{3}$ at $(\tfrac{2}{3}, \tfrac{1}{3})$.

The maximum is at a boundary point and the minimum at an interior point.

In many problems we are to maximize a function of three variables which are related by a *side condition*. We wish to find the maximum or minimum of

$$w = F(x, y, z)$$

given the side condition

$$g(x, y, z) = 0.$$

To work a problem of this type we use the side condition to get w as a function of just two independent variables and then proceed as before.

EXAMPLE 2   For a package to be mailed in the United States by parcel post, its length plus its girth (perimeter of cross section) must be at most 84 inches. Find the dimensions of the rectangular box of maximum volume which can be mailed by parcel post.

Step 1   Let x, y, and z be the dimensions of the box, with z the length. We wish to find the maximum of the volume

$$V = xyz$$

given the side condition

$$\text{length} + \text{girth} = z + 2x + 2y = 84.$$

We eliminate z using the side condition and express V as a function of x and y.

$$z = 84 - 2x - 2y,$$
$$V = xy(84 - 2x - 2y).$$

Since x, y, and z cannot be negative the domain is the closed triangle

$$0 \le x, \qquad 0 \le y, \qquad 0 \le 84 - 2x - 2y.$$

This is the same as the closed region

$$0 \le x \le 42, \qquad 0 \le y \le 42 - x.$$

The region is sketched in Figure 11.7.6.

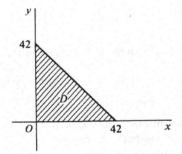

Figure 11.7.6

Step 2 $\dfrac{\partial V}{\partial x} = 84y - 4xy - 2y^2,$

$\dfrac{\partial V}{\partial y} = 84x - 2x^2 - 4xy.$

Step 3 $84y - 4xy - 2y^2 = 0,$
$84x - 2x^2 - 4xy = 0.$

Since $x > 0$ and $y > 0$ at all interior points, we have

$$84 - 4x - 2y = 0,$$
$$84 - 2x - 4y = 0.$$

There is one critical point

$$x = 14, \qquad y = 14,$$
$$V = (84 - 28 - 28) \cdot 14 \cdot 14 = 2(14)^3.$$

Step 4   On all three of the boundary lines

$$x = 0, \qquad y = 0, \qquad 84 - 2x - 2y = 0$$

we have $\qquad\qquad V = (84 - 2x - 2y)xy = 0.$

Therefore the maximum value of V on the boundary of D is 0.

CONCLUSION   The maximum of V is at $x = 14$, $y = 14$, where $V = 2(14)^3$ (Figure 11.7.7). The box has dimensions

$$x = 14, \qquad y = 14, \qquad z = 28.$$

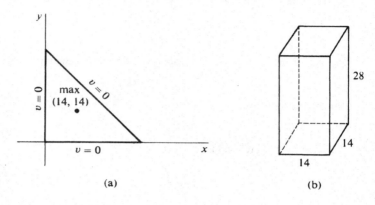

Figure 11.7.7 $\qquad\qquad\qquad$ (a) $\qquad\qquad\qquad\qquad\qquad\qquad\qquad$ (b)

We shall now develop a method for finding maxima and minima of functions defined on open regions.

A *bounded open region D* is a set of points given by strict inequalities

$$a < x < b, \qquad f(x) < y < g(x)$$

where f and g are continuous and $f(x) < g(x)$ on (a, b). A closed region with its boundary removed is a bounded open region.

We shall also consider *unbounded open regions*, which are given by strict inequalities where one or more of $a, b, f(x), g(x)$ are replaced by infinity symbols. For example, the following are unbounded open regions:

(1) $\quad -\infty < x < \infty, \qquad f(x) < y < g(x)$.
(2) $\quad 0 < x < \infty, \qquad 0 < y < \infty$.
(3) $\quad$ The whole plane $-\infty < x < \infty, \qquad -\infty < y < \infty$.

Unbounded open regions are pictured in Figure 11.7.8.

A smooth function whose domain is an open region may or may not have a maximum or minimum. Many problems have at most one critical point, and we shall concentrate on that case. The method can readily be extended to the case of two or more critical points. The Critical Point Theorem holds for open regions as well as closed regions. The corollary below shows how it can be used in maximum or minimum problems.

COROLLARY

Suppose the domain of the function $z = f(x, y)$ *is an open region D, and f is smooth on D.*

(i) *If f has no critical points it has no maximum or minimum.*

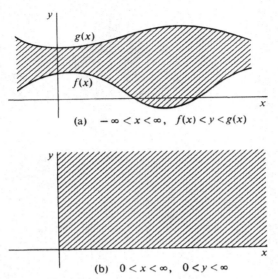

Figure 11.7.8 Unbounded Open Regions

(ii) *Let f have exactly one critical point (x_0, y_0). If f has a maximum or minimum, it occurs at (x_0, y_0).*

This corollary can be used to show certain functions do not have a maximum or minimum. If we are sure a function has a maximum or minimum, the corollary can be used to find it.

EXAMPLE 3 Show that the function $z = e^x \ln y$ has no maximum or minimum.

The domain is the open region

$$-\infty < x < \infty, \qquad 0 < y < \infty.$$

The partial derivatives are

$$\frac{\partial z}{\partial x} = e^x \ln y, \qquad \frac{\partial z}{\partial y} = \frac{e^x}{y}.$$

There are no critical points because $\partial z/\partial y$ is never zero. Therefore there is no maximum or minimum.

EXAMPLE 4 Show that the function $z = x^2 + 2y^2$ has no maximum.

The domain is the whole plane.

We have

$$\frac{\partial z}{\partial x} = 2x, \qquad \frac{\partial z}{\partial y} = 4y.$$

There is one critical point at $(0, 0)$. At this point, $z = 0$. This is not a maximum because, for example, $z = 3$ at $(1, 1)$. Hence z has no maximum.

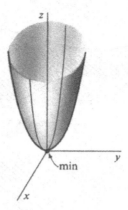

Figure 11.7.9

Notice that z has a minimum at $(0, 0)$ because $x^2 + 2y^2$ is always ≥ 0 (Figure 11.7.9).

EXAMPLE 5   Find the point on the plane $4x - 6y + 2z = 7$ which is nearest to the origin.

Step 1   The distance from the origin to (x, y, z) is $\sqrt{x^2 + y^2 + z^2}$. It is easier to work with the square of the distance, which has a minimum at the same point that the distance does. So we wish to find the minimum of

$$w = x^2 + y^2 + z^2$$

given that

$$4x - 6y + 2z = 7.$$

We eliminate z using the plane equation.

$$z = \tfrac{1}{2}(7 - 4x + 6y),$$
$$w = x^2 + y^2 + \tfrac{1}{4}(7 - 4x + 6y)^2.$$

The domain is the whole (x, y) plane.

Step 2 $\dfrac{\partial w}{\partial x} = 2x + 2 \cdot \dfrac{1}{4}(-4)(7 - 4x + 6y) = -14 + 10x - 12y,$

$\dfrac{\partial w}{\partial y} = 2y + 2 \cdot \dfrac{1}{4} \cdot 6(7 - 4x + 6y) = 21 - 12x + 20y.$

Step 3 $-14 + 10x - 12y = 0,$
$21 - 12x + 20y = 0.$

Solving for x and y we get one critical point

$$x = \tfrac{1}{2}, \qquad y = -\tfrac{3}{4}.$$

CONCLUSION   We know from geometry that there is a point on the plane which is closest to the origin (the point where a perpendicular line from the origin meets the plane). Therefore w has a minimum and it must be at the critical point

$$x = \tfrac{1}{2}, \qquad y = -\tfrac{3}{4}.$$

The value of z at this point is

$$z = \tfrac{1}{2}(7 - 4x + 6y) = \tfrac{1}{4}.$$

The answer is $(\tfrac{1}{2}, -\tfrac{3}{4}, \tfrac{1}{4})$. The plane is shown in Figure 11.7.10.

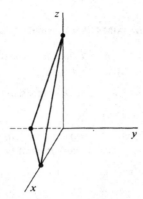

Figure 11.7.10

If we know a function has a maximum or minimum, we can find it simply by finding the critical point. But usually we are not sure whether a function has a maximum or minimum. Here is a method that can be used when a function has a unique critical point in an open region. It is based on the fact that the Extreme Value Theorem holds for closed regions of the hyperreal plane as well as the real plane (because of the Transfer Principle).

Given a real open region D we can find a hyperreal closed region E which contains the same real points as D (Figure 11.7.11).

For example, if D is the real region

$$a < x < b, \qquad f(x) < y < g(x),$$

we can take for E the hyperreal region

$$a + \varepsilon \le x \le b - \varepsilon, \qquad f(x) + \varepsilon \le y \le g(x) - \varepsilon$$

where ε is positive infinitesimal.

(a)   (b)

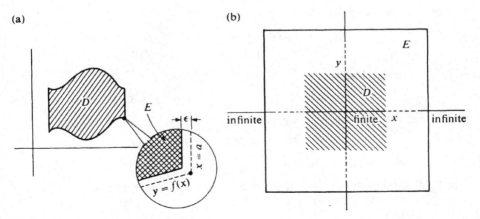

Figure 11.7.11 Hyperreal Closed Regions

If D is the whole real plane we can take for E the hyperreal region

$$-H \leq x \leq H, \qquad -H \leq y \leq H$$

where H is positive infinite.

METHOD FOR FINDING MAXIMA AND MINIMA ON AN OPEN REGION

When to Use $z = f(x, y)$ *is a smooth function whose domain is an open region D, and f has exactly one critical point.*

Step 1 *Set up the problem and sketch D if necessary.*

Step 2 *Compute $\partial z/\partial x$ and $\partial z/\partial y$.*

Step 3 *Find the critical point (x_0, y_0) and the value $f(x_0, y_0)$. If we already know there is a maximum (or minimum), it must be (x_0, y_0) and we can stop here.*

Step 4 *Find a hyperreal closed region E with the same real points as D.*

Step 5 *Compare $f(x_0, y_0)$ with the values of f on the boundary of E.*

CONCLUSION *f has a maximum at (x_0, y_0) if $f(x_0, y_0) \geq f(x, y)$ for every boundary point (x, y) of E. Otherwise f has no maximum.*

A similar rule holds for the minimum.

EXAMPLE 6   Find the maximum and minimum, if any, of the function

$$z = \frac{1}{(x + y)^2 + (x + 1)^2 + y^2}.$$

Step 1   The domain is the whole (x, y) plane because the denominator is always positive.

Step 2 $\dfrac{\partial z}{\partial x} = -[2(x + y) + 2(x + 1)][(x + y)^2 + (x + 1)^2 + y^2]^{-2},$

$\dfrac{\partial z}{\partial y} = -[2(x + y) + 2y][(x + y)^2 + (x + 1)^2 + y^2]^{-2}.$

Step 3   The partial derivatives are zero when

$$2(x + y) + 2(x + 1) = 0, \qquad 2(x + y) + 2y = 0,$$
or
$$2x + y + 1 = 0, \qquad x + 2y = 0.$$

The critical point is

$$x = -\tfrac{2}{3}, \qquad y = \tfrac{1}{3}, \quad \text{and} \quad z = 3.$$

Step 4   Let E be the hyperreal region

$$-H \leq x \leq H, \qquad -H \leq y \leq H$$

where H is positive infinite.

Step 5   At a boundary point of E where $x = \pm H$, $(x + 1)^2$ is infinite so z is infinitesimal. At a boundary point where $y = \pm H$, y^2 is infinite so again z is infinitesimal.

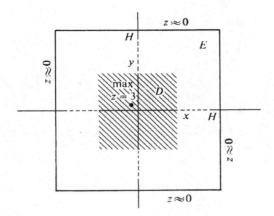

$z \approx 0$

H

E

$z \approx 0$

y

max
$z = 3$

D

x H

$z \approx 0$

$z \approx 0$

Figure 11.7.12

CONCLUSION z has a maximum of 3 at the critical point $(-\frac{2}{3}, \frac{1}{3})$. z has no minimum. The region E is sketched in Figure 11.7.12.

EXAMPLE 7   Find the dimensions of the box of volume one without a top which has the smallest area (if there is one). The box is sketched in Figure 11.7.13.

Step 1   Let x, y, and z be the dimensions of the box, with z the height. We want the minimum of the area

$$A = xy + 2xz + 2yz$$

given that

$$xyz = 1.$$

Eliminating z, we have

$$z = \frac{1}{xy},$$

$$A = xy + \frac{2}{y} + \frac{2}{x}.$$

The domain is the open region $x > 0$, $y > 0$ (see Figure 11.7.14).

Step 2 $\dfrac{\partial A}{\partial x} = y - \dfrac{2}{x^2}, \qquad \dfrac{\partial A}{\partial y} = x - \dfrac{2}{y^2}.$

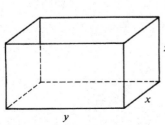

Figure 11.7.13

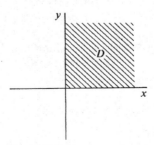

y

D

x

Figure 11.7.14

Step 3 $y - \dfrac{2}{x^2} = 0, \qquad x - \dfrac{2}{y^2} = 0.$

The critical point is $x = \sqrt[3]{2},\ y = \sqrt[3]{2},$

where $A = 2^{2/3} + 2 \cdot 2^{-1/3} + 2 \cdot 2^{-1/3} = 2^{2/3} + 2^{5/3}.$

Step 4 ·Take for E the hyperreal region $\varepsilon \le x \le H,\ \varepsilon \le y \le H$ where ε is positive infinitesimal and H is positive infinite.

Step 5   Let (x, y) be a boundary point of E. As we can see from Figure 11.7.15, there are four possible cases.

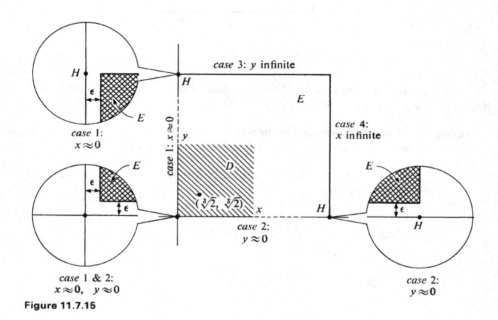

Figure 11.7.15

Case 1 x is infinitesimal. Then A is infinite because $2/x$ is.

Case 2 y is infinitesimal. A is infinite because $2/y$ is.

Case 3 x is not infinitesimal and y is infinite. A is infinite because xy is.

Case 4 y is not infinitesimal and x is infinite. A is infinite because xy is.

CONCLUSION A is infinite and hence greater than $2^{2/3} + 2^{5/3}$ on the boundary of E. Therefore A has a minimum at the critical point

$$x = \sqrt[3]{2}, \qquad y = \sqrt[3]{2}.$$

The box has dimensions

$$x = \sqrt[3]{2}, \qquad y = \sqrt[3]{2}, \qquad z = \frac{1}{xy} = \frac{1}{\sqrt[3]{4}}.$$

PROBLEMS FOR SECTION 11.7

In Problems 1–10, find the maxima and minima.

1      $x^2 + xy + y^2, \quad -1 \le x \le 1, \quad -1 \le y \le 1$

2      $-x^2 - 2y^2 + x - y + 2, \quad -1 \le x \le 1, \quad -1 \le y \le 1$

3      $x^2 + 2y^2 - 2x + 8y + 3, \quad -3 \le x \le 3, \quad -3 \le y \le x$

4      $x - xy + 2y, \quad -4 \le x \le 4, \quad -4 \le y \le x$

5      $x^2 - y^2 - 2x + 2y + 3, \quad 0 \le x \le 2, \quad 0 \le y \le 2x$

6      $xy + \dfrac{1}{x} + \dfrac{8}{y}, \quad \dfrac{1}{4} \le x \le 4, \quad 1 \le y \le 8$

7      $\sin x + \sin y, \quad 0 \le x \le \pi, \quad 0 \le y \le \pi$

8      $\sin x \sin y, \quad 0 \le x \le \pi, \quad 0 \le y \le \pi$

9      $x^2 + y^2 - y, \quad -1 \le x \le 1, \quad x^2 \le y \le 1$

10     $4 - x^2 - y^2, \quad -\sqrt{1 - x^2} \le y \le \sqrt{1 - x^2}$

In Problems 11–16, find the maximum and minimum subject to the given side conditions.

11     $z(x - y), \quad x + y + z = 1, \quad 0 \le x \le 1, \quad 0 \le y \le 1$

12     $xyz, \quad z = x + y, \quad -1 \le x \le 1, \quad -1 \le y \le 1$

13     $x + y + z, \quad z = x^2 + y^2, \quad z \le 1$

14     $x + y + z, \quad z = \sqrt{3 - x^2 - y^2}$

15     $x^2 + y^2 + z^2, \quad z = xy, \quad -1 \le x \le 1, \quad -1 \le y \le 1$

16     $xy + yz + xz, \quad xyz = 1, \quad \tfrac{1}{4} \le x \le 4, \quad \tfrac{1}{4} \le y \le 4$

In Problems 17–26, determine whether the maxima and minima exist, and if so, find them.

17     $x^2 + 4x + y^2$

18     $-x^2 - y^2 + 2x - 4y$

19     $1/xy, \quad 0 < x, \quad 0 < y$

20     $x^3 + 2x + y^3 - y^2$

21     $xy + \dfrac{1}{x} + \dfrac{8}{y}, \quad 0 < x, \quad 0 < y$

22     $x + 4y + \dfrac{1}{x} + \dfrac{1}{y}, \quad 0 < x, \quad 0 < y$

23     $\dfrac{1}{x^2 + y^2 + 1}$

24     $\dfrac{1}{\sqrt{1 - x^2 - y^2}}, \quad x^2 + y^2 < 1$

25     $x^2 - 4y^2$

26     $x^y, \quad 0 < x$

27   Find three positive numbers x, y, and z such that $x + y + z = 8$ and $x^2 yz$ is a maximum.

28   Find three positive numbers x, y, and z such that $x + y + z = 100$ and $x^2 y^2 z$ is a maximum.

29   A package can be sent overseas by the air mail small packet rate if its length plus girth is at most 36 inches. Find the dimensions of the rectangular solid of maximum volume which can be sent by the small packet rate.

30   Find the volume of the largest rectangular solid which can be inscribed in a sphere of radius one.

31   Find the volume of the largest rectangular solid with faces parallel to the coordinate planes which can be inscribed in the ellipsoid $x^2/4 + y^2 + z^2/9 = 1$.

32   A triangle with sides a, b, c and perimeter $p = a + b + c$ has area
$$A = \sqrt{2p(2p - a)(2p - b)(2p - c)}.$$
Find the triangle of maximum area with perimeter $p = 1$.

33   Find the point on the plane $x + 2y - z = 10$ which is nearest to the origin.

34   Find the point on the plane $x + y + z = 0$ which is nearest to the point $(1, 2, 3)$.

35   Find the points on the surface $xyz = 1$ which are nearest to the origin.

36   Find the point on the surface $z = xy + 1$ which is nearest to the origin.

37   Show that the rectangular solid with volume one and minimum surface area is the unit cube.

38   Show that the rectangular solid with surface area six and maximum volume is the unit cube.

39   A rectangular box with volume V in.^{3} is to be built with the sides and bottom made of material costing one cent per square inch, and the top costing two cents per square inch. Find the shape with the minimum cost.

40   A firm can produce and sell x units of one commodity and y units of another commodity for a profit of
$$P(x, y) = 100x + 200y - 10xy - x^2 - 500.$$
Due to limitations on plant capacity, $x \le 10$ and $y \le 5$. Find the values of x and y where the profit is a maximum.

41 x units of commodity one and y units of commodity two can be produced and sold at a profit of
$$P(x, y) = 400x + 500y - x^2 - y^2 - xy - 20000.$$
Find the values of x and y where the profit is a maximum.

42 x units of commodity one can be produced at a cost of
$$C_1(x) = 1000 + 5x,$$
and y units of commodity two can be produced at a cost of
$$C_2(y) = 2000 + 8y.$$
Moreover, x units of one and y units of two can be sold for a total revenue of
$$R(x, y) = 100\sqrt{x} + 200\sqrt{y} + 10\sqrt{xy}.$$
Find the values of x and y where the profit is a maximum.

43   Suppose that with x man hours of labor and y units of capital, $z = f(x, y)$ units of a commodity can be produced. The ratio z/x is called the average production per man hour. Show that $\partial z/\partial x = z/x$ when the average production per man hour is a maximum.

☐ **44**   (*Method of Least Squares*) A straight line is to be fit as closely as possible to the set of three experimentally observed points $(1, 6)$, $(2, 9)$, and $(3, 10)$. The line which *best fits* these points is the line $y = mx + b$ for which the sum of the squares of the errors,
$$E = [(m \cdot 1 + b) - 6]^2 + [(m \cdot 2 + b) - 9]^2 + [(m \cdot 3 + b) - 10]^2,$$
is a minimum. Find m and b such that E is a minimum.

11.8 HIGHER PARTIAL DERIVATIVES

Given a function $z = f(x, y)$ of two variables, the partial derivatives $f_x(x, y)$ and $f_y(x, y)$ may themselves be differentiated with respect to either x or y. Thus there are four possible second partial derivatives. Here they are.

Twice with respect to x: f_{xx},   *or* $\dfrac{\partial^2 z}{\partial x^2}$.

Twice with respect to y: f_{yy},   *or* $\dfrac{\partial^2 z}{\partial y^2}$.

First with respect to x and then with respect to y:

$$(f_x)_y = f_{xy}, \quad or \quad \frac{\partial}{\partial y}\!\left(\frac{\partial z}{\partial x}\right) = \frac{\partial^2 z}{\partial y\, \partial x}.$$

First with respect to y and then with respect to x:

$$(f_y)_x = f_{yx}, \quad or \quad \frac{\partial}{\partial x}\!\left(\frac{\partial z}{\partial y}\right) = \frac{\partial^2 z}{\partial x\, \partial y}.$$

Similar notation is used for three or more variables and for higher partial derivatives.

EXAMPLE 1   Find the four second partial derivatives of

$$z = e^x \sin y + xy^2.$$

$$\frac{\partial z}{\partial x} = e^x \sin y + y^2, \qquad \frac{\partial z}{\partial y} = e^x \cos y + 2xy.$$

$$\frac{\partial^2 z}{\partial x^2} = \frac{\partial}{\partial x}(e^x \sin y + y^2) = e^x \sin y,$$

$$\frac{\partial^2 z}{\partial y^2} = \frac{\partial}{\partial y}(e^x \cos y + 2xy) = -e^x \sin y + 2x.$$

$$\frac{\partial^2 z}{\partial y\, \partial x} = \frac{\partial}{\partial y}(e^x \sin y + y^2) = e^x \cos y + 2y,$$

$$\frac{\partial^2 z}{\partial x\, \partial y} = \frac{\partial}{\partial x}(e^x \cos y + 2xy) = e^x \cos y + 2y.$$

Notice that in this example the two mixed second partials $\partial^2 z/\partial y\, \partial x$ and $\partial^2 z/\partial x\, \partial y$ are equal. The following theorem shows that it is not just a coincidence.

THEOREM 1 (Equality of Mixed Partials)

Suppose that the first and second partial derivatives of $z = f(x, y)$ are continuous at (a, b). Then at (a, b),

$$\frac{\partial^2 z}{\partial y\, \partial x} = \frac{\partial^2 z}{\partial x\, \partial y}.$$

Discussion   This is a surprising theorem. $\partial^2 z/\partial y\, \partial x$ is the rate of change with respect to y of the slope $\partial z/\partial x$, while $\partial^2 z/\partial x\, \partial y$ is the rate of change with respect to x of the slope $\partial z/\partial y$. There is no simple intuitive way to see that these should be equal.

As a matter of fact, there *are* functions $f(x, y)$ whose mixed second partial derivatives exist but are not equal. One such example is the function

$$f(x, y) = \begin{cases} 0 & \text{if } (x, y) = (0, 0), \\ xy\dfrac{x^2 - y^2}{x^2 + y^2} & \text{if } (x, y) \neq (0, 0). \end{cases}$$

We have left the computation of the second partials of $f(x, y)$ as a problem. It turns out that at $(0, 0)$,

$$\partial^2 f / \partial x\, \partial y = 1, \qquad \partial^2 f / \partial y\, \partial x = -1.$$

How can this be in view of Theorem 1? The answer is that in this example the second partial derivatives exist but are not continuous at $(0, 0)$, so the theorem does not apply. We shall only rarely encounter functions whose second partial derivatives are not continuous, so in all ordinary problems it is true that the mixed partials are equal. We shall prove the theorem later. We now turn to some applications. Our first application concerns mixed third partial derivatives.

If the third partial derivatives of $z = f(x, y)$ are continuous, then

$$\frac{\partial^3 z}{\partial x\, \partial x\, \partial y} = \frac{\partial^3 z}{\partial x\, \partial y\, \partial x} = \frac{\partial^3 z}{\partial y\, \partial x\, \partial x},$$

so we write $\dfrac{\partial^3 z}{\partial^2 x\, \partial y}$ for each of them. Similarly,

$$\frac{\partial^3 z}{\partial x\, \partial y\, \partial y} = \frac{\partial^3 z}{\partial y\, \partial x\, \partial y} = \frac{\partial^3 z}{\partial y\, \partial y\, \partial x}$$

and we write $\dfrac{\partial^3 z}{\partial x\, \partial y^2}$ for each of them.

We prove the first equation as an illustration.

$$\frac{\partial^3 z}{\partial x\, \partial x\, \partial y} = \frac{\partial}{\partial x}\left(\frac{\partial^2 z}{\partial x\, \partial y}\right) = \frac{\partial}{\partial x}\left(\frac{\partial^2 z}{\partial y\, \partial x}\right) = \frac{\partial^3 z}{\partial x\, \partial y\, \partial x}.$$

EXAMPLE 2   Find the third partial derivatives of $z = e^{2x} \sin y$.

$$\frac{\partial z}{\partial x} = 2e^{2x} \sin y, \qquad \frac{\partial z}{\partial y} = e^{2x} \cos y,$$

$$\frac{\partial^2 z}{\partial x^2} = 4e^{2x} \sin y, \qquad \frac{\partial^2 z}{\partial y^2} = -e^{2x} \sin y,$$

$$\frac{\partial^2 z}{\partial y\, \partial x} = \frac{\partial^2 z}{\partial x\, \partial y} = 2e^{2x} \cos y,$$

$$\frac{\partial^3 z}{\partial x^3} = 8e^{2x} \sin y, \qquad \frac{\partial^3 z}{\partial y^3} = -e^{2x} \cos y,$$

$$\frac{\partial^3 z}{\partial x^2\, \partial y} = 4e^{2x} \cos y, \qquad \frac{\partial^3 z}{\partial x\, \partial y^2} = -2e^{2x} \sin y.$$

If a function has continuous second partial derivatives we may apply the Chain Rule to the first partial derivatives. For one independent variable,

$$\frac{d}{dt}\left(\frac{\partial z}{\partial x}\right) = \frac{\partial^2 z}{\partial x^2}\frac{dx}{dt} + \frac{\partial^2 z}{\partial y\,\partial x}\frac{dy}{dt},$$

$$\frac{d}{dt}\left(\frac{\partial z}{\partial y}\right) = \frac{\partial^2 z}{\partial x\,\partial y}\frac{dx}{dt} + \frac{\partial^2 z}{\partial y^2}\frac{dy}{dt}.$$

EXAMPLE 3 If $z = f(x, y)$ has continuous second partials, $x = r\cos\theta$, and $y = r\sin\theta$, find $\partial^2 z/\partial r^2$.

We use the Chain Rule three times.

$$\frac{\partial z}{\partial r} = \frac{\partial z}{\partial x}\frac{\partial x}{\partial r} + \frac{\partial z}{\partial y}\frac{\partial y}{\partial r} = \frac{\partial z}{\partial x}\cos\theta + \frac{\partial z}{\partial y}\sin\theta.$$

$$\frac{\partial^2 z}{\partial r^2} = \frac{\partial}{\partial r}\left(\cos\theta\frac{\partial z}{\partial x}\right) + \frac{\partial}{\partial r}\left(\sin\theta\frac{\partial z}{\partial y}\right)$$

$$= \cos\theta\frac{\partial}{\partial r}\left(\frac{\partial z}{\partial x}\right) + \sin\theta\frac{\partial}{\partial r}\left(\frac{\partial z}{\partial y}\right)$$

$$= \left(\frac{\partial^2 z}{\partial x^2}\frac{\partial x}{\partial r} + \frac{\partial^2 z}{\partial y\,\partial x}\frac{\partial y}{\partial r}\right)\cos\theta + \left(\frac{\partial^2 z}{\partial x\,\partial y}\frac{\partial x}{\partial r} + \frac{\partial^2 z}{\partial y^2}\frac{\partial y}{\partial r}\right)\sin\theta$$

$$= \frac{\partial^2 z}{\partial x^2}\cos^2\theta + \frac{\partial^2 z}{\partial y\,\partial x}\sin\theta\cos\theta + \frac{\partial^2 z}{\partial x\,\partial y}\cos\theta\sin\theta + \frac{\partial^2 z}{\partial y^2}\sin^2\theta$$

$$= \frac{\partial^2 z}{\partial x^2}\cos^2\theta + 2\frac{\partial^2 z}{\partial y\,\partial x}\sin\theta\cos\theta + \frac{\partial^2 z}{\partial y^2}\sin^2\theta.$$

By holding one variable fixed in Theorem 1, we get equalities of mixed partials for functions of three or more variables.

COROLLARY (Equality of Mixed Partials, Three Variables)

Suppose that the first and second partial derivatives of $w = f(x, y, z)$ are continuous at (a, b, c). Then at (a, b, c),

$$\frac{\partial^2 w}{\partial y\,\partial x} = \frac{\partial^2 w}{\partial x\,\partial y}, \qquad \frac{\partial^2 w}{\partial z\,\partial x} = \frac{\partial^2 w}{\partial x\,\partial z}, \qquad \frac{\partial^2 w}{\partial z\,\partial y} = \frac{\partial^2 w}{\partial y\,\partial z}.$$

PROOF OF THEOREM 1 The plan is to prove a corresponding result for average slopes and then use the Mean Value Theorem, which states that the average slope of a function on an interval is equal to the slope at some point in the interval.

Let Δx and Δy be positive infinitesimals. We hold Δx and Δy fixed. The first and second partial derivatives of $f(x, y)$ exist for (x, y) in the rectangle

$$a \le x \le a + \Delta x, \qquad b \le y \le b + \Delta y.$$

We shall use the following notation for average slopes in the x and y directions:

$$g(y) = \frac{f(a + \Delta x, y) - f(a, y)}{\Delta x}, \qquad h(x) = \frac{f(x, b + \Delta y) - f(x, b)}{\Delta y}.$$

Label the corners of the rectangle A, B, C, and D as in Figure 11.8.1.

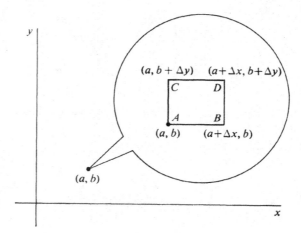

Figure 11.8.1

We first show that the following two quantities are equal:

$$\frac{\Delta^2 f}{\Delta y\, \Delta x} = \frac{g(b + \Delta y) - g(b)}{\Delta y}, \qquad \frac{\Delta^2 f}{\Delta x\, \Delta y} = \frac{h(a + \Delta x) - h(a)}{\Delta x}.$$

$\Delta^2 f/\Delta y\, \Delta x$ is the average slope in the y direction of the average slope in the x direction of f.

$$\frac{\Delta^2 f}{\Delta y\, \Delta x} = \frac{g(b + \Delta y) - g(b)}{\Delta y} = \frac{\dfrac{f(D) - f(C)}{\Delta x} - \dfrac{f(B) - f(A)}{\Delta x}}{\Delta y}$$

$$= \frac{f(D) - f(C) - f(B) + f(A)}{\Delta x\, \Delta y} = \frac{\dfrac{f(D) - f(B)}{\Delta y} - \dfrac{f(C) - f(A)}{\Delta y}}{\Delta x}$$

$$= \frac{h(a + \Delta x) - h(a)}{\Delta x} = \frac{\Delta^2 f}{\Delta x\, \Delta y}.$$

By the Mean Value Theorem,

$$\frac{\Delta^2 f}{\Delta y\, \Delta x} = \frac{g(b + \Delta y) - g(b)}{\Delta y} = g'(y_1),$$

where $b < y_1 < b + \Delta y$. Using the Mean Value Theorem again,

$$g'(y_1) = \frac{\partial \left[\dfrac{f(a + \Delta x, y) - f(a, y)}{\Delta x} \right]}{\partial y}(a, y_1)$$

$$= \frac{\dfrac{\partial f}{\partial y}(a + \Delta x, y_1) - \dfrac{\partial f}{\partial y}(a, y_1)}{\Delta x}$$

$$= \frac{\partial^2 f}{\partial x\, \partial y}(x_1, y_1),$$

where $a < x_1 < a + \Delta x$. Since $\partial^2 f/\partial x\, \partial y$ is continuous at (a, b),

$$st\left(\frac{\Delta^2 f}{\Delta y\, \Delta x}\right) = \frac{\partial^2 f}{\partial x\, \partial y}(a, b).$$

A similar computation gives

$$st\left(\frac{\Delta^2 f}{\Delta x\, \Delta y}\right) = \frac{\partial^2 f}{\partial y\, \partial x}(a, b).$$

Therefore
$$\frac{\partial^2 f}{\partial x\, \partial y}(a, b) = \frac{\partial^2 f}{\partial y\, \partial x}(a, b).$$

We conclude this section by stating a Second Derivative Test for maxima and minima of functions of two variables. In practice the test often fails except on small regions D. We therefore have emphasized the tests in the preceding section rather than the Second Derivative Test.

SECOND DERIVATIVE TEST

Suppose $z = f(x, y)$ has continuous first and second partial derivatives on a rectangle D, and (a, b) is a critical point of f in D.

(i)   *f has a minimum at (a, b) if*

$$\frac{\partial^2 z}{\partial x^2}\frac{\partial^2 z}{\partial y^2} - \left(\frac{\partial^2 z}{\partial x\, \partial y}\right)^2 > 0, \qquad \frac{\partial^2 z}{\partial x^2} > 0$$

at every point of D.

(ii) *f has a maximum at (a, b) if*

$$\frac{\partial^2 z}{\partial x^2}\frac{\partial^2 z}{\partial y^2} - \left(\frac{\partial^2 z}{\partial x\, \partial y}\right)^2 > 0, \qquad \frac{\partial^2 z}{\partial x^2} < 0$$

at every point of D.

(iii) *f has a saddle point at (a, b) if*

$$\frac{\partial^2 z}{\partial x^2}\frac{\partial^2 z}{\partial y^2} - \left(\frac{\partial^2 z}{\partial x\, \partial y}\right)^2 < 0 \qquad at\ (a, b).$$

For an indication of the proof of the Second Derivative Test, see Extra Problem 55 at the end of this chapter.

PROBLEMS FOR SECTION 11.8

In Problems 1–12, find all the second partial derivatives.

1	$z = x^2 + 2y^2$	2	$z = -3xy$
3	$z = ax^2 + bxy + cy^2$	4	$z = (ax + by + c)^n$
5	$z = xe^{x+y}$	6	$z = \cos(x + y) + \sin(x - y)$
7	$z = \ln(ax + by)$	8	$z = \sqrt{x^2 + y^2}$
9	$z = x^a y^b$	10	$w = xyz$

11 $w = \sqrt{x + y + z}$ **12** $w = z \cos x + z \sin y$

In Problems 13–16, find all the third partials.

13 $z = 2x^3 y^2 - 6x^2 y^3$ **14** $z = \sqrt{xy}$

15 $z = e^{ax + by}$ **16** $z = \cos x \sin y$

17   If $z = f(x, y)$, $x = r \cos \theta$, $y = r \sin \theta$, find $\partial^2 z / \partial \theta^2$.

18   If $z = f(x, y)$, $x = r \cos \theta$, $y = r \sin \theta$, find $\partial^2 z / \partial \theta \, \partial r$.

In Problems 19–24, find $\partial^2 z / \partial x^2$, $\partial^2 z / \partial y^2$, and $\partial^2 z / \partial x \, \partial y$.

19 $z = f(u), \quad u = ax + by$ **20** $z = f(u), \quad u = xy$

21 $z = g(x) + h(y)$ **22** $z = g(x)h(y)$

23 $z = u^n, \quad u = f(x, y)$ **24** $z = e^u, \quad u = f(x, y)$

In Problems 25–28 find $\partial^2 z / \partial s^2$, $\partial^2 z / \partial t^2$, and $\partial^2 z / \partial s \, \partial t$.

25 $z = ax + by, \quad x = f(s, t), \quad y = g(s, t)$

26 $z = xy, \quad x = f(s, t), \quad y = g(s, t)$

27 $z = f(x), \quad x = g(s) + h(t)$

28 $z = f(x, y), \quad x = g(s), \quad y = h(t)$

29   Suppose $z = f(x + at) + g(x - at)$ where f and g have continuous second derivatives. Show that z satisfies the *wave equation*

$$a^2 \frac{\partial^2 z}{\partial x^2} = \frac{\partial^2 z}{\partial t^2}.$$

30   Show that if

$$z = Ax^2 + Bxy + Cy^2 + Dx + Ey + F$$

then all the second partial derivatives of z are constant.

☐ **31**   Let $f(x, y)$ be the function

$$f(x, y) = \begin{cases} 0 & \text{if } (x, y) = (0, 0), \\ xy \dfrac{x^2 - y^2}{x^2 + y^2}, & \text{if } (x, y) \neq (0, 0), \end{cases}$$

Find the first and second partial derivatives of f. Show that
(a) $\partial^2 f / \partial x \, \partial y \neq \partial^2 f / \partial y \, \partial x$ at $(0, 0)$,
(b) $\partial^2 f / \partial x \, \partial y$ is not continuous at $(0, 0)$.
Hint: All the derivatives must be computed separately for the cases $(x, y) = (0, 0)$ and $(x, y) \neq (0, 0)$.

EXTRA PROBLEMS FOR CHAPTER 11

In Problems 1–4, make a contour map and sketch the surface.

1 $\frac{1}{9} x^2 + y^2 = z^2, -4 \leq z \leq 4$

2 $z = x^2 + \frac{1}{4} y^2, -4 \leq z \leq 4$

3 $z = x - y^2, -2 \leq x \leq 2, -2 \leq y \leq 2$

4 $z = \sqrt{xy}, -4 \leq x \leq 4, -4 \leq y \leq 4$

5   Find the largest set you can on which $f(x, y) = y + 1/x^2$ is continuous.

6   Find the largest set you can on which $f(x, y) = \sqrt{x^2 - y}/x$ is continuous.

7   Find the largest set you can on which $f(x, y) = \ln(1/x + 1/y)$ is continuous.

8   Find the largest set you can on which $f(x, y, z) = (\ln(x + y))/z$ is continuous.

9   Find the partial derivatives of $f(x, y) = ax - by$.

10 Find the partial derivatives of $f(x, y) = a_1 \sin(b_1 x) + a_2 \cos(b_2 y)$.

11 Find the partial derivatives of $z = \ln x/\ln y$.

12 Find the partial derivatives of $w = (x - y)e^z$.

13 Find the increment and total differential of $z = 1/x + 2/y$.

14 Find the increment and total differential of $z = \sqrt{x + y}$.

15 Find the tangent plane of $z = x^3 y + 4$ at $(2, 0)$.

16 Find the tangent plane of $z = \arcsin(xy)$ at $(3, \frac{1}{4})$.

17 Find dz/dt by the Chain Rule where $z = \log_{(2t + 1)}(3t + 2)$.

18 Find $\partial z/\partial s$ and $\partial z/\partial t$ where $z = x/y, x = e^{s+t}, y = as + bt$.

19 A particle moves in space so that $dx/dt = z \cos x$, $dy/dt = z \sin y$, $dz/dt = 1$. Find the rate of change of the distance from the origin when $x = 0, y = 0, z = 1$.

20 A company finds that it can produce x units of item 1 at a total cost of $x + 100\sqrt{x}$ dollars, and y units of item 2 at a total cost of $20y - \sqrt{y}$ dollars. Moreover, x units of item 1 and y units of item 2 can be sold for a total revenue of $10x + 30y - xy/100$ dollars. If z is the total profit (revenue minus cost), find $\partial z/\partial x$ and $\partial z/\partial y$, the partial marginal profit with respect to items 1 and 2.

21 Find the tangent line and slope of $x^4 + y^4 = 17$ at $(2, 1)$.

22 Find the tangent plane to the surface $x^4 + y^4 + z^2 = 18$ at $(1, 2, 1)$.

23 Find the maxima and minima of
$$z = x^2 + y^2 - 2x - 4y + 4, \qquad 0 \le x \le 3, \qquad x \le y \le 3.$$

24 Find the maxima and minima of
$$z = x + 4y + \frac{1}{x} + \frac{1}{y}, \qquad \frac{1}{4} \le x \le 4, \qquad \frac{1}{4} \le y \le 4.$$

25 Determine whether the surface $z = \log_x y, x > 1, y > 0$ has any maxima or minima.

26 Find the dimensions of the rectangular box of maximum volume such that the sum of the areas of the bottom and sides is one.

27 Find all second partial derivatives of $z = \arctan(xy)$.

28 Find all second partial derivatives of $w = (x^2 - y^2)z$.

29 Find $\partial^2 z/\partial r^2$ if $z = f(x, y), x = r \cosh \theta, y = r \sinh \theta$.

☐ **30** Let $f(x)$ be continuous for $a < x < b$. Prove that the function $F(u, v) = \int_u^v f(x) \, dx$ is continuous whenever u and v are in (a, b).

☐ **31** Prove that $f(x, y)$ is continuous at (a, b) if and only if the following ε, δ condition holds. For every real $\varepsilon > 0$ there is a real $\delta > 0$ such that whenever (x, y) is within δ of (a, b), $f(x, y)$ is within ε of $f(a, b)$.

☐ **32** Let
$$f(x, y) = \begin{cases} 1 & \text{if both } x \text{ and } y \text{ are rational,} \\ 0 & \text{otherwise.} \end{cases}$$
Prove that f is discontinuous at every point.

☐ **33** Prove that
$$\lim_{(x,y) \to (a,b)} f(x, y) = L$$
if and only if for every real $\varepsilon > 0$ there is a real $\delta > 0$ such that whenever (x, y) is different from but within δ of (a, b), $f(x, y)$ is within ε of L. (See Problems for Section 11.2.)

☐ **34** Prove that the following are equivalent.
(a) $f_x(x, y) = 0$ for all (x, y).
(b) The value of $f(x, y)$ depends only on y.

☐ **35**   Prove that the following are equivalent.
(a) $f_x(x, y) = 0$ and $f_y(x, y) = 0$ for all (x, y).
(b) f is a constant function.

☐ **36**   A function $z = f(x, y)$ is said to be *differentiable* at (x, y) if it satisfies the conclusion of the Increment Theorem. That is, whenever Δx and Δy are infinitesimal,

$$\Delta z = dz + \varepsilon_1 \, \Delta x + \varepsilon_2 \, \Delta y$$

for some infinitesimals ε_1 and ε_2 which depend on Δx and Δy. Prove the Chain Rule

$$\frac{dz}{dt} = \frac{\partial z}{\partial x} \frac{dx}{dt} + \frac{\partial z}{\partial y} \frac{dy}{dt}$$

assuming only that the functions $z = f(x, y)$, $x = g(t)$, and $y = h(t)$ are differentiable.

☐ **37**   Prove that the function $f(x, y) = |xy|$ is differentiable but not smooth at $(0, 0)$.

☐ **38**   A smooth function $z = f(x, y)$ is said to be *homogeneous of degree n* if

(1) $$f(tx, ty) = t^n f(x, y)$$

for all x, y, and t. Prove that if $z = f(x, y)$ is homogeneous of degree n then

$$x \frac{\partial z}{\partial x} + y \frac{\partial z}{\partial y} = nz.$$

Hint: Differentiate Equation 1 with respect to t and set $t = 1$.

☐ **39**   Suppose $f(x, y)$ has continuous second partial derivatives and that $\partial^2 f / \partial x \, \partial y$ is identically zero (i.e., zero at every point (x, y)). Prove that $f(x, y) = g(x) + h(y)$ for some functions g and h.

☐ **40**   Find all functions $f(x, y)$ all of whose second partial derivatives are identically zero.

MULTIPLE INTEGRALS

The first seven sections of this chapter develop the double and triple integral. They depend on Sections 11.1 and 11.2 on surfaces and continuous functions, but are independent of Chapter 10 on vectors.

Sections 8 through 10 of this chapter discuss the relationship between multiple integrals, line integrals, and surface integrals. Chapters 10 on vectors and 11 on partial derivatives are prerequisites.

12.1 DOUBLE INTEGRALS

The *double integral* is the analogue of the *single integral* (definite integral) suggested by Figure 12.1.1. Figure 12.1.1(a) shows the area A bounded by the interval $[a, b]$ and the curve $y = f(x)$, and corresponds to the single integral

$$A = \int_a^b f(x)\, dx.$$

Figure 12.1.1(b) shows the volume V bounded by the plane region D and the surface $z = f(x, y)$, and corresponds to the double integral

$$V = \iint_D f(x, y)\, dx\, dy.$$

Our development of double integrals will be similar to our development of single integrals in Chapter 4. Before going into detail, we give a brief intuitive preview.

Instead of closed intervals $[u, v]$ in the line, we deal with closed regions D in the plane. A *volume function* for $f(x, y)$ is a function B, which assigns a real number $B(D)$ to each closed region D, and has the following two properties: Addition Property and Cylinder Property.

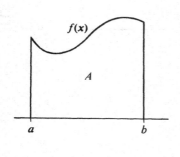

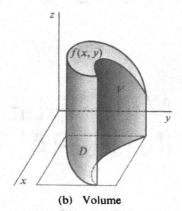

(a) Area   (b) Volume

Figure 12.1.1

ADDITION PROPERTY

If D is divided into two regions D_1 and D_2 which meet only on a common boundary curve, then

$$B(D) = B(D_1) + B(D_2).$$

(Intuitively, the volume over D is the sum of the volumes over D_1 and D_2.)
This property is illustrated in Figure 12.1.2(a).

CYLINDER PROPERTY

Let m and M be the minimum and maximum values of $f(x, y)$ on D and let A be the area of D. Then

$$mA \leq B(D) \leq MA.$$

(Intuitively, the volume over D is between the volumes of the cylinders over D of height m and M. This corresponds to the Rectangle Property for single integrals.)
This property is illustrated in Figure 12.1.2(b).

We shall see at the end of this section that the double integral

$$\iint_D f(x, y)\, dx\, dy$$

is the unique volume function for a continuous function $f(x, y)$. The double integral will be constructed using double Riemann sums, just as the single integral was constructed from single Riemann sums.

We now begin the construction of the double integral, starting with a careful discussion of closed regions in the plane.

A *closed region* in the (x, y) plane is a set D of real points (x, y) given by inequalities

$$a_1 \leq x \leq a_2, \qquad b_1(x) \leq y \leq b_2(x),$$

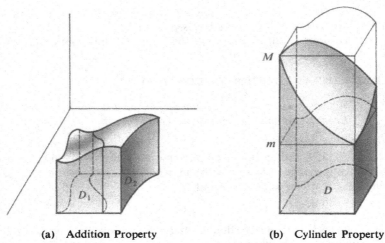

(a) Addition Property (b) Cylinder Property

Figure 12.1.2

where $b_1(x)$ and $b_2(x)$ are continuous and $b_1(x) \leq b_2(x)$ for x in $[a_1, a_2]$ (Figure 12.1.3(a)). The *boundary* of D is the set of points in D which are on the curves

$$x = a_1, \qquad x = a_2, \qquad y = b_1(x), \qquad y = b_2(x).$$

The simplest type of closed region is a *closed rectangle*

$$a_1 \leq x \leq a_2, \qquad b_1 \leq y \leq b_2,$$

shown in Figure 12.1.3(b).

Remark In this course we are restricting our attention to a very simple type of closed region, sometimes called a basic closed region. In advanced calculus and beyond, a much wider class of closed regions is studied.

An *open region* is a set of real points defined by strict inequalities of the form

$$c_1 < x < c_2, \qquad d_1(x) < y < d_2(x).$$

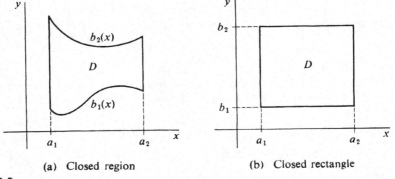

(a) Closed region (b) Closed rectangle

Figure 12.1.3

We shall usually be working with closed regions. So from now on when we use the word *region* alone we mean *closed region*.

To simplify our treatment we shall consider only continuous functions.

PERMANENT ASSUMPTION FOR CHAPTER 12

Whenever we refer to a function $f(x, y)$ and a region D, we assume that $f(x, y)$ is continuous on some open region containing D.

If $f(x, y) \geq 0$ on D, the double integral is intuitively the volume of the solid over D between the surfaces $z = 0$ and $z = f(x, y)$; i.e., the solid consisting of all points (x, y, z) where (x, y) is in D and

$$0 \leq z \leq f(x, y).$$

If $f(x, y) \leq 0$ on D the double integral is intuitively the negative of the volume of the solid under D between the surfaces $z = f(x, y)$ and $z = 0$. Thus volumes above the plane $z = 0$ are counted positively and volumes below $z = 0$ are counted negatively (Figure 12.1.4).

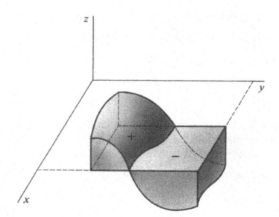

Figure 12.1.4

We now define the double Riemann sum and use it to give a precise definition of the double integral. We first consider the case where D is a rectangle

$$a_1 \leq x \leq a_2, \qquad b_1 \leq y \leq b_2,$$

shown in Figure 12.1.5.

Let Δx and Δy be positive real numbers. We partition the interval $[a_1, a_2]$ into subintervals of length Δx and $[b_1, b_2]$ into subintervals of length Δy. The partition points are

$$x_0 = a_1, \quad x_1 = a_1 + \Delta x, \quad x_2 = a_1 + 2\Delta x, \dots, x_n = a_1 + n\Delta x,$$
$$y_0 = b_1, \quad y_1 = b_1 + \Delta y, \quad y_2 = b_1 + 2\Delta y, \dots, y_p = b_1 + p\Delta y$$

where $\qquad x_n < a_2 \leq x_n + \Delta x, \qquad y_p < b_2 \leq y_p + \Delta y.$

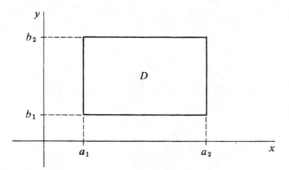

Figure 12.1.5

If Δx and Δy do not evenly divide $a_2 - a_1$ and $b_2 - b_1$, there will be little pieces left over at the end. We have partitioned the rectangle D into Δx by Δy subrectangles with partition points

$$(x_k, y_l), \qquad 0 \le k \le n, \qquad 0 \le l \le p,$$

as in Figure 12.1.6.

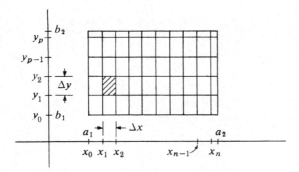

Figure 12.1.6

The *double Riemann sum* for a rectangle D is the sum

$$\sum\sum_D f(x, y)\, \Delta x\, \Delta y = \sum_{k=0}^{n} \sum_{l=0}^{p} f(x_k, y_l)\, \Delta x\, \Delta y.$$

This is the sum of the volume of the rectangular solids with base $\Delta x\, \Delta y$ and height $f(x_k, y_l)$.

As we can see from Figure 12.1.7,

$$\sum\sum_D f(x, y)\, \Delta x\, \Delta y$$

approximates the volume of the solid over D between $z = 0$ and $z = f(x, y)$.

Now let D be a general region

$$a_1 \le x \le a_2, \qquad b_1(x) \le y \le b_2(x).$$

The *circumscribed rectangle* of D is the rectangle

$$a_1 \le x \le a_2, \qquad B_1 \le y \le B_2,$$

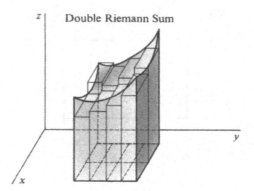

Figure 12.1.7

where B_1 = minimum value of $b_1(x)$,

B_2 = maximum value of $b_2(x)$.

It is shown in Figure 12.1.8.

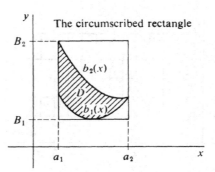

Figure 12.1.8

Given positive real numbers Δx and Δy, we partition the circumscribed rectangle of D into Δx by Δy subrectangles with partition points

$$(x_k, y_l), \qquad 0 \le k \le n, \qquad 0 \le l \le p.$$

DEFINITION

*The **double Riemann sum** over D is defined as the sum of the volumes of the rectangular solids with base $\Delta x \, \Delta y$ and height $f(x_k, y_l)$ corresponding to partition points (x_k, y_l) which belong to D. In symbols,*

$$\sum_D \sum f(x, y) \, \Delta x \, \Delta y = \sum_{(x_k, y_l) \text{ in } D} \sum f(x_k, y_l) \, \Delta x \, \Delta y.$$

Notice that in the double Riemann sum over D, we only use partition points (x_k, y_l) which belong to D (Figure 12.1.9).

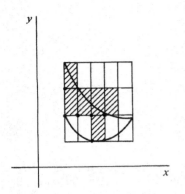

Figure 12.1.9   Double Riemann Sum

EXAMPLE 1   Find the double Riemann sum

$$\sum\sum_{D_1} x^2 y \, \Delta x \, \Delta y,$$

where D_1 is the square

$$0 \le x \le 1, \qquad 0 \le y \le 1,$$

and

$$\Delta x = \tfrac{1}{4}, \qquad \Delta y = \tfrac{1}{5}.$$

The partition of D_1 is shown in Figure 12.1.10 and the values of $x^2 y$ at the partition points are shown in the table.

$x^2 y$	$y_0 = 0$	$y_1 = \frac{1}{5}$	$y_2 = \frac{2}{5}$	$y_3 = \frac{3}{5}$	$y_4 = \frac{4}{5}$
$x_0 = 0$	0	0	0	0	0
$x_1 = \frac{1}{4}$	0	$\frac{1}{80}$	$\frac{2}{80}$	$\frac{3}{80}$	$\frac{4}{80}$
$x_2 = \frac{1}{2}$	0	$\frac{4}{80}$	$\frac{8}{80}$	$\frac{12}{80}$	$\frac{16}{80}$
$x_3 = \frac{3}{4}$	0	$\frac{9}{80}$	$\frac{18}{80}$	$\frac{27}{80}$	$\frac{36}{80}$

The double Riemann sum is

$$\sum\sum_{D_1} x^2 y \, \Delta x \, \Delta y$$
$$= (1 + 2 + 3 + 4 + 4 + 8 + 12 + 16 + 9 + 18 + 27 + 36)\tfrac{1}{80} \cdot \tfrac{1}{4} \cdot \tfrac{1}{5} = 0.0875.$$

A similar computation with $\Delta x = \tfrac{1}{10}, \Delta y = \tfrac{1}{10}$ gives

$$\sum\sum_{D_1} x^2 y \, \Delta x \, \Delta y = 0.12825.$$

EXAMPLE 2   Find the double Riemann sum

$$\sum\sum_{D_2} x^2 y \, \Delta x \, \Delta y,$$

where D_2 is the region

$$0 \le x \le 1, \qquad x^2 \le y \le \sqrt{x}$$

and

$$\Delta x = \tfrac{1}{4}, \qquad \Delta y = \tfrac{1}{5}.$$

The circumscribed rectangle of D_2 is the unit square. The partition and D_2 are shown in Figure 12.1.11 and the partition points which actually belong to D_2 are circled. The table shows the values of $x^2 y$ at the partition points which belong to D_2. It is a part of the table from Example 1.

$x^2 y$	$y_0 = 0$	$y_1 = \tfrac{1}{5}$	$y_2 = \tfrac{2}{5}$	$y_3 = \tfrac{3}{5}$	$y_4 = \tfrac{4}{5}$
$x_0 = 0$	0				
$x_1 = \tfrac{1}{4}$		$\tfrac{1}{80}$	$\tfrac{2}{80}$		
$x_2 = \tfrac{1}{2}$			$\tfrac{8}{80}$	$\tfrac{12}{80}$	
$x_3 = \tfrac{3}{4}$				$\tfrac{27}{80}$	$\tfrac{36}{80}$

The double Riemann sum is

$$\sum\sum_{D_2} x^2 y \, \Delta x \, \Delta y$$

$$= \left(\frac{1}{80} + \frac{2}{80} + \frac{8}{80} + \frac{12}{80} + \frac{27}{80} + \frac{36}{80}\right)\frac{1}{4}\cdot\frac{1}{5} = \frac{86}{80\cdot4\cdot5} = 0.05375.$$

A similar computation with $\Delta x = \tfrac{1}{10}, \Delta y = \tfrac{1}{10}$ gives

$$\sum\sum_{D_2} x^2 y \, \Delta x \, \Delta y = 0.04881.$$

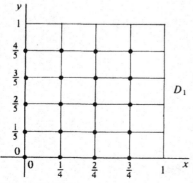

Figure 12.1.10

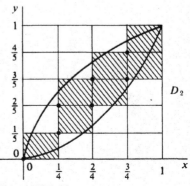

Figure 12.1.11

Given the function $f(x, y)$ and the region D, the double Riemann sum

$$\sum\sum_{D} f(x, y)\, \Delta x\, \Delta y$$

is a real function of Δx and Δy. When we replace Δx and Δy by positive infinitesimals dx and dy (Figure 12.1.12), we obtain (by the Function Axiom) the *infinite double Riemann sum*

$$\sum\sum_{D} f(x, y)\, dx\, dy.$$

The infinite double Riemann sum is in general a hyperreal number. Intuitively, it is equal to the sum of the volumes of infinitely many rectangular solids of infinitesimal base $dx\, dy$ and height $f(x_K, y_L)$. The double integral is defined as the standard part of the infinite double Riemann sum. The following lemma, based on our Permanent Assumption for Chapter 12, shows that this sum has a standard part.

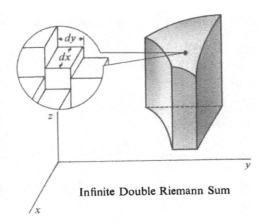

Infinite Double Riemann Sum

Figure 12.1.12

LEMMA

For any positive infinitesimals dx and dy, the double Riemann sum

$$\sum\sum_{D} f(x, y)\, dx\, dy$$

is a finite hyperreal number and thus has a standard part.

We omit the proof, which is similar to the proof that single Riemann sums are finite. We are now ready to define the double integral.

DEFINITION

*Given positive infinitesimals dx and dy, the **double integral** of a continuous function $f(x, y)$ over D is the standard part of the double Riemann sum·*

$$\iint_{D} f(x, y)\, dx\, dy = st\left(\sum\sum_{D} f(x, y)\, dx\, dy\right).$$

Here is a list of properties of the double integral. Each property is analogous to a property of the single integral given in Chapter Four and has a similar proof.

INDEPENDENCE OF *dx* AND *dy*

The value of the double integral $\iint_D f(x, y)\, dx\, dy$ does not depend on dx and dy. That is, if dx, $d_1 x$, dy, and $d_1 y$ are positive infinitesimals then

$$\iint_D f(x, y)\, dx\, dy = \iint_D f(x, y)\, d_1 x\, d_1 y.$$

This theorem shows that the value of the double integral depends only on the function f and the region D. From now on we shall usually use the simpler notation $dA = dx\, dy$ for the area of an infinitesimal dx by dy rectangle, and

$$\iint_D f(x, y)\, dA \qquad \text{for} \qquad \iint_D f(x, y)\, dx\, dy.$$

ADDITION PROPERTY

Let D be divided into two regions D_1 and D_2 which meet only on a common boundary as in Figure 12.1.13. Then

$$\iint_D f(x, y)\, dA = \iint_{D_1} f(x, y)\, dA + \iint_{D_2} f(x, y)\, dA.$$

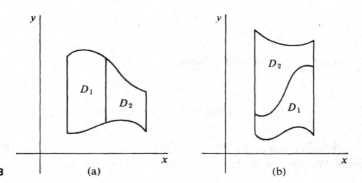

Figure 12.1.13   (a)   (b)

Interpreting the double integral as a volume, the Addition Property says that the volume of the solid over D is equal to the sum of the volume over D_1 and the volume over D_2, as shown in Figure 12.1.14.

A continuous function $z = f(x, y)$ always has a minimum and maximum value on a closed region D. The proof is similar to the one-variable case.

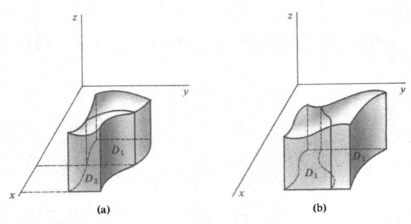

(a)   (b)

Figure 12.1.14 Addition Property

CYLINDER PROPERTY

Let m and M be the minimum and maximum values of $f(x, y)$ on D and let A be the area of D. Then

$$mA \leq \iint\limits_{D} f(x, y)\, dA \leq MA.$$

This corresponds to the Rectangle Property for single integrals. The solid with base D and constant height m is called the *inscribed cylinder*, and the solid with base D and height M is called the *circumscribed cylinder*. The inscribed cylinder and the circumscribed cylinder are shown in Figure 12.1.15. Intuitively, the volume of a cylinder is equal to the area of the base A times the height. Thus the Cylinder Property states that the volume of the solid is between the volumes of the inscribed and circumscribed cylinders.

Here are two consequences of the Cylinder Property.

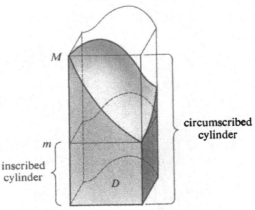

Figure 12.1.15 Cylinder Property

COROLLARY 1

The area of D is equal to the double integral of the constant function 1 over D, (Figure 12.1.16):

$$A = \iint_D dA.$$

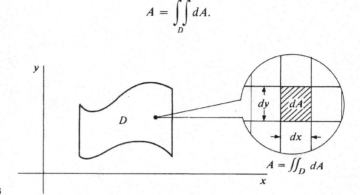

Figure 12.1.16

PROOF Both m and M are equal to 1, so $1 \cdot A \leq \iint_D dA \leq 1 \cdot A.$

COROLLARY 2

If $f(x, y) \geq 0$ on D then $\iint_D f(x, y) \, dA \geq 0$. If $f(x, y) \leq 0$ on D then $\iint_D f(x, y) \, dA \leq 0$.

To really be sure that the double integral corresponds to the volume, we need to know that it is the only operation that has the Addition and Cylinder Properties. To make this precise, we introduce the notion of a volume function.

We suppose $f(x, y)$ is continuous at every point of an open region D_0, and consider subregions D of D_0. A *volume function* for f is a function B which assigns a real number $B(D)$ to each subregion D of D_0 and has the Addition Property

$$B(D) = B(D_1) + B(D_2)$$

and the Cylinder Property

$$mA \leq B(D) \leq MA,$$

where m is the minimum and M the maximum value of f on D.

UNIQUENESS THEOREM

The double integral $\iint_D f(x, y) \, dA$ is the only volume function for f. That is, if B is a function which has the Addition and Cylinder Properties, then

$$B(D) = \iint_D f(x, y) \, dA \qquad \text{for every } D.$$

Given a continuous function f such that $f(x, y) \geq 0$ for all (x, y), the function

$$V(D) = \text{volume over } D$$

certainly has the Addition and Cylinder Properties. Thus we are justified in defining the volume as the double integral.

DEFINITION

Let $f(x, y) \geq 0$ for (x, y) in D. Then the volume over D between $z = 0$ and $z = f(x, y)$ is the double integral

$$V = \iint_D f(x, y)\, dA.$$

When $f(x, y)$ is the constant 1, we have

$$A = \iint_D dA = V.$$

That is, the area of D is equal to the volume of the cylinder with base D and height 1, as in Figure 12.1.17.

Given any unit of length (say meters), if the height is one meter then the area is in square meters and the volume has the same value but in cubic meters.

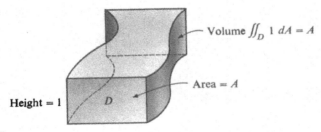

Height = 1

Volume $\iint_D 1\, dA = A$

Area = A

Figure 12.1.17

PROBLEMS FOR SECTION 12.1

Compute the following double Riemann sums.

1   $\displaystyle\sum_D\sum (3x + 4y)\,\Delta x\, \Delta y,\quad \Delta x = \tfrac{1}{4},\quad \Delta y = \tfrac{1}{4},\quad D: 0 \leq x \leq 1,\ 0 \leq y \leq 1$

2   $\displaystyle\sum_D\sum (4 + 2x - 5y)\,\Delta x\, \Delta y,\quad \Delta x = \tfrac{1}{2},\quad \Delta y = \tfrac{1}{5},\quad D: -2 \leq x \leq 2,\ -1 \leq y \leq 1$

3   $\displaystyle\sum_D\sum (x^2 + y^2)\,\Delta x\, \Delta y,\quad \Delta x = \tfrac{1}{2},\quad \Delta y = \tfrac{1}{2},\quad D: -2 \leq x \leq 2,\ -2 \leq y \leq 2$

4   $\displaystyle\sum_D\sum (1 + xy)\,\Delta x\, \Delta y,\quad \Delta x = \tfrac{1}{3},\quad \Delta y = \tfrac{1}{3},\quad D: 0 \leq x \leq 2,\ 0 \leq y \leq 1$

5   $\displaystyle\sum_D\sum \frac{x}{y}\,\Delta x\, \Delta y,\quad \Delta x = \tfrac{1}{4},\quad \Delta y = \tfrac{1}{5},\quad D: 1 \leq x \leq 2,\ 1 \leq y \leq 2$

6    $\sum\sum_{D} (\cos x + \sin y)\,\Delta x\,\Delta y, \quad \Delta x = \dfrac{\pi}{6}, \quad \Delta y = \dfrac{\pi}{6}, \quad D:-\dfrac{\pi}{2} \le x \le \dfrac{\pi}{2}, \quad 0 \le y \le \pi$

7    $\sum\sum_{D} (\cos x \sin y)\,\Delta x\,\Delta y, \quad \Delta x = \dfrac{\pi}{6}, \quad \Delta y = \dfrac{\pi}{6}, \quad D:-\dfrac{\pi}{2} \le x \le \dfrac{\pi}{2}, \quad 0 \le y \le \pi$

8    $\sum\sum_{D} xe^{y}\,\Delta x\,\Delta y, \quad \Delta x = \frac{1}{2}, \quad \Delta y = 1, \quad D:0 \le x \le 2, \quad -2 \le y \le 3$

9    $\sum\sum_{D} e^{2x-y}\,\Delta x\,\Delta y, \quad \Delta x = 1, \quad \Delta y = 1, \quad D:-2 \le x \le 2, \quad -2 \le y \le 2$

10   $\sum\sum_{D} (x + 2y)\,\Delta x\,\Delta y, \quad \Delta x = \frac{1}{4}, \quad \Delta y = \frac{1}{4}, \quad D:0 \le x \le 1, \quad 0 \le y \le x$

11   $\sum\sum_{D} (2 + x + 3y)\,\Delta x\,\Delta y, \quad \Delta x = \frac{1}{4}, \quad \Delta y = \frac{1}{4}, \quad D:0 \le x \le 1, \quad x \le y \le 1$

12   $\sum\sum_{D} (x^{2} + \sqrt{y})\,\Delta x\,\Delta y, \quad \Delta x = \frac{1}{5}, \quad \Delta y = \frac{1}{4}, \quad D:-1 \le x \le 1, \quad 0 \le y \le x^{2}$

13   $\sum\sum_{D} y \sin x\,\Delta x\,\Delta y, \quad \Delta x = \dfrac{\pi}{4}, \quad \Delta y = \dfrac{1}{4}, \quad D:0 \le x \le \pi, \quad \sin^{2}x \le y \le 2\sin x$

14   $\sum\sum_{D} (e^{x} + e^{y})\,\Delta x\,\Delta y, \quad \Delta x = 1, \quad \Delta y = 1, \quad D:-3 \le x \le 3, \quad -x \le y \le x$

15   $\sum\sum_{D} 4\,\Delta x\,\Delta y, \quad \Delta x = 1, \quad \Delta y = 1, \quad D:x^{2} + y^{2} \le 9$

16   $\sum\sum_{D} -10\,\Delta x\,\Delta y, \quad \Delta x = 1, \quad \Delta y = 1, \quad D:-3 \le x \le 3, \quad x^{2} \le y \le 18 - x^{2}$

☐ 17   Show that if D is a region with area A and c is constant, then $\iint_{D} c\,dA = cA$.

☐ 18   Prove the Constant Rule:

$$\sum\sum_{D} cf(x, y)\,\Delta x\,\Delta y = c\sum\sum_{D} f(x, y)\,\Delta x\,\Delta y,$$

$$\iint_{D} cf(x, y)\,dx\,dy = c\iint_{D} f(x, y)\,dx\,dy.$$

☐ 19   Prove the Sum Rule:

$$\sum\sum_{D} f(x, y) + g(x, y)\,\Delta x\,\Delta y = \sum\sum_{D} f(x, y)\,\Delta x\,\Delta y + \sum\sum_{D} g(x, y)\,\Delta x\,\Delta y,$$

$$\iint_{D} f(x, y) + g(x, y)\,dx\,dy = \iint_{D} f(x, y)\,dx\,dy + \iint_{D} g(x, y)\,dx\,dy.$$

12.2 ITERATED INTEGRALS

In this section we shall learn how to evaluate double integrals. A double integral can be evaluated by two single integrations. The Iterated Integral Theorem gives the key formula.

The *iterated integral*

$$\int_{a_1}^{a_2} \left[\int_{b_1(x)}^{b_2(x)} f(x, y)\,dy \right] dx$$

is an integral of an integral of $f(x, y)$. It is evaluated in two stages. First evaluate the inside integral

$$g(x) = \int_{b_1(x)}^{b_2(x)} f(x, y)\,dy$$

by ordinary definite integration, treating x as a constant. This gives us a function of x alone. Second, evaluate the outside integral

$$\int_{a_1}^{a_2} g(x)\,dx = \int_{a_1}^{a_2} \left[\int_{b_1(x)}^{b_2(x)} f(x, y)\,dy \right] dx$$

by a second definite integration.

We shall usually drop the brackets around the inside integral and write the iterated integral as

$$\int_{a_1}^{a_2} \int_{b_1(x)}^{b_2(x)} f(x, y)\,dy\,dx.$$

ITERATED INTEGRAL THEOREM

Let D be a region

$$a_1 \le x \le a_2, \qquad b_1(x) \le y \le b_2(x).$$

The double integral over D is equal to the iterated integral:

$$\iint_D f(x, y)\,dA = \int_{a_1}^{a_2} \int_{b_1(x)}^{b_2(x)} f(x, y)\,dy\,dx.$$

Discussion For a fixed x_0, $\int_{b_1(x_0)}^{b_2(x_0)} f(x_0, y)\,dy$ is the area of the cross section shown in Figure 12.2.1. The Iterated Integral Theorem states that the volume is equal to the integral of the areas of the cross sections.

The proof of the Iterated Integral Theorem is given at the end of this section. When using iterated integrals we must be sure that:

(1) $a_1 \le a_2$ and $b_1(x) \le b_2(x)$.
(2) The differentials dx and dy appear in the right order.
(3) The outer integral sign has constant limits.

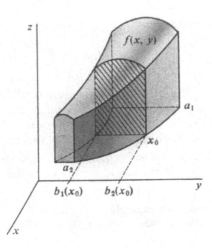

Figure 12.2.1

While the order of the differentials, $dx\,dy$ or $dy\,dx$, does not matter in a double integral, it is important in an iterated integral. The inside integral sign goes with the inside differential, and is performed first.

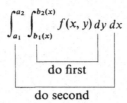

When the region D is a rectangle, there are two possible orders of integration, because all the boundaries are constant. Thus there are two different iterated integrals over a rectangle. Integrating first with respect to y we have

$$\int_{a_1}^{a_2}\int_{b_1}^{b_2} f(x, y)\,dy\,dx,$$

and integrating first with respect to x we have

$$\int_{b_1}^{b_2}\int_{a_1}^{a_2} f(x, y)\,dx\,dy.$$

Using the Iterated Integral Theorem twice, we see that both iterated integrals must equal the double integral.

$$\int_{a_1}^{a_2}\int_{b_1}^{b_2} f(x, y)\,dy\,dx = \iint_D f(x, y)\,dA,$$

$$\int_{b_1}^{b_2}\int_{a_1}^{a_2} f(x, y)\,dx\,dy = \iint_D f(x, y)\,dA.$$

Therefore the two iterated integrals are equal to each other. We have proved a corollary.

COROLLARY

The two iterated integrals over a rectangle are equal:

$$\int_{a_1}^{a_2}\int_{b_1}^{b_2} f(x, y)\,dy\,dx = \int_{b_1}^{b_2}\int_{a_1}^{a_2} f(x, y)\,dx\,dy.$$

Discussion   This corollary is the simplest form of a result known as Fubini's Theorem. Remember that by our Permanent Assumption, $f(x, y)$ is continuous on D. For an idea of the difficulties that arise when $f(x, y)$ is not assumed to be continuous, see Problem 49 at the end of this section.

There are also other regions besides rectangles over which we can integrate in either of two orders, such as Example 5 in this section.

In the following two examples we evaluate the double integrals which were approximated by double Riemann sums in the preceding section.

EXAMPLE 1 Evaluate

$$\iint_{D_1} x^2 y \, dA$$

where D_1 is the unit square

$$0 \le x \le 1, \qquad 0 \le y \le 1.$$

The limits of the outside integral are given by $0 \le x \le 1$, and those of the inside integral are given by $0 \le y \le 1$. The iterated integral is thus

$$\iint_{D_1} x^2 y \, dA = \int_0^1 \int_0^1 x^2 y \, dy \, dx.$$

The inside integral is

$$\int_0^1 x^2 y \, dy = \tfrac{1}{2} x^2 y^2 \Big]_{y=0}^{y=1} = \tfrac{1}{2} x^2.$$

Then

$$\iint_{D_1} x^2 y \, dA = \int_0^1 \tfrac{1}{2} x^2 \, dx = \tfrac{1}{6} x^3 \Big]_0^1 = \tfrac{1}{6} \sim 0.16667.$$

Since D_1 is a rectangle we may also integrate in the other order, and should get the same answer.

$$\iint_{D_1} x^2 y \, dA = \int_0^1 \int_0^1 x^2 y \, dx \, dy.$$

$$\int_0^1 x^2 y \, dx = \tfrac{1}{3} x^3 y \Big]_0^1 = \tfrac{1}{3} y.$$

$$\iint_{D_1} x^2 y \, dA = \int_0^1 \tfrac{1}{3} y \, dy = \tfrac{1}{6} y^2 \Big]_0^1 = \tfrac{1}{6} \sim 0.16667.$$

The Riemann sums in Section 12.1 were 0.0875, 0.12825.

EXAMPLE 2 Evaluate $\iint_{D_2} x^2 y \, dA$ where D_2 is the region in Figure 12.2.2:

$$0 \le x \le 1, \qquad x^2 \le y \le \sqrt{x}.$$

The limits on the outside integral are given by $0 \le x \le 1$, and those on the inside integral by $x^2 \le y \le \sqrt{x}$, so the iterated integral is

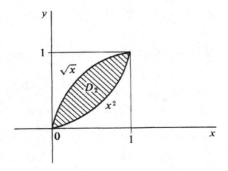

Figure 12.2.2

$$\iint_{D_2} x^2 y \, dA = \int_0^1 \int_{x^2}^{\sqrt{x}} x^2 y \, dy \, dx.$$

$$\int_{x^2}^{\sqrt{x}} x^2 y \, dy = \tfrac{1}{2} x^2 y^2 \Big]_{y=x^2}^{y=\sqrt{x}} = \tfrac{1}{2} x^3 - \tfrac{1}{2} x^6.$$

$$\iint_{D_2} x^2 y \, dA = \int_0^1 (\tfrac{1}{2} x^3 - \tfrac{1}{2} x^6) \, dx = \tfrac{1}{8} x^4 - \tfrac{1}{14} x^7 \Big]_0^1$$

$$= \tfrac{3}{56} \sim 0.05357.$$

The Riemann sums in Section 12.1 were 0.05375, 0.04881.

In many applications the region D is given verbally, and part of the problem is to find inequalities which describe D.

EXAMPLE 3   Let D be the region bounded by the curve $xy = 1$ and the line $y = \tfrac{5}{2} - x$. Find inequalities which describe D, and write down an iterated integral equal to $\iint_D f(x, y) \, dA$.

Step 1   Sketch the region D as in Figure 12.2.3.

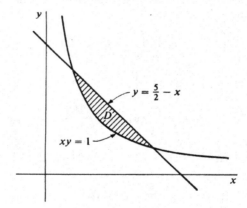

Figure 12.2.3

Step 2   The line and curve intersect where

$$x(\tfrac{5}{2} - x) = 1,$$
$$x^2 - \tfrac{5}{2} x + 1 = 0,$$
$$(x - \tfrac{1}{2})(x - 2) = 0,$$
$$x = \tfrac{1}{2}, \qquad x = 2.$$

For $1/2 \le x \le 2$, the curve $y = 1/x$ is below the line $y = 5/2 - x$. Therefore D is the region

$$\tfrac{1}{2} \le x \le 2, \qquad 1/x \le y \le \tfrac{5}{2} - x.$$

Step 3   The inequalities for x give the limits of the outside integral, and those for y give the limits of the inside integral. Thus

$$\iint_D f(x, y) \, dA = \int_{1/2}^2 \int_{1/x}^{(5/2)-x} f(x, y) \, dy \, dx.$$

EXAMPLE 4 Find the volume of the solid bounded by the surfaces $z = 0, z = y - x^2$, $y = 1$.

Step 1 Sketch the solid and the region D, as in Figure 12.2.4.

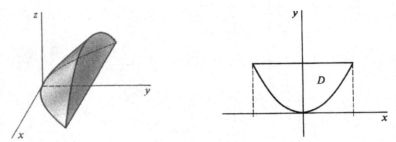

Figure 12.2.4

Step 2 Find the inequalities describing the region D.

This is the hardest step, and gives us the limits of integration. The surfaces $z = 0$ and $z = y - x^2$ intersect at the curve $y = x^2$. We see from the figure that D is the region between the curves $y = x^2$ and $y = 1$, so D is given by

$$-1 \le x \le 1, \qquad x^2 \le y \le 1.$$

Step 3 Set up the iterated integral and evaluate it.

$$V = \iint_D y - x^2 \, dA = \int_{-1}^{1} \int_{x^2}^{1} y - x^2 \, dy \, dx.$$

$$\int_{x^2}^{1} y - x^2 \, dy = \tfrac{1}{2}y^2 - x^2 y \Big]_{x^2}^{1}$$

$$= (\tfrac{1}{2} \cdot 1^2 - x^2 \cdot 1) - (\tfrac{1}{2}(x^2)^2 - x^2 \cdot x^2)$$

$$= \tfrac{1}{2} - x^2 + \tfrac{1}{2}x^4.$$

$$V = \int_{-1}^{1} \tfrac{1}{2} - x^2 + \tfrac{1}{2}x^4 \, dx = \tfrac{16}{30}.$$

Multiple integration problems can be solved by a three-step process as shown in Examples 3 and 4.

Step 1 Sketch the problem.

Step 2 Find the inequalities describing the region D.

Step 3 Set up the iterated integral and evaluate.

We can also integrate over a region in the (y, x) plane instead of the (x, y) plane. A region D in the (y, x) plane has the form

$$b_1 \le y \le b_2, \qquad a_1(y) \le x \le a_2(y),$$

as shown in Figure 12.2.5.

The double integral over D is equal to the iterated integral with dy on the outside and dx inside,

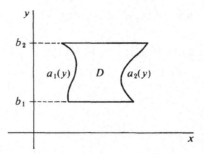

Figure 12.2.5

$$\iint_D f(x, y)\, dA = \int_{b_1}^{b_2} \int_{a_1(y)}^{a_2(y)} f(x, y)\, dx\, dy.$$

Some regions, such as rectangles and ellipses, may be regarded as regions in either the (x, y) plane or the (y, x) plane (Figure 12.2.6).

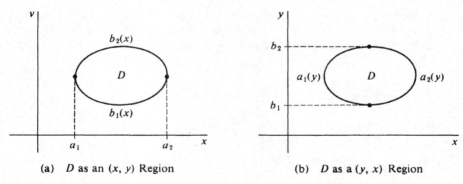

(a) D as an (x, y) Region   (b) D as a (y, x) Region

Figure 12.2.6

EXAMPLE 5   Let D be the region bounded by the curves

$$x = y^2, \qquad x = y + 2.$$

Evaluate the double integral $\iint_D xy\, dA$.

Step 1   The region D is sketched in Figure 12.2.7.

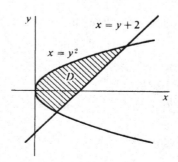

Figure 12.2.7

Step 2 Find inequalities for D. To do this we must find the points where the curves

$$x = y^2, \qquad x = y + 2$$

intersect. Solving for y and then x, we see that they intersect at

$$(1, -1), \qquad (4, 2).$$

We see from the figure that D is a region in either the (x, y) plane or the (y, x) plane. However, the boundary curves are simpler in the (y, x) plane. D is the region

$$-1 \le y \le 2, \qquad y^2 \le x \le y + 2.$$

Step 3 Set up the iterated integral and evaluate.

$$\iint_D xy \, dA = \int_{-1}^{2} \int_{y^2}^{y+2} xy \, dx \, dy.$$

$$\int_{y^2}^{y+2} xy \, dx = \tfrac{1}{2}x^2 y \Big]_{y^2}^{y+2}$$

$$= \tfrac{1}{2}(y + 2)^2 y - \tfrac{1}{2}(y^2)^2 y$$

$$= \tfrac{1}{2}y^3 + 2y^2 + 2y - \tfrac{1}{2}y^5.$$

$$\iint_D xy \, dA = \int_{-1}^{2} \tfrac{1}{2}y^3 + 2y^2 + 2y - \tfrac{1}{2}y^5 \, dy = \tfrac{135}{24}.$$

PROOF OF THE ITERATED INTEGRAL THEOREM For any region D, let $B(D)$ be the iterated integral over D. Our plan is to prove that B has the Addition and Cylinder Properties, so that by the Uniqueness Theorem $B(D)$ will equal the double integral.

PROOF OF ADDITION PROPERTY

Case 1 Let D be divided into D_1 and D_2 as in Figure 12.2.8(a). By the Addition Property for single integrals,

$$B(D) = \int_{a_1}^{a_2} \int_{b_1(x)}^{b_2(x)} f \, dy \, dx$$

$$= \int_{a_1}^{a_3} \int_{b_1(x)}^{b_2(x)} f \, dy \, dx + \int_{a_3}^{a_2} \int_{b_1(x)}^{b_2(x)} f \, dy \, dx$$

$$= B(D_1) + B(D_2).$$

Case 2 Let D be divided into D_1 and D_2 as in Figure 12.2.8(b). Then

$$B(D) = \int_{a_1}^{a_2} \int_{b_1(x)}^{b_2(x)} f \, dy \, dx$$

$$= \int_{a_1}^{a_2} \left[\int_{b_1(x)}^{b_3(x)} f \, dy + \int_{b_3(x)}^{b_2(x)} f \, dy \right] dx$$

$$= \int_{a_1}^{a_2} \int_{b_1(x)}^{b_3(x)} f \, dy \, dx + \int_{a_1}^{a_2} \int_{b_3(x)}^{b_2(x)} f \, dy \, dx$$

$$= B(D_1) + B(D_2).$$

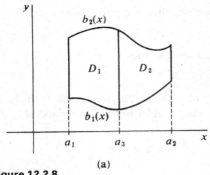
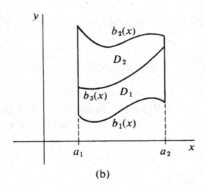

(a)   (b)

Figure 12.2.8

PROOF OF CYLINDER PROPERTY   Let m be the minimum value and M the maximum value of $f(x, y)$ on D For each fixed value of x,

$$\int_{b_1(x)}^{b_2(x)} m \, dy \leq \int_{b_1(x)}^{b_2(x)} f(x, y) \, dy.$$

Integrating from a_1 to a_2,

$$\int_{a_1}^{a_2}\int_{b_1(x)}^{b_2(x)} m \, dy \, dx \leq \int_{a_1}^{a_2}\int_{b_1(x)}^{b_2(x)} f \, dy \, dx = B(D).$$

But

$$\int_{a_1}^{a_2}\int_{b_1(x)}^{b_2(x)} m \, dy \, dx = \int_{a_1}^{a_2} m(b_2(x) - b_1(x)) \, dx$$

$$= m\int_{a_1}^{a_2} (b_2(x) - b_1(x)) \, dx = mA.$$

Therefore $mA \leq B(D)$.

By a similar argument, $B(D) \leq MA$.

Since B has both the Addition and Cylinder Properties,

$$B(D) = \iint_D f(x, y) \, dA.$$

The Constant, Sum, and Inequality Rules for double integrals follow easily from the corresponding rules for single integrals, using the Iterated Integral Theorem.

CONSTANT RULE

$$\iint_D cf(x, y) \, dA = c\iint_D f(x, y) \, dA.$$

SUM RULE

$$\iint_D f(x, y) + g(x, y) \, dA = \iint_D f(x, y) \, dA + \iint_D g(x, y) \, dA.$$

INEQUALITY RULE

> *If $f(x, y) \le g(x, y)$ for all (x, y) in D,*
>
> $$\iint_D f(x, y)\, dA \le \iint_D g(x, y)\, dA.$$

PROOF As an illustration we prove the Sum Rule.

$$\iint_D f + g\, dA = \int_{a_1}^{a_2} \int_{b_1(x)}^{b_2(x)} f + g\, dy\, dx$$

$$= \int_{a_1}^{a_2} \left[\int_{b_1(x)}^{b_2(x)} f\, dy + \int_{b_1(x)}^{b_2(x)} g\, dy \right] dx$$

$$= \int_{a_1}^{a_2} \int_{b_1(x)}^{b_2(x)} f\, dy\, dx + \int_{a_1}^{a_2} \int_{b_1(x)}^{b_2(x)} g\, dy\, dx$$

$$= \iint_D f\, dA + \iint_D g\, dA.$$

The Iterated Integral Theorem gives another proof that the area of D is equal to the double integral of 1 over D.

By definition of area between two curves,

$$A = \int_{a_1}^{a_2} (b_2(x) - b_1(x))\, dx.$$

Using iterated integrals,

$$\iint_D dA = \int_{a_1}^{a_2} \int_{b_1(x)}^{b_2(x)} dy\, dx$$

$$= \int_{a_1}^{a_2} (b_2(x) - b_1(x))\, dx = A.$$

PROBLEMS FOR SECTION 12.2

In Problems 1–16, evaluate the double integrals (compare these with the problems from Section 12.1).

1   $\displaystyle\iint_D (3x + 4y)\, dA, \quad D: 0 \le x \le 1, 0 \le y \le 1$

2   $\displaystyle\iint_D (4 + 2x - 5y)\, dA, \quad D: -2 \le x \le 2, -1 \le y \le 1$

3   $\displaystyle\iint_D (x^2 + y^2)\, dA, \quad D: -2 \le x \le 2, -2 \le y \le 2$

4   $\displaystyle\iint_D (1 + xy)\, dA, \quad D: 0 \le x \le 2, 0 \le y \le 1$

5     $\displaystyle\iint_D \frac{x}{y}\,dA,\quad D:1 \le x \le 2, 1 \le y \le 2$

6     $\displaystyle\iint_D (\cos x + \sin y)\,dA,\quad D: -\pi/2 \le x \le \pi/2, 0 \le y \le \pi$

7     $\displaystyle\iint_D (\cos x \sin y)\,dA,\quad D: -\pi/2 \le x \le \pi/2, 0 \le y \le \pi$

8     $\displaystyle\iint_D xe^y\,dA,\quad D:0 \le x \le 2, -2 \le y \le 3$

9     $\displaystyle\iint_D e^{2x-y}\,dA,\quad D: -2 \le x \le 2, -2 \le y \le 2$

10    $\displaystyle\iint_D (x + 2y)\,dA,\quad D:0 \le x \le 1, 0 \le y \le x$

11    $\displaystyle\iint_D (2 + x + 3y)\,dA,\quad D:0 \le x \le 1, x \le y \le 1$

12    $\displaystyle\iint_D (x^2 + \sqrt{y})\,dA,\quad D: -1 \le x \le 1, 0 \le y \le x^2$

13    $\displaystyle\iint_D y \sin x\,dA,\quad D:0 \le x \le \pi, \sin^2 x \le y \le 2 \sin x$

14    $\displaystyle\iint_D (e^x + e^y)\,dA,\quad D: -3 \le x \le 3, -x \le y \le x$

15    $\displaystyle\iint_D 4\,dA,\quad D:x^2 + y^2 \le 9$

16    $\displaystyle\iint_D -10\,dA,\quad D: -3 \le x \le 3, x^2 \le y \le 18 - x^2$

In Problems 17–24, evaluate the iterated integral. Then check your answer, by evaluating in the other order.

17    $\displaystyle\int_0^1 \int_0^1 (x^2 y - 3xy^2 + 5)\,dy\,dx$

18    $\displaystyle\int_0^1 \int_0^1 xy(2y + 1)\,dy\,dx$

19    $\displaystyle\int_3^6 \int_{-2}^8 dy\,dx$

20    $\displaystyle\int_2^4 \int_1^6 3x\,dy\,dx$

21    $\displaystyle\int_0^{\pi/2} \int_0^{\pi/2} \sin(x + y)\,dy\,dx$

22    $\displaystyle\int_{-1}^1 \int_0^2 \frac{y}{1 + x^2}\,dy\,dx$

23    $\displaystyle\int_0^3 \int_1^6 \sqrt{x + y}\,dy\,dx$

24    $\displaystyle\int_1^2 \int_0^1 \frac{1}{x + y}\,dy\,dx$

In Problems 25–30 evaluate the iterated integral.

25    $\displaystyle\int_0^1 \int_0^{e^x} dy\,dx$

26    $\displaystyle\int_0^\pi \int_{\sin x}^1 dy\,dx$

27   $\displaystyle\int_0^2 \int_0^{\sqrt{4-x^2}} y \, dy \, dx$ 28 $\displaystyle\int_0^1 \int_{-\sqrt{1-x^2}}^{\sqrt{1-x^2}} x^2 + y^2 \, dy \, dx$

29   $\displaystyle\int_0^3 \int_{y^2}^{3y} x^2 y \, dx \, dy$ 30 $\displaystyle\int_0^1 \int_0^y \frac{2}{\sqrt{1-x^2}} \, dx \, dy$

In Problems 31–38, find inequalities which describe the given region D, and write down an iterated integral equal to $\iint_D f(x, y) \, dA$.

31   The triangle with vertices $(0, 0)$, $(5, 0)$, $(0, 5)$.
32   The triangle with vertices $(1, -2)$, $(1, 4)$, $(5, 0)$.
33   The circle of radius 2 with center at the origin.
34   The bottom half of the circle of radius 1 with center at $(2, 3)$.
35   The region bounded by the parabola $y = 4 - x^2$ and the line $y = 3x$.
36   The region above the parabola $y = x^2$ and inside the circle $x^2 + y^2 = 1$.
37   The region bounded by the curves $x = \frac{1}{2}$ and $x = 1/(1 + y^2)$.
38   The region bounded by the curves $x = 12 + y^2$ and $x = y^4$.
39   Find the volume of the solid over the region $x^2 + y^2 \leq 1$ and between the surfaces $z = 0$, $z = x^2$.
40   Find the volume of the solid over the region
$$D : 1 \leq x \leq 2, \ x \leq y \leq x^2$$
and between the surfaces $z = 0$, $z = y/x$.
41   Find the volume of the solid between the surfaces $z = 0$, $z = 2 + 3x - y$, over the region $0 \leq x \leq 2, 0 \leq y \leq x$.
42   Find the volume of the solid between the surfaces $z = 0$, $z = \sqrt{y - x}$, over the region $0 \leq x \leq 1, \ x \leq y \leq 1$.
43   Find the volume of the solid bounded by the plane $z = 0$ and the paraboloid
$$z = 1 - \frac{x^2}{a^2} - \frac{y^2}{b^2}.$$
44   Find the volume of the solid bounded by the three coordinate planes and the plane $ax + by + cz = 1$, where a, b, and c are positive.

☐ 45   Show that
$$\int_{a_1}^{a_2} \int_{b_1}^{b_2} f(x) \, dy \, dx = (b_2 - b_1) \int_{a_1}^{a_2} f(x) \, dx.$$

☐ 46   Show that
$$\int_{a_1}^{a_2} \int_{b_1}^{b_2} f(x) + g(y) \, dy \, dx = (b_2 - b_1) \int_{a_1}^{a_2} f(x) \, dx + (a_2 - a_1) \int_{b_1}^{b_2} g(y) \, dy.$$

☐ 47   Show that
$$\int_{a_1}^{a_2} \int_{b_1}^{b_2} f(x)g(y) \, dy \, dx = \left(\int_{a_1}^{a_2} f(x) \, dx \right) \left(\int_{b_1}^{b_2} g(y) \, dy \right).$$

☐ 48   Show that
$$\int_a^b \int_{-g(x)}^{g(x)} y \, dy \, dx = 0.$$

☐ 49   Let
$$f(x, y) = \begin{cases} y & \text{if } x \text{ is rational,} \\ 1 - y & \text{if } x \text{ is irrational.} \end{cases}$$

Show that:

(a) $\displaystyle\int_0^1 \int_0^1 f(x, y) \, dy \, dx = \int_0^1 \frac{1}{2} \, dx = \frac{1}{2}.$

(b)   For each constant $y_0 \neq \frac{1}{2}$, the function $g(x) = f(x, y_0)$ is everywhere discontinuous, so that the iterated integral $\int_0^1 \int_0^1 f(x, y) \, dx \, dy$ is undefined.

12.3 INFINITE SUM THEOREM AND VOLUME

The double integral, like the single integral, has a number of applications to geometry and physics. The basic theorem which justifies these applications is the Infinite Sum Theorem. It shows how to get an integration formula by considering an infinitely small element of area.

An *element of area* is a rectangle ΔD whose sides are infinitesimal and parallel to the x and y axes. Given an element of area ΔD, we let

$$(x, y) = \text{lower left corner of } \Delta D,$$
$$\Delta x, \Delta y = \text{dimensions of } \Delta D,$$
$$\Delta A = \Delta x \, \Delta y = \text{area of } \Delta D.$$

ΔD is illustrated in Figure 12.3.1.

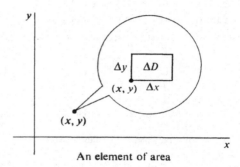

An element of area

Figure 12.3.1

INFINITE SUM THEOREM

Let $h(x, y)$ be continuous on an open region D_0 and let B be a function which assigns a real number $B(D)$ to each region D contained in D_0. Assume that

(i)   *B has the Addition Property* $B(D) = B(D_1) + B(D_2)$.

(ii) $B(D) \geq 0$ *for every D.*

(iii) *For every element of area* ΔD, $B(\Delta D) \approx h(x, y) \, \Delta A$ *(compared to* ΔA).

Then

$$B(D) = \iint_D h(x, y) \, dA.$$

We shall use the notation

$$\Delta B = B(\Delta D).$$

Given (i) and (ii), the theorem shows that if we always have

$$\Delta B \approx h(x, y) \, \Delta A \qquad \text{(compared to } \Delta A)$$

then

$$B(D) \approx \sum_D \sum h(x, y) \, \Delta A.$$

The proof is simplest in the case that D is a rectangle.

PROOF WHEN D IS A RECTANGLE   Choose positive infinitesimal Δx and Δy and partition D into elements of area ΔD (Figure 12.3.2). Since B has the Addition Property, $B(D)$ is the sum of the ΔB's. Let c be any positive real number. For each ΔD we have

$$\Delta B \approx h(x, y)\,\Delta A \qquad \text{(compared to } \Delta A\text{)},$$

$$\frac{\Delta B}{\Delta A} \approx h(x, y),$$

$$\frac{\Delta B}{\Delta A} - c < h(x, y) < \frac{\Delta B}{\Delta A} + c,$$

$$\Delta B - c\,\Delta A < h(x, y)\,\Delta A < \Delta B + c\,\Delta A.$$

Letting A be the area of D and adding up,

$$B(D) - cA < \sum\sum_{D} h(x, y)\,\Delta A < B(D) + cA.$$

Taking standard parts,

$$B(D) - cA \le \iint_{D} h(x, y)\,dA \le B(D) + cA,$$

so

$$B(D) = \iint_{D} h(x, y)\,dA.$$

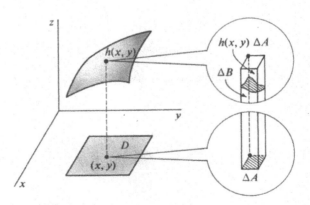

Figure 12.3.2

The proof in the general case is similar except that some of the elements of area ΔD will overlap the boundary of D and thus be only partly within D. (See Figure 12.3.3.) The method of proof is to change D to include all instead of part of each ΔD, use hypothesis (ii) to show that the new $B(D)$ is infinitely close to the old one, and then show as above that the new $B(D)$ is infinitely close to the double integral $\iint_{D} h(x, y)\,dA$.

In most applications of the Infinite Sum Theorem, hypotheses (i) and (ii) are automatic. To get a formula for $B(D)$ in practice, we take an element of area ΔD and

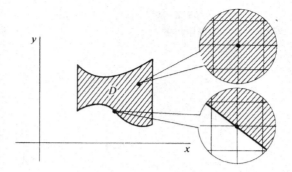

Figure 12.3.3

find an $h(x, y)$ such that

$$\Delta B \approx h(x, y)\Delta A \qquad \text{(compared to } \Delta A\text{)}.$$

Our first application is to the volume between two surfaces.

DEFINITION

Let $f(x, y) \leq g(x, y)$ for (x, y) in D and let E be the set of all points in space such that

$$(x, y) \text{ is in } D, \qquad f(x, y) \leq z \leq g(x, y).$$

The **volume** of E is

$$V = \iint\limits_{D} g(x, y) - f(x, y)\, dA.$$

V is called the volume over D between the surfaces $z = f(x, y)$ and $z = g(x, y)$ (Figure 12.3.4).

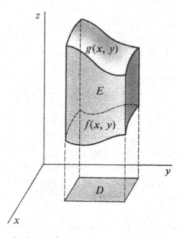

Figure 12.3.4   Volume between two surfaces

JUSTIFICATION The part ΔE of the solid E over an element of area ΔD is a rectangular solid with base ΔA and height $g(x, y) - f(x, y)$, except that the top and bottom surfaces are curved (Figure 12.3.5). Therefore the volume of ΔE is

$$\Delta V \approx (g(x, y) - f(x, y))\, \Delta A \qquad \text{(compared to } \Delta A\text{)}.$$

By the Infinite Sum Theorem,

$$V = \iint_D g(x, y) - f(x, y)\, dA.$$

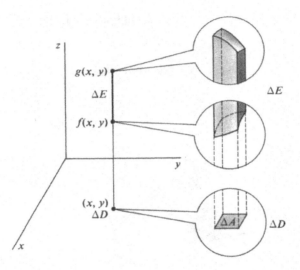

Figure 12.3.5

EXAMPLE 1 Find the volume of the solid

$$0 \le x \le 1, \cdot \quad 0 \le y \le x, \quad x + y \le z \le e^{x+y}.$$

Step 1 D is the triangle shown in Figure 12.3.6.

Step 2 D is the region $0 \le x \le 1, 0 \le y \le x$.

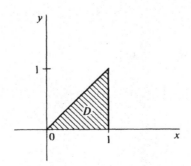

Figure 12.3.6

Step 3
$$V = \iint_D e^{x+y} - (x + y)\,dA$$

$$= \int_0^1 \int_0^x e^{x+y} - (x + y)\,dy\,dx.$$

$$\int_0^x e^{x+y} - (x + y)\,dy = e^{x+y} - xy - \tfrac{1}{2}y^2 \Big]_0^x = e^{2x} - e^x - \tfrac{3}{2}x^2.$$

$$V = \int_0^1 e^{2x} - e^x - \tfrac{3}{2}x^2\,dx = \tfrac{1}{2}e^2 - e.$$

EXAMPLE 2   Find the volume of the solid bounded by the four planes

$$x = 0, \qquad y = 0, \qquad z = x + y, \qquad z = 1 - x - y.$$

Step 1   Sketch the planes. We see from Figure 12.3.7 that $z = x + y$ is below $z = 1 - x - y$.

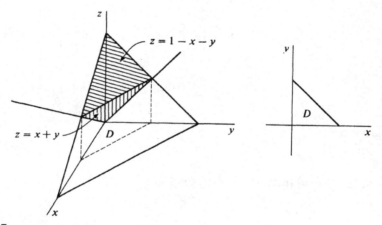

Figure 12.3.7

Step 2   Find inequalities for the region D. Since the two planes

$$z = x + y, \qquad z = 1 - x - y$$

meet at the line $2x + 2y = 1, \qquad y = \tfrac{1}{2} - x,$

D is the region $0 \le x \le \tfrac{1}{2}, \qquad 0 \le y \le \tfrac{1}{2} - x.$

Step 3
$$V = \iint_D (1 - x - y) - (x + y)\,dA = \iint_D 1 - 2x - 2y\,dA$$

$$= \int_0^{1/2} \int_0^{1/2-x} 1 - 2x - 2y\,dy\,dx.$$

$$\int_0^{1/2-x} 1 - 2x - 2y \, dy = y - 2xy - y^2 \Big]_0^{1/2-x}$$

$$= \tfrac{1}{2} - x - 2x(\tfrac{1}{2} - x) - (\tfrac{1}{2} - x)^2 = \tfrac{1}{4} - x + x^2.$$

$$V = \int_0^{1/2} \tfrac{1}{4} - x + x^2 \, dx = \tfrac{1}{24}.$$

EXAMPLE 3 Find the volume of the solid bounded by the plane $z = 2y$ and the paraboloid $z = 1 - 2x^2 - y^2$.

Step 1 The surfaces and the region D are sketched in Figure 12.3.8.

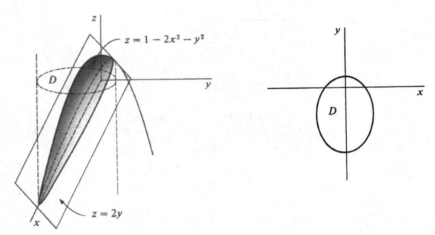

Figure 12.3.8

Step 2 The two surfaces intersect on the curve

$$2y = 1 - 2x^2 - y^2,$$

or solving for y, $y = -1 \pm \sqrt{2 - 2x^2}$.

Therefore D is the region

$$-1 \le x \le 1, \qquad -1 - \sqrt{2 - 2x^2} \le y \le -1 + \sqrt{2 - 2x^2}.$$

Step 3 We see from the figure that the plane is the lower surface and the paraboloid is the upper surface.

$$V = \iint_D (1 - 2x^2 - y^2) - 2y \, dA$$

$$= \int_{-1}^{1} \int_{-1-\sqrt{2-2x^2}}^{-1+\sqrt{2-2x^2}} (1 - 2x^2 - y^2 - 2y) \, dy \, dx.$$

$$\int_{-1-\sqrt{2-2x^2}}^{-1+\sqrt{2-2x^2}} (1 - 2x^2 - y^2 - 2y)\,dy = \int_{-\sqrt{2-2x^2}}^{\sqrt{2-2x^2}} (2 - 2x^2 - u^2)\,du$$

$$= \frac{8\sqrt{2}}{3}(1 - x^2)^{3/2}.$$

$$V = \int_{-1}^{1} \frac{8\sqrt{2}}{3}(1 - x^2)^{3/2}\,dx.$$

Put $x = \sin\theta$, $\sqrt{1 - x^2} = \cos\theta$, $dx = \cos\theta\,d\theta$ (Figure 12.3.9).

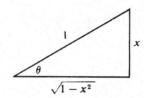

Figure 12.3.9

$$V = \frac{8\sqrt{2}}{3} \int_{-\pi/2}^{\pi/2} \cos^4\theta\,d\theta$$

$$= \frac{8\sqrt{2}}{3} \left[\frac{1}{4}\cos^3\theta\sin\theta + \frac{3}{4}\left(\frac{1}{2}\cos\theta\sin\theta + \frac{1}{2}\theta\right)\right]_{-\pi/2}^{\pi/2}$$

$$= \frac{8\sqrt{2}}{3} \cdot \frac{3}{4} \cdot \frac{1}{2}\pi = \sqrt{2}\pi.$$

Answer $V = \sqrt{2}\pi$.

PROBLEMS FOR SECTION 12.3

Find the volumes of the following solids in Problems 1–8.

1    $0 \le x \le 1$, $0 \le y \le 1$, $xy \le z \le 1$

2    $0 \le x \le 2$, $0 \le y \le 2$, $x^2 + y^2 \le z \le 8$

3    $0 \le x \le 1$, $1 \le y \le 2$, $x \le z \le y$

4    $0 \le x \le 4$, $0 \le y \le 1$, $x \le z \le xe^y$

5    $0 \le x \le 2$, $0 \le y \le x$, $y \le z \le x$

6    $1 \le x \le 4$, $x \le y \le 4$, $y \le x \le xy$

7    $-1 \le x \le 1$, $x^2 \le y \le 1$, $x\sqrt{y} \le z \le y$

8    $0 \le x \le \pi$, $-\sin x \le y \le \sin x$, $-\sin x \le z \le \sin x$

In Problems 9–16, find the volume of the solid bounded by the given surfaces.

9   The planes $y = 0$, $x + y = 2$, $z = -x$, $z = x$

10   The planes $x = 0$, $y = 0$, $2x + 3y + z = 4$, $6x + y - z = 8$

11 $\quad z = x^2 + y^2, \quad z = 4$	**12** $\quad z = x^2 + y^2 + 1, \quad z = 2x + 2y$
13 $\quad y = 0, \quad z = x^2 + y, \quad z = 1$	**14** $\quad x = 0, \quad x = y, \quad z^2 = 1 - y$
15 $\quad x^2 + y^2 = 9, \quad x^2 + z^2 = 9$	**16** $\quad z = x^2 + y^2, \quad z = 2 - x^2 - y^2$

17 Find the volume of the ellipsoid $\dfrac{x^2}{a^2} + \dfrac{y^2}{b^2} + \dfrac{z^2}{c^2} = 1$.

18 Find the volume of the solid bounded by the paraboloid $z = x^2/a^2 + y^2/b^2$ and the plane $z = c$, where c is positive.

12.4 APPLICATIONS TO PHYSICS

In this section we obtain double integrals for mass, center of mass, and moment of inertia.

DEFINITION

If a plane object fills a region D and has continuous density $\rho(x, y)$, its mass is

$$m = \iint_D \rho(x, y)\, dA.$$

On an element of area ΔD, the density is infinitely close to $\rho(x, y)$ (Figure 12.4.1). Therefore the mass is

$$\Delta m \approx \rho(x, y)\, \Delta A \qquad \text{(compared to } \Delta A\text{)}.$$

By the Infinite Sum Theorem $m = \iint_D \rho(x, y)\, dx\, dy$.

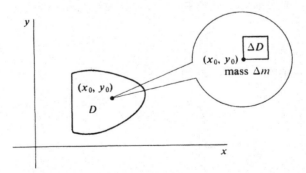

Figure 12.4.1

In Chapter 6 we were able to find the mass of a plane object whose density $\rho(x)$ depends only on x by a single integral,

$$m = \int_{a_1}^{a_2} \rho(x)(b_2(x) - b_1(x))\, dx.$$

Our new formula for mass reduces to the old formula in this case, for by the Iterated Integral Theorem,

$$
\begin{aligned}
m &= \iint_D \rho(x)\, dA \\
&= \int_{a_1}^{a_2} \int_{b_1(x)}^{b_2(x)} \rho(x)\, dy\, dx \\
&= \int_{a_1}^{a_2} \rho(x)(b_2(x) - b_1(x))\, dx.
\end{aligned}
$$

Now we can find the mass of a plane object whose density $\rho(x, y)$ depends on both x and y instead of on x alone.

EXAMPLE 1   Find the mass of an object in the shape of a unit square whose density is the sum of the distance from one edge and twice the distance from a second perpendicular edge.

Step 1   The region D is shown in Figure 12.4.2.

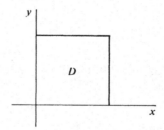

Figure 12.4.2

Step 2   Place the object so the first two edges are on the x and y axes. Then D is the region

$$0 \le x \le 1, \qquad 0 \le y \le 1.$$

Step 3   The density is $\rho(x, y) = y + 2x$.

$$m = \iint_D y + 2x\, dA = \int_0^1 \int_0^1 y + 2x\, dy\, dx.$$

$$\int_0^1 y + 2x\, dy = \tfrac{1}{2}y^2 + 2xy \Big]_0^1 = \tfrac{1}{2} + 2x.$$

$$m = \int_0^1 \tfrac{1}{2} + 2x\, dx = \tfrac{3}{2}.$$

DEFINITION

> *A plane object which fills a region D and has continuous density $\rho(x, y)$ has **moments** about the x and y axes given by*

$$M_x = \iint_D y\rho(x, y)\, dA.$$

$$M_y = \iint_D x\rho(x, y)\, dA.$$

M_x and M_y are sometimes called first moments to distinguish them from moments of inertia (which are called second moments).

*The **center of mass** of the object is the point $(\bar{x}, \bar{y})$ with coordinates*

$$\bar{x} = \frac{M_y}{m} = \frac{\displaystyle\iint_D x\rho(x, y)\, dA}{\displaystyle\iint_D \rho(x, y)\, dA},$$

$$\bar{y} = \frac{M_x}{m} = \frac{\displaystyle\iint_D y\rho(x, y)\, dA}{\displaystyle\iint_D \rho(x, y)\, dA}.$$

JUSTIFICATION The piece of the object on an element of area ΔD has mass

$$\Delta m \approx \rho(x, y)\, \Delta A \qquad \text{(compared to } \Delta A).$$

A point mass $\bar{m}$ at (x, y) has moments

$$\overline{M}_x = y\bar{m}, \qquad \overline{M}_y = x\bar{m}.$$

Therefore the piece of the object at ΔD has moments

$$\Delta M_x \approx y\, \Delta m \approx y\rho(x, y)\, \Delta A \qquad \text{(compared to } \Delta A),$$
$$\Delta M_y \approx x\, \Delta m \approx x\rho(x, y)\, \Delta A \qquad \text{(compared to } \Delta A).$$

The double integrals for M_x and M_y now follow from the Infinite Sum Theorem.

An object will balance on a pin at its center of mass (Figure 12.4.3). The center of mass is useful in finding the work done against gravity when moving the

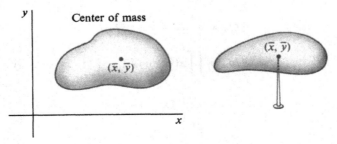

Figure 12.4.3

object. The work is the same as if the mass were all concentrated at the center of mass, and is given by

$$W = mgs$$

where s is the distance the center of mass is raised and g is constant.

EXAMPLE 2   A triangular plate bounded by the lines $x = 0$, $x = y$, $y = 1$ has density $\rho(x, y) = x + y$. Find the moments and center of mass.

Step 1   Sketch the region D, as in Figure 12.4.4.

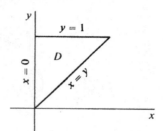

Figure 12.4.4

Step 2   We see from the figure that D is the region

$$0 \le x \le 1, \qquad x \le y \le 1.$$

Step 3   Set up and evaluate the iterated integrals for the mass m and moments M_x and M_y.

$$m = \iint_D x + y \, dA = \int_0^1 \int_x^1 x + y \, dy \, dx.$$

$$\int_x^1 x + y \, dy = x + \tfrac{1}{2} - \tfrac{3}{2}x^2.$$

$$m = \int_0^1 x + \tfrac{1}{2} - \tfrac{3}{2}x^2 \, dx = \tfrac{1}{2}.$$

$$M_x = \iint_D y(x + y) \, dA = \int_0^1 \int_x^1 yx + y^2 \, dy \, dx.$$

$$\int_x^1 yx + y^2 \, dy = \tfrac{1}{2}x + \tfrac{1}{3} - \tfrac{5}{6}x^3.$$

$$M_x = \int_0^1 \tfrac{1}{2}x + \tfrac{1}{3} - \tfrac{5}{6}x^3 \, dx = \tfrac{9}{24}.$$

$$M_y = \iint_D x(x + y) \, dA = \int_0^1 \int_x^1 x^2 + xy \, dy \, dx$$

$$\int_x^1 x^2 + xy \, dy = x^2 + \tfrac{1}{2}x - \tfrac{3}{2}x^3.$$

$$M_y = \int_0^1 x^2 + \tfrac{1}{2}x - \tfrac{3}{2}x^3 \, dx = \tfrac{5}{24}.$$

The answers are $M_x = \dfrac{9}{24}, \qquad M_y = \dfrac{5}{24}.$

$$\bar{x} = \frac{M_y}{m} = \frac{5/24}{1/2} = \frac{5}{12}.$$

$$\bar{y} = \frac{M_x}{m} = \frac{9/24}{1/2} = \frac{9}{12}.$$

The point $(\bar{x}, \bar{y})$ is shown in Figure 12.4.5.

EXAMPLE 3   The object in Example 2 is lying horizontally on the ground. Find the work required to stand the object up with the hypotenuse of the triangle on the ground (Figure 12.4.6).

We use the formula $W = mgs.$

From Example 2, $m = \frac{1}{2}$. We must find s.

s = minimum distance from $\left(\dfrac{5}{12}, \dfrac{9}{12}\right)$ to the line $x = y$.

s = minimum value of $z = \sqrt{\left(x - \dfrac{5}{12}\right)^2 + \left(x - \dfrac{9}{12}\right)^2}.$

$$z = \sqrt{2x^2 - \frac{28}{12}x + \frac{106}{144}}.$$

$$\frac{dz}{dx} = \left(4x - \frac{28}{12}\right)\frac{1}{2}z^{-1/2}.$$

$$\frac{dz}{dx} = 0 \quad \text{at} \quad 4x = \frac{28}{12}, \qquad x = \frac{7}{12}.$$

$$s = \sqrt{2\left(\frac{7}{12}\right)^2 - \frac{28}{12}\cdot\frac{7}{12} + \frac{106}{144}} = \frac{\sqrt{2}}{6}.$$

$$W = mgs = \frac{1}{2}\cdot g \cdot \frac{\sqrt{2}}{6} = \frac{\sqrt{2}}{12}g.$$

The *second moment*, or *moment of inertia*, of a point mass m about the origin is the mass times the square of the distance to the origin,

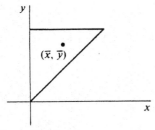

Figure 12.4.5

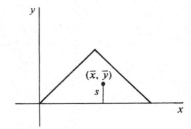

Figure 12.4.6

$$I = m(x^2 + y^2).$$

The moment of inertia is related to the kinetic energy of rotation. A mass m moving at speed v has kinetic energy

$$KE = \tfrac{1}{2} mv^2.$$

Hence if m is rotating about the origin with angular velocity ω radians per second, its speed is $v = \omega\sqrt{x^2 + y^2}$ and

$$KE = \tfrac{1}{2}m(\omega\sqrt{x^2 + y^2})^2 = \tfrac{1}{2}I\omega^2.$$

Thus moment of inertia is the rotational analogue of mass.

DEFINITION

Given a plane object on the region D with continuous density $\rho(x, y)$, the **moment of inertia** about the origin is

$$I = \iint_D \rho(x, y)(x^2 + y^2)\, dA.$$

JUSTIFICATION   On an element of volume ΔD, the moment of inertia is

$$\Delta I \approx (x^2 + y^2)\,\Delta m \approx \rho(x, y)(x^2 + y^2)\,\Delta A \qquad \text{(compared to } \Delta A\text{)}.$$

The integral for I follows by the Infinite Sum Theorem.

EXAMPLE 4   Find the moment of inertia about the origin of an object with constant density $\rho = 1$ which covers the square shown in Figure 12.4.7:

$$-\tfrac{1}{2} \le x \le \tfrac{1}{2}, \qquad -\tfrac{1}{2} \le y \le \tfrac{1}{2}.$$

$$I = \iint_D (x^2 + y^2)\, dA = \int_{-1/2}^{1/2} \int_{-1/2}^{1/2} x^2 + y^2 \; dy\, dx.$$

$$\int_{-1/2}^{1/2} x^2 + y^2 \; dy = x^2 y + \tfrac{1}{3}y^3 \Big]_{-1/2}^{1/2} = x^2 + \tfrac{1}{12}.$$

$$I = \int_{-1/2}^{1/2} x^2 + \tfrac{1}{12} \; dx = \tfrac{1}{6}.$$

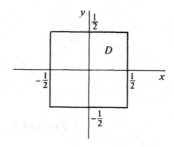

Figure 12.4.7

PROBLEMS FOR SECTION 12.4

In Problems 1–10, find (a) the mass, (b) the center of mass, (c) the moment of inertia about the origin, of the given plane object.

1      $-a \leq x \leq a, \quad -b \leq y \leq b, \quad \rho(x, y) = k$

2      $0 \leq x \leq a, \quad 0 \leq y \leq b, \quad \rho(x, y) = k$

3      $0 \leq x \leq 1, \quad x^2 \leq y \leq 1, \quad \rho(x, y) = k$

4      $0 \leq x \leq a, \quad 0 \leq y \leq bx, \quad \rho(x, y) = k$

5      $0 \leq x \leq 2, \quad x \leq y \leq 2x, \quad \rho(x, y) = x + y + 1$

6      $0 \leq x \leq 1, \quad 0 \leq y \leq x, \quad \rho(x, y) = x - y$

7      $0 \leq x \leq 1, \quad 0 \leq y \leq x^2, \quad \rho(x, y) = \sqrt{x} + \sqrt{y}$

8      $1 \leq x \leq 2, \quad x \leq y \leq x^2, \quad \rho(x, y) = 1/\sqrt{xy}$

9      $0 \leq x \leq 2, \quad e^{-x} \leq y \leq e^x, \quad \rho(x, y) = 1$

10     $-1 \leq x \leq 1, \quad 0 \leq y \leq 1/\sqrt{1 + x^2}, \quad \rho(x, y) = y$

11   Find the mass of an object in the shape of a unit square whose density is the sum of the four distances from the sides.

12   Find the mass of an object in the shape of a unit square whose density is the product of the distances from the four sides.

13   An object on the triangle $0 \leq x \leq 1, 0 \leq y \leq x$ has density equal to the distance from the hypotenuse $y = x$. Find the amount of work required to stand the object up (a) on one of the short sides, (b) on the hypotenuse.

14   An object in the shape of a unit square has density equal to the distance to the nearest side. Find the mass and the amount of work needed to stand the object up on a side.

15   An object on the plane region $-1 \leq x \leq 1, x^2 \leq y \leq 1$ has density $\rho(x, y) = 1 + x + \sqrt{y}$. Find the mass and the work needed to stand the object up on the flat side.

16   An object on the unit square $0 \leq x \leq 1, 0 \leq y \leq 1$ has density $\rho(x, y) = ax + by + c$. Find the mass and center of mass.

☐ 17   The moment of an object of density $\rho(x, y)$ in the region D about the vertical line $x = a$ is defined as

$$M_{y,x=a} = \iint_D (x - a)\rho(x, y)\, dA.$$

Show that

$$M_{y,x=a} = M_y - a \cdot m$$

where M_y is the moment about the y-axis and m is the mass.

☐ 18   The moment of inertia of an object in the region D of density $\rho(x, y)$ about the point $P(a, b)$ is defined as

$$I_P = \iint_D \rho(x, y)((x - a)^2 + (y - b)^2)\, dA.$$

Show that

$$I_P = I - 2aM_x - 2bM_y + m(a^2 + b^2)$$

where I is the moment of inertia about the origin, M_x and M_y are the first moments, and m is the mass.

12.5 DOUBLE INTEGRALS IN POLAR COORDINATES

A point with polar coordinates (θ, r) has rectangular coordinates

$$(x, y) = (r \cos \theta, r \sin \theta).$$

DEFINITION

A **polar region** is a region D in the (x, y) plane given by polar coordinate inequalities

$$\alpha \le \theta \le \beta, \qquad a(\theta) \le r \le b(\theta),$$

where $a(\theta)$ and $b(\theta)$ are continuous. To avoid overlaps, we also require that for all (θ, r) in D,

$$0 \le \theta \le 2\pi \quad \text{and} \quad 0 \le r.$$

The last requirement means that the limits α and β are between 0 and 2π, while the limits $a(\theta)$ and $b(\theta)$ are ≥ 0. Figure 12.5.1 shows a polar region.

The simplest polar regions are the *polar rectangles*

$$\alpha \le \theta \le \beta, \qquad a \le r \le b.$$

We see in Figure 12.5.2 that the θ boundaries are radii and the r boundaries are circular arcs.

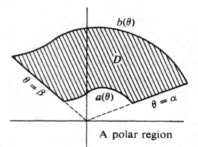

Figure 12.5.1   A polar region

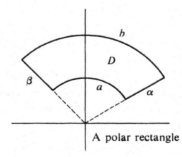

Figure 12.5.2   A polar rectangle

The polar rectangle

$$\alpha \le \theta \le \beta, \qquad 0 \le r \le b$$

is a sector of a circle of radius b (Figure 12.5.3(a)).

The polar rectangle

$$0 \le \theta \le 2\pi, \qquad 0 \le r \le b$$

is a whole circle of radius b (Figure 12.5.3(b)).

Less trivial examples of polar regions are the *circle with diameter from* $(0, 0)$ *to* $(0, b)$,

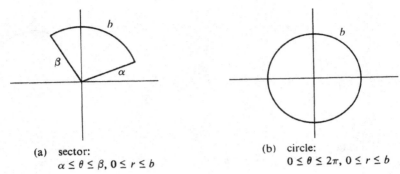

(a) sector:
$\alpha \le \theta \le \beta, 0 \le r \le b$

(b) circle:
$0 \le \theta \le 2\pi, 0 \le r \le b$

Figure 12.5.3

$$0 \le \theta \le \pi, \qquad 0 \le r \le b \sin \theta,$$

and the *cardioid*
$$0 \le \theta \le 2\pi, \qquad 0 \le r \le 1 + \cos \theta.$$

Both of these regions are shown in Figure 12.5.4.

We shall use the Infinite Sum Theorem to get a formula for the double integral over a polar region. In the proof we take for ΔD an infinitely small polar rectangle.

POLAR INTEGRATION FORMULA

Let D be the polar region
$$\alpha \le \theta \le \beta, \qquad a(\theta) \le r \le b(\theta).$$

The double integral of $f(x, y)$ over D is

$$\iint_D f(x, y)\, dA = \int_\alpha^\beta \int_{a(\theta)}^{b(\theta)} f(x, y) r\, dr\, d\theta$$

$$= \int_\alpha^\beta \int_{a(\theta)}^{b(\theta)} f(r \cos \theta, r \sin \theta) r\, dr\, d\theta.$$

Notice that in the iterated integral for a polar region we do not integrate $f(x, y)$ but the product of $f(x, y)$ and r. Intuitively, the extra r comes from the fact that a polar element of area is almost a rectangle of area $r\, \Delta\theta\, \Delta r$ (see Figure 12.5.6(b)).

PROOF We shall work with the rectangular (θ, r) plane. Let C be the region in the (θ, r) plane given by the inequalities

$$\alpha \le \theta \le \beta, \qquad a(\theta) \le r \le b(\theta).$$

Thus C has the same inequalities as D but they refer to the (θ, r) plane instead of the (x, y) plane. D and C are shown in Figure 12.5.5.

We must prove that

$$\iint_D f(x, y)\, dx\, dy = \iint_C f(x, y) r\, d\theta\, dr.$$

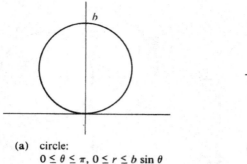

(a)   circle:
$0 \le \theta \le \pi, 0 \le r \le b \sin \theta$

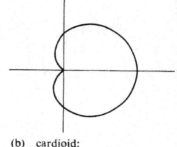

(b)   cardioid:
$0 \le \theta \le 2\pi, 0 \le r \le 1 + \cos \theta$

Figure 12.5.4

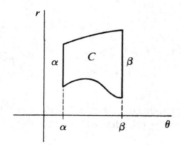

Figure 12.5.5

Our plan is to use the Infinite Sum Theorem in the (θ, r) plane. Assume first that $f(x, y) > 0$ for all (x, y) in D.

For any (θ, r) region C_1 corresponding to a polar region D_1 in the (x, y) plane, let

$$B(C_1) = \iint_{D_1} f(x, y) \, dx \, dy.$$

Then B has the Addition Property and is always ≥ 0. Consider an element of area ΔC in the (θ, r) plane with area $\Delta\theta \, \Delta r$. ΔC corresponds to a polar rectangle ΔD in the (x, y) plane. As we can see from Figure 12.5.6, ΔD is almost a rectangle with sides $r \, \Delta\theta$ and Δr and area $r \, \Delta\theta \, \Delta r$.

The volume over ΔD is almost a rectangular solid with base of area $r \, \Delta\theta \, \Delta r$ and height

$$f(x, y) = f(r \cos \theta, r \sin \theta).$$

Therefore $B(\Delta C) \approx f(x, y) r \, \Delta\theta \, \Delta r$ (compared to $\Delta\theta \, \Delta r$).

By the Infinite Sum Theorem

$$B(C) = \iint_C f(x, y) r \, d\theta \, dr,$$

and by definition $B(C) = \iint_D f(x, y) \, dx \, dy.$

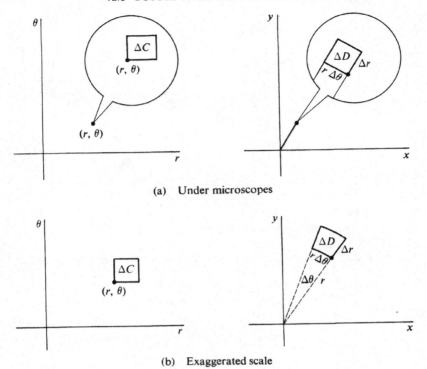

(a)   Under microscopes

(b)   Exaggerated scale

Figure 12.5.6

Finally we consider the case where $f(x, y)$ is not always positive. Pick a real constant $k > 0$ such that $f(x, y) + k$ is always positive for (x, y) in D. By the above proof,

$$\iint_D (f(x, y) + k)\, dx\, dy = \iint_C (f(x, y) + k) r\, d\theta\, dr,$$

$$\iint_D k\, dx\, dy = \iint_C kr\, d\theta\, dr.$$

When we use the Sum Rule and subtract the second equation from the first, we get

$$\iint_D f(x, y)\, dx\, dy = \iint_C f(x, y) r\, d\theta\, dr.$$

In a double integration problem where the region D is a circle or a sector of a circle, it is usually best to take the center as the origin and represent D as a polar rectangle.

EXAMPLE 1   Find the volume over the unit circle $x^2 + y^2 \le 1$ between the surfaces $z = 0$ and $z = x^2$.

Step 1   Sketch D and the solid, as in Figure 12.5.7.

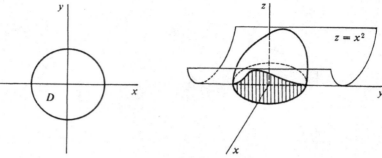

Figure 12.5.7

Step 2 D is the polar region $0 \le \theta \le 2\pi, 0 \le r \le 1$.

Step 3 $V = \iint_D x^2 \, dx \, dy = \int_0^{2\pi} \int_0^1 x^2 r \, dr \, d\theta = \int_0^{2\pi} \int_0^1 (r \cos \theta)^2 r \, dr \, d\theta$

$$= \int_0^{2\pi} \int_0^1 r^3 \cos^2 \theta \, dr \, d\theta$$

$$= \int_0^{2\pi} \tfrac{1}{4} \cos^2 \theta \, d\theta = \tfrac{1}{4}(\tfrac{1}{2} \sin \theta \cos \theta + \tfrac{1}{2}\theta)\bigg]_0^{2\pi} = \pi/4.$$

For comparison let us also work this problem in rectangular coordinates. We can see that it is easier to use polar coordinates.

D is the region $-1 \le x \le 1,$ $-\sqrt{1 - x^2} \le y \le \sqrt{1 - x^2}.$

$$V = \iint_D x^2 \, dx \, dy = \int_{-1}^1 \int_{-\sqrt{1-x^2}}^{\sqrt{1-x^2}} x^2 \, dy \, dx = \int_{-1}^1 2x^2 \sqrt{1 - x^2} \, dx.$$

We make the trigonometric substitution shown in Figure 12.5.8:

$$x = \sin \phi, \qquad \sqrt{1 - x^2} = \cos \phi, \qquad dx = \cos \phi \, d\phi.$$

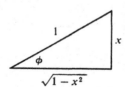

Figure 12.5.8

Then $\phi = -\pi/2$ at $x = -1$ and $\phi = \pi/2$ at $x = 1$, so

$$V = \int_{-\pi/2}^{\pi/2} 2 \sin^2 \phi \cos^2 \phi \, d\phi$$

$$= \int_{-\pi/2}^{\pi/2} 2\left(\frac{1 - \cos 2\phi}{2}\right)\left(\frac{1 + \cos 2\phi}{2}\right) d\phi$$

$$= \int_{-\pi/2}^{\pi/2} \tfrac{1}{2}(1 - \cos^2 2\phi) \, d\phi$$

$$= \int_{-\pi}^{\pi} \tfrac{1}{4}(1 - \cos^2 u) \, du$$

$$= \tfrac{1}{4}(u - \tfrac{1}{2}\cos u \sin u - \tfrac{1}{2}u)\Big]_{-\pi}^{\pi} = \pi/4.$$

EXAMPLE 2 Find the mass and center of mass of a flat plate in the shape of a semi-circle of radius one whose density is equal to the distance from the center of the circle.

Step 1 The region D is sketched in Figure 12.5.9.

Step 2 Take the origin at the center of the circle and the x-axis as the base of the semicircle. D is the polar region $0 \le \theta \le \pi$, $0 \le r \le 1$.

Step 3 The density is

$$\rho(x, y) = \sqrt{x^2 + y^2} = r.$$

$$m = \iint_D \sqrt{x^2 + y^2} \, dA = \int_0^\pi \int_0^1 r \cdot r \, dr \, d\theta$$

$$= \int_0^\pi \int_0^1 r^2 \, dr \, d\theta = \int_0^\pi \frac{1}{3} \, d\theta = \frac{\pi}{3}.$$

$$M_x = \iint_D y\sqrt{x^2 + y^2} \, dA = \int_0^\pi \int_0^1 r \sin \theta \cdot r \cdot r \, dr \, d\theta$$

$$= \int_0^\pi \int_0^1 r^3 \sin \theta \, dr \, d\theta = \int_0^\pi \tfrac{1}{4} \sin \theta \, d\theta = \tfrac{1}{2}.$$

$$M_y = \iint_D x\sqrt{x^2 + y^2} \, dA = \int_0^\pi \int_0^1 r \cos \theta \cdot r \cdot r \, dr \, d\theta$$

$$= \int_0^\pi \int_0^1 r^3 \cos \theta \, dr \, d\theta = \int_0^\pi \tfrac{1}{4} \cos \theta \, d\theta = 0.$$

Answer $m = \dfrac{\pi}{3}$, $\quad \bar{x} = \dfrac{M_y}{m} = 0$, $\quad \bar{y} = \dfrac{M_x}{m} = \dfrac{3}{2\pi} \sim 0.477.$

The point $(\bar{x}, \bar{y})$ is shown in Figure 12.5.10.

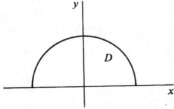

Figure 12.5.9

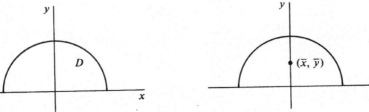

Figure 12.5.10

EXAMPLE 3   Find the moment of inertia of a circle of radius b and constant density ρ about the center of the circle.

Step 1   Draw the region D (Figure 12.5.11).

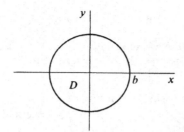

Figure 12.5.11

Step 2   Put the origin at the center, so D is the polar region

$$0 \leq \theta \leq 2\pi, \qquad 0 \leq r \leq b.$$

Step 3 $x^2 + y^2 = r^2$, so

$$I = \iint_D \rho \cdot (x^2 + y^2)\, dA = \int_0^{2\pi} \int_0^b \rho r^2 \cdot r\, dr\, d\theta$$

$$= \rho \int_0^{2\pi} \int_0^b r^3\, dr\, d\theta = \rho \int_0^{2\pi} \tfrac{1}{4}b^4\, d\theta$$

$$= \frac{\rho b^4 \pi}{2}.$$

PROBLEMS FOR SECTION 12.5

In Problems 1–16, find the volume using polar coordinates.

1    $x^2 + y^2 \leq 1, \quad 0 \leq z \leq 6$

2    $x^2 + y^2 \leq 1, \quad 0 \leq z \leq x^2 + y^2$

3    $x^2 + y^2 \leq 4, \quad 0 \leq z \leq x + 2$

4    $x^2 + y^2 \leq 4, \quad 0 \leq z \leq \sqrt{x^2 + y^2}$

5    $x^2 + y^2 \leq 9, \quad x^2 + y^2 \leq z \leq 9$

6    $x^2 + y^2 \leq 25, \quad 0 \leq z \leq e^{-x^2 - y^2}$

7    $1 \leq x^2 + y^2 \leq 4, \quad (x^2 + y^2)^{-1} \leq z \leq (x^2 + y^2)^{-1/2}$

8    $1 \leq x^2 + y^2 \leq 9, \quad 1/\sqrt{x^2 + y^2} \leq z \leq 1$

9    $0 \leq x \leq 1, \quad 0 \leq y \leq \sqrt{1 - x^2}, \quad 0 \leq z \leq x\sqrt{y}$

10    $-2 \leq x \leq 2, \quad 0 \leq y \leq \sqrt{4 - x^2}, \quad x \leq z \leq y + 2$

11    $\pi/4 \leq \theta \leq \pi/3, \quad 0 \leq r \leq 1, \quad 0 \leq z \leq r^2$

12    $0 \leq \theta \leq \pi/6, \quad 1 \leq r \leq 2, \quad 0 \leq z \leq \sqrt{9 - r^2}$

13    $0 \leq \theta \leq \pi, \quad 0 \leq r \leq 2\sin\theta, \quad 0 \leq z \leq r$

14    $0 \leq \theta \leq \pi/2, \quad 0 \leq r \leq \cos\theta, \quad r^3 \leq z \leq r^2$

15    $0 \leq \theta \leq 2\pi, \quad 0 \leq r \leq \theta, \quad 0 \leq z \leq r^2\theta^3 + 2r\theta$

16    $0 \leq \theta \leq 2\pi, \quad 0 \leq r \leq e^\theta, \quad 0 \leq z \leq \sqrt{r}$

17   Find the volume of the solid over the cardioid $r = 1 + \cos\theta$ between the plane $z = 0$ and the cone $z = r$.

18   Find the volume of the solid over the cardioid $r = 1 + \cos\theta$ between the paraboloids $z = r^2$ and $z = 8 - r^2$.

19   Find the volume of the solid over the circle $r = \sin\theta$ between the plane $z = 0$ and the hemisphere $z = \sqrt{1 - r^2}$.

20   Find the volume of the solid over the circle $r = 2\cos\theta$ between the plane $z = 0$ and the cone $z = 2 - r$.

21   Find the volume of the solid over the polar rectangle $\alpha \leq \theta \leq \beta, a \leq r \leq b$, between the plane $z = 0$ and the cone $z = r$.

22   Find the volume of the portion of the hemisphere $0 \leq z \leq \sqrt{1 - r^2}$ over the polar rectangle $\alpha \leq \theta \leq \beta, a \leq r \leq b$ (assuming $b \leq 1$).

23   A circular object of radius b has density equal to the distance from the outside of the circle. Find (a) the mass, (b) the moment of inertia about the origin.

24   A circular object of radius b has density equal to the cube of the distance from the center. Find (a) the mass, (b) the moment of inertia about the origin.

25   Find the moment of inertia about the origin of a circular ring $a \leq r \leq b, 0 \leq \theta \leq 2\pi$, of constant density k.

26   Find the moment of inertia of a circular object of radius b and constant density k about a point on its circumference. (The center can be put at $(0, b)$, so the object is on the polar region $0 \leq r \leq 2b \sin\theta, 0 \leq \theta \leq \pi$.)

27   An object has constant density k on the circular sector $0 \leq x \leq 1, 0 \leq y \leq \sqrt{1 - x^2}$. Find (a) the center of mass, (b) the moment of inertia about the origin.

28   An object of constant density k covers the cardioid $r \leq 1 + \cos\theta, 0 \leq \theta \leq 2\pi$. Find (a) the center of mass, (b) the moment of inertia about the origin.

29   An object of constant density k covers the region inside the circle $r = 2b \sin\theta$ and outside the circle $r = b$. Find (a) the center of mass, (b) the moment of inertia about the origin.

30   An object of constant density k covers the polar region
$$0 \leq \theta \leq \pi/2, \qquad 0 \leq r \leq b \sin 2\theta.$$
Find (a) the center of mass, (b) the moment of inertia about the origin.

☐ 31   (a) Use polar coordinates to evaluate $\int_{-\infty}^{\infty} \int_{-\infty}^{\infty} e^{-x^2 - y^2} \, dy \, dx$.

       (b) Show that $\int_{-\infty}^{\infty} \int_{-\infty}^{\infty} e^{-x^2 - y^2} \, dy \, dx = \left(\int_{-\infty}^{\infty} e^{-x^2} \, dx\right)\left(\int_{-\infty}^{\infty} e^{-y^2} \, dy\right)$.

       (c) Now evaluate the single integral $\int_{-\infty}^{\infty} e^{-x^2} \, dx$.

12.6 TRIPLE INTEGRALS

A *closed region in space*, or *solid region*, is a set E of points given by inequalities
$$a_1 \leq x \leq a_2, \qquad b_1(x) \leq y \leq b_2(x), \qquad c_1(x, y) \leq z \leq c_2(x, y)$$
where the functions $b_1(x), b_2(x)$ and $c_1(x, y), c_2(x, y)$ are continuous.

The *boundary* of E is the part of E on the following surfaces:
The planes $x = a_1, \quad x = a_2$.

The cylinders $y = b_1(x)$, $y = b_2(x)$.
The surfaces $z = c_1(x, y)$, $z = c_2(x, y)$.

The simplest type of closed region is a *rectangular solid*, or *rectangular box*,

$$a_1 \le x \le a_2, \qquad b_1 \le y \le b_2, \qquad c_1 \le z \le c_2.$$

Figure 12.6.1 shows a solid region and a rectangular box.

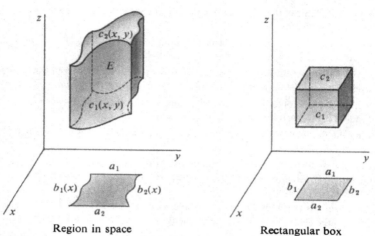

Region in space   Rectangular box

Figure 12.6.1

An *open region* in space is defined in a similar way but with strict inequalities. As in the two-dimensional case, the word *region* alone will mean *closed region*.

PERMANENT ASSUMPTION

Whenever we refer to a function $f(x, y, z)$ and a solid region E, we assume that $f(x, y, z)$ is continuous on some open region containing E.

The triple integral $$\iiint_E f(x, y, z)\, dx\, dy\, dz$$

is analogous to the double integral.

The first step in defining the triple integral is to form the *circumscribed rectangular box* of E (Figure 12.6.2). This is the rectangular box

$$a_1 \le x \le a_2, \qquad B_1 \le y \le B_2, \qquad C_1 \le z \le C_2,$$

where $B_1 = $ minimum value of $b_1(x)$,
              $B_2 = $ maximum value of $b_2(x)$,
              $C_1 = $ minimum value of $c_1(x, y)$,
              $C_2 = $ maximum value of $c_2(x, y)$.

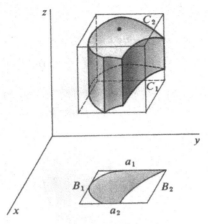

Figure 12.6.2   The circumscribed rectangular box

Our next step is to define the triple Riemann sum. Given positive real numbers Δx, Δy, and Δz, we partition the circumscribed rectangular box of E into rectangular boxes with sides Δx, Δy, and Δz (Figure 12.6.3). The partition points of this three-dimensional partition have the form

$$(x_k, y_l, z_m), \qquad 0 \le k \le n, \qquad 0 \le l \le p, \qquad 0 \le m \le q.$$

The *triple Riemann sum* of $f(x, y, z)\, \Delta x\, \Delta y\, \Delta z$ over E is defined as the sum

$$\sum\sum\sum_{E} f(x, y, z)\, \Delta x\, \Delta y\, \Delta z = \sum\sum\sum_{(x_k, y_l, z_m)\ \text{in}\ E} f(x_k, y_l, z_m)\, \Delta x\, \Delta y\, \Delta z.$$

When we replace Δx, Δy, Δz by positive infinitesimals dx, dy, dz we obtain an *infinite triple Riemann sum*

$$\sum\sum\sum_{E} f(x, y, z)\, dx\, dy\, dz.$$

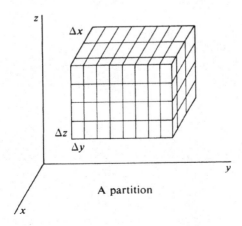

A partition

Figure 12.6.3

LEMMA

For all positive infinitesimals dx, dy, and dz, the triple Riemann sum

$$\sum\sum\sum_{E} f(x, y, z)\, dx\, dy\, dz$$

is a finite hyperreal number and therefore has a standard part.

We are now ready to define the triple integral (see Figure 12.6.4).

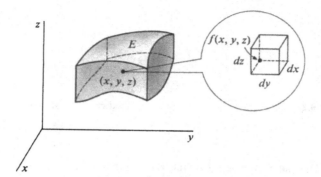

Figure 12.6.4

DEFINITION

*Given positive infinitesimal dx, dy, and dz, the **triple integral** of a continuous function $f(x, y, z)$ over E is*

$$\iiint_{E} f(x, y, z)\, dx\, dy\, dz = st\left(\sum\sum\sum_{E} f(x, y, z)\, dx\, dy\, dz\right).$$

We shall now briefly state some basic theorems on triple integrals, which are exactly like the corresponding theorems for double integrals.

INDEPENDENCE OF dx, dy, AND dz

The value of $\iiint_{E} f(x, y, z)\, dx\, dy\, dz$ does not depend on dx, dy, or dz.

We shall usually use the notation $dV = dx\, dy\, dz$ for the volume of an infinitesimal dx by dy by dz rectangular box, and write

$$\iiint_{E} f(x, y, z)\, dV \qquad \text{for} \qquad \iiint_{E} f(x, y, z)\, dx\, dy\, dz.$$

ADDITION PROPERTY

If E is divided into two regions E_1 and E_2 which meet only on a common boundary then

$$\iiint_{E} f(x, y, z)\, dV = \iiint_{E_1} f(x, y, z)\, dV + \iiint_{E_2} f(x, y, z)\, dV.$$

ITERATED INTEGRAL THEOREM

If E is the region

$$a_1 \leq x \leq a_2, \qquad b_1(x) \leq y \leq b_2(x), \qquad c_1(x, y) \leq z \leq c_2(x, y),$$

then

$$\iiint_E f(x, y, z) \, dV = \int_{a_1}^{a_2} \int_{b_1(x)}^{b_2(x)} \int_{c_1(x,y)}^{c_2(x,y)} f(x, y, z) \, dz \, dy \, dx.$$

If the region E is a rectangular box

$$a_1 \leq x \leq a_2, \qquad b_1 \leq y \leq b_2^{\cdot}, \qquad c_1 \leq z \leq c_2,$$

there are six different iterated integrals over E, corresponding to six different orders of integration. Here they are (in "alphabetical" order).

(1) $\displaystyle\int_{a_1}^{a_2} \int_{b_1}^{b_2} \int_{c_1}^{c_2} f(x, y, z) \, dz \, dy \, dx$ (2) $\displaystyle\int_{a_1}^{a_2} \int_{c_1}^{c_2} \int_{b_1}^{b_2} f(x, y, z) \, dy \, dz \, dx$

(3) $\displaystyle\int_{b_1}^{b_2} \int_{a_1}^{a_2} \int_{c_1}^{c_2} f(x, y, z) \, dz \, dx \, dy$ (4) $\displaystyle\int_{b_1}^{b_2} \int_{c_1}^{c_2} \int_{a_1}^{a_2} f(x, y, z) \, dx \, dz \, dy$

(5) $\displaystyle\int_{c_1}^{c_2} \int_{a_1}^{a_2} \int_{b_1}^{b_2} f(x, y, z) \, dy \, dx \, dz$ (6) $\displaystyle\int_{c_1}^{c_2} \int_{b_1}^{b_2} \int_{a_1}^{a_2} f(x, y, z) \, dx \, dy \, dz.$

The Iterated Integral Theorem shows that each of these iterated integrals is equal to the triple integral

$$\iiint_E f(x, y, z) \, dV.$$

EXAMPLE 1 Evaluate $\displaystyle\iiint_E xy^2z^3 \, dV$ where E is the rectangular box

$$0 \leq x \leq 2, \qquad 0 \leq y \leq 1, \qquad 0 \leq z \leq 4.$$

There are six iterated integrals which all have the same value. We compute one of them, and then another to check our answer.

FIRST SOLUTION $\displaystyle\iiint_E xy^2z^3 \, dV = \int_0^2 \int_0^1 \int_0^4 xy^2z^3 \, dz \, dy \, dx.$

The inside integral is

$$\int_0^4 xy^2z^3 \, dz = \frac{xy^2z^4}{4}\Bigg]_0^4 = 64xy^2.$$

The second integral is

$$\int_0^1 64xy^2 \, dy = \tfrac{64}{3}xy^3\Bigg]_0^1 = \tfrac{64}{3}x.$$

The final answer is

$$\int_0^2 \tfrac{64}{3}x \, dx = \tfrac{64}{6}x^2\Bigg]_0^2 = \tfrac{256}{6} = \tfrac{128}{3}.$$

SECOND SOLUTION $\displaystyle\iiint\limits_{E} xy^2z^3 \, dV = \int_0^4 \int_0^2 \int_0^1 xy^2z^3 \, dy \, dx \, dz.$

The inside integral is

$$\int_0^1 xy^2z^3 \, dy = \tfrac{1}{3}xy^3z^3 \Big]_0^1 = \tfrac{1}{3}xz^3.$$

The second integral is

$$\int_0^2 \tfrac{1}{3}xz^3 \, dx = \tfrac{1}{6}x^2z^3 \Big]_0^2 = \tfrac{4}{6}z^3.$$

The final answer is

$$\int_0^4 \tfrac{4}{6}z^3 \, dz = \tfrac{1}{6}z^4 \Big]_0^4 = \tfrac{256}{6} = \tfrac{128}{3}.$$

Triple integrals can be evaluated by iterated integrals.

EXAMPLE 2 Evaluate $\displaystyle\iiint\limits_{E} y + z \, dV$ where E is the region shown in Figure 12.6.5,

$$0 \le x \le \pi/2, \qquad 0 \le y \le \sin x, \qquad 0 \le z \le y \cos x.$$

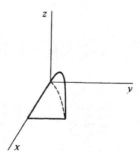

Figure 12.6.5

SOLUTION $\displaystyle\iiint\limits_{E} y + z \, dV = \int_0^{\pi/2} \int_0^{\sin x} \int_0^{y \cos x} y + z \, dz \, dy \, dx.$

We first evaluate the inside integral.

$$\int_0^{y \cos x} y + z \, dz = yz + \tfrac{1}{2}z^2 \Big]_0^{y \cos x} = y^2 \cos x + \tfrac{1}{2}y^2 \cos^2 x.$$

Now we evaluate the second integral.

$$\int_0^{\sin x} y^2 \cos x + \tfrac{1}{2}y^2 \cos^2 x \, dy = \tfrac{1}{3}y^3(\cos x + \tfrac{1}{2}\cos^2 x) \Big]_0^{\sin x}$$

$$= \tfrac{1}{3}\sin^3 x(\cos x + \tfrac{1}{2}\cos^2 x).$$

Finally we evaluate the outside integral.

$$\iiint_E y + z \, dV = \int_0^{\pi/2} \tfrac{1}{3} \sin^3 x (\cos x + \tfrac{1}{2} \cos^2 x) \, dx$$

$$= \int_0^{\pi/2} \tfrac{1}{3}(1 - \cos^2 x)(\cos x + \tfrac{1}{2} \cos^2 x) \sin x \, dx$$

$$= \int_1^0 -\tfrac{1}{3}(1 - u^2)(u + \tfrac{1}{2}u^2) \, du$$

$$= \int_0^1 \tfrac{1}{3}(u + \tfrac{1}{2}u^2 - u^3 - \tfrac{1}{2}u^4) \, du = \tfrac{19}{180}.$$

COROLLARY

The volume of a region E in space is equal to the triple integral of the constant 1 *over E* as illustrated in Figure 12.6.6,

$$V = \iiint_E dV.$$

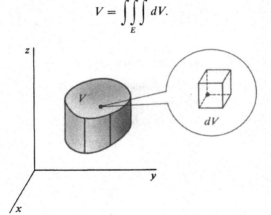

Figure 12.6.6

PROOF E is the solid over the plane region D given by

$$a_1 \leq x \leq a_2, \qquad b_1(x) \leq y \leq b_2(x)$$

between the surfaces $z = c_1(x, y)$ and $z = c_2(x, y)$. By definition of the volume between two surfaces,

$$V = \iint_D c_2(x, y) - c_1(x, y) \, dA.$$

Using the Iterated Integral Theorem,

$$\iiint_E dV = \int_{a_1}^{a_2} \int_{b_1(x)}^{b_2(x)} \int_{c_1(x,y)}^{c_2(x,y)} dz \, dy \, dx$$

$$= \int_{a_1}^{a_2} \int_{b_1(x)}^{b_2(x)} [c_2(x, y) - c_1(x, y)] \, dy \, dx = V.$$

We now come to the Infinite Sum Theorem for triple integrals, which is, again, the key result for applications.

We shall use Δx, Δy, and Δz for positive infinitesimals. By an *element of volume* we mean a rectangular box ΔE with sides Δx, Δy, and Δz (Figure 12.6.7). The volume of ΔE is

$$\Delta V = \Delta x \, \Delta y \, \Delta z.$$

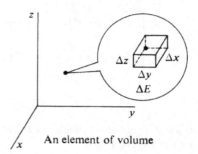

Figure 12.6.7   An element of volume

INFINITE SUM THEOREM

Let $h(x, y, z)$ be continuous on an open region E_0 and let B be a function which assigns a real number $B(E)$ to each region E contained in E_0. Assume that:

(i) B has the Addition Property.

(ii) $B(E) \geq 0$ for every E.

(iii) For every element of volume ΔE,

$$B(\Delta E) \approx h(x, y, z) \, \Delta V \qquad \text{(compared to } \Delta V\text{)}.$$

Then
$$B(E) = \iiint\limits_{E} h(x, y, z) \, dV.$$

Here are some applications of the triple Infinite Sum Theorem. Perhaps the simplest physical interpretation of the triple integral is mass as the triple integral of density.

DEFINITION

The **mass** of an object filling a solid region E with continuous density $\rho(x, y, z)$ is

$$m = \iiint\limits_{E} \rho(x, y, z) \, dV.$$

JUSTIFICATION   At every point of an element of volume ΔE the density is infinitely close to $\rho(x, y, z)$, so the element of mass is

$$\Delta m \approx \rho(x, y, z) \, \Delta V \qquad \text{(compared to } \Delta V\text{)}.$$

(See Figure 12.6.8.) By the Infinite Sum Theorem,

$$m = \iiint\limits_{E} \rho(x, y, z) \, \Delta V.$$

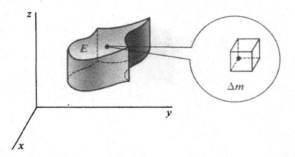

Figure 12.6.8

EXAMPLE 3   Find the mass of an object in the unit cube

$$0 \leq x \leq 1, \qquad 0 \leq y \leq 1, \qquad 0 \leq z \leq 1$$

with density $\rho(x, y, z) = x + y + z.$

$$
\begin{aligned}
m &= \iiint_E x + y + z \, dV \\
&= \int_0^1 \int_0^1 \int_0^1 x + y + z \, dz \, dy \, dx \\
&= \int_0^1 \int_0^1 x + y + \tfrac{1}{2} \, dy \, dx = \int_0^1 x + \tfrac{1}{2} + \tfrac{1}{2} \, dx = \tfrac{3}{2}.
\end{aligned}
$$

An object in space has a moment about each coordinate plane.

DEFINITION

If an object in space fills a region E and has continuous density $\rho(x, y, z)$, its moments about the coordinate planes are

$$M_{xy} = \iiint_E z\rho(x, y, z) \, dV.$$

$$M_{xz} = \iiint_E y\rho(x, y, z) \, dV.$$

$$M_{yz} = \iiint_E x\rho(x, y, z) \, dV.$$

The center of mass of the object is the point $(\bar{x}, \bar{y}, \bar{z})$, where m is mass and

$$\bar{x} = \frac{M_{yz}}{m}, \qquad \bar{y} = \frac{M_{xz}}{m}, \qquad \bar{z} = \frac{M_{xy}}{m}.$$

JUSTIFICATION   A point mass m has moment $M_{xy} = mz$ about the (x, y) plane (Figure 12.6.9). In an element of volume ΔE, the object has moment

$$\Delta M_{xy} \approx z \, \Delta m \approx z\rho(x, y, z) \, \Delta V \qquad \text{(compared to } \Delta V\text{).}$$

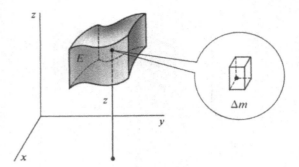

Figure 12.6.9

By the Infinite Sum Theorem,

$$M_{xy} = \iiint\limits_{E} z\rho(x, y, z)\, dV.$$

EXAMPLE 4 An object has constant density and the shape of a tetrahedron with vertices at the four points

$$(0, 0, 0), \qquad (1, 0, 0), \qquad (0, 1, 0), \qquad (0, 0, 1).$$

Find the center of mass.

Step 1 The region is sketched in Figure 12.6.10.

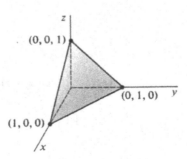

Figure 12.6.10

Step 2 The region E is the solid bounded by the coordinate planes and the plane $x + y + z = 1$ which passes through $(1, 0, 0)$, $(0, 1, 0)$, $(0, 0, 1)$. Solving for z, the plane is

$$z = 1 - x - y.$$

This plane meets the plane $z = 0$ at the line $1 - x - y = 0$, or $y = 1 - x$. Therefore E is the region

$$0 \le x \le 1, \qquad 0 \le y \le 1 - x, \qquad 0 \le z \le 1 - x - y.$$

Step 3   Let the density be $\rho = 1$.

$$m = \iiint\limits_{E} dV = \int_0^1 \int_0^{1-x} \int_0^{1-x-y} dz\, dy\, dx$$

$$= \int_0^1 \int_0^{1-x} 1 - x - y\, dy\, dx$$

$$= \int_0^1 \tfrac{1}{2}(1 - x)^2\, dx = \tfrac{1}{6}.$$

$$M_{yz} = \iiint\limits_{E} x\, dV$$

$$= \int_0^1 \int_0^{1-x} \int_0^{1-x-y} x\, dz\, dy\, dx$$

$$= \int_0^1 \int_0^{1-x} x(1 - x - y)\, dy\, dx$$

$$= \int_0^1 \tfrac{1}{2}x(1 - x)^2\, dx = \tfrac{1}{24}.$$

$$\bar{x} = \frac{M_{yz}}{m} = \frac{1/24}{1/6} = \frac{1}{4}.$$

Similarly
$$\bar{y} = \frac{1}{4}, \qquad \bar{z} = \frac{1}{4}.$$

The center of mass is

$$(\bar{x}, \bar{y}, \bar{z}) = \left(\frac{1}{4}, \frac{1}{4}, \frac{1}{4}\right).$$

An object in space has a moment of inertia about each coordinate axis. Intuitively, the moment of inertia about an axis is the analogue of mass for rotations about the axis.

DEFINITION

*If an object in space fills a region E and has continuous density $\rho(x, y, z)$, its **moments of inertia** about the coordinate axes are*

$$I_x = \iiint\limits_{E} (y^2 + z^2)\rho(x, y, z)\, dV,$$

$$I_y = \iiint\limits_{E} (x^2 + z^2)\rho(x, y, z)\, dV,$$

$$I_z = \iiint\limits_{E} (x^2 + y^2)\rho(x, y, z)\, dV.$$

JUSTIFICATION   A point mass m has a moment of inertia about the x-axis of

$$I_x = (y^2 + z^2)m.$$

On an element of volume ΔE, the object has moment of inertia

$$\Delta I_x \approx (y^2 + z^2)\,\Delta m \approx (y^2 + z^2)\rho(x, y, z)\,\Delta V \qquad \text{(compared to } \Delta V\text{)}.$$

The triple integral for I_x follows by the Infinite Sum Theorem.

EXAMPLE 5   Find the moments of inertia about the three axes of an object with constant density 1 filling the cube shown in Figure 12.6.11,

$$0 \le x \le a, \qquad 0 \le y \le a, \qquad 0 \le z \le a.$$

$$I_x = \iiint\limits_{E} y^2 + z^2 \, dV = \int_0^a \int_0^a \int_0^a y^2 + z^2 \, dz \, dy \, dx$$

$$= \int_0^a \int_0^a ay^2 + \tfrac{1}{3}a^3 \, dy \, dx = \int_0^a \tfrac{2}{3}a^4 \, dx = \tfrac{2}{3}a^5.$$

Similarly, $\qquad\qquad\qquad I_y = \tfrac{2}{3}a^5, \qquad I_z = \tfrac{2}{3}a^5.$

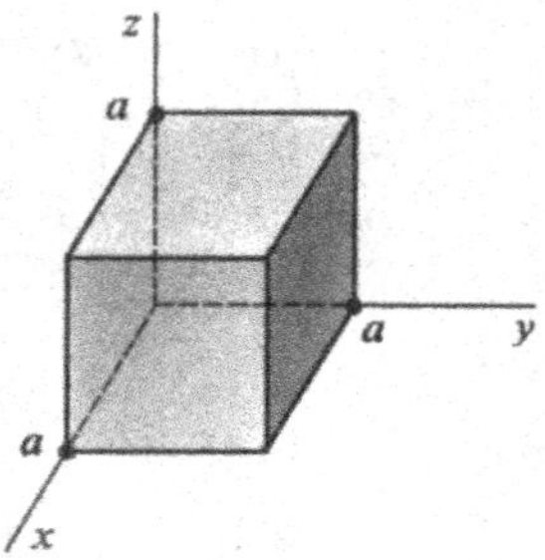

Figure 12.6.11

PROBLEMS FOR SECTION 12.6

In Problems 1–8, evaluate the iterated integral.

1   $\displaystyle\int_0^1 \int_2^4 \int_1^3 xyz \, dz \, dy \, dx$

2   $\displaystyle\int_1^2 \int_0^2 \int_{-1}^4 (x - 2y + 4z) \, dz \, dy \, dx$

3   $\displaystyle\int_0^4 \int_0^1 \int_0^1 (3y^2 + 6z^2) \, dz \, dy \, dx$

4   $\displaystyle\int_0^2 \int_1^2 \int_1^4 e^{x+z} \, dz \, dy \, dx$

5   $\displaystyle\int_0^1 \int_0^x \int_{xy}^1 (x^2 + yz) \, dz \, dy \, dx$

6   $\displaystyle\int_0^1 \int_x^1 \int_y^1 2x^2 z \, dz \, dy \, dx$

7   $\displaystyle\int_0^\pi \int_0^{\sin x} \int_0^{\sin x} \sqrt{yz} \, dz \, dy \, dx$

8   $\displaystyle\int_0^{\pi/2} \int_0^{\cos x} \int_0^{y \sin x} (x + 2z) \, dz \, dy \, dx$

In Problems 9–16, evaluate the triple integral.

9    $\iiint_E (x + 2y)\,dV,\quad E: 0 \le x \le 2, 1 \le y \le 3, 2 \le z \le 4$

10   $\iiint_E x^2 yz^3\,dV,\quad E: 0 \le x \le 1, 0 \le y \le 1, 0 \le z \le 1$

11   $\iiint_E (4xy + yz)\,dV,\quad E: 0 \le x \le 10, 0 \le y \le x^2, 0 \le z \le xy$

12   $\iiint_E \left(\frac{1}{x} + \frac{2}{y} + \frac{3}{z}\right)dV,\quad E: 1 \le x \le e, 1 \le y \le x, 1 \le z \le x$

13   $\iiint_E e^{x+2y+3z}\,dV,\quad E: -1 \le x \le 1, x \le y \le 1, x \le z \le y$

14   $\iiint_E xe^{y+z}\,dV,\quad E: 1 \le x \le 2, 0 \le y \le \ln x, 0 \le z \le y$

15   $\iiint_E \sqrt{x + y + z}\,dV,\quad E: 0 \le x \le 1, 0 \le y \le x, y \le z \le 2y$

16   $\iiint_E dV,\quad E: 0 \le x \le 1, x^2 \le y \le x, x^2 y \le z \le x\sqrt{y}$

In Problems 17–26, find (a) the mass, (b) the center of mass, (c) the moments of inertia about the three coordinate axes, of an object with density $\rho(x, y, z)$ filling the region E.

17    $E: 0 \le x \le 1, 0 \le y \le 1, 0 \le z \le 1 \qquad \rho(x, y, z) = x + 2y + 3z$
18    $E: 0 \le x \le 1, 0 \le y \le 1, 0 \le z \le 1 \qquad \rho(x, y, z) = x^2 + y^2 + z^2$
19    $E: 0 \le x \le 1, 0 \le y \le x, 0 \le z \le x + y \qquad \rho(x, y, z) = 2$
20    $E: -1 \le x \le 1, x^2 \le y \le 1, x^2 \le z \le 1 \qquad \rho(x, y, z) = z$
21    $E: 0 \le x \le 1, 0 \le y \le 1, \sqrt{xy} \le z \le 1 \qquad \rho(x, y, z) = xyz$
22    $E: 0 \le x \le 1, x \le y \le 1, x \le z \le y \qquad \rho(x, y, z) = 10$
23   E is the tetrahedron with vertices at $(0, 0, 0), (a, 0, 0), (0, b, 0), (0, 0, c), \rho(x, y, z) = k.$
24   E is the tetrahedron with vertices $(0, 0, 0), (1, 0, 0), (1, 1, 0), (1, 0, 1), \rho(x, y, z) = x + y + z.$
25   E is the rectangular box $0 \le x \le a, 0 \le y \le b, 0 \le z \le c, \rho(x, y, z) = k.$
26   E is the rectangular box $-a \le x \le a, -b \le y \le b, -c \le z \le c, \rho(x, y, z) = k.$

2.7 CYLINDRICAL AND SPHERICAL COORDINATES

In evaluating triple integrals it is sometimes easier to use cylindrical or spherical coordinates instead of rectangular coordinates.

A point (x, y, z) has *cylindrical coordinates* (θ, r, z) if

$$x = r \cos \theta, \qquad y = r \sin \theta, \qquad z = z.$$

That is, as we see in Figure 12.7.1, (θ, r) is a polar coordinate representation of (x, y), and z is the height above the (x, y) plane.

The name cylindrical coordinates is used because the graph of the cylindrical coordinate equation $r = $ constant is a circular cylinder as shown in Figure 12.7.2.

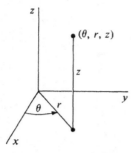

Cylindrical coordinates

Figure 12.7.1

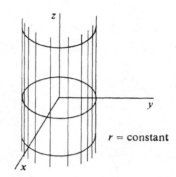

r = constant

Figure 12.7.2

DEFINITION

A **cylindrical region** is a region E in (x, y, z) space given by cylindrical coordinate inequalities

$$\alpha \le \theta \le \beta, \qquad a(\theta) \le r \le b(\theta), \qquad c_1(\theta, r) \le z \le c_2(\theta, r),$$

where all the functions are continuous. To avoid overlaps we also require that for (θ, r, z) in E,

$$0 \le \theta \le 2\pi \quad \text{and} \quad 0 \le r.$$

A cylindrical region is shown in Figure 12.7.3.
The simplest kind of cylindrical region is the *cylindrical box*

$$\alpha \le \theta \le \beta, \qquad a \le r \le b, \qquad c_1 \le z \le c_2.$$

This is a cylinder whose base is a polar rectangle and whose upper and lower faces are horizontal, as in Figure 12.7.4.

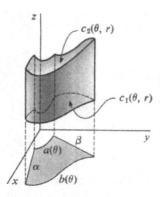

A cylindrical region

Figure 12.7.3

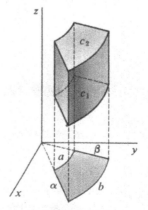

A cylindrical box

Figure 12.7.4

The cylinder box

$$0 \le \theta \le 2\pi, \qquad 0 \le r \le b, \qquad c_1 \le z \le c_2$$

is a cylinder whose base is a circle of radius b and whose top and bottom faces are horizontal (Figure 12.7.5).

The cylindrical box

$$0 \le \theta \le 2\pi, \qquad a \le r \le b, \qquad c_1 \le z \le c_2$$

is a circular pipe with inner radius a and outer radius b (Figure 12.7.6).

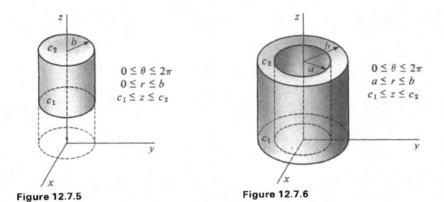

$$0 \le \theta \le 2\pi$$
$$0 \le r \le b$$
$$c_1 \le z \le c_2$$

Figure 12.7.5

$$0 \le \theta \le 2\pi$$
$$a \le r \le b$$
$$c_1 \le z \le c_2$$

Figure 12.7.6

To get a formula for the triple integral over a cylindrical region E, we use the Infinite Sum Theorem but take for ΔE an infinitely small cylindrical box instead of rectangular box.

CYLINDRICAL INTEGRATION FORMULA

Let E be the cylindrical region

$$\alpha \le \theta \le \beta, \qquad a(\theta) \le r \le b(\theta), \qquad c_1(\theta, r) \le z \le c_2(\theta, r).$$

The triple integral of $f(x, y, z)$ over E is

$$\iiint_E f(x, y, z)\, dV = \int_\alpha^\beta \int_{a(\theta)}^{b(\theta)} \int_{c_1(\theta, r)}^{c_2(\theta, r)} f(x, y, z)\, r\, dz\, dr\, d\theta.$$

To evaluate the triple integral we substitute

$$f(x, y, z) = f(r \cos \theta, r \sin \theta, z).$$

This is like the Polar Integration Formula but has an extra variable z. In the iterated integral we do not integrate $f(x, y, z)$ but the product of $f(x, y, z)$ and r.

PROOF Let C be the region in the rectangular (θ, r, z) space given by

$$\alpha \le \theta \le \beta, \qquad a(\theta) \le r \le b(\theta), \qquad c_1(\theta, r) \le z \le c_2(\theta, r).$$

The region C is shown in Figure 12.7.7.

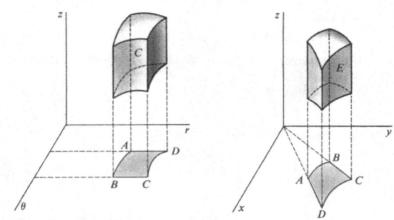

Figure 12.7.7

We must prove that

$$\iiint_E f(x, y, z)\, dx\, dy\, dz = \iiint_C f(x, y, z) r\, d\theta\, dr\, dz.$$

Assume first that $f(x, y, z) > 0$ on E. For any (θ, r, z) region C_1 corresponding to a cylindrical region E_1, define

$$B(C_1) = \iiint_{E_1} f(x, y, z)\, dx\, dy\, dz.$$

B has the Addition Property and is ≥ 0. An element of volume ΔC in the (θ, r, z) space has volume $\Delta\theta\, \Delta r\, \Delta z$. As we can see from Figure 12.7.8, ΔC corresponds to a cylindrical box ΔE. ΔE is almost a rectangular box with sides $r\, \Delta\theta$, Δr, and Δz, and volume $r\, \Delta\theta\, \Delta r\, \Delta z$.

At any point of ΔE, f has value infinitely close to

$$f(x, y, z) = f(r\cos\theta, r\sin\theta, z).$$

Therefore $B(\Delta C) \approx f(x, y, z) r\, \Delta\theta\, \Delta r\, \Delta z$ (compared to $\Delta\theta\, \Delta r\, \Delta z$).

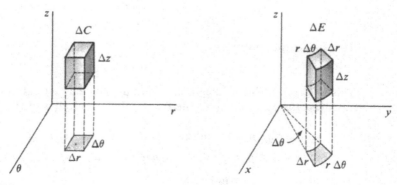

Figure 12.7.8

By the Infinite Sum Theorem

$$B(C) = \iiint_C f(x, y, z) r \, d\theta \, dr \, dr,$$

and by definition

$$B(C) = \iiint_E f(x, y, z) \, dx \, dy \, dz.$$

The general case where $f(x, y, z)$ is not always positive is dealt with as in the Polar Integration Formula proof.

When integrating over a solid region E whose base is a circle or polar rectangle, it is often easier to use cylindrical instead of rectangular coordinates.

EXAMPLE 1   Find the moment of inertia of a cylinder of height h, base a circle of radius b, and constant density 1, about its axis.

Step 1   Draw the region as in Figure 12.7.9.

Step 2   The problem is greatly simplified by a wise choice of coordinate axes. Let the z-axis be the axis of the cylinder and put the origin at the center of the base. Then the region E in rectangular coordinates is

$$-b \le x \le b, \qquad -\sqrt{b^2 - x^2} \le y \le \sqrt{b^2 - x^2}, \qquad 0 \le z \le h,$$

and in cylindrical coordinates is

$$0 \le \theta \le 2\pi, \quad 0 \le r \le b, \quad 0 \le z \le h.$$

Step 3   The problem looks easier in cylindrical coordinates.

$$x^2 + y^2 = r^2.$$

$$I_z = \iiint_E (x^2 + y^2) \, dV = \int_0^{2\pi} \int_0^b \int_0^h r^2 r \, dz \, dr \, d\theta$$

$$= \int_0^{2\pi} \int_0^b \int_0^h r^3 \, dz \, dr \, d\theta$$

$$= \int_0^{2\pi} \int_0^b r^3 h \, dr \, d\theta = \int_0^{2\pi} \frac{1}{4} b^4 h \, d\theta = \frac{\pi b^4 h}{2}.$$

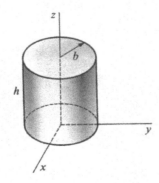

Figure 12.7.9

EXAMPLE 2 Find the center of mass of a cone of constant density with height h and base a circle of radius b.

Step 1 The region is sketched in Figure 12.7.10.

Step 2 Put the origin at the center of the base and let the z-axis be the axis of the cone. E is the cylindrical region

$$0 \le \theta \le 2\pi, \qquad 0 \le r \le b, \qquad 0 \le z \le h - \frac{h}{b}r.$$

Step 3 Let the density be 1.

$$m = \iiint_E dV = \int_0^{2\pi} \int_0^b \int_0^{h-hr/b} r \, dz \, dr \, d\theta$$

$$= \int_0^{2\pi} \int_0^b r\left(h - \frac{h}{b}r\right) dr \, d\theta$$

$$= \int_0^{2\pi} \left(\frac{1}{2}b^2 h - \frac{1}{3}b^3 \frac{h}{b}\right) d\theta = \int_0^{2\pi} \frac{1}{6}b^2 h \, d\theta = \frac{\pi b^2 h}{3}.$$

$$M_{xy} = \iiint_E z \, dV = \int_0^{2\pi} \int_0^b \int_0^{h-hr/b} zr \, dz \, dr \, d\theta$$

$$= \int_0^{2\pi} \int_0^b \frac{1}{2}r\left(h - \frac{hr}{b}\right)^2 dr \, d\theta$$

$$= \frac{1}{2}h^2 \int_0^{2\pi} \int_0^b r - \frac{2r^2}{b} + \frac{r^3}{b^2} \, dr \, d\theta$$

$$= \frac{1}{2}h^2 \int_0^{2\pi} \frac{1}{2}b^2 - \frac{2}{3}\cdot\frac{b^3}{b} + \frac{1}{4}\frac{b^4}{b^2} \, d\theta$$

$$= \frac{1}{2}h^2 \int_0^{2\pi} \frac{1}{12}b^2 \, d\theta = \frac{\pi b^2 h^2}{12}.$$

Since the cone is symmetric about the z-axis, $\bar{x} = 0$ and $\bar{y} = 0$.

$$\bar{z} = \frac{M_{xy}}{m} = \frac{1}{4}h, \qquad (\bar{x}, \bar{y}, \bar{z}) = \left(0, 0, \frac{1}{4}h\right).$$

The point $(\bar{x}, \bar{y}, \bar{z})$ is shown in Figure 12.7.11.

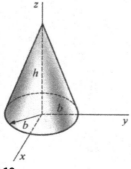

Figure 12.7.10

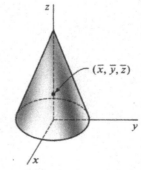

Figure 12.7.11

To express a point $P(x, y, z)$ in *spherical coordinates* we let ρ (rho) be the distance from the origin to P, let θ be the same angle as in cylindrical coordinates, and let ϕ be the angle between the positive z-axis and the line OP. Note that ϕ can always be chosen between 0 and π.

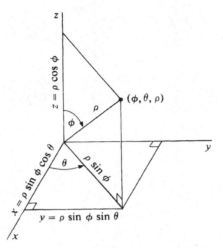

Figure 12.7.12

We see from Figure 12.7.12 that a point (x, y, z) has spherical coordinates (θ, ϕ, ρ) if

$$x = \rho \sin \phi \cos \theta, \qquad y = \rho \sin \phi \sin \theta, \qquad z = \rho \cos \phi.$$

The graph of the equation $\rho = $ constant is a sphere with center at the origin (hence the name spherical coordinates). The graph of $\phi = $ constant is a vertical cone with vertex at the origin. The graph of $\theta = $ constant is a half-plane through the z-axis. These surfaces are shown in Figure 12.7.13.

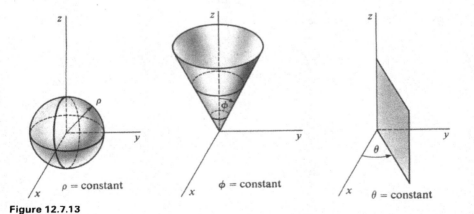

Figure 12.7.13

DEFINITION

>A **spherical region** E is a region in (x, y, z) space given by spherical coordinate inequalities

$$\alpha_1 \leq \theta \leq \alpha_2, \qquad \beta_1(\theta) \leq \phi \leq \beta_2(\theta), \qquad c_1(\theta, \phi) \leq \rho \leq c_2(\theta, \phi),$$

where all the functions are continuous. To avoid overlaps we also require that for (θ, ϕ, ρ) in E,

$$0 \leq \theta \leq 2\pi, \qquad 0 \leq \phi \leq \pi, \qquad 0 \leq \rho.$$

A *spherical box* is a spherical region of the simple form

$$\alpha_1 \leq \theta \leq \alpha_2, \qquad \beta_1 \leq \phi \leq \beta_2, \qquad c_1 \leq \rho \leq c_2.$$

The θ-boundaries are planes, the ϕ-boundaries are portions of cone surfaces, and the ρ-boundaries are portions of spherical surfaces. Figure 12.7.14 shows a spherical box.

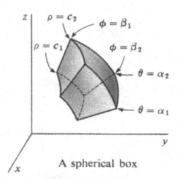

Figure 12.7.14

A spherical box

The spherical box

$$0 \leq \theta \leq 2\pi, \qquad 0 \leq \phi \leq \pi, \qquad 0 \leq \rho \leq c$$

is a sphere of radius c with center at the origin.

The spherical box

$$0 \leq \theta \leq 2\pi, \qquad 0 \leq \phi \leq \beta, \qquad 0 \leq \rho \leq c$$

is a cone whose vertex is at the origin and whose top is spherical instead of flat. (See Figure 12.7.15.)

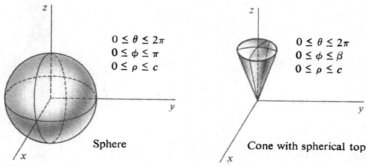

Sphere

Cone with spherical top

Figure 12.7.15

Another important example is the spherical region

$$0 \leq \theta \leq 2\pi, \qquad 0 \leq \phi \leq \pi/2, \qquad 0 \leq \rho \leq c \cos \phi,$$

which is a sphere of radius $\frac{1}{2}c$ whose center is on the z-axis at $z = \frac{1}{2}c$ (Figure 12.7.16).

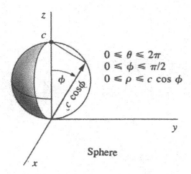

$$0 \leq \theta \leq 2\pi$$
$$0 \leq \phi \leq \pi/2$$
$$0 \leq \rho \leq c \cos \phi$$

Sphere

Figure 12.7.16

When integrating over a solid region E made up of spheres or cones, it is often easiest to use spherical coordinates.

SPHERICAL INTEGRATION FORMULA

Let E be a spherical region

$$\alpha_1 \leq \theta \leq \alpha_2, \qquad \beta_1(\theta) \leq \phi \leq \beta_2(\theta), \qquad c_1(\theta, \phi) \leq \rho \leq c_2(\theta, \phi).$$

The triple integral of $f(x, y, z)$ over E is

$$\iiint_E f(x, y, z) \, dV = \int_{\alpha_1}^{\alpha_2} \int_{\beta_1(\theta)}^{\beta_2(\theta)} \int_{c_1(\phi,\theta)}^{c_2(\phi,\theta)} f(x, y, z) \rho^2 \sin \phi \, d\rho \, d\phi \, d\theta.$$

In practice we make the substitution

$$f(x, y, z) = f(\rho \sin \phi \cos \theta, \rho \sin \phi \sin \theta, \rho \cos \phi)$$

before integrating.

PROOF Let C be the region in the rectangular (θ, ϕ, ρ) space which has the same inequalities as E. We prove

$$\iiint_E f(x, y, z) \, dx \, dy \, dz = \iiint_C f(x, y, z) \rho^2 \sin \phi \, d\theta \, d\phi \, d\rho.$$

As usual we let $f(x, y, z) > 0$ on E and put

$$B(C_1) = \iiint_{E_1} f(x, y, z) \, dx \, dy \, dz.$$

Consider an element of volume ΔC. As we see from Figure 12.7.17, ΔC corresponds to a spherical box ΔE. ΔE is almost a rectangular box with sides

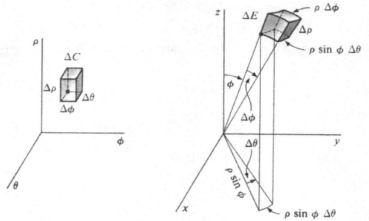

Figure 12.7.17   Spherical Element of Volume

$$\Delta\rho, \quad \rho\,\Delta\phi, \quad \rho\sin\phi\,\Delta\theta$$

and volume $\qquad\qquad \rho^2\sin\phi\,\Delta\theta\,\Delta\phi\,\Delta\rho.$

Thus $\qquad B(\Delta C) \approx f(x, y, z)\rho^2\sin\phi\,\Delta\theta\,\Delta\phi\,\Delta\rho \qquad$ (compared to $\Delta\theta\,\Delta\phi\,\Delta\rho$).

By the Infinite Sum Theorem

$$B(C) = \iiint_C f(x, y, z)\rho^2\sin\phi\,d\theta\,d\phi\,d\rho,$$

and by definition

$$B(C) = \iiint_E f(x, y, z)\,dx\,dy\,dz.$$

The triple integral for volume,

$$V = \iiint_E dV,$$

gives us iterated integral formulas for volume in rectangular, cylindrical, and spherical coordinates.

Rectangular $\qquad\qquad V = \int_{a_1}^{a_2} \int_{b_1(x)}^{b_2(x)} \int_{c_1(x,y)}^{c_2(x,y)} dz\,dy\,dx.$

Cylindrical $\qquad\qquad V = \int_{\alpha}^{\beta} \int_{a(\theta)}^{b(\theta)} \int_{c_1(\theta,r)}^{c_2(\theta,r)} r\,dz\,dr\,d\theta.$

Spherical $\qquad\qquad V = \int_{\alpha_1}^{\alpha_2} \int_{\beta_1(\theta)}^{\beta_2(\theta)} \int_{c_1(\theta,\phi)}^{c_2(\theta,\phi)} \rho^2\sin\phi\,d\rho\,d\phi\,d\theta.$

The rectangular formula is really equivalent to the double integral for the volume between two surfaces. Similarly, the cylindrical formula is equivalent to the double integral in polar coordinates for the volume between two surfaces.

On the other hand, the volume formula in spherical coordinates is something new which is useful for finding volumes of spherical regions.

EXAMPLE 3 Find the volume of the region above the cone $\phi = \beta$ and inside the sphere $\rho = c$.

The region, shown in Figure 12.7.18, is given by

$$0 \le \theta \le 2\pi, \qquad 0 \le \phi \le \beta, \qquad 0 \le \rho \le c.$$

$$V = \int_0^{2\pi} \int_0^\beta \int_0^c \rho^2 \sin \phi \, d\rho \, d\phi \, d\theta$$

$$= \int_0^{2\pi} \int_0^\beta \frac{c^3}{3} \sin \phi \, d\phi \, d\theta$$

$$= \int_0^{2\pi} (1 - \cos \beta) \frac{c^3}{3} \, d\theta = \frac{2\pi}{3} (1 - \cos \beta) c^3.$$

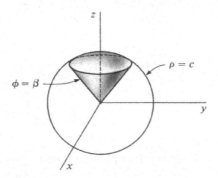

Figure 12.7.18

EXAMPLE 4 A sphere of diameter a passes through the center of a sphere of radius b, and $a > b$. Find the volume of the region inside the sphere of diameter a and outside the sphere of radius b.

Step 1 The region is sketched in Figure 12.7.19.

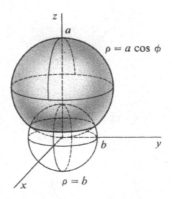

Figure 12.7.19

Step 2 We let the z-axis be the line through the two centers and put the origin at the center of the sphere of radius b. The two spheres have the spherical equations

$$\rho = a \cos \phi, \qquad \rho = b.$$

They intersect at $\cos \phi = \dfrac{b}{a}.$

Thus E is the region

$$0 \le \theta \le 2\pi, \qquad 0 \le \phi \le \arccos \frac{b}{a}, \qquad b \le \rho \le a \cos \phi.$$

Step 3 $V = \displaystyle\int_{0}^{2\pi} \int_{0}^{\arccos (b/a)} \int_{b}^{a \cos \phi} \rho^2 \sin \phi \, d\rho \, d\phi \, d\theta$

$$= \int_{0}^{2\pi} \int_{0}^{\arccos (b/a)} \frac{1}{3}(a^3 \cos^3 \phi - b^3) \sin \phi \, d\phi \, d\theta.$$

Put $u = \cos \phi, du = -\sin \phi \, d\phi$. Then

$$V = \int_{0}^{2\pi} \int_{1}^{b/a} -\frac{1}{3}(a^3 u^3 - b^3) \, du \, d\theta$$

$$= \frac{1}{3} \int_{0}^{2\pi} \int_{b/a}^{1} a^3 u^3 - b^3 \, du \, d\theta$$

$$= \frac{1}{3} \int_{0}^{2\pi} \frac{a^3}{4} - b^3 + \frac{3}{4} \frac{b^4}{a} \, d\theta = \frac{\pi}{6}\left(a^3 - 4b^3 + 3\frac{b^4}{a}\right).$$

EXAMPLE 5   Find the mass of a sphere of radius c whose density is equal to the distance from the surface. The sphere is shown in Figure 12.7.20.

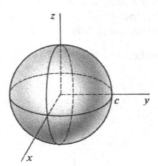

Figure 12.7.20

Put the center at the origin. The sphere is then given by

$$0 \le \theta \le 2\pi, \qquad 0 \le \phi \le \pi, \qquad 0 \le \rho \le c.$$

The density at (θ, ϕ, ρ) is

$$\text{density} = c - \rho.$$

The mass is

$$
\begin{aligned}
m &= \int_0^{2\pi} \int_0^{\pi} \int_0^c (c - \rho)\rho^2 \sin \phi \, d\rho \, d\phi \, d\theta \\
&= \int_0^{2\pi} \int_0^{\pi} \left(\frac{c}{3}c^3 - \frac{1}{4}c^4\right) \sin \phi \, d\phi \, d\theta \\
&= \int_0^{2\pi} \int_0^{\pi} \frac{1}{12}c^4 \sin \phi \, d\phi \, d\theta \\
&= \int_0^{2\pi} \frac{1}{6}c^4 \, d\theta = \frac{\pi}{3}c^4.
\end{aligned}
$$

PROBLEMS FOR SECTION 12.7

In Problems 1–6, evaluate the integral using cylindrical coordinates.

1    $\iiint_E \sqrt{x^2 + y^2} z \, dV$, E is the cylinder $x^2 + y^2 \leq 1, 0 \leq z \leq 2$

2    $\iiint_E x^2 + z \, dV$, E is the cylinder $x^2 + y^2 \leq 9, 0 \leq z \leq 6$

3    $\iiint_E x^2 + y^2 \, dV$, E is the cone $x^2 + y^2 \leq 1, 0 \leq z \leq 1 - \sqrt{x^2 + y^2}$

4    $\iiint_E 4 + \sqrt{z} \, dV$, E is the cone $x^2 + y^2 \leq 1, \sqrt{x^2 + y^2} \leq z \leq 1$

5    $\iiint_E (x + y)z \, dV$, E is the region $0 \leq x \leq 2, 0 \leq y \leq \sqrt{4 - x^2}, 0 \leq z \leq x^2 + y^2$

6    $\iiint_E \frac{z}{\sqrt{x^2 + y^2}} \, dV$, E is the region $1 \leq x^2 + y^2 \leq 4, 0 \leq z \leq |x|$

7   Find the mass of an object in the shape of a cylinder of radius b and height h whose density is equal to the distance from the axis.

8   Find the mass of an object in the shape of a cylinder of radius b and height h whose density is equal to the distance from the base.

9   Find the mass of an object in the shape of a cone of radius b and height h whose density is equal to the square of the distance from the axis.

10   Find the mass of an object in the shape of a cone of radius b and height h whose density is equal to the sum of the distance from the base and the distance from the axis.

11   Find the center of mass of an object of constant density filling the region above the paraboloid $z = x^2 + y^2$ and below the plane $z = 1$.

12   Find the center of mass of an object of constant density filling the region
$$
x^2 + y^2 \leq b, \qquad 0 \leq z \leq \sqrt{x^2 + y^2}.
$$

13   Find the moment of inertia of an object of constant density k in the cylinder $0 \leq r \leq b$, $-c \leq z \leq c$, about the x-axis.

14   Find the moment of inertia of an object of constant density k in the cylindrical shell $a \leq r \leq b, -c \leq z \leq c$, about the z-axis.

15   Find the moment of inertia of an object of constant density k in a cone of radius b and height h about its axis.

16   Find the moment of inertia of an object of constant density k in a cone of radius b and height h about a line through its apex and perpendicular to its axis.

In Problems 17–24, evaluate the integral using spherical coordinates.

17    $\iiint_E x^2 + y^2 + z^2 \, dV,$ E is the sphere $x^2 + y^2 + z^2 \le b^2$

18    $\iiint_E \sqrt{x^2 + y^2 + z^2} \, dV,$ E is the sphere $x^2 + y^2 + z^2 \le b^2$

19    $\iiint_E x^2 \, dV,$ E is the sphere $x^2 + y^2 + z^2 \le 1$

20    $\iiint_E z^2 \, dV,$ E is the sphere $x^2 + y^2 + z^2 \le 1$

21    $\iiint_E z \, dV,$ E is the sphere $\rho \le 2b \cos \phi$

22    $\iiint_E (x^2 + y^2 + z^2)^{3/2} \, dV,$ E is the intersection of the spheres $\rho \le 2b \cos \phi, \rho \le b$

23    $\iiint_E z\sqrt{x^2 + y^2 + z^2} \, dV,$ E is the region above the cone $\phi = \alpha$ and inside the sphere $\rho = b$

24    $\iiint_E \dfrac{1}{x^2 + y^2 + z^2} \, dV,$ E is the spherical shell $a \le \rho \le b$

25   Find the volume of the spherical shell $a \le \rho \le b$.

26   Find the volume of the spherical box $\alpha_1 \le \theta \le \alpha_2, \beta_1 \le \phi \le \beta_2, c_1 \le \rho \le c_2$.

27   Find the volume of the region above the cone $\phi = \beta$ and inside the sphere $\rho = b \cos \phi$.

28   Find the volume of the spherical region $0 \le \theta \le 2\pi, 0 \le \phi \le \pi, 0 \le \rho \le \sin \phi$.

29   Find the mass of an object in the shape of a sphere of radius c whose density is equal to the distance from the center.

30   Find the mass of a spherical shell $a \le \rho \le b$ whose density is equal to the reciprocal of the distance from the center.

31   Find the moment of inertia of a spherical object of radius b and constant density k about a diameter of the sphere.

32   Find the moment of inertia of a spherical shell $a \le \rho \le b$ of constant density k about any diameter.

33   A hole of radius a is bored through a sphere of radius b, and the surface of the hole passes through the center of the sphere, $a = \frac{1}{2}b$. Find the volume removed.

34   A hole of radius a is bored through a cone of height h and base of radius b, and the axis of the cone is on the surface of the hole ($a \le \frac{1}{2}b$). Find the volume removed.

35   Find the center of mass of a hemisphere of constant density and radius b.

36   Find the moment of inertia of an object of constant density k in the ellipsoid

$$\frac{x^2}{a^2} + \frac{y^2}{b^2} + \frac{z^2}{c^2} = 1$$

about the z-axis. *Hint*: Change variables $x_1 = x/a, y_1 = y/b, z_1 = z/c$ and use spherical coordinates.

EXTRA PROBLEMS FOR CHAPTER 12

1 Compute the Riemann sum
$$\sum\sum_D x^2 - y^2 \, \Delta x \, \Delta y, \Delta x = \tfrac{1}{2}, \Delta y = \tfrac{1}{2}, \quad D: -2 \le x \le 2, -2 \le y \le 2.$$

2 Compute the Riemann sum
$$\sum\sum_D x^2 - \sqrt{y} \, \Delta x \, \Delta y, \Delta x = \tfrac{1}{3}, \Delta y = \tfrac{1}{4}, \quad D: -1 \le x \le 1, 0 \le y \le 1 - x^2.$$

3 Evaluate $\displaystyle\iint_D x^2 - y^2 \, dA, \quad D: -2 \le x \le 2, -2 \le y \le 2.$

4 Evaluate $\displaystyle\iint_D x^2 - \sqrt{y} \, dA, \quad D: -1 \le x \le 1, 0 \le y \le 1 - x^2.$

5 Evaluate $\displaystyle\int_0^{\sqrt{2}/2} \int_{-\sqrt{1-2x^2}}^{\sqrt{1-2x^2}} x \, dy \, dx.$

6 Evaluate $\displaystyle\int_0^1 \int_{\sqrt{x}}^1 y e^x \, dy \, dx.$

7 Find the volume of the solid over the region $-1 \le x \le 1, 0 \le y \le 1 - x^2$ and between the surfaces $z = 0, z = 1 - y.$

8 Find the volume of the solid over the region $x^2 + y^2 = 4$ and between the surfaces $z = 0$ and $z = y^2 + x + 2.$

9 Find the volume of the solid $1 \le x \le 2, 0 \le y \le \ln x, y/x \le z \le 1/x.$

10 Find the volume of the solid $x^2 + y^2 \le 1, x^2 y^3 \le z \le 1.$

11 Find the volume of the solid bounded by the planes
$$x = 1, \qquad x = y, \qquad z = x + y, \qquad z = x + 2.$$

12 Find the volume of the solid bounded by the cylinders
$$x^2 + y^2 = 1, \quad x^2 + z^2 = 1.$$

13 Find the mass, center of mass, and moment of inertia about the origin of the plane object
$$0 \le x \le \pi, \qquad 0 \le y \le \sin x, \qquad \rho(x, y) = k.$$

14 Find the mass, center of mass, and moment of inertia about the origin of the plane object
$$0 \le x \le 1, \qquad x \le y \le 1, \qquad \rho(x, y) = x^2 y.$$

15 A circular disc filling the region $x^2 + y^2 \le r^2$ has density $\rho(x, y) = y^2.$ Find the mass, center of mass, and moment of inertia about the origin.

16 A semicircular object on the region
$$-r \le x \le r, \qquad 0 \le y \le \sqrt{r^2 - x^2}$$
has density $\rho(x, y) = y.$ Find the work required to stand the object up on its flat side.

17 Using polar coordinates, find the volume of the solid
$$x^2 + y^2 \le 9, \qquad y \le z \le x + 5.$$

18 Find the volume of the solid over the region $0 \le r \le 3 + \cos \theta$ between the plane $z = 0$ and the cone $z = r.$

19 Find the volume of the solid over the circle $0 \le r \le a$ between the plane $z = 0$ and the surface $z = 1/r.$

20 Find the mass and the moment of inertia about the origin of a semicircular object $0 \le r \le 1, 0 \le \theta \le \pi$ whose density is $\rho(r, \theta) = r\theta.$

21 A plane object covers the circle $0 \le r \le a$ and its density depends only on the distance r from the center, $\rho(r, \theta) = f(r).$ Show that the center of mass is at the origin.

22 Evaluate $\displaystyle\int_0^\pi \int_0^{\pi/2} \int_0^1 z \sin x + z \cos y \, dz \, dy \, dx.$

23 Evaluate $\int_0^1 \int_0^{x^2} \int_0^{y^2} x + y + z \, dz \, dy \, dx$.

24 Evaluate the triple integral

$$\iiint_E \frac{y+z}{x} \, dV, \quad E: 1 \leq x \leq 4, 1 \leq y \leq x, 1 \leq z \leq y.$$

25 An object has constant density k in the region

$$E: 0 \leq x \leq 1, 0 \leq y \leq 1 - x, 0 \leq z \leq xy.$$

Find its center of mass and its moments of inertia about the coordinate axes.

26 Use cylindrical coordinates to evaluate $\iiint_E z \, dV$, where E is the region inside the cylinder $x^2 + y^2 = 1$ which is above the plane $z = 0$ and within the sphere $x^2 + y^2 + z^2 = 9$.

27 An object of constant density k has the shape of a parabolic bowl

$$0 \leq \theta \leq 2\pi, \qquad 0 \leq r \leq b, \qquad r^2 \leq z \leq r^2 + c.$$

Find its center of mass and its moment of inertia about the z-axis.

28 Use spherical coordinates to evaluate the integral

$$\iiint_E x + y + z \, dV,$$

E is the spherical octant

$$x^2 + y^2 + z^2 \leq 1, \qquad 0 \leq x, \qquad 0 \leq y, \qquad 0 \leq z.$$

29 A spherical shell $a \leq \rho \leq b$ has density equal to the distance from the center. Find its mass and its moment of inertia about a diameter.

☐ **30** Prove that the double Riemann sum $\sum\sum_D f(x, y) \, dx \, dy$ is finite whenever $f(x, y)$ is continuous, D is a closed region, and dx, dy are positive infinitesimals.

☐ **31** Suppose a plane object is symmetric about the x-axis, that is, it covers a region D of the form

$$D: a \leq x \leq b, -g(x) \leq y \leq g(x)$$

and has density $\rho(x, y) = \rho(x, -y)$. Prove that the center of mass is on the x-axis.

☐ **32** The moment of inertia about the x-axis of a point in the plane of mass m is $I_x = my^2$. Use the Infinite Sum Theorem to show that the moment of inertia about the x-axis of a plane object with density $\rho(x, y)$ in the region D is $I_x = \iint_D \rho(x, y)y^2 \, dA$.

☐ **33** The kinetic energy of a point of mass m moving at speed v is $KE = \frac{1}{2}mv^2$. A rigid object of density $\rho(x, y)$ in the plane region D is rotating about the origin with angular velocity ω (so a point at distance d from the origin has speed ωd). Use the Infinite Sum Theorem to show that the kinetic energy of the object is

$$KE = \iint_D \tfrac{1}{2}\omega^2(x^2 + y^2)\rho(x, y) \, dA = \tfrac{1}{2}\omega^2 I.$$

☐ **34** Suppose a plane object is symmetric about the origin; that is, it fills a polar region $0 \leq r \leq g(\theta)$, $-\pi \leq \theta \leq \pi$, such that $g(\theta \pm \pi) = g(\theta)$, and its density has the property $\rho(r, \theta) = \rho(r, \theta \pm \pi)$. Show that the center of mass is at the origin.

☐ **35** Use the Infinite Sum Theorem to show that if D is a polar region of the form $a \leq r \leq b$, $\alpha(r) \leq \theta \leq \beta(r)$, then

$$\iint_D f(x, y) \, dA = \int_a^b \int_{\alpha(r)}^{\beta(r)} f(r, \theta)r \, d\theta \, dr.$$

VECTOR CALCULUS

3.1 DIRECTIONAL DERIVATIVES AND GRADIENTS

The partial derivatives $\partial z/\partial x$ and $\partial z/\partial y$ are the rates of change of $z = f(x, y)$ as the point (x, y) moves in the direction of the x-axis and the y-axis. We now consider the rate of change of z as the point (x, y) moves in other directions.

Let $P(a, b)$ be a point in the (x, y) plane and let

$$\mathbf{U} = \cos \alpha \mathbf{i} + \sin \alpha \mathbf{j}$$

be a unit vector, α is the angle from the x-axis to $\mathbf{U}$ (see Figure 13.1.1). The line through P with direction vector $\mathbf{U}$ has the vector equation

$$\mathbf{X} = \mathbf{P} + t\mathbf{U}$$

or in parametric form,

(1) $$x = a + t \cos \alpha, \qquad y = b + t \sin \alpha.$$

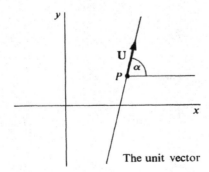

Figure 13.1.1

The unit vector

At $t = 0$ we have $x = a$ and $y = b$. If we intersect the surface $z = f(x, y)$ with the vertical plane through the line (Equation 1), we obtain the curve

$$z = f(a + t \cos \alpha, b + t \sin \alpha) = F(t).$$

The slope $dz/dt = F'(0)$ of this curve at $t = 0$ is called the *slope* or *derivative of f in the U direction* and is written $f_U(a, b)$ (Figure 13.1.2).

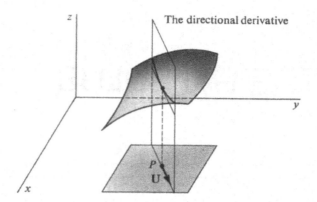

The directional derivative

Figure 13.1.2

Here is the precise definition.

DEFINITION

Given a function $z = f(x, y)$ and a unit vector $\mathbf{U} = \cos\alpha\mathbf{i} + \sin\alpha\mathbf{j}$, the *derivative of f in the U direction* is the limit

$$f_U(a, b) = \lim_{t \to 0} \frac{f(a + t\cos\alpha, b + t\sin\alpha) - f(a, b)}{t}.$$

$f_U(a, b)$ is called a *directional derivative* of f at (a, b).

The partial derivatives of $f(x, y)$ are equal to the derivatives of $f(x, y)$ in the **i** and **j** directions:

$$f_x(a, b) = \lim_{\Delta x \to 0} \frac{f(a + \Delta x, b) - f(a, b)}{\Delta x}$$

$$= \lim_{t \to 0} \frac{f(a + t\cos 0, b + t\sin 0) - f(a, b)}{t} = f_i(a, b).$$

$$f_y(a, b) = \lim_{\Delta y \to 0} \frac{f(a, b + \Delta y) - f(a, b)}{\Delta y}$$

$$= \lim_{t \to 0} \frac{f\left(a + t\cos\frac{\pi}{2}, b + t\sin\frac{\pi}{2}\right) - f(a, b)}{t} = f_j(a, b).$$

EXAMPLE 1   Find the derivative of $f(x, y) = xy + y^2$ in the direction of the unit vector

$$\mathbf{U} = \frac{\sqrt{3}}{2}\mathbf{i} + \frac{1}{2}\mathbf{j}.$$

$$f_{\mathbf{U}}(x, y) = \lim_{t \to 0} \frac{f\left(x + \dfrac{\sqrt{3}}{2}t, y + \dfrac{1}{2}t\right) - f(x, y)}{t}$$

$$= \lim_{t \to 0} \frac{\left(x + \dfrac{\sqrt{3}}{2}t\right)\left(y + \dfrac{1}{2}t\right) + \left(y + \dfrac{1}{2}t\right)^2 - (xy + y^2)}{t}$$

$$= \lim_{t \to 0} \frac{\dfrac{1}{2}xt + \dfrac{\sqrt{3}}{2}yt + \dfrac{\sqrt{3}}{4}t^2 + yt + \dfrac{1}{4}t^2}{t}$$

$$= \lim_{t \to 0} \frac{1}{2}x + \frac{\sqrt{3}}{2}y + \frac{\sqrt{3}}{4}t + y + \frac{1}{4}t$$

$$= \frac{1}{2}x + \left(\frac{\sqrt{3}}{2} + 1\right)y.$$

There is an easier way to find the directional derivatives of $f(x, y)$ using the partial derivatives. It is convenient to combine the partial derivatives into a vector called the gradient of f.

DEFINITION

The **gradient** of a function $z = f(x, y)$, denoted by **grad** z or **grad** f, is defined by

$$\mathbf{grad}\ z = \frac{\partial z}{\partial x}\mathbf{i} + \frac{\partial z}{\partial y}\mathbf{j}.$$

In functional notation,

$$\mathbf{grad}\ f = f_x(a, b)\mathbf{i} + f_y(a, b)\mathbf{j}.$$

Thus **grad** f is the vector valued function of two variables whose x and y components are the partial derivatives f_x and f_y (Figure 13.1.3). Sometimes the notation ∇f or ∇z is used for the gradient.

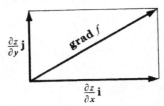

Figure 13.1.3

THEOREM 1

Suppose $z = f(x, y)$ is smooth at (a, b). Then for any unit vector $\mathbf{U} = \cos \alpha \mathbf{i} + \sin \alpha \mathbf{j}$, the directional derivative $f_{\mathbf{U}}(a, b)$ exists and

$$f_{\mathbf{U}}(a, b) = \mathbf{U} \cdot \mathbf{grad}\ f = \frac{\partial z}{\partial x}\cos \alpha + \frac{\partial z}{\partial y}\sin \alpha.$$

PROOF Let $\mathbf{U} = \cos \alpha \mathbf{i} + \sin \alpha \mathbf{j}$. Write x, y, and z as functions of t,

$$x = a + t \cos \alpha, \qquad y = a + t \sin \alpha,$$
$$z = f(a + t \cos \alpha, b + t \sin \alpha).$$

Then by the Chain Rule,

$$f_{\mathbf{U}}(a, b) = \frac{dz}{dt} = \frac{\partial z}{\partial x} \frac{dx}{dt} + \frac{\partial z}{\partial y} \frac{dy}{dt} = \frac{\partial z}{\partial x} \cos \alpha + \frac{\partial z}{\partial y} \sin \alpha.$$

EXAMPLE 2   Find the gradient of $f(x, y) = xy + y^2$ and use it to find the derivative in the direction of

$$\mathbf{U} = \frac{\sqrt{3}}{2}\mathbf{i} + \frac{1}{2}\mathbf{j}.$$
$$f_x(x, y) = y, \qquad f_y(x, y) = x + 2y.$$
$$\mathbf{grad}\, f(x, y) = y\mathbf{i} + (x + 2y)\mathbf{j}.$$
$$f_{\mathbf{U}}(x, y) = \frac{\sqrt{3}}{2}y + \frac{1}{2}(x + 2y) = \frac{1}{2}x + \left(\frac{\sqrt{3}}{2} + 1\right)y.$$

We can use Theorem 1 to give a geometric interpretation of the gradient vector. Let us assume that $f(x, y)$ is smooth at a point (a, b), and see what happens to the directional derivatives $f_{\mathbf{U}}(a, b)$ as the unit vector $\mathbf{U}$ varies. If both partial derivatives $f_x(a, b)$ and $f_y(a, b)$ are zero, then the gradient vector and hence all the directional derivatives are zero. Suppose the partial derivatives are not both zero, whence $\mathbf{grad}\, f \neq \mathbf{0}$. Then

$$f_{\mathbf{U}} = \mathbf{U} \cdot \mathbf{grad}\, f = |\mathbf{grad}\, f| \cos \theta$$

where θ is the angle between $\mathbf{U}$ and $\mathbf{grad}\, f$. Therefore $f_{\mathbf{U}}$ is a maximum when $\cos \theta = 1$ and $\theta = 0$, a minimum when $\cos \theta = -1$ and $\theta = \pi$, and zero when $\cos \theta = 0$ and $\theta = \pi/2$. We have proved the following corollary.

COROLLARY 1

Suppose $z = f(x, y)$ is smooth and $\mathbf{grad}\, f \neq \mathbf{0}$ at (a, b). Then the length of $\mathbf{grad}\, f$ is the largest directional derivative of f, and the direction of $\mathbf{grad}\, f$ is the direction of the largest directional derivative of f.

On a surface $z = f(x, y)$, the direction of the gradient vector is called the direction of *steepest ascent*, and the direction opposite the gradient vector is called the direction of *steepest descent* (Figure 13.1.4).

COROLLARY 2

Suppose $z = f(x, y)$ is smooth and $\partial z / \partial y \neq 0$ at (a, b). Then $\mathbf{grad}\, f$ is normal (perpendicular) to the level curve at (a, b). That is, $\mathbf{grad}\, f$ is perpendicular to the tangent line of the level curve (Figure 13.1.5).

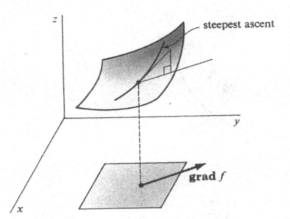

steepest ascent

grad f

Figure 13.1.4

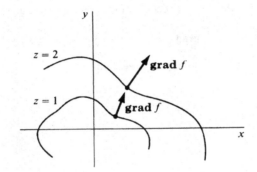

$z = 2$

grad f

$z = 1$

grad f

Figure 13.1.5

PROOF By the Implicit Function Theorem, the level curve

$$f(x, y) - f(a, b) = 0$$

has the tangent line

$$\frac{\partial z}{\partial x}(x - a) + \frac{\partial z}{\partial y}(y - b) = 0.$$

(a, b) is on this line. Let (x_0, y_0) be any other point on the line. Then

$$\mathbf{D} = (x_0 - a)\mathbf{i} + (y_0 - b)\mathbf{j}$$

is a direction vector of the line, and

$$\mathbf{D} \cdot \mathbf{grad}\, f = (x_0 - a)\frac{\partial z}{\partial x} + (y_0 - b)\frac{\partial z}{\partial y} = 0.$$

Thus **grad** f is perpendicular to the direction vector **D**.

Water always flows down a hill in the direction of steepest descent. Thus on a topographic map, the course of a river must always be perpendicular to the level curves, as in Figure 13.1.6.

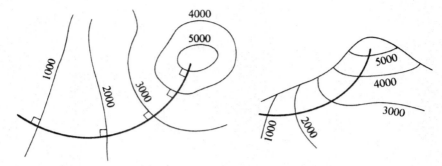

Figure 13.1.6

EXAMPLE 3   A ball is placed at rest on the surface $z = 2x^2 - 3y^2$ at the point $(2, 1, 5)$ (Figure 13.1.7). Which direction will the ball roll?

The ball will roll in direction of steepest descent, given by $-\textbf{grad}\ z$.

$$\textbf{grad}\ z = \frac{\partial z}{\partial x}\textbf{i} + \frac{\partial z}{\partial y}\textbf{j} = 4x\textbf{i} - 6y\textbf{j} = 8\textbf{i} - 6\textbf{j}.$$

$$-\textbf{grad}\ z = -8\textbf{i} + 6\textbf{j}.$$

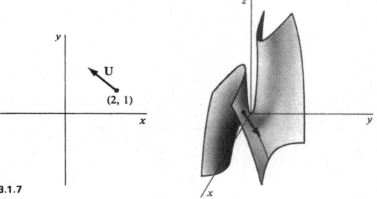

Figure 13.1.7

The unit vector in this direction is

$$\textbf{U} = \frac{-8\textbf{i} + 6\textbf{j}}{\sqrt{8^2 + 6^2}} = -\frac{8}{10}\textbf{i} + \frac{6}{10}\textbf{j}$$

Directional derivatives and gradients for functions of three variables are similar to the case of two variables.

DEFINITION

Given a real function $w = f(x, y, z)$ and a unit vector

$$\textbf{U} = \cos \alpha \textbf{i} + \cos \beta \textbf{j} + \cos \gamma \textbf{k}$$

in space, the derivative of f in the direction **U** *and the gradient of f at* (a, b, c)
are defined as follows.

$$f_{\mathbf{U}}(a, b, c) = \lim_{t \to 0} \frac{f(a + t\cos\alpha, b + t\cos\beta, c + t\cos\gamma) - f(a, b, c)}{t},$$

$$\mathbf{grad}\, w = \frac{\partial w}{\partial x}\mathbf{i} + \frac{\partial w}{\partial y}\mathbf{j} + \frac{\partial w}{\partial z}\mathbf{k}.$$

THEOREM 2

Suppose $w = f(x, y, z)$ *is smooth at* (a, b, c). *Then for any unit vector*

$$\mathbf{U} = \cos\alpha\mathbf{i} + \cos\beta\mathbf{j} + \cos\gamma\mathbf{k},$$

the directional derivative $f_{\mathbf{U}}(a, b, c)$ *exists and*

$$f_{\mathbf{U}}(a, b, c) = \mathbf{U} \cdot \mathbf{grad}\, f = \frac{\partial w}{\partial x}\cos\alpha + \frac{\partial w}{\partial y}\cos\beta + \frac{\partial w}{\partial z}\cos\gamma.$$

Corollaries 1 and 2 also hold for functions of three variables. In Corollary 2,
grad f is normal to the tangent plane of the level surface $f(x, y, z) - f(a, b, c) = 0$
at (a, b, c).

EXAMPLE 4   Given the function

$$w = z\cos x + z\sin y$$

at the point $(0, 0, 3)$, find the gradient vector and the derivative in the direction
of

$$\mathbf{U} = \frac{2}{3}\mathbf{i} - \frac{1}{3}\mathbf{j} + \frac{2}{3}\mathbf{k}.$$

$$\begin{aligned}
\mathbf{grad}\, w &= \frac{\partial w}{\partial x}\mathbf{i} + \frac{\partial w}{\partial y}\mathbf{j} + \frac{\partial w}{\partial z}\mathbf{k} \\
&= -z\sin x\mathbf{i} + z\cos y\mathbf{j} + (\cos x + \sin y)\mathbf{k} \\
&= -3\cdot\sin 0\mathbf{i} + 3\cdot\cos 0\mathbf{j} + (\cos 0 + \sin 0)\mathbf{k} \\
&= 3\mathbf{j} + \mathbf{k}.
\end{aligned}$$

$$f_{\mathbf{U}}(0, 0, 3) = \mathbf{U} \cdot \mathbf{grad}\, w = \frac{2}{3}\cdot 0 - \frac{1}{3}\cdot 3 + \frac{2}{3}\cdot 1 = -\frac{1}{3}.$$

EXAMPLE 5   Find a unit vector normal to the surface

$$z = x^2 + 2y^2 + 1$$

at $(1, 2, 10)$ shown in Figure 13.1.8.

Let $\qquad\qquad f(x, y, z) = -z + x^2 + 2y^2 + 1.$

By Corollary 2, **grad** f is normal to the given surface $-z + x^2 + 2y^2 + 1 = 0$.
We compute

$$\mathbf{grad}\, f = 2x\mathbf{i} + 4y\mathbf{j} - \mathbf{k}.$$

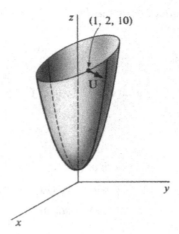

Figure 13.1.8

At $(1, 2, 10)$, **grad** $f = 2\mathbf{i} + 8\mathbf{j} - \mathbf{k}$. The required unit vector is found by dividing **grad** f by its length,

$$\mathbf{U} = \frac{2\mathbf{i} + 8\mathbf{j} - \mathbf{k}}{\sqrt{2^2 + 8^2 + 1^2}} = \frac{2\mathbf{i} + 8\mathbf{j} - \mathbf{k}}{\sqrt{69}}.$$

PROBLEMS FOR SECTION 13.1

In Problems 1–14, find the gradient vector, **grad** f, and the directional derivative $f_{\mathbf{U}}$.

1 $f(x, y) = x^2 + y^2, \quad \mathbf{U} = \dfrac{\mathbf{i} + \mathbf{j}}{\sqrt{2}}$

2 $f(x, y) = x^2 + y^2, \quad \mathbf{U} = \dfrac{\mathbf{i} - \mathbf{j}}{\sqrt{2}}$

3 $f(x, y) = x^2 y^3, \quad \mathbf{U} = \dfrac{3\mathbf{i} - 4\mathbf{j}}{5}$

4 $f(x, y) = x^2 y^3, \quad \mathbf{U} = \dfrac{3\mathbf{i} + 4\mathbf{j}}{5}$

5 $f(x, y) = \cos x \sin y, \quad \mathbf{U} = \dfrac{\mathbf{i} + 2\mathbf{j}}{\sqrt{5}}$

6 $f(x, y) = e^{ax+by}, \quad \mathbf{U} = \dfrac{a\mathbf{i} + b\mathbf{j}}{\sqrt{a^2 + b^2}}$

7 $f(x, y) = \sqrt{x^2 + y^2}, \quad \mathbf{U} = \dfrac{\mathbf{i} - \mathbf{j}}{\sqrt{2}}$

8 $f(x, y) = \sqrt{x^2 - y^2}, \quad \mathbf{U} = \dfrac{4\mathbf{i} - 3\mathbf{j}}{5}$

9 $f(x, y, z) = xyz, \quad \mathbf{U} = \dfrac{\mathbf{i} + 2\mathbf{j} - 2\mathbf{k}}{3}$

10 $f(x, y, z) = x^2 + y^2 + z^2, \quad \mathbf{U} = \dfrac{\mathbf{i} + \mathbf{j} + \mathbf{k}}{\sqrt{3}}$

11 $f(x, y, z) = \dfrac{1}{x} + \dfrac{2}{y} + \dfrac{3}{z}, \quad \mathbf{U} = \dfrac{\mathbf{i} - \mathbf{k}}{\sqrt{2}}$

12 $f(x, y, z) = \dfrac{1}{x} + \dfrac{2}{y} + \dfrac{3}{z}, \quad \mathbf{U} = \dfrac{\mathbf{i} - \mathbf{j} + \mathbf{k}}{\sqrt{3}}$

13 $f(x, y, z) = \sqrt{x^2 + y^2 + z^2}, \quad \mathbf{U} = \cos \alpha \mathbf{i} + \cos \beta \mathbf{j} + \cos \gamma \mathbf{k}$

14 $f(x, y, z) = Ax + By + Cz, \quad \mathbf{U} = \cos \alpha \mathbf{i} + \cos \beta \mathbf{j} + \cos \gamma \mathbf{k}$

15   Find the derivative of $z = \sqrt{x}/y$ at the point $(1, 1)$ in the direction $\mathbf{U} = (\mathbf{i} - 3\mathbf{j})/\sqrt{10}$.

16   Find the derivative of $z = 1/(x + y)$ at the point $(2, 3)$ in the direction $\mathbf{U} = (-\mathbf{i} - \mathbf{j})/\sqrt{2}$.

17   Find the derivative of $z = 2x^2 + xy - y^2$ at the point $(2, 1)$ in the direction $\mathbf{U} = a\mathbf{i} + b\mathbf{j}$.

18   Find the derivative of $w = \sqrt{xyz}$ at $(1, 1, 1)$ in the direction $\mathbf{U} = (2\mathbf{i} + \mathbf{j} + 2\mathbf{k})/3$.

19   Find the derivative of $w = \sqrt{4 - x^2 - y^2 - z^2}$ at $(1, 1, 1)$ in the direction $\mathbf{U} = (\mathbf{i} - \mathbf{j} + \mathbf{k})/\sqrt{3}$.

20   Find the direction of steepest ascent on the surface $z = 2x^2 + 3y^2$ at the point $(1, -1)$.

21   Find the direction of steepest descent on the surface $z = \sqrt{4 - x^2 - y^2}$ at the point $(1, 1)$.

22   Find a unit vector normal to the sphere $x^2 + y^2 + z^2 = 9$ at the point $(1, 2, 2)$.

23   Find a unit vector normal to the ellipsoid $\frac{1}{4}x^2 + y^2 + \frac{1}{2}z^2 = 3$ at the point $(2, 1, 3)$.

☐ **24**   Given a unit vector $\mathbf{U} = a\mathbf{i} + b\mathbf{j}$ and a function $z = f(x, y)$ with continuous second partials, find a formula for the second directional derivative $f_{\mathbf{UU}}(x, y)$, i.e., the derivative of $f_{\mathbf{U}}(x, y)$ in the direction $\mathbf{U}$.

☐ **25**   Given unit vectors $\mathbf{U} = u_1\mathbf{i} + u_2\mathbf{j}$ and $\mathbf{V} = v_1\mathbf{i} + v_2\mathbf{j}$, and a function $z = f(x, y)$ with continuous second partials, find a formula for the mixed second directional derivative $(f_{\mathbf{U}})_{\mathbf{V}}(x, y)$.

3.2. LINE INTEGRALS

There are two ways to generalize the integral to functions of two or more variables. One way is the line integral, which we shall study in this section. The other is the multiple integral, which was studied in Chapter 12.

The line integral can be motivated by the notion of *work* in physics. The work done by a constant force vector $\mathbf{F}$ acting along a directed line segment from A to B is the inner product

$$W = \mathbf{F} \cdot \mathbf{S}$$

where $\mathbf{S}$ is the vector from A to B (Figure 13.2.1).

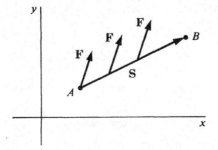

Figure 13.2.1

If the force vector

$$F(x, y) = P(x, y)\mathbf{i} + Q(x, y)\mathbf{j}$$

varies with x and y and acts along a curve C instead of a straight line S, the work turns out to be the line integral of $\mathbf{F}$ along the curve C (Figure 13.2.2).

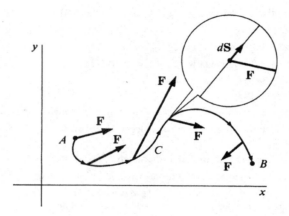

Figure 13.2.2

The intuitive idea of the line integral is an integral

$$W = \int_C \mathbf{F} \cdot d\mathbf{S}$$

of infinitesimal bits of work

$$dW = \mathbf{F} \cdot d\mathbf{S}$$

along infinitesimal pieces $d\mathbf{S}$ of the curve C. We now give a precise definition.
An *open rectangle* is a region of the plane of the form

$$a_1 < x < a_2, \qquad b_1 < y < b_2$$

where the a's and b's are either real numbers or infinity symbols (Figure 13.2.3).

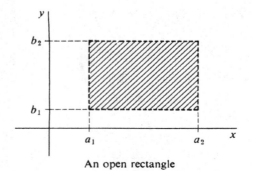

An open rectangle

Figure 13.2.3

DEFINITION

> A **smooth curve** from A to B is a curve C given by parametric equations
>
> $$x = g(s), \quad y = h(s) \quad 0 \le s \le L,$$
>
> where: $\quad A = (g(0), h(0)), \quad\quad B = (g(L), h(L)),$
>
> $L = $ length of curve,
>
> $s = $ length of the curve from A to (x, y),
>
> dx/ds and dy/ds are continuous for $0 \le s \le L$.

We call A the *initial point* and B the *terminal point* of C. A smooth curve from A to B is also called a *directed curve*, and is drawn with arrows.

Given s and an infinitesimal change $\Delta s = ds$, we let,

$$\Delta x = g(s + \Delta s) - g(s), \quad dx = g'(s)\, ds,$$
$$\Delta y = h(s + \Delta s) - h(s), \quad dy = h'(s)\, ds,$$
$$\Delta \mathbf{S} = \Delta x\mathbf{i} + \Delta y\mathbf{j}, \quad d\mathbf{S} = dx\mathbf{i} + dy\mathbf{j}.$$

Thus $\Delta \mathbf{S}$ is the vector from the point (x, y) to $(x + \Delta x, y + \Delta y)$ on C, and $d\mathbf{S}$ is an infinitesimal vector tangent to C at (x, y) (Figure 13.2.4).

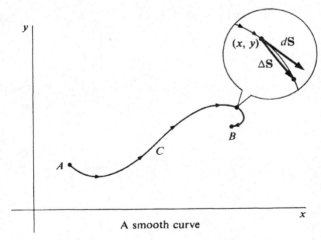

A smooth curve

Figure 13.2.4

DEFINITION

> Let $\quad\quad \mathbf{F}(x, y) = P(x, y)\mathbf{i} + Q(x, y)\mathbf{j}$
>
> be a continuous vector valued function on an open rectangle D and let C be a smooth curve in D. The **line integral** of **F** along C,
>
> $$\int_C \mathbf{F} \cdot d\mathbf{S} = \int_C P\, dx + Q\, dy,$$
>
> is defined as the definite integral

$$\int_0^L \left(P\frac{dx}{ds} + Q\frac{dy}{ds} \right) ds.$$

Notice that the inner product of **F** and *d***S** is

$$\mathbf{F} \cdot d\mathbf{S} = (P\mathbf{i} + Q\mathbf{j}) \cdot (dx\mathbf{i} + dy\mathbf{j}) = P\,dx + Q\,dy.$$

This is why we use both notations $\int_C \mathbf{F} \cdot d\mathbf{S}$ and $\int_C P\,dx + Q\,dy$ for the line integral.

DEFINITION

The **work** done by a continuous force vector $\mathbf{F}(x, y)$ along a smooth curve C is given by the line integral

$$W = \int_C \mathbf{F} \cdot d\mathbf{S}.$$

JUSTIFICATION We can justify this definition by using the Infinite Sum Theorem from Chapter 6. Let $W(u, v)$ be the work done along C from $s = u$ to $s = v$ (Figure 13.2.5). Then $W(u, v)$ has the Addition Property, because the work done from u to v plus the work done from v to w is the work done from u to w. On an infinitesimal piece of C from s to $s + \Delta s$, the work done is

$$\Delta W \approx \mathbf{F}(x, y) \cdot \Delta \mathbf{S} \approx \mathbf{F}(x, y) \cdot d\mathbf{S} \qquad \text{(compared to } \Delta s\text{)}.$$

But $\qquad \mathbf{F}(x, y) \cdot d\mathbf{S} = P\,dx + Q\,dy = \left(P\frac{dx}{ds} + Q\frac{dy}{ds} \right) ds.$

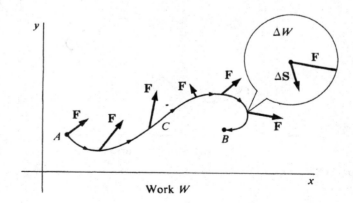

Figure 13.2.5   Work W

By the Infinite Sum Theorem,

$$W = \int_0^L \left(P\frac{dx}{ds} + Q\frac{dy}{ds} \right) ds = \int_C \mathbf{F} \cdot d\mathbf{S}.$$

The next theorem is useful for evaluating line integrals. It shows that any other parameter t can be used in place of the length s of the curve. Figure 13.2.6 illustrates the four parts of this theorem.

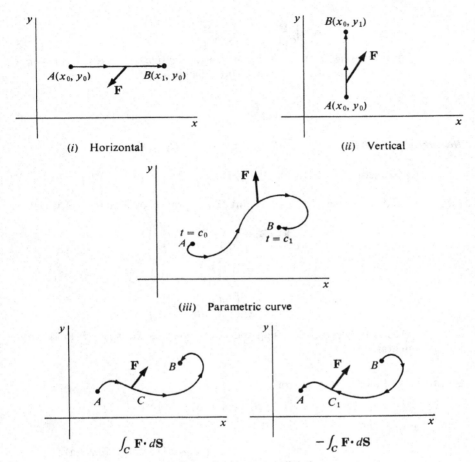

(i) Horizontal

(ii) Vertical

(iii) Parametric curve

$\int_C \mathbf{F} \cdot d\mathbf{S}$

$-\int_C \mathbf{F} \cdot d\mathbf{S}$

(iv) Reversing the curve direction

Figure 13.2.6

THEOREM

Let $\int_C \mathbf{F} \cdot d\mathbf{S}$ be a line integral.

(i) If C is a horizontal directed line segment $x_0 \leq x \leq x_1, y = y_0$, then

$$\int_C \mathbf{F} \cdot d\mathbf{S} = . \int_{x_0}^{x_1} P(x, y_0)\, dx.$$

(ii) If C is a vertical directed line segment $x = x_0, y_0 \leq y \leq y_1$, then

$$\int_C \mathbf{F} \cdot d\mathbf{S} = \int_{y_0}^{y_1} Q(x_0, y)\, dy.$$

(iii) If C is traced by a parametric curve

$$x = g(t), \qquad y = h(t), \qquad c_0 \leq t \leq c_1$$

where dx/dt and dy/dt are continuous, then

$$\int_C \mathbf{F} \cdot d\mathbf{S} = \int_{c_0}^{c_1} \left(P\frac{dx}{dt} + Q\frac{dy}{dt} \right) dt.$$

(iv) *Reversing the curve direction changes the sign of a line integral. That is, if C_1 is the curve C with its direction reversed, then*

$$\int_{C_1} \mathbf{F} \cdot d\mathbf{S} = - \int_C \mathbf{F} \cdot d\mathbf{S}.$$

Remark The integrals $\displaystyle\int_{x_0}^{x_1} P(x, y_0)\,dx, \qquad \int_{y_0}^{y_1} Q(x_0, y)\,dy$

are sometimes called *partial integrals*.

PROOF (i) and (ii) are special cases of (iii). (iii) is proved by a change of variables,

$$\int_C \mathbf{F} \cdot d\mathbf{S} = \int_0^L \left(P\frac{dx}{ds} + Q\frac{dy}{ds} \right) ds$$

$$= \int_{c_0}^{c_1} \left(P\frac{dx}{ds} + Q\frac{dy}{ds} \right) \frac{ds}{dt}\,dt$$

$$= \int_{c_0}^{c_1} \left(P\frac{dx}{dt} + Q\frac{dy}{dt} \right) dt.$$

(iv) is true because reversing the limits changes the sign of an ordinary integral.

EXAMPLE 1   Find the line integral of

$$\mathbf{F}(x, y) = \sin x \cos y\mathbf{i} + e^{xy}\mathbf{j}$$

along

(a)   The horizontal line $C_1 : 0 \le x \le \pi$, $y = \pi/3$ (Figure 13.2.7(a)).

(b)   The vertical line $C_2 : 0 \le y \le 1$, $x = 2$ (Figure 13.2.7(b)).

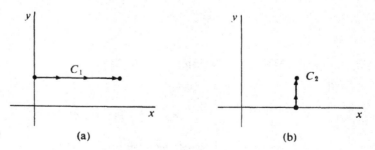

Figure 13.2.7   (a)   (b)

We use partial integrals.

(a) $\displaystyle\int_{C_1} \mathbf{F} \cdot d\mathbf{S} = \int_0^{\pi} \sin x \cos\frac{\pi}{3}\,dx$

$$= \int_0^{\pi} \tfrac{1}{2}\sin x\,dx = -\tfrac{1}{2}\cos x \Big]_0^{\pi} = -\tfrac{1}{2}(-1 - 1) = 1.$$

(b) $\int_{C_2} \mathbf{F} \cdot d\mathbf{S} = \int_0^1 e^{2y} \, dy = \frac{1}{2}e^{2y} \Big]_0^1 = \frac{1}{2}(e^2 - 1).$

Given two points A and B, there are infinitely many different smooth curves C from A to B. In general the value of a line integral will be different for different curves from A to B.

EXAMPLE 2   Let the force vector $\mathbf{F}$ be $\mathbf{F} = -y\mathbf{i} + x\mathbf{j}$.

$\mathbf{F}$ is perpendicular to the position vector $x\mathbf{i} + y\mathbf{j}$ but has the same length as $x\mathbf{i} + y\mathbf{j}$. Find the work done by $\mathbf{F}$ along the following curves, shown in Figure 13.2.8, from $(0, 0)$ to $(1, 1)$:

(a) C_1: The line $y = x, 0 \le x \le 1$.
(b) C_2: The parabola $y = x^2, 0 \le x \le 1$.
(c) C_3: The curve $y = \sqrt[3]{x}, 0 \le x \le 1$.

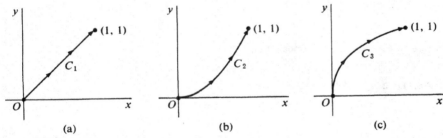

(a)   (b)   (c)

Figure 13.2.8

(a)   Put $x = t, y = t$.

$$W_1 = \int_{C_1} \mathbf{F} \cdot d\mathbf{S} = \int_{C_1} -y \, dx + x \, dy$$
$$= \int_0^1 (-t + t) \, dt = 0.$$

The work is zero because the force $\mathbf{F}$ is perpendicular to $d\mathbf{S}$ along C_1.

(b)   Put $x = t, y = t^2$.

$$W_2 = \int_{C_2} -y \, dx + x \, dy$$
$$= \int_0^1 (-t^2 + t \cdot 2t) \, dt = \int_0^1 t^2 \, dt = \frac{1}{3}.$$

(c)   Put $x = t^3, y = t$.

$$W_3 = \int_{C_3} -y \, dx + x \, dy = \int_0^1 (-t \cdot 3t^2 + t^3) \, dt$$
$$= \int_0^1 -2t^3 \, dt = -\frac{1}{2}.$$

A *piecewise smooth* curve is a curve C that can be broken into finitely many smooth pieces $C_1, C_2, \ldots, C_n$ where the terminal point of one piece is the initial point of the next (Figure 13.2.9). For example, a curve formed by two or more sides of a rectangle or a polygon is piecewise smooth. The *line integral* of $\mathbf{F}(x, y)$ over a piecewise smooth curve C is defined as the sum

$$\int_C \mathbf{F} \cdot d\mathbf{S} = \int_{C_1} \mathbf{F} \cdot d\mathbf{S} + \int_{C_2} \mathbf{F} \cdot d\mathbf{S} + \cdots + \int_{C_n} \mathbf{F} \cdot d\mathbf{S}.$$

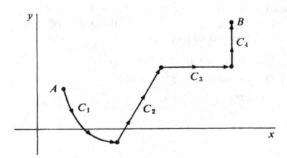

Figure 13.2.9   A Piecewise Smooth Curve from A to B

EXAMPLE 3   Find the line integral

$$\int_C xy\,dx + x^2y\,dy$$

where C is the rectangular curve from $(2, 5)$ to $(4, 5)$ to $(4, 6)$.

We see in Figure 13.2.10 that C is a piecewise smooth curve made up of a horizontal piece

$$C_1 : 2 \leq x \leq 4, \qquad y = 5$$

and a vertical piece

$$C_2 : x = 4, \qquad 5 \leq y \leq 6.$$

The line integral is the sum of two partial integrals,

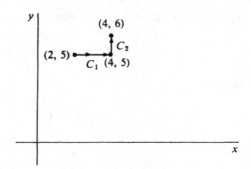

Figure 13.2.10

$$\int_C \mathbf{F} \cdot d\mathbf{S} = \int_{C_1} \mathbf{F} \cdot d\mathbf{S} + \int_{C_2} \mathbf{F} \cdot d\mathbf{S} = \int_2^4 x \cdot 5\, dx + \int_5^6 4^2 \cdot y\, dy$$

$$= 5 \cdot \tfrac{1}{2}x^2 \bigg]_2^4 + 16 \cdot \tfrac{1}{2}y^2 \bigg]_5^6 = 30 + 88 = 118.$$

A *simple closed curve* is a piecewise smooth curve whose initial and terminal points are equal and that does not cross or retrace its path. Examples of simple closed curves are the perimeters of a circle, a triangle, and a rectangle. The value of a line integral around a simple closed curve C depends on whether the length s is measured clockwise or counterclockwise, but does not depend on the initial point (Figure 13.2.11). The clockwise and counterclockwise line integrals of $\mathbf{F}$ around a simple closed curve C are denoted by

$$\oint_C \mathbf{F} \cdot d\mathbf{S}, \qquad \oint_C \mathbf{F} \cdot d\mathbf{S}.$$

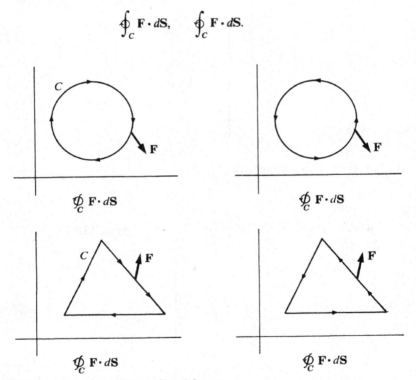

Figure 13.2.11 Integrals around Simple Closed Curves

THEOREM 2

If C is a simple closed curve, then

$$\oint_C \mathbf{F} \cdot d\mathbf{S} = -\oint_C \mathbf{F} \cdot d\mathbf{S}$$

and the values do not depend on the initial point of C.

PROOF The equation in Theorem 2 holds because reversing the direction of the curve changes the sign of the line integral. Suppose C has the initial point A, and its direction is clockwise. Let A_1 be any other point on C, and let C_1 and C_2 be as in Figure 13.2.12.

With the initial point A,

$$\oint_C \mathbf{F} \cdot d\mathbf{S} = \int_{C_1} \mathbf{F} \cdot d\mathbf{S} + \int_{C_2} \mathbf{F} \cdot d\mathbf{S}.$$

With the initial point A_1,

$$\oint_C \mathbf{F} \cdot d\mathbf{S} = \int_{C_2} \mathbf{F} \cdot d\mathbf{S} + \int_{C_1} \mathbf{F} \cdot d\mathbf{S}.$$

These are equal as required.

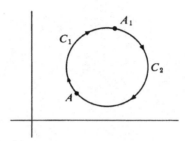

Figure 13.2.12

EXAMPLE 4   Find the line integral

$$\oint_C -y\,dx + x\,dy$$

where C is the circle $x^2 + y^2 = 4$, shown in Figure 13.2.13.

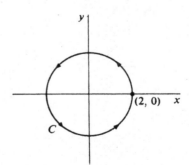

Figure 13.2.13

We may start at any point of C. Take $(2, 0)$ as the initial point. Then C has the parametric equations

$$x = 2\cos\theta, \qquad y = 2\sin\theta, \qquad 0 \le \theta \le 2\pi.$$

As θ goes from 0 to 2π, (x, y) goes around C once counterclockwise as required.

$$\oint_C -y\,dx + x\,dy = \int_0^{2\pi} \left(-y\frac{dx}{d\theta} + x\frac{dy}{d\theta} \right) d\theta$$

$$= \int_0^{2\pi} (-2\sin\theta(-2\sin\theta) + 2\cos\theta(2\cos\theta))\,d\theta$$

$$= \int_0^{2\pi} 4\sin^2\theta + 4\cos^2\theta\,d\theta = \int_0^{2\pi} 4\,d\theta = 8\pi.$$

Line integrals in space are developed in a similar way. Instead of an open rectangle we work in an *open rectangular solid*. A *smooth curve C* in space has three parametric equations with continuous derivatives,

$$x = g(s), \qquad y = h(s), \qquad z = l(s), \qquad 0 \le s \le L.$$

Given a continuous vector valued function

$$\mathbf{F}(x, y, z) = P(x, y, z)\mathbf{i} + Q(x, y, z)\mathbf{j} + R(x, y, z)\mathbf{k}$$

and a smooth curve C in space, we define the line integral of **F** along C, in symbols,

$$\int_C \mathbf{F} \cdot d\mathbf{S} = \int_C P\,dx + Q\,dy + R\,dz,$$

as

$$\int_0^L \left(P\frac{dx}{ds} + Q\frac{dy}{ds} + R\frac{dz}{ds} \right) ds.$$

EXAMPLE 5   Find the line integral

$$\int_C (x + y)\,dx + \frac{z}{x}\,dy + xy\,dz$$

along the spiral C given by

$$x = \cos t, \qquad y = \sin t, \qquad z = 2t, \qquad 0 \le t \le \frac{\pi}{2}.$$

The line integral is

$$\int_0^{\pi/2} (\cos t + \sin t)\,d(\cos t) + \frac{2t}{\cos t}\,d(\sin t) + (\cos t \sin t)\,d(2t)$$

$$= \int_0^{\pi/2} (-\cos t \sin t - \sin^2 t + 2t + 2\cos t \sin t)\,dt$$

$$= -\frac{1}{2}\sin^2 t - \left(\frac{1}{2}t - \frac{1}{2}\sin t \cos t \right) + t^2 + \sin^2 t \Big]_0^{\pi/2}$$

$$= \frac{1}{2}\sin^2 t - \frac{1}{2}t + \frac{1}{2}\sin t \cos t + t^2 \Big]_0^{\pi/2}$$

$$= \frac{1}{2} - \frac{\pi}{4} + \frac{\pi^2}{4}.$$

PROBLEMS FOR SECTION 13.2

Evaluate the following line integrals.

1 $\int_C xe^y\,dx + x^2y\,dy, \quad C: 0 \le x \le 2, y = 3$

2 $\int_C xe^y\,dx + x^2y\,dy, \quad C: 0 \le y \le 4, x = 4$

3 $\int_C xe^y\,dx + x^2y\,dy, \quad C: x = 3t, y = t^2, 0 \le t \le 1$

4 $\int_C xe^y\,dx + x^2y\,dy, \quad C: x = e^t, y = e^t, -1 \le t \le 1$

5 $\int_C (\cos x\mathbf{i} + \sin y\mathbf{j}) \cdot d\mathbf{S}, \quad C: x = t, y = t, 0 \le t \le 1$

6 $\int_C \left(\dfrac{\mathbf{i}}{xy} + \dfrac{\mathbf{j}}{x+y}\right) \cdot d\mathbf{S}, \quad C$ is the rectangular curve from $(1, 1)$ to $(3, 1)$ to $(3, 6)$.

7 $\int_C \left(\dfrac{\mathbf{i}}{xy} + \dfrac{\mathbf{j}}{x+y}\right) \cdot d\mathbf{S}, \quad C: x = 2t, y = 5t, 1 \le t \le 4$

8 $\int_C \left(\dfrac{\mathbf{i}}{xy} + \dfrac{\mathbf{j}}{x+y}\right) \cdot d\mathbf{S}, \quad C: x = t, y = t^2, 1 \le t \le 4$

9 $\oint_C y\,dx - x\,dy$ and $\oint_C y\,dx - x\,dy, \quad C: x^2 + y^2 = 1$

10 $\oint_C x^2y\,dx + xy^2\,dy, \quad C: x^2 + y^2 = 4$

11 $\oint_C (x + y)\,dx - 3xy\,dy, \quad C: x^2 + y^2 = 4$

12 $\oint_C (e^x \cos y\mathbf{i} + e^x \sin y\mathbf{j}) \cdot d\mathbf{S}, \quad C$ is the square with vertices $(0, 0), (1, 0), (1, 1), (0, 1)$.

13 $\oint_C (\sqrt{xy}\mathbf{i} + x^2y^2\mathbf{j}) \cdot d\mathbf{S}, \quad C$ is the triangle with vertices $(0, 0), (1, 1), (1, 0)$.

14 $\int_C yz\,dx + xz\,dy + xy\,dz, \quad C: x = t, y = t^2, z = t^3, 0 \le t \le 1.$

15 $\int_C yz\,dx + xz\,dy + xy\,dz, \quad C: x = \cos t, y = \sin t, z = \tan t, 0 \le t \le \pi/4.$

16 $\int_C (x\mathbf{i} + y\mathbf{j} + z\mathbf{k}) \cdot d\mathbf{S}, \quad C$ is the rectangular curve from $(0, 0, 0)$ to $(1, 0, 0)$ to $(1, 1, 0)$ to $(1, 1, 1)$.

17 Find the work done by the force $\mathbf{F} = (x\mathbf{i} + y\mathbf{j})/(x^2 + y^2)$ acting along a straight line from $(1, 1)$ to $(2, 5)$.

18 Find the work done by the force $\mathbf{F} = (\mathbf{i}/(y + 1)) - (\mathbf{j}/(x + 1))$ acting along the parabola $x = t, y = t^2, 0 \le t \le 1$.

19 Find the work done by the force $\mathbf{F} = x^2\mathbf{i} + y^2\mathbf{j} + z^2\mathbf{k}$ acting along a straight line from $(0, 0, 0)$ to $(3, 6, 10)$.

20 Find the work done by the force $\mathbf{F} = y\mathbf{i} + z\mathbf{j} + x\mathbf{k}$ along the curve $x = \sqrt{t}, y = 1/\sqrt{t}, z = t, 1 \le t \le 4$.

3.3 INDEPENDENCE OF PATH

For functions of one variable, the Fundamental Theorem of Calculus shows that the integral is the opposite of the derivative. In this section we shall see that the line integral is the opposite of the gradient.

By a *vector field* we mean a vector valued function

$$\mathbf{F}(x, y) = P(x, y)\mathbf{i} + Q(x, y)\mathbf{j}$$

where P and Q are smooth functions on an open rectangle D.

For example, if $f(x, y)$ has continuous second partials on D then its gradient **grad** f is a vector field.

Many vector fields are found in physics. Examples are gravitational force fields and magnetic force fields, in which a force vector $\mathbf{F}(x, y)$ is associated with each point (x, y). Another example is the flow velocity $\mathbf{V}(x, y)$ of a fluid. A vector field in economics is the demand vector

$$\mathbf{D}(x, y) = D_1(x, y)\mathbf{i} + D_2(x, y)\mathbf{j},$$

where $D_1(x, y)$ is the demand for commodity one and $D_2(x, y)$ is the demand for commodity two at the prices x for commodity one and y for commodity two. All of the examples above have analogues for three variables and three dimensions (and the demand vector for n commodities has n variables and n dimensions).

DEFINITION

$f(x, y)$ is a ***potential function*** of the vector field $P\mathbf{i} + Q\mathbf{j}$ if the gradient of f is $P\mathbf{i} + Q\mathbf{j}$.

Not every vector field has a potential function. Theorem 1 below shows which vector fields have potential functions, and Theorem 2 tells how to find a potential function when there is one.

Using the equality of mixed partials, we see that *if the vector field $P\mathbf{i} + Q\mathbf{j}$ has a potential function, then $\partial P/\partial y = \partial Q/\partial x$.* If f is a potential function of $P\mathbf{i} + Q\mathbf{j}$, we have

$$\mathbf{grad}\, f = \frac{\partial f}{\partial x}\mathbf{i} + \frac{\partial f}{\partial y}\mathbf{j} = P\mathbf{i} + Q\mathbf{j},$$

$$\frac{\partial P}{\partial y} = \frac{\partial^2 f}{\partial y\, \partial x} = \frac{\partial^2 f}{\partial x\, \partial y} = \frac{\partial Q}{\partial x}.$$

EXAMPLE 1   The vector field $-y\mathbf{i} + x\mathbf{j}$ has no potential function, because

$$\frac{\partial P}{\partial y} = \frac{\partial(-y)}{\partial y} = -1, \qquad \frac{\partial Q}{\partial x} = \frac{\partial x}{\partial x} = 1.$$

THEOREM 1

A vector field $P\mathbf{i} + Q\mathbf{j}$ has a potential function if and only if $\dfrac{\partial P}{\partial y} = \dfrac{\partial Q}{\partial x}.$

We have already proved one direction. We postpone the proof of the other direction until later.

By definition, **grad** $f = P\mathbf{i} + Q\mathbf{j}$ if and only if $df = P\,dx + Q\,dy$. In general, an expression $P\,dx + Q\,dy$ is called a *differential form*. A differential form is called an *exact differential* if it is equal to the total differential df of some function $f(x, y)$.

Using this terminology, Theorem 1 states that: $P\,dx + Q\,dy$ *is an exact differential if and only if* $\partial P/\partial y = \partial Q/\partial x$.

EXAMPLE 2   Test for existence of a potential function:

$$x^2 y\mathbf{i} + \sin x \cos y\mathbf{j}.$$

$$\frac{\partial P}{\partial y} = \frac{\partial(x^2 y)}{\partial y} = x^2,$$

$$\frac{\partial Q}{\partial x} = \frac{\partial(\sin x \cos y)}{\partial x} = \cos x \cos y.$$

There is no potential function.

EXAMPLE 3   Test for existence of a potential function:

$$3x^2 y^2\mathbf{i} + (y^2 + 2x^3 y)\mathbf{j}.$$

$$\frac{\partial P}{\partial y} = \frac{\partial(3x^2 y^2)}{\partial y} = 6x^2 y,$$

$$\frac{\partial Q}{\partial x} = \frac{\partial(y^2 + 2x^3 y)}{\partial x} = 6x^2 y.$$

There is a potential function.

THEOREM 2 (Path Independence Theorem)

Let $P\mathbf{i} + Q\mathbf{j}$ *be a vector field such that* $\partial P/\partial y = \partial Q/\partial x$ *and let A and B be two points of D.*

(i) *Let f be a potential function for* $P\mathbf{i} + Q\mathbf{j}$. *For any piecewise smooth curve C from A to B,*

$$\int_C P\,dx + Q\,dy = f(B) - f(A).$$

Since the line integral in this case depends only on the points A and B and not on the curve C (Figure 13.3.1), we write

$$\int_A^B P\,dx + Q\,dy = \int_C P\,dx + Q\,dy.$$

(ii) *g is a potential function for* $P\mathbf{i} + Q\mathbf{j}$ *if and only if g has the form*

$$g(x, y) = \int_A^{(x,y)} P\,dx + Q\,dy + K$$

for some constant K.

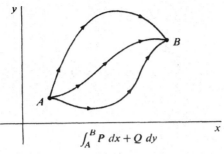

$$\int_A^B P\,dx + Q\,dy$$

Figure 13.3.1   Independence of path

Theorem 2 is important in physics. A vector field of forces which has a potential function is called a *conservative force field*. The negative of a potential function for a conservative force field is called a *potential energy function*. Gravity, static electricity, and magnetism are conservative force fields. Part (i) of the theorem shows that the work done by a conservative force field along a curve depends only on the initial and terminal points of the curve and is equal to the decrease in potential energy.

Mathematically, Theorem 2 is like the Fundamental Theorem of Calculus. It shows that the line integral of **grad** f along any curve from A to B is equal to the change in the value of f from A to B. When $A = B$, we have an interesting consequence:

If $f(x, y)$ has continuous second partials then the line integral of the gradient of f around a simple closed curve is zero,

$$\oint_C \mathbf{grad}\, f \cdot d\mathbf{S} = 0.$$

Using part (ii), we can find a potential function $f(x, y)$ for a vector field $P\mathbf{i} + Q\mathbf{j}$ in three steps.

When to Use $$\frac{\partial P}{\partial y} = \frac{\partial Q}{\partial x} \text{ on } D.$$

Step 1 Choose an initial point $A(a, b)$ in D.

Step 2 Choose and sketch a piecewise smooth curve C from A to an arbitrary point $X(x_0, y_0)$.

Step 3 Compute $f(x_0, y_0)$ by evaluating the line integral

$$f(x_0, y_0) = \int_C P\,dx + Q\,dy.$$

We postpone the proof of Theorem 2 to the end of this section.

EXAMPLE 3 (Continued)   Find a potential function for the vector field

$$3x^2 y^2 \mathbf{i} + (y^2 + 2x^3 y)\mathbf{j}.$$

We have already shown that $\dfrac{\partial P}{\partial y} = \dfrac{\partial Q}{\partial x}$.

Step 1   Pick $(0, 0)$ for the initial point.

Step 2   Let C be the rectangular curve from $(0, 0)$ to $(0, y_0)$ to (x_0, y_0), shown in Figure 13.3.2.

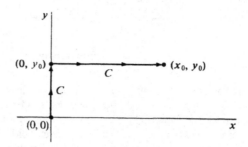

Figure 13.3.2

Step 3   A potential function is

$$f(x_0, y_0) = \int_C 3x^2 y^2 \, dx + (y^2 + 2x^3 y) \, dy$$

$$= \int_0^{y_0} (y^2 + 2 \cdot 0^3 y) \, dy + \int_0^{x_0} 3x^2 y_0^2 \, dx$$

$$= \tfrac{1}{3} y_0^3 + x_0^3 y_0^2.$$

$$f(x, y) = \tfrac{1}{3} y^3 + x^3 y^2.$$

As a check we may compute **grad** f.

$$\mathbf{grad}\, f = \frac{\partial f}{\partial x} \mathbf{i} + \frac{\partial f}{\partial y} \mathbf{j} = 3x^2 y^2 \mathbf{i} + (y^2 + 2x^3 y)\mathbf{j}.$$

We can get the same answer by choosing another curve in Step 2.

FIRST ALTERNATE SOLUTION

Step 2   Let C_1 be the rectangular curve from $(0, 0)$ to $(x_0, 0)$ to (x_0, y_0), shown in Figure 13.3.3.

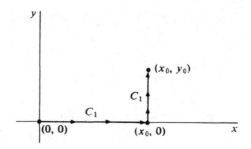

Figure 13.3.3

Step 3 $f(x_0, y_0) = \int_{C_1} 3x^2 y^2 \, dx + (y^2 + 2x^3 y) \, dy$

$= \int_0^{x_0} 3x^2 \cdot 0^2 \, dx + \int_0^{y_0} (y^2 + 2x_0^3 y) \, dy$

$= 0 + (\tfrac{1}{3} y_0^3 + x_0^3 y_0^2),$

$f(x, y) = \tfrac{1}{3} y^3 + x^3 y^2.$

SECOND ALTERNATE SOLUTION

Step 2 Let C_2 be the straight line from $(0, 0)$ to (x_0, y_0), shown in Figure 13.3.4. It has parametric equations

$$x = tx_0, \qquad y = ty_0, \qquad 0 \le t \le 1.$$

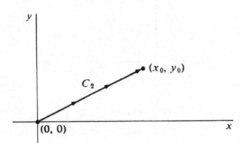

Figure 13.3.4

Step 3 $f(x_0, y_0) = \int_{C_2} 3x^2 y^2 \, dx + (y^2 + 2x^3 y) \, dy$

$= \int_0^1 [3(tx_0)^2 (ty_0)^2] x_0 + [(ty_0)^2 + 2(tx_0)^3 (ty_0)] y_0 \, dt$

$= \int_0^1 3x_0^3 y_0^2 t^4 + t^2 y_0^3 + 2t^4 x_0^3 y_0^2 \, dt$

$= \tfrac{3}{5} x_0^3 y_0^2 + \tfrac{1}{3} y_0^3 + \tfrac{2}{5} x_0^3 y_0^2 = x_0^3 y_0^2 + \tfrac{1}{3} y_0^3,$

$f(x, y) = x^3 y^2 + \tfrac{1}{3} y^3.$

EXAMPLE 4   An object at the origin $(0, 0)$ has a gravity force field with magnitude proportional to $1/(x^2 + y^2)$ and the direction of $-x\mathbf{i} - y\mathbf{j}$. Show that this force field is conservative and find a potential function.

The force vector is

$$\mathbf{F}(x, y) = \left(\frac{-x\mathbf{i} - y\mathbf{j}}{\sqrt{x^2 + y^2}} \right) \frac{k}{x^2 + y^2}$$

$$= -kx(x^2 + y^2)^{-3/2} \mathbf{i} - ky(x^2 + y^2)^{-3/2} \mathbf{j},$$

for some constant k. $\mathbf{F}(x, y)$ is undefined at $(0, 0)$ but is a vector field on the open rectangle $0 < x$.

$$\frac{\partial P}{\partial y} = -kx(2y)\left(-\frac{3}{2}\right)(x^2 + y^2)^{-5/2} = 3kxy(x^2 + y^2)^{-5/2}.$$

$$\frac{\partial Q}{\partial x} = -ky(2x)\left(-\frac{3}{2}\right)(x^2 + y^2)^{-5/2} = 3kxy(x^2 + y^2)^{-5/2}.$$

Therefore **F** is conservative.

Step 1   Take the initial point $(1, 0)$.

Step 2   Let C be the rectangular curve from $(1, 0)$ to $(1, y_0)$ to (x_0, y_0), shown in Figure 13.3.5.

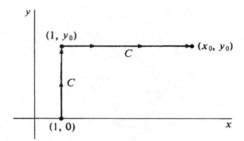

Figure 13.3.5

Step 3 $f(x_0, y_0) = \displaystyle\int_0^{y_0} -ky(1 + y^2)^{-3/2}\, dy + \int_1^{x_0} -kx(x^2 + y_0^2)^{-3/2}\, dx$

$$= k(1 + y^2)^{-1/2}\Big]_0^{y_0} + k(x^2 + y_0^2)^{-1/2}\Big]_1^{x_0}$$

$$= k(1 + y_0)^{-1/2} - k + k(x_0^2 + y_0^2)^{-1/2} - k(1 + y_0^2)^{-1/2}$$

$$= k(x_0^2 + y_0^2)^{-1/2} + \text{constant},$$

$$f(x, y) = \frac{k}{\sqrt{x^2 + y^2}} + \text{constant}.$$

Any choice of the constant will give a potential function. The same method works on the open rectangle $x < 0$.

An *exact differential equation* is an equation of the form

$$P(x, y)\, dx + Q(x, y)\, dy = 0,$$

where $\partial P/\partial y = \partial Q/\partial x$. Exact differential equations can be solved using Theorem 2.

EXAMPLE 5   Solve the differential equation

$$(x^2 + \sin y)\, dx + (x + 1)\cos y\, dy = 0.$$

First we test for exactness.

$$\frac{\partial(x^2 + \sin y)}{\partial y} = \cos y, \qquad \frac{\partial((x + 1)\cos y)}{\partial x} = \cos y.$$

Next we find a function with the given total differential. That is, we find a potential function for the vector field

$$(x^2 + \sin y)\mathbf{i} + (x + 1)\cos y\mathbf{j}.$$

Step 1   Take $(0, 0)$ for the initial point.

Step 2   Let C be the rectangular curve from $(0, 0)$ to $(0, y_0)$ to (x_0, y_0), shown in Figure 13.3.6.

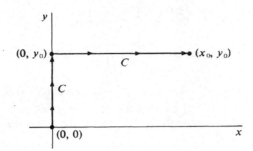

Figure 13.3.6

Step 3 $f(x_0, y_0) = \displaystyle\int_0^{y_0} (0 + 1) \cos y \, dy + \int_0^{x_0} (x^2 + \sin y_0) \, dx$

$\qquad\qquad = \sin y_0 + \tfrac{1}{3}x_0^3 + x_0 \sin y_0,$

$\quad f(x, y) = \sin y + \tfrac{1}{3}x^3 + x \sin y.$

Step 4 $f(x, y)$ is a constant k because $df = 0$. The general solution is

$$\tfrac{1}{3}x^3 + x \sin y + \sin y = k,$$

$$\sin y = \frac{k - \tfrac{1}{3}x^3}{x + 1},$$

$$y = \arcsin \frac{k - \tfrac{1}{3}x^3}{x + 1}, \quad k \text{ constant.}$$

We conclude this section with the proofs of Theorems 1 and 2. The proof of Theorem 1 uses a lemma about derivatives of partial integrals.

LEMMA

Suppose $P(x, y)$ is smooth on an open rectangle containing the point (a, b). Then

$$\frac{\partial}{\partial x} \int_a^x P(t, y) \, dt = P(x, y),$$

$$\frac{\partial}{\partial y} \int_a^x P(t, y) \, dt = \int_a^x \frac{\partial P}{\partial y}(t, y) \, dt.$$

PROOF The first formula follows at once from the Fundamental Theorem of Calculus. For the second formula, let Δy be a nonzero infinitesimal and let

$$z = \int_a^x P(t, y) \, dt.$$

As y changes to $y + \Delta y$, we have

$$\frac{\Delta z}{\Delta y} = \frac{\int_a^x P(t, y + \Delta y)\, dt - \int_a^x P(t, y)\, dt}{\Delta y}$$

$$= \int_a^x \frac{P(t, y + \Delta y) - P(t, y)}{\Delta y}\, dt \approx \int_a^x \frac{\partial P}{\partial y}(t, y)\, dt.$$

Taking standard parts, $\qquad \dfrac{\partial z}{\partial y} = \displaystyle\int_a^x \dfrac{\partial P}{\partial y}(t, y)\, dt.$

PROOF OF THEOREM 1   We must find a potential function for $P\mathbf{i} + Q\mathbf{j}$.

Assume $\partial P/\partial y = \partial Q/\partial x$. Pick a point (a, b) in D, and let $f(x_0, y_0)$ be the line integral of $P\mathbf{i} + Q\mathbf{j}$ on the rectangular curve C from (a, b) to (a, y_0) to (x_0, y_0) (Figure 13.3.7). Thus

$$f(x_0, y_0) = \int_C P\, dx + Q\, dy$$

$$= \int_b^{y_0} Q(a, y_0)\, dy + \int_a^{x_0} P(x, y_0)\, dx.$$

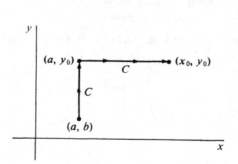

Figure 13.3.7

By the Lemma,

$$\frac{\partial f}{\partial x} = \frac{\partial}{\partial x}\left(\int_b^y Q(a, y)\, dy + \int_a^x P(x, y)\, dx \right) = P(x, y).$$

$$\frac{\partial f}{\partial y} = \frac{\partial}{\partial y}\left(\int_b^y Q(a, y)\, dy + \int_a^x P(x, y)\, dx \right)$$

$$= Q(a, y) + \int_a^x \frac{\partial P}{\partial y}(x, y)\, dx$$

$$= Q(a, y) + \int_a^x \frac{\partial Q}{\partial x}(x, y)\, dx$$

$$= Q(a, y) + [Q(x, y) - Q(a, y)] = Q(x, y).$$

Thus $\qquad \dfrac{\partial f}{\partial x} = P(x, y), \qquad \dfrac{\partial f}{\partial y} = Q(x, y),$

and $\qquad\qquad df = P\, dx + Q\, dy.$

PROOF OF THEOREM 2

(i) Let C have the parametric equations

$$x = g(t), \qquad y = h(t), \qquad c_1 \leq t \leq c_2.$$

Then $A = (g(c_1), h(c_1))$ and $B = (g(c_2), h(c_2))$. By the Chain Rule,

$$\frac{dz}{dt} = \frac{\partial z}{\partial x}\frac{dx}{dt} + \frac{\partial z}{\partial y}\frac{dy}{dt} = Pg'(t) + Qh'(t).$$

Then
$$\int_C P(x, y)\,dx + Q(x, y)\,dy = \int_{c_1}^{c_2} Pg'(t) + Qh'(t)\,dt$$

$$= \int_{c_1}^{c_2} \frac{dz}{dt}\,dt$$

$$= f(g(c_2), h(c_2)) - f(g(c_1), h(c_1))$$

$$= f(B) - f(A).$$

A similar computation works for piecewise smooth curves. This proves (i).

(ii) Define $f(x, y)$ by

$$f(X) = \int_A^X P\,dx + Q\,dy,$$

where $A = (a, b)$, $X = (x, y)$. Let C be the rectangular curve from (a, b) to (a, y) to (x, y). Then

$$f(x, y) = \int_C P\,dx + Q\,dy.$$

We already showed in the proof of Theorem 1 that this function $f(x, y)$ is a potential function for $P\mathbf{i} + Q\mathbf{j}$. To complete the proof we note that the following are equivalent.

$$\mathbf{grad}\ g = P\mathbf{i} + Q\mathbf{j},$$

$$\frac{\partial g}{\partial x} = \frac{\partial f}{\partial x} \quad \text{and} \quad \frac{\partial g}{\partial y} = \frac{\partial f}{\partial y},$$

$$\frac{\partial(g - f)}{\partial x} = 0 \quad \text{and} \quad \frac{\partial(g - f)}{\partial y} = 0,$$

$g - f$ depends only on y and only on x,

$g(x, y) - f(x, y) = $ constant,

$$g(x, y) = \int_A^X P\,dx + Q\,dy + \text{constant}.$$

Theorems 1 and 2 also hold for three variables. For three variables a vector field has the form

$$\mathbf{F}(x, y, z) = P(x, y, z)\mathbf{i} + Q(x, y, z)\mathbf{j} + R(x, y, z)\mathbf{k}.$$

Theorem 1 for three variables reads as follows.

THEOREM 1 (Three Variables)

A vector field $P\mathbf{i} + Q\mathbf{j} + R\mathbf{k}$ *has a potential function if and only if*

$$\frac{\partial P}{\partial y} = \frac{\partial Q}{\partial x}, \qquad \frac{\partial P}{\partial z} = \frac{\partial R}{\partial x}, \qquad \frac{\partial Q}{\partial z} = \frac{\partial R}{\partial y}.$$

Theorem 2 is modified in the same way.

PROBLEMS FOR SECTION 13.3

Test the following vector fields for existence of a potential function and find the potential function when there is one.

1    $(2x + y^2)\mathbf{i} + (x^2 + 2y)\mathbf{j}$

2    $x^3\mathbf{i} - y^4\mathbf{j}$

3    $y\mathbf{i} + 2x\mathbf{j}$

4    $xe^y\mathbf{i} + ye^x\mathbf{j}$

5    $\sqrt{x^2 + y^2}(\mathbf{i} + \mathbf{j})$ $(x > 0, y > 0)$

6    $y \cos x\mathbf{i} + y \sin x\mathbf{j}$

7    $y \cos x\mathbf{i} + \sin x\mathbf{j}$

8    $e^{x+y}(\mathbf{i} + \mathbf{j})$

9    $-2\mathbf{i} + 6\mathbf{j}$

10    $y\sqrt{x^2 + y^2}\mathbf{i} + x\sqrt{x^2 + y^2}\mathbf{j}$ $(x > 0, y > 0)$

11    $x^2y^3\mathbf{i} + xy^4\mathbf{j}$

12    $\dfrac{y}{x}\mathbf{i} + \dfrac{x}{y}\mathbf{j}$ $(x > 0, y > 0)$

13    $(3x + 5y)\mathbf{i} + (5x - 2y)\mathbf{j}$

14    $\dfrac{y^2}{x}\mathbf{i} + 2y \ln x\mathbf{j}$ $(x > 0)$

15    $\sinh x \cosh y\mathbf{i} + \cosh x \sinh y\mathbf{j}$

16    $\sqrt{y/x}\mathbf{i} + \sqrt{x/y}\mathbf{j}$ $(x > 0, y > 0)$

17   Show that every vector field of the form $P(x)\mathbf{i} + Q(y)\mathbf{j}$ has a potential function.

18   Show that every vector field of the form $f(x + y)(\mathbf{i} + \mathbf{j})$ has a potential function.

19   Show that every vector field of the form $f(x^2 + y^2)(x\mathbf{i} + y\mathbf{j})$ has a potential function.

20   Show that every vector field of the form $f(xy)(y\mathbf{i} + x\mathbf{j})$ has a potential function.

21   Show that the sum of two conservative force fields is conservative.

In Problems 22–31 solve the given exact differential equation.

22    $e^x\,dx + \sin y\,dy = 0$

23    $(3x + 4y)\,dx + (4x - 2y)\,dy = 0$

24    $(x^3 + 2xy + y^2)\,dx + (x^2 + 2xy + y^3)\,dy = 0$

25    $(\sqrt{x} + \sqrt{y})\,dx + (x/2\sqrt{y})\,dy = 0$

26      $2x \sin y \, dx + (y + x^2 \cos y) \, dy = 0$

27      $\dfrac{y}{x^2 + 1} \, dx + (y^2 + \arctan x) \, dy = 0$

28      $(ax + by) \, dx + (bx + cy) \, dy = 0$

29      $\sin x \sin y \, dx - \cos x \cos y \, dy = 0$

30      $\dfrac{\arcsin y}{x} \, dx + \dfrac{\ln x}{\sqrt{1 - y^2}} \, dy = 0$

31      $(x + \sqrt{x + y}) \, dx + (y + \sqrt{x + y}) \, dy = 0$

32   Find a function $Q(x, y)$ such that $\sqrt{xy^3} \, dx + Q(x, y) \, dy$ is an exact differential.

33   Find a function $P(x, y)$ such that $P(x, y) \, dx + \sin^2 x \cos y \, dy$ is an exact differential.

34   The gravity force field of a point mass in three dimensions has magnitude proportional to $1/(x^2 + y^2 + z^2)$ and the direction of $-x\mathbf{i} - y\mathbf{j} - z\mathbf{k}$. Show that the force field is conservative.

3.4 GREEN'S THEOREM

Green's Theorem gives a relationship between double integrals and line integrals. It is a two-dimensional analogue of the Fundamental Theorem of Calculus,

$$F(b) - F(a) = \int_a^b F'(x) \, dx,$$

and shows that the line integral of $\mathbf{F}(x, y)$ around the boundary of a plane region D is equal to a certain double integral over D.

Let D be a plane region

$$a_1 \leq x \leq a_2, \qquad b_1(x) \leq y \leq b_2(x).$$

The directed curve which goes around the boundary of D in the counterclockwise direction is denoted by ∂D and is called the *boundary* of D (Figure 13.4.1).

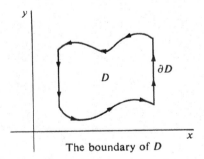

Figure 13.4.1   The boundary of D

If $b_1(x)$ and $b_2(x)$ have continuous derivatives, ∂D will be a piecewise smooth curve and thus a simple closed curve (see Section 13.1).

GREEN'S THEOREM

Let $P(x, y)$ and $Q(x, y)$ be smooth functions on a region D with a piecewise smooth boundary. Then

$$\oint_{\partial D} P \, dx + Q \, dy = \iint_D \frac{\partial Q}{\partial x} - \frac{\partial P}{\partial y} \, dA,$$

$$\oint_{\partial D} - Q \, dx + P \, dy = \iint_D \frac{\partial P}{\partial x} + \frac{\partial Q}{\partial y} \, dA.$$

(See Figure 13.4.2.)

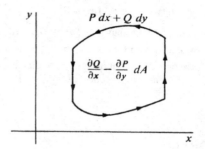

Figure 13.4.2

The second formula follows at once from the first formula by replacing P by $-Q$ and Q by P. We shall prove the theorem only in the simplest case, where D is a rectangle.

PROOF FOR D A RECTANGLE D is shown in Figure 13.4.3.

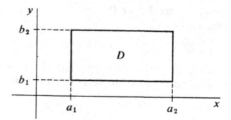

Figure 13.4.3

The line integral around ∂D is a sum of four partial integrals,

$$\oint_{\partial D} P \, dx + Q \, dy = \int_{a_1}^{a_2} P(x, b_1) \, dx + \int_{b_1}^{b_2} Q(a_2, y) \, dy$$

$$+ \int_{a_2}^{a_1} P(x, b_2) \, dx + \int_{b_2}^{b_1} Q(a_1, y) \, dy$$

$$= \int_{b_1}^{b_2} Q(a_2, y) - Q(a_1, y) \, dy - \int_{a_1}^{a_2} P(x, b_2) - P(x, b_1) \, dx.$$

By the Fundamental Theorem of Calculus,

$$Q(a_2, y) - Q(a_1, y) = \int_{a_1}^{a_2} \frac{\partial Q}{\partial x} \, dx,$$

$$P(x, b_2) - P(x, b_1) = \int_{b_1}^{b_2} \frac{\partial P}{\partial y} \, dy.$$

Therefore

$$\oint_{\partial D} P\,dx + Q\,dy = \int_{b_1}^{b_2}\int_{a_1}^{a_2} \frac{\partial Q}{\partial x}\,dx\,dy - \int_{a_1}^{a_2}\int_{b_1}^{b_2} \frac{\partial P}{\partial y}\,dy\,dx$$

$$= \int_{a_1}^{a_2}\int_{b_1}^{b_2} \frac{\partial Q}{\partial x} - \frac{\partial P}{\partial y}\,dx\,dy = \iint_D \frac{\partial Q}{\partial x} - \frac{\partial P}{\partial y}\,dA.$$

We may apply Green's theorem to evaluate a line integral by double integration, or to evaluate a double integral by line integration.

EXAMPLE 1   Compute the line integral

$$\oint_{\partial D} x^2 y\,dx + (x + y)\,dy$$

by Green's Theorem, where D is the rectangle shown in Figure 13.4.4,

$$0 \le x \le 2, \qquad 0 \le y \le 1.$$

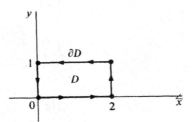

Figure 13.4.4

By Green's Theorem,

$$\oint_{\partial D} x^2 y\,dx + (x + y)\,dy = \iint_D \frac{\partial(x + y)}{\partial x} - \frac{\partial(x^2 y)}{\partial y}\,dA$$

$$= \iint_D (1 - x^2)\,dA = \int_0^2 \int_0^1 1 - x^2\,dy\,dx$$

$$= \int_0^2 1 - x^2\,dx = -\tfrac{2}{3}.$$

As a check, we also compute the line integral directly.

$$\oint_D x^2 y\,dx + (x + y)\,dy = \int_0^2 x^2 \cdot 0\,dx + \int_0^1 2 + y\,dy$$

$$+ \int_2^0 x^2 \cdot 1\,dx + \int_1^0 0 + y\,dy$$

$$= 0 + \tfrac{5}{2} - \tfrac{8}{3} - \tfrac{1}{2} = -\tfrac{2}{3}.$$

EXAMPLE 2   Evaluate by Green's Theorem the line integral

$$\oint_{\partial D} \frac{y}{x + 1}\,dx + 2xy\,dy$$

where D is the region bounded by the curve $y = x^2$ and the line $y = x$, shown in Figure 13.4.5.

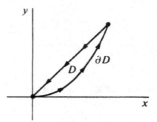

Figure 13.4.5

D is the region $\qquad 0 \le x \le 1, \qquad x^2 \le y \le x.$

$$\oint_{\partial D} \frac{y}{x + 1}\, dx + 2xy\, dy = \iint_D \frac{\partial(2xy)}{\partial x} - \frac{\partial(y/x + 1)}{\partial y}\, dA$$

$$= \iint_D 2y - \frac{1}{x + 1}\, dA$$

$$= \int_0^1 \int_{x^2}^x 2y - \frac{1}{x + 1}\, dy\, dx$$

$$= \int_0^1 x^2 - x^4 - \frac{x}{x + 1} + \frac{x^2}{x + 1}\, dx$$

$$= 2 \ln 2 - \tfrac{41}{30}.$$

As a corollary to Green's Theorem we get a formula for the area of D.

COROLLARY

> If D has a piecewise smooth boundary, then the area of D is
>
> $$A = \oint_{\partial D} x\, dy = \oint_{\partial D} - y\, dx.$$

PROOF By Green's Theorem,

$$\oint_{\partial D} x\, dy = \iint_D \frac{\partial x}{\partial x} - \frac{\partial 0}{\partial y}\, dA = \iint_D dA = A,$$

$$\oint_{\partial D} - y\, dx = \iint_D \frac{\partial 0}{\partial x} - \frac{\partial(-y)}{\partial y}\, dA = \iint_D dA = A.$$

EXAMPLE 3   Use Green's Theorem to find the area of the ellipse shown in Figure 13.4.6,

$$\frac{x^2}{a^2} + \frac{y^2}{b^2} \le 1.$$

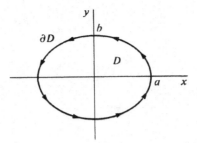

Figure 13.4.6

The boundary of the ellipse is the parametric curve

$$x = a \cos t, \qquad y = b \sin t, \qquad 0 \le t \le 2\pi.$$

By the corollary,

$$A = \oint_{\partial D} x \, dy = \int_0^{2\pi} x \frac{dy}{dt} \, dt = \int_0^{2\pi} (a \cos t)(b \cos t) \, dt$$

$$= ab \int_0^{2\pi} \cos^2 t \, dt = \pi ab.$$

Green's theorem has a vector form which is convenient for physical applications. We define two new functions obtained from a vector field, the curl and the divergence.

DEFINITION

Given a vector field $\mathbf{F}(x, y) = P(x, y)\mathbf{i} + Q(x, y)\mathbf{j}$ *in the plane.*

*The **curl** of* $\mathbf{F}$ *is* $\qquad\qquad$ $\operatorname{curl} \mathbf{F} = \dfrac{\partial Q}{\partial x} - \dfrac{\partial P}{\partial y}.$

*The **divergence** of* $\mathbf{F}$ *is* $\qquad$ $\operatorname{div} \mathbf{F} = \dfrac{\partial P}{\partial x} + \dfrac{\partial Q}{\partial y}.$

On the boundary ∂D, the differential forms $P \, dx + Q \, dy$ and $-Q \, dx + P \, dy$ may be written in the vector form

$$P \, dx + Q \, dy = \mathbf{F} \cdot \mathbf{T} \, ds,$$
$$-Q \, dx + P \, dy = \mathbf{F} \cdot \mathbf{N} \, ds,$$

where $\qquad\qquad$ $\mathbf{T} = $ unit tangent vector to ∂D,
$$\mathbf{T} \, ds = dx\mathbf{i} + dy\mathbf{j},$$

and $\qquad\qquad$ $\mathbf{N} = $ unit outward normal vector to ∂D,
$$\mathbf{N} \, ds = dy\mathbf{i} - dx\mathbf{j}.$$

$\mathbf{T}$ and $\mathbf{N}$ are shown in Figure 13.4.7.

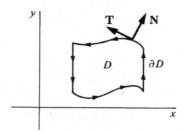

Figure 13.4.7

Substituting the vector notation into the original form of Green's Theorem, we get the following.

GREEN'S THEOREM (Vector Form)

Given a vector field $\mathbf{F}(x, y) = P\mathbf{i} + Q\mathbf{j}$ *on a region D with a piecewise smooth boundary,*

$$\oint_{\partial D} \mathbf{F} \cdot \mathbf{T} \, ds = \iint_{D} \text{curl } \mathbf{F} \, dA,$$

$$\oint_{\partial D} \mathbf{F} \cdot \mathbf{N} \, ds = \iint_{D} \text{div } \mathbf{F} \, dA.$$

The physical meaning of Green's theorem can be explained in terms of the flow of a fluid (a liquid or gas). Let the vector field $\mathbf{F}(x, y)$ represent the rate and direction of fluid flow at a point (x, y) in the plane. Consider a plane region D and element of area ΔD containing (x, y) (Figure 13.4.8).

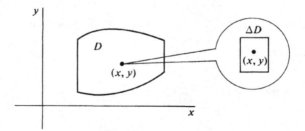

Figure 13.4.8

We first explain the formula

$$\oint_{\partial D} \mathbf{F} \cdot \mathbf{T} \, ds = \iint_{D} \text{curl } \mathbf{F} \, dA.$$

The line integral $\oint_{\partial D} \mathbf{F} \cdot \mathbf{T} \, ds$ of the flow component in the direction tangent to the boundary is called the *circulation* of $\mathbf{F}$ *around* ∂D. Green's Theorem states that the circulation of $\mathbf{F}$ around the boundary of D equals the integral of the curl of $\mathbf{F}$ over D (Figure 13.4.9).

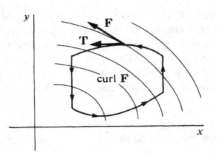

Figure 13.4.9

When we apply Green's theorem to an element of area ΔD we get

$$\oint_{\partial \Delta D} \mathbf{F} \cdot \mathbf{T} \, ds \approx \operatorname{curl} \mathbf{F} \, \Delta A \qquad \text{(compared to } \Delta A\text{)}.$$

Thus the curl of $\mathbf{F}$ at (x, y) is equal to the circulation per unit area at (x, y).

If curl $\mathbf{F}$ is identically zero, the fluid flow $\mathbf{F}$ is called *irrotational*. By the Exactness Criterion, $\mathbf{F}$ is irrotational if and only if $P \, dx + Q \, dy$ is an exact differential. The circulation of an irrotational field around any ∂D is zero.

Next we explain the formula

$$\oint_{\partial D} \mathbf{F} \cdot \mathbf{N} \, ds = \iint_D \operatorname{div} \mathbf{F} \, dA.$$

The line integral $\oint_{\partial D} \mathbf{F} \cdot \mathbf{N} \, ds$ of the flow component in the direction of the outward normal vector is called the *flux across* ∂D. The flux is the net rate at which fluid is flowing from inside D across the boundary and is therefore equal to the rate of decrease of the mass inside D. Green's Theorem states that the flux of $\mathbf{F}$ across the boundary of D equals the integral of the divergence of $\mathbf{F}$ over D (Figure 13.4.10).

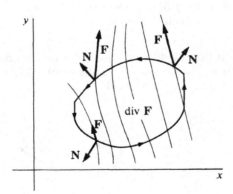

Figure 13.4.10

When we apply this to ΔD we get

$$\oint_{\partial \Delta D} \mathbf{F} \cdot \mathbf{N} \, ds \approx \operatorname{div} \mathbf{F} \, \Delta A \qquad \text{(compared to } \Delta A\text{)}.$$

Therefore the divergence of **F** at (x, y) is the net rate of flow of fluid away from (x, y), and is equal to the rate of decrease in density at (x, y). Positive divergence means that the density is decreasing, and negative divergence means that the density is increasing.

If div **F** is identically zero, the fluid flow is called *solenoidal*, or *incompressible*. By the Exactness Criterion, **F** is incompressible if and only if $-Q\,dx + P\,dy$ is an exact differential. The flux of an incompressible field across any ∂D is zero.

EXAMPLE 4   A fluid is rotating about the origin with angular velocity ω radians per second. Find the curl and divergence of the velocity field $\mathbf{F}(x, y)$.

As we can see from Figure 13.4.11, the velocity at a point (x, y) is

$$\mathbf{F}(x, y) = \omega(-y\mathbf{i} + x\mathbf{j}) = -\omega y\mathbf{i} + \omega x\mathbf{j}.$$

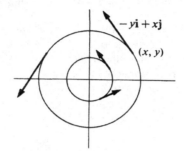

Figure 13.4.11

Then

$$\text{curl } \mathbf{F} = \frac{\partial(\omega x)}{\partial x} - \frac{\partial(-\omega y)}{\partial y} = 2\omega,$$

$$\text{div } \mathbf{F} = \frac{\partial(-\omega y)}{\partial x} + \frac{\partial(\omega x)}{\partial y} = 0.$$

Thus a purely rotating fluid is incompressible and its curl at every point is equal to twice the angular velocity.

EXAMPLE 5   A fluid is flowing directly away from the origin at a rate equal to a constant b times the distance from the origin (Figure 13.4.12). Find the curl and divergence of the flow field.

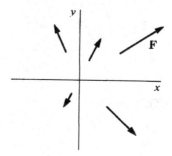

Figure 13.4.12

We have $\qquad F(x, y) = b\dfrac{x\mathbf{i} + y\mathbf{j}}{\sqrt{x^2 + y^2}}\sqrt{x^2 + y^2} = bx\mathbf{i} + by\mathbf{j}.$

$$\text{curl } F = \frac{\partial(by)}{\partial x} - \frac{\partial(bx)}{\partial y} = 0.$$

$$\text{div } F = \frac{\partial(bx)}{\partial x} + \frac{\partial(by)}{\partial y} = 2b.$$

The fluid flow field is irrotational and the divergence at every point is $2b$.

PROBLEMS FOR SECTION 13.4

In Problems 1–12, find the line integral by Green's Theorem.

1. $\quad \oint_{\partial D} 2y\, dx + 3x\, dy, \quad D: 0 \le x \le 1, 0 \le y \le 1$

2. $\quad \oint_{\partial D} xy\, dx + xy\, dy, \quad D: 0 \le x \le 1, 0 \le y \le 1$

3. $\quad \oint_{\partial D} e^{2x+3y}\, dx + e^{xy}\, dy, \quad D: -2 \le x \le 2, -1 \le y \le 1$

4. $\quad \oint_{\partial D} y\cos x\, dx + y\sin x\, dy, \quad D: 0 \le x \le \pi/2, 1 \le y \le 2$

5. $\quad \oint_{\partial D} x^2 y\, dx + xy^2\, dy, \quad D: 0 \le x \le 1, 0 \le y \le x$

6. $\quad \oint_{\partial D} x\sqrt{y}\, dx + \sqrt{x+y}\, dy, \quad D: 1 \le x \le 2, 2x \le y \le 4$

7. $\quad \oint_{\partial D} (x/y)\, dx + (2 + 3x)\, dy, \quad D: 1 \le x \le 2, 1 \le y \le x^2$

8. $\quad \oint_{\partial D} \sin y\, dx + \sin x\, dy, \quad D: 0 \le x \le \pi/2, x \le y \le \pi/2$

9. $\quad \oint_{\partial D} x \ln y\, dx, \quad D: 1 \le x \le 2, e^x \le y \le e^{x^2}$

10. $\quad \oint_{\partial D} \sqrt{1 + x^2}\, dy, \quad D: -1 \le x \le 1, x^2 \le y \le 1$

11. $\quad \oint_{\partial D} x^2 y\, dx - xy^2\, dy, \quad D: x^2 + y^2 \le 1 \qquad$ *Hint*: Use polar coordinates.

12. $\quad \oint_{\partial D} y^3\, dx + 2x^3\, dy, \quad D: x^2 + y^2 \le 4$

In Problems 13–18, find (a) curl F, (b) $\oint_{\partial D} F \cdot T\, ds$, (c) div F, (d) $\oint_{\partial D} F \cdot N\, ds$.

13. $\quad F(x, y) = xy\mathbf{i} - xy\mathbf{j}, \quad D: 0 \le x \le 1, 0 \le y \le 1$
14. $\quad F(x, y) = ax^2\mathbf{i} + by^2\mathbf{j}, \quad D: 0 \le x \le 1, 0 \le y \le 1$
15. $\quad F(x, y) = ay^2\mathbf{i} + bx^2\mathbf{j}, \quad D: 0 \le x \le 1, 0 \le y \le x$
16. $\quad F(x, y) = \sin x \cos y\mathbf{i} + \cos x \sin y\mathbf{j}, \quad D: 0 \le x \le \pi/2, 0 \le y \le x$
17. $\quad F(x, y) = y\mathbf{i} - x\mathbf{j}, \quad D: x^2 + y^2 \le 1$
18. $\quad F(x, y) = x\mathbf{i} + y\mathbf{j}, \quad D: x^2 + y^2 \le 1$

19   Use Green's Theorem to find the area inside the curve $r = a + \cos\theta$, $(a \geq 1)$.

20   Use Green's Theorem to find the area inside the ellipse $x^2/a^2 + y^2/b^2 = 1$ and above the line $y = c$ $(0 < c < b)$.

☐ **21**   Show that if D has a piecewise smooth boundary, the area of D is $A = \frac{1}{2}\oint_{\partial D} -y\,dx + x\,dy$.

☐ **22**   Show that for any continuous function $f(t)$ and constants a, b, c,

$$\oint_{\partial D} af(x^2 + y^2)\,dx + bf(x^2 + y^2)\,dy = 0$$

where D is the circle $x^2 + y^2 \leq c^2$.

☐ **23**   Find the value of the line integral

$$\oint_{\partial D} (a_1x + b_1y)\,dx + (a_2x + b_2y)\,dy$$

where D is a region with area A.

☐ **24**   Show that any vector field of the form

$$\mathbf{F}(x, y) = xf(x^2 + y^2)\mathbf{i} + yf(x^2 + y^2)\mathbf{j}$$

is irrotational.

☐ **25**   Show that any vector field of the form

$$\mathbf{F}(x, y) = yf(x^2 + y^2)\mathbf{i} - xf(x^2 + y^2)\mathbf{j}$$

is incompressible.

☐ **26**   Show that any vector field of the form

$$\mathbf{F}(x, y) = f(x)\mathbf{i} + g(y)\mathbf{j}$$

is irrotational.

13.5 SURFACE AREA AND SURFACE INTEGRALS

In Chapter 6 we were able to find the area of a surface of revolution by a single integral. To find the area of a smooth surface in general (Figure 13.5.1), we need a double integral.

We call a function $f(x, y)$, or a surface $z = f(x, y)$, *smooth* if both partial derivatives of f are continuous.

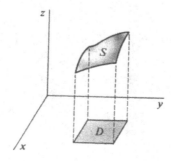

Figure 13.5.1

DEFINITION

> The area of a smooth surface
> $$z = f(x, y), \qquad (x, y) \text{ in } D$$
> is
> $$S = \iint_D \sqrt{\left(\frac{\partial z}{\partial x}\right)^2 + \left(\frac{\partial z}{\partial y}\right)^2 + 1} \, dx \, dy.$$

JUSTIFICATION   Let $S(D_1)$ be the area of the part of the surface with (x, y) in D_1. $S(D_1)$ has the Addition Property, and $S(D_1) \geq 0$. Consider the piece of the surface ΔS above an element of area ΔD (Figure 13.5.2). ΔS is infinitely close to the piece of the tangent plane above ΔD, which is a parallelogram with sides

$$\mathbf{U} = \Delta x \mathbf{i} + \frac{\partial z}{\partial x} \Delta x \mathbf{k}, \qquad \mathbf{V} = \Delta y \mathbf{j} + \frac{\partial z}{\partial y} \Delta y \mathbf{k}.$$

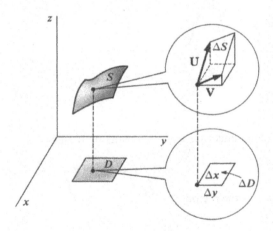

Figure 13.5.2

The quickest way to find the area of this parallelogram is to use the vector product formula (Section 10.4, Problem 39),

$$\text{Area} = |\mathbf{U} \times \mathbf{V}|.$$

Then
$$\text{Area} = \begin{Vmatrix} \mathbf{i} & \mathbf{j} & \mathbf{k} \\ \Delta x & 0 & \dfrac{\partial z}{\partial x} \Delta x \\ 0 & \Delta y & \dfrac{\partial z}{\partial y} \Delta y \end{Vmatrix}$$

$$= \left| -\Delta y \frac{\partial z}{\partial x} \Delta x \mathbf{i} - \Delta x \frac{\partial z}{\partial y} \Delta y \mathbf{j} + \Delta x \, \Delta y \mathbf{k} \right|$$

$$= \sqrt{\left(\frac{\partial z}{\partial x}\right)^2 + \left(\frac{\partial z}{\partial y}\right)^2 + 1} \, \Delta x \, \Delta y.$$

Therefore $\Delta S \approx \sqrt{\left(\dfrac{\partial z}{\partial x}\right)^2 + \left(\dfrac{\partial z}{\partial y}\right)^2 + 1}\ \Delta x\ \Delta y$ (compared to $\Delta x\ \Delta y$),

and by the Infinite Sum Theorem,

$$S = \iint_D \sqrt{\left(\dfrac{\partial z}{\partial x}\right)^2 + \left(\dfrac{\partial z}{\partial y}\right)^2 + 1}\ dx\ dy.$$

EXAMPLE 1   Find the area of the triangle cut from the plane $2x + 3y + z = 1$ by the coordinate planes.

Step 1   Sketch the region as in Figure 13.5.3.

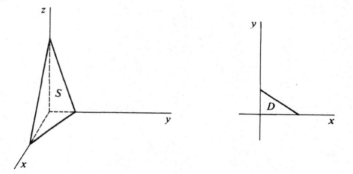

Figure 13.5.3

Step 2   The plane intersects the (x, y) plane on the line

$$2x + 3y = 1, \qquad y = \frac{1 - 2x}{3}.$$

Thus D is the region

$$0 \le x \le \tfrac{1}{2}, \qquad 0 \le y \le \frac{1 - 2x}{3}.$$

Step 3   On the surface,

$$z = 1 - 2x - 3y, \qquad \frac{\partial z}{\partial x} = -2, \qquad \frac{\partial z}{\partial y} = -3.$$

Then $S = \displaystyle\iint_D \sqrt{\left(\dfrac{\partial z}{\partial x}\right)^2 + \left(\dfrac{\partial z}{\partial y}\right)^2 + 1}\ dx\ dy$

$\qquad = \displaystyle\iint_D \sqrt{4 + 9 + 1}\ dx\ dy = \sqrt{14} \iint_D dx\ dy$

$\qquad = \sqrt{14} \displaystyle\int_0^{1/2} \int_0^{(1-2x)/3} dy\ dx = \sqrt{14} \int_0^{1/2} \frac{1 - 2x}{3}\ dx = \frac{\sqrt{14}}{12}.$

EXAMPLE 2   Find the area of the portion of the hyperbolic paraboloid $z = x^2 - y^2$ which is inside the cylinder $x^2 + y^2 = 1$.

Step 1 Sketch the region (Figure 13.5.4).

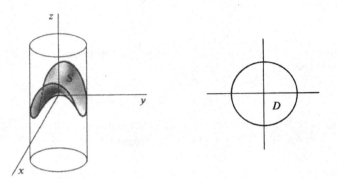

Figure 13.5.4

Step 2 D is the region

$$-1 \le x \le 1, \qquad -\sqrt{1 - x^2} \le y \le \sqrt{1 - x^2},$$

or in polar coordinates,

$$0 \le \theta \le 2\pi, \qquad 0 \le r \le 1.$$

Step 3 $\dfrac{\partial z}{\partial x} = 2x, \qquad \dfrac{\partial z}{\partial y} = -2y.$

Then

$$S = \iint_D \sqrt{\left(\frac{\partial z}{\partial x}\right)^2 + \left(\frac{\partial z}{\partial y}\right)^2 + 1}\, dx\, dy$$

$$= \iint_D \sqrt{4x^2 + 4y^2 + 1}\, dx\, dy.$$

It is easier to use polar coordinates, where

$$\sqrt{4x^2 + 4y^2 + 1} = \sqrt{4r^2 + 1}.$$

$$S = \int_0^{2\pi} \int_0^1 \sqrt{4r^2 + 1}\, r\, dr\, d\theta.$$

Put

$$u = 4r^2 + 1, \quad du = 8r\, dr,$$

$$S = \int_0^{2\pi} \int_1^5 \frac{1}{8}\sqrt{u}\, du\, d\theta$$

$$= \int_0^{2\pi} \frac{1}{12}(5^{3/2} - 1)\, d\theta = \frac{\pi}{6}(5^{3/2} - 1).$$

The line integral has an analogue for surfaces called the surface integral. The form of the line integral which is most easily generalized to surfaces is the vector form

$$\int_C \mathbf{F} \cdot \mathbf{N}\, ds = \int_C -Q\, dx + P\, dy$$

where $\mathbf{N}$ is the unit normal vector of C. This is convenient because surfaces also have unit normal vectors.

Before stating the definition we motivate it with a fluid flow interpretation. Remember that in the plane the line integral

$$\int_C \mathbf{F} \cdot \mathbf{N} \, ds$$

is equal to the flux, or net rate of fluid flow across the curve C in the direction of the normal vector $\mathbf{N}$.

Consider a fluid flow field

$$\mathbf{F}(x, y, z) = P\mathbf{i} + Q\mathbf{j} + R\mathbf{k}$$

and a surface S in space. Call one side of S positive and the other side negative, and at each point of S let $\mathbf{N}$ be the unit normal vector on the positive side of S. The *surface integral*

$$\iint_S \mathbf{F} \cdot \mathbf{N} \, dS$$

will be the flux, or net rate of fluid flow across the surface S from the negative to the positive side (Figure 13.5.5).

With this interpretation in mind we shall define the surface integral and then justify the definition. First we need the notion of an oriented surface.

DEFINITION

> An *oriented surface* S is a smooth surface
>
> $$z = g(x, y)$$
>
> over a plane region D with a piecewise smooth boundary, together with an *orientation* that designates one side of the surface as positive and the other side as negative. (See Figure 13.5.6.)

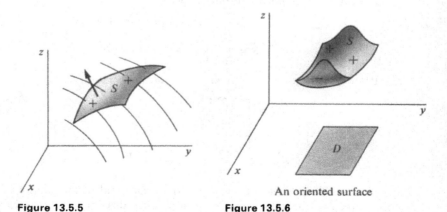

Figure 13.5.5   Figure 13.5.6

An oriented surface

DEFINITION

> Let S be an oriented surface $z = g(x, y)$ over D and let
>
> $$\mathbf{F}(x, y, z) = P\mathbf{i} + Q\mathbf{j} + R\mathbf{k}$$

be a vector field defined on S. The **surface integral** of $\mathbf{F}$ over S is defined by

$$\iint_S \mathbf{F} \cdot \mathbf{N}\, dS = \pm \iint_D -P\frac{\partial z}{\partial x} - Q\frac{\partial z}{\partial y} + R\, dA,$$

$+$ if the top side of S is positive, $-$ if the top side of S is negative.

Thus a change in orientation of S changes the sign of the surface integral.

JUSTIFICATION We show that this definition corresponds to the intuitive concept of flux, or net rate of fluid flow, across a surface. Suppose S is oriented so that the top surface of S is positive.

Let $B(D)$ be the flux across the part of S over a region D. Consider an element of area ΔD and let ΔS be the area of S over ΔD. Then ΔS is almost a piece of the tangent plane. The component of fluid flow perpendicular to ΔS is given by the scalar product $\mathbf{F} \cdot \mathbf{N}$ where $\mathbf{N}$ is the unit normal vector on the top side of ΔS (Figure 13.5.7). Thus the flux across ΔS is

$$\Delta B \approx \mathbf{F} \cdot \mathbf{N} \, \Delta S \qquad \text{(compared to } \Delta A\text{)}.$$

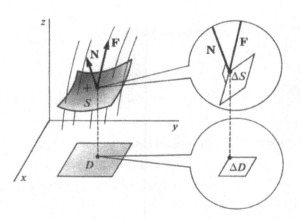

Figure 13.5.7

This suggests the surface integral notation

$$\iint_S \mathbf{F} \cdot \mathbf{N}\, dS.$$

Let us find $\mathbf{F}$, $\mathbf{N}$, and ΔS. The vector $\mathbf{F}$ at (x, y, z) is

(1) $$\mathbf{F}(x, y, z) = P\mathbf{i} + Q\mathbf{j} + R\mathbf{k}.$$

From Section 13.1, one normal vector at (x, y, z) is

$$-\frac{\partial z}{\partial x}\mathbf{i} - \frac{\partial z}{\partial y}\mathbf{j} + \mathbf{k}.$$

The unit normal vector $\mathbf{N}$ on the top side of ΔS has positive $\mathbf{k}$ component and length one, so

(2)
$$\mathbf{N} = \frac{\dfrac{\partial z}{\partial x}\mathbf{i} - \dfrac{\partial z}{\partial y}\mathbf{j} + \mathbf{k}}{\sqrt{\left(\dfrac{\partial z}{\partial x}\right)^2 + \left(\dfrac{\partial z}{\partial y}\right)^2 + 1}}.$$

From our study of surface areas,

(3)
$$\Delta S \approx \sqrt{\left(\frac{\partial z}{\partial x}\right)^2 + \left(\frac{\partial z}{\partial y}\right)^2 + 1}\ \Delta A \qquad \text{(compared to } \Delta A\text{).}$$

When we substitute Equations 1–3 into $\mathbf{F} \cdot \mathbf{N}\ \Delta S$, the radicals cancel out and we have

$$\Delta B \approx \left(-P\frac{\partial z}{\partial x} - Q\frac{\partial z}{\partial y} + R\right)\Delta A \qquad \text{(compared to } \Delta A\text{).}$$

Using the Infinite Sum Theorem we get the surface integral formula

$$B(D) = \iint_D -P\frac{\partial z}{\partial x} - Q\frac{\partial z}{\partial y} + R\, dA.$$

EXAMPLE 3   Evaluate the surface integral

$$\iint_S \mathbf{F} \cdot \mathbf{N}\, dS,$$

where S is the surface $z = e^{x-y}$ over the region D given by

$$0 \le x \le 1, \qquad x \le y \le 1,$$

S is oriented with the top side positive, and

$$\mathbf{F}(x, y, z) = 2\mathbf{i} + \mathbf{j} + z^2\mathbf{k}.$$

The region is sketched in Figure 13.5.8. The first step is to find $\partial z/\partial x$ and $\partial z/\partial y$.

$$\frac{\partial z}{\partial x} = e^{x-y}, \qquad \frac{\partial z}{\partial y} = -e^{x-y}.$$

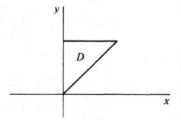

Figure 13.5.8

By definition of surface integral,

$$\iint\limits_{S} \mathbf{F} \cdot \mathbf{N} \, dS = \iint\limits_{D} -2\frac{\partial z}{\partial x} - \frac{\partial z}{\partial y} + z^2 \, dA$$

$$= \iint\limits_{D} -2e^{x-y} + e^{x-y} + e^{2x-2y} \, dA$$

$$= \int_{0}^{1} \int_{x}^{1} -e^{x-y} + e^{2x-2y} \, dy \, dx$$

$$= \int_{0}^{1} -\tfrac{1}{2} + e^{x-1} - \tfrac{1}{2}e^{2x-2} \, dx = \tfrac{1}{4} - e^{-1} + \tfrac{1}{4}e^{-2}.$$

The same surface integral with S oriented with the top side negative has minus the above value.

PROBLEMS FOR SECTION 13.5

1   Find the area of the triangle cut from the plane $x + 2y + 4z = 10$ by the coordinate planes.

2   Find the area cut from the plane $2x + 4y + z = 0$ by the cylinder $x^2 + y^2 = 1$.

3   Find the area of the surface of the paraboloid $z = x^2 + y^2$ below the plane $z = 1$.

4   Find the area of the surface of the cone $z = \sqrt{x^2 + y^2}$ below the plane $z = 2$.

5   Find the surface area of the part of the sphere $x^2 + y^2 + z^2 = a^2$ which lies in the first octant; i.e., $x \geq 0$, $y \geq 0$, $z \geq 0$.

6   Find the surface area of the part of the sphere $x^2 + y^2 + z^2 = a^2$ which is above the circle $x^2 + y^2 \leq b^2$ $(b \leq a)$.

7   Find the surface area cut from the hyperboloid $z = x^2 - y^2$ by the cylinder $x^2 + y^2 = a^2$.

8   Find the area cut from the surface $z = xy$ by the cylinder $x^2 + y^2 = a^2$.

9   Find the surface area of the part of the sphere $r^2 + z^2 = a^2$ above the circle $r = a \cos \theta$.

10   Find the surface area of the part of the cone $z = cr$ above the circle $r = a \cos \theta$.

11   Find the area of the part of the plane $z = ax + by + c$ over a region D of area A.

12   Find the surface area of the part of the cone $z = c\sqrt{x^2 + y^2}$ over a region D of area A.

13   Find the surface area of the part of the cylinder $x^2 + z^2 = a^2$ cut out by the cylinder $x^2 + y^2 \leq a^2$.

14   Find the surface area of the part of the cylinder $x^2 + z^2 = a^2$ above and below the square $-b \leq x \leq b$, $-b \leq y \leq b$ $(b \leq a)$.

15   Evaluate the surface integral

$$\iint\limits_{S} (2\mathbf{i} - 3\mathbf{j} + 4\mathbf{k}) \cdot \mathbf{N} \, dS,$$

where S is the surface $z = x^2 + y^2$, $-1 \leq x \leq 1$, $-1 \leq y \leq 1$, oriented with the top side positive.

16   Evaluate the surface integral

$$\iint\limits_{S} (x\mathbf{i} + y\mathbf{j} + 3\mathbf{k}) \cdot \mathbf{N} \, dS$$

where S is the surface $z = 3x - 5y$ over the rectangle $1 \leq x \leq 2$, $0 \leq y \leq 2$, oriented with the top side positive.

17   Evaluate the surface integral

$$\iint_S (x\mathbf{i} + y\mathbf{j} - 2\mathbf{k}) \cdot \mathbf{N} \, dS$$

where S is the surface $z = 1 - x^2 - y^2$, $x^2 + y^2 \le 1$, oriented with the top side positive.

18   Evaluate the surface integral

$$\iint_S (xy\mathbf{i} + yz\mathbf{j} + zx\mathbf{k}) \cdot \mathbf{N} \, dS$$

where S is the surface $z = x + y^2 + 2$, $0 \le x \le 1$, $x \le y \le 1$, oriented with the top side positive.

19   Evaluate the surface integral

$$\iint_S (e^x\mathbf{i} + e^y\mathbf{j} + z\mathbf{k}) \cdot \mathbf{N} \, dS$$

where S is the surface $z = xy$, $0 \le x \le 1$, $-x \le y \le x$, oriented with the top side positive.

20   Evaluate the surface integral

$$\iint_S xz\mathbf{i} + yz\mathbf{j} + z\mathbf{k}$$

where S is the surface $z = \sqrt{a^2 - x^2 - y^2}$, $x^2 + y^2 \le b$, oriented with the top side positive $(b < a)$.

☐ 21   Show that if S is a horizontal surface $z = c$ over a region D, oriented with the top side positive, then the surface integral over S is

$$\iint_S (P(x, y, z)\mathbf{i} + Q(x, y, z)\mathbf{j} + R(x, y, z)\mathbf{k}) \cdot \mathbf{N} \, dS = \iint_D R(x, y, c) \, dA.$$

13.6 THEOREMS OF STOKES AND GAUSS

Both Stokes' Theorem and Gauss' Theorem are three-dimensional generalizations of Green's Theorem. To state these theorems we need the notions of curl and divergence in three dimensions. The curl of a vector field in the plane is a scalar field, while the curl of a vector field in space is another vector field. However, the divergence in both cases is scalar.

DEFINITION

Given a vector field $\mathbf{F}(x, y, z) = P\mathbf{i} + Q\mathbf{j} + R\mathbf{k}$

in space. The **curl** of $\mathbf{F}$ is the new vector field

$$\text{curl } \mathbf{F} = \left(\frac{\partial R}{\partial y} - \frac{\partial Q}{\partial z}\right)\mathbf{i} + \left(\frac{\partial P}{\partial z} - \frac{\partial R}{\partial x}\right)\mathbf{j} + \left(\frac{\partial Q}{\partial x} - \frac{\partial P}{\partial y}\right)\mathbf{k}.$$

This can be remembered by writing the curl as a "determinant"

$$\text{curl } \mathbf{F} = \begin{vmatrix} \mathbf{i} & \mathbf{j} & \mathbf{k} \\ \dfrac{\partial}{\partial x} & \dfrac{\partial}{\partial y} & \dfrac{\partial}{\partial z} \\ P & Q & R \end{vmatrix}.$$

*The **divergence** of* **F** *is the real valued function*

$$\operatorname{div} \mathbf{F} = \frac{\partial P}{\partial x} + \frac{\partial Q}{\partial y} + \frac{\partial R}{\partial z}.$$

EXAMPLE 1 Find the curl and divergence of the vector field

$$\mathbf{F}(x, y, z) = xy\mathbf{i} + yz\mathbf{j} + zx\mathbf{k}.$$

$$\operatorname{curl} \mathbf{F} = \begin{vmatrix} \mathbf{i} & \mathbf{j} & \mathbf{k} \\ \dfrac{\partial}{\partial x} & \dfrac{\partial}{\partial y} & \dfrac{\partial}{\partial z} \\ xy & yz & zx \end{vmatrix}$$

$$= \left(\frac{\partial(zx)}{\partial y} - \frac{\partial(yz)}{\partial z} \right)\mathbf{i} + \left(\frac{\partial(xy)}{\partial z} - \frac{\partial(zx)}{\partial x} \right)\mathbf{j} + \left(\frac{\partial(yz)}{\partial x} - \frac{\partial(xy)}{\partial y} \right)\mathbf{k}$$

$$= -y\mathbf{i} - z\mathbf{j} - x\mathbf{k}.$$

$$\operatorname{div} \mathbf{F} = \frac{\partial(xy)}{\partial x} + \frac{\partial(yz)}{\partial y} + \frac{\partial(xz)}{\partial z} = y + z + x.$$

Two interesting identities are given in the next theorem.

THEOREM 1

Assume the function $f(x, y, z)$ *and vector field* $\mathbf{F}(x, y, z)$ *have continuous second partials. Then*

$$\operatorname{curl}(\operatorname{grad} f) = \mathbf{0}, \qquad \operatorname{div}(\operatorname{curl} \mathbf{F}) = 0.$$

PROOF We use the equality of mixed partials.

$$\operatorname{grad} f = \frac{\partial f}{\partial x}\mathbf{i} + \frac{\partial f}{\partial y}\mathbf{j} + \frac{\partial f}{\partial z}\mathbf{k}.$$

$$\operatorname{curl}(\operatorname{grad} f) = \begin{vmatrix} \mathbf{i} & \mathbf{j} & \mathbf{k} \\ \dfrac{\partial}{\partial x} & \dfrac{\partial}{\partial y} & \dfrac{\partial}{\partial z} \\ \dfrac{\partial f}{\partial x} & \dfrac{\partial f}{\partial y} & \dfrac{\partial f}{\partial z} \end{vmatrix}$$

$$= \left(\frac{\partial^2 f}{\partial y\,\partial z} - \frac{\partial^2 f}{\partial z\,\partial y} \right)\mathbf{i} + \left(\frac{\partial^2 f}{\partial z\,\partial x} - \frac{\partial^2 f}{\partial x\,\partial z} \right)\mathbf{j} + \left(\frac{\partial^2 f}{\partial x\,\partial y} - \frac{\partial^2 f}{\partial y\,\partial x} \right)\mathbf{k}$$

$$= \mathbf{0}.$$

The other proof is similar and is left as a problem.

Stokes' Theorem relates a surface integral over S to a line integral over the boundary of S. It corresponds to Green's Theorem in the form

$$\oint_{\partial D} \mathbf{F} \cdot \mathbf{T} \, ds = \iint_D \operatorname{curl} \mathbf{F} \, dA.$$

Let S be an oriented surface over a region D. The *boundary* of S, ∂S, is the simple closed space curve whose direction depends on the orientation of S as shown in Figure 13.6.1.

The notation

$$\oint_{\partial S} \mathbf{F} \cdot \mathbf{T} \, ds \quad \text{or} \quad \oint_{\partial S} P \, dx + Q \, dy + R \, dz,$$

denotes the line integral around ∂S in the direction determined by the orientation of S.

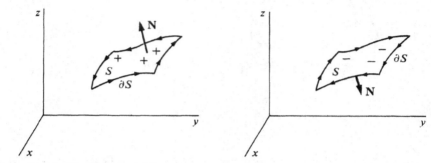

Figure 13.6.1 The Boundary of S

STOKES' THEOREM

Given a vector field $\mathbf{F}(x, y, z)$ *on an oriented surface* S,

$$\oint_{\partial S} \mathbf{F} \cdot \mathbf{T} \, ds = \iint_S \text{curl } \mathbf{F} \cdot \mathbf{N} \, dS.$$

(See Figure 13.6.2.)

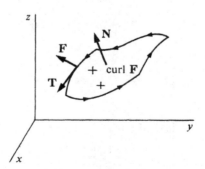

Figure 13.6.2

To put this equation in scalar form, let

$$\mathbf{F} = P\mathbf{i} + Q\mathbf{j} + R\mathbf{k}, \qquad \text{curl } \mathbf{F} = H\mathbf{i} + L\mathbf{j} + M\mathbf{k}.$$

Then
$$\mathbf{F} \cdot \mathbf{T} \, ds = P \, dx + Q \, dy + R \, dz,$$

and if S is oriented with the top side positive,

$$\text{curl } \mathbf{F} \cdot \mathbf{N} \, dS = \left(-H\frac{\partial z}{\partial x} - L\frac{\partial z}{\partial y} + M \right) dA.$$

Thus Stokes' Theorem has the scalar form

$$\oint_{\partial S} P\,dx + Q\,dy + R\,dz = \iint_{D}\left(-H\frac{\partial z}{\partial x} - L\frac{\partial z}{\partial y} + M\right) dA.$$

Stokes' Theorem has two corollaries which are analogous to the Path Independence Theorem.

COROLLARY 1

If $f(x, y, z)$ has continuous second partials, then the line integral of grad f around the boundary of any oriented surface is zero,

$$\oint_{\partial S} \mathbf{grad}\, f \cdot \mathbf{T}\, ds = 0.$$

(See Figure 13.6.3.)

PROOF $\mathbf{curl}(\mathbf{grad}\, f) = 0$, so

$$\oint_{\partial S} \mathbf{grad}\, f \cdot \mathbf{T}\, ds = \iint_{S} \mathbf{curl}(\mathbf{grad}\, f) \cdot \mathbf{N}\, dS = \iint_{S} 0\, dS = 0.$$

COROLLARY 2

The surface integral of curl F *over an oriented surface depends only on the boundary of the surface. That is, if $\partial S_1 = \partial S_2$ then*

$$\iint_{S_1} \mathbf{curl}\, \mathbf{F} \cdot \mathbf{N}_1\, dS_1 = \iint_{S_2} \mathbf{curl}\, \mathbf{F} \cdot \mathbf{N}_2\, dS_2.$$

(See Figure 13.6.4.)

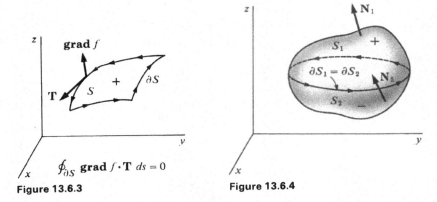

Figure 13.6.3   Figure 13.6.4

PROOF By Stokes' Theorem, both surface integrals are equal to the line integral

$$\oint_{\partial S_1} \mathbf{F} \cdot \mathbf{T}\, ds = \oint_{\partial S_2} \mathbf{F} \cdot \mathbf{T}\, ds.$$

For fluid flows, Stokes' Theorem states that the circulation of fluid around the boundary of an oriented surface S is equal to the surface integral of the curl over S.

We shall not prove Stokes' Theorem, but will illustrate it in the following examples.

EXAMPLE 2   Let S_1 be the portion of the plane

$$z = 2x + 2y - 1$$

and S_2 the portion of the paraboloid

$$z = x^2 + y^2$$

bounded by the curve where the plane and paraboloid intersect. Orient both surfaces with the top side positive, so they have the same boundary

$$C = \partial S_1 = \partial S_2.$$

Let
$$F(x, y, z) = z\mathbf{i} + x\mathbf{j} + y\mathbf{k}.$$

Evaluate the integrals

(a) $\displaystyle \iint_{S_1} \operatorname{curl} \mathbf{F} \cdot \mathbf{N}_1 \, dS_1$.

(b) $\displaystyle \iint_{S_2} \operatorname{curl} \mathbf{F} \cdot \mathbf{N}_2 \, dS_2$.

(c) $\displaystyle \oint_C \mathbf{F} \cdot \mathbf{T} \, ds$.

By Stokes' Theorem, all three answers are equal, but we compute them separately as a check.

The regions are drawn in Figure 13.6.5. First we find the plane region D over which S_1 and S_2 are defined. The two surfaces intersect at

$$2x + 2y - 1 = x^2 + y^2,$$
$$(x - 1)^2 + (y - 1)^2 = 1.$$

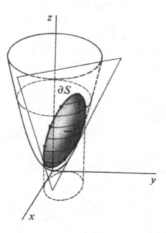

Figure 13.6.5

So D is the unit circle with center at $(1, 1)$ shown in Figure 13.6.6; that is,

$$0 \le x \le 2, \qquad 1 - \sqrt{1 - (x - 1)^2} \le y \le 1 + \sqrt{1 - (x - 1)^2}.$$

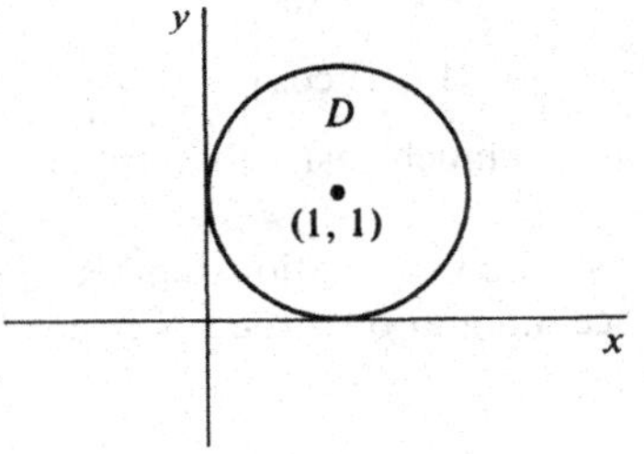

Figure 13.6.6

Next we compute **curl F**.

$$\text{curl } \mathbf{F} = \begin{vmatrix} \mathbf{i} & \mathbf{j} & \mathbf{k} \\ \dfrac{\partial}{\partial x} & \dfrac{\partial}{\partial y} & \dfrac{\partial}{\partial z} \\ z & x & y \end{vmatrix} = \mathbf{i} + \mathbf{j} + \mathbf{k}.$$

(a) On the surface $z = 2x + 2y - 1$,

$$\frac{\partial z}{\partial x} = 2, \qquad \frac{\partial z}{\partial y} = 2.$$

Thus $\displaystyle\iint_{S_1} \text{curl } \mathbf{F} \cdot \mathbf{N}_1 \, dS_1 = \iint_D -2 - 2 + 1 \, dA = -3 \iint_D dA = -3\pi.$

(b) On the surface $z = x^2 + y^2$,

$$\frac{\partial z}{\partial x} = 2x, \qquad \frac{\partial z}{\partial y} = 2y.$$

Thus $\displaystyle\iint_{S_2} \text{curl } \mathbf{F} \cdot \mathbf{N}_2 \, dS_2 = \iint_D -2x - 2y + 1 \, dA$

$$= \int_0^2 \int_{1 - \sqrt{1 - (x-1)^2}}^{1 + \sqrt{1 - (x-1)^2}} -2x - 2y + 1 \, dy \, dx$$

$$= \int_0^2 -2xy - y^2 + y \Big]_{1 - \sqrt{1 - (x-1)^2}}^{1 + \sqrt{1 - (x-1)^2}} \, dx$$

$$= \int_0^2 -4x\sqrt{1 - (x - 1)^2} - 2\sqrt{1 - (x - 1)^2} \, dx$$

$$= -3\pi.$$

(c) The boundary curve $C = \partial S_1 = \partial S_2$ is a space curve on the plane $z = 2x + 2y - 1$ and over the circle

$$(x - 1)^2 + (y - 1)^2 = 1.$$

Thus C has the parametric equations

$$x = 1 + \cos\theta, \quad y = 1 + \sin\theta, \quad z = 2\cos\theta + 2\sin\theta + 3, \quad 0 \le \theta \le 2\pi.$$

Then $\qquad dx = -\sin\theta \, d\theta, \qquad dy = \cos\theta \, d\theta,$

$$dz = (-2\sin\theta + 2\cos\theta) \, d\theta.$$

$$\oint_C \mathbf{F} \cdot \mathbf{T}\, ds = \oint_C z\, dx + x\, dy + y\, dz$$
$$= \int_0^{2\pi} [(2\cos\theta + 2\sin\theta + 3)(-\sin\theta) + (1 + \cos\theta)\cos\theta$$
$$+ (1 + \sin\theta)(-2\sin\theta + 2\cos\theta)]\, d\theta$$
$$= \int_0^{2\pi} (1 + 3\cos\theta - 5\sin\theta - 5\sin^2\theta)\, d\theta = -3\pi.$$

Notice that (a) was much easier than (b) or (c).

Gauss' Theorem shows a relationship between a triple integral over a region E in space and a surface integral over the boundary of E. It corresponds to Green's Theorem in the form

$$\oint_{\partial D} \mathbf{F} \cdot \mathbf{N}\, ds = \iint_D \operatorname{div} \mathbf{F}\, dA.$$

Before stating Gauss' Theorem, we must explain what is meant by the surface integral over the boundary of a solid region E. In general, the boundary of E is made up of six surfaces corresponding to the six faces of a cube (Figure 13.6.7). Sometimes one or more faces will degenerate to a line or a point.

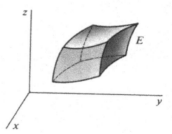

Figure 13.6.7

The top and bottom faces of E are (x, y) surfaces, that is, they are given by equations $z = c(x, y)$. However, the left and right faces of E are (x, z) surfaces $y = b(x, z)$, while the front and back faces of E are (y, z) surfaces of the form $x = a(y, z)$. Surface integrals over oriented (x, z) and (y, z) surfaces are defined exactly as for (x, y) surfaces except that the variables are interchanged.

In the following discussion E is a solid region all of whose faces are smooth surfaces.

DEFINITION

*The **boundary** of E, ∂E, is the union of the six faces of E oriented so that the outside surfaces are positive. The **surface integral** of a vector field $\mathbf{F}(x, y, z)$ over ∂E,*

$$\iint_{\partial E} \mathbf{F} \cdot \mathbf{N}\, dS,$$

is the sum of the surface integrals of $\mathbf{F}$ over the six faces of E. (See Figure 13.6.8.)

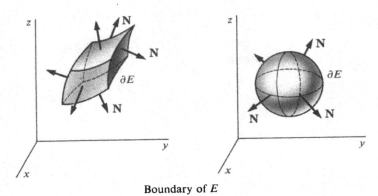

Boundary of E

Figure 13.6.8

We are now ready to state Gauss' Theorem.

GAUSS' THEOREM

Given a vector field $\mathbf{F}(x, y, z)$ *and a solid region* E,

$$\iint_{\partial E} \mathbf{F} \cdot \mathbf{N} \, dS = \iiint_{E} \operatorname{div} \mathbf{F} \, dV.$$

This equation may also be written in the form

$$\iint_{\partial E} \mathbf{F} \cdot \mathbf{N} \, dS = \iiint_{E} \frac{\partial P}{\partial x} + \frac{\partial Q}{\partial y} + \frac{\partial R}{\partial z} \, dV.$$

Gauss' Theorem is sometimes called the *Divergence Theorem.*

For fluid flow, Gauss' Theorem states that the outward rate of flow across the boundary of E is equal to the integral of the divergence over E (Figure 13.6.9). As in the two-dimensional case, the divergence is the rate at which the density is decreasing.

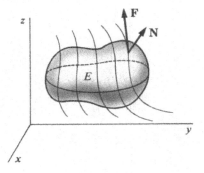

Figure 13.6.9

The following corollary is another analogue of the Path Independence Theorem.

COROLLARY 3

If $\mathbf{F}(x, y, z)$ has continuous second partials, the surface integral of **curl F** over the boundary of E is zero. In symbols,

$$\iint_{\partial E} \text{curl } \mathbf{F} \cdot \mathbf{N} \, dS = 0.$$

PROOF Since div(**curl F**) = 0,

$$\iint_{\partial E} \text{curl } \mathbf{F} \cdot \mathbf{N} \, dS = \iiint_{E} \text{div}(\text{curl } \mathbf{F}) \, dV = \iiint_{E} 0 \, dV = 0.$$

EXAMPLE 3 Use Gauss' Theorem to evaluate the surface integral

$$\iint_{\partial E} \mathbf{F} \cdot \mathbf{N} \, dS,$$

where $\qquad \mathbf{F}(x, y, z) = e^x \mathbf{i} + e^y \mathbf{j} + xyz \mathbf{k}$

and E is the unit cube in Figure 13.6.10.

$$0 \leq x \leq 1, \qquad 0 \leq y \leq 1, \qquad 0 \leq z \leq 1.$$

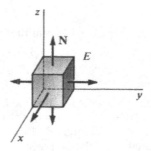

Figure 13.6.10

By Gauss' Theorem,

$$\iint_{\partial E} \mathbf{F} \cdot \mathbf{N} \, dS = \iiint_{E} \text{div } \mathbf{F} \, dV$$

$$= \iiint_{E} e^x + e^y + xy \, dV$$

$$= \int_{0}^{1} \int_{0}^{1} \int_{0}^{1} e^x + e^y + xy \, dz \, dy \, dx$$

$$= \int_{0}^{1} \int_{0}^{1} e^x + e^y + xy \, dy \, dx$$

$$= \int_{0}^{1} e^x + e - 1 + \tfrac{1}{2}x \, dx$$

$$= 2e - \tfrac{7}{4}.$$

PROBLEMS FOR SECTION 13.6

In Problems 1–6, find the curl and divergence of the vector field.

1      $F(x, y, z) = x^2 i + y^2 j + z^2 k$

2      $F(x, y, z) = x \cos z i + y \sin z j + z k$

3      $F(x, y, z) = (x + y + z)i + (y + z)j + z k$

4      $F(x, y, z) = yz i + xz j + xy k$

5      $F(x, y, z) = x e^{y+z} i + y e^{x+z} j + z e^{x+y} k$

6      $F(x, y, z) = y i + x j + k$

7   Prove that for every vector field $F(x, y, z)$ with continuous second partials, div(**curl** F) = 0.

8   Given a function $f(x, y, z)$ with continuous second partials, show that

$$\text{div}(\textbf{grad } f) = \frac{\partial^2 f}{\partial x^2} + \frac{\partial^2 f}{\partial y^2} + \frac{\partial^2 f}{\partial z^2}.$$

9   Use Stokes' Theorem to evaluate the surface integral $\iint_S \textbf{curl } F \cdot \textbf{N} \, dS$ where S is the portion of the paraboloid $z = 1 - x^2 - y^2$ above the (x, y) plane and $F(x, y, z) = xy^2 i - x^2 yj + xyz k$. ($S$ is oriented with the top side positive.)

10   Use Stokes' Theorem to evaluate the line integral

$$\oint_{\partial S} (y i + z j - x k) \cdot \textbf{T} \, ds$$

where S is the portion of the plane $z = 2x + 5y$ inside the cylinder $x^2 + y^2 = 1$ oriented with the top side positive.

11   Use Stokes' Theorem to evaluate the line integral

$$\oint_{\partial S} (ax + by + cz)(i + j + k) \cdot \textbf{T} \, ds$$

where S is the portion of the plane $z = px + qy + r$ over a region D of area A, oriented with the top side positive.

12   Use Stokes' Theorem to show that the line integral

$$\oint_{\partial S} (P(x)i + Q(y)j + R(z)k) \cdot \textbf{T} \, ds = 0$$

for any oriented surface S.

13   Use Gauss' Theorem to compute the surface integral

$$\iint_{\partial E} (x^2 i + y^2 j + z^2 k) \cdot \textbf{N} \, dS$$

where E is the rectangular box $0 \leq x \leq a, 0 \leq y \leq b, 0 \leq z \leq c$.

14   Use Gauss' Theorem to compute the surface integral

$$\iint_{\partial E} (2xy i + 3xy j + z e^{x+y} k) \cdot \textbf{N} \, dS$$

where E is the rectangular box $0 \leq x \leq 1, 0 \leq y \leq 1, 0 \leq z \leq 1$.

15   Use Gauss' Theorem to evaluate

$$\iint_{\partial E} (x i + 2y j + 3z k) \cdot \textbf{N} \, dS$$

where E is the region $0 \leq x \leq 1, 0 \leq y \leq x, 0 \leq z \leq x + y$.

16   Use Gauss' Theorem to evaluate

$$\int\int_{\partial E}(x^3\mathbf{i} + y^3\mathbf{j} + z^3\mathbf{k})\cdot\mathbf{N}\,dS$$

where E is the sphere $x^2 + y^2 + z^2 \leq 4$.

17   Use Gauss' Theorem to evaluate

$$\int\int_{\partial E}(\sqrt{x^2 + y^2 + z^2})(\mathbf{i} + \mathbf{j} + \mathbf{k})\cdot\mathbf{N}\,dS$$

where E is the hemisphere $0 \leq z \leq \sqrt{1 - x^2 - y^2}$.

18   Use Gauss' Theorem to evaluate

$$\int\int_{\partial E}(xy^2\mathbf{i} + yz\mathbf{j} + x^2z\mathbf{k})\cdot\mathbf{N}\,dS$$

where S is the cylinder $x^2 + y^2 \leq 1, 0 \leq z \leq 4$.

19   Use Gauss' Theorem to evaluate

$$\int\int_{\partial E}(x\cos^2 z\mathbf{i} + y\sin^2 z\mathbf{j} + \sqrt{x^2 + y^2}\,z\mathbf{k})\cdot\mathbf{N}\,dS$$

where E is the part of the cone $z = 1 - \sqrt{x^2 + y^2}$ above the (x, y) plane.

EXTRA PROBLEMS FOR CHAPTER 13

1   Find the derivative of $z = \cos x + \sin y$ in the direction of the unit vector $\mathbf{U} = \cos \alpha\mathbf{i} + \sin \alpha\mathbf{j}$.

2   Find **grad** f and f_U if

$$f(x, y) = \cosh x \sinh y, \qquad \mathbf{U} = \frac{-\mathbf{i} + \mathbf{j}}{\sqrt{2}}.$$

3   Find **grad** f and f_U if

$$f(x, y) = e^{xy}, \qquad \mathbf{U} = \cos \alpha\mathbf{i} + \sin \alpha\mathbf{j}.$$

4   Find the derivative of $z = \ln(x^2 + y^2)$ at the point $(-1, 1)$ in the direction of the unit vector $\mathbf{U} = a\mathbf{i} + b\mathbf{j}$.

5   Find a unit vector normal to the surface $z = xy$ at the point $(2, 3, 6)$.

6   Evaluate the line integral

$$\int_C (\cos x\mathbf{i} - \sin y\mathbf{j})\cdot d\mathbf{S}$$

where C is the curve $x = t^2, y = t^3, 0 \leq t \leq 1$.

7   Evaluate the line integral

$$\int_C \left(\frac{\mathbf{i}}{xy^2} + \frac{\mathbf{j}}{2x + y}\right)\cdot d\mathbf{S}$$

where C is the rectangular curve from $(1, 2)$ to $(4, 2)$ to $(4, 4)$.

8   Evaluate the line integral

$$\int_C (x\mathbf{i} + y\mathbf{j} + z\mathbf{k})\cdot d\mathbf{S}$$

where C is the line $x = 2t, y = 3t, z = -t, 0 \leq t \leq 1$.

9   Find the work done by the force $\mathbf{F} = y^2\mathbf{i} + x^2\mathbf{j}$ acting once counterclockwise around the circle $x^2 + y^2 = 1$.

10   Find a potential function for $y \cosh x\mathbf{i} + \sinh x\mathbf{j}$.

11   Find a potential function for $(y \ln y + \ln x)\mathbf{i} + (x \ln y + x)\mathbf{j}$.

12   Solve the differential equation
$$(2x - 6x^2y + y^3)\,dx + (-2x^3 + 3xy^2 + 1)\,dy = 0.$$

13   Solve the differential equation $e^{-y} \sin x\,dx + (e^{-y} \cos x + 3y)\,dy$.

14   Use Green's Theorem to evaluate the line integral
$$\oint_{\partial D} \sin x \sin y\,dx + \cos x \cos y\,dy,$$
$$D: \pi/6 \le x \le \pi/3, \pi/6 \le y \le \pi/3.$$

15   Use Green's Theorem to evaluate the line integral
$$\oint_{\partial D} 2xy^2\,dx + 3x^2y^3\,dy, \quad D: 0 \le x \le 1, x^2 \le y \le 2x.$$

16   Use Green's Theorem to find the area of the region bounded by the parametric curve
$$x = a \cos^3 \theta, \quad y = a \sin^3 \theta, \quad 0 \le \theta \le 2\pi.$$

17   Find the area of the part of the surface $z = x^2 + y$ which lies over the triangular region $0 \le x \le 1, 0 \le y \le x$.

18   Find the area of the part of the surface $z = xy$ which is inside the cylinder $x^2 + y^2 = 4$.

19   Evaluate the surface integral
$$\iint_S (x\mathbf{i} + y\mathbf{j} + z\mathbf{k}) \cdot \mathbf{N}\,dS,$$
where S is the upper half of the sphere $x^2 + y^2 + z^2 = 1$, oriented with the top side positive.

20   Find the curl and divergence of the vector field
$$\mathbf{F}(x, y) = xe^y\mathbf{i} + ye^x\mathbf{j}.$$

21   Find the curl and divergence of the vector field
$$\mathbf{F}(x, y, z) = xyz\mathbf{i} + xy^2z^3\mathbf{j} + x^2yz\mathbf{k}.$$

22   Use Gauss' Theorem to evaluate the surface integral
$$\iint_{\partial E} (xy^2\mathbf{i} + yz^2\mathbf{j} + x^2y\mathbf{k}) \cdot \mathbf{N}\,dS$$
where E is the region $x^2 + y^2 \le 1, x^2 + y^2 \le z \le 1$.

☐ **23**   The gravitational force of a point mass m_1 acting on another point mass m_2 has the direction of the vector $\mathbf{D}$ from m_2 to m_1 and has magnitude proportional to the inverse square of the distance $|\mathbf{D}|$. Thus
$$\mathbf{F} = \frac{cm_1m_2\mathbf{D}}{|\mathbf{D}|^3}$$
where c is constant. Use the Infinite Sum Theorem to show that the gravitational force of an object with density $h(x, y, z)$ in a region E on a point mass m at (a, b, c) is
$$\mathbf{F} = P\mathbf{i} + Q\mathbf{j} + R\mathbf{k},$$

where
$$P = \iiint_E \frac{cmh(x, y, z)(x - a)}{[(x - a)^2 + (y - b)^2 + (z - c)^2]^{3/2}}\,dV,$$

$$Q = \iiint_E \frac{cmh(x, y, z)(y - b)}{[(x - a)^2 + (y - b)^2 + (z - c)^2]^{3/2}}\,dV,$$

$$R = \iiint_E \frac{cmh(x, y, z)(z - c)}{[(x - a)^2 + (y - b)^2 + (z - c)^2]^{3/2}}\,dV.$$

☐ 24   Suppose $z = f(x, y)$ is differentiable at (a, b). Prove that the directional derivatives $f_U(a, b)$ exist for all U. (See also extra Problem 36 in Chapter 11.)

☐ 25   Let $U = \cos \alpha i + \sin \alpha j$. Suppose that $z = f(x, y)$ has continuous second partial derivatives. Prove that the second directional derivative of f in the direction U is given by

$$f_{UU}(x, y) = \frac{\partial^2 f}{\partial x^2} \cos^2 \alpha + 2\frac{\partial^2 f}{\partial x\, \partial y} \cos \alpha \sin \alpha + \frac{\partial^2 f}{\partial y^2} \sin^2 \alpha.$$

☐ 26   Second Derivative Test for two variables. Suppose
(a) $f(x, y)$ has an interior critical point (a, b) in a rectangle D.

(b) Throughout D, $\dfrac{\partial^2 f}{\partial x^2}, \dfrac{\partial^2 f}{\partial y^2}, \dfrac{\partial^2 f}{\partial x\, \partial y}$ are continuous and

$$\frac{\partial^2 f}{\partial x^2} > 0, \qquad \frac{\partial^2 f}{\partial y^2} > 0, \qquad \frac{\partial^2 f}{\partial x^2} \frac{\partial^2 f}{\partial y^2} - \left(\frac{\partial^2 f}{\partial x \partial y}\right)^2 > 0.$$

Prove that f has a minimum in D at (a, b). *Hint:* Use the preceding problem to show that all the second directional derivatives $f_{UU}(x, y)$ are positive so that the surface $z = f(x, y)$ has a minimum in every direction at (a, b). In the case $\cos \alpha \sin \alpha > 0$, use the inequality

$$0 \le \left(\sqrt{\frac{\partial^2 f}{\partial x^2}} \cos \alpha - \sqrt{\frac{\partial^2 f}{\partial y^2}} \sin \alpha \right)^2,$$

and use a similar inequality when $\cos \alpha \sin \alpha < 0$.

☐ 27   Given a sphere of mass m_1 and constant density, and a point mass m_2 outside the sphere at distance D from the center. Show that the gravitational force on m_2 is the same as it would be if all the mass of the sphere were concentrated at the center. That is, F points toward the center and has magnitude

$$|F| = \frac{cm_1 m_2}{D^2}.$$

Hint: For simplicity let the center of the sphere be at the origin and let m_2 be at the point $(0, 0, D)$ on the z-axis. Let the sphere have radius b and density h, so

$$h = m_1/\text{volume} = 3m_1/4\pi b^3, \qquad b < D.$$

By symmetry the i and j components of the force are zero. Use spherical coordinates to find the k component,

$$R = \iiint_E \frac{cm_2 h \cdot (z - D)}{[x^2 + y^2 + (z - D)^2]^{3/2}} \, dV$$

$$= \int_0^{2\pi} \int_0^b \int_0^\pi \frac{cm_2 h(\rho \cos \phi - D)\rho^2 \sin \phi}{[\rho^2 + D^2 - 2D\rho \cos \phi]^{3/2}} \, d\phi \, d\rho \, d\theta.$$

☐ 28   A region D in the plane has a piecewise smooth boundary ∂D and area A. Use Green's Theorem to show that an object with constant density k in D has center of mass

$$\bar{x} = \frac{1}{2A} \oint_{\partial D} x^2 \, dy, \qquad \bar{y} = -\frac{1}{2A} \oint_{\partial D} y^2 \, dx.$$

☐ 29   Show that the object in the preceding exercise has moment of inertia about the origin

$$I = \frac{k}{3} \oint_{\partial D} - y^3 \, dx + x^3 \, dy.$$

☐ 30   Use the Infinite Sum Theorem to show that the mass of a film of density $\rho(x, y)$ per unit area on a surface $z = f(x, y)$, (x, y) in D, is

$$m = \int\int_D \sqrt{\left(\frac{\partial z}{\partial x}\right)^2 + \left(\frac{\partial z}{\partial y}\right)^2 + 1} \;\; \rho(x, y) \, dx \, dy.$$

☐ **31**   Show that the volume of a region E is equal to the surface integral

$$V = \tfrac{1}{3} \iint\limits_{\partial E} (x\mathbf{i} + y\mathbf{j} + z\mathbf{k}) \cdot \mathbf{N} \, dS.$$

☐ **32**   Show that the gravity force field of a mass m at the origin,

$$\mathbf{F}(x, y, z) = \frac{m}{x^2 + y^2 + z^2} \frac{x\mathbf{i} + y\mathbf{j} + z\mathbf{k}}{\sqrt{x^2 + y^2 + z^2}},$$

is irrotational (except at the origin). Use Stokes' Theorem to show that

$$\oint_{\partial S} \mathbf{F}(x, y, z) \cdot \mathbf{T} \, ds = 0$$

where S is any oriented surface not containing the origin.

☐ **33**   Show that for any smooth closed curve C around the origin,

$$\oint_C \frac{-y}{x^2 + y^2} \, dx + \frac{x}{x^2 + y^2} \, dy = 2\pi.$$

Assume for simplicity that C has the parametric equation

$$C : r = f(\theta), 0 \leq \theta \leq 2\pi \qquad \text{where } 0 < f(\theta), f(0) = f(2\pi).$$

DIFFERENTIAL EQUATIONS

14.1 EQUATIONS WITH SEPARABLE VARIABLES

A *first order differential equation* is an equation involving an independent variable t, a dependent variable y, and the derivative dy/dt. In many applications, the independent variable t is time. A first order differential equation can be put in the following form, where $f(t, y)$ is continuous in both t and y.

FIRST ORDER DIFFERENTIAL EQUATION

(1)
$$\frac{dy}{dt} = f(t, y).$$

For instance, if t is time and y is the position of a particle at time t, the differential equation (1) gives the velocity of the particle in terms of time and position. A differential equation gives information about an unknown function $y(t)$. The *general solution* of a first order differential equation is the family of all functions $y(t)$ that satisfy the equation. Each function in this family is called a *particular solution* of the differential equation. In most cases, the family of functions will depend in some way on a constant C, and the graphs of these functions will form a family of curves that fill up the (t, y) plane but do not touch each other, as in Figure 14.1.1.

Some examples of first order differential equations were solved in Section 8.6. For instance, it was shown that the general solution of the differential equation

$$\frac{dy}{dt} = y(1 - y)$$

is

$$y(t) = \frac{1}{1 + Ce^{-t}}, \qquad y(t) = 0.$$

There is one particular solution for each value of the constant C, and one additional particular solution $y(t) = 0$. The graph of this general solution is shown in Figure 14.1.2.

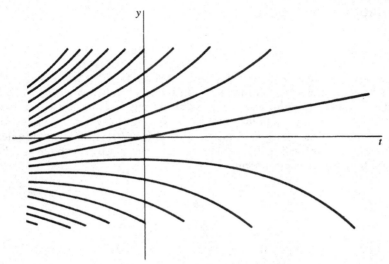

Figure 14.1.1

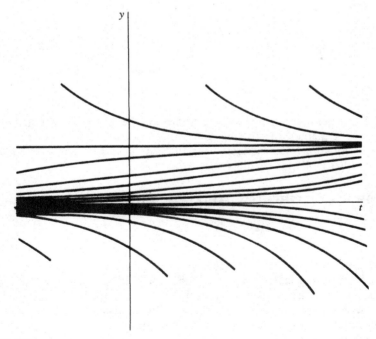

Figure 14.1.2

In most applications, a first order differential equation will describe a process that starts at some initial time t_0. In order to determine a particular solution, we need both the differential equation and the value of $y(t)$ at the initial time t_0. A *first order initial value problem* is a pair of equations consisting of a first order differential equation and an initial value.

FIRST ORDER INITIAL VALUE PROBLEM

(2) $$\frac{dy}{dt} = f(t, y), \qquad y(t_0) = y_0.$$

An initial value problem usually has just one solution, which will be a particular solution of the differential equation. This can be seen intuitively as follows. Figure 14.1.3 shows a moving point controlled by an infinitesimal driver with a steering wheel. At $t = t_0$, the moving point starts at (t_0, y_0). At each $t > t_0$, the infinitesimal driver measures his position (t, y), computes the value of $f(t, y)$, and turns the steering wheel so that the slope will be $f(t, y)$. The curve traced out by this point will be the solution of the initial value problem.

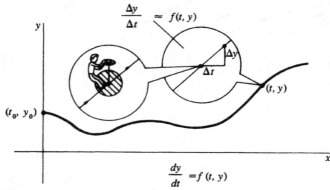

Figure 14.1.3

The general problem of solving a differential equation is difficult. In Section 8.6, we gave a method for solving differential equations of an especially simple type, called equations with separable variables. As a starting point in this chapter, we take another look at these differential equations.

FIRST ORDER DIFFERENTIAL EQUATION WITH SEPARABLE VARIABLES

(3) $$\frac{dy}{dt} = g(t)h(y).$$

In an equation with separable variables, dy/dt is a product of a function of t and a function of y. In particular, $dy/dt = g(t)$ is a differential equation with separable variables in which $h(y)$ is the constant 1. Similarly, $dy/dt = h(y)$ is a differential equation with separable variables in which $g(t)$ is the constant 1. An equation with separable variables can be solved by separating the variables and integrating both sides of the equation.

METHOD FOR SOLVING A DIFFERENTIAL EQUATION WITH SEPARABLE VARIABLES

$$\frac{dy}{dt} = g(t)h(y).$$

Step 1 *Find all points y_1 where $h(y_1) = 0$. For each such point, the constant function $y(t) = y_1$ is a particular solution.*

Step 2 *Separate the variables by dividing by $h(y)$ and multiplying by dt,*

$$\frac{1}{h(y)} \, dy = g(t) \, dt,$$

then integrate both sides of the equation. That is, find antiderivatives of each side.

$$K'(y) = \frac{1}{h(y)}, \qquad G'(t) = g(t),$$

so that $$K(y) = G(t) + C.$$

If possible, solve for y as a function of t.

Step 3 *The general solution is the family of all solutions found in Steps 1 and 2. It will usually depend on a constant C.*

Step 4 *If an initial value $y(t_0) = y_0$ is given, use it to find the constant C and the particular solution of the initial value problem.*

Remark The cases $h(y) = 0$ and $h(y) \neq 0$ must be done separately in Steps 1 and 2, because the division by $h(y)$ in Step 2 cannot be done when $h(y) = 0$.

The general solution of a differential equation $dy/dt = g(t)$, where dy/dt is a function of t alone, is just the indefinite integral

$$y = \int g(t) \, dt = G(t) + C.$$

In this case, C is the familiar constant of integration, which is added to a particular solution. For example, the general solution of the differential equation $dy/dt = 1/t$ is $y = \ln |t| + C$.

In the examples that follow, the constant C appears in a more complicated manner.

EXAMPLE 1 Solve the initial value problem

$$\frac{dy}{dt} = -2y, \qquad y(1) = -5.$$

Step 1 $-2y = 0$ when $y = 0$. Thus the constant $y(t) = 0$ is a particular solution.

Step 2 Separate the variables and integrate both sides.

$$\frac{dy}{y} = -2 \, dt,$$
$$\ln |y| = -2t + B,$$
$$|y| = e^B e^{-2t},$$
$$y = Ce^{-2t},$$

where $C = e^B$ if $y > 0$, and $C = -e^B$ if $y < 0$.

Step 3   General solution:

$$y(t) = Ce^{-2t},$$

where C is any constant (C can be 0 from Step 1).

Step 4   Substitute 1 for t and -5 for y, and solve for C.

$$-5 = Ce^{-2\cdot 1}, \qquad -5 = Ce^{-2}, \qquad C = -5e^2.$$

Particular solution:

$$y(t) = -5e^2 e^{-2t} = -5e^{(2-2t)}.$$

The graph of the solution is shown in Figure 14.1.4.

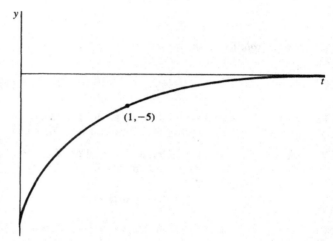

Figure 14.1.4   Example 1

EXAMPLE 2   Find the general solution of the differential equation

$$\frac{dy}{dt} = y^2 \sin t.$$

Then find the particular solution with $y(0) = \tfrac{1}{2}$.

Step 1 $y^2 = 0$ when $y = 0$. Thus $y(t) = 0$ is a constant solution.

Step 2
$$y^{-2}\,dy = \sin t\,dt,$$
$$-y^{-1} = -\cos t + C,$$
$$y = (\cos t - C)^{-1}.$$

Step 3   General solution:

$$y(t) = 0 \qquad \text{and} \qquad y(t) = (\cos t - C)^{-1}.$$

Step 4
$$y(0) = \tfrac{1}{2} = (\cos 0 - C)^{-1} = (1 - C)^{-1},$$
$$2 = 1 - C, \qquad C = -1.$$

Particular solution:

$$y(t) = (\cos t + 1)^{-1}.$$

The particular solution to Example 2 is illustrated in Figure 14.1.5. It is defined only for $-\pi \leq t < \pi$ and approaches ∞ as t approaches π. It is said to have an *explosion* at $t = \pi$.

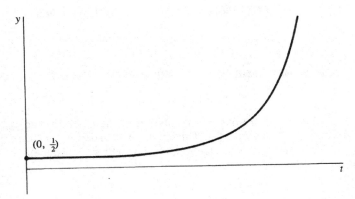

Figure 14.1.5 Example 2

To avoid errors, the general solution can be checked by differentiating. The solution for Example 2 is checked as follows.

$$y(t) = (\cos t - C)^{-1}.$$
$$\frac{dy}{dt} = -(\cos t - C)^{-2}(-\sin t)$$
$$= (\cos t - C)^{-2}(\sin t)$$
$$= y^2 \sin t,$$

as required.

PROBLEMS FOR SECTION 14.1

In Problems 1–12, find the general solution of the given differential equation.

1	$y' = t \cdot \sin(t^2)$	**2**	$y' = e^{-3t}$
3	$y' = e^{-y}$	**4**	$y' = y^3$
5	$y' = y^2 - 1$	**6**	$y' = ty(y + 1)$
7	$y' = y^2 t$	**8**	$y' = \dfrac{2t + 1}{2y - 1}$
9	$y' = (1 + y^2)e^t$	**10**	$y' = \sqrt{1 - y^2} \cos t$
11	$y' = y \tan t$	**12**	$y' = y \sin t$

In Problems 13–18, solve the initial value problem.

13	$y' = y^2 t^3, \quad y(1) = 2$	**14**	$y' = t\sqrt{y}, \quad y(0) = 3$

15 $\quad y' = \dfrac{\ln t}{y}, \quad y(1) = -2$ $\qquad$ **16** $\qquad y' = ty - y + 2t - 2, \quad y(0) = 0$

17 $\quad y' = (y^2 - 3y + 2)\sqrt{t}, \quad y(1) = 2$

18 $\quad y' = e^y(y - 4)\sin t, \quad y(2) = 4$

14.2 FIRST ORDER HOMOGENEOUS LINEAR EQUATIONS

In this section we study the following special type of differential equation.

FIRST ORDER HOMOGENEOUS LINEAR DIFFERENTIAL EQUATION

(1) $\qquad\qquad\qquad y' + p(t)y = 0.$

It is understood that t varies over some interval in the real line, and $p(t)$ is a continuous function of t in the interval. The equation is called *linear* because y and y' occur only linearly and *homogeneous* because the right side of the equation is zero. The equation

$$y' = ky, \quad \text{or} \quad y' - ky = 0,$$

for exponential growth (see Section 8.6) is an example. The first order homogeneous linear differential equation (1) has separable variables, because it can be written as

$$\frac{dy}{dt} = -p(t)y.$$

Its solution is given by the next formula.

METHOD FOR SOLVING FIRST ORDER HOMOGENEOUS LINEAR DIFFERENTIAL EQUATION (1)

The general solution is

$$y(t) = Ce^{-P(t)},$$

where $P(t)$ is an antiderivative of $p(t)$. That is,

$$y(t) = Ce^{-\int p(t)\,dt}.$$

This formula is obtained by the procedure described in Section 14.1 for differential equations with separable variables, as follows. First write the equation in the form

$$\frac{dy}{dt} = -p(t)y.$$

Step 1 There is a constant solution $y(t) = 0$.

Step 2 Separate the variables and integrate:

$$y^{-1}\,dy = -p(t)\,dt.$$

$$\ln|y| = -\int p(t)\,dt + B.$$

Now solve for y.

$$|y| = e^{-\int p(t)\,dt + B},$$
$$y = Ce^{-\int p(t)\,dt},$$

where $C = e^B$ if $y > 0$, and $C = -e^B$ if $y < 0$.

Step 3 Combining Steps 1 and 2, we get the general solution

$$y(t) = Ce^{-\int p(t)\,dt}.$$

Remark The case $C = 0$ gives the constant solution $y(t) = 0$ of Step 1.

Discussion The constant of integration in the indefinite integral

$$\int p(t)\,dt$$

will be absorbed in the constant C.

The particular solution for the initial value $y(t_0) = y_0$ is found by substituting and computing C. Notice that any two particular solutions of the same homogeneous linear differential equation differ only by a constant factor. If $x(t)$ is any nonzero particular solution, then the general solution is $Cx(t)$.

EXAMPLE 1

(a) Find the general solution of the equation $y' + y\cos t = 0$.
(b) Find the particular solution with initial value $y(0) = \frac{1}{2}$.
(c) Find the particular solution with initial value $y(2) = \frac{1}{2}$.

SOLUTION

(a) First evaluate the integral

$$\int \cos t\,dt = \sin t + B.$$

General solution:

$$y(t) = Ce^{-\sin t}.$$

(b) First substitute and solve for C.

$$y(0) = \frac{1}{2} = Ce^{-\sin 0} = Ce^0 = C.$$

Particular solution:

$$y(t) = \frac{1}{2}e^{-\sin t}.$$

(c) Substitute and solve for C.

$$y(2) = \frac{1}{2} = Ce^{-\sin 2},$$
$$C = \frac{1}{2}e^{\sin 2} = 1.2413.$$

Particular solution:

$$y(t) = 1.2413e^{-\sin t}.$$

The solution to this example is shown in Figure 14.2.1.

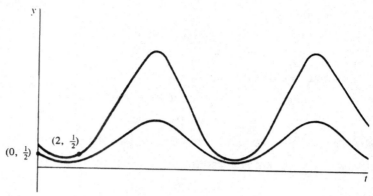

Figure 14.2.1   Example 1

EXAMPLE 2

(a)   Find the general solution of the equation $ty' + 3y = 0$ for $t > 0$.
(b)   Find the particular solution with the initial value $y(1) = 2$.

SOLUTION

(a)   We first put the equation into the homogeneous linear form (1) by dividing by t:

$$y' + 3t^{-1}y = 0.$$

Next evaluate the integral,

$$\int 3t^{-1}\, dt = 3\ln t + B.$$

The constant of integration B is absorbed into the constant C, and the general solution is

$$y(t) = Ce^{-3\ln t} = Ct^{-3}.$$

(b)   The particular solution with initial value $y(1) = 2$ is

$$y(t) = 2t^{-3}.$$

The solution to this example is shown in Figure 14.2.2.

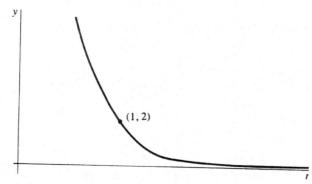

Figure 14.2.2   Example 2

The coefficient $3t^{-1}$ in the equation

$$y' + 3t^{-1}y = 0$$

of Example 2 is discontinuous at $t = 0$. However, it is continuous on the interval $t > 0$ and on the interval $t < 0$. An initial value at a positive time $t_0 > 0$ will determine a particular solution only for the interval $t > 0$, while an initial value at a negative time $t_0 < 0$ will determine a particular solution for the interval $t < 0$. Each interval must be solved separately. The next example is like Example 2 but on the negative time interval.

EXAMPLE 3

(a) Find the general solution of the equation $y' + 3t^{-1}y = 0$ from Example 2, but for the interval $t < 0$ instead of $t > 0$.

(b) Find the particular solution of the initial value problem with $y(-2) = 1$.

SOLUTION

(a) This time we integrate with a negative t,

$$\int 3t^{-1}\,dt = 3\ln|t| + B = 3\ln(-t) + B.$$

The general solution for $t < 0$ is thus

$$y(t) = Ce^{-3\ln(-t)} = C(-t)^{-3} = -Ct^{-3},$$

or $$y(t) = At^{-3},$$

where A is the constant $-C$.

(b) The particular solution with the initial value $y(-2) = 1$ is found by solving for A:

$$1 = A(-2)^{-3}, \qquad A = -8, \qquad y(t) = -8t^{-3}.$$

The solution to this example is shown in Figure 14.2.3.

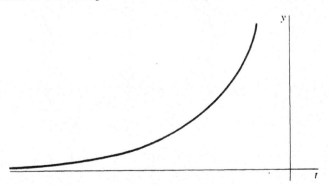

Figure 14.2.3   Example 3

For some purposes, it is useful to describe the solution of a differential equation using a definite integral from some point a to t, instead of using an indefinite integral. In the definite integral form, the general solution of the linear homogeneous differential equation

(1)
$$y' + p(t)y = 0$$
is
$$y(t) = Ce^{-\int_a^t p(s)\,ds},$$

where a is any point in the interval I, and C is any real number. This formula is helpful in a problem where one cannot evaluate the integral of $p(t)$ exactly and must use a numerical approximation. The formula holds because the integral

$$\int_a^t p(s)\,ds$$

is an antiderivative of $p(t)$ by the Fundamental Theorem of Calculus. The choice of the endpoint a does not matter because a change in the value of a will be absorbed by a change in the value of the constant C. If we are given an initial value $y(t_0) = y_0$, the particular solution can again be found by substituting and solving for C. If we take $a = t_0$, the constant C will be equal to y_0,

$$y_0 = y(a) = Ce^{-\int_a^a p(s)\,ds} = Ce^0 = C.$$

Thus the particular solution of the initial value problem (1) with $y(a) = y_0$ is given by

$$y(t) = y_0 e^{-\int_a^t p(s)\,ds}.$$

PROBLEMS FOR SECTION 14.2

In Problems 1–4, find the general solution of the given differential equation.

1	$y' + 5y = 0$	**2**	$y' - 2y = 0$
3	$y' + \dfrac{y}{1 + t^2} = 0$	**4**	$y' + t^2 y = 0$

In Problems 5–12, find the particular solution of the initial value problem.

5	$y' + y = 0,$ $y(0) = 4$	**6**	$y' - 3y = 0,$ $y(1) = -2$
7	$y' + y \sin t = 0,$ $y(\pi) = 1$	**8**	$y' + ye^t = 0,$ $y(0) = e$
9	$y' + y\sqrt{1 + t^4} = 0,$ $y(0) = 0$		
10	$y' + y \cos(e^t) = 0,$ $y(0) = 0$		
11	$ty' - 2y = 0,$ $y(1) = 4,$ $t > 0$		
12	$t^2 y' + y = 0,$ $y(1) = -2,$ $t > 0$		
13	$t^3 y' = 2y,$ $y(1) = 1,$ $t > 0$		
14	$t^3 y' = 2y,$ $y(1) = 0,$ $t > 0$		

15   A function $y(t)$ is a solution of the differential equation

$$y' + ky = 0$$

for some constant k. Given that $y(0) = 100$, and $y(2) = 4$, find k and find y as a function of t.

16   A function $y(t)$ is a solution of the differential equation

$$y' + t^k y = 0$$

for some constant k. Given that $y(0) = 1$, and $y(1) = e^{-1/3}$, find k and find y as a function of t.

17   A bacterial culture grows at a rate proportional to its population. If it has a population of one million at time $t = 0$ hours and 1.5 million at time $t = 1$ hour, find its population as a function of t.

18   A radioactive element decays with a half-life of 6 years. Starting with 10 lb of the element at time $t = 0$ years, find the amount of the element as a function of t.

14.3 FIRST ORDER LINEAR EQUATIONS

We shall now give a method for solving a differential equation of the following type.

FIRST ORDER LINEAR DIFFERENTIAL EQUATION

(1) $$y' + p(t)y = f(t).$$

Both $p(t)$ and $f(t)$ are continuous functions of t, where t varies over some interval I in the real line. When $f(t)$ is the constant function with value 0, the equation is a homogeneous linear differential equation of the type studied in Section 14.2.

First order linear differential equations arise in models of population growth with immigration. Suppose a population $y(t)$ has a net birthrate of $b(t)$ and net immigration rate of $f(t)$. The net birthrate $b(t)$ is the excess of births over deaths per unit of population in one unit of time. In a small period of time of length Δt, the difference of births and deaths is $b(t) \cdot y(t) \cdot \Delta t$, and the net immigration is $f(t) \cdot \Delta t$. Then the population will be a solution of the differential equation $y' = b(t)y + f(t)$, which is the same as equation (1) with $p(t) = -b(t)$.

The size of a bank account that earns interest and also changes due to deposits and withdrawals can be described by a first order linear differential equation. If the account earns interest at the rate of $r(t)$ at time t, and the net deposit per unit of time is $f(t)$, then the account size $y(t)$ will be a solution of the differential equation (1) with $p(t) = -r(t)$.

The next theorem will be helpful in solving an equation of the type (1).

THEOREM 1

Suppose that $y(t)$ is a particular solution of the first order linear differential equation

(1) $$y' + p(t)y = f(t),$$

and $x(t)$ is a nonzero particular solution of the corresponding homogeneous equation

(2) $$x' + p(t)x = 0.$$

Then the general solution of the original equation (1) is

$$y(t) + Cx(t).$$

We already know from Section 14.2 how to solve the homogeneous linear equation (2). So if we can find one particular solution of the linear equation (1), we can use Theorem 1 to find the general solution. We postpone the proof of Theorem 1 to the end of this section.

A particular solution of a linear equation (1) can be found by the method called *variation of constants*. Start with a particular solution $x(t)$ of the corresponding homogeneous equation (2). For any constant C, $Cx(t)$ is also a solution of (2). Now replace the constant C by a variable $v(t)$, and see what happens. Let $y(t) = v(t)x(t)$. We shall compute the left side of equation (1), $y' + p(t)y$. If it turns out to be equal to $f(t)$, then $y(t)$ will be a particular solution of (1) as required. We carry out the computations using the Product Rule for derivatives.

$$y = vx.$$
$$y' + py = (vx)' + pvx$$
$$= v'x + vx' + pvx$$
$$= v'x + v(x' + px).$$

Since x is a solution of the homogeneous equation (2),

$$x' + px = 0.$$

Therefore
$$y' + py = v'x.$$

Thus if we can find a function $v(t)$ such that

$$v'(t)x(t) = f(t),$$

then
$$y(t) = v(t)x(t)$$

is a particular solution of the linear equation (1).

Putting all the ideas together, we have a method for solving a first order linear differential equation.

METHOD FOR SOLVING A FIRST ORDER LINEAR DIFFERENTIAL EQUATION

(1)
$$y' + p(t)y = f(t).$$

The corresponding homogeneous linear differential equation is

(2)
$$x' + p(t)x = 0.$$

Step 1 *Find a nonzero particular solution $x(t)$ of the corresponding homogeneous linear differential equation (2). By the method of Section 14.2, we may take*

$$x(t) = e^{-\int p(t)\, dt}.$$

Step 2 *Find a function $v(t)$ whose derivative is given by*

$$v'(t) = \frac{f(t)}{x(t)}.$$

This is done by integration,

$$v(t) = \int \frac{f(t)}{x(t)}\, dt.$$

Step 3 *The general solution of (1) is*

$$y(t) = v(t)x(t) + Cx(t).$$

Step 4 *If an initial value is given, the particular solution for the initial value problem is found by substituting and solving for the constant C.*

Discussion Step 2 gives us a function $v(t)$ for which $v'(t)x(t) = f(t)$. Therefore, by our previous discussion, $v(t)x(t)$ is a particular solution of the linear equation (1).

Step 3 is then justified by Theorem 1.

EXAMPLE 1   Find the general solution of the equation
$$y' + 3t^{-1}y = t^2, \qquad t > 0.$$

Then find the particular solution with the initial value $y(1) = \frac{1}{2}$.

Step 1   The corresponding homogeneous equation is
$$x' + 3t^{-1}x = 0.$$

From Example 2 in Section 14.2, a particular solution is
$$x = t^{-3}.$$

Step 2
$$v' = \frac{t^2}{t^{-3}} = t^5,$$

$$v = \left(\frac{1}{6}\right)t^6.$$

Step 3   The general solution is
$$y = vx + Cx,$$
or
$$y = (\tfrac{1}{6})t^3 + Ct^{-3}.$$

Step 4
$$y(1) = \tfrac{1}{2} = (\tfrac{1}{6})1^3 + C1^{-3},$$

$$C = \tfrac{1}{3}.$$

The required particular solution (Figure 14.3.1) is
$$y = (\tfrac{1}{6})t^3 + (\tfrac{1}{3})t^{-3}.$$

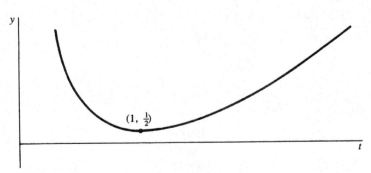

Figure 14.3.1   Example 1

EXAMPLE 2   A population has a net birthrate of 2% per year and a net annual immigration rate of $100{,}000 \sin t$. At time $t = 0$ years, the population is $1{,}000{,}000$. Find the population as a function of t.

The given verbal problem can be expressed as the initial value problem

$$y' = 0.02y + 100{,}000 \sin t, \qquad y(0) = 1{,}000{,}000.$$

We first put the equation in the usual form with all the y terms on the left side,

$$y' - 0.02y = 100{,}000 \sin t, \qquad y(0) = 1{,}000{,}000.$$

Step 1   The corresponding homogeneous equation is

$$x' - 0.02x = 0.$$

The particular solution is

$$x(t) = e^{0.02t}.$$

Step 2
$$v' = \frac{100{,}000 \sin t}{e^{0.02t}} = 100{,}000 \sin t e^{-0.02t}.$$

v can now be found by integration by parts. With $u = \sin t$ and $dw = e^{-0.02t}\, dt$, we have $w = -50e^{-0.02t}$ and

$$\int \sin t e^{-0.02t}\, dt = -50 \sin t e^{-0.02t} + 50 \int \cos t e^{-0.02t}\, dt.$$

Similarly,

$$\int \cos t e^{-0.02t}\, dt = -50 \cos t e^{-0.02t} - 50 \int \sin t e^{-0.02t}\, dt.$$

Combining the last two equations and solving for the integral of $\sin t e^{-0.02t}$, we get

$$\int \sin t e^{-0.02t}\, dt = \frac{-1}{2501} e^{-0.02t}[50 \sin t + 2500 \cos t] + K,$$

$$v(t) = \frac{-100{,}000}{2501} e^{-0.02t}[50 \sin t + 2500 \cos t].$$

Step 3   The general solution is $y = vx + Cx$, or

$$y(t) = -\frac{100{,}000}{2501}[50 \sin t + 2500 \cos t] + Ce^{0.02t}$$

Step 4   Substitute at $t = 0$.

$$1{,}000{,}000 = -\frac{100{,}000}{2501}[50 \sin 0 + 2500 \cos 0] + Ce^{0}$$

$$= -\frac{100{,}000}{2501}[2500] + C,$$

$$C = 1{,}099{,}960.$$

The particular solution (Figure 14.3.2) is then

$$y(t) = -\frac{100{,}000}{2501}[50 \sin t + 2500 \cos t] + 1{,}099{,}960e^{0.02t}.$$

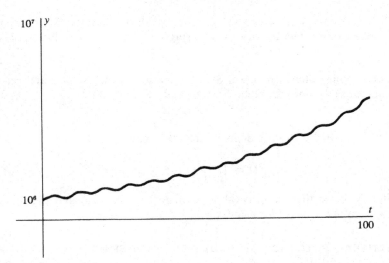

Figure 14.3.2   Example 2

EXAMPLE 3   Find the general solution of the equation

$$y' - sy = Ke^{rt},$$

where r, s, and K are constants.

Step 1   The corresponding homogeneous equation is

$$x' - sx = 0.$$

It has the particular solution

$$x(t) = e^{st}.$$

Step 2 $$v'(t) = \frac{Ke^{rt}}{e^{st}} = Ke^{(r-s)t}.$$

There are two cases, $r \neq s$ and $r = s$.

Step 3 (Case 1) $r \neq s$.

$$v(t) = \int Ke^{(r-s)t}\, dt,$$

$$v(t) = \frac{K}{r-s} e^{(r-s)t}.$$

The general solution $y = vx + Cx$ in this case is

$$y(t) = \frac{K}{r-s} e^{rt} + Ce^{st}.$$

Step 3 (Case 2) $r = s$. In this case $v'(t) = K$, and $v(t) = Kt$. The general solution in this case is

$$y(t) = Kte^{st} + Ce^{st}.$$

We now return to the general first order linear differential equation (1). Using definite integrals, we can get a single formula for the solution of equation

(1)
$$y' + p(t)y = f(t)$$

by combining Steps 1 to 4. For Step 1, choose an initial point a, and get a particular solution of the corresponding homogeneous equation,

$$x(t) = e^{-\int_a^t p(s)\, ds}.$$

For Step 2, write $v(t)$ as a definite integral from a to t,

$$v(t) = \int_a^t \frac{f(s)}{x(s)}\, ds = \int_a^t f(s)e^{\int_a^s p(r)\, dr}\, ds.$$

Step 3 shows that the general solution is $y = vx + Cx$, and the final formula is found by substituting for v and x.

GENERAL SOLUTION OF EQUATION (1), DEFINITE INTEGRAL FORM

$$y(t) = e^{-\int_a^t p(r)\, dr}\left[\int_a^t f(s)e^{\int_a^s p(r)\, dr}\, ds + C\right].$$

By taking $t = a$ in the above equation, we find that $C = y(a)$. Thus the particular solution of equation (1) with the initial condition $y(a) = y_0$ is obtained by replacing C by y_0. This formula is useful when one or both of the integrals cannot be evaluated. In a simple problem, it is better to use Steps 1 to 4, which break the solution process into smaller parts.

In the following example, we are able to evaluate the first integral but not the second, so the solution is left in a form with one definite integral.

EXAMPLE 4   Find the general solution of the equation

$$y' + y\cos t = t.$$

Step 1   From Example 1 in Section 14.2, the corresponding homogeneous equation has the particular solution

$$x = e^{-\sin t}.$$

Step 2   The function $v(t)$ is expressed by an integral.

$$v' = \frac{t}{e^{-\sin t}} = te^{\sin t},$$

$$v = \int_a^t se^{\sin s}\, ds.$$

We cannot evaluate the integral, so we leave it in this form. It does not matter which value is chosen for the lower endpoint in the integral, so we take the lower endpoint zero.

Step 3   The general solution is

$$y = vx + Cx,$$

or
$$y(t) = e^{-\sin t}\int_0^t se^{\sin s}\, ds + Ce^{-\sin t}.$$

EXAMPLE 5 Find the solution of the initial value problem

$$y' + y \ln(2 + \cos t) = t, \qquad y(1) = 4.$$

We are not able to evaluate the integral of $\ln(2 + \cos t)$, so we shall use the definite integral form of the solution. With the initial point $a = 1$, the solution is

$$y(t) = e^{-\int_1^t \ln(2 + \cos r)\, dr} \left[\int_1^t s e^{\int_1^s \ln(2 + \cos r)\, dr} \, ds + 4 \right].$$

We conclude this section with a proof of Theorem 1. The proof uses the Principle of Superposition.

PRINCIPLE OF SUPERPOSITION (First Order)

Suppose $x(t)$ and $y(t)$ are solutions of the two first order linear differential equations

$$x' + p(t)x = f(t),$$
$$y' + p(t)y = g(t).$$

Then for any constants A and B, the function

$$u(t) = Ax(t) + By(t)$$

is a solution of the linear differential equation

$$u' + p(t)u = Af(t) + Bg(t).$$

Notice that all three differential equations have the same $p(t)$. The Principle of Superposition follows from the Constant and Sum Rules for derivatives:

$$\begin{aligned}
u' + p(t)u &= (Ax + By)' + p(t)(Ax + By) \\
&= Ax' + By' + Ap(t)x + Bp(t)y \\
&= A(x' + p(t)x) + B(y' + p(t)y) \\
&= Af(t) + Bg(t).
\end{aligned}$$

PROOF OF THEOREM 1 We are given that y and x are solutions of

(1) $$y' + p(t)y = f(t)$$

and

(2) $$x' + p(t)x = 0.$$

We must prove that a function $u(t)$ is a solution of

(3) $$u' + p(t)u = f(t)$$

if and only if $u = y + Cx$ for some constant C.

Assume first that $u = y + Cx$. By the Principle of Superposition,

$$u' + p(t)u = f(t) + C \cdot 0 = f(t),$$

so u is a solution of (3).

Now assume that u is a solution of (3). Using the Principle of Superposition again,

$$(u - y)' + p(t)(u - y) = f(t) - f(t) = 0.$$

Thus $u - y$ is a solution of the homogeneous linear equation (2). The general solution of equation (2) is Cx. Therefore for some constant C,

$$u - y = Cx \quad \text{and} \quad u = y + Cx.$$

PROBLEMS FOR SECTION 14.3

In Problems 1–10, find the general solution of the given differential equation.

1	$y' + 4y = 8$	2	$y' - 2y = 6$
3	$y' + ty = 5t$	4	$y' + e^t y = -2e^t$
5	$y' - y = t^2$	6	$2y' + y = t$
7	$ty' - 2y = 1/t, \quad t > 0$	8	$ty' + y = \sqrt{t}, \quad t > 0$
9	$y' \cos t + y \sin t = 1, \quad -\pi/2 < t < \pi/2$		
10	$y' + y \sec t = \tan t, \quad -\pi/2 < t < \pi/2$		

In Problems 11–14, find the general solution using the definite integral form when the integral cannot be evaluated.

11	$y' + y \sin t = t$	12	$y' + yt^2 = \tan t, \quad -\pi/2 < t < \pi/2$
13	$y' + y \cos(e^t) = 1$	14	$y' + ye^{1/t} = 2e^t, \quad t > 0$

15   A population has a net birthrate of 2.5% per year and a net annual immigration equal to $10,000t - 40,000$, where t is measured in years. At time $t = 0$, the population is $y(0) = 100,000$. Find the population as a function of t.

16   Work Problem 15 if the net annual immigration is $1,000(\cos t - 1)$.

17   A bank account earns interest at the rate of 10% per year, and money is deposited continuously into the account at the rate of $5t^2$ dollars per year. The earnings due to interest are also left in the account. If the account had $5000 at time $t = 0$ years, find the amount in the account at time $t = 10$ years.

18   Work Problem 17 if there are no deposits but money is withdrawn continuously from the account at the rate of $5t^2$ dollars per year.

19   Use differential equations to prove the capital accumulation formula in Section 8.4. The formula says that if money is deposited continuously in an account at the rate of $f(t)$ dollars per year, and the account earns interest at the annual rate r, and there are zero dollars in the account at time $t = a$, then the value of the account at time $t = b$ will be

$$y(b) = \int_a^b f(t)e^{r(b-t)} \, dt.$$

14.4 EXISTENCE AND APPROXIMATION OF SOLUTIONS

This section deals with arbitrary first order differential equations. It is optional and therefore can be omitted if desired. Most first order differential equations cannot be solved explicitly. However, it is possible to approximate a solution by a method similar to the Riemann sum for the definite integral. The Euler approximation starts by dividing the interval $[a, \infty)$ into small subintervals of length Δt. When Δt is real, it gives an approximate solution that can be computed numerically. When Δt is infinitesimal, it leads to a precise solution and is useful because it shows that a solution exists.

Throughout this section, we shall work with a first order differential equation with an initial value

(1)
$$y' = f(t, y), \qquad y(a) = y_0.$$

We assume once and for all that $f(t, y)$ is continuous for all t and y.

DEFINITION

> *Let Δt be positive, and partition the interval $[a, \infty)$ into subintervals of length Δt. The **Euler approximation** for the initial value problem (1) is the function $Y(t)$, $a \le t$, defined as follows. Start the graph of $Y(t)$ at the point (a, y_0). Then move from (a, y_0) to $(a + \Delta t, Y(a + \Delta t))$ along a straight line with slope $f(a, Y(a))$. Once the value $Y(t)$ is computed for a partition point $t = a + k\,\Delta t$, move from $(t, Y(t))$ to the next partition point $(t + \Delta t, Y(t + \Delta t))$ along a straight line with slope $f(t, Y(t))$.*

The graph of $Y(t)$ is the broken line shown in Figure 14.4.1. Each piece has the slope required by the differential equation (1) at the beginning of the subinterval. If Δt is small, then since $f(t, y)$ is continuous, the slope of $Y(t)$ should be close to the correct slope. Thus we would expect $Y(t)$ to be close to a solution of (1).

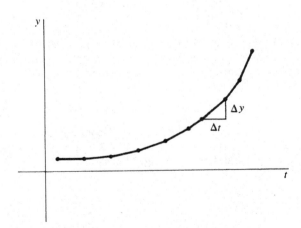

Figure 14.4.1

The values of $Y(t)$ at the partition points can be computed by an iteration that can easily be carried out on a computer. The first three values are

$$Y(a) = y_0,$$
$$Y(a + \Delta t) = y_0 + f(a, y_0)\,\Delta t,$$
$$Y(a + 2\,\Delta t) = Y(a + \Delta t) + f(a + \Delta t, Y(a + \Delta t))\,\Delta t.$$

Given the value $Y(t)$ for a partition point $t = a + k\,\Delta t$, the next value $Y(t + \Delta t)$ is given by the rule

$$Y(t + \Delta t) = Y(t) + f(t, Y(t)) \, \Delta t.$$

Using the sigma notation, the $(k + 1)^{st}$ value of $Y(t)$ can be written as

$$Y(a + k \, \Delta t) = y_0 + \sum_{n=0}^{k-1} f(a + n \, \Delta t, Y(a + n \, \Delta t)) \, \Delta t.$$

This equation may also be written in the manner of a Riemann sum with $b = a + k \, \Delta t$:

$$Y(b) = y_0 + \sum_a^b f(t, Y(t)) \, \Delta t.$$

In the simple case

$$y'(t) = f(t),$$

the Euler approximation is just y_0 plus the Riemann sum,

$$Y(b) = y_0 + \sum_a^b f(t) \, \Delta t,$$

which is approximately equal to y_0 plus the integral

$$y(b) = y_0 + \int_a^b f(t) \, dt.$$

EXAMPLE 1   Compute the Euler approximation to the initial value problem

$$y' = t - y^2, \qquad y(0) = 0$$

for $0 \le t \le 1$, with $\Delta t = 0.2$.

Notice that the differential equation is not linear because of the y^2, and the variables are not separable, so we cannot solve the equation by the methods of the preceding sections. Given $Y(t)$, the next value $Y(t + \Delta t)$ is computed by the rule

$$Y(t + \Delta t) = Y(t) + (t - Y(t)^2) \, \Delta t.$$

We record the values in a table. The third column gives the change in $Y(t)$. The graph of $Y(t)$, shown in Figure 14.4.2, is obtained by connecting the points $(t, Y(t))$ in the table by straight lines.

$$\Delta t = 0.2$$

t	$Y(t)$	$Y(t + \Delta t) - Y(t) = (t - Y(t)^2) \, \Delta t$
0.0	0.0	0.0
0.2	0.0	0.04
0.4	0.04	0.0797
0.6	0.1197	0.1171
0.8	0.2368	0.1488
1.0	0.3856	

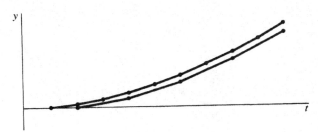

Figure 14.4.2   Example 1

Work the same problem with $\Delta t = 0.1$.

$$\Delta t = 0.1$$

t	$Y(t)$	$Y(t + \Delta t) - Y(t) = (t - Y(t)^2)\,\Delta t$
0.0	0.0	0.0
0.1	0.0	0.01
0.2	0.01	0.02
0.3	0.03	0.0299
0.4	0.0599	0.0396
0.5	0.0995	0.0490
0.6	0.1486	0.0578
0.7	0.2063	0.0657
0.8	0.2721	0.0726
0.9	0.3447	0.0781
1.0	0.4228	

We now consider Euler approximations with infinitesimal Δt. These approximations cannot be computed directly but are useful in showing that a differential equation has a solution.

The Euler approximation $Y(t)$ depends on both t and the increment size Δt. Now let Δt be positive infinitesimal. By the Transfer Principle, $Y(t + \Delta t)$ is still given by the rule

$$Y(t + \Delta t) - Y(t) = f(t, Y(t))\,\Delta t.$$

Intuitively, the graph of $Y(t)$ as a function of t is formed from infinitesimal line segments, and the segment from t to $t + \Delta t$ has slope $f(t, Y(t))$, as in Figure 14.4.3.

The next theorem shows that the Euler approximation for infinitesimal Δt is infinitely close to a solution of the initial value problem.

EXISTENCE THEOREM

Let Δt be positive infinitesimal and let $Y(t)$ be the Euler approximation of the initial value problem

(1)
$$y' = f(t, y), \qquad y(a) = y_0$$

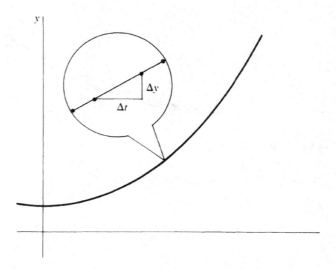

Figure 14.4.3

with increment Δt. Let b be a real number greater than a, and suppose that $Y(t)$ is finite for all t between a and b. Then for real numbers t in the interval $[a, b]$, the function $y(t)$ given by

$$y(t) = st(Y(t))$$

is a solution of the initial value problem (1).

Discussion   The theorem shows that the initial value problem (1) has a solution as long as $Y(t)$ remains finite. The solution is found by taking the standard part of $Y(t)$. When $Y(t)$ becomes infinite, we say that an *explosion* occurs (see Example 2 in Section 14.1 and Example 3 in this section).

PROOF OF THE EXISTENCE THEOREM   At $t = a$, $y(a) = st(Y(a)) = y_0$. Let M be the largest value of $|f(t, Y(t))|$ for t, a partition point between a and b. Then M is finite. Since $Y(t)$ never changes by more than $M \Delta t$ from one partition point to the next, we always have

$$|Y(t) - Y(s)| \le M|t - s|.$$

Taking standard parts, we see that for real s and t in the interval $[a, b]$,

$$|y(t) - y(s)| \le M|t - s|.$$

By the Transfer Principle, this also holds for all hyperreal s and t between a and b. Then for any $a \le t \le b$,

$$Y(t) \approx Y(st(t)) \approx y(st(t)) \approx y(t)$$

and hence, because $f(t, z)$ is continuous in z,

$$f(t, Y(t)) \approx f(t, y(t)).$$

Let $h(t)$ be the real function

$$h(t) = f(t, y(t)).$$

Since Y is an Euler approximation,

$$Y(t) = y_0 + \sum_{s=a}^{t} f(s, Y(s)) \, \Delta t$$

for each real point t between a and b. But $h(s) = f(s, y(s)) \approx f(s, Y(s))$, so

$$Y(t) \approx y_0 + \sum_{s=a}^{t} h(s) \, \Delta t.$$

This is just the Riemann sum of h. Taking standard parts, we get the integral of h:

$$y(t) = y_0 + \int_{a}^{t} h(s) \, ds.$$

Finally, by the Fundamental Theorem of Calculus,

$$y'(t) = h(t) = f(t, y(t)).$$

Thus $y(t)$ is a solution of (1) as required.

To apply the Existence Theorem, we need a way of checking that $Y(t)$ is finite. Here is a convenient criterion.

LEMMA

Let $Y(t)$ be an Euler approximation of the initial value problem (1) *with infinitesimal Δt, and let M and b be finite.*

(i) *If $|f(t, y)| \leq M$ for all $a \leq t \leq b$ and all y, then $Y(t)$ is finite for all $a \leq t \leq b$.*

(ii) *If $|f(t, y)| \leq M$ for all $a \leq t \leq b$ and all y within $M \cdot (t - a)$ of y_0, then $Y(t)$ is finite for all $a \leq t \leq b$.*

PROOF (i) Since $Y(t)$ cannot change by more than $M \, \Delta t$ from one partition point to the next, we have

$$|Y(t) - y_0| \leq M \cdot (t - a) \leq M \cdot (b - a).$$

$M \cdot (b - a)$ is finite, so $Y(t)$ is finite.

The proof of (ii) is similar.

Discussion The lemma is illustrated in Figure 14.4.4. Choose a positive real number M. Part (i) of the lemma says that if $f(t, y)$ is between $-M$ and M, everywhere in the vertical strip between $t = a$ and $t = b$, then $Y(t)$ is finite for $a \leq t \leq b$. Part (ii) of the lemma says that if $f(t, y)$ is between $-M$ and M, everywhere in the shaded triangle, then $Y(t)$ is finite for $a \leq t \leq b$. Part (ii) is stronger because the shaded triangle is a subset of the vertical strip. The proof shows that $Y(t)$ stays within the shaded rectangle for $a \leq t \leq b$.

The lemma and the Existence Theorem combined show that if we can find an M such that $f(t, y)$ is continuous and $f(t, y)$ is between $-M$ and M everywhere in the shaded triangle, then the initial value problem (1) has a solution $y(t)$ for $a \leq t \leq b$. The proof also shows that the solution $y(t)$ is within the shaded triangle for $a \leq t \leq b$.

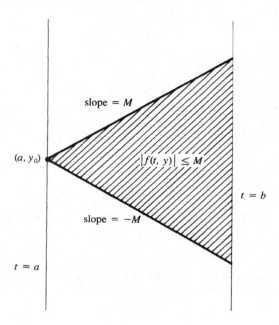

Figure 14.4.4

EXAMPLE 2 Show that the initial value problem

$$y' = t - y^2, \qquad y(0) = 0$$

from Example 1 has a solution for $0 \le t \le 1$.

Let Δt be infinitesimal and form the Euler approximation $Y(t)$ with increment Δt. Apply the lemma with $M = 1$, $b = 1$. In this example,

$$|f(t, y)| = |y^2 - t| \le 1$$

whenever $0 \le t \le 1$ and $-1 \le y \le 1$. Therefore, by the lemma, $Y(t)$ is finite for $0 \le t \le 1$. By the Existence Theorem, the standard part of $Y(t)$ is a solution for $0 \le t \le 1$.

Here is another theorem that shows that in most cases the solution is unique and is close to the Euler approximations for small real increments Δt. In this theorem, we shall write $Y_{\Delta t}(t)$ instead of $Y(t)$ to keep track of the fact that $Y(t)$ depends on Δt as well as on t.

UNIQUENESS THEOREM

Assume the hypotheses of the Existence Theorem and also that $f(t, y)$ is smooth; that is, the partial derivatives of f are continuous. Then the initial value problem (1) has only one solution $y(t)$ for t in $[a, b]$. Furthermore, the Euler approximations $Y_{\Delta t}(t)$ approach $y(t)$ as the real number Δt approaches zero; that is,

$$\lim_{\Delta t \to 0^+} Y_{\Delta t}(t) = y(t)$$

for each t in $[a, b]$.

We shall not give the proof. The Uniqueness Theorem tells us two important things about differential equations in which $f(t, y)$ is smooth. First, it tells us that a particular solution of such a differential equation will depend only on the initial condition. Thus if an experiment is accurately described by a differential equation with $f(t, y)$ smooth, then repeated trials of the experiment with the same initial condition will give the same outcome. Second, it tells us that the Euler approximations will approach the solution of the differential equation as Δt approaches zero. Thus we can get better and better approximations of the solution by taking Δt small.

EXAMPLE 2 (Continued)   The function $f(t, y) = t - y^2$ is smooth. The Uniqueness Theorem shows that the initial value problem of Example 1 has just one solution $y(t)$ for $0 \le t \le 1$. Moreover, the Euler approximations $Y(t)$ get close to $y(t)$ as the real increment Δt approaches zero. Thus the approximations computed in Example 1 really are approaching the solution.

We conclude with an example of an explosion and an example with more than one solution.

EXAMPLE 3 (An Explosion)   The initial value problem

$$y' = y^2, \qquad y(0) = 1$$

may be solved by separation of variables:

$$y^{-2} \, dy = dt,$$

$$\frac{-1}{y(t)} = t + C,$$

$$y(t) = -(t + C)^{-1} \qquad \qquad \text{(general solution),}$$

$$1 = -(0 + C)^{-1}, \qquad C = -1,$$

$$y(t) = (1 - t)^{-1} \qquad \qquad \text{(particular solution).}$$

The graph, shown in Figure 14.4.5, approaches infinity as t approaches 1 from the left. The function $y(t) = (1 - t)^{-1}$ is a solution for $0 \le t < 1$, and the solution has an explosion at $t = 1$.

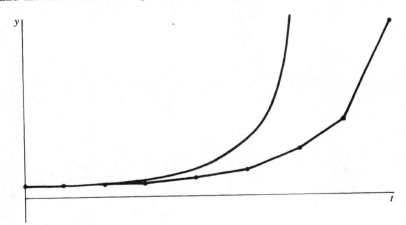

Figure 14.4.5   Example 3

The Euler approximation $Y(t)$ with a real increment Δt can be computed even for t greater than 1, but will approach $y(t)$ only for $0 \le t < 1$. For infinitesimal Δt, the Euler approximation will be finite and infinitely close to $y(t)$ when t is in the real interval $[0, 1)$. $Y(t)$ will keep on increasing and will be infinite for all t with standard part ≥ 1.

Continuing the example, compute the Euler approximation $Y(t)$ for $\Delta t = 0.2$ and $0 \le t \le 2$, and compare the values with the solution $y(t) = (1 - t)^{-1}$ for $0 \le t < 1$. The results are shown in the next table and are graphed in Figure 14.4.5.

$$\Delta t = 0.2$$

t	$y(t)$	$Y(t)$	$Y(t)^2 \, \Delta t$
0.0	1.0	1.0	0.2
0.2	1.25	1.2	0.288
0.4	1.6667	1.4488	0.4428
0.6	2.5	1.9309	0.7456
0.8	5.0	2.6764	1.4327
1.0	∞	4.1091	3.3770
1.2		7.4861	11.2084
1.4		18.6945	69.8966
1.6		88.5910	1569.6736
1.8		1658.26	549968.3
2.0		551627.6	

EXAMPLE 4 (Nonuniqueness)   The initial value problem

$$y' = 3y^{2/3}, \qquad y(0) = 0$$

has infinitely many solutions. The graphs split apart, as shown in Figure 14.4.6.

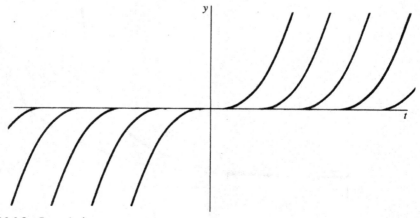

Figure 14.4.6   Example 4

One solution is the constant solution $y(t) = 0$. A second solution is found by separating variables:

$$\tfrac{1}{3}y^{-2/3}\,dy = dt,$$
$$y^{1/3} = t + C, \qquad C = 0$$
$$y = t^3.$$

This solution can be checked by differentiation:

$$y' = 3t^2 = 3(t^3)^{2/3} = 3y^{2/3}.$$

The other solutions go along the line $y(t) = 0$ in some interval and branch off the line $y(t) = 0$ to the right and left of the interval. The full list of solutions is:

$$y(t) = \begin{cases} (t-a)^3 & \text{for } t \le a \\ 0 & \text{for } a < t < b \\ (t-b)^3 & \text{for } b \le t \end{cases}$$

Here a is either a real number or $-\infty$, b is either a real number or $+\infty$, and $a \le 0 \le b$. In the case that $a = -\infty$ and $b = +\infty$, the solution is the constant function $y(t) = 0$.

These solutions all have the same initial value $y(0) = 0$. The Uniqueness Theorem does not apply in this example because the function $f(t, y) = 3y^{2/3}$ has no derivative at $y = 0$, so that $f(t, y)$ is not smooth at $y = 0$.

PROBLEMS FOR SECTION 14.4

In Problems 1–4, compute the Euler approximation to the given initial value problem with $\Delta t = 0.1$ for $0 \le t \le 1$.

1 $y' = t/y, \qquad y(0) = 1$ **2** $y' = t + y^3, \qquad y(0) = 1$

3 $y' = \cos(t + y), \qquad y(0) = 1$ **4** $y' = t \ln y, \qquad y(0) = 2$

5   Show that the initial value problem

$$y' = 1 + ty^2, \qquad y(0) = 0$$

has a unique solution $y(t)$ for $0 \le t \le \tfrac{1}{2}$.

6   Show that the initial value problem

$$y' = t + \frac{y}{2} + \frac{y^2}{4}, \qquad y(0) = 0$$

has a unique solution $y(t)$ for $0 \le t \le \tfrac{1}{2}$.

7   Show that the initial value problem

$$y' = \arctan(t + e^y), \qquad y(0) = 4$$

has a unique solution $y(t)$ for $0 \le t < \infty$.

8   Show that the initial value problem

$$y' = t + \exp(-y^2), \qquad y(0) = 1$$

has a unique solution $y(t)$ for $0 \le t < \infty$.

9   Show that the initial value problem

$$y' = y^{1/3}, \qquad y(0) = 0$$

has infinitely many solutions for $0 \le t < \infty$.

☐ **10**   Show that the initial value problem

$$y' = t(|1 - y^2|)^{1/2}, \qquad y(0) = 1$$

has infinitely many solutions for $0 \le t < \infty$.

☐ **11**   Suppose that $f(t, y)$ is continuous for all t and y. Prove that for each point (a, y_0), the initial value problem

$$y' = \cos (f(t, y)), \qquad y(a) = y_0$$

has a solution $y(t)$, $a \le t < \infty$.

☐ **12**   Suppose that $f(t, y)$ and $g(t)$ are continuous for all t and y and that $|f(t, y)| \le g(t)$ for all t and y. Prove that for each point (a, y_0) the initial value problem

$$y' = f(t, y), \qquad y(a) = y_0$$

has a solution $y(t)$ for $a \le t < \infty$.

☐ **13**   Suppose that $f(t, y)$ is continuous for all t and y. Prove that for each point (a, y_0) there is a number $b > a$ such that the initial value problem

$$y' = f(t, y), \qquad y(a) = y_0$$

has a solution $y(t)$, $a \le t \le b$.

14.5 COMPLEX NUMBERS

This section begins with a review of the complex numbers. Complex numbers are useful in the solution of second order differential equations. The starting point is the imaginary number i, which is the square root of -1. The *complex number system* is an extension of the real number system that is formed by adding the number i and keeping the usual rules for sums and products. The set of *complex numbers*, or *complex plane*, is the set of all numbers of the form

$$z = x + iy$$

where x and y are real numbers. The number x is called the *real part* of z, and y is called the *imaginary part* of z. A complex number z can be represented by a point in the plane, with the real part drawn on the horizontal axis and the imaginary part on the vertical axis, as in Figure 14.5.1. The *sum* of two complex numbers is computed in the same way as the sum of two vectors,

$$(a + ib) + (c + id) = (a + c) + i(b + d).$$

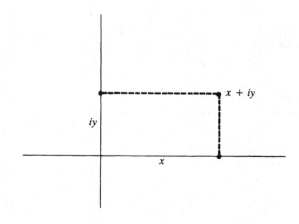

Figure 14.5.1

The *product* of two complex numbers is computed using the basic rule $i^2 = -1$ and the rules of algebra:

$$(a + ib) \cdot (c + id) = ac + ibc + iad + i^2 bd = (ac - bd) + i(bc + ad).$$

EXAMPLE 1   Compute the product of $3 + i6$ and $7 - i$.

$$(3 + i6) \cdot (7 - i) = (3 \cdot 7 - 6 \cdot (-1)) + i(6 \cdot 7 + 3 \cdot (-1)) = 27 + i39.$$

The *complex conjugate* $\bar{z}$ of z is formed by changing the sign of the imaginary part of z:

$$\overline{a + ib} = a - ib.$$

The product of a complex number with its conjugate is always a nonnegative real number, computed as follows.

(1) $(a + ib)(a - ib) = a^2 - iab + iab + b^2 = a^2 + b^2.$

The quotient of two complex numbers can be computed by multiplying the numerator and denominator by the conjugate of the denominator, as follows.

$$\frac{a + ib}{c + id} = \frac{(a + ib)(c - id)}{(c + id)(c - id)} = \frac{ac + bd}{c^2 + d^2} + i\,\frac{-ad + bc}{c^2 + d^2}\,.$$

EXAMPLE 2   Compute the quotient $(1 + i)/(1 - i2)$.

$$\frac{1 + i}{1 - i2} = \frac{(1 + i) \cdot (1 + i2)}{(1 - i2) \cdot (1 + i2)} = \frac{(1 \cdot 1 - 1 \cdot 2) + i(1 \cdot 2 + 1 \cdot 1)}{1 + 4}$$

$$= \frac{-1 + i3}{5} = -\frac{1}{5} + i\frac{3}{5}.$$

In the real number system, a positive real number b has two square roots, $\sqrt{b}$ and $-\sqrt{b}$, and negative real numbers have no square roots. In the complex number system, a negative real number $-b$ has two imaginary square roots, $i\sqrt{b}$ and $-i\sqrt{b}$. The quadratic formula gives the roots of any second degree polynomial in the complex number system.

QUADRATIC FORMULA

The roots of the polynomial

$$az^2 + bz + c \qquad \text{where } a \neq 0$$

in the complex number system are given by

$$z = \frac{-b \pm \sqrt{b^2 - 4ac}}{2a}.$$

The number $b^2 - 4ac$ is called the *discriminant*. If a, b, and c are real, there are three cases:

Case 1   If $b^2 - 4ac > 0$, there are two real roots.

Case 2   If $b^2 - 4ac = 0$, there is one real root.

Case 3   If $b^2 - 4ac < 0$, there are two complex roots, which are complex conjugates of each other.

EXAMPLE 3   Find the roots of the polynomial $z^2 + z + 2$ in the complex number system.

$$z = \frac{-1 \pm \sqrt{1 - 4 \cdot 1 \cdot 2}}{2} = \frac{-1 \pm \sqrt{-7}}{2} = -\frac{1}{2} \pm i\frac{\sqrt{7}}{2}.$$

It is often useful to represent a complex number in polar form. A point (x, y) in the plane has polar coordinates (r, θ) where $x = r \cos \theta$, $y = r \sin \theta$. The complex number $x + iy$ may be written in the *polar form*

$$x + iy = r(\cos \theta + i \sin \theta).$$

The coordinates r and θ can always be chosen so that $r \geq 0$ and $-\pi < \theta \leq \pi$. The number r, which is the distance of the point (x, y) from the origin, is the *absolute value* of the complex number $x + iy$:

$$r = |x + iy| = (x^2 + y^2)^{1/2}.$$

The formula (1) for the product of a complex number and its conjugate may now be written in the short form

$$z\bar{z} = |z|^2.$$

The real number θ is an angle in radians and is called the *argument* of $x + iy$. The argument can be computed by using the formula

$$\tan \frac{y}{x} = \theta,$$

and then choosing θ in the correct quadrant. The polar form of a complex number is illustrated in Figure 14.5.2. This figure is sometimes called the *Argand diagram* of the complex number. The complex number with absolute value one and argument θ is sometimes called *cis* θ:

$$cis\ \theta = \cos \theta + i \sin \theta.$$

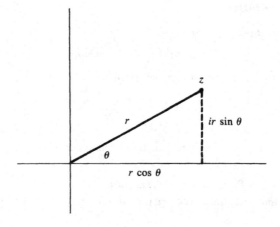

Figure 14.5.2

In Figure 14.5.3, we see that in the complex plane, *cis θ* is on the unit circle at an angle θ counterclockwise from the *x*-axis. Using the symbol *cis θ*, the polar form can be written

$$x + iy = r \, cis \, \theta.$$

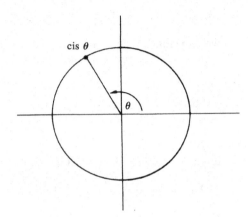

Figure 14.5.3

EXAMPLE 4   Write the complex number $z = -2 + i2$ in polar form.

The absolute value of *z* is $|z| = (2^2 + (-2)^2)^{1/2} = \sqrt{8}$. To find the argument θ, we use $\tan \theta = 2/(-2) = -1$. Since *z* is in the second quadrant (*x* negative and *y* positive), θ must be $3\pi/4$. Thus

$$z = \sqrt{8} \, cis \, \frac{3\pi}{4}.$$

cis θ is helpful in computing products, quotients, and powers of complex numbers. Using the addition formulas for sines and cosines, we can prove the product formula

(2) $$(r \, cis \, \theta) \cdot (s \, cis \, \phi) = rs \, cis \, (\theta + \phi).$$

In words, this formula states: *To multiply two complex numbers, multiply the absolute values and add the arguments.* There is a similar formula for quotients:

(3) $$\frac{r \, cis \, \theta}{s \, cis \, \phi} = \frac{r}{s} \, cis \, (\theta - \phi).$$

To divide two complex numbers, divide the absolute values and subtract the arguments.

EXAMPLE 5   Using the polar form, find the quotient $(1 + i)/(1 - i)$.

In polar form,

$$1 + i = \sqrt{2} \, cis \, \frac{\pi}{4}, \qquad 1 - i = \sqrt{2} \, cis \left(-\frac{\pi}{4}\right).$$

$$\frac{1 + i}{1 - i} = \frac{\sqrt{2}}{\sqrt{2}} \, cis \left(\frac{\pi}{4} - \frac{-\pi}{4}\right) = cis \left(\frac{\pi}{2}\right)$$

$$= \cos \frac{\pi}{2} + i \sin \frac{\pi}{2} = i.$$

Using the product formula (2) n times, we get a formula for the nth power of a complex number,

(4) $$(r\ cis\ \theta)^n = r^n\ cis\ (n\theta).$$

This formula in the case $r = 1$ is called *De Moivre's Formula*,

$$(\cos \theta + i \sin \theta)^n = \cos (n\theta) + i \sin (n\theta).$$

We can see from the power formula (4) that the complex number $r\ cis\ \theta$ has the square root $\sqrt{r}\ cis\ (\theta/2)$. In fact, each complex number except zero has two square roots,

(5) $$(r\ cis\ \theta)^{1/2} = \pm\sqrt{r}\ cis\ \frac{\theta}{2}.$$

EXAMPLE 6   Find the square roots of i.

By the computation in Example 3, the polar form of i is $i = cis\ (\pi/2)$.

$$i^{1/2} = \pm\sqrt{1}\ cis\ \frac{\pi}{4} = \pm \left(\cos \frac{\pi}{4} + i \sin \frac{\pi}{4} \right)$$

$$= \pm \left(\frac{\sqrt{2}}{2} + i \frac{\sqrt{2}}{2} \right).$$

The two square roots of i are shown in Figure 14.5.4.

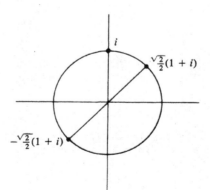

Figure 14.5.4

We now turn to complex exponents, which are useful in the study of differential equations. In order to give a meaning to an exponent e^z, we consider infinite series of complex numbers. The sum of an infinite series of complex numbers is defined by summing the real and imaginary parts separately. If $z_n = x_n + iy_n$, and the series $\sum x_n$ and $\sum y_n$ both converge, the sum of the series $\sum z_n$ is defined by the formula

$$\sum_{n=0}^{\infty} z_n = \sum_{n=0}^{\infty} x_n + i \sum_{n=0}^{\infty} y_n.$$

In Chapter 9, we found that for real numbers z the exponent e^z is given by the power series

$$e^z = 1 + z + \frac{z^2}{2!} + \frac{z^3}{3!} + \frac{z^4}{4!} + \cdots.$$

When z is a complex number, this formula is taken as the definition of e^z. It can be shown that the power series converges for every z and that the exponential rule $e^{u+z} = e^u e^z$ holds for complex exponents. In the case that z is a purely imaginary number $z = iy$, the power series takes the form

$$e^{iy} = 1 + iy - \frac{y^2}{2!} - i\frac{y^3}{3!} + \frac{y^4}{4!} + i\frac{y^5}{5!} - \cdots$$

$$= \left[1 - \frac{y^2}{2!} + \frac{y^4}{4!} - \cdots\right] + i\left[y - \frac{y^3}{3!} + \frac{y^5}{5!} - \cdots\right].$$

Using the power series for $\cos y$ and $\sin y$, we obtain *Euler's Formula*:

$$e^{iy} = \cos y + i \sin y = \text{cis } y.$$

When z is a complex number $z = x + iy$, the exponent e^z is given by the formula

$$e^{x+iy} = e^x e^{iy} = e^x(\cos y + i \sin y).$$

EXAMPLE 7   Find $e^{-2+i\pi/3}$.

$$e^{-2+i\pi/3} = e^{-2}\left[\cos\frac{\pi}{3} + i\sin\frac{\pi}{3}\right] = e^{-2}\left(\frac{1}{2} + i\frac{\sqrt{3}}{2}\right).$$

In Chapter 8, the hyperbolic cosine and hyperbolic sine were defined in terms of e^x by the equations

$$\cosh x = \frac{e^x + e^{-x}}{2}, \qquad \sinh x = \frac{e^x - e^{-x}}{2}.$$

Euler's Formula leads to similar equations for the cosine and sine.

$$e^{iy} = \cos y + i \sin y, \qquad e^{-iy} = \cos y - i \sin y,$$

$$\cos y = \frac{e^{iy} + e^{-iy}}{2}, \qquad \sin y = \frac{e^{iy} - e^{-iy}}{i2}.$$

In the next section we will make use of *complex valued functions*, that is, functions $f(t)$ that assign a complex number $z = f(t)$ to each real number t. The derivative of a complex valued function is obtained by differentiating the real and complex parts separately. Thus, if $h(t) = f(t) + ig(t)$, where g and h are real functions, then $h'(t) = f'(t) + ig'(t)$.

For example, if $h(t) = e^{rt}$, where $r = a + ib$ is a complex constant, then

$$h(t) = e^{at}(\cos bt + i \sin bt),$$

$$h'(t) = ae^{at}(\cos bt + i \sin bt) + be^{at}(-\sin bt + i \cos bt)$$

$$= (a + ib)e^{at}(\cos bt + i \sin bt) = re^{rt}.$$

Summing up, the usual rule $(e^{rt})' = re^{rt}$ still holds when r is a complex constant. We can also consider complex valued differential equations. The example we shall need is the homogeneous linear differential equation

$$z' + rz = 0,$$

where r is a complex constant. The general solution of this equation is

$$z(t) = Ce^{-rt},$$

where C is a complex constant. This solution can be checked by differentiation as before.

EXAMPLE 8   Solve the complex initial value problem

$$z' + (3 + i4)z = 0, \qquad z(0) = e^{1+i2}.$$

The general solution is

$$z(t) = Ce^{-(3+i4)t}.$$

Substituting the initial value at $t = 0$, $e^{1+i2} = C$. The particular solution is then

$$z(t) = e^{1+i2}e^{-(3+i4)t} = e^{1-3t+i(2-4t)}.$$

The solution may also be written in polar form using Euler's Formula,

$$z(t) = e^{1-3t} \, cis \, (2 - 4t).$$

PROBLEMS FOR SECTION 14.5

In Problems 1–6, put the complex number in the form $a + ib$.

1	$(7 - i4) + (3 + i2)$	2	$(4 - i6) - (8 - i)$
3	$(-4 + i2) \cdot (1 - i2)$	4	$(3 + i) \cdot (-2 - i6)$
5	$(1 - i2)/(3 + i)$	6	$(7 + i3)/(2 - i5)$

In Problems 7–12, find the roots of the given equation.

7	$z^2 - 8z + 16 = 0$	8	$z^2 + 6z + 9 = 0$
9	$z^2 + 25 = 0$	10	$z^2 + 100 = 0$
11	$z^2 + 2z + 5 = 0$	12	$z^2 + z + 3 = 0$

In Problems 13–18, put the complex number into the polar form $r \, cis \, \theta$.

13	$i5$	14	$-i3$
15	$-3 - i3$	16	$-4 + i4$
17	$\sqrt{3} - i$	18	$2 + i2\sqrt{3}$

In Problems 19–24, use the polar form to simplify the given expression.

19	$(2 + i2)/(-3 + i3)$	20	$(-4 - i4)/(5 - i5)$
21	$(\sqrt{3} + i)/(1 - i)$	22	$(1 + i)/(1 - i\sqrt{3})$
23	$(1 - i)^5$	24	$(1 + i\sqrt{3})^6$

In Problems 25–28, compute both square roots of the given complex number using the polar form.

25	$1 + i$	26	$-1 + i$
27	$-i4$	28	$-1 - i\sqrt{3}$

In Problems 29–32, put the given exponent in the form $a + ib$.

29	$e^{-3+i\pi/2}$	30	$e^{2-i\pi}$

31	$e^{1-i\pi/4}$	**32**	$e^{-i\pi/6}$

In Problems 33–36, find the derivative.

33	$z = e^{(5-i3)t}$	**34**	$z = e^{(4+i2)t}$
35	$z = e^{(2-i7)+(3+i2)t}$	**36**	$z = e^{5+(-1+i4)t}$

In Problems 37–40, find the general solution of the complex differential equation.

37	$z' + (2 - i3)z = 0$	**38**	$z' - i4z = 0$
39	$z' + (-3 + i5)z = 0$	**40**	$z' + (-2 - i)z = 0$

In Problems 41–44, solve the given complex initial value problem.

41	$z' + (-2 + i)z = 0,$	$z(0) = e^i$
42	$z' + (3 + i4)z = 0,$	$z(0) = e^{2+i4}$
43	$z' + (-2 + i)z = 0,$	$z(0) = 4$
44	$z' + (3 + i4)z = 0,$	$z(0) = -1$

45 Show that for any complex number z, $z + \bar{z}$ is a real number.

46 Prove that the conjugate of a complex number $r\,cis\,\theta$ is $r\,cis\,(-\theta)$.

47 Prove that for any nonzero complex number z, $z/\bar{z} = cis\,(2\theta)$, where θ is the argument of z.

48 Prove that for any two complex numbers u and z, the sum of the conjugates of u and z is equal to the conjugate of the sum of u and z, and similarly for products. In symbols,
$$\bar{u} + \bar{z} = \overline{u+z}, \qquad (\bar{u}) \cdot (\bar{z}) = \overline{u \cdot z}.$$

49 Prove that for any two complex numbers u and z,
$$|u + z| \le |u| + |z|.$$

50 Use De Moivre's Formula with $n = 2$,
$$\cos(2\theta) + i\sin(2\theta) = (\cos\theta + i\sin\theta)^2,$$
to obtain expressions for $\cos(2\theta)$ and $\sin(2\theta)$ in terms of $\cos\theta$ and $\sin\theta$.

51 Use De Moivre's Formula with $n = 3$,
$$\cos(3\theta) + i\sin(3\theta) = (\cos\theta + i\sin\theta)^3,$$
to obtain expressions for $\cos(3\theta)$ and $\sin(3\theta)$ in terms of $\cos\theta$ and $\sin\theta$.

52 Find the solution of the initial value problem
$$z' + (a + ib)z = 0, \qquad z(0) = e^{c+id}$$
where $a, b, c,$ and d are real numbers.

53 Show that every solution of the differential equation $z' + ibz = 0$ has constant absolute value (where b is a real number).

14.6 SECOND ORDER HOMOGENEOUS LINEAR EQUATIONS

A *second order differential equation* is an equation involving an independent variable t, a dependent variable y, and the first two derivatives y' and y''. The general solution of a second order differential equation will usually involve two constants. Two initial values are needed to determine a particular solution. A *second order initial value problem* is a second order differential equation together with initial values for y and y'. This section gives a solution method for second order equations of the following very simple type.

SECOND ORDER HOMOGENEOUS LINEAR DIFFERENTIAL EQUATION WITH CONSTANT COEFFICIENTS

(1)
$$ay'' + by' + cy = 0,$$

where a, b, and c are real constants and a, $c \neq 0$.

Discussion If $a = 0$, the equation is a first order differential equation $by' + cy = 0$. If $c = 0$, the change of variables $u = y'$ turns the given equation (1) into a first order differential equation $au' + bu = 0$. In each of these cases, the equation can be solved by the method of Section 14.1 or 14.2. (After finding u, y can be found by integration because $y' = u$.)

In Section 14.2 we found that the first order homogeneous linear differential equation with constant coefficients, $y' + cy = 0$, has the solution $y = e^{-ct}$. To get an idea of what to expect in the second order case, let us try to find a solution of equation (1) of the form $y = e^{rt}$ where r is a constant. Differentiating and substituting into equation (1), we see that

$$a(e^{rt})'' + b(e^{rt})' + ce^{rt} = ar^2 e^{rt} + bre^{rt} + ce^{rt}$$
$$= (ar^2 + br + c)e^{rt}.$$

This shows that $y(t) = e^{rt}$ is a solution of equation (1) if and only if

$$ar^2 + br + c = 0.$$

We should therefore expect that the solutions of the equation (1) will be built up from the functions $y(t) = e^{rt}$ where r is a root of the polynomial $az^2 + bz + c$. We shall state the rule for finding the general solution of the equation (1) now and prove the rule later on.

METHOD FOR SOLVING A SECOND ORDER HOMOGENEOUS DIFFERENTIAL EQUATION WITH CONSTANT COEFFICIENTS

(1)
$$ay'' + by' + cy = 0, \qquad a \neq 0.$$

Step 1 Form the **characteristic polynomial**

$$az^2 + bz + c.$$

Find its roots by using the quadratic equation or by factoring.

Step 2 The general solution is described by three cases.

Case 1 Two distinct real roots: $z = r, z = s$.

$$y = Ae^{rt} + Be^{st}.$$

Case 2 One real root: $z = r$.

$$y = Ae^{rt} + Bte^{rt}.$$

Case 3 Two complex conjugate roots: $z = \alpha \pm i\beta$.

$$y = e^{\alpha t}[A \cos(\beta t) + B \sin(\beta t)].$$

Step 3 If initial values for y and y' are given, solve for A and B, and substitute to

*obtain the particular solution. The two initial values will specify the position
and velocity at one time:*

$$y = y_0 \quad \text{and} \quad y' = v_0 \quad \text{at } t = t_0.$$

Discussion The general solution in Case 3 is sometimes written in the same form
as Case 1 by using complex exponents,

$$y = Ce^{rt} + De^{st},$$

where
$$r = \alpha + i\beta, \quad s = \alpha - i\beta,$$

$$C = \tfrac{1}{2}(A - iB), \quad D = \tfrac{1}{2}(A + iB).$$

To show that the two forms of the solution are really the same, use the
complex exponent formula

$$e^{\alpha + i\beta} = e^{\alpha}(\cos \beta + i \sin \beta)$$

from the preceding section.

EXAMPLE 1 Find the general solution of

$$y'' - \omega^2 y = 0, \quad \omega \neq 0.$$

Step 1 The characteristic polynomial is $z^2 - \omega^2$. It has two real roots, $z = \omega$ and
$z = -\omega$.

Step 2 The general solution is

$$y = Ae^{\omega t} + Be^{-\omega t},$$

where ω is constant.

EXAMPLE 2 Find the solution of the initial value problem

$$y'' - y' - 2y = 0, \quad y(0) = 5, \quad y'(0) = 0.$$

Step 1 The characteristic polynomial $z^2 - z - 2$ has two real roots, $z = -1$
and $z = 2$.

Step 2 The general solution is

$$y = Ae^{-t} + Be^{2t}.$$

Step 3 The initial value $y(0) = 5$ gives the equation

$$5 = A + B.$$

To get a second equation, we differentiate the general solution and substitute
the initial value for $y'(0)$.

$$y' = -Ae^{-t} + 2Be^{2t}, \quad 0 = -A + 2B.$$

The solution of the two equations for A and B is

$$A = \frac{10}{3}, \quad B = \frac{5}{3}.$$

The particular solution of the initial value problem, shown in Figure 14.6.1, is

$$y = \left(\frac{10}{3}\right)e^{-t} + \left(\frac{5}{3}\right)e^{2t}.$$

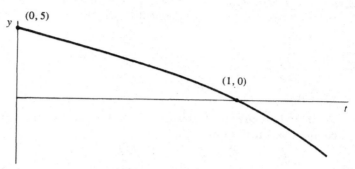

Figure 14.6.1   Example 2

EXAMPLE 3   Find the solution of the initial value problem

$$y'' - 4y' + 4y = 0, \qquad y(0) = -3, \qquad y'(0) = 1.$$

Step 1   The characteristic polynomial $z^2 - 4z + 4$ has one real root, $z = 2$.

Step 2   The general solution is

$$y = Ae^{2t} + Bte^{2t}.$$

Step 3   Substitute 0 for t and -3 for y.

$$-3 = Ae^0 + B \cdot 0 \cdot e^0, \qquad A = -3.$$

Compute y' for the general solution.

$$y' = 2Ae^{2t} + 2Bte^{2t} + Be^{2t}.$$

Substitute 0 for t and 1 for y'.

$$1 = 2Ae^0 + 2B \cdot 0 \cdot e^0 + Be^0 = 2A + B, \qquad B = 7.$$

The particular solution, shown in Figure 14.6.2, is

$$y = -3e^{2t} + 7te^{2t}.$$

EXAMPLE 4   Find the solution of

$$2y'' + 18y = 0, \qquad y(0) = 2, \qquad y'(0) = 15.$$

Step 1   The characteristic polynomial is $2z^2 + 18$, and its roots are $z = \pm i3$.

Step 2   The general solution is

$$y = A \cos(3t) + B \sin(3t).$$

Step 3   Substitute 0 for t and 2 for y.

$$2 = A \cos 0 + B \sin 0 = A.$$

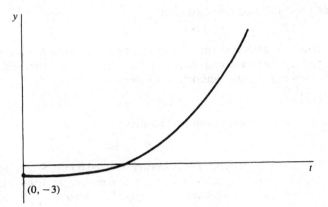

Figure 14.6.2 Example 3

Compute $y'(t)$ for the general solution.

$$y' = -3A \sin(3t) + 3B \cos(3t).$$

Substitute 0 for t and 15 for y'.

$$15 = -3A \sin 0 + 3B \cos 0 = -3 \cdot 0 + 3 \cdot B, \qquad B = 5.$$

The particular solution, shown in Figure 14.6.3, is

$$y = 2 \cos(3t) + 5 \sin(3t).$$

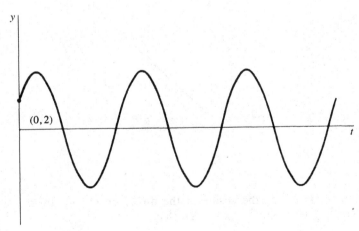

Figure 14.6.3 Example 4

GRAPHS OF SOLUTIONS OF SECOND ORDER HOMOGENEOUS LINEAR EQUATIONS

Our next topic is graphs of solutions of second order homogeneous linear equations. Several cases arise, including simple, damped, and overdamped oscillations.

Consider the second order homogeneous linear differential equation

(1) $$ay'' + by' + c = 0.$$

We shall concentrate on the case that

$$a > 0, \qquad b \geq 0, \qquad c > 0.$$

(We can always make a positive by changing all the signs if a is negative. The cases with negative b or c are considered in the problem set at the end of the section.) The equation has the characteristic polynomial

$$az^2 + bz + c.$$

Let d be the discriminant of this polynomial,

$$d = b^2 - 4ac.$$

Simple Oscillation This type of solution arises when $b = 0$. Since a and c are assumed to be positive, the discriminant $d = -4ac$ is negative, and the characteristic polynomial has two purely imaginary roots, $\pm i\beta$. The general solution of equation (1) is

$$y(t) = A \cos(\beta t) + B \sin(\beta t).$$

It is helpful to put this equation in a different form. The point (A, B) is on the circle with center at the origin and radius $C = (A^2 + B^2)^{1/2}$. There is thus an angle θ for which

$$A = C \cos \theta, \qquad B = C \sin \theta,$$

as in Figure 14.6.4. The angle θ can be computed as follows:

$$\frac{B}{A} = \frac{C \sin \theta}{C \cos \theta} = \tan \theta, \qquad \theta = \arctan\left(\frac{B}{A}\right).$$

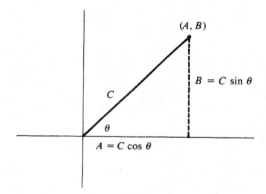

Figure 14.6.4

Using the formula for the cosine of the difference of two angles, $\cos(\phi - \theta) = \cos(\phi) \cos(\theta) + \sin(\phi) \sin(\theta)$, we find that

$$y(t) = C \cos(\beta t) \cos(\theta) + C \sin(\beta t) \sin(\theta) = C \cos(\beta t - \theta),$$

so that

$$y(t) = C \cos(\beta t - \theta).$$

The number C is called the *amplitude*, because the cosine curve oscillates between C and $-C$. The number β is called the *frequency*, because the curve will complete β cycles each 2π units of time. The number $2\pi/\beta$ is called the *period*, because each cycle is $2\pi/\beta$ units long. The angle θ is called the *phase shift*. Thus the graph of each particular solution is a cosine wave with amplitude C, period $2\pi/\beta$, and phase shift θ, as illustrated in Figure 14.6.5.

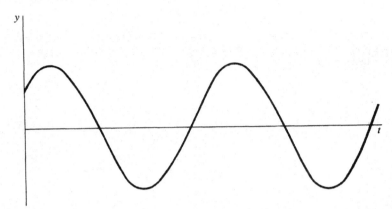

Figure 14.6.5

Damped Oscillation   This case arises when b is positive and the discriminant is negative, $b > 0$ and $d < 0$. Here b is in the range $0 < b < \sqrt{4ac}$. The roots of the characteristic polynomial are complex conjugates, $\alpha \pm i\beta$. The real part $\alpha = -b/2a$ is negative. The general solution is

$$y(t) = e^{\alpha t}[A \cos(\beta t) + B \sin(\beta t)].$$

Each particular solution oscillates with period $2\pi/\beta$, but the amplitude dies down exponentially as t increases, as in Figure 14.6.6.

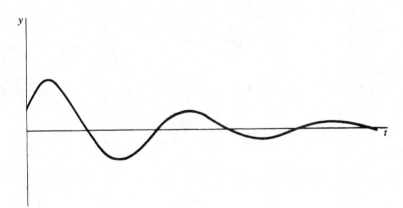

Figure 14.6.6

As in the case of simple oscillation, the solution may be written in the form

$$y(t) = e^{\alpha t}C \cos(\beta t - \theta),$$

where $C = (A^2 + B^2)^{1/2}$ and θ is a constant angle. The amplitude at time t will then be $e^{\alpha t}C$, which is decreasing because α is negative.

Critical Damping   This case arises when b is positive and the discriminant is zero, so' that $b = \sqrt{4ac}$. The characteristic polynomial has one negative real root, $r = -b/2a$. The general solution is

$$y(t) = Ae^{rt} + Bte^{rt}.$$

Each particular solution will approach 0 as t approaches infinity and will never complete one oscillation. A solution can cross the x-axis once, but never more than once. See Figure 14.6.7.

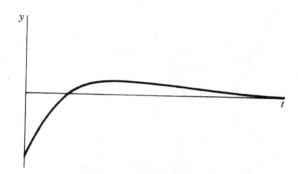

Figure 14.6.7

Overdamping   This case arises when b is positive and the discriminant is positive, so that $b > \sqrt{4ac}$. The characteristic polynomial has two real roots, r and s. Since a, b, and c are positive, the characteristic polynomial cannot have any positive or zero roots. Therefore both roots r and s are negative. The general solution is

$$y(t) = Ae^{rt} + Be^{st}.$$

Again, each particular solution approaches zero as t approaches infinity and will never complete one oscillation, as in Figure 14.6.7.

The differential equations of this section provide simple models for a variety of physical systems that oscillate, such as mass-spring systems and electrical networks. When a horizontal spring of natural length L is compressed a distance x, it exerts a force of approximately $F = -kx$. k is called the *spring constant* and depends on the particular spring. The negative sign indicates that the force is in the opposite direction from x, as in Figure 14.6.8. A mass m is attached to the end of the spring. From Newton's Law,

$$F = ma = mx'',$$

we obtain a second order differential equation for the position $x(t)$,

$$mx'' = -kx,$$

or

$$mx'' + kx = 0.$$

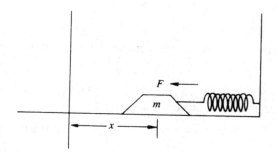

Figure 14.6.8

Both constants k and m are positive, and the solution is the simple oscillation

$$x(t) = A \cos (\beta t) + B \sin (\beta t)$$

where $\beta = (k/m)^{1/2}$.

A mass-spring system immersed in oil (such as an automobile shock absorber) is subject to a damping force $bx'(t)$, which is proportional to the velocity $x'(t)$ but in the opposite direction. This additional force will slow down the motion of the spring and lead to a damped oscillation. The force is approximately

$$F = -bx' - kx$$

and thus satisfies the differential equation

$$mx'' + bx' + kx = 0.$$

When the damping constant b is between 0 and $(4mk)^{1/2}$, the solution will be a damped oscillation. The greater the value of b, the more quickly the oscillation will be damped down. When b is equal to $(4mk)^{1/2}$, the solution will be critically damped; when b is greater than $(4mk)^{1/2}$, the solution will be overdamped.

EXAMPLE 5 Suppose a mass-spring system

$$mx'' + bx' + kx$$

has spring constant $k = 5$, damping constant $b = 4$, and mass $m = 1$. At time $t = 0$, the position is $x(0) = 1$ and the velocity is $x'(0) = 2$. Find the position $x(t)$ as a function of time.

The differential equation is

$$x'' + 4x' + 5x = 0.$$

Step 1 The characteristic polynomial $z^2 + 4z + 5$ has roots $-2 - i$ and $-2 + i$. (These roots can be found using the quadratic equation.)

Step 2 The general solution is

$$x(t) = e^{-2t}[A \cos t + B \sin t].$$

Step 3 Find A and B using the given initial values.

$$1 = e^0[A \cos 0 + B \sin 0], \qquad A = 1.$$

Compute $x'(t)$ and substitute to find B.

$$x'(t) = -2e^{-2t}[A \cos t + B \sin t] + e^{-2t}[-A \sin t + B \cos t],$$
$$2 = -2e^0[\cos 0 + B \sin 0] + e^0[-\sin 0 + B \cos 0],$$
$$B = 4.$$

The particular solution is

$$x(t) = e^{-2t}[\cos t + 4 \sin t].$$

This is a damped oscillation with period 2π.

Let us now justify the solution method given at the beginning of this section. We may take the coefficient of y'' to be one and consider the second order differential equation

(2) $$y'' + by' + cy = 0.$$

Let r and s be the roots of the characteristic polynomial, so that

$$z^2 + bz + c = (z - r)(z - s) = z^2 - rz - sz + rs.$$

If r and s are distinct (either two different real numbers or complex conjugates), we must show that the general solution is

$$y = Ae^{rt} + Be^{st}.$$

If $r = s$, we must show that the general solution is

$$y = Ae^{rt} + Bte^{rt}.$$

The plan is to break the second order equation (2) into a pair of first order differential equations. It is useful to use the symbol D for the first derivative and D^2 for the second derivative with respect to t. Thus

$$Dy = y', \qquad D^2y = y''.$$

The differential equation (2) can then be written in the form

(3) $$(D^2 + bD + c)y = 0.$$

We now wish to "factor" the expression $D^2 + bD + c$ as if it were a polynomial. It is to be understood that $(D - r)(D - s)y$ means $(D - r)u$ where u is the function $(D - s)y = y' - sy$. Thus, using the Sum and Product Rules for derivatives,

$$(D - r)(D - s)y = (y' - sy)' - r(y' - sy) = y'' - sy' - ry' + rsy$$
$$= y'' + by' + cy = (D^2 + bD + c)y.$$

This shows that the second order equation (3) is equivalent to the pair of first order equations

(4) $$(D - r)u = 0,$$

(5) $$(D - s)y = u.$$

Equation (4) is a homogeneous linear equation whose general solution is

$$u = Ke^{rt}.$$

Equation (5) may now be put in the form

(6) $$y' - sy = Ke^{rt}.$$

This first order linear equation was solved in Example 3 in Section 14.3. In the case $r \neq s$, the general solution came out to be

$$y = \frac{K}{r - s}e^{rt} + Be^{st}.$$

Putting $A = K/(r - s)$, we get the general solution

$$y = Ae^{rt} + Be^{st}.$$

In case $r = s$, the general solution is

$$y = Ate^{st} + Be^{st},$$

where $A = K$. In each case we have the required formula for the general solution of the original equation (2).

PROBLEMS FOR SECTION 14.6

In Problems 1–8, find the general solution of the given differential equation.

1	$y'' + y' - 6y = 0$	2	$y'' - 4y' + 3y = 0$
3	$y'' - 10y' + 25y = 0$	4	$y'' + 2\sqrt{2}y' + 2y = 0$
5	$y'' + 16y = 0$	6	$y'' + 2y = 0$
7	$y'' - 2y' + 2y = 0$	8	$y'' - 6y + 13 = 0$

In Problems 9–16, find the particular solution of the initial value problem.

9    $y'' + 6y' + 5y = 0$, $y(0) = 1$, $y'(0) = 0$

10    $y'' - y' - 12y = 0$, $y(0) = 0$, $y'(0) = 14$

11    $y'' + 12y' + 36y = 0$, $y(0) = 5$, $y'(0) = -10$

12    $y'' - 8y' + 16y = 0$, $y(0) = -3$, $y'(0) = 4$

13    $y'' + 5y = 0$, $y(0) = -2$, $y'(0) = 5$

14    $y'' + y = 0$, $y(\pi/4) = 0$, $y'(\pi/4) = 2$

15    $y'' + 12y' + 37y = 0$, $y(0) = 4$, $y'(0) = 0$

16    $y'' + 6y' + 18y = 0$, $y(0) = 0$, $y'(0) = 6$

In Problems 17–20, solve the initial value problem and find the amplitude, frequency, and phase shift of the solution.

17    $y'' + 4y = 0$, $y(0) = \sqrt{3}$, $y'(0) = 2$

18    $y'' + 100y = 0$, $y(0) = 5$, $y'(0) = 50$

19    $y'' + 2y' + 10y = 0$, $y(0) = 1$, $y'(0) = 1$

20    $y'' - 8y' + 25y = 0$, $y(0) = 3$, $y'(0) = 0$

21   A mass-spring system $mx'' + bx' + kx = 0$ has spring constant $k = 29$, damping constant $b = 4$, and mass $m = 1$. At time $t = 0$, the position is $x(0) = 2$ and the velocity is $x'(0) = 1$. Find the position $x(t)$ as a function of time.

22   A mass-spring system $mx'' + bx' + kx = 0$ has spring constant $k = 24$, damping constant $b = 12$, and mass $m = 3$. At time $t = 0$, the position is $x(0) = 0$ and the velocity is $x'(0) = -1$. Find the position $x(t)$ as a function of time.

23   Show that if $y(t)$ is a solution of a differential equation $ay'' + by' + cy = 0$, such that $y(t_0) = 0$ and $y'(t_0) = 0$ at some time t_0, then $y(t) = 0$ for all t.

☐ 24   In the differential equation $ay'' + by' + cy = 0$, suppose that a is positive and c is negative. Show that the characteristic equation has one positive real root and one negative real root, so that the general solution has the form $y = Ae^{rt} + Be^{st}$ where r is positive and s is negative.

☐ 25   In the differential equation $ay'' + by' + cy = 0$, suppose that a and c are positive and b is negative. Show that there are three cases for the general solution, depending on the sign of the discriminant d:

Case 1   If d is positive, the general solution has the form $y = Ae^{rt} + Be^{st}$ where r and s are positive.

Case 2   If d is zero, the general solution has the form $y = Ae^{rt} + Bte^{rt}$ where r is positive.

Case 3   If d is negative, the general solution has the form $y = e^{\alpha t}[A \cos (\beta t) + B \sin (\beta t)]$ where α is positive, so that the graph is an oscillation whose amplitude is increasing instead of decreasing.

14.7 SECOND ORDER LINEAR EQUATIONS

This section contains a method for solving nonhomogeneous second order differential equations. As in the previous section, we deal only with linear equations with constant coefficients. We consider equations of the following type.

SECOND ORDER LINEAR DIFFERENTIAL EQUATION WITH CONSTANT COEFFICIENTS

(1)
$$ay'' + by' + cy = f(t)$$

where a, b, and c are real constants and $a, c \neq 0$.

A differential equation of this form describes a mass-spring system where an outside force $f(t)$ is applied to the mass. The function $f(t)$ is called the *forcing term*.

As before, if $a = 0$ the equation is a first order linear differential equation in y, and if $c = 0$ it is a first order linear differential equation in y'. In each of these cases, the equation should be solved by the methods of Section 14.3. Hereafter we assume $a, c \neq 0$.

To get started, let us review the first theorem on first order linear differential equations. Theorem 1 in Section 14.3 states that the general solution of a first order linear differential equation is the sum

$$y(t) + Bx(t),$$

where $y(t)$ is a particular solution of the given equation and $x(t)$ is a particular solution of the corresponding homogeneous equation. Here is a similar theorem for second order equations.

THEOREM 1

Suppose that $y(t)$ *is a particular solution of the second order linear differential equation*

(1)
$$ay'' + by' + cy = f(t),$$

and $Ax_1(t) + Bx_2(t)$ *is the general solution of the corresponding homogeneous linear differential equation*

(2)
$$ax'' + bx' + cx = 0.$$

Then the general solution of the original equation (1) *is*

$$y(t) + Ax_1(t) + Bx_2(t).$$

As in the first order case, this theorem is proved using the Principle of Superposition.

PRINCIPLE OF SUPERPOSITION (Second Order)

Suppose $x(t)$ *and* $y(t)$ *are solutions of the two second order linear differential equations*

$$ax'' + bx' + cx = f(t),$$
$$ay'' + by' + cy = g(t).$$

Then for any constants A and B, the function

$$u = Ax + By$$

is a solution of the linear differential equation

$$au'' + bu' + cu = Af(t) + Bg(t).$$

Theorem 1 breaks the problem of finding the general solution of the equation (1) into two simpler problems.

First problem: Find the general solution of the corresponding linear homogeneous equation

$$ax'' + bx' + c = 0.$$

Second problem: Find some particular solution of the given equation

$$ay'' + by' + cy = f(t).$$

The first problem was solved in the preceding section. We now present a method for solving the second problem. This method is sometimes called the method of *judicious guessing*, or the method of *undetermined coefficients*. The method works only when the forcing term $f(t)$ is a fairly simple function, of the form

(3) $$p(t)e^{\alpha t}\cos(\beta t) + q(t)e^{\alpha t}\sin(\beta t),$$

where $p(t)$ and $q(t)$ are polynomials. However, when it works it is a very efficient method of solution. Often $f(t)$ will be of an even simpler form, such as a polynomial alone, or a single exponential or trigonometric function. In the case of a homogeneous equation, where $f(t) = 0$, the zero function $y(t) = 0$ is a particular solution. The idea for solving a linear equation is to guess that the differential equation has a particular solution, which looks like the forcing term $f(t)$ but has different constant coefficients. By working backwards, it is possible to find the unknown constants and discover a particular solution. We illustrate the method with several examples.

EXAMPLE 1 A mass of one gram is suspended from a vertical spring with spring constant $k = 100$, as in Figure 14.7.1. At time $t = 0$, the mass is at position $y(0) = 2$ cm and has velocity $y'(0) = 50$ cm/sec. Find the equation of motion of the mass. It is understood that there is no damping, and the origin is at the point where the spring is at its natural length.

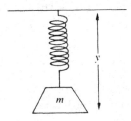

Figure 14.7.1

In this problem there are two forces, the force of the spring and the force of gravity. The force of gravity is a constant and is equal to mg dynes, always in the downward direction. The system is described by a second order linear differential equation with a constant forcing term,

$$my'' + ky = -mg.$$

In this case $k = 100$, $m = 1$, and $g = 980$, so the differential equation is

(4)
$$y'' + 100y = -980.$$

To solve the problem, we first find some particular solution of the differential equation (4), then use Theorem 1 to find the general solution, and finally substitute to find the particular solution for the given initial values $y(0) = 2$ and $y'(0) = 50$.

Since the forcing term is a constant $f(t) = -980$, we guess that the differential equation (4) has a particular solution, which is a constant. In this example it is easy to see by inspection that the constant function

$$u(t) = \frac{980}{100} = -9.8$$

is a particular solution of the differential equation (4). By the method of the preceding section, the characteristic polynomial $z^2 + 100$ has roots $\pm i10$, and the corresponding homogeneous differential equation $x'' + 100x = 0$ has the general solution

$$A \cos(10t) + B \sin(10t).$$

According to Theorem 1, the general solution of the original differential equation (4) is the sum of the particular solution of the original equation and the general solution of the homogeneous equation. Thus the general solution of equation (4) is

$$y(t) = A \cos(10t) + B \sin(10t) - 9.8.$$

Use the initial value $y(0) = 2$ to find A.

$$2 = A \cos(0) + B \sin(0) - 9.8 = A - 9.8, \qquad A = 11.8.$$

Now compute $y'(t)$, and substitute the given initial value $y'(0) = 50$ to find B.

$$y'(t) = -10A \sin(10t) + 10B \cos(10t).$$
$$50 = -10A \sin(0) + 10B \cos(0) = 10B, \qquad B \doteq 5.0.$$

The required particular solution is thus

$$y(t) = 11.8 \cos(10t) + 5.0 \sin(10t) - 9.8,$$

shown in Figure 14.7.2.

In the remaining examples we shall concentrate on the first part of the problem, finding some particular solution of the given differential equation. In each case we could then find the general solution by solving the corresponding homogeneous equation and applying Theorem 1 as we did in Example 1.

EXAMPLE 2 Find a particular solution of the differential equation

$$y'' - y' - 6y = 5 + 18t^2.$$

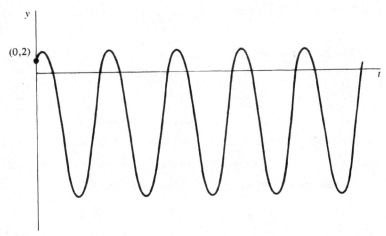

Figure 14.7.2   Example 1

Since $f(t)$ is a polynomial of degree two, we guess that some particular solution $y(t)$ is a polynomial of degree two,

$$y(t) = K + Lt + Mt^2.$$

First we find the first and second derivatives of $y(t)$.

$$y' = L + 2Mt, \qquad y'' = 2M.$$

Next we substitute these derivatives into the given differential equation.

$$\begin{aligned}
y'' - y' - 6y &= 2M - (L + 2Mt) - 6(K + Lt + Mt^2) \\
&= (2M - L - 6K) + (-2M - 6L)t - 6Mt^2 \\
&= 5 + 18t^2.
\end{aligned}$$

In the last equation the coefficients for each power of t must be equal. There are three equations, for units, t, and t^2.

$$\begin{aligned}
\text{units:} \quad & 2M - L - 6K = 5 \\
t: \quad & -2M - 6L = 0 \\
t^2: \quad & -6M = 18.
\end{aligned}$$

We can now solve the three equations for the three unknowns K, L, and M.

$$K = -2, \qquad L = 1, \qquad M = -3.$$

The required particular solution is then

$$y(t) = -2 + t - 3t^2.$$

It can be shown that whenever the forcing term $f(t)$ is a polynomial of degree n, the differential equation (1) will have a particular solution that is a polynomial of degree n. When $f(t)$ is a polynomial of degree n, the guess $y(t)$ should be a polynomial of degree n with unknown coefficients,

$$y(t) = A_0 + A_1 t + \cdots + A_n t^n.$$

EXAMPLE 3 Find a particular solution of the differential equation

$$y'' + 7y' + 10y = e^{3t}.$$

We guess that there is a particular solution that is a constant times e^{3t},

$$y(t) = Me^{3t}.$$

The first two derivatives of $y(t)$ are

$$y'(t) = 3Me^{3t}, \qquad y''t = 9Me^{3t}.$$

Substitute these into the original differential equation.

$$9Me^{3t} + 21Me^{3t} + 10Me^{3t} = e^{3t}.$$

Cancel the e^{3t}, and solve for the unknown constant M.

$$9M + 21M + 10M = 1, \qquad M = \frac{1}{40} = 0.025.$$

The required particular solution is

$$y(t) = 0.025e^{3t}.$$

Here is the rule for guessing a particular solution of the differential equation of the form (1) when the forcing term $f(t)$ is an exponential function $f(t) = e^{kt}$. We first should find the roots of the characteristic polynomial $az^2 + bz + c$. *If k is not a root of the characteristic polynomial, there is a particular solution of the form $y(t) = Me^{kt}$ (as in Example 3 above). If k is a single root of the characteristic polynomial, there is a particular solution of the form $y(t) = Mte^{kt}$. If k is a double root of the characteristic polynomial, there is a particular solution of the form $y(t) = Mt^2e^{kt}$.*

EXAMPLE 4 Find a particular solution of the differential equation

$$y'' + 7y' + 10y = e^{-2t}.$$

The characteristic polynomial $z^2 + 7z + 10$ has roots -2 and -5. Since -2 is a single root of the characteristic polynomial, our guess at a particular solution should be

$$y(t) = Mte^{-2t}.$$

The first two derivatives of $y(t)$ are

$$y'(t) = Me^{-2t} - 2Mte^{-2t},$$
$$y''(t) = -4Me^{-2t} + 4Mte^{-2t}.$$

Now substitute into the original differential equation.

$$e^{-2t} = y'' + 7y' + 10y$$
$$= -4Me^{-2t} + 4Mte^{-2t} + 7Me^{-2t} - 14Mte^{-2t} + 10Mte^{-2t}$$
$$= 3Me^{-2t}.$$

Then $M = 1/3$, and the required particular solution is

$$y(t) = \left(\frac{1}{3}\right)te^{-2t}.$$

In this example the simpler guess Le^{-2t} would not have worked. The trouble is that Le^{-2t} is a solution of the corresponding homogeneous equation, so it cannot also be a solution of the original differential equation. To see what happens, let us try to use the method with the guess $u(t) = Le^{-2t}$. Computing the first two derivatives and substituting, we get

$$u'(t) = -2Le^{-2t}, \qquad u''(t) = 4Le^{-2t},$$
$$e^{-2t} = u'' + 7u' + 10u = 4Le^{-2t} - 14Le^{-2t} + 10Le^{-2t}.$$

The right side of the above equation adds up to zero, so we cannot solve for the unknown constant L.

In a physical system, the forcing term is often a simple oscillation that can be represented by a function of the form $f(t) = G \cos(\omega t) + H \sin(\omega t)$. Here is the rule for guessing a particular solution of the differential equation (1) when the forcing term is $f(t) = G \cos(\omega t) + H \sin(\omega t)$. *If $z = i\omega$ is not a root of the characteristic polynomial, then the differential equation will have a particular solution of the form $y(t) = K \cos(\omega t) + L \sin(\omega t)$. On the other hand, if $z = i\omega$ is a root of the characteristic polynomial, then there is a particular solution of the form $y(t) = Kt \cos(\omega t) + Lt \sin(\omega t)$.*

EXAMPLE 5 Find a particular solution of the differential equation

$$y'' + 16y = -\sin(4t).$$

The characteristic polynomial has roots $\pm i4$. Then $\cos(4t)$ and $\sin(4t)$ are already solutions of the homogeneous equation, so our guess must have an extra factor of t. The guess for a solution is then

$$y(t) = Kt \cos(4t) + Lt \sin(4t).$$

Compute the first two derivatives of $y(t)$.

$$y'(t) = K[-4t \sin(4t) + \cos(4t)] + L[4t \cos(4t) + \sin(4t)],$$
$$y''(t) = K[-16t \cos(4t) - 8 \sin(4t)] + L[-16t \sin(4t) + 8 \cos(4t)].$$

Now substitute into the original differential equation.

$$K[(-16t + 16t)\cos(4t) - 8 \sin(4t)]$$
$$+ L[(-16t + 16t)\sin(4t) + 8 \cos(4t)] = -\sin(4t).$$
$$-8K \sin(4t) + 8L \cos(4t) = -\sin(4t).$$

From the sine terms we get $-8K = -1$, so $K = \frac{1}{8}$. From the cosine terms we get $8L = 0$, so $L = 0$. The particular solution is therefore

$$y(t) = 0.125t \cos(4t).$$

In Example 5, the particular solution oscillates more and more wildly as t approaches infinity, as shown in Figure 14.7.3. This happens because the forcing term $\cos(4t)$ has the same frequency as the solutions of the homogeneous equation, $A \cos(4t) + B \sin(4t)$. In this case the forcing term causes the oscillation to build up. This phenomenon is called *resonance*.

If, instead, the forcing term in Example 5 had a different frequency, $-\sin(\omega t)$ where ω is not equal to 4, then the particular solution of the differential equation would be a simple oscillation of the form $K \cos(\omega t) + L \sin(\omega t)$, whose amplitude does not change with time.

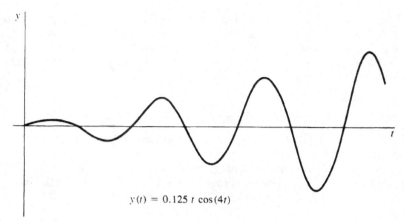

$$y(t) = 0.125\, t \cos(4t)$$

Figure 14.7.3   Exercise 5

EXAMPLE 6   Find a particular solution of the differential equation

$$y'' + 6y' + 25y = \cos(4t).$$

The characteristic polynomial $z^2 + 6z + 25$ has roots $-3 \pm i4$. Since $i4$ itself is not a root, the proper guess is

$$y(t) = K\cos(4t) + L\sin(4t).$$

Both a cosine term and a sine term are required, because the derivative of a sine is a cosine. Compute the first two derivatives of $y(t)$.

$$y'(t) = -4K\sin(4t) + 4L\cos(4t),$$
$$y''(t) = -16K\cos(4t) - 16L\sin(4t).$$

Now substitute into the original differential equation.

$$\cos(4t) = [-16K\cos(4t) - 16L\sin(4t)] + 6[-4K\sin(4t) + 4L\cos(4t)]$$
$$+ 25[K\cos(4t) + L\sin(4t)]$$
$$= (-16K + 24L + 25K)\cos(4t) + (-16L - 24K + 25L)\sin(4t).$$

Both the $\sin(4t)$ coefficients and the $\cos(4t)$ coefficients must be the same on both sides of the equation. Thus we have two equations in the two unknowns K and L.

$$\sin(4t) \text{ terms: } \quad 0 = -24K + 9L.$$
$$\cos(4t) \text{ terms: } \quad 1 = 9K + 24L.$$

Solve for K and L.

$$K = \frac{1}{73}, \qquad L = \frac{8}{219}.$$

The required particular solution is

$$y(t) = \left(\frac{1}{73}\right)\cos(4t) + \left(\frac{8}{219}\right)\sin(4t).$$

In Example 6, the particular solution is a simple oscillation, while the general solution of the corresponding homogeneous equation is a damped oscillation. The general solution is their sum:

$$y(t) = e^{-3t}[A \cos (4t) + B \sin (4t)] + \left[\left(\frac{1}{73}\right) \cos (4t) + \left(\frac{8}{219}\right) \sin (4t)\right].$$

The first term, $e^{-3t}[A \cos (4t) + B \sin (4t)]$, approaches zero as $t \to \infty$ and is called the *transient part of the solution*. The second term, $[(1/73) \cos (4t) + (8/219) \sin (4t)]$, is called the *steady part of the solution*. The constants A and B, which depend on the initial conditions, appear only in the transient part and not in the steady state part of the solution. No matter what the initial conditions are, every particular solution will approach the steady state part of the solution as $t \to \infty$. The effect of the initial conditions dies out as $t \to \infty$,

The same thing happens in any mass-spring system with damping where the forcing term is a simple oscillation. Let us consider a mass-spring system

(5) $$my'' + by' + ky = \cos (\omega t),$$

where m, b, and k are positive and the forcing term $\cos (\omega t)$ has frequency ω. A particular solution can be found of the form

$$K \cos (\omega t) + L \sin (\omega t).$$

The constants K and L can be computed as in Example 6. The general solution of equation (5) is

$$y(t) = e^{-\alpha t}[A \cos (\beta t) + B \sin (\beta t)] + [K \cos (\omega t) + L \sin (\omega t)].$$

As in Example 6, the first term approaches 0 as $t \to \infty$ and is called the *transient part of the solution*, and the second term is called the *steady state part of the solution*. Again, every particular solution of the mass-spring system will approach the steady state part of the solution as $t \to \infty$.

PROBLEMS FOR SECTION 14.7

In Problems 1–12, find a particular solution of the given differential equation.

1	$y'' - 10y' + 25y = \cos t$	2	$y'' + 2\sqrt{2}y' + 2y = 10$
3	$y'' + 16y = 8t^2 + 3t - 4$	4	$y'' + 2y = \cos (5t) + \sin (5t)$
5	$y'' - 2y' + 2y = e^{2t}$	6	$y'' - 6y + 13 = 1 + 2t + e^{-t}$
7	$y'' + y' - 6y = e^{-3t}$	8	$y'' - 4y' + 3y = e^{3t}$
9	$y'' + 16y = \cos (4t)$	10	$y'' + 9y = 3 \sin (3t)$
11	$y'' + 12y' + 36y = 6e^{-6t}$	12	$y'' - 8y' + 16y = -2e^{4t}$

In Problems 13–16, find the general solution of the given differential equation.

13	$y'' + 6y' + 5y = 4$	14	$y'' - y' - 12y = t$
15	$y'' + 5y = 8 \sin (2t)$	16	$y'' - 4y = 4e^{2t}$

In Problems 17–20, find the particular solution of the initial value problem.

17    $y'' - y = 3t + 5$, $y(0) = 0$, $y'(0) = 0$

18 $y'' + 9y = 4t,$ $y(0) = 0,$ $y'(0) = 0$

19 $y'' + 12y' + 37y = 10e^{-4t},$ $y(0) = 4,$ $y'(0) = 0$

20 $y'' + 6y' + 18y = \cos t - \sin t,$ $y(0) = 0,$ $y'(0) = 2$

21   A mass-spring system $mx'' + bx' + kx = 689 \cos(2t)$ has an external force of $689 \cos(2t)$ dynes, spring constant $k = 29$, damping constant $b = 4$, and mass $m = 1$ gm. Find the general solution for the motion of the spring and the steady state part of the solution.

22   A mass-spring system $mx'' + bx' + kx = 2 \sin t$ has an external force of $2 \sin t$ dynes, spring constant $k = 24$, damping constant $b = 12$, and mass $m = 3$ gm. Find the general solution for the motion of the spring and the steady state part of the solution.

☐ **23**   In the mass-spring system

(5) $my'' + by' + ky = \cos(\omega t),$

where m, b, and k are positive, show that the steady state part of the solution has amplitude

$$\sqrt{K^2 + L^2} = \frac{1}{\sqrt{(k - m\omega^2)^2 + b^2}}.$$

☐ **24**   In Problem 23, show that the frequency ω in the forcing term for which the steady state has the largest amplitude is

$$\omega = \sqrt{\frac{k}{m}},$$

and the largest amplitude is $1/b$. This frequency ω is called the *resonant frequency*.

25   Using Problem 24, find the resonant frequency for the mass-spring system

$$y'' + 6y' + 25y = \cos(\omega t).$$

EXTRA PROBLEMS FOR CHAPTER 14

1   Find the general solution of $y' = y^2 \cos t$.

2   Find the general solution of $y' = \sqrt{y}$.

3   Solve $y' = e^y/t, y(1) = 2$.

4   Solve $y' = t^3/(y + 1), y(0) = 1$.

5   Find the general solution of $y' + 10ty = 0$.

6   Find the general solution of $y' + e^{-t}y = 0$.

7   Solve $y' + 6y = 0, y(0) = 1$.

8   Solve $y' - y\sqrt{t} = 0, y(0) = 2$.

9   Find the general solution of $y' + 3y = 2t$.

10   Find the general solution of $y' - ty = t^2$.

11   A population has a net birthrate of 2% per year and a constant net immigration rate of 50,000 per year. At time $t = 0$, the population is one million. Find the population y as a function of t.

12   Repeat Problem 11 for a net immigration rate of $-50,000$ per year (that is, emigration exceeds immigration by 50,000 per year).

13   Show that the initial value problem $y' = \cos(y^2 + t), y(0) = 1$, has a unique solution for $0 \le t < \infty$.

14   Show that the initial value problem $y' = 1/(2 + \sin y), y(0) = 1$, has a unique solution for $0 \le t < \infty$.

15   Find the general solution of $y'' - 5y' + 4y = 0$.

16   Find the general solution of $y'' + 400y = 0$.

17   Find the general solution of $y'' - 4y' + 8y = 0$.

18   Find the general solution of $y'' - 14y' + 49y = 0$.

19   Solve $y'' + 4y' - 5y = 0$, $y(0) = 0$, $y'(0) = 1$.

20   Solve $y'' - 20y' + 100y = 0$, $y(0) = 1$, $y'(0) = 0$.

21   A mass-spring system $mx'' + bx' + kx = 0$ has mass $m = 2$ gm and constants $b = 6$ and $k = 5$. At time $t = 0$, its position is $x(0) = 10$ and its velocity is $x'(0) = 0$. Find its position x as a function of t.

22   Work Problem 21 if the system is subjected to constant external force of 3 dynes.

23   Find the general solution of $y'' - 5y' + 4y = 2 + t$.

24   Find the general solution of $y'' + 400y = e^t$.

25   Find the general solution of $y'' - 4y' + 8y = \cos t$.

26   Find the general solution of $y'' - 14y' + 49y = t^2$.

27   Solve $y'' + 4y' - 5y = 26 \sin t$, $y(0) = 0$, $y'(0) = 0$.

28   Solve $y'' - 20y' + 100y = e^{10t}$, $y(0) = 0$, $y'(0) = 0$.

EPILOGUE

How does the infinitesimal calculus as developed in this book relate to the traditional (or ε, δ) calculus? To get the proper perspective we shall sketch the history of the calculus.

Many problems involving slopes, areas, and volumes, which we would today call calculus problems, were solved by the ancient Greek mathematicians. The greatest of them was Archimedes (287–212 B.C.). Archimedes anticipated both the infinitesimal and the ε, δ approach to calculus. He sometimes discovered his results by reasoning with infinitesimals, but always published his proofs using the "method of exhaustion," which is similar to the ε, δ approach.

Calculus problems became important in the early 1600's with the development of physics and astronomy. The basic rules for differentiation and integration were discovered in that period by informal reasoning with infinitesimals. Kepler, Galileo, Fermat, and Barrow were among the contributors.

In the 1660's and 1670's Sir Isaac Newton and Gottfried Wilhelm Leibniz independently "invented" the calculus. They took the major step of recognizing the importance of a collection of isolated results and organizing them into a whole.

Newton, at different times, described the derivative of y (which he called the "fluxion" of y) in three different ways, roughly

(1) The ratio of an infinitesimal change in y to an infinitesimal change in x. (The infinitesimal method.)

(2) The limit of the ratio of the change in y to the change in x, $\Delta y/\Delta x$, as Δx approaches zero. (The limit method.)

(3) The velocity of y where x denotes time. (The velocity method.)

In his later writings Newton sought to avoid infinitesimals and emphasized the methods (2) and (3).

Leibniz rather consistently favored the infinitesimal method but believed (correctly) that the same results could be obtained using only real numbers. He regarded the infinitesimals as "ideal" numbers like the imaginary numbers. To justify them he proposed his law of continuity: "In any supposed transition, ending in any terminus, it is permissible to institute a general reasoning, in which the terminus may also be included."[1] This "law" is far too imprecise by present standards. But it was a remarkable forerunner of the Transfer Principle on which modern infinitesimal calculus is based. Leibniz was on the right track, but 300 years too soon!

The notation developed by Leibniz is still in general use today, even though it was meant to suggest the infinitesimal method: dy/dx for the derivative (to suggest an infinitesimal change in y divided by an infinitesimal change in x), and $\int_a^b f(x)\,dx$ for the integral (to suggest the sum of infinitely many infinitesimal quantities $f(x)\,dx$).

All three approaches had serious inconsistencies which were criticized most effectively by Bishop Berkeley in 1734. However, a precise treatment of the calculus was beyond the state of the art at the time, and the three intuitive descriptions (1)–(3) of the derivative competed with each other for the next two hundred years. Until sometime after 1820, the infinitesimal method (1) of Leibniz was dominant on the European continent, because of its intuitive appeal and the convenience of the Leibniz notation. In England the velocity method (3) predominated; it also has intuitive appeal but cannot be made rigorous.

In 1821 A. L. Cauchy published a forerunner of the modern treatment of the calculus based on the limit method (2). He defined the integral as well as the derivative in terms of limits, namely

$$\int_a^b f(x)\,dx = \lim_{\Delta x \to 0^+} \sum_a^b f(x)\,\Delta x.$$

He still used infinitesimals, regarding them as variables which approach zero. From that time on, the limit method gradually became the dominant approach to calculus, while infinitesimals and appeals to velocity survived only as a manner of speaking. There were two important points which still had to be cleared up in Cauchy's work, however. First, Cauchy's definition of limit was not sufficiently clear; it still relied on the intuitive use of infinitesimals. Second, a precise definition of the real number system was not yet available. Such a definition required a better understanding of the concepts of set and function which were then evolving.

A completely rigorous treatment of the calculus was finally formulated by Karl Weierstrass in the 1870's. He introduced the ε, δ condition as the definition of limit. At about the same time a number of mathematicians, including Weierstrass, succeeded in constructing the real number system from the positive integers. The problem of constructing the real number system also led to development of set theory by Georg Cantor in the 1870's. Weierstrass' approach has become the traditional or "standard" treatment of calculus as it is usually presented today. It begins with the ε, δ condition as the definition of limit and goes on to develop the calculus entirely in terms of the real number system (with no mention of infinitesimals). However, even when calculus is presented in the standard way, it is customary to argue informally in terms of infinitesimals, and to use the Leibniz notation which suggests infinitesimals.

From the time of Weierstrass until very recently, it appeared that the limit method (2) had finally won out and the history of elementary calculus was closed. But in 1934, Thoralf Skolem constructed what we here call the hyperintegers and proved that the analogue of the Transfer Principle holds for them. Skolem's construction (now called the ultraproduct construction) was later extended to a wide class of structures, including the construction of the hyperreal numbers from the real numbers.

---

[1] See Kline, p. 385.

The name "hyperreal" was first used by E. Hewitt in a paper in 1948. The hyperreal numbers were known for over a decade before they were applied to the calculus.

Finally in 1961 Abraham Robinson discovered that the hyperreal numbers could be used to give a rigorous treatment of the calculus with infinitesimals. The presentation of the calculus which was given in this book is based on Robinson's treatment (but modified to make it suitable for a first course).

Robinson's calculus is in the spirit of Leibniz' old method of infinitesimals. There are major differences in detail. For instance, Leibniz defined the derivative as the ratio $\Delta y/\Delta x$ where Δx is infinitesimal, while Robinson defines the derivative as the *standard part* of the ratio $\Delta y/\Delta x$ where Δx is infinitesimal. This is how Robinson avoids the inconsistencies in the old infinitesimal approach. Also, Leibniz' vague law of continuity is replaced by the precisely formulated Transfer Principle.

The reason Robinson's work was not done sooner is that the Transfer Principle for the hyperreal numbers is a type of axiom that was not familiar in mathematics until recently. It arose in the subject of model theory, which studies the relationship between axioms and mathematical structures. The pioneering developments in model theory were not made until the 1930's, by Gödel, Malcev, Skolem, and Tarski; and the subject hardly existed until the 1950's.

Looking back we see that the method of infinitesimals was generally preferred over the method of limits for over 150 years after Newton and Leibniz invented the calculus, because infinitesimals have greater intuitive appeal. But the method of limits was finally adopted around 1870 because it was the first mathematically precise treatment of the calculus. Now it is also possible to use infinitesimals in a mathematically precise way. Infinitesimals in Robinson's sense have been applied not only to the calculus but to the much broader subject of analysis. They have led to new results and problems in mathematical research. Since Skolem's infinite hyperintegers are usually called nonstandard integers, Robinson called the new subject "nonstandard analysis." (He called the real numbers "standard" and the other hyperreal numbers "nonstandard." This is the origin of the name "standard part.")

The starting point for this course was a pair of intuitive pictures of the real and hyperreal number systems. These intuitive pictures are really only rough sketches that are not completely trustworthy. In order to be sure that the results are correct, the calculus must be based on mathematically precise descriptions of these number systems, which fill in the gaps in the intuitive pictures. There are two ways to do this. The quickest way is to list the mathematical properties of the real and hyperreal numbers. These properties are to be accepted as basic and are called *axioms*. The second way of mathematically describing the real and hyperreal numbers is to start with the positive integers and, step by step, construct the integers, the rational numbers, the real numbers, and the hyperreal numbers. This second method is better because it shows that there really is a structure with the desired properties. At the end of this epilogue we shall briefly outline the construction of the real and hyperreal numbers and give some examples of infinitesimals.

We now turn to the first way of mathematically describing the real and hyperreal numbers. We shall list two groups of axioms in this epilogue, one for the real numbers and one for the hyperreal numbers. The axioms for the hyperreal numbers will just be more careful statements of the Extension Principle and Transfer Principle of Chapter 1. The axioms for the real numbers come in three sets: the Algebraic Axioms, the Order Axioms, and the Completeness Axiom. All the familiar facts about the real numbers can be proved using only these axioms.

I. ALGEBRAIC AXIOMS FOR THE REAL NUMBERS

A   Closure laws   *0 and 1 are real numbers. If a and b are real numbers, then so are $a + b$, ab, and $-a$. If a is a real number and $a \neq 0$, then $1/a$ is a real number.*

B   Commutative laws $a + b = b + a$ $ab = ba.$

C   Associative laws $a + (b + c) = (a + b) + c$ $a(bc) = (ab)c.$

D   Identity laws $0 + a = a$ $1 \cdot a = a.$

E   Inverse laws $a + (-a) = 0$ *If* $a \neq 0$, $a \cdot \dfrac{1}{a} = 1.$

F   Distributive law $a \cdot (b + c) = ab + ac.$

DEFINITION

The ***positive integers*** *are the real numbers* $1, 2 = 1 + 1,\ 3 = 1 + 1 + 1,$ $4 = 1 + 1 + 1 + 1,$ *and so on.*

II. ORDER AXIOMS FOR THE REAL NUMBERS

A $0 < 1.$

B   Transitive law   *If* $a < b$ *and* $b < c$ *then* $a < c.$

C   Trichotomy law   *Exactly one of the relations* $a < b$, $a = b$, $b < a$, *holds.*

D   Sum law   *If* $a < b$, *then* $a + c < b + c.$

E   Product law   *If* $a < b$ *and* $0 < c$, *then* $ac < bc.$

F   Root axiom   *For every real number* $a > 0$ *and every positive integer* n, *there is a real number* $b > 0$ *such that* $b^n = a.$

III. COMPLETENESS AXIOM

Let A be a set of real numbers such that whenever x and y are in A, any real number between x and y is in A. Then A is an interval.

THEOREM

An increasing sequence $\langle S_n \rangle$ *either converges or diverges to* $\infty.$

PROOF   Let T be the set of all real numbers x such that $x \leq S_n$ for some n. T is obviously nonempty.

Case 1 T is the whole real line. If H is infinite we have $x \leq S_H$ for all real numbers x. So S_H is positive infinite and $\langle S_n \rangle$ diverges to ∞.

Case 2 T is not the whole real line. By the Completeness Axiom, T is an interval $(-\infty, b]$ or $(-\infty, b)$. For each real $x < b$, we have

$$x \leq S_n \leq S_{n+1} \leq S_{n+2} \leq \cdots \leq b$$

for some n. It follows that for infinite H, $S_H \leq b$ and $S_H \approx b$. Therefore $\langle S_n \rangle$ converges to b.

We now take up the second group of axioms, which give the properties of the hyperreal numbers. There will be two axioms, called the Extension Axiom and the Transfer Axiom, which correspond to the Extension Principle and Transfer Principle of Section 1.5. We first state the Extension Axiom.

I*. EXTENSION AXIOM

(a) *The set R of real numbers is a subset of the set R* of hyperreal numbers.*

(b) *There is given a relation $<^*$ on R^*, such that the order relation $<$ on R is a subset of $<^*$, $<^*$ is transitive ($a < {}^*b$ and $b <{}^* c$ implies $a <{}^* c$), and $<^*$ satisfies the Trichotomy Law: for all a, b in R^*, exactly one of $a <{}^* b$, $a = b$, $b <{}^* a$ holds.*

(c) *There is a hyperreal number ε such that $0 <{}^* \varepsilon$ and $\varepsilon <{}^* r$ for each positive real number r.*

(d) *For each real function f, there is given a hyperreal function f^* with the same number of variables, called the natural extension of f.*

Part (c) of the Extension Axiom states that there is at least one positive infinitesimal. Part (d) gives us the natural extension for each real function. The Transfer Axiom will say that this natural extension has the same properties as the original function.

Recall that the Transfer Principle of Section 1.5 made use of the intuitive idea of a real statement. Before we can state the Transfer Axiom, we must give an exact mathematical explanation of the notion of a real statement. This will be done in several steps, first introducing the concepts of a real expression and a formula.

We begin with the concept of a real *expression*, or *term*, built up from variables and real constants using real functions. Real expressions can be built up as follows:

(1) A real constant standing alone is a real expression.

(2) A variable standing alone is a real expression.

(3) If e is a real expression, and f is a real function of one variable, then $f(e)$ is a real expression. Similarly, if $e_1, \ldots, e_n$ are real expressions, and g is a real function of n variables, then $g(e_1, \ldots, e_n)$ is a real expression.

Step (3) can be used repeatedly to build up longer expressions. Here are some examples of real expressions, where x and y are variables:

$$2, \quad x + y, \quad |x - 4|, \quad \sin(\pi y^2), \quad \frac{\sqrt{x} + \sqrt{y}}{\sqrt{3}}, \quad g(x, f(0)), \quad 1/0.$$

By a *formula*, we mean a statement of one of the following kinds, where d and e are real expressions:

(1) An *equation* between two real expressions, $d = e$.

(2) An *inequality* between two real expressions, $d < e, d \leq e, d > e, d \geq e$, or $d \neq e$.

(3)   A statement of the form "e is defined" or "e is undefined."

Here are some examples of formulas:

$$x + y = 5,$$
$$f(x) = \frac{1 - x^2}{1 + x},$$
$$g(x, y) < f(t),$$
$$f(x, x) \text{ is undefined.}$$

If each variable in a formula is replaced by a real number, the formula will be either true or false. Ordinarily, a formula will be true for some values of the variables and false for others. For example, the formula $x + y = 5$ will be true when $(x, y) = (4, 1)$ and false when $(x, y) = (7, -2)$.

DEFINITION

*A **real statement** is either a nonempty finite set of formulas T or a combination involving two nonempty finite sets of formulas S and T that states that "whenever every formula in S is true, every formula in T is true."*

We shall give several comments and examples to help make this definition clear. Sometimes, instead of writing "whenever every formula in S is true, every formula in T is true" we use the shorter form "if S then T" for a real statement. Each of the Algebraic Axioms for the Real Numbers is a real statement. The commutative laws, associative laws, identity laws, and distributive laws are real statements. For example, the commutative laws are the pair of formulas

$$a + b = b + a, \qquad ab = ba,$$

which involve the two variables a and b. The closure laws may be expressed as four real statements:

$$a + b \text{ is defined,}$$
$$ab \text{ is defined,}$$
$$-a \text{ is defined,}$$
$$\text{if } a \neq 0, \text{ then } 1/a \text{ is defined.}$$

The inverse laws consist of two more real statements. The Trichotomy Law is part of the Extension Axiom, and all of the other Order Axioms for the Real Numbers are real statements. However, the Completeness Axiom is not a real statement, because it is not built up from equations and inequalities between terms.

A typical example of a real statement is the inequality for exponents discussed in Section 8.1:

$$\text{if } a \geq 0, \text{ and } q \geq 1, \text{ then } (a + 1)^q \geq aq + 1.$$

This statement is true for all real numbers a and q.

A formula can be given a meaning in the hyperreal number system as well as in the real number system. Consider a formula with the two variables x and y. When x and y are replaced by particular real numbers, the formula will be either true or false in the real number system. To give the formula a meaning in the hyperreal number system, we replace each real function by its natural extension and replace the real order relation $<$ by the hyperreal relation $<^*$. When x and y are replaced by hyperreal numbers, each real function f is replaced by its natural extension f^*, and the real order relation $<$ is replaced by $<^*$, the formula will be either true or false in the hyperreal number system.

For example, the formula $x + y = 5$ is true in the hyperreal number system when $(x, y) = (2 - \varepsilon, 3 + \varepsilon)$, but false when $(x, y) = (2 + \varepsilon, 3 + \varepsilon)$, if ε is nonzero.

We are now ready to state the Transfer Axiom.

II*. TRANSFER AXIOM

Every real statement that holds for all real numbers holds for all hyperreal numbers.

It is possible to develop the whole calculus course as presented in this book from these axioms for the real and hyperreal numbers. By the Transfer Axiom, all the Algebraic Axioms for the Real Numbers also hold true for the hyperreal numbers. In other words, we can transfer every Algebraic Axiom for the real numbers to the hyperreal numbers. We can also transfer every Order Axiom for the real numbers to the hyperreal numbers. The Trichotomy Law is part of the Extension Axiom. Each of the other Order Axioms is a real statement and thus carries over to the hyperreal numbers by the Transfer Axiom. Thus we can make computations with the hyperreal numbers in the same way as we do for the real numbers.

There is one fact of basic importance that we state now as a theorem.

THEOREM (Standard Part Principle)

For every finite hyperreal number b, there is exactly one real number r that is infinitely close to b.

PROOF We first show that there cannot be more than one real number infinitely close to b. Suppose r and s are real numbers such that $r \approx b$ and $s \approx b$. Then $r \approx s$, and since r and s are real, r must be equal to s. Thus there is at most one real number infinitely close to b.

We now show that there is a real number infinitely close to b. Let A be the set of all real numbers less than b. Then any real number between two elements of A is an element of A. By the Completeness Axiom for the real numbers, A is an interval. Since the hyperreal number b is finite. A must be an interval of the form $(-\infty, r)$ or $(-\infty, r]$ for some real number r. Every real number $s < r$ belongs to A, so $s < b$. Also, every real number $t > r$ does not belong to A, so $t \geq b$. This shows that r is infinitely close to b.

It was pointed out earlier that the Completeness Axiom does not qualify as a real statement. For this reason, the Transfer Principle cannot be used to transfer the Completeness Axiom to the hyperreal numbers. In fact, the Completeness Axiom is *not* true for the hyperreal numbers. By a *closed hyperreal interval*, we mean a set of hyperreal numbers of the form $[a, b]$, the set of all hyperreal numbers x for which $a \leq x \leq b$, where a and b are hyperreal constants. Open and half-open hyperreal intervals are defined in a similar way. When we say that the Completeness Axiom is not true for the hyperreal numbers, we mean that there actually are sets A of hyperreal numbers such that:

(a)　Whenever x and y are in A, any hyperreal number between x and y is in A.

(b)　A is not a hyperreal interval.

Here are two quite familiar examples.

EXAMPLE 1　The set A of all infinitesimals has property (a) above but is not a hyperreal interval. It has property (a) because any hyperreal number that is between two infinitesimals is itself infinitesimal. We show that A is not a hyperreal interval. A cannot be of the form $[a, \infty)$ or (a, ∞) because every infinitesimal is less than 1. A cannot be of the form $[a, b]$ or $(a, b]$, because if b is positive infinitesimal, then $2 \cdot b$ is a larger infinitesimal. A cannot be of the form $[a, b)$ or (a, b), because if b is positive and not infinitesimal, then $b/2$ is less than b but still positive and not infinitesimal.

The set B of all finite hyperreal numbers is another example of a set that has property (a) above but is not an interval.

Here are some examples that may help to illustrate the nature of the hyperreal number system and the use of the Transfer Axiom.

EXAMPLE 2　Let f be the real function given by the equation

$$f(x) = \sqrt{1 - x^2}.$$

Its graph is the unit semicircle with center at the origin. The following two real statements hold for all real numbers x:

$$\text{whenever } 1 - x^2 \geq 0, \qquad f(x) = \sqrt{1 - x^2};$$
$$\text{whenever } 1 - x^2 < 0, \qquad f(x) \text{ is undefined.}$$

By the Transfer Axiom, these real statements also hold for all hyperreal numbers x. Therefore the natural extension f^* of f is given by the same equation

$$f^*(x) = \sqrt{1 - x^2}.$$

The domain of f^* is the set of all hyperreal numbers between -1 and 1. The hyperreal graph of f^*, shown in Figure E.1, can be drawn on paper by drawing the real graph of $f(x)$ and training an infinitesimal microscope on certain key points.

EXAMPLE 3　Let f be the identity function on the real numbers, $f(x) = x$. By the Transfer Axiom, the equation $f(x) = x$ is true for all hyperreal x. Thus the natural extension f^* of f is defined, and $f^*(x) = x$ for all hyperreal x. Figure E.2 shows the hyperreal graph of f^*. Under a microscope, it has a 45° slope.

Here is an example of a hyperreal function that is not the natural extension of a real function.

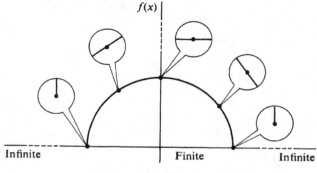

Hyperreal graph of $f(x) = \sqrt{1 - x^2}$

Figure E.1

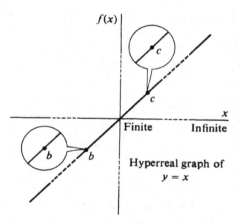

Hyperreal graph of
$y = x$

Figure E.2

EXAMPLE 4   One hyperreal function, which we have already studied in some detail, is the *standard part function* $st(x)$. For real numbers the standard part function has the same values as the identity function,

$$st(x) = x \quad \text{for all real } x.$$

However, the hyperreal graph of $st(x)$, shown in Figure E.3, is very different from the hyperreal graph of the identity function f^*. The domain of the standard part function is the set of all finite numbers, while f^* has domain R^*. Thus for infinite x, $f^*(x) = x$, but $st(x)$ is undefined. If x is finite but not real, $f^*(x) = x$ but $st(x) \neq x$. Under the microscope, an infinitesimal piece of the graph of the standard part function is horizontal, while the identity function has a 45° slope.

The standard part function is not the natural extension of the identity function, and hence is not the natural extension of any real function.

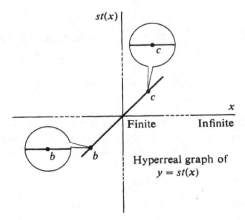

Figure E.3

The standard part function is the only hyperreal function used in this course except for natural extensions of real functions.

We conclude with a few words about the construction of the real and the hyperreal numbers. Before Weierstrass, the rational numbers were on solid ground but the real numbers were something new. Before one could safely use the axioms for the real numbers, it had to be shown that the axioms did not lead to a contradiction. This was done by starting with the rational numbers and constructing a structure which satisfied all the axioms for the real numbers. Since anything proved from the axioms is true in this structure, the axioms cannot lead to a contradiction.

The idea is to construct *real* numbers out of Cauchy sequences of *rational* numbers.

DEFINITION

A Cauchy Sequence is a sequence $\langle a_1, a_2, \ldots \rangle$ of numbers such that for every real $\varepsilon > 0$ there is an integer n_ε such that the numbers

$$\langle a_{n_\varepsilon}, a_{n_\varepsilon + 1}, a_{n_\varepsilon + 2}, \ldots \rangle$$

are all within ε of each other.

Two Cauchy sequences

$$\langle a_1, a_2, \ldots \rangle, \qquad \langle b_1, b_2, \ldots \rangle$$

of rational numbers are called *Cauchy equivalent*, in symbols $\langle a_1, a_2, \ldots \rangle \equiv \langle b_1, b_2, \ldots \rangle$, if the difference sequence

$$\langle a_1 - b_1, a_2 - b_2, \ldots \rangle$$

converges to zero. (Intuitively this means that the two sequences have the same limit.)

PROPERTIES OF CAUCHY EQUIVALENCE

(1)   If $\langle a_1, a_2, \ldots \rangle \equiv \langle a_1', a_2', \ldots \rangle$ and $\langle b_1, b_2, \ldots \rangle \equiv \langle b_1', b_2', \ldots \rangle$

then the sum sequences are equivalent,

$$\langle a_1 + b_1, a_2 + b_2, \ldots \rangle \equiv \langle a'_1 + b'_1, a'_2 + b'_2, \ldots \rangle.$$

(2)   Under the same hypotheses, the product sequences are equivalent,

$$\langle a_1 \cdot b_1, a_2 \cdot b_2, \ldots \rangle \equiv \langle a'_1 \cdot b'_1, a'_2 \cdot b'_2, \ldots \rangle.$$

(3)   If $a_n = b_n$ for all but finitely many n, then

$$\langle a_1, a_2, \ldots \rangle \equiv \langle b_1, b_2, \ldots \rangle.$$

The set of real numbers is then defined as the set of all equivalence classes of Cauchy sequences of rational numbers. A rational number r corresponds to the equivalence class of the constant sequence $\langle r, r, r, \ldots \rangle$. The sum of the equivalence class of $\langle a_1, a_2, \ldots \rangle$ and the equivalence class of $\langle b_1, b_2, \ldots \rangle$ is defined as the equivalence class of the sum sequence

$$\langle a_1 + b_1, a_2 + b_2, \ldots \rangle.$$

The product is defined in a similar way. It can be shown that all the axioms for the real numbers hold for this structure.

Today the real numbers are on solid ground and the hyperreal numbers are a new idea. Robinson used the ultraproduct construction of Skolem to show that the axioms for the hyperreal numbers (for example, as used in this book) do not lead to a contradiction. The method is much like the construction of the real numbers from the rationals. But this time the real number system is the starting point. We construct *hyperreal* numbers out of arbitrary (not just Cauchy) sequences of *real* numbers.

By an *ultraproduct equivalence* we mean an equivalence relation $\equiv$ on the set of all sequences of real numbers which have the properties of Cauchy equivalence (1)–(3) and also

(4)   If each a_n belongs to the set $\{0, 1\}$ then $\langle a_1, a_2, \ldots \rangle$ is equivalent to exactly one of the constant sequences $\langle 0, 0, 0, \ldots \rangle$ or $\langle 1, 1, 1, \ldots \rangle$.

Given an ultraproduct equivalence relation, the set of hyperreal numbers is defined as the set of all equivalence classes of sequences of real numbers. A real number r corresponds to the equivalence class of the constant sequence $\langle r, r, r, \ldots \rangle$. Sums and products are defined as for Cauchy sequences. The natural extension f^* of a real function $f(x)$ is defined so that the image of the equivalence class of $\langle a_1, a_2, \ldots \rangle$ is the equivalence class of $\langle f(a_1), f(a_2), \ldots \rangle$. It can be proved that ultraproduct equivalence relations exist, and that all the axioms for the real and hyperreal numbers hold for the structure defined in this way.

When hyperreal numbers are constructed as equivalence classes of sequences of real numbers, we can give specific examples of infinite hyperreal numbers. The equivalence class of

$$\langle 1, 2, 3, \ldots, n, \ldots \rangle$$

is a positive infinite hyperreal number. The equivalence class of

$$\langle 1, 4, 9, \ldots, n^2, \ldots \rangle$$

is larger, and the equivalence class of

$$\langle 1, 2, 4, \ldots, 2^n, \ldots \rangle$$

is a still larger infinite hyperreal number.

We can also give examples of nonzero infinitesimals. The equivalence classes

of

$$\langle 1, 1/2, 1/3, \ldots, 1/n, \ldots \rangle,$$

$$\langle 1, 1/4, 1/9, \ldots, n^{-2}, \ldots \rangle,$$

and

$$\langle 1, 1/2, 1/4, \ldots, 2^{-n}, \ldots \rangle$$

are progressively smaller positive infinitesimals.

The mistake of Leibniz and his contemporaries was to identify all the infinitesimals with zero. This leads to an immediate contradiction because dy/dx becomes $0/0$. In the present treatment the equivalence classes of

$$\langle 1, 1/2, 1/3, \ldots, 1/n, \ldots \rangle$$

and

$$\langle 0, 0, 0, \ldots, 0, \ldots \rangle$$

are different hyperreal numbers. They are not equal but merely have the same standard part, zero. This avoids the contradiction and once again makes infinitesimals a mathematically sound method.

For more information about the ideas touched on in this epilogue, see the instructor's supplement, *Foundations of Infinitesimal Calculus*, which has a self-contained treatment of ultraproducts and the hyperreal numbers.

FOR FURTHER READING ON THE HISTORY OF THE CALCULUS SEE:

The History of the Calculus and its Conceptual Development; Carl C. Boyer, Dover, New York, 1959.

Mathematical Thought from Ancient to Modern Times; Morris Kline, Oxford Univ. Press, New York, 1972.

Non-standard Analysis; Abraham Robinson, North-Holland, Amsterdam, London, 1966.

FOR ADVANCED READING ON INFINITESIMAL ANALYSIS SEE NON-STANDARD ANALYSIS BY ABRAHAM ROBINSON AND:

Lectures on Non-standard Analysis; M. Machover and J. Hirschfeld, Springer-Verlag, Berlin, Heidelberg, New York, 1969.

Victoria Symposium on Nonstandard Analysis; A. Hurd and P. Loeb, Springer-Verlag, Berlin, Heidelberg, New York, 1973.

Studies in Model Theory; M. Morley, Editor, Mathematical Association of America, Providence, 1973.

Applied Nonstandard Analysis: M. Davis, Wiley, New York, 1977.

Introduction to the Theory of Infinitesimals: K. D. Stroyan and W. A. J. Luxemburg, Academic Press, New York and London, 1976.

Foundations of Infinitesimal Stochastic Analysis: K. D. Stroyan and J. M. Bayod, North-Holland Publ. Co., in press.

Appendix

TABLES

Table I   Trigonometric Functions

Degrees	Radians	Sin	Tan	Cot	Cos		
0	.000	.000	.000		1.000	1.571	90
1	.017	.017	.017	57.29	1.000	1.553	89
2	.035	.035	.035	28.64	.999	1.536	88
3	.052	.052	.052	19.081	.999	1.518	87
4	.070	.070	.070	14.301	.998	1.501	86
5	.087	.087	.087	11.430	.996	1.484	85
6	.105	.105	.105	9.514	.995	1.466	84
7	.122	.122	.123	8.144	.993	1.449	83
8	.140	.139	.141	7.115	.990	1.431	82
9	.157	.156	.158	6.314	.988	1.414	81
10	.175	.174	.176	5.671	.985	1.396	80
11	.192	.191	.194	5.145	.982	1.379	79
12	.209	.208	.213	4.705	.978	1.361	78
13	.227	.225	.231	4.331	.974	1.344	77
14	.244	.242	.249	4.011	.970	1.326	76
15	.262	.259	.268	3.732	.966	1.309	75
16	.279	.276	.287	3.487	.961	1.292	74
17	.297	.292	.306	3.271	.956	1.274	73
18	.314	.309	.325	3.078	.951	1.257	72
19	.332	.326	.344	2.904	.946	1.239	71
20	.349	.342	.364	2.747	.940	1.222	70
21	.367	.358	.384	2.605	.934	1.204	69
22	.384	.375	.404	2.475	.927	1.187	68
23	.401	.391	.424	2.356	.921	1.169	67
24	.419	.407	.445	2.246	.914	1.152	66
		Cos	Cot	Tan	Sin	Radians	Degrees

(Table I is continued on the next page.)

Table I   Trigonometric Functions (continued)

Degrees	Radians	Sin	Tan	Cot	Cos		
25	.436	.423	.466	2.144	.906	1.134	65
26	.454	.438	.488	2.050	.899	1.117	64
27	.471	.454	.510	1.963	.891	1.100	63
28	.489	.469	.532	1.881	.883	1.082	62
29	.506	.485	.554	1.804	.875	1.065	61
30	.524	.500	.577	1.732	.866	1.047	60
31	.541	.515	.601	1.664	.857	1.030	59
32	.559	.530	.625	1.600	.848	1.012	58
33	.576	.545	.649	1.540	.839	.995	57
34	.593	.559	.675	1.483	.829	.977	56
35	.611	.574	.700	1.428	.819	.960	55
36	.628	.588	.727	1.376	.809	.942	54
37	.646	.602	.754	1.327	.799	.925	53
38	.663	.616	.781	1.280	.788	.908	52
39	.681	.629	.810	1.235	.777	.890	51
40	.698	.643	.839	1.192	.766	.873	50
41	.716	.656	.869	1.150	.755	.855	49
42	.733	.669	.900	1.111	.743	.838	48
43	.750	.682	.933	1.072	.731	.820	47
44	.768	.695	.966	1.036	.719	.803	46
45	.785	.707	1.000	1.000	.707	.785	45
		Cos	Cot	Tan	Sin	Radians	Degrees

Table II   Greek Alphabet

A α	Alpha	N ν	Nu
B β	Beta	Ξ ξ	Xi, Si
Γ γ	Gamma	O o	Omicron
Δ δ	Delta	Π π	Pi
E ε	Epsilon	P ρ	Rho
Z ζ	Zeta	Σ σ	Sigma
H η	Eta	T τ	Tau
Θ θ	Theta	Υ υ	Upsilon
I ι	Iota	Φ φ	Phi
K κ	Kappa	X χ	Chi
Λ λ	Lambda	Ψ ψ	Psi
M μ	Mu	Ω ω	Omega

Table III Exponential Functions

x	e^x	e^{-x}		x	e^x	e^{-x}
0.0	1.00	1.00		3.0	20.1	.050
0.1	1.11	.905		3.1	22.2	.045
0.2	1.22	.819		3.2	24.5	.041
0.3	1.35	.741		3.3	27.1	.037
0.4	1.49	.670		3.4	30.0	.033
0.5	1.65	.607		3.5	33.1	.030
0.6	1.82	.549		3.6	36.6	.027
0.7	2.01	.497		3.7	40.4	.025
0.8	2.23	.449		3.8	44.7	.022
0.9	2.46	.407		3.9	49.4	.020
1.0	2.72	.368		4.0	54.6	.018
1.1	3.00	.333		4.1	60.3	.017
1.2	3.32	.301		4.2	66.7	.015
1.3	3.67	.273		4.3	73.7	.014
1.4	4.06	.247		4.4	81.5	.012
1.5	4.48	.223		4.5	90.0	.011
1.6	4.95	.202		4.6	99.5	.010
1.7	5.47	.183		4.7	110	.0091
1.8	6.05	.165		4.8	122	.0082
1.9	6.69	.150		4.9	134	.0074
2.0	7.39	.135		5.0	148	.0067
2.1	8.17	.122		5.1	164	.0061
2.2	9.02	.111		5.2	181	.0055
2.3	9.97	.100		5.3	200	.0050
2.4	11.0	.091		5.4	221	.0045
2.5	12.2	.082		5.5	245	.0041
2.6	13.5	.074		5.6	270	.0037
2.7	14.9	.067		5.7	299	.0033
2.8	16.4	.061		5.8	330	.0030
2.9	18.2	.055		5.9	365	.0027
				6.0	403	.0025

Table IV Natural Logarithms

n	.0	.1	.2	.3	.4	.5	.6	.7	.8	.9
0*		7.697	8.391	8.796	9.084	9.307	9.489	9.643	9.777	9.895
1	0.000	0.095	0.182	0.262	0.336	0.405	0.470	0.531	0.588	0.642
2	0.693	0.742	0.788	0.833	0.875	0.916	0.956	0.993	1.030	1.065
3	1.099	1.131	1.163	1.194	1.224	1.253	1.281	1.308	1.335	1.361
4	1.386	1.411	1.435	1.459	1.482	1.504	1.526	1.548	1.569	1.589
5	1.609	1.629	1.649	1.668	1.686	1.705	1.723	1.740	1.758	1.775
6	1.792	1.808	1.825	1.841	1.856	1.872	1.887	1.902	1.917	1.932
7	1.946	1.960	1.974	1.988	2.001	2.015	2.028	2.041	2.054	2.067
8	2.079	2.092	2.104	2.116	2.128	2.140	2.152	2.163	2.175	2.186
9	2.197	2.208	2.219	2.230	2.241	2.251	2.262	2.272	2.282	2.293
10	2.303	2.313	2.322	2.332	2.342	2.351	2.361	2.370	2.380	2.389

* Subtract 10 if $n < 1$; for example, $\ln 0.3 \approx 8.796 - 10 = -1.204$.

Table V Powers and Roots

n	n^2	$\sqrt{n}$	n^3	$\sqrt[3]{n}$	n	n^2	$\sqrt{n}$	n^3	$\sqrt[3]{n}$
1	1	1.000	1	1.000	51	2,601	7.141	132,651	3.708
2	4	1.414	8	1.260	52	2,704	7.211	140,608	3.733
3	9	1.732	27	1.442	53	2,809	7.280	148,877	3.756
4	16	2.000	64	1.587	54	2,916	7.348	157,464	3.780
5	25	2.236	125	1.710	55	3,025	7.416	166,375	3.803
6	36	2.449	216	1.817	56	3,136	7.483	175,616	3.826
7	49	2.646	343	1.913	57	3,249	7.550	185,193	3.849
8	64	2.828	512	2.000	58	3,364	7.616	195,112	3.871
9	81	3.000	729	2.080	59	3,481	7.681	205,379	3.893
10	100	3.162	1,000	2.154	60	3,600	7.746	216,000	3.915
11	121	3.317	1,331	2.224	61	3,721	7.810	226,981	3.936
12	144	3.464	1,728	2.289	62	3,844	7.874	238,328	3.958
13	169	3.606	2,197	2.351	63	3,969	7.937	250,047	3.979
14	196	3.742	2,744	2.410	64	4,096	8.000	262,144	4.000
15	225	3.873	3,375	2.466	65	4,225	8.062	274,625	4.021
16	256	4.000	4,096	2.520	66	4,356	8.124	287,496	4.041
17	289	4.123	4,913	2.571	67	4,489	8.185	300,763	4.062
18	324	4.243	5,832	2.621	68	4,624	8.246	314,432	4.082
19	361	4.359	6,859	2.668	69	4,761	8.307	328,509	4.102
20	400	4.472	8,000	2.714	70	4,900	8.367	343,000	4.121
21	441	4.583	9,261	2.759	71	5,041	8.426	357,911	4.141
22	484	4.690	10,648	2.802	72	5,184	8.485	373,248	4.160
23	529	4.796	12,167	2.844	73	5,329	8.544	389,017	4.179
24	576	4.899	13,824	2.884	74	5,476	8.602	405,224	4.198
25	625	5.000	15,625	2.924	75	5,625	8.660	421,875	4.217
26	676	5.099	17,576	2.962	76	5,776	8.718	438,976	4.236
27	729	5.196	19,683	3.000	77	5,929	8.775	456,533	4.254
28	784	5.292	21,952	3.037	78	6,084	8.832	474,552	4.273
29	841	5.385	24,389	3.072	79	6,241	8.888	493,039	4.291
30	900	5.477	27,000	3.107	80	6,400	8.944	512,000	4.309
31	961	5.568	29,791	3.141	81	6,561	9.000	531,441	4.327
32	1,024	5.657	32,768	3.175	82	6,724	9.055	551,368	4.344
33	1,089	5.745	35,937	3.208	83	6,889	9.110	571,787	4.362
34	1,156	5.831	39,304	3.240	84	7,056	9.165	592,704	4.380
35	1,225	5.916	42,875	3.271	85	7,225	9.220	614,125	4.397
36	1,296	6.000	46,656	3.302	86	7,396	9.274	636,056	4.414
37	1,369	6.083	50,653	3.332	87	7,569	9.327	658,503	4.431
38	1,444	6.164	54,872	3.362	88	7,744	9.381	681,472	4.448
39	1,521	6.245	59,319	3.391	89	7,921	9.434	704,969	4.465
40	1,600	6.325	64,000	3.420	90	8,100	9.487	729,000	4.481
41	1,681	6.403	68,921	3.448	91	8,281	9.539	753,571	4.498
42	1,764	6.481	74,088	3.476	92	8,464	9.592	778,688	4.514
43	1,849	6.557	79,507	3.503	93	8,649	9.644	804,357	4.531
44	1,936	6.633	85,184	3.530	94	8,836	9.695	830,584	4.547
45	2,025	6.708	91,125	3.557	95	9,025	9.747	857,375	4.563
46	2,116	6.782	97,336	3.583	96	9,216	9.798	884,736	4.579
47	2,209	6.856	103,823	3.609	97	9,409	9.849	912,673	4.595
48	2,304	6.928	110,592	3.634	98	9,604	9.899	941,192	4.610
49	2,401	7.000	117,649	3.659	99	9,801	9.950	970,299	4.626
50	2,500	7.071	125,000	3.684	100	10,000	10.000	1,000,000	4.642

ANSWERS TO SELECTED PROBLEMS

Section 1.1

1 5　　**3** $\sqrt{13}$　　**5** 13

7

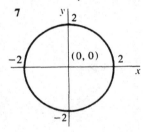

9

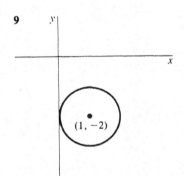

11

13 $(x - 3)^2 + y^2 = 4$
15 $(x - 1)^2 + (y - \sqrt{3})^2 = 4, (x - 1)^2 + (y + \sqrt{3})^2 = 4$
17 $(x - 2)^2 + (y - 4)^2 = 25$

Section 1.2

	-1	$-\frac{1}{2}$	0	$\frac{1}{2}$	1
1	$-1/3$	$-1/6$	0	$1/6$	$1/3$
3	-6	$3/8$	2	$9/8$	0
5	1	$1/\sqrt{2}$	0	$*$	$*$
7	2	1	1	1	2

9 yes

11 $f(2) = 7, f(t) = 1 + t + t^2, f(t + \Delta t) = 1 + (t + \Delta t) + (t + \Delta t)^2,$
$f(1 + t + t^2) = 1 + (1 + t + t^2) + (1 + t + t^2)^2, f(g(t)) = 1 + g(t) + (g(t))^2$

13 $f(t) = t\sqrt{t}, f(t + \Delta t) = (t + \Delta t)^{3/2}, f(t^2) = t^3, f(\sqrt{t}) = t^{3/4}, f(g(t)) = (g(t))^{3/2}$

15 $4\,\Delta x$

17 $\dfrac{1}{x^2 + 2x\,\Delta x + \Delta x^2} - \dfrac{1}{x^2} = -\dfrac{(2x + \Delta x)\,\Delta x}{x^2(x^2 + 2x\,\Delta x + \Delta x^2)}$

19 $\sqrt{x + \Delta x} - \sqrt{x} = \dfrac{\Delta x}{\sqrt{x + \Delta x} + \sqrt{x}}$

21 everywhere except $x = 1$ and $x = -1$ **23** $x \geq 0$ **25** $(-1, 1)$

Section 1.3

1 $m = 1, y = x + 1$ **3** vertical, $x = -4$ **5** $m = -\frac{1}{3}, y = (-\frac{1}{3})x + 1$
7 $m = 0, y = 3$ **9** $y = 2x - 3$ **11** $y = -(\frac{1}{4})x - \frac{7}{2}$ **13** $y = 5x$ **15** $y = 4$
17 $v = 2, y = 2t$ **19** $v = -\frac{1}{2}, y = -(\frac{1}{2})t + \frac{9}{2}$ **21** $v = 0, y = 4$ **23** $y + 3t + 2$

25
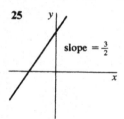
slope $= \frac{3}{2}$

27
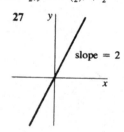
slope $= 2$

29
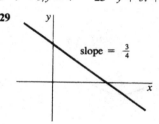
slope $= \frac{3}{4}$

33 $(-c/a, 0), (0, -c/b)$

Section 1.4 (no problem set)

Section 1.5

1 infinitesimal **3** infinite **5** infinite **7** infinite **9** infinitesimal
11 infinitesimal **13** infinite **15** infinite **17** finite **19** finite **21** infinite
23 infinitesimal **25** infinite **27** finite **29** finite **31** finite **33** infinite
35 infinitesimal **37** infinite **39** finite

41 (a) $\varepsilon > \varepsilon^2$ (b) $1/\varepsilon^3 < 1/\varepsilon^4$ (c) $H < H^2$ (d) $\varepsilon < \sqrt{\varepsilon}$ (e) $H > \sqrt{H}$ (f) $\sqrt{H} > \sqrt[3]{H}$

43 (a) $a = 0$ and $b \neq 0$ (b) $a \neq 0$ and $b \neq 0$, or $a = b = 0$ (c) $a \neq 0$ and $b = 0$

Section 1.6

1 2 **3** $2/5$ **5** x^{12} **7** 0 **9** 2 **11** 3 **13** 0 **15** $1/2$ **17** $1/2$
19 $x + y$ **21** $3x^2$ **23** 10 **25** $-1/6$ **27** 1 **29** $1/2$ **31** $-1/16$

33 $\dfrac{1}{1 + \sqrt{2}}$ or $\sqrt{2} - 1$

Extra Problems for Chapter 1.

1 $\sqrt{122}$ **3** $-5/3$ **5** $y = -2x + 13$ **7** $(x-1)^2 + (y-3)^2 = 5$

9 finite   **11** infinite   **13** infinite   **15** infinitesimal   **17** 48

19 4/7   **21** $-15/4$ **23** $\dfrac{1}{\sqrt{x + \Delta x}} - \dfrac{1}{\sqrt{x}}$ **25** $A = 6V^{2/3}$

Section 2.1

1 $2x$ **3** $-4x$ **5** 4   **7** $12t^2$ **9** $5/(2\sqrt{u})$ **11** $\frac{3}{2}\sqrt{x}$ **13** $-2t^{-3}$
15 $-3y^{-2} + 4$ **17** a **19** $a/(2\sqrt{ax + b})$ **21** $2(3 - 2x)^{-2}$ **23** 1/4   **25** $4t^3$
27 $600, $0, -$1000$ **29** $-3x^2$ if $x \le 0$, $3x^2$ if $x > 0$.

Section 2.2

1 $\Delta y = 2x\,\Delta x + \Delta x^2$, $dy = 2x\,\Delta x$, $\Delta y = dy + \Delta x^2$

3 $\Delta y = \dfrac{2\,\Delta x}{\sqrt{x + \Delta x} + \sqrt{x}}$, $dy = \dfrac{\Delta x}{\sqrt{x}}$, $\Delta y = dy + \left[\dfrac{2}{\sqrt{x + \Delta x} + \sqrt{x}} - \dfrac{1}{\sqrt{x}}\right]\Delta x$

5 $\Delta y = -\dfrac{\Delta x}{x(x + \Delta x)}$, $dy = -\dfrac{\Delta x}{x^2}$, $\Delta y = dy + \left[\dfrac{1}{x^2} - \dfrac{1}{x(x + \Delta x)}\right]\Delta x$

7 $\Delta y = \left(1 + \dfrac{1}{x(x + \Delta x)}\right)\Delta x$, $dy = \left(1 + \dfrac{1}{x^2}\right)\Delta x$, $\Delta y = dy + \left[\dfrac{1}{x(x + \Delta x)} - \dfrac{1}{x^2}\right]\Delta x$

9 $\Delta y = 4x\,\Delta x + 2\,\Delta x^2$, $\Delta z = 3x^2\,\Delta x + 3x\,\Delta x^2 + \Delta x^3$, $dy = 4x\,dx$, $dz = 3x^2\,dx$ **11** $2\,dx$

13 $dx/(2\sqrt{x + 1})$ **15** $a\,dx$ **17** $(-2/x^2)\,dx$ **19** $-\frac{1}{2}x^{-3/2}\,dx$

21 $d(y + z) = \left(\dfrac{1}{2\sqrt{x}} + 3\right)dx$, $d\left(\dfrac{y}{z}\right) = -\dfrac{1}{6}x^{-3/2}$ **23** $y = 4x - 4$ **25** $y = 0$

27 $y = 3x - 4$ **29** $y = -32x - 48$ **31** $y = (2x_0)x - x_0^2$

Section 2.3

1 $6x + 5$ **3** $5(x + 8)^4$ **5** $-3(4 - t)^2$ **7** $6x(x^2 + 5)^2$ **9** $-12x(6 - 2x^2)^2$
11 $24x^2(1 - 4x^3)^{-3}$ **13** $5(1 + x^{-2})$ **15** $-8(4x - 1)(2x^2 - x + 3)^{-3}$
17 $t^{-2}(1 + t^{-1})^{-2}$ or $(t + 1)^{-2}$ **19** $3(8t - 2)(4t^2 - 2t + 1)^{-2}$ **21** $-3x^2 + 5x - 1$
23 $6(3t^2 + 1)(2t - 4)^2 + 6t(2t - 4)^3$ or $6(2t - 4)^2(5t^2 - 4t + 1)$ **25** $-2(x - 1)^{-2}$
27 $4x(x^2 + 1)^{-2}$ **29** $(s^2 - 6s + 7)(s - 3)^{-2}$ **31** $-5(3x - 4)^{-2}$ **33** 0
35 $6(x^2 + 1)(2x^2 - 1) + 12x(x^2 + 1)(2x + 3) + 6x(2x^2 - 1)(2x + 3)$
37 $-(4x - 5)(2x + 1)^{-2}(x - 3)^{-2}$ **39** $2(2x + 1)^{-2}[(2x + 1)^{-1} + 3]^{-2}$
41 $4x(2x + 1)^3(x^2 + 1) + 6(2x + 1)^2(x^2 + 1)^2$ or $2(2x + 1)^2(x^2 + 1)(7x^2 + 2x + 3)$

43 $\dfrac{dy}{dx} = \dfrac{du}{dx} - \dfrac{dv}{dx}$ **45** $\dfrac{dy}{dx} = 4\dfrac{du}{dx} + 2v\dfrac{dv}{dx}$ **47** $\dfrac{dy}{dx} = -\left[u\dfrac{dv}{dx} + v\dfrac{du}{dx}\right](uv)^{-2}$

49 $y = 6x - 2$ **51** $b = -2, c = 4$

Section 2.4

1 $(9y^2 + 2)^{-1}$ **3** $-\frac{1}{4}y^{-1}$ **5** $-\frac{1}{2}y^{-1}(y^2 + 2)^2$ or $-1/(2x^2y)$ **7** $\frac{4}{3}x^{1/3}$

9 $-\dfrac{1}{\sqrt{x}(\sqrt{x} - 1)^2}$ **11** $\frac{2}{3}x^{-2/3} + \frac{8}{3}x^{-1/3} + 6$ **13** $-2(\frac{5}{3}x^{2/3} - 1)(x^{5/3} - x)^{-3}$

15 $(y^{-2/3} + 2)^{-1}$ **17** $y = \dfrac{1}{k}(x - c), \dfrac{dy}{dx} = \dfrac{1}{k}$ **19** $y = \sqrt{\dfrac{x - 1}{2}}, \dfrac{dy}{dx} = \dfrac{1}{2\sqrt{2}\sqrt{x - 1}}$

21 $y = (x + 3)^{1/4}, \dfrac{dy}{dx} = \dfrac{1}{4}(x + 3)^{-1/4}$

23 $y = \left[\dfrac{\sqrt{4x-3}-1}{2}\right]^{1/2}, \dfrac{dy}{dx} = (4y^3 + 2y)^{-1}$ or $\dfrac{dy}{dx} = \dfrac{1}{2}\left[\dfrac{\sqrt{4x-3}-1}{2}\right]^{-1/2}(4x-3)^{-1/2}$

25 $y = \left[\dfrac{\sqrt{1+8x}-1}{4}\right]^2 = \dfrac{1+4x-\sqrt{1+8x}}{8}, \dfrac{dy}{dx} = \left(\dfrac{1}{2\sqrt{y}}+2\right)^{-1}$

or $\dfrac{dy}{dx} = \dfrac{1}{2} - \dfrac{1}{2\sqrt{1+8x}}$

Section 2.5

1 $-2\sin\theta\cos\theta$ **3** $2\cos x - 3\sin x$ **5** $\sin z(\cos z)^{-2}$ **7** $n\sin^{n-1}\theta\cos\theta$
9 $t\cos t + \sin t$ **11** $e^x + xe^x$ **13** $(2\ln x)/x$ **15** $e^x/x + e^x\ln x$
17 $-e^v\sqrt{v} + (1-e^v)/2\sqrt{v}$ **19** $x^n/x + nx^{n-1}\ln x$
21 $y - 1/2 = (\sqrt{3}/2)(x - \pi/6)$ **23** $y = x - x/e$

Section 2.6

1 $\frac{1}{2}(x+2)^{-1/2}$ **3** $-\frac{1}{2}(5-x)^{-1/2}$ **5** $-\frac{3}{2}(2+3x)^{-3/2}$ **7** $2(6x+1)^{-2/3}$
9 $x(x^2+1)^{-1/2}$ **11** $3\cos(3x)$ **13** $-2x^{-3}\cos(3x)$ **15** $4e^{4x}$
17 $-\sin xe^{\cos x}$ **19** $-e^x\sin(e^x)$ **21** $-14(1-4x)^9$
23 $-2x\cos(1-x^2) + 2\cos(2x-1)$

25 $\dfrac{e^{\sqrt{\sin x}}\cos x}{2\sqrt{\sin x}}$ **27** $-\frac{1}{3}x(1+\sqrt{x^2-1})^{-4/3}(x^2-1)^{-1/2}$

29 $-12[1 + (x^3+1)^{-1}]^3 x^2 (x^3+1)^{-2}$
31 $\frac{1}{3}[\sqrt{x^2-1} + \sqrt{x^2+1}]^{-2/3}[x(x^2-1)^{-1/2} + x(x^2+1)^{-1/2}]$
33 $6x\cos(2x-1) + 3\sin(2x-1)$

35 $-\dfrac{\cos(3t)}{\sin(3t)}$ **37** $\dfrac{2\cos(2t)}{\cos t}$

39 $\dfrac{2t}{t+1}$ **41** $\dfrac{\sqrt{t^2-4}}{3t\sqrt{t+2}}$

43 $(t+1)^{2/3}(t+2)^{-2/3}$ **45** $(9/2)t$

Section 2.7

1 $2x^{-3}$ **3** $-10(x+1)^{-3}$ **5** $-(1/4)x^{-3/2} + (3/4)x^{-5/2}$ **7** $(3/4)t^{-1/2}$
9 $-\sin x$ **11** $-AB^2\sin(Bx)$ **13** a^2e^{ax} **15** $-x^{-2}$
17 $-2(t^2+1)^{-2} + 8t^2(t^2+1)^{-3}$ **19** $-14(x+2)^{-3}$
21 $-\frac{1}{2}x(x+1)^{-3/2} + (x+1)^{-1/2}$ or $(\frac{3}{4}x+1)(x+1)^{-3/2}$
23 $\frac{3}{4}t^{1/2}(t+3)^{-5/2} - \frac{1}{2}t^{-1/2}(t+3)^{-3/2} - \frac{1}{4}t^{-3/2}(t+3)^{-1/2}$ or $-(3t^{-1/2} + \frac{9}{4}t^{-3/2})(t+3)^{-5/2}$
25 $-6t^{-4}$ **27** $3\dfrac{d^2u}{dx^2}$ **29** $2u\dfrac{d^2u}{dx^2} + 2\left(\dfrac{du}{dx}\right)^2$

Section 2.8

1 $-(y/x)$ **3** $-(x/y)^2$ **5** $\dfrac{-y}{2y+x}$ **7** $-(y/x)^3$ **9** $-\dfrac{2x+3y}{3x+2y}$ **11** $\dfrac{5x^4}{2y-1}$

13 $\dfrac{y}{2y-x}$ **15** $\dfrac{6xy-y^2-1}{2xy-3x^2}$ **17** $\dfrac{y\cos(xy)}{1-x\cos(xy)}$ **19** $-\dfrac{1}{2\sin y\cos y}$

21 $\dfrac{y}{1-2y}$ **23** ye^x **25** $2(6x^2+4xy-3)^{-1}$

27   at $(1, 2)$, $y' = -4/5$; at $(-1, 3)$, $y' = -1/5$

29   at $(2, 1)$, $y' = 2$; at $(2, -1)$, $y' = -2$; at $(\sqrt{3}, 0)$, vertical

31 $-2/3$ **33** e **35** $\dfrac{2x}{y}, \dfrac{2y^2 - 4x^2}{y^3}$

Extra Problems for Chapter 2

1 $12x^2 - 2$ **3**   8   **5** $\Delta y = \dfrac{1}{(x + \Delta x)^3} - \dfrac{1}{x^3} = -\dfrac{3x^2 + 3x\,\Delta x + \Delta x^2}{x^3(x + \Delta x)^3}\,\Delta x$,

$dy = -3x^{-4}\,\Delta x$ **7** $(2x - 2x^{-3})\,dx$ **9** $y = -x$ **11** $-9x^2 - 5$

13 $9t^2 + 8t - 15$ **15** $(-2v^4 + 10v^3 - 3v^2 - 16v + 20)(v^3 - 4)^{-2}$

17 $\tfrac{1}{2}x^{-1/2} + 6x^{1/2}$ **19** $\tfrac{1}{3}x^{-2/3} + \tfrac{1}{4}x^{-5/4}$ **21** $(2y + \tfrac{1}{2}y^{-1/2})^{-1}$ or $\dfrac{2\sqrt{y}}{4y\sqrt{y} + 1}$

23 $\dfrac{-3}{2\sqrt{1 - 3x}}$ **25** $-\tfrac{5}{2}(5x + 4)^{-3/2}$ **27** $-(3 + \tfrac{1}{2}t^{-1/2})(t^{-2} + 2t)^{-1}$

29 $-4(4x - 1)^{-3/2}$ **31** $v = \dfrac{3t + 6}{2\sqrt{t + 3}}, a = \dfrac{3t + 12}{4(t + 3)^{3/2}}$ **33** $-\dfrac{3y^3 + 6x^2 y}{9xy^2 + 2x^3}$

35 $f'(x) = \begin{cases} 2x & \text{if } |x| > 1 \\ \text{undefined} & \text{if } |x| = 1 \\ -2x & \text{if } |x| < 1 \end{cases}$

Section 3.1

1 $p = 4\sqrt{A}$ **3** $V = S^{3/2}/(6\sqrt{\pi})$ **5** $z = \sqrt{x^4 - x^2 + 1}$ **7** $V = (s - 2x)^2 x$

9 $s = 3y/(10 - y)$, $0 < y < 10$ **11** $b/16$ **13** $A = (1 + t)(w - 2t)$

15 $P(x) = x(100 - \sqrt{x}) - 10x$ or $P(x) = x(90 - \sqrt{x})$

Section 3.2

1 $100 \text{ cm}^2/\text{sec}$ **3** $-2 \text{ in.}^2/\text{sec}$ **5** $10\sqrt{61} \text{ mph}$ **7** $10/3 \text{ in.}^2/\text{sec}$

9 $40(3g)^{1/2} \text{ ft/sec}$ where $g = 32$, or $160\sqrt{6} \text{ ft/sec}$ **11** -120 mph

13 $\dfrac{800 \cdot 400}{\sqrt{(800)^2 + 6^2}} = \dfrac{160,000}{\sqrt{160,009}} \text{ mph}$ **15** -2 ft/sec

17 $1,000,000t - 300,000\sqrt{t} \text{ people/year}$ **19**   \$60 per person per year

21 $(55/144)\pi \text{ in.}^2/\text{min}$ **23** -0.01 **25** $2 - (x/500)$ **27** -100 radians/hr

29   22

Section 3.3

1   53   **3**   does not exist   **5**   does not exist   **7** $-1/(4\sqrt{2})$ **9** $1/3$

11   does not exist   **13**   does not exist   **15**   1   **17** $(1 + \sqrt{2})^{1/2}$ **19**   1

21 $-1/2$ **23** $2x$ **25** $1/(2\sqrt{t})$ **27** $3x^2$ **29** -2 **31**   does not exist

Section 3.4

1 R **3** > -2 **5** R **7**   all x except $x = 0, x = -1$ **9** $-2 < x < 2$

11   all x except $x = (-1 \pm \sqrt{5})/2$ and $x = -1$

13   all x except $x = 2, x = 3$ **15** $x < 1$

Section 3.5

1 0, min   **3** 0, min   **5** 0, neither   **7** $-1/3$, min   **9** 0, min   **11** 0, max
13 0, neither   **15** $-1/(2^{1/3})$, min   **17** 3/2, min   **19** 0, max   **21** π, neither
23 0, max   **25** 0, min   **27** -1, min   **29** 1, min   **31** 3, min   **33** 0, max
35 1/2, max   **37** $1/\sqrt{17}$ at $(4/17, 1/17)$ **39** $1 - m$ at $x = 1$
41 max of 1/4 at $x = 2$, min of $-1/4$ at $x = -2$
43 max of $1/(4 \cdot 3^{1/4})$ at $x = 1/(3^{1/4})$, min of $-1/(4 \cdot 3^{1/4})$ at $x = -1/(3^{1/4})$

Section 3.6

1 $x = 20/3, y = 40/3$ **3** $x = 50, y = 1$ **5** $x = 3/4, y = 1/4, x^3y = 27/256$
7 width $= 10\sqrt{3}$ in., length $= 40/\sqrt{3}$ in.   **9** base $=$ height $= 2/\sqrt{2}$
11 side of square $= 10/3$ in., height of triangle $= 10/3$ in.   **13** area $= r^2$
15 base of radius 2/3, height 1   **17** base of radius $4/(3\pi)$, height 4/3   **19** 50
21 $x = 50,000, p = \$5$ **23** 256 sec   **25** 1/2   **27** base 4, height 2
29 $1/\sqrt{17}$ at $x = 4/17, y = 1/17$ **31** height $(3/2) \cdot$ side of base
33 $r = (2\pi)^{-1/3}, h = (4/\pi)^{1/3}$ **35** $x = (a + b)/2$
37 $A = 2/r, dA/dr$ is never zero, $0 < r < \infty$ **39** $10\sqrt[3]{2}$ hours   **41** 62,500

Section 3.7

1

x	$f(x)$	$f'(x)$	$f''(x)$		
-2	6	-4	$+$	decr.	$\cup$
0	2	0	$+$	min.	$\cup$
2	6	4	$+$	incr.	$\cup$

3

x	$f(x)$	$f'(x)$	$f''(x)$		
-2	8	-6	$+$	decr.	$\cup$
1	-1	0	$+$	min.	$\cup$
2	0	2	$+$	incr.	$\cup$

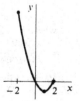

5

x	$f(x)$	$f'(x)$	$f''(x)$		
0	3	-4	$+$	decr.	$\cup$
1	1	0	$+$	min.	$\cup$
2	3	4	$+$	incr.	$\cup$

7

x	$f(x)$	$f'(x)$	$f''(x)$		
-2	16	-32	$+$	decr.	$\cup$
0	0	0	$+$	min.	$\cup$
2	16	32	$+$	incr.	$\cup$

9

x	$f(x)$	$f'(x)$	$f''(x)$		
-2	-6	9	$-$	incr.	$\cap$
$-\frac{1}{3}$	$-\frac{7}{27}$	$\frac{2}{3}$	0	incr.	infl.
2	14	17	$+$	incr.	$\cup$

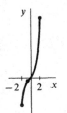
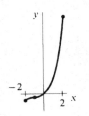

11

x	$f(x)$	$f'(x)$	$f''(x)$		
-2	$-\frac{2}{3}$	1	$-$	incr.	$\cap$
-1	$-\frac{1}{3}$	0	0	horiz.	infl.
2	$8\frac{2}{3}$	9	$+$	incr.	$\cup$

13

x	$f(x)$	$f'(x)$	$f''(x)$		
-4	2	-64	$+$	decr.	$\cup$
-3	-25	0	$+$	min.	$\cup$
-2	-14	16	0	incr.	infl.
-1	-1	8	$-$	incr.	$\cap$
0	2	0	0	horiz.	infl.
1	7	16	$+$	incr.	$\cup$
2	50	96	$+$	incr.	$\cup$

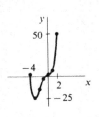

15

x	$f(x)$	$f'(x)$	$f''(x)$		
-2	-4	12	$-$	incr.	$\cap$
-1	$\frac{1}{2}$	0	$-$	max.	$\cap$
$-1/\sqrt{3}$	$\frac{5}{18}$	$-\frac{4}{3}\sqrt{3}$	0	decr.	infl.
0	0	0	$+$	min.	$\cup$
$1/\sqrt{3}$	$\frac{5}{18}$	$\frac{4}{3}\sqrt{3}$	0	incr.	infl.
1	$\frac{1}{2}$	0	$-$	max.	$\cap$
2	-4	-12	$-$	decr.	$\cap$

17

x	$f(x)$	$f'(x)$	$f''(x)$		
-4	$-\frac{1}{4}$	$-\frac{1}{16}$	$-$	decr.	$\cap$
-1	-1	-1	$-$	decr.	$\cap$
$-\frac{1}{4}$	-4	-16	$-$	decr.	$\cap$
$\frac{1}{4}$	4	-16	$+$	decr.	$\cup$
1	1	-1	$+$	decr.	$\cup$
4	$\frac{1}{4}$	$-\frac{1}{16}$	$+$	decr.	$\cup$

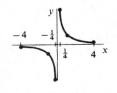

19

x	$f(x)$	$f'(x)$	$f''(x)$		
-2	$\frac{1}{4}$	$\frac{1}{4}$	$+$	incr.	$\cup$
-1	1	2	$+$	incr.	$\cup$
$-\frac{1}{2}$	4	16	$+$	incr.	$\cup$
$\frac{1}{2}$	4	-16	$+$	decr.	$\cup$
1	1	-2	$+$	decr.	$\cup$
2	$\frac{1}{4}$	$-\frac{1}{4}$	$+$	decr.	$\cup$

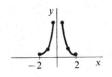

21

x	$f(x)$	$f'(x)$	$f''(x)$		
0	-1	2	$-$	incr.	$\frown$
1	0	$\frac{1}{2}$	$-$	incr.	$\frown$
10	$\frac{9}{11}$	$\frac{2}{121}$	$-$	incr.	$\frown$

23

x	$f(x)$	$f'(x)$	$f''(x)$		
-4	$\frac{1}{17}$	.03	$+$	incr.	$\cup$
$-1/\sqrt{3}$	$\frac{3}{4}$	1.9	0	incr.	infl.
0	1	0	$-$	max.	$\frown$
$1/\sqrt{3}$	$\frac{3}{4}$	-1.9	0	decr.	infl.
4	$\frac{1}{17}$	$-.03$	$+$	decr.	$\cup$

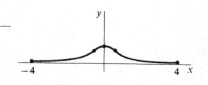

25

x	$f(x)$	$f'(x)$	$f''(x)$		
-2	$\frac{4}{5}$	$-\frac{2}{25}$	$-$	decr.	$\frown$
$-1/\sqrt{3}$	$\frac{1}{4}$	$-\sqrt{3}/2$	0	decr.	infl.
0	0	0	$+$	min.	$\cup$
$1/\sqrt{3}$	$\frac{1}{4}$	$\sqrt{3}/2$	0	incr.	infl.
2	$\frac{4}{5}$	$\frac{2}{25}$	$-$	incr.	$\frown$

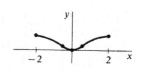

27

x	$f(x)$	$f'(x)$	$f''(x)$		
$\frac{1}{4}$	$\frac{1}{2}$	1	$-$	incr.	$\frown$
1	1	$\frac{1}{2}$	$-$	incr.	$\frown$
4	2	$\frac{1}{4}$	$-$	incr.	$\frown$

29

x	$f(x)$	$f'(x)$	$f''(x)$		
$\frac{1}{4}$	2	-4	$+$	decr.	$\cup$
1	1	$-\frac{1}{2}$	$+$	decr.	$\cup$
4	$\frac{1}{2}$	$-\frac{1}{16}$	$+$	decr.	$\cup$

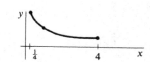

31

x	$f(x)$	$f'(x)$	$f''(x)$		
-2	$\sqrt{5}$	$2/\sqrt{5}$	$-$	incr.	$\frown$
0	3	0	$-$	max.	$\frown$
2	$\sqrt{5}$	$-2/\sqrt{5}$	$-$	decr.	$\frown$

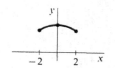

33

x	$f(x)$	$f'(x)$	$f''(x)$		
0	0	1	0	incr.	infl.
$\pi/4$	$\frac{1}{2}$	0	$-$	max.	$\frown$
$\pi/2$	0	-1	0	decr.	infl.
$3\pi/4$	$-\frac{1}{2}$	0	$+$	min.	$\cup$
π	0	1	0	incr.	infl.

repeats

35

x	$f(x)$	$f'(x)$	$f''(x)$		
0	0	$\frac{3}{2}$	0	incr.	infl.
π	3	0	$-$	max.	$\cap$
2π	0	$-\frac{3}{2}$	0	decr.	infl.

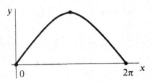

37

x	$f(x)$	$f'(x)$	$f''(x)$		
$-\pi/3$	$-\sqrt{3}$	4	$-$	incr.	$\cap$
0	0	1	0	incr.	infl.
$\pi/3$	$\sqrt{3}$	4	$+$	incr.	$\cup$

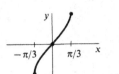

39

x	$f(x)$	$f'(x)$	$f''(x)$		
-2	e^2	$-e^2$	$+$	decr.	$\cup$
0	1	-1	$+$	decr.	$\cup$
2	e^{-2}	$-e^{-2}$	$+$	decr.	$\cup$

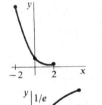

41

x	$f(x)$	$f'(x)$	$f''(x)$		
$1/e$	-1	e	$-$	incr.	$\cap$
1	0	1	$-$	incr.	$\cap$
e	1	$1/e$	$-$	incr.	$\cap$

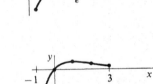

43

x	$f(x)$	$f'(x)$	$f''(x)$		
-1	$-e$	$2e$	$-$	incr.	$\cap$
0	0	1	$-$	incr.	$\cap$
1	e^{-1}	0	$-$	max.	$\cap$
2	$2e^{-2}$	$-e^{-2}$	0	decr.	infl.
3	$3e^{-3}$	$-2e^{-3}$	$+$	decr.	$\cup$

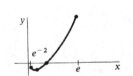

45

x	$f(x)$	$f'(x)$	$f''(x)$		
e^{-2}	$-2e^{-2}$	-1	e^2	decr.	$\cup$
e^{-1}	$-e^{-1}$	0	e	min.	$\cup$
1	0	1	1	incr.	$\cup$
e	e	2	e^{-1}	incr.	$\cup$

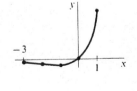

47

x	$f(x)$	$f'(x)$	$f''(x)$		
-3	$-3e^{-3}$	$-2e^{-3}$	$-$	decr.	$\cap$
-2	$-2e^{-2}$	$-e^{-2}$	0	decr.	infl.
-1	$-e^{-1}$	0	$+$	min.	$\cup$
0	0	1	$+$	incr.	$\cup$
1	e	$2e$	$+$	incr.	$\cup$

49

x	$f(x)$	$f'(x)$	$f''(x)$		
$\frac{1}{4}$	$4e^{\frac{1}{4}}$	$-12e^{\frac{1}{4}}$	$+$	decr.	$\cup$
1	e	0	$+$	min.	$\cup$
4	$e^4/4$	$\frac{3}{16}e^4$	$+$	incr.	$\cup$

Section 3.8

1 $f(0) = 1, f(1) = -1$ **3** $f(4) = 0.236, f(9) = -2.838$
5 $f(0) = 0.586, f(1) = -0.732$ **7** $f(0) = -1, f(1) = 1$ **9** $f(0) = 1, f(1) = -1$
11 $f(0) = -1, f(1) = 1$ **13** $f(0) = 0.9, f(1) = -0.1$ **15** $f(1) = -1, f(e) = 0.632$
17 yes **19** yes **21** no **23** yes, $f'(16/7) = 0$ **25** no **27** yes **29** no
31 one **33** two **35** 1/2 **37** 8/27 **39** 1 **41** $\sqrt{3} - 1$
43 $f'(x) = 3x^2 - 3$ has no zeros in $(-1, 1)$

Extra Problems for Chapter 3

1 $A = 6V^{2/3}$ **3** $x = 300t$ **5** $s = 5x/4$ **7** $16/\pi$ in./sec **9** 0.01 in./sec
11 -3 **13** $-1/4$ **15** $-1 < x < 1$ **17** $-2 < x < -1, 1 < x < 2$ **19** none
21 max $= 5$ at $x = 1$, min $= 3$ at $x = 2$ **23** max $= 4$ at $x = 0$, min $= 1$ at $x = \pm1$
25 square of side $\sqrt{2}$ **27** $8b^3/27a^3$ **29** min $= -32$ at $x = -2$, no max
31 base $=$ height $= 1$ **33** three zeros

33

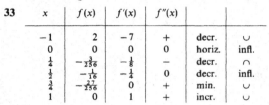

x	$f(x)$	$f'(x)$	$f''(x)$		
-1	2	-7	$+$	decr.	$\cup$
0	0	0	0	horiz.	infl.
$\frac{1}{4}$	$-\frac{3}{256}$	$-\frac{1}{8}$	$-$	decr.	$\cap$
$\frac{1}{2}$	$-\frac{1}{16}$	$-\frac{1}{4}$	0	decr.	infl.
$\frac{3}{4}$	$-\frac{27}{256}$	0	$+$	min.	$\cup$
1	0	1	$+$	incr.	$\cup$

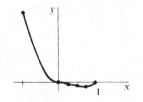

37 none **39** $f(-1) = \sqrt[3]{7} - 2 < 0, f(0) = 1 > 0$

Section 4.1

1 2 **3** 5/4 **5** 11/8 **7** 11 **9** $-1/4$ **11** 136 **13** $901/280 = 3.2$
15 $177/512 = 0.35$ **17** $1 + \sqrt{2} + \sqrt{3}(\pi - 3) = 2.7$ **19** 1.90 **21** 1.55
23 3.18

Section 4.2

1 $(16/3)x^{3/2}$ **3** $t^3 + t$ **5** $4t - t^3$ **7** $-(7/2)s^{-2}$ **9** $(1/3)(x - 6)^3$
11 $(2/5)y^{5/2}$ **13** $x^2/2, x \geq 0; -x^2/2, x < 0$ **15** 2/3 **17** 12 **19** 0
21 32/3 **23** 14 **25** 12 **27** 9.9 **29** 16/3 **31** 1/6 **33** 2

Section 4.3

1 $x + x^2 + x^3 + C$ **3** $(3/2)t^8 - (1/2)t^6 + (2/3)t^3 + t + C$ **5** $(2/3)t^{3/2} + 2t^{1/2} + C$
7 $(1/6)(2x - 3)^3 + C$ **9** $(1/3)z^3 + 2z - z^{-1} + C$ **11** $5 \sin x + C$
13 $x + \ln x + C$ **15** $x + 2\ln x - x^{-1} + C$ **17** $12t - (4/3)t^{3/2} - t^2 + C$
19 $-4y^{-1} - 6y^{-1/2} + 2\sqrt{y} + C$ **21** $(1/3)ax^3 + (1/2)bx^2 + cx + C$ **23** 0
25 68/15 **27** 0 **29** $3\ln 2$ **31** $-\ln 3$ **33** $(4/3)t^3 - t + 2$
35 $-2\cos t + 12$ **37** $v = (1/2)t^2, y = (1/6)t^3 + 1$
39 $v = t^3 + 1, y = (1/4)t^4 + t + 2$
41 $v = -(1/2)t^{-2} + 3/2, y = (1/2)t^{-1} + (3/2)t - 2$ **43** $(a), (d), (f), (g), (i), (l)$
45 $y = t^3 + 2t + 1$

Section 4.4

1 $-\dfrac{1}{2(2x + 1)} + C$ **3** $-\dfrac{1}{28}(3 - 4z)^7 + C$ **5** $-(2/3)(1 - t^2)^{2/3} + C$

7 $(1/30)(4 + 5x^2)^3 + C$ **9** $-(1/3)\cos(3x) + C$ **11** $-(3/2)\cos(4x - 1) + C$
13 $(1/2)\sin^2\theta + C$ **15** $-(1/4)\cos^4\theta + C$ **17** $-(1/2)\cos(x^2 + 1) + C$
19 $-\cos(\ln x) + C$ **21** $(2/3)(\sin t)^{3/2} + C$ **23** $(1/2)e^{2x} + C$
25 $ae^x - be^{-x} + C$ **27** $\frac{1}{2}e^{x^2} + C$ **29** $(b/a)e^{ax} + C$ **31** $e^{\sin\theta} + C$
33 $\ln(x + 2) + C$ **35** $\ln(e^x + 1) + C$ **37** $x - \ln(x + 1) + C$
39 $2\ln(1 + \sqrt{x}) + C$ **41** $-3t - 19\ln(5 - t) + C$ **43** $(1/6)(x^4 + 5)^{3/2} + C$
45 $(1/3)(2 + y^2)^{3/2} + C$ **47** $-(1 - u^2)^{1/2} + C$ **49** $(2/3)(3s + 2)^{1/2} + C$
51 $-2(1 + x^{-1})^{1/2} + C$ **53** $-(1/15)(3 + 5x^{-2})^{3/2} + C$ **55** $-(2/3)(3 - \sqrt{x})^3 + C$
57 $-2z^{-1} - (1/2)z^{-2} + C$ **59** $(2/3)(x^3 + 4)^{1/2} + C$ **61** $(1/2)(1 + x^4)^{1/2} + C$
63 $(2/5)(t + 1)^{5/2} - (2/3)(t + 1)^{3/2} + C$ **65** $(8/3)(1 - s)^{-3} - (1 - s)^{-2} + C$
67 $-(1/2)(y^2 + 1)^{-1} + (1/4)(y^2 + 1)^{-2} + C$
69 $(1/24)(4x + 1)^{3/2} - (1/8)(4x + 1)^{1/2} + C$ or $(1/12)(2x - 1)(4x + 1)^{1/2} + C$
71 $-(2/27)(1 - 3u)^{3/2} + (2/45)(1 - 3u)^{5/2} + C$ **73** $(1/6)(4x + 1)^{3/2} - (4x + 1)^{1/2} + C$
75 $-(1/4)\ln(1 - x^4) + C$ **77** $(1/4)y^4 - (1/2)y^2 + (1/2)\ln(1 + y^2) + C$ or
$(1/4)(1 + y^2)^2 - (1 + y^2) + (1/2)\ln(1 + y^2) + C$
79 $(2/3)\ln(3u + 2) + 3/(3u + 2) + C$ **81** $x - 4\sqrt{x} + 8\ln(2 + \sqrt{x}) + C$
83 $\ln(\sin\theta) + C$ **85** $(1/b)\ln(a + bx) + C$ **87** $-\ln(1 + \cos\theta) + C$
89 $(1/2)(\ln x)^2 + C$ **91** $1/2$ **93** $e - e^{-1}$ **95** $\ln 2$ or $(1/2)\ln 4$ **97** $1/2$
99 $(2/3)(3\sqrt{3} - 1)$ **101** $242/5$ **103** 100 **105** $(1/6)(17\sqrt{17} - 27)$ **107** 0
109 $(\ln 7 - \ln 4)/3$ **111** $2/3$ **113** $93/35$ **115** 0 **117** $f(g(x)) + C$

Section 4.5

1 $56/3$ **3** 2 **5** $(10\sqrt{5}/3) - 6$ **7** $32/3$ **9** $500/3$ **11** 2 **13** 4
15 2 **17** $e^2 - 3$ **19** $2(e - e^{-1})$ **21** $2 - \ln 3$ **23** $(3/2) - \ln 2$ **25** $1/5$
27 $128\sqrt{6}/5$ **29** $4/3$ **31** $32/3$ **33** $1/6$ **35** $19/15$ **37** $49/15$
39 $23/12$ **41** $3/4$ **43** $1 - (4/3)\cdot 2^{3/4} + (5/6)\cdot 2^{3/5} = 0.02$ **45** $11/4$
47 $(3 + \sqrt{5})/2$ **49** $\sqrt[3]{54}$ or $3\sqrt[3]{2}$ **51** $c = (1/2)^{2/3}$ or $c = 2^{-2/3}$

Section 4.6

1 In all cases, sum = 2, error = 0.
3 (a) $\Delta x = 0.25$: sum = 1.0426 $\Delta x = 0.1$: sum = 1.0652
   (b) $\Delta x = 0.25$: sum = 1.0594 $\Delta x = 0.1$: sum = 1.0700
   In all cases, there is no error estimate.
5 (a) $\Delta x = 0.25$: sum = 0.3229, error $\leq 1/384$
   $\Delta x = 0.1$: sum = 0.3220, error $\leq 1/2400$
   (b) $\Delta x = 0.25$: sum = 0.3217108, error $\leq 1/15,360$
   $\Delta x = 0.1$: sum = 0.3217505, error $\leq 1/600,000$
7 (a) $\Delta x = 0.5$: sum = 2.8968, error $\leq 1/48$
   $\Delta x = 0.1$: sum = 2.9012, error $\leq 1/1200$
   (b) $\Delta x = 0.5$: sum = 2.9013, error $\leq 1/960$
   $\Delta x = 0.1$: sum = 2.901388, error $\leq 1/600,000$
9 (a) $\Delta x = 0.25$: sum = 1.0968, error $\leq \sqrt{2}/96$
   $\Delta x = 0.1$: sum = 1.0906, error $\leq \sqrt{2}/600$
   (b) $\Delta x = 0.25$: sum = 1.089413
   $\Delta x = 0.1$: sum = 1.089430
11 (a) $\Delta x = 1$: sum = 2.0214, error ≤ 1
   $\Delta x = 0.1$: sum = 1.9447, error ≤ 0.01
   (b) $\Delta x = 1$: sum = 1.9587, error ≤ 0.8
   $\Delta x = 0.1$: sum = 1.945913, error ≤ 0.0008
13 (a) $\Delta x = 3$: sum = 1.8793, error $\leq 45/8$
   $\Delta x = 0.1$: sum = 1.6685, error $\leq 5/800$
   (b) $\Delta x = 3$: sum = 1.738132
   $\Delta x = 0.1$: sum = 1.668231

15 (a) $\Delta x = \pi/2$: sum $= 1.5708$, error $\leq \pi^3/48$ or 0.65
$\Delta x = \pi/10$: sum $= 1.9835$, error $\leq \pi^3/1200$ or 0.026
(b) $\Delta x = \pi/2$: sum $= 2.0944$, error $\leq \pi^5/2880$ or 0.1
$\Delta x = \pi/10$: sum $= 2.00011$, error $\leq \pi^5/1,800,000$ or 0.00017
17 (a) $\Delta x = 0.25$: sum $= 1.7272$ error $\leq e/192$ or 0.014
$\Delta x = 0.1$: sum $= 1.7197$, error $\leq e/1200$ or 0.002
(b) $\Delta x = 0.25$: sum $= 1.718319$, error $\leq e/46,080$ or 0.00006
$\Delta x = 0.1$: sum $= 1.718283$, error $\leq e/1,800,000$ or 0.0000015
19 (a) $\Delta x = 0.25$: sum $= 0.3837$, error $\leq 1/192$
$\Delta x = 0.1$: sum $= 0.3859$, error $\leq 1/1200$
(b) $\Delta x = 0.25$: sum $= 0.386260$, error $\leq 1/46,080$
$\Delta x = 0.1$: sum $= 0.386293$, error $\leq 1/1,800,000$

Extra Problems for Chapter 4

1 0.6025 **3** 7.875 **5** $1/3$ **7** $76 + 40\sqrt{5}$ **9** $946\frac{2}{3}$ **11** 2
13 $2x + x^2/2 - x^3 + C$ **15** $-(1/4)(x^2 - 1)^{-2} + C$ **17** $-(1/3)(1 - 3u^2)^{1/2} + C$
19 $(1/3)[(2t + 1)^{3/2} - (2t - 1)^{3/2}] + C$ **21** $(2/5)u^{5/2} - (4/3)u^{3/2} + C$
23 $2\sin(x/2) + C$ **25** $-e^{-t} + C$ **27** $40/3$ **29** $(e^4 - 1)/4$ **31** $(x^3 + 2)^{1/2}$
33 $-\sqrt{u}(u - 1)^{1/2}$ **35** $(1/2)x^2 - x + b + 1/2$ **37** $3x^2 - 6x + 5$
39 $F(x) = 1$ if $x < 0$, $(x^2/2) + 1$ if $x \geq 0$ **43** $\pi ab/2$ **45** $xf(x) + \int_0^x f(t)\,dt$

Section 5.1

1 3 **3** ∞ **5** $1/3$ **7** $5/3$ **9** $1/\sqrt{3}$ **11** ∞ **13** ∞ **15** 0
17 $-\infty$ **19** ∞ **21** 0 **23** $1/2$ **25** $-1/4$ **27** 0 **29** $-5/6$ **31** 0
33 does not exist   **35** ∞ **37** does not exist   **39** does not exist   **41** ∞
43 0 **45** $-\infty$ **47** $1/2$ **49** ∞ **51** $-3/2$ **53** 0 **55** 1 **57** 0
59 does not exist   **61** 0 **63** $-\infty$ **65** does not exist   **67** ∞

Section 5.2

1 $1/6$ **3** $1/16$ **5** 0 **7** $-1/4$ **9** $1/2$ **11** -1 **13** 5 **15** ∞
17 2 **19** 3 **21** 1 **23** 2 **25** 1 **27** 1 **29** 1 **31** 1 **33** $1/2$
35 0 **37** $1/2$ **39** $1/2$ **41** $4/\sqrt{2} = \sqrt{8}$ **43** 2 **45** ∞ **47** $3/2$
49 5 **51** does not exist

Section 5.3

1 (a) A, B, F, H (b) B, F, H (c) B, F (d) A, B, C, D, F, G, H
(e) B, D, F, G, H (f) B, D, F

3

x	$f(x)$	$f'(x)$	$f''(x)$		
$\lim_{x \to -\infty}$	∞	$-\infty$			
0	0	-2	$+$	decr.	$\cup$
1	-1	0	$+$	min.	$\cup$
2	0	2	$+$	incr.	$\cup$
$\lim_{x \to \infty}$	∞	∞			

5

x	$f(x)$	$f'(x)$	$f''(x)$		
$\lim_{x\to-\infty}$	∞	$-\infty$			
-2	$5\frac{1}{3}$	-6	$+$	decr.	$\cup$
0	0	0	$+$	min.	$\cup$
2	$2\frac{2}{3}$	2	0	incr.	infl.
4	$5\frac{1}{3}$	0	$-$	max.	$\cap$
6	0	-6	$-$	decr.	$\cap$
$\lim_{x\to\infty}$	$-\infty$	$-\infty$			

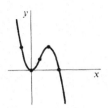

7

x	$f(x)$	$f'(x)$	$f''(x)$		
$\lim_{x\to-\infty}$	$-\infty$	∞			
-2	-12	20	$-$	incr.	$\cap$
0	0	0	0	horiz.	infl.
1	$\frac{3}{4}$	2	$+$	incr.	$\cup$
2	4	4	0	incr.	infl.
3	$6\frac{3}{4}$	0	$-$	max.	$\cap$
4	0	-16	$-$	decr.	$\cap$
$\lim_{x\to\infty}$	$-\infty$	$-\infty$			

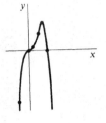

9

x	$f(x)$	$f'(x)$	$f''(x)$		
$\lim_{x\to-\infty}$	0	0			
1	1	1	$+$	incr.	$\cup$
$\lim_{x\to2^-}$	∞	∞			
$\lim_{x\to2^+}$	$-\infty$	∞			
3	-1	1	$-$	incr.	$\cap$
$\lim_{x\to\infty}$	0	0			

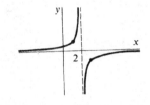

11

x	$f(x)$	$f'(x)$	$f''(x)$		
$\lim_{x\to-\infty}$	0	0			
-1	1	2	$+$	incr.	$\cup$
$\lim_{x\to0^-}$	∞	∞			
$\lim_{x\to0^+}$	∞	∞			
1	1	-2	$+$	decr.	$\cup$
$\lim_{x\to\infty}$	0	0			

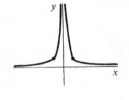

13

x	$f(x)$	$f'(x)$	$f''(x)$		
$\lim_{x\to0^+}$	0	∞			
1	1	$\frac{1}{2}$	$-$	incr.	$\cap$
$\lim_{x\to\infty}$	∞	0			

15

x	$f(x)$	$f'(x)$	$f''(x)$			
$\lim_{x \to 0^+}$	∞	$-\infty$				
1	1	$-\frac{1}{2}$	$+$	decr.	$\cap$	
$\lim_{x \to \infty}$	0	0				

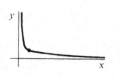

17

x	$f(x)$	$f'(x)$	$f''(x)$			
$\lim_{x \to -\infty}$	$-\infty$	0				
-1	-1	$\frac{1}{3}$	$+$	incr.	$\cup$	
$\lim_{x \to 0^-}$	0	∞				
$\lim_{x \to 0^+}$	0	∞				
1	1	$\frac{1}{3}$	$-$	incr.	$\cap$	
$\lim_{x \to \infty}$	∞	0				

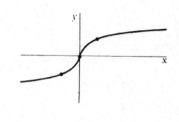

19

x	$f(x)$	$f'(x)$	$f''(x)$			
$\lim_{x \to -\infty}$	1	0				
-2	3	2	$-$	incr.	$\cap$	
$\lim_{x \to -1^-}$	∞	∞				
$\lim_{x \to -1^+}$	$-\infty$	∞				
0	-1	2	$-$	incr.	$\cap$	
$\lim_{x \to \infty}$	1	0				

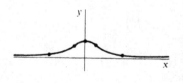

21

x	$f(x)$	$f'(x)$	$f''(x)$			
$\lim_{x \to -\infty}$	0	0				
-2	$\frac{1}{5}$	$\frac{4}{25}$	$+$	incr.	$\cup$	
$-1/\sqrt{3}$	$\frac{3}{4}$	1.9	0	incr.	infl.	
0	1	0	$-$	max.	$\cap$	
$1/\sqrt{3}$	$\frac{3}{4}$	-1.9	0	decr.	infl.	
2	$\frac{1}{5}$	$-\frac{4}{25}$	$+$	decr.	$\cup$	
$\lim_{x \to \infty}$	0	0				

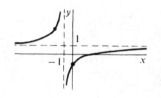

23

x	$f(x)$	$f'(x)$	$f''(x)$			
$\lim_{x \to -\infty}$	1	0				
-2	$\frac{4}{5}$	$-\frac{2}{25}$	$-$	decr.	$\cap$	
$-1/\sqrt{3}$	$\frac{1}{4}$	$-\sqrt{3}/2$	0	decr.	infl.	
0	0	0	$+$	min.	$\cup$	
$1/\sqrt{3}$	$\frac{1}{4}$	$\sqrt{3}/2$	0	incr.	infl.	
2	$\frac{4}{5}$	$\frac{2}{25}$	$-$	incr.	$\cap$	
$\lim_{x \to \infty}$	1	0				

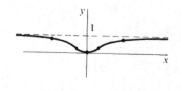

25

x	$f(x)$	$f'(x)$	$f''(x)$		
$\lim\limits_{x\to-\infty}$	0	0			
-2	$-\frac{2}{3}$	$-\frac{5}{9}$	$-$	decr.	$\cap$
$\lim\limits_{x\to-1^-}$	$-\infty$	$-\infty$			
$\lim\limits_{x\to-1^+}$	∞	$-\infty$			
$-\frac{1}{2}$	$\frac{2}{3}$	$-\frac{20}{9}$	$+$	decr.	$\cup$
0	0	-1	0	decr.	infl.
$\frac{1}{2}$	$-\frac{2}{3}$	$-\frac{20}{9}$	$-$	decr.	$\cap$
$\lim\limits_{x\to1^-}$	$-\infty$	$-\infty$			
$\lim\limits_{x\to1^+}$	∞	$-\infty$			
2	$\frac{2}{3}$	$-\frac{5}{9}$	$+$	decr.	$\cup$
∞	0	0			

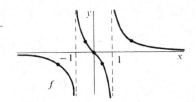

27

x	$f(x)$	$f'(x)$	$f''(x)$		
$\lim\limits_{x\to-\infty}$	∞	0			
-1	1	$-\frac{2}{3}$	$-$	decr.	$\cap$
$\lim\limits_{x\to0^-}$	0	$-\infty$		min. (cusp)	
$\lim\limits_{x\to0^+}$	0	∞			
1	1	$\frac{2}{3}$	$-$	incr.	$\cap$
$\lim\limits_{x\to\infty}$	∞	0			

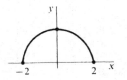

29

x	$f(x)$	$f'(x)$	$f''(x)$		
$\lim\limits_{x\to-2^+}$	0	∞			
0	2	0	$-$	max.	$\cap$
$\lim\limits_{x\to2^-}$	0	$-\infty$			

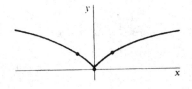

31

x	$f(x)$	$f'(x)$	$f''(x)$		
$\lim\limits_{x\to-1^+}$	1	$-\infty$			
0	0	0	$+$	min.	$\cup$
$\lim\limits_{x\to1^-}$	1	∞			

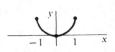

33

x	$f(x)$	$f'(x)$	$f''(x)$		
$\lim_{x \to 0^+}$	∞	$-\infty$			
$\pi/4$	$\sqrt{2}$	$-\sqrt{2}$	$+$	decr.	$\cup$
$\pi/2$	1	0	$+$	min.	$\cup$
$3\pi/4$	$\sqrt{2}$	$\sqrt{2}$	$+$	incr.	$\cup$
$\lim_{x \to \pi^-}$	∞	∞			
$\lim_{x \to \pi^+}$	$-\infty$	∞			
$5\pi/4$	$-\sqrt{2}$	$\sqrt{2}$	$-$	incr.	$\cap$
$3\pi/2$	-1	0	$-$	max.	$\cap$
$7\pi/4$	$-\sqrt{2}$	$-\sqrt{2}$	$-$	decr.	$\cap$
$\lim_{x \to 2\pi^-}$	$-\infty$	$-\infty$			

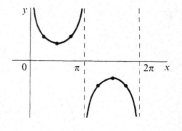

35

x	$f(x)$	$f'(x)$	$f''(x)$		
0	0	1	0	incr.	infl.
$\pi/4$	1,	2	$+$	incr.	$\cup$
$\lim_{x \to \pi/2^-}$	∞	∞			
$\lim_{x \to \pi/2^+}$	$-\infty$	∞			
$3\pi/4$	-1	2	$-$	incr.	$\cap$
π	0	1	0	incr.	infl.
$5\pi/4$	1	2	$+$	incr.	$\cup$
$\lim_{x \to 3\pi/2^-}$	∞	∞			
$\lim_{x \to 3\pi/2^+}$	$-\infty$	∞			
$7\pi/4$	-1	2	$-$	incr.	$\cap$
2π	0	1	0	incr.	infl.

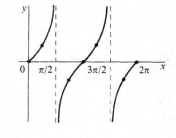

37

x	$f(x)$	$f'(x)$	$f''(x)$		
0	1	-1	$+$	decr.	$\cup$
$\pi/4$	$\sqrt{2}/2$	0	$+$	min.	$\cup$
$\pi/2$	1	1	$+$	incr.	$\cup$
$\lim_{x \to 3\pi/4^-}$	∞	∞			
$\lim_{x \to 3\pi/4^+}$	$-\infty$	∞			
π	-1	1	$-$	incr.	$\cap$
$5\pi/4$	$-\sqrt{2}/2$	0	$-$	max.	$\cap$
$3\pi/2$	-1	-1	$-$	decr.	$\cap$
$\lim_{x \to 7\pi/4^-}$	$-\infty$	$-\infty$			
$\lim_{x \to 7\pi/4^+}$	∞	$-\infty$			
2π	1	-1	$+$	decr.	$\cup$

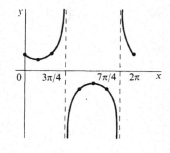

39

x	$f(x)$	$f'(x)$	$f''(x)$		
$\lim\limits_{x\to-\infty}$	$-\infty$	1			
-3	$-\sqrt5$	$3/\sqrt5$	$+$	incr.	$\cup$
$\lim\limits_{x\to-2^-}$	0	∞			
$\lim\limits_{x\to2^+}$	0	$-\infty$			
3	$-\sqrt5$	$-3/\sqrt5$	$+$	decr.	$\cup$
$\lim\limits_{x\to\infty}$	$-\infty$	-1			

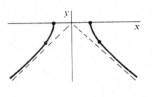

41

x	$f(x)$	$f(x)$	$f''(x)$		
$\lim\limits_{x\to-\infty}$	0	0			
-2	$1/\sqrt3$	$2/(3\sqrt3)$	$+$	incr.	$\cup$
$\lim\limits_{x\to-1^-}$	∞	∞			
$\lim\limits_{x\to1^+}$	∞	$-\infty$			
2	$1/\sqrt3$	$-2/(3\sqrt3)$	$+$	decr.	$\cup$
$\lim\limits_{x\to\infty}$	0	0			

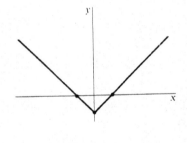

43

x	$f(x)$	$f'(x)$	$f''(x)$		
$\lim\limits_{x\to-\infty}$	∞	-1			
-1	0	-1	0	decr.	linear
$\lim\limits_{x\to0^-}$	-1	-1		min. (corner)	
$\lim\limits_{x\to0^+}$	-1	1			
1	0	1	0	incr.	linear
$\lim\limits_{x\to\infty}$	∞	1			

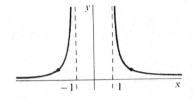

45

x	$f(x)$	$f'(x)$	$f''(x)$		
$\lim\limits_{x\to-\infty}$	∞	-2			
0	1	-2	0	decr.	linear
$\lim\limits_{x\to\frac12^-}$	0	-2		min. (corner)	
$\lim\limits_{x\to\frac12^+}$	0	2			
1	1	2	0	incr.	linear
$\lim\limits_{x\to\infty}$	∞	2			

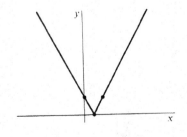

47

x	$f(x)$	$f'(x)$	$f''(x)$		
$\lim\limits_{x \to -\infty}$	$-\infty$	1			
0	2	1	0	incr.	linear
$\lim\limits_{x \to 2^-}$	4	1		corner	
$\lim\limits_{x \to 2^+}$	4	3			
4	10	3	0	incr.	linear
$\lim\limits_{x \to \infty}$	∞	3			

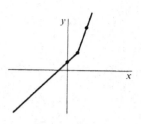

49

x	$f(x)$	$f'(x)$	$f''(x)$		
$\lim\limits_{x \to -\infty}$	∞	$-\infty$			
-2	5	-5	2	decr.	$\cup$
$\lim\limits_{x \to -1^-}$	1	-3		corner	
$\lim\limits_{x \to -1^+}$	1	-1			
$-\frac{1}{2}$	$\frac{3}{4}$	0	2	min.	$\cup$
0	1	1	2	incr.	$\cup$
1	3	3	2	incr.	$\cup$
$\lim\limits_{x \to \infty}$	∞	∞			

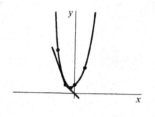

51

x	$f(x)$	$f'(x)$	$f''(x)$		
$\lim\limits_{x \to -\infty}$	∞	0			
-1	1	$-\frac{1}{2}$	$-$	decr.	$\cap$
$\lim\limits_{x \to 0^-}$	0	$-\infty$		min. (cusp)	
$\lim\limits_{x \to 0^+}$	0	∞			
1	1	$\frac{1}{2}$	$-$	incr.	$\cap$
$\lim\limits_{x \to \infty}$	∞	0			

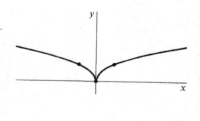

53

x	$f(x)$	$f'(x)$	$f''(x)$		
$\lim\limits_{x \to -\infty}$	$-\infty$	1			
-1	-2	1	0	incr.	linear
$\lim\limits_{x \to 0^-}$	-1	1		jump	
$\lim\limits_{x \to 0^+}$	1	1			
1	2	1	0	incr.	linear
$\lim\limits_{x \to \infty}$	∞	1			

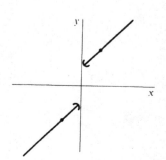

55

x	$f(x)$	$f'(x)$	$f''(x)$		
$\lim\limits_{x \to -\infty}$	$-\infty$	1			
-1	$-\sqrt{2}$	$1/\sqrt{2}$	$-$	incr.	$\cap$
$\lim\limits_{x \to 0^-}$	-1	0		jump	
$\lim\limits_{x \to 0^+}$	1	0			
1	$\sqrt{2}$	$1/\sqrt{2}$	$+$	incr.	$\cup$
$\lim\limits_{x \to \infty}$	∞	1			

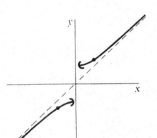

Section 5.4

1

3

5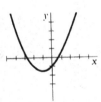

7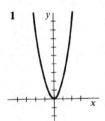

9

11

13

15 focus $(1/4, 0)$, directrix $x = -1/4$ **17** focus $(3/2, 1/2)$, directrix $x = 2$

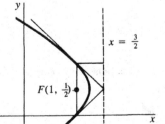

19 $x = (1/2)y^2 - (1/2)$ **21** $y = -(1/4)(x^2 + 2x + 1)$

Section 5.5

1

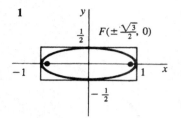

$F(\pm \frac{\sqrt{3}}{2}, 0)$

3

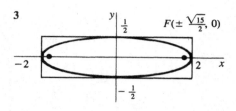

$F(\pm \frac{\sqrt{15}}{2}, 0)$

5

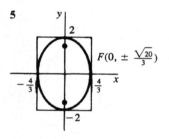

$F(0, \pm \frac{\sqrt{20}}{3})$

7

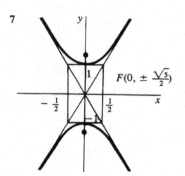

$F(0, \pm \frac{\sqrt{5}}{2})$

9

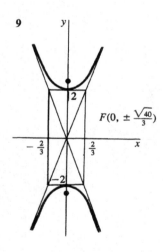

$F(0, \pm \frac{\sqrt{40}}{3})$

11

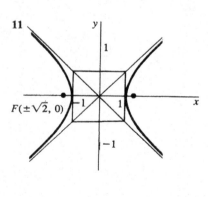

$F(\pm\sqrt{2}, 0)$

Section 5.6

1 hyperbola   **3** parabola   **5** ellipse   **7**

9

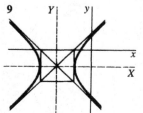

11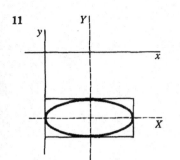

13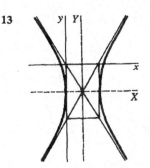

15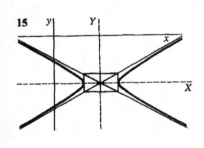

17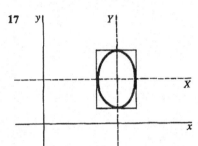

Section 5.7

1 $\alpha = 45°, X^2 - Y^2 + 8 = 0$ **3** $\alpha = 45°, -2X^2 + 6Y^2 = 1$
5 $\alpha = 30°, 2X^2 - 2Y^2 = 7$ **7** $\alpha = 22.5°, 1.207X^2 - 0.207Y^2 = 3$
9 $\alpha = -15°, 4.232X^2 + 0.768Y^2 = 5$

Section 5.8

Any smaller value of $\delta > 0$ or larger value of B is also correct.

1 $\delta = 10^{-3}$ **3** $\delta = 2 - (0.51)^{-1} \sim 0.039$ **5** $\delta = 10^{-4}$
7 $\delta = 1 - \sqrt{1 - 10^{-6}} \sim 5 \times 10^{-7}$ **9** $\delta = 10^{-2}$ **11** $\delta = 10^{-3}$ **13** $\delta = 10^{-4}$
15 $\delta = 1 - \sqrt{0.99} \sim 5 \times 10^{-3}$ **17** $B = 99/4 = 24.75$ **19** $B = \dfrac{5 + \sqrt{8025}}{4} \sim 23.7$
21 $B = \dfrac{10^4 - 1}{5} \sim 2 \times 10^3$
23 For all $\varepsilon > 0$ there exists $\delta > 0$ such that whenever $c - \delta < x < c, |f(x) - L| < \varepsilon$.
25 For all $A > 0$ there exists $B > 0$ such that whenever $x > B, f(x) < -A$.

Section 5.9

1 1.42332   **3** 0.63688   **5** 1.22074   **7** 2.93923   **9** 0.38197   **11** 0.56714
13 1.10606   **15** 1.30633

Section 5.10

1 $f'(1) = 2$, error $\leq \Delta x$ **3** $f'(4) = -(1/8)$, error $\leq (3/128) \Delta x$
5 $f'(3) = -1/9$, error $\leq (1/27) \Delta x$ **7** $f'(0) = 1$, error $\leq 0.5 \Delta x$
9 $f'(\pi/3) = -\sqrt{3}$, error $\leq 2 \Delta x$ **11** $f'(1) = 1$, error $\leq \Delta x/2$

13 $f'(1) = e$, error $\leq (1/2)e\,\Delta x$

15 $f'(100) = 1/20$, error $\leq (1/8) \cdot 99^{-3/2}\,\Delta x$ or $0.000127\,\Delta x$

17 $f'(2) = (2/\sqrt{5})$, error $\leq 1/(4\sqrt{2})\,\Delta x$ **19** $f'(1) = \sqrt{2} + 2^{-3/2}$, error $\leq \Delta x/2$

21 $\sqrt{65} \sim 8\frac{1}{16}$, error $\leq 1/4096$ **23** $(0.301)^4$ or 0.008208, error $\leq 6 \times 10^{-7}$

25 $1/97$ or $10^{-2} + 3 \times 10^{-4} = 0.0103$, error $\leq 10^{-5}$

27 $\sqrt{1.02} + \sqrt[3]{1.02} \sim 2\frac{1}{60}$, error $\leq \frac{17}{18} \times 10^{-4}$ **29** $(1.003)^5$ or 1.015, error $\leq 10^{-4}$

31 $\sin(\pi/3 + 0.004)$ or $\sqrt{3}/2 + 0.002$, error $\leq 8 \times 10^{-6}$

33 $\tan(0.005)$ or 0.005, error $\leq 10^{-6}$ **35** $e^{0.002}$ or 1.002, error $\leq 3 \times 10^{-6}$

37 $\ln(1.006)$ or 0.006, error $\leq 18 \times 10^{-6}$

Extra Problems for Chapter 5

1 0   **3** 0   **5** ∞ **7** 0   **9** 3/2

11

x	$f(x)$	$f'(x)$	$f''(x)$		
$\lim\limits_{x \to -\infty}$	$-\infty$	1			
-1	0	2	$+$	incr.	$\cup$
$\lim\limits_{x \to 0^-}$	∞	∞			
$\lim\limits_{x \to 0^+}$	$-\infty$	∞			
1	0	2	$-$	incr.	$\cap$
$\lim\limits_{x \to \infty}$	∞	1			

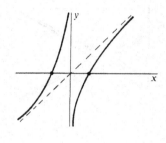

13

x	$f(x)$	$f'(x)$	$f''(x)$		
$\lim\limits_{x \to -\infty}$	0	0			
0	$\frac{1}{2}$	$\frac{3}{4}$	$+$	incr.	$\cup$
$\lim\limits_{x \to 1^-}$	∞	∞			
$\lim\limits_{x \to 1^+}$	$-\infty$	∞			
$\frac{3}{2}$	-4	0	$-$	max.	$\cap$
$\lim\limits_{x \to 2^-}$	$-\infty$	$-\infty$			
$\lim\limits_{x \to 2^+}$	∞	$-\infty$			
3	$\frac{1}{2}$	$-\frac{3}{4}$	$+$	decr.	$\cup$
$\lim\limits_{x \to \infty}$	0	0			

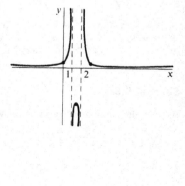

15

x	$f(x)$	$f'(x)$	$f''(x)$		
$\lim\limits_{x \to -\infty}$	∞	-2			
-2	4	-2	0	decr.	linear
$\lim\limits_{x \to -1^-}$	2	-2		corner	
$\lim\limits_{x \to -1^+}$	2	0			
0	2	0	0	constant	linear
$\lim\limits_{x \to 1^-}$	2	0		corner	
$\lim\limits_{x \to 1^+}$	2	2			
2	4	2	0	incr.	linear
$\lim\limits_{x \to \infty}$	∞	2			

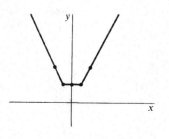

17
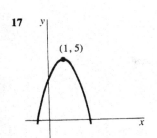

19
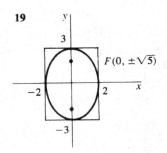

21
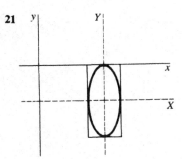

23 $\alpha = 45°, X^2 - Y^2 = 18$ **25** $\delta \le 0.155$ **27** 1.7866   **29** $(3/4096)\,\Delta x$
31 $25 - 2/15 = 24.8666$, error $\le 1/(9 \cdot (124)^{4/3})$

Section 6.1

1 1/3   **3** $\sqrt{3}/12$ **5** $\pi/24$ **7** $2cr^2/3$

Section 6.2

1 (a) $\pi/5$ (b) $\pi/2$ **3** (a) 8π (b) $128\pi/5$ **5** (a) $\pi/3$ (b) $\pi/3$
7 (a) $4\pi/3$ (b) $(2/3)\pi(2\sqrt{2} - 1)$ **9** (a) $31\pi/160$ (b) π **11** (a) $4\pi/45$ (b) $\pi/3$
13 (a) 27π (b) 18π **15** (a) $2\pi/35$ (b) $\pi/10$ **17** (a) $\pi/3$ (b) $\pi/9$
19 $20\pi/3$ (b) $29\pi/6$ **21** (a) $16\pi/3$ (b) $184\pi/15$ **23** 2π **25** $(\pi^2/4) - (\pi/2)$
27 $(\pi/2)(e^2 - 1)$ **29** $(\pi/6)(e^2 - 1)$ **31** $\pi \ln 2$ **33** $\pi(3 - \ln 4)$ **35** 2π
37 2π **39** $\pi(e - 1)$ **41** $2\pi(e^{-1} - e^{-4})$ **43** $2\pi \ln 2$ **45** $(\pi/2) \ln 7$

47 $\frac{4}{3}\pi(r^2 - a^2)^{3/2}$ **49** $\dfrac{1}{3}\pi r^2 h - \pi a^2 h + \dfrac{2\pi h a^3}{3r}$

51 (a) $\pi\left(\dfrac{b^2 a}{3} + \dfrac{2}{3}r^3 - r^2 a + \dfrac{a^3}{3}\right)$ or $\dfrac{2}{3}\pi r^2(r - a)$

   (b) $\frac{2}{3}\pi(ba^2 + b^3)$ or $\frac{2}{3}\pi[a^2\sqrt{r^2 - a^2} + (r^2 - a^2)^{3/2}]$ **53** $2\pi^2 cr^2$

Section 6.3

1 $\frac{2}{3}(6\sqrt{6} - 3\sqrt{3})$ **3** $\frac{8}{3}((3\sqrt{3}/2\sqrt{2}) - 1)$ **5** $\frac{8}{27}(10\sqrt{10} - 1)$ **7** 379/12

9 387/20   **11** 3/2   **13** 36   **15** $\frac{4}{3}(2^{5/2} - 1)$ **17** $2\sqrt{3}/3$

19 $8 \cdot 10\sqrt{10} - 37\sqrt{37}$ **21** 2π **23** $4\sqrt{2}$ **25** $\dfrac{t^2}{\sqrt{1 + 4t^2}}$

27 $\displaystyle\int_0^4 \sqrt{1 + (4x - 1)^2}\,dx$ **29** $\displaystyle\int_1^2 \sqrt{4 + \dfrac{1}{4t}}\,dt$

31 $\displaystyle\int_1^2 \sqrt{1 + \frac{1}{4x}}\,dx \sim \frac{1}{4}\left(\frac{\sqrt{5}}{4} + \sqrt{\frac{3}{2}} + \sqrt{\frac{7}{4}} + \sqrt{2} + \frac{3}{4}\right) \sim 1.32$

33 $\displaystyle\int_1^5 \sqrt{1 + x^{-4}}\,dx \sim \frac{\sqrt{2}}{2} + \frac{\sqrt{17}}{4} + \frac{\sqrt{82}}{9} + \frac{\sqrt{257}}{16} + \frac{\sqrt{626}}{50} \sim 4.246$

Section 6.4

1 $\dfrac{\pi}{6}(17\sqrt{17} - 1)$ **3** $\dfrac{2\pi}{81}\left(\dfrac{2}{5}\cdot 10^{5/2} - \dfrac{2}{3}\cdot 10^{3/2} + \dfrac{4}{15}\right) \sim 8.2$ **5** $256\pi/15$

7 $1575\pi/8$ **9** $\dfrac{2\pi}{3}(10\sqrt{10} - 2\sqrt{2})$ **11** $\pi(2\sqrt{2} - 1)$ **13** $\dfrac{\pi}{9}(2\sqrt{2} - 1)$

15 $47\pi/16$ **17** $7\pi/9$ **19** $16\pi\sqrt{5}$ **21** $2\pi ra$ **23** (a) $\displaystyle\int_0^1 2\pi x\sqrt{1 + 25x^8}\,dx$

(b) $\displaystyle\int_0^1 2\pi x^5\sqrt{1 + 25x^8}\,dx$ **25** (a) $\displaystyle\int_1^{10} 2\pi(t^2 + t)\sqrt{8t^2 + 4t + 1}\,dt$

(b) $\displaystyle\int_1^{10} 2\pi(t^2 - 1)\sqrt{8t^2 + 4t + 1}\,dt$ **27** $\displaystyle\int_0^1 \pi x^2\sqrt{1 + x^2}\,dx \sim 1.357$

Section 6.5

1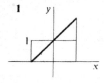

3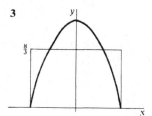

5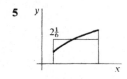

7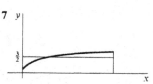

9 $3 - (2/\sqrt{3})$ **11** -6 **13** 0 **15** $2/\pi$ **17** $1/\pi$ **19** $(1/2)(e - e^{-1})$
21 $(1/3)\ln 4$ **23** 1 **25** $8/9$ **27** $6\sqrt{15}/25$ **29** $13/2$ **31** $116/45$
33 (a) 2 (b) $32/7$

Section 6.6

1 (a) 4 (b) $M_x = 0, M_y = 4$ (c) $(1, 0)$ **3** (a) $128/3$ (b) $M_x = 512/5, M_y = 0$
(c) $(0, 12/5)$ **5** (a) $6k$ (b) $M_x = 8k, M_y = 2k$ (c) $(\frac{1}{3}, \frac{4}{3})$ **7** (a) $128/15$
(b) $M_x = 1024/105, M_y = 0$ (c) $(0, \frac{8}{7})$ **9** (a) $\frac{1}{2}$ (b) $M_x = \frac{1}{8}, M_y = \frac{1}{3}$ (c) $(\frac{2}{3}, \frac{1}{4})$

11 (a) $2\sqrt{2} - 2$ (b) $M_x = 1 - \dfrac{\sqrt{2}}{2}, M_y = \dfrac{4\sqrt{2} - 2}{3}$

(c) $\left(\dfrac{2\sqrt{2} - 1}{3\sqrt{2} - 3}, \dfrac{2 - \sqrt{2}}{4\sqrt{2} - 4}\right)$ or $\left(1 + \dfrac{\sqrt{2}}{3}, \dfrac{\sqrt{2}}{4}\right)$

13 (a) $4/3$ (b) $M_x = 1/2, M_y = 0$ (c) $(0, 3/8)$
15 (a) $128/3$ (b) $M_x = 1024/15, M_y = 512/5$ (c) $(\frac{12}{5}, \frac{8}{5})$

17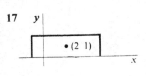

19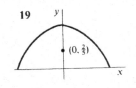

21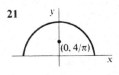

23

25 2/3   **27** $e^2 - 1$ **29** (2 ln 2, 7/24)   **31** (8/15, 16/105)   **33** (1/2, 2/5)
35 80 ft lbs   **37** 320/3 ft lbs   **39** 50 ft lbs   **41** 215 ft lbs   **43** 9k/400

Section 6.7

1 1/2   **3** diverges to ∞ **5** diverges to $-\infty$ **7** diverges to ∞ **9** 1/2
11 diverges to ∞ **13** 3   **15** 6   **17** $-3/2$ **19** 3   **21** diverges to ∞
23 diverges to ∞ **25** diverges   **27** 1/2   **29** diverges   **31** $4\sqrt{2}$ **33** ∞
37 4   **39** 2/3   **41** (a) π (b) ∞ **43** (a) ∞ (b) 8π **45** 4/3
47 (a) $11\pi/9$ (b) $16\pi/15$ **49** (a) $\dfrac{\pi}{6}(5\sqrt{5} - 1)$ (b) $\displaystyle\int_0^1 2\pi x\sqrt{1 + \dfrac{1}{4x}}\,dx$
51 (a) $2\pi ra$ (b) $2\pi r^2 - 2\pi r\sqrt{r^2 - a^2}$

Extra Problems for Chapter 6

1 $8r^3/3$ **3** $2r^3/3$ **5** (a) $11\pi/3$ (b) $(16\pi/3) - 2\pi\sqrt{3}$ **7** (a) 16π (b) $16\pi/3$
9 $\pi p/(p + 1)$ **11** $\frac{1}{27}(46\sqrt{46} - 10\sqrt{10})$ **13** 5   **15** $(b - a)\sqrt{A^2 + C^2}$
17 $\dfrac{\pi}{3A}[(8A^2 + 4AB + B^2)^{3/2} - (4A^2 + B^2)^{3/2}]$ **19** $\dfrac{b^{p+1} - 1}{(b - 1)(p + 1)}$ **21** $(p + 1)^{-1/p}$
23 $(\frac{1}{2}, \frac{2}{3})$ **25** $(\frac{5}{7}, 0)$ **27** 15 ft lbs   **29** diverges to ∞ **31** $-5/4$
33 diverges to ∞ **35** 8   **37** $a^2h/3$ **39** $m = \pi\rho,\ M_x = 2\rho,\ M_y = 0,\ (\bar{x}, \bar{y}) = (0, 2/\pi)$
43 $\pi\rho r^4/4$ **45** $w\rho b^2/2$

Section 7.1

7 $(\pi/4) + k\pi$ **9** none   **13** 0   **15** 0   **17** 0   **19** $k\pi$

Section 7.2

1 $5\cos(5x)$ **3** $6\theta\cos(3\theta^2)$ **5** $4\sec^2(4\theta - 3)$ **7** $a\cos\theta - b\sin\theta$
9 $-\dfrac{1}{2\sqrt{x}}\sin(\sqrt{x})$ **11** $\cos\theta\sec^2(\sin\theta)$ **13** $\dfrac{3\csc(3t)\cot(3t)}{(2 + \csc(3t))^2}$ **15** $\dfrac{1}{2\sin y\cos y}$
17 3/2   **19** ∞ **21** 2   **23** 1

25

x	$f(x)$	$f'(x)$	$f''(x)$		
0	0	3	0	incr.	infl.
$\pi/4$	$3/\sqrt{2}$	$3/\sqrt{2}$	$-$	incr.	$\cap$
$\pi/2$	3	0	$-$	max.	$\cap$
$3\pi/4$	$3/\sqrt{2}$	$-3/\sqrt{2}$	$-$	decr.	$\cap$
π	0	-3	0	decr.	infl.
$5\pi/4$	$-3/\sqrt{2}$	$-3/\sqrt{2}$	$+$	decr.	$\cup$
$3\pi/2$	-3	0	$+$	min.	$\cup$
$7\pi/4$	$-3/\sqrt{2}$	$3/\sqrt{2}$	$+$	incr.	$\cup$
2π	0	3	0	incr.	infl.

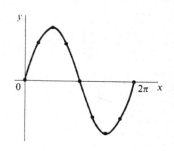

27

x	$f(x)$	$f'(x)$	$f''(x)$		
0	0	0	$+$	min.	$\cup$
$\pi/4$	$1/2$	1	0	incr.	infl.
$\pi/2$	1	0	$-$	max.	$\cap$
$3\pi/4$	$1/2$	-1	0	decr.	infl.
π	0	0	$+$	min.	$\cup$
$5\pi/4$	$1/2$	1	0	incr.	infl.
$3\pi/2$	1	0	$-$	max.	$\cap$
$7\pi/4$	$1/2$	-1	0	decr.	infl.
2π	0	0	$+$	min.	$\cup$

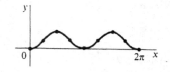

29

x	$f(x)$	$f'(x)$	$f''(x)$		
0	$-1/\sqrt{2}$	$1/\sqrt{2}$	$+$	incr.	$\cup$
$\pi/4$	0	1	0	incr.	infl.
$\pi/2$	$1/\sqrt{2}$	$1/\sqrt{2}$	$-$	incr.	$\cap$
$3\pi/4$	1	0	$-$	max.	$\cap$
π	$1/\sqrt{2}$	$-1/\sqrt{2}$	$-$	decr.	$\cap$
$5\pi/4$	0	-1	0	decr.	infl.
$3\pi/2$	$-1/\sqrt{2}$	$-1/\sqrt{2}$	$+$	decr.	$\cup$
$7\pi/4$	-1	0	$+$	min.	$\cup$
2π	$-1/\sqrt{2}$	$1/\sqrt{2}$	$+$	incr.	$\cup$

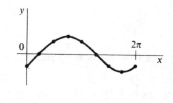

31

x	$f(x)$	$f'(x)$	$f''(x)$		
0	0	1	0	incr.	infl.
$\pi/4$	1	2	$+$	incr.	$\cup$
$\lim\limits_{x \to \pi/2^-}$	∞	∞			
$\lim\limits_{x \to \pi/2^+}$	$-\infty$	∞			
$3\pi/4$	-1	2	$-$	incr.	$\cap$
π	0	1	0	incr.	infl.
$5\pi/4$	1	2	$+$	incr.	$\cup$
$\lim\limits_{x \to 3\pi/2^-}$	∞	∞			
$\lim\limits_{x \to 3\pi/2^+}$	$-\infty$	∞			
$7\pi/4$	-1	2	$-$	incr.	$\cap$
2π	0	1	0	incr.	infl.

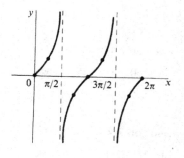

33

x	$f(x)$	$f'(x)$	$f''(x)$		
$\lim\limits_{x \to 0^+}$	∞	$-\infty$			
$\pi/4$	2	-4	$+$	decr.	$\cup$
$\pi/2$	1	0	$+$	min.	$\cup$
$3\pi/4$	2	4	$+$	incr.	$\cup$
$\lim\limits_{x \to \pi^-}$	∞	∞			
$\lim\limits_{x \to \pi^+}$	∞	$-\infty$			
$5\pi/4$	2	-4	$+$	decr.	$\cup$
$3\pi/2$	1	0	$+$	min.	$\cup$
$7\pi/4$	2	4	$+$	incr.	$\cup$
$\lim\limits_{x \to 2\pi^-}$	∞	∞			

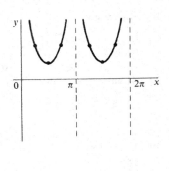

37 $-\frac{1}{2}\cos(2t) + C$ **39** $\frac{1}{3}\sec^3 x + C$ **41** $2\sin(\sqrt{x}) + C$ **43** $-\frac{1}{5}\csc(5\theta) + C$
45 $2\sqrt{\sec x + 1} + C$ **47** $\tan\theta - \sec\theta + C$ **49** 2   **51** $(3 - \sqrt{3})/2$ **53** ∞
55 $-\frac{3}{10}$ radians/sec   **57** 6   **59** $1/2$ **61** 2π **63** $\sqrt{2}$ **65** $\pi\sqrt{2}$

Section 7.3

1 $\pi/3$ **3** $-\pi/4$ **5** $\pi/3$ **7** $\sqrt{1 - x^2}$ **9** $\dfrac{\pi}{2} - x$

15 $\dfrac{5}{|5x - 2|\sqrt{(5x - 2)^2 - 1}}$ **17** $\dfrac{2x}{\sqrt{1 - x^4}}$ **19** $\arcsin t + \dfrac{t}{\sqrt{1 - t^2}}$ **21** $\dfrac{1 - x}{\sqrt{1 - x^2}}$

23 $\arcsin x$ **25** $-\dfrac{1}{2\sqrt{x}(x + 1)}$ **27** $-\pi/2$ **29** 1   **31** $\frac{1}{3}\arctan(x/3) + C$

33 $\arcsin(2x - 1) + C$ **35** $\operatorname{arcsec}(2x) + C$ **37** $\frac{1}{2}\arcsin(x^2) + C$
39 $2\operatorname{arcsec}(\sqrt{x}) + C$ or $2\arctan(\sqrt{x - 1}) + C$ **41** $\frac{1}{2}(\arcsin x)^2 + C$ **43** $\pi/3$
45 $\pi/6$ **47** π/a **49** $\pi/2$

Section 7.4

1 $x\sin x + \cos x + C$ **3** $t^2\sin t + 2t\cos t - 2\sin t + C$
5 $-\frac{1}{2}t\cos(2t - 1) + \frac{1}{4}\sin(2t - 1) + C$
7 $-\dfrac{x^2}{4}\cos(4x) + \dfrac{x}{8}\sin(4x) + \dfrac{1}{32}\cos(4x) + C$
9 $\frac{1}{4}x^4\operatorname{arcsec} x - \frac{1}{12}(x^2 - 1)^{3/2} - \frac{1}{4}\sqrt{x^2 - 1} + C$
11 $-2\sqrt{x}\cos(\sqrt{x}) + 2\sin(\sqrt{x}) + C$ **13** $(x + 1)\arctan(\sqrt{x}) - \sqrt{x} + C$
15 $x^2\sqrt{x^2 - 1} - \frac{2}{3}(x^2 - 1)^{3/2} + C$ or $\frac{1}{3}(x^2 - 1)^{3/2} + \sqrt{x^2 - 1} + C$ or
$\quad (\frac{1}{3}x^2 + \frac{2}{3})\sqrt{x^2 - 1} + C$
17 $-\frac{1}{4}x\cos(2x) + \frac{1}{8}\sin(2x) + C$ or $\frac{1}{2}x\sin^2 x - \frac{1}{4}x + \frac{1}{4}\sin x\cos x + C$
19 $\frac{2}{3}\sin^3\theta + C$ or $\frac{1}{3}[\cos\theta\sin(2\theta) - 2\sin\theta(\cos(2\theta)] + C$
21 $\frac{1}{8}\cos(4x) - \frac{1}{12}\cos(6x) + C$ or $\frac{5}{24}\sin x\sin(5x) + \frac{1}{24}\cos x\cos(5x) + C$
23 $\frac{1}{2}\sin(t^2) - \frac{1}{2}t^2\cos(t^2) + C$ **25** $\dfrac{1}{x}\cos\left(\dfrac{1}{x}\right) - \sin\left(\dfrac{1}{x}\right) + C$

27 $(t^2 + 4)^{3/2}\left(\dfrac{t^2}{5} - \dfrac{8}{15}\right) + C$ or $\frac{1}{3}t^2(t^2 + 4)^{3/2} - \dfrac{2}{15}(t^2 + 4)^{5/2} + C$ or

$\quad -\dfrac{4}{3}(t^2 + 4)^{3/2} + \dfrac{1}{5}(t^2 + 4)^{5/2} + C$ **29** $(\pi/2) - 1$ **31** $\pi/2$ **33** $1/2$

35 $\dfrac{2}{3}\pi - \dfrac{\sqrt{3}}{2}$

Section 7.5

1 $(1/\cos t) + \cos t + C$ **3** $-\cot x - x + C$ **5** $\frac{1}{4}\cos^3 x \sin x + \frac{3}{8}\sin x \cos x + \frac{3}{8}x + C$
  or $\frac{3}{8}x + \frac{1}{4}\sin(2x) + \frac{1}{32}\sin(4x) + C$ **7** $\frac{1}{6}\tan^6 x + \frac{1}{4}\tan^4 x + C$ or
  $\frac{1}{6}\sec^6 x - \frac{1}{4}\sec^4 x + C$ **9** $\frac{1}{3}\sin^3 x - \frac{1}{5}\sin^5 x + C$ **11** $-\frac{1}{3}\cot^3 \theta + C$

13 $\frac{2}{9}(\tan x)^{9/2} + \frac{2}{5}(\tan x)^{5/2} + C$ **15** $\tan\theta - \cot\theta + C$ or $-2\cot(2\theta) + C$

17 $-\cot\theta + \csc\theta + C$ **19** $4/3$ **21** $\dfrac{\pi}{4} - \dfrac{2}{3}$ **23** $3\pi/8$

25 $\cos(\sqrt{x})\sin(\sqrt{x}) + \sqrt{x} + C$ or $\frac{1}{2}\sin(2\sqrt{x}) + \sqrt{x} + C$

27 $-x\cos x + \frac{1}{3}x\cos^3 x + \frac{2}{3}\sin x + \frac{1}{9}\sin^3 x + C$

29 $\frac{1}{4}x\sin^3 x + \frac{1}{3}\cos x - \frac{1}{9}\cos^3 x + C$ **31** $\frac{1}{9}\tan^9 \theta + \frac{2}{7}\tan^7 \theta + \frac{1}{5}\tan^5 \theta + C$

33 $\frac{1}{2}\sec x \tan x + \frac{1}{2}\int \sec x\, dx$ **35** $\frac{1}{2}\sec x \tan x - \frac{1}{2}\int \sec x\, dx$

37 $-\frac{1}{4}\csc^4 x \cos x + \frac{1}{8}\csc^2 x \cos x - \frac{1}{8}\int \csc x\, dx$ **39** $\int \sec x\, dx + \int \csc x\, dx$

41 $\int x^n \sin x\, dx = -x^n \cos x + n\int x^{n-1}\cos x\, dx = -x^n \cos x + nx^{n-1}\sin x - n(n-1)\int x^{n-2}\sin x\, dx$

43   (a) $\pi^2/16$ (b) $\pi^2/4$

Section 7.6

1 $\frac{1}{2}\arcsin(2x) + C$ **3** $\frac{1}{3}(9 + x^2)^{3/2} - 9(9 + x^2)^{1/2} + C$ or $\sqrt{9 + x^2}(\frac{1}{3}x^2 - 6) + C$

5 $\dfrac{x}{4\sqrt{4 - x^2}} + C$ **7** $\arccos\left(\dfrac{\cos\theta}{\sqrt{2}}\right) + C$ or $-\arcsin\left(\dfrac{\cos\theta}{\sqrt{2}}\right) + C$

9 $-\dfrac{1}{x} - \arctan x + C$ **11** $2\arcsin(x/2) - \frac{1}{2}x\sqrt{4 - x^2} + C$

13 $\frac{1}{8}\arcsin x - \frac{1}{4}x(1 - x^2)^{3/2} + \frac{1}{8}x\sqrt{1 - x^2} + C$

15 $2\arcsin\left(\dfrac{x - 2}{2}\right) + \dfrac{1}{2}(x - 2)\sqrt{4x - x^2} + C$ **17** $\frac{1}{48}(4x^2 - 1)^{3/2} + \frac{1}{16}\sqrt{4x^2 - 1} + C$
  or $\sqrt{4x^2 - 1}\,(\frac{1}{24} - \frac{1}{12}x^2) + C$ **19** $\frac{1}{3}(a^2 - x^2)^{3/2} - a^2\sqrt{a^2 - x^2} + C$
  or $-\frac{1}{3}\sqrt{a^2 - x^2}(2a^2 + x^2) + C$ **21** $\frac{1}{2}\sqrt{x^4 - 1} - \frac{1}{2}\operatorname{arcsec}(x^2) + C$

23 $\sqrt{a^2 + x^2} + \dfrac{a^2}{\sqrt{a^2 + x^2}} + C$ **25** $\pi/2$ **27** $1/9$ **29** ∞

31 $(\frac{1}{2}x^2 - \frac{1}{4})\arcsin x + \frac{1}{4}x\sqrt{1 - x^2} + C$ **33** $\frac{1}{3}x^3 \arcsin x + \frac{1}{3}\sqrt{1 - x^2} - \frac{1}{9}(1 - x^2)^{3/2} + C$

35 $-\dfrac{\arcsin x}{2x^2} - \dfrac{\sqrt{1 - x^2}}{2x} + C$ **37** $\dfrac{2\pi^2}{3\sqrt{3}} + \dfrac{\pi}{2}$

Section 7.7

1

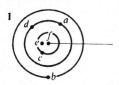

3 $r = \dfrac{2}{\sin\theta - 5\cos\theta}$ **5** $r = 2\sec\theta$ **7** $r\sin\theta = r^2\cos^2\theta + 1$

9 $r\sin\theta = 3r^2\cos^2\theta - 2r\cos\theta$ or $r = \dfrac{\sin\theta + 2\cos\theta}{3\cos^2\theta}$ **11** $r = \sin\theta$

13   **15**   **17**

 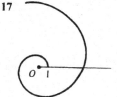

19

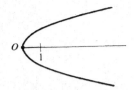

21 $x = \sin(3\theta)\cos\theta,\ y = \sin(3\theta)\sin\theta$ **23** $x = \theta^2\cos\theta,\ y = \theta^2\sin\theta$

Section 7.8

1 θ **3** $-\cot\theta$ **5** $-\dfrac{1 + \cot\theta}{\sin\theta}$

7

θ	r	$dr/d\theta$	$\tan\psi$	$\lvert r\rvert$
0	1	1	1	incr.
$\pi/4$	$\sqrt{2}$	0		max.
$\pi/2$	1	-1	-1	decr.
$3\pi/4$	0	$-\sqrt{2}$		crosses O
π	-1	-1	1	incr.
$5\pi/4$	$-\sqrt{2}$	0		max.
$3\pi/2$	-1	1	-1	decr.
$7\pi/4$	0	$\sqrt{2}$		crosses O
2π	1	1	1	incr.

9

θ	r	$dr/d\theta$	$\tan\psi$	$\lvert r\rvert$
0	1.5	1	1.5	incr.
$\pi/4$	2.2	0.7	3.1	incr.
$\pi/2$	2.5	0		max.
$3\pi/4$	2.2	-0.7	-3.1	decr.
π	1.5	-1	-1.5	decr.
$5\pi/4$	0.8	-0.7	-1.1	decr.
$3\pi/2$	0.5	0		min.
$7\pi/4$	0.8	0.7	1.1	incr.
2π	1.5	1	1.5	incr.

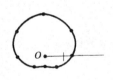

11

θ	r	$dr/d\theta$	$\tan\psi$	$\lvert r\rvert$
0	1.5	0		max.
$\pi/2$	0.5	-1	-0.5	decr.
$2\pi/3$	0	-0.9		crosses O
$3\pi/4$	-0.2	-0.7	0.3	incr.
π	-0.5	0		max.
$5\pi/4$	-0.2	0.7	-0.3	decr.
$4\pi/3$	0	0.9		crosses O
$3\pi/2$	0.5	1	0.5	incr.
2π	1.5	0		max.

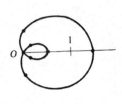

13

θ	r	$dr/d\theta$	$\tan\psi$	$\|r\|$
0	0	$\frac{1}{3}$		crosses O
π	0.9	$\frac{1}{6}$	5	incr.
1.5π	1	0		max.
2π	0.9	$-\frac{1}{6}$	-5	decr.
3π	0	$-\frac{1}{3}$		crosses O
4π	-0.9	$-\frac{1}{6}$	5	incr.
4.5π	-1	0		max.
5π	-0.9	$\frac{1}{6}$	-5	decr.
6π	0	$\frac{1}{3}$		crosses O

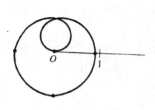

15

θ	r	$dr/d\theta$	$\tan\psi$	$\|r\|$
0	4	0		max.
$\pi/8$	2.5	-6	-0.4	decr.
$2\pi/8$	1	0		min.
$3\pi/8$	2.5	6	0.4	incr.
$4\pi/8$	4	0		max.
$5\pi/8$	2.5	-6	-0.4	decr.
$6\pi/8$	1	0		min.
$7\pi/8$	2.5	6	0.4	incr.
π	4	0		max.
repeats				

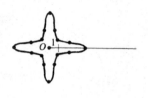

17

θ	r	$dr/d\theta$	$\tan\psi$	$\|r\|$
0	0	1		crosses O
$\pi/4$	1	2	$\frac{1}{2}$	incr.
$\displaystyle\lim_{\theta\to\pi/2^-}$	∞			$x \to 1$
$\displaystyle\lim_{\theta\to\pi/2^+}$	$-\infty$			$x \to 1$
$3\pi/4$	-1	2	$-\frac{1}{2}$	decr.
π	0	1		crosses O

repeats, with $x \to -1$ as $\theta \to 3\pi/2$

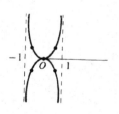

19

θ	r	$dr/d\theta$	$\tan\psi$	$\|r\|$
0	2	0		min.
$\pi/4$	2.4	2.4	1	incr.
$\displaystyle\lim_{\theta\to\pi/2^-}$	∞			$x \to 1$
$\displaystyle\lim_{\theta\to\pi/2^+}$	$-\infty$			$x \to 1$
$3\pi/4$	-0.4	0.4	-1	decr.
π	0	0		cusp at O
$5\pi/4$	-0.4	-0.4	1	incr.
$\displaystyle\lim_{\theta\to 3\pi/2^-}$	$-\infty$			$x \to 1$
$\displaystyle\lim_{\theta\to 3\pi/2^+}$	∞			$x \to 1$
$7\pi/4$	2.4	-2.4	-1	decr.
2π	2	0		min.

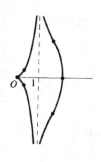

21

θ	r	$dr/d\theta$	$\tan\psi$	$|r|$
0	1	-1	-1	decr.
$\pi/4$	0.6	-0.3	-2	decr.
$\pi/2$	0.5	0		min.
$3\pi/4$	0.6	0.3	2	incr.
π	1	1	1	incr.
$5\pi/4$	3.4	8	0.4	incr.
$\lim\limits_{\theta\to 3\pi/2^-}$	∞		$x\to-\infty$	
$\lim\limits_{\theta\to 3\pi/2^+}$	∞		$x\to\infty$	
$7\pi/4$	3.4	-8	-0.4	decr.
2π	1	-1	-1	decr.

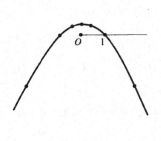

23

θ	r	$dr/d\theta$	$\tan\psi$	$|r|$
$\lim\limits_{\theta\to 0^+}$	∞		$y\to\pi$	
$\pi/4$	4	$-16/\pi$	$-\pi/4$	decr.
$\pi/2$	2	$-4/\pi$	$-\pi/2$	decr.
π	1	$-1/\pi$	$-\pi$	decr.
$3\pi/2$	$\frac{2}{3}$	$-\frac{4}{9}\pi$	$-3\pi/2$	decr.
2π	$\frac{1}{2}$	$-\frac{1}{4}\pi$	-2π	decr.
3π	$\frac{1}{3}$	$-\frac{1}{9}\pi$	-3π	decr.
4π	$\frac{1}{4}$	$-\frac{1}{16}\pi$	-4π	decr.

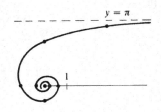

25

θ	r	$dr/d\theta$	$\tan\psi$	$|r|$
$\lim\limits_{\theta\to 0^+}$	∞		$y\to O$	
$\pi/4$	2	$-4/\pi$	$-\pi/2$	decr.
$\pi/2$	$\sqrt{2}$	$-\sqrt{2}/\pi$	$-\pi$	decr.
π	1	$-\frac{1}{\pi}\pi$	-2π	decr.
$3\pi/2$	$\sqrt{\frac{2}{3}}$	$-$	$-$	decr.
2π	$1/\sqrt{2}$	$-$	$-$	decr.
3π	$1/\sqrt{3}$	$-$	$-$	decr.
4π	$\frac{1}{2}$	$-$	$-$	decr.

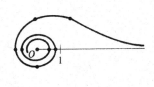

27 x: max of $\sqrt{32/27}$ at $\theta = \arcsin(1/\sqrt{3})$, $2\pi - \arcsin(1/\sqrt{3})$; min of $-\sqrt{32/27}$ at $\theta = \pi + \arcsin(1/\sqrt{3})$, $\pi - \arcsin(1/\sqrt{3})$
y: max of 2 at $\theta = \pi/2$; min of -2 at $\theta = 3\pi/2$

29 x: max of $\frac{5}{2}$ at $\theta = 0$; min of $-\frac{9}{16}$ at $\theta = \arccos(-\frac{3}{4})$, $2\pi - \arccos(-\frac{3}{4})$

y: max of $\sin\theta_0(\frac{3}{2} + \cos\theta_0)$ at $\theta = \theta_0 = \arccos\left(\dfrac{\sqrt{41}-3}{8}\right)$; min of $-\sin\theta_0\left(\dfrac{3}{2} + \cos\theta_0\right)$

at $\theta = 2\pi - \arccos\left(\dfrac{\sqrt{41}-3}{8}\right)$

31 $\left(\dfrac{1}{2}, \dfrac{\pi}{12}\right), \left(\dfrac{1}{2}, \dfrac{5\pi}{12}\right), \left(\dfrac{1}{2}, \dfrac{13\pi}{12}\right), \left(\dfrac{1}{2}, \dfrac{17\pi}{12}\right), \left(\dfrac{1}{2}, \dfrac{7\pi}{12}\right), \left(\dfrac{1}{2}, \dfrac{11\pi}{12}\right), \left(\dfrac{1}{2}, \dfrac{19\pi}{12}\right), \left(\dfrac{1}{2}, \dfrac{23\pi}{12}\right)$

Section 7.9

1 πa^2 **3** 1   **5** $2 - \dfrac{\pi}{2}$ **7** $\dfrac{3\pi}{16}$ **9** $\dfrac{\pi}{4} - \dfrac{3\sqrt{3}}{8}$ **11** $\dfrac{\pi}{12} + \dfrac{\sqrt{3}}{16}$ **13** $\dfrac{3}{2} - \dfrac{\pi}{4}$

15 $\dfrac{\pi}{3} - \dfrac{\sqrt{3}}{4}$ **17** π^2 **19** $\dfrac{7\pi}{12} - \sqrt{3}$ **21** $\dfrac{1}{2}\displaystyle\int_a^b (g(\theta))^2 - (f(\theta))^2\, d\theta$

Section 7.10

1 14π **3** 2   **5** $\dfrac{2048 - 493\sqrt{17}}{15} \sim 1.02$ **7** 4   **9** 4

11 $\displaystyle\int_0^{2\pi} \sqrt{\sin^2(2\theta) + 4\cos^2(2\theta)}\, d\theta$ **13** $\displaystyle\int_0^b \sqrt{\theta^2 + 1}\, d\theta$ **17** $\pi\sqrt{a^2 + b^2}\left(a + \dfrac{b\pi}{2}\right)$

19 $\pi\sqrt{2}$ **21** $\dfrac{\pi}{5}(8 - \sqrt{2})$

Extra Problems for Chapter 7

1 $1 + \cos x$ **3** $\dfrac{\cos\theta}{2\sqrt{\theta}} - \sqrt{\theta}\sin\theta$ **5** $4/3$ **7** $-\sin(\cos\theta) + C$ **9** 8

11 $-600/29$ radians/hr   **13** $-\dfrac{1}{2\sqrt{x}\sqrt{1-x}}$ **15** $\dfrac{1}{1+t^2} - 1$ or $-\dfrac{t^2}{1+t^2}$ **17** 1

19 $\operatorname{arcsec}(x-1) + C$ **21** $\tfrac{1}{3}(x^2+1)^{3/2} - (x^2+1)^{1/2} + C = x^2(x^2+1)^{1/2}$
$-\tfrac{2}{3}(x^2+1)^{3/2} + C = (x^2+1)^{1/2}(\tfrac{1}{3}x^2 - \tfrac{2}{3}) + C$ **23** $2\sin 1 + 2\cos 1 - 2$

25 $\pi\left(1 - \dfrac{\pi}{4}\right)$ **27** $\dfrac{\sec^9\theta}{9} - \dfrac{2\sec^7\theta}{7} + \dfrac{\sec^5\theta}{5} + C$

29 $-\dfrac{\sqrt{2-x^2}}{x} - \arcsin\left(\dfrac{x}{\sqrt{2}}\right) + C$

31 (a)   (b)

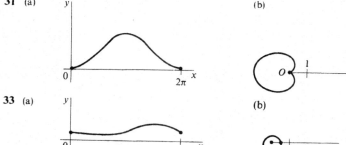

33 (a) y (b)

35 $19\pi/8$ **37** $(3\pi/2) - 4$ **39** $4\sqrt{2}\pi/5$

Section 8.1

9 0   **11** ∞ **13** 0   **15** ∞ **17** 27   **19** ∞ **21** 1/6   **23** $1/(2\pi)$

Section 8.2

1 x **3** $-x^2$ **5** x/y^2 **7** $\tfrac{2}{3}\log_b x$ **9** $\tfrac{1}{2}(\log_b x + \log_b y)$ **11** 2   **13** 1/2
15 $x = \log_5 3$ **17** $x = -\tfrac{2}{3}$ **19** $x = 5^{1/3}$ **21** $x = 2, x = 3$
23 $x = 2^{1/(1 - \log_3 2)} = 3^{1/(\log_2 3 - 1)}$ **25** $-\infty$ **27** $-\infty$ **29** $-\infty$

Section 8.3

1 $3e^{3x+4}$ **3** $-4^{-x}\ln 4$ **5** $e^t\cos(e^t)$ **7** $2^{t^2+1}t\ln 2$ **9** $e^t e^{(e^t)}$

11 $\dfrac{\ln 3}{2\sqrt{x}}3^{\sqrt{x}}$ **13** $-\dfrac{e^{x+y}}{e^{x+y} + \sin y}$ **15** $\dfrac{1}{2\sqrt{e^t(t^{-1} - t^{-2})}}$ or $\dfrac{t^2}{2\sqrt{e^t(t-1)}}$ **17** ∞

19 1   **21** $-\infty$ **23** e^e **25** e^e

27

x	$f(x)$	$f'(x)$	$f''(x)$		
$\lim\limits_{x \to -\infty}$	0	0			
-2	1/4	$(\ln 2)/4$	$+$	incr.	$\cup$
-1	1/2	$(\ln 2)/2$	$+$	incr.	$\cup$
0	1	$\ln 2$	$+$	incr.	$\cup$
1	2	$2 \ln 2$	$+$	incr.	$\cup$
2	4	$4 \ln 2$	$+$	incr.	$\cup$
$\lim\limits_{x \to \infty}$	∞	∞			

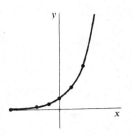

29

x	$f(x)$	$f'(x)$	$f''(x)$		
$\lim\limits_{x \to -\infty}$	0	0			
-3	$-3e^{-3}$	$-2e^{-3}$	$-$	decr.	$\cap$
-2	$-2e^{-2}$	$-e^{-2}$	0	decr.	infl.
-1	$-e^{-1}$	0	$+$	min.	$\cup$
0	0	1	$+$	incr.	$\cup$
1	e	$2e$	$+$	incr.	$\cup$
$\lim\limits_{x \to \infty}$	∞	∞			

31

x	$f(x)$	$f'(x)$	$f''(x)$		
$\lim\limits_{x \to -\infty}$	0	0			
-2	e^{-8}	$12e^{-8}$	$+$	incr.	$\cup$
-1	e^{-1}	$3e^{-1}$	$+$	incr.	$\cup$
$-(\frac{2}{3})^{1/3}$	$e^{-2/3}$	$+$	0	incr.	infl.
$-\frac{1}{2}$	$e^{-1/8}$	$\frac{3}{4}e^{-1/8}$	$-$	incr.	$\cap$
0	1	0	0	horiz.	infl.
1	e	$3e$	$+$	incr.	$\cup$
$\lim\limits_{x \to \infty}$	∞	∞			

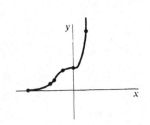

33

x	$f(x)$	$f'(x)$	$f''(x)$		
$\lim\limits_{x \to -\infty}$	1	0			
-2	0.88	-0.1	$-$	decr.	$\cap$
-1	0.73	-0.2	$-$	decr.	$\cap$
0	0.5	-0.25	0	decr.	infl.
1	0.27	-0.2	$+$	decr.	$\cup$
2	0.12	-0.1	$+$	decr.	$\cup$
$\lim\limits_{x \to \infty}$	0	0			

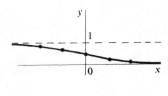

35 $\frac{1}{2}e^{2x} + C$ **37** $-\frac{1}{2}e^{-x^2} + C$ **39** $\frac{1}{3}(1 + e^{2x})^{3/2} + C$ **41** $xe^x - e^x + C$

43 $\frac{1}{2}e^x(\sin x - \cos x) + C$ **45** $\dfrac{e^{10} - 1}{5}$ **47** ∞

49 $\dfrac{1}{r^2}$ if $r > 0$, ∞ if $r \le 0$ **51** (a) $\dfrac{\pi}{2}(e^2 - 1)$ (b) 2π **53** $\sqrt{2}(e^{2\pi} - 1)$

Section 8.4

1 $3 \cosh (3x)$ **3** $-\tanh x \operatorname{sech} x$ **5** 1   **7** 0

9

x	$f(x)$	$f'(x)$	$f''(x)$			
$\lim\limits_{x \to -\infty}$	-1	0				
-1	-0.76	0.4	$+$	incr.	$\cup$	
0	0	1	0	incr.	infl.	
1	0.76	0.4	$-$	incr.	$\cap$	
$\lim\limits_{x \to \infty}$	1	0				

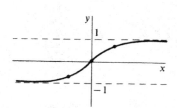

11

x	$f(x)$	$f'(x)$	$f''(x)$			
$\lim\limits_{x \to -\infty}$	0	0				
-2	0.27	0.26	$+$	incr.	$\cup$	
-0.88	0.71	0.5	0	incr.	infl.	
0	1	1	$-$	max.	$\cap$	
0.88	0.71	-0.5	0	decr.	infl.	
2	0.27	-0.26	$+$	decr.	$\cup$	
$\lim\limits_{x \to \infty}$	0	0				

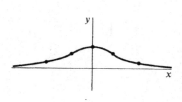

13 $\dfrac{\sinh^2 x}{2} + C$ or $\dfrac{\cosh^2 x}{2} + C$ **15** $x \cosh x - \sinh x + C$

17 $\frac{1}{2}x \sinh x \cosh x + \frac{1}{4}x^2 - \frac{1}{4}\sinh^2 x + C$ or $\frac{1}{4}x \sinh (2x) - \frac{1}{8}\cosh (2x) + \frac{1}{4}x^2 + C$ **19** ∞

23 (a) $\dfrac{\pi}{8}(4 + e^2 - e^{-2})$ (b) $2\pi(1 - e^{-1})$ **25** $50{,}000(e^2 - 1) \sim 320{,}000$ dollars

27 (a) $10^6(90e^{0.2} - 110) \sim -74{,}000$ dollars   (b) $10^6(90e^{0.3} - 120) \sim 1{,}490{,}000$ dollars

Section 8.5

1 $(3/x)(\ln x)^2$ **3** $-\tan x$ **5** $\ln t$ **7** $t^{-2}(1 - \ln t)$ **9** $\dfrac{1}{x \ln 2}$

11 $\dfrac{1}{y} + \dfrac{3}{2(3y + 1)}$ **13** $y - \dfrac{y}{x}$ **15** $\dfrac{1}{x + y - 1}$ **17** 0   **19** 0   **21** $\ln a$

23 1   **25** 1

27

x	$f(x)$	$f'(x)$	$f''(x)$		
$\lim\limits_{x \to 0^+}$	$-\infty$	∞			
$\frac{1}{2}$	-0.3	$\frac{4}{3}$	$-$	incr.	$\cap$
1	0	0	$-$	max.	$\cap$
$\frac{3}{2}$	0.3	$-\frac{4}{3}$	$-$	decr.	$\cap$
$\lim\limits_{x \to 2^-}$	$-\infty$	$-\infty$			

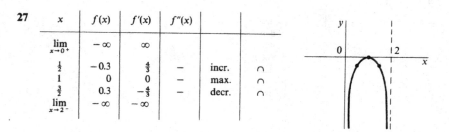

29 $\frac{1}{2}\ln |2x + 3| + C$ **31** $2x + 2 \ln |x - 1| + C$ **33** $\frac{1}{2}(\ln x)^2 + C$ **35** $\ln |\ln t| + C$

37 $\ln x(\ln (\ln x) - 1) + C$ **39** $\dfrac{x^{n+1}}{n + 1}\ln x - \dfrac{x^{n+1}}{(n + 1)^2} + C$

41 $x(\ln x)^3 - 3x(\ln x)^2 + 6x \ln x - 6x + C$ **43** $\sin (\ln x) + C$ **45** $10 - \ln (11)$

47 1   **49** $-\infty$ **51** ∞ **53** (a) $\pi(e-2)$ (b) $\dfrac{\pi}{2}(e^2+1)$

55 $\sqrt{e^2+1} - \sqrt{2} + 1 + \ln\left(\dfrac{1+\sqrt{2}}{1+\sqrt{e^2+1}}\right)$

Section 8.6

1 $y = -\dfrac{2}{x^2+C}$ **3** $y = \pm\sqrt{\tfrac{2}{3}x^3 + C}$ **5** $y = -\ln\left(\dfrac{C-x^2}{2}\right)$

7 $y = \ln(e^x + C)$ **9** $y = \dfrac{1}{\pm\sqrt{2\cos x + C}}$ **11** $y = \dfrac{x^3}{3} + \dfrac{x^2}{2} + Ax + B$

13 $y = Ax + B$ **15** $y = A\sinh(\sqrt{3}x) + B\cosh(\sqrt{3}x)$ or $y = Ae^{\sqrt{3}x} + Be^{-\sqrt{3}x}$

17 $y = 10^7 e^{(0.01)t}$ **19** $\dfrac{6\ln 10}{\ln 2}$ min ~ 19.93 min   **21** $10^5 \times (1.15)^4 \sim 174{,}900$

23 $t = \dfrac{\ln(0.5)}{\ln(0.9)} \sim 6.576$ years   **25** $y = e^t + e^{-t}$ **27** $y = \cos t - 2\sin t$

29 $y = 10^{6.5}\sqrt{t}$ **31** $y = \dfrac{10^8}{1 + 9e^{-t}}$ **33** (b) $y_0 = L/2$

35 (a) $y = 10^7 - 6 \times 10^6 2^{-t}$ (b) $y = 10^7 + 3 \times 10^6 3^{-t}$

37 $y^2 - x^2 = C$ or $y = \pm\sqrt{x^2 + C}$

Section 8.7

1 $\left(\dfrac{3x-2}{4x+3}\right)\left[\dfrac{3}{3x-2} - \dfrac{4}{4x+3}\right]$

3 $\dfrac{(x^2+1)\sqrt{3x+4}}{(2x-3)\sqrt{x^2-4}}\left[\dfrac{2x}{x^2+1} + \dfrac{3}{2(3x+4)} - \dfrac{2}{2x-3} - \dfrac{x}{x^2-4}\right]$

5 $(x-1)^{x^2+1}\left[2x\ln(x-1) + \dfrac{x^2+1}{x-1}\right]$ **7** $e^{(e^x)}e^x$ **9** $\sqrt[t]{t}\left[\dfrac{1}{t^2} - \dfrac{1}{t^2}\ln t\right]$

11

x	$f(x)$	$f'(x)$	$f''(x)$		
$\lim\limits_{x\to 0^+}$	1	$-\infty$			
$1/10$	$10^{-0.1}$	$-$	$+$	decr.	$\cup$
$1/e$	$e^{-1/e}$	0	$+$	min.	$\cup$
1	1	1	$+$	incr.	$\cup$
$\lim\limits_{x\to\infty}$	∞	∞			

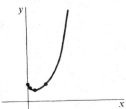

15 $\ln|\sin\theta| + C$ **17** $\tfrac{1}{3}\ln|\sec(3\theta)| + C$

19 $\tfrac{1}{4}\sec^4\theta\sin\theta + \tfrac{3}{8}\sec^2\theta\sin\theta + \tfrac{3}{8}\ln|\sec\theta + \tan\theta| + C$ **21** $\ln|\tan x| + C$

23 $\ln|\sec x + \tan x| - \sin x + C$ **25** $\tfrac{1}{2}x\sqrt{x^2-1} - \tfrac{1}{2}\ln|x + \sqrt{x^2-1}| + C$

27 $\tfrac{1}{2}\ln\left|\dfrac{x}{\sqrt{4-x^2}+2}\right| + C$ or $\tfrac{1}{4}\ln\left|\dfrac{2-\sqrt{4-x^2}}{2+\sqrt{4-x^2}}\right| + C$ **29** $\tfrac{1}{4}\ln\left|\dfrac{x-2}{x+2}\right| + C$

31 $\sqrt{x^2-1} - \operatorname{arcsec} x + C$ or $\sqrt{x^2-1} - \arctan\sqrt{x^2-1} + C$

33 $\dfrac{x}{4}(x^2+1)^{3/2} - \dfrac{x}{8}(x^2+1)^{1/2} - \dfrac{1}{8}\ln(x + \sqrt{x^2+1}) + C$ or

$x(x^2+1)^{1/2}\left(\dfrac{1}{4}x^2 + \dfrac{1}{8}\right) - \dfrac{1}{8}\ln(x + \sqrt{x^2+1}) + C$

35 $\quad -\dfrac{\arcsin x}{x} - \ln\left|\dfrac{1+\sqrt{1-x^2}}{x}\right| + C$ or $-\dfrac{\arcsin x}{x} + \ln\left|\dfrac{1-\sqrt{1-x^2}}{x}\right| + C$

37 $\quad x\,\text{arccsc}\,x + \ln(x + \sqrt{x^2-1}) + C,\, x \ge 1;\, x\,\text{arccsc}\,x - \ln|x + \sqrt{x^2-1}| + C,\, x \le -1$

39 $\quad \sqrt{5} + \frac{1}{2}\ln(2 + \sqrt{5}) \sim 2.96$ **41** $\quad \frac{1}{2}a\sqrt{1+a^2} + \frac{1}{2}\ln(a + \sqrt{1+a^2})$

Section 8.8

1 $\quad \frac{1}{2}\ln|2x-7| + C$ **3** $\quad \frac{1}{12}\ln|x-4| - \frac{1}{12}\ln|x| + C$ **5** $\quad -\frac{1}{5}\ln|x-1| + \frac{11}{5}\ln|x+4| + C$

7 $\quad \dfrac{x^2}{2} - 3x + \dfrac{1}{4}\ln|x| + \dfrac{51}{4}\ln|x+4| + C$ **9** $\quad -\dfrac{1}{2(x+1)^2} + C$

11 $\quad \ln|x-1| - \dfrac{1}{x-1} - \dfrac{1}{2(x-1)^2} + C$ **13** $\quad -\ln|x| + \frac{1}{2}\ln|x+1| + \frac{1}{2}\ln|x-1| + C$

15 $\quad x + \ln|x| - \dfrac{1}{x} + C$ **17** $\quad x - 4\arctan(x/4) + C$ **19** $\quad \frac{5}{4}\ln|x-3| - \frac{1}{4}\ln|x+1| + C$

21 $\quad \frac{1}{3}x^3 - x + \arctan x + C$ **23** $\quad \arctan x - (1/\sqrt{2})\arctan(x/\sqrt{2}) + C$

25 $\quad -x^{-1} - \arctan x + C$ **27** $\quad \dfrac{1}{3}\ln|x+1| - \dfrac{1}{6}\ln|x^2-x+1| + \dfrac{1}{\sqrt{3}}\arctan\left(\dfrac{2x-1}{\sqrt{3}}\right) + C$

29 $\quad \dfrac{x^2}{2} - \dfrac{3}{4}\ln|x+1| + \dfrac{5}{4}\ln|x-1| - \dfrac{1}{4}\ln|x^2+1| + \arctan x + C$

31 $\quad \ln|x| - \dfrac{1}{2}\ln|x^2+1| - \dfrac{\arctan x}{x} + C$

Section 8.9

1 $\quad -3\cos x + 4\sin x + C$ **3** $\quad \frac{3}{4}(x^2-1)^{2/3} + C$ **5** $\quad \frac{1}{3}[(x+2)^{3/2} + x^{3/2}] + C$

7 $\quad 2(1 + \sqrt{x}) - 2\ln(1 + \sqrt{x}) + C$ or $2\sqrt{x} - 2\ln(1 + \sqrt{x}) + C$

9 $\quad -(\sqrt{4x+1}/x) + C$ **11** $\quad -2\cos(\sqrt{x}) + C$

13 $\quad \frac{2}{7}(x-3)^{7/2} + \frac{12}{5}(x-3)^{5/2} + 6(x-3)^{3/2} + C$

15 $\quad \frac{1}{3}[(3x+4)\ln(3x+4) - (3x+4)] + C$ or $\frac{1}{3}(3x+4)\ln(3x+4) - x + C$

17 $\quad x\tan x - \ln|\sec x| - (x^2/2) + C$ **19** $\quad -\frac{1}{6}\cos(3x^2+1) + C$

21 $\quad 2x\ln|x| - 3x + (x+1)\ln(x+1) + C$ **23** $\quad \cos\theta - \cos\theta\ln(\cos\theta) + C$

25 $\quad \ln(2 - \cos\theta) + C$ **27** $\quad \frac{1}{4}(e^x+1)^4 + C$ **29** $\quad 2\sin(\sqrt{x}) - 2\sqrt{x}\cos(\sqrt{x}) + C$

31 $\quad \ln(1 + \cosh x) + C$ **33** $\quad -\dfrac{\ln x}{1+x} + \ln\left(\dfrac{x}{1+x}\right) + C$

35 $\quad \frac{4}{5}(1 - \cos x)^{5/2} - \frac{2}{7}(1 - \cos x)^{7/2} + C$ **37** $\quad \dfrac{1}{\sqrt{2}}\ln\left|\dfrac{x}{\sqrt{2} + \sqrt{2+x^2}}\right| + C$

39 $\quad 4\arcsin\left(\dfrac{\sqrt{x}}{2}\right) + \sqrt{x}\sqrt{4-x} - \dfrac{1}{2}\sqrt{x}(4-x)^{3/2} + C$ or

$\qquad \dfrac{1}{2}(x-2)\sqrt{x}\sqrt{4-x} + 2\arcsin\left(\dfrac{x-2}{2}\right) + C$

41 $\quad \frac{1}{3}[(5x-2)\arcsin(5x-2) + \sqrt{1-(5x-2)^2}] + C$ **43** $\quad \frac{1}{2}e^x(\sin x + \cos x) + C$

45 $\quad (x^3/3) - x + 2\arctan x + C$ **47** $\quad \frac{1}{2}x^2\,\text{arcsec}(x^2) - \frac{1}{2}\ln(x^2 + \sqrt{x^4-1}) + C$

49 $\quad 2x\ln x + \dfrac{4x-1}{8}\ln(4x-1) - \dfrac{5}{2}x + C$ **51** $\quad 2\ln|2x + \sqrt{4x^2-1}| - \dfrac{\sqrt{4x^2-1}}{x} + C$

53 $\quad (x+1)\arctan(\sqrt{x}) - \sqrt{x} + C$ **55** $\quad \frac{1}{8}\ln|\sec(4x^2+7) + \tan(4x^2+7)| + C$

57 $\quad -\frac{1}{243}\sqrt{1-9x^2}(2+9x^2) + C$ **59** $\quad (\sqrt{x^2-3}/3x) + C$

61 $\quad 3x^{2/3}\sin(\sqrt[3]{x}) + 6(\sqrt[3]{x})\cos(\sqrt[3]{x}) - 6\sin(\sqrt[3]{x}) + C$

63 $\quad \frac{1}{3}x^3(1-x^2)^{-3/2} + x(1-x^2)^{-1/2} + C$ **65** $\quad \dfrac{x}{5}(\cos^2(\ln x) + \sin(2\ln x) + 2) + C$

Extra Problems for Chapter 8

1 $-\infty$ **3** $-e^{\cos\theta}\sin\theta$ **5** $-3\operatorname{csch}^3 x\coth x$ **7** $-\dfrac{1}{\ln 3}\cos(3^x)+C$

9 $\frac{1}{2}e^x\sqrt{1-e^{2x}}+\frac{1}{2}\arcsin(e^x)+C$ or $\frac{1}{2}e^x\sqrt{1-e^{2x}}-\frac{1}{2}\arccos(e^x)+C$

11 $\frac{1}{4}e^{2x}-\frac{1}{2}x+C$ **13** $\dfrac{8x}{x^2-1}$ **15** $\dfrac{3}{3x+2}+\dfrac{5}{5x-4}-\dfrac{2}{2x-1}-\dfrac{2x}{x^2+1}$ **17** e^2

19 $(1/a)\ln|x|-(b/a)\ln|a+bx|+C$ **21** ∞ **23** $y=\pm\sqrt{ax^2}+C$

25 $P=CV^{-1/k}$ **27** air temp. $=60°$, $y=60+80\cdot 2^{-t/10}$

29 $(4t+1)^t(t-3)^{2t+1}\left\{\ln\left[(t-3)^2(4t+1)\right]+\dfrac{4t}{4t+1}+\dfrac{2t+1}{t-3}\right\}$ **31** $\ln|\cosh x|+C$

33 $\frac{2}{5}(x+1)^{5/2}+C$ **35** $\pi(2\sqrt{2}+\ln|\sqrt{2}+1|-\ln|\sqrt{2}-1|)$ or $2\pi(\sqrt{2}+\ln|\sqrt{2}+1|)$

37 $e^{0.03}\sim 1.03$, error ≤ 0.0005 **39** $\ln 6\sim\frac{1}{2}+\frac{1}{2}+\frac{1}{3}+\frac{1}{4}+\frac{1}{5}+\frac{1}{12}=\frac{28}{15}$, error $\le\frac{1}{2}$

41 $\bar x=\dfrac{2\ln 2-\frac{3}{4}}{2\ln 2-1},\ \bar y=\dfrac{(\ln 2-1)^2}{2\ln 2-1}$

43 $\pi[e\sqrt{1+e^2}-\sqrt{2}+\ln(e+\sqrt{1+e^2})-\ln(1+\sqrt{2})]$

Section 9.1

1 2^{-n} **3** $(-1)^n n$ **5** $2-(1/2^{n-1})$ **7** $2^{2^{(n-1)}}$ **9** diverges   **11** 1
13 0   **15** diverges   **17** diverges   **19** $1/e$ **21** 0   **23** -1 **25** diverges
27 diverges   **29** 0

Section 9.2

1 $\frac{3}{2}(1-(1/3)^n)$, converges to $\frac{3}{2}$ **3** $4(1-(3/4)^n)$, converges to 4

5 $1-\dfrac{1}{(n+1)!}$, converges to 1   **7** $1-\dfrac{1}{n+1}$, converges to 1

9 n even: $-\dfrac{n}{2}$, n odd: $\dfrac{n}{2}+\dfrac{1}{2}$, diverges   **11** $1-(n+1)^{-2}$, converges to 1

13 $\dfrac{1}{2}\left(1-\dfrac{1}{2n+1}\right)$, converges to $\dfrac{1}{2}$ **15** diverges because $\lim\limits_{n\to\infty}a_n=\frac{1}{2}$

17 diverges because for infinite H, $S_{2H}\neq S_{2\cdot 2H}$
19 diverges because for infinite H, $S_{2H}\neq S_H$ **21** A: 2/3, B: 1/3
23 A: 9/19, B: 6/19; C: 4/19

Section 9.3

1 1/42   **3** 11/4   **5** 5/48   **7** 216/5   **9** 80/9   **11** $5\frac{43}{99}$
13 $492.315+(41/999{,}000)$ **15** 1/7

Section 9.4

1 diverges   **3** converges   **5** converges   **7** converges   **9** diverges
11 diverges   **13** converges   **15** diverges   **17** converges   **19** converges
21 diverges   **23** converges   **25** converges   **27** diverges   **29** diverges
31 converges   **33** diverges   **35** converges   **37** converges

Section 9.5

1 diverges   **3** diverges   **5** converges   **7** diverges   **9** diverges
11 diverges   **13** converges   **15** converges   **17** converges   **19** diverges
21 0.90   **23** 0.37

Section 9.6

1 diverges   **3** diverges   **5** absolutely converges   **7** diverges   **9** diverges
11 diverges   **13** conditionally converges   **15** absolutely converges
17 conditionally converges   **19** diverges   **21** diverges   **23** no information
25 converges   **27** diverges   **29** converges   **31** converges   **33** no information
35 converges   **37** converges   **39** converges   **41** no information   **43** converges

Section 9.7

1 1   **3** 0   **5** e **7** $4/e^2$ **9** 1   **11** 1   **13** 0   **15** 1/3   **17** ∞
19 ∞ **21** 0   **23** $\sqrt[3]{5}$ **25** 1   **27** $(-1, 1)$ **29** $(-1, 1]$ **31** $[-1/2, 1/2)$
33 $(-\infty, \infty)$ **35** $(-3, -1]$ **37** $(-\infty, \infty)$ **39** $(-1, 1)$ **41** $(-5/4, 5/4)$
43 $(-\sqrt{5}, \sqrt{5})$ **45** $(-\infty, \infty)$

Section 9.8

1 $f'(x) = \sum_{n=1}^{\infty} n\,10^n x^{n-1}, \int_0^x f(t)\,dt = \sum_{n=0}^{\infty} \frac{10^n}{n+1} x^{n+1}$

3 $f'(x) = \sum_{n=1}^{\infty} n^{-2} x^{n-1}, \int_0^x f(t)\,dt = \sum_{n=1}^{\infty} \frac{n^{-3}}{n+1} x^{n+1}$

5 $f'(x) = \sum_{n=1}^{\infty} (n+1) x^{n-1}, \int_0^x f(t)\,dt = \sum_{n=1}^{\infty} \frac{1}{n} x^{n+1}$

7 $f'(x) = \sum_{n=1}^{\infty} \frac{n!}{n^{n-1}} x^{n-1}, \int_0^x f(t)\,dt = \sum_{n=1}^{\infty} \frac{n!}{n^n(n+1)} x^{n+1}$

9 $f'(x) = \sum_{n=1}^{\infty} 2nx^{2n-1}, \int_0^x f(t)\,dt = \sum_{n=0}^{\infty} \frac{1}{2n+1} x^{2n+1}$ **11** $\sum_{n=0}^{\infty} (-1)^n 3^n x^n, r = \frac{1}{3}$

13 $\sum_{n=0}^{\infty} \frac{(-1)^n 4^{2n+1} x^{4n+2}}{2n+1}, r = \frac{1}{2}$ **15** $\sum_{n=0}^{\infty} \frac{(-1)^n 2^{n+1} x^{n+2}}{n+1}, r = \frac{1}{2}$

17 $\sum_{n=0}^{\infty} \frac{(-1)^n 4^n x^n}{n!}, r = \infty$ **19** $\sum_{n=0}^{\infty} \frac{3^{2n+1} x^{2n+1}}{(2n+1)!}, r = \infty$

21 $\sum_{n=0}^{\infty} \frac{(-1)^n 2^{n+1} x^{2n+3}}{(n+1)(2n+3)}, r = \frac{1}{\sqrt{2}}$ **23** $\sum_{n=0}^{\infty} \frac{x^{3n+1}}{n!(3n+1)}, r = \infty$

25 $\sum_{n=0}^{\infty} -\frac{x^{n+3}}{(n+1)(n+3)}, r = 1$ **27** $\sum_{n=0}^{\infty} \frac{(-1)^n x^{n+1}}{(n+1)^2}, r = 1$

29 $\sum_{n=1}^{\infty} (-1)^{n+1} 2nx^{2n-1}, r = 1$ **31** $\sum_{n=0}^{\infty} (-1)^n 2 \cdot x^{4n+1}, r = 1$

33 $\sum_{n=0}^{\infty} \frac{(-1)^n(1 + 2^{2n+1})x^{2n+1}}{2n+1}, r = \frac{1}{2}$

Section 9.9

1 0.18232   **3** 0.7788008   **5** 0.4854019   **7** 0.3293740   **9** 1.003009
11 1.098614   **13** $f(x) = x + x^2 + \cdots + x^{n+1} + E, |E| \le 2|x|^{n+2}, f(\frac{1}{2}) = 1$
15 $f(x) = 1 + x^2 + \cdots + x^{2n} + E, |E| \le \frac{4}{3} x^{2n+2}, f(\frac{1}{2}) = \frac{4}{3}$
17 $f(x) = -x - (x^2/2^2) - \cdots - (x^n/n^2) + E, |E| \le 2|x|^{m+1}/(m+1)^2, f(\frac{1}{2}) \sim 0.58$

Section 9.10

1 $f(x) = 1 - \frac{x^2}{2!} + \frac{x^4}{4!} - \cdots + \frac{(-1)^n x^{2n}}{(2n)!} + \frac{(-1)^{n+1}(\cos t)x^{2n+2}}{(2n+2)!}, f\left(\frac{1}{2}\right) \sim 0.8776$

3 $f(x) = 2x - \dfrac{2^3 x^3}{3!} + \cdots + \dfrac{(-1)^{n+1} 2^{2n+1} x^{2n+1}}{(2n+1)!} + \dfrac{(-1)^n \cos(2t)(2x)^{2n+3}}{(2n+3)!}, f\left(\dfrac{1}{2}\right) \sim 0.8415$

5 $f(x) = x - \dfrac{4x^3}{3!} + \dfrac{4^2 x^5}{5!} - \cdots + \dfrac{(-1)^n 4^n x^{2n+1}}{(2n+1)!} + \dfrac{(-1)^{n+1} 4^{n+1}(\cos^2 t - \sin^2 t)x^{2n+3}}{(2n+3)!},$

$f(\tfrac{1}{2}) \sim 0.4207$

7 $f(x) = 2^{-3} + 3 \cdot 2^{-6} x + \cdots + \dfrac{(-1)^n(1 \cdot 3 \cdots (2n+1))2^{-3(n+1)} x^n}{n!}$

$+ \dfrac{(-1)^{n+1}(1 \cdot 3 \cdots (2n+3))(4+t)^{(-5/2-n)} x^{n+1}}{2^{n+1}(n+1)!}, f\left(\dfrac{1}{2}\right) \sim 0.1048$

9 $x + \dfrac{x^3}{3}, 0.346$ **11** $x + \dfrac{x^3}{6}, 0.340$ **13** $x - \dfrac{x^2}{2}, 0.28$ **15** $x + \dfrac{x^3}{3}, 0.346$

17 $\dfrac{x^2}{2} - \dfrac{x^3}{6}, 0.049$

19 $e^x = e^2 + e^2(x-2) + \dfrac{e^2(x-2)^2}{2!} + \cdots + \dfrac{e^2(x-2)^n}{n!} + \dfrac{e^t(x-2)^{n+1}}{(n+1)!}$

21 $x^p = 1 + p(x-1) + \dfrac{p(p-1)(x-1)^2}{2} + \cdots + \dfrac{p(p-1)\cdots(p-n+1)(x-1)^n}{n!}$

$+ \dfrac{p(p-1)\cdots(p-n)t^{p-n-1}(x-1)^{n+1}}{(n+1)!}$

Section 9.11

1 96   **3** 6   **5** 1008   **7** $\sum_{n=0}^{\infty} x^n/(2^n n!), r = \infty$

9 $1 + x + \sum_{n=2}^{\infty} \dfrac{(-1)^{n+1}(1 \cdot 3 \cdot 5 \cdots (2n-3))x^n}{n!}, r = \dfrac{1}{2}$

11 $\sum_{n=0}^{\infty} \dfrac{(-1)^n x^n}{(2n)!}$, converges to $\cos\sqrt{x}$ for $x \ge 0$ **13** $\sum_{n=0}^{\infty} \dfrac{(-1)^n x^{2n}}{(2n+1)!}, r = \infty$

15 $1 - \dfrac{x^2}{2} + \sum_{n=2}^{\infty} - \dfrac{(1 \cdot 3 \cdot 5 \cdots (2n-3))x^{2n}}{2^n n!}, r = 1$ **17** $\sum_{n=0}^{\infty} \dfrac{(-1)^n x^{6n+4}}{(6n+4)(2n+1)!}, r = \infty$

19 $\sum_{n=0}^{\infty} \dfrac{x^{4n+1}}{(4n+1)(2n+1)!}, r = \infty$ **21** $x + \sum_{n=1}^{\infty} \dfrac{\frac{1}{3}(\frac{1}{3}-1)\cdots(\frac{1}{3}-n+1)x^{2n+1}}{n!(2n+1)}, r = 1$

23 $x + \sum_{n=1}^{\infty} \dfrac{1 \cdot 3 \cdots (2n-1)x^{2n+1}}{2^n n!(2n+1)^2}, r = 1$ **25** $\sum_{n=1}^{\infty} \dfrac{(-1)^{n+1}(x-1)^n}{n}, r = 1$

Extra Problems for Chapter 9

1 converges to 1   **3** converges to 1   **5** diverges to ∞ **7** diverges
9 converges to $28\frac{4}{15}$ **11** converges to 42   **13** diverges   **15** diverges
17 converges   **19** converges   **21** diverges   **23** converges   **25** converges
27 converges   **29** diverges   **31** 1/2   **33** 1   **35** $\sqrt{e}$
37 $f'(x) = \sum_{n=1}^{\infty} n^{a+1}(n+1)^b 2^n x^{n-1}, \int_0^x f(t)\,dt = \sum_{n=1}^{\infty} n^a(n+1)^{b-1} 2^n x^{n+1}, r = \frac{1}{2}$

39 $\sum_{n=0}^{\infty} \dfrac{(-1)^n x^{4n+1}}{(2n+1)(4n+1)}, r = 1$ **41** -0.006 **43** 0.646

45 $x + \sum_{n=1}^{\infty} \dfrac{2^n(-1)^n(2 \cdot 5 \cdot 8 \cdots (3n-1))x^{2n+1}}{3^n(2n+1)n!}, r = \dfrac{1}{\sqrt{2}}$ **47** e^{50}

Section 10.1

1 $i + 2j$ **3** $-3i - 4j$ **5** $(2, -4)$ **7** $(-1, 12)$ **9** $A + B = -3i + j$
11 $A + B + C = j$ **13** $3A = 3i - 6j$ **15** $B - A = -5i + 5j$

17 $\mathbf{A} - 2\mathbf{B} + 3\mathbf{C} = 18\mathbf{i} - 8\mathbf{j}$ **19** $|\mathbf{B}| = 5$ **21** $|\mathbf{A} - \mathbf{B}| = 5\sqrt{2}$ **23** $|6\mathbf{A}| = 6\sqrt{5}$
25 $\frac{13}{4}\mathbf{i} - \frac{5}{4}\mathbf{j}$ **27** $-\frac{4}{5}\mathbf{i} + \frac{3}{5}\mathbf{j}, (-\frac{4}{5}, \frac{3}{5})$ **29** $\arccos(-(2\sqrt{5})/5)$ **31** $\arccos(-\frac{4}{5})$
33 $7\mathbf{i} + 3\mathbf{j}$ **35** $8\mathbf{i} - 2\mathbf{j}$ **37** $-4\mathbf{i} - 6\mathbf{j}$ **39** 5   **41** $4\mathbf{i} + 3\mathbf{j}$ **43** $10\mathbf{i} + 2\mathbf{j}$
45 $-2\sqrt{2}\mathbf{i} + 2\sqrt{2}\mathbf{j}$ **47** $6\mathbf{i} + 8\mathbf{j}$ **49** $2\mathbf{i} - \mathbf{j}, 2\mathbf{i} + 2\mathbf{j}, -5\mathbf{i} + 2\mathbf{j}, \mathbf{i} - 3\mathbf{j}$

Section 10.2

1 $\mathbf{X} = 3\mathbf{i} - \mathbf{j} + t(-\mathbf{i} + \mathbf{j})$ **3** $\mathbf{X} = 3\mathbf{i} + 4\mathbf{j} + t(-2\mathbf{i} + 5\mathbf{j})$ **5** $\mathbf{X} = \mathbf{i} + 4\mathbf{j} + t(\mathbf{i} - 5\mathbf{j})$
7 $\mathbf{X} = 2\mathbf{i} + 5\mathbf{j} + t\mathbf{j}$ **9** $\mathbf{X} = 2\mathbf{j} + t(\mathbf{i} + 5\mathbf{j})$ **11** $\mathbf{X} = 3\mathbf{j} + t\mathbf{i}$
13 $\mathbf{X} = 6\mathbf{i} + 5\mathbf{j} + t(\mathbf{i} - 3\mathbf{j})$ **15** $y = -2x + 10$ **17** $y = 3$ **19** no   **21** yes
23 yes   **25** $(-2, 3)$ **27** $(2, 10)$ **29** $(4, 3)$

Section 10.3

1 $5\mathbf{i} - \mathbf{j} + 7\mathbf{k}$ **3** $-\mathbf{k}$ **5** $(0, 0, 0)$ **7** $3\mathbf{i} + \mathbf{j} - 4\mathbf{k}$ **9** $\mathbf{i} + 5\mathbf{j} - 8\mathbf{k}$
11 $-2\mathbf{i} - 3\mathbf{j} + 6\mathbf{k}$ **13** 3   **15** $\sqrt{26}$ **17** $\arccos(\frac{8}{21})$ **19** 0
21 $\frac{1}{3}\mathbf{i} - \frac{2}{3}\mathbf{j} + \frac{2}{3}\mathbf{k}, (\frac{1}{3}, -\frac{2}{3}, \frac{2}{3})$ **23** $-3\mathbf{i} + 3\mathbf{j} + 3\sqrt{2}\mathbf{k}$ **25** $\frac{2}{3}, -\frac{2}{3}$ **27** $\frac{3}{10}\mathbf{i} - \frac{1}{2}\mathbf{j} + \frac{1}{10}\mathbf{k}$
29 $125\mathbf{i} + 250\mathbf{j} + 625\mathbf{k}$ **31** $\mathbf{X} = \mathbf{i} + 3\mathbf{j} + \mathbf{k} + t(\mathbf{i} - \mathbf{k})$
33 $\mathbf{X} = -\mathbf{i} + 4\mathbf{j} + 3\mathbf{k} + t(-\mathbf{i} - 7\mathbf{j} + 3\mathbf{k})$ **35** $\mathbf{X} = t(-3\mathbf{i} + 4\mathbf{k})$

Section 10.4

1 (a) 20 (b) $\mathbf{A} \parallel \mathbf{B}$ (c) 1   **3** (a) 0 (b) $\mathbf{A} \perp \mathbf{B}$ (c) 0
5 (a) 24 (b) neither (c) 12/37   **7** (a) -92 (b) neither (c) $-92/117$
9 (a) 0 (b) $\mathbf{A} \perp \mathbf{B}$ (c) 0   **11** (a) $8\sqrt{5}$ (b) $\mathbf{A} \parallel \mathbf{B}$ (c) 1   **13** 41
15 $3\mathbf{i} + 3\mathbf{j} + 3\mathbf{k}$ **17** $200\sqrt{2}$ **19** $\mathbf{i} - \mathbf{j}$ **21** $-2\mathbf{i} - 2\mathbf{j} - 4\mathbf{k}$ **23** $-\mathbf{i} + \mathbf{j} + \mathbf{k}$
23 $-\mathbf{i} - \mathbf{j} + \mathbf{k}$ **27** $\frac{4}{5}\mathbf{i} + \frac{3}{5}\mathbf{j}$ **29** $\mathbf{i} - \mathbf{j}$ **31** $\arccos(\frac{1}{3})$ **33** $\pi/3$

Section 10.5

1
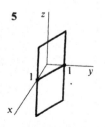

3
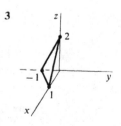

5
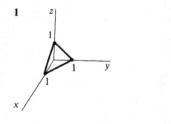

7
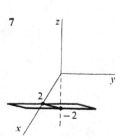

9 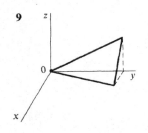   **11**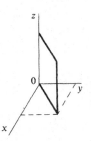

13 (a) $\mathbf{i} - 3\mathbf{j} + 6\mathbf{k}$ (b) $\mathbf{i} + 2\mathbf{j}$ (c) $-3\mathbf{i} + 4\mathbf{j} + \mathbf{k}$ (d) $\mathbf{i} + 6\mathbf{k}$ (e) $\mathbf{j}$ (f) $-\mathbf{j} + \mathbf{k}$
15 $y + 2z = -9$ **17** $x + y + 2z = 0$ **19** $y - z = -1$
21 $4x + 20y - 5z = 20$ **23** $x + 2y + z = 9$ **25** $2x + y + 3z = 13$
27 $x + y - 2z = 20$ **29** $x - y - 3z = 3$ **31** $24x - 19y + 4z = 57$ **33** parallel
35 neither   **37** $\mathbf{X} = 5\mathbf{i} + 3\mathbf{j} - \mathbf{k} + t(\mathbf{i} - \mathbf{j} + 3\mathbf{k})$ **39** $\mathbf{X} = -\mathbf{i} + \mathbf{k} + t(3\mathbf{i} - \mathbf{j} - 2\mathbf{k})$
41 $\mathbf{X} = \mathbf{j} - \mathbf{k} + t(-\mathbf{i} + \mathbf{j} + \mathbf{k})$ **43** $(3, 1, 1)$ **45** $(-\frac{4}{3}, 0, \frac{8}{3})$ **47** $(-2, 1, 2)$
49 $(1, 1, 3)$

Section 10.6

1 $\mathbf{X} = 3t\mathbf{i} + 9t^2\mathbf{j}$ **3** $\mathbf{X} = \dfrac{1}{\sqrt{5}}t^3\mathbf{i} + \dfrac{2}{\sqrt{5}}t^3\mathbf{j}$ **5** $\mathbf{X} = (1/\sqrt{t^2 + 1})(t\mathbf{i} + \mathbf{j})$

7 $\mathbf{X} = \dfrac{\cos t + \cos (3t)}{2}\mathbf{i} + \dfrac{\sin t + \sin (3t)}{2}\mathbf{j}$

9 $\mathbf{X} = (4 \cos (t/3) - \cos (4t/3))\mathbf{i} + (4 \sin (t/3) - \sin (4t/3))\mathbf{j}$
11 $\mathbf{X} = (\cos t + t \sin t)\mathbf{i} + (\sin t - t \cos t)\mathbf{j}$ **13** $\mathbf{X} = \cos t\mathbf{i} + \sin t\mathbf{j} + t^2\mathbf{k}$

15 $\mathbf{X} = \mathbf{i} + \left(2 + \dfrac{t^2}{\sqrt{2t^4 - 2t^2 + 1}}\right)\mathbf{j} + \left(1 + \dfrac{t^2 - 1}{\sqrt{2t^4 - 2t^2 + 1}}\right)\mathbf{k}$

17 $\mathbf{X} = \frac{1}{2}(\cos t + \cos (2t))\mathbf{i} + \frac{1}{2} \sin t\mathbf{j} + \frac{1}{2} \sin (2t)\mathbf{k}$
19 $\mathbf{X} = (-\frac{5}{3} + \frac{2}{3} \cos t - \frac{1}{3} \sin t)\mathbf{i} + (-\frac{5}{3} - \frac{1}{3} \cos t + \frac{2}{3} \sin t)\mathbf{j} + (\frac{13}{3} - \frac{1}{3} \cos t - \frac{1}{3} \sin t)\mathbf{k}$
21 $\mathbf{X} = (t - t \cos^2 t - 2t \sin t \cos t)\mathbf{i} + (2t - t \sin t \cos t - 2t \sin^2 t)\mathbf{j} + 3t\mathbf{k}$

23 $\mathbf{P}(t) = 2t\mathbf{i} + \dfrac{1}{2t(t + 1)}\mathbf{j} + (t + 1)\mathbf{k}$

Section 10.7

1 $5 \cos t\mathbf{i} - 5 \sin t\mathbf{j}$ **3** $-e^t \sin (e^t)\mathbf{i} + e^t \cos (e^t)\mathbf{j}$ **5** $-6\mathbf{i} + (6/t)\mathbf{j} - 6e^t\mathbf{k}$
7 $2 \cos 2t$ **9** $t/\sqrt{1 + t^2}$ **11** $\mathbf{i} + \mathbf{j} + \mathbf{k} + t(\mathbf{i} + 2\mathbf{j} + 3\mathbf{k})$
13 $\mathbf{V} = 2\mathbf{i} + 3\mathbf{j} - 4\mathbf{k}, |\mathbf{V}| = \sqrt{29}, \mathbf{A} = \mathbf{0}$
15 $\mathbf{V} = -\sin t\mathbf{i} + \cos t\mathbf{j} + \mathbf{k}, |\mathbf{V}| = \sqrt{2}, \mathbf{A} = - \cos t\mathbf{i} - \sin t\mathbf{j}$
17 $\mathbf{V} = -e^t \sin (e^t)\mathbf{i} + e^t \cos (e^t)\mathbf{j}, |\mathbf{V}| = e^t,$
  $\mathbf{A} = (-e^t \sin e^t - e^{2t} \cos e^t)\mathbf{i} + (e^t \cos e^t - e^{2t} \sin e^t)\mathbf{j}$
19 $\mathbf{V} = 2t\mathbf{i} + 4t\mathbf{j} - 2t\mathbf{k}, |\mathbf{V}| = 2\sqrt{6}|t|, \mathbf{A} = 2\mathbf{i} + 4\mathbf{j} - 2\mathbf{k}$
21 $\mathbf{S} = t \cos t\mathbf{i} + t \sin t\mathbf{j}, \mathbf{V} = (-t \sin t + \cos t)\mathbf{i} + (t \cos t + \sin t)\mathbf{j}, |\mathbf{V}| = \sqrt{1 + t^2},$
  $\mathbf{A} = (-t \cos t - 2 \sin t)\mathbf{i} + (-t \sin t + 2 \cos t)\mathbf{j}$
23 $\mathbf{S} = \cos (t^2)\mathbf{i} + \sin (t^2)\mathbf{j}, \mathbf{V} = -2t \sin (t^2)\mathbf{i} + 2t \cos (t^2)\mathbf{j}, |\mathbf{V}| = |2t|,$
  $\mathbf{A} = (-2 \sin (t^2) - 4t^2 \cos (t^2))\mathbf{i} + (2 \cos (t^2) - 4t^2 \sin (t^2))\mathbf{j}$
25 $\mathbf{V} = (e^x/\sqrt{1 + e^{2x}})\mathbf{i} + (e^{2x}/\sqrt{1 + e^{2x}})\mathbf{j}, |\mathbf{V}| = e^x,$
  $\mathbf{A} = [-e^{3x}(1 + e^{2x})^{-3/2} + e^x(1 + e^{2x})^{-1/2}]\mathbf{i} + [-e^{4x}(1 + e^{2x})^{-3/2} + 2e^{2x}(1 + e^{2x})^{-1/2}]\mathbf{j}$
27 2π **29** 5 **31** $2\sqrt{3} + \sqrt{2} \ln (\sqrt{3} + \sqrt{2})$ **33** $\frac{3}{2} + \ln 2$
35 $\mathbf{F}(t) = (\frac{1}{2}t^2 + \frac{1}{2})\mathbf{i} + (\frac{1}{3}t^3 + \frac{5}{3})\mathbf{j} + (\frac{1}{4}t^4 + \frac{11}{4})\mathbf{k}$

37 $\mathbf{F}(t) = \ln |t - 1|\mathbf{i} + (\ln |t - 2| - \ln 2)\mathbf{j} + (\ln |t - 3| - \ln 3)\mathbf{k}$
39 $\mathbf{S} = (t - \sin t)\mathbf{i} + (1 - \cos t)\mathbf{j} + \frac{1}{2}t^2\mathbf{k}$

Section 10.8

1 infinitesimal   **3** infinite   **5** finite   **7** infinite   **9** infinite
11 infinitesimal (zero)   **13** infinite   **15** infinite   **17** infinitesimal   **19** infinite
21 $-\sin x\mathbf{i} + \cos x\mathbf{j}$ **23** $10\mathbf{i} - 20\mathbf{j} + 5\mathbf{k}$ **25** $2\mathbf{i} + 3\mathbf{j} - \mathbf{k}$ **27** 0
29 $(1/|\mathbf{A}|)\mathbf{A} \cdot \mathbf{U}$ if $\mathbf{A} \neq \mathbf{0}$, 1 if $\mathbf{A} = \mathbf{0}$ **31** (a) no (b) no   **33** (a) yes (b) no
35 (a) yes (b) no   **37** (a) no (b) no   **39** (a) yes (b) no

Extra Problems for Chapter 10

1 $5\mathbf{i} - 12\mathbf{j}$ **3** $-3\mathbf{i} - 3\mathbf{j}$ **5** $9\mathbf{i} + 11\mathbf{j}$ **7** $\mathbf{X} = -\mathbf{i} + \mathbf{j} + t(4\mathbf{i} - 3\mathbf{j})$ **9** $(2, 5/2)$
11 $(1/\sqrt{105}, -10/\sqrt{105}, 2/\sqrt{105})$ **13** $100\mathbf{i} - 500\mathbf{j} + 300\mathbf{k}$
15 $\mathbf{X} = \mathbf{i} + 4\mathbf{j} + 3\mathbf{k} + t\mathbf{k}$ **17** $\mathbf{A} \perp \mathbf{B}$ **19** -20 **21** $-\mathbf{i} + 3\mathbf{j} - \mathbf{k}$
23

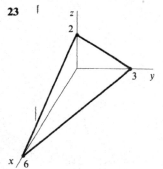

25 $-x - y + 2z = 0$ **27** $(-1, -2, 4)$
29 $\mathbf{S} = \sin t \cos t\mathbf{i} + \sin^2 t\mathbf{j} + \cos t\mathbf{k}$, $\mathbf{S} = -\sin t \cos t\mathbf{i} - \sin^2 t\mathbf{j} + \cos t\mathbf{k}$
31 $\mathbf{V} = \cos (2t)\mathbf{i} + \sin (2t)\mathbf{j} - \sin t\mathbf{k}$ or $\mathbf{V} = -\cos (2t)\mathbf{i} - \sin (2t)\mathbf{j} - \sin t\mathbf{k}$, $|\mathbf{V}| = \sqrt{1 + \sin^2 t}$,
  $\mathbf{A} = -2 \sin (2t)\mathbf{i} + 2 \cos (2t)\mathbf{j} - \cos t\mathbf{k}$ or $\mathbf{A} = 2 \sin (2t)\mathbf{i} - 2 \cos (2t)\mathbf{j} - \cos t\mathbf{k}$
33 $\mathbf{i} + \frac{1}{2}\mathbf{j} + \frac{1}{3}\mathbf{k} + t(-\mathbf{i} - \frac{1}{4}\mathbf{j} - \frac{1}{9}\mathbf{k})$
35 $\mathbf{S} = (-\cos e^t + \cos 1)\mathbf{i} + (\sin e^t - \sin 1)\mathbf{j} + (e^t - 1)\mathbf{k}$

Section 11.1

1

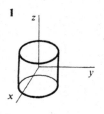

3

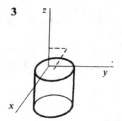

5

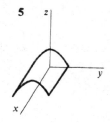

7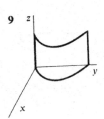

9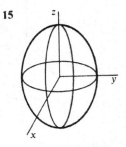

11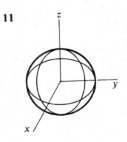

13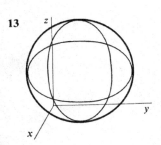

15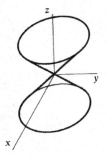

17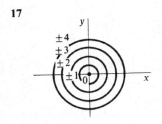

19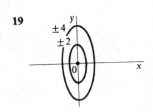

21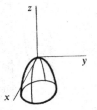

23

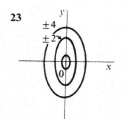

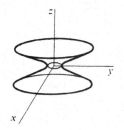

25

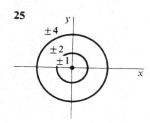

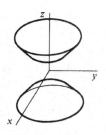

27

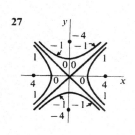

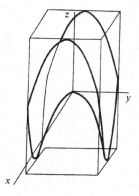

29

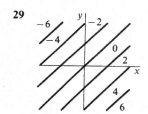

31

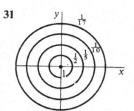

33

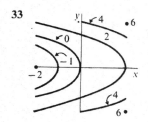

35

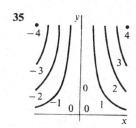

37

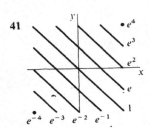

39

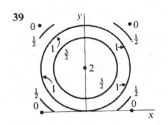

41

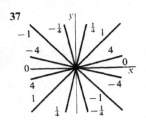

43

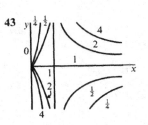

Section 11.2

1 all (x, y) **3** all (x, y) **5** $y \neq -x$ **7** $y \neq -2$ **9** $x \neq 2$ and $y \neq -1$
11 $x + y > 0$ **13** $x > y$ **15** $x > 0$ **17** $x^2 > y$ **19** $x > 0$ and $y > 0$
21 $x^2 > y$ **23** $x > 0$ and $y > 0$ and $x \neq 1/y$ **25** $x > 0$ and $y > 0$
27 $x \neq 4y$ and $y^2 > x$ **29** $\cos x + y > 0$ **31** $x \neq y$ **33** $x > y$ and $y \neq z$
35 $(x, y, z) \neq (0, 0, 0)$ **37** $x + y > 0$ and $z \neq 0$ **43** e **45** ∞

Section 11.3

1 $\partial z/\partial x = 4, \partial z/\partial y = -3$ **3** $\partial z/\partial x = y^2 + 3x^2y, \partial z/\partial y = 2xy + x^3$

5 $\dfrac{\partial z}{\partial x} = -\dfrac{2x}{(x^2 + y^2)^2}, \dfrac{\partial z}{\partial y} = -\dfrac{2y}{(x^2 + y^2)^2}$ **7** $f_x(x, y) = y, f_y(x, y) = x$

9 $f_x(x, y) = a, f_y(x, y) = b$ **11** $f_x(x, y) = 2xe^{x^2 - y^2}, f_y(x, y) = -2ye^{x^2 - y^2}$

13 $f_x(x, y) = \dfrac{1}{2\sqrt{x + 2y}}, f_y(x, y) = \dfrac{1}{\sqrt{x + 2y}}$ **15** $\partial z/\partial x = yx^{y-1}, \partial z/\partial y = x^y \ln x$

17 $\partial z/\partial x = 1/x, \partial z/\partial y = 1/y$

19 $\dfrac{\partial z}{\partial x} = -\dfrac{\ln y}{x(\ln x)^2}, \dfrac{\partial z}{\partial y} = \dfrac{1}{y \ln x}$ **21** $\dfrac{\partial z}{\partial x} = \dfrac{2xy}{\sqrt{1 - x^4y^2}}, \dfrac{\partial z}{\partial y} = \dfrac{x^2}{\sqrt{1 - x^4y^2}}$

23 $\dfrac{\partial w}{\partial x} = \dfrac{x}{\sqrt{x^2 + y^2 + z^2}}, \dfrac{\partial w}{\partial y} = \dfrac{y}{\sqrt{x^2 + y^2 + z^2}}, \dfrac{\partial w}{\partial z} = \dfrac{z}{\sqrt{x^2 + y^2 + z^2}}$

25 $f_x(x, y, z) = a, f_y(x, y, z) = b, f_z(x, y, z) = c$
27 $\partial w/\partial x = -z \sin x, \partial w/\partial y = z \cos y, \partial w/\partial z = \cos x + \sin y$ **29** $f_x(1, 2) = 4, f_y(1, 2) = 4$

31 $f_x(-1, 1) = -1, f_y(-1, 1) = 1$ **33** $\dfrac{\partial z}{\partial x}(0, 2) = 2, \dfrac{\partial z}{\partial y}(0, 2) = 0$

35 $\dfrac{\partial z}{\partial x}(1, 0) = e, \dfrac{\partial z}{\partial y}(1, 0) = 0$ **37** $\dfrac{\partial z}{\partial x}(2, 3) = -\dfrac{4}{961}, \dfrac{\partial z}{\partial y}(2, 3) = -\dfrac{27}{961}$

39 $f_x(1, 2, 3) = 2, f_y(1, 2, 3) = 4, f_z(1, 2, 3) = 6$ **41** $1/\sqrt{5}$ **43** $1/3$ **45** $b = 2, c = 1$
47 $C_x(x, y) = 3 - \frac{1}{2}\sqrt{y/x}, C_y(x, y) = 4 - \frac{1}{2}\sqrt{x/y}$

Section 11.4

1 $\Delta z = 3\,\Delta x - 2\,\Delta y, dz = 3\,dx - 2\,dy$

3 $\Delta z = 2xy^2\,\Delta x + 2x^2 y\,\Delta y + y^2\,\Delta x^2 + 4xy\,\Delta x\,\Delta y + x^2\,\Delta y^2 + 2x\,\Delta x\,\Delta y^2 + 2y\,\Delta x^2\,\Delta y + \Delta x^2\,\Delta y^2, dz = 2xy^2\,dx + 2x^2 y\,dy$

5 $\Delta z = -\dfrac{x\,\Delta y + y\,\Delta x + \Delta x\,\Delta y}{(x + \Delta x)(y + \Delta y)xy}, dz = -\dfrac{dx}{x^2 y} - \dfrac{dy}{xy^2}$

7 $\Delta z = e^{3x-4y}(e^{3\,\Delta x - 4\,\Delta y} - 1), dz = e^{3x-4y}(3\,dx - 4\,dy)$

9 $\Delta z = \cos(x + \Delta x)\sin(y + \Delta y) - \cos x \sin y, dz = -\sin x \sin y\,dx + \cos x \cos y\,dy$

11 $\Delta z = x\ln(1 + (\Delta y/y)) + \Delta x\ln(y + \Delta y), dz = \ln y\,dx + (x/y)\,dy$

13 $\Delta w = \Delta x + 2\,\Delta y + 3\,\Delta z, dw = dx + 2\,dy + 3\,dz$

15 $\Delta w = y\,\Delta x + (x + z)\,\Delta y + y\,\Delta z + \Delta x\,\Delta y + \Delta y\,\Delta z, dw = y\,dx + (x + z)\,dy + y\,dz$

17 $\Delta z = dz + \Delta x\,\Delta x + \Delta y\,\Delta y$ **19** $\Delta z = dz + (y\,\Delta x + 2x\,\Delta y + \Delta x\,\Delta y)\,\Delta x + 0\,\Delta y$

21 $\Delta z = dz + \dfrac{\Delta y}{y(y + \Delta y)}\Delta x + \dfrac{x\,\Delta y}{y^2(y + \Delta y)}\Delta y$ **23** $z = 4x + 4y - 6$

25 $z = 4x + 4y - 2$ **27** $z = \frac{1}{2}x + \frac{1}{2}y + 1$ **29** $z = 6e^3 x + e^3 y - 8e^3$

31 $z = \dfrac{\sqrt{2}}{4}x + \dfrac{\sqrt{6}}{4}y + \dfrac{\sqrt{6}}{4} - \pi\left(\dfrac{\sqrt{2}}{12} + \dfrac{\sqrt{6}}{16}\right)$ **33** $z = x - 2$ **35** $z = 1$

37 $x - 2y + 2z = 9$ **39** $x + y - z = 1$

Section 11.5

1 $2e^{2t} + 2e^{-2t}$ **3** $-\dfrac{\cos(t/a) + \cos(t/b)}{[a\sin(t/a) + b\sin(t/b)]^2}$ **5** $\dfrac{2y}{x\ln y} - \dfrac{2x\ln x}{y(\ln y)^2}$

7 $(t + 1)^{1/t}\left[\dfrac{1}{t(t + 1)} - \dfrac{\ln(t + 1)}{t^2}\right]$ **9** $(\sin t)^{\cos t}\left[\dfrac{\cos^2 t}{\sin t} - \sin t\ln(\sin t)\right]$

11 $\dfrac{2t}{(t^2 - 1)\ln(t^2 + 1)} - \dfrac{2t\ln(t^2 - 1)}{(t^2 + 1)[\ln(t^2 + 1)]^2}$ **13** $3\sqrt{1 - t^4} - 2\sqrt{1 - t^3}$

15 $\dfrac{\partial z}{\partial s} = 3s^2\cos^3 t, \dfrac{\partial z}{\partial t} = -3s^2\sin t\cos^2 t$ **17** $\dfrac{\partial z}{\partial s} = \dfrac{2s}{s^2 - t^2}, \dfrac{\partial z}{\partial t} = -\dfrac{2t}{s^2 - t^2}$

19 $\dfrac{\partial z}{\partial s} = -\dfrac{a}{(s + t)^2} - \dfrac{b}{(s - t)^2}, \dfrac{\partial z}{\partial t} = -\dfrac{a}{(s + t)^2} + \dfrac{b}{(s - t)^2}$ **21** $b\dfrac{\partial z}{\partial x} = abf'(ax + by) = a\dfrac{\partial z}{\partial y}$

25 $\dfrac{dw}{dt} = e^t\cos(\sqrt{t}) - e^{-t}\sin(\sqrt{t}) + \dfrac{1}{2\sqrt{t}}(-e^t\sin(\sqrt{t}) + e^{-t}\cos(\sqrt{t}))$

27 $\dfrac{dz}{dt} = \dfrac{x(dx/dt) + y(dy/dt)}{\sqrt{x^2 + y^2}}$ or $\dfrac{dz}{dt} = \dfrac{x}{z}\dfrac{dx}{dt} + \dfrac{y}{z}\dfrac{dy}{dt}$

29 $\dfrac{dz}{dt} = x^y\left(\dfrac{y}{x}\dfrac{dx}{dt} + \ln x\dfrac{dy}{dt}\right)$ or $\dfrac{dz}{dt} = \dfrac{zy}{x}\dfrac{dx}{dt} + z\ln x\dfrac{dy}{dt}$

31 $\dfrac{\partial z}{\partial s} = af'(u), \dfrac{\partial z}{\partial t} = bf'(u)$ **33** $\dfrac{\partial z}{\partial s} = z\dfrac{\partial u}{\partial s}, \dfrac{\partial z}{\partial t} = z\dfrac{\partial u}{\partial t}$

35 $\dfrac{\partial z}{\partial s} = g'(s)h(t), \dfrac{\partial z}{\partial t} = g(s)h'(t)$ **37** -2 **39** -3 **41** 2 **43** $1/25$

Section 11.6

1 $\dfrac{\partial z}{\partial x} = 3, \dfrac{dz}{dx} = 3 - 4e^x$ **3** $\dfrac{\partial z}{\partial x} = -\sin x, \dfrac{dz}{dx} = -\sin x + 3\cos(3x)$

5 $\dfrac{\partial z}{\partial x} = yx^{y-1}, \dfrac{dz}{dx} = x^x(1 + \ln x)$ **7** $\dfrac{\partial z}{\partial x} = \dfrac{y}{1 + x^2 y^2}, \dfrac{dz}{dx} = \dfrac{e^{-x}(1 - x)}{1 + x^2 e^{-2x}}$ **9** $\dfrac{y + x}{y - x}$

11 $-\dfrac{2x + 2y^3}{6xy^2 + 1}$ **13** $-\dfrac{1 + y\cos(xy)}{x\cos(xy)}$ **15** slope $= -\frac{3}{4}, 4y = -3x + 14$

17 slope $= 5, y = 5x - 6$ **19** slope $= -1, y = -x + 2$ **21** slope $= 1, y = x$

23 $\dfrac{\partial w}{\partial x}(x, y) = 15, \dfrac{\partial w}{\partial y}(x, y) = -34$

25 $\dfrac{\partial w}{\partial x}(x, y) = \dfrac{10x - 6y}{\sqrt{10x^2 - 12xy + 5y^2}}, \dfrac{\partial w}{\partial y}(x, y) = \dfrac{-6x + 5y}{\sqrt{10x^2 - 12xy + 5y^2}}$

27 $6x - 5y + 4z = -21$ **29** $x + y + z = 3$ **31** $x + y + z = 0$

33 $\dfrac{\partial z}{\partial x} = -\dfrac{xy}{z}, \dfrac{\partial z}{\partial y} = -\dfrac{x^2}{2z}$ **35** $\dfrac{\partial z}{\partial x} = \dfrac{\cos(xy)}{\sin(yz)}, \dfrac{\partial z}{\partial y} = \dfrac{x\cos(xy) - z\sin(yz)}{y\sin(yz)}$

37 $\dfrac{\partial z}{\partial x} = -\dfrac{z}{3x}, \dfrac{\partial z}{\partial y} = -\dfrac{2z}{3y}$ **39** $\dfrac{dz}{dx} = y - \dfrac{x}{1 + \sqrt{x}}$

41 $\dfrac{dw}{dx} = \dfrac{\partial w}{\partial x} + \dfrac{\partial w}{\partial y}\dfrac{dy}{dx} + \dfrac{\partial w}{\partial z}\dfrac{dz}{dx}$

Section 11.7

1 max $= 3$ at $(-1, -1)$ and $(1, 1)$; min $= 0$ at $(0, 0)$

3 max $= 48$ at $(3, 3)$; min $= -6$ at $(1, -2)$ **5** max $= 4$ at $(2, 1)$; min $= -5$ at $(2, 4)$

7 max $= 2$ at $(\pi/2, \pi/2)$; min $= 0$ at $(0, 0), (0, \pi), (\pi, 0), (\pi, \pi)$

9 max $= 1$ at $(-1, 1)$ and $(1, 1)$; min $= -\frac{1}{4}$ at $(0, \frac{1}{2})$

11 max $= \frac{1}{4}$ at $(\frac{1}{2}, 0, \frac{1}{2})$ and $(\frac{1}{2}, 1, -\frac{1}{2})$; min $= -\frac{1}{4}$ at $(0, \frac{1}{2}, \frac{1}{2})$ and $(1, \frac{1}{2}, -\frac{1}{2})$

13 max $= 1 + \sqrt{2}$ at $(\sqrt{2}/2, \sqrt{2}/2, 1)$; min $= -\frac{1}{2}$ at $(-\frac{1}{2}, -\frac{1}{2}, \frac{1}{2})$

15 max $= 3$ at $(-1, -1, 1), (-1, 1, 1), (1, -1, 1), (1, 1, 1)$; min $= 0$ at $(0, 0, 0)$

17 no max; min $= -4$ at $(-2, 0)$ **19** no max; no min

21 no max; min $= 6$ at $(\frac{1}{2}, 4)$ **23** max $= 1$ at $(0, 0)$; no min   **25** no max, no min

27 $x = 4, y = 2, z = 2$ **29** 6 in. $\times$ 6 in. $\times$ 12 in.   **31** $2\sqrt{3}/3$ **33** $(\frac{5}{3}, \frac{10}{3}, -\frac{5}{3})$

35 $(1, 1, 1), (1, -1, -1), (-1, 1, -1), (-1, -1, 1)$

39 $\frac{2}{3}(\frac{9}{4}V)^{1/3}$ in. $\times$ $\frac{2}{3}(\frac{9}{4}V)^{1/3}$ in. $\times$ $(\frac{9}{4}V)^{1/3}$ in.   **41** $x = 100, y = 200, P(100, 200) = 50{,}000$

Section 11.8

1 $\dfrac{\partial^2 z}{\partial x^2} = 2, \dfrac{\partial^2 z}{\partial y^2} = 4, \dfrac{\partial^2 z}{\partial x \partial y} = 0$ **3** $\dfrac{\partial^2 z}{\partial x^2} = 2a, \dfrac{\partial^2 z}{\partial y^2} = 2c, \dfrac{\partial^2 z}{\partial x \partial y} = b$

5 $\dfrac{\partial^2 z}{\partial x^2} = e^{x+y}(x + 2), \dfrac{\partial^2 z}{\partial y^2} = xe^{x+y}, \dfrac{\partial^2 z}{\partial x \partial y} = e^{x+y}(x + 1)$

7 $\dfrac{\partial^2 z}{\partial x^2} = -\dfrac{a^2}{(ax + by)^2}, \dfrac{\partial^2 z}{\partial y^2} = -\dfrac{b^2}{(ax + by)^2}, \dfrac{\partial^2 z}{\partial x \partial y} = -\dfrac{ab}{(ax + by)^2}$

9 $\dfrac{\partial^2 z}{\partial x^2} = a(a - 1)x^{a-2}y^b, \dfrac{\partial^2 z}{\partial y^2} = b(b - 1)x^a y^{b-2}, \dfrac{\partial^2 z}{\partial x \partial y} = abx^{a-1}y^{b-1}$

11 $\dfrac{\partial^2 w}{\partial x^2} = \dfrac{\partial^2 w}{\partial y^2} = \dfrac{\partial^2 w}{\partial z^2} = \dfrac{\partial^2 w}{\partial x \partial y} = \dfrac{\partial^2 w}{\partial x \partial z} = \dfrac{\partial^2 w}{\partial y \partial z} = -\dfrac{1}{4}(x + y + z)^{-3/2}$

13 $\dfrac{\partial^3 z}{\partial x^3} = 12y^3, \dfrac{\partial^3 z}{\partial y^3} = -36x^2, \dfrac{\partial^3 z}{\partial x^2 \partial y} = 24xy - 36y^2, \dfrac{\partial^3 z}{\partial x \partial y^2} = 12x^2 - 72xy$

15 $\dfrac{\partial^3 z}{\partial x^3} = a^3 e^{ax+by}, \dfrac{\partial^3 z}{\partial y^3} = b^3 e^{ax+by}, \dfrac{\partial^3 z}{\partial x^2 \partial y} = a^2 b e^{ax+by}, \dfrac{\partial^2 z}{\partial x \partial y^2} = ab^2 e^{ax+by}$

17 $\dfrac{\partial^2 z}{\partial \theta^2} = r^2 \sin^2 \theta \dfrac{\partial^2 f}{\partial x^2} - 2r^2 \sin \theta \dfrac{\partial^2 f}{\partial x \partial y} + r^2 \cos^2 \theta \dfrac{\partial^2 f}{\partial y^2} - r \cos \theta \dfrac{\partial f}{\partial x} - r \sin \theta \dfrac{\partial f}{\partial y}$

19 $\dfrac{\partial^2 z}{\partial x^2} = a^2 f''(u), \dfrac{\partial^2 z}{\partial y^2} = b^2 f''(u), \dfrac{\partial^2 z}{\partial x \partial y} = abf''(u)$ **21** $\dfrac{\partial^2 z}{\partial x^2} = g''(x), \dfrac{\partial^2 z}{\partial y^2} = h''(y), \dfrac{\partial^2 z}{\partial x \partial y} = 0$

23 $\dfrac{\partial^2 z}{\partial x^2} = n(n - 1)u^{n-2}\left(\dfrac{\partial u}{\partial x}\right)^2 + nu^{n-1}\dfrac{\partial^2 u}{\partial x^2}, \dfrac{\partial^2 z}{\partial y^2} = n(n - 1)u^{n-2}\left(\dfrac{\partial u}{\partial y}\right)^2 + nu^{n-1}\dfrac{\partial^2 u}{\partial y^2},$

$\dfrac{\partial^2 z}{\partial x \partial y} = n(n - 1)u^{n-2}\left(\dfrac{\partial u}{\partial x}\right)\left(\dfrac{\partial u}{\partial y}\right) + nu^{n-1}\dfrac{\partial^2 u}{\partial x \partial y}$

25 $\dfrac{\partial^2 z}{\partial s^2} = a\dfrac{\partial^2 x}{\partial s^2} + b\dfrac{\partial^2 y}{\partial s^2}, \dfrac{\partial^2 z}{\partial t^2} = a\dfrac{\partial^2 x}{\partial t^2} + b\dfrac{\partial^2 y}{\partial t^2}, \dfrac{\partial^2 z}{\partial s \partial t} = a\dfrac{\partial^2 x}{\partial s \partial t} + b\dfrac{\partial^2 y}{\partial s \partial t}$

27 $\dfrac{\partial^2 z}{\partial s^2} = f'(x)[(g'(s))^2 + g'(s)g''(s)], \dfrac{\partial^2 z}{\partial t^2} = f'(x)[(h'(t))^2 + h'(t)h''(t)], \dfrac{\partial^2 z}{\partial s \partial t} = f'(x)g'(s)h'(t)$

Extra Problems for Chapter 11

1

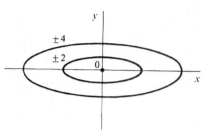

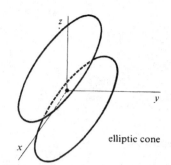

elliptic cone

3

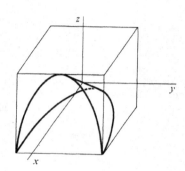

5 $x \neq 0$ **7** $1/x + 1/y > 0$ **9** $f_x(x, y) = a, f_y(x, y) = b$

11 $\dfrac{\partial z}{\partial x} = \dfrac{1}{x \ln y}, \dfrac{\partial z}{\partial y} = -\dfrac{\ln x}{y(\ln y)^2}$ **13** $\Delta z = -\dfrac{\Delta x}{x(x + \Delta x)} - \dfrac{2 \Delta y}{y(y + \Delta y)}, dz = -\dfrac{dx}{x^2} - \dfrac{2 \, dy}{y^2}$

15 $-8y + z = 4$ **17** $\dfrac{dz}{dt} = -\dfrac{2 \log_{(2t+1)} (3t + 2)}{(2t + 1) \ln (2t + 1)} + \dfrac{3}{(3t + 2) \ln (2t + 1)}$ **19** 1

21 $y = -8x + 17$, slope $= -8$ **23** max $= 4$ at $(0, 0)$ and $(3, 3)$; min $= -1$ at $(1, 2)$

25 no max; no min   **27** $\dfrac{\partial^2 z}{\partial x^2} = -\dfrac{2xy^3}{(1 + x^2 y^2)^2}, \dfrac{\partial^2 z}{\partial y^2} = -\dfrac{2x^3 y}{(1 + x^2 y^2)^2}, \dfrac{\partial^2 z}{\partial x \partial y} = -\dfrac{1 - x^2 y^2}{(1 + x^2 y^2)^2}$

29 $\dfrac{\partial^2 z}{\partial r^2} = \cosh^2 \theta \dfrac{\partial^2 f}{\partial x^2} + 2 \sinh \theta \cosh \theta \dfrac{\partial^2 f}{\partial x \partial y} + \sinh^2 \theta \dfrac{\partial^2 f}{\partial y^2}$

Section 12.1

1 $168/64 = 2.625$ **3** 44   **5** 1.025   **7** $(\pi/6)^2(7 + 4\sqrt{3}) \sim 3.8$
9 $e^{-5} + e^{-4} + 2e^{-3} + 2e^{-2} + 2e^{-1} + 2 + 2e + 2e^2 + e^3 + e^4 \sim 97.52$
11 $150/64 \sim 2.34$ **13** $((11 + 7\sqrt{2})\pi)/32 \sim 2.05$ **15** 108

Section 12.2

1 $7/2 = 3.5$ **3** $128/3 = 42\frac{2}{3}$ **5** $\frac{3}{2} \ln 2 \sim 1.04$ **7** 4
9 $\frac{1}{2}(e^6 - e^2 - e^{-2} + e^{-6}) \sim 197.9$ **11** $13/6 \sim 2.17$ **13** $32/15 \sim 2.13$
15 $36\pi \sim 113.1$ **17** $14/3$ **19** 30   **21** 2   **23** $\frac{4}{15}(212 - 36\sqrt{6})$ **25** $e - 1$
27 $8/3$ **29** $3^8/40$
31 $0 \leq x \leq 5, 0 \leq y \leq 5 - x$, or $0 \leq y \leq 5, 0 \leq x \leq 5 - y$; $\int_0^5 \int_0^{5-x} f(x, y) \, dy \, dx$, or
$\int_0^5 \int_0^{5-y} f(x, y) \, dx \, dy$

33 $-2 \leq x \leq 2, -\sqrt{4 - x^2} \leq y \leq \sqrt{4 - x^2}, \int_{-2}^{2} \int_{-\sqrt{4-x^2}}^{\sqrt{4-x^2}} f(x, y) \, dy \, dx$

35 $-4 \leq x \leq 1, 3x \leq y \leq 4 - x^2, \int_{-4}^{1} \int_{3x}^{4-x^2} f(x, y) \, dy \, dx$

37 $-1 \leq y \leq 1, 1/2 \leq x \leq 1/(1 + y^2), \int_{-1}^{1} \int_{1/2}^{1/(1+y^2)} f(x, y) \, dx \, dy$ 39 $\pi/4$ 41 32/3

43 $\frac{1}{2}\pi ab$

Section 12.3

1 3/4   3 1   5 4/3   7 16/15   9 8/3   11 8π 13 8/15   15 144

17 $\frac{4}{3}\pi abc$

Section 12.4

1 (a) $4abk$ (b) $(0, 0)$ (c) $\frac{4}{3}(ba^3 + ab^3)k$ 3 (a) $\frac{2}{3}k$ (b) $(\frac{3}{8}, \frac{3}{5})$ (c) $\frac{44}{105}k$

5 (a) 26/3 (b) $(19/13, 151/65)$ (c) 944/15

7 (a) $\frac{2}{7} + \frac{1}{6}$ (b) $\left(\frac{\frac{2}{9} + \frac{2}{15}}{\frac{2}{7} + \frac{1}{6}}, \frac{\frac{1}{11} + \frac{1}{15}}{\frac{2}{7} + \frac{1}{6}}\right)$ (c) $\frac{2}{11} + \frac{1}{9} + \frac{2}{45} + \frac{1}{28}$

9 (a) $e^2 + e^{-2} - 2$ (b) $\left(\frac{e^2 + 3e^{-2} - 2}{e^2 + e^{-2} - 2}, \frac{e^4 + e^{-4} - 2}{4(e^2 + e^{-2} - 2)}\right)$

    (c) $\frac{1}{9}e^6 + 2e^2 - \frac{2}{9} - 10e^{-2} + \frac{1}{9}e^{-6}$ 11 2   13 (a) $(\sqrt{2}/48)g$ (b) $g/24$

15 $m = \frac{8}{3}, W = \frac{16}{15}g$

Section 12.5

1 6π 3 8π 5 $\frac{81}{2}\pi$ 7 $2\pi(1 - \ln 2)$ 9 4/21   11 $\pi/48$

13 32/9   15 $8\pi^8 + \frac{64}{15}\pi^5$ 17 $5\pi/3$ 19 $\pi/3 - 4/9$ 21 $\frac{1}{3}(\beta - \alpha)(b^3 - a^3)$

23 (a) $\pi b^3/3$ (b) $\pi b^5/10$ 25 $(\pi k/2)(b^4 - a^4)$ 27 (a) $\bar{x} = 4/(3\pi), \bar{y} = 4/(3\pi)$ (b) $\pi k/8$

29 (a) $\bar{x} = 0, \bar{y} = \frac{3\sqrt{3} + 8\pi}{6\sqrt{3} + 4\pi}b$ (b) $\left(\frac{5}{6}\pi + \frac{7\sqrt{3}}{8}\right)b^4 k$ 31 (a) π (c) $\sqrt{\pi}$

Section 12.6

1 12   3 12   5 13/56   7 16/27   9 40

11 $\frac{763}{594} \times 10^9 = (\frac{4}{27} + \frac{100}{88}) \times 10^9 \sim 1.3 \times 10^9$

13 $\frac{e^6}{24} - \frac{e^4}{15} + \frac{e^{-2}}{24} - \frac{e^{-6}}{60}$ 15 $\frac{52}{63} - \frac{12\sqrt{3}}{35}$

17 (a) 3 (b) $(\frac{19}{36}, \frac{5}{9}, \frac{7}{12})$ (c) $I_z = \frac{7}{4}, I_y = \frac{7}{3}, I_x = \frac{29}{12}$

19 (a) 1 (b) $(\frac{3}{4}, \frac{5}{12}, \frac{7}{12})$ (c) $I_z = \frac{5}{6}, I_y = \frac{11}{10}, I_x = \frac{11}{15}$

21 (a) $\frac{5}{18}$ (b) $(\frac{3}{20}, \frac{3}{20}, \frac{65}{972})$ (c) $I_z = \frac{5}{16}, I_y = \frac{73}{960}, I_x = \frac{73}{960}$

23 (a) $\frac{1}{6}abck$ (b) $\left(\frac{a}{4}, \frac{b}{4}, \frac{c}{4}\right)$ (c) $I_z = \frac{abck}{60}(a^2 + b^2), I_y = \frac{abck}{60}(a^2 + c^2), I_x = \frac{abck}{60}(b^2 + c^2)$

25 (a) $abck$ (b) $\left(\frac{a}{2}, \frac{b}{2}, \frac{c}{2}\right)$ (c) $I_z = \frac{abck}{3}(a^2 + b^2), I_y = \frac{abck}{3}(a^2 + c^2), I_x = \frac{abck}{3}(b^2 + c^2)$

Section 12.7

1 $\frac{4\pi}{3}$ 3 $\frac{\pi}{10}$ 5 $\frac{128}{7}$ 7 $\frac{2\pi hb^3}{3}$ 9 $\frac{\pi hb^4}{10}$ 11 $\left(0, 0, \frac{2}{3}\right)$

13 $\frac{b^4 ck\pi}{2} + \frac{2b^2 c^3 k\pi}{3}$ 15 $\frac{\pi hb^4 k}{10}$ 17 $\frac{4\pi b^5}{5}$ 19 $\frac{4\pi}{15}$ 21 $\frac{4\pi b^4}{3}$

23 $\frac{\pi b^5}{10}(1 - \cos(2\alpha))$ or $\frac{\pi b^5}{5}\sin^2\alpha$ 25 $\frac{4}{3}\pi(b^3 - a^3)$ 27 $\frac{\pi b^3}{6}(1 - \cos^4\beta)$ 29 πc^4

31 $\frac{8\pi kb^5}{15}$ 33 $b^3\left(\frac{2\pi}{3} - \frac{8}{9}\right)$ 35 $\frac{3}{8}b$ from the center

Extra Problems for Chapter 12

1 0 **3** 0 **5** 1/3 **7** 4/5 **9** $\frac{1}{2}(\ln 2)^2 - \frac{1}{6}(\ln 2)^3$ **11** 1/6
13 $m = 2k, (\bar{x}, \bar{y}) = (\pi/2, \pi/8), I_0 = (\pi^2 - \frac{32}{9})k$ **15** $m = \pi r^4/4, (\bar{x}, \bar{y}) = (0, 0), I_0 = \pi r^6/6$
17 45π **19** $2\pi a$ **23** $311/3960$

25 $(\bar{x}, \bar{y}, \bar{z}) = (\frac{1}{5}, \frac{1}{5}, \frac{1}{30}), I_z = \dfrac{k}{60}, I_y = I_x = \dfrac{29k}{3360}$

27 $(\bar{x}, \bar{y}, \bar{z}) = \left(0, 0, \dfrac{b^2 + c}{2}\right), I_z = \dfrac{\pi b^4 ck}{2}$ **29** $m = \pi(b^4 - a^4), I = \dfrac{4\pi}{9}(b^6 - a^6)$

Section 13.1

1 $\text{grad} f = 2x\mathbf{i} + 2y\mathbf{j}, f_{\mathbf{U}} = \sqrt{2}x + \sqrt{2}y$ **3** $\text{grad} f = 2xy^3\mathbf{i} + 3x^2y^2\mathbf{j}, f_{\mathbf{U}} = \frac{6}{5}xy^3 - \frac{12}{5}x^2y^2$
5 $\text{grad} f = -\sin x \sin y\mathbf{i} + \cos x \cos y\mathbf{j}, f_{\mathbf{U}} = (1/\sqrt{5})(-\sin x \sin y + 2\cos x \cos y)$
7 $\text{grad} f = \dfrac{x}{\sqrt{x^2 + y^2}}\mathbf{i} + \dfrac{y}{\sqrt{x^2 + y^2}}\mathbf{j}, f_{\mathbf{U}} = \dfrac{x - y}{\sqrt{2x^2 + 2y^2}}$
9 $\text{grad} f = yz\mathbf{i} + xz\mathbf{j} + xy\mathbf{k}, f_{\mathbf{U}} = \frac{1}{3}(yz + 2xz - 2xy)$
11 $\text{grad} f = -x^{-2}\mathbf{i} - 2y^{-2}\mathbf{j} - 3z^{-2}\mathbf{k}, f_{\mathbf{U}} = (1/\sqrt{2})(-x^{-2} + 3z^{-2})$
13 $\text{grad} f = \dfrac{x\mathbf{i} + y\mathbf{j} + z\mathbf{k}}{\sqrt{x^2 + y^2 + z^2}}, f_{\mathbf{U}} = \dfrac{x \cos \alpha + y \cos \beta + z \cos \gamma}{\sqrt{x^2 + y^2 + z^2}}$ **15** $7\sqrt{10}/20$
17 $9a/\sqrt{a^2 + b^2}$ **19** $-1/\sqrt{3}$ **21** $(1/\sqrt{2})(\mathbf{i} + \mathbf{j})$ **23** $\frac{1}{7}(3\mathbf{i} + 6\mathbf{j} + 2\mathbf{k})$
25 $(f_{\mathbf{U}})_{\mathbf{V}} = \dfrac{\partial^2 z}{\partial x^2}u_1 v_1 + \dfrac{\partial^2 z}{\partial x \partial y}(u_1 v_2 + u_2 v_1) + \dfrac{\partial^2 z}{\partial y^2}u_2 v_2$

Section 13.2

1 $2e^3$ **3** $\frac{9}{2}e - \frac{3}{2}$ **5** $\sin 1 - \cos 1 + 1$ **7** $\frac{3}{20} + \frac{5}{7}\ln 4$ **9** $-2\pi, 2\pi$
11 -4π **13** $\frac{11}{30}$ **15** $\frac{1}{2}$ **17** $\frac{1}{2}\ln(\frac{29}{2})$ **19** $1243/3$

Section 13.3

1 no potential function **3** no potential function **5** no potential function
7 $y \sin x + C$ **9** $-2x + 6y + C$ **11** no potential function
13 $\frac{3}{2}x^2 + 5xy - y^2 + C$ **15** $\cosh x \cosh y + C$ **23** $\frac{3}{2}x^2 + 4xy - y^2 = C$
25 $\frac{2}{3}x^{3/2} + x\sqrt{y} = C$ **27** $\frac{1}{3}y^3 + y \arctan x = C$ **29** $\cos x \sin y = C$
31 $(x^2/2) + (y^2/2) + (2/3)(x + y)^{3/2} = C$ **33** $P(x, y) = 2 \sin x \cos x \sin y$

Section 13.4

1 1 **3** $\frac{1}{2}(-e^7 + e^{-1} + e - e^{-7})$ **5** $-\frac{1}{6}$ **7** $\frac{11}{2} - \ln 2$ **9** $-17/12$
11 $-\pi/2$ **13** (a) $-x - y$ (b) -1 (c) $y - x$ (d) 0
15 (a) $2bx - 2ay$ (b) $(2b - a)/3$ (c) 0 (d) 0 **17** (a) -2 (b) -2π (c) 0 (d) 0
19 $\pi a^2 + \pi/2$

Section 13.5

1 $25\sqrt{21}/4$ **3** $(\pi/6)(5\sqrt{5} - 1)$ **5** $\pi a^2/2$ **7** $(\pi/6)[(1 + 4a^2)^{3/2} - 1]$
9 $a^2(\pi - 2)$ **11** $A\sqrt{1 + a^2 + b^2}$ **13** $8a^2$ **15** 16 **17** $-\pi$ **19** $-2e^{-1}$

Section 13.6

1 $\text{curl } \mathbf{F} = \mathbf{0}, \text{div } \mathbf{F} = 2x + 2y + 2z$ **3** $\text{curl } \mathbf{F} = -\mathbf{i} + \mathbf{j} - \mathbf{k}, \text{div } \mathbf{F} = 3$
5 $\text{curl } \mathbf{F} = (ze^{x+y} - ye^{x+z})\mathbf{i} + (xe^{y+z} - ze^{x+y})\mathbf{j} + (ye^{x+z} - xe^{y+z})\mathbf{k},$
$\text{div } \mathbf{F} = e^{y+z} + e^{x+z} + e^{x+y}$ **9** 0 **11** $A(p(c - b) + q(a - c) + (a - b))$
13 $abc(a + b + c)$ **15** 3 **17** $\pi/3$ **19** $\pi/2$

Extra Problems for Chapter 13

1 $-\sin x \cos \alpha + \cos y \sin \alpha$ **3** $\operatorname{grad} f = ye^{xy}\mathbf{i} + xe^{xy}\mathbf{j}$, $f_U = e^{xy}(y \cos \alpha + x \sin \alpha)$

5 $\dfrac{1}{\sqrt{14}}(-2\mathbf{i} - 3\mathbf{j} + \mathbf{k})$ **7** $\ln\left(\dfrac{6\sqrt{2}}{5}\right)$ **9** 0   **11** $x \ln x + xy \ln y - x$

13 $\frac{3}{2}y^2 - e^{-y} \cos x = C$ **15** $\dfrac{131}{60}$ **17** $\dfrac{\sqrt{6}}{2} - \dfrac{\sqrt{2}}{6}$ **19** 2π

21 $\operatorname{curl} \mathbf{F} = (x^2 z - 3xy^2 z^2)\mathbf{i} + (xy - 2xyz)\mathbf{j} + (y^2 z^3 - xz)\mathbf{k}$, $\operatorname{div} \mathbf{F} = yz + 2xyz^3 + x^2 y$

Section 14.1

1 $y = -1/2 \cos(t^2) + C$ **3** $y = \ln(t + C)$ **5** $y = \dfrac{1 + Ce^{2t}}{1 - Ce^{2t}}, y = 1, y = -1$

7 $y = -\dfrac{1}{0.5t^2 + C}, y = 0$ **9** $y = \tan(e^t + C)$ **11** $y = C \sec t, y = 0$

13 $y = \left(\dfrac{t^2}{4} + 9\right)^{1/2}$ **15** $y = -\sqrt{2(t \ln t - t) + 6}$ **17** $y = 2$

Section 14.2

1 $y = C \cdot e^{-5t}$ **3** $y = C \cdot e^{\arctan t}$ **5** $y = 4 \cdot e^{-t}$ **7** $y = e \cdot e^{\cos t}$ **9** $y = 0$
11 $y = 4t^2$ **13** $y = e^{1 - t - t^2}$ **15** $y = 100 \cdot e^{-t \cdot \ln 5}$ **17** $y = 1{,}000{,}000 \cdot e^{t \cdot \ln 1.5}$

Section 14.3

1 $y = 2 + Ce^{-4t}$ **3** $y = 5 + Ce^{-(1/2)t^2}$ **5** $y = -(t^2 + 2t + 2) \cdot Ce^{-4t}$

7 $y = \dfrac{1}{3}t^{-1} + Ct^2$ **9** $y = \sin t + C \cos t$ **11** $y = e^{-\cos t}\left(\displaystyle\int_0^t s e^{\cos s}\, ds + C\right)$

13 $y = e^{-\int_0^t \cos(e^r)\, dr}\left[\displaystyle\int_0^t e^{\int_0^s \cos(e^r)\, dr}\, ds + C\right]$

15 $-[400{,}000t + 15{,}840{,}000] + 15{,}940{,}000e^{0.025t}$
17 $15{,}000e - 25{,}000$ dollars, or $15,774.23

Section 14.4

t	0.0	0.1	0.2	0.3	0.4	0.5	0.6	0.7	0.8	0.9	1.0
1 $Y(t)$	1.0	1.0	1.01	1.030	1.059	1.097	1.142	1.195	1.253	1.317	1.386
3 $Y(t)$	1.0	1.054	1.095	1.122	1.137	1.140	1.133	1.117	1.093	1.061	1.023

5 Apply the lemma with $M = 4$.   **7** Apply the lemma with $M = \pi/2$.
9 $y(t) = 0$ for $0 \le t \le b$, $y(t) = [(2/3)(t - b)]^{3/2}$ for $b < t < \infty$.

Section 14.5

1 $10 - i2$ **3** $i10$ **5** $0.1 - i(0.7)$ **7** 4 **9** $\pm i5$ **11** $-1 \pm i2$
13 $5 \operatorname{cis}(\pi/2)$ **15** $3\sqrt{2} \operatorname{cis}(-3\pi/4)$ **17** $2 \operatorname{cis}(-\pi/6)$ **19** $-i2/3$
21 $2 \operatorname{cis}(-\pi/12)$ **23** $-4 + i4$ **25** $\pm 2^{1/4} \operatorname{cis}(\pi/8)$ **27** $\pm 2 \operatorname{cis}(-\pi/4)$,
  or $\pm\sqrt{2}(1 - i)$ **29** ie^{-3} **31** $e\sqrt{2}/2 - ie\sqrt{2}/2$ **33** $(5 - i3)e^{(5 - i3)t}$
35 $(3 + i2)e^{(2 - i7) + (3 + i2)t}$ **37** $Ce^{(-2 + i3)t}$ **39** $Ce^{(3 - i5)t}$
41 $e^{t + (2 - i)t}$, or $e^{2t} \operatorname{cis}(1 - t)$ **43** $4e^{(2 - i)t}$, or $4e^{2t} \operatorname{cis}(-t)$

Section 14.6

1 $Ae^{2t} + Be^{-3t}$ **3** $Ae^{5t} + Bte^{5t}$ **5** $A\cos(4t) + B\sin(4t)$
7 $e^t[A\cos t + B\sin t]$ **9** $-\frac{1}{4}e^{-5t} + \frac{5}{4}e^{-t}$ **11** $5e^{-6t} + 20te^{-6t}$
13 $-2\cos(\sqrt{5}t) + \sqrt{5}\sin(\sqrt{5}t)$ **15** $e^{-6t}[4\cos t + 24\sin t]$
17 $2\cos(2t - \pi/6)$, amplitude $= 2$, frequency $= 2$, phase shift $= \pi/6$
19 $e^{-2t}\sqrt{2}\cos(3t - \pi/4)$, amplitude $= \sqrt{2}e^{-2t}$, frequency $= 3$, phase shift $= \pi/4$
21 $e^{-2t}[2\cos(5t) + \sin(5t)]$

Section 14.7

1 $(24/676)\cos t - (10/676)\sin t$ **3** $(1/2)t^2 + (3/16)t - (3/8)$ **5** $0.5e^{2t}$
7 $-0.2te^{-3t}$ **9** $0.125t\sin(4t)$ **11** $3t^2e^{-6t}$ **13** $Ae^{-3t} + Be^{-2t} + 0.8$
15 $A\cos(\sqrt{5}t) + B\sin(\sqrt{5}t) + 2\sin(2t)$ **17** $4e^t + e^{-t} - 3t - 5$
19 $e^{-6t}[2\cos t + 20\sin t] + 2e^{-4t}$
21 $e^{-2t}[A\cos(5t) + B\sin(5t)] + 25\cos(2t) + 8\sin(2t)$, steady state $= 25\cos(2t) + 8\sin(2t)$
25 5

Extra Problems for Chapter 14

1 $y = -1/(\sin t + C)$ **3** $y = -\ln(-\ln|t| + e^{-2})$ **5** $y = Ce^{-5t^2}$ **7** $y = e^{-6t}$
9 $y = (2/3)t - 2/9 + Ce^{-3t}$ **11** $y = 3,500,000e^{0.02t} - 2,500,000$
15 $y = Ae^{4t} + Be^t$ **17** $y = e^{2t}[A\cos(2t) + B\sin(2t)]$ **19** $y = -(1/6)e^{-5t} + (1/6)e^t$
21 $y = e^{-(3/2)t}[10\cos(t/2) + 30\sin(t/2)]$ **23** $y = Ae^{4t} + Be^t + 13/16 + t/4$
25 $y = -2\cos t - 3\sin t - (1/6)e^{-5t} + (13/6)e^t$

INDEX

A CATALOG OF SELECTED
DOVER BOOKS
IN SCIENCE AND MATHEMATICS

Mathematics–Bestsellers

HANDBOOK OF MATHEMATICAL FUNCTIONS: with Formulas, Graphs, and Mathematical Tables, Edited by Milton Abramowitz and Irene A. Stegun. A classic resource for working with special functions, standard trig, and exponential logarithmic definitions and extensions, it features 29 sets of tables, some to as high as 20 places. 1046pp. 8 x 10 1/2. 0-486-61272-4

ABSTRACT AND CONCRETE CATEGORIES: The Joy of Cats, Jiri Adamek, Horst Herrlich, and George E. Strecker. This up-to-date introductory treatment employs category theory to explore the theory of structures. Its unique approach stresses concrete categories and presents a systematic view of factorization structures. Numerous examples. 1990 edition, updated 2004. 528pp. 6 1/8 x 9 1/4. 0-486-46934-4

MATHEMATICS: Its Content, Methods and Meaning, A. D. Aleksandrov, A. N. Kolmogorov, and M. A. Lavrent'ev. Major survey offers comprehensive, coherent discussions of analytic geometry, algebra, differential equations, calculus of variations, functions of a complex variable, prime numbers, linear and non-Euclidean geometry, topology, functional analysis, more. 1963 edition. 1120pp. 5 3/8 x 8 1/2. 0-486-40916-3

INTRODUCTION TO VECTORS AND TENSORS: Second Edition–Two Volumes Bound as One, Ray M. Bowen and C.-C. Wang. Convenient single-volume compilation of two texts offers both introduction and in-depth survey. Geared toward engineering and science students rather than mathematicians, it focuses on physics and engineering applications. 1976 edition. 560pp. 6 1/2 x 9 1/4. 0-486-46914-X

AN INTRODUCTION TO ORTHOGONAL POLYNOMIALS, Theodore S. Chihara. Concise introduction covers general elementary theory, including the representation theorem and distribution functions, continued fractions and chain sequences, the recurrence formula, special functions, and some specific systems. 1978 edition. 272pp. 5 3/8 x 8 1/2. 0-486-47929-3

ADVANCED MATHEMATICS FOR ENGINEERS AND SCIENTISTS, Paul DuChateau. This primary text and supplemental reference focuses on linear algebra, calculus, and ordinary differential equations. Additional topics include partial differential equations and approximation methods. Includes solved problems. 1992 edition. 400pp. 7 1/2 x 9 1/4. 0-486-47930-7

PARTIAL DIFFERENTIAL EQUATIONS FOR SCIENTISTS AND ENGINEERS, Stanley J. Farlow. Practical text shows how to formulate and solve partial differential equations. Coverage of diffusion-type problems, hyperbolic-type problems, elliptic-type problems, numerical and approximate methods. Solution guide available upon request. 1982 edition. 414pp. 6 1/8 x 9 1/4. 0-486-67620-X

VARIATIONAL PRINCIPLES AND FREE-BOUNDARY PROBLEMS, Avner Friedman. Advanced graduate-level text examines variational methods in partial differential equations and illustrates their applications to free-boundary problems. Features detailed statements of standard theory of elliptic and parabolic operators. 1982 edition. 720pp. 6 1/8 x 9 1/4. 0-486-47853-X

LINEAR ANALYSIS AND REPRESENTATION THEORY, Steven A. Gaal. Unified treatment covers topics from the theory of operators and operator algebras on Hilbert spaces; integration and representation theory for topological groups; and the theory of Lie algebras, Lie groups, and transform groups. 1973 edition. 704pp. 6 1/8 x 9 1/4. 0-486-47851-3

Browse over 9,000 books at www.doverpublications.com

A SURVEY OF INDUSTRIAL MATHEMATICS, Charles R. MacCluer. Students learn how to solve problems they'll encounter in their professional lives with this concise single-volume treatment. It employs MATLAB and other strategies to explore typical industrial problems. 2000 edition. 384pp. 5 3/8 x 8 1/2.   0-486-47702-9

NUMBER SYSTEMS AND THE FOUNDATIONS OF ANALYSIS, Elliott Mendelson. Geared toward undergraduate and beginning graduate students, this study explores natural numbers, integers, rational numbers, real numbers, and complex numbers. Numerous exercises and appendixes supplement the text. 1973 edition. 368pp. 5 3/8 x 8 1/2.   0-486-45792-3

A FIRST LOOK AT NUMERICAL FUNCTIONAL ANALYSIS, W. W. Sawyer. Text by renowned educator shows how problems in numerical analysis lead to concepts of functional analysis. Topics include Banach and Hilbert spaces, contraction mappings, convergence, differentiation and integration, and Euclidean space. 1978 edition. 208pp. 5 3/8 x 8 1/2.   0-486-47882-3

FRACTALS, CHAOS, POWER LAWS: Minutes from an Infinite Paradise, Manfred Schroeder. A fascinating exploration of the connections between chaos theory, physics, biology, and mathematics, this book abounds in award-winning computer graphics, optical illusions, and games that clarify memorable insights into self-similarity. 1992 edition. 448pp. 6 1/8 x 9 1/4.   0-486-47204-3

SET THEORY AND THE CONTINUUM PROBLEM, Raymond M. Smullyan and Melvin Fitting. A lucid, elegant, and complete survey of set theory, this three-part treatment explores axiomatic set theory, the consistency of the continuum hypothesis, and forcing and independence results. 1996 edition. 336pp. 6 x 9.   0-486-47484-4

DYNAMICAL SYSTEMS, Shlomo Sternberg. A pioneer in the field of dynamical systems discusses one-dimensional dynamics, differential equations, random walks, iterated function systems, symbolic dynamics, and Markov chains. Supplementary materials include PowerPoint slides and MATLAB exercises. 2010 edition. 272pp. 6 1/8 x 9 1/4.   0-486-47705-3

ORDINARY DIFFERENTIAL EQUATIONS, Morris Tenenbaum and Harry Pollard. Skillfully organized introductory text examines origin of differential equations, then defines basic terms and outlines general solution of a differential equation. Explores integrating factors; dilution and accretion problems; Laplace Transforms; Newton's Interpolation Formulas, more. 818pp. 5 3/8 x 8 1/2.   0-486-64940-7

MATROID THEORY, D. J. A. Welsh. Text by a noted expert describes standard examples and investigation results, using elementary proofs to develop basic matroid properties before advancing to a more sophisticated treatment. Includes numerous exercises. 1976 edition. 448pp. 5 3/8 x 8 1/2.   0-486-47439-9

THE CONCEPT OF A RIEMANN SURFACE, Hermann Weyl. This classic on the general history of functions combines function theory and geometry, forming the basis of the modern approach to analysis, geometry, and topology. 1955 edition. 208pp. 5 3/8 x 8 1/2.   0-486-47004-0

THE LAPLACE TRANSFORM, David Vernon Widder. This volume focuses on the Laplace and Stieltjes transforms, offering a highly theoretical treatment. Topics include fundamental formulas, the moment problem, monotonic functions, and Tauberian theorems. 1941 edition. 416pp. 5 3/8 x 8 1/2.   0-486-47755-X

Mathematics–Logic and Problem Solving

PERPLEXING PUZZLES AND TANTALIZING TEASERS, Martin Gardner. Ninety-three riddles, mazes, illusions, tricky questions, word and picture puzzles, and other challenges offer hours of entertainment for youngsters. Filled with rib-tickling drawings. Solutions. 224pp. 5 3/8 x 8 1/2. 0-486-25637-5

MY BEST MATHEMATICAL AND LOGIC PUZZLES, Martin Gardner. The noted expert selects 70 of his favorite "short" puzzles. Includes The Returning Explorer, The Mutilated Chessboard, Scrambled Box Tops, and dozens more. Complete solutions included. 96pp. 5 3/8 x 8 1/2. 0-486-28152-3

THE LADY OR THE TIGER?: and Other Logic Puzzles, Raymond M. Smullyan. Created by a renowned puzzle master, these whimsically themed challenges involve paradoxes about probability, time, and change; metapuzzles; and self-referentiality. Nineteen chapters advance in difficulty from relatively simple to highly complex. 1982 edition. 240pp. 5 3/8 x 8 1/2. 0-486-47027-X

SATAN, CANTOR AND INFINITY: Mind-Boggling Puzzles, Raymond M. Smullyan. A renowned mathematician tells stories of knights and knaves in an entertaining look at the logical precepts behind infinity, probability, time, and change. Requires a strong background in mathematics. Complete solutions. 288pp. 5 3/8 x 8 1/2.
0-486-47036-9

THE RED BOOK OF MATHEMATICAL PROBLEMS, Kenneth S. Williams and Kenneth Hardy. Handy compilation of 100 practice problems, hints and solutions indispensable for students preparing for the William Lowell Putnam and other mathematical competitions. Preface to the First Edition. Sources. 1988 edition. 192pp. 5 3/8 x 8 1/2. 0-486-69415-1

KING ARTHUR IN SEARCH OF HIS DOG AND OTHER CURIOUS PUZZLES, Raymond M. Smullyan. This fanciful, original collection for readers of all ages features arithmetic puzzles, logic problems related to crime detection, and logic and arithmetic puzzles involving King Arthur and his Dogs of the Round Table. 160pp. 5 3/8 x 8 1/2. 0-486-47435-6

UNDECIDABLE THEORIES: Studies in Logic and the Foundation of Mathematics, Alfred Tarski in collaboration with Andrzej Mostowski and Raphael M. Robinson. This well-known book by the famed logician consists of three treatises: "A General Method in Proofs of Undecidability," "Undecidability and Essential Undecidability in Mathematics," and "Undecidability of the Elementary Theory of Groups." 1953 edition. 112pp. 5 3/8 x 8 1/2. 0-486-47703-7

LOGIC FOR MATHEMATICIANS, J. Barkley Rosser. Examination of essential topics and theorems assumes no background in logic. "Undoubtedly a major addition to the literature of mathematical logic." — *Bulletin of the American Mathematical Society.* 1978 edition. 592pp. 6 1/8 x 9 1/4. 0-486-46898-4

INTRODUCTION TO PROOF IN ABSTRACT MATHEMATICS, Andrew Wohlgemuth. This undergraduate text teaches students what constitutes an acceptable proof, and it develops their ability to do proofs of routine problems as well as those requiring creative insights. 1990 edition. 384pp. 6 1/2 x 9 1/4. 0-486-47854-8

FIRST COURSE IN MATHEMATICAL LOGIC, Patrick Suppes and Shirley Hill. Rigorous introduction is simple enough in presentation and context for wide range of students. Symbolizing sentences; logical inference; truth and validity; truth tables; terms, predicates, universal quantifiers; universal specification and laws of identity; more. 288pp. 5 3/8 x 8 1/2. 0-486-42259-3

Mathematics–Algebra and Calculus

VECTOR CALCULUS, Peter Baxandall and Hans Liebeck. This introductory text offers a rigorous, comprehensive treatment. Classical theorems of vector calculus are amply illustrated with figures, worked examples, physical applications, and exercises with hints and answers. 1986 edition. 560pp. 5 3/8 x 8 1/2.   0-486-46620-5

ADVANCED CALCULUS: An Introduction to Classical Analysis, Louis Brand. A course in analysis that focuses on the functions of a real variable, this text introduces the basic concepts in their simplest setting and illustrates its teachings with numerous examples, theorems, and proofs. 1955 edition. 592pp. 5 3/8 x 8 1/2.   0-486-44548-8

ADVANCED CALCULUS, Avner Friedman. Intended for students who have already completed a one-year course in elementary calculus, this two-part treatment advances from functions of one variable to those of several variables. Solutions. 1971 edition. 432pp. 5 3/8 x 8 1/2.   0-486-45795-8

METHODS OF MATHEMATICS APPLIED TO CALCULUS, PROBABILITY, AND STATISTICS, Richard W. Hamming. This 4-part treatment begins with algebra and analytic geometry and proceeds to an exploration of the calculus of algebraic functions and transcendental functions and applications. 1985 edition. Includes 310 figures and 18 tables. 880pp. 6 1/2 x 9 1/4.   0-486-43945-3

BASIC ALGEBRA I: Second Edition, Nathan Jacobson. A classic text and standard reference for a generation, this volume covers all undergraduate algebra topics, including groups, rings, modules, Galois theory, polynomials, linear algebra, and associative algebra. 1985 edition. 528pp. 6 1/8 x 9 1/4.   0-486-47189-6

BASIC ALGEBRA II: Second Edition, Nathan Jacobson. This classic text and standard reference comprises all subjects of a first-year graduate-level course, including in-depth coverage of groups and polynomials and extensive use of categories and functors. 1989 edition. 704pp. 6 1/8 x 9 1/4.   0-486-47187-X

CALCULUS: An Intuitive and Physical Approach (Second Edition), Morris Kline. Application-oriented introduction relates the subject as closely as possible to science with explorations of the derivative; differentiation and integration of the powers of x; theorems on differentiation, antidifferentiation; the chain rule; trigonometric functions; more. Examples. 1967 edition. 960pp. 6 1/2 x 9 1/4.   0-486-40453-6

ABSTRACT ALGEBRA AND SOLUTION BY RADICALS, John E. Maxfield and Margaret W. Maxfield. Accessible advanced undergraduate-level text starts with groups, rings, fields, and polynomials and advances to Galois theory, radicals and roots of unity, and solution by radicals. Numerous examples, illustrations, exercises, appendixes. 1971 edition. 224pp. 6 1/8 x 9 1/4.   0-486-47723-1

AN INTRODUCTION TO THE THEORY OF LINEAR SPACES, Georgi E. Shilov. Translated by Richard A. Silverman. Introductory treatment offers a clear exposition of algebra, geometry, and analysis as parts of an integrated whole rather than separate subjects. Numerous examples illustrate many different fields, and problems include hints or answers. 1961 edition. 320pp. 5 3/8 x 8 1/2.   0-486-63070-6

LINEAR ALGEBRA, Georgi E. Shilov. Covers determinants, linear spaces, systems of linear equations, linear functions of a vector argument, coordinate transformations, the canonical form of the matrix of a linear operator, bilinear and quadratic forms, and more. 387pp. 5 3/8 x 8 1/2.   0-486-63518-X

STANDARD PARTS

In the following b, c are finite (possibly infinitesimal).

$st(b + c) = st(b) + st(c)$

$st(bc) = st(b)st(c)$

$st(\sqrt[n]{b}) = \sqrt[n]{st(b)}$ if $b > 0, n > 0$

$b \approx st(b)$

$b = st(b)$ if and only if b is real

$st(\varepsilon) = 0, st(H)$ is undefined

$st(b - c) = st(b) - st(c)$

$st(b/c) = st(b)/st(c)$ if $st(c) \neq 0$

$st(b^c) = st(b)^{st(c)}$ if $st(b) > 0$

$b \approx c$ if and only if $st(b) = st(c)$

if $b \leq c$, then $st(b) \leq st(c)$

TRIGONOMETRIC IDENTITIES

$\sin^2 \theta + \cos^2 \theta = 1$

$\cot^2 \theta + 1 = \csc^2 \theta$

$\cos(-\theta) = \cos\theta$

$\cos(\pi/2 - \theta) = \sin\theta$

$\cos(\theta + \phi) = \cos\theta\cos\phi - \sin\theta\sin\phi$

$\tan^2 \theta + 1 = \sec^2 \theta$

$\sin(-\theta) = -\sin\theta$

$\sin(\pi/2 - \theta) = \cos\theta$

$\sin(\theta + \phi) = \sin\theta\cos\phi + \cos\theta\sin\phi$

RULES OF EXPONENTS AND LOGARITHMS

Assume $a > 0, b > 0$.

$1^x = 1$

$a^{x+y} = a^x a^y$

$a^{xy} = (a^x)^y$

$a^{\log_a x} = x$

$\log_a(xy) = \log_a x + \log_a y$

$\log_a(x^y) = y \log_a x$

$\log_b y = \log_a y / \log_a b$

$a^0 = 1$

$a^{x-y} = a^x/a^y$

$a^x b^x = (ab)^x$

$\log_a(a^x) = x$

$\log_a(x/y) = \log_a x - \log_a y$

$b^x = a^{x \log_a b}$

TABLE OF DERIVATIVES

$du = (du/dx)\,dx$ (Chain Rule)

$d(u + v) = du + dv$ (Sum Rule)

$d\left(\dfrac{u}{v}\right) = \dfrac{v\,du - u\,dv}{v^2}$ (Quotient Rule)

$d(\ln u) = du/u$

$d(\sin u) = \cos u\,du$

$d(\tan u) = \sec^2 u\,du$

$d(\sec u) = \tan u \sec u\,du$

$d(\arcsin u) = \dfrac{du}{\sqrt{1 - u^2}}$

$d(\arctan u) = \dfrac{du}{1 + u^2}$

$d(\text{arcsec}\,u) = \dfrac{du}{|u|\sqrt{u^2 - 1}}$

$d(ku) = k\,du$ (Constant Rule)

$d(uv) = u\,dv + v\,du$ (Product Rule)

$d(u^r) = ru^{r-1}\,du$ (Power Rule)

$d(e^u) = e^u\,du$

$d(\cos u) = -\sin u\,du$

$d(\cot u) = -\csc^2 u\,du$

$d(\csc u) = -\cot u \csc u\,du$

$d(\arccos u) = -\dfrac{du}{\sqrt{1 - u^2}}$

$d(\text{arccot}\,u) = -\dfrac{du}{1 + u^2}$

$d(\text{arccsc}\,u) = -\dfrac{du}{|u|\sqrt{u^2 - 1}}$